The Handloader's Manual Of Cartridge Conversions

John J. Donnelly and Judy Donnelly

Reviews, design, and illustrations by Hughes Design, LLC.

Skyhorse Publishing

Skyhorse Publishing books may be purchased in bulk at special discounts for sales promotion, corporate gifts, fund-raising, or educational purposes. Special editions can also be created to specifications. For details, contact the Special Sales Department, Skyhorse Publishing, 307 West 36th Street, 11th Floor, New York, NY 10018 or info@skyhorsepublishing.com.

www.skyhorsepublishing.com

10 9 8 7 6 5 4

Library of Congress Cataloging-in-Publication Data

Donnelly, John J. (John James), 1946-
The handloader's manual of cartridge conversions / John J. Donnelly & Judy Donnelly.
p. cm.
Includes index.
ISBN 978-1-61608-238-3 (pbk. : alk. paper)
1. Handloading of ammunition--Handbooks, manuals, etc. I. Donnelly, Judy. II. Title.
TS538.3.D66 2011
683.4'06--dc22

2011002800

Printed in the United States of America

ACKNOWLEDGEMENTS: Numerous individuals and companies have helped me prepare this book, and it is only fitting that they receive proper recognition along with my sincere thanks. Robert E. Weise, my publisher, guided me through the intricacies of assembling the manuscript and has been most helpful, patient and tolerant. My mother, Arleen, deserves a hearty cheer for her much needed assistance, as does the rest of my family for putting up with the whole business. Paul Eastus was kind enough to provide many of the fine drawings that embellish these pages.

DISCLAIMER: Any and all loading information contained in this book should be used with caution. Since they have no knowledge or control over the reader's choice of guns, components, and loadings, the author, publisher, and the various firearms and components manufacturers mentioned herein assume no responsibility for the use of this information. Futhermore, while all sizes, dimensions and loadings included in the conversion pages (Parts I and II) have been checked for accuracy, neither the author nor the publisher can assume any liability for any incorrect information contained therein.

FOREWORD

For riflemen interested in reloading, creating their own cartridges or duplicating the wildcat loads of other enthusiastic shooters, John J. Donnelly's Handbook of Cartridge Conversions is the ultimate reference. It's not very often that a master of a given topic decides to put his life's work into print, and handloaders around the world should give a collective cheer for Donnelly's efforts.

I began reloading in the early 1960s. My father had set up a bench in the attic above our third-floor apartment where we spent many evening and weekend hours creating hot loads for our .220 Swift, .30/06 and other rifles. Dad did most of the critical work (resizing, case trimming, powder measuring and bullet seating), while I sorted bullets (by weight), checked primers (for bulging anvils) and polished the brass cases after each operation. We even melted lead wheel weights on the kitchen stove to make our own bullets, two at a time with our Lyman "nutcracker" hand tools. Using only the cheapest, most primitive reloading tools available at the time and a second-hand copy of a Speer reloading manual, we would spend hours cooking up exciting new loads, three cartridges per "recipe," and then head for the shooting range on Saturdays to test our creations. Some of our loads were super accurate, some were too "hot" and some were borderline dangerous - one load for the .220 Swift resulted in the bullets literally evaporating a few yards from the muzzle, a result of too much powder and too light a bullet spinning too fast. We were probably a grain or two of powder from serious disaster, and when we had to replace the (Model 70 Supergrade) Swift's stainless steel barrel the second time due to chamber erosion caused by excessive heat and pressure, we quit experimenting and went back to the rather tame loads listed in the manual.

Oh, how we would have loved to have found a copy of John Donnelly's book! Not only were we unaware of the variety of reloading tools and techniques that were available, we had no idea that one could create new cartridge designs out of existing calibers or, miracle of miracles, build cartridges (even guns) of our own creation. He had already done all the hard work and was foresighted enough to write it all down for the benefit of future "gun nuts." Anyone who has the least inclination to reload his own cartridges should own a copy of this book, the culmination of a lifetime of one who has truly been there, done that!

For safety's sake, understand that the loads, conversions and techniques included here are Donnelly's and should be duplicated and approached with care and caution. Do not tinker with his measurements, loads or calculations. Read every powder label and check each measurement twice before you proceed. Always wear standard safety gear (goggles and gloves) and conduct each operation with utmost caution.

If you enjoy rifle shooting, creating your own ammunition is the next logical step. Welcome to the club!

—Stephen D. Carpenteri

This book is for Judy, my brightest star.

—J. J. Donnelley, 1987

INTRODUCTION

One of the major influences in my career has been the late George C. Nonte's book, *The Home Guide to Cartridge Conversions*. I've been gleaning information from its dog-eared pages ever since 1972, and it has helped me immeasurably in building an internationally known custom ammunition business, which I've named Ballistek.

It is my intention to repay a debt to Major Nonte with this book. Although I can never thank George personally for his assistance and insight (he passed away in 1978), I hope that somehow he knows that I have added to his work and passed on our accumulated knowledge to others.

Before I am accused of plagiarism, let me state flatly that I have NOT copied any of the Major's information. Similarities do exist, simply because there is only one logical way to present this information. But while the form may be similar, the information contained therein reflects today's technology and tool availability, plus my own years of experience.

Actually, I wrote this book for myself! As a custom loader, I use this data every working day, and I had grown tired of constantly searching for this or that piece of information. My ammunition library consists of numerous hardbound books, several thousand magazines, tons of notes, and a card file index to pull everything together. Whenever I worked on a cartridge, I had to pull a master card listing nearly every literary citation for that cartridge during the past 12 to 15 years. Each book, magazine or note then had to be located before the necessary data could be extracted. Cumbersome was the most polite description my long-suffering wife could offer to describe my research system. Things progressed to the point where the need for this book became obvious, if not desperate.

Today, there's growing interest in obsolete, foreign, metric and wildcat ammunition. For reasons of economics and greater accuracy, we are also witnessing an increasing number of handloaders. These factors combined cannot survive for very long without spurring interest in more advanced handloading. This book was written for all those handloaders as well.

My background as a manufacturing engineer and tool designer will be evident in many of the more specialized forming techniques you'll run across in this book. The methods contained herein often reflect my experience in the highly technical fields of fixture design, process engineering, carbide tooling technology, and chemical engineering. Many of these operations are obviously inappropriate for the novice handloader. This manual further assumes that the reader has mastered basic handloading principles and will not be baffled by such terms as "full-length size," "flashhole," and "extractor groove." In short, this book is not meant to be an "Introduction to Handloading." On the other hand, I have made every effort to "keep it simple," or at least as simple as possible consistent with the effective completion of whatever task is at hand.

Reloaders will also benefit from the information provided concerning tools and metalworking techniques. For example, many of my forming operations require the use of a lathe. I also believe that most reloaders will find this book to be both interesting and challenging.

Most of the cartridges detailed in this book are long obsolete, and the weapons themselves are old and worn. When fabricating ammunition for these pieces, I remind the reader that minute-of-angle accuracy is not always possible. It is my intention to offer one method of producing shootable ammo. Although I have not personally fired many of these cartridges, I've relied on feedback from shooters who have. These loads, which perform quite well, often duplicate the velocity and internal pressure of the original round. I cannot, however, be held responsible for the use of any loading or forming information. When it comes to converting cartridges, you are pretty much on your own. In case of an accident, you have no one to blame but yourself!

The information contained in this manual is, to the best of my knowledge, true and accurate. It would be virtually impossible, however, to produce a work like this—one that is so highly dependent on numbers—without a few errors here and there.

So have fun reloading—and work safe.

—J. J. Donnelley, 1987

CONTENTS

The Handloader's Manual Of Cartridge Conversions

PART ONE Methods in the Art of Handloading

1 Cartridge Conversion: History and Nomenclature

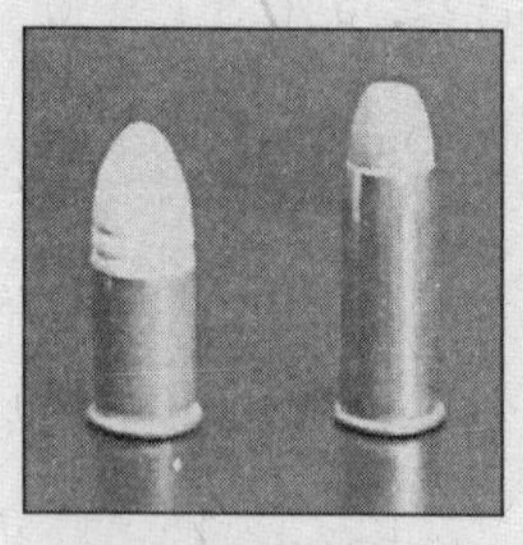

These two cartridges date from the mid-19th century when cartridges were moving from inside priming to centerfire. On the left is a .56-56 Spencer. The one on the right is a .50-60 Peabody.

While it is not within the scope or intent of this work to present the long chain of technological milestones that have coalesced into our modern centerfire brass cartridge cases, a brief history may be in order. During the Civil War, most firearms in America featured either percussion or rimfire ignition. By the end of the war, the percussion firearm was already passing into history as bigger and better rimfire cartridges appeared on the market. Notable among these were the .56 caliber Spencer, the .54 Ballard, the .52 Sharps & Hankins and the .44 Henry.

Following the Civil War, experiments with centerfire priming dealt a death blow to the large bore rimfires. The U.S. Army tested several inside-primed cartridges and used the Benet inside cup primed round for several years before Colonel Hiram Berdan's exterior primer, invented in 1866, was adopted. This primer is still used in the military ammunition of many nations.

At the same time, the British were working on ammunition for the Snider conversion of the Enfield muzzleloading rifle. Under the direction of Colonel Edward Boxer, whose cartridge used a primer different from that of Berdan's, this model carried its own anvil. Berdan's primer, on the other hand, incorporated the anvil as part of the brass case in the center of the primer pocket, with two flashholes, one on either side of the anvil. U. S. ammunition producers subsequently adopted Boxer's idea, in part because it made the reloading of ammunition a much more practical matter.

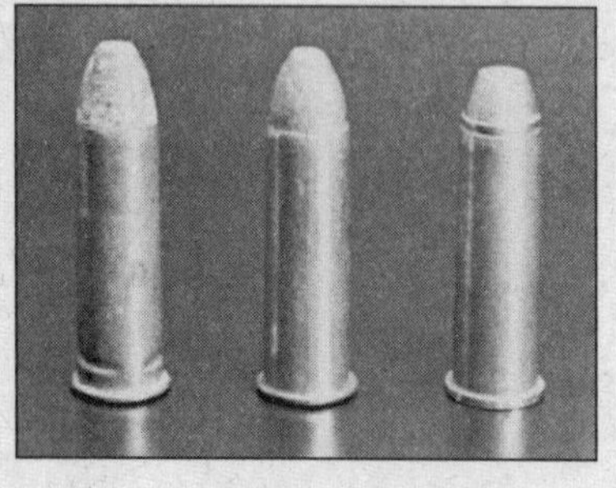

The cartridge shown on the left is an early rimfire version of the .50-70 Government. In the center is a Remington centerfire; version and on the right is a .50-70 Government cartridge made in 1975 from Dixie Gun Works brass.

By the early 1870s, the march of American settlers westward was in full swing, and thus a good, available supply of reliable ammunition became a necessity. It soon became apparent that Berdan-primed ammunition was too difficult to reload on the open plains, especially with screaming hordes of belligerent warriors dashing about. Moreover, not many of our frontier ancestors perceived much merit in lugging tons of ammo west in their Conestoga wagons. Enter the Boxer primer.

After adoption of the .45-70 Springfield cartridge in 1873, inside primed ammunition was used for a few years, but eventually the Boxer-primed cartridge replaced it. Reloading apparatus soon became a part of every infantry company's supply kit, thus marking the advent of reloading in America.

Britain—and for that matter most of Europe—did not consider field reloading

DISCLAIMER

Any and all loading information contained in this book should be used with caution. Since they have no knowledge or control over the reader's choice of guns, components, and loadings, the author, publisher, and the various firearms and components manufacturers mentioned herein assume no responsibility for the use of this information.

Furthermore, while all sizes, dimensions and loadings included in the conversion pages (Parts I and II) have been checked for accuracy, neither the author nor the publisher can assume any liability for any incorrect information contained therein.

a necessary pastime for their troops, however, and adopted the cheaper Berdan primer for their metallic cartridgess. Surprisingly few changes have taken place—in both cases and primers—since 1875 except that the materials in current use are vastly superio and more reliable.

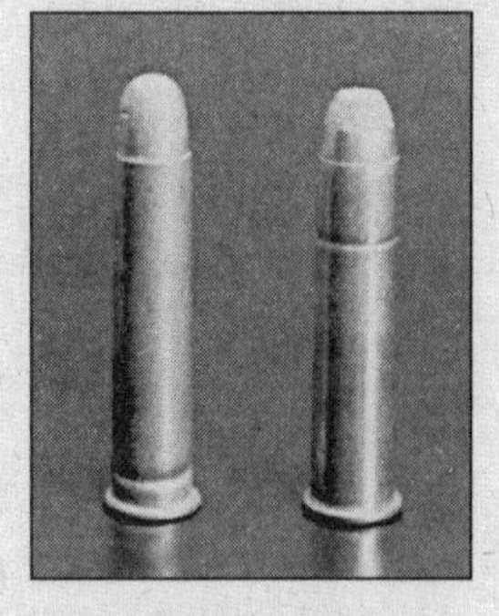

On the left is an early .45-70 cartridge utilizing inside priming (rimfire)and to its right a contemporary .45-70 centerfire made by Winchester.

MODERN PRODUCTION

Modern brass cases are produced under exacting specifications. While each case manufacturer has its own recipe for brass, the basic formulas of copper and zinc remain unchanged. The exact ratio affects the forming characteristics and ultimate strength of the case considerably; hence, each manufacturer uses the blend it believes is most superior.

Cartridge cases begin life as flat sheets rolled into long coils. These brass sheets are first heated and then rolled to an exact thickness. The sheets pass through blanking presses, which punch disks of the proper diameter to minimize waste . The disks then enter a series of annealing, lubrication and drawing processes.

A drawing die is little more than a polished hole in a tool steel plate. Poised above this hole is the male portion of the die, which is a highly polished tool steel finger. The outside diameter of this finger is slightly smaller than the inside diameter of the female die. A disk of annealed brass is centered above the female cavity. When the press cycles, the male portion enters the female cavity, pulling the brass disk with it.

After drawing, the disk, which now resembles a small cup, passes into another draw die. The second die set contains a female cavity with a smaller inside diameter and a male pin with a larger outside diameter. Drawing is performed several times until a long, cylindrical tube of brass with a solid bottom is produced. Later in the process, a small nipple is added to the base of one of the female dies, thus forming the primer pocket and headstamp.

The brass tubes move to another machine that plunge-cuts the rim diameter and the extractor groove. After turning, the flashhole is punched, followed by annealing, subsequent neck forming (if required) and final trimming to the required length.

Some suppliers offer selections of rare and exotic cases, including the .577 Nitro Express shown here. Most cases are "basic," meaning they must be formed and trimmed before use.

Quality control instruments and personnel monitor each step in this procedure. The chances of a missed operation are therefore quite slim. A finished round of ammunition without a flashhole connecting the primer to the powder, for example, would be dangerous indeed. During 20 years of handloading ammunition, the author has found one case—in a box of new, unprimed .38 Special brass—that did not sport a flashhole. This case is still around as a reminder that large manufacturers are not infallible, and that new cases should always be closely scrutinized before use.

It does not require 20 years on a factory assembly line to realize that the above procedure is complex and expensive. Many modern machines of this type may take days to tool up, and the added expense of tool maintenance and constant upkeep can only be justified by extended production runs producing millions of cases.

WHY THEY'RE GONE

In 1902, Winchester Repeating Arms introduced the .33 WCF cartridge and simultaneously announced the availability of its companion M86 lever rifle. This combination became very popular mostly because the 200-grain factory bullet exited the muzzle at some 2,200 fps, delivering slightly more than a ton of energy. In 1936, Winchester unveiled its much stronger M71 rifle and .348 WCF cartridge, whose 200-grain factory loading produced almost 1-1/2 tons of energy at 2,325 fps. Owners of the M86 quickly began trading for M71s and .33 WCF ammo sales plummeted. The cartridge was discontinued in 1941.

Although World War II certainly had some influence on this disappearance, the bottom line indicated that sales of .33 WCF ammo had dropped off in favor of the more powerful .348. Moreover, Winchester could not justify the set-up and tooling costs for the .33, and thus this once great cartridge became obsolete. A few years later, the .358 WCF appeared, causing the .348 to pass into oblivion as well.

At left is an original Winchester .33 WCF cartridge. The case on the right is a .45-70 Government reformed to .33 WCF.

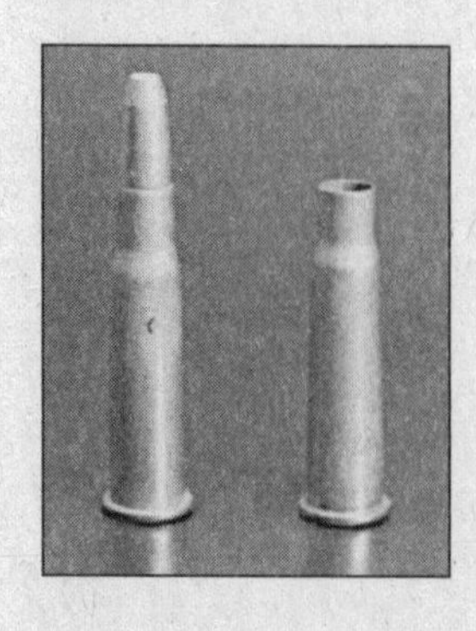

A cartridge can become obsolete in other ways. In some instances, a manufacturer may conjure up a new cartridge and firearm, place it on the market, and then discontinue it after only a year or so because of general lack of interest. An example of this would be the .38-45 Stevens. For whatever reason, a host of cartridges has been introduced and later been discontinued over the years. It follows that many firearms are also produced for these cartridges, but they do not evaporate when the ammo is discontinued. Some individuals opt to rechamber, rebarrel, or rebore these relics in an effort to remain on the firing line. This is unfortunate and should only be considered when every possibility of obtaining shootable ammo has been exhausted because the collectible value of such guns is severely diminished.

In this book we shall concentrate on those methods and tools that allow fabrication by the handloader of long extinct cases from commercially available brass. By the time you read this, several more familiar cartridges will have disappeared, notably the .303 Savage and .30 Rem. A point is reached where it is no longer profitable to manufacture these and other cartridges, and they will then pass into the realm of the custom loader and cartridge collector. Meanwhile, newer cartridges are constantly being developed.

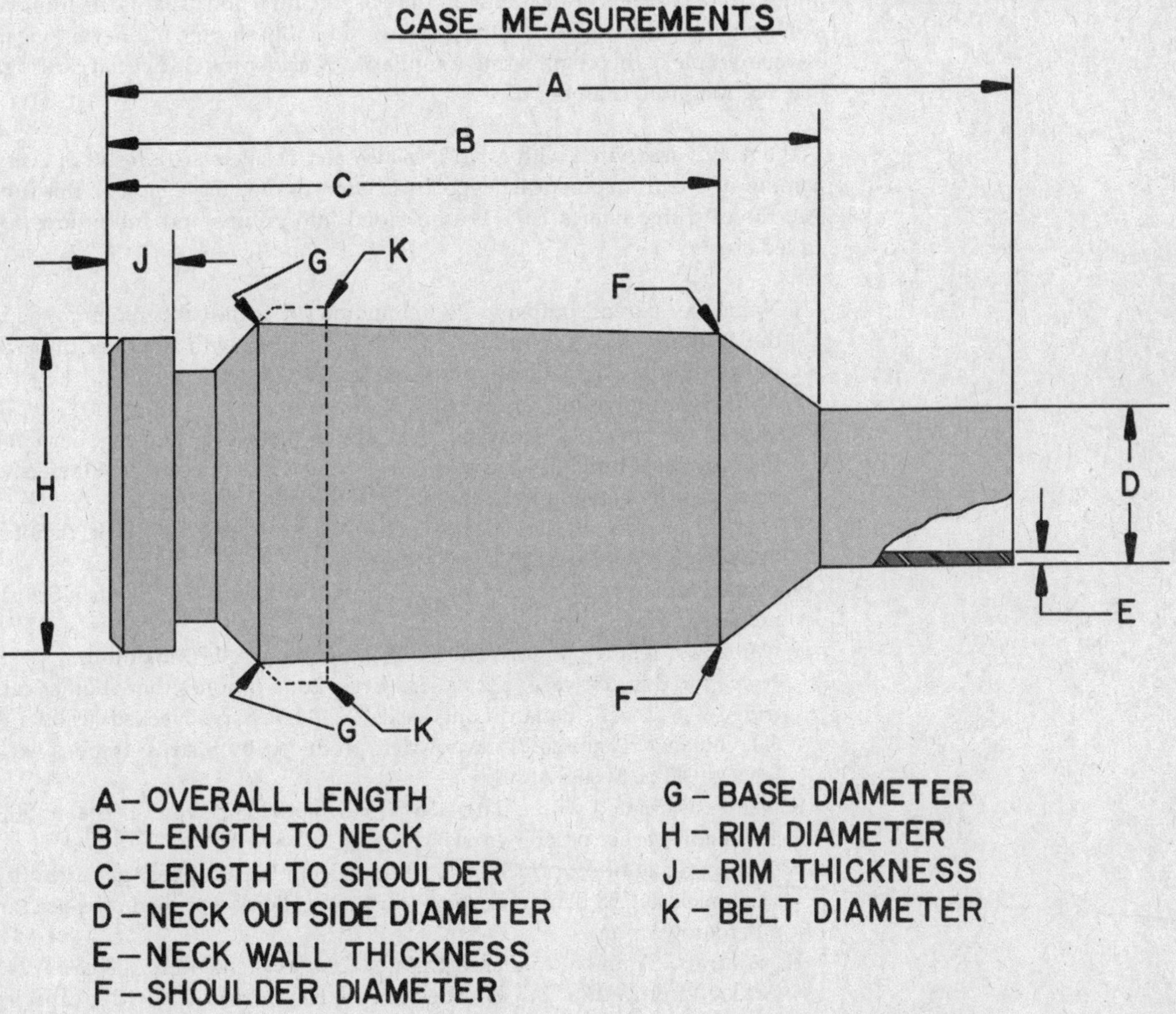

CASE MEASUREMENTS

A – OVERALL LENGTH
B – LENGTH TO NECK
C – LENGTH TO SHOULDER
D – NECK OUTSIDE DIAMETER
E – NECK WALL THICKNESS
F – SHOULDER DIAMETER
G – BASE DIAMETER
H – RIM DIAMETER
J – RIM THICKNESS
K – BELT DIAMETER

CASE NOMENCLATURE

Most reloaders are familiar with the basic terms that apply to cartridge brass. For the purpose of later discussions, however, a review of all terms in current use is provided, including standard case measurements. We shall also review the common types of existing cases and the manner in which they headspace. The "target" notation indicates that portion of the case that must position itself accurately against the mating chamber surface so as to establish the headspace dimension. The dimension from the target to the head is critical; a tolerance of plus-or minus- .001 must be held. The remaining distance – i.e., case length minus headspace—is critical only in the positive direction. That is, the dimension from the target to the mouth can be considerably shorter, but never longer. As an example of this chambering irregularity, consider the .38 Special cartridge in a .357 Magnum chamber.

The names associated with cartridges new and obsolete are a result of every naming system from practical to ego. In an effort to shed more light on this subject, the cartridge names have been divided into groups and sub-groups for added clarity:

1. Named for parent company. .17 Remington (.17 is the caliber,Remington is the designer). .25-20 Winchester (.25 is the caliber, with 20 grains of black powder in the original load, produced by Winchester).
2. Named for individual designer. (.257 Roberts).
3. Named for particular weapon. .30 Carbine (.30 is the caliber; "Carbine" refers to the M1 military carbine). .30-40 Krag (.30 is the caliber; 40 refers to black powder content; Krag-Jorgensen is the rifle).
4. Named for weapon action type. .351 WSL (caliber is .35; WSL denotes "Winchester Self-loading").
5. Named for issuing government. .303 British (caliber is .303; British: British military).
6. Named for relative performance. .357 Magnum or .357 Maximum.
7. Named to describe case. .32 Ballard Extra Long (implies that another cartridge—i.e., the .32 Ballard Long – exists) .40-50 Sharps Necked (caliber is .40; contains 50 grains black powder; produced by Sharps; implies existence of .40-50 Straight Sharps).
8. Names associated with a date. .38-50 Maynard 1882 (suggests that another version, earlier or later, exists).
9. Named for ignition principle. .38 Long Centerfire (implies that a rimfire version exists) .58 Berdan Carbine (this case is Berdan primed or is used in a Berdan weapon).
10. Wildcats. 7mm-08 (caliber is 7mm or .284; projectile in a .308 case) .243 Ackley Improved (a .243 Winchester case has increased powder capacity due to Ackley's improvement).

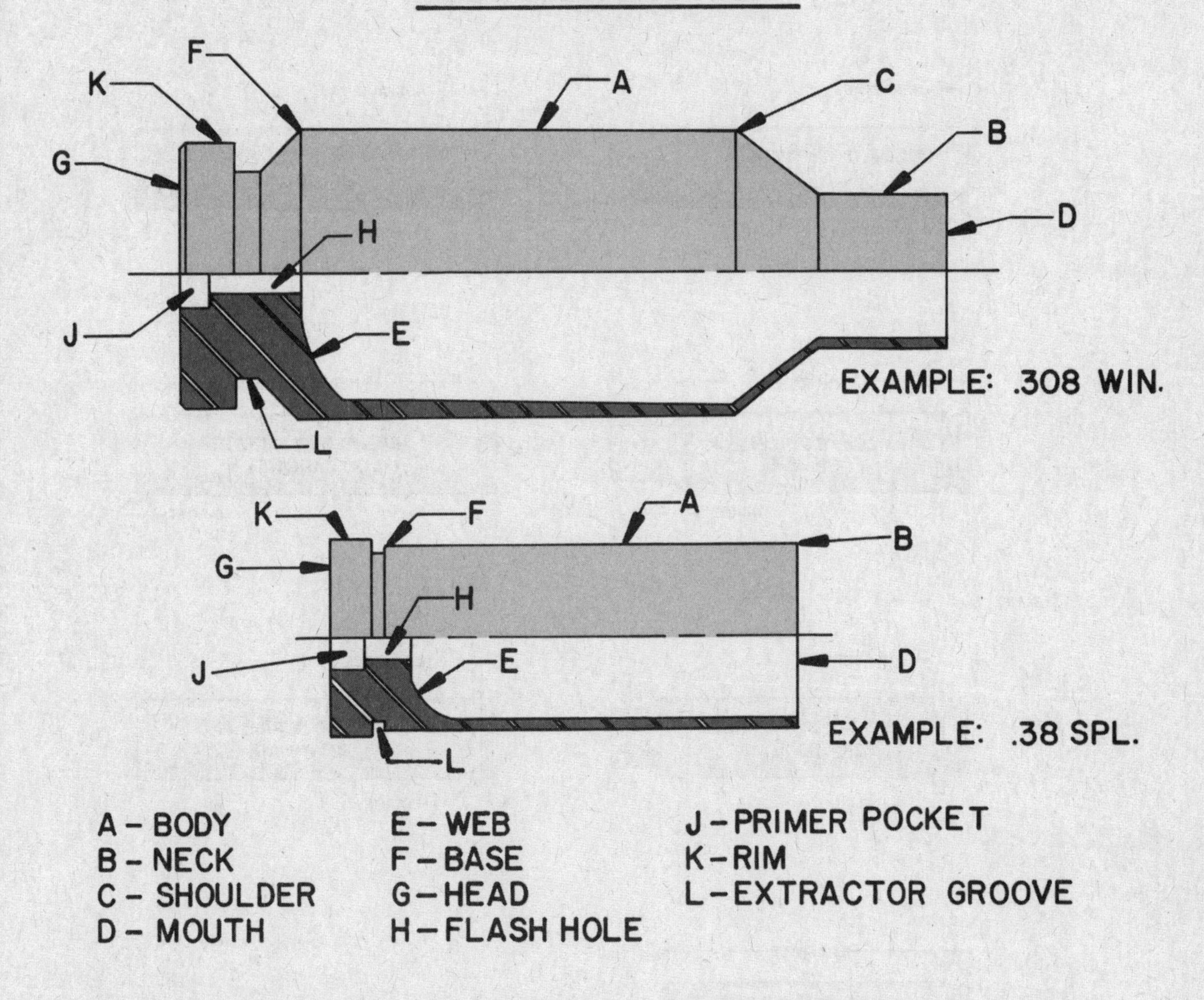

11. Names selected at random. .218 Bee, or .22 Hornet (in which "Bee"and "Hornet" are merely arbitrary terms). .40-70 Peabody "What Cheer" (refers to a rifle range in Rhode Island known by that name).
12. Metric. 7 x 57mm Mauser (caliber is 7mm or .284; case is 57mm or 2.244" long). 6.5 x 57R (caliber is 6.5mm or .264; case is 57mm or 2.244" long; "R" means rimmed). 8 x 57JS ("J" denotes .318 dia.; case is rimless). "J" denotes the German infantry cartridge with its smaller diameter (.318). "JS" is slightly larger (.323). 8 x 57JR (same as above, except case is rimmed). 8 x 57 (.323 diameter assumed. Rimless).

The 12 major categories listed above are provided merely in an attempt to illustrate how cartridges are named. Shooters who spend a great amount of time handloading will soon become accustomed to these and other strange letters and symbols.

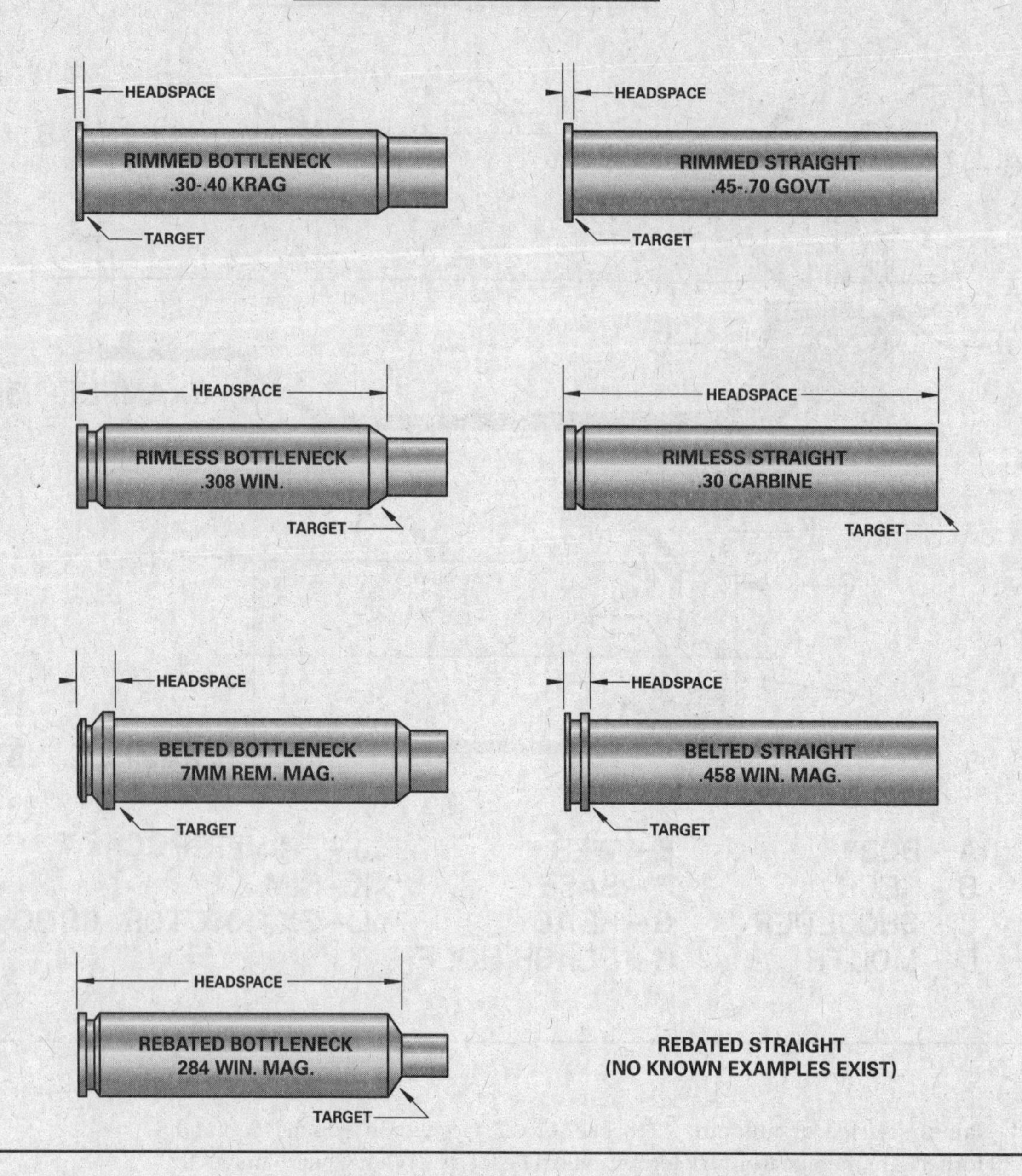

To illustrate further why it is so difficult to place each cartridge in its proper niche, let's examine some additional (but common) aberrations:

- The .218 Bee uses .224 diameter projectiles, as do the .220 Swift, .222 Remington, .223 Remington, and .225 Winchester .
- The 6.5 Remington Magnum uses .264 diameter bullets (which actually measure 6.70mm in diameter). Why isn't it called the 6.7 Remington Magnum? That's a good—but unanswerable—question.
- The .270 Winchester and .270 Weatherby both use .277 diameter bullets.
- The 7 x 57 Mauser cartridge (see above) uses a .284 diameter bullet and is actually 2.235″ in length; to be consistent, the cartridge should be labeled: "7.2 x 56.8."

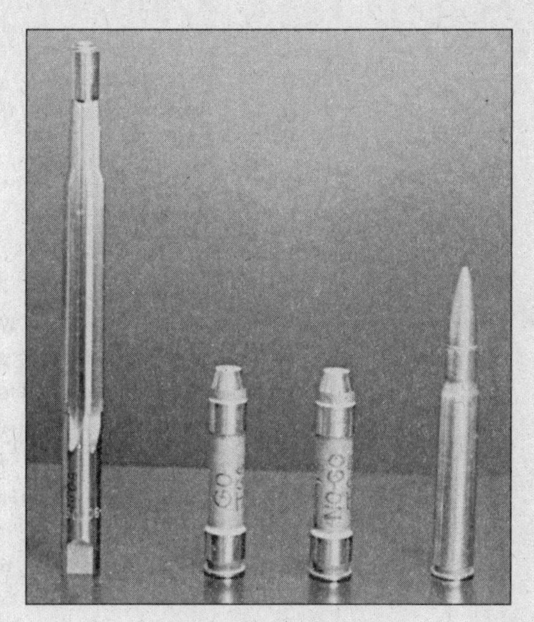

To cut a rifle chamber, these tools are essential: piloted reamer (left), Chamber GO and NO-GO guages (center), and a sample case (.30-06 shown here on right).

- The .303 Savage uses a .308 diameter bullet, while the .303 British uses a .311 diameter bullet.
- The .256 (6,5) Newton uses a .264 diameter bullet, which measures 6.7 mm in diameter.
- In U.S. nomenclature, .40 can indicate anything from .403 diameter to .424, while .44 can mean anything from .422 to .450 in diameter.

There are no absolutes in cartridge nomenclature. These examples are only part of the mystery. Metric and British ammunition compound the confusion. Delving into cartridge nomenclature is much like the first day on a new job—it seems impossible that you could ever memorize everyone's name, but you will in time. The same applies to handloading.

This discussion of cartridge names requires one more clarification. In referring to the group called "Names at Random." this may be an accurate description in terms of "Bee" and "Hornet" but such is not always the case. Often the tags applied to such cartridges have a legitimate, if cloudy, validity. When undertaking the fabrication of any cartridge for the first time, it is strongly rec-

A&M Atkinson & Marquart, Arizona gunsmiths.
Ackley Parker O. Ackley, gunsmith and experimenter.
AMP Auto Mag. Pistol.
AR Auto Rim (rimmed version of automatic for use in revolvers).
Auto Implies use in a semi or full-automatic weapon.
B&D Bain & Davis, gunsmiths.
Barnes Frank C. Barnes, experimenter and writer.
Belted Belted base case (i.e. 7mm Remington Magnum).
Bergmann Theodore Bergmann, weapon & ammunition designer.
Borchardt Hugo Borchardt, weapon designer.
BR Benchrest.
Donaldson Harvey Donaldson, ammunition experimenter.
DWM Deutsche Waffen und Munitionsfabriken.Extra heavy walled cases
Everlasting (for longer life in reloading).
Express Magnum (-ish), special design.
Flanged Rimmed (e.g., .30-30 WCF).
H&H Holland & Holland, British weapon manufacturers.
Harvey Jim Harvey, Connecticut ammunition experimenter.
Herrett Steve Herrett, stockmaker, experimenter.
Johnson Johnson—Melvin M. Johnson, marine weapon designer.
Kilbourne Kilbourne—Lysle Kilbourne, experimenter.
Kynoch Kynoch—British ammunition manufacturer.
Maynard Maynard—Maynard Co., weapon manufacturer (circa 1882).
MAS Manufacture d' Arms de St. Etienne (French arsenal).

MMJ	See Johnson.
Newton	Charles Newton, weapon designer & manufacturer (circa 1915).
Nitro Express	Generally indicates smokeless powder is used.
Norma	A.B. Norma Projektilfabrik, Swedish munitions manufacturer.
OKH	Charles O'Neil, Elmer Keith, Don Hopkins.
Page	Warren Page, experimenter, shooter, writer.
Peabody	H.L. Peabody, weapon designer (Boston, Mass.).
PPC	Pindell Palm Corporation.
RCBS	Fred Huntington, Huntington Die Specialties. (founder of RCBS; also Rockchucker & Super RC).
Rigby	J. Rigby & Co., English weapon manufacturer.
Roberts	Ned H. Roberts, ballistic experimenter (circa 1925).
Sharps	Sharps Rifle Manufacturing Company (circa 1870).
Springfield	Springfield Armory (Springfield, Mass.).
SS	Single Shot.
S&W	Smith & Wesson.
Vom Hofe	E.A. Vom Hofe, Austrian weapon manufacturer.
WCF	Winchester Centerfire.
Webley	British Arms Manufacturer.
Wesson	Frank Wesson, weapon designer & manufacturer (circa 1870).
Whelen	Col. Townsend Whelen, hunter, writer, experimenter.
WSL	Winchester Self-loading.

ommended that you begin with a clean workbench and a note pad. First, select the appropriate starter case and sketch out a rough plan of attack. As each operation is performed, note your success or failure. Keep these notes as clear and concise as possible, including details of any lubrication applied or any extra cleaning or polishing required. Gradually, these notes will grow into a record that includes all the information required to produce that same ammunition months or years later without repeating the negative results of the first attempt. Unfortunately, there is no standard form with which to record this work; all you need, actually, is a yellow legal pad on which you have carefully noted these points:

- Original case
- Trimming
- Precleaning
- F/L sizing
- Annealing
- Fireforming
- Pre-trimming
- Cleaning
- Forming
- Special notes
- Seating
- Lathe work (including tools)
- Crimping

You should also keep careful records on powder, primers, lengths and diameters, along with dates and other pertinent information.

CARTRIDGE:	ORIGINAL CASE:

PRE-TRIMMING:	TRIMMING:
CLEANING:	PRE-CLEANING:
ANNEALING:	FIREFORMING:
F/L SIZING:	FORMING:
SEATING:	CRIMPING:

LATHE WORK (INCLUDING TOOLS):

DIMENSIONAL SKETCH:	SPECIAL NOTES:

2 Hints for the Handloader's Shop

A number of factors should be considered when selecting a location for reloading work. Because you undoubtedly have more than a casual interest in reloading, the kitchen and living room in your home should be ruled out. The most advantageous location is an outbuilding, a place removed from the family living quarters so that your machinery won't bother the rest of the family nor will they disturb your concentration at critical moments in the forming-loading operation.

Basements, utility rooms, spare bedrooms and garages can be satisfactory given some careful thought and preparation. The floor of the workshop should be hardwood, although tile and cement are also acceptable. Wood is preferred, though, because it will not damage tools and dies that may fall accidentally to the floor. Carpet, tile and linoleum are acceptable, but you must use care when handling solvents, for they will destroy these materials. Cement is acceptable, too, but only as a last resort and provided rubber mats are laid down to reduce fatigue. Also, be sure to apply two or three coats of concrete sealer to cut down on dust and reduce clean-up time. Never consider a dirt floor. Moisture, fine dust and grit will inevitably seep into your dies and onto the cutting edges of your tools.

Plan for all the shelving space you can fit into the area, while windows and wall obstructions should be kept to a minimum. Pegboards and wood paneling are ideal for organizing tools. Any space not used for shelving offers good potential for hanging odds and ends. Dedicated handloaders are frugal to the extreme: they save everything! Do not select your workroom site according to the supplies and equipment now in your possession; expect to double your collection as the years pass by.

Ceiling configuration is not important so long as no dirt or dust can settle on tools or cartridge cases just before sizing. If the ceiling of your shop consists of joists and flooring from the room above, you might consider installing a cheap ceiling made of particle board or rigid foam.

Provide the doors and windows of your work room with strong locks. This is especially important in a house filled with small children. Your tools and cartridge paraphernalia will draw kids like steel filings to a magnet, so be sure your shop is safe and secure at all times.

These photos show three sides of the author's prototype ammunition shop. This is the area where experimental work on cartridge conversions takes place

Ammo storage boxes and workbench are shown at upper right.

Side view of reloader and more storage area. A lone window admits plenty of light and air.

Lathe and drill press occupy the west wall .

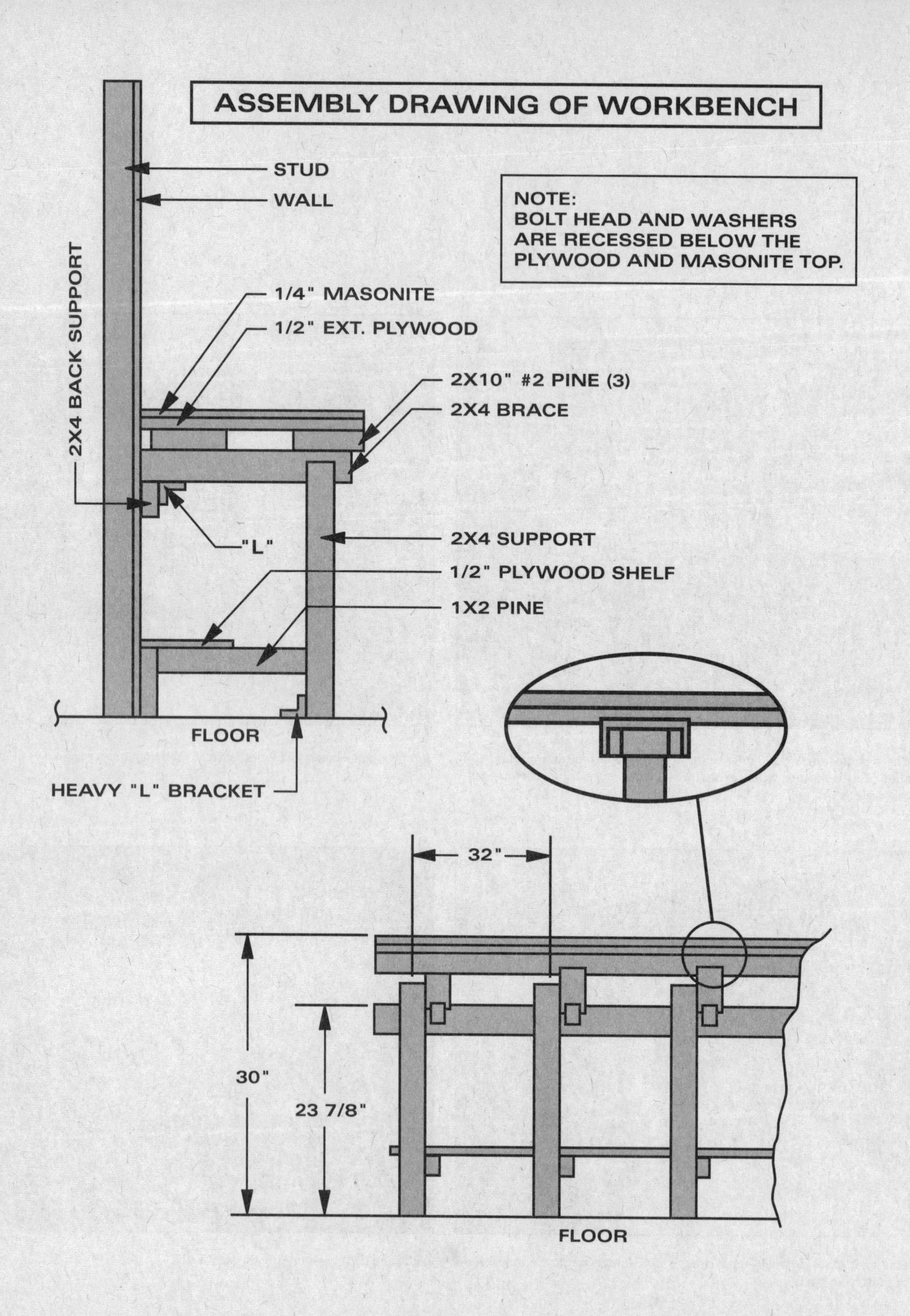
ASSEMBLY DRAWING OF WORKBENCH
STUD
WALL
NOTE:
BOLT HEAD AND WASHERS
ARE RECESSED BELOW THE
PLYWOOD AND MASONITE TOP.
2X4 BACK SUPPORT
1/4" MASONITE
1/2" EXT. PLYWOOD
2X10" #2 PINE (3)
2X4 BRACE
"L"
2X4 SUPPORT
1/2" PLYWOOD SHELF
1X2 PINE
FLOOR
HEAVY "L" BRACKET
32"
30"
23 7/8"
FLOOR

THE WORKBENCH: A HANDLOADER'S HOME AWAY FROM HOME

The handloader's single most important item of furniture is his work bench. One of the most efficient and versatile work bench designs is described and illustrated at left. It mounts to any type wall and may be adjusted to any length and height.

The easiest place to start is the 2x4 back support. Establish the 23.875" dimension and snap a chalk line level with the floor. Locate the wall studs and drill two 1/4-inch holes through the back support at each stud. Anchor the back support using 5-inch lag screws with washers. Cut the 2x4' braces to 27 inches. Position the end of the brace over the back support and attach it using heavy "L" brackets and 1-1/2-inch lags. Use a piece of scrap lumber and a clamp to support the end. Repeat for each brace.

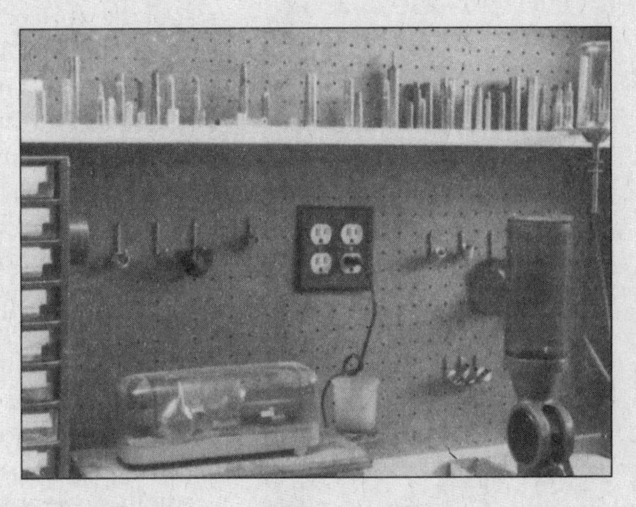

Because many modern handloading tools are electric, several outlets are needed, as shown here.

Next, cut the support legs to 26.5 inches. Position the support to the same side of the brace as the wall stud. Adjust up or down until level and hold in place with a clamp. Drill two 1/4-inch holes through the support and brace and secure with 3 1/4-inch carriage bolts. Repeat for each brace. Lay the three 2x10s in their final position. Snap a chalk line across all three, centered above the braces. Mark two locations for each 2x10 and counterbore a 1-inch diameter hole about 1/2 inch deep. This will recess the washer and bolt head below the 2x10 and provide a flat surface for the plywood. Repeat at each location, dropping a 5 1/2-inch carriage bolt (with washer) through each hole as soon as it is drilled. When all of the bolts are in place, apply washers and nuts and tighten. Cut the 1/2-inch plywood approximately 1/2 inch shorter than the dimensions from the wall to the outer edge of the bench. Sweep the 2 x10s clean and nail the plywood in place using 1 1/2-inch finish nails 6 inches on center. Cut the 1/4 inch Masonite approximately 1/4 inch longer than the dimension from the wall to the outer edge of the plywood. Position the Masonite, smooth side up over the plywood and against the back wall. Nail with panel nails 5 inchews on center. Use a router to dress the masonite flush with the plywood and attach a suitable piece of door molding for a smooth outer edge. Anchor the legs to the floor using masonry anchors, bracket and lags.

This bench will last many years with proper care. When the masonite top gets too scarred and pitted, it may be easily removed and replaced. The primary asset of this bench is its ruggedness: it can withstand any handloading operation.

You may also want to consider building a smaller auxiliary bench for less rigorous work. Using the same construction technique, design this bench according to your needs, such as mounting a vise and for such general tasks as painting, silver soldering, drilling and tapping, assembly work and general tinkering. This secondary bench can be invaluable if only because it enables you to work on other projects while reloading operations are underway.

Beneath this bench you might want to add a shelf made of 1/2-inch plywood and supported by 1x 2 stringers. This shelf extends only 12 inches from the rear and does not interfere with leg room.

The concentrated weight of a largee collection of lead bullets can sag long g shelf sections. A high, narrow shelf module, as illustrated,, will solve that problem..

ADDING THE EXTRAS

No matter how many electric outlets exist or where they are located, you will eventually want to add more. Place them on the wall about 3 feet on center at the rear of the work benches. A few under-bench outlets are also convenient for items such as tumblers. Fluorescent lights hung in pairs in 4-foot lengths are ideal for workshop use. Hang as many as necessary to eliminate dark spots in the shop. These fixtures, which hang by chains from ceiling hooks, are easily adjusted to clear shelves and may be relocated in a matter of minutes. A single control switch near the main door offers the most convenient method for controlling these lights.

You cannot do precision work when you are shivering with cold or dripping with sweat. For hot climates, you may want to consider installing a small window air conditioner or dehumidifier. For cold weather, heat can be supplied by the existing system or from a portable electric space heater. You may be doing considerable cleaning with volatile solvents, so the use of wood stoves or kerosene heaters is not advised. You may also want to install a small venting fan. Simply cut a hole high on the wall in one corner of your shop and install an exhaust fan.

Another essential ingredient in your workshop is water, which finds constant application in the more advanced forms of handloading; i.e., quenching after annealing brass, hardening tools, and washing hands before priming. An inexpensive fiberglass mop sink, the kind commonly used in mud and laundry rooms, will work perfectly.

Two other items have proven their usefulness in many handloading shops: a fire extinguisher and an alarm system. Local fire authorities will help you select these items.

SHELVING AND STORAGE

The old adage "A place for everything and everything in its place" certainly holds true for the handloader and his unwieldy assortment of paraphernalia. You probably already had a good supply of tools on hand when you began handloading, and you will definitely add more as time goes on.

Storage areas will always be important, so address this subject at the outset. Boxes for new brass, powder, dies, molds, sizers, top punches, trimmer pilots, lubricants, primers, spent cases, instruments, gauges, ammo boxes, small tools, books, and magazines will demand some of the available space.

The most expensive shelving is made from 1x6 #2 pine. Modular units, each designed to fit a particular wall location, are easy to make and assemble using 1 1/2-inch #10 flat head wood screws. The hangers are 5-foot pieces of 1x2 pine braced as shown to prevent sagging. A similar, but shorter, floor-standing unit may be constructed in the same manner for heavier items. Modular shelf units need not be painted or stained and they can be easily relocated as necessary..

For larger items, consider using metal shelf kits, which are available in most hardware stores. These kits are ready to assemble, are very sturdy, and offer wider shelf space for bulkier items.

If more protected storage areas are needed for your more delicate tools, instruments, and gauges, drawers may be installed under the bench. Especially useful are the compartmented plastic drawer units, which are ideal for storing screws and nuts. These bin-type drawers are also handy for holding sample cartridges, sizer/top-punch combination, and other tools.

This typical shelf module is sturdy and easily moved.

THE LAST WORD

Assuming your work area is complete (benches are in place, electricity installed, lights hung and shelves mounted) it's time to turn your work space into a handloader's shop. Although you may already have many of the tools you'll need, it may be wise not to bolt any pieces of equipment in place immediately, but instead use clamps temporarily. You can then make changes after you've had a chance to work with each tool and are satisfied with the setup. Only when you are assured that each press, powder measure, trimmer, etc., is located exactly where you want it should these items be permanently bolted to the bench.

Tooling

Some handloading presses are quite sophisticated, such as this straight-line seating die made by J. Dewey Gun Co. It is too slow however, for anything but reloading for benchrest shooters.

HANDLOADING PRESSES

The handloading press is the centerpiece of the handloader's bench. There is a variety of presses available but only a few basic types, and each one has its use in custom handloading.

At the bottom of this lengthy list is the simple hand-type tool offered by Lee and other suppliers. These hand-held loading tools were originally designed for the hobby reloader with little or no work space at his disposal. Similar tools are made now for benchrest shooters and others who demand super accuracy. A fine but distinct line separates the two categories. The science of matching a benchrest rifle to a custom, hand-held neck sizer cannot be accomplished with normal tools. Further, these tools are not available in any but common calibers; they cannot reasonably be used in forming (e.g., .243 from .30-06), and they are deathly slow. It is safe to say they are of limited use to the custom loader.

The next step up the ladder is what might be called, for lack of a better term, the "occasional loader's" press. These tools are first-rate for those who reload a few boxes of a few different calibers each year. These presses will never be used for anything more than full-length sizing .300 Winchester cases, nor will filing, reaming, tapping or bullet pulling abuse them. Several forms of press design are included in this category, such as the "O", "T", and "C" frames, plus two notable exceptions—the CO-AX inline press and the Huntington Compact Tool.

Most manufacturers actively address this market, and rightly so, because all average work can easily be accomplished with any representative press. Prime examples are the RCBS Rock Chucker Supreme and Reloader Special-5, the Hornady Lock-N-Load, the Forster CO-AX, the Lyman Crusher II, plus other similar presses manufactured by C-H Tool & Die, Redding Reloading, Lee, Midway and Dillon Precision.

Of all the presses mentioned above, by far the most rigid are those featuring the "O" type frame. In this configuration, the case is retained by a shellholder inside and at the base of the "O" while the die is located on top, outside the "O."The tremendous pressure that is generated in base swaging, for

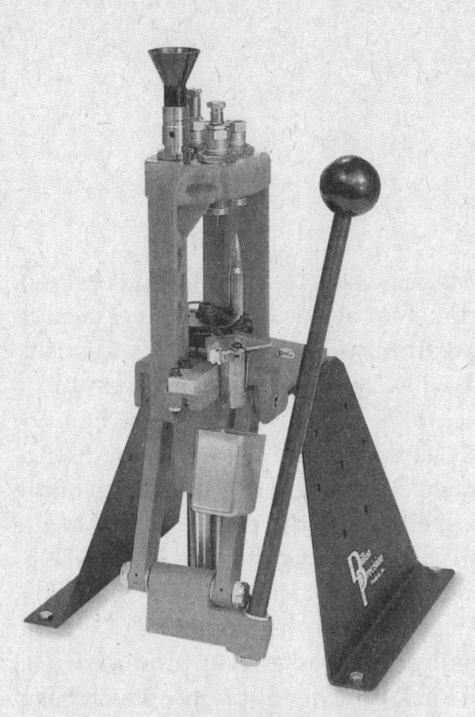

Dillon's Advanced Turret 500 press offers a universal shellplate and interchangeable toolhead. The AT 500 uses standard 7/8 x 14 inch dies and is shown here with a Dillon "strong mount" for bench-top mounting..

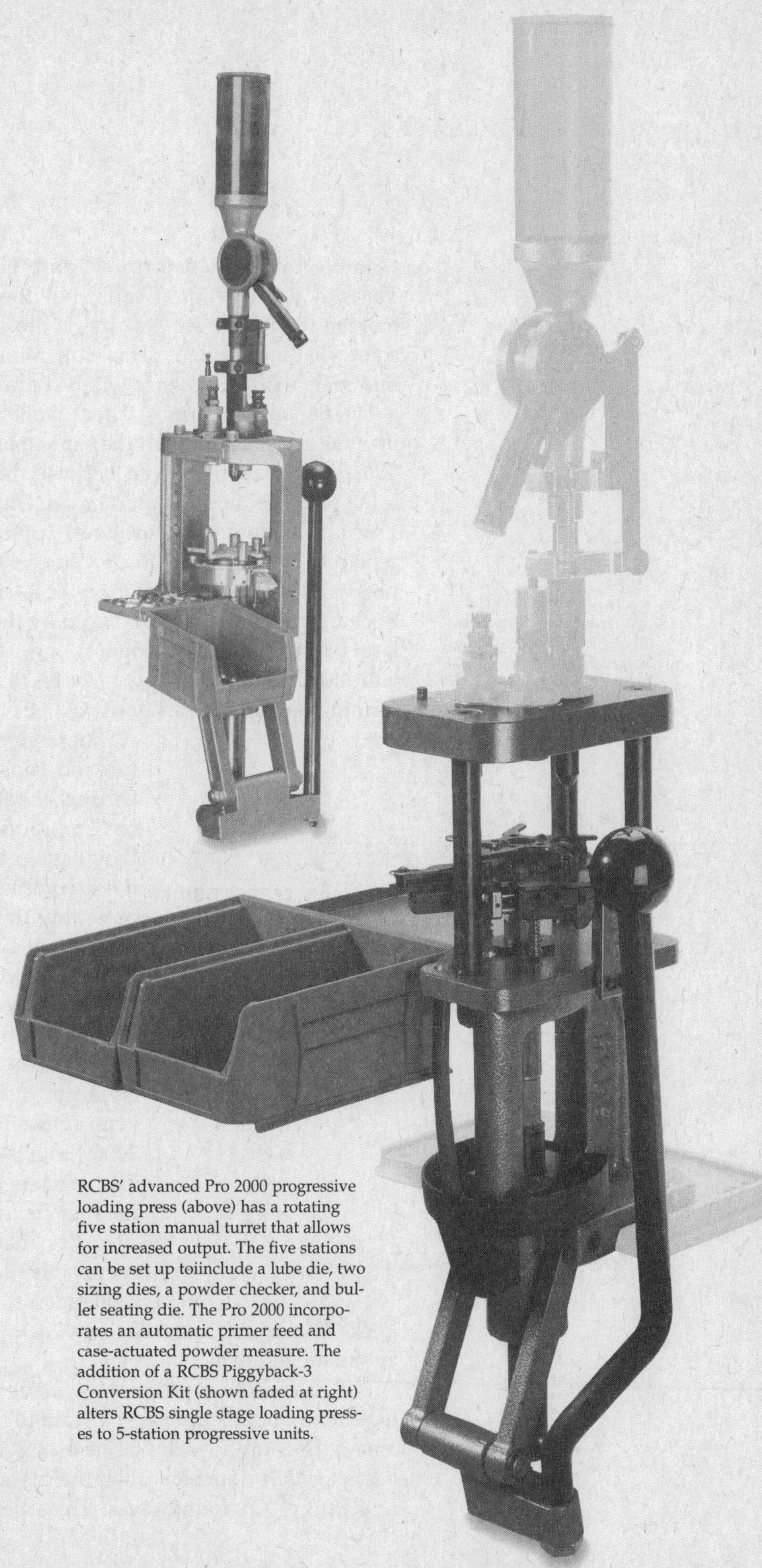

RCBS' advanced Pro 2000 progressive loading press (above) has a rotating five station manual turret that allows for increased output. The five stations can be set up toiinclude a lube die, two sizing dies, a powder checker, and bullet seating die. The Pro 2000 incorporates an automatic primer feed and case-actuated powder measure. The addition of a RCBS Piggyback-3 Conversion Kit (shown faded at right) alters RCBS single stage loading presses to 5-station progressive units.

This CH Tool & Die heavy duty "O" frame press can withstand considerable pressures.

example, can easily deform "T" and "C" type presses (see illustrations below). The only deformation is caused by the inherent elasticity of the steel used in making the frame. An "O" frame press distributes the effect of this elasticity somewhat more evenly. This elongation, when applied with hand pressure, is minimal; and even then it will be parallel to the die/case/ram centerline.

The Bonanza B-1 press, later available as the Forster Co-Ax press, distributes these forces in much the same manner. Some slight characteristics of the "C" frame exist, however, because the shellholder/die combo is not in the same plane as the steel guide rods. Until RCBS released its "Big Max", the B-1 was considered the strongest top-quality press available. The Big Max, replaced by the RCBS Rock Chucker, was a super, heavy-duty "O" frame press and it remains one of the best single stage presses ever produced. The Rock Chucker was later replaced by the Rock Chucker Supreme, which has a larger frame to accommodate long cartridges like those in the Remington Ultra Mag family. This type of press is highly recommended for most of the forming detailed in this book.

These drawings illustrate typical loading press deflections. The "O" frame design consistently yields the most stable and rigid press..

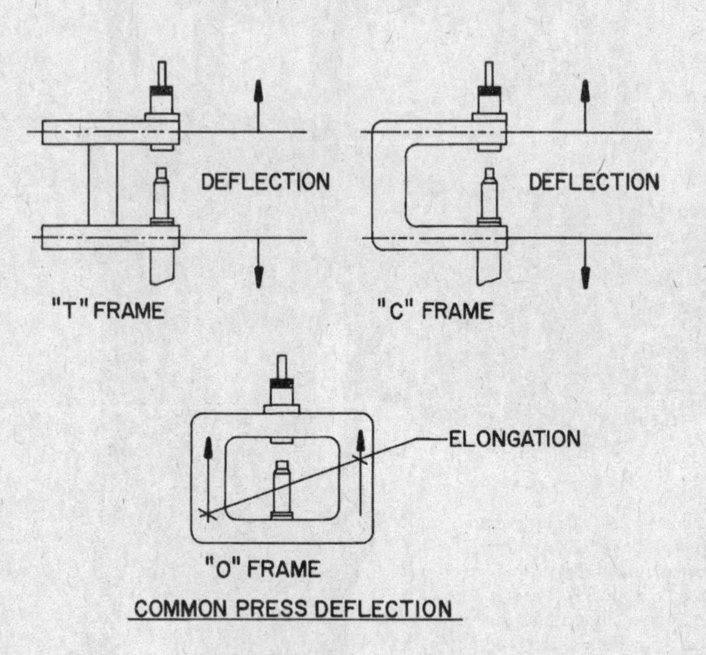

Progressive presses seem to have been around forever, mostly in the shotshell loading genre. During the last few years, many manufacturers have introduced presses for metallic loading, and they have proven useful for repetitive work. You'll not accomplish any heavy forming in a progressive press, but they are ideal for normal sizing and bullet seating. The Dillon RL-550 model has all the features required in a progressive, including a removable tool head, auto powdering, and auto prime (but not auto advance). Lee Precision's Progressive 1000 is also highly recommended, with its auto advance, auto prime, auto powder charges and even a magazine holder for the cases. The RCBS Pro 2000 Progressive features the APS priming system. The primers are in plastic strips that feed into the loading press, not unlike a machine gun belt feed. The dies mount in five top stations and are in a removable block for a quick change without losing adjustments. Indexing is automatic. The Hornady Lock 'n Load AP features their Lock 'n Load quick change die system and automatic indexing.

The reason these progressive loaders are mentioned here is that they do have their place. Suppose you want to manufacture 300 rounds of 8x50R Lebel ammo. Beginning with 300 new .348 WCF cases, you must first reduce the bases to 0.534 diameter. Then the cases are formed, using the 8x50R form die in a heavy "O" frame press. This operation will finish swage the cartridge

Among many long obselete, yet still requested cartridges is the 8 x 50R French Lebel. At left is an original 8 x 50 with a French arsenal headstamp. The center cartridge is of Remington manufacture made around 1930 and the case on the right was formed from Winchester .348 brass.

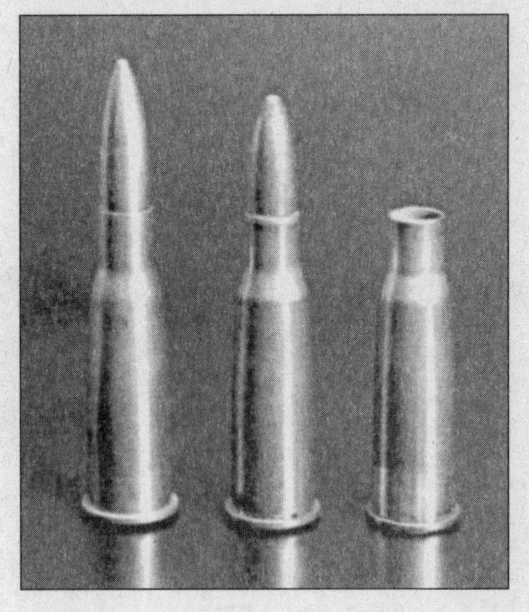

base to assure the extra rigidity required. The next step is to full-length size and neck expand, and the quickest way to do that is to mount both dies in a progressive press.

How many and what kinds of presses to use is a matter of personal preference. A very busy custom loading shop could use, for example, one "Big Max" for forming and swaging; a Pacific Pro-7 for lighter, multiple operations; a Bonanza B-1 for bullet seating and crimping; and an RCBS A3 Rock Chucker for neck expanding, interior dimension (I.D.) neck reaming, and bullet needs. Each has specific capabilities geared to perform certain operations.

POWDER SCALES

In the past, most scales contained both good and bad features. The bad features have mostly been eliminated. Most of the scales available today are essentially the same, although any serious reloader should use an electronic scale. While not necessarily more accurate than a balance beam, electronic scales are much faster for most operations. Here are a few points to consider when choosing a scale:

1. Scale Capacity: Many of the older British and metric cartridges utilize 600 and 700 grain bullets, and you may want to sort these bullets for one reason or another. Select a scale, therefore, with a minimum 1,000-grain capacity.
2. Grain vs. Gram: Virtually all reloading terminology utilizes the grain (437.5 gr = 1 oz). The gram (which equals 15.43 grains) is far too cumbersome.
3. Accuracy: Plus or minus .1 grain when checked against a calibrated standard.
4. Bearings: (Applies to balance beam scales only.) The knife edges that support the beam must contact a super hard material. Agate is in common use because of its low friction and long wearing properties.
5. Dampened: (Applies to balance beam scales only.) To stop the beam from oscillating as quickly as possible, a magnet is used to arrest motion.
6. Ease of Reading: Purchase a scale that can be read without the use of magnification. Graduations or screens should be sharp and clear.
7. Convenient to Use: For handloaders with large, clumsy fingers,a scale requiring delicate moves of a poise member is not advisable.(Applies primarily to balance beam scales.)
8. Base: To help steady the scale, the base should be quite heavy; with three "feet" for easier leveling.
9. Cover: A suitable cover should be provided to protect the scale from dust when not in use.

The RCBS PowderMaster electronic powder dispenser uses infrared light to accurately weigh and calibrate powder measurements. The dispenser delivers charges with ±0.1 grain accuracy.

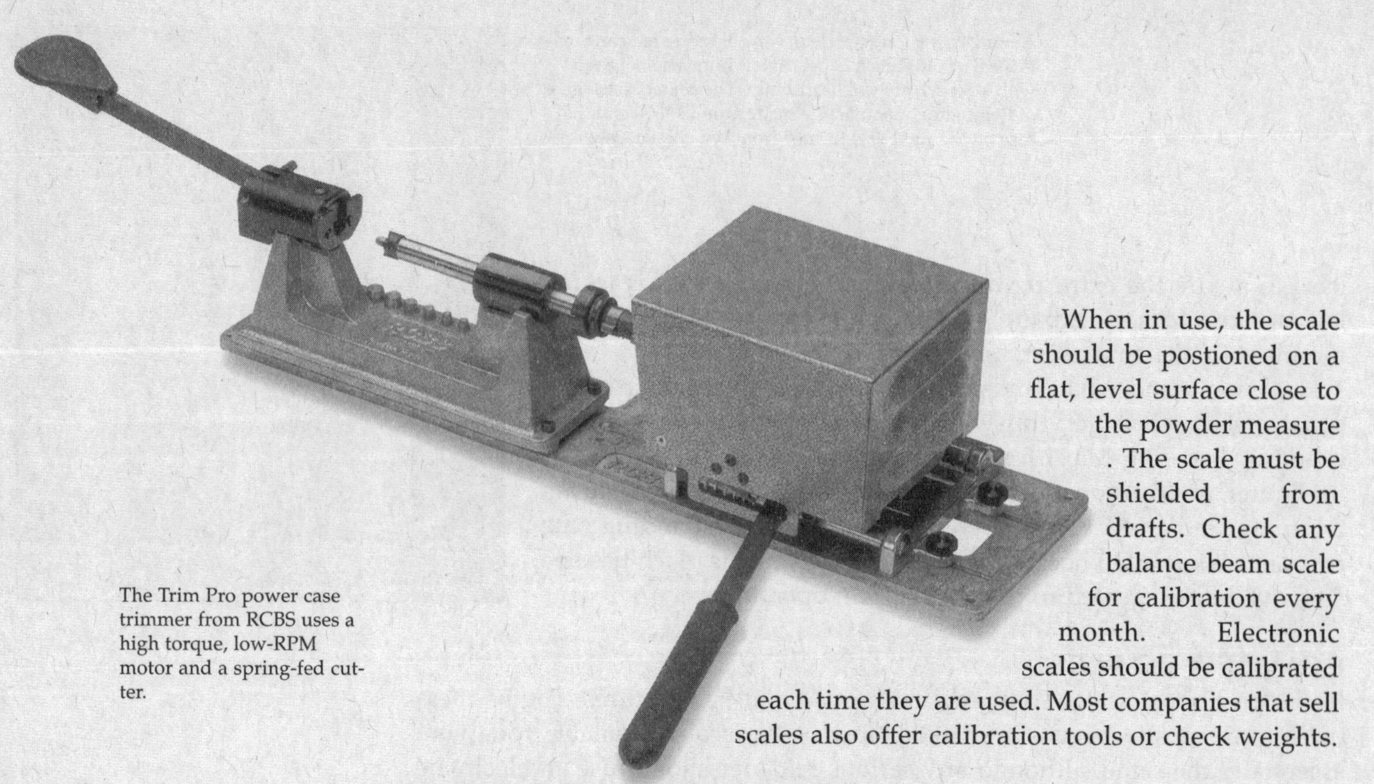

The Trim Pro power case trimmer from RCBS uses a high torque, low-RPM motor and a spring-fed cutter.

When in use, the scale should be postioned on a flat, level surface close to the powder measure . The scale must be shielded from drafts. Check any balance beam scale for calibration every month. Electronic scales should be calibrated each time they are used. Most companies that sell scales also offer calibration tools or check weights.

THE CASE LENGTH TRIMMER

Case trimming is the process by which the case is reduced in length by a few thousandths of an inch up to 50 percent or more. (radical length reduction is more accurately termed "case cutting," which is discussed in Chapter 5). You cannot reasonably contemplate reducing a case from, say, 3.25" to 2.10" with a basic rotary case mouth trimmer.

In the basic hand-held case length trimmer, the case is held in a shellholder, which, in turn, grips the rim of the case. The actual milling trimmer consists of a metal spindle, which projects through the case mouth and bottoms out against a metal piece at the base. Mounted on this spindle is a four-flute cutter, pre-set at the correct length. The reloader then applies rotary force to the cutter with his fingers. These simple hand trimmers are not practical for the custom loader, however, because they are available in only a few common lengths and are not adjustable.

This drill press-type case length trimmer by Forster sits upright, below the drill press spindle. The drill chuck holds the cutter and pilot.

By far the most common type of trimmer utilizes the principles of a miniature lathe. These trimmers mount directly to the workbench, or to a board held in place with a clamp. Each trimmer has its own method of holding the case, either by the extractor groove or by the rim's outside diameter. Each uses interchangeable pilots to correspond to the inside diameter of the case mouth. In use, a collar is adjusted on the trimmer spindle, bottoming out against the tool body at a predetermined distance. Most of these trimmers feature both fine and coarse length adjustments and all are acceptable for removing up to 0.10 inches of brass.

The milling cutters are manufactured from tool steel and are easily replaced when dull. RCBS, Lyman, Forster, Redding, and a few others manufacture lathe type trimmers and many offer motorized versions. A few also offer features such as an attachment for cleaning primer pockets or tools to chamfer case mouths inside and out.

Another trimmer possibility is the type that clamps (or bolts) to the drill table and holds the case as mentioned above, using the drill spindle to rotate the cutter. If you choose this method of trimming, remember to run the drill at the lowest possible speed, and apply a light lubricant to the cutter.

POWDER MEASURES

The rule of thumb for powder measures is to buy the best you can afford, and view with suspicion those that advocate the charging of cases with anything resembling a scoop. This may be acceptable for novice reloaders and under certain adverse survival conditions, but it is fraught with potentially disastrous complications.

Powder measures are designed to throw consistently accurate powder charges. The powder should pour uniformly from measuring cylinder into case, to eliminate the risk of an overload.

Many powder measures are fitted with baffles located just above the powder drum. In theory, these baffles help maintain constant pressure on the powder as it feeds into the charging drum. One device that has been used with success is a small electric motor bolted directly to the powder measure stand. Using a steel pulley with a mounting hole that is slightly off-center, this motor produces a vibrating motion which, in turn, settles the powder and maintains a more constant powder density. The powder measure, by the way, should have its own bench area set aside for optimal convenience. Allow plenty of room for cartridge boards.

Most manufacturers warn against the use of black powder in their powder measures, and for good reason. These tools usually involve metal-to-metal contact between the measure and the charge drum. A spark generated by this contact could produce some potentially lethal fireworks. Black powder handloaders should use a hand measure or the Hornady Lock 'n Load Powder Measure designed for black powder and Pyrodex (and safety).

The neck expander die and plug rod by RCBS is used to expand the neck of a case to permit accurate seating of bullets.

DIES

Dies fall into several categories, notably full-length, form, trim, base swaging, neck sizing and reaming. The same companies that produce loading presses and other major handloading equipment also manufacture dies. For custom loaders who deal in oddball cartridges, a company like Huntington Die Specialties is recommended. Its list of standard custom dies is extensive and even includes chambering reamers for highly obscure wildcat cartridges.

The case pictured here is being filed to proper length in a trim die. The top of the die is specially hardened to resist the action of the file.

In any event, all handloaders require a set of full-length dies for final forming. In many instances, a seater with the same geometry is not needed, but the full-length die is always required for cartridge uniformity and safety. All dies should be kept in their original dust-proof boxes for longer life. Some handloaders build wooden holders so their dies remain in plain view, but this setup encourages dirt to collect in the dies.

A better suggestion: Once you're finished using a die, break it down into its component parts, clean each part with acetone, and then dip the parts in lacquer thinner diluted with a light (10 percent)) mineral oil solution. The solvent will evaporate, leaving a fine film of oil. Be sure to wipe off the oil before reusing the die.

Forming dies are used to compress brass cases to smaller dimensions. Most forming takes place in the neck area of the case. In a typical form set, each individual die is stamped to note its place in the forming process; i.e., Form #1, Form #2, Form #3, etc. These dies are almost always used in the order noted, or else the case will collapse. Individual form dies from any given set can be used on different cases for many intermediate forming applications. Consider, for example, the Jurras/Contender series of cartridges formed from .500 Nitro Express brass. The cartridges are all essentially the same, varying only in projectile diameter; i.e., .375, .411, .416, .458, .475, and .510.

A hand tap driver is used to rotate .510″ diameter reamer in a .50-90 Winchester reaming die.

Some die manufacturers will also quote prices on individual forming sets designed for each cartridge. Thus, in forming the smaller .375 case from the .500 N.E. parent, the brass would gradually pass through most of the intermediate diameters. You might, therefore, purchase the full-length die for each cartridge and only the form set for the .375. To make the .416 case, for example, you would use the form dies, in order, until you reached the closest dimension possible to the required .416. As you collect form die sets, you may also find it advantageous to keep a record of each individual die, including the outside diameter of the neck produced.

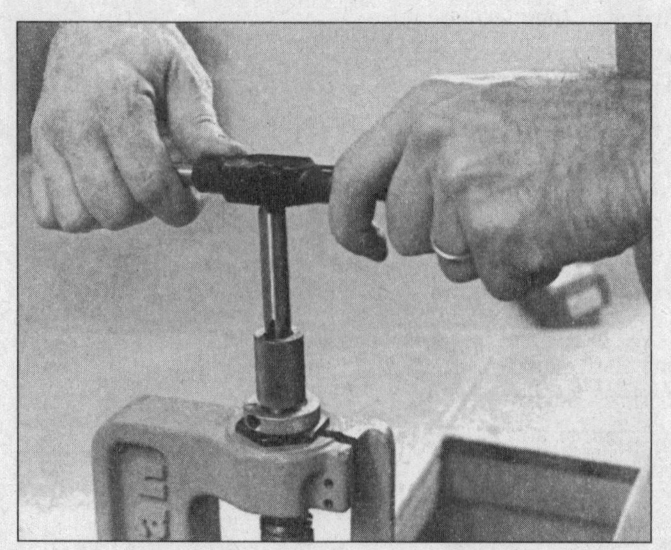

The trim die is often the last die used in a forming process and is often included in the form set. These special dies are produced to the exact length required for the finished cartridge. A trim die, in use, is installed exactly like a full-length or forming die. The case is placed in the shell-holder and pushed up and into the die body. The excess brass, which protrudes from the top of the die, is then removed, first with a fine-toothed (32-pitch) hacksaw blade and then with a file set flush with the tip of the die. Don't worry about damage to the die; the top surface

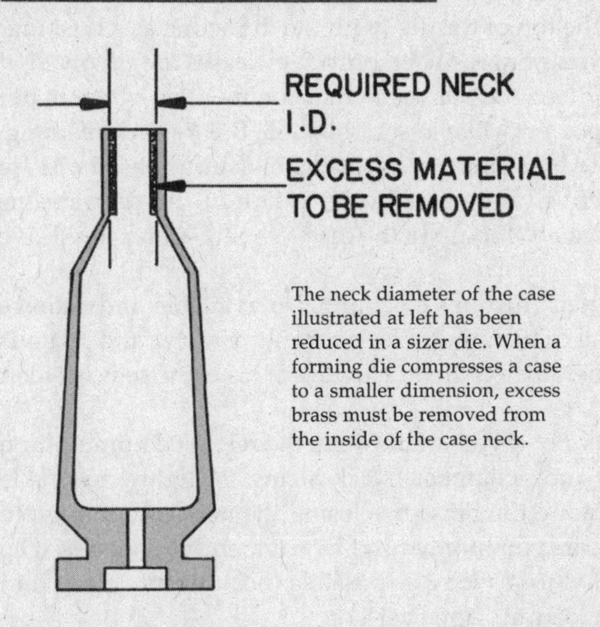

The neck diameter of the case illustrated at left has been reduced in a sizer die. When a forming die compresses a case to a smaller dimension, excess brass must be removed from the inside of the case neck.

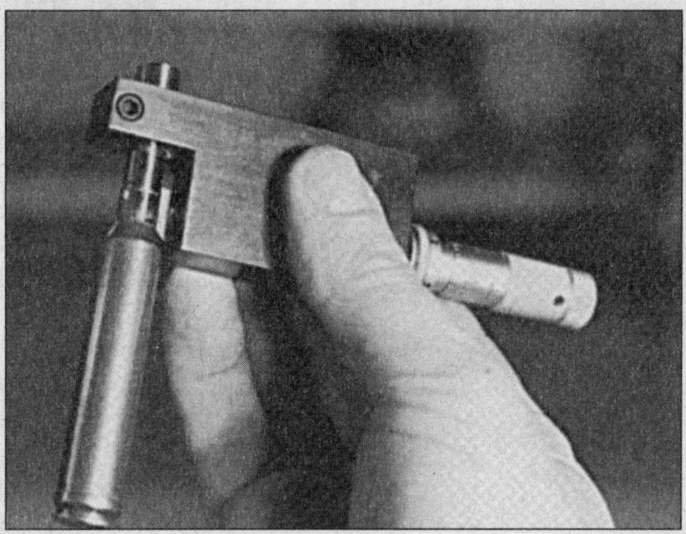

Hand turning the outside neck dimension (top right) with a Marquart turning tool.

The Robert Hart deluxe turning tool (above) is shown at work on the outside of a .300 Winchester case.

is especially hardened to resist the effects of sawing and filing. If scratch marks are noted on this top surface, return the die to the manufacturer for replacement. In most instances, your length trimmer will prove more suitable for this final length adjustment; the trim die is quite useful for hack-sawing large excesses because it acts as a rough gauge. Neck reaming dies are used for removing excess brass from the inside diameter of the case mouth.

Imagine you are forming a batch of .44 AMP brass from surplus .30-06 military brass. This common forming operation utilizes nothing more than the .44 AMP from a .30-06 form die cut to a length of 1.298. The problem is the thicker case wall (roughly 0.022) of the .30-06 body must be re-positioned to the AMP's neck area. Because the walls at the neck of the AMP measure approximately 0.429 in diameter, the outside neck diameter should measure 0.429+(2 x .014) or, 0.457. When the heavier walls are reduced to this outside diameter (0.457), the inside diameter becomes 0.457 –(2 x .0214) or 0.414. Thus, 0.015 (.429 -.414) must be removed from the inside neck diameter of the newly formed case. A .44 AMP neck reaming die, which is included in the form kit (along with a suitable reamer), will do the job easily. Reaming dies are also available, apart from the forming kit, and are also useful as fire-forming chambers (see also Chapter 5).

With the reamer die installed in the press, and the case positioned within the die, the reamer enters from the top of the die (with the die acting as a bushing) and rotates to remove the excess brass. When only a few cases are involved, it may be more convenient to drive the reamer with a common tap wrench, but for continuous operation a power drill is recommended. Before each reaming, lubricate the tip of the reamer by dipping it in cutting oil (sulfur-based oils are messy but work quite effectively). Be certain the inside of the die remains clean, for a brass chip could easily end up in the case's shoulder area and leave a sizable dent.

Various types of outside neck turning tools are also available and achieve the same results as inside neck reaming. Neck turning is slow and tedious work, but the results are generally excellent. However, this exercise is best left to the benchrest crowd.

Neck sizer dies are useful only if you are positive the reloaded ammunition will be fired again in the same chamber. Neck sizing is highly advisable because it prevents overworking the brass by closing the neck just enough to secure the bullet. These dies are commonly used by competitive shooters who wish to utilize a batch of brass for as long as possible (neck sizers do not find much application in custom loading, however).

Special dies appear in a number of forms. The most common—called carbide sizers—are used to size (full-length) straight-walled cases (usually pistol cases such as .357 Magnum, .44 Magnum, etc.). Carbide dies eliminate the need for lubing the cases before sizing but the cases must be kept as clean as possible. Carbide dies, which are actually rings of carbide pressed into regular die bodies, are also useful in high-volume, progressive loading presses. They are not widely utilized for bottle necked cases or normal reloading, however, because of their higher cost. The carbide ring used in these dies is one of the hardest man-made substances available, but it is very brittle.

Another type of special die, or series of dies, is used to swage the base of a case to a smaller diameter. For example, there is a set for swaging .308 Winchester brass to a base diameter of 0.442". These cases are then used to form 7.62 x 39 M43 Russian brass. This swaging takes four separate steps to complete and requires a strong press, such as the Big Max. When fitted with special accessory handles, this press can perform the operation with ease.

RCBS and other offer custom dies used to swage-form a magnum-type belt on an otherwise beltless case. An easier method, however, is to lathe turn (from solid brass or silver solder) a ring in place, turning it to the required belt dimensions.

In any event, reloading dies are the most important tools a cartridge converter can own. Other tools may be substituted to achieve passing results, but this is seldom true with dies.

RCBS Carbide Pistol Dies. The smooth, diamond-lapped tungsten-carbide inner ring eliminates the need to lubricate cases.

RCBS Case Neck Turning Accessories.

RCBS X-Dies. A precision mandrel limits growth of the case eliminating the need to trim after each sizing.

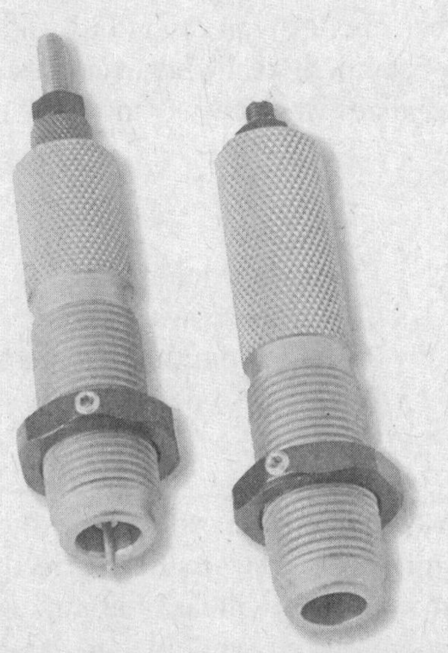

RCBS Precision Dies. Basic die sets for reloading most cartridge types.

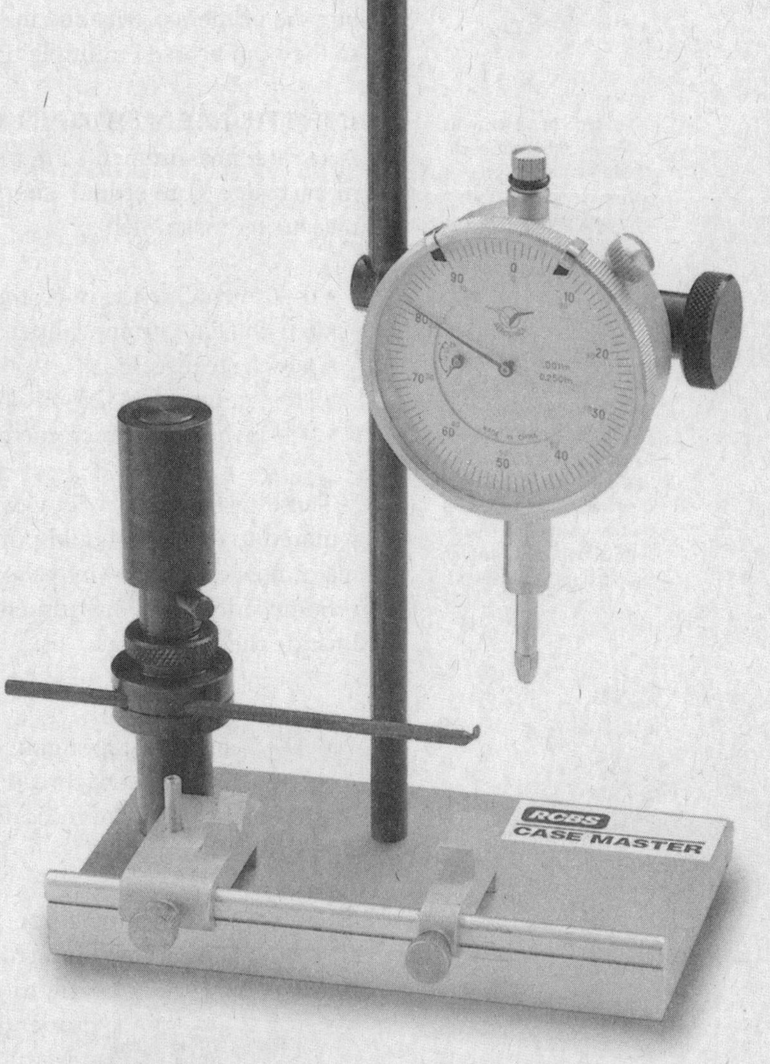

RCBS CaseMaster Gauguing Tool. Measures several important cartridge and case dimensions.

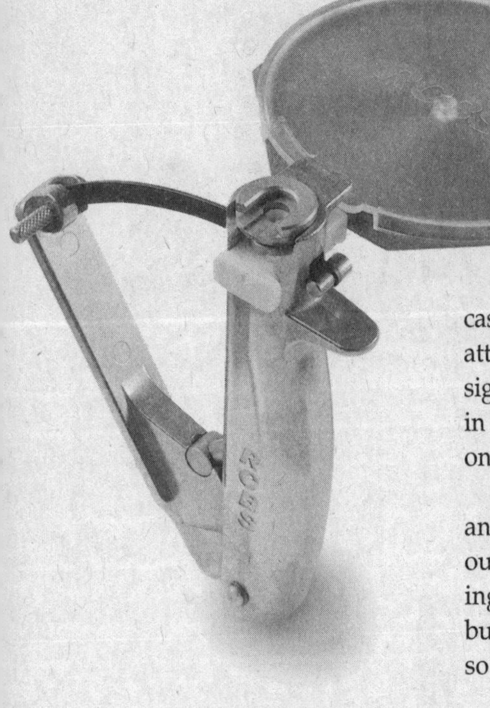

The RCBS hand-priming tool uses the same shell holders as RCBS presses. A safety device separates the seating operation from the primer tray to eliminate the risk of tray detonation.

PRIMING TOOLS

It should come as no surprise to learn that primers must be inserted in the case before you can do much shooting! Simple hand priming tools exist, along with attachments for most loading presses, to do the job reasonably well, but there is a significant drawback: the primers must be picked up with the fingers and inserted in the priming tool. In so doing, oily and sweaty fingers can deposit contaminants on the prime and render it useless, especially after long periods of storage.

In recent years there have been many new priming tools introduced. Hornady and Lee both have hand-held tools with primer reservoirs that feed the tool without the need to touch the primers. RCBS has hand held and bench mounted priming tools that use their strip fed APS (Automatic Priming System) strips. You can buy the primers pre-loaded in strips or they also offer a tool to recharge the strips so they can be used multiple times.

INSTRUMENTS AND GAUGES

Accurate measurements are important in handloading and this is especially so in cartridge converting. Every handloader should have the following basic measuring instruments:

- 0—1 inch Outside micrometer graduated to 0.0001 inch.
- 6 inch Dial vernier caliper graduated to 0.001 inch.
- 6 inch steel scale
- Hole gauge set (.125- to .500-inch range).
- 0—-inch tubing micrometer graduated to 0.0001 inch.

Dial calipers, produced in stainless steel or plastic; incorporate an easy-to-read dial gauge with .001-inch graduations for case and bullet measurements. Calipers measure four dimensions -- outer, iinner, depth, and step..

These instruments will accomplish most of the measuring you will ever be required to do. You should purchase the best models you can afford, and plastic instruments should be bypassed in favor of stainless steel whenever possible.

Some additional instruments for specialized measuring and tool making include the following:

- .1—.5-inch I.D. taper gauge
- English and metric thread gauges
- Magnetic base dial indicator
- Anvil micrometer
- Precision machine level
- Cast iron surface plate
- 0—6-inch depth micrometer
- Blade micrometer

All of these tools are used, one way or another, in the manufacture of both common and exotic cartridge cases. Learn to use each of these instruments correctly, take proper care of them, and they will last a lifetime.

Many other gauges are available to assist the handloader. While none of them are essential, they can often prove helpful under certain conditions and circumstances. A partial list includes the following:

- Primer pocket depth gauge
- Bullet roundness (concentricity) gauge
- Cartridge run-out gauge
- Cartridge head-squareness gauge

Many more gauges and instruments exist, but most are geared to the needs of the benchrest shooter who may spend hours on each box of ammunition. If you have a strong interest in increased accuracy, check the manufacturers and magazines listed in the back of this book.

The RCBS automatic bench priming tool accurately feeds primers through an auto primer feed tube one by one. A single-stage lever system permits sensitive and precise primer seating.

HAND TOOLS

The following partial listing of common, everyday hand tools should be on hand before you attempt the conversions included in Part II of this book:

- Vise
- Tack hammer
- Numbered drill set
- Letter drill set
- Blade screwdriver set
- Tap set (2-56 to 9/16-18)
- Number stamp set
- Tap handles
- Oxy/gas torch
- Assorted open-end wrenches
- "V" block set
- Reamers
- Countersinks
- Jeweler's screwdrivers
- Small anvil
- Hacksaw
- Electric hand drill
- Fractional drill set
- Allen wrench set
- Phillips screwdriver set
- File assortment
- Letter stamp set
- Nail hammer
- Propane torch
- Assorted pliers
- Pocket scriber
- 6-inch square
- Counterbores
- Bench grinder
- Arbor press

POWER TOOLS

For the advanced handloader, the acquisition of a floor or bench power drill press and a lathe is highly recommended. Buy the best drill press you can afford, preferably one with a 0—1/2-inch chuck, variable speed drive (4 or 5 speeds will do), and a reversing switch. If possible, add a small, two-direction (axis) milling table, along with an adjustable drill press vice. This set-up allows fine to medium milling in brass and aluminum, plus the fabricating of many tools and other items that might otherwise have to be purchased.

Locate the drill press in a convenient, out-of-the-way corner of your workroom if possible. If you choose a bench-mounted press, bolt the press to the bench and then bolt the bench to the floor. This is especially important if you intend to mill, where absolute rigidity is a must.

As you read through the cartridge conversions contained in this book (see Part II) you will find many references to lathe work. Phrases such as "Turn the rim to...." or, "Turn off the belt....," or, "Make from solid brass," all indicate the use of a lathe. If you are to follow the case preparation instructions accurately, you will need a lathe or have access to one. You might be able to thin the rims of .45 Auto Rimmed brass with a file in order to produce an emergency supply of .45 Webley ammo, but it would be like asking a carpenter to whittle out a Chippendale sideboard from an oak tree—possible, but excruciatingly difficult. You'll find that most case modifying operations require some degree of concentricity not commonly attainable by hand.

Among the best lathes available for case conversions is the Unimat lathe and its wide range of accessories. Marketed in the U.S. by Hobby Products Corporation, this unit is available as part of Hobby's EMCO Compact 5 system. The EMCO line of machine tools (manufactured in Austria) includes the Unimat and the Compact 5, which features a 14-inch swing between centers and requires only 20x40 inches of bench space. Many of the illustrations in this book show conversion work being done with the Compact 5.

In addition to the usual lathe functions of turning and facing, this lathe will thread (inch and metric), bore, taper turn, and drill. It can also be used, with the addition of certain accessories, as a milling machine. For those who plan to do a

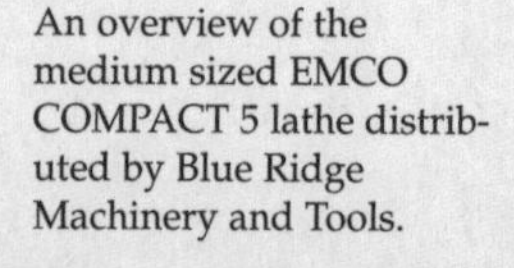

An overview of the medium sized EMCO COMPACT 5 lathe distributed by Blue Ridge Machinery and Tools.

lot of cartridge conversion work, or who simply want a medium-sized lathe, the Compact 5 should be seriously considered.

Basic Cartridge Identification

CALIBER DETERMINATION

Imagine this scene at your local gunshop: "I'd like some bullets for my gun," says the customer. "Yes, sir," replies the owner, "and what caliber and bullet weight would you be interested in?" "What do you mean?" exclaims the customer, assuming his needs must be obvious from the lever rifle he is holding. "I mean," calmly explains the custom loader, "do you want bullets, or do you want loaded cartridges?" "Bullets, cartridges, whatever you want to call 'em. I need them for this here rifle!" The loader, who is beginning to get the picture, then asks, "What cartridge is your rifle chambered for, sir?" "Damned if I know! Bought her this morning at a yard sale!"

That may sound like an improbable scenario, but it has occurred many times at one gunshop or another all over the country. Too often, gun owners who seek ammo for their weapons do not have the slightest idea what cartridges are required. Granted, it is often possible to determine the chambering from markings on the weapon, but many weapons have been produced in this country and abroad for which ammo has long since disappeared. Such firearms must be rechambered, rebored or otherwise altered to accept more common ammo. In other instances, chambers are reworked to accept their owners' own wildcat designs. It is often necessary to start with a firearm of unknown caliber and determine what cartridge can be safely used, a task best accomplished by using two simple gunsmithing techniques: bore slugging and chamber casting.

BORE SLUGGING

Slugging the bore of a firearm is a highly accurate method of determining what diameter bullet is required for a specific rifle. Knowing the bullet diameter will help determine final cartridge dimensions.

To start this operation, select a lead bullet with a diameter slightly larger than the bore. To determine the bore diameter at the muzzle, use the inside diameter measuring fingers of a dial vernier caliper. Be certain that both fingers contact the larger diameter.

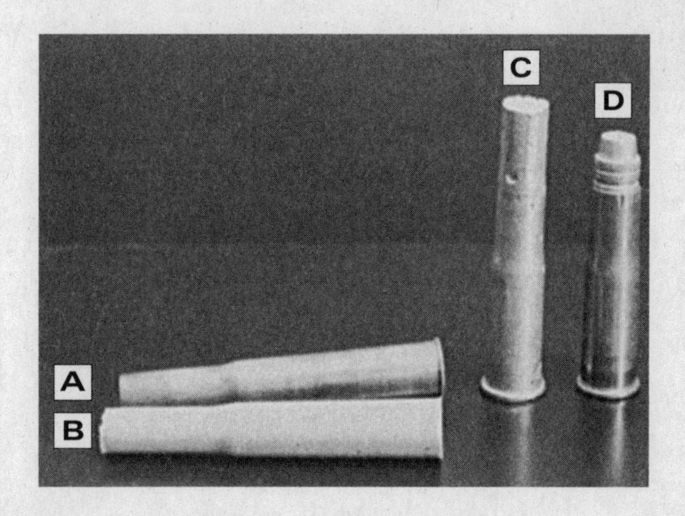

CHAMBER CASTINGS AND CASES.
.450 Nitro Express case: (a) and acrylic plastic chamber casting (b). Unknown Sharps or Winchester: Alloy 158 chamber casting (c) and cartridge formed from chamber-cast dimensions (d) .

Next, select a bullet mold that will cast a slug slightly larger in diameter than the bore. If the bore measures, say, 0.446 inches, then any .45 caliber mold (0.451 to 0.458) will do the trick. A long rifle slug is easier to measure, by the way, than a shorter pistol bullet.

Now, cast a couple of bullets from pure, soft lead. Or, use a small section of brass tubing (lighly oiled for ease of release) as a mold and melt the proper amount of lead with a propane torch. If necessary, turn this slug in your lathe to the necessary oversize diameter.

The slug may be driven through the bore in either direction. While it is usually easier to start in the chamber and drive toward the muzzle, this may be difficult in the case of lever action carbines, semi-autos, etc. Before slugging, push an oiled patch through the bore a few times to add lubrication. Start the slug with a plastic (or phenolic) mallet and continue with a stiff, non-aluminum cleaning rod once the bullet has entered the bore completely. Drive gently and completely through the length of the bore, paying close attention to the force required to move the slug along. Should you encounter greater resistance in some sections than in others, you can assume the bore has constrictions and swells caused by any number of past events and conditions. Once the slug is out, it can be used to determine the nominal bore diameter and the caliber bullet needed for average shooting. Keep the lead slug (along with the chamber cast) in a jar or bag with notations for future reference.

CHAMBER CASTING

Bore diameter may not always be enough to allow a positive caliber identificaiton. There are at least 75 different cartridges that show a 0.308 diameter bore, and yet no two are interchangeable! Therefore, a chamber cast must be prepared for final identification.

Chamber casts have been made from lead, linotype metal, sulfur and plastic compounds. Lead and linotype have unreliable—and usually unknown—shrinkage characteristics. Sulfur is somewhat dangerous and messy to melt and the casts are brittle. Plastic compounds tend to warp. The best material for

The RCBS Precision Mic micrometer measures chamber headspace and bullet seating depth to within .001 of an inch. The Precision Mic measures from the case shoulder datum point to the base.

chamber casting is low temperature melting point alloys, called fusible alloys, and the best of these is cerrosafe, specifically a variation called Alloy 158. This inexpensive bismuth alloy (available from Small Parts, Inc. at www.smallparts.com and other suppliers) melts at 158° Fahrenheit, well below the boiling point of water, and it is reusable.

To prepare the chamber cast, the rifle must be held in a vise or some other suitable stabilizing device with the chamber in a vertical position, muzzle down. Clean the chamber—first with a solvent and brush, and then follow with a light coating of gun oil. Use a cleaning rod to position a piece of cotton or other fabric approximately one inch below the chamber. This will act as a dam to hold back the casting material.

Using a hair dryer, direct hot air onto the barrel in the chamber area. Meanwhile, melt the alloy in a water bath (do not heat the fusible alloy directly or the metal will be destroyed). Once the chamber area is warm and the alloy has melted, the alloy can be poured directly into the chamber, filling it to the top. Do not add any excess as it will only have to be removed later. Now wait about one hour until the cast has cooled to the ambient temperature. Most fusible alloys expand slightly as they cool. This expansion will disappear when the cast cools, but not before. Do not try to remove the cast before one hour is up; otherwise, you may damage the weapon and the cast.

Once cooled, the cast is easily tapped out with a cleaning rod. You now have an exact reproduction of the chamber and should be able to make an accurate identification.

At this point, the reader is directed to the conversion tables in Part II of this book. There you should find the required cartridge. Use the known bullet diameter as a starting point and compare the dimensions of the cast with those listed for each possible cartridge candidate. If you cannot find a cartridge that conforms exactly to your cast, check cartridges with dimensions that are slightly smaller than the cast. If you cannot find your cartridge, consider the possibility of using an existing form die, or fabricate one yourself. This will depend on the quantity of ammo required. As a last resort, you can use your chamber cast as a starting point in negotiations with a custom die maker.

5 Case Conversions

The art and craft of cartridge conversion is a fascinating and challenging business, however, a few practical words of advice are in order. First, after reviewing the conversion for your particular cartridge, be sure to refer back to the text (Part I) to make certain that each step in the process is fully understood. Do you have all the tools, gauges and dies that are required? Nothing is more irritating than to complete a rim turning operation only to discover that you never ordered the smaller shellholder. If special lathe arbors are required, why not build them well ahead of time? Proceed through each operation slowly and keep meticulous records for future reference. Also, keep the project rifle close at hand to facilitate testing of the conversion as you progress.

Readers who have only recently discovered the joys and challenges of handloading are cautioned not to begin by turning .500 Nitro Express cases from solid brass and making your own swaged bullets. Move ahead to more complicated conversions only after you have become thoroughly familiar with less involved operations.

An easy way to get started with cartridge converting would be to purchase a box of .270 Winchester brass and make up a batch of .30-06 ammo. To do this you must to familiarize yourself with tapered expanders, annealing and length trimming. Once mastered, you'll be proclaimed an ordnance genius as you produce .35 Whelen ammo for your handloading buddies using the same technique.

LENGTH ALTERATIONS

The most common modification involving a cartridge case is length reduction, which can range from only a few thousandths of an inch (called trimming) to the removal of a considerable amount of brass (called cutting).

Trimming is best accomplished with a lathe-type tool. You will need the correct pilot and collet (if one is required), but the rest is easy and quick. As an example, let's assume you are forming 7mm TCU cases from .223 Remington brass. The .223 is simply passed over a .283-inch diameter tapered expander, full-length sized, and then trimmed to 1.760 inches. The

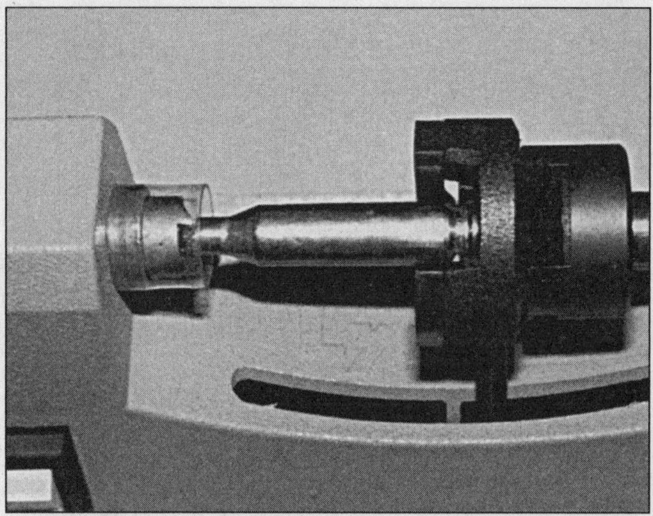

trimming will remove only a few thousandths of an inch of brass and may not remove any in some cases. Should this occur, readjust the cutter, a few thousandths at a time, until a full cut is made on the shortest case of the lot. Now finish trimming the entire batch.

Another simple method of trimming is accomplished with a trim die. When the die is properly adjusted in the press, the parent case will project slightly above the die. The top surface of this die is very hard, so a file should be used to remove the excess (if the amount of brass projecting above the die is considerable, a fine hacksaw should be used, followed by the file). Be sure to keep the inside of the die clean and free of file swarf.

When a large amount of brass must be removed, a different approach is called for. Suppose you want to form 8mm Nambu brass from .30 Remington cases. The Remington case is 1.170 inch or longer than the Nambu, which means there is far too much brass to remove for a trimmer. Because the Remington must be cut off at the thicker portion of the body, the process will be extensive with a trim die. If you want to make 500 rounds, knowing that each case must be inside neck reamed and full-length sized. The only practical answer may be to design your own case cutter.

A case cutter is essentially a lathe with a bed that has been replaced by polished stainless steel guide rods. Upon these two rods ride linear bearings that support an aluminum plate. This plate moves the length of the rods and is controlled by a hand-screw feed. The aluminum plate supports another series of guide rods and linear bearings, which in turn support a motorized spindle. This can best be described as a two-axis, compound table. The long axis moves parallel with the lathe spindle centerline, while the short (top) axis moves perpendicular (see photograph above).

Top left: With the case cutting machine shown here, cases are cut with thin (.060) abrasive disks or even thinner (.025)) jeweler saw blades.

Top right: A Lyman Power Trimmer is shown here trimming a case. The cutter revolves inside a plastic guard, preventing chips from flying up into operators face.

Above: A brass cutting machine. The case is mounted on a spindle, which passes through the center of a large, round, bearing cylinder. The circular cutting blade (center) is driven at high speed (7-9000 rpm).

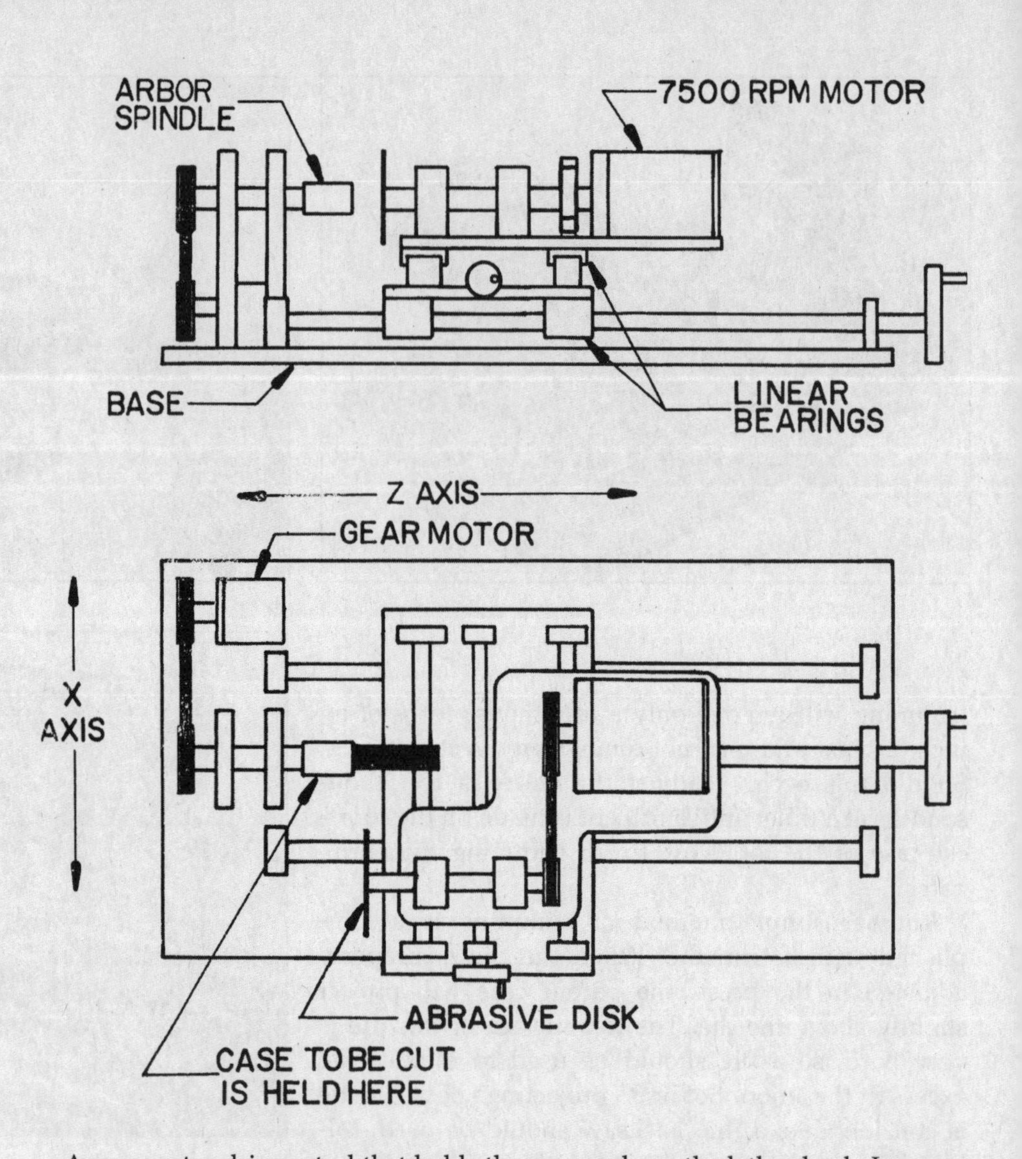

A gear motor-driven stud that holds the case replaces the lathe chuck. In operation, the case revolves clockwise at about 100 rpm. The cutter is an abrasive disk or steel saw blade—2 inches in diameter and .020- to .065- inches thick—revolving counterclockwise at 7,500 rpm. The case length is read directly at the dial indicator, while the in-and-out movement of the abrasive wheel cuts the case. The secret of this method lies in the way in which the case is held for cutting. To begin, the flashhole must be drilled through with a No. 44 drill, allowing a 2-56 threaded stud to pass through it. This drilling also increases the diameter of the hole from 0.080- to 0.086-inches. For drilling, use an electric motor hooked directly to a small Jacobs chuck. The arbor is made from hardened and ground 0-1 tool steel. Its outside diameter slip-fits into the gear motor-driven spindle.

This close-up view of the brass cutting machine (see p.41) shows the arbor and blade in greater detail. The case rotates at about 60 rpm.

After drilling, be sure to chamfer the flashhole, inside and out. This will remove any burrs that might interfere with the key that prevents the case from turning. The machine used in this method of cutting is not important; getting the job done is what matters. Handloaders often need to improvise their own methods.

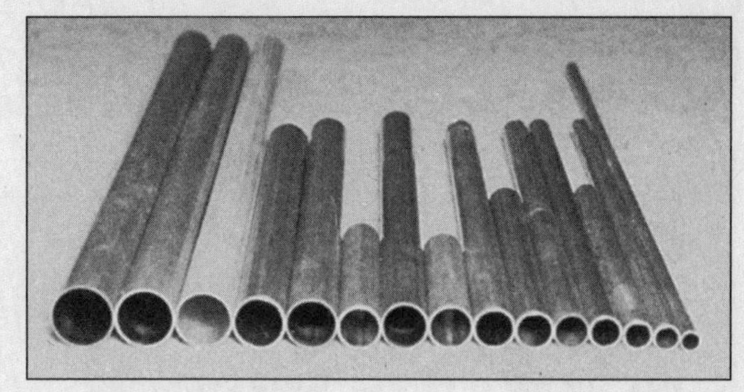

A selection of thin wall (.013") brass tubing. Larger tubing is 12/32"; smallest is .25" O.D.

CASE EXTENSIONS

Often you will find a parent case with the precise head and rim dimensions, but the length is shorter than needed. This should pose no problem as long as the headspace requirements are met.

Suppose you have a rimmed, straight case, such as a .38/90 Winchester. Cases are easily made by taper expanding .30-30 or .32 Winchester Special brass to accept .375-inch diameter bullets. The loaded round is fired in the .38/90 to form a case that is 0.32-inch too short, but otherwise perfect. Before making full-length cases, therefore, it is wise to investigate the use of shorter, readily available brass.

Assume you are in need of a certain amount of brass for use with a .45-120 Borchardt rifle. This can be done quite easily using factory .45-70 Government cases that are run partially into a .41 S&W Magnum carbide sizer die with a base diameter of 0.433-inch. The .45-70 case now sports a neck with an outside diameter of 0.433 inches, but it need not be longer than .375 inch.

The next step is to obtain a quantity of thin wall brass tubing with a thickness of approximately 0.15 inches. Cut the tubing into 1 1/2-inch sections with the ends faced in the lathe for squareness. Using steel wool, carefully clean the neck of the .45-70s as well as the inside of the tubing section. The outside diameter of the neck and the inside of the tubing are now tinned evenly using flux and a special 30,000 psi silver solder. The tube may then be dropped onto the necked case. Re-apply heat from a propane torch until the solder flows, meanwhile holding the head in a heat sink (or water) to keep the base cool.

The silver solder will form a very strong bond, and after cleaning in the lathe with 240 grit silicon carbide paper it will look fine. After fabrication, the cases may be cut to the proper length, full-length sized and then loaded with black powder and paper patched bullets (cases made in this manner may only be used for black powder, Pyrodex and other low pressure loads).

BASE DIAMETER REDUCTION

Forming 8x50R Lebel cases is a prime example of how a lathe can be used in custom loading. The parent case is the .348 Winchester with a base diameter of 0.553 inches. The base diameter of the required Lebel is 0.534 inches, so some brass must be removed. The .348 case is first run into the full-length 8x50R die as far as possible. This partially reduces the body of the case but is not effective on the solid base. Remove the case from the die and turn the base to the required diameter. The cases are held in the lathe much the same as in the cutting machine. Simply place them in the arbor with a three-jaw chuck. Thinning case bases in this manner, by the way, does not weaken the case. Once the case has been run into the full-length die, wherever the case is thin and can adjust inwards, it will. Also, since it cannot be easily swaged at the base, any additional sizing is imped-

Shown at left is an original 8x50R Lebel. Center is a .348 Winchester base that has been turned and reformed. Case at right is the original .348 Winchester.

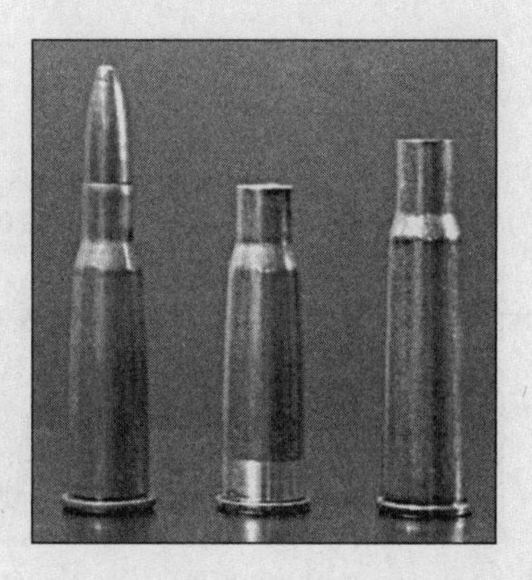

ed. As the case is turned, most of the brass is removed from the heaviest web area.

While it is certainly possible to use high-speed steel tool bits that are individually ground for each job, there is a better method. The answer lies in the use of disposable, indexable carbide inserts, along with the proper tool holders to support them. Kennametal, Inc., lists in its catalog 3/8-inch and 1/2-inch square tool holders and a variety of inserts. Their performance has been thoroughly tested and approved for all operations described in this book. These carbide tools may seem expensive in the beginning, but they will pay for themselves many times over the long haul.

A case (above) is mounted on the arbor prior to base turning. A short screw mandrel is located just inside the case mouth.

CASE SPINNING

Another method for lengthening a case is spinning. A case is placed over a smaller diameter mandrel and rotated at high speed. Pressure is applied to the outside case wall and then towards the case mouth. This method is considered legitimate for forming sheet metal and for industrial applications, but it is of little use to the custom loader. For one thing, only minute increases in length can be accomplished before the remaining walls become too thin. And for another, the typical shop lacks the consistency needed for this method. No two cases are guaranteed to come out alike.

HEAD ALTERATIONS

Turning the base of a case may reduce the head diameter satisfactorily, but often it is necessary to increase this diameter. Here again, thin-walled brass tubing comes into play. Generally, the added ring head need not measure more than 3/8- to 1/2-inch long. All that's necessary is to hold the case in the center of the chamber. After firing, the brass in front of the ring will expand to full chamber dimensions, enough to hold the ring firmly in place. If no standard diameter tubing is available in the exact diameter required, the ring may be carefully soldered to the body with lead/tin solder and then turned to the correct diameter.

A .348 Winchester is lathe turned to the proper diameter after reforming to 8 x 50R Lebel.

RIM DIAMETER

Suppose you want to produce 11mm French Revolver ammo from .44 Magnum brass. You must first reduce the rim diameter of the .44 Magnum from .515 inch to .490. Next, install the large primer pocket arbor and extend the arbor out from the chuck in preparation for turning to diameter and back chamfering. Use a neutral hand tool with a 50° diamond insert.

This drawing (top) illustrates how to increase diameter using the sleeving method. Case will fireform (forward of sleever) to full body diameter. Drawing below represents the setup after fireforming.

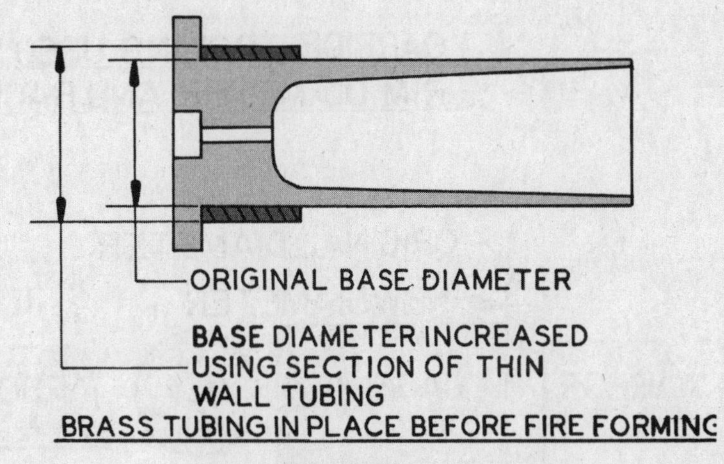

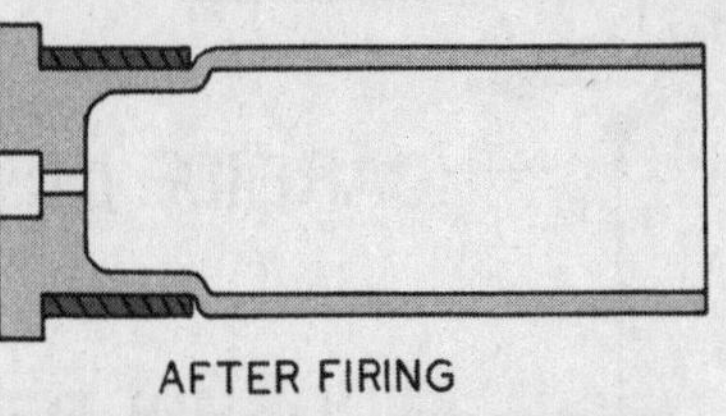

With the case rotating, move the tool toward the chuck, taking a very light cut, and then check the diameter. Adjust the tool for the prescribed diameter (.490 inch) and make another cut. You should arrive at .490 inch with one pass. Once the tool completes the turning operation, it is simply fed toward the arbor to cut the chamfer on the rim.

This method should produce a perfect case and no additional filing or polishing is required.

In some instances, where a good deal of brass must be removed from a case, it is better to use a rough-and-finish cut. Adjust the tool to rough-cut the rim .005- to .010-inch oversize. The tool will loaf through the cut and leave a smooth finish.

Sometimes it is necessary to increase a rim diameter, but this is rare. Normally, a rim that is considerably undersized will work fine so long as it provides some support. To increase the diameter, you must turn a ring with the inside diameter equal to the case rim and an outside diameter somewhat greater than required. This ring is pushed onto the case rim and soldered in place. After soldering, the new rim may be turned and chamfered as described above.

Be careful when soldering or in any way heating the head of a case. The best procedure is to stand the case, with the ring in place, on a hot plate. Turn the heat on and hold the solder in place. The solder will flow when the head reaches the melting point of the solder. At the instant the solder flows, turn off the heat. Heating tends to anneal the brass in the head and can easily render it too soft for safe use. Allow the case to air cool — *Do not quench in water*! Slow air-cooling optimizes the hardness of brass, whereas forced cooling causes it to anneal. This method of increasing rim diameter, by the way, is recommended only for cases that are used for black powder loads.

An extractor groove is shown being re-cut with a carbide form tool. A carbide blank is ground to precise dimensions and angle of groove.

Many other types of rim alterations exist. For example, when making 9mm Browning Long cartridges from .38 Special cases, the rim must first be turned to .404 inches, thus removing most of the rim. Change tools at this point and use a common grooving insert modified to cut a new extractor groove. This insert is not a standard item, so you will have to perform the modification yourself.

The best way to gauge the depth of a new extractor groove is by using a shellholder as a "go" gauge. When the proper shellholder fits, the groove is deep enough. A blade micrometer also comes in handy for this work. If you hap-

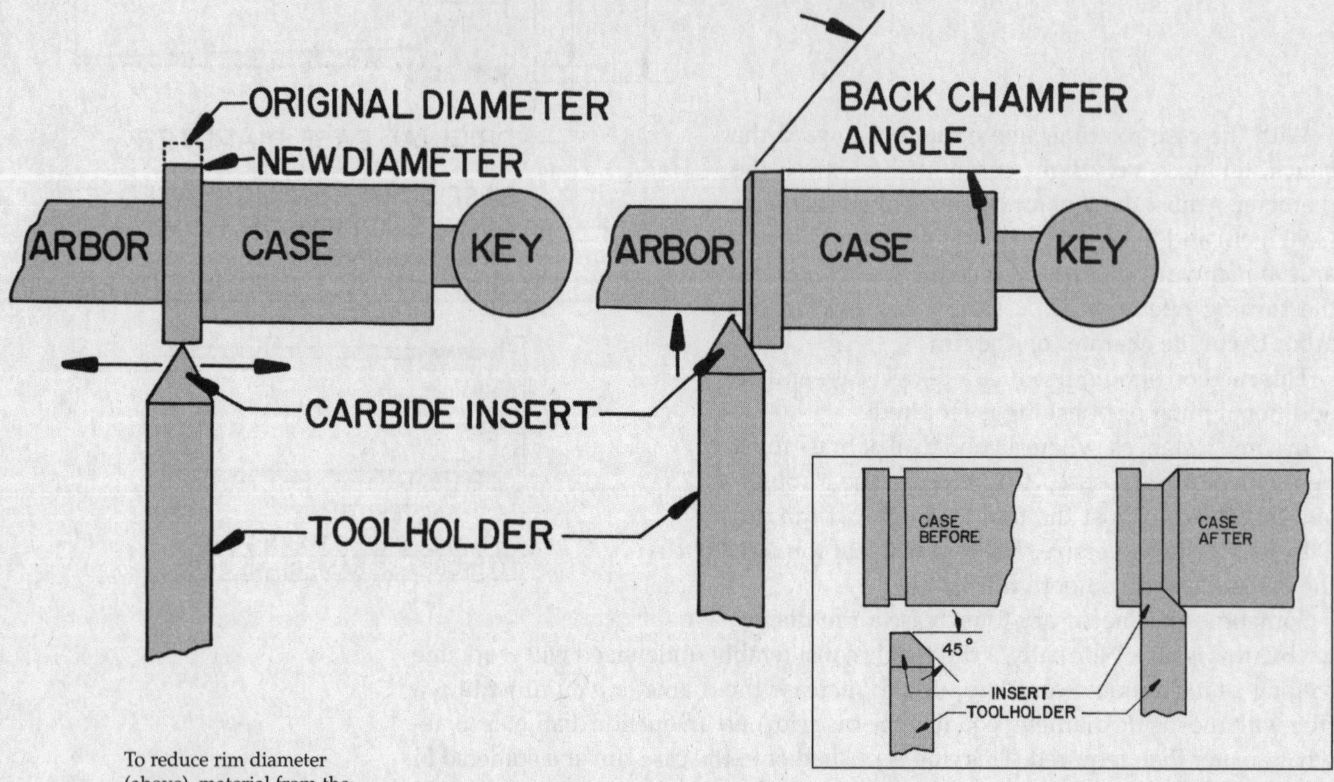

To reduce rim diameter (above), material from the rim's outer diameter is removed. To back-chamfer (inset), the rim is touched with the same tool to create a slightly angled surface.

pen to have the firearm available, it can also be used as a gauge. The extractor groove should be cut no more than is necessary for it to be functional.

RIM THINNING

This operation, commonly referred to as "thinning the rim," uses a similar grooving insert, but requires no alteration. With this method, for example, .45 Webley brass may be converted from .45 Auto rimmed cases. The rim must first be cut to the same depth as the existing extractor groove. Remember to thin the rim by removing brass forward of the case head. Removing material from the head itself will result in shallow primer pockets, which are quite difficult to re-cut.

BELTED CASES

Belted cases present their own special problems because the belt must be added or removed. Removing an existing belt is a relatively easy prime lathe operation. For example, .280 Ross Rimless brass is prepared from .300 H&H Magnum brass by holding the magnum case on an arbor (in the lathe) and removing the belt to the head diameter of the .280 Ross (0.522 inch). A small portion of the belt will remain, but the head will be as required for proper chambering. Normal full-length sizing completes this conversion.

Adding a belt is far more difficult. Suppose you want to make several boxes of .240 Belted Rimless Nitro Express brass from .270 Winchester cases on a home-made belt-forming die. The die in this instance conforms to the base and belt diameter only, not the entire cartridge. Here the die is installed in the press and

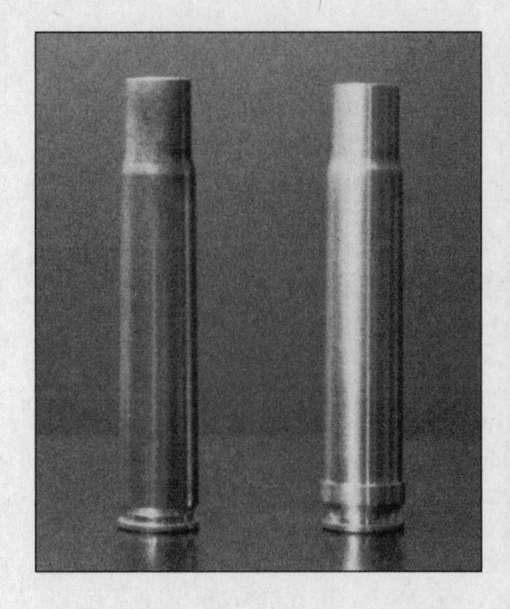

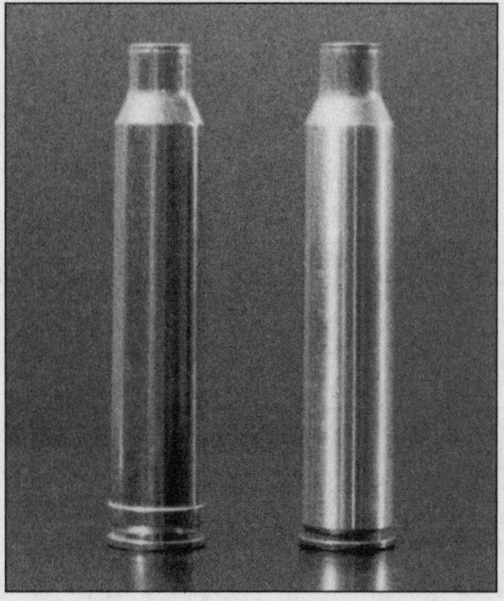

Far left; at left is a .35 Whelen case, and on the right is the same case with a soldered-in-place belt (see text for a full discussion of methods).

Left; on the left is a standard .300 Winchester case; at right is the same case with the belt removed, flush with the rim.

the case carefully lubricated. If an original case is available, the width of the belt can be approximated: keep adjusting the die and swaging the case. The smooth edges of the die will force brass towards the head to form a rough base. After all bases are formed, the belts can be trued for diameter in the lathe. The resulting belt will not be as well formed as a factory case, but it will suffice for headspacing. It may be simpler to solder a ring in place and lathe turn the belt, as described in Chapter 3.

When undergoing any new case forming, it is always preferable to have on hand an original cartridge for reference. To obtain one of the more exotic cartridges, contact one of the better-known cartridge collector/dealers. The few dollars spent for a master cartridge can save a lot of aggravation later on.

NECK ALTERATIONS

Form dies are generally used to form one case into a smaller or shorter one. The die has a chamber, much like the chamber in a firearm, with dimensions that produce the outside dimensions of the newly formed case. All the excess brass that has been squeezed in during the process must then be removed in the neck area before a bullet can be seated. If this excess is not removed, the bullet will expand the neck area and increase the neck diameter. The chamber may not accept this oversized neck, but there is always the chance that the cartridge will be driven completely into the chamber, thus greatly increasing chamber pressures—perhaps dangerously.

The solution is to remove this excess brass in one of two ways. Inside neck reaming is the quickest and most common method using a suitable reaming die and reamer. Bench-type case length trimmers are also used for this purpose, but the results are not quite as good. When reaming with a trimmer, the outside wall of the case remains unsupported and will thus expand slightly—but definitely—as the reamer cuts its way through.

The second method for removing excess brass—outside neck turning—is an outgrowth of that used by benchrest shooters, and while it works effectively, it does take more time. Outside neck turners are available as hand-held tools or as modifications to bench-mounted case length trimmers. None of the procedures in this book require outside neck turning.

To increase rim thickness, solder or cement (with cyanoacrylate) metal or hard plastic rings at low temperature up to existing rim to desired thickness. Here a brass washer has been soldered in place and turned to the proper diameter and thickness.

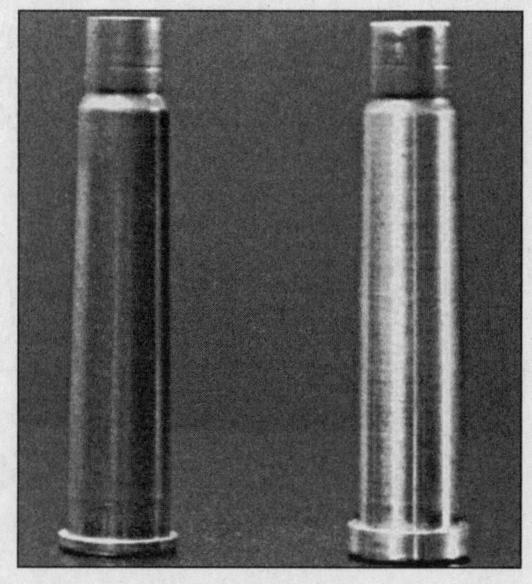

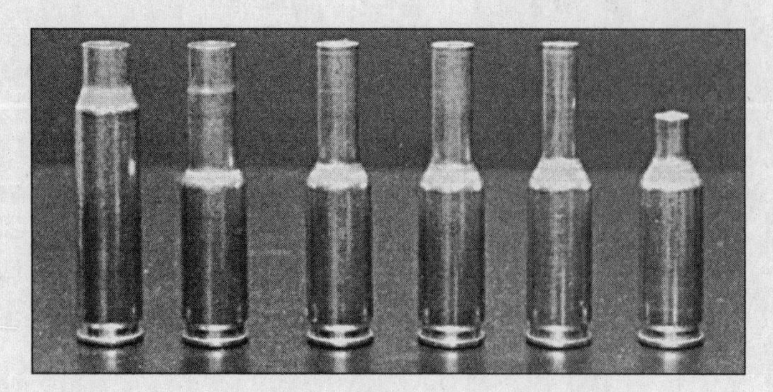

Forming a .22 benchrest case from Remington BR brass, the following stages take place (from left to right): original case; shoulder set back; first neck reduction; second neck reduction; third neck reduction; case is run into the trim die; excess is cut and filed away; I.D. neck is reamed and full-length size (far right).

NECK EXPANDERS

A great many wildcat cartridges are the result of taking one case and necking it up to a new diameter. For example, nothing could be easier than expanding .30-06 necks to .357" -inch diameter to produce .35 Whelen cases. The secret to increasing neck diameter lies in the tapered expander. By trying to force the blunt expander button of the .35 Whelen die into the .30-06 case, you will succeed only in producing a badly bulged, non-concentric neck. The .35 expander is simply too blunt to center itself properly in the neck of the .30-06. What is required is a gradually tapered expander.

Homemade expanders can be made from type 316 stainless steel, 9/16-18 threaded rod. A set of these rods might start at .150 diameter and proceed to .510. The tapered portion of each expander is highly polished to reduce drag, but a few drops of light oil should be applied to the inside of the case neck before expansion. When using this set, expand to the next closest diameter above that of the finished round. The .30-06 case mentioned above would therefore be expanded to .308, .315, .330, .345 and finally .360. The case is now full-length sized in the Whelen die, bringing the inside neck diameter to .357—exactly as it should be. Expanding slightly above the diameter and then backing down helps to form a better neck and shoulder angle on the finished case.

A full set of expanders will easily produce straight cases from .30-40 Krag brass for making .405 Winchester ammo. Nearly any expander die body from a three-die set can be used to hold the homemade expanders. Even larger expanders may be fashioned from 5/8-18 threaded rod and held in the expander die of a .577 Snider. All of the expanders described above are available from any good die manufacturer, but the ability to create special expanders with odd diameters is an important asset for the serious cartridge converter.

ANNEALING

In forming brass cases up or down in caliber, you will soon realize that brass cannot be pushed around easily. Suppose you have a new box of .30-06 brass and want to expand the necks to .37 caliber. You probably won't get much farther than .35 caliber without cracking and splitting the brass. That's because of the crystalline structure of brass, which is not very ductile. This problem is easily resolved, however, by means of annealing, or the heating of a given material—in this case, brass—to facilitate forming.

Some of the methods used in annealing cartridge brass involve the use of lead pots and molten lead, while others advocate placing the cases in a pan of water and applying heat with a torch. A simpler method is to use a Burnz-O-Matic torch, a pan of cold water and

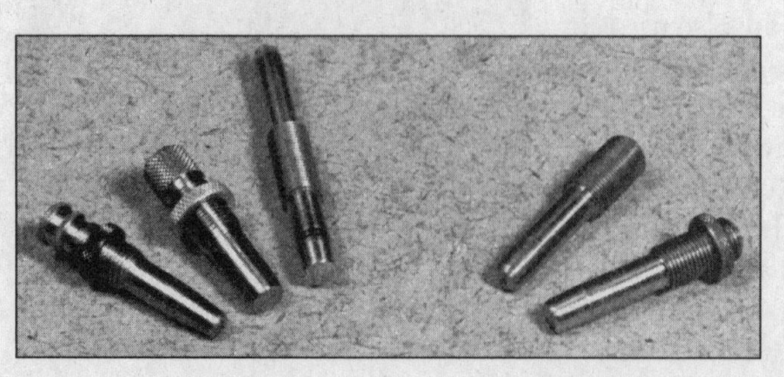

These case neck expanders include both commercial and homemade varieties. With a small lathe and supply of 9/16-18" stainless threaded rods, virtually any expander can be fabricated.

The case on this annealing machine revolves around slowly in a gas flame at about 60 rpm. Case is then knocked into a coffee tin partially filled with water.

common pliers (with a plastic coating over the handles to protect your hands). Adjust the torch until the pencil-point flame is 1 1/2 inches from the tip. Holding the case with the pliers at the base (just below the rim), slowly rotate the case mouth in the flame until it begins to glow red, and then drop it in the cold water.

The pliers used in this process should have a circular, toothed nut gripper located below the flat ends. That's where you should hold the case. The pliers also act as a heat sink to keep the base from becoming too hot. Otherwise, the annealed heads will become much too soft for safe shooting. In cases that are ready to be formed, the brass should be hard at the base, medium hard at the case mid-point and soft at the neck.

Nothing destroys annealing faster than cold working. As the annealed .30-06 is passed over a .30 to .35 expander, the brass will stretch. When a radical change is called for in a case neck—i.e., from .30 to .41 caliber—two annealings may be required, once at .30 caliber and again at .35 caliber. All this is best learned from experience. You may lose a case or two, but by keeping careful notes you should not have that same experience a second time.

It is also possible to anneal at the wrong time. An annealed .30-30 Winchester case, if passed over a .30 to .35 expander, will have a tendency to fold up on itself like an accordion. That's because the brass is too soft in the neck area to support the neck properly when pressure is applied downward.

Never anneal all cases in a batch of brass until you have ascertained the exact method of forming. Some cases may require one neck expansion in the cold state followed by annealing, and then a second necking. Always experiment first.

FIREFORMING

As in annealing, several methods are recommended for proper fireforming. One, called hydraulic forming, is too messy and slow. Cases formed by this method are never perfect or alike.

A better way to fireform a given case is to first purchase a trim die for the finished cartridge. The parent case is then primed, charged with a suitable load of fast-burning pistol powder (10 to 15 grains of Bullseye), and the remainder of the case filled with cornmeal. The meal is packed tightly in the case, using a rod that fits the case neck properly, to a point 1/8 inch from the mouth. Fill this 1/8-inch portion with a few drops of common wood glue. The cases are allowed to dry, and then lubricated and inserted into the trim die.

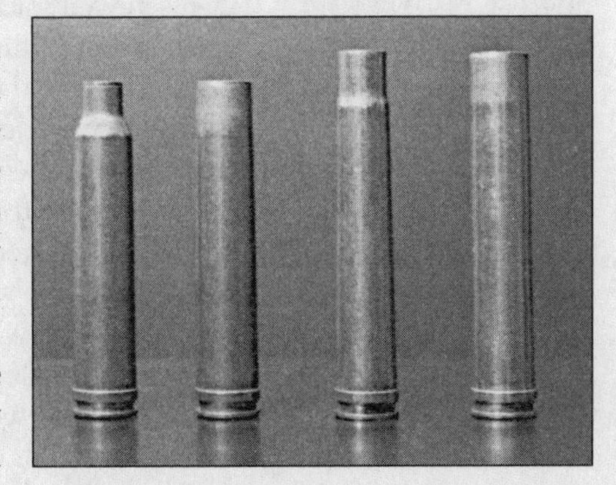

This drawing illustrates how to increase diameter using the sleeving method. Case will fireform to full body diameter.

Adjust the case within the die using a plastic mallet until the head protrudes 0.125 inch below the die mouth. You can easily devise a special "gun" to fire this load. It's nothing more than a

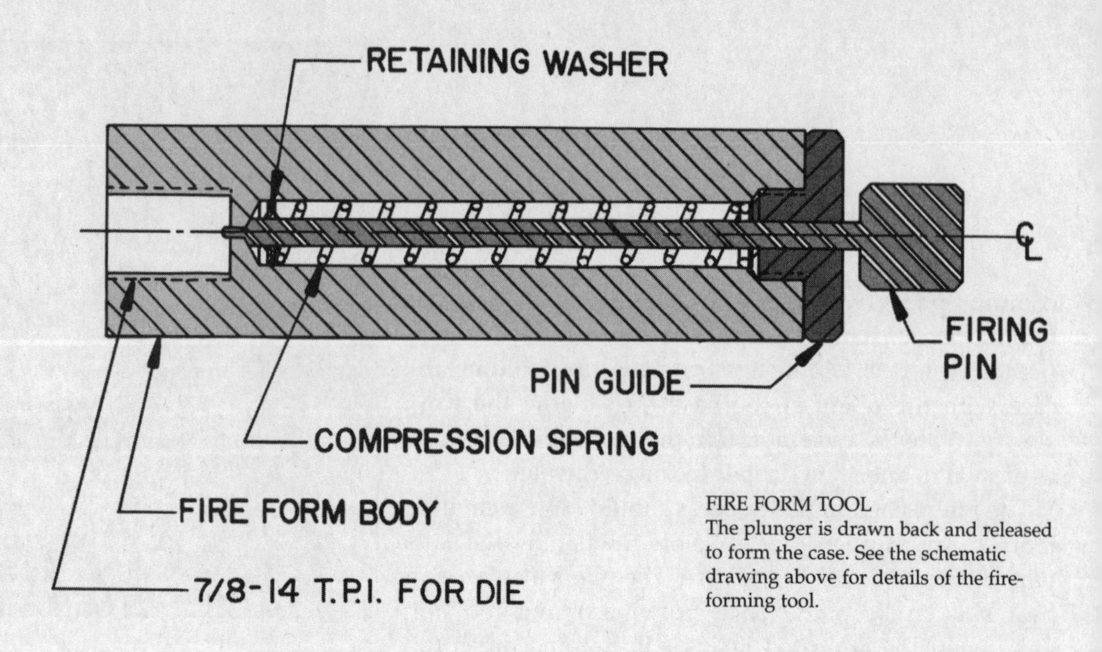

FIRE FORM TOOL
The plunger is drawn back and released to form the case. See the schematic drawing above for details of the fire-forming tool.

Here a Huntington Compact Tool is used as teh basis for fireforming a tool. A .300 Winchester case is filled to the case mouth with 10 grains of Bullseye and corn meal. The meal is tamped down and the neck is plugged with several drops of paraffin.

bar of steel threaded at one end to accept the 7/8-14 thread of the trim die. Running through the center of this "gun" is a spring-loaded firing pin. The die is then screwed into the gun until the case head bottoms out. With the gun held in one hand and pointed at the ground, pull the plunger back and release. Upon firing, the glue and cornmeal will be blown free of the case, but not before the case has been expanded to fill the die exactly. The report will sound much like an actual round, but the cornmeal projectile is quite harmless beyond five feet.

The finished formed case is now tapped out of the die with a metal rod. Each case will be perfectly formed and ready to clean and reload. Some extreme cases will require two such formings. An example is the .450/348 Ackley Improved. It is better to use two moderate loads than one heavy load. A load that is too strong can punch a hole in the neck area of the case near the die's lube vent hole, and render the case useless.

The beauty of fireforming is that it removes all evidence of the old neck. This is not always true with plain neck expansion. Fireforming is most commonly used to blow out a case to an improved configuration. Repeated mention of this operation will be found in the conversions section (Part II).

6 Special Cases

Once you've fabricated your first case from scratch, you'll begin to feel like an accomplished custom loader. Occasionally, you'll be required to dig into your new bag of tricks and perform a bit of special magic. Here are a few techniques to use in preparing odd or exotic cases.

THE REBODIED CASE

Marrying an existing case head to a new case body results in a rebodied case. Soldering is involved in preparing these cases, which means they must be used only with black powder or light Pyrodex loads. This point cannot be stressed too strongly. Soldered cases will crack or separate at 40,000-plus psi of chamber pressure.

After years of forming many wildcat and obsolete cartridges, the author had assembled a collection of ruined .348 Winchester cases (i.e., the necks had been split or the bodies had partially folded up), but the case heads were still in good condition. These spoiled cases should be cut off at about 1/2 inch to save the head. The head is then turned on a lathe to the proper diameter so that it can be slip-fit into a brass tube, which serves as the body of the new case. A 3 1/4-inch length of 17/32-inch (outside diameter) tubing is then placed on the turned diameter. Next, insert a drop of flux and a coil of low temperature silver solder into each case and place them upright on the hot plate. Turn on the heat and wait for the solder to flow, and then turn off the heat and allow the case to air-cool to room temperature. Now run these cases into a full-length .45-125 Winchester die and trim them to 3.25 inches.

Small Parts, Inc., offers thin-walled brass tubing for these applications. The sizes begin at 5/8 inch outside diameter with a .015 inch wall thickness. The next smaller size is 19/32 inch (outside diameter) also with a wall thickness of .015 inches. The sizes progress downward so that they can be telescoped together. For example, 17/32-inch O.D. tubing may be used for a .45-125 body; the next smaller size—1/2-inch O.D.—will then drop right into the case. This technique can be used to good advantage in making cases with stronger (i.e., thicker) walls near the base and thin walls at the mouth. The case thus becomes a three-piece assembly.

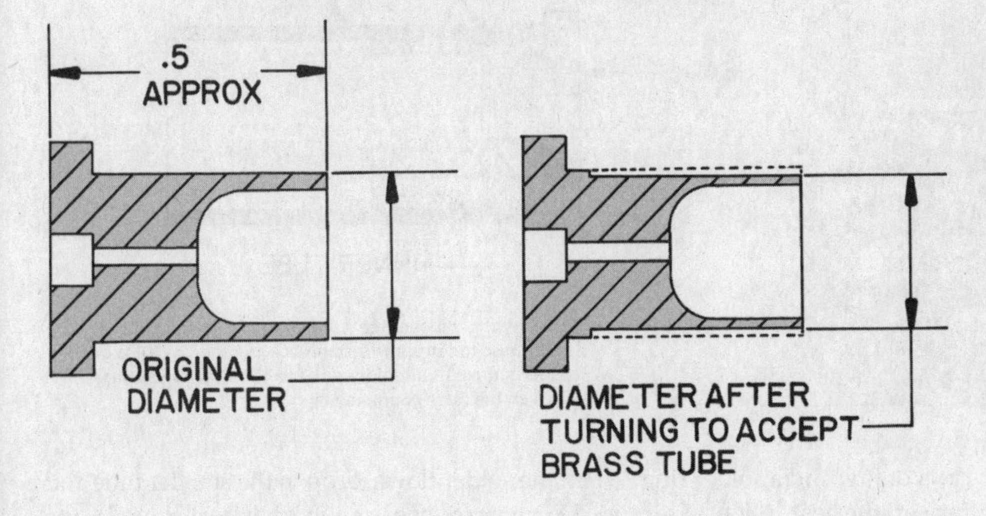

The base of a .380 Winchester case (illustrated in the diagram above) has been reduced on a lathe so that a section of brass tubing can be soldered on to create a rebodied case.

BUILT-UP CASES

When no existing heads are available, a variation of the construction method described above can be used to make cases from scratch, using brass tubing supplied by Small Parts, Inc., and solid brass purchased locally as round stock (called "1/2 hard brass," which indicates good machinability). Several hundred .577 basic cases (for .577 Snider and .577/450 Martini Henry) were prepared in the author's shop this way and all functioned properly when the appropriate silver solder was used.

To manufacture .577 basic cases, begin first with brass round stock slightly larger in diameter than the rim of the .577 (3/4-inch will do fine for a 0.748-inch diameter rim). Put a section of this 3/4-inch diameter brass in the lathe chuck and turn the major diameter to 0.748 inches. The .577 will use 5/8 O.D. tubing for the body, so turn next a 1/2-inch portion to accept the tubing (0.595-inch dia.). Cut a large chamfer on the front edge of the head piece to act as a solder well, and then cut the turned piece off long enough to leave about 0.075 inches of rim thickness. Trim the 0.595-inch diameter piece off to a finished rim thickness of 0.055 inch. Center drill the head face (primer pocket side) and drill completely through with a No. 44 (.086-inch dia.) drill. Use a No. 4 (.201-inch dia.) drill to start the primer pocket. Finish the pocket with a Sinclair Primer Pocket Uniformer. Deburr completely, clean with solvent, and then soak the pieces in vinegar (acetic acid) to etch the brass for soldering. Next, cut one piece of 5/8-inch tubing about 3 1/2 inches long and another piece of 17/32 tubing 1-inch long. Square both edges of each piece, clean with solvent and soak the tube section in vinegar.

When you are ready to assemble the case, dip the head/rim in water and again in alcohol. Polish the inside diameter of the larger tube with steel wool, rinse in solvent and alcohol, and allow to dry. Rinse the smaller tube—first in water and then in alcohol—and dry. Place the larger tube on the head and stand the whole business on the hot plate. Turn the heat on and add a

In preparing a .348 Winchester case head for rebodying, the straight portion of the case is lathe turned to accept the proper tubing.

A carbide bit (above) is used to turn the case body's largest diameter.

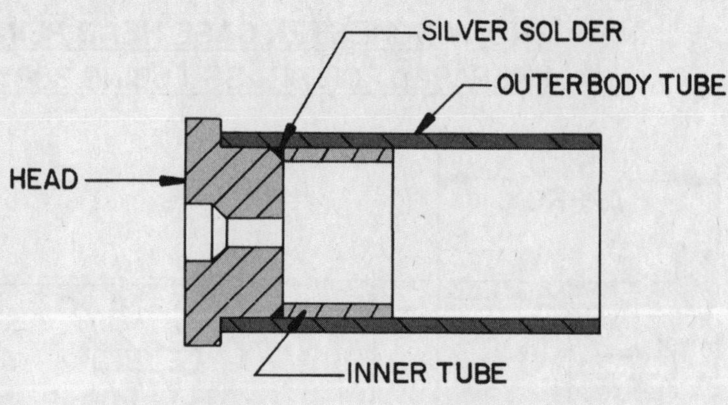

The drawing (above right) illustrates the components of a tubing case (or, in some instances, a rebodied case). When properly fluxed, the solder will "wet" all mating surfaces between the three components.

drop of flux and a solder ring. When the solder flows, drop in the smaller tube and turn off the heat. Allow to air cool. Do not force-cool or quench in water, or else the head will become too soft.

Tubing cases of this type must be neck-annealed before forming. If you have a bad solder joint, the case will pull away from the head while forming, so check the joints first. Any solder running into the flashhole or primer pocket can be removed with a No. 44 drill and primer pocket uniformer. Remember, these cases must be soldered with low temperature silver solder and they are only safe for black powder loads.

LATHE TURNED CASE

Solid cases that are turned from solid brass bar stock should be considered only when no other means of producing cases is available. A great deal of work is involved in this process and it can be quite expensive. For example, .600 Nitro Express brass is now all but impossible to find. When cases or ammunition can be located, they are either in bad shape or are too costly.

To manufacture your own .600 Nitro Express cases, begin with several sections (depending on the number of cases required, in this instance 25) of 1-inch diameter round stock brass (half hard), each section measuring 4 inches in length. With about 3/4- to 7/8-inch of this stock held in the lathe chuck, face and center drill (with a 1/4-inch diameter drill) to a full diameter depth of 2.625 inches. Next, drill the piece with 3/8-, 1/2-, and 23/32-inch diameter drills, in that order, to bottoming depth. Use a small diameter boring bar to counterbore the case mouth to 0.648 inches inside diameter and 3/4 inch in depth. The outside diameter is then turned—using the power feed and in cuts of no more than .005 inch per side—to .696-inch diameter for a length of 3 inches. Use a high spindle speed and a slow feed to produce a good finish.

Turn the case around in the lathe, chucked so that it just touches the full 1-inch diameter. Remove all but 3/16 inch of the larger diameter with a hacksaw. Turn this larger diameter to .800 inch, and then face the piece, leaving the rim about .065 inch thick. Center drill the face, drill No. 44 (.084-inch) through, and drill No. 4 to slightly less than proper depth for a large rifle primer. Complete the primer pocket with a Sinclair Uniformer, and gently break the edges with a fine file. This case will require annealing and full-length sizing. Because the walls of this case are extra heavy, the powder capacity is somewhat reduced; but with modern black

Shown on the left is an original .577/450 Martini Henry cartridge. The case on the right was made from 21/32" and 5/8" diameter tubing and a lathe turned case.

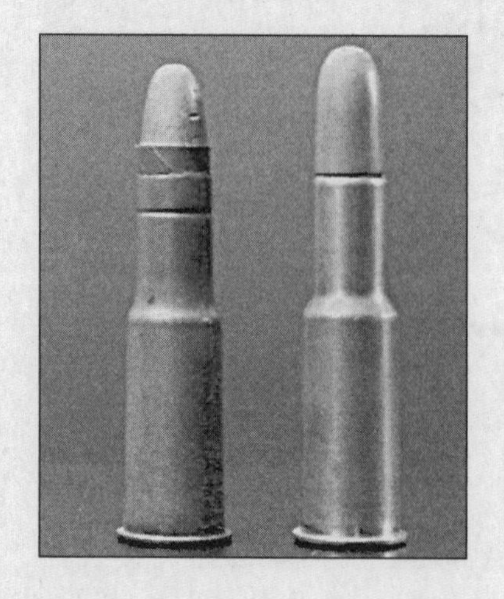

powder and magnum rifle primers this is of small importance. Bottleneck cases may be prepared in the same way, although they require annealing and careful forming.

Using a lathe to fabricate cases from solid brass stock is the epitome of the custom loader's craft. With proper dies, any case ever made can be reproduced. Those who are fortunate enough to own sophisticated lathes and taper turning attachments can even forget about the proper dies, for they can easily duplicate the taper of any case with their lathes to produce shootable ammo, including the following:

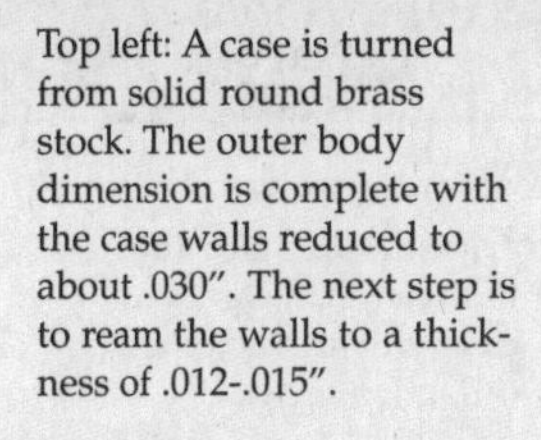

- .35-40 Maynard 1873
- .35-30 Maynard 1873
- .40-90 Ballard
- .44-90 Sharps BN
- .44-100 Wesson
- .577 Basic
- .500/465 N.E.
- .475 #2 N.E.
- 8x72R S&S
- 9.3x82R
- .600 Nitro Express

Top left: A case is turned from solid round brass stock. The outer body dimension is complete with the case walls reduced to about .030″. The next step is to ream the walls to a thickness of .012-.015″.

Top right: After the body has been turned and the inner dimension drilled, the case is cut off leaving enough stock to form the rim.

Above: The primer pocket is drilled to a controlled depth using a No. 4 drill.

Readers should note the following facts concerning silver soldering of tubing cases. First, any abnormal heat applied to a cartridge case, particularly in the head area, can be detrimental. Heat modifies the crystalline structure of brass and renders it softer. The actual degree of softness is roughly proportional to the temperature applied and the duration of application. Heating a case in a gas flame to dull redness, followed by immediate water quenching, will cause the case to become "dead soft." Cases that are dipped in a molten lead bath will retain some degree of hardness, but not enough to be noticed in any case forming operation.

The silver solder used for tubing cases as described in this book—Brookstone catalog No. S-01590—has a melting point of 430°F. (221°C.), and a tensile strength approaching 25,000 psi. Heat the case slowly on a hot plate to the point where the solder begins to flow. Follow by air cooling to create the strongest possible joint with the least amount of annealing.

Cases soldered in this manner are excellent for black powder and Pyrodex loads and, in most instances, light to moderate charges of, say, IMR 4198. Do not make this type of case for a 7mm Remington Magnum!

The Handloading Process

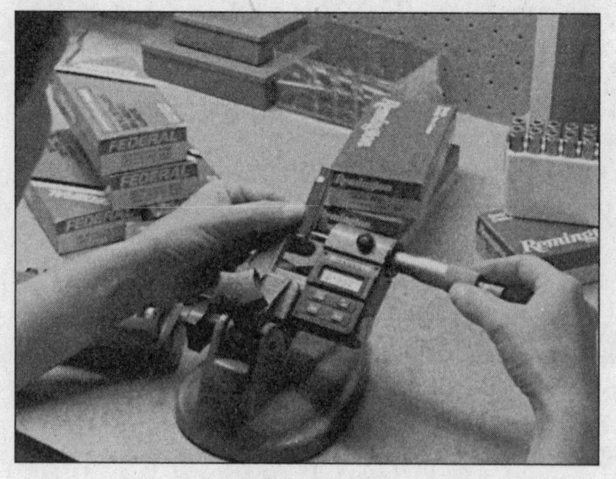

A handloader measures the head diameter of a case with an electronic 0-1" micrometer.

INSPECTION

Once the cases are formed and ready for full-length sizing, they are physically complete and ready to be put to use. One last operation must be performed, however, and it should be considered as an integral part of the case manufacturing procedure: Inspection!

The time to discover that the case necks need to be reamed is not while you are seating bullets into charged cases! The best way to inspect cases is to sit down at a well-lit, clean workbench and evaluate each case. Pay particular attention to any features that may have been overworked such as the case necks. If you did extensive necking, up or down, or cut a new extractor groove, check these areas carefully. Using the proper measuring instruments be sure that all base diameters and rims are in tolerance. Is the case length correct? Are the primer pockets the proper depth? Now is the time to make these adjustments if necessary.

CLEANING

Dirty cases should never be placed into your dies. Dents, scratches and other weak spots will result. Case cleaning is best accomplished in two stages. Cases that have been full-length sized will be coated with sizing lube. This lube must first be removed. The simplest way to do that is to place the cases in a half-filled bottle of lacquer thinner or acetone for a minute or so. Next, stand the cases (on their mouths) in wooden cartridge blocks to drain and dry. (Do not prime or load any case that retains the slightest odor of solvent and, of course, do not smoke or allow any open flames in the area where you are using solvents!)

Solvents should be reused only for as long as the dry cases do not feel oily to the touch. Otherwise, the solvent should be discarded. If a water-based case lubricant is used, you must still clean the cases. Plain hot water will do the trick.

An outside micrometer (below), fitted with blades that are only .030" thick, is used to determine the diameter of extractor grooves.

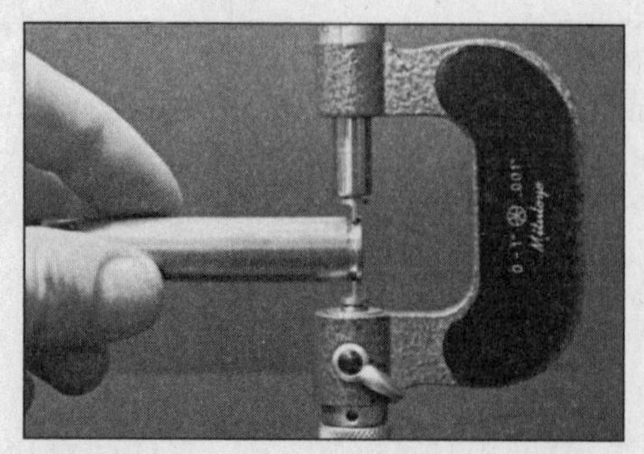

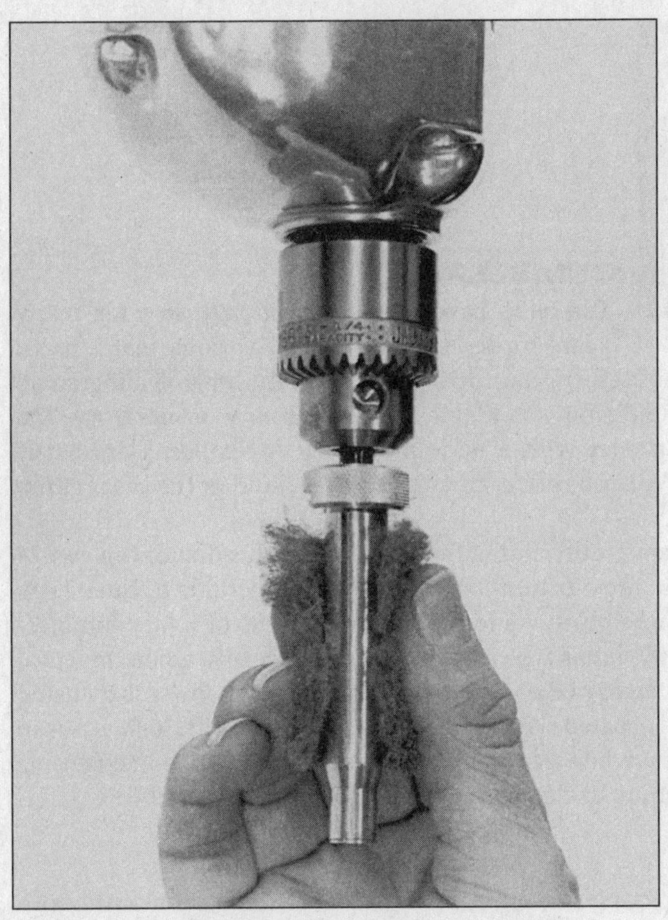

In this slow but effective method, the case exterior is cleaned using an electric drill and steel wool.

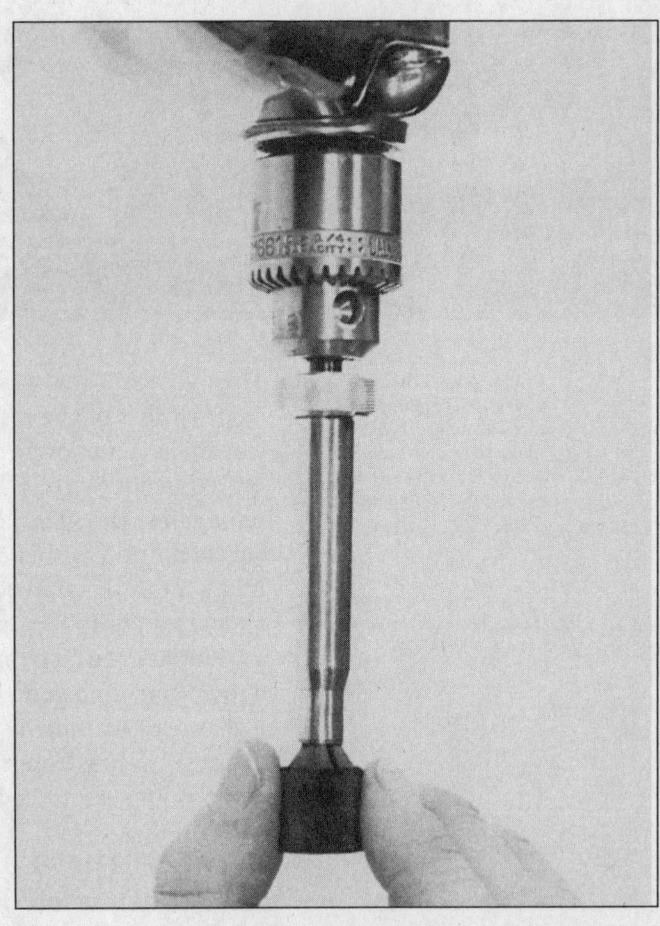

This arrangement provides power for the small Lee-type hand-held chamfering tool. The drill must turn slowly to avoid chatter that may damage the case.

Dillon's CV-2001 case cleaner uses vibratory action to clean and polish up to 1,300 .38 or 550 .30-06 cases per hour. The tumbler cleans cases prior to sizing and removes sizing lube from sized cases.

The Sidewinder rotary case tumbler from RCBS cleans brass cases using either wet or dry media. The drum may be fitted with an extra perforated lid that separates the cleaning media from the cleaned cases.

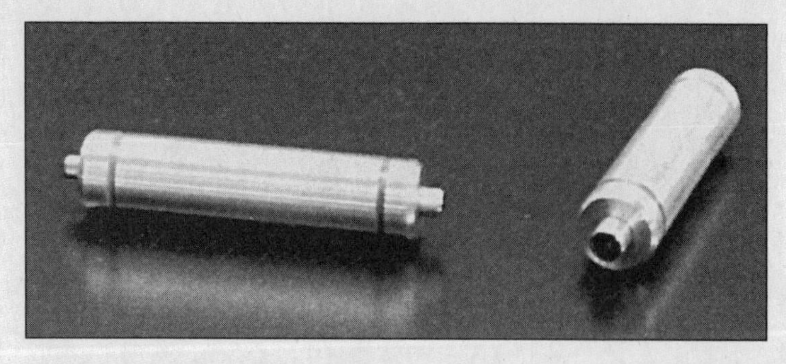

Primer pocket diameter GO/NO-GO gauges are mandatory for those who make their own case heads. Bands at each end are red (NO-GO) and green (GO).

TUMBLERS

Tumblers have been used in industry for many years to de-burr and polish various metal parts. They've also seen wide used by the rock/gem fraternity for polishing small stones. Most of the tumblers now available are either rotary or vibratory. You can make your own tumbler with a wide-mouthed, one-gallon glass bottle coated inside with RTV silicon rubber (to deaden the sound of the cases clinking against the glass).

When fitted with a heavy-duty, industrial gear-motor, this device can run 24 hours a day. If you operate your tumbler over extended periods of time, keep an eye on the motor temperature. As for the relative merits of rotary tumblers vs. vibratory tumblers, the latter type is far superior, especially when time is at a premium. One good way to make your rotary tumbler work faster is by using a charge of walnut shells coated with rouge. This will clean 140 .30-06 cases in about six hours. Common white rice may also be used but it's quite inexpensive and requires about 24 hours to clean any quantity of cases thoroughly.

CHAMFERING

The last step before priming is a light chamfering of the case mouth, both inside and outside diameters. Chamfering is required on the outside of the case so that it slides smoothly and uniformly into the weapon chamber. A large clump of brass, if left on the case neck, will cause additional gripping pressure on the bullet and increased chamber pressure as well.

Chamfering the inside diameter helps to seat the bullet and allows a smooth, even crimp. Be sure to chamfer just enough to remove any excessive brass. A good tool to use for chamfering the inside diameter is a thin-handled Ex-Acto knife with a No. 11 blade.

This RCBS deburring tool can be used by hand or fitted to a lathe. The pointed end removes burrs and bevels the inside of the case. The open end can be fitted over the case mouth to remove exterior burrs.

BULLET SEATING

Bullet seating offers no special problems for the custom loader, other than the occasional modification of an existing seater. These seaters, which are manufactured by the large die makers, are designed to conform to the ogival shape of the more common commercial and cast bullets. If the bullet geometry is changed, the seater can damage the meplate of the projectile (the blunt end of the tip of a projectile). The author first discovered this problem while making ammunition for an 11.75mm Montenegrin revolver. The brass cases were easily formed from .45-70 Government brass, but the bullet was another story. A suitable projectile was eventually produced by casting a standard 0.446-inch diameter RCBS .44-370-FN bullet. This slug, when lathe turned to a length of .75 inch with an SWC configuration, presented definite problems in seating. All existing seaters were either too small, the wrong configuration, or they damaged the bullet by rolling over the lead edges.

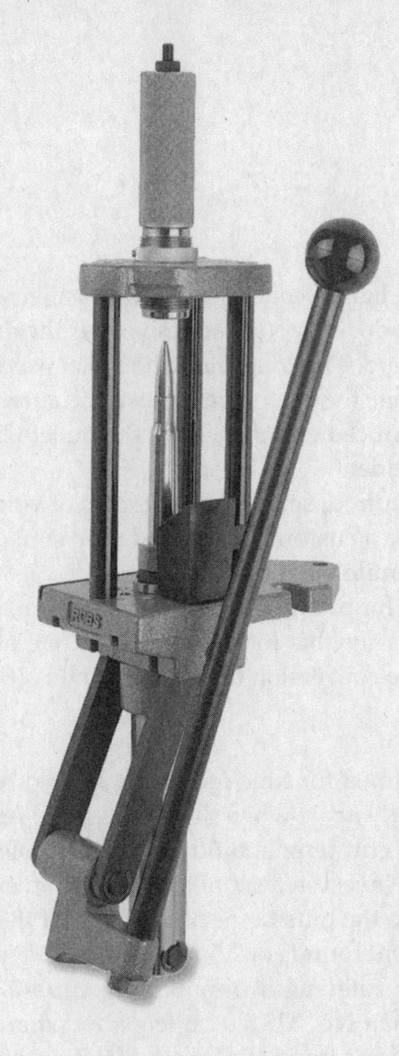

The AmmoMaster .50 BMG Pack by RCBS fitted to the AmmoMaster Single Stage Press is provided with the dies and accessory items needed to reload .50 Browning machine gun cartridges. The press has a massive solid steel ram and sufficient height to accept the big .50 caliber case.

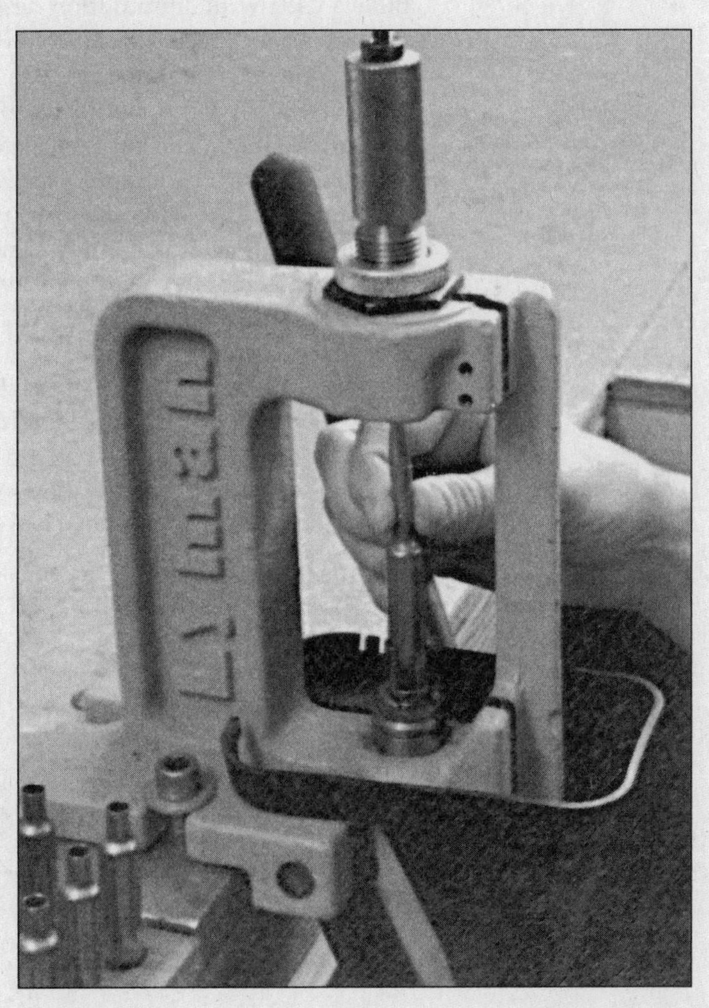

Commercial bullets are seated with Lyman's "O"-frame "Orange Crusher" press. Note the wooden blocks used to hold powder-charged cases.

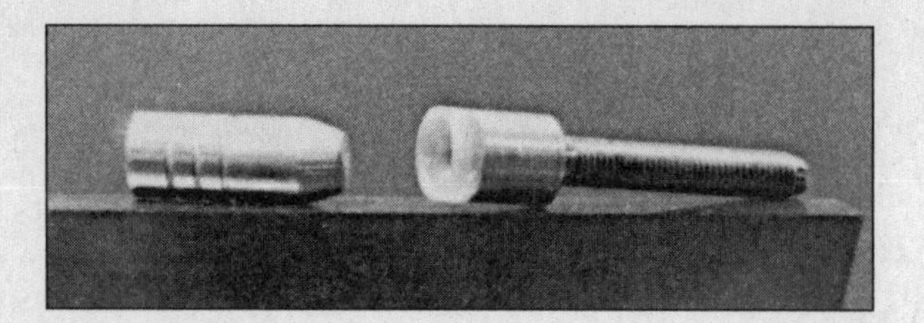

Standard seating plugs can be modified to accommodate odd bullet meplats by filling the seater with epoxy and allowing it to cure around the bullet. The bullet is coated with silicone or teflon for easy release.

This problem was solved by first lightly coating a finished bullet with Teflon grease (case sizing lube will also work). An epoxy adhesive was then mixed and applied to the seating cup of a standard .44 SWC seater. The bullet was forced into the epoxy and held in alignment while the epoxy was allowed to cure. After a little trimming, a seater plug was produced that duplicates the bullet . Bullet seating was then completed without incident.

If a more permanent seater is required, send a few samples of your bullet to a die manufacturer, who will create a custom seater made for you, or you can make your own on the lathe from mild steel or brass.

Seating depth is generally critical for proper operation of a firearm action. You will have to rely on published data for this information, and one of the most complete sources you can find is the conversion tables in this book (Part II).

CRIMPING

Bullets used with ammunition destined for tube-feed rifles should be crimped in place. Otherwise, the ammo can fly apart when the rifle is fired. Armed with a complete set of full-length dies, crimping should pose no problem for the handloader. If you are not ready to invest in a set of dies, locate an existing die with a base diameter that is close to the outside neck diameter of the cartridge being crimped. For example, excellent rounds of 7.5mm Swedish Nagant ammo may be produced without form or full-length dies. Simply cut a .32-20 WCF case to 0.89 inch and chamfer. Lyman No. 323470 bullets are cast and the rear portion of each bullet is cut off, leaving a projectile length of 0.52 inch. Size these bullets to 0.321-inch diameter. All that remains is to seat the bullets and crimp the brass by "bumping" the case into the bottom of an 8mm French Lebel revolver full-length sizer. This will roll the brass in neatly to form a perfect crimp. If you don't have the Lebel die, any other die with a corresponding base diameter can be used with equally good results.

Bullet selection generally allows the loader some latitude. There are really no hard and fast rules about what can or cannot be used. If the projectile is approx-

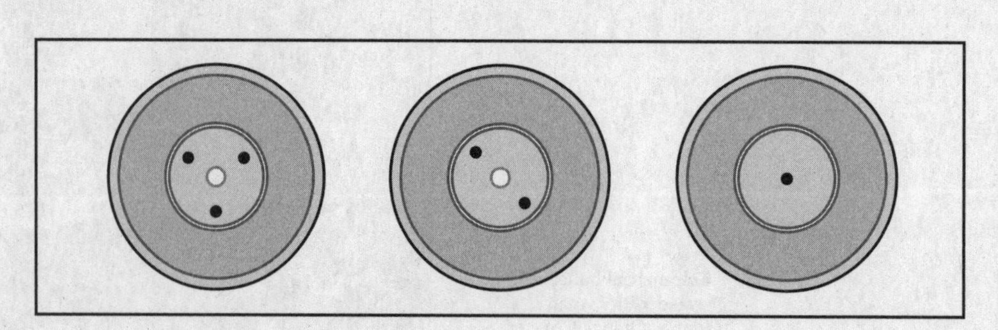

Shown above are three common systems for transferring the primer's flash to the powder. The 3-hole Berdan (left), 2-hole Berda (center), and the single-hole Boxer (right). The Boxer primer is the predominant method used in the U.S.

imately the correct weight and geometry, it can be made to function in the weapon at hand. Experimenting is, after all, half the fun of creating exotic ammo—so long as you KEEP IT SAFE!

8

Primers and Their Pockets

CLEANING AND INSPECTION

Cleaning of primer pockets should be a routine procedure at any handloading bench. All priming compound residue must be removed from the pocket before seating a new primer. An accumulation of residue could prevent proper, flush seating of new primers. Pockets are cleaned most conveniently with a small, flat screwdriver blade, or any commercial tool sold for this purpose.

If the flashhole has not been drilled oversize, it's an excellent idea to push an 0.080-inch diameter drill through each hole. This will remove residue and assure hole-to-hole uniformity. The primer pocket diameter must be closely controlled to ensure a tight primer fit. If you intend to produce any quantity of custom cases from scratch, fabricate or buy a "GO/NO GO" pocket diameter gauge. When using this instrument, the 0.2085-inch member should enter tightly, and the 0.2095-inch member should not enter at all.

A primer pocket is cleaned with a Lyman Power Trimmer.

Any cases with pockets that are oversized should be discarded immediately. There is no effective way to re-swage pockets to a smaller diameter, and you should never attempt to cement loose primers in place.

Check the pocket depth with a micrometer or a gauge made expressly for that purpose. A shallow pocket is easily cut deeper with a Sinclair "Uniformer." Removing small amounts of brass from the case head can occasionally rework pockets that are too deep. Be careful not to thin the rim beyond proper headspacing limits. This operation may remove the headstamp in some instances, so remember to record any information that is pertinent to duplicating an operation in the future.

MILITARY PRIMER POCKETS

Automatic weapons produce a good deal of vibration while firing, and occasionally a primer will vibrate free of the case. This can jam the action, so military ammo-makers anchor their primers in place with a crimp or stake.

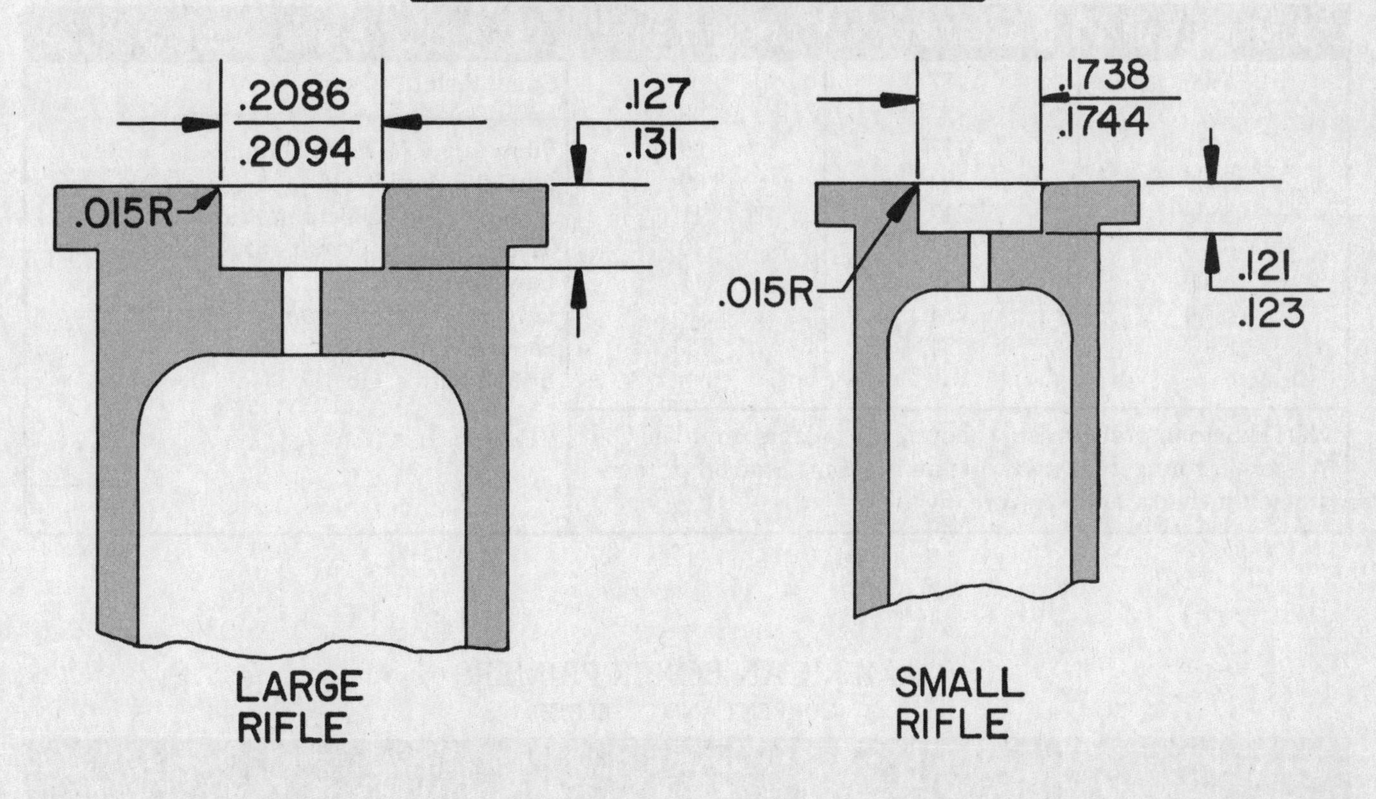

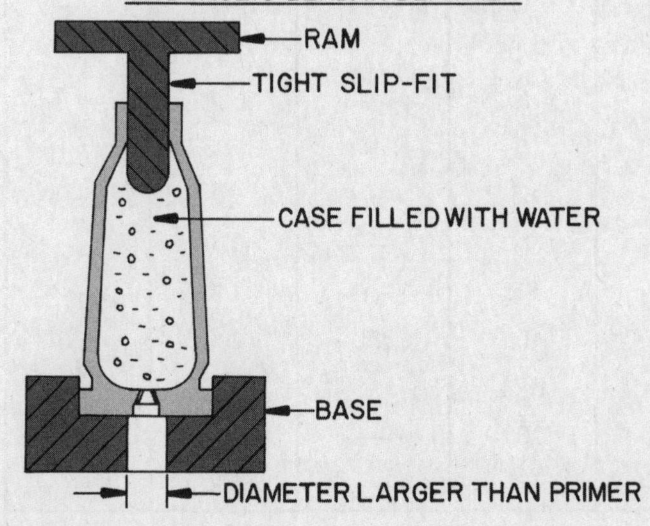

Berdan cases may be deprimed with water and a tight-fitting ram, as illustrated here. A sharp blow on the ram will displace the primer.

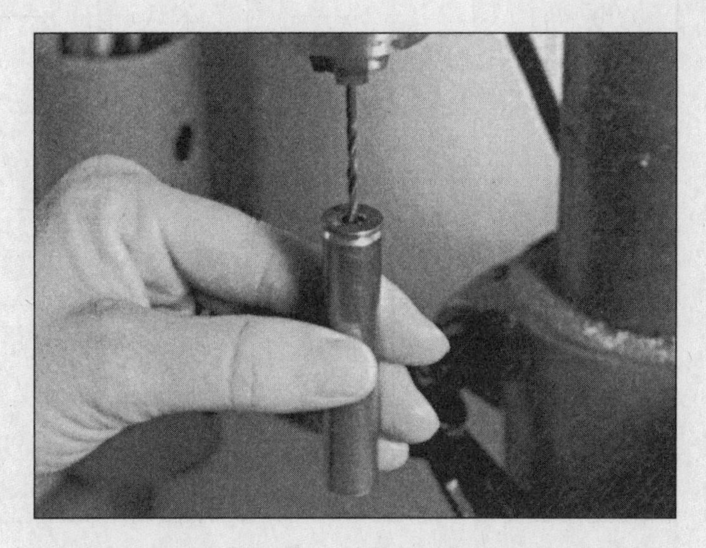

An 0.008-inch diameter drill is used to open up flash holes. This removes any priming compound residue and assures uniform flash hole diameters.

RWS SINOXID BERDAN PRIMERS

PRIMER NUMBER	DIAMETER MIN.	HEIGHT MIN.	RECOMMENDATION
4506	.177	.09	Small Pistol
4520	.177	.08	Small Rifle
4521	.177	.09	9mm Luger (4, 5mm Berdan)
5005	.197	.09	Non-standard Large
5608	.217	.11	Standard NATO Berdan Primer
5620	.217	.10	Most Common Berdan Rifle Primer
6000	.250	.11	Large Berdan Rifle
6504	.254	.09	Larger 6, 45mm Berdan
6507	.254	.133	Same as ELEY #172 for the large British Nitros, etc.

NOTE: Berdan primers listed above are available from The Old Western Scrounger (at www.ows-ammo.com). Sinoxid primers are non-mercuric and non-corrosive.

AMERICAN BOXER PRIMERS

(CURRENT AND OBSELETE)

	SMALL PISTOL	SMALL PISTOL MAG.	LARGE PISTOL	LARGE PISTOL MAG.	SMALL RIFLE	SMALL RIFLE MAG.	LARGE RIFLE	LARGE RIFLE MAG.
REM	1 1/2	-------	2 1/2	-------	6 1/2	-------	9 1/2	9 1/2M
RWS	4031	4047	5337	-------	4033	-------	5341	5333
WIN	1 1/2 -108	-------	7-111	-------	6 1/2 -116	-------	8 1/2 -120	-------
Western	1 1/2	-------	7	-------	6 1/2	-------	8 1/2 81/2G	-------
Peters	15	-------	20x	-------	65	-------	12	-------
Herters	1 1/2	-------	111	-------	6 1/2	-------	120	-------
Alcan	SP	-------	LP	-------	SR	-------	LR	-------
CCI	500	550	300	350	400 BR4	450	200 BR2	250
Federal	100	-------	150	155	200	-------	210	215
Norma	SP	-------	LP	-------	SR	-------	LR	-------

NOTE: Current Boxer Primer Pocket Dimensions (in inches):

Large: Diameter - .2086 (+ .0008/-0)
Depth - .129 (+ .002/-.002)
Small: Diameter - .1738 (+ .0006/-0)
Depth - .121 (+ .002/-.002)

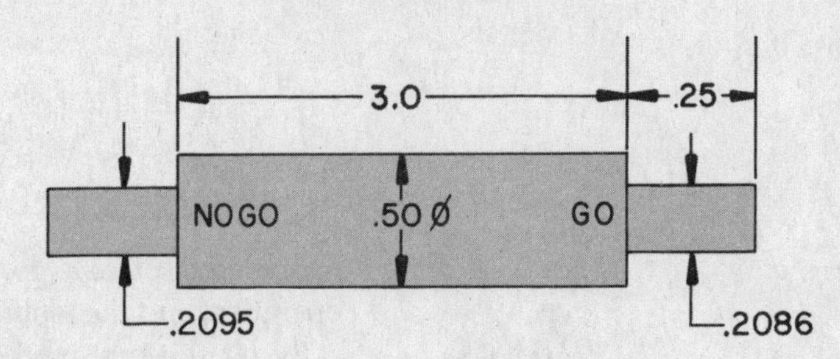

Dimensions for a large U.S. Boxer primer GO/NO-GO gauge. The body length of the guage is optional.

Crimping is found on several types of ammunition, notably .30-06, 7.62 NATO, .30 Carbine, .45 ACP, and 9mm Parabellum. Military crimped primers are often quite difficult to remove and may damage the depriming rod of your die set. To prepare crimped cases for reloading, it's a good idea to drive the spent military primer out first with a mallet and a simple, homemade depriming rod. A good deprimer to try is a piece of cold-rolled steel, about 3/16 inch in diameter, with a standard depriming pin that is press-fitted into a hole at one end. Allow as little of the pin to protrude as necessary to keep the pin from bending.

When removing live primers from military (or any other) ammo, you must first "kill" the primer. This can be done effectively by boiling the cases for 10 minutes in soapy water and soaking them overnight in acetone or lacquer thinner, and then reboiling them in soap, followed by a clean water rinse. Be sure to wear safety glasses when depriming any case, no matter what method is used.

Once the primer has been removed, the crimp may also be removed. It usually appears as a complete 360° rolled lip around the primer, or a series of "tits" that are mushroomed out towards the center of the primer.

One common fix for crimped primers is to reswage the pockets, forcing the brass crimp back and out of the way. This method may work fine, but it's much slower than simply milling the crimp away. Try using an 82° included angle carbide countersink for this operation. The countersink is first turned in the drill chuck, and the case is then simply pushed against the tool. Be careful not to remove too much metal. Lightly chase this chamfer with the point of a sharp knife and the job should be complete.

BERDAN PRIMERS

Unfortunately, Berdan primers are still with us, mostly in European cases. This was once a major problem for reloaders, but Berdan primers no longer offer special problems and updating requires only a few special tools.

One popular method for depriming Berdan cases is through the use of hydraulics. The case is first positioned over a hole in a metal plate to clear the primer. The case is then filled halfway to the neck with water, glycerin or oil. A rod (usually custom-made) is tightly fitted in the neck and a series of sharp blows are then made with a hammer. Because the liquid will not compress and

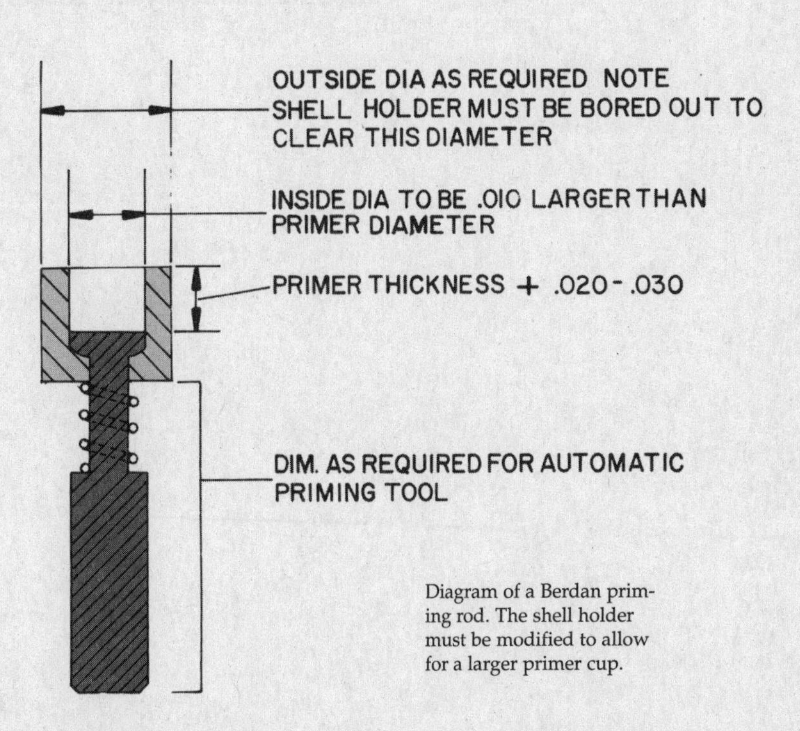

Diagram of a Berdan priming rod. The shell holder must be modified to allow for a larger primer cup.

has no other place to go, it forces the Berdan primer out. This is obviously a messy operation, and the tooling-up process takes a considerable amount of time.

By far the best method for depriming Berdan cases is with the Lachmiller Berdan depriming tool marketed by RCBS. The Lachmiller tool is similar to a metal can opener, which first pierces and then lifts the primer with ease.

One last consideration in seating Berdan primers is the modification of the shellholder to allow the larger primer cup to pass through its center. This is easily accomplished by reaming (in the lathe) and lightly chamfering.

The RCBS Ram Priming tool seats all Boxer-type rifle and pistol primers. It can be pre-set for a positive stop and works with any press with a 7\8x14 thread and RCBS-type removable shell holder.

PRIMER POCKET ADAPTERS

A friend who owned a bolt action Mauser rifle chambered for 11.15x60R ammo asked if his Berdan primers could be converted to Boxer. A local screw machine house provided a number of stainless steel primer adapters with no problem. The 11.15x60R cases were then chucked in the lathe and the Berdan primer pockets were removed by drilling through with an "F" (.257-inch) diameter drill. Each hole was then tapped 5/16-32 UNF and provided with a .410 diameter counter-bore 0.030 inch deep (see accompanying drawing).

The cases and adapters were carefully degreased and a light coat of hydraulic Loc-Tite adhesive applied to the female threads. The adapters were threaded home and the adhesive allowed to cure. The combination of very fine thread and proper sealing produced an excellent case that did not leak even medium pressure smokeless powder loads. This may be an extreme example, but it does illustrate that nearly any conversion is possible.

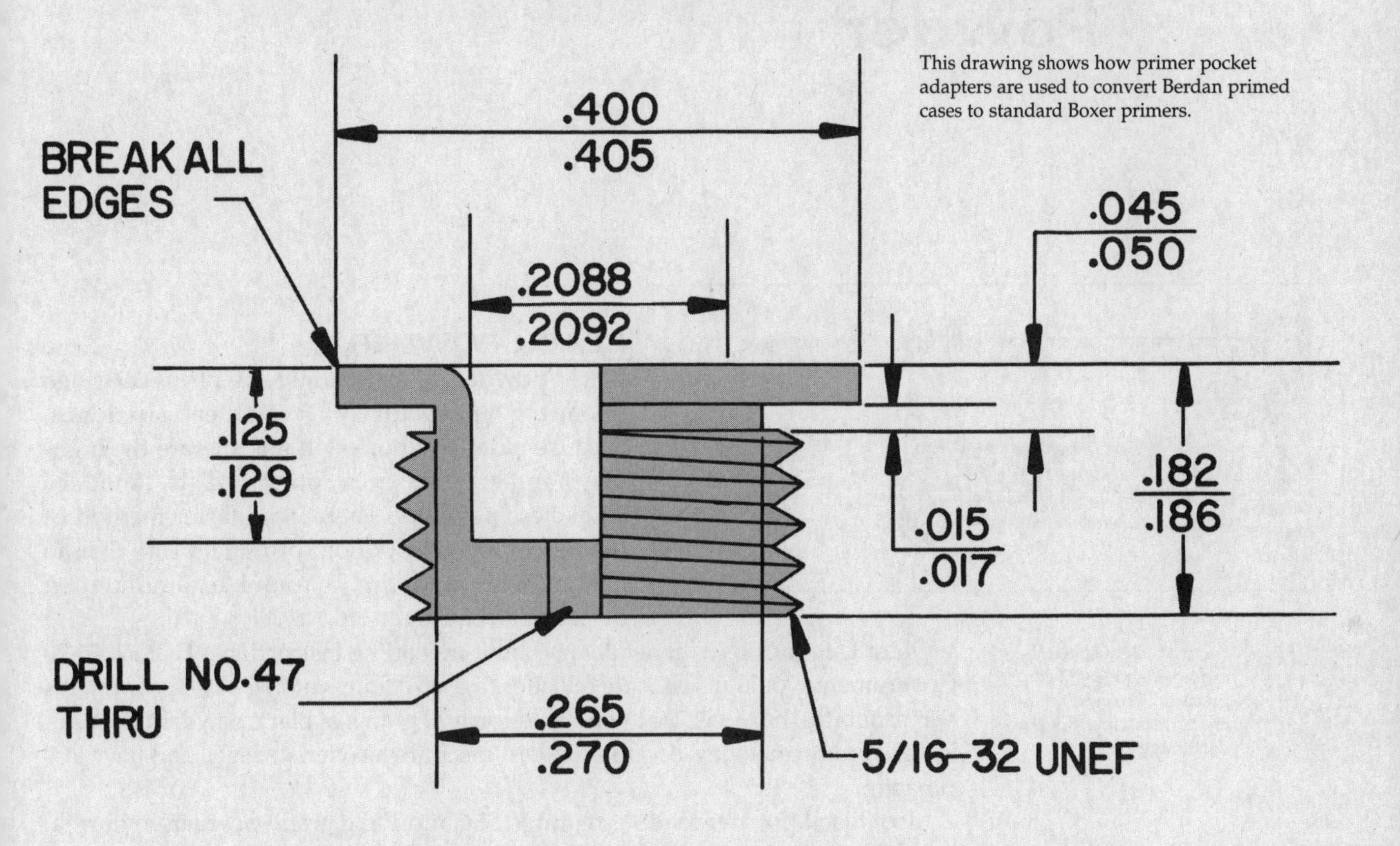

This drawing shows how primer pocket adapters are used to convert Berdan primed cases to standard Boxer primers.

Methods in the Art of Handloading

9 Gun Powder

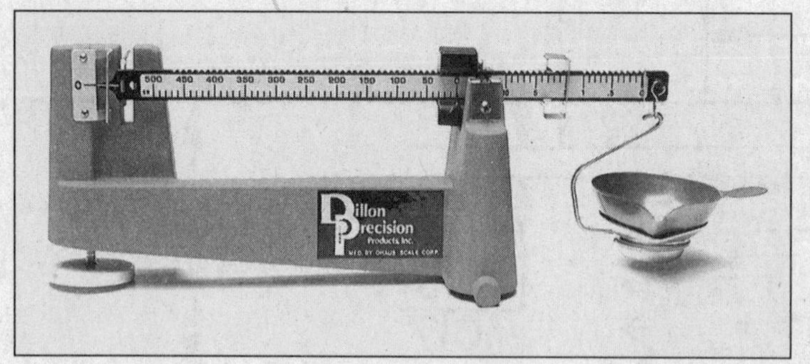

Beam scales, such as this example made by Dillon Precision, are used to check the accuracy of powder charges.

BLACK POWDER

Black powder is commonly used in cartridge reloading, particularly for older cartridges, because older weapons will not tolerate the higher chamber pressures produced by modern smokeless powders. There is no faster method of field-stripping a trapdoor Springfield rifle than to load it with, and fire, a round of ammunition designed for a Ruger No.1 rifle!

Older U.S. cartridges generally had built-in loading instructions. That is, .45-70 Government would instruct the reloader to use 70 grains of FFg black powder. It's very doubtful, however, that you could force 70 grains of black powder into a .45-70 case produced today, because modern cases are heavier, stronger, and have less capacity.

Most black powder loading requires FFg and FFFg grade powder. In normal use, there's not much difference between these two powders and they can be used interchangeably. For cases smaller than .30-30, FFFFg or FFFg are recommended (generally, the smaller the case, the finer the grade of powder). The table below (left) indicates the proper uses for each grade.

It is no longer possible to load modern black powder cartridge cases to their original black powder load, so a rule of thumb is to use just enough powder to touch the bullet when fully seated.

Black powder should *never* be thrown from a powder measure that has two contacting and movable metal parts (other than brass). The chance of creating a spark may be remote, but should half a pound of black powder ever blow up in your face, you'll wish you had heeded this warning. Black powder should be thrown from a dipper or, even better, a hand-held, plastic volumetric measure. This measure may be adjusted to throw a predetermined volume of powder. Heavier charges are thrown with two pulls of the trigger, and so on. This measure is especially effective with black powder and Pyrodex, Black Mag, Triple 7.

Black Powder	Approx. Grain Size	Use
FFFFg	.016-.018	Small ctg; blanks
FFFg	.037-.039	Med. ctg; muzzle loads
FFg	.057-.059	Med. ctg; lg. pistol
Fg	.068-.070	Large ctg
Cannon	.130-.135	1" & large bore

PYRODEX

Pyrodex is the trade name for the synthetic black powder manufactured by Hodgdon Powder Company. This material has the same applications as regular black powder (use "CTG grade for cased ammunition). Pyrodex is used by volume. Simply fill the case up to the base of the seated bullet. When loading with Pyrodex, use magnum primers for better ignition.

SMOKELESS POWDER

Most loading manuals today contain lengthy discussions about smokeless powder, and the reader is referred to these sources for more extensive coverage. The best substitute for black powder is DuPont's smokeless IMR 4198. A standard conversion for this powder is as follows:

IMR 4198 = (BP) (.29)

Translation: To determine the weight of IMR 4198 for a specific cartridge, multiply the original weight of black powder by .29, which is a constant. For example, when reloading .45-125 Winchester ammo without using black powder, apply this formula: (125) (.29) = 36.25

In this case, 36.3 grains of IMR 4198 is the suggested smokeless powder load for use with the original weight bullet (in this case, 300 grains).

In using IMR 4198, you enter the rather dangerous world of "reduced loads," which means reduced in volume. A comparative charge of IMR 4198 will fill only a small portion of the case. When such a load is held in a horizontal position, the powder tends to lay flat in the bottom of the case, below the flashhole. Upon ignition, the flame may blast through the flashhole and ignite the powder in the center of the case. The scientific principle involved here is best left to the ballistics experts, but it's enough to know that abnormal things can happen—including weapons that explode in your face! If you are going to experiment with black powder load conversions, do so in small increments and *never exceed cartridge manufacturer's recommendations!*

One safe technique is to pack the powder against the flashhole with a cotton ball, a wad of tissue or a tuft of Dacron pillow filling material (called Kapoc). You can also use plain cardboard wads to hold the powder in place, but it's still advisable to add some packing to help hold the wad.

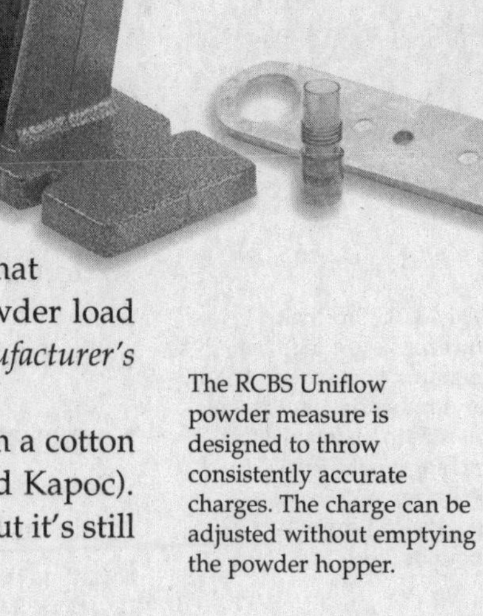

The RCBS Uniflow powder measure is designed to throw consistently accurate charges. The charge can be adjusted without emptying the powder hopper.

CORDITE

Cordite is a double-based, nitroglycerine-based propellant common to British cartridges. It can be replaced safely by an equivalent weight of IMR 3031, but as a starting load only. Heavier loads should be approached with caution.

Projectiles

The RCBS Pro-Melt furnace uses a high temperature heating element and industrial grade thermostat to maintain a temperature range of 650 to 850 degrees F. The furnace's lead capacity is 22 pounds.

LEAD BULLETS

Once a special or exotic cartridge has been fully formed, it's time to think about making bullets for it.

Projectiles fall conveniently into two categories: cast lead and jacketed. Lead and jacketed bullets are available for most calibers. The selection of one bullet over another is a matter of individual preference for downrange accuracy and performance.

In duplicating many of the obsolete cartridges, the reloader will also encounter obsolete diameters. Even common calibers can cause problems. The .44-40 WCF, .44 Ballard Long, .44-90 Remington Straight, .44-77 Sharps, and .44-100 Maynard 1873, all are .44's; but the actual bullet diameters are .428, .439, .442, .446, and .450 respectively.

It is worth the time and effort to write to each of the current mold manufacturers and ask for their most recent catalogs. Chances are good that one of them carries the mold required for your cartridge. Otherwise, you should begin thinking about lathe turning or "bumping" bullets. To do this, you must also have on hand a proper diameter bullet sizer as well as the mold. Top punches can be fabricated, but a sizer would be difficult to manufacture at the home shop level.

MODIFIED LEAD PROJECTILES

A lead bullet that has the weight required but is slightly too large in diameter is easily reduced to the proper diameter. First, hold the base of the bullet in the lathe chuck and test to see whether the bullet revolves true (using a dial indicator, if possible). Turn the front portion to the required diameter, re-chuck on the turned portion and turn the base to the same diameter. Bullets that are .458-inch diameter can be easily turned to .446-inch diameter (for 11mm cartridges) in the same manner. Turning lead is not difficult; just remember to

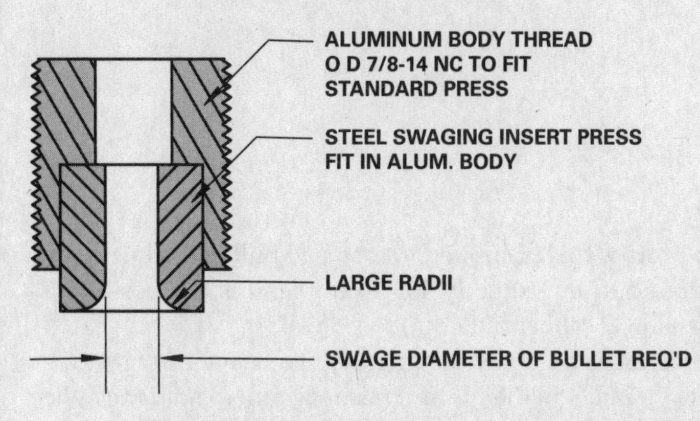

Bullet swaging dies are used to reduce the diameter of a cast lead bullet. The aluminum body fits into a reloading press and a ram is used to push the bullet through the die with an extension rack.

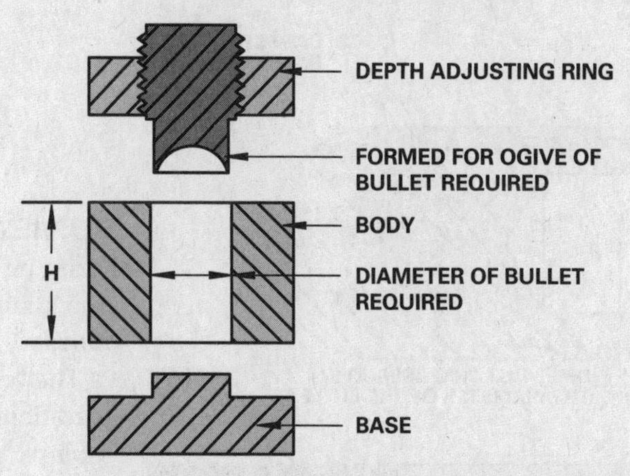

This bullet "bumping" die (above), which fits into a handloading (or arbor) press, compresses the bullet within the die body causing the bullet to mushroom out to a slightly greater dimension.

use a tool big with plenty of rake. Also, keep the spindle speed high and the feed low.

BOAT-TAILING

Many cast lead bullets are produced with perfectly square bases; i.e., they have no recesses for a gas check. Such bullets can be somewhat difficult to seat, especially if the case mouth can't be belled. The simple solution is to chuck the bullet on the forward half of the full diameter, and cut a simple 30° boattail. This not only aids in bullet seating, it may actually improve the aerodynamics of the bullet. These should be no significant reduction in projectile weight.

CUTTING BULLETS

If you can't find the correct weight bullet for your .30 caliber revolver ammo, what about a .30 caliber lead rifle bullet cut off to produce the required bullet weight? Do the cutting in the lathe with a very fine jeweler's saw, and then dress the base with a medium file to produce a small chamfer.

SPECIAL SIZERS

If only a few projectiles are required, and it's not feasible to use the lathe for diameter reduction, construct your own sizer using a section of brass tubing (or one of the cases for which the bullets are being made). Stand the bullets in a shallow container and pour in whatever amount of melted lubricant is required to cover the bullets above the uppermost grease groove. When the lube is hard, push the tubing (or case) down over the bullet. Upon lifting the tubing, the lubed bullets should come up with it. Gently knock the bullet out of the tube with a dowel.

SPECIAL MOLDS

If you decide that you must have a mold, sizer, and top punch for your own bullet design, contact the mold manufacturer of your choice. Include a good sketch, including all dimensions, of the bullet you wish to cast.

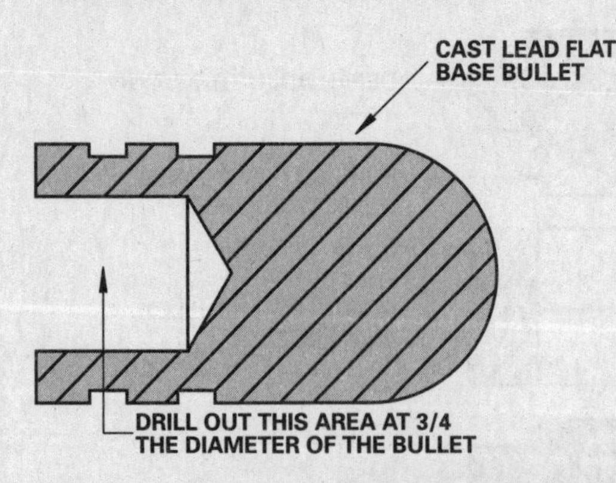

When "Miniéing" a lead bullet, the bullet must first be center-drilled. If not, the drill will "walk" in the soft lead and, most likely, drill oversize.

BUMPING

Bumping is a process that is applied to cast lead bullets for increasing their diameters—but only slightly. For example, it is not possible to bump a 150 grain .30 caliber bullet up to .58 caliber.

The bumping die is normally designed to fit a heavy-duty press. In operation, insert a bullet in the die and raise the ram to hold the bullet in place. With the ram at its highest position, adjust the die top until it just touches the top of the bullet. Lower the ram slightly and turn the die top down, one-half turn only. Raise the ram to "bump" the bullet. Examine and measure after each bump until the bullet is uniformly expanded to the desired diameter. This type of die can also be used in an arbor press, if slightly re-designed. You'll probably have to make these dies yourself from aluminum or brass (for longer production runs, use mild steel). Ream or turn the I.D. to the required dimension.

"MINIÉING"

Inspector Claude Minié was a French weapons inspector whose hollow base principle is used at the custom loader's convenience to make allowances for slightly undersized projectiles. Let's assume that the bore of your 11mm Beaumont is badly worn and slugging reveals a nominal diameter of .460 inch. The .457-inch diameter flat base bullets are not "taking" (expanding to fill) what little rifling remains and, as a result, accuracy is poor. Place the bullet in the lathe and drill a 25/64-inch diameter hole in the base about .25-inch deep. This hole provides a skirt that expands into the rifling once the powder ignites, extracting whatever spin is available.

JACKETED BULLET MODIFICATIONS

Perhaps the best way to illustrate a jacketed bullet modification is to make a projectile for the 7.35mm Carcano. Jacketed bullets at .300 diameter are all but nonexistent. Bullets for the Carcano are easily produced by selecting .30 caliber (.308-inch) 220-grain round nose bullets. These bullets are held in the lathe by the base and trued with a dial indicator to eliminate runout. Turn the bullets to .300-inch diameter, cut them off evenly to 150 grains (approx.), and boattail the bullet slightly. This operation is recommended only for very slight reductions. Otherwise, the jacket becomes too thin (.308 inch to .300 inch is considered maximum for a .30 caliber bullet).

SWAGING

Home-swaged bullets are as good as, or better than, the bullets you can buy from any bullet manufacturer. Armed with the proper equipment, you can, in fact, produce bullets of your own design at a special weight and for a special purpose. The truth is, next time you purchase .416 inch or .475 inch diameter bullets, chances are good some fellow just like yourself made them—with home swaging equipment!

When a large quantity of one particular projectile is desired, swaging is the only way to proceed. This process consists of forming a gilded metal jacket around a lead core. The cores are produced by cutting specially sized lead wire to the correct length (for the intended bullet); or lead cores may be cast, much as you'd cast a lead bullet. You'll need a supply of jackets, too, which must be purchased from a jacket manufacturer.

To swage bullets, you'll need a swaging die and a press heavy enough to do the swaging (swaging dies are available for most swaging presses). You'll also need a large quantity of one particular bullet, however, to justify the original outlay for swaging equipment.

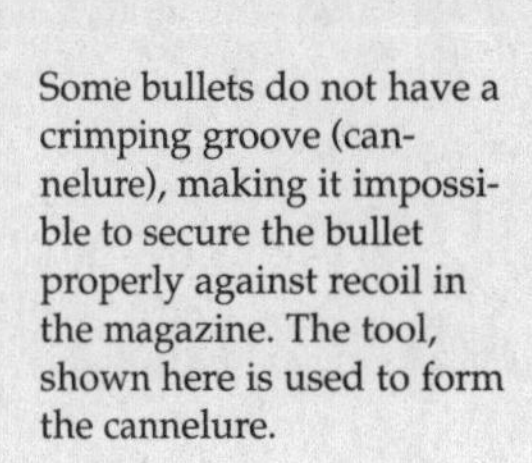

Some bullets do not have a crimping groove (cannelure), making it impossible to secure the bullet properly against recoil in the magazine. The tool, shown here is used to form the cannelure.

A BRIEF SUMMARY

More than 900 cartridge conversions are included in the following section of this book (Part II). You may view these upcoming pages as 900 separate sets of manufacturing instructions, or as 900 facets of the same "diamond."

As long as we are free to own and enjoy firearms in this country, there will always be a need for small arms ordnance technicians and the exotic ammunition they produce. Pick up any contemporary gun magazine and inevitably you'll find a letter to the editor asking something along this line: "Where can I find a long, rimmed cartridge in the style of the great British Nitro's, but one for which the starter case is more readily available?"

Using this book as a reference, that reader can look it up himself. In this instance, he'll find just what he's looking for: a 3 1/4 inch, .45 caliber basic case necked to .416-inch diameter. All he needs to do now is design the cartridge, build a set of dies, fabricate some test ammo and supply his gunsmith with a drawing for chambering reamers.

The original concept for this process is nearly 100 years olds. More importantly, the reader will have the satisfaction of owning a custom rifle and ammo combination that is tuned exactly to his wants and needs. And that's what the cartridge conversion business is all about.

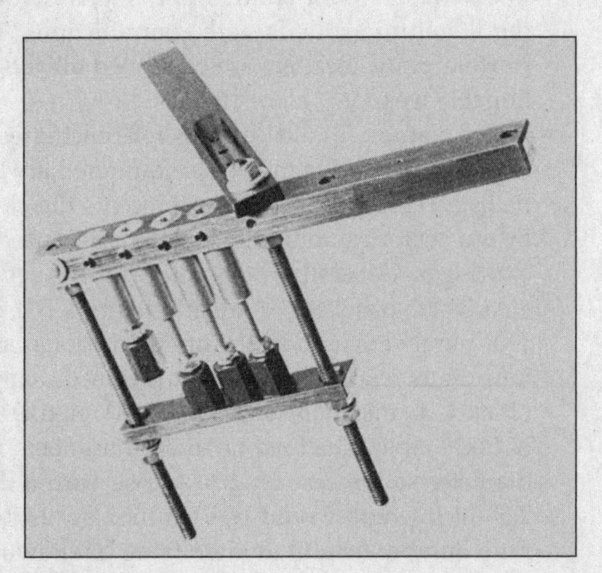

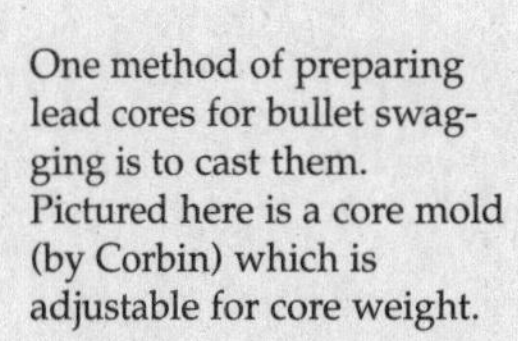

One method of preparing lead cores for bullet swagging is to cast them. Pictured here is a core mold (by Corbin) which is adjustable for core weight.

PART TWO Metallic Cartridge Conversions

Introduction

The purpose of the following conversions are to give the experimental handloader a reasonable starting point in the manufacture of specific cases. You will note that most conversions finish with the use of a full-length sizer die. Anyone attempting to produce ammo for a given weapon is advised to have the firearm handy and should take the time to obtain a suitable chamber cast to establish accurate chamber and bore dimensions. The die manufacturer can use these dimensions to produce a set of accurate, full-length dies. In many instances, a cast will not be necessary. Correspondence with a manufacturer should be enough to secure the proper dies.

Listed below is a key to the data listed on the conversion pages:

- Cartridge Name: The title given to each cartridge is that which is in widest use or, in the case of obscure cartridges, is most descriptive. Thus, 8mm Mauser is listed as 8x57 Mauser.
- Diameter: The projectile diameter listed is that which gives the most accurate, reliable performance. Many bullet diameters are somewhat hazy, and the literature abounds with contradictions. The listed diameter is simly a starting point. Readers are cautioned always to slug the bore of any questionable weapon.
- Other Names: To aid in cross-referencing each cartridge, other names by which the cartridge may be encountered are listed.
- Ballistek No.: This number represents the cartridge identification assigned by our custom ammunition business; it is used to avoid confusion with other cartridges. Generally, the first three digits represent the projectile diameter.
- NAI No.: Northwest Arizona Industries, Inc. is Ballistek's mother company. This number is used to classify the various cartridges by geometry. The various digits relate to the ratio of the head diameter to: (1) length to shoulder, (2) neck diameter, (3) neck length, (4) relative shoulder angle and (5) amount of body taper. The final three digit numbers represent the ratio of the head diameter to the case length. A case with a .500-inch head diameter and a 2.50-inch length would be identified as (2.50/5), or 5.00. You may find this ratio number helpful in identifying unknown cases.

DISCLAIMER

Any and all loading information contained in this book should be used with caution. Since they have no knowledge or control over the reader's choice of guns, components, and loadings, the author, publisher, and the various firearms and components manufacturers mentioned herein assume no responsibility for the use of this information.

Furthermore, while all sizes, dimensions and loadings included in the conversion pages (Parts I and II) have been checked for accuracy, neither the author nor the publisher can assume any liability for any incorrect information contained therein.

This numbering system was developed by the author as a method of categorizing metallic cartridge cases. Certain elements of this number were used to computer-select the drawings that accompany each conversion. While there are several thousand NAI number combinations, it was possible to prepare a rather finite selection of drawings and allow the computer to select that which best reflected the essence of a particular case. For this reason, the reader will note that some drawings do not appear exactly as described—but they do show the basic geometry of the case and where the various dimensions apply to it.

Here's how this numbering system works: A typical number might be: BEN 22344/4.625. The various elements of this number work out as follows:

1. Alpha
   RMS = Rimmed straight
   RMB = Rimmed bottleneck
   RXS = Rimless straight
   RXB = Rimless bottleneck
   BES = Belted straight
   BEN = Belted bottleneck
   RBS = Rebated straight
   RBB = Rebated bottleneck
2. Position One (000N0000/0.000). Body Taper.
   1 = Less than 1 degree
   2 = Greater than one degree, less than 2 degrees.
   3 = Greater than 2 degrees.
3. Position Two (000 0N000/0.000). Length to shoulder.
   1 = No shoulder (straight case).
   2 = Long length to shoulder. Greater than 3.2 head diameters long.
   3 = Moderate length to shoulder. Greater than 2.5 head diameters long, less than 3.199.4 = Short length to shoulder. Less than 2.499 head diameters long.
4. Position Three (000 00N00/0.000). Neck diameter.
   1 = No neck (straight case).

2 = Large neck. Greater than .63 head diameters in diameter.
3 = Moderate neck. Between .60 and .629 head diameters in diameter.
4 = Small neck. Less than .599 head diameters in diameter.

5. Position Four (000000N0/0.000. Neck Length.
   1 = no neck (straight case)
   2 = long neck length. Greater than 1.6 projectile diameters long.
   3 = Moderate neck length. Between 1.4 and 1.59 projectile diameters long.
   4 = Short neck length. Less than 1.39 projectile diameters long.
6. Position Five (000 0000n/0.000). Shoulder angle.
   1 = 6 to 10 degree shoulder.
   2 = 11 to 20 degree shoulder.
   3 = 21 to 30 degree shoulder.
   4 = Greater than 31 degree shoulder.
   5 = No shoulder (straight case).
7. Position Six, Seven & Eight (000 00000/N.NNN).
   Overall length in terms of head diameters long. A case with an overall length of 2.55 inches and a head diameter of .470 inch would be represented as (2.55/.470) 5.420. In other words, this case is 5.42 head diameters long.

There may one day be a system of cataloguing cartridge cases that will leave no question as to what a particular case looks like. Meanwhile, this system will allow anyone to construct a reasonably accurate, proportional drawing of the case in question. The reason for this whole exercise, beyond that of assembling drawings for this book, is to create a computer program that will identify all unknown cases, including those sent in by individuals who do not know what they have or are not certain of a particular chambering. Through the use of precision measuring instruments, certain dimensions can be programmed into the computer, which will, in turn, identify the cartridge.

- Data Source: The listed source provides the reader with a secondary source of information for more complete coverage of historical or actual loading data.
- Historical Data: This listing indicates who developed the cartridge and approximately when. Sadly, much of this information has never been recorded; thus, there are cartridges for which no such information can be supplied.
- Notes: A listing of manufacturing notes and other data pertinent to each cartridge.
- Loading Data: This book is concerned with the mechanics of building a particular cartridge case and is not intended to be a loading manual. At least one load for each cartridge has been presented here for use as a rough guide only. Many of these loads come from other sources and are, in such instances, attributed to the proper individual. Other loads are those developed in the author's shop or by his computer.

Neither the author nor the publisher is responsible for any use of this data. Loads taken from Ackley's books are notoriously hot, and the velocity figures given make one wonder what barrel lengths were used. Our own computer-generated loads operate only on the premise of new, strong weapons and perfect brass cases. Too many variables enter the picture when one tries to allow for the age and quality of a specific rifle (or case).

CASE PREPARATION

- Make From: The case referred to is that which forms into the required case with the least amount of work. If you are making, say, .243 Winchester cases, start with .270 Winchester cases. The .30-06 will work equally well; however, the .270 case requires one less neck reduction operation. It's almost always preferable to form cases down as opposed to expanding them. A .35 Ackley Magnum can be made from .300 Holland & Holland brass by expanding the neck, but it's much easier to start with .375 Holland & Holland and squeeze the whole business down.
- Shellholder: The RCBS shellholder is listed for the finished case and can thus be used as a rough gauge in making radical case-head changes. In a conversion that begins with a .348 Winchester case and calls for a No. 4 shellholder, the shellholder will help to verify the required lathe work.

The balance of the Case Preparation section concerns the procedure for making the case and is based on actual experience in the author's own shop. You may discover shortcuts and/or better methods, in which case the publisher would enjoy hearing from you about these valid improvements.

PHYSICAL DATA

The author does not have samples of all cases listed. Some were obtained from cartridge collectors for the sole purpose of taking measurements. Other dimensions were taken from the literature or were deduced by manipulation of known data. A great many of the cases listed herein were designed and manufactured before the days of quality control and precision measuring instruments; thus, variations exist. Any errors should be brought to the publisher's attention.

- Shoulder Angle and Body Angle: These two figures, which came from the computer, are only as accurate as the data that's available. There will most likely be discrepancies in the shoulder angle figures. Many cartridges are presented as "having a 30-degree shoulder," or some other equally precise figure. When the actual case dimensions are fed into the computer, however, a slightly different angle may appear. In such cases, the error probably lies within the original statement.
- Case Capacity: This data, which also comes from the computer, refers to the volumetric capacity of each case expressed in cubic centimeters (cc's) and in

theoretical grains of water. The data was obtained from known external case dimensions. Again, this data is for reference only—the computer does not "know" that a military match 7.62mm NATO case has a smaller capacity than a commercial .308 Winchester case.

Regardless of accuracy, these figures are useful for comparing one case with another. The figures given here are 100 percent capacity; that is, level with the case mouth. To use this information properly, the reader must determine the volume of that portion of the bullet that enters the case (volume = 3.1416 x radius squared x length), and subtract that figure from the capacity listed.

- Dimensional Case Drawing: As mentioned, the NAI number was used to classify ammunition cases by various case dimensional parameters and ratios. These numbers suggested the closest master drawing, which is pictured. This drawing is meant to show relative form only. Do not scale these drawings or in any way try to take measurements directly from them. You should find all that information listed under "Physical Data." The letters "A" through "J" in the drawings refer, of course, to the corresponding letters provided under the heading, "Physical Data."
- Abbreviations: The following abbreviations are used throughout Part II.

ACP—Automatic Colt Pistol
APPROX—Approximately
AR—As required
A/R –Auto rimmed
AUTO—Automatic
B/C—Backchamfer
B'EYE—Bullseye
B/N—Bottleneck
BP—Black powder
BR—Benchrest
BT—Boattail
CAL—Caliber
C/D—Centerdrill
CF—Centerfire
CTG—Cartridge
DIA—Diameter
DRL—Drill
EXP—Expand
FAB—Fabricate
FF—Fireform
F/H—Flashhole
FL—Full-length
FMJ—Full metal jacket
FP—Flat point
GA—Gauge
GC—Gas check
GG—Grease groove
GR—Grain
H&H—Holland & Holland
HJ—Half Jacket
HP—Hollow point
IMP—Improved
JRN—Jacketed round nose
LUB—Lubricated
MAG—Magnum
MAX—Maximum
MC—Metal case
MM—Millimeter
NE—Nitro Express
P/P—Paper Patched
PP—Primer pocket
REM—Remington
RN—Round nose
S/H—Shellholder
SJ—Short Jacket
SL—Self-loading
SOL—Solvent
SP—Spire point
SPL—Special
SS—Single shot
ST—Straight
S&W—Smith & Wesson
SWC—Semi-wadcutter
WIN—Winchester
WC—Wadcutter

ENGLISH AND AMERICAN CARTRIDGES

CARTRIDGE: .14 Jones

OTHER NAMES:	
	DIA: .140
	BALLISTEK NO: 140A
	NAI NO: RMB 14344/2.668

DATA SOURCE: Ackley Vol.2 Pg.109

HISTORICAL DATA: By A. Jones in 1920

NOTES: A .14 Velo Dog also exists.

LOADING DATA:

BULLET WT./TYPE	POWDER WT./TYPE	VELOCITY ('/SEC)	SOURCE
No loading data is available for this cartridge.			

CASE PREPARATION: SHELLHOLDER (RCBS): 29

MAKE FROM: .22 Long Rifle. Pull factory bullet and remove powder. Run into Jones form die. Square case mouth and chamfer. Seat .140-inch dia. bullet. I do not have loading data but, you can start with about 75% of the original powder.

PHYSICAL DATA (INCHES):

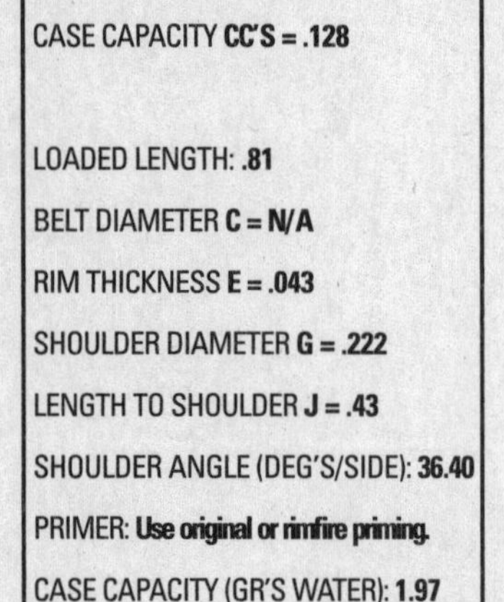

CASE TYPE: **Rimmed Bottleneck**

CASE LENGTH **A = .595**

HEAD DIAMETER **B = .223**

RIM DIAMETER **D = .274**

NECK DIAMETER **F = .163**

NECK LENGTH **H = .125**

SHOULDER LENGTH **K = .039**

BODY ANGLE (DEG'S/SIDE): **.125**

CASE CAPACITY **CC'S = .128**

LOADED LENGTH: **.81**

BELT DIAMETER **C = N/A**

RIM THICKNESS **E = .043**

SHOULDER DIAMETER **G = .222**

LENGTH TO SHOULDER **J = .43**

SHOULDER ANGLE (DEG'S/SIDE): **36.40**

PRIMER: **Use original or rimfire priming.**

CASE CAPACITY (GR'S WATER): **1.97**

DIMENSIONAL DRAWING:

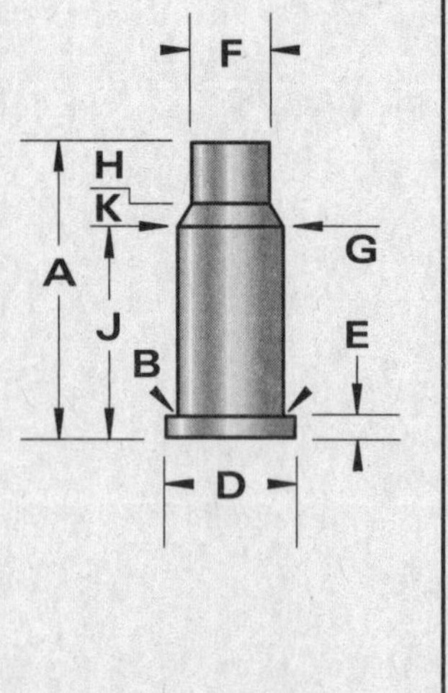

-NOT ACTUAL SIZE-
-DO NOT SCALE-

CARTRIDGE: .14/221 Walker

OTHER NAMES:	
	DIA: .140
	BALLISTEK NO: 140B
	NAI NO: RXB 24423/3.763

DATA SOURCE: NAI/Ballistek

HISTORICAL DATA: By Walker Machine Tool Co.

NOTES:

LOADING DATA:

BULLET WT./TYPE	POWDER WT./TYPE	VELOCITY ('/SEC)	SOURCE
20/–	15.0/3031	3643	Walker

CASE PREPARATION: SHELLHOLDER (RCBS): 10

MAKE FROM: .221 Fireball. Anneal case neck. Run case into form die. Trim to length. F/L size and chamfer.

PHYSICAL DATA (INCHES):

CASE TYPE: **Rimless Bottleneck**

CASE LENGTH **A = 1.415**

HEAD DIAMETER **B = .376**

RIM DIAMETER **D = .378**

NECK DIAMETER **F = .164**

NECK LENGTH **H = .188**

SHOULDER LENGTH **K = .177**

BODY ANGLE (DEG'S/SIDE): **.876**

CASE CAPACITY **CC'S = 1.28**

LOADED LENGTH: **1.78**

BELT DIAMETER **C = N/A**

RIM THICKNESS **E = .045**

SHOULDER DIAMETER **G = .350**

LENGTH TO SHOULDER **J = 1.05**

SHOULDER ANGLE (DEG'S/SIDE): **27.72**

PRIMER: **S/R**

CASE CAPACITY (GR'S WATER): **19.80**

DIMENSIONAL DRAWING:

-NOT ACTUAL SIZE-
-DO NOT SCALE-

CARTRIDGE: .17 Ackley Bee

OTHER NAMES:
.17/218 Improved Bee

DIA: .172

BALLISTEK NO: 172D

NAI NO: RMB 23424/3.868

DATA SOURCE: NAI/Ballistek

HISTORICAL DATA: By P.O. Ackley in 1944

NOTES: Ideal case for .172 inch dia. bore.

LOADING DATA:

BULLET WT./TYPE	POWDER WT./TYPE	VELOCITY ('/SEC)	SOURCE
25/Spire	14.0/IMR4198	3535	Ackley

CASE PREPARATION: SHELLHOLDER (RCBS): 1

MAKE FROM: .218 Bee. Anneal case neck. Form case in form die (or in F/L die with expander removed [you'll have some loss]). Trim to length and chamfer. Fireform in chamber.

PHYSICAL DATA (INCHES):

CASE TYPE: **Rimmed Bottleneck**

CASE LENGTH **A = 1.35**

HEAD DIAMETER **B = .349**

RIM DIAMETER **D = .408**

NECK DIAMETER **F = .195**

NECK LENGTH **H = .243**

SHOULDER LENGTH **K = .093**

BODY ANGLE (DEG'S/SIDE): **.527**

CASE CAPACITY **CC'S = 1.09**

LOADED LENGTH: **1.67**

BELT DIAMETER **C = N/A**

RIM THICKNESS **E = .065**

SHOULDER DIAMETER **G = .334**

LENGTH TO SHOULDER **J = 1.014**

SHOULDER ANGLE (DEG'S/SIDE): **35.77**

PRIMER: **S/R**

CASE CAPACITY (GR'S WATER): **16.89**

DIMENSIONAL DRAWING:

F H K G A J E B D

-NOT ACTUAL SIZE-
-DO NOT SCALE-

CARTRIDGE: .17 Ackley Hornet

OTHER NAMES:
.17/22 Hornet Improved

DIA: .172

BALLISTEK NO: 172K

NAI NO: RMB 12324/4.682

DATA SOURCE: Ackley Vol.1 Pg.257

HISTORICAL DATA: By P.O. Ackley

NOTES:

LOADING DATA:

BULLET WT./TYPE	POWDER WT./TYPE	VELOCITY ('/SEC)	SOURCE
25/Spire	11.0/4227	3390	Ackley

CASE PREPARATION: SHELLHOLDER (RCBS): 2

MAKE FROM: .22 Hornet. Anneal case neck. I suggest a form die, however, cases can be formed in the F/L die with some loss. Trim to length. Chamfer. F/L size.

PHYSICAL DATA (INCHES):

CASE TYPE: **Rimmed Bottleneck**

CASE LENGTH **A = 1.40**

HEAD DIAMETER **B = .299**

RIM DIAMETER **D = .350**

NECK DIAMETER **F = .193**

NECK LENGTH **H = .225**

SHOULDER LENGTH **K = .060**

BODY ANGLE (DEG'S/SIDE): **.346**

CASE CAPACITY **CC'S = .888**

LOADED LENGTH: **1.80**

BELT DIAMETER **C = N/A**

RIM THICKNESS **E = .062**

SHOULDER DIAMETER **G = .288**

LENGTH TO SHOULDER **J = 1.109**

SHOULDER ANGLE (DEG'S/SIDE)**35.62**

PRIMER: **S/R**

CASE CAPACITY (GR'S WATER): **13.71**

DIMENSIONAL DRAWING:

F H K G A J E B D

-NOT ACTUAL SIZE-
-DO NOT SCALE-

CARTRIDGE: .17 Bumblebee

OTHER NAMES:

DIA: .172

BALLISTEK NO: 172Q

NAI NO: RMB 24433/2.628

DATA SOURCE: Handloader #98, pg. 40

HISTORICAL DATA:

NOTES:

LOADING DATA:

BULLET WT./TYPE	POWDER WT./TYPE	VELOCITY ('/SEC)	SOURCE
25/Spire	7.0/IMR4227	2245	Waters

CASE PREPARATION: SHELLHOLDER (RCBS): 1

MAKE FROM: .218 Bee. Anneal case. Run into form die. Trim to length and chamfer. F/L size. Clean.

PHYSICAL DATA (INCHES):

CASE TYPE: **Rimmed Bottleneck**

CASE LENGTH: **A = .920**

HEAD DIAMETER: **B = .350**

RIM DIAMETER: **D = .408**

NECK DIAMETER: **F = .197**

NECK LENGTH: **H = .195**

SHOULDER LENGTH: **K = .136**

BODY ANGLE (DEG'S/SIDE): **.736**

CASE CAPACITY **CC'S = .597**

LOADED LENGTH: **1.34**

BELT DIAMETER: **C = N/A**

RIM THICKNESS: **E = .063**

SHOULDER DIAMETER: **G = .34**

LENGTH TO SHOULDER: **J = .589**

SHOULDER ANGLE (DEG'S/SIDE): **27.73**

PRIMER: **S/R**

CASE CAPACITY (GR'S WATER): **9.22**

DIMENSIONAL DRAWING:

F H K A G J B E D

-NOT ACTUAL SIZE-
-DO NOT SCALE-

CARTRIDGE: .17 Hempalina

OTHER NAMES:
.17 Javalina

DIA: .172

BALLISTEK NO: 172L

NAI NO: RXB 23423/4.090

DATA SOURCE: NAI/Ballistek

HISTORICAL DATA:

NOTES:

LOADING DATA:

BULLET WT./TYPE	POWDER WT./TYPE	VELOCITY ('/SEC)	SOURCE
25/Spire	15.5/IMR4198	3510	JJD

CASE PREPARATION: SHELLHOLDER (RCBS): 10

MAKE FROM: .222 Remington. Form dies are required. Run .222 case into form die #1. Run into .17 Mach IV form #3. Trim to length. Chamfer. F/L size. Clean.

PHYSICAL DATA (INCHES):

CASE TYPE: **Rimless Bottleneck**

CASE LENGTH: **A = 1.53**

HEAD DIAMETER: **B = .374**

RIM DIAMETER: **D = .375**

NECK DIAMETER: **F = .201**

NECK LENGTH: **H = .218**

SHOULDER LENGTH: **K = .152**

BODY ANGLE (DEG'S/SIDE): **.626**

CASE CAPACITY **CC'S = 1.41**

LOADED LENGTH: **1.922**

BELT DIAMETER: **C = N/A**

RIM THICKNESS: **E = .042**

SHOULDER DIAMETER: **G = .353**

LENGTH TO SHOULDER: **J = 1.16**

SHOULDER ANGLE (DEG'S/SIDE): **26.56**

PRIMER: **S/R**

CASE CAPACITY (GR'S WATER): **21.77**

DIMENSIONAL DRAWING:

F H K G A J E B D

-NOT ACTUAL SIZE-
-DO NOT SCALE-

CARTRIDGE: .17 Mach IV

OTHER NAMES:

DIA: .172

BALLISTEK NO: 172C

NAI NO: RXB 23433/3.723

DATA SOURCE: Hornady Manual 3rd pg.58

HISTORICAL DATA: Designed by the O'Brien Rifle Company.

NOTES:

LOADING DATA:

BULLET WT./TYPE	POWDER WT./TYPE	VELOCITY ('/SEC)	SOURCE
25/HP	14.0/IMR4227	3600	Horn.

CASE PREPARATION: SHELLHOLDER (RCBS): 10

MAKE FROM: .221 Remington Fireball. A form die set is required. Run case into form #1 followed by form #2. Trim to length. F/L size. Chamfer neck.

PHYSICAL DATA (INCHES):

CASE TYPE: **Rimless Bottleneck**

CASE LENGTH: **A = 1.400**

HEAD DIAMETER: **B = .376**

RIM DIAMETER: **D = .378**

NECK DIAMETER: **F = .199**

NECK LENGTH: **H = .203**

SHOULDER LENGTH: **K = .139**

BODY ANGLE (DEG'S/SIDE): **.534**

CASE CAPACITY **CC'S = 1.315**

LOADED LENGTH: **1.77**

BELT DIAMETER: **C = N/A**

RIM THICKNESS: **E = .045**

SHOULDER DIAMETER: **G = .360**

LENGTH TO SHOULDER: **J = 1.058**

SHOULDER ANGLE (DEG'S/SIDE): **30.07**

PRIMER: **S/R**

CASE CAPACITY (GR'S WATER): **20.31**

DIMENSIONAL DRAWING:

-NOT ACTUAL SIZE-
-DO NOT SCALE-

CARTRIDGE: .17 Remington

OTHER NAMES:

DIA: .172

BALLISTEK NO: 172B

NAI NO: RXB 22423/4.764

DATA SOURCE: Hornady Manual 3rd pg.60

HISTORICAL DATA: By Remington in 1971

NOTES:

LOADING DATA:

BULLET WT./TYPE	POWDER WT./TYPE	VELOCITY ('/SEC)	SOURCE
25/Spire	22.5/BL-C2	3900	JJD

CASE PREPARATION: SHELLHOLDER (RCBS): 10

MAKE FROM: Factory or .223 Rem. Run the .223 case into the .17 Rem. die (with the expander removed). You may have trouble with the neck so, experiment with annealing. Purchase a form die set, if possible - the job is much quicker.

PHYSICAL DATA (INCHES):

CASE TYPE: **Rimless Bottleneck**

CASE LENGTH: **A = 1.796**

HEAD DIAMETER: **B = .377**

RIM DIAMETER: **D = .378**

NECK DIAMETER: **F = .199**

NECK LENGTH: **H = .260**

SHOULDER LENGTH: **K = .185**

BODY ANGLE (DEG'S/SIDE): **.522**

CASE CAPACITY **CC'S = 1.76**

LOADED LENGTH: **2.17**

BELT DIAMETER: **C = N/A**

RIM THICKNESS: **E = .045**

SHOULDER DIAMETER: **G = .356**

LENGTH TO SHOULDER: **J = 1.351**

SHOULDER ANGLE (DEG'S/SIDE): **22.99**

PRIMER: **S/R**

CASE CAPACITY (GR'S WATER): **27.23**

DIMENSIONAL DRAWING:

-NOT ACTUAL SIZE-
-DO NOT SCALE-

CARTRIDGE: .17/218 Bee

OTHER NAMES:
.17 Bee

DIA: .172

BALLISTEK NO: 172M

NAI NO: RMB
13424/3.839

DATA SOURCE: NAI/Ballistek

HISTORICAL DATA:

NOTES:

LOADING DATA:

BULLET WT./TYPE	POWDER WT./TYPE	VELOCITY ('/SEC)	SOURCE
Use loading data for .17 Ackley Bee reduced for starting loads.			

CASE PREPARATION: SHELLHOLDER (RCBS): 1

MAKE FROM: .218 Bee. Anneal case neck. Run into form die (required!). Trim case to length, F/L size and chamfer. Fireform in chamber.

PHYSICAL DATA (INCHES):

CASE TYPE: **Rimmed Bottleneck**

CASE LENGTH: **A = 1.34**

HEAD DIAMETER: **B = .349**

RIM DIAMETER: **D = .408**

NECK DIAMETER: **F = .201**

NECK LENGTH: **H = .224**

SHOULDER LENGTH: **K = .096**

BODY ANGLE (DEG'S/SIDE): **.314**

CASE CAPACITY **CC'S = 1.06**

LOADED LENGTH: **1.70**

BELT DIAMETER: **C = N/A**

RIM THICKNESS: **E = .063**

SHOULDER DIAMETER: **G = .340**

LENGTH TO SHOULDER: **J = 1.02**

SHOULDER ANGLE (DEG'S/SIDE): **35.90**

PRIMER: **S/R**

CASE CAPACITY (GR'S WATER): **16.37**

DIMENSIONAL DRAWING:

-NOT ACTUAL SIZE-
-DO NOT SCALE-

CARTRIDGE: .17/22-250 Remington

OTHER NAMES:
.17 Super Eyebunger

DIA: .172

BALLISTEK NO: 172H

NAI NO: RXB
32433/4.149

DATA SOURCE: NAI/Ballistek

HISTORICAL DATA:

NOTES: Overbore for .172" dia.

LOADING DATA:

BULLET WT./TYPE	POWDER WT./TYPE	VELOCITY ('/SEC)	SOURCE
25/Spire	34.0/IMR4350	4250(!)	Ackley

CASE PREPARATION: SHELLHOLDER (RCBS): 3

MAKE FROM: .243 Winchester. (Note: if .243 brass is used, it must first be F/L sized back to .243.) Run into .22-250 F/L die with expander removed. Run into .17/22-250 form die. Trim to length and chamfer. Ream neck I.D. Chamfer and F/L size.

PHYSICAL DATA (INCHES):

CASE TYPE: **Rimless Bottleneck**

CASE LENGTH: **A = 1.917**

HEAD DIAMETER: **B = .462**

RIM DIAMETER: **D = .466**

NECK DIAMETER: **F = .204**

NECK LENGTH: **H = .206**

SHOULDER LENGTH: **K = .197**

BODY ANGLE (DEG'S/SIDE): **1.22**

CASE CAPACITY **CC'S = 2.70**

LOADED LENGTH: **2.311**

BELT DIAMETER: **C = N/A**

RIM THICKNESS: **E = .043**

SHOULDER DIAMETER: **G = .406**

LENGTH TO SHOULDER: **J = 1.514**

SHOULDER ANGLE (DEG'S/SIDE): **27.14**

PRIMER: **L/R**

CASE CAPACITY (GR'S WATER): **41.81**

DIMENSIONAL DRAWING:

-NOT ACTUAL SIZE-
-DO NOT SCALE-

CARTRIDGE: .17/222 Remington

OTHER NAMES:	
	DIA: .172
	BALLISTEK NO: 172E
	NAI NO: RXB 22423/4.582

DATA SOURCE: Ackley Vol.1 Pg.258

HISTORICAL DATA:

NOTES:

LOADING DATA:

BULLET WT./TYPE	POWDER WT./TYPE	VELOCITY ('/SEC)	SOURCE
25/Spire	16.0/IMR4198	3510	Ackley

CASE PREPARATION: SHELLHOLDER (RCBS): 10

MAKE FROM: .222 Remington. A form die is required. Anneal case neck. Form case in form die. Trim to length. F/L size. Case will fireform in chamber.

PHYSICAL DATA (INCHES):

CASE TYPE: **Rimless Bottleneck**

CASE LENGTH: **A = 1.723**

HEAD DIAMETER: **B = .376**

RIM DIAMETER: **D = .378**

NECK DIAMETER: **F = .201**

NECK LENGTH: **H = .24**

SHOULDER LENGTH: **K = .206**

BODY ANGLE (DEG'S/SIDE): **.558**

CASE CAPACITY **CC'S = 1.63**

LOADED LENGTH: **2.195**

BELT DIAMETER: **C = N/A**

RIM THICKNESS: **E = .045**

SHOULDER DIAMETER: **G = .355**

LENGTH TO SHOULDER: **J = 1.277**

SHOULDER ANGLE (DEG'S/SIDE): **20.49**

PRIMER: **S/R**

CASE CAPACITY (GR'S WATER): **25.25**

DIMENSIONAL DRAWING:

F
H
K
G
A
J
E
B
D

-NOT ACTUAL SIZE-
-DO NOT SCALE-

CARTRIDGE: .17/223 Remington

OTHER NAMES:	
	DIA: .172
	BALLISTEK NO: 172F
	NAI NO: RXB 22434/4.681

DATA SOURCE: NAI/Ballistek

HISTORICAL DATA: By Harrington and Richardson about 1968

NOTES:

LOADING DATA:

BULLET WT./TYPE	POWDER WT./TYPE	VELOCITY ('/SEC)	SOURCE
25/HP	19.0/IMR4198	3940	Wooters

CASE PREPARATION: SHELLHOLDER (RCBS): 10

MAKE FROM: .223 Remington. Anneal case. Form in form die. Trim to length. F/L size. Chamfer. Case will finish form in chamber.

PHYSICAL DATA (INCHES):

CASE TYPE: **Rimless Bottleneck**

CASE LENGTH: **A = 1.76**

HEAD DIAMETER: **B = .376**

RIM DIAMETER: **D = .378**

NECK DIAMETER: **F = .203**

NECK LENGTH: **H = .205**

SHOULDER LENGTH: **K = .125**

BODY ANGLE (DEG'S/SIDE): **.419**

CASE CAPACITY **CC'S = 1.74**

LOADED LENGTH: **2.15**

BELT DIAMETER: **C = N/A**

RIM THICKNESS: **E = .045**

SHOULDER DIAMETER: **G = .358**

LENGTH TO SHOULDER: **J = 1.43**

SHOULDER ANGLE (DEG'S/SIDE): **31.79**

PRIMER: **S/R**

CASE CAPACITY (GR'S WATER): **26.95**

DIMENSIONAL DRAWING:

F
H
K
G
A
J
E
B
D

-NOT ACTUAL SIZE-
-DO NOT SCALE-

CARTRIDGE: .17/224 Weatherby

OTHER NAMES:

DIA: .172

BALLISTEK NO: 172T

NAI NO: BEN 22423/4.626

DATA SOURCE: NAI/Ballistek

HISTORICAL DATA:

NOTES:

LOADING DATA:

BULLET WT./TYPE	POWDER WT./TYPE	VELOCITY ('/SEC)	SOURCE
25/Spire	34.0/H335	4150	JJD

CASE PREPARATION: **SHELLHOLDER (RCBS): 27**

MAKE FROM: .224 Weatherby. Form die required. Anneal case and form in form die. Trim to length and chamfer. F/L size. Fireform in chamber.

PHYSICAL DATA (INCHES):

CASE TYPE: **Belted Bottleneck**

CASE LENGTH: **A = 1.92**

HEAD DIAMETER: **B = .415**

RIM DIAMETER: **D = .429**

NECK DIAMETER: **F = .196**

NECK LENGTH: **H = .250**

SHOULDER LENGTH: **K = .190**

BODY ANGLE (DEG'S/SIDE): **.559**

CASE CAPACITY **CC'S = 2.41**

LOADED LENGTH: **A/R**

BELT DIAMETER: **C = .429**

RIM THICKNESS: **E = .049**

SHOULDER DIAMETER: **G = .390**

LENGTH TO SHOULDER: **J = 1.48**

SHOULDER ANGLE (DEG'S/SIDE): **N/A**

PRIMER: **L/R**

CASE CAPACITY (GR'S WATER): **37.25**

DIMENSIONAL DRAWING:

F
H
K
G
A
J
E
B
D
C

-NOT ACTUAL SIZE-
-DO NOT SCALE-

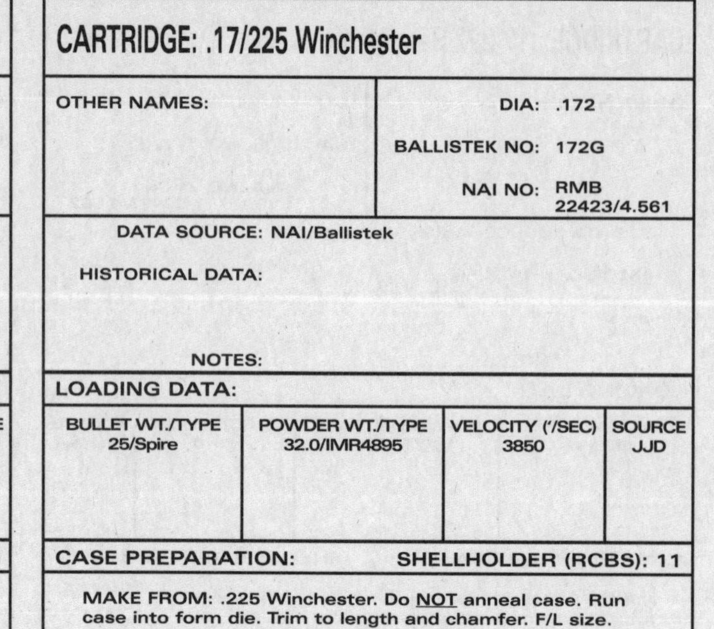

CARTRIDGE: .17/225 Winchester

OTHER NAMES:

DIA: .172

BALLISTEK NO: 172G

NAI NO: RMB 22423/4.561

DATA SOURCE: NAI/Ballistek

HISTORICAL DATA:

NOTES:

LOADING DATA:

BULLET WT./TYPE	POWDER WT./TYPE	VELOCITY ('/SEC)	SOURCE
25/Spire	32.0/IMR4895	3850	JJD

CASE PREPARATION: **SHELLHOLDER (RCBS): 11**

MAKE FROM: .225 Winchester. Do NOT anneal case. Run case into form die. Trim to length and chamfer. F/L size.

PHYSICAL DATA (INCHES):

CASE TYPE: **Rimmed Bottleneck**

CASE LENGTH: **A = 1.925**

HEAD DIAMETER: **B = .422**

RIM DIAMETER: **D = .473**

NECK DIAMETER: **F = .206**

NECK LENGTH: **H = .259**

SHOULDER LENGTH: **K = .176**

BODY ANGLE (DEG'S/SIDE): **.555**

CASE CAPACITY **CC'S = 2.32**

LOADED LENGTH: **2.33**

BELT DIAMETER: **C = N/A**

RIM THICKNESS: **E = .05**

SHOULDER DIAMETER: **G = .397**

LENGTH TO SHOULDER: **J = .149**

SHOULDER ANGLE (DEG'S/SIDE): **28.48**

PRIMER: **L/R**

CASE CAPACITY (GR'S WATER): **35.89**

DIMENSIONAL DRAWING:

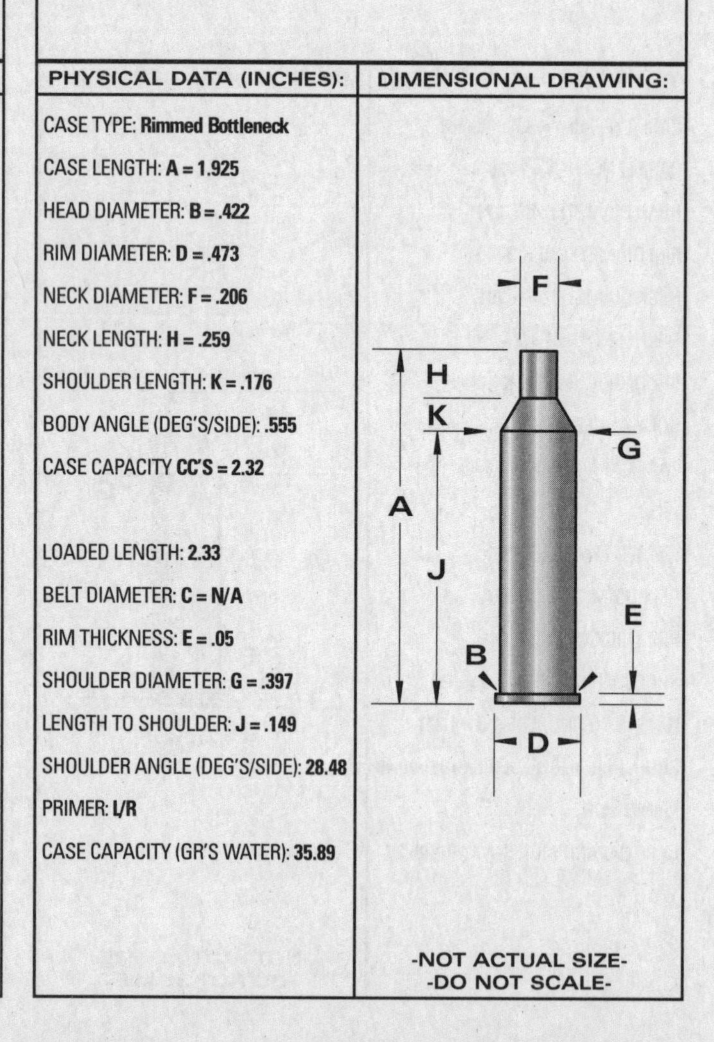

-NOT ACTUAL SIZE-
-DO NOT SCALE-

CARTRIDGE: .17/30 Carbine

OTHER NAMES:
.17 Pee Wee

DIA: .172

BALLISTEK NO: 172N

NAI NO: RXB
34433/3.558

DATA SOURCE: NAI/Ballistek

HISTORICAL DATA: R. Schuetz version.

NOTES:

LOADING DATA:

BULLET WT./TYPE	POWDER WT./TYPE	VELOCITY ('/SEC)	SOURCE
25/Spire	10.5/2400	3789	Ackley

CASE PREPARATION: SHELLHOLDER (RCBS): 17

MAKE FROM: .30 Carbine. Anneal case neck and form in form die. Trim to length. Chamfer. F/L size.

PHYSICAL DATA (INCHES):

CASE TYPE: **Rimless Bottleneck**

CASE LENGTH: **A = 1.267**

HEAD DIAMETER: **B = .356**

RIM DIAMETER: **D = .360**

NECK DIAMETER: **F = .204**

NECK LENGTH: **H = .197**

SHOULDER LENGTH: **K = .101**

BODY ANGLE (DEG'S/SIDE): **1.23**

CASE CAPACITY **CC'S = .935**

LOADED LENGTH: **1.64**

BELT DIAMETER: **C = N/A**

RIM THICKNESS: **E = .05**

SHOULDER DIAMETER: **G = .323**

LENGTH TO SHOULDER: **J = .969**

SHOULDER ANGLE (DEG'S/SIDE): **30.50**

PRIMER: **S/R**

CASE CAPACITY (GR'S WATER): **14.43**

DIMENSIONAL DRAWING:

-NOT ACTUAL SIZE-
-DO NOT SCALE-

CARTRIDGE: .20/222 Remington

OTHER NAMES:

DIA: .204

BALLISTEK NO: 204C

NAI NO: RXB
22423/4.654

DATA SOURCE: Ackley Vol.1 Pg.260

HISTORICAL DATA:

NOTES:

LOADING DATA:

BULLET WT./TYPE	POWDER WT./TYPE	VELOCITY ('/SEC)	SOURCE
45/Spire	20.0/IMR4198	4350	Ackley

CASE PREPARATION: SHELLHOLDER (RCBS): 10

MAKE FROM: .222 Remington. Form neck down in .20-222 trim die. Trim to length and chamfer. F/L size. Use 5mm lead bullets pulled from 5mm Rimfire ammo (see 5mm/223 Rem #204B)

PHYSICAL DATA (INCHES):

CASE TYPE: **Rimless Bottleneck**

CASE LENGTH: **A = 1.75**

HEAD DIAMETER: **B = .376**

RIM DIAMETER: **D = .378**

NECK DIAMETER: **F = .226**

NECK LENGTH: **H = .290**

SHOULDER LENGTH: **K = .140**

BODY ANGLE (DEG'S/SIDE): **.741**

CASE CAPACITY **CC'S = 1.77**

LOADED LENGTH: **2.22**

BELT DIAMETER: **C = N/A**

RIM THICKNESS: **E = .045**

SHOULDER DIAMETER: **G = .347**

LENGTH TO SHOULDER: **J = 1.32**

SHOULDER ANGLE (DEG'S/SIDE): **23.37**

PRIMER: **S/R**

CASE CAPACITY (GR'S WATER): **27.38**

DIMENSIONAL DRAWING:

-NOT ACTUAL SIZE-
-DO NOT SCALE-

CARTRIDGE: .20/222 Remington Mag

OTHER NAMES:

DIA: .204

BALLISTEK NO: 204D

NAI NO: RXB 22423/4.906

DATA SOURCE: NAI/Ballistek

HISTORICAL DATA:

NOTES:

LOADING DATA:

BULLET WT./TYPE	POWDER WT./TYPE	VELOCITY ('/SEC)	SOURCE
45/Spire	20.2/IMR4198	3400	JJD

CASE PREPARATION: SHELLHOLDER (RCBS): 10

MAKE FROM: .222 Remington Magnum. Form die is required but, this case can use the form die for .20/222 Rem. or .20/223 Rem. Otherwise, prepare in the same manner. Bullets are also obtained the same way.

PHYSICAL DATA (INCHES):

CASE TYPE: **Rimless Bottleneck**

CASE LENGTH: **A = 1.84**

HEAD DIAMETER: **B = .375**

RIM DIAMETER: **D = .378**

NECK DIAMETER: **F = .225**

NECK LENGTH: **H = .260**

SHOULDER LENGTH: **K = .160**

BODY ANGLE (DEG'S/SIDE): **.54**

CASE CAPACITY **CC'S = 1.95**

LOADED LENGTH: **A/R**

BELT DIAMETER: **C = N/A**

RIM THICKNESS: **E = .045**

SHOULDER DIAMETER: **G = .352**

LENGTH TO SHOULDER: **J = 1.42**

SHOULDER ANGLE (DEG'S/SIDE): **21.64**

PRIMER: **S/R**

CASE CAPACITY (GR'S WATER): **30.16**

DIMENSIONAL DRAWING:

-NOT ACTUAL SIZE-
-DO NOT SCALE-

CARTRIDGE: .204 Ruger

OTHER NAMES:

DIA: .204

BALLISTEK NO:

NAI NO:

DATA SOURCE: Hornady

HISTORICAL DATA: Created jointly with Hornady and Ruger in 2003. Based on the .222 Magnum case.

NOTES: Developed to provide a long range, flat shooting alternative to .22 centerfire cartridges for varmint hunting. Highest muzzle velocity of any current factory loaded hunting cartridge.

LOADING DATA:

BULLET WT./TYPE	POWDER WT./TYPE	VELOCITY ('/SEC)	SOURCE
33/Hornady V-Max		4225	Factory Load

CASE PREPARATION: SHELLHOLDER (RCBS): 10

MAKE FROM: Can be made by necking down .222 Magnum and fire forming. Then ream neck. Trim to length.

PHYSICAL DATA (INCHES):

CASE TYPE: **Rimless Bottleneck**

CASE LENGTH: **A = 1.8500**

HEAD DIAMETER: **B = .3763**

RIM DIAMETER: **D = .3780**

NECK DIAMETER: **F = .231**

NECK LENGTH: **H = .200**

SHOULDER LENGTH: **K = .1116**

BODY ANGLE (DEG'S/SIDE):

CASE CAPACITY **CC'S =**

LOADED LENGTH: **2.260**

BELT DIAMETER: **C = N/A**

RIM THICKNESS: **E = .0450**

SHOULDER DIAMETER: **G = .3599**

LENGTH TO SHOULDER: **J = 1.5386**

SHOULDER ANGLE (DEG'S/SIDE): **30.0**

PRIMER: **S/R**

CASE CAPACITY (GR'S WATER): **32**

DIMENSIONAL DRAWING:

-NOT ACTUAL SIZE-
-DO NOT SCALE-

CARTRIDGE: .218 Bee

OTHER NAMES:

DIA: .224

BALLISTEK NO: 224A

NAI NO: RMB 24342/3.667

DATA SOURCE: Hornady Manual 3rd Pg.64

HISTORICAL DATA: By Win. in 1938

NOTES:

LOADING DATA:

BULLET WT./TYPE	POWDER WT./TYPE	VELOCITY ('/SEC)	SOURCE
45/Spire	12.0/IMR4227	2800	Horn.

CASE PREPARATION: SHELLHOLDER (RCBS): 1

MAKE FROM: Factory or .32-20 Win. or .25-20 Win. (for .32-20, anneal neck). F/L size in .218 Bee die. Trim to length. Chamfer mouth.

PHYSICAL DATA (INCHES):

CASE TYPE: **Rimmed Bottleneck**

CASE LENGTH: **A = 1.280**

HEAD DIAMETER: **B = .349**

RIM DIAMETER: **D = .408**

NECK DIAMETER: **F = .242**

NECK LENGTH: **H = .200**

SHOULDER LENGTH: **K = .215**

BODY ANGLE (DEG'S/SIDE): **.689**

CASE CAPACITY **CC'S = 1.106**

LOADED LENGTH: **1.63**

BELT DIAMETER: **C = N/A**

RIM THICKNESS: **E = .065**

SHOULDER DIAMETER: **G = .333**

LENGTH TO SHOULDER: **J = .865**

SHOULDER ANGLE (DEG'S/SIDE): **11.94**

PRIMER: **S/R**

CASE CAPACITY (GR'S WATER): **17.07**

DIMENSIONAL DRAWING:

F
H
K
A
G
E
J
B
D

-NOT ACTUAL SIZE-
-DO NOT SCALE-

CARTRIDGE: .218 ICL Bobcat

OTHER NAMES:
.218 ICL Improved Hornet

DIA: .224

BALLISTEK NO: 224BQ

NAI NO: RMB 22232/4.615

DATA SOURCE: Ackley Vol.1 Pg.267

HISTORICAL DATA:

NOTES:

LOADING DATA:

BULLET WT./TYPE	POWDER WT./TYPE	VELOCITY ('/SEC)	SOURCE
40/Spire	14.0/4227	2675	Ackley

CASE PREPARATION: SHELLHOLDER (RCBS): 12

MAKE FROM: .218 Bee. Fireform factory ammo in ICL chamber.

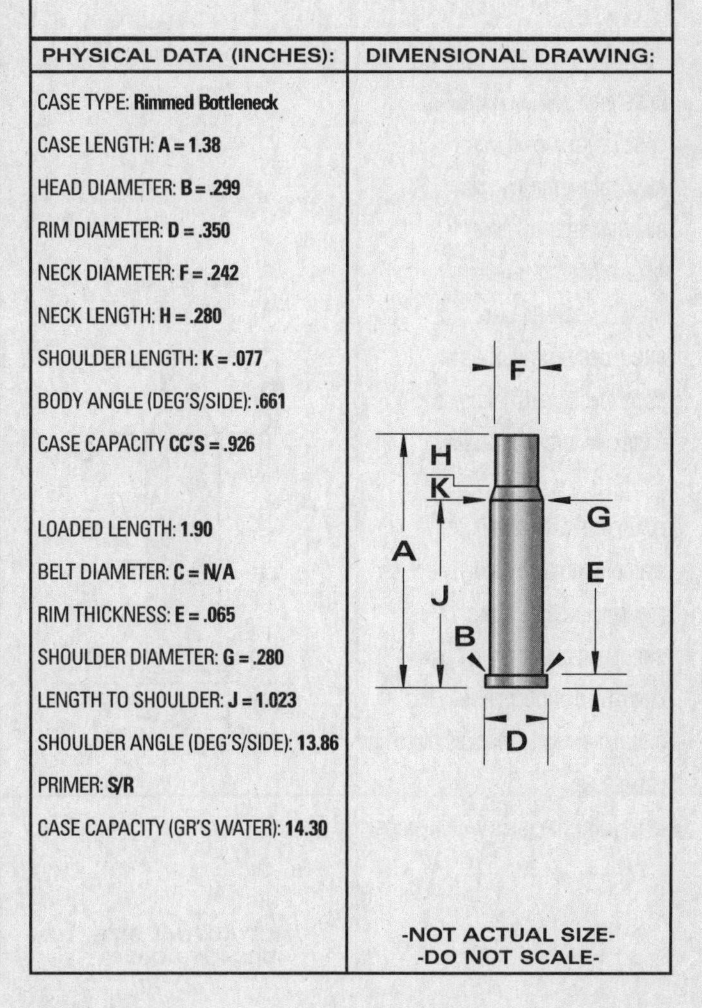

PHYSICAL DATA (INCHES):

CASE TYPE: **Rimmed Bottleneck**

CASE LENGTH: **A = 1.38**

HEAD DIAMETER: **B = .299**

RIM DIAMETER: **D = .350**

NECK DIAMETER: **F = .242**

NECK LENGTH: **H = .280**

SHOULDER LENGTH: **K = .077**

BODY ANGLE (DEG'S/SIDE): **.661**

CASE CAPACITY **CC'S = .926**

LOADED LENGTH: **1.90**

BELT DIAMETER: **C = N/A**

RIM THICKNESS: **E = .065**

SHOULDER DIAMETER: **G = .280**

LENGTH TO SHOULDER: **J = 1.023**

SHOULDER ANGLE (DEG'S/SIDE): **13.86**

PRIMER: **S/R**

CASE CAPACITY (GR'S WATER): **14.30**

DIMENSIONAL DRAWING:

-NOT ACTUAL SIZE-
-DO NOT SCALE-

CARTRIDGE: .219 Donaldson Wasp

OTHER NAMES:

DIA: .224

BALLISTEK NO: 224C

NAI NO: RMB 23423/4.296

DATA SOURCE: Hornady Manual 3rd Pg.83

HISTORICAL DATA: By Harvey Donaldson about 1940

NOTES:

LOADING DATA:

BULLET WT./TYPE	POWDER WT./TYPE	VELOCITY ('/SEC)	SOURCE
55/Spire	26.6/IMR3031	3400	Horn.

CASE PREPARATION: SHELLHOLDER (RCBS): 6

MAKE FROM: .30-30 Winchester. Do not anneal! Run into form die #1, form die #2 and F/L die with expander removed. Cut to length and I.D. neck ream. Clean. F/L size. Chamfer.

PHYSICAL DATA (INCHES):

CASE TYPE: **Rimmed Bottleneck**

CASE LENGTH: **A = 1.813**

HEAD DIAMETER: **B = .422**

RIM DIAMETER: **D = .506**

NECK DIAMETER: **F = .250**

NECK LENGTH: **H = .316**

SHOULDER LENGTH: **K = .134**

BODY ANGLE (DEG'S/SIDE): **.394**

CASE CAPACITY **CC'S = 2.39**

LOADED LENGTH: **2.21**

BELT DIAMETER: **C = N/A**

RIM THICKNESS: **E = .063**

SHOULDER DIAMETER: **G = .406**

LENGTH TO SHOULDER: **J = 1.363**

SHOULDER ANGLE (DEG'S/SIDE): **30.20**

PRIMER: **L/R**

CASE CAPACITY (GR'S WATER): **36.85**

DIMENSIONAL DRAWING:

-NOT ACTUAL SIZE-
-DO NOT SCALE-

CARTRIDGE: .219 ICL Wolverine

OTHER NAMES:

DIA: .224

BALLISTEK NO: 224R

NAI NO: RMB 22424/4.774

DATA SOURCE: Ackley Vol.1 Pg.278

HISTORICAL DATA:

NOTES:

LOADING DATA:

BULLET WT./TYPE	POWDER WT./TYPE	VELOCITY ('/SEC)	SOURCE
50/Spire	30.0/IMR4895	3575	Ackley

CASE PREPARATION: SHELLHOLDER (RCBS): 2

MAKE FROM: .30-30 Win. Form normal. .219 Zipper case (#224B) and fire this ammo in ICL chamber. Best to anneal the case, a second time, after forming the Zipper. The ICL shoulder is rough on brass!

PHYSICAL DATA (INCHES):

CASE TYPE: **Rimmed Bottleneck**

CASE LENGTH: **A = 2.01**

HEAD DIAMETER: **B = .421**

RIM DIAMETER: **D = .506**

NECK DIAMETER: **F = .252**

NECK LENGTH: **H = .298**

SHOULDER LENGTH: **K = .082**

BODY ANGLE (DEG'S/SIDE): **.821**

CASE CAPACITY **CC'S = 2.57**

LOADED LENGTH: **2.58**

BELT DIAMETER: **C = N/A**

RIM THICKNESS: **E = .063**

SHOULDER DIAMETER: **G = .380**

LENGTH TO SHOULDER: **J = 1.63**

SHOULDER ANGLE (DEG'S/SIDE): **37.97**

PRIMER: **L/R**

CASE CAPACITY (GR'S WATER): **39.68**

DIMENSIONAL DRAWING:

-NOT ACTUAL SIZE-
-DO NOT SCALE-

CARTRIDGE: .219 Stingray

OTHER NAMES:

DIA: .224

BALLISTEK NO: 224BH

NAI NO: RMB 23444/3.881

DATA SOURCE: Ackley Vol.2 Pg.125

HISTORICAL DATA: By M. Marquez

NOTES:

LOADING DATA:

BULLET WT./TYPE	POWDER WT./TYPE	VELOCITY ('/SEC)	SOURCE
50/Spire	25.6/IMR4895	—	JJD

CASE PREPARATION: SHELLHOLDER (RCBS): 2

MAKE FROM: .30-30 Win. Form a .219 Zipper (#224B) and fireform in the Stingray chamber. Not really much difference.

PHYSICAL DATA (INCHES):

CASE TYPE: **Rimmed Bottleneck**

CASE LENGTH: **A = 1.63**

HEAD DIAMETER: **B = .420**

RIM DIAMETER: **D = .506**

NECK DIAMETER: **F = .253**

NECK LENGTH: **H = .220**

SHOULDER LENGTH: **K = .11**

BODY ANGLE (DEG'S/SIDE): **.52**

CASE CAPACITY **CC'S = 2.09**

LOADED LENGTH: **2.15**

BELT DIAMETER: **C = N/A**

RIM THICKNESS: **E = .06**

SHOULDER DIAMETER: **G = .400**

LENGTH TO SHOULDER: **J = 1.30**

SHOULDER ANGLE (DEG'S/SIDE): **33.70**

PRIMER: **L/R**

CASE CAPACITY (GR'S WATER): **32.29**

DIMENSIONAL DRAWING:

-NOT ACTUAL SIZE-
-DO NOT SCALE-

CARTRIDGE: .219 Zipper

OTHER NAMES:

DIA: .224

BALLISTEK NO: 224B

NAI NO: RMB 33422/4.592

DATA SOURCE: Hornady Manual 3rd Pg.86

HISTORICAL DATA: By Win. in 1937

NOTES: No longer a commercial cartridge.

LOADING DATA:

BULLET WT./TYPE	POWDER WT./TYPE	VELOCITY ('/SEC)	SOURCE
50/Spire	27.5/IMR3031	3500	JJD

CASE PREPARATION: SHELLHOLDER (RCBS): 6

MAKE FROM: .30-30 Win. or .32 Win. Spl. Anneal brass. Run into form die #1 and form die #2. F/L size with expander removed. Trim to length. F/L size. Chamfer.

PHYSICAL DATA (INCHES):

CASE TYPE: **Rimmed Bottleneck**

CASE LENGTH: **A = 1.938**

HEAD DIAMETER: **B = .422**

RIM DIAMETER: **D = .506**

NECK DIAMETER: **F = .252**

NECK LENGTH: **H = .317**

SHOULDER LENGTH: **K = .262**

BODY ANGLE (DEG'S/SIDE): **1.41**

CASE CAPACITY **CC'S = 2.24**

LOADED LENGTH: **2.40**

BELT DIAMETER: **C = N/A**

RIM THICKNESS: **E = .063**

SHOULDER DIAMETER: **G = .365**

LENGTH TO SHOULDER: **J = 1.359**

SHOULDER ANGLE (DEG'S/SIDE): **12.17**

PRIMER: **L/R**

CASE CAPACITY (GR'S WATER): **34.63**

DIMENSIONAL DRAWING:

-NOT ACTUAL SIZE-
-DO NOT SCALE-

CARTRIDGE: .219 Zipper Improved (Ackley)

OTHER NAMES:

DIA: .224

BALLISTEK NO: 224Q

NAI NO: RMB 22433/4.620

DATA SOURCE: Ackley Vol.1 Pg.277

HISTORICAL DATA: By P.O. Ackley in 1938

NOTES:

LOADING DATA:

BULLET WT./TYPE	POWDER WT./TYPE	VELOCITY ('/SEC)	SOURCE
50/Spire	33.0/IMR3031	3782	Ackley

CASE PREPARATION: SHELLHOLDER (RCBS): 2

MAKE FROM: .30-30 Win. Form die set required. Form in form die #1 and form die #2. Trim to length. F/L size. Fireform in chamber. Note: form dies can be normal Zipper form dies; "Improvement" will be done upon fireforming.

PHYSICAL DATA (INCHES):

CASE TYPE: **Rimmed Bottleneck**

CASE LENGTH: **A = 1.95**

HEAD DIAMETER: **B = .422**

RIM DIAMETER: **D = .506**

NECK DIAMETER: **F = .253**

NECK LENGTH: **H = .277**

SHOULDER LENGTH: **K = .173**

BODY ANGLE (DEG'S/SIDE): **.418**

CASE CAPACITY **CC'S = 2.58**

LOADED LENGTH: **2.39**

BELT DIAMETER: **C = N/A**

RIM THICKNESS: **E = .06**

SHOULDER DIAMETER: **G = .403**

LENGTH TO SHOULDER: **J = 1.50**

SHOULDER ANGLE (DEG'S/SIDE): **23.43**

PRIMER: **L/R**

CASE CAPACITY (GR'S WATER): **39.82**

DIMENSIONAL DRAWING:

-NOT ACTUAL SIZE-
-DO NOT SCALE-

CARTRIDGE: .22 Benchrest

OTHER NAMES:
.22 Remington BR

DIA: .224

BALLISTEK NO: 224BA

NAI NO: RXB 24433/3.241

DATA SOURCE: Rifle #63, Pg.32

HISTORICAL DATA: .22 version of Rem. case by J. Stekl.

NOTES: Generally considered a competition ctg.

LOADING DATA:

BULLET WT./TYPE	POWDER WT./TYPE	VELOCITY ('/SEC)	SOURCE
55/Spire	29.8/IMR4064	–	JJD

CASE PREPARATION: SHELLHOLDER (RCBS): 3

MAKE FROM: Rem. BR. A form set is a must! Do not anneal cases. Run through form die set (four dies) and trim to length. Chamfer. F/L size.

PHYSICAL DATA (INCHES):

CASE TYPE: **Rimless Bottleneck**

CASE LENGTH: **A = 1.52**

HEAD DIAMETER: **B = .469**

RIM DIAMETER: **D = .473**

NECK DIAMETER: **F = .246**

NECK LENGTH: **H = .260**

SHOULDER LENGTH: **K = .185**

BODY ANGLE (DEG'S/SIDE): **.360**

CASE CAPACITY **CC'S = 2.41**

LOADED LENGTH: **1.99**

BELT DIAMETER: **C = N/A**

RIM THICKNESS: **E = .052**

SHOULDER DIAMETER: **G = .458**

LENGTH TO SHOULDER: **J = 1.075**

SHOULDER ANGLE (DEG'S/SIDE): **29.81**

PRIMER: **L/R**

CASE CAPACITY (GR'S WATER): **37.29**

DIMENSIONAL DRAWING:

-NOT ACTUAL SIZE-
-DO NOT SCALE-

CARTRIDGE: .22 Harvey Kay-Chuk

OTHER NAMES:

DIA: .224

BALLISTEK NO: 224P

NAI NO: RMB 12245/4.591

DATA SOURCE: COTW 4th Pg.141

HISTORICAL DATA: By J. Harvey about 1956

NOTES:

LOADING DATA:

BULLET WT./TYPE	POWDER WT./TYPE	VELOCITY ('/SEC)	SOURCE
40/RN	5.0/Unique	1650	Barnes

CASE PREPARATION: SHELLHOLDER (RCBS): 12

MAKE FROM: .22 Hornet. Fire .22 Hornet ammo in Kay-Chuk chamber.

PHYSICAL DATA (INCHES):

CASE TYPE: **Rimmed Bottleneck**

CASE LENGTH: **A = 1.35**

HEAD DIAMETER: **B = .294**

RIM DIAMETER: **D = .347**

NECK DIAMETER: **F = .243**

NECK LENGTH: **H = .184**

SHOULDER LENGTH: **K = N/A**

BODY ANGLE (DEG'S/SIDE): **.025**

CASE CAPACITY **CC'S = .966**

LOADED LENGTH: **1.60**

BELT DIAMETER: **C = N/A**

RIM THICKNESS: **E = .065**

SHOULDER DIAMETER: **G = .293**

LENGTH TO SHOULDER: **J = 1.335**

SHOULDER ANGLE (DEG'S/SIDE): **N/A**

PRIMER: **S/R**

CASE CAPACITY (GR'S WATER): **14.91**

DIMENSIONAL DRAWING:

-NOT ACTUAL SIZE-
-DO NOT SCALE-

CARTRIDGE: .22 Hornet

OTHER NAMES:
5.6 x 35R Vierling

DIA: .224

BALLISTEK NO: 224D

NAI NO: RMB 24221/4.475

DATA SOURCE: Hornady Manual 3rd Pg.62

HISTORICAL DATA: By Win. in 1933. Based on the .22 Win. Centerfire case.

NOTES: Developed at Springfield Arsenal

LOADING DATA:

BULLET WT./TYPE	POWDER WT./TYPE	VELOCITY ('/SEC)	SOURCE
50/Spire	11.5/IMR4227	2585	JJD

CASE PREPARATION: SHELLHOLDER (RCBS): 12

MAKE FROM: Factory. No other case available from which the .22 Hornet can be formed. Cases can be made from solid brass.

PHYSICAL DATA (INCHES):

CASE TYPE: **Rimmed Bottleneck**

CASE LENGTH: **A = 1.338**

HEAD DIAMETER: **B = .299**

RIM DIAMETER: **D = .350**

NECK DIAMETER: **F = .242**

NECK LENGTH: **H = .323**

SHOULDER LENGTH: **K = .190**

BODY ANGLE (DEG'S/SIDE): **.96**

CASE CAPACITY **CC'S = .86**

LOADED LENGTH: **1.80**

BELT DIAMETER: **C = N/A**

RIM THICKNESS: **E = .065**

SHOULDER DIAMETER: **G = .278**

LENGTH TO SHOULDER: **J = .825**

SHOULDER ANGLE (DEG'S/SIDE): **5.41**

PRIMER: **S/R**

CASE CAPACITY (GR'S WATER): **13.27**

DIMENSIONAL DRAWING:

-NOT ACTUAL SIZE-
-DO NOT SCALE-

CARTRIDGE: .22 ICL Gopher

OTHER NAMES:

DIA: .224

BALLISTEK NO: 224AA

NAI NO: RMB 22232/4.381

DATA SOURCE: Ackley Vol.1 Pg.262

HISTORICAL DATA: By A. Juenke

NOTES:

LOADING DATA:

BULLET WT./TYPE	POWDER WT./TYPE	VELOCITY ('/SEC)	SOURCE
40/Spire	10.0/4227	2515	Ackley/ Juenke

CASE PREPARATION: SHELLHOLDER (RCBS): 12

MAKE FROM: .22 Hornet. Hornet ammo is fired in ICL chamber.

PHYSICAL DATA (INCHES):

CASE TYPE: **Rimmed Bottleneck**

CASE LENGTH: **A = 1.31**

HEAD DIAMETER: **B = .299**

RIM DIAMETER: **D = .350**

NECK DIAMETER: **F = .242**

NECK LENGTH: **H = .225**

SHOULDER LENGTH: **K = .067**

BODY ANGLE (DEG'S/SIDE): **.665**

CASE CAPACITY **CC'S = .88**

LOADED LENGTH: **1.83**

BELT DIAMETER: **C = N/A**

RIM THICKNESS: **E = .065**

SHOULDER DIAMETER: **G = .280**

LENGTH TO SHOULDER: **J = 1.018**

SHOULDER ANGLE (DEG'S/SIDE): **15.83**

PRIMER: **S/R**

CASE CAPACITY (GR'S WATER): **13.56**

DIMENSIONAL DRAWING:

F H K G A J E B D

-NOT ACTUAL SIZE-
-DO NOT SCALE-

CARTRIDGE: .22 Jet Improved (Ackley)

OTHER NAMES:
.22 Sabre
.22 Super Jet

DIA: .224

BALLISTEK NO: 224V

NAI NO: RMB 14343/3.328

DATA SOURCE: Ackley Vol.1 Pg.265

HISTORICAL DATA: By D. Cotterman

NOTES: 30° shoulder

LOADING DATA:

BULLET WT./TYPE	POWDER WT./TYPE	VELOCITY ('/SEC)	SOURCE
40/—	17.5/4198	2975	Ackley/ Helbig

CASE PREPARATION: SHELLHOLDER (RCBS): 6

MAKE FROM: .22 Jet. Fireform .22 Jet ammo in "Improved" chamber OR form the .22 Jet from non-plated .357 Mag cases and fireform.

PHYSICAL DATA (INCHES):

CASE TYPE: **Rimmed Bottleneck**

CASE LENGTH: **A = 1.265**

HEAD DIAMETER: **B = .380**

RIM DIAMETER: **D = .440**

NECK DIAMETER: **F = .248**

NECK LENGTH: **H = .192**

SHOULDER LENGTH: **K = .107**

BODY ANGLE (DEG'S/SIDE): **.336**

CASE CAPACITY **CC'S = 1.30**

LOADED LENGTH: **1.74**

BELT DIAMETER: **C = N/A**

RIM THICKNESS: **E = .06**

SHOULDER DIAMETER: **G = .371**

LENGTH TO SHOULDER: **J = .966**

SHOULDER ANGLE (DEG'S/SIDE): **29.88**

PRIMER: **S/P**

CASE CAPACITY (GR'S WATER): **20.16**

DIMENSIONAL DRAWING:

F H K G A E J B D

-NOT ACTUAL SIZE-
-DO NOT SCALE-

CARTRIDGE: .22 K-Hornet

OTHER NAMES:
.22 Kilbourn Hornet

DIA: .224

BALLISTEK NO: 224AF

NAI NO: RMB 22234/4.645

DATA SOURCE: Sierra Manual 1985 Pg.92

HISTORICAL DATA: By L. Kilbourn in 1940

NOTES: Use only new Hornet brass to form!

LOADING DATA:

BULLET WT./TYPE	POWDER WT./TYPE	VELOCITY ('/SEC)	SOURCE
45/Spire	12.0/IMR4227	2600	Sierra
50/—	14.0/IMR4198	2690	Ackley

CASE PREPARATION: SHELLHOLDER (RCBS): 12

MAKE FROM: .22 Hornet. Cases are produced by firing Hornet ammo in the Kilbourn chamber. Cases can also be made by fireforming, with corn meal, in the proper trim die.

PHYSICAL DATA (INCHES):

CASE TYPE: **Rimmed Bottleneck**

CASE LENGTH: **A = 1.389**

HEAD DIAMETER: **B = .299**

RIM DIAMETER: **D = .350**

NECK DIAMETER: **F = .242**

NECK LENGTH: **H = .279**

SHOULDER LENGTH: **K = .025**

BODY ANGLE (DEG'S/SIDE): **.42**

CASE CAPACITY **CC'S = .959**

LOADED LENGTH: **1.75**

BELT DIAMETER: **C = N/A**

RIM THICKNESS: **E = .065**

SHOULDER DIAMETER: **G = .286**

LENGTH TO SHOULDER: **J = 1.085**

SHOULDER ANGLE (DEG'S/SIDE): **41.34**

PRIMER: **S/R**

CASE CAPACITY (GR'S WATER): **14.80**

DIMENSIONAL DRAWING:

-NOT ACTUAL SIZE-
-DO NOT SCALE-

CARTRIDGE: .22 Krag Short

OTHER NAMES:
.22-250 Rimmed

DIA: .224

BALLISTEK NO: 224Y

NAI NO: RMB 22443/4.170

DATA SOURCE: Handloader #71 Pg.34

HISTORICAL DATA: This version is by S. Hopkins.

NOTES: Other versions exist.

LOADING DATA:

BULLET WT./TYPE	POWDER WT./TYPE	VELOCITY ('/SEC)	SOURCE
52/Spire	33.5/IMR4895	3593	Hopkins

CASE PREPARATION: SHELLHOLDER (RCBS): 7

MAKE FROM: .30-40 Krag. Trim case to 1.93". Run into form die to reduce neck diameter. Note: .250 Savage die may be used in place of form die. Run into .22-250 F/L die to size. Fireform in chamber.

PHYSICAL DATA (INCHES):

CASE TYPE: **Rimmed Bottleneck**

CASE LENGTH: **A = 1.91**

HEAD DIAMETER: **B = .458**

RIM DIAMETER: **D = .545**

NECK DIAMETER: **F = .253**

NECK LENGTH: **H = .20**

SHOULDER LENGTH: **K = .15**

BODY ANGLE (DEG'S/SIDE): **.905**

CASE CAPACITY **CC'S = 2.93**

LOADED LENGTH: **A/R**

BELT DIAMETER: **C = N/A**

RIM THICKNESS: **E = .063**

SHOULDER DIAMETER: **G = .415**

LENGTH TO SHOULDER: **J = 1.56**

SHOULDER ANGLE (DEG'S/SIDE): **28.37**

PRIMER: **L/R**

CASE CAPACITY (GR'S WATER): **45.16**

DIMENSIONAL DRAWING:

-NOT ACTUAL SIZE-
-DO NOT SCALE-

CARTRIDGE: .22 KSS

OTHER NAMES:	
.22 Krag Short-Short	DIA: .224 BALLISTEK NO: 224S NAI NO: RMB 24423/3.286

DATA SOURCE: Rifle #55 Pg.12

HISTORICAL DATA: By L.E. Wilson in 1955

NOTES: Similar to .22 PPC

LOADING DATA:

BULLET WT./TYPE	POWDER WT./TYPE	VELOCITY ('/SEC)	SOURCE
No data available: Use .22 PPC as a starting point. Page 98			

CASE PREPARATION: SHELLHOLDER (RCBS): 3

MAKE FROM: .30-40 Krag. A three-die form set is required OR size case in .308x1.5" die. 7mm PPC die, 6mm PPC die and .22 KSS trim die. F/L size and trim to length. Anneal neck between 7mm and 6mm neck reduction. Chamfer.

PHYSICAL DATA (INCHES):

CASE TYPE: **Rimmed Bottleneck**

CASE LENGTH: **A = 1.505**

HEAD DIAMETER: **B = .458**

RIM DIAMETER: **D = .473**

NECK DIAMETER: **F = .243**

NECK LENGTH: **H = .310**

SHOULDER LENGTH: **K = .160**

BODY ANGLE (DEG'S/SIDE): **.823**

CASE CAPACITY **CC'S = 2.18**

LOADED LENGTH: **A/R**

BELT DIAMETER: **C = N/A**

RIM THICKNESS: **E = .052**

SHOULDER DIAMETER: **G = .434**

LENGTH TO SHOULDER: **J = 1.035**

SHOULDER ANGLE (DEG'S/SIDE): **30.83**

PRIMER: **L/R**

CASE CAPACITY (GR'S WATER): **33.78**

DIMENSIONAL DRAWING:

F H K G A E J B D

-NOT ACTUAL SIZE-
-DO NOT SCALE-

CARTRIDGE: .22 Mashburn Bee

OTHER NAMES:	
	DIA: .224 BALLISTEK NO: 224AB NAI NO: RMB 23344/3.839

DATA SOURCE: Ackley Vol.1 Pg.265

HISTORICAL DATA: By A. Mashburn

NOTES:

LOADING DATA:

BULLET WT./TYPE	POWDER WT./TYPE	VELOCITY ('/SEC)	SOURCE
50/Spire	17.3/4198	3300	Ackley

CASE PREPARATION: SHELLHOLDER (RCBS): 12

MAKE FROM: .218 Bee. Fire Hornet ammo in Mashburn chamber or fireform in Mashburn trim die (with corn meal).

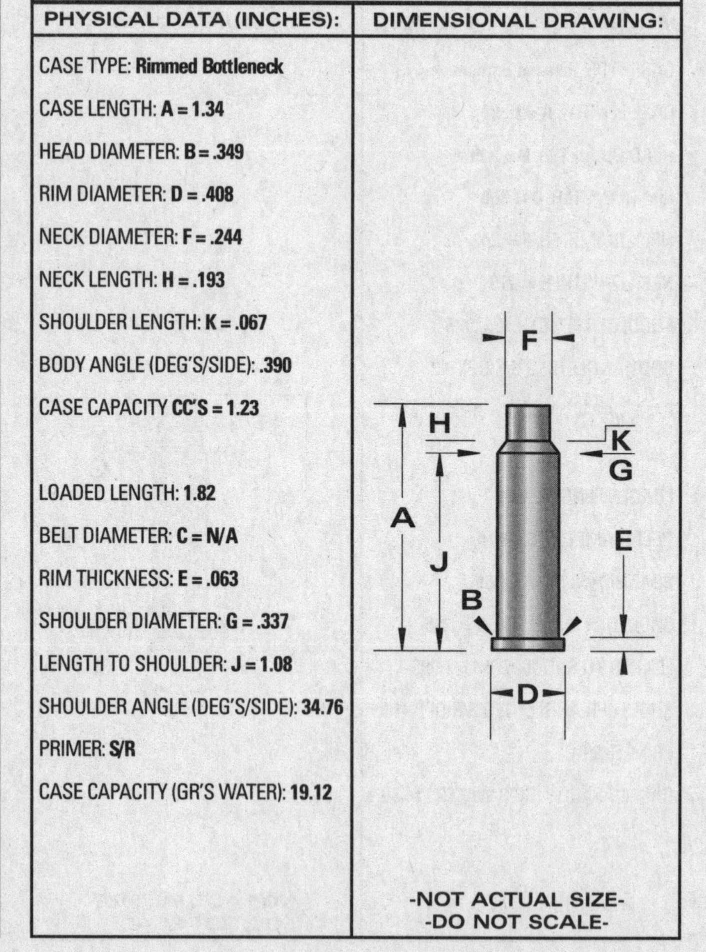

PHYSICAL DATA (INCHES):

CASE TYPE: **Rimmed Bottleneck**

CASE LENGTH: **A = 1.34**

HEAD DIAMETER: **B = .349**

RIM DIAMETER: **D = .408**

NECK DIAMETER: **F = .244**

NECK LENGTH: **H = .193**

SHOULDER LENGTH: **K = .067**

BODY ANGLE (DEG'S/SIDE): **.390**

CASE CAPACITY **CC'S = 1.23**

LOADED LENGTH: **1.82**

BELT DIAMETER: **C = N/A**

RIM THICKNESS: **E = .063**

SHOULDER DIAMETER: **G = .337**

LENGTH TO SHOULDER: **J = 1.08**

SHOULDER ANGLE (DEG'S/SIDE): **34.76**

PRIMER: **S/R**

CASE CAPACITY (GR'S WATER): **19.12**

DIMENSIONAL DRAWING:

-NOT ACTUAL SIZE-
-DO NOT SCALE-

CARTRIDGE: .22 Maynard Extra Long

OTHER NAMES:
.22 Maynard XL

DIA: .228

BALLISTEK NO: 228S

NAI NO: RMS
11115/4.643

DATA SOURCE: COTW 4th, Pg.88

HISTORICAL DATA:

NOTES: Original ctg. used very small (#0) primer.

LOADING DATA:

BULLET WT./TYPE	POWDER WT./TYPE	VELOCITY ('/SEC)	SOURCE
45/Lead	7.0/FFFg	1100	JJD

CASE PREPARATION: SHELLHOLDER (RCBS): 29

MAKE FROM: Solid. If you need this one, you'll have to turn from solid brass. Use 5/16" dia. stock. Anneal neck area but, F/L sizing is not needed.

PHYSICAL DATA (INCHES):

CASE TYPE: **Rimmed Straight**
CASE LENGTH: **A = 1.17**
HEAD DIAMETER: **B = .252**
RIM DIAMETER: **D = .310**
NECK DIAMETER: **F = .252**
NECK LENGTH: **H = N/A**
SHOULDER LENGTH: **K = N/A**
BODY ANGLE (DEG'S/SIDE): **.385**
CASE CAPACITY **CC'S = .517**

LOADED LENGTH: **1.40**
BELT DIAMETER: **C = N/A**
RIM THICKNESS: **E = .063**
SHOULDER DIAMETER: **G = N/A**
LENGTH TO SHOULDER: **J = N/A**
SHOULDER ANGLE (DEG'S/SIDE): **N/A**
PRIMER: **S/R**
CASE CAPACITY (GR'S WATER): **7.98**

DIMENSIONAL DRAWING:

-NOT ACTUAL SIZE-
-DO NOT SCALE-

CARTRIDGE: .22 Newton

OTHER NAMES:
.228 Newton

DIA: .228

BALLISTEK NO: 228A

NAI NO: RXB
22423/4.744

DATA SOURCE: COTW 4th Pg.140

HISTORICAL DATA: By Charles Newton about 1912.

NOTES:

LOADING DATA:

BULLET WT./TYPE	POWDER WT./TYPE	VELOCITY ('/SEC)	SOURCE
70/Spire	40.0/4064	3300	Nonte
90/—	38.0/4350	3100	Barnes

CASE PREPARATION: SHELLHOLDER (RCBS): 3

MAKE FROM: .30-06 Springfield. Anneal case neck. Set shoulder back in .308 die to the 1.72" dim. Run into 7mm/308 Win. or 7mm BR sizer to reduce neck dia. Run into .243 Win. sizer. Cut case to length. I.D. neck ream. F/L size. (Note: if you need a bunch of these cases, purchase a form die set!)

PHYSICAL DATA (INCHES):

CASE TYPE: **Rimless Bottleneck**
CASE LENGTH: **A = 2.225**
HEAD DIAMETER: **B = .469**
RIM DIAMETER: **D = .470**
NECK DIAMETER: **F = .256**
NECK LENGTH: **H = .302**
SHOULDER LENGTH: **K = .203**
BODY ANGLE (DEG'S/SIDE): **.923**
CASE CAPACITY **CC'S = 3.53**

LOADED LENGTH: **2.85**
BELT DIAMETER: **C = N/A**
RIM THICKNESS: **E = .05**
SHOULDER DIAMETER: **G = .42**
LENGTH TO SHOULDER: **J = 1.72**
SHOULDER ANGLE (DEG'S/SIDE): **21.99**
PRIMER: **L/R**
CASE CAPACITY (GR'S WATER): **54.55**

DIMENSIONAL DRAWING:

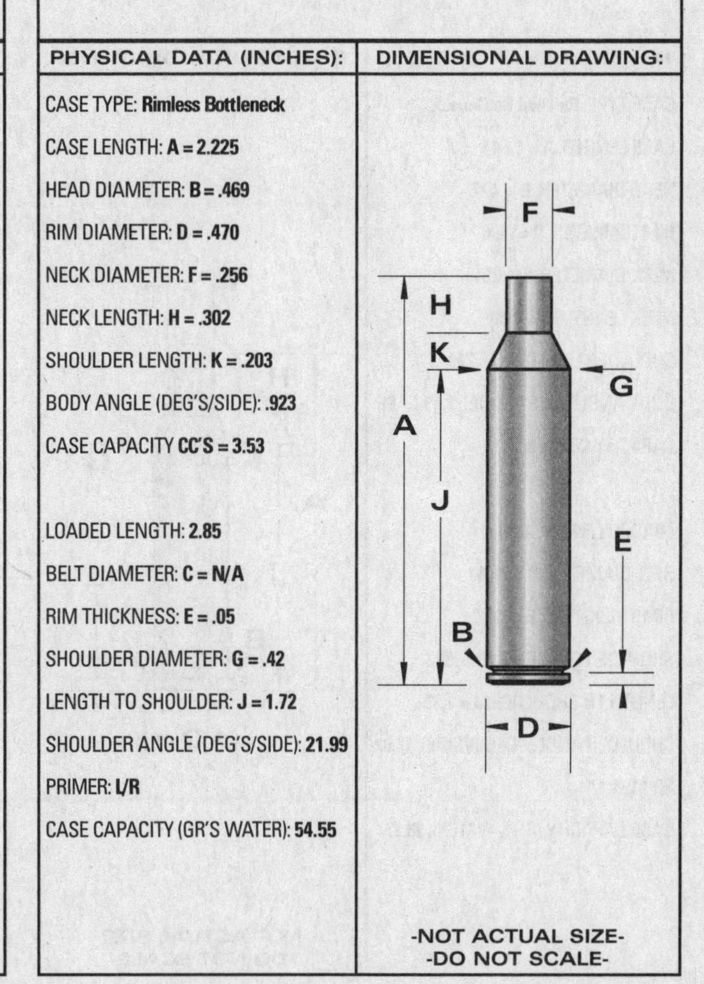

-NOT ACTUAL SIZE-
-DO NOT SCALE-

CARTRIDGE: .22 Niedner Magnum

OTHER NAMES:

DIA: .224

BALLISTEK NO: 224U

NAI NO: RMB 33432/4.360

DATA SOURCE: Handloader #71 Pg.14

HISTORICAL DATA:

NOTES: Reported as 17° shoulder.

LOADING DATA:

BULLET WT./TYPE	POWDER WT./TYPE	VELOCITY ('/SEC)	SOURCE
50/Spire	27.0/IMR4064	—	Waters

CASE PREPARATION: **SHELLHOLDER (RCBS):** 2

MAKE FROM: .30-30 Win. or .25-35. Purchase a form die set OR modify a .219 Zipper to establish the shoulder at 1.35" (for use with .25-35 brass, only). Fireform the case in the chamber or in a trim die, with corn meal and 6-7 grs. of Bullseye.

PHYSICAL DATA (INCHES):

CASE TYPE: **Rimmed Bottleneck**

CASE LENGTH: **A = 1.84**

HEAD DIAMETER: **B = .422**

RIM DIAMETER: **D = .506**

NECK DIAMETER: **F = .251**

NECK LENGTH: **H = .250**

SHOULDER LENGTH: **K = .240**

BODY ANGLE (DEG'S/SIDE): **1.49**

CASE CAPACITY **CC'S = 2.16**

LOADED LENGTH: **2.30**

BELT DIAMETER: **C = N/A**

RIM THICKNESS: **E = .062**

SHOULDER DIAMETER: **G = .362**

LENGTH TO SHOULDER: **J = 1.35**

SHOULDER ANGLE (DEG'S/SIDE): **13.02**

PRIMER: **L/R**

CASE CAPACITY (GR'S WATER): **33.37**

DIMENSIONAL DRAWING:

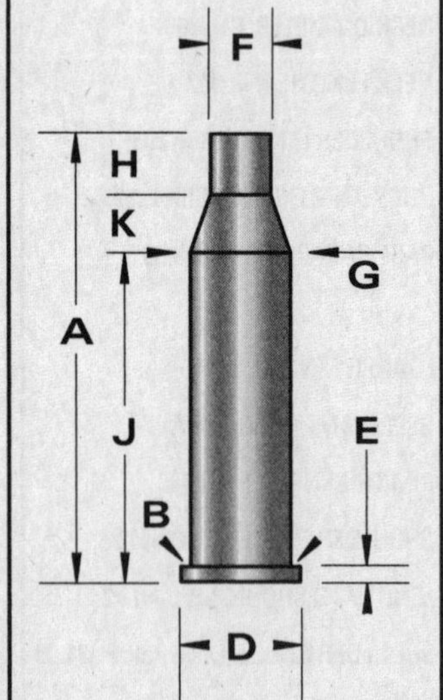

-NOT ACTUAL SIZE-
-DO NOT SCALE-

CARTRIDGE: .22 PPC

OTHER NAMES:

DIA: .224

BALLISTEK NO: 224T

NAI NO: RXB 14423/3.443

DATA SOURCE: Hornady 3rd Pg.78

HISTORICAL DATA: By Ferris Pindell in 1974

NOTES: Generally used as a benchrest ctg.

LOADING DATA:

BULLET WT./TYPE	POWDER WT./TYPE	VELOCITY ('/SEC)	SOURCE
53/Spire	29.0/BL-C2	3400	Hom.

CASE PREPARATION: **SHELLHOLDER (RCBS):** 6

MAKE FROM: .220 Russian (Sako). Anneal neck and shoulder area of case. F/L size in PPC die.

PHYSICAL DATA (INCHES):

CASE TYPE: **Rimless Bottleneck**

CASE LENGTH: **A = 1.515**

HEAD DIAMETER: **B = .440**

RIM DIAMETER: **D = .440**

NECK DIAMETER: **F = .246**

NECK LENGTH: **H = .285**

SHOULDER LENGTH: **K = .155**

BODY ANGLE (DEG'S/SIDE): **.327**

CASE CAPACITY **CC'S = 2.11**

LOADED LENGTH: **2.07**

BELT DIAMETER: **C = N/A**

RIM THICKNESS: **E = .06**

SHOULDER DIAMETER: **G = .430**

LENGTH TO SHOULDER: **J = 1.075**

SHOULDER ANGLE (DEG'S/SIDE): **30.69**

PRIMER: **S/R**

CASE CAPACITY (GR'S WATER): **32.65**

DIMENSIONAL DRAWING:

-NOT ACTUAL SIZE-
-DO NOT SCALE-

CARTRIDGE: .22 Remington Jet

OTHER NAMES:
.22 Jet
.22 Centerfire Magnum

DIA: .223

BALLISTEK NO: 223A

NAI NO: RMB 34341/3.389

DATA SOURCE: Hornady Manual 3rd Pg.365

HISTORICAL DATA: By Rem. and S&W in 1961.

NOTES:

LOADING DATA:

BULLET WT./TYPE	POWDER WT./TYPE	VELOCITY ('/SEC)	SOURCE
45/Spire	13.2/IMR4227	2400	Horn.

CASE PREPARATION: SHELLHOLDER (RCBS): 6

MAKE FROM: .357 Magnum or factory. Form dies required. Form in die #1 and die #2. Trim to length. F/L size. Chamfer. Do not use plated cases!

PHYSICAL DATA (INCHES):

CASE TYPE: **Rimmed Bottleneck**
CASE LENGTH: **A = 1.288**
HEAD DIAMETER: **B = .380**
RIM DIAMETER: **D = .439**
NECK DIAMETER: **F = .251**
NECK LENGTH: **H = .2**
SHOULDER LENGTH: **K = .49**
BODY ANGLE (DEG'S/SIDE): **1.00**
CASE CAPACITY **CC'S = 1.12**
LOADED LENGTH: **1.66**
BELT DIAMETER: **C = N/A**
RIM THICKNESS: **E = .06**
SHOULDER DIAMETER: **G = .366**
LENGTH TO SHOULDER: **J = .598**
SHOULDER ANGLE (DEG'S/SIDE): **6.69**
PRIMER: **S/P**
CASE CAPACITY (GR'S WATER): **17.39**

DIMENSIONAL DRAWING:

F H K A J B G E D

-NOT ACTUAL SIZE-
-DO NOT SCALE-

CARTRIDGE: .22 Savage High Power

OTHER NAMES:
5,6 x 52R

DIA: .228

BALLISTEK NO: 228B

NAI NO: RMB 32422/4.897

DATA SOURCE: Hornady Manual 3rd Pg.104

HISTORICAL DATA: By Savage in 1912

NOTES: Designed by Charles Newton

LOADING DATA:

BULLET WT./TYPE	POWDER WT./TYPE	VELOCITY ('/SEC)	SOURCE
70/Spire	26.5/IMR3031	3000	Horn.

CASE PREPARATION: SHELLHOLDER (RCBS): 2

MAKE FROM: .30-30 Win. A form die set is required. Anneal case. Run into form dies #1, #2 and #3, in order. Trim to 2.05" and chamfer. F/L size. Case will finish fireform in chamber.

PHYSICAL DATA (INCHES):

CASE TYPE: **Rimmed Bottleneck**
CASE LENGTH: **A = 2.047**
HEAD DIAMETER: **B = .418**
RIM DIAMETER: **D = .492**
NECK DIAMETER: **F = .252**
NECK LENGTH: **H = .409**
SHOULDER LENGTH: **K = .259**
BODY ANGLE (DEG'S/SIDE): **1.336**
CASE CAPACITY **CC'S = 2.37**
LOADED LENGTH: **2.51**
BELT DIAMETER: **C = N/A**
RIM THICKNESS: **E = .063**
SHOULDER DIAMETER: **G = .363**
LENGTH TO SHOULDER: **J = 1.379**
SHOULDER ANGLE (DEG'S/SIDE): **12.09**
PRIMER: **L/R**
CASE CAPACITY (GR'S WATER): **36.70**

DIMENSIONAL DRAWING:

F H K A G J E B D

-NOT ACTUAL SIZE-
-DO NOT SCALE-

CARTRIDGE: .22 Savage HP Improved

OTHER NAMES:

DIA: .228

BALLISTEK NO: 228M

NAI NO: RMB 22433/4.890

DATA SOURCE: NAI/Ballistek

HISTORICAL DATA: Early Ackley development.

NOTES:

LOADING DATA:

BULLET WT./TYPE	POWDER WT./TYPE	VELOCITY ('/SEC)	SOURCE
70/Spire	35.0/4320	3480	Ackley

CASE PREPARATION: SHELLHOLDER (RCBS): 2

MAKE FROM: .30-30 Win. Form .22 Savage High Power (#228A). Load this case with a light load and fireform in the "improved" chamber.

PHYSICAL DATA (INCHES):

CASE TYPE: **Rimmed Bottleneck**

CASE LENGTH: **A = 2.035**

HEAD DIAMETER: **B = .416**

RIM DIAMETER: **D = .492**

NECK DIAMETER: **F = .253**

NECK LENGTH: **H = .275**

SHOULDER LENGTH: **K = .150**

BODY ANGLE (DEG'S/SIDE): **.528**

CASE CAPACITY **CC'S = 2.70**

LOADED LENGTH: **2.51**

BELT DIAMETER: **C = N/A**

RIM THICKNESS: **E = .063**

SHOULDER DIAMETER: **G = .390**

LENGTH TO SHOULDER: **J = 1.61**

SHOULDER ANGLE (DEG'S/SIDE): **24.54**

PRIMER: **L/R**

CASE CAPACITY (GR'S WATER): **41.70**

DIMENSIONAL DRAWING:

-NOT ACTUAL SIZE-
-DO NOT SCALE-

CARTRIDGE: .22 Stark

OTHER NAMES:
.224 Stark

DIA: .224

BALLISTEK NO: 224W

NAI NO: RMB 24444/2.822

DATA SOURCE: Ackley Vol.2 Pg.119

HISTORICAL DATA: By J. Stark

NOTES:

LOADING DATA:

BULLET WT./TYPE	POWDER WT./TYPE	VELOCITY ('/SEC)	SOURCE
50/Spire	21.0/4198	3150	Ackley

CASE PREPARATION: SHELLHOLDER (RCBS): 18

MAKE FROM: .44 Mag. You must have form dies! Anneal case and run through form dies #1, #2 and #3, in order. Trim to length. F/L size and chamfer. Fireform in chamber. This case is quite difficult to produce.

PHYSICAL DATA (INCHES):

CASE TYPE: **Rimmed Bottleneck**

CASE LENGTH: **A = 1.29**

HEAD DIAMETER: **B = .457**

RIM DIAMETER: **D = .514**

NECK DIAMETER: **F = .252**

NECK LENGTH: **H = .213**

SHOULDER LENGTH: **K = .147**

BODY ANGLE (DEG'S/SIDE): **.431**

CASE CAPACITY **CC'S = 1.82**

LOADED LENGTH: **1.71**

BELT DIAMETER: **C = N/A**

RIM THICKNESS: **E = .06**

SHOULDER DIAMETER: **G = .446**

LENGTH TO SHOULDER: **J = .93**

SHOULDER ANGLE (DEG'S/SIDE): **33.42**

PRIMER: **L/P**

CASE CAPACITY (GR'S WATER): **28.04**

DIMENSIONAL DRAWING:

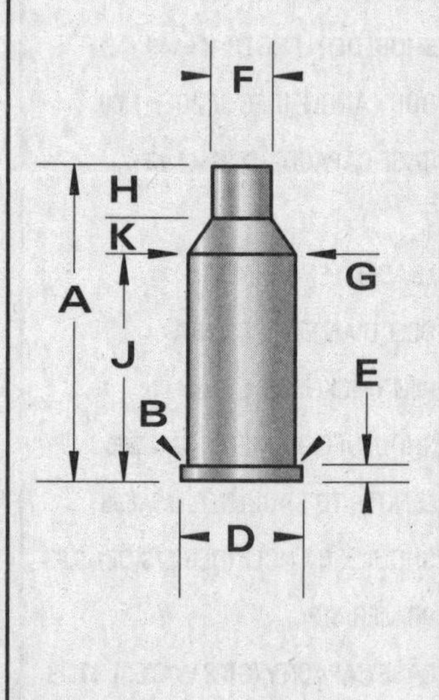

-NOT ACTUAL SIZE-
-DO NOT SCALE-

CARTRIDGE: .22 Wasp

OTHER NAMES:

DIA: .224

BALLISTEK NO: 224BU

NAI NO: RMB 13424/4.329

DATA SOURCE: Ackley Vol.2 Pg.124

HISTORICAL DATA: Another Australian design.

NOTES:

LOADING DATA:

BULLET WT./TYPE	POWDER WT./TYPE	VELOCITY ('/SEC)	SOURCE
50/Spire	37.5/IMR4350	—	JJD

CASE PREPARATION: SHELLHOLDER (RCBS): 7

MAKE FROM: .303 British. Form die set required. Anneal case neck. Form in form dies #1 and #2. Trim to length. Chamfer. F/L size.

PHYSICAL DATA (INCHES):

CASE TYPE: **Rimmed Bottleneck**

CASE LENGTH: **A = 1.97**

HEAD DIAMETER: **B = .455**

RIM DIAMETER: **D = .540**

NECK DIAMETER: **F = .250**

NECK LENGTH: **H = .373**

SHOULDER LENGTH: **K = .147**

BODY ANGLE (DEG'S/SIDE): **.206**

CASE CAPACITY **CC'S = 3.07**

LOADED LENGTH: **2.50**

BELT DIAMETER: **C = N/A**

RIM THICKNESS: **E = .063**

SHOULDER DIAMETER: **G = .446**

LENGTH TO SHOULDER: **J = 1.45**

SHOULDER ANGLE (DEG'S/SIDE): **33.69**

PRIMER: **L/R**

CASE CAPACITY (GR'S WATER): **47.37**

DIMENSIONAL DRAWING:

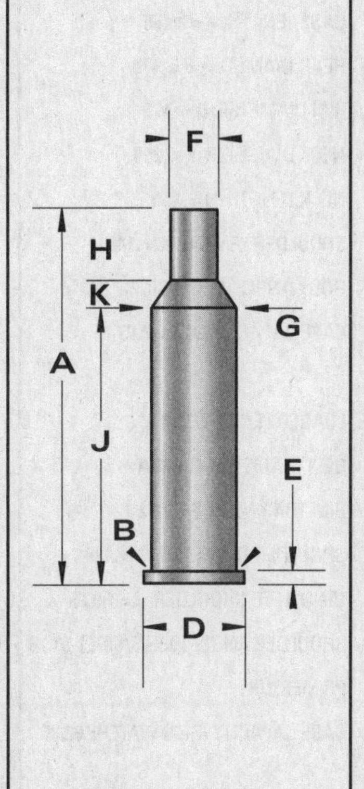

-NOT ACTUAL SIZE-
-DO NOT SCALE-

CARTRIDGE: .22 Winchester Centerfire

OTHER NAMES:
.22 WCF

DIA: .228

BALLISTEK NO: 228W

NAI NO: RMB 24221/4.712

DATA SOURCE: COTW 4th Pg.88

HISTORICAL DATA: By Win. in 1885

NOTES:

LOADING DATA:

BULLET WT./TYPE	POWDER WT./TYPE	VELOCITY ('/SEC)	SOURCE
45/Lead	4.0/Unique	1500	Barnes

CASE PREPARATION: SHELLHOLDER (RCBS): 12

MAKE FROM: .22 Hornet. Trim Hornet case to length. Load with 2 grs of Bullseye and corn meal and fireform in .22 WCF trim die. Clean and F/L size.

PHYSICAL DATA (INCHES):

CASE TYPE: **Rimmed Bottleneck**

CASE LENGTH: **A = 1.39**

HEAD DIAMETER: **B = .295**

RIM DIAMETER: **D = .342**

NECK DIAMETER: **F = .241**

NECK LENGTH: **H = .383**

SHOULDER LENGTH: **K = .202**

BODY ANGLE (DEG'S/SIDE): **.804**

CASE CAPACITY **CC'S = .925**

LOADED LENGTH: **1.61**

BELT DIAMETER: **C = N/A**

RIM THICKNESS: **E = .055**

SHOULDER DIAMETER: **G = .278**

LENGTH TO SHOULDER: **J = .805**

SHOULDER ANGLE (DEG'S/SIDE): **5.23**

PRIMER: **S/R**

CASE CAPACITY (GR'S WATER): **14.28**

DIMENSIONAL DRAWING:

F
H
K
G
A
J
E
B
D

-NOT ACTUAL SIZE-
-DO NOT SCALE-

CARTRIDGE: .22-15-60 Stevens

OTHER NAMES:

DIA: .226

BALLISTEK NO: 226B

NAI NO: RMS
11115/7.585

DATA SOURCE: COTW 4th Pg.88

HISTORICAL DATA: By Stevens in 1896 for M44 S.S. rifle.

NOTES:

LOADING DATA:

BULLET WT./TYPE	POWDER WT./TYPE	VELOCITY ('/SEC)	SOURCE
60/Lead	3.4/Unique	1070	Barnes

CASE PREPARATION: SHELLHOLDER (RCBS): 12

MAKE FROM: Solid brass. There is no case which will form into the .22-15. Lathe turn from 3/8" dia. brass.

PHYSICAL DATA (INCHES):

CASE TYPE: **Rimmed Straight**
CASE LENGTH: **A = 2.01**
HEAD DIAMETER: **B = .265**
RIM DIAMETER: **D = .342**
NECK DIAMETER: **F = .243**
NECK LENGTH: **H = N/A**
SHOULDER LENGTH: **K = N/A**
BODY ANGLE (DEG'S/SIDE): **.325**
CASE CAPACITY **CC'S = 1.14**

LOADED LENGTH: **2.26**
BELT DIAMETER: **C = N/A**
RIM THICKNESS: **E = .072**
SHOULDER DIAMETER: **G = N/A**
LENGTH TO SHOULDER: **J = N/A**
SHOULDER ANGLE (DEG'S/SIDE): **N/A**
PRIMER: **S/R**
CASE CAPACITY (GR'S WATER): **17.64**

DIMENSIONAL DRAWING:

F
A
E
B
D

-NOT ACTUAL SIZE-
-DO NOT SCALE-

CARTRIDGE: .22/06 Easling

OTHER NAMES:

DIA: .224

BALLISTEK NO: 224BN

NAI NO: RXB
22424/5.244

DATA SOURCE: Ackley Vol.2 Pg.137

HISTORICAL DATA: By S. Easling

NOTES:

LOADING DATA:

BULLET WT./TYPE	POWDER WT./TYPE	VELOCITY ('/SEC)	SOURCE
89/Spire	59.0/H870	3455	Ackley

CASE PREPARATION: SHELLHOLDER (RCBS): 3

MAKE FROM: .25-06 Rem. Anneal case neck. Size in .243 Win. F/L die to reduce neck dia. (do not set shoulder back!). Size in .22-250 F/L die until bolt will close on chambered case. Fireform, in chamber, with light loads.

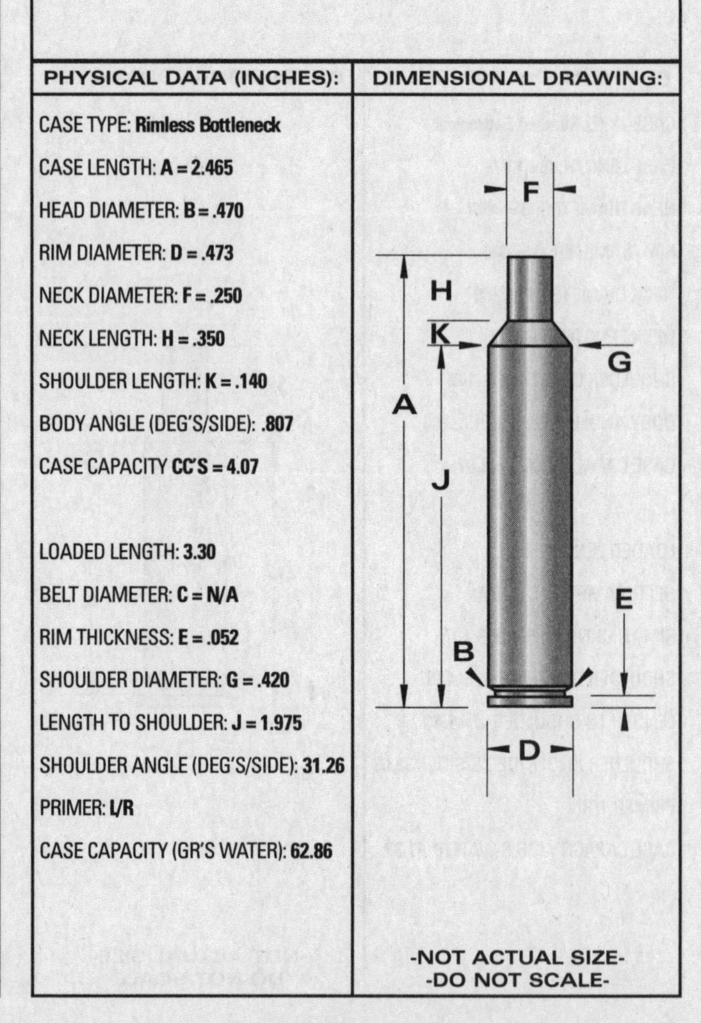

PHYSICAL DATA (INCHES):

CASE TYPE: **Rimless Bottleneck**
CASE LENGTH: **A = 2.465**
HEAD DIAMETER: **B = .470**
RIM DIAMETER: **D = .473**
NECK DIAMETER: **F = .250**
NECK LENGTH: **H = .350**
SHOULDER LENGTH: **K = .140**
BODY ANGLE (DEG'S/SIDE): **.807**
CASE CAPACITY **CC'S = 4.07**

LOADED LENGTH: **3.30**
BELT DIAMETER: **C = N/A**
RIM THICKNESS: **E = .052**
SHOULDER DIAMETER: **G = .420**
LENGTH TO SHOULDER: **J = 1.975**
SHOULDER ANGLE (DEG'S/SIDE): **31.26**
PRIMER: **L/R**
CASE CAPACITY (GR'S WATER): **62.86**

DIMENSIONAL DRAWING:

-NOT ACTUAL SIZE-
-DO NOT SCALE-

CARTRIDGE: .22/06 Short

OTHER NAMES:

DIA: .224

BALLISTEK NO: 224AG

NAI NO: RXB 14424/3.676

DATA SOURCE: Rifle #55 Pg.12

HISTORICAL DATA: By C. Graves.

NOTES: Similar to .22 PPC

LOADING DATA:

BULLET WT./TYPE	POWDER WT./TYPE	VELOCITY ('/SEC)	SOURCE
50/Spire	31.5/IMR4895	3570	JJD

CASE PREPARATION: SHELLHOLDER (RCBS): 3

MAKE FROM: .22-250 Rem. Use form die or .22 PPC F/L die to set shoulder back. Annealing may be required. I.D. neck ream if seated bullet bulges neck beyond .251" dia. F/L size and chamfer.

PHYSICAL DATA (INCHES):

CASE TYPE: **Rimless Bottleneck**
CASE LENGTH: **A = 1.728**
HEAD DIAMETER: **B = .470**
RIM DIAMETER: **D = .473**
NECK DIAMETER: **F = .250**
NECK LENGTH: **H = .355**
SHOULDER LENGTH: **K = .163**
BODY ANGLE (DEG'S/SIDE): **.312**
CASE CAPACITY **CC'S = 2.71**
LOADED LENGTH: **A/R**
BELT DIAMETER: **C = N/A**
RIM THICKNESS: **E = .053**
SHOULDER DIAMETER: **G = .459**
LENGTH TO SHOULDER: **J = 1.21**
SHOULDER ANGLE (DEG'S/SIDE): **32.66**
PRIMER: **L/R**
CASE CAPACITY (GR'S WATER): **41.86**

DIMENSIONAL DRAWING:

F H K G A J E B D

-NOT ACTUAL SIZE-
-DO NOT SCALE-

CARTRIDGE: .22/243 Winchester

OTHER NAMES:

DIA: .224

BALLISTEK NO: 224AM

NAI NO: RXB 23433/4.212

DATA SOURCE: Handloader #21 Pg.38

HISTORICAL DATA: By P. Middlested in 1964

NOTES: Reported as 30° shoulder.

LOADING DATA:

BULLET WT./TYPE	POWDER WT./TYPE	VELOCITY ('/SEC)	SOURCE
72/–	45.0/4831	3450	Mason

CASE PREPARATION: SHELLHOLDER (RCBS): 3

MAKE FROM: .22-250 Rem. Fireform in chamber. This wildcat was developed in the pre-.22-250 days. It is nothing more than a .22-250 with a 30° (or so) shoulder. Offers no advantage over the .22-250.

PHYSICAL DATA (INCHES):

CASE TYPE: **Rimless Bottleneck**
CASE LENGTH: **A = 1.98**
HEAD DIAMETER: **B = .470**
RIM DIAMETER: **D = .473**
NECK DIAMETER: **F = .242**
NECK LENGTH: **H = .240**
SHOULDER LENGTH: **K = .234**
BODY ANGLE (DEG'S/SIDE): **.661**
CASE CAPACITY **CC'S = 3.44**
LOADED LENGTH: **2.56**
BELT DIAMETER: **C = N/A**
RIM THICKNESS: **E = .052**
SHOULDER DIAMETER: **G = .440**
LENGTH TO SHOULDER: **J = 1.50**
SHOULDER ANGLE (DEG'S/SIDE): **22.41**
PRIMER: **L/R**
CASE CAPACITY (GR'S WATER): **53.19**

DIMENSIONAL DRAWING:

F H K G A J E B D

-NOT ACTUAL SIZE-
-DO NOT SCALE-

CARTRIDGE: .22/250 Ackley Improved

OTHER NAMES:

DIA: .224

BALLISTEK NO: 224AN

NAI NO: RXB 23433/4.105

DATA SOURCE: Ackley Vol.1 Pg.282

HISTORICAL DATA:

NOTES: This is the 28° (reported) shoulder version.

LOADING DATA:

BULLET WT./TYPE	POWDER WT./TYPE	VELOCITY ('/SEC)	SOURCE
50/Spire	34.0/3031	3650	JJD

CASE PREPARATION: SHELLHOLDER (RCBS): 3

MAKE FROM: .22-250 Rem. Fire ammo in "improved" chamber.

PHYSICAL DATA (INCHES):

CASE TYPE: **Rimless Bottleneck**

CASE LENGTH: **A = 1.917**

HEAD DIAMETER: **B = .467**

RIM DIAMETER: **D = .473**

NECK DIAMETER: **F = .253**

NECK LENGTH: **H = .225**

SHOULDER LENGTH: **K = .192**

BODY ANGLE (DEG'S/SIDE): **.771**

CASE CAPACITY **CC'S = 3.05**

LOADED LENGTH: **2.43**

BELT DIAMETER: **C = N/A**

RIM THICKNESS: **E = .049**

SHOULDER DIAMETER: **G = .432**

LENGTH TO SHOULDER: **J = 1.50**

SHOULDER ANGLE (DEG'S/SIDE): **24.99**

PRIMER: **L/R**

CASE CAPACITY (GR'S WATER): **47.18**

DIMENSIONAL DRAWING:

-NOT ACTUAL SIZE-
-DO NOT SCALE-

CARTRIDGE: .22-250 Remington

OTHER NAMES:
.22 Varminter
.220 Wotkyns Original Swift (WOS)

DIA: .224

BALLISTEK NO: 224J

NAI NO: RXB 33433/4.085

DATA SOURCE: Hornady 3rd Pg.95

HISTORICAL DATA: By Rem. in 1965

NOTES: Original design by Capt. Wotkyns.

LOADING DATA:

BULLET WT./TYPE	POWDER WT./TYPE	VELOCITY ('/SEC)	SOURCE
55/Spire	35.0/IMR4064	3575	JJD

CASE PREPARATION: SHELLHOLDER (RCBS): 3

MAKE FROM: Factory or .250 Savage. Anneal case neck and full-length size in .22-250 die. Trim to length. .243 Winchester brass may also be used in the same manner.

PHYSICAL DATA (INCHES):

CASE TYPE: **Rimless Bottleneck**

CASE LENGTH: **A = 1.912**

HEAD DIAMETER: **B = .468**

RIM DIAMETER: **D = .472**

NECK DIAMETER: **F = .253**

NECK LENGTH: **H = .249**

SHOULDER LENGTH: **K = .223**

BODY ANGLE (DEG'S/SIDE): **1.155**

CASE CAPACITY **CC'S = 2.89**

LOADED LENGTH: **2.35**

BELT DIAMETER: **C = N/A**

RIM THICKNESS: **E = .05**

SHOULDER DIAMETER: **G = .418**

LENGTH TO SHOULDER: **J = 1.44**

SHOULDER ANGLE (DEG'S/SIDE): **20.3**

PRIMER: **L/R**

CASE CAPACITY (GR'S WATER): **44.62**

DIMENSIONAL DRAWING:

-NOT ACTUAL SIZE-
-DO NOT SCALE-

CARTRIDGE: .22-250 Remington Improved (Shannon)

OTHER NAMES:

DIA: .224

BALLISTEK NO: 224BP

NAI NO: RXB 23433/4.068

DATA SOURCE: Ackley Vol.2 Pg.136

HISTORICAL DATA: By J. Shannon.

NOTES:

LOADING DATA:

BULLET WT./TYPE	POWDER WT./TYPE	VELOCITY ('/SEC)	SOURCE
63/—	41.0/4350	—	Ackley

CASE PREPARATION: SHELLHOLDER (RCBS): 3

MAKE FROM: .22-250 Rem. Anneal case neck and fireform (with corn meal) in a Shannon trim die. It may be possible to fireform this ctg. in the Shannon Imp. chamber but, I have never done so. Trim to length and chamfer.

PHYSICAL DATA (INCHES):

CASE TYPE: **Rimless Bottleneck**

CASE LENGTH: **A = 1.925**

HEAD DIAMETER: **B = .467**

RIM DIAMETER: **D = .473**

NECK DIAMETER: **F = .251**

NECK LENGTH: **H = .27**

SHOULDER LENGTH: **K = .133**

BODY ANGLE (DEG'S/SIDE): **.680**

CASE CAPACITY **CC'S = 3.11**

LOADED LENGTH: **A/R**

BELT DIAMETER: **C = N/A**

RIM THICKNESS: **E = .049**

SHOULDER DIAMETER: **G = .44**

LENGTH TO SHOULDER: **J = 1.522**

SHOULDER ANGLE (DEG'S/SIDE): **35**

PRIMER: **L/R**

CASE CAPACITY (GR'S WATER): **48.06**

DIMENSIONAL DRAWING:

-NOT ACTUAL SIZE-
-DO NOT SCALE-

CARTRIDGE: .22/284 Winchester

OTHER NAMES:

DIA: .224

BALLISTEK NO: 224AO

NAI NO: RBB 22434/4.330

DATA SOURCE: Ackley Vol.2 Pg.137

HISTORICAL DATA:

NOTES: Overbore for .224" dia.!

LOADING DATA:

BULLET WT./TYPE	POWDER WT./TYPE	VELOCITY ('/SEC)	SOURCE
50/—	53.0/4350	4040(?)	Ackley

CASE PREPARATION: SHELLHOLDER (RCBS): 3

MAKE FROM: .284 Winchester. Use form die set. Form #1 and form #2. Trim to length. F/L size. I do not know of any way to avoid purchasing the form dies due to the large (.477" dia.) shoulder dia. of this ctg.

PHYSICAL DATA (INCHES):

CASE TYPE: **Rebated Bottleneck**

CASE LENGTH: **A = 2.165**

HEAD DIAMETER: **B = .500**

RIM DIAMETER: **D = .473**

NECK DIAMETER: **F = .251**

NECK LENGTH: **H = .262**

SHOULDER LENGTH: **K = .098**

BODY ANGLE (DEG'S/SIDE): **.410**

CASE CAPACITY **CC'S = 4.34**

LOADED LENGTH: **2.70**

BELT DIAMETER: **C = N/A**

RIM THICKNESS: **E = .053**

SHOULDER DIAMETER: **G = .477**

LENGTH TO SHOULDER: **J = 1.805**

SHOULDER ANGLE (DEG'S/SIDE): **49.06**

PRIMER: **L/R**

CASE CAPACITY (GR'S WATER): **67.11**

DIMENSIONAL DRAWING:

-NOT ACTUAL SIZE-
-DO NOT SCALE-

CARTRIDGE: .22/30-30 Imp. (Ackley)

OTHER NAMES:

DIA: .224

BALLISTEK NO: 224AP

NAI NO: RMB 22424/4.845

DATA SOURCE: Ackley Vol.1 Pg.278

HISTORICAL DATA:

NOTES: Reported as 40° shoulder.

LOADING DATA:

BULLET WT./TYPE	POWDER WT./TYPE	VELOCITY ('/SEC)	SOURCE
45/—	38.0/4895	3972	Ackley

CASE PREPARATION: SHELLHOLDER (RCBS): 2

MAKE FROM: .30-30 Win. Use .219 Zipper form set but allow for greater length to shoulder and case length. F/L size in Zipper die. Fireform in chamber.

PHYSICAL DATA (INCHES):

CASE TYPE: **Rimmed Bottleneck**

CASE LENGTH: **A = 2.04**

HEAD DIAMETER: **B = .421**

RIM DIAMETER: **D = .506**

NECK DIAMETER: **F = .253**

NECK LENGTH: **H = .312**

SHOULDER LENGTH: **K = .098**

BODY ANGLE (DEG'S/SIDE): **.621**

CASE CAPACITY **CC'S = 2.64**

LOADED LENGTH: **2.55**

BELT DIAMETER: **C = N/A**

RIM THICKNESS: **E = .063**

SHOULDER DIAMETER: **G = .390**

LENGTH TO SHOULDER: **J = 1.63**

SHOULDER ANGLE (DEG'S/SIDE): **34.95**

PRIMER: **L/R**

CASE CAPACITY (GR'S WATER): **40.83**

DIMENSIONAL DRAWING:

-NOT ACTUAL SIZE-
-DO NOT SCALE-

CARTRIDGE: .22/3000 Lovell

OTHER NAMES:
R-2 Lovell

DIA: .224

BALLISTEK NO: 224AD

NAI NO: RMB 22332/5.061

DATA SOURCE: Ackley Vol.1 Pg.267

HISTORICAL DATA: By H. Lovell about 1934.

NOTES:

LOADING DATA:

BULLET WT./TYPE	POWDER WT./TYPE	VELOCITY ('/SEC)	SOURCE
50/—	17.0/4198	3050	Ackley

CASE PREPARATION: SHELLHOLDER (RCBS):

MAKE FROM: You'll have to make this case from solid brass. Recently, we have made some cases by using 5/16" dia. tubing on a .223 Rem. case head. These cases require a form die as F/L sizing will crush the case.

Note: Jim Bell as told me that he plans a run of .22 Single Shot brass soon. Let's hope so!

PHYSICAL DATA (INCHES):

CASE TYPE: **Rimmed Bottleneck**

CASE LENGTH: **A = 1.65**

HEAD DIAMETER: **B = .326**

RIM DIAMETER: **D = .390**

NECK DIAMETER: **F = .247**

NECK LENGTH: **H = .270**

SHOULDER LENGTH: **K = .12**

BODY ANGLE (DEG'S/SIDE): **.891**

CASE CAPACITY **CC'S = 1.25**

LOADED LENGTH: **2.12**

BELT DIAMETER: **C = N/A**

RIM THICKNESS: **E = .06**

SHOULDER DIAMETER: **G = .293**

LENGTH TO SHOULDER: **J = 1.26**

SHOULDER ANGLE (DEG'S/SIDE): **10.85**

PRIMER: **S/R**

CASE CAPACITY (GR'S WATER): **19.36**

DIMENSIONAL DRAWING:

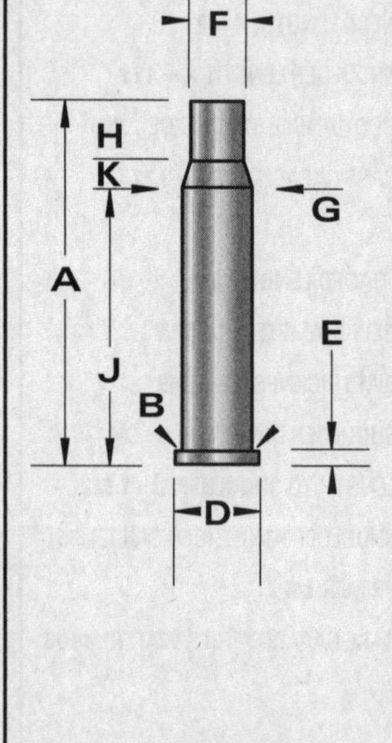

-NOT ACTUAL SIZE-
-DO NOT SCALE-

CARTRIDGE: .22/303 Sprinter

OTHER NAMES:

DIA: .224

BALLISTEK NO: 224AR

NAI NO: RMB 32423/4.605

DATA SOURCE: Ackley Vol.1 Pg.280

HISTORICAL DATA: Australian design.

NOTES:

LOADING DATA:

BULLET WT./TYPE	POWDER WT./TYPE	VELOCITY ('/SEC)	SOURCE
50/Spire	30.0/3031	3375	Ackley

CASE PREPARATION: SHELLHOLDER (RCBS): 7

MAKE FROM: .303 British. Anneal case neck. Reduce neck dia. in 7mm/08 die followed by .243 Win. die (or, better, obtain the form set). Trim to proper length and F/L size. Chamfer.

PHYSICAL DATA (INCHES):

CASE TYPE: **Rimmed Bottleneck**

CASE LENGTH: **A = 2.100**

HEAD DIAMETER: **B = .456**

RIM DIAMETER: **D = .540**

NECK DIAMETER: **F = .254**

NECK LENGTH: **H = .316**

SHOULDER LENGTH: **K = .124**

BODY ANGLE (DEG'S/SIDE): **1.58**

CASE CAPACITY **CC'S = 2.83**

LOADED LENGTH: **2.55**

BELT DIAMETER: **C = N/A**

RIM THICKNESS: **E = .063**

SHOULDER DIAMETER: **G = .375**

LENGTH TO SHOULDER: **J = 1.66**

SHOULDER ANGLE (DEG'S/SIDE): **26.00**

PRIMER: **L/R**

CASE CAPACITY (GR'S WATER): **43.78**

DIMENSIONAL DRAWING:

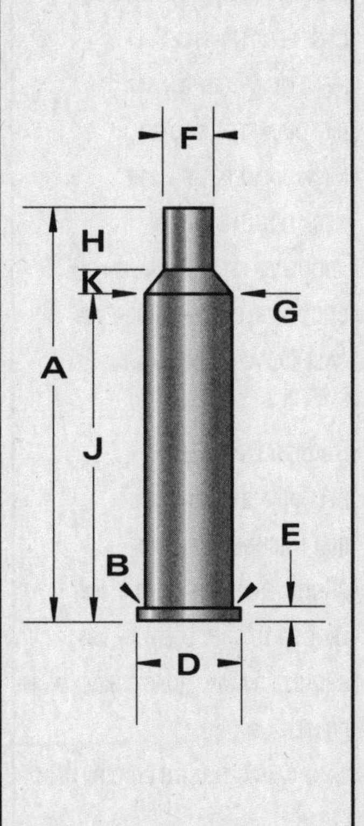

-NOT ACTUAL SIZE-
-DO NOT SCALE-

CARTRIDGE: .22/303 Varmint-R

OTHER NAMES:
.22 Varmint-R

DIA: .224

BALLISTEK NO: 224AC

NAI NO: RMB 22423/4.464

DATA SOURCE: Ackley Vol.1 Pg.279

HISTORICAL DATA: By G. Crandall about 1937

NOTES: Other versions exist.

LOADING DATA:

BULLET WT./TYPE	POWDER WT./TYPE	VELOCITY ('/SEC)	SOURCE
55/Spire	30.0/3031	3380	Ackley

CASE PREPARATION: SHELLHOLDER (RCBS): 7

MAKE FROM: .303 British. Form die set required. Anneal case and form in form set. Trim length, F/L size.

PHYSICAL DATA (INCHES):

CASE TYPE: **Rimmed Bottleneck**

CASE LENGTH: **A = 2.031**

HEAD DIAMETER: **B = .455**

RIM DIAMETER: **D = .540**

NECK DIAMETER: **F = .254**

NECK LENGTH: **H = .325**

SHOULDER LENGTH: **K = .144**

BODY ANGLE (DEG'S/SIDE): **.967**

CASE CAPACITY **CC'S = 2.92**

LOADED LENGTH: **2.48**

BELT DIAMETER: **C = N/A**

RIM THICKNESS: **E = .063**

SHOULDER DIAMETER: **G = .409**

LENGTH TO SHOULDER: **J = 1.562**

SHOULDER ANGLE (DEG'S/SIDE): **28.28**

PRIMER: **L/R**

CASE CAPACITY (GR'S WATER): **45.07**

DIMENSIONAL DRAWING:

F H K G A J E B D

-NOT ACTUAL SIZE-
-DO NOT SCALE-

CARTRIDGE: .22/350 Remington Mag.

OTHER NAMES:

DIA: .224

BALLISTEK NO: 224AS

NAI NO: BEN 22433/4.356

DATA SOURCE: Ackley Vol.2 Pg.140

HISTORICAL DATA:

NOTES: Badly overbore for .224" dia.

LOADING DATA:

BULLET WT./TYPE	POWDER WT./TYPE	VELOCITY ('/SEC)	SOURCE
55/Spire	59.0/4831	4263	Hutton

CASE PREPARATION: SHELLHOLDER (RCBS): 4

MAKE FROM: .350 Rem. Mag. Best to obtain form set but, case can be made by first annealing the magnum case and necking, in order, in the following dies: .33 Win., .284 Win., 6.5 Rem. Mag., 6mm/284 Win. and, finally, the .22/350 F/L die. Anneal 2nd time between 6mm/284 and F/L die. Of course, you can cheat and use .257 Weatherby brass. Anneal neck and F/L size with expander removed. Trim to length, F/L size, chamfer and fireform, in chamber.

PHYSICAL DATA (INCHES):

CASE TYPE: **Belted Bottleneck**

CASE LENGTH: **A = 2.235**

HEAD DIAMETER: **B = .513**

RIM DIAMETER: **D = .532**

NECK DIAMETER: **F = .250**

NECK LENGTH: **H = .240**

SHOULDER LENGTH: **K = .265**

BODY ANGLE (DEG'S/SIDE): **.861**

CASE CAPACITY **CC'S = 4.43**

LOADED LENGTH: **2.74**

BELT DIAMETER: **C = .532**

RIM THICKNESS: **E = .049**

SHOULDER DIAMETER: **G = .467**

LENGTH TO SHOULDER: **J = 1.73**

SHOULDER ANGLE (DEG'S/SIDE): **22.26**

PRIMER: **L/R Mag**

CASE CAPACITY (GR'S WATER): **68.44**

DIMENSIONAL DRAWING:

-NOT ACTUAL SIZE-
-DO NOT SCALE-

CARTRIDGE: .22/6.5 Mannlicher Schoenauer

OTHER NAMES:

DIA: .224

BALLISTEK NO: 224AQ

NAI NO: RXB 22423/4.787

DATA SOURCE: Ackley Vol.2 Pg.133

HISTORICAL DATA:

NOTES:

LOADING DATA:

BULLET WT./TYPE	POWDER WT./TYPE	VELOCITY ('/SEC)	SOURCE
55/Spire	34.0/4320	3278	Ackley

CASE PREPARATION: SHELLHOLDER (RCBS): 9

MAKE FROM: 6,5 M/S. Anneal case neck. Run into .243 Win. die to reduce neck dia. F/L size in .22/6.5mm die. Trim to length and chamfer. Fireform in chamber.

PHYSICAL DATA (INCHES):

CASE TYPE: **Rimless Bottleneck**

CASE LENGTH: **A = 2.14**

HEAD DIAMETER: **B = .447**

RIM DIAMETER: **D = .453**

NECK DIAMETER: **F = .250**

NECK LENGTH: **H = .308**

SHOULDER LENGTH: **K = .174**

BODY ANGLE (DEG'S/SIDE): **.785**

CASE CAPACITY **CC'S = 3.14**

LOADED LENGTH: **2.62**

BELT DIAMETER: **C = N/A**

RIM THICKNESS: **E = .055**

SHOULDER DIAMETER: **G = .407**

LENGTH TO SHOULDER: **J = 1.657**

SHOULDER ANGLE (DEG'S/SIDE): **24.28**

PRIMER: **L/R**

CASE CAPACITY (GR'S WATER): **48.54**

DIMENSIONAL DRAWING:

-NOT ACTUAL SIZE-
-DO NOT SCALE-

CARTRIDGE: .22/6/40

OTHER NAMES:

DIA: .224

BALLISTEK NO: 224CF

NAI NO: RMB 24224/3.451

DATA SOURCE: Handloader #102 Pg.38

HISTORICAL DATA: By G. Sengel for Contender pistols.

NOTES:

LOADING DATA:

BULLET WT./TYPE	POWDER WT./TYPE	VELOCITY ('/SEC)	SOURCE
40/Spire	3.0/Unique	1472	Sengel

CASE PREPARATION: SHELLHOLDER (RCBS): 12

MAKE FROM: .22 Hornet. Use a metal rod to push the Hornet cases into a .222 Rem. F/L sizer. Watch the location of the shoulder and make certain that the de-capping pin is removed. Use a punch to tap the case out of the die. I.D. neck ream. Sengel annealed his cases and then O.D. neck turned! Fireform cases. If you need a goodly supply of these cases, I'd purchase a form and F/L die set.

PHYSICAL DATA (INCHES):

CASE TYPE: **Rimmed Bottleneck**
CASE LENGTH: **A = 1.025**
HEAD DIAMETER: **B = .297**
RIM DIAMETER: **D = .350**
NECK DIAMETER: **F = .242**
NECK LENGTH: **H = .350**
SHOULDER LENGTH: **K = .030**
BODY ANGLE (DEG'S/SIDE): **.45**
CASE CAPACITY **CC'S = .613**

LOADED LENGTH: **1.22**
BELT DIAMETER: **C = N/A**
RIM THICKNESS: **E = .064**
SHOULDER DIAMETER: **G = .290**
LENGTH TO SHOULDER: **J = .645**
SHOULDER ANGLE (DEG'S/SIDE): **38.66**
PRIMER: **S/R**
CASE CAPACITY (GR'S WATER): **9.46**

DIMENSIONAL DRAWING:

-NOT ACTUAL SIZE-
-DO NOT SCALE-

CARTRIDGE: .220 Russian

OTHER NAMES: 5,6 x 39mm

DIA: .224

BALLISTEK NO: 224CA

NAI NO: RXB 34433/3.455

DATA SOURCE: NAI/Ballistek

HISTORICAL DATA:

NOTES: Starting case for PPC's.

LOADING DATA:

BULLET WT./TYPE	POWDER WT./TYPE	VELOCITY ('/SEC)	SOURCE
52/Spire	24.9/IMR3031	–	JJD

CASE PREPARATION: SHELLHOLDER (RCBS): 6

MAKE FROM: Factory (Sako) or 6,5 x 54 M/S. Anneal case and form in two die form set. Cut case to length. Turn rim to .441" dia. and back chamfer. F/L size to form and swage base. Trim to length. Fireform in chamber.

PHYSICAL DATA (INCHES):

CASE TYPE: **Rimless Bottleneck**
CASE LENGTH: **A = 1.517**
HEAD DIAMETER: **B = .439**
RIM DIAMETER: **D = .441**
NECK DIAMETER: **F = .242**
NECK LENGTH: **H = .253**
SHOULDER LENGTH: **K = .198**
BODY ANGLE (DEG'S/SIDE): **1.22**
CASE CAPACITY **CC'S = 2.03**

LOADED LENGTH: **A/R**
BELT DIAMETER: **C = N/A**
RIM THICKNESS: **E = .056**
SHOULDER DIAMETER: **G = .402**
LENGTH TO SHOULDER: **J = 1.066**
SHOULDER ANGLE (DEG'S/SIDE): **22.00**
PRIMER: **S/R**
CASE CAPACITY (GR'S WATER): **31.48**

DIMENSIONAL DRAWING:

-NOT ACTUAL SIZE-
-DO NOT SCALE-

CARTRIDGE: .220 Swift (Shannon)

OTHER NAMES:

DIA: .224

BALLISTEK NO: 224AT

NAI NO: RMB 22423/5.078

DATA SOURCE: Ackley Vol.2 Pg.136

HISTORICAL DATA: By J. Shannon

NOTES:

LOADING DATA:

BULLET WT./TYPE	POWDER WT./TYPE	VELOCITY ('/SEC)	SOURCE
50/Spire	35.0/IMR4895	3560	JJD

CASE PREPARATION: SHELLHOLDER (RCBS): 11

MAKE FROM: .220 Swift. Fire factory ammo in Shannon chamber or, with corn meal, in a Shannon trim die. F/L size and chamfer.

PHYSICAL DATA (INCHES):

CASE TYPE: **Rimmed Bottleneck**

CASE LENGTH: **A = 2.21**

HEAD DIAMETER: **B = .445**

RIM DIAMETER: **D = .473**

NECK DIAMETER: **F = .262**

NECK LENGTH: **H = .31**

SHOULDER LENGTH: **K = .123**

BODY ANGLE (DEG'S/SIDE): **.919**

CASE CAPACITY **CC'S = 3.07**

LOADED LENGTH: **2.83**

BELT DIAMETER: **C = N/A**

RIM THICKNESS: **E = .05**

SHOULDER DIAMETER: **G = .405**

LENGTH TO SHOULDER: **J = 1.777**

SHOULDER ANGLE (DEG'S/SIDE): **30**

PRIMER: **L/R**

CASE CAPACITY (GR'S WATER): **46.3**

DIMENSIONAL DRAWING:

-NOT ACTUAL SIZE-
-DO NOT SCALE-

CARTRIDGE: .220 Swift

OTHER NAMES:

DIA: .224

BALLISTEK NO: 224E

NAI NO: RMB 22423/4.955

DATA SOURCE: Hornady Manual 3rd Pg.99

HISTORICAL DATA: By Win. in 1935

NOTES: About max case capacity for .224" dia. bore

LOADING DATA:

BULLET WT./TYPE	POWDER WT./TYPE	VELOCITY ('/SEC)	SOURCE
60/Spire	34.0/IMR4895	3460	JJD

CASE PREPARATION: SHELLHOLDER (RCBS): 11

MAKE FROM: Factory or .270 Win. Swage base to .445" dia. F/L size (with expander removed). Inside neck ream. F/L size. Trim to length. Chamfer.

PHYSICAL DATA (INCHES):

CASE TYPE: **Rimmed Bottleneck**

CASE LENGTH: **A = 2.205**

HEAD DIAMETER: **B = .445**

RIM DIAMETER: **D = .473**

NECK DIAMETER: **F = .260**

NECK LENGTH: **H = .300**

SHOULDER LENGTH: **K = .183**

BODY ANGLE (DEG'S/SIDE): **.809**

CASE CAPACITY **CC'S = 2.99**

LOADED LENGTH: **2.70**

BELT DIAMETER: **C = N/A**

RIM THICKNESS: **E = .049**

SHOULDER DIAMETER: **G = .402**

LENGTH TO SHOULDER: **J = 1.722**

SHOULDER ANGLE (DEG'S/SIDE): **21.20**

PRIMER: **L/R**

CASE CAPACITY (GR'S WATER): **46.25**

DIMENSIONAL DRAWING:

-NOT ACTUAL SIZE-
-DO NOT SCALE-

CARTRIDGE: .220 Swift Improved (Ackley)

OTHER NAMES:

DIA: .224

BALLISTEK NO: 224AK

NAI NO: RMB 12424/5.140

DATA SOURCE: Ackley Vol.1 Pg.285

HISTORICAL DATA: By P.O. Ackley

NOTES:

LOADING DATA:

BULLET WT./TYPE	POWDER WT./TYPE	VELOCITY ('/SEC)	SOURCE
50/Spire	16.5/4895	3555	Ackley

CASE PREPARATION: SHELLHOLDER (RCBS): 11

MAKE FROM: .220 Swift. Fire .220 ammo in "improved" chamber or fireform in trim die.

PHYSICAL DATA (INCHES):

CASE TYPE: **Rimmed Bottleneck**

CASE LENGTH: **A = 2.19**

HEAD DIAMETER: **B = .426**

RIM DIAMETER: **D = .473**

NECK DIAMETER: **F = .260**

NECK LENGTH: **H = .320**

SHOULDER LENGTH: **K = .110**

BODY ANGLE (DEG'S/SIDE): **.073**

CASE CAPACITY **CC'S = 3.07**

LOADED LENGTH: **2.80**

BELT DIAMETER: **C = N/A**

RIM THICKNESS: **E = .05**

SHOULDER DIAMETER: **G = .430**

LENGTH TO SHOULDER: **J = 1.76**

SHOULDER ANGLE (DEG'S/SIDE): **37.69**

PRIMER: **L/R**

CASE CAPACITY (GR'S WATER): **47.45**

DIMENSIONAL DRAWING:

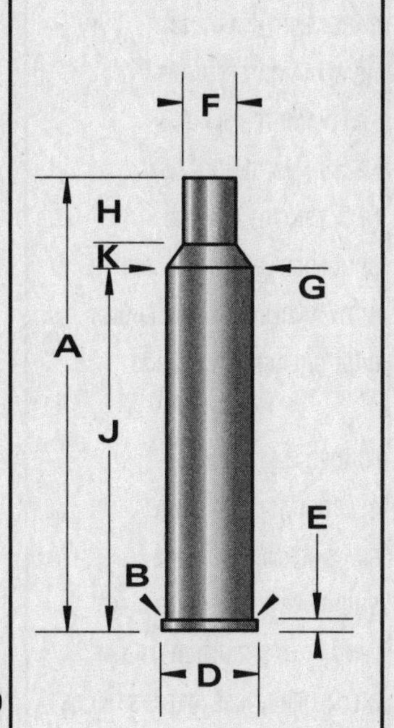

-NOT ACTUAL SIZE-
-DO NOT SCALE-

CARTRIDGE: .220 Weatherby Rocket

OTHER NAMES:

DIA: .224

BALLISTEK NO: 224BS

NAI NO: RMB 12423/5.089

DATA SOURCE: Ackley Vol.2 Pg.134

HISTORICAL DATA: By R. Weatherby

NOTES: This case has semi-venturi-type shoulder.

LOADING DATA:

BULLET WT./TYPE	POWDER WT./TYPE	VELOCITY ('/SEC)	SOURCE
50/Spire	40.0/3031	4000 (max)	Ackley

CASE PREPARATION: SHELLHOLDER (RCBS): 11

MAKE FROM: .220 Swift. Fire Swift ammo in Rocket chamber.

PHYSICAL DATA (INCHES):

CASE TYPE: **Rimmed Bottleneck**

CASE LENGTH: **A = 2.265**

HEAD DIAMETER: **B = .445**

RIM DIAMETER: **D = .473**

NECK DIAMETER: **F = .260**

NECK LENGTH: **H = .325**

SHOULDER LENGTH: **K = .140**

BODY ANGLE (DEG'S/SIDE): **.304**

CASE CAPACITY **CC'S = 3.31**

LOADED LENGTH: **2.78**

BELT DIAMETER: **C = N/A**

RIM THICKNESS: **E = .049**

SHOULDER DIAMETER: **G = .428**

LENGTH TO SHOULDER: **J = 1.80**

SHOULDER ANGLE (DEG'S/SIDE): **30.69**

PRIMER: **L/R**

CASE CAPACITY (GR'S WATER): **51.13**

DIMENSIONAL DRAWING:

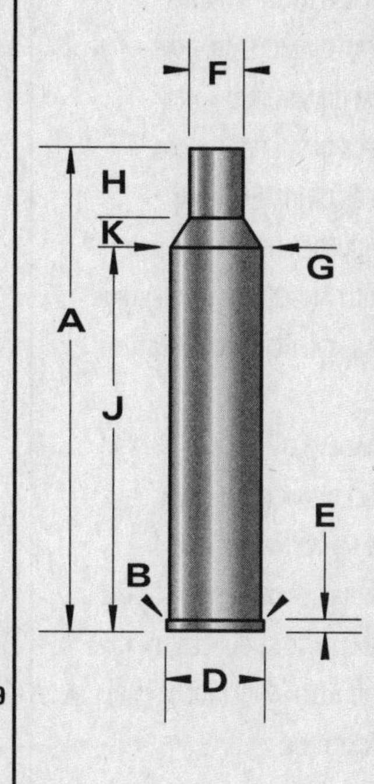

-NOT ACTUAL SIZE-
-DO NOT SCALE-

CARTRIDGE: .220 Wilson Arrow

OTHER NAMES:
.220 Wotkyns-Wilson Arrow
.220 Arrow

DIA: .224

BALLISTEK NO: 224CB

NAI NO: RMB 22423/4.955

DATA SOURCE: Handloader #102 Pg.24

HISTORICAL DATA: By Wilson & Wotkyns about 1940

NOTES: Dim's vary somewhat!

LOADING DATA:

BULLET WT./TYPE	POWDER WT./TYPE	VELOCITY ('/SEC)	SOURCE
55/HPBT	38.0/IMR4895	3755	Simpson

CASE PREPARATION: SHELLHOLDER (RCBS): 11

MAKE FROM: .220 Swift. Size the Swift cases in the Arrow F/L die. Trim, if necessary, to square case mouth. Chamfer.

PHYSICAL DATA (INCHES):

CASE TYPE: **Rimmed Bottleneck**

CASE LENGTH: **A = 2.205**

HEAD DIAMETER: **B = .445**

RIM DIAMETER: **D = .473**

NECK DIAMETER: **F = .260**

NECK LENGTH: **H = .360**

SHOULDER LENGTH: **K = .122**

BODY ANGLE (DEG'S/SIDE): **.808**

CASE CAPACITY **CC'S = 2.97**

LOADED LENGTH: **2.70**

BELT DIAMETER: **C = N/A**

RIM THICKNESS: **E = .05**

SHOULDER DIAMETER: **G = .402**

LENGTH TO SHOULDER: **J = 1.723**

SHOULDER ANGLE (DEG'S/SIDE): **30.19**

PRIMER: **L/R**

CASE CAPACITY (GR'S WATER): **45.80**

DIMENSIONAL DRAWING:

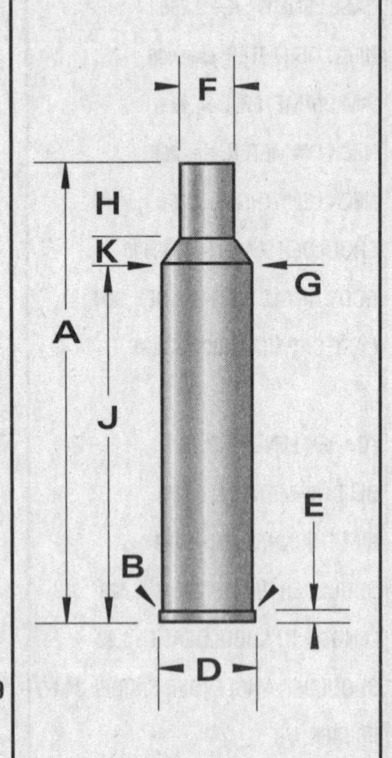

-NOT ACTUAL SIZE-
-DO NOT SCALE-

CARTRIDGE: .220/303 British (Shannon)

OTHER NAMES:

DIA: .224

BALLISTEK NO: 224BR

NAI NO: RMB 22424/5.089

DATA SOURCE: Ackley Vol.2 Pg.134

HISTORICAL DATA: By J. Shannon

NOTES: ALso see .22/303 Springer (#224AR)

LOADING DATA:

BULLET WT./TYPE	POWDER WT./TYPE	VELOCITY ('/SEC)	SOURCE
50/Spire	40.0/IMR4895	3825	JJD

CASE PREPARATION: SHELLHOLDER (RCBS): 7

MAKE FROM: .303 British. Anneal case mouth. Run neck portion into .308 F/L die to reduce dia. and repeat with 7mm/08 die, any short 6,5mm die, .243 Win die and .22-250 die. Repeat until rough shoulder is formed that will chamber in weapon. F/L size and fireform (with light loads) in chamber. You could also obtain a form set which does a much neater job of the whole business!

PHYSICAL DATA (INCHES):

CASE TYPE: **Rimmed Bottleneck**

CASE LENGTH: **A = 2.22**

HEAD DIAMETER: **B = .45**

RIM DIAMETER: **D = .540**

NECK DIAMETER: **F = .252**

NECK LENGTH: **H = .315**

SHOULDER LENGTH: **K = .055**

BODY ANGLE (DEG'S/SIDE): **.885**

CASE CAPACITY **CC'S = 3.33**

LOADED LENGTH: **2.815**

BELT DIAMETER: **C = N/A**

RIM THICKNESS: **E = .063**

SHOULDER DIAMETER: **G = .405**

LENGTH TO SHOULDER: **J = 1.85**

SHOULDER ANGLE (DEG'S/SIDE): **35**

PRIMER: **L/R**

CASE CAPACITY (GR'S WATER): **51.5**

DIMENSIONAL DRAWING:

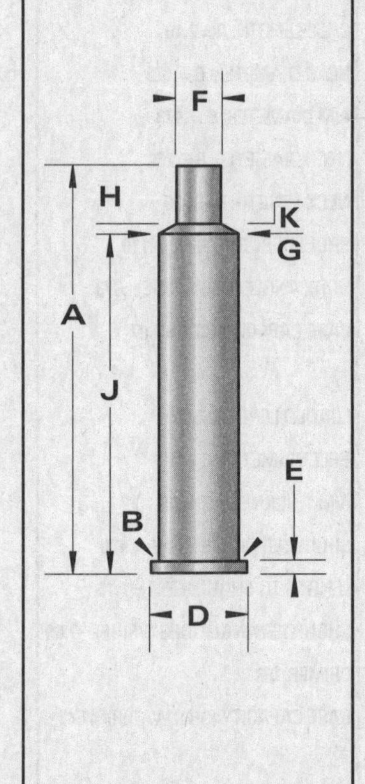

-NOT ACTUAL SIZE-
-DO NOT SCALE-

CARTRIDGE: .221 Remington Fireball

OTHER NAMES: .221 Fireball

DIA: .224

BALLISTEK NO: 224N

NAI NO: RXB 23343/3.713

DATA SOURCE: Hornady 3rd Pg.318

HISTORICAL DATA: By Rem. in 1963

NOTES: Designed for XP-100 pistol.

LOADING DATA:

BULLET WT./TYPE	POWDER WT./TYPE	VELOCITY ('/SEC)	SOURCE
53/HP	15.4/IMR4227	2500	Horn.

CASE PREPARATION: SHELLHOLDER (RCBS): 10

MAKE FROM: Factory or .223 Rem. Size in Fireball die with expander removed. Trim to length and chamber.

PHYSICAL DATA (INCHES):

CASE TYPE: **Rimless Bottleneck**

CASE LENGTH: **A = 1.400**

HEAD DIAMETER: **B = .377**

RIM DIAMETER: **D = .378**

NECK DIAMETER: **F = .253**

NECK LENGTH: **H = .203**

SHOULDER LENGTH: **K = .126**

BODY ANGLE (DEG'S/SIDE): **.526**

CASE CAPACITY **CC'S = 1.38**

LOADED LENGTH: **1.84**

BELT DIAMETER: **C = N/A**

RIM THICKNESS: **E = .045**

SHOULDER DIAMETER: **G = .361**

LENGTH TO SHOULDER: **J = 1.071**

SHOULDER ANGLE (DEG'S/SIDE): **23.19**

PRIMER: **S/R**

CASE CAPACITY (GR'S WATER): **21.34**

DIMENSIONAL DRAWING:

-NOT ACTUAL SIZE-
-DO NOT SCALE-

CARTRIDGE: .222 Eichhorn Lynx

OTHER NAMES: 5,7 x 43mm Eichhorn

DIA: .224

BALLISTEK NO: 224AJ

NAI NO: RXB 22334/4.533

DATA SOURCE: Ackley Vol.1 Pg.272

HISTORICAL DATA: By E. Eichhorn

NOTES: Reported as a 40° shoulder

LOADING DATA:

BULLET WT./TYPE	POWDER WT./TYPE	VELOCITY ('/SEC)	SOURCE
50/Spire	23.5/BL-C2	2640	Ackley

CASE PREPARATION: SHELLHOLDER (RCBS): 10

MAKE FROM: .222 Rem. Mag. Trim cases to 1.79". Use light load to fireform in the Eichhorn chamber.

PHYSICAL DATA (INCHES):

CASE TYPE: **Rimless Bottleneck**

CASE LENGTH: **A = 1.70**

HEAD DIAMETER: **B = .375**

RIM DIAMETER: **D = .378**

NECK DIAMETER: **F = .253**

NECK LENGTH: **H = .25**

SHOULDER LENGTH: **K = .03**

BODY ANGLE (DEG'S/SIDE): **.352**

CASE CAPACITY **CC'S = 1.77**

LOADED LENGTH: **2.31**

BELT DIAMETER: **C = N/A**

RIM THICKNESS: **E = .045**

SHOULDER DIAMETER: **G = .36**

LENGTH TO SHOULDER: **J = 1.42**

SHOULDER ANGLE (DEG'S/SIDE): **60.71**

PRIMER: **S/R**

CASE CAPACITY (GR'S WATER): **27.43**

DIMENSIONAL DRAWING:

-NOT ACTUAL SIZE-
-DO NOT SCALE-

CARTRIDGE: .222 Remington Improved

OTHER NAMES:
.222 Improved (Kilbourn)
.222 "K" Improved

DIA: .224

BALLISTEK NO: 224AH

NAI NO: RXB 22324/4.668

DATA SOURCE: Ackley Vol.1 Pg.269

HISTORICAL DATA:

NOTES: Several versions exist.

LOADING DATA:

BULLET WT./TYPE	POWDER WT./TYPE	VELOCITY ('/SEC)	SOURCE
50/—	23.0/H380	3024	Ackley

CASE PREPARATION: SHELLHOLDER (RCBS): 10

MAKE FROM: .222 Rem. Fire .222 Rem. ammo in "K" chamber or fireform in "K" trim die.

PHYSICAL DATA (INCHES):

CASE TYPE: Rimless Bottleneck
CASE LENGTH: A = 1.76
HEAD DIAMETER: B = .377
RIM DIAMETER: D = .378
NECK DIAMETER: F = .254
NECK LENGTH: H = .327
SHOULDER LENGTH: K = .058
BODY ANGLE (DEG'S/SIDE): .414
CASE CAPACITY CC'S = 1.79
LOADED LENGTH: 2.325
BELT DIAMETER: C = N/A
RIM THICKNESS: E = .045
SHOULDER DIAMETER: G = .360
LENGTH TO SHOULDER: J = 1.375
SHOULDER ANGLE (DEG'S/SIDE): 42.42
PRIMER: S/R
CASE CAPACITY (GR'S WATER): 27.62

DIMENSIONAL DRAWING:

F H K G A J E B D

-NOT ACTUAL SIZE-
-DO NOT SCALE-

CARTRIDGE: .222 Remington

OTHER NAMES:

DIA: .224

BALLISTEK NO: 224F

NAI NO: RXB 22323/4.509

DATA SOURCE: Hornady Manual 3rd Pg.67

HISTORICAL DATA: By Rem. in 1950

NOTES:

LOADING DATA:

BULLET WT./TYPE	POWDER WT./TYPE	VELOCITY ('/SEC)	SOURCE
53/SpireHP	26.0/W748	3100	JJD

CASE PREPARATION: SHELLHOLDER (RCBS): 10

MAKE FROM: Factory or .222 Rem. Mag. You should have no trouble finding these cases, however, the Magnum .222s can be F/L sized and trimmed to length if necessary.

PHYSICAL DATA (INCHES):

CASE TYPE: Rimless Bottleneck
CASE LENGTH: A = 1.700
HEAD DIAMETER: B = .377
RIM DIAMETER: D = .378
NECK DIAMETER: F = .253
NECK LENGTH: H = .313
SHOULDER LENGTH: K = .129
BODY ANGLE (DEG'S/SIDE): .514
CASE CAPACITY CC'S = 1.70
LOADED LENGTH: 2.20
BELT DIAMETER: C = N/A
RIM THICKNESS: E = .045
SHOULDER DIAMETER: G = .358
LENGTH TO SHOULDER: J = 1.258
SHOULDER ANGLE (DEG'S/SIDE): 22.14
PRIMER: S/R
CASE CAPACITY (GR'S WATER): 26.19

DIMENSIONAL DRAWING:

F H K G A J E B D

-NOT ACTUAL SIZE-
-DO NOT SCALE-

CARTRIDGE: .222 Remington Magnum

OTHER NAMES:

DIA: .224

BALLISTEK NO: 224G

NAI NO: RXB 22332/4.933

DATA SOURCE: Hornady Manual 3rd Pg.75

HISTORICAL DATA: By Rem. in 1958

NOTES: Long version of .222 Remington

LOADING DATA:

BULLET WT./TYPE	POWDER WT./TYPE	VELOCITY ('/SEC)	SOURCE
55/Spire	25.6/BL-C2	3200	Horn.

CASE PREPARATION: SHELLHOLDER (RCBS): 10

MAKE FROM: .223 Remington. Taper expand the case neck to .250" dia. F/L size in .222 Magnum die to create pseudo-shoulder to support case in fireforming. Form with corn meal over 3 grains of Bullseye in trim die or with normal load in actual chamber. With any luck, you should be able to find factory cases!

PHYSICAL DATA (INCHES):

CASE TYPE: **Rimless Bottleneck**

CASE LENGTH: **A = 1.850**

HEAD DIAMETER: **B = .375**

RIM DIAMETER: **D = .378**

NECK DIAMETER: **F = .253**

NECK LENGTH: **H = .235**

SHOULDER LENGTH: **K = .161**

BODY ANGLE (DEG'S/SIDE): **.388**

CASE CAPACITY **CC'S = 1.933**

LOADED LENGTH: **2.33**

BELT DIAMETER: **C = N/A**

RIM THICKNESS: **E = .045**

SHOULDER DIAMETER: **G = .358**

LENGTH TO SHOULDER: **J = 1.454**

SHOULDER ANGLE (DEG'S/SIDE): **23.06**

PRIMER: **S/R**

CASE CAPACITY (GR'S WATER): **29.84**

DIMENSIONAL DRAWING:

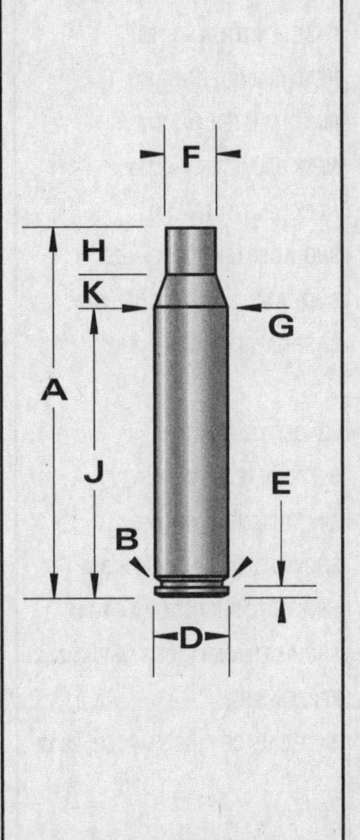

-NOT ACTUAL SIZE-
-DO NOT SCALE-

CARTRIDGE: .222 Remington Magnum Improved

OTHER NAMES:

DIA: .224

BALLISTEK NO: 224AI

NAI NO: RXB 12324/4.933

DATA SOURCE: Ackley Vol.1 Pg.271

HISTORICAL DATA: Ackley version.

NOTES:

LOADING DATA:

BULLET WT./TYPE	POWDER WT./TYPE	VELOCITY ('/SEC)	SOURCE
Increase loading data for .222 Rem Mag by 5% (per Ackley).			

CASE PREPARATION: SHELLHOLDER (RCBS): 10

MAKE FROM: .222 Rem. Mag. Fireform magnum ammo in "improved" chamber.

PHYSICAL DATA (INCHES):

CASE TYPE: **Rimless Bottleneck**

CASE LENGTH: **A = 1.85**

HEAD DIAMETER: **B = .375**

RIM DIAMETER: **D = .378**

NECK DIAMETER: **F = .253**

NECK LENGTH: **H = .280**

SHOULDER LENGTH: **K = .10**

BODY ANGLE (DEG'S/SIDE): **.315**

CASE CAPACITY **CC'S = 1.96**

LOADED LENGTH: **2.397**

BELT DIAMETER: **C = N/A**

RIM THICKNESS: **E = .045**

SHOULDER DIAMETER: **G = .360**

LENGTH TO SHOULDER: **J = 1.56**

SHOULDER ANGLE (DEG'S/SIDE): **45.00**

PRIMER: **S/R**

CASE CAPACITY (GR'S WATER): **30.34**

DIMENSIONAL DRAWING:

F
H
K
G
A
J
E
B
D

-NOT ACTUAL SIZE-
-DO NOT SCALE-

CARTRIDGE: .222 Rimmed

OTHER NAMES:
.222 (R) Remington

DIA: .224

BALLISTEK NO: 224BM

NAI NO: RMB 12333/4.933

DATA SOURCE: Ackley Vol.2 Pg.119

HISTORICAL DATA: Australian design

NOTES: Rimmed version of standard .222 Rem.

LOADING DATA:

BULLET WT./TYPE	POWDER WT./TYPE	VELOCITY ('/SEC)	SOURCE
50/Spire	19.0/N200	3100	Ackley

CASE PREPARATION: **SHELLHOLDER (RCBS): 6**

MAKE FROM: .357 max. Anneal case neck and F/L size in .222R die. This case will have a very short neck but, enough to hold a bullet. It is possible to re-head a .38 Spl. or .357 Mag. case head with a section of .375" dia tubing or solder a tubing extension on a .38 Spl. case. Use only very light loads for any cases made in this manner.

PHYSICAL DATA (INCHES):

CASE TYPE: **Rimmed Bottleneck**

CASE LENGTH: **A = 1.85**

HEAD DIAMETER: **B = .375**

RIM DIAMETER: **D = .450**

NECK DIAMETER: **F = .253**

NECK LENGTH: **H = .265**

SHOULDER LENGTH: **K = .125**

BODY ANGLE (DEG'S/SIDE): **.341**

CASE CAPACITY **CC'S = 1.93**

LOADED LENGTH: **A/R**

BELT DIAMETER: **C = N/A**

RIM THICKNESS: **E = .06**

SHOULDER DIAMETER: **G = .360**

LENGTH TO SHOULDER: **J = 1.46**

SHOULDER ANGLE (DEG'S/SIDE): **23.17**

PRIMER: **S/R**

CASE CAPACITY (GR'S WATER): **29.88**

DIMENSIONAL DRAWING:

-NOT ACTUAL SIZE-
-DO NOT SCALE-

CARTRIDGE: .223 Remington

OTHER NAMES:
5,6mm NATO M193
.223 Armalite

DIA: .224

BALLISTEK NO: 224H

NAI NO: RXB 22343/4.668

DATA SOURCE: Hornady Manual 3rd Pg.71

HISTORICAL DATA: Commercial by Rem. in 1958

NOTES: Current U.S. Military ctg.

LOADING DATA:

BULLET WT./TYPE	POWDER WT./TYPE	VELOCITY ('/SEC)	SOURCE
53/Spire	26.2/BL-C2	3250	JJD

CASE PREPARATION: **SHELLHOLDER (RCBS): 10**

MAKE FROM: Factory or .222 Rem. Mag. Size magnum case in .223 F/L die. Square case mouth. Chamfer.

PHYSICAL DATA (INCHES):

CASE TYPE: **Rimless Bottleneck**

CASE LENGTH: **A = 1.760**

HEAD DIAMETER: **B = .377**

RIM DIAMETER: **D = .378**

NECK DIAMETER: **F = .253**

NECK LENGTH: **H = .191**

SHOULDER LENGTH: **K = .125**

BODY ANGLE (DEG'S/SIDE): **.622**

CASE CAPACITY **CC'S = .1.82**

LOADED LENGTH: **2.30**

BELT DIAMETER: **C = N/A**

RIM THICKNESS: **E = .045**

SHOULDER DIAMETER: **G = .350**

LENGTH TO SHOULDER: **J = 1.444**

SHOULDER ANGLE (DEG'S/SIDE): **21.21**

PRIMER: **S/R**

CASE CAPACITY (GR'S WATER): **28.12**

DIMENSIONAL DRAWING:

-NOT ACTUAL SIZE-
-DO NOT SCALE-

CARTRIDGE: .223 Winchester Super Short Magnum

OTHER NAMES:

DIA: .224

BALLISTEK NO:

NAI NO:

DATA SOURCE: Most current loading manuals.

HISTORICAL DATA: Developed by Winchester in 2003 to fit in new super Short rifle actions introduced by Browning and Winchester. The case is .43-inch shorter than the Winchester Short Magnum.

LOADING DATA:

BULLET WT./TYPE	POWDER WT./TYPE	VELOCITY ('/SEC)	SOURCE
60/Nosler Partition	38.5/Varget	3,656	Hodgdon

CASE PREPARATION: SHELL HOLDER (RCBS): 43

MAKE FROM: Factory. Can be made from .404 Jeffery. Shorten and die form to fit chamber. Fireform body to get the proper diameter, shoulder angle and head space. Then turn the rim to correct dimension. Ream neck if necessary. Square mouth and chamfer and deburr.

PHYSICAL DATA (INCHES):

CASE TYPE: **Rebated Bottleneck**

CASE LENGTH **A = 1.670**

HEAD DIAMETER **B = .550**

RIM DIAMETER **D = .535**

NECK DIAMETER **F = .272**

NECK LENGTH **H = .264**

SHOULDER LENGTH **K = .256**

BODY ANGLE (DEG'S/SIDE):

CASE CAPACITY **CC'S =**

LOADED LENGTH: **2.360**

BELT DIAMETER **C =**

RIM THICKNESS **E = .54**

SHOULDER DIAMETER **G = .5444**

LENGTH TO SHOULDER **J = .1.150**

SHOULDER ANGLE (DEG'S/SIDE): **28**

PRIMER: **L/R**

CASE CAPACITY (GR'S WATER): **54.6**

DIMENSIONAL DRAWING:

F H K G A J E B D

-NOT ACTUAL SIZE-
-DO NOT SCALE-

CARTRIDGE: .224 Clark

OTHER NAMES:

DIA: .224

BALLISTEK NO: 224BL

NAI NO: RXB 12423/4.749

DATA SOURCE: Handloader #107 Pg.12

HISTORICAL DATA: By K. Clark in 1962 for heavy .224s

NOTES:

LOADING DATA:

BULLET WT./TYPE	POWDER WT./TYPE	VELOCITY ('/SEC)	SOURCE
82/Spire	53.0/H450	3513	Clark/ Ackley

CASE PREPARATION: SHELLHOLDER (RCBS): 3

MAKE FROM: .257 Roberts. Size cases in Clark die to reduce neck dia. (with expander removed). I.D. neck ream & chamfer. F/L size and fireform in chamber.

PHYSICAL DATA (INCHES):

CASE TYPE: **Rimless Bottleneck**

CASE LENGTH: **A = 2.237**

HEAD DIAMETER: **B = .471**

RIM DIAMETER: **D = .473**

NECK DIAMETER: **F = .225**

NECK LENGTH: **H = .304**

SHOULDER LENGTH: **K = .205**

BODY ANGLE (DEG'S/SIDE): **.299**

CASE CAPACITY **CC'S = 4.59**

LOADED LENGTH: **3.075**

BELT DIAMETER: **C = N/A**

RIM THICKNESS: **E = .05**

SHOULDER DIAMETER: **G = .455**

LENGTH TO SHOULDER: **J = 1.728**

SHOULDER ANGLE (DEG'S/SIDE): **29.25**

PRIMER: **L/R**

CASE CAPACITY (GR'S WATER): **70.94**

DIMENSIONAL DRAWING:

F H K G A J E B D

-NOT ACTUAL SIZE-
-DO NOT SCALE-

CARTRIDGE: .224 Critser Comet

OTHER NAMES:

DIA: .224

BALLISTEK NO: 224BK

NAI NO: RMB 34334/2.951

DATA SOURCE: Ackley Vol.1 Pg.510

HISTORICAL DATA: By C. Critser

NOTES: Reported as 53° shoulder.

LOADING DATA:

BULLET WT./TYPE	POWDER WT./TYPE	VELOCITY ('/SEC)	SOURCE
45/Spire	6.5/2400	1740	Ackley

CASE PREPARATION: SHELLHOLDER (RCBS): 1

MAKE FROM: .218 Bee. Run case into form die. Trim to length and chamfer. F/L size and fireform in chamber.

PHYSICAL DATA (INCHES):

CASE TYPE: **Rimmed Bottleneck**

CASE LENGTH: **A = 1.03**

HEAD DIAMETER: **B = .349**

RIM DIAMETER: **D = .408**

NECK DIAMETER: **F = .241**

NECK LENGTH: **H = .252**

SHOULDER LENGTH: **K = .017**

BODY ANGLE (DEG'S/SIDE): **1.32**

CASE CAPACITY **CC'S = .816**

LOADED LENGTH: **1.50**

BELT DIAMETER: **C = N/A**

RIM THICKNESS: **E = .065**

SHOULDER DIAMETER: **G = .323**

LENGTH TO SHOULDER: **J = .761**

SHOULDER ANGLE (DEG'S/SIDE): **67.47**

PRIMER: **S/P**

CASE CAPACITY (GR'S WATER): **12.59**

DIMENSIONAL DRAWING:

F
H
K
G
A
J
E
B
D

-NOT ACTUAL SIZE-
-DO NOT SCALE-

CARTRIDGE: .224 Donaldson Ace

OTHER NAMES:

DIA: .224

BALLISTEK NO: 224AV

NAI NO: RMB 22424/4.287

DATA SOURCE: Rifle #24 Pg.30

HISTORICAL DATA: By H. Donaldson to replace the .219 Donaldson Wasp.

NOTES:

LOADING DATA:

BULLET WT./TYPE	POWDER WT./TYPE	VELOCITY ('/SEC)	SOURCE
52/HPBT	30.0/3031	3600	Czarnota

CASE PREPARATION: SHELLHOLDER (RCBS): 11

MAKE FROM: .225 Win. One form die is required. Anneal case and form. Trim to length and F/L size. Fireform in chamber.

PHYSICAL DATA (INCHES):

CASE TYPE: **Rimmed Bottleneck**

CASE LENGTH: **A = 1.805**

HEAD DIAMETER: **B = .421**

RIM DIAMETER: **D = .473**

NECK DIAMETER: **F = .260**

NECK LENGTH: **H = .312**

SHOULDER LENGTH: **K = .118**

BODY ANGLE (DEG'S/SIDE): **.365**

CASE CAPACITY **CC'S = 2.22**

LOADED LENGTH: **A/R**

BELT DIAMETER: **C = N/A**

RIM THICKNESS: **E = .049**

SHOULDER DIAMETER: **G = .406**

LENGTH TO SHOULDER: **J = 1.375**

SHOULDER ANGLE (DEG'S/SIDE): **31.74**

PRIMER: **L/R**

CASE CAPACITY (GR'S WATER): **34.22**

DIMENSIONAL DRAWING:

F
H
K
G
A
J
E
B
D

-NOT ACTUAL SIZE-
-DO NOT SCALE-

CARTRIDGE: .224 Durham Jet

OTHER NAMES:

DIA: .224

BALLISTEK NO: 224AX

NAI NO: RXB 22424/4.351

DATA SOURCE: Ackley Vol.1 Pg.286

HISTORICAL DATA: By C. Durham

NOTES:

LOADING DATA:

BULLET WT./TYPE	POWDER WT./TYPE	VELOCITY ('/SEC)	SOURCE
60/Spire	46.0/H380	3980	Ackley

CASE PREPARATION: SHELLHOLDER (RCBS): 3

MAKE FROM: .243 Win. A form die is required because of the .446" dia. shoulder dim. Anneal case and form. Trim to length and F/L size. Chamfer.

PHYSICAL DATA (INCHES):

CASE TYPE: **Rimless Bottleneck**

CASE LENGTH: **A = 2.045**

HEAD DIAMETER: **B = .470**

RIM DIAMETER: **D = .473**

NECK DIAMETER: **F = .254**

NECK LENGTH: **H = .300**

SHOULDER LENGTH: **K = .117**

BODY ANGLE (DEG'S/SIDE): **.481**

CASE CAPACITY **CC'S = 3.38**

LOADED LENGTH: **2.68**

BELT DIAMETER: **C = N/A**

RIM THICKNESS: **E = .05**

SHOULDER DIAMETER: **G = .446**

LENGTH TO SHOULDER: **J = 1.628**

SHOULDER ANGLE (DEG'S/SIDE): **39.36**

PRIMER: **L/R**

CASE CAPACITY (GR'S WATER): **52.30**

DIMENSIONAL DRAWING:

F H K G A J E B D

-NOT ACTUAL SIZE-
-DO NOT SCALE-

CARTRIDGE: .224 ICL Benchrester

OTHER NAMES:

DIA: .224

BALLISTEK NO: 224AW

NAI NO: RXB 12425/3.276

DATA SOURCE: Ackley Vol.1 Pg.272

HISTORICAL DATA:

NOTES: This cartridge is quite hard to form!

LOADING DATA:

BULLET WT./TYPE	POWDER WT./TYPE	VELOCITY ('/SEC)	SOURCE
45/Spire	28.0/3031	3721	Ackley

CASE PREPARATION: SHELLHOLDER (RCBS): 3

MAKE FROM: .250 Savage. .22 BR form dies can be used to neck down and set shoulder back but, a ICL F/L die is required for final forming. May be possible to fireform in the ICL chamber but, I have never done so.

PHYSICAL DATA (INCHES):

CASE TYPE: **Rimless Bottleneck**

CASE LENGTH: **A = 1.54**

HEAD DIAMETER: **B = .470**

RIM DIAMETER: **D = .473**

NECK DIAMETER: **F = .254**

NECK LENGTH: **H = .300**

SHOULDER LENGTH: **K = .110**

BODY ANGLE (DEG'S/SIDE): **.215**

CASE CAPACITY **CC'S = 3.02**

LOADED LENGTH: **2.075**

BELT DIAMETER: **C = N/A**

RIM THICKNESS: **E = .05**

SHOULDER DIAMETER: **G = .457**

LENGTH TO SHOULDER: **J = 1.93**

SHOULDER ANGLE (DEG'S/SIDE): **53.36**

PRIMER: **L/R**

CASE CAPACITY (GR'S WATER): **46.63**

DIMENSIONAL DRAWING:

F H K G A J E B D

-NOT ACTUAL SIZE-
-DO NOT SCALE-

CARTRIDGE: .224 ICL Marmot

OTHER NAMES:

DIA: .224

BALLISTEK NO: 224BJ

NAI NO: RMB 22424/4.898

DATA SOURCE: Ackley Vol.1 Pg.287

HISTORICAL DATA: Improved Swift by ICL.

NOTES:

LOADING DATA:

BULLET WT./TYPE	POWDER WT./TYPE	VELOCITY ('/SEC)	SOURCE
55/Spire	36.0/4895	3755	Ackley

CASE PREPARATION: SHELLHOLDER (RCBS): 11

MAKE FROM: .220 Swift. Modify a standard .220 Swift die by grinding away base of die until dimension from case base to front of shoulder is 1.840. Use this die to set Swift's shoulder back. Fireform in ICL chamber to produce blown-out shoulder.

PHYSICAL DATA (INCHES):

CASE TYPE: **Rimmed Bottleneck**

CASE LENGTH: **A = 2.18**

HEAD DIAMETER: **B = .445**

RIM DIAMETER: **D = .473**

NECK DIAMETER: **F = .258**

NECK LENGTH: **H = .340**

SHOULDER LENGTH: **K = .114**

BODY ANGLE (DEG'S/SIDE): **.657**

CASE CAPACITY **CC'S = 3.05**

LOADED LENGTH: **2.73**

BELT DIAMETER: **C = N/A**

RIM THICKNESS: **E = .049**

SHOULDER DIAMETER: **G = .410**

LENGTH TO SHOULDER: **J = 1.726**

SHOULDER ANGLE (DEG'S/SIDE): **33.69**

PRIMER: **L/R**

CASE CAPACITY (GR'S WATER): **47.15**

DIMENSIONAL DRAWING:

-NOT ACTUAL SIZE-
-DO NOT SCALE-

CARTRIDGE: .224 Weatherby

OTHER NAMES:

DIA: .224

BALLISTEK NO: 224BC

NAI NO: BEN 22432/4.626

DATA SOURCE: Most current handloading manuals.

HISTORICAL DATA: By Roy Weatherby and introduced in 1963.

NOTES: Cartridge was approved by SAAMI on 6/26/02. Dimensions are now standard. Older rifles and loading dies may vary slightly. Use as Weatherby double radius shoulder.

LOADING DATA:

BULLET WT./TYPE	POWDER WT./TYPE	VELOCITY ('/SEC)	SOURCE
50/Hornady	34.5/H335	3900	Hornady #6
55/Hornady	32.7/IMR-4895	3700	

CASE PREPARATION: SHELLHOLDER (RCBS): 27

MAKE FROM: Use factory only. Prior suggestions to create a case by sweating on a belt on a 5,6mm case might be dangerous given the high pressure levels for this cartridge.

PHYSICAL DATA (INCHES):

CASE TYPE: **Belted Bottleneck**

CASE LENGTH: **A = 1.923**

HEAD DIAMETER: **B = .4125**

RIM DIAMETER: **D = .4295**

NECK DIAMETER: **F = .252**

NECK LENGTH: **H = .248**

SHOULDER LENGTH: **K = .188**

BODY ANGLE (DEG'S/SIDE): **.493**

CASE CAPACITY **CC'S = 2.54**

LOADED LENGTH: **2.360**

BELT DIAMETER: **C = .429**

RIM THICKNESS: **E = .050**

SHOULDER DIAMETER: **G = .3956**

LENGTH TO SHOULDER: **J = 1.487**

SHOULDER ANGLE (DEG'S/SIDE):

Front radius: **R .151 + .005**

Rear radius: **R .130**

PRIMER: **L/R**

CASE CAPACITY (GR'S WATER): **39.19**

DIMENSIONAL DRAWING:

-NOT ACTUAL SIZE-
-DO NOT SCALE-

CARTRIDGE: .225 Winchester

OTHER NAMES:

DIA: .224

BALLISTEK NO: 224L

NAI NO: RMB
12433/4.573

DATA SOURCE: Hornady Manual 3rd Pg.89

HISTORICAL DATA: By Win. in 1964

NOTES:

LOADING DATA:

BULLET WT./TYPE	POWDER WT./TYPE	VELOCITY ('/SEC)	SOURCE
60/Spire	32.0/IMR4895	3400	JJD

CASE PREPARATION: SHELLHOLDER (RCBS): 11

MAKE FROM: .30-30 Winchester. Anneal case. .219 Zipper of .219 Don. Wasp progressive necking dies may be used to slowly reduce neck dia. Watch shoulder location! F/L size in .225 Win. die. Trim. Fireform in chamber. Reduce rim dia. as required. I have never had to thin the 30-30s rim for .225 Win., but this may depend on the actual weapon.

PHYSICAL DATA (INCHES):

CASE TYPE: **Rimmed Bottleneck**

CASE LENGTH: **A = 1.93**

HEAD DIAMETER: **B = .422**

RIM DIAMETER: **D = .473**

NECK DIAMETER: **F = .26**

NECK LENGTH: **H = .243**

SHOULDER LENGTH: **K = .157**

BODY ANGLE (DEG'S/SIDE): **.344**

CASE CAPACITY **CC'S = 2.48**

LOADED LENGTH: **2.43**

BELT DIAMETER: **C = N/A**

RIM THICKNESS: **E = .05**

SHOULDER DIAMETER: **G = .406**

LENGTH TO SHOULDER: **J = 1.53**

SHOULDER ANGLE (DEG'S/SIDE): **24.94**

PRIMER: **L/R**

CASE CAPACITY (GR'S WATER): **38.30**

DIMENSIONAL DRAWING:

F
H
K
G
A
J
E
B
D

-NOT ACTUAL SIZE-
-DO NOT SCALE-

CARTRIDGE: .226 Barnes QT

OTHER NAMES:

DIA: .226

BALLISTEK NO: 226A

NAI NO: RXB
22424/4.767

DATA SOURCE: Ackley Vol.1 Pg.292

HISTORICAL DATA: By F. Barnes. Designed for fast (quick twist) rifles.

NOTES: .226" diameter is correct!

LOADING DATA:

BULLET WT./TYPE	POWDER WT./TYPE	VELOCITY ('/SEC)	SOURCE
125/Spire	47.0/MRP	–	JJD

CASE PREPARATION: SHELLHOLDER (RCBS): 11

MAKE FROM: .257 Roberts. Reduce case neck in form die. F/L size. Trim to length, chamfer and fireform in the chamber.

PHYSICAL DATA (INCHES):

CASE TYPE: **Rimless Bottleneck**

CASE LENGTH: **A = 2.25**

HEAD DIAMETER: **B = .472**

RIM DIAMETER: **D = .473**

NECK DIAMETER: **F = .257**

NECK LENGTH: **H = .308**

SHOULDER LENGTH: **K = .142**

BODY ANGLE (DEG'S/SIDE): **.465**

CASE CAPACITY **CC'S = 3.80**

LOADED LENGTH: **3.31**

BELT DIAMETER: **C = N/A**

RIM THICKNESS: **E = .05**

SHOULDER DIAMETER: **G = .446**

LENGTH TO SHOULDER: **J = 1.80**

SHOULDER ANGLE (DEG'S/SIDE): **33.64**

PRIMER: **L/R**

CASE CAPACITY (GR'S WATER): **58.66**

DIMENSIONAL DRAWING:

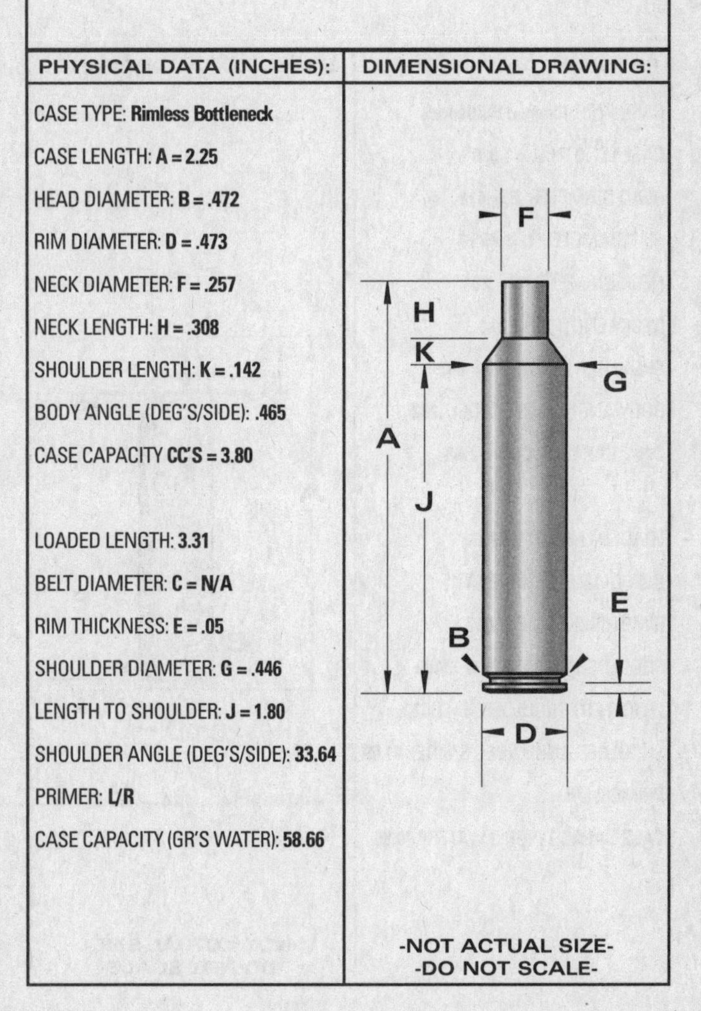

-NOT ACTUAL SIZE-
-DO NOT SCALE-

CARTRIDGE: .226 JDJ

OTHER NAMES:

DIA: .224

BALLISTEK NO: 224CJ

NAI NO: RMB 12444/4.600

DATA SOURCE: SSK Industries

HISTORICAL DATA: By J.D. Jones

NOTES:

LOADING DATA:

BULLET WT./TYPE	POWDER WT./TYPE	VELOCITY ('/SEC)	SOURCE
60/—	33.0/4064	2842	SSK

CASE PREPARATION: SHELLHOLDER (RCBS): 11

MAKE FROM: .225 Winchester. Fireform factory case with a 10% reduced load.

PHYSICAL DATA (INCHES):

CASE TYPE: **Rimmed Bottleneck**

CASE LENGTH: **A = 1.923**

HEAD DIAMETER: **B = .418**

RIM DIAMETER: **D = .465**

NECK DIAMETER: **F = .254**

NECK LENGTH: **H = .214**

SHOULDER LENGTH: **K = .089**

BODY ANGLE (DEG'S/SIDE): **.182**

CASE CAPACITY **CC'S = 2.63**

LOADED LENGTH: **2.520**

BELT DIAMETER: **C = N/A**

RIM THICKNESS: **E = .048**

SHOULDER DIAMETER: **G = .409**

LENGTH TO SHOULDER: **J = 1.620**

SHOULDER ANGLE (DEG'S/SIDE): **41.09**

PRIMER: **L/R**

CASE CAPACITY (GR'S WATER): **40.86**

DIMENSIONAL DRAWING:

-NOT ACTUAL SIZE-
-DO NOT SCALE-

CARTRIDGE: .228 Ackley Magnum

OTHER NAMES:

DIA: .228

BALLISTEK NO: 228C

NAI NO: RXB 22423/4.872

DATA SOURCE: Ackley Vol.1 Pg.291

HISTORICAL DATA: By P.O. Ackley in 1938

NOTES: Overbore for .228" dia.

LOADING DATA:

BULLET WT./TYPE	POWDER WT./TYPE	VELOCITY ('/SEC)	SOURCE
70/Spire	44.0/4350	3451	Ackley

CASE PREPARATION: SHELLHOLDER (RCBS): 3

MAKE FROM: .25-06 Rem. Anneal case neck. Use of a form die is suggested but, it is possible to reduce the neck dia. in a .243 Win. F/L die (do not size fully!) and then F/L size in the Ackley die. Trim case and chamfer.

PHYSICAL DATA (INCHES):

CASE TYPE: **Rimless Bottleneck**

CASE LENGTH: **A = 2.29**

HEAD DIAMETER: **B = .470**

RIM DIAMETER: **D = .473**

NECK DIAMETER: **F = .260**

NECK LENGTH: **H = .320**

SHOULDER LENGTH: **K = .180**

BODY ANGLE (DEG'S/SIDE): **.558**

CASE CAPACITY **CC'S = 3.73**

LOADED LENGTH: **2.91**

BELT DIAMETER: **C = N/A**

RIM THICKNESS: **E = .052**

SHOULDER DIAMETER: **G = .439**

LENGTH TO SHOULDER: **J = 1.79**

SHOULDER ANGLE (DEG'S/SIDE): **26.43**

PRIMER: **L/R**

CASE CAPACITY (GR'S WATER): **57.71**

DIMENSIONAL DRAWING:

-NOT ACTUAL SIZE-
-DO NOT SCALE-

CARTRIDGE: .228 Belted Express (Ackley)

OTHER NAMES:

DIA: .228

BALLISTEK NO: 228R

NAI NO: BEN 22423/4.977

DATA SOURCE: Ackley Vol.1 Pg.289

HISTORICAL DATA: Ackley design.

NOTES:

LOADING DATA:

BULLET WT./TYPE	POWDER WT./TYPE	VELOCITY ('/SEC)	SOURCE
70/Spire	44.0/4350	3415	Ackley

CASE PREPARATION: SHELLHOLDER (RCBS): 3

MAKE FROM: .25-06 Rem. Either swage the base to .452" dia. and silver solder a belt in place or obtain a set of belt swagging die. Complete the belt. Cut case to 2.3" and anneal neck. Chamfer. Form in F/L die with expander removed. Trim. F/L size.

PHYSICAL DATA (INCHES):

CASE TYPE: **Belted Bottleneck**
CASE LENGTH: **A = 2.25**
HEAD DIAMETER: **B = .452**
RIM DIAMETER: **D = .473**
NECK DIAMETER: **F = .262**
NECK LENGTH: **H = .328**
SHOULDER LENGTH: **K = .162**
BODY ANGLE (DEG'S/SIDE): **.385**
CASE CAPACITY **CC'S = 3.40**

LOADED LENGTH: **2.83**
BELT DIAMETER: **C = .470**
RIM THICKNESS: **E = .05**
SHOULDER DIAMETER: **G = .431**
LENGTH TO SHOULDER: **J = 1.76**
SHOULDER ANGLE (DEG'S/SIDE): **27.54**
PRIMER: **L/R**
CASE CAPACITY (GR'S WATER): **52.49**

DIMENSIONAL DRAWING:

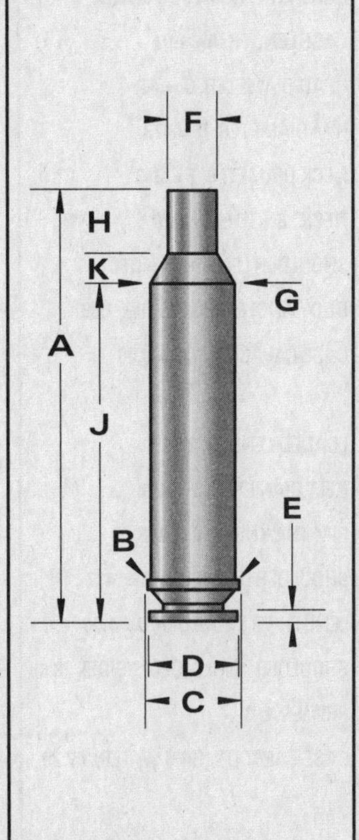

-NOT ACTUAL SIZE-
-DO NOT SCALE-

CARTRIDGE: .228 Hawk

OTHER NAMES:
.228/6mm Hawk

DIA: .228

BALLISTEK NO: 228D

NAI NO: RXB 22423/4.781

DATA SOURCE: Handloader #27 Pg.26

HISTORICAL DATA: By W. Schwartz in 1970

NOTES:

LOADING DATA:

BULLET WT./TYPE	POWDER WT./TYPE	VELOCITY ('/SEC)	SOURCE
84/Spire	42.0/4831	3090	Waters

CASE PREPARATION: SHELLHOLDER (RCBS): 3

MAKE FROM: 6mm. (.244) Remington. Anneal case neck (may not always be required) and F/L size the 6mm case in the Hawk die. Use a light load to fireform, in the chamber.

PHYSICAL DATA (INCHES):

CASE TYPE: **Rimless Bottleneck**
CASE LENGTH: **A = 2.233**
HEAD DIAMETER: **B = .467**
RIM DIAMETER: **D = .472**
NECK DIAMETER: **F = .261**
NECK LENGTH: **H = .342**
SHOULDER LENGTH: **K = .221**
BODY ANGLE (DEG'S/SIDE): **.643**
CASE CAPACITY **CC'S = 3.46**

LOADED LENGTH: **2.98**
BELT DIAMETER: **C = N/A**
RIM THICKNESS: **E = .048**
SHOULDER DIAMETER: **G = .434**
LENGTH TO SHOULDER: **J = 1.67**
SHOULDER ANGLE (DEG'S/SIDE): **21.37**
PRIMER: **L/R**
CASE CAPACITY (GR'S WATER): **53.49**

DIMENSIONAL DRAWING:

-NOT ACTUAL SIZE-
-DO NOT SCALE-

CARTRIDGE: .228 Krag (Ackley)

OTHER NAMES:
.228 Ackley Krag

DIA: .228

BALLISTEK NO: 228G

NAI NO: RMB
32433/4.683

DATA SOURCE: Ackley Vol.1 Pg.291

HISTORICAL DATA:

NOTES:

LOADING DATA:

BULLET WT./TYPE	POWDER WT./TYPE	VELOCITY ('/SEC)
70/—	38.7/4350	—

CASE PREPARATION: **SHELLHOLDER (RCBS): 7**

MAKE FROM: .30-40 Krag. Forming dies required. Form in form dies #1 and #2. Trim to length. Chamfer and F/L size.

PHYSICAL DATA (INCHES):

CASE TYPE: **Rimmed Bottleneck**

CASE LENGTH: **A = 2.145**

HEAD DIAMETER: **B = .458**

RIM DIAMETER: **D = .545**

NECK DIAMETER: **F = .258**

NECK LENGTH: **H = .278**

SHOULDER LENGTH: **K = .167**

BODY ANGLE (DEG'S/SIDE): **1.20**

CASE CAPACITY **CC'S = 3.10**

LOADED LENGTH: **2.58**

BELT DIAMETER: **C = N/A**

RIM THICKNESS: **E = .063**

SHOULDER DIAMETER: **G = .395**

LENGTH TO SHOULDER: **J = 1.70**

SHOULDER ANGLE (DEG'S/SIDE): **22.30**

PRIMER: **L/R**

CASE CAPACITY (GR'S WATER): **47.90**

DIMENSIONAL DRAWING:

F
H
K
G
A
J
E
B
D

-NOT ACTUAL SIZE-
-DO NOT SCALE-

CARTRIDGE: .228/22-250 R&M

OTHER NAMES:

DIA: .228

BALLISTEK NO: 228U

NAI NO: RXB
32443/4.089

DATA SOURCE: Ackley Vol.2 Pg.141

HISTORICAL DATA: By R&M Chronograph Service.

NOTES:

LOADING DATA:

BULLET WT./TYPE	POWDER WT./TYPE	VELOCITY ('/SEC)	SOURCE
70/Spire	38.7/H414	—	JJD

CASE PREPARATION: **SHELLHOLDER (RCBS): 3**

MAKE FROM: .22-250 Rem. Taper expand to take .228" dia. bullets. Use light load to fireform in R&M chamber. Trim, chamfer and F/L size.

PHYSICAL DATA (INCHES):

CASE TYPE: **Rimless Bottleneck**

CASE LENGTH: **A = 1.91**

HEAD DIAMETER: **B = .467**

RIM DIAMETER: **D = .473**

NECK DIAMETER: **F = .253**

NECK LENGTH: **H = .220**

SHOULDER LENGTH: **K = .150**

BODY ANGLE (DEG'S/SIDE): **1.06**

CASE CAPACITY **CC'S = 3.06**

LOADED LENGTH: **2.65**

BELT DIAMETER: **C = N/A**

RIM THICKNESS: **E = .049**

SHOULDER DIAMETER: **G = .417**

LENGTH TO SHOULDER: **J = 1.54**

SHOULDER ANGLE (DEG'S/SIDE): **28.66**

PRIMER: **L/R**

CASE CAPACITY (GR'S WATER): **47.29**

DIMENSIONAL DRAWING:

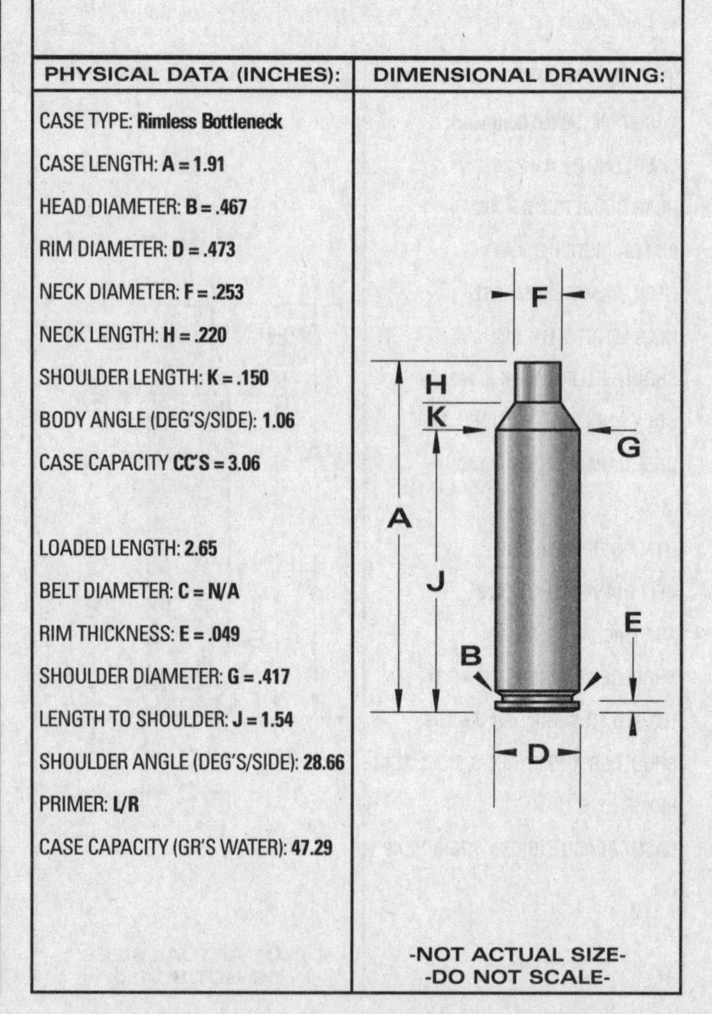

-NOT ACTUAL SIZE-
-DO NOT SCALE-

CARTRIDGE: .228/225 Winchester

OTHER NAMES:

DIA: .228

BALLISTEK NO: 228H

NAI NO: RMB 22433/4.562

DATA SOURCE: NAI/Ballistek

HISTORICAL DATA:

NOTES:

LOADING DATA:

BULLET WT./TYPE	POWDER WT./TYPE	VELOCITY ('/SEC)	SOURCE
70/Spire	24.0/IMR4198	2900	JJD

CASE PREPARATION: **SHELLHOLDER (RCBS):** 11

MAKE FROM: .225 Win. Run case neck over .224 to .230" dia tapered expander. Trim to length. Chamfer. F/L size.

PHYSICAL DATA (INCHES):

CASE TYPE: **Rimmed Bottleneck**

CASE LENGTH: **A = 1.93**

HEAD DIAMETER: **B = .423**

RIM DIAMETER: **D = .473**

NECK DIAMETER: **F = .264**

NECK LENGTH: **H = .243**

SHOULDER LENGTH: **K = .157**

BODY ANGLE (DEG'S/SIDE): **.366**

CASE CAPACITY **CC'S = 2.49**

LOADED LENGTH: **2.46**

BELT DIAMETER: **C = N/A**

RIM THICKNESS: **E = .049**

SHOULDER DIAMETER: **G = .406**

LENGTH TO SHOULDER: **J = 1.53**

SHOULDER ANGLE (DEG'S/SIDE): **24.33**

PRIMER: **L/R**

CASE CAPACITY (GR'S WATER): **38.51**

DIMENSIONAL DRAWING:

F

H

K

G

A

J

E

B

D

-NOT ACTUAL SIZE-
-DO NOT SCALE-

CARTRIDGE: .228/257 Roberts

OTHER NAMES:

DIA: .228

BALLISTEK NO: 228J

NAI NO: RXB 23422/4.734

DATA SOURCE: NAI/Ballistek

HISTORICAL DATA:

NOTES:

LOADING DATA:

BULLET WT./TYPE	POWDER WT./TYPE	VELOCITY ('/SEC)	SOURCE
70/Spire	41.0/IMR3031	3240	JJD

CASE PREPARATION: **SHELLHOLDER (RCBS):** 11

MAKE FROM: .257 Roberts. Reduce neck dia. in .243 Win. F/L die. Anneal. Run into .228/257 F/L die with expander removed. Trim to length and chamfer. F/L size.

PHYSICAL DATA (INCHES):

CASE TYPE: **Rimless Bottleneck**

CASE LENGTH: **A = 2.23**

HEAD DIAMETER: **B = .471**

RIM DIAMETER: **D = .473**

NECK DIAMETER: **F = .252**

NECK LENGTH: **H = .320**

SHOULDER LENGTH: **K = .385**

BODY ANGLE (DEG'S/SIDE): **.886**

CASE CAPACITY **CC'S = 3.55**

LOADED LENGTH: **2.60**

BELT DIAMETER: **C = N/A**

RIM THICKNESS: **E = .049**

SHOULDER DIAMETER: **G = .430**

LENGTH TO SHOULDER: **J = 1.525**

SHOULDER ANGLE (DEG'S/SIDE): **13.01**

PRIMER: **L/R**

CASE CAPACITY (GR'S WATER): **54.87**

DIMENSIONAL DRAWING:

-NOT ACTUAL SIZE-
-DO NOT SCALE-

CARTRIDGE: .228/270 R&M

OTHER NAMES:

DIA: .228

BALLISTEK NO: 228T

NAI NO: RXB 22432/5.425

DATA SOURCE: Ackley Vol.2 Pg.141

HISTORICAL DATA:

NOTES:

LOADING DATA:

BULLET WT./TYPE	POWDER WT./TYPE	VELOCITY ('/SEC)	SOURCE
70/Spire	64.0/MRP	3515	JJD

CASE PREPARATION: SHELLHOLDER (RCBS): 3

MAKE FROM: .25-06 Rem. Anneal case neck. Run case neck into .243 Win. die to reduce neck dia. (not full length!). F/L size in R&M die. Square case mouth and chamfer. Fireform.

PHYSICAL DATA (INCHES):

CASE TYPE: **Rimless Bottleneck**

CASE LENGTH: **A = 2.55**

HEAD DIAMETER: **B = .470**

RIM DIAMETER: **D = .473**

NECK DIAMETER: **F = .259**

NECK LENGTH: **H = .264**

SHOULDER LENGTH: **K = .296**

BODY ANGLE (DEG'S/SIDE): **.704**

CASE CAPACITY **CC'S = .4.17**

LOADED LENGTH: **3.32**

BELT DIAMETER: **C = N/A**

RIM THICKNESS: **E = .052**

SHOULDER DIAMETER: **G = .426**

LENGTH TO SHOULDER: **J = 1.99**

SHOULDER ANGLE (DEG'S/SIDE): **15.75**

PRIMER: **L/R**

CASE CAPACITY (GR'S WATER): **64.44**

DIMENSIONAL DRAWING:

-NOT ACTUAL SIZE-
-DO NOT SCALE-

CARTRIDGE: .228/284 R&M

OTHER NAMES:

DIA: .228

BALLISTEK NO: 228V

NAI NO: RBB 22444/4.380

DATA SOURCE: Ackley Vol.2 Pg.141

HISTORICAL DATA: By R&M Chronograph Service

NOTES:

LOADING DATA:

BULLET WT./TYPE	POWDER WT./TYPE	VELOCITY ('/SEC)	SOURCE
90/Spire	50.0/IMR4831	3340	JJD

CASE PREPARATION: SHELLHOLDER (RCBS): 3

MAKE FROM: .284 Winchester. Neck size in .270 Win die (neck only!). Further reduce neck in any short .25 cal. sizer and in any short .243 cal. sizer. Anneal as needed. Trim to length, chamfer and F/L size in R&M die. Fireform.

PHYSICAL DATA (INCHES):

CASE TYPE: **Rebated Bottleneck**

CASE LENGTH: **A = 2.19**

HEAD DIAMETER: **B = .500**

RIM DIAMETER: **D = .473**

NECK DIAMETER: **F = .251**

NECK LENGTH: **H = .225**

SHOULDER LENGTH: **K = .145**

BODY ANGLE (DEG'S/SIDE): **.548**

CASE CAPACITY **CC'S = 4.47**

LOADED LENGTH: **A/R**

BELT DIAMETER: **C = N/A**

RIM THICKNESS: **E = .054**

SHOULDER DIAMETER: **G = .469**

LENGTH TO SHOULDER: **J = 1.82**

SHOULDER ANGLE (DEG'S/SIDE): **36.93**

PRIMER: **L/R**

CASE CAPACITY (GR'S WATER): **69.07**

DIMENSIONAL DRAWING:

-NOT ACTUAL SIZE-
-DO NOT SCALE-

CARTRIDGE: .228/300 Holland & Holland Magnum

OTHER NAMES:	
	DIA: .228
	BALLISTEK NO: 228L
	NAI NO: BEN 22422/5.555

DATA SOURCE: NAI/Ballistek

HISTORICAL DATA:

NOTES:

LOADING DATA:

BULLET WT./TYPE	POWDER WT./TYPE	VELOCITY ('/SEC)	SOURCE
100/RN	45.9/W785	—	JJD

CASE PREPARATION: SHELLHOLDER (RCBS): 4

MAKE FROM: .300 H&H Mag. Form dies are required. Anneal case and form in form dies. I.D. neck ream. Trim to length. F/L size & chamfer. Fireform in the chamber.

PHYSICAL DATA (INCHES):

CASE TYPE: **Belted Bottleneck**
CASE LENGTH **A = 2.85**
HEAD DIAMETER **B = .513**
RIM DIAMETER **D = .532**
NECK DIAMETER **F = .258**
NECK LENGTH **H = .30**
SHOULDER LENGTH **K = .440**
BODY ANGLE (DEG'S/SIDE): **.85**
CASE CAPACITY **CC'S = 5.46**

LOADED LENGTH: **A/R**
BELT DIAMETER **C = .532**
RIM THICKNESS **E = .05**
SHOULDER DIAMETER **G = .456**
LENGTH TO SHOULDER **J = 2.11**
SHOULDER ANGLE (DEG'S/SIDE): **12.68**
PRIMER: **L/R Mag**
CASE CAPACITY (GR'S WATER): **84.31**

DIMENSIONAL DRAWING:

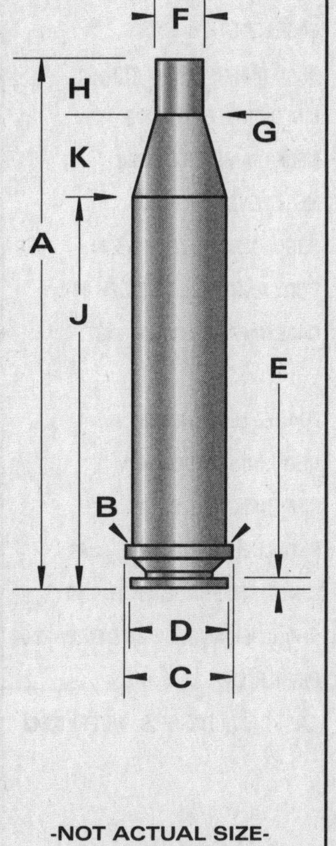

-NOT ACTUAL SIZE-
-DO NOT SCALE-

CARTRIDGE: .228/6.5 Mannlicher Schoenauer

OTHER NAMES:	
	DIA: .228
	BALLISTEK NO: 228I
	NAI NO: RXB 22424/4.698

DATA SOURCE: NAI/Ballistek

HISTORICAL DATA:

NOTES:

LOADING DATA:

BULLET WT./TYPE	POWDER WT./TYPE	VELOCITY ('/SEC)	SOURCE
70/Spire	28.3/IMR4064	—	JJD

CASE PREPARATION: SHELLHOLDER (RCBS): 9

MAKE FROM: 6,5x54 M/S. Anneal case neck and F/L size. Sometimes a form die is required and other times it is not. Trim case to length, chamfer and fireform.

PHYSICAL DATA (INCHES):

CASE TYPE: **Rimless Bottleneck**
CASE LENGTH **A = 2.10**
HEAD DIAMETER **B = .447**
RIM DIAMETER **D = .453**
NECK DIAMETER **F = .253**
NECK LENGTH **H = .305**
SHOULDER LENGTH **K = .145**
BODY ANGLE (DEG'S/SIDE): **.375**
CASE CAPACITY **CC'S = 3.29**

LOADED LENGTH: **A/R**
BELT DIAMETER **C = N/A**
RIM THICKNESS **E = .056**
SHOULDER DIAMETER **G = .428**
LENGTH TO SHOULDER **J = 1.65**
SHOULDER ANGLE (DEG'S/SIDE): **31.11**
PRIMER: **L/R**
CASE CAPACITY (GR'S WATER): **50.74**

DIMENSIONAL DRAWING:

F
H
K
G
A
J
E
B
D

-NOT ACTUAL SIZE-
-DO NOT SCALE-

CARTRIDGE: .230 Ackley

OTHER NAMES:

DIA: .230

BALLISTEK NO: 230A

NAI NO: RXB 12423/4.840

DATA SOURCE: Ackley Vol.1 Pg.293

HISTORICAL DATA: Ackley design

NOTES:

LOADING DATA:

BULLET WT./TYPE	POWDER WT./TYPE	VELOCITY ('/SEC)	SOURCE
75/Spire	45.0/4350	3355	Barnes

CASE PREPARATION: SHELLHOLDER (RCBS): 3

MAKE FROM: .25-06 Rem. Form die required. Anneal. Form in form die and run into trim die. Trim & chamfer. F/L size.

PHYSICAL DATA (INCHES):

CASE TYPE: **Rimless Bottleneck**

CASE LENGTH **A = 2.275**

HEAD DIAMETER **B = .470**

RIM DIAMETER **D = .473**

NECK DIAMETER **F = .268**

NECK LENGTH **H = .300**

SHOULDER LENGTH **K = .167**

BODY ANGLE (DEG'S/SIDE): **.267**

CASE CAPACITY **CC'S = 3.74**

LOADED LENGTH: **A/R**

BELT DIAMETER **C = N/A**

RIM THICKNESS **E = .052**

SHOULDER DIAMETER **G = .455**

LENGTH TO SHOULDER **J = 1.808**

SHOULDER ANGLE (DEG'S/SIDE): **29.24**

PRIMER: **L/R**

CASE CAPACITY (GR'S WATER): **57.78**

DIMENSIONAL DRAWING:

F
H
K
G
A
J
E
B
D

-NOT ACTUAL SIZE-
-DO NOT SCALE-

CARTRIDGE: .230 LLF

OTHER NAMES:

DIA: .230

BALLISTEK NO: 230B

NAI NO: RXB 22433/4.425

DATA SOURCE: Ackley Vol.1 Pg.294

HISTORICAL DATA: By the LLF Gun Shop

NOTES: .230" is the correct diameter.

LOADING DATA:

BULLET WT./TYPE	POWDER WT./TYPE	VELOCITY ('/SEC)	SOURCE
70/Spire	48.0/4350	3733	Ackley

CASE PREPARATION: SHELLHOLDER (RCBS): 3

MAKE FROM: .243 Winchester. Size the .243 case in the LLF die to reduce neck dia. Trim to length & chamfer.

PHYSICAL DATA (INCHES):

CASE TYPE: **Rimless Bottleneck**

CASE LENGTH **A = 2.08**

HEAD DIAMETER **B = .470**

RIM DIAMETER **D = .473**

NECK DIAMETER **F = .260**

NECK LENGTH **H = .250**

SHOULDER LENGTH **K = .230**

BODY ANGLE (DEG'S/SIDE): **.49**

CASE CAPACITY **CC'S = 3.46**

LOADED LENGTH: **2.65**

BELT DIAMETER **C = N/A**

RIM THICKNESS **E = .05**

SHOULDER DIAMETER **G = .446**

LENGTH TO SHOULDER **J = 1.60**

SHOULDER ANGLE (DEG'S/SIDE): **22.02**

PRIMER: **L/R**

CASE CAPACITY (GR'S WATER): **53.43**

DIMENSIONAL DRAWING:

F
H
K
G
A
J
E
B
D

-NOT ACTUAL SIZE-
-DO NOT SCALE-

CARTRIDGE: .240 Belted Rimless Nitro Express

OTHER NAMES:
.240 Apex

DIA: .245

BALLISTEK NO: 245B

NAI NO: BEN 22422/5.558

DATA SOURCE: COTW 4th Pg.224

HISTORICAL DATA: By Holland & Holland about 1923

NOTES:

LOADING DATA:

BULLET WT./TYPE	POWDER WT./TYPE	VELOCITY ('/SEC)	SOURCE
100/Spire	39.0/4350	2890	Barnes

CASE PREPARATION: SHELLHOLDER (RCBS): 3

MAKE FROM: .25-06 Rem. Swage base to .448" dia. and silver solder a magnum-type belt in place OR use a set of belt-forming swage dies. Anneal the case neck and form in the .240 die, with expander removed. Trim to length & chamfer. F/L size.

PHYSICAL DATA (INCHES):

CASE TYPE: **Belted Bottleneck**
CASE LENGTH **A = 2.49**
HEAD DIAMETER **B = .451**
RIM DIAMETER **D = .468**
NECK DIAMETER **F = .461**
NECK LENGTH **H = .360**
SHOULDER LENGTH **K = .207**
BODY ANGLE (DEG'S/SIDE): **.748**
CASE CAPACITY **CC'S = 3.68**

LOADED LENGTH: **3.21**
BELT DIAMETER **C = .475**
RIM THICKNESS **E = .05**
SHOULDER DIAMETER **G = .403**
LENGTH TO SHOULDER **J = 1.923**
SHOULDER ANGLE (DEG'S/SIDE): **17.30**
PRIMER: **L/R**
CASE CAPACITY (GR'S WATER): **56.92**

DIMENSIONAL DRAWING:

F
H
K
G
A
J
E
B
D
C

-NOT ACTUAL SIZE-
-DO NOT SCALE-

CARTRIDGE: .240 Cobra

OTHER NAMES:

DIA: .243

BALLISTEK NO: 243L

NAI NO: RMB 12424/4.972

DATA SOURCE: NAI/Ballistek

HISTORICAL DATA: By Custom Gunsmith Service about 1948.

NOTES:

LOADING DATA:

BULLET WT./TYPE	POWDER WT./TYPE	VELOCITY ('/SEC)	SOURCE
85/Spire	42.0/4895	3690	CGS

CASE PREPARATION: SHELLHOLDER (RCBS): 11

MAKE FROM: .220 Swift. Taper expand the Swift case to about .250" I.D. F/L size in Cobra die set and fireform (with light loads) in the chamber.

PHYSICAL DATA (INCHES):

CASE TYPE: **Rimmed Bottleneck**
CASE LENGTH **A = 2.198**
HEAD DIAMETER **B = .442**
RIM DIAMETER **D = .469**
NECK DIAMETER **F = .278**
NECK LENGTH **H = .335**
SHOULDER LENGTH **K = .123**
BODY ANGLE (DEG'S/SIDE): **.130**
CASE CAPACITY **CC'S = 3.29**

LOADED LENGTH: **A/R**
BELT DIAMETER **C = N/A**
RIM THICKNESS **E = .05**
SHOULDER DIAMETER **G = .435**
LENGTH TO SHOULDER **J = 1.74**
SHOULDER ANGLE (DEG'S/SIDE): **32.54**
PRIMER: **L/R**
CASE CAPACITY (GR'S WATER): **50.76**

DIMENSIONAL DRAWING:

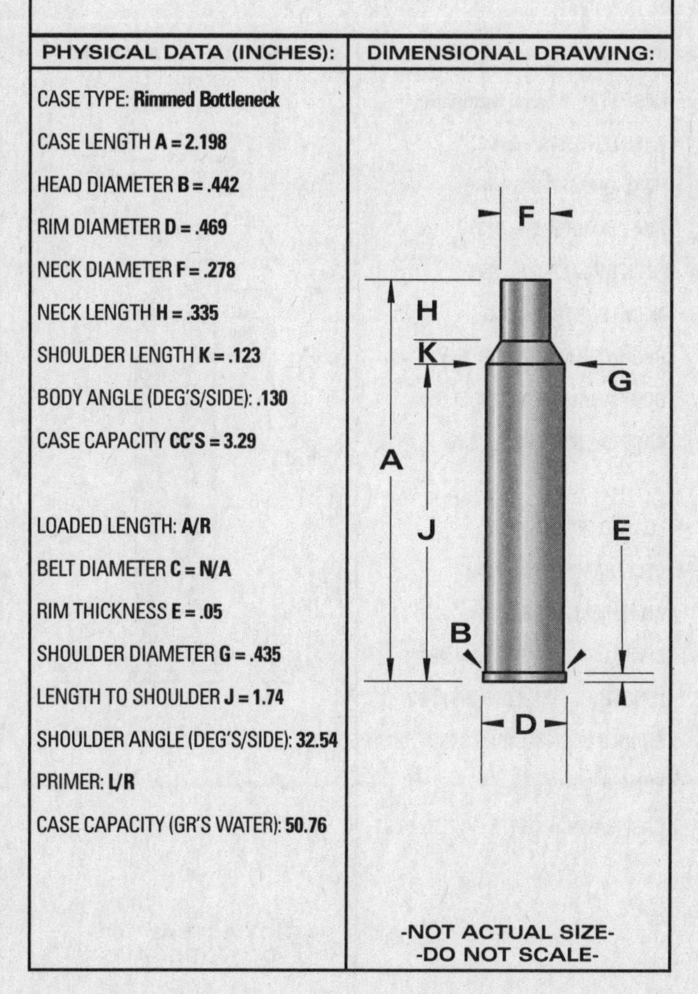

CARTRIDGE: .240 Flanged Nitro Express

OTHER NAMES:

DIA: .245

BALLISTEK NO: 245A

NAI NO: RMB 22423/5.571

DATA SOURCE: COTW 4th Pg.223

HISTORICAL DATA: By H&H about 1920.

NOTES:

LOADING DATA:

BULLET WT./TYPE	POWDER WT./TYPE	VELOCITY ('/SEC)	SOURCE
100/Spire	42.0/4350	2900	Barnes

CASE PREPARATION: SHELLHOLDER (RCBS): 4

MAKE FROM: 9,3 x 74R. Turn rim to .513" dia. F/L size case with expander removed. This will cause a bulge just forward of the rim. Turn head dia. (the bulge) to .448" diameter. Form in form die (anneal if necessary). Trim to length and F/L size.

PHYSICAL DATA (INCHES):

CASE TYPE: **Rimmed Bottleneck**

CASE LENGTH **A = 2.496**

HEAD DIAMETER **B = .448**

RIM DIAMETER **D = .513**

NECK DIAMETER **F = .274**

NECK LENGTH **H = .345**

SHOULDER LENGTH **K = .163**

BODY ANGLE (DEG'S/SIDE): **.736**

CASE CAPACITY **CC'S = 3.73**

LOADED LENGTH: **3.25**

BELT DIAMETER **C = N/A**

RIM THICKNESS **E = .06**

SHOULDER DIAMETER **G = .402**

LENGTH TO SHOULDER **J = 1.988**

SHOULDER ANGLE (DEG'S/SIDE): **21.44**

PRIMER: **L/R**

CASE CAPACITY (GR'S WATER): **57.63**

DIMENSIONAL DRAWING:

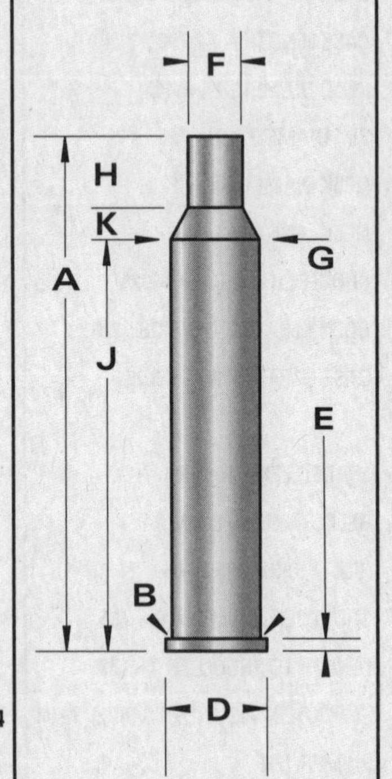

-NOT ACTUAL SIZE-
-DO NOT SCALE-

CARTRIDGE: .240 Madame

OTHER NAMES:

DIA: .243

BALLISTEK NO: 243M

NAI NO: RXB 22344/4.083

DATA SOURCE: Ackley Vol.1 Pg.295

HISTORICAL DATA: By Mr. Foerster.

NOTES:

LOADING DATA:

BULLET WT./TYPE	POWDER WT./TYPE	VELOCITY ('/SEC)	SOURCE
70/—	28.0/3031	3110	Ackley

CASE PREPARATION: SHELLHOLDER (RCBS): 19

MAKE FROM: .25 Remington. Anneal case neck and size in Madame sizer with expander removed. Trim to length & chamfer. F/L size and fireform in chamber.

PHYSICAL DATA (INCHES):

CASE TYPE: **Rimless Bottleneck**

CASE LENGTH **A = 1.715**

HEAD DIAMETER **B = .420**

RIM DIAMETER **D = .421**

NECK DIAMETER **F = .263**

NECK LENGTH **H = .205**

SHOULDER LENGTH **K = .099**

BODY ANGLE (DEG'S/SIDE): **.473**

CASE CAPACITY **CC'S = 2.44**

LOADED LENGTH: **2.325**

BELT DIAMETER **C = N/A**

RIM THICKNESS **E = .052**

SHOULDER DIAMETER **G = .400**

LENGTH TO SHOULDER **J = 1.41**

SHOULDER ANGLE (DEG'S/SIDE): **34.41**

PRIMER: **L/R**

CASE CAPACITY (GR'S WATER): **37.76**

DIMENSIONAL DRAWING:

F H K G A J B E D

-NOT ACTUAL SIZE-
-DO NOT SCALE-

CARTRIDGE: .240 Mashburn Falcon

OTHER NAMES:	
	DIA: .243
	BALLISTEK NO: 243P
	NAI NO: RXB 22434/5.319

DATA SOURCE: Ackley Vol.1 Pg.321

HISTORICAL DATA: By Mashburn Arms

NOTES:

LOADING DATA:

BULLET WT./TYPE	POWDER WT./TYPE	VELOCITY ('/SEC)	SOURCE
100/Spire	46.5/4350	2850	JJD

CASE PREPARATION: SHELLHOLDER (RCBS): 3

MAKE FROM: .25-06 Rem. Anneal shoulder area. Size in .243 Win. die until bolt will close on chambered round. Trim to length. Fireform.

PHYSICAL DATA (INCHES):

CASE TYPE: **Rimless Bottleneck**

CASE LENGTH **A = 2.500**

HEAD DIAMETER **B = .470**

RIM DIAMETER **D = .473**

NECK DIAMETER **F = .264**

NECK LENGTH **H = .293**

SHOULDER LENGTH **K = .142**

BODY ANGLE (DEG'S/SIDE): **.385**

CASE CAPACITY **CC'S = 4.68**

LOADED LENGTH: **3.03**

BELT DIAMETER **C = N/A**

RIM THICKNESS **E = .05**

SHOULDER DIAMETER **G = .445**

LENGTH TO SHOULDER **J = 2.065**

SHOULDER ANGLE (DEG'S/SIDE): **32.51**

PRIMER: **L/R**

CASE CAPACITY (GR'S WATER): **72.26**

DIMENSIONAL DRAWING:

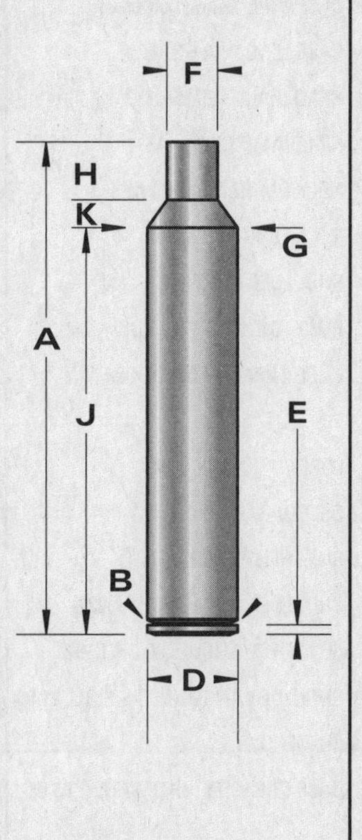

-NOT ACTUAL SIZE-
-DO NOT SCALE-

CARTRIDGE: .240 Page Souper Pooper

OTHER NAMES:	
.240 PSP	DIA: .243
	BALLISTEK NO: 243G
	NAI NO: RXB 22424/4.841

DATA SOURCE: Ackley Vol.1 Pg.314

HISTORICAL DATA: By W. Page

NOTES:

LOADING DATA:

BULLET WT./TYPE	POWDER WT./TYPE	VELOCITY ('/SEC)	SOURCE
75/Spire	51.0/4831	3410	Ackley

CASE PREPARATION: SHELLHOLDER (RCBS): 3

MAKE FROM: .25-06 Rem. Anneal case neck. Reduce neck dia. and set back shoulder with .243 Win. die until case will chamber. Fireform with light load. Form dies are helpful (and, available) in producing this case.

PHYSICAL DATA (INCHES):

CASE TYPE: **Rimless Bottleneck**

CASE LENGTH **A = 2.28**

HEAD DIAMETER **B = .471**

RIM DIAMETER **D = .472**

NECK DIAMETER **F = .270**

NECK LENGTH **H = .353**

SHOULDER LENGTH **K = .127**

BODY ANGLE (DEG'S/SIDE): **.376**

CASE CAPACITY **CC'S = 3.98**

LOADED LENGTH: **A/R**

BELT DIAMETER **C = N/A**

RIM THICKNESS **E = .048**

SHOULDER DIAMETER **G = .450**

LENGTH TO SHOULDER **J = 1.80**

SHOULDER ANGLE (DEG'S/SIDE): **35.32**

PRIMER: **L/R**

CASE CAPACITY (GR'S WATER): **61.49**

DIMENSIONAL DRAWING:

F H K G A J E B D

-NOT ACTUAL SIZE-
-DO NOT SCALE-

CARTRIDGE: .240 Super Varminter

OTHER NAMES:

DIA: .243

BALLISTEK NO: 243N

NAI NO: RXB 22432/5.362

DATA SOURCE: Ackley Vol.1 Pg.320

HISTORICAL DATA: By J. Gebby

NOTES:

LOADING DATA:

BULLET WT./TYPE	POWDER WT./TYPE	VELOCITY ('/SEC)	SOURCE
100/Spire	45.0/4350	2997	Ackley

CASE PREPARATION: SHELLHOLDER (RCBS): 3

MAKE FROM: .25-06 Rem. Anneal case and size neck only in a .243 Win. F/L die. F/L size and fireform or just fireform as long as weapon will chamber round. Use light load for forming.

PHYSICAL DATA (INCHES):

CASE TYPE: **Rimless Bottleneck**

CASE LENGTH **A = 2.52**

HEAD DIAMETER **B = .470**

RIM DIAMETER **D = .473**

NECK DIAMETER **F = .264**

NECK LENGTH **H = .297**

SHOULDER LENGTH **K = .253**

BODY ANGLE (DEG'S/SIDE): **.453**

CASE CAPACITY **CC'S = 4.59**

LOADED LENGTH: **2.27**

BELT DIAMETER **C = N/A**

RIM THICKNESS **E = .05**

SHOULDER DIAMETER **G = .442**

LENGTH TO SHOULDER **J = 1.97**

SHOULDER ANGLE (DEG'S/SIDE): **19.38**

PRIMER: **L/R**

CASE CAPACITY (GR'S WATER): **70.90**

DIMENSIONAL DRAWING:

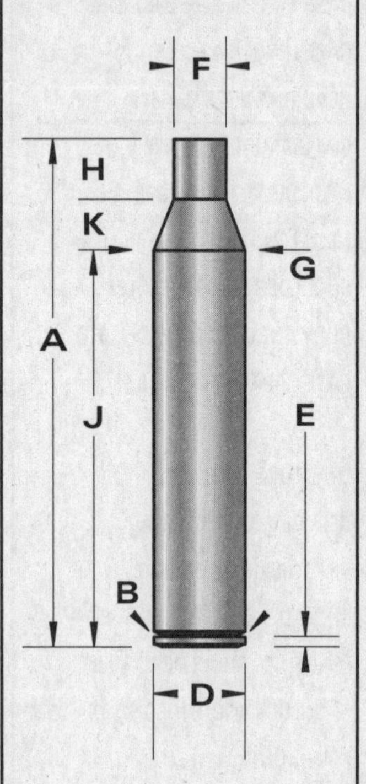

-NOT ACTUAL SIZE-
-DO NOT SCALE-

CARTRIDGE: .240 Weatherby Magnum

OTHER NAMES:

DIA: .243

BALLISTEK NO: 243F

NAI NO: BEN 12423/5.518

DATA SOURCE: Hornady 3rd Pg.119

HISTORICAL DATA: By R. Weatherby in 1968

NOTES: About max. case cap. for .243" dia. bore.

LOADING DATA:

BULLET WT./TYPE	POWDER WT./TYPE	VELOCITY ('/SEC)	SOURCE
100/Spire	50.0/IMR4831	3275	JJD

CASE PREPARATION: SHELLHOLDER (RCBS): 3

MAKE FROM: Factory. No case lends itself to forming the .240 Weatherby. I would not try tuning cases from solid; the pressure could be too much.

PHYSICAL DATA (INCHES):

CASE TYPE: **Belted Bottleneck**

CASE LENGTH **A = 2.500**

HEAD DIAMETER **B = .453**

RIM DIAMETER **D = .473**

NECK DIAMETER **F = .272**

NECK LENGTH **H = .308**

SHOULDER LENGTH **K = .200**

BODY ANGLE (DEG'S/SIDE): **.336**

CASE CAPACITY **CC'S = 4.08**

LOADED LENGTH: **3.08**

BELT DIAMETER **C = .473**

RIM THICKNESS **E = .05**

SHOULDER DIAMETER **G = .432**

LENGTH TO SHOULDER **J = 1.992**

SHOULDER ANGLE (DEG'S/SIDE): **21.80**

PRIMER: **L/R**

CASE CAPACITY (GR'S WATER): **63.08**

DIMENSIONAL DRAWING:

F
H
K
G
E
A
J
B
D
C

-NOT ACTUAL SIZE-
-DO NOT SCALE-

CARTRIDGE: .242 Belted Rimless Holland & Holland

OTHER NAMES:	
	DIA: .243
	BALLISTEK NO: 243J
	NAI NO: BEN 22424/5.302

DATA SOURCE: Ackley Vol.1 Pg.325

HISTORICAL DATA: Very specialized and very hot ctg!

NOTES: Sublimely overbore

LOADING DATA:

BULLET WT./TYPE	POWDER WT./TYPE	VELOCITY ('/SEC)	SOURCE
105/Spire	66.0/4831	3197	Ackley

CASE PREPARATION: SHELLHOLDER (RCBS): 4

MAKE FROM: .300 H&H Mag. or 7mm Rem. Mag. Size in 7mm Rem. Mag sizer (.300 H&H). Size in .264 Mag sizer. Anneal. Trim to 2.7". F/L size in .242 die. Chamfer.

PHYSICAL DATA (INCHES):

CASE TYPE: **Belted Bottleneck**
CASE LENGTH **A = 2.72**
HEAD DIAMETER **B = .513**
RIM DIAMETER **D = .532**
NECK DIAMETER **F = .264**
NECK LENGTH **H = .320**
SHOULDER LENGTH **K = .050**
BODY ANGLE (DEG'S/SIDE): **.826**
CASE CAPACITY **CC'S = 5.82**

LOADED LENGTH: **A/R**
BELT DIAMETER **C = .532**
RIM THICKNESS **E = .05**
SHOULDER DIAMETER **G = .451**
LENGTH TO SHOULDER **J = 2.35**
SHOULDER ANGLE (DEG'S/SIDE): **31.86**
PRIMER: **L/R**
CASE CAPACITY (GR'S WATER): **89.83**

DIMENSIONAL DRAWING:

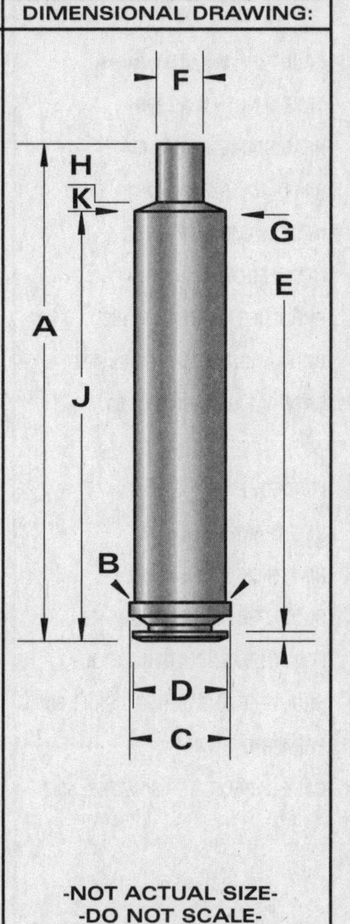

-NOT ACTUAL SIZE-
-DO NOT SCALE-

CARTRIDGE: .242 Rimless Nitro Exp.

OTHER NAMES:	
	DIA: .249
	BALLISTEK NO: 249A
	NAI NO: RXB 22442/5.118

DATA SOURCE: COTW 4th Pg.224

HISTORICAL DATA: By Vickers about 1922.

NOTES:

LOADING DATA:

BULLET WT./TYPE	POWDER WT./TYPE	VELOCITY ('/SEC)	SOURCE
100/Spire	39.0/4350	2890	Barnes

CASE PREPARATION: SHELLHOLDER (RCBS): 3

MAKE FROM: 7 x 64 Brenneke. Anneal case and form in the .242 F/L die with the expander removed. Trim & chamfer. F/L size. Fireform.

PHYSICAL DATA (INCHES):

CASE TYPE: **Rimless Bottleneck**
CASE LENGTH **A = 2.380**
HEAD DIAMETER **B = .465**
RIM DIAMETER **D = .466**
NECK DIAMETER **F = .281**
NECK LENGTH **H = .212**
SHOULDER LENGTH **K = .146**
BODY ANGLE (DEG'S/SIDE): **.971**
CASE CAPACITY **CC'S = 3.75**

LOADED LENGTH: **3.20**
BELT DIAMETER **C = N/A**
RIM THICKNESS **E = .05**
SHOULDER DIAMETER **G = .405**
LENGTH TO SHOULDER **J = 2.022**
SHOULDER ANGLE (DEG'S/SIDE): **23**
PRIMER: **L/R**
CASE CAPACITY (GR'S WATER): **57.94**

DIMENSIONAL DRAWING:

-NOT ACTUAL SIZE-
-DO NOT SCALE-

CARTRIDGE: .243 Epps Improved

OTHER NAMES:

DIA: .243

BALLISTEK NO: 243AY

NAI NO: RXB 22444/4.336

DATA SOURCE: Ackley Vol.1 Pg.306

HISTORICAL DATA: By E. Epps.

NOTES:

LOADING DATA:

BULLET WT./TYPE	POWDER WT./TYPE	VELOCITY ('/SEC)	SOURCE
75/Spire	42.0/IMR4320	3170	JJD

CASE PREPARATION: SHELLHOLDER (RCBS): 3

MAKE FROM: .243 Winchester. Fire factory ammo in Epps chamber.

PHYSICAL DATA (INCHES):

CASE TYPE: **Rimless Bottleneck**

CASE LENGTH **A = 2.038**

HEAD DIAMETER **B = .470**

RIM DIAMETER **D = .473**

NECK DIAMETER **F = .273**

NECK LENGTH **H = .240**

SHOULDER LENGTH **K = .128**

BODY ANGLE (DEG'S/SIDE): **.389**

CASE CAPACITY **CC'S = 3.53**

LOADED LENGTH: **2.74**

BELT DIAMETER **C = N/A**

RIM THICKNESS **E = .05**

SHOULDER DIAMETER **G = .450**

LENGTH TO SHOULDER **J = 1.67**

SHOULDER ANGLE (DEG'S/SIDE): **34.66**

PRIMER: **L/R**

CASE CAPACITY (GR'S WATER): **54.43**

DIMENSIONAL DRAWING:

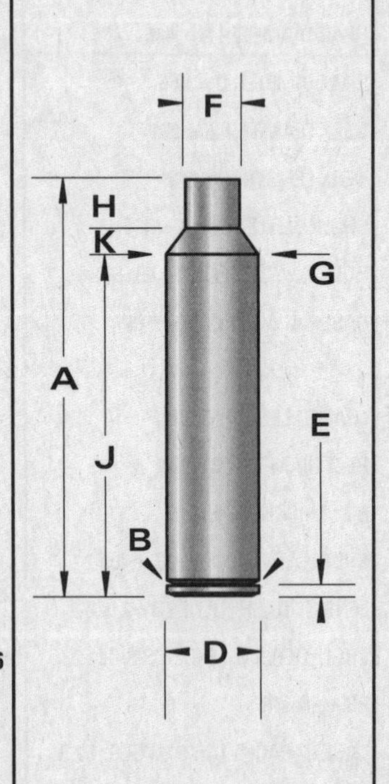

-NOT ACTUAL SIZE-
-DO NOT SCALE-

CARTRIDGE: .243 JS

OTHER NAMES:

DIA: .243

BALLISTEK NO: 243AZ

NAI NO: BEN 22423/5.393

DATA SOURCE: Ackley Vol.1 Pg.318

HISTORICAL DATA: By J. Shannon.

NOTES:

LOADING DATA:

BULLET WT./TYPE	POWDER WT./TYPE	VELOCITY ('/SEC)	SOURCE
85/Spire	50.0/4350	3600	Ackley

CASE PREPARATION: SHELLHOLDER (RCBS): 3

MAKE FROM: .25-06 Rem. A magnum-type belt must be swaged or soldered to the base of the .25-06 case. Anneal case neck, F/L size and chamfer. Other than the "scientific" value, this case is seldom worth the effort.

PHYSICAL DATA (INCHES):

CASE TYPE: **Belted Bottleneck**

CASE LENGTH **A = 2.375**

HEAD DIAMETER **B = .452**

RIM DIAMETER **D = .47**

NECK DIAMETER **F = .277**

NECK LENGTH **H = .34**

SHOULDER LENGTH **K = .182**

BODY ANGLE (DEG'S/SIDE): **.407**

CASE CAPACITY **CC'S = 4.09**

LOADED LENGTH: **3.01**

BELT DIAMETER **C = .47**

RIM THICKNESS **E = .05**

SHOULDER DIAMETER **G = .445**

LENGTH TO SHOULDER **J = 1.873**

SHOULDER ANGLE (DEG'S/SIDE): **30**

PRIMER: **L/R**

CASE CAPACITY (GR'S WATER): **63.2**

DIMENSIONAL DRAWING:

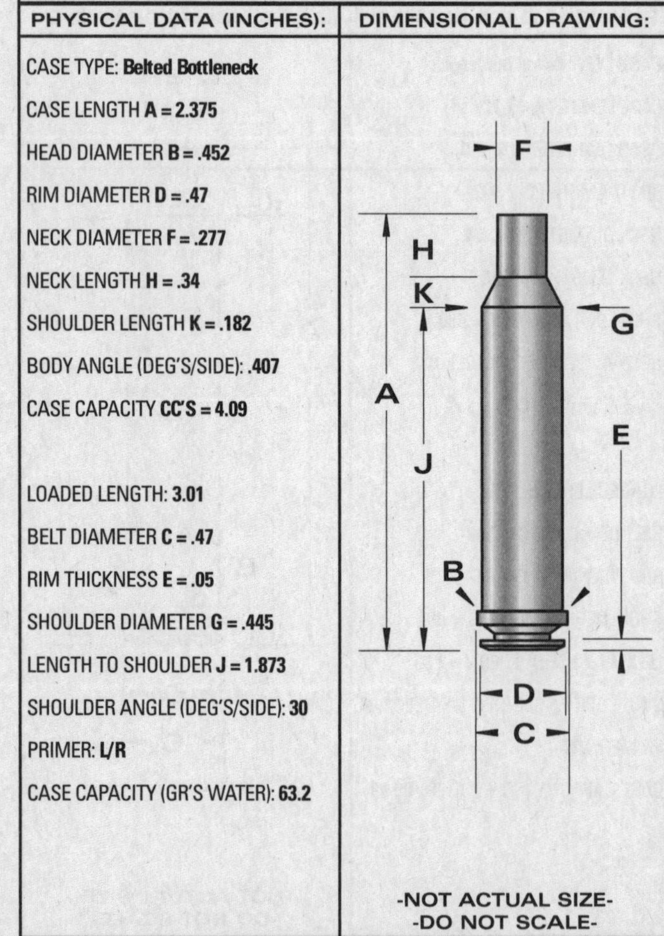

-NOT ACTUAL SIZE-
-DO NOT SCALE-

CARTRIDGE: .243 R.K.B.

OTHER NAMES:

DIA: .243

BALLISTEK NO: 243S

NAI NO: RMB 12424/4.814

DATA SOURCE: Ackley Vol.1 Pg.318

HISTORICAL DATA: By J. Shannon.

NOTES:

LOADING DATA:

BULLET WT./TYPE	POWDER WT./TYPE	VELOCITY ('/SEC)	SOURCE
85/Spire	43.0/4895	3500	Ackley

CASE PREPARATION: SHELLHOLDER (RCBS): 7

MAKE FROM: .30-40 Krag. Anneal neck. Use form die or size neck in any short 6,5mm die (e.g. 6,5x55). F/L size, trim & chamfer. Fireform.

PHYSICAL DATA (INCHES):

CASE TYPE: **Rimmed Bottleneck**

CASE LENGTH **A = 2.196**

HEAD DIAMETER **B = .454**

RIM DIAMETER **D = .528**

NECK DIAMETER **F = .277**

NECK LENGTH **H = .330**

SHOULDER LENGTH **K = .125**

BODY ANGLE (DEG'S/SIDE): **.339**

CASE CAPACITY **CC'S = 3.62**

LOADED LENGTH: **2.80**

BELT DIAMETER **C = N/A**

RIM THICKNESS **E = .058**

SHOULDER DIAMETER **G = .442**

LENGTH TO SHOULDER **J = 1.695**

SHOULDER ANGLE (DEG'S/SIDE): **34.21**

PRIMER: **L/R**

CASE CAPACITY (GR'S WATER): **55.90**

DIMENSIONAL DRAWING:

-NOT ACTUAL SIZE-
-DO NOT SCALE-

CARTRIDGE: .243 RCBS

OTHER NAMES:

DIA: .243

BALLISTEK NO: 243AT

NAI NO: RXB 22433/4.383

DATA SOURCE: Ackley Vol.1 Pg.305

HISTORICAL DATA: By F. Huntington.

NOTES:

LOADING DATA:

BULLET WT./TYPE	POWDER WT./TYPE	VELOCITY ('/SEC)	SOURCE
100/Spire	43.0/4350	3150	Hunting-ton

CASE PREPARATION: SHELLHOLDER (RCBS): 3

MAKE FROM: .243 Win. Factory ammo may be fired in the "improved" chamber.

PHYSICAL DATA (INCHES):

CASE TYPE: **Rimless Bottleneck**

CASE LENGTH **A = 2.06**

HEAD DIAMETER **B = .470**

RIM DIAMETER **D = .473**

NECK DIAMETER **F = .274**

NECK LENGTH **H = .300**

SHOULDER LENGTH **K = .172**

BODY ANGLE (DEG'S/SIDE): **.475**

CASE CAPACITY **CC'S = 3.41**

LOADED LENGTH: **2.75**

BELT DIAMETER **C = N/A**

RIM THICKNESS **E = .05**

SHOULDER DIAMETER **G = .447**

LENGTH TO SHOULDER **J = 1.588**

SHOULDER ANGLE (DEG'S/SIDE): **26.70**

PRIMER: **L/R**

CASE CAPACITY (GR'S WATER): **52.59**

DIMENSIONAL DRAWING:

-NOT ACTUAL SIZE-
-DO NOT SCALE-

CARTRIDGE: .243 Rockchucker

OTHER NAMES:

DIA: .243

BALLISTEK NO: 243I

NAI NO: RXB 22423/4.786

DATA SOURCE: Ackley Vol.1 Pg.317

HISTORICAL DATA: By F. Huntington in early 1950s.

NOTES:

LOADING DATA:

BULLET WT./TYPE	POWDER WT./TYPE	VELOCITY ('/SEC)	SOURCE
75/Spire	51.5/IMR4831	3200	JJD

CASE PREPARATION: SHELLHOLDER (RCBS): 11

MAKE FROM: .257 Roberts. Run Roberts case into .243 Rockchucker trim die and trim to length (forming is done in trim die). F/L size & chamfer.

Note: RCBS trim die length is 2.223".

PHYSICAL DATA (INCHES):

CASE TYPE: **Rimless Bottleneck**

CASE LENGTH **A = 2.24**

HEAD DIAMETER **B = .468**

RIM DIAMETER **D = .473**

NECK DIAMETER **F = .275**

NECK LENGTH **H = .350**

SHOULDER LENGTH **K = .140**

BODY ANGLE (DEG'S/SIDE): **.665**

CASE CAPACITY **CC'S = 3.58**

LOADED LENGTH: **2.90**

BELT DIAMETER **C = N/A**

RIM THICKNESS **E = .05**

SHOULDER DIAMETER **G = .432**

LENGTH TO SHOULDER **J = 1.75**

SHOULDER ANGLE (DEG'S/SIDE): **29.28**

PRIMER: **L/R**

CASE CAPACITY (GR'S WATER): **55.34**

DIMENSIONAL DRAWING:

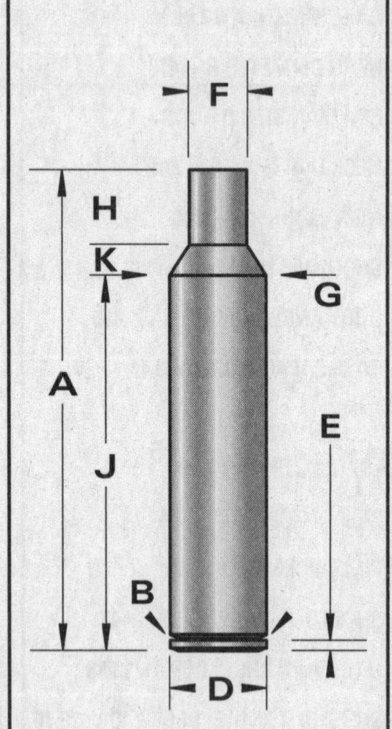

-NOT ACTUAL SIZE-
-DO NOT SCALE-

CARTRIDGE: .243 Winchester

OTHER NAMES:
6mm/308 Winchester

DIA: .243

BALLISTEK NO: 243H

NAI NO: RXB 12442/4.351

DATA SOURCE: Hornady 3rd Pg.108

HISTORICAL DATA: By Win. in 1955.

NOTES:

LOADING DATA:

BULLET WT./TYPE	POWDER WT./TYPE	VELOCITY ('/SEC)	SOURCE
80/Spire	38.0/IMR3031	3200	JJD

CASE PREPARATION: SHELLHOLDER (RCBS): 3

MAKE FROM: Factory or .270 Win. Anneal neck. F/L size with expander removed. Trim to length. Chamfer. F/L Size. Inside neck ream. Clean.

PHYSICAL DATA (INCHES):

CASE TYPE: **Rimless Bottleneck**

CASE LENGTH **A = 2.045**

HEAD DIAMETER **B = .470**

RIM DIAMETER **D = .473**

NECK DIAMETER **F = .276**

NECK LENGTH **H = .240**

SHOULDER LENGTH **K = .252**

BODY ANGLE (DEG'S/SIDE): **.317**

CASE CAPACITY **CC'S = 3.42**

LOADED LENGTH: **2.65**

BELT DIAMETER **C = N/A**

RIM THICKNESS **E = .055**

SHOULDER DIAMETER **G = .455**

LENGTH TO SHOULDER **J = 1.553**

SHOULDER ANGLE (DEG'S/SIDE): **19.55**

PRIMER: **L/R**

CASE CAPACITY (GR'S WATER): **52.81**

DIMENSIONAL DRAWING:

F
H
K
G
A
E
J
B
D

-NOT ACTUAL SIZE-
-DO NOT SCALE-

CARTRIDGE: .243 Winchester Improved Ackley

OTHER NAMES:

DIA: .243

BALLISTEK NO: 243Q

NAI NO: RXB 22434/4.308

DATA SOURCE: Ackley Vol.1 Pg.303

HISTORICAL DATA:

NOTES:

LOADING DATA:

BULLET WT./TYPE	POWDER WT./TYPE	VELOCITY ('/SEC)	SOURCE
90/Spire	39.0/4895	3225	Ackley

CASE PREPARATION: SHELLHOLDER (RCBS): 3

MAKE FROM: .243 Win. Fire factory ammo in "improved" chamber or fireform case (with corn meal) in the proper trim die. Trim to length and F/L size.

PHYSICAL DATA (INCHES):

CASE TYPE: **Rimless Bottleneck**

CASE LENGTH **A = 2.025**

HEAD DIAMETER **B = .470**

RIM DIAMETER **D = .473**

NECK DIAMETER **F = .260**

NECK LENGTH **H = .266**

SHOULDER LENGTH **K = .109**

BODY ANGLE (DEG'S/SIDE): **.711**

CASE CAPACITY **CC'S = 3.64**

LOADED LENGTH: **2.66**

BELT DIAMETER **C = N/A**

RIM THICKNESS **E = .05**

SHOULDER DIAMETER **G = .434**

LENGTH TO SHOULDER **J = 1.65**

SHOULDER ANGLE (DEG'S/SIDE): **38.59**

PRIMER: **L/R**

CASE CAPACITY (GR'S WATER): **56.31**

DIMENSIONAL DRAWING:

-NOT ACTUAL SIZE-
-DO NOT SCALE-

CARTRIDGE: .243 Win. Reynolds Spec.

OTHER NAMES:

DIA: .243

BALLISTEK NO: 243BA

NAI NO: RXB 22444/4.362

DATA SOURCE: Ackley Vol.1 Pg.307

HISTORICAL DATA: By R. Reynolds.

NOTES:

LOADING DATA:

BULLET WT./TYPE	POWDER WT./TYPE	VELOCITY ('/SEC)	SOURCE
100/Spire	47.0/4350	3290	Ackley

CASE PREPARATION: SHELLHOLDER (RCBS): 3

MAKE FROM: .243 Win. Fire factory ammo in Reynolds chamber or fireform, with corn meal, in a Reynolds trim die. F/L size, trim & chamfer.

PHYSICAL DATA (INCHES):

CASE TYPE: **Rimless Bottleneck**

CASE LENGTH **A = 2.05**

HEAD DIAMETER **B = .470**

RIM DIAMETER **D = .473**

NECK DIAMETER **F = .268**

NECK LENGTH **H = .240**

SHOULDER LENGTH **K = .089**

BODY ANGLE (DEG'S/SIDE): **.490**

CASE CAPACITY **CC'S = 3.65**

LOADED LENGTH: **2.80**

BELT DIAMETER **C = N/A**

RIM THICKNESS **E = .05**

SHOULDER DIAMETER **G = .444**

LENGTH TO SHOULDER **J = 1.72**

SHOULDER ANGLE (DEG'S/SIDE): **44.35**

PRIMER: **L/R**

CASE CAPACITY (GR'S WATER): **56.38**

DIMENSIONAL DRAWING:

-NOT ACTUAL SIZE-
-DO NOT SCALE-

CARTRIDGE: .243 Winchestser Super Short Magnum

OTHER NAMES:

DIA: .243

BALLISTEK NO:

NAI NO:

DATA SOURCE: Most current loading manuals.

HISTORICAL DATA: Developed by Winchester in 2003 to fit in new Super Short rifle actions introduced by Browning and Winchester. The case is .43-inch shorter than the Winchester Short Magnum.

LOADING DATA:

BULLET WT./TYPE	POWDER WT./TYPE	VELOCITY ('/SEC)	SOURCE
55/Ballistic Silvertip	Factory Load	4,060	Win
95/Ballistic Silvertip	Factory Load	3,250	Win

CASE PREPARATION: SHELL HOLDER (RCBS): 43

MAKE FROM: Factory. Can be made from .404 Jeffery. Shorten and die form to fit chamber. Fireform body to get the proper diameter, shoulder angle and head space. Then turn the rim to correct dimension. Ream neck if necessary. Square mouth and chamfer.

PHYSICAL DATA (INCHES):

CASE TYPE: **Rebated Bottleneck**

CASE LENGTH **A = 1.670**

HEAD DIAMETER **B = .5550**

RIM DIAMETER **D = .535**

NECK DIAMETER **F = .2870**

NECK LENGTH **H = .278**

SHOULDER LENGTH **K = .2421**

BODY ANGLE (DEG'S/SIDE):

CASE CAPACITY **CC'S =**

LOADED LENGTH: **2.360**

BELT DIAMETER **C =**

RIM THICKNESS **E = .54**

SHOULDER DIAMETER **G = .5444**

LENGTH TO SHOULDER **J = 1.1499**

SHOULDER ANGLE (DEG'S/SIDE): **28**

PRIMER: **L/R**

CASE CAPACITY (GR'S WATER): **55.0**

DIMENSIONAL DRAWING:

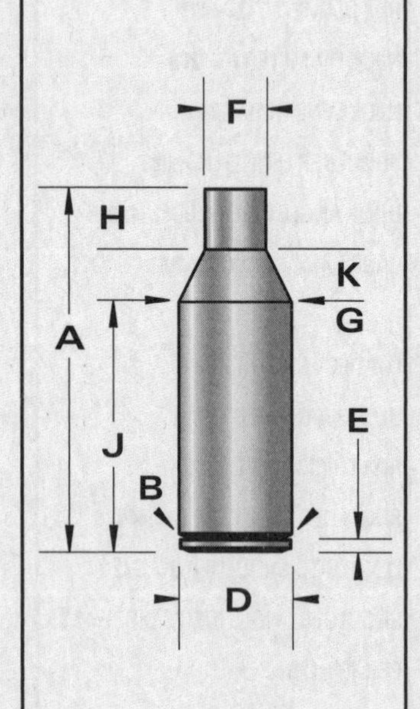

-NOT ACTUAL SIZE-
-DO NOT SCALE-

CARTRIDGE: .244 Durham Magnum

OTHER NAMES:

DIA: .243

BALLISTEK NO: 243AR

NAI NO: BEN 22424/4.386

DATA SOURCE: Ackley Vol.1 Pg.323

HISTORICAL DATA:

NOTES:

LOADING DATA:

BULLET WT./TYPE	POWDER WT./TYPE	VELOCITY ('/SEC)	SOURCE
70/Spire	57.2/4350	3850	JJD

CASE PREPARATION: SHELLHOLDER (RCBS): 4

MAKE FROM: .264 Win. Mag. Anneal case mouth & neck. Run case into F/L die with expander removed. Trim & chamfer. F/L size. If seated bullet bulges neck beyond the .275" dia, I.D. neck ream.

PHYSICAL DATA (INCHES):

CASE TYPE: **Belted Bottleneck**

CASE LENGTH **A = 2.25**

HEAD DIAMETER **B = .513**

RIM DIAMETER **D = .532**

NECK DIAMETER **F = .275**

NECK LENGTH **H = .336**

SHOULDER LENGTH **K = .078**

BODY ANGLE (DEG'S/SIDE): **.753**

CASE CAPACITY **CC'S = 4.40**

LOADED LENGTH: **3.08**

BELT DIAMETER **C = .532**

RIM THICKNESS **E = .05**

SHOULDER DIAMETER **G = .470**

LENGTH TO SHOULDER **J = 1.836**

SHOULDER ANGLE (DEG'S/SIDE): **51.34**

PRIMER: **L/R Mag**

CASE CAPACITY (GR'S WATER): **67.96**

DIMENSIONAL DRAWING:

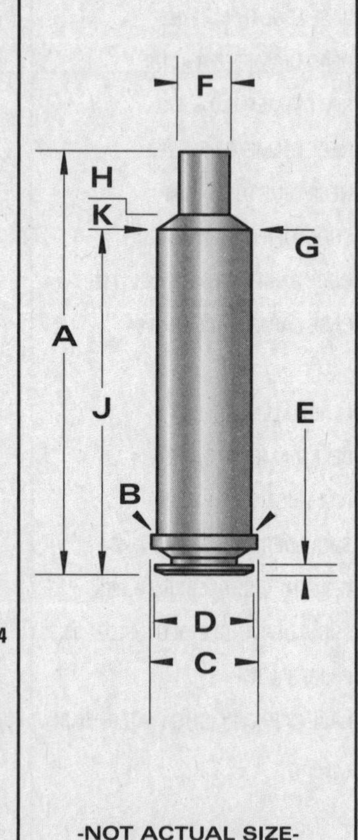

-NOT ACTUAL SIZE-
-DO NOT SCALE-

CARTRIDGE: .244 Remington Improved (Ackley)

OTHER NAMES:

DIA: .243

BALLISTEK NO: 243BB

NAI NO: RXB 12423/4.989

DATA SOURCE: Ackley Vol.1 Pg.315

HISTORICAL DATA:

NOTES:

LOADING DATA:

BULLET WT./TYPE	POWDER WT./TYPE	VELOCITY ('/SEC)	SOURCE
75/Spire	41.0/3031	3492	Ackley

CASE PREPARATION: SHELLHOLDER (RCBS): 3

MAKE FROM: 6mm Remington. Fire factory ammo in "improved" chamber. This version has a 26° shoulder and very little body taper. Another version, with a 40° shoulder, also exists and is formed in the same manner.

PHYSICAL DATA (INCHES):

CASE TYPE: **Rimless Bottleneck**

CASE LENGTH **A = 2.35**

HEAD DIAMETER **B = .471**

RIM DIAMETER **D = .472**

NECK DIAMETER **F = .273**

NECK LENGTH **H = .380**

SHOULDER LENGTH **K = .245**

BODY ANGLE (DEG'S/SIDE): **.056**

CASE CAPACITY **CC'S = 4.09**

LOADED LENGTH: **A/R**

BELT DIAMETER **C = N/A**

RIM THICKNESS **E = .048**

SHOULDER DIAMETER **G = .468**

LENGTH TO SHOULDER **J = 1.725**

SHOULDER ANGLE (DEG'S/SIDE): **25.70**

PRIMER: **L/R**

CASE CAPACITY (GR'S WATER): **64.25**

DIMENSIONAL DRAWING:

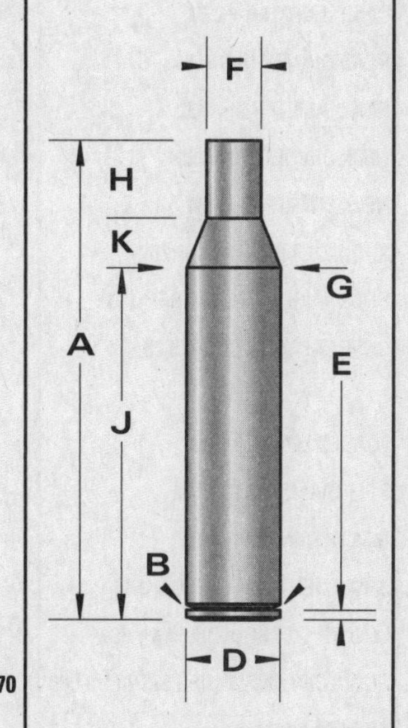

-NOT ACTUAL SIZE-
-DO NOT SCALE-

CARTRIDGE: .244 Remington Improved (Mashburn)

OTHER NAMES:

DIA: .243

BALLISTEK NO: 243BC

NAI NO: RXB 22424/4.755

DATA SOURCE: Ackley Vol.1 Pg.316

HISTORICAL DATA:

NOTES:

LOADING DATA:

BULLET WT./TYPE	POWDER WT./TYPE	VELOCITY ('/SEC)	SOURCE
100/Spire	39.0/4895	3140	Ackley

CASE PREPARATION: SHELLHOLDER (RCBS): 3

MAKE FROM: 6mm Rem. Fire factory ammo in the Mashburn chamber.

PHYSICAL DATA (INCHES):

CASE TYPE: **Rimless Bottleneck**

CASE LENGTH **A = 2.24**

HEAD DIAMETER **B = .471**

RIM DIAMETER **D = .473**

NECK DIAMETER **F = .259**

NECK LENGTH **H = .370**

SHOULDER LENGTH **K = .135**

BODY ANGLE (DEG'S/SIDE): **.690**

CASE CAPACITY **CC'S = 3.99**

LOADED LENGTH: **A/R**

BELT DIAMETER **C = N/A**

RIM THICKNESS **E = .048**

SHOULDER DIAMETER **G = .434**

LENGTH TO SHOULDER **J = 1.735**

SHOULDER ANGLE (DEG'S/SIDE): **32.94**

PRIMER: **L/R**

CASE CAPACITY (GR'S WATER): **61.69**

DIMENSIONAL DRAWING:

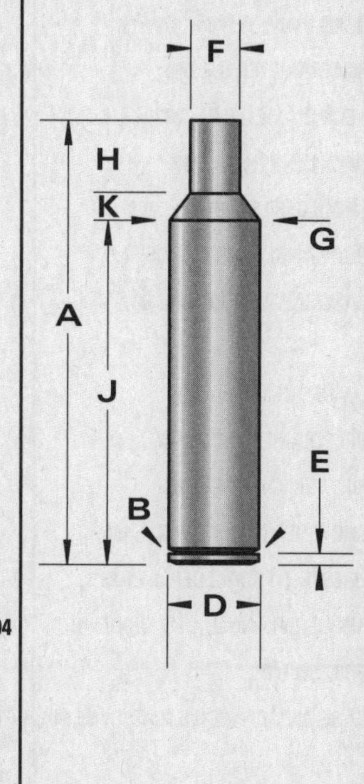

-NOT ACTUAL SIZE-
-DO NOT SCALE-

CARTRIDGE: .246 Purdey Flanged

OTHER NAMES:
.246 Purdey Express

DIA: .253

BALLISTEK NO: 253A

NAI NO: RMB 32442/4.704

DATA SOURCE: COTW 4th Pg.225

HISTORICAL DATA: Circa 1921

NOTES:

LOADING DATA:

BULLET WT./TYPE	POWDER WT./TYPE	VELOCITY ('/SEC)	SOURCE
70/RN Lead	4.5/Unique	1000	JJD
100/Spire	—/—	2950	Factory

CASE PREPARATION: SHELLHOLDER (RCBS): 4

MAKE FROM: .444 Marlin. Best bet is to obtain a form set, however, the Marlin's neck can be annealed and reduced in diameter in a .348 WCF die, a .33 WCF die and a .308 die. Watch shoulder location; this is pretty tricky stuff. Re-anneal and F/L size. Neck I.D.'s may need reaming.

PHYSICAL DATA (INCHES):

CASE TYPE: **Rimmed Bottleneck**

CASE LENGTH **A = 2.23**

HEAD DIAMETER **B = .474**

RIM DIAMETER **D = .544**

NECK DIAMETER **F = .283**

NECK LENGTH **H = .180**

SHOULDER LENGTH **K = .200**

BODY ANGLE (DEG'S/SIDE): **1.26**

CASE CAPACITY **CC'S = 3.58**

LOADED LENGTH: **2.98**

BELT DIAMETER **C = N/A**

RIM THICKNESS **E = .062**

SHOULDER DIAMETER **G = .401**

LENGTH TO SHOULDER **J = 1.85**

SHOULDER ANGLE (DEG'S/SIDE): **16.43**

PRIMER: **L/R**

CASE CAPACITY (GR'S WATER): **55.34**

DIMENSIONAL DRAWING:

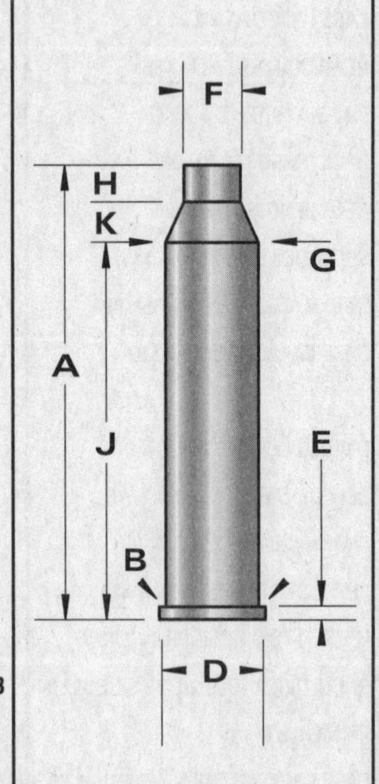

-NOT ACTUAL SIZE-
-DO NOT SCALE-

CARTRIDGE: .25 Ackley Magnum

OTHER NAMES:

DIA: .257

BALLISTEK NO: 257Q

NAI NO: BEN 12423/5.009

DATA SOURCE: Ackley Vol.1 Pg.349

HISTORICAL DATA:

NOTES:

LOADING DATA:

BULLET WT./TYPE	POWDER WT./TYPE	VELOCITY ('/SEC)	SOURCE
100/Spire	60.0/4350	3368	Ackley

CASE PREPARATION: SHELLHOLDER (RCBS): 5

MAKE FROM: .264 Win. Mag. Anneal case neck and F/L size in Ackley die. Trim to length and chamfer.

PHYSICAL DATA (INCHES):

CASE TYPE: **Belted Bottleneck**

CASE LENGTH **A = 2.57**

HEAD DIAMETER **B = .513**

RIM DIAMETER **D = .532**

NECK DIAMETER **F = .294**

NECK LENGTH **H = .404**

SHOULDER LENGTH **K = .202**

BODY ANGLE (DEG'S/SIDE): **.113**

CASE CAPACITY **CC'S = 5.26**

LOADED LENGTH: **3.38**

BELT DIAMETER **C = .532**

RIM THICKNESS **E = .05**

SHOULDER DIAMETER **G = .506**

LENGTH TO SHOULDER **J = 1.964**

SHOULDER ANGLE (DEG'S/SIDE): **27.68**

PRIMER: **L/R Mag**

CASE CAPACITY (GR'S WATER): **81.16**

DIMENSIONAL DRAWING:

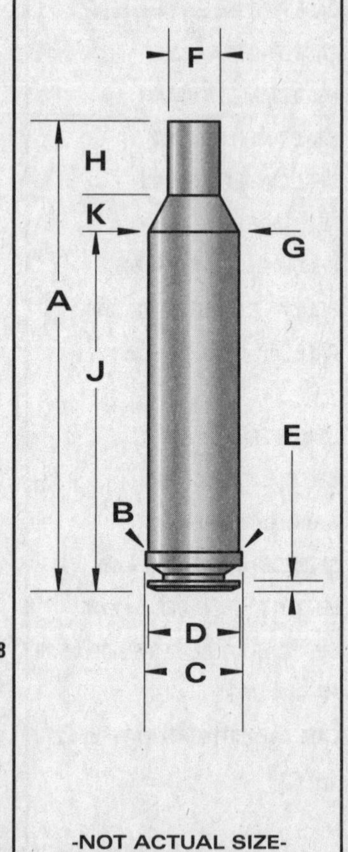

-NOT ACTUAL SIZE-
-DO NOT SCALE-

CARTRIDGE: .25 ACP

OTHER NAMES:
6,35mm Browning
6,35 x 15.5mm

DIA: .251

BALLISTEK NO: 250A

NAI NO: RMS
11115/2.440

DATA SOURCE: Hornady 3rd Pg.321

HISTORICAL DATA: By John Browning in 1905.

NOTES:

LOADING DATA:

BULLET WT./TYPE	POWDER WT./TYPE	VELOCITY ('/SEC)	SOURCE
50/FMJ	1.2/B'eye	800	Horn.

CASE PREPARATION: SHELLHOLDER (RCBS): 29

MAKE FROM: Factory. There is no other case from which .25 ACP cases can be made. It would be possible to turn cases from solid brass, but not worth the effort since commercial cases are easy to find.

PHYSICAL DATA (INCHES):

CASE TYPE: **Rimmed Straight**
CASE LENGTH **A = .615**
HEAD DIAMETER **B = .274**
RIM DIAMETER **D = .302**
NECK DIAMETER **F = .273**
NECK LENGTH **H = N/A**
SHOULDER LENGTH **K = N/A**
BODY ANGLE (DEG'S/SIDE): **.05**
CASE CAPACITY **CC'S = .228**

LOADED LENGTH: **.90**
BELT DIAMETER **C = N/A**
RIM THICKNESS **E = .041**
SHOULDER DIAMETER **G = N/A**
LENGTH TO SHOULDER **J = N/A**
SHOULDER ANGLE (DEG'S/SIDE): **N/A**
PRIMER: **S/P**
CASE CAPACITY (GR'S WATER): **3.52**

DIMENSIONAL DRAWING:

-NOT ACTUAL SIZE-
-DO NOT SCALE-

CARTRIDGE: .25 Gibbs

OTHER NAMES:

DIA: .257

BALLISTEK NO: 257U

NAI NO: RXB
12434/5.319

DATA SOURCE: Ackley Vol.2 Pg.154

HISTORICAL DATA: By R. Gibbs.

NOTES:

LOADING DATA:

BULLET WT./TYPE	POWDER WT./TYPE	VELOCITY ('/SEC)	SOURCE
100/Spire	63.0/4831	3525	Ackley

CASE PREPARATION: SHELLHOLDER (RCBS): 3

MAKE FROM: .25-06 Rem. F/L size the '06 case in the Gibbs die. Fireform in the chamber with light loads. Trim, chamfer and reload.

PHYSICAL DATA (INCHES):

CASE TYPE: **Rimless Bottleneck**
CASE LENGTH **A = 2.50**
HEAD DIAMETER **B = .470**
RIM DIAMETER **D = .473**
NECK DIAMETER **F = .282**
NECK LENGTH **H = .290**
SHOULDER LENGTH **K = .066**
BODY ANGLE (DEG'S/SIDE): **.118**
CASE CAPACITY **CC'S = 4.84**

LOADED LENGTH: **3.13**
BELT DIAMETER **C = N/A**
RIM THICKNESS **E = .05**
SHOULDER DIAMETER **G = .462**
LENGTH TO SHOULDER **J = 2.144**
SHOULDER ANGLE (DEG'S/SIDE): **53.74**
PRIMER: **L/R**
CASE CAPACITY (GR'S WATER): **74.69**

DIMENSIONAL DRAWING:

-NOT ACTUAL SIZE-
-DO NOT SCALE-

CARTRIDGE: .25 Hornet

OTHER NAMES:

DIA: .257

BALLISTEK NO: 257T

NAI NO: RMS 21115/4.615

DATA SOURCE: Handloader #111 Pg.36

HISTORICAL DATA:

NOTES:

LOADING DATA:

BULLET WT./TYPE	POWDER WT./TYPE	VELOCITY ('/SEC)	SOURCE
75/Lead	3.5/Unique	1343	Sengel

CASE PREPARATION: SHELLHOLDER (RCBS): 12

MAKE FROM: .22 Hornet. Taper expand Hornet's neck to accept .25 cal bullets and fireform in the chamber OR fireform with corn meal, in the chamber or in a trim die. Trim & chamfer. You should have a F/L die for reloading.

PHYSICAL DATA (INCHES):

CASE TYPE: **Rimmed Straight**

CASE LENGTH **A = 1.38**

HEAD DIAMETER **B = .299**

RIM DIAMETER **D = .350**

NECK DIAMETER **F = .273**

NECK LENGTH **H = 0**

SHOULDER LENGTH **K = N/A**

BODY ANGLE (DEG'S/SIDE): **.566**

CASE CAPACITY **CC'S = .946**

LOADED LENGTH: **A/R**

BELT DIAMETER **C = N/A**

RIM THICKNESS **E = .065**

SHOULDER DIAMETER **G = N/A**

LENGTH TO SHOULDER **J = N/A**

SHOULDER ANGLE (DEG'S/SIDE): **.285**

PRIMER: **S/R**

CASE CAPACITY (GR'S WATER): **14.61**

DIMENSIONAL DRAWING:

-NOT ACTUAL SIZE-
-DO NOT SCALE-

CARTRIDGE: .25 ICL Magnum

OTHER NAMES:

DIA: .257

BALLISTEK NO: 257AV

NAI NO: BEN 12424/4.873

DATA SOURCE: Ackley Vol.2 Pg.155

HISTORICAL DATA:

NOTES:

LOADING DATA:

BULLET WT./TYPE	POWDER WT./TYPE	VELOCITY ('/SEC)	SOURCE
100/Spire	63.0/4350	3595	Ackley

CASE PREPARATION: SHELLHOLDER (RCBS): 4

MAKE FROM: .264 Win. Mag. Size .264 case in ICL F/L die, trim and chamfer. Fireform in ICL chamber.

PHYSICAL DATA (INCHES):

CASE TYPE: **Belted Bottleneck**

CASE LENGTH **A = 2.500**

HEAD DIAMETER **B = .513**

RIM DIAMETER **D = .532**

NECK DIAMETER **F = .279**

NECK LENGTH **H = .336**

SHOULDER LENGTH **K = .094**

BODY ANGLE (DEG'S/SIDE): **.260**

CASE CAPACITY **CC'S = 5.67**

LOADED LENGTH: **3.18**

BELT DIAMETER **C = .532**

RIM THICKNESS **E = .05**

SHOULDER DIAMETER **G = .496**

LENGTH TO SHOULDER **J = 2.07**

SHOULDER ANGLE (DEG'S/SIDE): **49.09**

PRIMER: **L/R Mag**

CASE CAPACITY (GR'S WATER): **87.59**

DIMENSIONAL DRAWING:

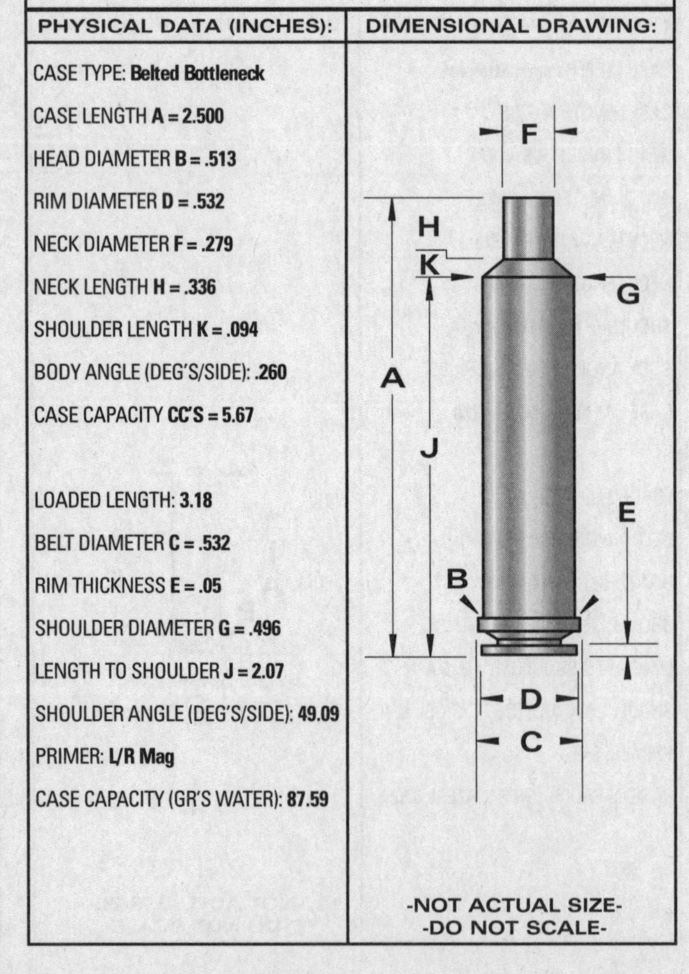

-NOT ACTUAL SIZE-
-DO NOT SCALE-

CARTRIDGE: .5 Krag (Short)

OTHER NAMES:
.25 Short Krag (Ackley)

DIA: .257

BALLISTEK NO: 257R

NAI NO: RMB 22423/5.022

DATA SOURCE: Ackley Vol.1 Pg.340

HISTORICAL DATA:

NOTES: Several versions of a .25 Short Krag exist.

LOADING DATA:

BULLET WT./TYPE	POWDER WT./TYPE	VELOCITY ('/SEC)	SOURCE
100/Spire	43.0/4064	3265	Ackley

CASE PREPARATION: SHELLHOLDER (RCBS): 7

MAKE FROM: .30-40 Krag. Anneal case and neck size in .257 Roberts die until bolt will close on case. Use a light load to fireform, in chamber.

PHYSICAL DATA (INCHES):

CASE TYPE: **Rimmed Bottleneck**
CASE LENGTH **A = 2.300**
HEAD DIAMETER **B = .458**
RIM DIAMETER **D = .545**
NECK DIAMETER **F = .283**
NECK LENGTH **H = .357**
SHOULDER LENGTH **K = .143**
BODY ANGLE (DEG'S/SIDE): **.358**
CASE CAPACITY **CC'S = 3.86**

LOADED LENGTH: **3.03**
BELT DIAMETER **C = N/A**
RIM THICKNESS **E = .063**
SHOULDER DIAMETER **G = .438**
LENGTH TO SHOULDER **J = 1.80**
SHOULDER ANGLE (DEG'S/SIDE): **28.45**
PRIMER: **L/R**
CASE CAPACITY (GR'S WATER): **59.58**

DIMENSIONAL DRAWING:

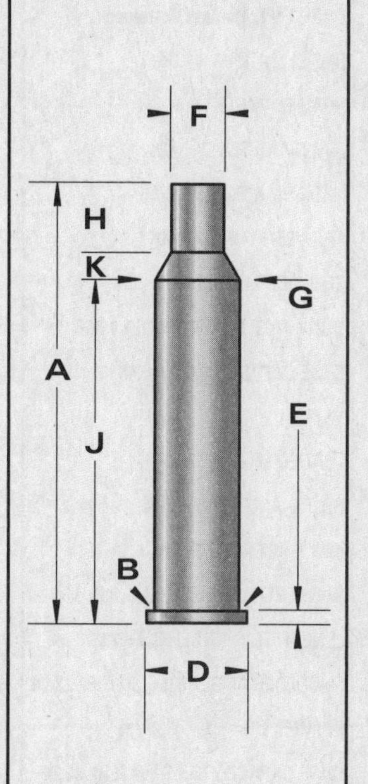

-NOT ACTUAL SIZE-
-DO NOT SCALE-

CARTRIDGE: .25 Krag

OTHER NAMES:
.25/30-40 Krag

DIA: .257

BALLISTEK NO: 257A

NAI NO: RMB 22423/5.022

DATA SOURCE: COTW 4th Pg.144

HISTORICAL DATA: Very early wildcat.

NOTES:

LOADING DATA:

BULLET WT./TYPE	POWDER WT./TYPE	VELOCITY ('/SEC)	SOURCE
100/Spire	32.0/3031	2550	Barnes

CASE PREPARATION: SHELLHOLDER (RCBS): 7

MAKE FROM: .30-40 Krag. Anneal case neck and F/L size in .25 Krag die. Trim and chamfer.

PHYSICAL DATA (INCHES):

CASE TYPE: **Rimmed Bottleneck**
CASE LENGTH **A = 2.300**
HEAD DIAMETER **B = .458**
RIM DIAMETER **D = .545**
NECK DIAMETER **F = .290**
NECK LENGTH **H = .350**
SHOULDER LENGTH **K = .114**
BODY ANGLE (DEG'S/SIDE): **.630**
CASE CAPACITY **CC'S = 3.57**

LOADED LENGTH: **2.95**
BELT DIAMETER **C = N/A**
RIM THICKNESS **E = .063**
SHOULDER DIAMETER **G = .422**
LENGTH TO SHOULDER **J = 1.836**
SHOULDER ANGLE (DEG'S/SIDE): **30.06**
PRIMER: **L/R**
CASE CAPACITY (GR'S WATER): **55.17**

DIMENSIONAL DRAWING:

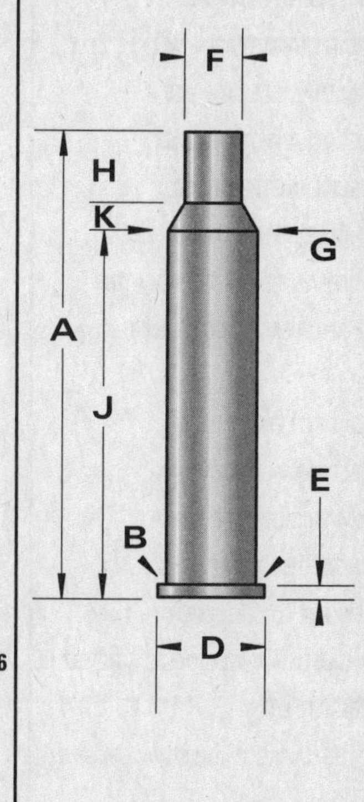

-NOT ACTUAL SIZE-
-DO NOT SCALE-

CARTRIDGE: .25 Krag Improved

OTHER NAMES:

DIA: .257

BALLISTEK NO: 257S

NAI NO: RMB 12424/5.032

DATA SOURCE: Rifle #67 Pg.30

HISTORICAL DATA: From the early 1900s.

NOTES:

LOADING DATA:

BULLET WT./TYPE	POWDER WT./TYPE	VELOCITY ('/SEC)	SOURCE
117/JSP	52.5/H4831	3100	Gates

CASE PREPARATION: SHELLHOLDER (RCBS): 7

MAKE FROM: .30-40 Krag. Cases can be formed as described under the .25 Krag (Short) (#257R) and fire-formed or the form set can be obtained from RCBS.

PHYSICAL DATA (INCHES):

CASE TYPE: **Rimmed Bottleneck**

CASE LENGTH **A = 2.300**

HEAD DIAMETER **B = .457**

RIM DIAMETER **D = .545**

NECK DIAMETER **F = .282**

NECK LENGTH **H = .390**

SHOULDER LENGTH **K = .105**

BODY ANGLE (DEG'S/SIDE): **.285**

CASE CAPACITY **CC'S = 3.88**

LOADED LENGTH: **2.98**

BELT DIAMETER **C = N/A**

RIM THICKNESS **E = .063**

SHOULDER DIAMETER **G = .441**

LENGTH TO SHOULDER **J = 1.804**

SHOULDER ANGLE (DEG'S/SIDE): **37.13**

PRIMER: **L/R**

CASE CAPACITY (GR'S WATER): **59.99**

DIMENSIONAL DRAWING:

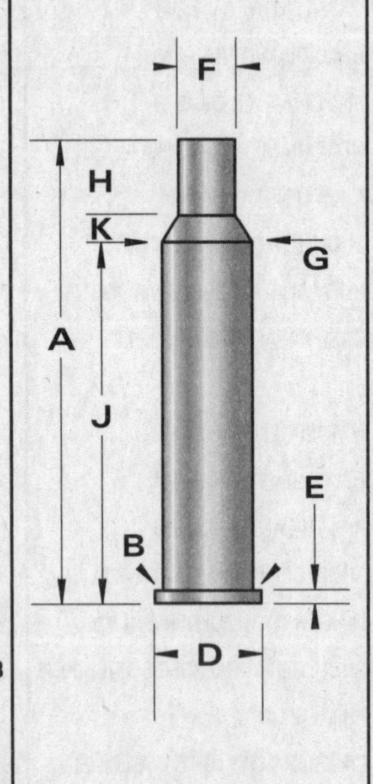

-NOT ACTUAL SIZE-
-DO NOT SCALE-

CARTRIDGE: .25 Remington

OTHER NAMES:

DIA: .257

BALLISTEK NO: 257C

NAI NO: RXB 32323/4.892

DATA SOURCE: COTW 4th Pg.92

HISTORICAL DATA: By Remington in 1906.

NOTES: Rimless version of .25-35.

LOADING DATA:

BULLET WT./TYPE	POWDER WT./TYPE	VELOCITY ('/SEC)	SOURCE
87/Spire	30.0/IMR4895	2800	Barnes

CASE PREPARATION: SHELLHOLDER (RCBS): 19

MAKE FROM: Factory or .30 Rem. Taper neck expand to .310" dia. F/L size in .30 Rem. die. Square case mouth. Chamfer.

PHYSICAL DATA (INCHES):

CASE TYPE: **Rimless Bottleneck**

CASE LENGTH **A = 2.04**

HEAD DIAMETER **B = .417**

RIM DIAMETER **D = .421**

NECK DIAMETER **F = .280**

NECK LENGTH **H = .449**

SHOULDER LENGTH **K = .091**

BODY ANGLE (DEG'S/SIDE): **1.366**

CASE CAPACITY **CC'S = 2.48**

LOADED LENGTH: **2.53**

BELT DIAMETER **C = N/A**

RIM THICKNESS **E = .06**

SHOULDER DIAMETER **G = .355**

LENGTH TO SHOULDER **J = 1.50**

SHOULDER ANGLE (DEG'S/SIDE): **22.39**

PRIMER: **L/R**

CASE CAPACITY (GR'S WATER): **38.26**

DIMENSIONAL DRAWING:

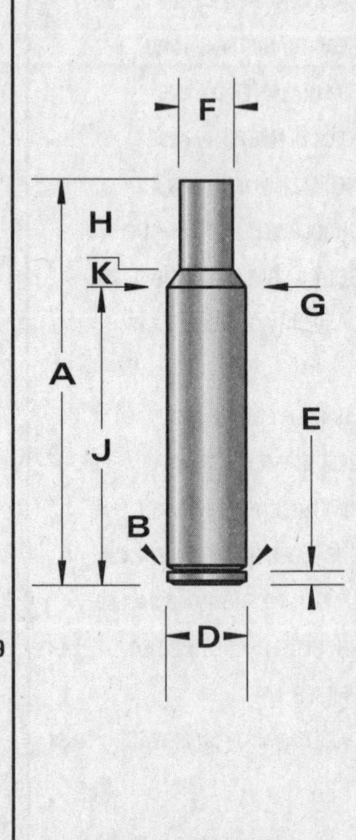

-NOT ACTUAL SIZE-
-DO NOT SCALE-

CARTRIDGE: .25 Souper

OTHER NAMES:
.25/308 Winchester

DIA: .257

BALLISTEK NO: 257V

NAI NO: RXB 12433/4.468

DATA SOURCE: Ackley Vol.1 Pg.338

HISTORICAL DATA: Design may be by Warren Page.

NOTES:

LOADING DATA:

BULLET WT./TYPE	POWDER WT./TYPE	VELOCITY ('/SEC)	SOURCE
100/Spire	39.0/4320	3000	Ackley

CASE PREPARATION: SHELLHOLDER (RCBS): 3

MAKE FROM: .243 Winchester. Taper neck expand .243 case to .260" dia. F/L size, trim & chamfer.

PHYSICAL DATA (INCHES):

CASE TYPE: **Rimless Bottleneck**

CASE LENGTH **A = 2.100**

HEAD DIAMETER **B = .470**

RIM DIAMETER **D = .473**

NECK DIAMETER **F = .289**

NECK LENGTH **H = .260**

SHOULDER LENGTH **K = .230**

BODY ANGLE (DEG'S/SIDE): **.203**

CASE CAPACITY **CC'S = 3.63**

LOADED LENGTH: **2.88**

BELT DIAMETER **C = N/A**

RIM THICKNESS **E = .052**

SHOULDER DIAMETER **G = .460**

LENGTH TO SHOULDER **J = 1.61**

SHOULDER ANGLE (DEG'S/SIDE): **20.39**

PRIMER: **L/R**

CASE CAPACITY (GR'S WATER): **56.04**

DIMENSIONAL DRAWING:

F H K G A E J B D

-NOT ACTUAL SIZE-
-DO NOT SCALE-

CARTRIDGE: .25-06 Ackley Improved

OTHER NAMES:

DIA: .257

BALLISTEK NO: 257W

NAI NO: RXB 12424/5.311

DATA SOURCE: Ackley Vol.1 Pg.345

HISTORICAL DATA:

NOTES:

LOADING DATA:

BULLET WT./TYPE	POWDER WT./TYPE	VELOCITY ('/SEC)	SOURCE
117/Spire	50.0/4350	3051	Hutton

CASE PREPARATION: SHELLHOLDER (RCBS): 3

MAKE FROM: .25-06 Rem. Fireform factory ammo in "improved" chamber. Ackley cautions that light loads should not be used for fireforming, with this ctg. The light charge may not fully "blow-out" the shoulders and could cause a head space problem. I have only formed this case by fireforming, in a trim die, with 10 grs. of Bullseye, a case full of corn meal and a wax wad. No head space problems were noticed.

PHYSICAL DATA (INCHES):

CASE TYPE: **Rimless Bottleneck**

CASE LENGTH **A = 2.496**

HEAD DIAMETER **B = .470**

RIM DIAMETER **D = .473**

NECK DIAMETER **F = .282**

NECK LENGTH **H = .338**

SHOULDER LENGTH **K = .154**

BODY ANGLE (DEG'S/SIDE): **.154**

CASE CAPACITY **CC'S = 4.70**

LOADED LENGTH: **3.40**

BELT DIAMETER **C = N/A**

RIM THICKNESS **E = .05**

SHOULDER DIAMETER **G = .460**

LENGTH TO SHOULDER **J = 2.05**

SHOULDER ANGLE (DEG'S/SIDE): **39.49**

PRIMER: **L/R**

CASE CAPACITY (GR'S WATER): **72.60**

DIMENSIONAL DRAWING:

F H K G A J E B D

-NOT ACTUAL SIZE-
-DO NOT SCALE-

CARTRIDGE: .25-06 Remington

OTHER NAMES:
.25/30-06 Springfield
.25 Neidner

DIA: .257

BALLISTEK NO: 257B

NAI NO: RXB 22432/5.306

DATA SOURCE: Hornady 3rd Pg.140

HISTORICAL DATA: Early wildcat. Offered by Rem. in 1969.

NOTES:

LOADING DATA:

BULLET WT./TYPE	POWDER WT./TYPE	VELOCITY ('/SEC)	SOURCE
120/Spire	42.5/IMR4895	2850	JJD

CASE PREPARATION: SHELLHOLDER (RCBS): 3

MAKE FROM: .30-06 Spgf. Full length size in .25-06 die. Square case mouth. .270 Win. may also be used but, will require more length trimming.

PHYSICAL DATA (INCHES):

CASE TYPE: **Rimless Bottleneck**

CASE LENGTH **A = 2.494**

HEAD DIAMETER **B = .470**

RIM DIAMETER **D = .473**

NECK DIAMETER **F = .290**

NECK LENGTH **H = .308**

SHOULDER LENGTH **K = .246**

BODY ANGLE (DEG'S/SIDE): **.477**

CASE CAPACITY **CC'S = 4.21**

LOADED LENGTH: **3.25**

BELT DIAMETER **C = N/A**

RIM THICKNESS **E = .05**

SHOULDER DIAMETER **G = .441**

LENGTH TO SHOULDER **J = 1.94**

SHOULDER ANGLE (DEG'S/SIDE): **17.06**

PRIMER: **L/R**

CASE CAPACITY (GR'S WATER): **65.03**

DIMENSIONAL DRAWING:

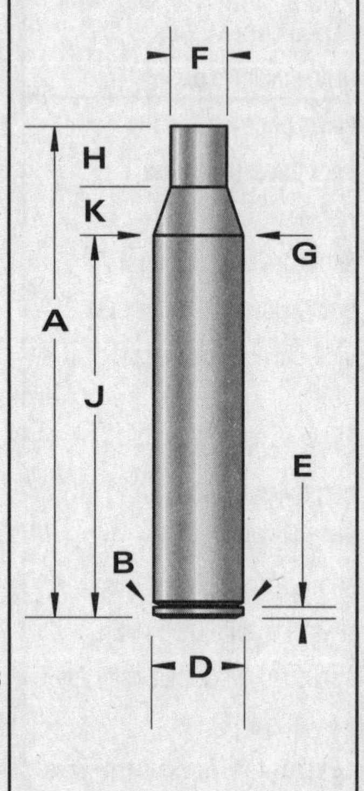

-NOT ACTUAL SIZE-
-DO NOT SCALE-

CARTRIDGE: .25-06 Vickery

OTHER NAMES:

DIA: .257

BALLISTEK NO: 257AW

NAI NO: RXB 22423/5.206

DATA SOURCE: Ackley Vol.1 Pg.344

HISTORICAL DATA: By W.F. Vickery

NOTES:

LOADING DATA:

BULLET WT./TYPE	POWDER WT./TYPE	VELOCITY ('/SEC)	SOURCE
100/Spire	55.0/4831	3014	Ackley

CASE PREPARATION: SHELLHOLDER (RCBS): 3

MAKE FROM: .25-06 Rem. Fireform, with corn meal, in Vickery trim die and F/L size OR anneal .270 Win. case, size in Vickery F/L die, trim and chamfer. It may be possible to fireform .25-06 factory ammo in the Vickery chamber but, I have not done so.

PHYSICAL DATA (INCHES):

CASE TYPE: **Rimless Bottleneck**

CASE LENGTH **A = 2.447**

HEAD DIAMETER **B = .470**

RIM DIAMETER **D = .473**

NECK DIAMETER **F = .283**

NECK LENGTH **H = .350**

SHOULDER LENGTH **K = .167**

BODY ANGLE (DEG'S/SIDE): **.513**

CASE CAPACITY **CC'S = 4.29**

LOADED LENGTH: **3.24**

BELT DIAMETER **C = N/A**

RIM THICKNESS **E = .05**

SHOULDER DIAMETER **G = .439**

LENGTH TO SHOULDER **J = 1.93**

SHOULDER ANGLE (DEG'S/SIDE): **25.03**

PRIMER: **L/R**

CASE CAPACITY (GR'S WATER): **66.16**

DIMENSIONAL DRAWING:

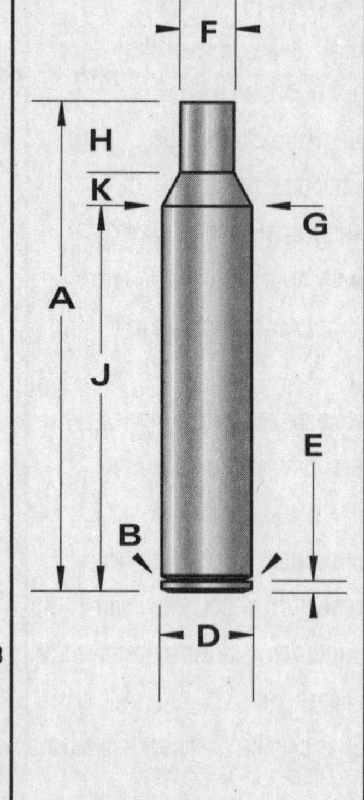

-NOT ACTUAL SIZE-
-DO NOT SCALE-

CARTRIDGE: 25-10 Halsted

OTHER NAMES:
25-10 Center Fire
25 Stevens Centerfire

DIA: .257

BALLISTEK NO:

NAI NO:

DATA SOURCE: Willard Halsted correspondence

HISTORICAL DATA: C.F. Cartridge developed to use in rifles converted from rimfire 25 to reloadable centerfire cartridges.

NOTES:

LOADING DATA:

BULLET WT./TYPE	POWDER WT./TYPE	VELOCITY ('/SEC)	SOURCE
65-67 RNFP	2.5 grains/unique	1156	Halsted
Reccommended bullet is Lyman 25720 (67 grain ver.)			

CASE PREPARATION: SHELL HOLDER (RCBS): 12

MAKE FROM: .22 Hornet
Reduce dia in .278 die, turn base to .280, reduce rim to .333 dia and .050 thk., anneal neck and shoulder, expand with .257 tapered plug expander. Full length resize in .278 die slightly relieved at base.

PHYSICAL DATA (INCHES):

CASE TYPE: **Rimmed Straight**
CASE LENGTH **A = 1.125**
HEAD DIAMETER **B = .278(.280 max)**
RIM DIAMETER **D = .333(.340 max)**
NECK DIAMETER **F = .276**
NECK LENGTH **H = N/A**
SHOULDER LENGTH **K = N/A**
BODY ANGLE (DEG'S/SIDE): **N/A**
CASE CAPACITY **CC'S = 4.29**

LOADED LENGTH: **1.395**
BELT DIAMETER **C = N/A**
RIM THICKNESS **E = .050**
SHOULDER DIAMETER **G = N/A**
LENGTH TO SHOULDER **J = N/A**
SHOULDER ANGLE (DEG'S/SIDE):
PRIMER: **S/P (w/ slight hammers) or S/R**
CASE CAPACITY (GR'S WATER): **66.16**

DIMENSIONAL DRAWING:

-NOT ACTUAL SIZE-
-DO NOT SCALE-

CARTRIDGE: .25-20 Single Shot

OTHER NAMES:

DIA: .257

BALLISTEK NO: 257N

NAI NO: RMB 22225/5.174

DATA SOURCE: COTW 4th Pg.90

HISTORICAL DATA: By F. Rabbeth about 1882

NOTES: Case is used for .22/3000 (R-2) Lovell.

LOADING DATA:

BULLET WT./TYPE	POWDER WT./TYPE	VELOCITY ('/SEC)	SOURCE
86/FN	8.0/2400	1470	Barnes

CASE PREPARATION: SHELLHOLDER (RCBS): 10

MAKE FROM: There is, to my regret, no case that will form into the .25-20 SS. I have had good luck making cases from solid, 3/8" dia. brass and, also, from 5/16" dia. thin wall tubing on turned, brass heads. It is also possible to re-body a .223 Rem. case head with 5/16 tubing. Recently, Jim Bell informed me that his Brass Extrusion Labs was going to offer this case, in the near future. I hope so!

PHYSICAL DATA (INCHES):

CASE TYPE: **Rimmed Bottleneck**
CASE LENGTH **A = 1.630**
HEAD DIAMETER **B = .315**
RIM DIAMETER **D = .378**
NECK DIAMETER **F = .275**
NECK LENGTH **H = .338**
SHOULDER LENGTH **K = .229**
BODY ANGLE (DEG'S/SIDE): **.630**
CASE CAPACITY **CC'S = 1.312**

LOADED LENGTH: **1.90**
BELT DIAMETER **C = N/A**
RIM THICKNESS **E = .06**
SHOULDER DIAMETER **G = .296**
LENGTH TO SHOULDER **J = 1.06**
SHOULDER ANGLE (DEG'S/SIDE): **2.63**
PRIMER: **S/R**
CASE CAPACITY (GR'S WATER): **20.25**

DIMENSIONAL DRAWING:

-NOT ACTUAL SIZE-
-DO NOT SCALE-

CARTRIDGE: .25-20 Winchester

OTHER NAMES:
.25-20 WCF
.25/32-20 WCF

DIA: .257

BALLISTEK NO: 257P

NAI NO: RMB 24222/3.810

DATA SOURCE: Hornady Manual 3rd Pg.123

HISTORICAL DATA: By Win. in 1893

NOTES:

LOADING DATA:

BULLET WT./TYPE	POWDER WT./TYPE	VELOCITY ('/SEC)	SOURCE
60/RN	11.0/2400	2070	JJD

CASE PREPARATION: SHELLHOLDER (RCBS): 10

MAKE FROM: .32-20 Win. Full-length size, square case mouth and chamfer.

PHYSICAL DATA (INCHES):

CASE TYPE: **Rimmed Bottleneck**

CASE LENGTH **A = 1.33**

HEAD DIAMETER **B = .349**

RIM DIAMETER **D = .408**

NECK DIAMETER **F = .273**

NECK LENGTH **H = .345**

SHOULDER LENGTH **K = .131**

BODY ANGLE (DEG'S/SIDE): **.657**

CASE CAPACITY **CC'S = 1.22**

LOADED LENGTH: **1.50**

BELT DIAMETER **C = N/A**

RIM THICKNESS **E = .065**

SHOULDER DIAMETER **G = .334**

LENGTH TO SHOULDER **J = .854**

SHOULDER ANGLE (DEG'S/SIDE): **13.10**

PRIMER: **L/R**

CASE CAPACITY (GR'S WATER): **18.85**

DIMENSIONAL DRAWING:

-NOT ACTUAL SIZE-
-DO NOT SCALE-

CARTRIDGE: .25-21 Stevens

OTHER NAMES:

DIA: .257

BALLISTEK NO: 257AX

NAI NO: RMS 11115/6.833

DATA SOURCE: COTW 4th Pg.91

HISTORICAL DATA: By Capt. Carpenter about 1897.

NOTES: Straight version of .25-20 Win. S/S

LOADING DATA:

BULLET WT./TYPE	POWDER WT./TYPE	VELOCITY ('/SEC)	SOURCE
86/Lead	5.0/Unique	1500	Barnes

CASE PREPARATION: SHELLHOLDER (RCBS): 12

MAKE FROM: .22 Hornet. Only way to form this case would be to silver solder a home-made tubing extension onto the Hornet case. Also, cases can be made by turning from solid brass.

PHYSICAL DATA (INCHES):

CASE TYPE: **Rimmed Straight**

CASE LENGTH **A = 2.05**

HEAD DIAMETER **B = .300**

RIM DIAMETER **D = .376**

NECK DIAMETER **F = .280**

NECK LENGTH **H = N/A**

SHOULDER LENGTH **K = N/A**

BODY ANGLE (DEG'S/SIDE): **.286**

CASE CAPACITY **CC'S = 1.46**

LOADED LENGTH: **2.30**

BELT DIAMETER **C = N/A**

RIM THICKNESS **E = .05**

SHOULDER DIAMETER **G = N/A**

LENGTH TO SHOULDER **J = N/A**

SHOULDER ANGLE (DEG'S/SIDE): **N/A**

PRIMER: **S/R**

CASE CAPACITY (GR'S WATER): **22.58**

DIMENSIONAL DRAWING:

-NOT ACTUAL SIZE-
-DO NOT SCALE-

CARTRIDGE: .25-25 Stevens

OTHER NAMES:

DIA: .257

BALLISTEK NO: 257AY

NAI NO: RMS 21115/7.337

DATA SOURCE: COTW 4th Pg.91

HISTORICAL DATA: By Capt. Carpenter in 1895.

NOTES:

LOADING DATA:

BULLET WT./TYPE	POWDER WT./TYPE	VELOCITY ('/SEC)	SOURCE
86/Lead	9.0/IMR4198	1350	JJD

CASE PREPARATION: SHELLHOLDER (RCBS): 16

MAKE FROM: Make case from solid brass or re-body a .223 Rem. case with 5/16 tubing. Trim and chamfer. F/L size.

PHYSICAL DATA (INCHES):

CASE TYPE: **Rimmed Straight**

CASE LENGTH **A = 2.37**

HEAD DIAMETER **B = .299**

RIM DIAMETER **D = .376**

NECK DIAMETER **F = .282**

NECK LENGTH **H = N/A**

SHOULDER LENGTH **K = N/A**

BODY ANGLE (DEG'S/SIDE): **.513**

CASE CAPACITY **CC'S = 1.86**

LOADED LENGTH: **2.63**

BELT DIAMETER **C = N/A**

RIM THICKNESS **E = .052**

SHOULDER DIAMETER **G = N/A**

LENGTH TO SHOULDER **J = N/A**

SHOULDER ANGLE (DEG'S/SIDE): **N/A**

PRIMER: **S/R**

CASE CAPACITY (GR'S WATER): **28.77**

DIMENSIONAL DRAWING:

F A E B D

-NOT ACTUAL SIZE-
-DO NOT SCALE-

CARTRIDGE: .25-35 Ackley Improved

OTHER NAMES:

DIA: .257

BALLISTEK NO: 247AS

NAI NO: RMB 12324/4.883

DATA SOURCE: Ackley Vol.1 Pg.331

HISTORICAL DATA:

NOTES:

LOADING DATA:

BULLET WT./TYPE	POWDER WT./TYPE	VELOCITY ('/SEC)	SOURCE
75/—	39.0/3031	3600	Ackley

CASE PREPARATION: SHELLHOLDER (RCBS): 2

MAKE FROM: .30-30 Win. Anneal case neck and form with a .25-35 Win. form die. F/L size .25-35, trim and chamfer. Fire these "strong" cases in the Ackley chamber.

PHYSICAL DATA (INCHES):

CASE TYPE: **Rimmed Bottleneck**

CASE LENGTH **A = 2.056**

HEAD DIAMETER **B = .421**

RIM DIAMETER **D = .506**

NECK DIAMETER **F = .284**

NECK LENGTH **H = .350**

SHOULDER LENGTH **K = .076**

BODY ANGLE (DEG'S/SIDE): **.06**

CASE CAPACITY **CC'S = 2.96**

LOADED LENGTH: **2.85**

BELT DIAMETER **C = N/A**

RIM THICKNESS **E = .06**

SHOULDER DIAMETER **G = .418**

LENGTH TO SHOULDER **J = 1.63**

SHOULDER ANGLE (DEG'S/SIDE): **41.39**

PRIMER: **L/R**

CASE CAPACITY (GR'S WATER): **45.72**

DIMENSIONAL DRAWING:

F H K G A J E B D

-NOT ACTUAL SIZE-
-DO NOT SCALE-

CARTRIDGE: .25-35 ICL Coyote

OTHER NAMES:	DIA: .257 BALLISTEK NO: 257X NAI NO: RMB 12324/4.786

DATA SOURCE: Ackley Vol.1 Pg.332

HISTORICAL DATA:

NOTES:

LOADING DATA:

BULLET WT./TYPE	POWDER WT./TYPE	VELOCITY ('/SEC)	SOURCE
100/RN	32.5/IMR4350	2500	JJD

CASE PREPARATION: SHELLHOLDER (RCBS): 2

MAKE FROM: .25-35 Win. Fire factory ammo in ICL chamber OR make from .30-30 Win. by F/L sizing to .25-35 Win. and fireforming in the ICL chamber or in a proper trim die.

PHYSICAL DATA (INCHES):

CASE TYPE: **Rimmed Bottleneck**

CASE LENGTH **A = 2.02**

HEAD DIAMETER **B = .422**

RIM DIAMETER **D = .506**

NECK DIAMETER **F = .281**

NECK LENGTH **H = .450**

SHOULDER LENGTH **K = .12**

BODY ANGLE (DEG'S/SIDE): **.091**

CASE CAPACITY **CC'S = 2.88**

LOADED LENGTH: **A/R**

BELT DIAMETER **C = N/A**

RIM THICKNESS **E = .062**

SHOULDER DIAMETER **G = .426**

LENGTH TO SHOULDER **J = 1.45**

SHOULDER ANGLE (DEG'S/SIDE): **31.13**

PRIMER: **L/R**

CASE CAPACITY (GR'S WATER): **44.57**

DIMENSIONAL DRAWING:

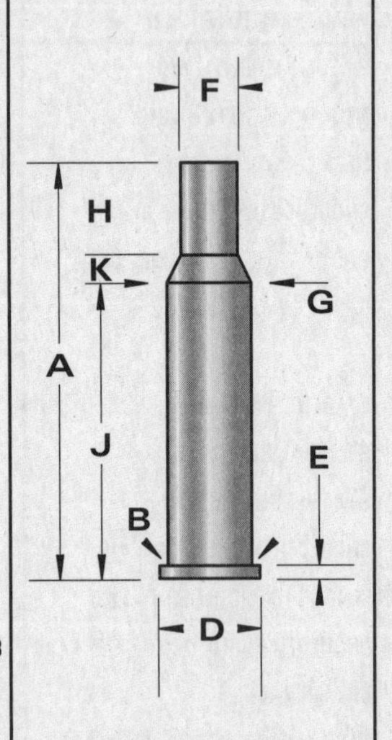

-NOT ACTUAL SIZE-
-DO NOT SCALE-

CARTRIDGE: .25-35 Winchester

OTHER NAMES: .25-35 WCF 6,5 x 52R	DIA: .257 BALLISTEK NO: 257D NAI NO: RMB 32321/4.841

DATA SOURCE: Hornady 3rd Pg.126

HISTORICAL DATA: By Win. in 1895

NOTES:

LOADING DATA:

BULLET WT./TYPE	POWDER WT./TYPE	VELOCITY ('/SEC)	SOURCE
117/RN	24.0/IMR3031	2175	JJD

CASE PREPARATION: SHELLHOLDER (RCBS): 2

MAKE FROM: .30-30 Win or .32 Win Spl. F/L size in .25-35 WCF die. Trim to length. Chamfer. Watch neck wall thickness - I.D. ream if necessary.

PHYSICAL DATA (INCHES):

CASE TYPE: **Rimmed Bottleneck**

CASE LENGTH **A = 2.043**

HEAD DIAMETER **B = .422**

RIM DIAMETER **D = .506**

NECK DIAMETER **F = .281**

NECK LENGTH **H = .428**

SHOULDER LENGTH **K = .140**

BODY ANGLE (DEG'S/SIDE): **2.17**

CASE CAPACITY **CC'S = 2.31**

LOADED LENGTH: **2.60**

BELT DIAMETER **C = N/A**

RIM THICKNESS **E = .063**

SHOULDER DIAMETER **G = .325**

LENGTH TO SHOULDER **J = 1.475**

SHOULDER ANGLE (DEG'S/SIDE): **8.93**

PRIMER: **L/R**

CASE CAPACITY (GR'S WATER): **35.68**

DIMENSIONAL DRAWING:

-NOT ACTUAL SIZE-
-DO NOT SCALE-

CARTRIDGE: .25-36 Marlin

OTHER NAMES:
.25-37 Marlin

DIA: .257

BALLISTEK NO: 257E

NAI NO: RMB 32325/5.084

DATA SOURCE: COTW 4th Pg.92

HISTORICAL DATA: By Marlin in 1895.

NOTES: Similar to .25-35 Win.

LOADING DATA:

BULLET WT./TYPE	POWDER WT./TYPE	VELOCITY ('/SEC)	SOURCE
117/Spire	25.5/IMR4895	2300	Nonte

CASE PREPARATION: SHELLHOLDER (RCBS): 2

MAKE FROM: .30-30 Win. Size .30-30 case in Marlin die which should slightly swage base. Trim to length. I.D. neck ream only if necessary. Chamfer.

PHYSICAL DATA (INCHES):

CASE TYPE: **Rimmed Bottleneck**

CASE LENGTH **A = 2.12**

HEAD DIAMETER **B = .417**

RIM DIAMETER **D = .500**

NECK DIAMETER **F = .287**

NECK LENGTH **H = .54**

SHOULDER LENGTH **K = .126**

BODY ANGLE (DEG'S/SIDE): **1.37**

CASE CAPACITY **CC'S = 2.04**

LOADED LENGTH: **2.50**

BELT DIAMETER **C = N/A**

RIM THICKNESS **E = .063**

SHOULDER DIAMETER **G = .357**

LENGTH TO SHOULDER **J = 1.454**

SHOULDER ANGLE (DEG'S/SIDE): **15.52**

PRIMER: **L/R**

CASE CAPACITY (GR'S WATER): **31.52**

DIMENSIONAL DRAWING:

-NOT ACTUAL SIZE-
-DO NOT SCALE-

CARTRIDGE: .25/222 Copperhead

OTHER NAMES:

DIA: .257

BALLISTEK NO: 257Z

NAI NO: RXB 22324/4.414

DATA SOURCE: Rifle #22 Pg.34

HISTORICAL DATA: By John Wootters.

NOTES:

LOADING DATA:

BULLET WT./TYPE	POWDER WT./TYPE	VELOCITY ('/SEC)	SOURCE
60/Spire	19.5/H4227	2955	Wootters

CASE PREPARATION: SHELLHOLDER (RCBS): 10

MAKE FROM: .222 Rem. Taper expand neck to .260" dia. F/L size in .25/222 die. Trim & chamfer.

PHYSICAL DATA (INCHES):

CASE TYPE: **Rimless Bottleneck**

CASE LENGTH **A = 1.66**

HEAD DIAMETER **B = .376**

RIM DIAMETER **D = .378**

NECK DIAMETER **F = .286**

NECK LENGTH **H = .335**

SHOULDER LENGTH **K = .108**

BODY ANGLE (DEG'S/SIDE): **.903**

CASE CAPACITY **CC'S = 1.65**

LOADED LENGTH: **A/R**

BELT DIAMETER **C = N/A**

RIM THICKNESS **E = .045**

SHOULDER DIAMETER **G = .341**

LENGTH TO SHOULDER **J = 1.264**

SHOULDER ANGLE (DEG'S/SIDE): **31.38**

PRIMER: **S/R**

CASE CAPACITY (GR'S WATER): **25.59**

DIMENSIONAL DRAWING:

-NOT ACTUAL SIZE-
-DO NOT SCALE-

CARTRIDGE: .25/224 Weatherby

OTHER NAMES:

DIA: .257

BALLISTEK NO: 257AA

NAI NO: BEN 22334/4.708

DATA SOURCE: Ackley Vol.2 Pg.153

HISTORICAL DATA:

NOTES:

LOADING DATA:

BULLET WT./TYPE	POWDER WT./TYPE	VELOCITY ('/SEC)	SOURCE
100/Spire	31.0/4895	2992	Ackley

CASE PREPARATION: **SHELLHOLDER (RCBS): 27**

MAKE FROM: .224 Weatherby. Taper expand the Weatherby's neck to about .260" dia. Square case mouth and chamfer. F/L size.

PHYSICAL DATA (INCHES):

CASE TYPE: **Belted Bottleneck**

CASE LENGTH **A = 1.954**

HEAD DIAMETER **B = .415**

RIM DIAMETER **D = .429**

NECK DIAMETER **F = .282**

NECK LENGTH **H = .305**

SHOULDER LENGTH **K = .034**

BODY ANGLE (DEG'S/SIDE): **.545**

CASE CAPACITY **CC'S = 2.62**

LOADED LENGTH: **2.58**

BELT DIAMETER **C = .429**

RIM THICKNESS **E = .05**

SHOULDER DIAMETER **G = .389**

LENGTH TO SHOULDER **J = 1.565**

SHOULDER ANGLE (DEG'S/SIDE): **N/A**

PRIMER: **L/R**

CASE CAPACITY (GR'S WATER): **40.44**

DIMENSIONAL DRAWING:

-NOT ACTUAL SIZE-
-DO NOT SCALE-

CARTRIDGE: .25/270 ICL Ram

OTHER NAMES:

DIA: .257

BALLISTEK NO: 257AZ

NAI NO: RXB 12424/5.276

DATA SOURCE: Ackley Vol.1 Pg.346

HISTORICAL DATA:

NOTES:

LOADING DATA:

BULLET WT./TYPE	POWDER WT./TYPE	VELOCITY ('/SEC)	SOURCE
117/Spire	61.0/4831	3450	Ackley

CASE PREPARATION: **SHELLHOLDER (RCBS): 3**

MAKE FROM: .270 Win. Size neck in .25-06 Rem. F/L die until bolt will just close on case. Fireform. Do not start with .25-06 case because of headspace problems. This is a good case to fireform in the proper trim die.

PHYSICAL DATA (INCHES):

CASE TYPE: **Rimless Bottleneck**

CASE LENGTH **A = 2.48**

HEAD DIAMETER **B = .470**

RIM DIAMETER **D = .473**

NECK DIAMETER **F = .319**

NECK LENGTH **H = .375**

SHOULDER LENGTH **K = .105**

BODY ANGLE (DEG'S/SIDE): **.318**

CASE CAPACITY **CC'S = 3.506**

LOADED LENGTH: **A/R**

BELT DIAMETER **C = N/A**

RIM THICKNESS **E = .05**

SHOULDER DIAMETER **G = .450**

LENGTH TO SHOULDER **J = 2.000**

SHOULDER ANGLE (DEG'S/SIDE): **31.95**

PRIMER: **L/R**

CASE CAPACITY (GR'S WATER): **54.12**

DIMENSIONAL DRAWING:

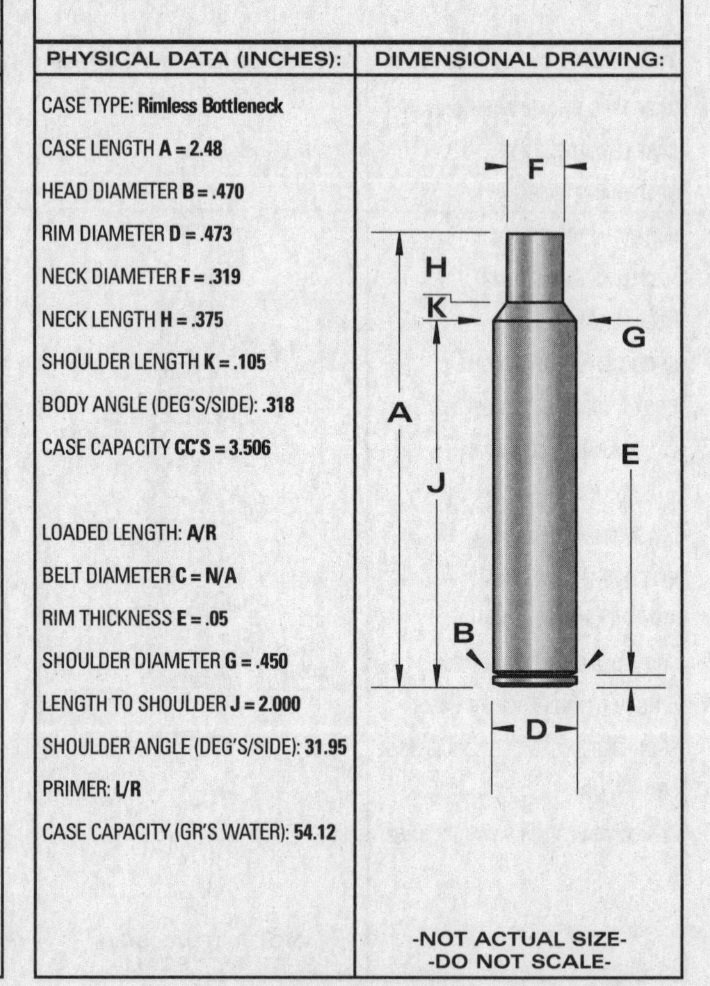

-NOT ACTUAL SIZE-
-DO NOT SCALE-

CARTRIDGE: .25/284 Winchester

OTHER NAMES:

DIA: .257

BALLISTEK NO: 257AB

NAI NO: RBB 22434/4.388

DATA SOURCE: Handloader #26 Pg.30

HISTORICAL DATA:

NOTES:

LOADING DATA:

BULLET WT./TYPE	POWDER WT./TYPE	VELOCITY ('/SEC)	SOURCE
117/Spire	51.5/4350	3150	Ackley

CASE PREPARATION: SHELLHOLDER (RCBS): 3

MAKE FROM: .284 Winchester. Form case in form/trim die, trim to length and F/L size. Neck I.D. may need reaming.

PHYSICAL DATA (INCHES):

CASE TYPE: **Rebated Bottleneck**

CASE LENGTH **A = 2.194**

HEAD DIAMETER **B = .500**

RIM DIAMETER **D = .473**

NECK DIAMETER **F = .284**

NECK LENGTH **H = .290**

SHOULDER LENGTH **K = .104**

BODY ANGLE (DEG'S/SIDE): **.608**

CASE CAPACITY **CC'S = 4.33**

LOADED LENGTH: **2.93**

BELT DIAMETER **C = N/A**

RIM THICKNESS **E = .054**

SHOULDER DIAMETER **G = .466**

LENGTH TO SHOULDER **J = 1.80**

SHOULDER ANGLE (DEG'S/SIDE): **35.18**

PRIMER: **L/R**

CASE CAPACITY (GR'S WATER): **66.85**

DIMENSIONAL DRAWING:

F H K G A E J B D

-NOT ACTUAL SIZE-
-DO NOT SCALE-

CARTRIDGE: .25/303 British

OTHER NAMES:

DIA: .257

BALLISTEK NO: 257AD

NAI NO: RMB 22333/4.802

DATA SOURCE: Ackley Vol.1 Pg.341

HISTORICAL DATA:

NOTES:

LOADING DATA:

BULLET WT./TYPE	POWDER WT./TYPE	VELOCITY ('/SEC)	SOURCE
100/—	33.0/4895	2750	Ackley

CASE PREPARATION: SHELLHOLDER (RCBS): 7

MAKE FROM: .30-40 British. Neck size in a 7mm/08 F/L die (do not set the shoulder back). Trim to length, chamfer and F/L size.

PHYSICAL DATA (INCHES):

CASE TYPE: **Rimmed Bottleneck**

CASE LENGTH **A = 2.185**

HEAD DIAMETER **B = .455**

RIM DIAMETER **D = .540**

NECK DIAMETER **F = .290**

NECK LENGTH **H = .284**

SHOULDER LENGTH **K = .151**

BODY ANGLE (DEG'S/SIDE): **.794**

CASE CAPACITY **CC'S = 3.29**

LOADED LENGTH: **2.85**

BELT DIAMETER **C = N/A**

RIM THICKNESS **E = .063**

SHOULDER DIAMETER **G = .412**

LENGTH TO SHOULDER **J = 1.75**

SHOULDER ANGLE (DEG'S/SIDE): **21.99**

PRIMER: **L/R**

CASE CAPACITY (GR'S WATER): **50.86**

DIMENSIONAL DRAWING:

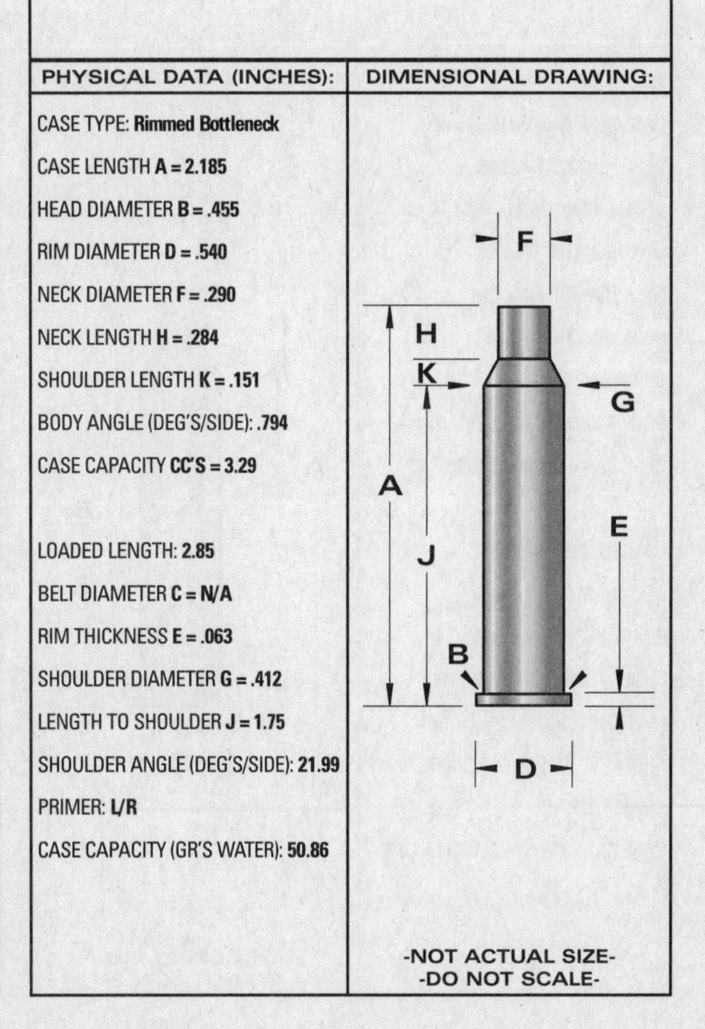

-NOT ACTUAL SIZE-
-DO NOT SCALE-

CARTRIDGE: .25/303 Epps

OTHER NAMES:

DIA: .257

BALLISTEK NO: 257BA

NAI NO: RMB 12423/5.011

DATA SOURCE: Ackley Vol.2 Pg.153

HISTORICAL DATA: By E. Epps.

NOTES:

LOADING DATA:

BULLET WT./TYPE	POWDER WT./TYPE	VELOCITY ('/SEC)	SOURCE
90/Spire	48.2/H414	—	JJD

CASE PREPARATION: SHELLHOLDER (RCBS): 7

MAKE FROM: .303 British. Reduce neck dia. in 7mm/08 F/L die. F/L size in Epps die. Trim and chamfer.

PHYSICAL DATA (INCHES):

CASE TYPE: **Rimmed Bottleneck**

CASE LENGTH **A = 2.285**

HEAD DIAMETER **B = .456**

RIM DIAMETER **D = .540**

NECK DIAMETER **F = .281**

NECK LENGTH **H = .336**

SHOULDER LENGTH **K = .139**

BODY ANGLE (DEG'S/SIDE): **.267**

CASE CAPACITY **CC'S = 3.92**

LOADED LENGTH: **2.875**

BELT DIAMETER **C = N/A**

RIM THICKNESS **E = .063**

SHOULDER DIAMETER **G = .441**

LENGTH TO SHOULDER **J = 1.81**

SHOULDER ANGLE (DEG'S/SIDE): **29.92**

PRIMER: **L/R**

CASE CAPACITY (GR'S WATER): **60.50**

DIMENSIONAL DRAWING:

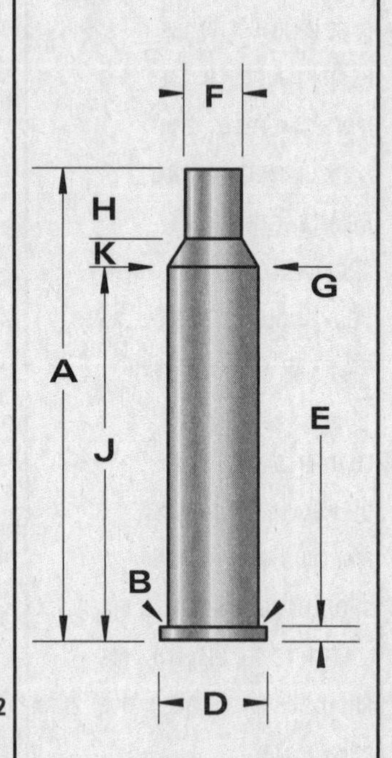

-NOT ACTUAL SIZE-
-DO NOT SCALE-

CARTRIDGE: .25/350 Remington Mag.

OTHER NAMES:

DIA: .257

BALLISTEK NO: 257AF

NAI NO: BEN 22443/4.288

DATA SOURCE: Ackley Vol.2 Pg.155

HISTORICAL DATA:

NOTES: Overbore!

LOADING DATA:

BULLET WT./TYPE	POWDER WT./TYPE	VELOCITY ('/SEC)	SOURCE
117/Spire	58.0/IMR4831	3400	JJD

CASE PREPARATION: SHELLHOLDER (RCBS): 4

MAKE FROM: 6.5 Rem. Mag. Size 6,5mm case in .25/350 F/L die. Trim & chamfer. This cartridge is as worthless as it is difficult to make!

PHYSICAL DATA (INCHES):

CASE TYPE: **Belted Bottleneck**

CASE LENGTH **A = 2.200**

HEAD DIAMETER **B = .513**

RIM DIAMETER **D = .532**

NECK DIAMETER **F = .280**

NECK LENGTH **H = .220**

SHOULDER LENGTH **K = .230**

BODY ANGLE (DEG'S/SIDE): **.591**

CASE CAPACITY **CC'S = 4.72**

LOADED LENGTH: **2.92**

BELT DIAMETER **C = .532**

RIM THICKNESS **E = .05**

SHOULDER DIAMETER **G = .481**

LENGTH TO SHOULDER **J = 1.75**

SHOULDER ANGLE (DEG'S/SIDE): **23.60**

PRIMER: **L/R Mag**

CASE CAPACITY (GR'S WATER): **72.84**

DIMENSIONAL DRAWING:

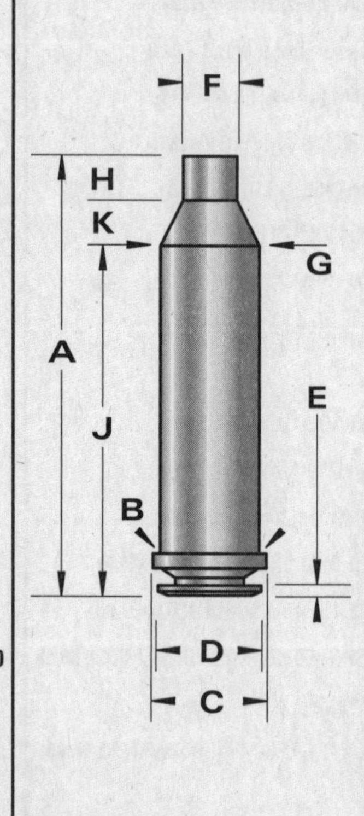

-NOT ACTUAL SIZE-
-DO NOT SCALE-

CARTRIDGE: .250 Bennett Magnum

OTHER NAMES:

DIA: .257

BALLISTEK NO: 257BB

NAI NO: BEN 32433/4.717

DATA SOURCE: Ackley Vol.1 Pg.349

HISTORICAL DATA: By B. Bennett.

NOTES:

LOADING DATA:

BULLET WT./TYPE	POWDER WT./TYPE	VELOCITY ('/SEC)	SOURCE
100/Spire	42.0/IMR4064	3038	JJD

CASE PREPARATION: SHELLHOLDER (RCBS): 4

MAKE FROM: .300 H&H Mag. Form in form die set. Trim to length & chamfer. F/L size and fireform in Bennett chamber.

PHYSICAL DATA (INCHES):

CASE TYPE: **Belted Bottleneck**

CASE LENGTH **A = 2.42**

HEAD DIAMETER **B = .513**

RIM DIAMETER **D = .532**

NECK DIAMETER **F = .282**

NECK LENGTH **H = .308**

SHOULDER LENGTH **K = .182**

BODY ANGLE (DEG'S/SIDE): **1.24**

CASE CAPACITY **CC'S = 4.69**

LOADED LENGTH: **3.08**

BELT DIAMETER **C = .532**

RIM THICKNESS **E = .05**

SHOULDER DIAMETER **G = .438**

LENGTH TO SHOULDER **J = 1.93**

SHOULDER ANGLE (DEG'S/SIDE): **23.19**

PRIMER: **L/R Mag**

CASE CAPACITY (GR'S WATER): **72.45**

DIMENSIONAL DRAWING:

-NOT ACTUAL SIZE-
-DO NOT SCALE-

CARTRIDGE: .250 Durham Magnum

OTHER NAMES:

DIA: .257

BALLISTEK NO: 257AG

NAI NO: BEN 32433/4.522

DATA SOURCE: Ackley Vol.1 Pg.348

HISTORICAL DATA:

NOTES:

LOADING DATA:

BULLET WT./TYPE	POWDER WT./TYPE	VELOCITY ('/SEC)	SOURCE
117/Spire	55.0/4350	3330	Ackley

CASE PREPARATION: SHELLHOLDER (RCBS): 4

MAKE FROM: .264 Win. Mag. Anneal case neck. Set back shoulder in 6,5 Rem. Mag. sizer. F/L size in Durham die. Trim & chamfer. This is a good ctg. to have a form set for!

PHYSICAL DATA (INCHES):

CASE TYPE: **Belted Bottleneck**

CASE LENGTH **A = 2.32**

HEAD DIAMETER **B = .513**

RIM DIAMETER **D = .532**

NECK DIAMETER **F = .282**

NECK LENGTH **H = .310**

SHOULDER LENGTH **K = .157**

BODY ANGLE (DEG'S/SIDE): **1.057**

CASE CAPACITY **CC'S = 4.60**

LOADED LENGTH: **2.94**

BELT DIAMETER **C = .532**

RIM THICKNESS **E = .05**

SHOULDER DIAMETER **G = .452**

LENGTH TO SHOULDER **J = 1.853**

SHOULDER ANGLE (DEG'S/SIDE): **28.43**

PRIMER: **L/R Mag**

CASE CAPACITY (GR'S WATER): **71.08**

DIMENSIONAL DRAWING:

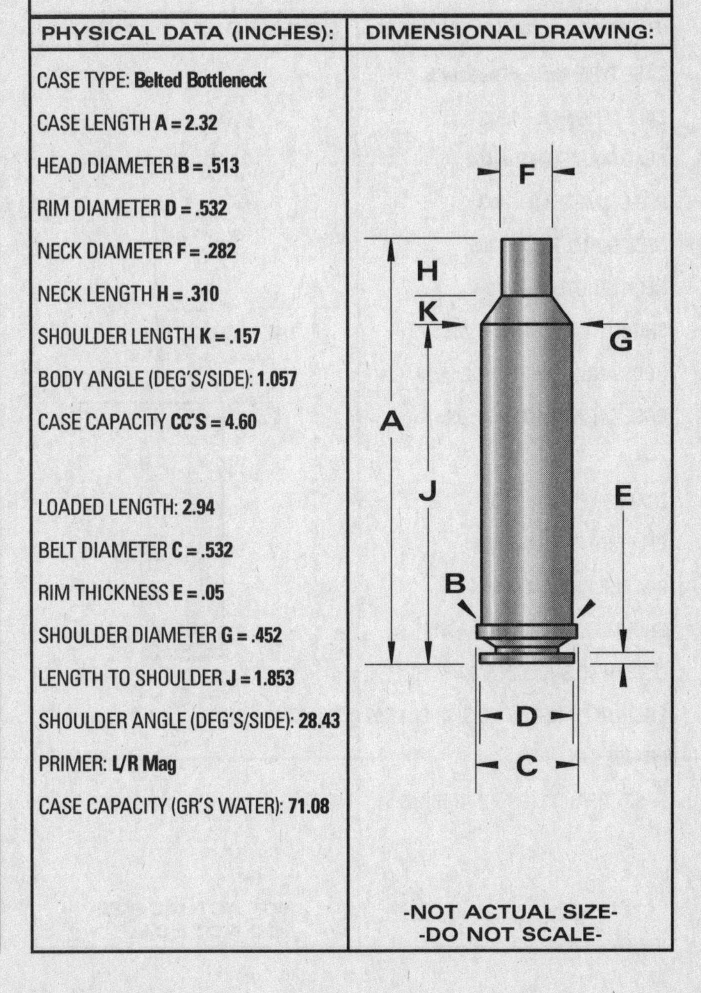

-NOT ACTUAL SIZE-
-DO NOT SCALE-

CARTRIDGE: .250-3000 Savage

OTHER NAMES:
.250 Savage

DIA: .257

BALLISTEK NO: 257H

NAI NO: RXB 33432/4.085

DATA SOURCE: Hornady 3rd Pg.128

HISTORICAL DATA: Developed by C. Newton in 1915.

NOTES: Commercial by Rem(?) in 1915.

LOADING DATA:

BULLET WT./TYPE	POWDER WT./TYPE	VELOCITY ('/SEC)	SOURCE
75/Spire	32.6/IMR3031	3100	Horn.

CASE PREPARATION: SHELLHOLDER (RCBS): 3

MAKE FROM: .270 Win. Anneal case. Set shoulder back with .308 die. Run into .250 Savage F/L die with expander removed. Trim to length. I.D. neck ream if required (and, it normally is). Chamfer. F/L size.

PHYSICAL DATA (INCHES):

CASE TYPE: **Rimless Bottleneck**

CASE LENGTH **A = 1.912**

HEAD DIAMETER **B = .468**

RIM DIAMETER **D = .473**

NECK DIAMETER **F = .285**

NECK LENGTH **H = .275**

SHOULDER LENGTH **K = .231**

BODY ANGLE (DEG'S/SIDE): **1.18**

CASE CAPACITY **CC'S = 2.95**

LOADED LENGTH: **2.40**

BELT DIAMETER **C = N/A**

RIM THICKNESS **E = .05**

SHOULDER DIAMETER **G = .418**

LENGTH TO SHOULDER **J = 1.405**

SHOULDER ANGLE (DEG'S/SIDE): **15.99**

PRIMER: **L/R**

CASE CAPACITY (GR'S WATER): **45.54**

DIMENSIONAL DRAWING:

-NOT ACTUAL SIZE-
-DO NOT SCALE-

CARTRIDGE: .250/3000 Savage Improved (Ackley)

OTHER NAMES:

DIA: .257

BALLISTEK NO: 257AT

NAI NO: RXB 12434/4.226

DATA SOURCE: Ackley Vol.1 Pg.334

HISTORICAL DATA:

NOTES: Almost perfect case capacity for .257" dia. bore.

LOADING DATA:

BULLET WT./TYPE	POWDER WT./TYPE	VELOCITY ('/SEC)	SOURCE
100/Spire	42.0/4350	3129	Ackley

CASE PREPARATION: SHELLHOLDER (RCBS): 3

MAKE FROM: .250/3000 Savage. Fire the factory ammo in the "improved" chamber OR fireform the case, with Bullseye and corn meal, in the proper trim die.

PHYSICAL DATA (INCHES):

CASE TYPE: **Rimless Bottleneck**

CASE LENGTH **A = 1.978**

HEAD DIAMETER **B = .468**

RIM DIAMETER **D = .473**

NECK DIAMETER **F = .288**

NECK LENGTH **H = .278**

SHOULDER LENGTH **K = .130**

BODY ANGLE (DEG'S/SIDE): **.293**

CASE CAPACITY **CC'S = 3.36**

LOADED LENGTH: **2.71**

BELT DIAMETER **C = N/A**

RIM THICKNESS **E = .049**

SHOULDER DIAMETER **G = .454**

LENGTH TO SHOULDER **J = 1.57**

SHOULDER ANGLE (DEG'S/SIDE): **32.55**

PRIMER: **L/R**

CASE CAPACITY (GR'S WATER): **51.92**

DIMENSIONAL DRAWING:

-NOT ACTUAL SIZE-
-DO NOT SCALE-

CARTRIDGE: .255 Rook Rifle

OTHER NAMES:	DIA: .255
	BALLISTEK NO: 255A
	NAI NO: RMB 24232/3.343

DATA SOURCE: COTW 4th Pg.226

HISTORICAL DATA: By Jeffrey for Rook rifles.

NOTES:

LOADING DATA:

BULLET WT./TYPE	POWDER WT./TYPE	VELOCITY ('/SEC)	SOURCE
70/RN	9.0/IMR4227	1200	JJD

CASE PREPARATION: SHELLHOLDER (RCBS): 1

MAKE FROM: .25-20 Win. Trim case to length, chamfer and F/L size in Rook die. Fireform. Some weapons may require that you turn the .25-20's rim to the proper dia.

PHYSICAL DATA (INCHES):

CASE TYPE: **Rimmed Bottleneck**
CASE LENGTH **A = 1.15**
HEAD DIAMETER **B = .344**
RIM DIAMETER **D = .401**
NECK DIAMETER **F = .274**
NECK LENGTH **H = .318**
SHOULDER LENGTH **K = .134**
BODY ANGLE (DEG'S/SIDE): **.920**
CASE CAPACITY **CC'S = .953**

LOADED LENGTH: **1.43**
BELT DIAMETER **C = N/A**
RIM THICKNESS **E = .065**
SHOULDER DIAMETER **G = .328**
LENGTH TO SHOULDER **J = .698**
SHOULDER ANGLE (DEG'S/SIDE): **11.39**
PRIMER: **S/R**
CASE CAPACITY (GR'S WATER): **14.70**

DIMENSIONAL DRAWING:

F
H
K
G
A
J
B
E
D

-NOT ACTUAL SIZE-
-DO NOT SCALE-

CARTRIDGE: .256 Gibbs Magnum

OTHER NAMES: .256 Magnum	DIA: .264
	BALLISTEK NO: 264AK
	NAI NO: RXB 22432/4.587

DATA SOURCE: COTW 4th Pg.226

HISTORICAL DATA: By Gibbs in 1913.

NOTES: Similar to 6,5x55mm.

LOADING DATA:

BULLET WT./TYPE	POWDER WT./TYPE	VELOCITY ('/SEC)	SOURCE
150/Spire	42.7/IMR4350	–	JJD

CASE PREPARATION: SHELLHOLDER (RCBS): 3

MAKE FROM: .270 Win. Size the .270 case in the Gibbs F/L die with the expander removed. Trim & chamfer. Neck I.D. may require reaming. Nonte reports bore dia. as .266" but, .264" projectiles work fine.

PHYSICAL DATA (INCHES):

CASE TYPE: **Rimless Bottleneck**
CASE LENGTH **A = 2.17**
HEAD DIAMETER **B = .473**
RIM DIAMETER **D = .476**
NECK DIAMETER **F = .298**
NECK LENGTH **H = .275**
SHOULDER LENGTH **K = .252**
BODY ANGLE (DEG'S/SIDE): **.913**
CASE CAPACITY **CC'S = 3.49**

LOADED LENGTH: **3.05**
BELT DIAMETER **C = N/A**
RIM THICKNESS **E = .052**
SHOULDER DIAMETER **G = .427**
LENGTH TO SHOULDER **J = 1.643**
SHOULDER ANGLE (DEG'S/SIDE): **14.35**
PRIMER: **L/R**
CASE CAPACITY (GR'S WATER): **53.87**

DIMENSIONAL DRAWING:

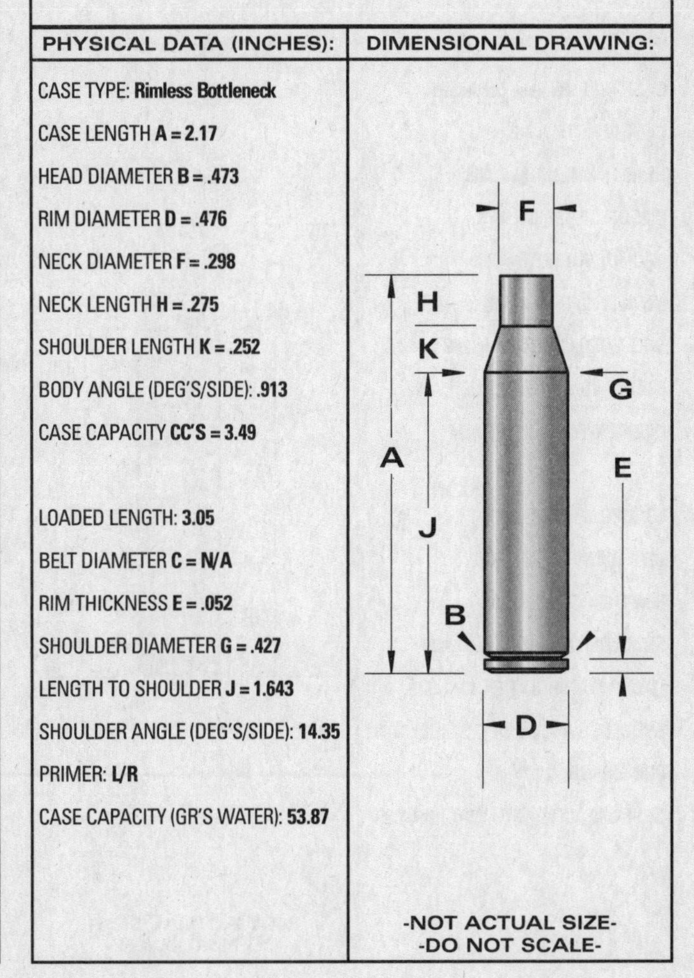

-NOT ACTUAL SIZE-
-DO NOT SCALE-

CARTRIDGE: .256 Newton

OTHER NAMES:

DIA: .264

BALLISTEK NO: 264A

NAI NO: RXB 22323/5.25

DATA SOURCE: NAI/Ballistek

HISTORICAL DATA: By Charles Newton in 1913.

NOTES:

LOADING DATA:

BULLET WT./TYPE	POWDER WT./TYPE	VELOCITY ('/SEC)	SOURCE
140/Spire	57.0/4831	2890	Barnes

CASE PREPARATION: SHELLHOLDER (RCBS): 3

MAKE FROM: .270 Win. Anneal case. Trim to 2.5" F/L size in Newton die. Trim to length. Chamfer & clean.

PHYSICAL DATA (INCHES):

CASE TYPE: **Rimless Bottleneck**

CASE LENGTH **A = 2.457**

HEAD DIAMETER **B = .468**

RIM DIAMETER **D = .473**

NECK DIAMETER **F = .289**

NECK LENGTH **H = .372**

SHOULDER LENGTH **K = .182**

BODY ANGLE (DEG'S/SIDE): **.756**

CASE CAPACITY **CC'S = 4.14**

LOADED LENGTH: **3.40**

BELT DIAMETER **C = N/A**

RIM THICKNESS **E = .054**

SHOULDER DIAMETER **G = .423**

LENGTH TO SHOULDER **J = 1.903**

SHOULDER ANGLE (DEG'S/SIDE): **20.21**

PRIMER: **L/R**

CASE CAPACITY (GR'S WATER): **63.99**

DIMENSIONAL DRAWING:

F H K G A E J B D

-NOT ACTUAL SIZE-
-DO NOT SCALE-

CARTRIDGE: .256 Winchester Mag.

OTHER NAMES:

DIA: .257

BALLISTEK NO: 257AU

NAI NO: RMB 24343/3.371

DATA SOURCE: Hornady 3rd Pg.124

HISTORICAL DATA: By Win. in 1961.

NOTES:

LOADING DATA:

BULLET WT./TYPE	POWDER WT./TYPE	VELOCITY ('/SEC)	SOURCE
75/HP	14.2/IMR4227	2400	Horn.

CASE PREPARATION: SHELLHOLDER (RCBS): 6

MAKE FROM: .357 Magnum. Anneal case neck (carefully!). F/L size. Trim to length.

PHYSICAL DATA (INCHES):

CASE TYPE: **Rimmed Bottleneck**

CASE LENGTH **A = 1.281**

HEAD DIAMETER **B = .380**

RIM DIAMETER **D = .440**

NECK DIAMETER **F = .285**

NECK LENGTH **H = .208**

SHOULDER LENGTH **K = .083**

BODY ANGLE (DEG'S/SIDE): **.435**

CASE CAPACITY **CC'S = 1.32**

LOADED LENGTH: **1.78**

BELT DIAMETER **C = N/A**

RIM THICKNESS **E = .060**

SHOULDER DIAMETER **G = .368**

LENGTH TO SHOULDER **J = .99**

SHOULDER ANGLE (DEG'S/SIDE): **26.56**

PRIMER: **S/P**

CASE CAPACITY (GR'S WATER): **20.49**

DIMENSIONAL DRAWING:

F H K G A J E B D

-NOT ACTUAL SIZE-
-DO NOT SCALE-

CARTRIDGE: .257 Arch

OTHER NAMES:

DIA: .257

BALLISTEK NO: 257AL

NAI NO: RXB 12334/4.809

DATA SOURCE: Ackley Vol.1 Pg.340

HISTORICAL DATA: By Dr. E. Arch for very long projectiles.

NOTES:

LOADING DATA:

BULLET WT./TYPE	POWDER WT./TYPE	VELOCITY ('/SEC)	SOURCE
117/Spire	52.5/4831	3174	Ackley

CASE PREPARATION: SHELLHOLDER (RCBS): 2

MAKE FROM: 6.5 x 55 Swedish. F/L size the 6,5 case in the Arch die. Fireform, trim, chamfer and reload.

PHYSICAL DATA (INCHES):

CASE TYPE: **Rimless Bottleneck**

CASE LENGTH **A = 2.15**

HEAD DIAMETER **B = .474**

RIM DIAMETER **D = .476**

NECK DIAMETER **F = .286**

NECK LENGTH **H = .275**

SHOULDER LENGTH **K = .125**

BODY ANGLE (DEG'S/SIDE): **.203**

CASE CAPACITY **CC'S = 3.92**

LOADED LENGTH: **3.20**

BELT DIAMETER **C = N/A**

RIM THICKNESS **E = .056**

SHOULDER DIAMETER **G = .463**

LENGTH TO SHOULDER **J = 1.75**

SHOULDER ANGLE (DEG'S/SIDE): **35.29**

PRIMER: **L/R**

CASE CAPACITY (GR'S WATER): **60.64**

DIMENSIONAL DRAWING:

-NOT ACTUAL SIZE-
-DO NOT SCALE-

CARTRIDGE: .257 Baker Magnum

OTHER NAMES:

DIA: .257

BALLISTEK NO: 257AO

NAI NO: BEN 22433/4.922

DATA SOURCE: Ackley Vol.2 Pg.156

HISTORICAL DATA: By H. Baker.

NOTES:

LOADING DATA:

BULLET WT./TYPE	POWDER WT./TYPE	VELOCITY ('/SEC)	SOURCE
120/Spire	81.5/H570	3480	Ackley

CASE PREPARATION: SHELLHOLDER (RCBS): 4

MAKE FROM: .264 Win. Mag. Anneal case neck and F/L size in Baker die. Trim and chamfer.

PHYSICAL DATA (INCHES):

CASE TYPE: **Belted Bottleneck**

CASE LENGTH **A = 2.525**

HEAD DIAMETER **B = .513**

RIM DIAMETER **D = .532**

NECK DIAMETER **F = .291**

NECK LENGTH **H = .270**

SHOULDER LENGTH **K = .185**

BODY ANGLE (DEG'S/SIDE): **.352**

CASE CAPACITY **CC'S = 5.30**

LOADED LENGTH: **3.18**

BELT DIAMETER **C = .532**

RIM THICKNESS **E = .049**

SHOULDER DIAMETER **G = .490**

LENGTH TO SHOULDER **J = 2.07**

SHOULDER ANGLE (DEG'S/SIDE): **28.27**

PRIMER: **L/R Mag**

CASE CAPACITY (GR'S WATER): **81.93**

DIMENSIONAL DRAWING:

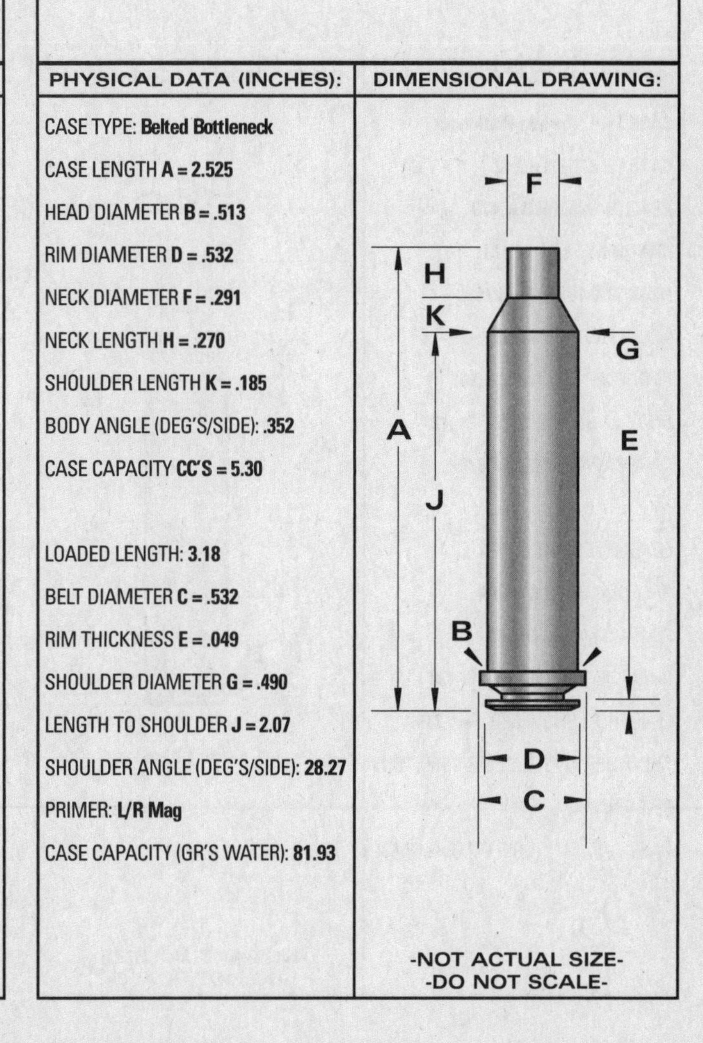

-NOT ACTUAL SIZE-
-DO NOT SCALE-

CARTRIDGE: .257 Big Horn

OTHER NAMES:

DIA: .257

BALLISTEK NO: 257BC

NAI NO: RXB 22423/5.149

DATA SOURCE: Ackley Vol.1 Pg.343

HISTORICAL DATA: By J. Ochocki

NOTES:

LOADING DATA:

BULLET WT./TYPE	POWDER WT./TYPE	VELOCITY ('/SEC)	SOURCE
87/Spire	45.0/4320	3325	Ackley

CASE PREPARATION: SHELLHOLDER (RCBS): 3

MAKE FROM: .25-06 Rem. Size Rem. case in .228 Ackley Mag. (#228c) F/L die to set shoulder back. Taper expand back up to .260" dia and F/L size in Big Horn die. This leaves a very long neck.

PHYSICAL DATA (INCHES):

CASE TYPE: **Rimless Bottleneck**
CASE LENGTH **A = 2.42**
HEAD DIAMETER **B = .470**
RIM DIAMETER **D = .473**
NECK DIAMETER **F = .279**
NECK LENGTH **H = .503**
SHOULDER LENGTH **K = .141**
BODY ANGLE (DEG'S/SIDE): **.727**
CASE CAPACITY **CC'S = 4.05**
LOADED LENGTH: **3.19**
BELT DIAMETER **C = N/A**
RIM THICKNESS **E = .05**
SHOULDER DIAMETER **G = .430**
LENGTH TO SHOULDER **J = 1.776**
SHOULDER ANGLE (DEG'S/SIDE): **28.16**
PRIMER: **L/R**
CASE CAPACITY (GR'S WATER): **62.63**

DIMENSIONAL DRAWING:

-NOT ACTUAL SIZE-
-DO NOT SCALE-

CARTRIDGE: .257 Condor

OTHER NAMES:

DIA: .257

BALLISTEK NO: 257AN

NAI NO: BEN 22434/4.678

DATA SOURCE: Ackley Vol.1 Pg.352

HISTORICAL DATA: By Dr. R. Somovia for long & heavy bullets.

NOTES:

LOADING DATA:

BULLET WT./TYPE	POWDER WT./TYPE	VELOCITY ('/SEC)	SOURCE
160/HP	56.0/4831	2730	Ackley

CASE PREPARATION: SHELLHOLDER (RCBS): 4

MAKE FROM: .300 Win. Mag. Use form set to form 7x61 Sharpe & Hart case (#284F). Form the S&H case in the Condor form die. Trim, chamfer and F/L size.

PHYSICAL DATA (INCHES):

CASE TYPE: **Belted Bottleneck**
CASE LENGTH **A = 2.400**
HEAD DIAMETER **B = .413**
RIM DIAMETER **D = .532**
NECK DIAMETER **F = .288**
NECK LENGTH **H = .300**
SHOULDER LENGTH **K = .120**
BODY ANGLE (DEG'S/SIDE): **.997**
CASE CAPACITY **CC'S = 4.67**
LOADED LENGTH: **3.30**
BELT DIAMETER **C = .532**
RIM THICKNESS **E = .05**
SHOULDER DIAMETER **G = .451**
LENGTH TO SHOULDER **J = 1.98**
SHOULDER ANGLE (DEG'S/SIDE): **34.18**
PRIMER: **L/R Mag**
CASE CAPACITY (GR'S WATER): **72.11**

DIMENSIONAL DRAWING:

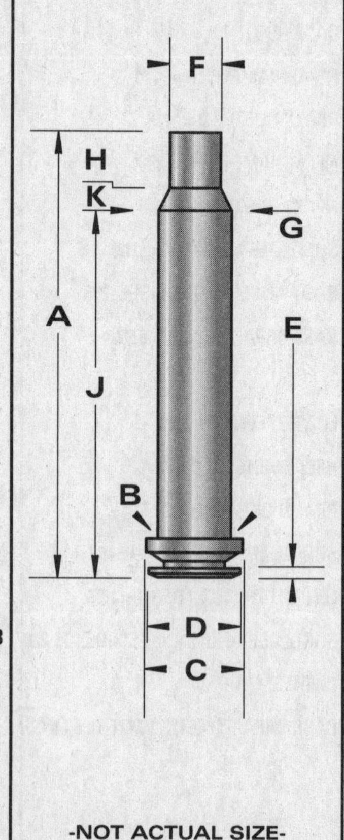

-NOT ACTUAL SIZE-
-DO NOT SCALE-

CARTRIDGE: .257 Critser Magnum

OTHER NAMES:

DIA: .257

BALLISTEK NO: 257AM

NAI NO: BEN 22444/5.068

DATA SOURCE: Ackley Vol.1 Pg.351

HISTORICAL DATA: By C. Critser.

NOTES:

LOADING DATA:

BULLET WT./TYPE	POWDER WT./TYPE	VELOCITY ('/SEC)	SOURCE
120/Spire	65.0/4350	3400	Ackley

CASE PREPARATION: SHELLHOLDER (RCBS): 4

MAKE FROM: .300 H&H. Anneal neck and size in 7mm Rem. Mag. die. Size again in .264 Win. Mag. die (do not trim neck!). Size in Critser F/L die and fireform in chamber or trim die (with corn meal). Trim to length.

PHYSICAL DATA (INCHES):

CASE TYPE: **Belted Bottleneck**

CASE LENGTH **A = 2.600**

HEAD DIAMETER **B = .513**

RIM DIAMETER **D = .532**

NECK DIAMETER **F = .288**

NECK LENGTH **H = .250**

SHOULDER LENGTH **K = .110**

BODY ANGLE (DEG'S/SIDE): **.365**

CASE CAPACITY **CC'S = 5.67**

LOADED LENGTH: **3.20**

BELT DIAMETER **C = .532**

RIM THICKNESS **E = .05**

SHOULDER DIAMETER **G = .487**

LENGTH TO SHOULDER **J = 2.24**

SHOULDER ANGLE (DEG'S/SIDE): **42.13**

PRIMER: **L/R Mag**

CASE CAPACITY (GR'S WATER): **87.65**

DIMENSIONAL DRAWING:

-NOT ACTUAL SIZE-
-DO NOT SCALE-

CARTRIDGE: .257 Durham Jet

OTHER NAMES:

DIA: .257

BALLISTEK NO: 257AK

NAI NO: RXB 12434/4.255

DATA SOURCE: Ackley Vol.1 Pg.339

HISTORICAL DATA:

NOTES:

LOADING DATA:

BULLET WT./TYPE	POWDER WT./TYPE	VELOCITY ('/SEC)	SOURCE
100/Spire	48.5/4350	3330	Ackley

CASE PREPARATION: SHELLHOLDER (RCBS): 3

MAKE FROM: .243 Winchester. Taper expand neck to .260" dia. Trim and size F/L in Durham die. Fireform in chamber or in trim die.

PHYSICAL DATA (INCHES):

CASE TYPE: **Rimless Bottleneck**

CASE LENGTH **A = 2.000**

HEAD DIAMETER **B = .470**

RIM DIAMETER **D = .473**

NECK DIAMETER **F = .290**

NECK LENGTH **H = .300**

SHOULDER LENGTH **K = .112**

BODY ANGLE (DEG'S/SIDE): **.206**

CASE CAPACITY **CC'S = 3.41**

LOADED LENGTH: **A/R**

BELT DIAMETER **C = N/A**

RIM THICKNESS **E = .05**

SHOULDER DIAMETER **G = .460**

LENGTH TO SHOULDER **J = 1.588**

SHOULDER ANGLE (DEG'S/SIDE): **37.19**

PRIMER: **L/R**

CASE CAPACITY (GR'S WATER): **52.66**

DIMENSIONAL DRAWING:

-NOT ACTUAL SIZE-
-DO NOT SCALE-

CARTRIDGE: .257 ICL Whitetail

OTHER NAMES:

DIA: .257

BALLISTEK NO: 257BE

NAI NO: RXB 12424/4.692

DATA SOURCE: Ackley Vol.1 Pg338

HISTORICAL DATA:

NOTES:

LOADING DATA:

BULLET WT./TYPE	POWDER WT./TYPE	VELOCITY ('/SEC)	SOURCE
117/Spire	48.0/4350	3040	Ackley

CASE PREPARATION: SHELLHOLDER (RCBS): 3

MAKE FROM: .257 Roberts. Fire factory ammo in ICL chamber.

PHYSICAL DATA (INCHES):

CASE TYPE: **Rimless Bottleneck**

CASE LENGTH **A = 2.21**

HEAD DIAMETER **B = .471**

RIM DIAMETER **D = .473**

NECK DIAMETER **F = .290**

NECK LENGTH **H = .330**

SHOULDER LENGTH **K = .080**

BODY ANGLE (DEG'S/SIDE): **.161**

CASE CAPACITY **CC'S = 3.88**

LOADED LENGTH: **2.77**

BELT DIAMETER **C = N/A**

RIM THICKNESS **E = .05**

SHOULDER DIAMETER **G = .462**

LENGTH TO SHOULDER **J = 1.80**

SHOULDER ANGLE (DEG'S/SIDE): **47.07**

PRIMER: **L/R**

CASE CAPACITY (GR'S WATER): **59.96**

DIMENSIONAL DRAWING:

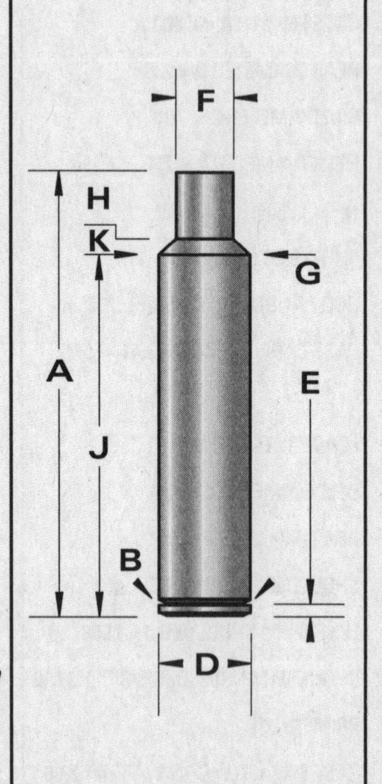

-NOT ACTUAL SIZE-
-DO NOT SCALE-

CARTRIDGE: .257 JDJ

OTHER NAMES:

DIA: .257

BALLISTEK NO:

NAI NO:

DATA SOURCE:

HISTORICAL DATA: Design to give added range and power to the Thompson/Center Contender pistol.

NOTES:

LOADING DATA:

BULLET WT./TYPE	POWDER WT./TYPE	VELOCITY ('/SEC)	SOURCE
75/HP	30/H-322	2310	SSK
75/HP	37/W748	2645	SSK
100/SP	34/W748	1405	SSK
117/SP	35/IMR 4350	2195	SSK

CASE PREPARATION: SHELL HOLDER (RCBS): 11

MAKE FROM: 225 Winchester

PHYSICAL DATA (INCHES):

CASE TYPE: **Rimmed Bottleneck**

CASE LENGTH **A = 1.905**

HEAD DIAMETER **B = .421**

RIM DIAMETER **D = .473**

NECK DIAMETER **F = .288**

NECK LENGTH **H = .190**

SHOULDER LENGTH **K = .065**

BODY ANGLE (DEG'S/SIDE): **40°**

CASE CAPACITY **CC'S**

LOADED LENGTH: **2.81**

BELT DIAMETER **C = N/A**

RIM THICKNESS **E = .049**

SHOULDER DIAMETER **G = .415**

LENGTH TO SHOULDER **J = .1.64**

SHOULDER ANGLE (DEG'S/SIDE):

PRIMER: **L/R**

CASE CAPACITY (GR'S WATER):

DIMENSIONAL DRAWING:

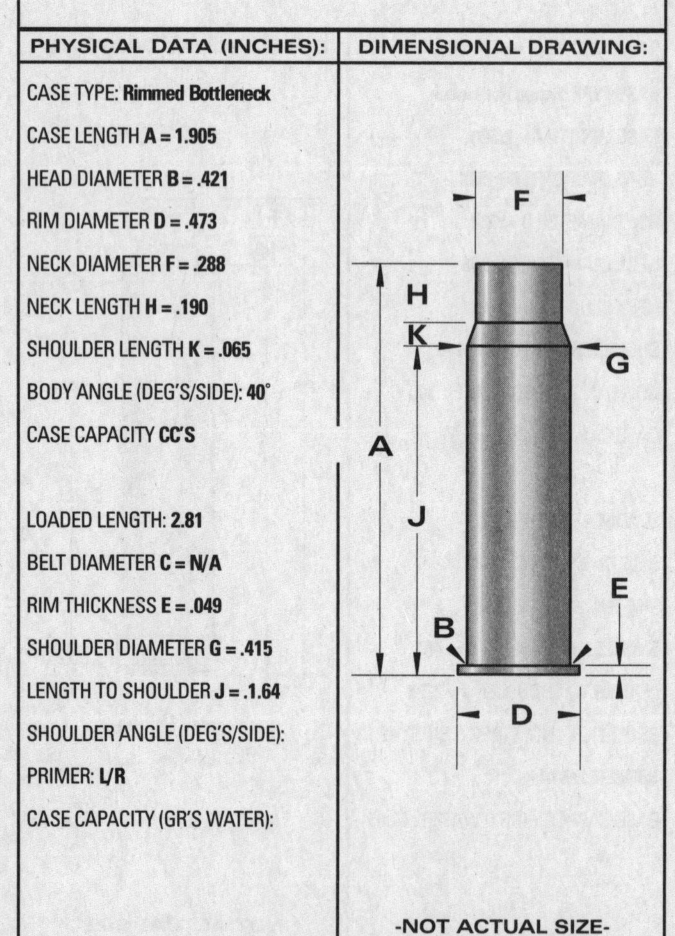

-NOT ACTUAL SIZE-
-DO NOT SCALE-

CARTRIDGE: .257 Roberts

OTHER NAMES:

DIA: .257

BALLISTEK NO: 257J

NAI NO: RXB 22432/4.755

DATA SOURCE: Hornady 3rd Pg132

HISTORICAL DATA: Designed by Ned Roberts.

NOTES: Released by Rem. in 1934.

LOADING DATA:

BULLET WT./TYPE	POWDER WT./TYPE	VELOCITY ('/SEC)	SOURCE
117/RN	36.0/IMR4895	2715	JJD

CASE PREPARATION: SHELLHOLDER (RCBS): 11

MAKE FROM: 7 x 57 Mauser. Anneal case neck. Size in Roberts die with expander removed. Trim to length. I.D. neck ream. Chamfer. F/L size.

PHYSICAL DATA (INCHES):

CASE TYPE: **Rimless Bottleneck**
CASE LENGTH **A = 2.221**
HEAD DIAMETER **B = .467**
RIM DIAMETER **D = .467**
NECK DIAMETER **F = .284**
NECK LENGTH **H = .281**
SHOULDER LENGTH **K = .205**
BODY ANGLE (DEG'S/SIDE): **.821**
CASE CAPACITY **CC'S = 3.39**

LOADED LENGTH: **2.99**
BELT DIAMETER **C = N/A**
RIM THICKNESS **E = .048**
SHOULDER DIAMETER **G = .423**
LENGTH TO SHOULDER **J = 1.735**
SHOULDER ANGLE (DEG'S/SIDE): **18.73**
PRIMER: **L/R**
CASE CAPACITY (GR'S WATER): **52.43**

DIMENSIONAL DRAWING:

-NOT ACTUAL SIZE-
-DO NOT SCALE-

CARTRIDGE: .257 Roberts Improved

OTHER NAMES:

DIA: .257

BALLISTEK NO: 257AJ

NAI NO: RXB 12423/4.734

DATA SOURCE: Hornady 3rd Pg.136

HISTORICAL DATA: This is, most likely, Ackley's version.

NOTES: Several variations exist.

LOADING DATA:

BULLET WT./TYPE	POWDER WT./TYPE	VELOCITY ('/SEC)	SOURCE
100/Spire	50.5/N204	3200	Horn.

CASE PREPARATION: SHELLHOLDER (RCBS): 11

MAKE FROM: Factory .257 Roberts ammo can be fired in the improved chamber or, cases can be made from .270 Win. Anneal neck and set back with .308 dia. F/L size with expander removed. Trim to length and F/L size.

PHYSICAL DATA (INCHES):

CASE TYPE: **Rimless Bottleneck**
CASE LENGTH **A = 2.23**
HEAD DIAMETER **B = .471**
RIM DIAMETER **D = .473**
NECK DIAMETER **F = .290**
NECK LENGTH **H = .323**
SHOULDER LENGTH **K = .178**
BODY ANGLE (DEG'S/SIDE): **.224**
CASE CAPACITY **CC'S = 3.84**

LOADED LENGTH: **2.99**
BELT DIAMETER **C = N/A**
RIM THICKNESS **E = .05**
SHOULDER DIAMETER **G = .459**
LENGTH TO SHOULDER **J = 1.729**
SHOULDER ANGLE (DEG'S/SIDE): **25.39**
PRIMER: **LR**
CASE CAPACITY (GR'S WATER): **59.35**

DIMENSIONAL DRAWING:

-NOT ACTUAL SIZE-
-DO NOT SCALE-

CARTRIDGE: .257 Weatherby

OTHER NAMES:

DIA: .257

BALLISTEK NO: 257K

NAI NO: BEN 12423/4.970

DATA SOURCE: Most current handloading manuals.

HISTORICAL DATA: Created by Roy Weatherby and introduced in 1944.

NOTES: Cartridge was approved by SAAMI on 1/12/94. Dimensions are now standard. Older rifles and loading dies may vary slightly. Use as Weatherby double radius shoulder.

LOADING DATA:

BULLET WT./TYPE	POWDER WT./TYPE	VELOCITY ('/SEC)	SOURCE
100/Hornady	75/RL-25	3500	Hornady
120/Hornady	69.4/RL-25	3200	#6

CASE PREPARATION: SHELLHOLDER (RCBS): 4

MAKE FROM: Factory. Or .300 H&H or .300 WBY. Anneal neck. F/L size. Trim to length. Fireform in chamber to create final shoulder. Check neck wall thickness and turn if needed.

PHYSICAL DATA (INCHES):

CASE TYPE: **Belted Bottleneck**

CASE LENGTH **A = 2.549**

HEAD DIAMETER **B = .5117**

RIM DIAMETER **D = .5315**

NECK DIAMETER **F = .285**

NECK LENGTH **H = .319**

SHOULDER LENGTH **K = .218**

BODY ANGLE (DEG'S/SIDE): **.349**

CASE CAPACITY **CC'S = 5.55**

LOADED LENGTH: **3.170**

BELT DIAMETER **C = .530**

RIM THICKNESS **E = .050**

SHOULDER DIAMETER **G = .492**

LENGTH TO SHOULDER **J = 2.012**

SHOULDER ANGLE (DEG'S/SIDE):

Front radius: **R .1513 + .005**

Rear radius: **R .130 - .005**

PRIMER: **L/R Mag**

CASE CAPACITY (GR'S WATER): **85.63**

DIMENSIONAL DRAWING:

-NOT ACTUAL SIZE-
-DO NOT SCALE-

CARTRIDGE: .26 Epps

OTHER NAMES:

DIA: .264

BALLISTEK NO: 264AC

NAI NO: RXB 22443/5.383

DATA SOURCE: Ackley Vol.1 Pg.363

HISTORICAL DATA: By E. Epps

NOTES:

LOADING DATA:

BULLET WT./TYPE	POWDER WT./TYPE	VELOCITY ('/SEC)	SOURCE
140/RN	55.0/4831	3126	Ackley

CASE PREPARATION: SHELLHOLDER (RCBS): 3

MAKE FROM: .270 Win. F/L size .270 case in .26 Epps die. Trim, chamfer and fireform, in chamber.

PHYSICAL DATA (INCHES):

CASE TYPE: **Rimless Bottleneck**

CASE LENGTH **A = 2.53**

HEAD DIAMETER **B = .470**

RIM DIAMETER **D = .473**

NECK DIAMETER **F = .293**

NECK LENGTH **H = .330**

SHOULDER LENGTH **K = .125**

BODY ANGLE (DEG'S/SIDE): **.534**

CASE CAPACITY **CC'S = 4.42**

LOADED LENGTH: **3.53**

BELT DIAMETER **C = N/A**

RIM THICKNESS **E = .049**

SHOULDER DIAMETER **G = .435**

LENGTH TO SHOULDER **J = 2.075**

SHOULDER ANGLE (DEG'S/SIDE): **29.59**

PRIMER: **L/R**

CASE CAPACITY (GR'S WATER): **68.20**

DIMENSIONAL DRAWING:

-NOT ACTUAL SIZE-
-DO NOT SCALE-

CARTRIDGE: .26 Rimless Nitro Express (BSA)

OTHER NAMES: .26 BSA

DIA: .267

BALLISTEK NO: 267A

NAI NO: BEN 12425/4.658

DATA SOURCE: NAI/Ballistek

HISTORICAL DATA: By BSA in early 1930's.

NOTES:

LOADING DATA:

BULLET WT./TYPE	POWDER WT./TYPE	VELOCITY ('/SEC)	SOURCE
140/Spire	49.0/4350	2700	Barnes

CASE PREPARATION: SHELLHOLDER (RCBS): 4

MAKE FROM: .264 Winchester Mag. Set back .264's shoulder in .26 BSA F/L die with expander removed. Trim & chamfer. Neck ream if necessary. F/L size.

PHYSICAL DATA (INCHES):

CASE TYPE: **Belted Bottleneck**

CASE LENGTH **A = 2.39**

HEAD DIAMETER **B = .513**

RIM DIAMETER **D = .530**

NECK DIAMETER **F = .305**

NECK LENGTH **H = .287**

SHOULDER LENGTH **K = .153**

BODY ANGLE (DEG'S/SIDE): **.868**

CASE CAPACITY **CC'S = 4.55**

LOADED LENGTH: **A/R**

BELT DIAMETER **C = .530**

RIM THICKNESS **E = .049**

SHOULDER DIAMETER **G = .460**

LENGTH TO SHOULDER **J = 1.95**

SHOULDER ANGLE (DEG'S/SIDE): **28.85**

PRIMER: **L/R Mag**

CASE CAPACITY (GR'S WATER): **70.22**

DIMENSIONAL DRAWING:

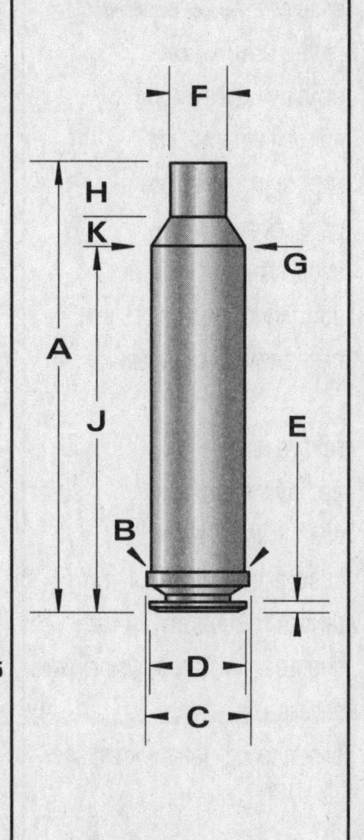

-NOT ACTUAL SIZE-
-DO NOT SCALE-

CARTRIDGE: .260 A.A.R.

OTHER NAMES:

DIA: .264

BALLISTEK NO: 264AD

NAI NO: RXB 22434/4.607

DATA SOURCE: Ackley Vol.1 Pg. 358

HISTORICAL DATA: By Apex Rifle Co.

NOTES: AAR = All Around Rifle

LOADING DATA:

BULLET WT./TYPE	POWDER WT./TYPE	VELOCITY ('/SEC)	SOURCE
140/Spire	48.0/4350	2767	Apex/ Ackley

CASE PREPARATION: SHELLHOLDER (RCBS): 11

MAKE FROM: .257 Roberts. Taper expand Roberts case to .270" dia. F/L size in AAR die, trim and chamfer.

PHYSICAL DATA (INCHES):

CASE TYPE: **Rimless Bottleneck**

CASE LENGTH **A = 2.17**

HEAD DIAMETER **B = .471**

RIM DIAMETER **D = .473**

NECK DIAMETER **F = .290**

NECK LENGTH **H = .320**

SHOULDER LENGTH **K = .115**

BODY ANGLE (DEG'S/SIDE): **.638**

CASE CAPACITY **CC'S = 3.77**

LOADED LENGTH: **3.00**

BELT DIAMETER **C = N/A**

RIM THICKNESS **E = .05**

SHOULDER DIAMETER **G = .436**

LENGTH TO SHOULDER **J = 1.77**

SHOULDER ANGLE (DEG'S/SIDE): **32.38**

PRIMER: **L/R**

CASE CAPACITY (GR'S WATER): **58.31**

DIMENSIONAL DRAWING:

-NOT ACTUAL SIZE-
-DO NOT SCALE-

CARTRIDGE: .260 Remington

OTHER NAMES:	DIA: .264
	BALLISTEK NO:
	NAI NO:

DATA SOURCE: Most current loading manuals.

HISTORICAL DATA: Created by gun writer Jim Carmichael. Introduced by Remington as a factory load in 1997.

NOTES:

LOADING DATA:

BULLET WT./TYPE	POWDER WT./TYPE	VELOCITY ('/SEC)	SOURCE
140/Hornady SST	19/RL	2,700	Hornady #6

CASE PREPARATION: SHELL HOLDER (RCBS): 3

MAKE FROM:

PHYSICAL DATA (INCHES):

CASE TYPE: **Rimless Bottleneck**
CASE LENGTH **A = 2.035**
HEAD DIAMETER **B = .470**
RIM DIAMETER **D = .473**
NECK DIAMETER **F = .297**
NECK LENGTH **H = .259**
SHOULDER LENGTH **K = .216**
BODY ANGLE (DEG'S/SIDE):
CASE CAPACITY **CC'S =**
LOADED LENGTH: **2.80**
BELT DIAMETER **C =**
RIM THICKNESS **E = .054**
SHOULDER DIAMETER **G = .454**
LENGTH TO SHOULDER **J = 1.560**
SHOULDER ANGLE (DEG'S/SIDE): **20**
PRIMER: **L/R**
CASE CAPACITY (GR'S WATER):

DIMENSIONAL DRAWING:

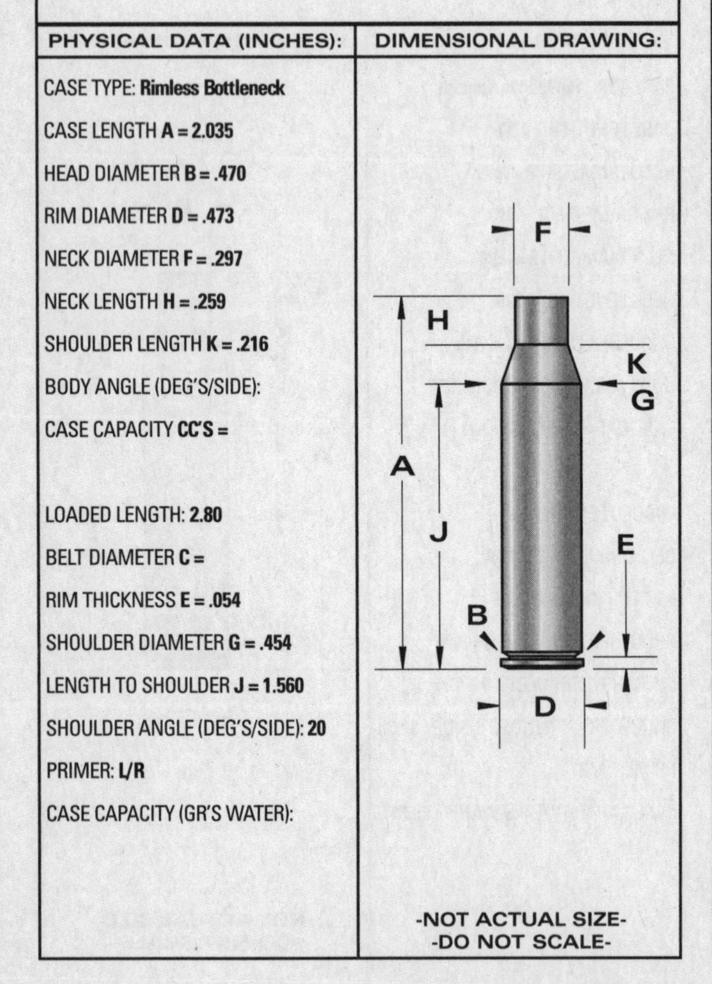

CARTRIDGE: .263 Express

OTHER NAMES: 6,5/308 Winchester	DIA: .264
	BALLISTEK NO: 264M
	NAI NO: RXB 12433/4.340

DATA SOURCE: Ackley Vol.1 Pg.356

HISTORICAL DATA: By Ken Waters

NOTES:

LOADING DATA:

BULLET WT./TYPE	POWDER WT./TYPE	VELOCITY ('/SEC)	SOURCE
140/—	43.0/4350	2850	Waters

CASE PREPARATION: SHELLHOLDER (RCBS): 3

MAKE FROM: .243 Winchester. Taper expand to about .270" dia. Trim & chamfer. F/L size.

PHYSICAL DATA (INCHES):

CASE TYPE: **Rimless Bottleneck**
CASE LENGTH **A = 2.04**
HEAD DIAMETER **B = .470**
RIM DIAMETER **D = .473**
NECK DIAMETER **F = .293**
NECK LENGTH **H = .300**
SHOULDER LENGTH **K = .18**
BODY ANGLE (DEG'S/SIDE): **.337**
CASE CAPACITY **CC'S = 3.51**
LOADED LENGTH: **A/R**
BELT DIAMETER **C = N/A**
RIM THICKNESS **E = .05**
SHOULDER DIAMETER **G = .454**
LENGTH TO SHOULDER **J = 1.56**
SHOULDER ANGLE (DEG'S/SIDE): **24.09**
PRIMER: **L/R**
CASE CAPACITY (GR'S WATER): **54.23**

DIMENSIONAL DRAWING:

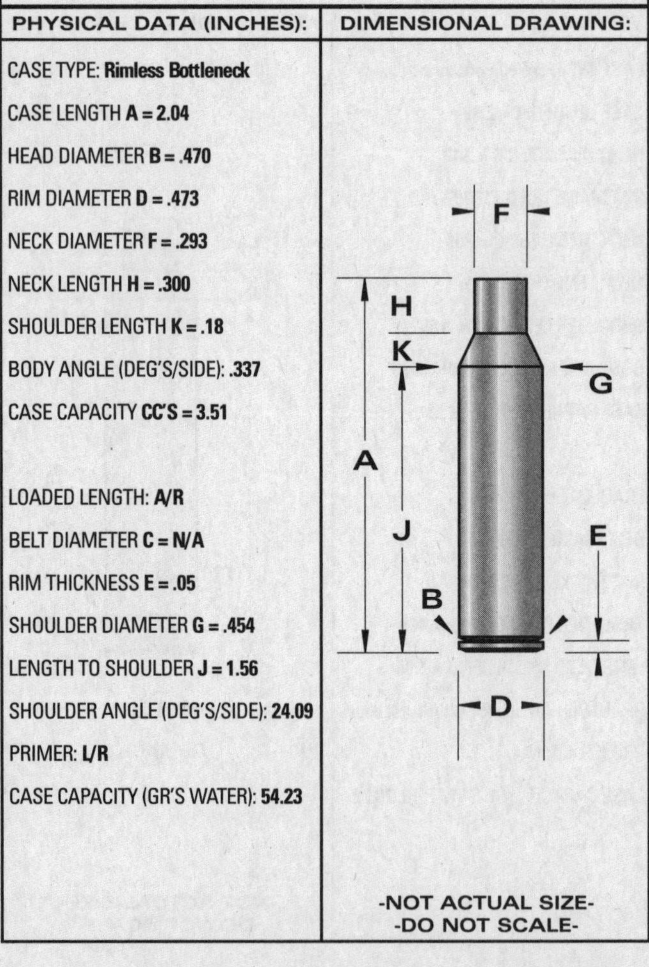

CARTRIDGE: .263 Sabre

OTHER NAMES:

DIA: .264

BALLISTEK NO: 264AF

NAI NO: RXB 22424/5.298

DATA SOURCE: Ackley Vol.1 Pg.365

HISTORICAL DATA: By J. Shannon.

NOTES:

LOADING DATA:

BULLET WT./TYPE	POWDER WT./TYPE	VELOCITY ('/SEC)	SOURCE
120/Spire	58.5/4350	3400	Ackley

CASE PREPARATION: SHELLHOLDER (RCBS): 3

MAKE FROM: .25-06 Rem. Taper expand neck to .27 cal. F/L size in Sabre die. It may be possible to taper expand until the neck will just hold a .264" dia. bullet and fireform in the Sabre chamber.

PHYSICAL DATA (INCHES):

CASE TYPE: **Rimless Bottleneck**

CASE LENGTH **A = 2.54**

HEAD DIAMETER **B = .467**

RIM DIAMETER **D = .47**

NECK DIAMETER **F = .297**

NECK LENGTH **H = .354**

SHOULDER LENGTH **K = .101**

BODY ANGLE (DEG'S/SIDE): **.421**

CASE CAPACITY **CC'S = 4.30**

LOADED LENGTH: **3.43**

BELT DIAMETER **C = N/A**

RIM THICKNESS **E = .050**

SHOULDER DIAMETER **G = .439**

LENGTH TO SHOULDER **J = 2.085**

SHOULDER ANGLE (DEG'S/SIDE): **35**

PRIMER: **L/R**

CASE CAPACITY (GR'S WATER): **66.4**

DIMENSIONAL DRAWING:

-NOT ACTUAL SIZE-
-DO NOT SCALE-

CARTRIDGE: .264 Brooks

OTHER NAMES:
6,5/257 Weatherby
6,5 Brooks

DIA: .264

BALLISTEK NO: 264T

NAI NO: BEN 22423/4.970

DATA SOURCE: Ackley Vol.1 Pg.368

HISTORICAL DATA: By G. Brooks.

NOTES: Weatherby-type shoulders

LOADING DATA:

BULLET WT./TYPE	POWDER WT./TYPE	VELOCITY ('/SEC)	SOURCE
140/RN	63.0/4350	3125	Ackley

CASE PREPARATION: SHELLHOLDER (RCBS): 4

MAKE FROM: .257 Weatherby. Taper expand the Weatherby case to .270" dia. Trim and F/L size in the Brooks die.

PHYSICAL DATA (INCHES):

CASE TYPE: **Belted Bottleneck**

CASE LENGTH **A = 2.54**

HEAD DIAMETER **B = .511**

RIM DIAMETER **D = .530**

NECK DIAMETER **F = .288**

NECK LENGTH **H = .335**

SHOULDER LENGTH **K = .165**

BODY ANGLE (DEG'S/SIDE): **.622**

CASE CAPACITY **CC'S = 5.35**

LOADED LENGTH: **3.22**

BELT DIAMETER **C = .530**

RIM THICKNESS **E = .05**

SHOULDER DIAMETER **G = .471**

LENGTH TO SHOULDER **J = 2.04**

SHOULDER ANGLE (DEG'S/SIDE): **N/A**

PRIMER: **L/R Mag**

CASE CAPACITY (GR'S WATER): **82.71**

DIMENSIONAL DRAWING:

-NOT ACTUAL SIZE-
-DO NOT SCALE-

CARTRIDGE: .264 Connell Short Magnum

OTHER NAMES:

DIA: .264

BALLISTEK NO: 264AW

NAI NO: BEN 23433/4.093

DATA SOURCE: Ackley Vol.1 Pg.367

HISTORICAL DATA: By C. Connell.

NOTES:

LOADING DATA:

BULLET WT./TYPE	POWDER WT./TYPE	VELOCITY ('/SEC)	SOURCE
140/Spire	54.0/4350	3095	Connell/ Ackley

CASE PREPARATION: SHELLHOLDER (RCBS): 4

MAKE FROM: 6,5 Rem. Mag. Size the 6,5 Rem. case in the Connell F/L die. Trim to length, chamfer and fireform in chamber. This case is so similar to the 6.5 Rem. Mag., in the first place, that there is little need to go this route!

PHYSICAL DATA (INCHES):

CASE TYPE: **Belted Bottleneck**

CASE LENGTH **A = 2.100**

HEAD DIAMETER **B = .513**

RIM DIAMETER **D = .532**

NECK DIAMETER **F = .293**

NECK LENGTH **H = .270**

SHOULDER LENGTH **K = .170**

BODY ANGLE (DEG'S/SIDE): **.804**

CASE CAPACITY **CC'S = 4.19**

LOADED LENGTH: **2.86**

BELT DIAMETER **C = .532**

RIM THICKNESS **E = .05**

SHOULDER DIAMETER **G = .472**

LENGTH TO SHOULDER **J = 1.66**

SHOULDER ANGLE (DEG'S/SIDE): **27.76**

PRIMER: **L/R Mag**

CASE CAPACITY (GR'S WATER): **64.73**

DIMENSIONAL DRAWING:

-NOT ACTUAL SIZE-
-DO NOT SCALE-

CARTRIDGE: .264 Durham Magnum

OTHER NAMES:

DIA: .264

BALLISTEK NO: 264AG

NAI NO: BEN 32434/4.425

DATA SOURCE: Ackley Vol.1 Pg.367

HISTORICAL DATA:

NOTES:

LOADING DATA:

BULLET WT./TYPE	POWDER WT./TYPE	VELOCITY ('/SEC)	SOURCE
140/—	56.0/4350	3130	Ackley

CASE PREPARATION: SHELLHOLDER (RCBS): 4

MAKE FROM: .264 Win. Mag. Use a F/L Durham die to set the .264's shoulder back. Square case mouth and fireform.

PHYSICAL DATA (INCHES):

CASE TYPE: **Belted Bottleneck**

CASE LENGTH **A = 2.27**

HEAD DIAMETER **B = .513**

RIM DIAMETER **D = .530**

NECK DIAMETER **F = .294**

NECK LENGTH **H = .325**

SHOULDER LENGTH **K = .091**

BODY ANGLE (DEG'S/SIDE): **1.05**

CASE CAPACITY **CC'S = 4.40**

LOADED LENGTH: **3.06**

BELT DIAMETER **C = .530**

RIM THICKNESS **E = .05**

SHOULDER DIAMETER **G = .452**

LENGTH TO SHOULDER **J = 1.852**

SHOULDER ANGLE (DEG'S/SIDE): **40.96**

PRIMER: **L/R Mag**

CASE CAPACITY (GR'S WATER): **67.86**

DIMENSIONAL DRAWING:

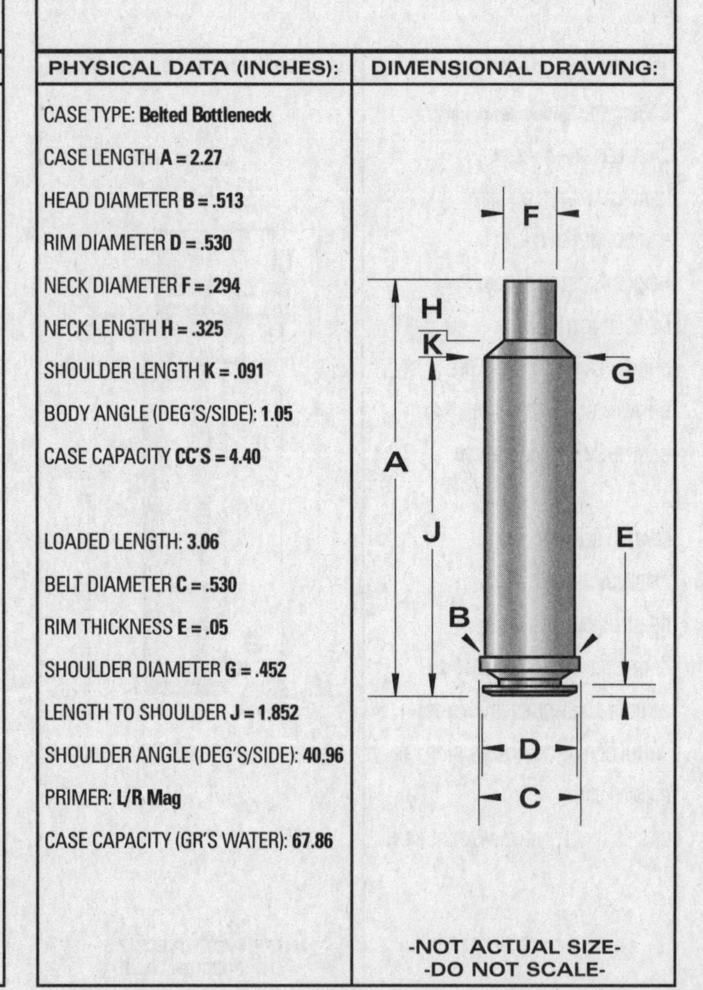

-NOT ACTUAL SIZE-
-DO NOT SCALE-

CARTRIDGE: .264 Hollis

OTHER NAMES:	DIA: .264 BALLISTEK NO: 264BA NAI NO: RXB 22434/4.703

DATA SOURCE: Ackley Vol.2 Pg.156

HISTORICAL DATA: By E. Follis

NOTES: Similar to .260 A.A.R.

LOADING DATA:

BULLET WT./TYPE	POWDER WT./TYPE	VELOCITY ('/SEC)	SOURCE
140/Spire	48.0/4350	2767	Ackley

CASE PREPARATION: SHELLHOLDER (RCBS): 11

MAKE FROM: .257 Roberts. Taper expand case to .270" dia. Trim to length and chamfer. F/L size in Hollis die and fireform. Do not fireform this case (without sizing) in the chamber as the shoulder will not headspace. We have fire-formed this case, in a trim die, with corn meal, with good results.

PHYSICAL DATA (INCHES):

CASE TYPE: **Rimless Bottleneck**
CASE LENGTH **A = 2.215**
HEAD DIAMETER **B = .471**
RIM DIAMETER **D = .473**
NECK DIAMETER **F = .295**
NECK LENGTH **H = .330**
SHOULDER LENGTH **K = .095**
BODY ANGLE (DEG'S/SIDE): **.360**
CASE CAPACITY **CC'S = 3.85**

LOADED LENGTH: **3.00**
BELT DIAMETER **C = N/A**
RIM THICKNESS **E = .049**
SHOULDER DIAMETER **G = .451**
LENGTH TO SHOULDER **J = 1.79**
SHOULDER ANGLE (DEG'S/SIDE): **39.38**
PRIMER: **L/R**
CASE CAPACITY (GR'S WATER): **59.54**

DIMENSIONAL DRAWING:

-NOT ACTUAL SIZE-
-DO NOT SCALE-

CARTRIDGE: .264 Williams

OTHER NAMES:	DIA: .264 BALLISTEK NO: 264AE NAI NO: RXB 22433/5.323

DATA SOURCE: Ackley Vol.1 Pg.366

HISTORICAL DATA: By W. Williams in 1951.

NOTES:

LOADING DATA:

BULLET WT./TYPE	POWDER WT./TYPE	VELOCITY ('/SEC)	SOURCE
140/Spire	52.0/4350	3000	Ackley

CASE PREPARATION: SHELLHOLDER (RCBS): 3

MAKE FROM: .30-06 Spgf. Size '06 case in Williams die and fireform in the chamber. Trim, chamfer, F/L size and reload. Since this shoulder is being "blown-forward", this is an excellent case to fireform in a trim die.

PHYSICAL DATA (INCHES):

CASE TYPE: **Rimless Bottleneck**
CASE LENGTH **A = 2.502**
HEAD DIAMETER **B = .470**
RIM DIAMETER **D = .473**
NECK DIAMETER **F = .290**
NECK LENGTH **H = .290**
SHOULDER LENGTH **K = .142**
BODY ANGLE (DEG'S/SIDE): **.413**
CASE CAPACITY **CC'S = 4.57**

LOADED LENGTH: **A/R**
BELT DIAMETER **C = N/A**
RIM THICKNESS **E = .05**
SHOULDER DIAMETER **G = .443**
LENGTH TO SHOULDER **J = 2.07**
SHOULDER ANGLE (DEG'S/SIDE): **28.31**
PRIMER: **L/R**
CASE CAPACITY (GR'S WATER): **70.53**

DIMENSIONAL DRAWING:

-NOT ACTUAL SIZE-
-DO NOT SCALE-

CARTRIDGE: .264 Winchester Magnum

OTHER NAMES:

DIA: .264

BALLISTEK NO: 264B

NAI NO: BEN 12443/4.873

DATA SOURCE: Hornady Manual 3rd Pg.165

HISTORICAL DATA: By Win. in 1958.

NOTES: Overbore for .264" dia.

LOADING DATA:

BULLET WT./TYPE	POWDER WT./TYPE	VELOCITY ('/SEC)	SOURCE
140/Spire	60.0/IMR4831	3050	JJD

CASE PREPARATION: SHELLHOLDER (RCBS): 4

MAKE FROM: .300 H&H Mag. Run Holland's case into .264 die. Trim to length. Case will fireform in chamber.

PHYSICAL DATA (INCHES):

CASE TYPE: **Belted Bottleneck**

CASE LENGTH **A = 2.500**

HEAD DIAMETER **B = .513**

RIM DIAMETER **D = .532**

NECK DIAMETER **F = .298**

NECK LENGTH **H = .254**

SHOULDER LENGTH **K = .206**

BODY ANGLE (DEG'S/SIDE): **.342**

CASE CAPACITY **CC'S = 5.28**

LOADED LENGTH: **3.35**

BELT DIAMETER **C = .532**

RIM THICKNESS **E = .05**

SHOULDER DIAMETER **G = .491**

LENGTH TO SHOULDER **J = 2.04**

SHOULDER ANGLE (DEG'S/SIDE): **25.10**

PRIMER: **L/R Mag**

CASE CAPACITY (GR'S WATER): **81.51**

DIMENSIONAL DRAWING:

F H K G A J E B D C

-NOT ACTUAL SIZE-
-DO NOT SCALE-

CARTRIDGE: .264 x 61 Shannon

OTHER NAMES:
6.5/7 x 61 Sharpe & Hart

DIA: .264

BALLISTEK NO: 264BB

NAI NO: BEN 22424/4.873

DATA SOURCE: Ackley Vol.2 Pg.163

HISTORICAL DATA: By J. Shannon.

NOTES:

LOADING DATA:

BULLET WT./TYPE	POWDER WT./TYPE	VELOCITY ('/SEC)	SOURCE
140/Spire	63.9/MRP	–	JJD

CASE PREPARATION: SHELLHOLDER (RCBS): 4

MAKE FROM: .300 Win. Mag. Form the Win. case in a 7x61 S&H form die. F/L size in 7x61 S&H die. Trim & chamfer. Anneal neck and F/L size in Shannon die. It may be easier to just obtain a .264x61 form set and F/L die.

PHYSICAL DATA (INCHES):

CASE TYPE: **Belted Bottleneck**

CASE LENGTH **A = 2.47**

HEAD DIAMETER **B = .511**

RIM DIAMETER **D = .532**

NECK DIAMETER **F = .299**

NECK LENGTH **H = .369**

SHOULDER LENGTH **K = .118**

BODY ANGLE (DEG'S/SIDE): **.51**

CASE CAPACITY **CC'S = 4.90**

LOADED LENGTH: **3.325**

BELT DIAMETER **C = .532**

RIM THICKNESS **E = .05**

SHOULDER DIAMETER **G = .476**

LENGTH TO SHOULDER **J = 1.987**

SHOULDER ANGLE (DEG'S/SIDE): **35**

PRIMER: **L/R Mag**

CASE CAPACITY (GR'S WATER): **75.7**

DIMENSIONAL DRAWING:

F H K G A J E B D C

-NOT ACTUAL SIZE-
-DO NOT SCALE-

CARTRIDGE: .270 Ackley Magnum

OTHER NAMES:

DIA: .277

BALLISTEK NO: 277G

NAI NO: BEN 22423/4.804

DATA SOURCE: Ackley Vol.1 Pg.382

HISTORICAL DATA:

NOTES:

LOADING DATA:

BULLET WT./TYPE	POWDER WT./TYPE	VELOCITY ('/SEC)	SOURCE
150/Spire	52.0/4895	2821	Ackley

CASE PREPARATION: SHELLHOLDER (RCBS): 4

MAKE FROM: 7mm Rem. Mag. F/L size Rem. case. Square case mouth and fireform in Ackley chamber or trim die. F/L size and reload.

PHYSICAL DATA (INCHES):

CASE TYPE: **Belted Bottleneck**

CASE LENGTH **A = 2.460**

HEAD DIAMETER **B = .512**

RIM DIAMETER **D = .532**

NECK DIAMETER **F = .316**

NECK LENGTH **H = .400**

SHOULDER LENGTH **K = .140**

BODY ANGLE (DEG'S/SIDE): **.466**

CASE CAPACITY **CC'S = 4.80**

LOADED LENGTH: **3.24**

BELT DIAMETER **C = .532**

RIM THICKNESS **E = .05**

SHOULDER DIAMETER **G = .484**

LENGTH TO SHOULDER **J = 1.92**

SHOULDER ANGLE (DEG'S/SIDE): **30.96**

PRIMER: **L/R Mag**

CASE CAPACITY (GR'S WATER): **74.07**

DIMENSIONAL DRAWING:

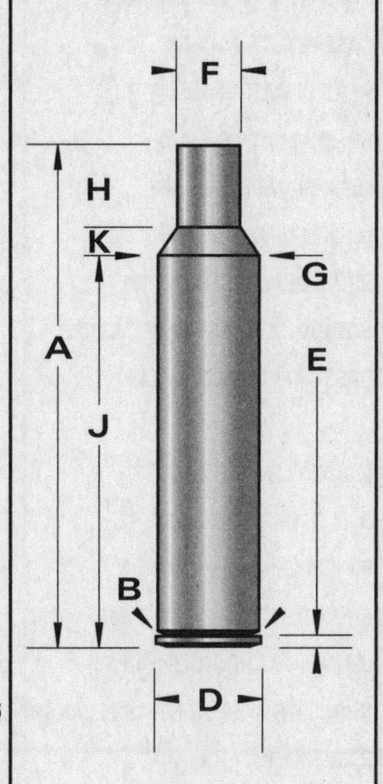

-NOT ACTUAL SIZE-
-DO NOT SCALE-

CARTRIDGE: .270 Ingram

OTHER NAMES:
.270/223 Remington

DIA: .277

BALLISTEK NO: 277W

NAI NO: RXB 22242/4.680

DATA SOURCE: Handloader #86 Pg.39

HISTORICAL DATA: By P. Briggs as Contender ctg. in 1979.

NOTES:

LOADING DATA:

BULLET WT./TYPE	POWDER WT./TYPE	VELOCITY ('/SEC)	SOURCE
110/SHP	22.0/RL-7	2104	Briggs

CASE PREPARATION: SHELLHOLDER (RCBS): 10

MAKE FROM: .223 Remington. Taper expand the .223 case to .280" dia. and F/L size in the Ingram die. Trim & chamfer. Case will fireform in the chamber.

PHYSICAL DATA (INCHES):

CASE TYPE: **Rimless Bottleneck**

CASE LENGTH **A = 1.76**

HEAD DIAMETER **B = .376**

RIM DIAMETER **D = .378**

NECK DIAMETER **F = .304**

NECK LENGTH **H = .210**

SHOULDER LENGTH **K = .095**

BODY ANGLE (DEG'S/SIDE): **.456**

CASE CAPACITY **CC'S = 1.96**

LOADED LENGTH: **A/R**

BELT DIAMETER **C = N/A**

RIM THICKNESS **E = .045**

SHOULDER DIAMETER **G = .356**

LENGTH TO SHOULDER **J = 1.455**

SHOULDER ANGLE (DEG'S/SIDE): **30.30**

PRIMER: **S/R**

CASE CAPACITY (GR'S WATER): **30.37**

DIMENSIONAL DRAWING:

F H K G A E J B D

-NOT ACTUAL SIZE-
-DO NOT SCALE-

CARTRIDGE: .270 JDJ

OTHER NAMES:

DIA: .277

BALLISTEK NO: 277Y

NAI NO: RMB
12344/4.520

DATA SOURCE: SSK Industries

HISTORICAL DATA: By J.D. Jones

NOTES:

LOADING DATA:

BULLET WT./TYPE	POWDER WT./TYPE	VELOCITY ('/SEC)	SOURCE
130/Nosler	36.0/3031	2410	SSK

CASE PREPARATION: SHELLHOLDER (RCBS): 11

MAKE FROM: .225 Winchester. Taper expand neck to hold .277" dia. bullets and fireform with a 10% reduced load.

PHYSICAL DATA (INCHES):

CASE TYPE: **Rimmed Bottleneck**

CASE LENGTH **A = 1.894**

HEAD DIAMETER **B = .419**

RIM DIAMETER **D = .469**

NECK DIAMETER **F = .304**

NECK LENGTH **H = .185**

SHOULDER LENGTH **K = .069**

BODY ANGLE (DEG'S/SIDE): **.199**

CASE CAPACITY **CC'S = 2.78**

LOADED LENGTH: **2.745**

BELT DIAMETER **C = N/A**

RIM THICKNESS **E = .049**

SHOULDER DIAMETER **G = .409**

LENGTH TO SHOULDER **J = 1.64**

SHOULDER ANGLE (DEG'S/SIDE): **37.26**

PRIMER: **L/R**

CASE CAPACITY (GR'S WATER): **42.97**

DIMENSIONAL DRAWING:

-NOT ACTUAL SIZE-
-DO NOT SCALE-

CARTRIDGE: .270 Savage

OTHER NAMES:
.270 Evans
.270 Titus

DIA: .277

BALLISTEK NO: 277H

NAI NO: RXB
23344/3.970

DATA SOURCE: Ackley Vol.1 Pg.376

HISTORICAL DATA:

NOTES: Several versions exist.

LOADING DATA:

BULLET WT./TYPE	POWDER WT./TYPE	VELOCITY ('/SEC)	SOURCE
130/Spire	39.0/4064	2763	Ackley

CASE PREPARATION: SHELLHOLDER (RCBS): 3

MAKE FROM: .270 Win. F/L size the .270 case in the Savage die with the expander removed. Trim to length and chamfer. I.D. neck ream. F/L size again. I have not had any luck necking the .300 Savage to this dia., not sufficient support for neck.

PHYSICAL DATA (INCHES):

CASE TYPE: **Rimless Bottleneck**

CASE LENGTH **A = 1.87**

HEAD DIAMETER **B = .471**

RIM DIAMETER **D = .473**

NECK DIAMETER **F = .308**

NECK LENGTH **H = .245**

SHOULDER LENGTH **K = .094**

BODY ANGLE (DEG'S/SIDE): **.538**

CASE CAPACITY **CC'S = 3.19**

LOADED LENGTH: **2.52**

BELT DIAMETER **C = N/A**

RIM THICKNESS **E = .049**

SHOULDER DIAMETER **G = .446**

LENGTH TO SHOULDER **J = 1.53**

SHOULDER ANGLE (DEG'S/SIDE): **35.99**

PRIMER: **L/R**

CASE CAPACITY (GR'S WATER): **49.35**

DIMENSIONAL DRAWING:

-NOT ACTUAL SIZE-
-DO NOT SCALE-

CARTRIDGE: .270 Weatherby Magnum

OTHER NAMES:

DIA: .277

BALLISTEK NO: 277B

NAI NO: BEN 12423/4.980

DATA SOURCE: Hornady 3rd Pg.172

HISTORICAL DATA: Another Roy Weatherby development.

NOTES:

LOADING DATA:

BULLET WT./TYPE	POWDER WT./TYPE	VELOCITY ('/SEC)	SOURCE
150/Spire	59.5/IMR4350	2900	Horn.

CASE PREPARATION: SHELLHOLDER (RCBS): 4

MAKE FROM: .300 H&H Mag. Anneal case neck. F/L size in Weatherby die. Case will fireform upon ignition.

PHYSICAL DATA (INCHES):

CASE TYPE: **Belted Bottleneck**

CASE LENGTH **A = 2.545**

HEAD DIAMETER **B = .511**

RIM DIAMETER **D = .530**

NECK DIAMETER **F = .301**

NECK LENGTH **H = .360**

SHOULDER LENGTH **K = .223**

BODY ANGLE (DEG'S/SIDE): **.341**

CASE CAPACITY **CC'S = 5.53**

LOADED LENGTH: **3.28**

BELT DIAMETER **C = .530**

RIM THICKNESS **E = .05**

SHOULDER DIAMETER **G = .490**

LENGTH TO SHOULDER **J = 1.962**

SHOULDER ANGLE (DEG'S/SIDE): **N/A**

PRIMER: **L/R Mag**

CASE CAPACITY (GR'S WATER): **85.34**

DIMENSIONAL DRAWING:

F H K G A J E B D

-NOT ACTUAL SIZE-
-DO NOT SCALE-

CARTRIDGE: .270 Winchester

OTHER NAMES:

DIA: .277

BALLISTEK NO: 277A

NAI NO: RXB 22322/5.404

DATA SOURCE: Hornady 3rd Pg.168

HISTORICAL DATA: By Win. in 1925

NOTES:

LOADING DATA:

BULLET WT./TYPE	POWDER WT./TYPE	VELOCITY ('/SEC)	SOURCE
150/Spire	52.0/IMR4350	2775	JJD

CASE PREPARATION: SHELLHOLDER (RCBS): 3

MAKE FROM: Factory or .30-06 Spgf. Anneal neck. Full length size in .270 die. Square case mouth and chamfer. Case will be short but, otherwise, fine.

PHYSICAL DATA (INCHES):

CASE TYPE: **Rimless Bottleneck**

CASE LENGTH **A = 2.54**

HEAD DIAMETER **B = .470**

RIM DIAMETER **D = .473**

NECK DIAMETER **F = .308**

NECK LENGTH **H = .383**

SHOULDER LENGTH **K = .208**

BODY ANGLE (DEG'S/SIDE): **.475**

CASE CAPACITY **CC'S = 4.36**

LOADED LENGTH: **3.35**

BELT DIAMETER **C = N/A**

RIM THICKNESS **E = .049**

SHOULDER DIAMETER **G = .441**

LENGTH TO SHOULDER **J = 1.949**

SHOULDER ANGLE (DEG'S/SIDE): **17.73**

PRIMER: **L/R**

CASE CAPACITY (GR'S WATER): **67.39**

DIMENSIONAL DRAWING:

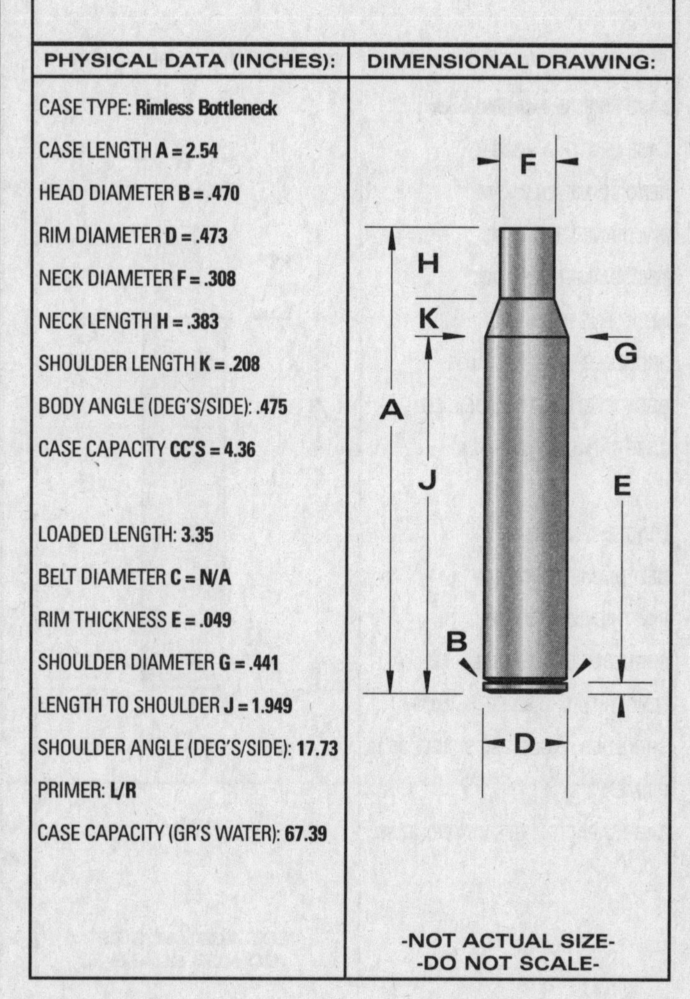

-NOT ACTUAL SIZE-
-DO NOT SCALE-

CARTRIDGE: .270 Winchester Improved (Ackley)

OTHER NAMES:

DIA: .277

BALLISTEK NO: 277T

NAI NO: RXB 22324/5.383

DATA SOURCE: Ackley Vol.1 Pg.382

HISTORICAL DATA:

NOTES:

LOADING DATA:

BULLET WT./TYPE	POWDER WT./TYPE	VELOCITY ('/SEC)	SOURCE
150/Spire	54.0/4350	2900	JJD

CASE PREPARATION: SHELLHOLDER (RCBS): 3

MAKE FROM: .270 Winchester. Fire factory ammo in the "improved" chamber or, with corn meal, in the "Imp" trim die. Trim and F/L size.

PHYSICAL DATA (INCHES):

CASE TYPE: **Rimless Bottleneck**

CASE LENGTH **A = 2.53**

HEAD DIAMETER **B = .47**

RIM DIAMETER **D = .473**

NECK DIAMETER **F = .308**

NECK LENGTH **H = .400**

SHOULDER LENGTH **K = .095**

BODY ANGLE (DEG'S/SIDE): **.437**

CASE CAPACITY **CC'S = 4.40**

LOADED LENGTH: **A/R**

BELT DIAMETER **C = N/A**

RIM THICKNESS **E = .05**

SHOULDER DIAMETER **G = .442**

LENGTH TO SHOULDER **J = 2.035**

SHOULDER ANGLE (DEG'S/SIDE): **35.19**

PRIMER: **L/R**

CASE CAPACITY (GR'S WATER): **67.98**

DIMENSIONAL DRAWING:

-NOT ACTUAL SIZE-
-DO NOT SCALE-

CARTRIDGE: .270 Winchester Short Magnum

OTHER NAMES:

DIA: .277

BALLISTEK NO:

NAI NO:

DATA SOURCE: Most current loading manuals.

HISTORICAL DATA: Developed by Winchester in 2002 to extract the most performance possible from a .270-caliber cartridge chambered in current short rifle designs.

NOTES:

LOADING DATA:

BULLET WT./TYPE	POWDER WT./TYPE	VELOCITY ('/SEC)	SOURCE
130/Nosler	63.0/IMR-4350	3,294	Nosler #5
150/Nosler	60.0/H4831sc	2,966	Nosler #5

CASE PREPARATION: SHELL HOLDER (RCBS): 43

MAKE FROM: Factory. Can be made from .404 Jeffery. Shorten and die form to fit chamber. Fireform body to get the proper diameter, shoulder angle and head space. Then turn the rim to correct dimension. Ream neck if necessary. Square mouth and chamfer.

PHYSICAL DATA (INCHES):

CASE TYPE: **Rebated Bottleneck**

CASE LENGTH **A = 2.100**

HEAD DIAMETER **B = .5550**

RIM DIAMETER **D = .535**

NECK DIAMETER **F = .3140**

NECK LENGTH **H = .2765**

SHOULDER LENGTH **K = .1595**

BODY ANGLE (DEG'S/SIDE):

CASE CAPACITY **CC'S =**

LOADED LENGTH: **2.860**

BELT DIAMETER **C =**

RIM THICKNESS **E = .54**

SHOULDER DIAMETER **G = .5381**

LENGTH TO SHOULDER **J = 1.664**

SHOULDER ANGLE (DEG'S/SIDE): **35**

PRIMER: **L/R or L/R Mag**

CASE CAPACITY (GR'S WATER): **77.9**

DIMENSIONAL DRAWING:

-NOT ACTUAL SIZE-
-DO NOT SCALE-

CARTRIDGE: .270/257 Roberts

OTHER NAMES:

DIA: .277

BALLISTEK NO: 277V

NAI NO: RXB 22332/4.734

DATA SOURCE: NAI/Ballistek

HISTORICAL DATA:

NOTES: Several versions exist.

LOADING DATA:

BULLET WT./TYPE	POWDER WT./TYPE	VELOCITY ('/SEC)	SOURCE
150/Spire	44.0/4350	—	JJD

CASE PREPARATION: SHELLHOLDER (RCBS): 11

MAKE FROM: .257 Roberts. Taper expand the Roberts case to .280" dia. F/L size, trim and chamfer. Fireform in chamber.

PHYSICAL DATA (INCHES):

CASE TYPE: **Rimless Bottleneck**

CASE LENGTH A = **2.23**

HEAD DIAMETER B = **.471**

RIM DIAMETER D = **.473**

NECK DIAMETER F = **.308**

NECK LENGTH H = **.315**

SHOULDER LENGTH K = **.185**

BODY ANGLE (DEG'S/SIDE): **.786**

CASE CAPACITY CC'S = **3.69**

LOADED LENGTH: **A/R**

BELT DIAMETER C = **N/A**

RIM THICKNESS E = **.05**

SHOULDER DIAMETER G = **.429**

LENGTH TO SHOULDER J = **1.73**

SHOULDER ANGLE (DEG'S/SIDE): **18.11**

PRIMER: **L/R**

CASE CAPACITY (GR'S WATER): **57.05**

DIMENSIONAL DRAWING:

F H K G A E J B D

-NOT ACTUAL SIZE-
-DO NOT SCALE-

CARTRIDGE: .270/257 Roberts Improved

OTHER NAMES:

DIA: .277

BALLISTEK NO: 277Q

NAI NO: RXB 12333/4.734

DATA SOURCE: Ackley Vol.1 Pg378

HISTORICAL DATA:

NOTES:

LOADING DATA:

BULLET WT./TYPE	POWDER WT./TYPE	VELOCITY ('/SEC)	SOURCE
150/Spire	39.0/4064	2541	Ackley

CASE PREPARATION: SHELLHOLDER (RCBS): 11

MAKE FROM: 7 x 57mm Mauser. Size Mauser case F/L in "improved" die with the expander removed. Trim and chamfer. Fireform in chamber.

PHYSICAL DATA (INCHES):

CASE TYPE: **Rimless Bottleneck**

CASE LENGTH A = **2.230**

HEAD DIAMETER B = **.471**

RIM DIAMETER D = **.473**

NECK DIAMETER F = **.309**

NECK LENGTH H = **.318**

SHOULDER LENGTH K = **.182**

BODY ANGLE (DEG'S/SIDE): **.206**

CASE CAPACITY CC'S = **3.94**

LOADED LENGTH: **A/R**

BELT DIAMETER C = **N/A**

RIM THICKNESS E = **.05**

SHOULDER DIAMETER G = **.460**

LENGTH TO SHOULDER J = **1.73**

SHOULDER ANGLE (DEG'S/SIDE): **22.53**

PRIMER: **L/R**

CASE CAPACITY (GR'S WATER): **60.83**

DIMENSIONAL DRAWING:

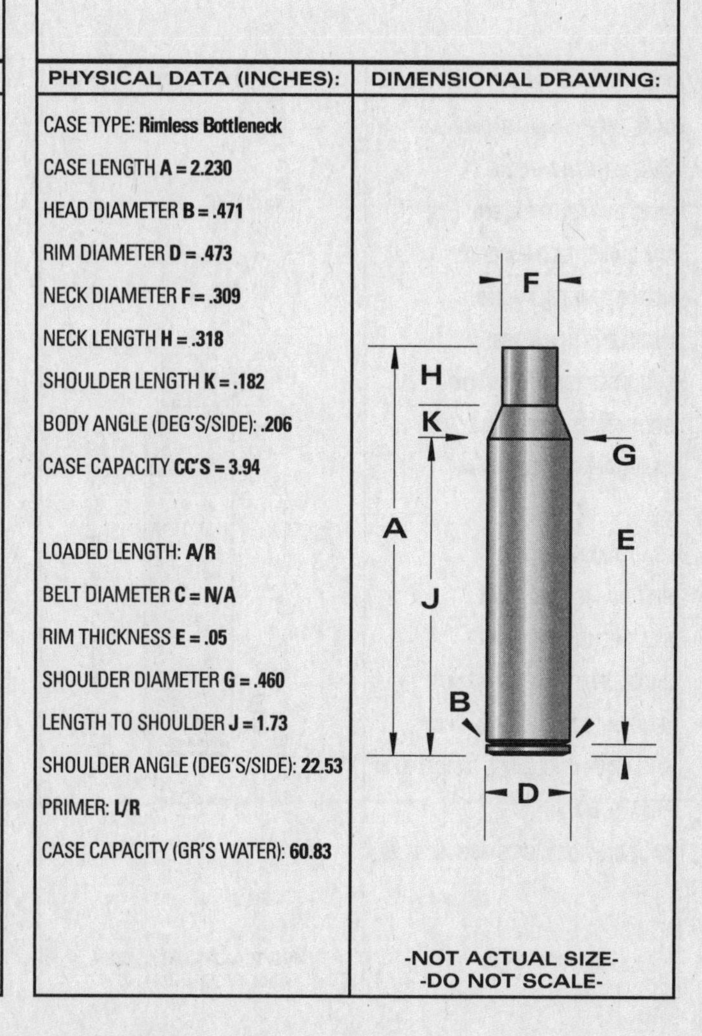

-NOT ACTUAL SIZE-
-DO NOT SCALE-

CARTRIDGE: .270/284 Winchester

OTHER NAMES:	DIA: .277
	BALLISTEK NO: 277C
	NAI NO: RBB 22434/4.400

DATA SOURCE: Ackley Vol.2 Pg168

HISTORICAL DATA:

NOTES:

LOADING DATA:

BULLET WT./TYPE	POWDER WT./TYPE	VELOCITY ('/SEC)	SOURCE
150/RN	53.0/4350	2900	Ackley

CASE PREPARATION: SHELLHOLDER (RCBS): 3

MAKE FROM: .284 Winchester. Size .284 case in F/L die. Square case mouth and chamfer.

PHYSICAL DATA (INCHES):

CASE TYPE: **Rebated Bottleneck**

CASE LENGTH **A = 2.200**

HEAD DIAMETER **B = .500**

RIM DIAMETER **D = .473**

NECK DIAMETER **F = .299**

NECK LENGTH **H = .285**

SHOULDER LENGTH **K = .115**

BODY ANGLE (DEG'S/SIDE): **.447**

CASE CAPACITY **CC'S = 4.61**

LOADED LENGTH: **2.89**

BELT DIAMETER **C = N/A**

RIM THICKNESS **E = .05**

SHOULDER DIAMETER **G = .475**

LENGTH TO SHOULDER **J = 1.80**

SHOULDER ANGLE (DEG'S/SIDE): **37.42**

PRIMER: **L/R**

CASE CAPACITY (GR'S WATER): **71.26**

DIMENSIONAL DRAWING:

-NOT ACTUAL SIZE-
-DO NOT SCALE-

CARTRIDGE: .270/300 Weatherby Magnum

OTHER NAMES:	DIA: .277
	BALLISTEK NO: 277R
	NAI NO: BEN 12433/5.479

DATA SOURCE: Ackley Vol.2 Pg170

HISTORICAL DATA:

NOTES: Overbore!

LOADING DATA:

BULLET WT./TYPE	POWDER WT./TYPE	VELOCITY ('/SEC)	SOURCE
150/Spire	79.0/IMR4831	–	JJD

CASE PREPARATION: SHELLHOLDER (RCBS): 4

MAKE FROM: .300 Weatherby. F/L size, trim to length & chamfer.

PHYSICAL DATA (INCHES):

CASE TYPE: **Belted Bottleneck**

CASE LENGTH **A = 2.800**

HEAD DIAMETER **B = .511**

RIM DIAMETER **D = .530**

NECK DIAMETER **F = .304**

NECK LENGTH **H = .315**

SHOULDER LENGTH **K = .165**

BODY ANGLE (DEG'S/SIDE): **.283**

CASE CAPACITY **CC'S = 6.26**

LOADED LENGTH: **A/R**

BELT DIAMETER **C = .530**

RIM THICKNESS **E = .05**

SHOULDER DIAMETER **G = .490**

LENGTH TO SHOULDER **J = 2.32**

SHOULDER ANGLE (DEG'S/SIDE): **N/A**

PRIMER: **L/R Mag**

CASE CAPACITY (GR'S WATER): **96.62**

DIMENSIONAL DRAWING:

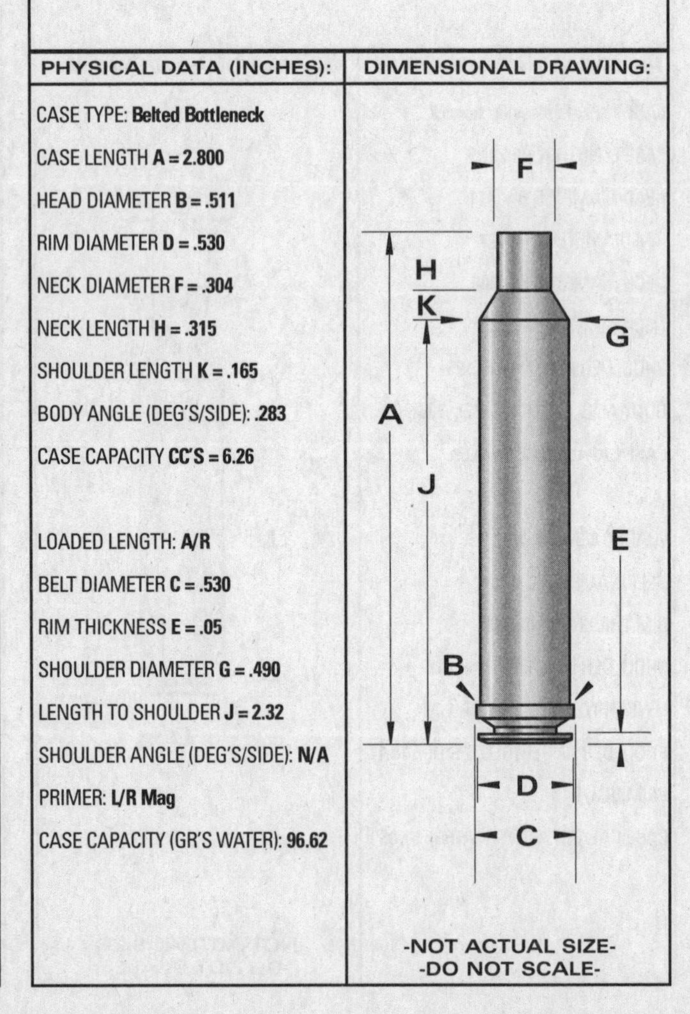

-NOT ACTUAL SIZE-
-DO NOT SCALE-

CARTRIDGE: .270/308 Winchester

OTHER NAMES:

DIA: .277

BALLISTEK NO: 277N

NAI NO: RXB 12343/4.287

DATA SOURCE: Rifle #59 Pg30

HISTORICAL DATA:

NOTES:

LOADING DATA:

BULLET WT./TYPE	POWDER WT./TYPE	VELOCITY ('/SEC)	SOURCE
130/SBT	46.0/4350	3041	Huber

CASE PREPARATION: SHELLHOLDER (RCBS): 3

MAKE FROM: .308 Winchester. F/L size the .308 case in the .270/308 die. Square case mouth and chamfer.

PHYSICAL DATA (INCHES):

CASE TYPE: **Rimless Bottleneck**

CASE LENGTH **A = 2.015**

HEAD DIAMETER **B = .470**

RIM DIAMETER **D = .473**

NECK DIAMETER **F = .308**

NECK LENGTH **H = .257**

SHOULDER LENGTH **K = .198**

BODY ANGLE (DEG'S/SIDE): **.337**

CASE CAPACITY **CC'S = 3.49**

LOADED LENGTH: **2.83**

BELT DIAMETER **C = N/A**

RIM THICKNESS **E = .05**

SHOULDER DIAMETER **G = .454**

LENGTH TO SHOULDER **J = 1.56**

SHOULDER ANGLE (DEG'S/SIDE): **20.24**

PRIMER: **L/R**

CASE CAPACITY (GR'S WATER): **53.90**

DIMENSIONAL DRAWING:

-NOT ACTUAL SIZE-
-DO NOT SCALE-

CARTRIDGE: .270/308 Winchester Improved

OTHER NAMES:

DIA: .277

BALLISTEK NO: 277P

NAI NO: RXB 12333/4.362

DATA SOURCE: Ackley Vol.2 Pg166

HISTORICAL DATA:

NOTES:

LOADING DATA:

BULLET WT./TYPE	POWDER WT./TYPE	VELOCITY ('/SEC)	SOURCE
150/—	46.0/4350	2817	Sorensen

CASE PREPARATION: SHELLHOLDER (RCBS): 3

MAKE FROM: .308 Winchester. Size .308 case in either .270/308 or .270/308 "Imp" die. Square case mouth and fireform.

PHYSICAL DATA (INCHES):

CASE TYPE: **Rimless Bottleneck**

CASE LENGTH **A = 2.01**

HEAD DIAMETER **B = .470**

RIM DIAMETER **D = .473**

NECK DIAMETER **F = .309**

NECK LENGTH **H = .300**

SHOULDER LENGTH **K = .109**

BODY ANGLE (DEG'S/SIDE): **.204**

CASE CAPACITY **CC'S = 3.58**

LOADED LENGTH: **A/R**

BELT DIAMETER **C = N/A**

RIM THICKNESS **E = .051**

SHOULDER DIAMETER **G = .460**

LENGTH TO SHOULDER **J = 1.60**

SHOULDER ANGLE (DEG'S/SIDE): **34.46**

PRIMER: **L/R**

CASE CAPACITY (GR'S WATER): **55.24**

DIMENSIONAL DRAWING:

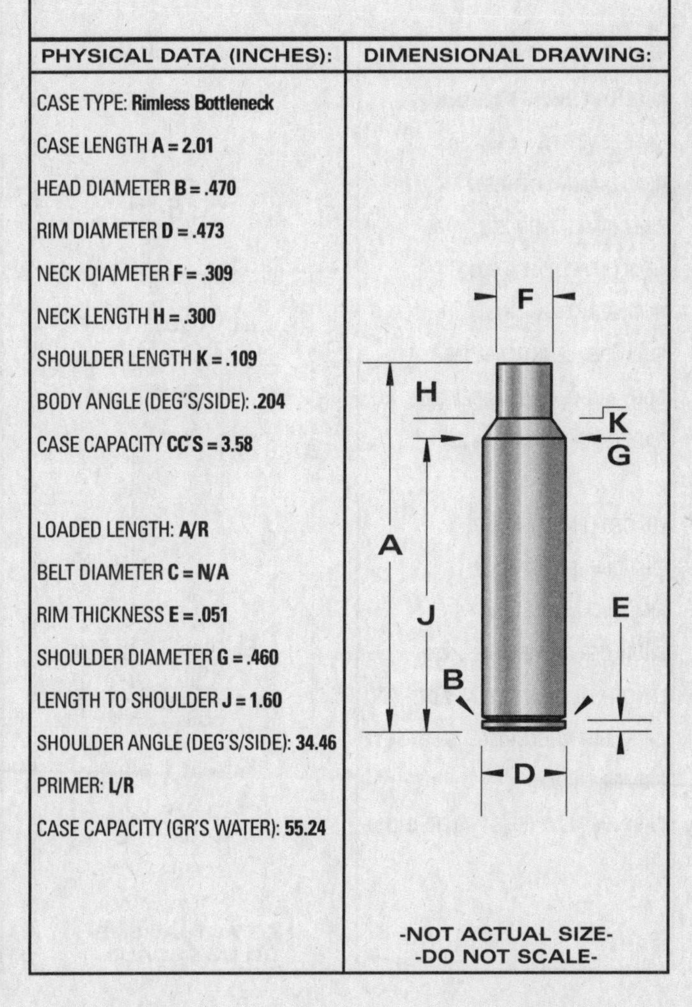

-NOT ACTUAL SIZE-
-DO NOT SCALE-

CARTRIDGE: .270/338 Winchester Magnum

OTHER NAMES:

DIA: .277

BALLISTEK NO: 277D

NAI NO: BEN 12434/4.854

DATA SOURCE: NAI/Ballistek

HISTORICAL DATA:

NOTES:

LOADING DATA:

BULLET WT./TYPE	POWDER WT./TYPE	VELOCITY ('/SEC)	SOURCE
150/Spire	66.6/MRP	—	JJD

CASE PREPARATION: SHELLHOLDER (RCBS): 4

MAKE FROM: 7mm Rem. Mag. Size Rem. case in .270/338 F/L die. Trim & chamfer.

PHYSICAL DATA (INCHES):

CASE TYPE: **Belted Bottleneck**
CASE LENGTH **A = 2.490**
HEAD DIAMETER **B = .513**
RIM DIAMETER **D = .532**
NECK DIAMETER **F = .309**
NECK LENGTH **H = .315**
SHOULDER LENGTH **K = .135**
BODY ANGLE (DEG'S/SIDE): **.327**
CASE CAPACITY **CC'S = 5.32**

LOADED LENGTH: **A/R**
BELT DIAMETER **C = .532**
RIM THICKNESS **E = .05**
SHOULDER DIAMETER **G = .492**
LENGTH TO SHOULDER **J = 2.04**
SHOULDER ANGLE (DEG'S/SIDE): **34.12**
PRIMER: **L/R Mag**
CASE CAPACITY (GR'S WATER): **82.19**

DIMENSIONAL DRAWING:

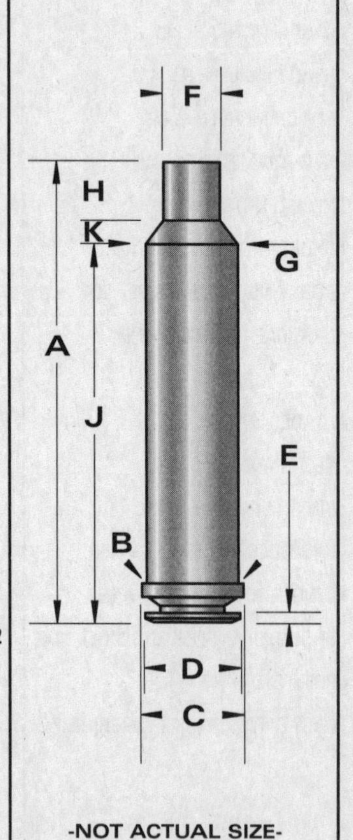

-NOT ACTUAL SIZE-
-DO NOT SCALE-

CARTRIDGE: .275 Bennett Magnum

OTHER NAMES:

DIA: .284

BALLISTEK NO: 284BE

NAI NO: BEN 22424/4.971

DATA SOURCE: Ackley Vol.1 Pg406

HISTORICAL DATA: By B. Bennett

NOTES:

LOADING DATA:

BULLET WT./TYPE	POWDER WT./TYPE	VELOCITY ('/SEC)	SOURCE
160/Spire	61.0/IMR4831	—	JJD

CASE PREPARATION: SHELLHOLDER (RCBS): 4

MAKE FROM: .300 H&H Mag. F/L size the H&H case in the Bennett die with the expander removed. Trim to length & chamfer. F/L size and fireform in the chamber.

PHYSICAL DATA (INCHES):

CASE TYPE: **Belted Bottleneck**
CASE LENGTH **A = 2.55**
HEAD DIAMETER **B = .513**
RIM DIAMETER **D = .532**
NECK DIAMETER **F = .316**
NECK LENGTH **H = .375**
SHOULDER LENGTH **K = .095**
BODY ANGLE (DEG'S/SIDE): **.990**
CASE CAPACITY **CC'S = 4.96**

LOADED LENGTH: **3.40**
BELT DIAMETER **C = .532**
RIM THICKNESS **E = .05**
SHOULDER DIAMETER **G = .448**
LENGTH TO SHOULDER **J = 2.08**
SHOULDER ANGLE (DEG'S/SIDE): **34.79**
PRIMER: **L/R Mag**
CASE CAPACITY (GR'S WATER): **76.56**

DIMENSIONAL DRAWING:

F H K G A J E B D C

-NOT ACTUAL SIZE-
-DO NOT SCALE-

CARTRIDGE: .275 Flanged Magnum (Holland & Holland)

OTHER NAMES:
.275 H&H Rimmed Magnum

DIA: .284

BALLISTEK NO: 284AK

NAI NO: RMB 22443/4.902

DATA SOURCE: COTW 4th Pg227

HISTORICAL DATA: By Holland & Holland in 1912

NOTES: Rimmed version of .275 Belted.

LOADING DATA:

BULLET WT./TYPE	POWDER WT./TYPE	VELOCITY ('/SEC)	SOURCE
160/Spire	51.0/4064	2860	Barnes

CASE PREPARATION: **SHELLHOLDER (RCBS): 4**

MAKE FROM: .375 Flanged (B.E.L.L.) Anneal case. Cut case to 2.55". Chamfer and F/L size. Two form dies are generally required although we have neck sized in a .33 WCF and .30-30 die, before F/L sizing.

PHYSICAL DATA (INCHES):

CASE TYPE: **Rimmed Bottleneck**

CASE LENGTH **A = 2.500**

HEAD DIAMETER **B = .510**

RIM DIAMETER **D = .585**

NECK DIAMETER **F = .318**

NECK LENGTH **H = .277**

SHOULDER LENGTH **K = .126**

BODY ANGLE (DEG'S/SIDE): **.906**

CASE CAPACITY **CC'S = 4.88**

LOADED LENGTH: **3.26**

BELT DIAMETER **C = N/A**

RIM THICKNESS **E = .055**

SHOULDER DIAMETER **G = .450**

LENGTH TO SHOULDER **J = 2.097**

SHOULDER ANGLE (DEG'S/SIDE): **27.65**

PRIMER: **L/R**

CASE CAPACITY (GR'S WATER): **75.26**

DIMENSIONAL DRAWING:

-NOT ACTUAL SIZE-
-DO NOT SCALE-

CARTRIDGE: .275 Holland & Holland Belted

OTHER NAMES:
.275 Belted Rimless Nitro

DIA: .284

BALLISTEK NO: 284M

NAI NO: BEN 22443/4.873

DATA SOURCE: COTW 4th Pg227

HISTORICAL DATA:

NOTES:

LOADING DATA:

BULLET WT./TYPE	POWDER WT./TYPE	VELOCITY ('/SEC)	SOURCE
160/Spire	51.0/4064	2860	Barnes

CASE PREPARATION: **SHELLHOLDER (RCBS): 4**

MAKE FROM: .300 H&H Mag. F/L size in the .275 die with expander removed. Trim to length and chamfer. Fireform.

PHYSICAL DATA (INCHES):

CASE TYPE: **Belted Bottleneck**

CASE LENGTH **A = 2.500**

HEAD DIAMETER **B = .513**

RIM DIAMETER **D = .532**

NECK DIAMETER **F = .319**

NECK LENGTH **H = .280**

SHOULDER LENGTH **K = .123**

BODY ANGLE (DEG'S/SIDE): **.875**

CASE CAPACITY **CC'S = 4.93**

LOADED LENGTH: **3.26**

BELT DIAMETER **C = .532**

RIM THICKNESS **E = .05**

SHOULDER DIAMETER **G = .455**

LENGTH TO SHOULDER **J = 2.097**

SHOULDER ANGLE (DEG'S/SIDE): **28.93**

PRIMER: **L/R Mag**

CASE CAPACITY (GR'S WATER): **76.06**

DIMENSIONAL DRAWING:

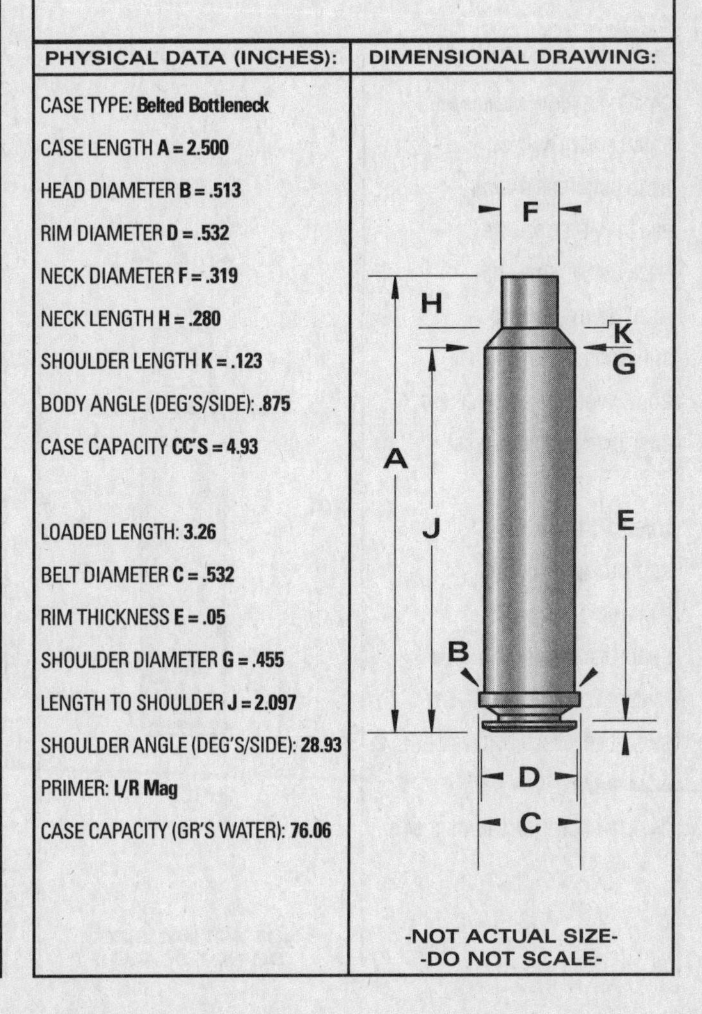

-NOT ACTUAL SIZE-
-DO NOT SCALE-

CARTRIDGE: .275 Rimless (Rigby)

OTHER NAMES:
.275 High Velocity

DIA: .284

BALLISTEK NO: 284AL

NAI NO: RXB
22332/4.725

DATA SOURCE: COTW 4th Pg227

HISTORICAL DATA: By Rigby in 1907.

NOTES: Similar to 7x57 Mauser.

LOADING DATA:

BULLET WT./TYPE	POWDER WT./TYPE	VELOCITY ('/SEC)	SOURCE
175/RN	42.7/H414	—	JJD

CASE PREPARATION: SHELLHOLDER (RCBS): 30

MAKE FROM: 9,3 x 62mm Mauser. Anneal case neck and F/L size with expander removed. Trim and chamfer. F/L size again. 7,7mm Jap brass will also work in the same manner, as will .30-06.

PHYSICAL DATA (INCHES):

CASE TYPE: **Rimless Bottleneck**

CASE LENGTH **A = 2.24**

HEAD DIAMETER **B = .475**

RIM DIAMETER **D = .475**

NECK DIAMETER **F = .325**

NECK LENGTH **H = .320**

SHOULDER LENGTH **K = .150**

BODY ANGLE (DEG'S/SIDE): **.839**

CASE CAPACITY **CC'S = 3.55**

LOADED LENGTH: **3.02**

BELT DIAMETER **C = N/A**

RIM THICKNESS **E = .05**

SHOULDER DIAMETER **G = .428**

LENGTH TO SHOULDER **J = 1.77**

SHOULDER ANGLE (DEG'S/SIDE): **19.12**

PRIMER: **L/R**

CASE CAPACITY (GR'S WATER): **54.82**

DIMENSIONAL DRAWING:

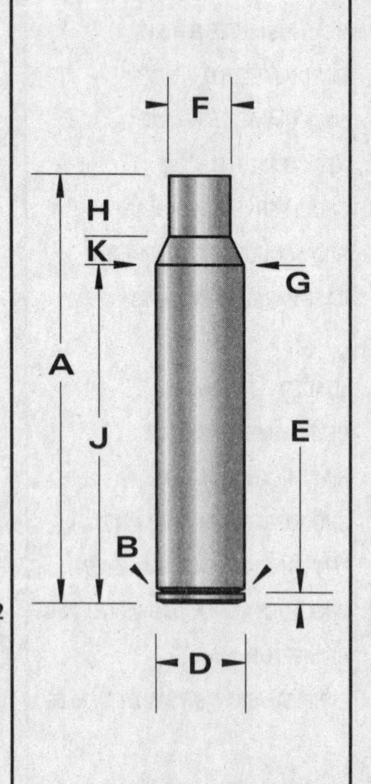

-NOT ACTUAL SIZE-
-DO NOT SCALE-

CARTRIDGE: .276 B-J Express

OTHER NAMES:

DIA: .284

BALLISTEK NO: 284BF

NAI NO: BEN
22434/4.795

DATA SOURCE: Ackley Vol.1 Pg399

HISTORICAL DATA: By Barnes & Johnson.

NOTES:

LOADING DATA:

BULLET WT./TYPE	POWDER WT./TYPE	VELOCITY ('/SEC)	SOURCE
160/Spire	65.4/W785	—	JJD

CASE PREPARATION: SHELLHOLDER (RCBS): 4

MAKE FROM: .300 H&H Mag. F/L size the H&H case in the B-J die with the expander removed. Trim and chamfer. F/L size and fireform in the chamber.

PHYSICAL DATA (INCHES):

CASE TYPE: **Belted Bottleneck**

CASE LENGTH **A = 2.46**

HEAD DIAMETER **B = .513**

RIM DIAMETER **D = .532**

NECK DIAMETER **F = .317**

NECK LENGTH **H = .287**

SHOULDER LENGTH **K = .165**

BODY ANGLE (DEG'S/SIDE): **.404**

CASE CAPACITY **CC'S = 5.23**

LOADED LENGTH: **3.39**

BELT DIAMETER **C = .532**

RIM THICKNESS **E = .05**

SHOULDER DIAMETER **G = .487**

LENGTH TO SHOULDER **J = 2.045**

SHOULDER ANGLE (DEG'S/SIDE): **33.58**

PRIMER: **L/R Mag**

CASE CAPACITY (GR'S WATER): **80.70**

DIMENSIONAL DRAWING:

F
H
K
G
A
J
E
B
D
C

-NOT ACTUAL SIZE-
-DO NOT SCALE-

CARTRIDGE: .276 Carlson Magnum

OTHER NAMES:

DIA: .284

BALLISTEK NO: 284P

NAI NO: BEN 22424/4.775

DATA SOURCE: Ackley Vol.1 Pg399

HISTORICAL DATA: By R. Carlson.

NOTES:

LOADING DATA:

BULLET WT./TYPE	POWDER WT./TYPE	VELOCITY ('/SEC)	SOURCE
160/—	60.0/4350	3052	Ackley

CASE PREPARATION: SHELLHOLDER (RCBS): 4

MAKE FROM: 7mm Rem. Mag. F/L size Rem. case in Carlson die. Trim, chamfer and fireform.

PHYSICAL DATA (INCHES):

CASE TYPE: **Belted Bottleneck**

CASE LENGTH **A = 2.45**

HEAD DIAMETER **B = .513**

RIM DIAMETER **D = .532**

NECK DIAMETER **F = .317**

NECK LENGTH **H = .360**

SHOULDER LENGTH **K = .120**

BODY ANGLE (DEG'S/SIDE): **.534**

CASE CAPACITY **CC'S = 5.02**

LOADED LENGTH: **3.27**

BELT DIAMETER **C = .532**

RIM THICKNESS **E = .05**

SHOULDER DIAMETER **G = .480**

LENGTH TO SHOULDER **J = 1.97**

SHOULDER ANGLE (DEG'S/SIDE): **34.18**

PRIMER: **L/R Mag**

CASE CAPACITY (GR'S WATER): **77.48**

DIMENSIONAL DRAWING:

-NOT ACTUAL SIZE-
-DO NOT SCALE-

CARTRIDGE: .277 Brooks Short Magnum

OTHER NAMES:

DIA: .277

BALLISTEK NO: 277S

NAI NO: BEN 23444/3.898

DATA SOURCE: Ackley Vol.1 Pg379

HISTORICAL DATA: By G. Brooks.

NOTES: Ctg. has Weatherby-type shoulders.

LOADING DATA:

BULLET WT./TYPE	POWDER WT./TYPE	VELOCITY ('/SEC)	SOURCE
150/Spire	55.0/4831	—	JJD

CASE PREPARATION: SHELLHOLDER (RCBS): 4

MAKE FROM: 7mm Rem. Mag. Anneal case neck and set back shoulder in Brooks F/L die with expander removed. Trim to length. I.D. neck ream. Chamfer and F/L size. Fireform in chamber.

PHYSICAL DATA (INCHES):

CASE TYPE: **Belted Bottleneck**

CASE LENGTH **A = 2.00**

HEAD DIAMETER **B = .513**

RIM DIAMETER **D = .532**

NECK DIAMETER **F = .292**

NECK LENGTH **H = .230**

SHOULDER LENGTH **K = .145**

BODY ANGLE (DEG'S/SIDE): **.864**

CASE CAPACITY **CC'S = 4.39**

LOADED LENGTH: **2.83**

BELT DIAMETER **C = .532**

RIM THICKNESS **E = .05**

SHOULDER DIAMETER **G = .470**

LENGTH TO SHOULDER **J = 1.625**

SHOULDER ANGLE (DEG'S/SIDE): **N/A**

PRIMER: **L/R Mag**

CASE CAPACITY (GR'S WATER): **67.77**

DIMENSIONAL DRAWING:

-NOT ACTUAL SIZE-
-DO NOT SCALE-

CARTRIDGE: .277 ICL Flying Saucer

OTHER NAMES:

DIA: .277
BALLISTEK NO: 277U
NAI NO: RXB 22324/4.734

DATA SOURCE: Ackley Vol.1 Pg 379

HISTORICAL DATA:

NOTES:

LOADING DATA:

BULLET WT./TYPE	POWDER WT./TYPE	VELOCITY ('/SEC)	SOURCE
110/—	48.0/4831	2631	Ackley

CASE PREPARATION: **SHELLHOLDER (RCBS): 11**

MAKE FROM: .257 Roberts. Taper expand the Roberts brass to .277 cal. and fireform in the ICL chamber.

PHYSICAL DATA (INCHES):

CASE TYPE: **Rimless Bottleneck**
CASE LENGTH **A = 2.23**
HEAD DIAMETER **B = .471**
RIM DIAMETER **D = .473**
NECK DIAMETER **F = .310**
NECK LENGTH **H = .364**
SHOULDER LENGTH **K = .046**
BODY ANGLE (DEG'S/SIDE): **.495**
CASE CAPACITY **CC'S = 3.79**

LOADED LENGTH: **2.93**
BELT DIAMETER **C = N/A**
RIM THICKNESS **E = .049**
SHOULDER DIAMETER **G = .443**
LENGTH TO SHOULDER **J = 1.82**
SHOULDER ANGLE (DEG'S/SIDE): **55.32**
PRIMER: **L/R**
CASE CAPACITY (GR'S WATER): **58.58**

DIMENSIONAL DRAWING:

-NOT ACTUAL SIZE-
-DO NOT SCALE-

CARTRIDGE: .280 Halger

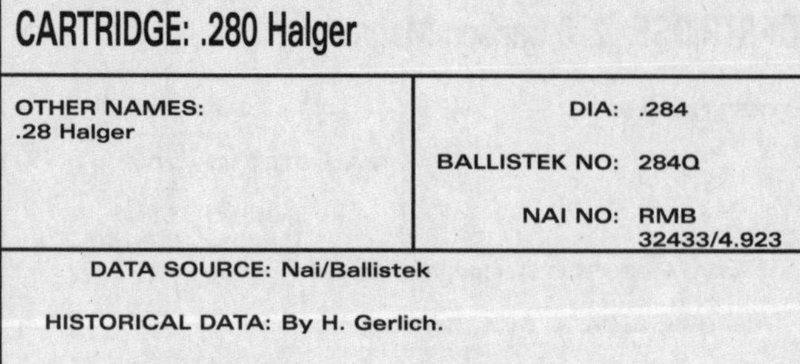

OTHER NAMES:
.28 Halger

DIA: .284
BALLISTEK NO: 284Q
NAI NO: RMB 32433/4.923

DATA SOURCE: Nai/Ballistek

HISTORICAL DATA: By H. Gerlich.

NOTES: Similar to .280 Ross

LOADING DATA:

BULLET WT./TYPE	POWDER WT./TYPE	VELOCITY ('/SEC)	SOURCE
140/Spire	56.0/IMR4895	3000	JJD

CASE PREPARATION: **SHELLHOLDER (RCBS): 13**

MAKE FROM: .280 Ross (B.E.L.L.) Size the Ross case in the Halger die. Trim and chamfer. Fireform.

PHYSICAL DATA (INCHES):

CASE TYPE: **Rimmed Bottleneck**
CASE LENGTH **A = 2.58**
HEAD DIAMETER **B = .524**
RIM DIAMETER **D = .555**
NECK DIAMETER **F = .317**
NECK LENGTH **H = .300**
SHOULDER LENGTH **K = .105**
BODY ANGLE (DEG'S/SIDE): **1.51**
CASE CAPACITY **CC'S = 4.90**

LOADED LENGTH: **3.50**
BELT DIAMETER **C = N/A**
RIM THICKNESS **E = .05**
SHOULDER DIAMETER **G = .420**
LENGTH TO SHOULDER **J = 2.175**
SHOULDER ANGLE (DEG'S/SIDE): **26.12**
PRIMER: **L/R**
CASE CAPACITY (GR'S WATER): **75.75**

DIMENSIONAL DRAWING:

-NOT ACTUAL SIZE-
-DO NOT SCALE-

CARTRIDGE: .28-30 Stevens

OTHER NAMES:
.28-30-120 Stevens

DIA: .285

BALLISTEK NO: 285A

NAI NO: RMS 21245/7.031

DATA SOURCE: COTW 4th Pg93

HISTORICAL DATA: By Steven Arms in 1900.

NOTES:

LOADING DATA:

BULLET WT./TYPE	POWDER WT./TYPE	VELOCITY ('/SEC)	SOURCE
135/FN	17.0/4198	1500	Barnes

CASE PREPARATION: SHELLHOLDER (RCBS): 1

MAKE FROM: There is no "good" way to make this case other than to turn from solid 7/16" dia. brass. Turn case, trim to length, chamfer, anneal and F/L size. It may be possible to re-body a .224 Weatherby case head but, I have not done so.

PHYSICAL DATA (INCHES):

CASE TYPE: **Rimmed Straight**
CASE LENGTH **A = 2.51**
HEAD DIAMETER **B = .357**
RIM DIAMETER **D = .412**
NECK DIAMETER **F = .309**
NECK LENGTH **H = N/A**
SHOULDER LENGTH **K = N/A**
BODY ANGLE (DEG'S/SIDE): **.561**
CASE CAPACITY **CC'S = 2.50**

LOADED LENGTH: **2.82**
BELT DIAMETER **C = N/A**
RIM THICKNESS **E = .06**
SHOULDER DIAMETER **G = N/A**
LENGTH TO SHOULDER **J = N/A**
SHOULDER ANGLE (DEG'S/SIDE): **N/A**
PRIMER: **S/R**
CASE CAPACITY (GR'S WATER): **38.61**

DIMENSIONAL DRAWING:

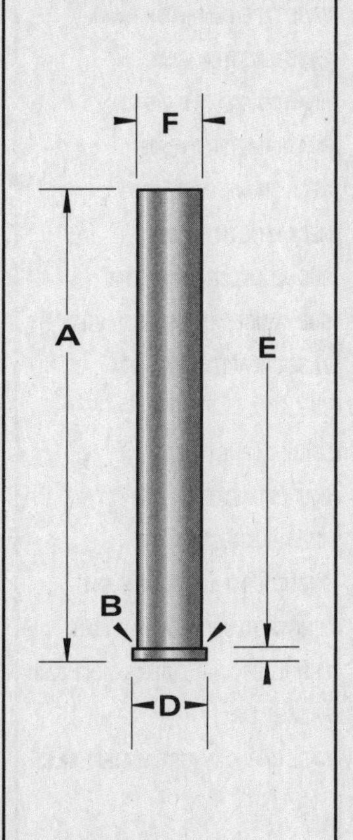

-NOT ACTUAL SIZE-
-DO NOT SCALE-

CARTRIDGE: .280 DuBiel

OTHER NAMES:
.276 DuBiel
7mm DuBiel

DIA: .284

BALLISTEK NO: 284L

NAI NO: BEN 32421/5.594

DATA SOURCE: Nai/Ballistek

HISTORICAL DATA: By J. DuBiel.

NOTES:

LOADING DATA:

BULLET WT./TYPE	POWDER WT./TYPE	VELOCITY ('/SEC)	SOURCE
150/—	57.0/4064	3300	DuBiel

CASE PREPARATION: SHELLHOLDER (RCBS): 4

MAKE FROM: .300 H&H Mag. F/L size the H&H case in the DuBiel die. Trim & chamfer. Fireform in the proper trim die or chamber.

PHYSICAL DATA (INCHES):

CASE TYPE: **Belted Bottleneck**
CASE LENGTH **A = 2.87**
HEAD DIAMETER **B = .513**
RIM DIAMETER **D = .532**
NECK DIAMETER **F = .313**
NECK LENGTH **H = .386**
SHOULDER LENGTH **K = .370**
BODY ANGLE (DEG'S/SIDE): **1.01**
CASE CAPACITY **CC'S = 5.55**

LOADED LENGTH: **3.64**
BELT DIAMETER **C = .532**
RIM THICKNESS **E = .05**
SHOULDER DIAMETER **G = .445**
LENGTH TO SHOULDER **J = 2.114**
SHOULDER ANGLE (DEG'S/SIDE): **10.11**
PRIMER: **L/R Mag**
CASE CAPACITY (GR'S WATER): **85.75**

DIMENSIONAL DRAWING:

F
H
G
K
A
J
E
B
D
C

-NOT ACTUAL SIZE-
-DO NOT SCALE-

CARTRIDGE: .280 Flanged Nitro Express

OTHER NAMES:

DIA: .287

BALLISTEK NO: 287B

NAI NO: RMB 32432/4.504

DATA SOURCE: COTW 4th Pg228

HISTORICAL DATA: By Lancaster about 1910

NOTES:

LOADING DATA:

BULLET WT./TYPE	POWDER WT./TYPE	VELOCITY ('/SEC)	SOURCE
160/—	55.0/4350	2650	Barnes

CASE PREPARATION: SHELLHOLDER (RCBS): 5

MAKE FROM: .280 Flanged (B.E.L.L.) F/L size case in .280 die. Trim & chamfer.

PHYSICAL DATA (INCHES):

CASE TYPE: **Rimmed Bottleneck**

CASE LENGTH **A = 2.41**

HEAD DIAMETER **B = .535**

RIM DIAMETER **D = .607**

NECK DIAMETER **F = .316**

NECK LENGTH **H = .330**

SHOULDER LENGTH **K = .190**

BODY ANGLE (DEG'S/SIDE): **1.89**

CASE CAPACITY **CC'S = 4.66**

LOADED LENGTH: **3.62**

BELT DIAMETER **C = N/A**

RIM THICKNESS **E = .058**

SHOULDER DIAMETER **G = .423**

LENGTH TO SHOULDER **J = 1.89**

SHOULDER ANGLE (DEG'S/SIDE): **15.72**

PRIMER: **L/R**

CASE CAPACITY (GR'S WATER): **71.96**

DIMENSIONAL DRAWING:

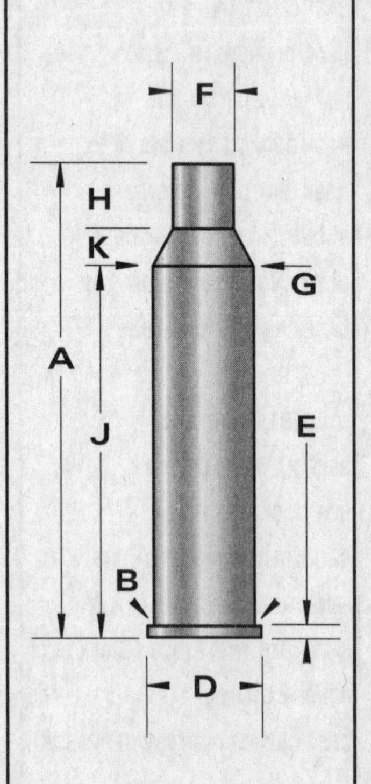

-NOT ACTUAL SIZE-
-DO NOT SCALE-

CARTRIDGE: .280 Jeffery Rimless

OTHER NAMES:

DIA: .288

BALLISTEK NO: 288A

NAI NO: RXB 22434/4.539

DATA SOURCE: COTW 4th Pg228

HISTORICAL DATA: Circa 1915

NOTES:

LOADING DATA:

BULLET WT./TYPE	POWDER WT./TYPE	VELOCITY ('/SEC)	SOURCE
140/Spire	60.0/4350	3000	Barnes

CASE PREPARATION: SHELLHOLDER (RCBS): 7

MAKE FROM: .404 Jeffery (B.E.L.L.) Anneal case and form in form die set. Trim to length, chamfer and F/L size.

PHYSICAL DATA (INCHES):

CASE TYPE: **Rimless Bottleneck**

CASE LENGTH **A = 2.46**

HEAD DIAMETER **B = .542**

RIM DIAMETER **D = .538**

NECK DIAMETER **F = .317**

NECK LENGTH **H = .350**

SHOULDER LENGTH **K = .145**

BODY ANGLE (DEG'S/SIDE): **.616**

CASE CAPACITY **CC'S = 5.74**

LOADED LENGTH: **3.38**

BELT DIAMETER **C = N/A**

RIM THICKNESS **E = .04**

SHOULDER DIAMETER **G = .504**

LENGTH TO SHOULDER **J = 1.965**

SHOULDER ANGLE (DEG'S/SIDE): **32.81**

PRIMER: **L/R**

CASE CAPACITY (GR'S WATER): **88.63**

DIMENSIONAL DRAWING:

F
H
K
G
A
J
E
B
D

-NOT ACTUAL SIZE-
-DO NOT SCALE-

CARTRIDGE: .280 Remington

OTHER NAMES:
7mm Express Remington

DIA: .284

BALLISTEK NO: 284A

NAI NO: RXB
22332/5.404

DATA SOURCE: Hornady Manual 3rd Pg184

HISTORICAL DATA: By Rem. in 1957.

NOTES: Renamed 7mm Express, by Rem., in 1980.

LOADING DATA:

BULLET WT./TYPE	POWDER WT./TYPE	VELOCITY ('/SEC)	SOURCE
175/RN	47.0/IMR4350	2400	Lyman

CASE PREPARATION: SHELLHOLDER (RCBS): 3

MAKE FROM: Factory or .30-06 Spgf. Anneal case neck and F/L size. Square case mouth & chamfer.

PHYSICAL DATA (INCHES):

CASE TYPE: **Rimless Bottleneck**

CASE LENGTH **A = 2.54**

HEAD DIAMETER **B = .470**

RIM DIAMETER **D = .473**

NECK DIAMETER **F = .315**

NECK LENGTH **H = .341**

SHOULDER LENGTH **K = .200**

BODY ANGLE (DEG'S/SIDE): **.462**

CASE CAPACITY **CC'S = 4.44**

LOADED LENGTH: **3.33**

BELT DIAMETER **C = N/A**

RIM THICKNESS **E = .049**

SHOULDER DIAMETER **G = .441**

LENGTH TO SHOULDER **J = 1.999**

SHOULDER ANGLE (DEG'S/SIDE): **17.48**

PRIMER: **L/R**

CASE CAPACITY (GR'S WATER): **68.58**

DIMENSIONAL DRAWING:

-NOT ACTUAL SIZE-
-DO NOT SCALE-

CARTRIDGE: .280 Remington Improved

OTHER NAMES:

DIA: .284

BALLISTEK NO: 284AC

NAI NO: RXB
22324/5.297

DATA SOURCE: Ackley Vol.1 Pg395

HISTORICAL DATA:

NOTES:

LOADING DATA:

BULLET WT./TYPE	POWDER WT./TYPE	VELOCITY ('/SEC)	SOURCE
175/—	58.0/4350	2850	Ackley

CASE PREPARATION: SHELLHOLDER (RCBS): 3

MAKE FROM: .280 Rem. or .30-06 Spgf. Fire factory .280 ammo in the "improved" chamber or F/L size the '06 case, trim and chamfer.

PHYSICAL DATA (INCHES):

CASE TYPE: **Rimless Bottleneck**

CASE LENGTH **A = 2.49**

HEAD DIAMETER **B = .470**

RIM DIAMETER **D = .473**

NECK DIAMETER **F = .314**

NECK LENGTH **H = .400**

SHOULDER LENGTH **K = .100**

BODY ANGLE (DEG'S/SIDE): **.448**

CASE CAPACITY **CC'S = 4.36**

LOADED LENGTH: **3.25**

BELT DIAMETER **C = N/A**

RIM THICKNESS **E = .049**

SHOULDER DIAMETER **G = .442**

LENGTH TO SHOULDER **J = 1.99**

SHOULDER ANGLE (DEG'S/SIDE): **32.62**

PRIMER: **L/R**

CASE CAPACITY (GR'S WATER): **67.39**

DIMENSIONAL DRAWING:

-NOT ACTUAL SIZE-
-DO NOT SCALE-

CARTRIDGE: .280 Ross

OTHER NAMES:

DIA: .287

BALLISTEK NO: 287A

NAI NO: RMB 32433/4.914

DATA SOURCE: Handloader #45 Pg29

HISTORICAL DATA: By F. Jones about 1906.

NOTES: Case referred to as "rimless" but, actually, a slight rim is present.

LOADING DATA:

BULLET WT./TYPE	POWDER WT./TYPE	VELOCITY ('/SEC)	SOURCE
140/Spire	58.0/4895	3170	Barnes

CASE PREPARATION: **SHELLHOLDER (RCBS):** 13

MAKE FROM: .280 Ross (B.E.L.L.) Anneal case and F/L size. Trim to length. .300 H&H cases will also work if the belt is turned to .525" dia. anneal case F/L size and trim to length.

.287" dia. bullets are available from a few of the custom swaggers.

PHYSICAL DATA (INCHES):

CASE TYPE: **Rimmed Bottleneck**

CASE LENGTH **A = 2.58**

HEAD DIAMETER **B = .525**

RIM DIAMETER **D = .556**

NECK DIAMETER **F = .322**

NECK LENGTH **H = .300**

SHOULDER LENGTH **K = .105**

BODY ANGLE (DEG'S/SIDE): **1.54**

CASE CAPACITY **CC'S = 5.01**

LOADED LENGTH: **3.50**

BELT DIAMETER **C = N/A**

RIM THICKNESS **E = .05**

SHOULDER DIAMETER **G = .419**

LENGTH TO SHOULDER **J = 2.176**

SHOULDER ANGLE (DEG'S/SIDE): **25.90**

PRIMER: **L/R Mag**

CASE CAPACITY (GR'S WATER): **77.38**

DIMENSIONAL DRAWING:

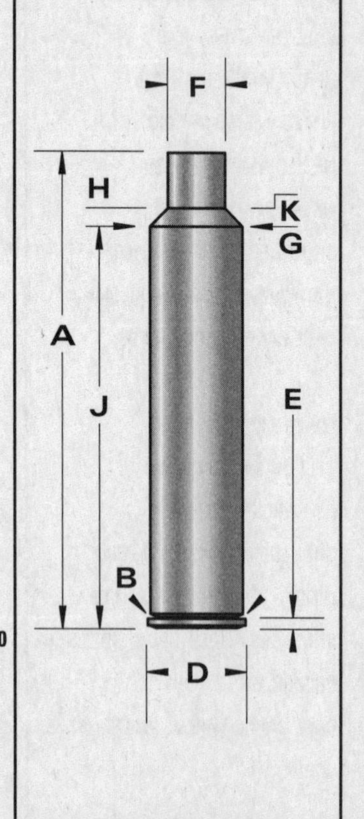

-NOT ACTUAL SIZE-
-DO NOT SCALE-

CARTRIDGE: .280/30 British

OTHER NAMES:
7mm NATO
.280 British

DIA: .284

BALLISTEK NO: 284BO

NAI NO: RXB 24343/3.234

DATA SOURCE: COTW 4th Pg200

HISTORICAL DATA: British mil. experimental ctg about 1946.

NOTES:

LOADING DATA:

BULLET WT./TYPE	POWDER WT./TYPE	VELOCITY ('/SEC)	SOURCE
150/Spire	28.5/IMR3031	–	JJD

CASE PREPARATION: **SHELLHOLDER (RCBS):** 3

MAKE FROM: .308 Winchester. Anneal neck and shoulder area. F/L size in .280/30 die with expander removed. Trim to length and chamfer. I.D. neck ream. Rechamfer and F/L size.

PHYSICAL DATA (INCHES):

CASE TYPE: **Rimless Bottleneck**

CASE LENGTH **A = 1.52**

HEAD DIAMETER **B = .470**

RIM DIAMETER **D = .473**

NECK DIAMETER **F = .313**

NECK LENGTH **H = .265**

SHOULDER LENGTH **K = .145**

BODY ANGLE (DEG'S/SIDE): **.692**

CASE CAPACITY **CC'S = 2.43**

LOADED LENGTH: **2.54**

BELT DIAMETER **C = N/A**

RIM THICKNESS **E = .049**

SHOULDER DIAMETER **G = .448**

LENGTH TO SHOULDER **J = 1.11**

SHOULDER ANGLE (DEG'S/SIDE): **24.96**

PRIMER: **L/R**

CASE CAPACITY (GR'S WATER): **37.51**

DIMENSIONAL DRAWING:

F
H
K
G
A
J
E
B
D

-NOT ACTUAL SIZE-
-DO NOT SCALE-

CARTRIDGE: .284 Williams

OTHER NAMES:

DIA: .284

BALLISTEK NO: 284BA

NAI NO: BEN 22423/4.873

DATA SOURCE: Ackley Vol.1 Pg403

HISTORICAL DATA: By W. Williams in 1945.

NOTES:

LOADING DATA:

BULLET WT./TYPE	POWDER WT./TYPE	VELOCITY ('/SEC)	SOURCE
160/—	62.0/4350	2850	Williams

CASE PREPARATION: SHELLHOLDER (RCBS): 4

MAKE FROM: 7mm Rem. Mag. Size the Rem. case in the Williams F/L die to set the shoulder back. Square case mouth and chamfer. Fireform.

PHYSICAL DATA (INCHES):

CASE TYPE: **Belted Bottleneck**

CASE LENGTH **A = 2.500**

HEAD DIAMETER **B = .513**

RIM DIAMETER **D = .532**

NECK DIAMETER **F = .315**

NECK LENGTH **H = .360**

SHOULDER LENGTH **K = .180**

BODY ANGLE (DEG'S/SIDE): **.293**

CASE CAPACITY **CC'S = 5.11**

LOADED LENGTH: **3.28**

BELT DIAMETER **C = .532**

RIM THICKNESS **E = .049**

SHOULDER DIAMETER **G = .495**

LENGTH TO SHOULDER **J = 1.96**

SHOULDER ANGLE (DEG'S/SIDE): **26.56**

PRIMER: **L/R Mag**

CASE CAPACITY (GR'S WATER): **78.81**

DIMENSIONAL DRAWING:

-NOT ACTUAL SIZE-
-DO NOT SCALE-

CARTRIDGE: .284 Winchester

OTHER NAMES:

DIA: .284

BALLISTEK NO: 284C

NAI NO: RBB 22334/4.34

DATA SOURCE: Hornady Manual 3rd Pg182

HISTORICAL DATA: By Win. in 1963

NOTES: Only American rebated case

LOADING DATA:

BULLET WT./TYPE	POWDER WT./TYPE	VELOCITY ('/SEC)	SOURCE
154/Spire	46.8/IMR4064	2800	Horn.

CASE PREPARATION: SHELLHOLDER (RCBS): 3

MAKE FROM: .45 Basic. Turn rim flush with base (.500" dia.) and cut new extractor groove. Anneal case. Size in .40-65 Win and .33 Win die to slowly reduce neck dia. Watch shoulder location! Anneal again. F/L size in .284 die with expander removed. Trim to 2.2". Inside neck ream. F/L size, trim and chamfer. Generally, however, factory cases are available.

PHYSICAL DATA (INCHES):

CASE TYPE: **Rebated Bottleneck**

CASE LENGTH **A = 2.17**

HEAD DIAMETER **B = .500**

RIM DIAMETER **D = .473**

NECK DIAMETER **F = .320**

NECK LENGTH **H = .285**

SHOULDER LENGTH **K = .110**

BODY ANGLE (DEG'S/SIDE): **454**

CASE CAPACITY **CC'S = 4.185**

LOADED LENGTH: **2.95**

BELT DIAMETER **C = N/A**

RIM THICKNESS **E = .054**

SHOULDER DIAMETER **G = .475**

LENGTH TO SHOULDER **J = 1.775**

SHOULDER ANGLE (DEG'S/SIDE): **35.16**

PRIMER: **L/R**

CASE CAPACITY (GR'S WATER): **64.59**

DIMENSIONAL DRAWING:

-NOT ACTUAL SIZE-
-DO NOT SCALE-

CARTRIDGE: .285 O.K.H.

OTHER NAMES: 7mm/06 OKH	
DIA:	.284
BALLISTEK NO:	284BG
NAI NO:	RXB 22332/5.298

DATA SOURCE: Ackley Vol.1 Pg392

HISTORICAL DATA: By O'Neil, Keith & Hopkins

NOTES:

LOADING DATA:

BULLET WT./TYPE	POWDER WT./TYPE	VELOCITY ('/SEC)	SOURCE
160/Spire	43.0/4064	2616	Ackley

CASE PREPARATION: SHELLHOLDER (RCBS): 3

MAKE FROM: .280 Remington. F/L size the Rem. case in the OKH die. Trim to length and chamfer. So similar to the .280 Rem. (#284A) that you may as well stay with the factory ctg.

PHYSICAL DATA (INCHES):

CASE TYPE: **Rimless Bottleneck**
CASE LENGTH **A = 2.49**
HEAD DIAMETER **B = .470**
RIM DIAMETER **D = .473**
NECK DIAMETER **F = .304**
NECK LENGTH **H = .350**
SHOULDER LENGTH **K = .225**
BODY ANGLE (DEG'S/SIDE): **.534**
CASE CAPACITY **CC'S = 4.59**

LOADED LENGTH: **3.40**
BELT DIAMETER **C = N/A**
RIM THICKNESS **E = .05**
SHOULDER DIAMETER **G = .438**
LENGTH TO SHOULDER **J = 1.915**
SHOULDER ANGLE (DEG'S/SIDE): **16.58**
PRIMER: **L/R**
CASE CAPACITY (GR'S WATER): **70.82**

DIMENSIONAL DRAWING:

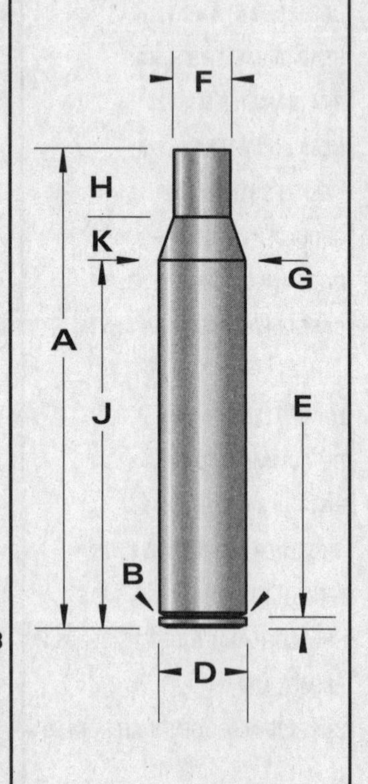

-NOT ACTUAL SIZE-
-DO NOT SCALE-

CARTRIDGE: .288 Barnes Supreme

OTHER NAMES:	
DIA:	.284
BALLISTEK NO:	284BL
NAI NO:	BEN 12424/5.536

DATA SOURCE: Ackley Vol.1 Pg408

HISTORICAL DATA: By F. Barnes

NOTES:

LOADING DATA:

BULLET WT./TYPE	POWDER WT./TYPE	VELOCITY ('/SEC)	SOURCE
140/Spire	55.0/IMR4831	–	JJD

CASE PREPARATION: SHELLHOLDER (RCBS): 4

MAKE FROM: .300 H&H Mag. Anneal case and F/L size in the Barnes die. Trim to length and chamfer. Fireform in the chamber. Cases can also be formed by fireforming, with corn meal, in the proper trim die.

PHYSICAL DATA (INCHES):

CASE TYPE: **Belted Bottleneck**
CASE LENGTH **A = 2.84**
HEAD DIAMETER **B = .513**
RIM DIAMETER **D = .532**
NECK DIAMETER **F = .315**
NECK LENGTH **H = .360**
SHOULDER LENGTH **K = .080**
BODY ANGLE (DEG'S/SIDE): **.169**
CASE CAPACITY **CC'S = 6.39**

LOADED LENGTH: **A/R**
BELT DIAMETER **C = .532**
RIM THICKNESS **E = .049**
SHOULDER DIAMETER **G = .50**
LENGTH TO SHOULDER **J = 2.40**
SHOULDER ANGLE (DEG'S/SIDE): **49.14**
PRIMER: **L/R Mag**
CASE CAPACITY (GR'S WATER): **98.68**

DIMENSIONAL DRAWING:

-NOT ACTUAL SIZE-
-DO NOT SCALE-

CARTRIDGE: .293/230 Morris Long

OTHER NAMES:

DIA: .225

BALLISTEK NO: 225B

NAI NO: RMB 34242/2.721

DATA SOURCE: COTW 4th Pg223

HISTORICAL DATA: Ca. 1880 by R. Morris.

NOTES: Practice round for .577/450 Martini Henry chamber insert

LOADING DATA:

BULLET WT./TYPE	POWDER WT./TYPE	VELOCITY ('/SEC)	SOURCE
43/Lead	4.0/2400	1200(?)	Barnes

CASE PREPARATION: SHELLHOLDER (RCBS): 12

MAKE FROM: .22 Hornet. Conversion same as .293/230 Morris Short (#225A). Note: Length to shoulder dim. is different (.57).

PHYSICAL DATA (INCHES):

CASE TYPE: **Rimmed Bottleneck**

CASE LENGTH **A = .800**

HEAD DIAMETER **B = .294**

RIM DIAMETER **D = .346**

NECK DIAMETER **F = .240**

NECK LENGTH **H = .180**

SHOULDER LENGTH **K = .05**

BODY ANGLE (DEG'S/SIDE): **1.54**

CASE CAPACITY **CC'S = .438**

LOADED LENGTH: **1.01**

BELT DIAMETER **C = N/A**

RIM THICKNESS **E = .06**

SHOULDER DIAMETER **G = .274**

LENGTH TO SHOULDER **J = .57**

SHOULDER ANGLE (DEG'S/SIDE): **18.77**

PRIMER: **S/R**

CASE CAPACITY (GR'S WATER): **6.76**

DIMENSIONAL DRAWING:

F H K A G J B D E

-NOT ACTUAL SIZE-
-DO NOT SCALE-

CARTRIDGE: .293/230 Morris Short

OTHER NAMES:

DIA: .225

BALLISTEK NO: 225A

NAI NO: RMB 34242/1.972

DATA SOURCE: COTW 4th Pg223

HISTORICAL DATA: Ca. 1880 by R. Morris.

NOTES: As practice round for .577/450 Martini Henry insert

LOADING DATA:

BULLET WT./TYPE	POWDER WT./TYPE	VELOCITY ('/SEC)	SOURCE
43/Lead	3.0/Unique	900	Barnes

CASE PREPARATION: SHELLHOLDER (RCBS): 12

MAKE FROM: .22 Hornet. Run Hornet case into .25 ACP die just far enough to establish shoulder at .35" dim. Neck I.D. may need reaming. Really not necessary to F/L size. Use .224" dia. 40 gr. bullets.

PHYSICAL DATA (INCHES):

CASE TYPE: **Rimmed Bottleneck**

CASE LENGTH **A = .58**

HEAD DIAMETER **B = .294**

RIM DIAMETER **D = .346**

NECK DIAMETER **F = .240**

NECK LENGTH **H = .18**

SHOULDER LENGTH **K = .05**

BODY ANGLE (DEG'S/SIDE): **3.81**

CASE CAPACITY **CC'S = .244**

LOADED LENGTH: **.83**

BELT DIAMETER **C = N/A**

RIM THICKNESS **E = .06**

SHOULDER DIAMETER **G = .274**

LENGTH TO SHOULDER **J = .35**

SHOULDER ANGLE (DEG'S/SIDE): **18.77**

PRIMER: **S/R**

CASE CAPACITY (GR'S WATER): **3.77**

DIMENSIONAL DRAWING:

F H K A G J B D E

-NOT ACTUAL SIZE-
-DO NOT SCALE-

CARTRIDGE: .297/250 Rook Rifle

OTHER NAMES:

DIA: .250

BALLISTEK NO: 250B

NAI NO: RMB 14241/2.779

DATA SOURCE: COTW 4th Pg225

HISTORICAL DATA: By Holland & Holland before 1900

NOTES:

LOADING DATA:

BULLET WT./TYPE	POWDER WT./TYPE	VELOCITY ('/SEC)	SOURCE
70/Lead RN	4.5/Unique	1000	JJD

CASE PREPARATION: SHELLHOLDER (RCBS): 12

MAKE FROM: .22 Hornet. Turn Hornet's rim to .343 (if necessary). Cut off to .84" and chamfer. F/L size in Rook die OR reduce neck dia. in base of .25 ACP F/L die (careful of shoulder location!). Use light load to fireform. .25 ACP proj's work fine.

PHYSICAL DATA (INCHES):

CASE TYPE: **Rimmed Bottleneck**

CASE LENGTH **A = .82**

HEAD DIAMETER **B = .295**

RIM DIAMETER **D = .343**

NECK DIAMETER **F = .267**

NECK LENGTH **H = .200**

SHOULDER LENGTH **K = .079**

BODY ANGLE (DEG'S/SIDE): **.084**

CASE CAPACITY **CC'S = .504**

LOADED LENGTH: **1.06**

BELT DIAMETER **C = N/A**

RIM THICKNESS **E = .04**

SHOULDER DIAMETER **G = .294**

LENGTH TO SHOULDER **J = .54**

SHOULDER ANGLE (DEG'S/SIDE): **9.57**

PRIMER: **S/R**

CASE CAPACITY (GR'S WATER): **7.79**

DIMENSIONAL DRAWING:

F H K A G J B D E

-NOT ACTUAL SIZE-
-DO NOT SCALE-

CARTRIDGE: .30 Borchardt

OTHER NAMES:
7,65mm Borchardt

DIA: .307

BALLISTEK NO: 307B

NAI NO: RXB 24242/2.571

DATA SOURCE: COTW 4th Pg165

HISTORICAL DATA: About 1893 by H. Borchardt.

NOTES:

LOADING DATA:

BULLET WT./TYPE	POWDER WT./TYPE	VELOCITY ('/SEC)	SOURCE
85/RN	5.0/B'Eye	1200	JJD

CASE PREPARATION: SHELLHOLDER (RCBS): 16

MAKE FROM: .38 ACP (or Super ACP). Turn rim to .390" dia. F/L size. .38 Spl. brass will also work by turning rim, cutting to length and sizing. The case made from .38 ACP will have a very short neck; that made from .38 Spl. will be to correct length. I have not had good luck with the shorter neck holding bullets during recoil.

PHYSICAL DATA (INCHES):

CASE TYPE: **Rimless Bottleneck**

CASE LENGTH **A = .99**

HEAD DIAMETER **B = .385**

RIM DIAMETER **D = .390**

NECK DIAMETER **F = .331**

NECK LENGTH **H = .175**

SHOULDER LENGTH **K = .054**

BODY ANGLE (DEG'S/SIDE): **.767**

CASE CAPACITY **CC'S = 1.06**

LOADED LENGTH: **1.34**

BELT DIAMETER **C = N/A**

RIM THICKNESS **E = .05**

SHOULDER DIAMETER **G = .370**

LENGTH TO SHOULDER **J = .76**

SHOULDER ANGLE (DEG'S/SIDE): **19.52**

PRIMER: **S/P**

CASE CAPACITY (GR'S WATER): **16.34**

DIMENSIONAL DRAWING:

F H K G A J E B D

-NOT ACTUAL SIZE-
-DO NOT SCALE-

CARTRIDGE: .30 #1 Ackley Short Magnum

OTHER NAMES:

DIA: .308

BALLISTEK NO: 308AX

NAI NO: BEN 22323/4.838

DATA SOURCE: Ackley Vol.1 Pg430

HISTORICAL DATA: Designed for short actioned rifles.

NOTES:

LOADING DATA:

BULLET WT./TYPE	POWDER WT./TYPE	VELOCITY ('/SEC)	SOURCE
180/Spire	58.0/4064	2940	Ackley

CASE PREPARATION: **SHELLHOLDER (RCBS): 4**

MAKE FROM: .300 H&H Mag. Single form die required. Form case and cut off to 2.5". F/L size, trim to final length and chamfer. Case will fireform in chamber.

PHYSICAL DATA (INCHES):

CASE TYPE: **Belted Bottleneck**

CASE LENGTH **A = 2.482**

HEAD DIAMETER **B = .513**

RIM DIAMETER **D = .532**

NECK DIAMETER **F = .340**

NECK LENGTH **H = .390**

SHOULDER LENGTH **K = .142**

BODY ANGLE (DEG'S/SIDE): **.409**

CASE CAPACITY **CC'S = 5.24**

LOADED LENGTH: **A/R**

BELT DIAMETER **C = .532**

RIM THICKNESS **E = .049**

SHOULDER DIAMETER **G = .488**

LENGTH TO SHOULDER **J = 1.95**

SHOULDER ANGLE (DEG'S/SIDE): **27.52**

PRIMER: **L/R Mag**

CASE CAPACITY (GR'S WATER): **80.88**

DIMENSIONAL DRAWING:

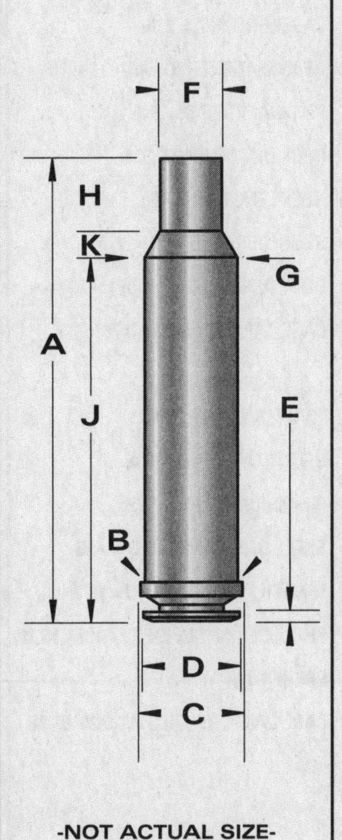

-NOT ACTUAL SIZE-
-DO NOT SCALE-

CARTRIDGE: .30 #2 Ackley Short Magnum

OTHER NAMES:

DIA: .308

BALLISTEK NO: 308BL

NAI NO: BEN 22323/5.146

DATA SOURCE: Ackley Vol.1 Pg431

HISTORICAL DATA:

NOTES:

LOADING DATA:

BULLET WT./TYPE	POWDER WT./TYPE	VELOCITY ('/SEC)	SOURCE
180/—	54.0/4064	2810	Ackley

CASE PREPARATION: **SHELLHOLDER (RCBS): 4**

MAKE FROM: .300 H&H Mag. F/L size the H&H case, in the "improved" die, with the expander removed. Trim & chamfer. F/L size and fireform in the Ackley #2 chamber. .30 #1 Ackley dies may be used if adjusted to properly locate the shoulder.

PHYSICAL DATA (INCHES):

CASE TYPE: **Belted Bottleneck**

CASE LENGTH **A = 2.64**

HEAD DIAMETER **B = .513**

RIM DIAMETER **D = .532**

NECK DIAMETER **F = .340**

NECK LENGTH **H = .390**

SHOULDER LENGTH **K = .188**

BODY ANGLE (DEG'S/SIDE): **.353**

CASE CAPACITY **CC'S = 5.65**

LOADED LENGTH: **A/R**

BELT DIAMETER **C = .532**

RIM THICKNESS **E = .04**

SHOULDER DIAMETER **G = .490**

LENGTH TO SHOULDER **J = 2.062**

SHOULDER ANGLE (DEG'S/SIDE): **21.75**

PRIMER: **L/R Mag**

CASE CAPACITY (GR'S WATER): **87.15**

DIMENSIONAL DRAWING:

F H K G A J E B D C

-NOT ACTUAL SIZE-
-DO NOT SCALE-

CARTRIDGE: .30 Carbine

OTHER NAMES:
.30 M1 Carbine
7,62 x 3/33mm

DIA: .308
BALLISTEK NO: 308A
NAI NO: RXS 21245/3.623

DATA SOURCE: Hornady Manual 3rd Pg207

HISTORICAL DATA: U.S. Mil. Ctg. ca. 1940.

NOTES:

LOADING DATA:

BULLET WT./TYPE	POWDER WT./TYPE	VELOCITY ('/SEC)	SOURCE
110/RN	15.0/IMR4227	1845	Lyman

CASE PREPARATION: **SHELLHOLDER (RCBS):** 17

MAKE FROM: Factory or .25-20 Win. Turn rim to .360" dia. and cut new extractor groove. Fireform (with tissue paper or cornmeal) in .30 Carbine trim die. Clean. F/L size, length trim and chamfer. Best to just find the factory cases!

PHYSICAL DATA (INCHES):

CASE TYPE: **Rimless Straight**
CASE LENGTH **A = 1.29**
HEAD DIAMETER **B = .356**
RIM DIAMETER **D = .360**
NECK DIAMETER **F = .336**
NECK LENGTH **H = N/A**
SHOULDER LENGTH **K = N/A**
BODY ANGLE (DEG'S/SIDE): **.462**
CASE CAPACITY **CC'S = 1.169**

LOADED LENGTH: **1.68**
BELT DIAMETER **C = N/A**
RIM THICKNESS **E = .05**
SHOULDER DIAMETER **G = N/A**
LENGTH TO SHOULDER **J = N/A**
SHOULDER ANGLE (DEG'S/SIDE): **N/A**
PRIMER: **S/R**
CASE CAPACITY (GR'S WATER): **18.048**

DIMENSIONAL DRAWING:

-NOT ACTUAL SIZE-
-DO NOT SCALE-

CARTRIDGE: .30 Flanged Nitro (Purdey)

OTHER NAMES:

DIA: .308
BALLISTEK NO: 308BH
NAI NO: RMB 22322/5.164

DATA SOURCE: COTW 4th Pg 230

HISTORICAL DATA:

NOTES: Similar to .30-40 Krag.

LOADING DATA:

BULLET WT./TYPE	POWDER WT./TYPE	VELOCITY ('/SEC)	SOURCE
150/RN	45.0/4320	2700	Barnes

CASE PREPARATION: **SHELLHOLDER (RCBS):** 7

MAKE FROM: .30-40 Krag. Size the Krag case in the Purdey F/L die. Case will be slightly short but, otherwise, fine. Fireform.

PHYSICAL DATA (INCHES):

CASE TYPE: **Rimmed Bottleneck**
CASE LENGTH **A = 2.36**
HEAD DIAMETER **B = .457**
RIM DIAMETER **D = .545**
NECK DIAMETER **F = .338**
NECK LENGTH **H = .480**
SHOULDER LENGTH **K = .110**
BODY ANGLE (DEG'S/SIDE): **.766**
CASE CAPACITY **CC'S = 3.76**

LOADED LENGTH: **2.97**
BELT DIAMETER **C = N/A**
RIM THICKNESS **E = .062**
SHOULDER DIAMETER **G = .415**
LENGTH TO SHOULDER **J = 1.77**
SHOULDER ANGLE (DEG'S/SIDE): **19.20**
PRIMER: **L/R**
CASE CAPACITY (GR'S WATER): **57.98**

DIMENSIONAL DRAWING:

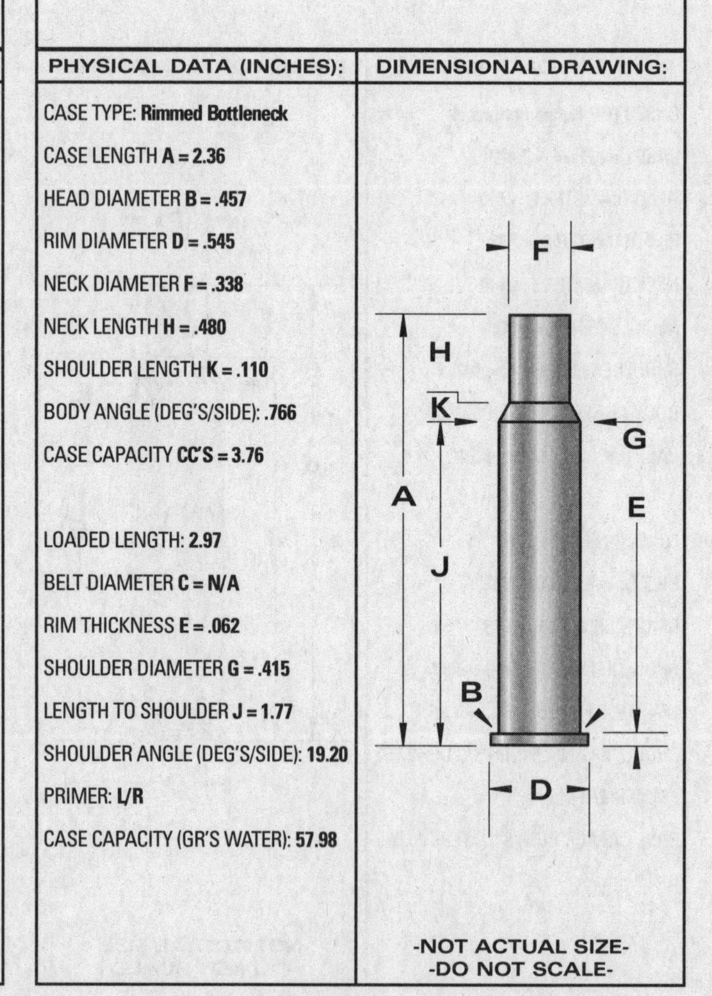

-NOT ACTUAL SIZE-
-DO NOT SCALE-

CARTRIDGE: .30 Gibbs

OTHER NAMES:

DIA: .308

BALLISTEK NO: 308AG

NAI NO: RXB 12343/5.319

DATA SOURCE: Ackley Vol.2 Pg188

HISTORICAL DATA:

NOTES:

LOADING DATA:

BULLET WT./TYPE	POWDER WT./TYPE	VELOCITY ('/SEC)	SOURCE
150/Spire	62.6/4831	2850	O'Connor

CASE PREPARATION: SHELLHOLDER (RCBS): 3

MAKE FROM: .30-06 Spgf. Taper expand the '06 case to about .315" dia. F/L size in the Gibbs die. Square case mouth and fireform in the Gibbs chamber.

PHYSICAL DATA (INCHES):

CASE TYPE: **Rimless Bottleneck**

CASE LENGTH **A = 2.500**

HEAD DIAMETER **B = .470**

RIM DIAMETER **D = .471**

NECK DIAMETER **F = .339**

NECK LENGTH **H = .225**

SHOULDER LENGTH **K = .130**

BODY ANGLE (DEG'S/SIDE): **.338**

CASE CAPACITY **CC'S = 4.65**

LOADED LENGTH: **3.34**

BELT DIAMETER **C = N/A**

RIM THICKNESS **E = .05**

SHOULDER DIAMETER **G = .447**

LENGTH TO SHOULDER **J = 2.145**

SHOULDER ANGLE (DEG'S/SIDE): **42.55**

PRIMER: **L/R**

CASE CAPACITY (GR'S WATER): **71.71**

DIMENSIONAL DRAWING:

-NOT ACTUAL SIZE-
-DO NOT SCALE-

CARTRIDGE: .30 Herrett

OTHER NAMES:

DIA: .308

BALLISTEK NO: 308X

NAI NO: RMB 23243/3.812

DATA SOURCE: Hornady Manual 3rd Pg370

HISTORICAL DATA: By Herrett & Milek in 1972.

NOTES:

LOADING DATA:

BULLET WT./TYPE	POWDER WT./TYPE	VELOCITY ('/SEC)	SOURCE
125/Spire	22.9/IMR4227	2200	Sierra
150/Spire	21.5/IMR4227	1900	Horn.

CASE PREPARATION: SHELLHOLDER (RCBS): 2

MAKE FROM: .30-30 Win. Not necessary to anneal. Run case into form die. Cut case to 1.62". F/L size and trim to length. Chamfer & clean.

PHYSICAL DATA (INCHES):

CASE TYPE: **Rimmed Bottleneck**

CASE LENGTH **A = 1.605**

HEAD DIAMETER **B = .421**

RIM DIAMETER **D = .506**

NECK DIAMETER **F = .329**

NECK LENGTH **H = .290**

SHOULDER LENGTH **K = .067**

BODY ANGLE (DEG'S/SIDE): **.41**

CASE CAPACITY **CC'S = 2.34**

LOADED LENGTH: **2.32**

BELT DIAMETER **C = N/A**

RIM THICKNESS **E = .062**

SHOULDER DIAMETER **G = .406**

LENGTH TO SHOULDER **J = 1.248**

SHOULDER ANGLE (DEG'S/SIDE): **29.88**

PRIMER: **L/R**

CASE CAPACITY (GR'S WATER): **36.13**

DIMENSIONAL DRAWING:

-NOT ACTUAL SIZE-
-DO NOT SCALE-

CARTRIDGE: .30 Howell

OTHER NAMES:	
	DIA: .308
	BALLISTEK NO: 308AH
	NAI NO: RMB 24424/3.617

DATA SOURCE: Ackley Vol.1 Pg428

HISTORICAL DATA: By J. Howell.

NOTES:

LOADING DATA:

BULLET WT./TYPE	POWDER WT./TYPE	VELOCITY ('/SEC)	SOURCE
165/Spire	63.0/4350	3150	Ackley

CASE PREPARATION: SHELLHOLDER (RCBS): 5

MAKE FROM: .348 Winchester. Case forming dies are required. Anneal case and form in 3-die set. Trim to length. Chamfer and F/L size. Some cases require I.D. neck reaming and others do not. Check! Fireform in chamber to fill out shoulders.

PHYSICAL DATA (INCHES):

CASE TYPE: **Rimmed Bottleneck**
CASE LENGTH **A = 2.00**
HEAD DIAMETER **B = .553**
RIM DIAMETER **D = .610**
NECK DIAMETER **F = .337**
NECK LENGTH **H = .420**
SHOULDER LENGTH **K = .110**
BODY ANGLE (DEG'S/SIDE): **.406**
CASE CAPACITY **CC'S = 4.70**

LOADED LENGTH: **2.77**
BELT DIAMETER **C = N/A**
RIM THICKNESS **E = .07**
SHOULDER DIAMETER **G = .535**
LENGTH TO SHOULDER **J = 1.47**
SHOULDER ANGLE (DEG'S/SIDE): **41.98**
PRIMER: **L/R**
CASE CAPACITY (GR'S WATER): **72.60**

DIMENSIONAL DRAWING:

-NOT ACTUAL SIZE-
-DO NOT SCALE-

CARTRIDGE: .30 ICL Grizzly

OTHER NAMES:	
	DIA: .308
	BALLISTEK NO: 308BB
	NAI NO: BEN 22334/5.546

DATA SOURCE: Ackley Vol.1 Pg441

HISTORICAL DATA:

NOTES:

LOADING DATA:

BULLET WT./TYPE	POWDER WT./TYPE	VELOCITY ('/SEC)	SOURCE
180/—	78.0/4350	3330	Ackley

CASE PREPARATION: SHELLHOLDER (RCBS): 4

MAKE FROM: .300 Winchester Mag. Fireform factory ammo in the ICL chamber.

PHYSICAL DATA (INCHES):

CASE TYPE: **Belted Bottleneck**
CASE LENGTH **A = 2.610**
HEAD DIAMETER **B = .513**
RIM DIAMETER **D = .532**
NECK DIAMETER **F = .340**
NECK LENGTH **H = .340**
SHOULDER LENGTH **K = .085**
BODY ANGLE (DEG'S/SIDE): **.684**
CASE CAPACITY **CC'S = 5.95**

LOADED LENGTH: **A/R**
BELT DIAMETER **C = .532**
RIM THICKNESS **E = .05**
SHOULDER DIAMETER **G = .460**
LENGTH TO SHOULDER **J = 2.185**
SHOULDER ANGLE (DEG'S/SIDE): **35.22**
PRIMER: **L/R**
CASE CAPACITY (GR'S WATER): **82.81**

DIMENSIONAL DRAWING:

-NOT ACTUAL SIZE-
-DO NOT SCALE-

CARTRIDGE: .30 Johnson Special

OTHER NAMES:

DIA: .309

BALLISTEK NO: 308BP

NAI NO: RMS 31115/3.895

DATA SOURCE: Handloader #45 Pg30

HISTORICAL DATA: By C. Johnson for lead projectiles.

NOTES:

LOADING DATA:

BULLET WT./TYPE	POWDER WT./TYPE	VELOCITY ('/SEC)	SOURCE
158/Lead	16.0/IMR4198	1700	Johnson

CASE PREPARATION: SHELLHOLDER (RCBS): 2

MAKE FROM: .30-30 Win. Anneal case and F/L size in a .32-40 Win. die. Cut case to 1.70" and chamfer. Reduce neck dia. in a .32 S&W long die (watch shoulder location). Seat bullets with an arbor press. I do not believe any mfg. has ever made dies for this case. Johnson also turned the case rims but, I don't know why!

PHYSICAL DATA (INCHES):

CASE TYPE: **Rimmed Straight**

CASE LENGTH **A = 1.64**

HEAD DIAMETER **B = .421**

RIM DIAMETER **D = .473**

NECK DIAMETER **F = .335**

NECK LENGTH **H = N/A**

SHOULDER LENGTH **K = N/A**

BODY ANGLE (DEG'S/SIDE): **1.55**

CASE CAPACITY **CC'S = 1.97**

LOADED LENGTH: **A/R**

BELT DIAMETER **C = N/A**

RIM THICKNESS **E = .06**

SHOULDER DIAMETER **G = N/A**

LENGTH TO SHOULDER **J = N/A**

SHOULDER ANGLE (DEG'S/SIDE): **N/A**

PRIMER: **L/R**

CASE CAPACITY (GR'S WATER): **30.50**

DIMENSIONAL DRAWING:

F
H
A
E
B
D

-NOT ACTUAL SIZE-
-DO NOT SCALE-

CARTRIDGE: .30 Kurz

OTHER NAMES:
.30/30-06 Short (Kurz)

DIA: .308

BALLISTEK NO: 308Y

NAI NO: RXB 34343/2.742

DATA SOURCE: Ackley Vol.2 Pg185

HISTORICAL DATA:

NOTES:

LOADING DATA:

BULLET WT./TYPE	POWDER WT./TYPE	VELOCITY ('/SEC)	SOURCE
110/RN	25.0/4198	–	JJD

CASE PREPARATION: SHELLHOLDER (RCBS): 3

MAKE FROM: .308 Win. Form die set required. Anneal shoulder area of case. Set back shoulder in form die. Cut to 1.3" & chamfer. I.D. neck ream. Trim to length, F/L size and chamfer.

PHYSICAL DATA (INCHES):

CASE TYPE: **Rimless Bottleneck**

CASE LENGTH **A = 1.289**

HEAD DIAMETER **B = .470**

RIM DIAMETER **D = .473**

NECK DIAMETER **F = .333**

NECK LENGTH **H = .181**

SHOULDER LENGTH **K = .123**

BODY ANGLE (DEG'S/SIDE): **1.02**

CASE CAPACITY **CC'S = 2.09**

LOADED LENGTH: **1.70**

BELT DIAMETER **C = N/A**

RIM THICKNESS **E = .05**

SHOULDER DIAMETER **G = .442**

LENGTH TO SHOULDER **J = .985**

SHOULDER ANGLE (DEG'S/SIDE): **23.89**

PRIMER: **L/R**

CASE CAPACITY (GR'S WATER): **32.39**

DIMENSIONAL DRAWING:

F
H
K
G
A
J
E
B
D

-NOT ACTUAL SIZE-
-DO NOT SCALE-

CARTRIDGE: .30 Lever Power

OTHER NAMES:

DIA: .308

BALLISTEK NO: 308BM

NAI NO: RMB 12334/4.329

DATA SOURCE: Ackley Vol.1 Pg412

HISTORICAL DATA: By F. Wade.

NOTES:

LOADING DATA:

BULLET WT./TYPE	POWDER WT./TYPE	VELOCITY ('/SEC)	SOURCE
150/—	41.0/4064	2670	Ackley

CASE PREPARATION: SHELLHOLDER (RCBS): 7

MAKE FROM: .303 British. Turn rim dia. to .520" dia. Size in .30 L/P F/L die with expander removed. Trim to length & chamfer. F/L size. Fireform.

PHYSICAL DATA (INCHES):

CASE TYPE: **Rimmed Bottleneck**

CASE LENGTH **A = 1.97**

HEAD DIAMETER **B = .455**

RIM DIAMETER **D = .520**

NECK DIAMETER **F = .339**

NECK LENGTH **H = .325**

SHOULDER LENGTH **K = .045**

BODY ANGLE (DEG'S/SIDE): **.286**

CASE CAPACITY **CC'S = 3.28**

LOADED LENGTH: **A/R**

BELT DIAMETER **C = N/A**

RIM THICKNESS **E = .065**

SHOULDER DIAMETER **G = .441**

LENGTH TO SHOULDER **J = 1.60**

SHOULDER ANGLE (DEG'S/SIDE): **48.57**

PRIMER: **L/R**

CASE CAPACITY (GR'S WATER): **50.67**

DIMENSIONAL DRAWING:

-NOT ACTUAL SIZE-
-DO NOT SCALE-

CARTRIDGE: .30 Luger

OTHER NAMES:
7,65mm Luger (Parabellum)
7,65mm Bergmanns SMG

DIA: .308

BALLISTEK NO: 308V

NAI NO: RXB 34242/2.188

DATA SOURCE: NAI Ballistek

HISTORICAL DATA: By DWM about 1900.

NOTES:

LOADING DATA:

BULLET WT./TYPE	POWDER WT./TYPE	VELOCITY ('/SEC)	SOURCE
100/RN	4.8/Unique	1210	Barnes

CASE PREPARATION: SHELLHOLDER (RCBS): 1

MAKE FROM: 9mm Win. Mag. Trim case to length and F/L size in .30 Luger die. Re-trim and chamfer.

PHYSICAL DATA (INCHES):

CASE TYPE: **Rimless Bottleneck**

CASE LENGTH **A = .849**

HEAD DIAMETER **B = .388**

RIM DIAMETER **D = .392**

NECK DIAMETER **F = .330**

NECK LENGTH **H = .155**

SHOULDER LENGTH **K = .02**

BODY ANGLE (DEG'S/SIDE): **1.39**

CASE CAPACITY **CC'S = .862**

LOADED LENGTH: **1.13**

BELT DIAMETER **C = N/A**

RIM THICKNESS **E = .042**

SHOULDER DIAMETER **G = .368**

LENGTH TO SHOULDER **J = .610**

SHOULDER ANGLE (DEG'S/SIDE): **12.44**

PRIMER: **S/P**

CASE CAPACITY (GR'S WATER): **13.30**

DIMENSIONAL DRAWING:

-NOT ACTUAL SIZE-
-DO NOT SCALE-

CARTRIDGE: .30 Newton

OTHER NAMES:
.30 Adolph Express

DIA: .308

BALLISTEK NO: 308Z

NAI NO: RXB
12333/4.799

DATA SOURCE: NAI/Ballistek

HISTORICAL DATA: By C. Newton for F. Adolph in 1912.

NOTES:

LOADING DATA:

BULLET WT./TYPE	POWDER WT./TYPE	VELOCITY ('/SEC)	SOURCE
180/Spire	73.0/4831	2890	Barnes

CASE PREPARATION: SHELLHOLDER (RCBS): 28

MAKE FROM: 8 x 68S. F/L size the 8x68S case in the Newton die. Trim to length and chamfer. Fireform in the chamber.

PHYSICAL DATA (INCHES):

CASE TYPE: **Rimless Bottleneck**

CASE LENGTH **A = 2.515**

HEAD DIAMETER **B = .524**

RIM DIAMETER **D = .519**

NECK DIAMETER **F = .340**

NECK LENGTH **H = .310**

SHOULDER LENGTH **K = .185**

BODY ANGLE (DEG'S/SIDE): **.220**

CASE CAPACITY **CC'S = 5.78**

LOADED LENGTH: **3.29**

BELT DIAMETER **C = N/A**

RIM THICKNESS **E = .05**

SHOULDER DIAMETER **G = .51**

LENGTH TO SHOULDER **J = 2.02**

SHOULDER ANGLE (DEG'S/SIDE): **24.67**

PRIMER: **L/R**

CASE CAPACITY (GR'S WATER): **89.20**

DIMENSIONAL DRAWING:

F H K G A J E B D

-NOT ACTUAL SIZE-
-DO NOT SCALE-

CARTRIDGE: .30 Remington

OTHER NAMES:

DIA: .308

BALLISTEK NO: 308B

NAI NO: RXB
22223/4.869

DATA SOURCE: NAI/Ballistek

HISTORICAL DATA: By Rem. in 1906.

NOTES:

LOADING DATA:

BULLET WT./TYPE	POWDER WT./TYPE	VELOCITY ('/SEC)	SOURCE
150/RN	34.0/3031	2407	Barnes

CASE PREPARATION: SHELLHOLDER (RCBS): 19

MAKE FROM: .30-30 Win. Turn rim to .422" dia. and re-cut extractor groove. F/L size and trim to length. .25 Rem. brass may also be taper expanded to .31 cal. and F/L sized. .225 Win. may be used by turning the rim to dia., cutting a new extractor groove, taper expanding and F/L sizing.

PHYSICAL DATA (INCHES):

CASE TYPE: **Rimless Bottleneck**

CASE LENGTH **A = 2.05**

HEAD DIAMETER **B = .421**

RIM DIAMETER **D = .422**

NECK DIAMETER **F = .332**

NECK LENGTH **H = .472**

SHOULDER LENGTH **K = .08**

BODY ANGLE (DEG'S/SIDE): **.441**

CASE CAPACITY **CC'S = 2.96**

LOADED LENGTH: **2.54**

BELT DIAMETER **C = N/A**

RIM THICKNESS **E = .049**

SHOULDER DIAMETER **G = .401**

LENGTH TO SHOULDER **J = 1.498**

SHOULDER ANGLE (DEG'S/SIDE): **23.32**

PRIMER: **L/R**

CASE CAPACITY (GR'S WATER): **45.75**

DIMENSIONAL DRAWING:

F H K G A E J B D

-NOT ACTUAL SIZE-
-DO NOT SCALE-

CARTRIDGE: .30 Smith

OTHER NAMES:	DIA: .308
	BALLISTEK NO: 308AF
	NAI NO: BEN 22323/4.873

DATA SOURCE: Ackley Vol.2 Pg190

HISTORICAL DATA:

NOTES:

LOADING DATA:

BULLET WT./TYPE	POWDER WT./TYPE	VELOCITY ('/SEC)	SOURCE
180/Spire	69.0/4350	3162	Ackley

CASE PREPARATION: SHELLHOLDER (RCBS): 4

MAKE FROM: .300 H&H Mag. F/L size case with expander removed. Trim to length & chamfer. F/L size. Fireform in chamber.

PHYSICAL DATA (INCHES):

CASE TYPE: **Belted Bottleneck**

CASE LENGTH **A = 2.500**

HEAD DIAMETER **B = .513**

RIM DIAMETER **D = .532**

NECK DIAMETER **F = .339**

NECK LENGTH **H = .405**

SHOULDER LENGTH **K = .155**

BODY ANGLE (DEG'S/SIDE): **.378**

CASE CAPACITY **CC'S = 5.31**

LOADED LENGTH: **3.35**

BELT DIAMETER **C = .532**

RIM THICKNESS **E = .05**

SHOULDER DIAMETER **G = .490**

LENGTH TO SHOULDER **J = 1.94**

SHOULDER ANGLE (DEG'S/SIDE): **25.97**

PRIMER: **L/R Mag**

CASE CAPACITY (GR'S WATER): **81.94**

DIMENSIONAL DRAWING:

F H K G A J E B D C

-NOT ACTUAL SIZE-
-DO NOT SCALE-

CARTRIDGE: .30 Streaker

OTHER NAMES:	DIA: .308
	BALLISTEK NO: 308CA
	NAI NO: RMB 24243/3.481

DATA SOURCE: Handloader #77 Pg36

HISTORICAL DATA: Cal. .30 revolver ctg. about 1978 by B. Rayzak.

NOTES:

LOADING DATA:

BULLET WT./TYPE	POWDER WT./TYPE	VELOCITY ('/SEC)	SOURCE
110/JHP	22.0/IMR4227	1821	Rayzak

CASE PREPARATION: SHELLHOLDER (RCBS): 2

MAKE FROM: .30-30 Win. Cut cases 1.5" long and chamfer the O.D. F/L size in Streaker die, with expander removed. Trim to final length, chamfer and I.D. neck ream. Clean and F/L size.

PHYSICAL DATA (INCHES):

CASE TYPE: **Rimmed Bottleneck**

CASE LENGTH **A = 1.469**

HEAD DIAMETER **B = .422**

RIM DIAMETER **D = .506**

NECK DIAMETER **F = .330**

NECK LENGTH **H = .290**

SHOULDER LENGTH **K = .067**

BODY ANGLE (DEG'S/SIDE): **.439**

CASE CAPACITY **CC'S = 2.08**

LOADED LENGTH: **A/R**

BELT DIAMETER **C = N/A**

RIM THICKNESS **E = .063**

SHOULDER DIAMETER **G = .408**

LENGTH TO SHOULDER **J = 1.112**

SHOULDER ANGLE (DEG'S/SIDE): **30.20**

PRIMER: **L/R**

CASE CAPACITY (GR'S WATER): **32.19**

DIMENSIONAL DRAWING:

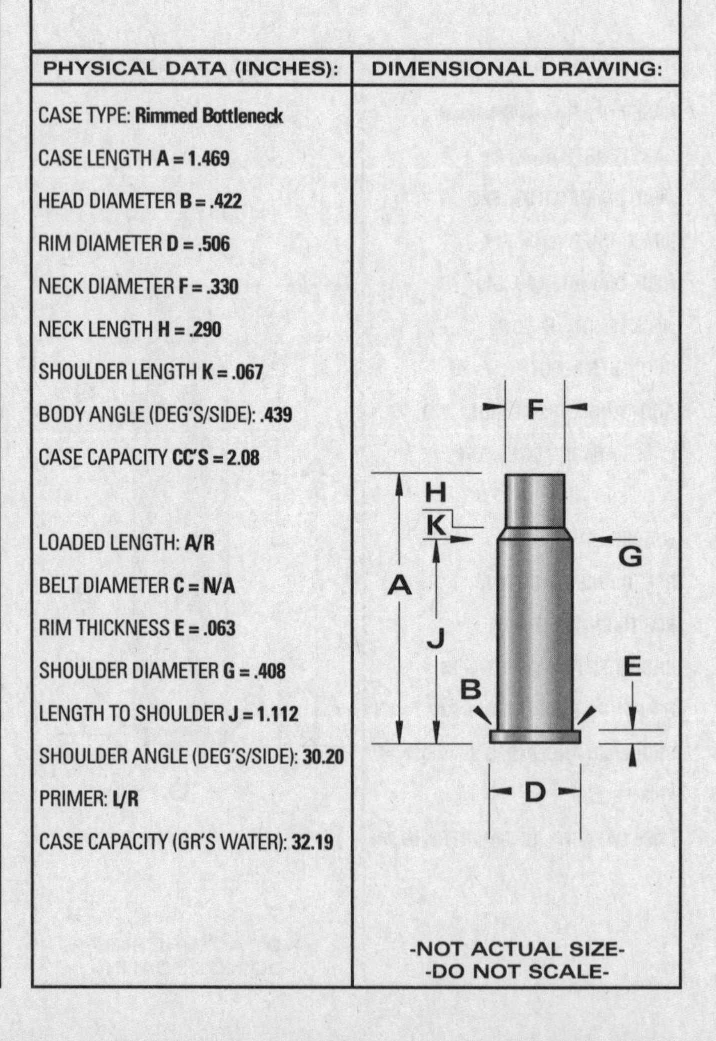

-NOT ACTUAL SIZE-
-DO NOT SCALE-

CARTRIDGE: .30-06 ICL Caribou

OTHER NAMES:

DIA: .308

BALLISTEK NO: 308BN

NAI NO: RXB 12324/5.297

DATA SOURCE: Ackley Vol.1 Pg427

HISTORICAL DATA:

NOTES:

LOADING DATA:

BULLET WT./TYPE	POWDER WT./TYPE	VELOCITY ('/SEC)	SOURCE
180/Spire	60.0/4350	2950	Ackley

CASE PREPARATION: SHELLHOLDER (RCBS): 3

MAKE FROM: .30-06 Spgf. Taper expand the '06 case to about .315" dia. F/L size in the ICL die. Square case mouth & chamfer. Fireform in the ICL chamber.

PHYSICAL DATA (INCHES):

CASE TYPE: **Rimless Bottleneck**

CASE LENGTH **A = 2.49**

HEAD DIAMETER **B = .470**

RIM DIAMETER **D = .473**

NECK DIAMETER **F = .339**

NECK LENGTH **H = .420**

SHOULDER LENGTH **K = .05**

BODY ANGLE (DEG'S/SIDE):**.283**

CASE CAPACITY **CC'S = 4.52**

LOADED LENGTH: **A/R**

BELT DIAMETER **C = N/A**

RIM THICKNESS **E = .050**

SHOULDER DIAMETER **G = .452**

LENGTH TO SHOULDER **J = 2.02**

SHOULDER ANGLE (DEG'S/SIDE): **48.49**

PRIMER: **L/R**

CASE CAPACITY (GR'S WATER): **69.83**

DIMENSIONAL DRAWING:

F
H
K
G
A
J
E
B
D

-NOT ACTUAL SIZE-
-DO NOT SCALE-

CARTRIDGE: .30-06 Improved (Ackley)

OTHER NAMES:

DIA: .308

BALLISTEK NO: 308AO

NAI NO: RXB 22333/5.276

DATA SOURCE: Ackley Vol.1 Pg427

HISTORICAL DATA:

NOTES:

LOADING DATA:

BULLET WT./TYPE	POWDER WT./TYPE	VELOCITY ('/SEC)	SOURCE
180/Spire	55.0/4350	2720	Ackley

CASE PREPARATION: SHELLHOLDER (RCBS): 3

MAKE FROM: .30-06 Spgf. Fire factory '06 ammo in the "improved" chamber.

PHYSICAL DATA (INCHES):

CASE TYPE: **Rimless Bottleneck**

CASE LENGTH **A = 2.48**

HEAD DIAMETER **B = .470**

RIM DIAMETER **D = .473**

NECK DIAMETER **F = .338**

NECK LENGTH **H = .380**

SHOULDER LENGTH **K = .099**

BODY ANGLE (DEG'S/SIDE): **.477**

CASE CAPACITY **CC'S = 4.43**

LOADED LENGTH: **A/R**

BELT DIAMETER **C = N/A**

RIM THICKNESS **E = .05**

SHOULDER DIAMETER **G = .440**

LENGTH TO SHOULDER **J = 2.00**

SHOULDER ANGLE (DEG'S/SIDE): **27.02**

PRIMER: **L/R**

CASE CAPACITY (GR'S WATER): **68.34**

DIMENSIONAL DRAWING:

F
H
K
G
A
J
E
B
D

-NOT ACTUAL SIZE-
-DO NOT SCALE-

CARTRIDGE: .30-30 Improved (Ackley)

OTHER NAMES:

DIA: .308

BALLISTEK NO: 308AP

NAI NO: RMB 22233/4.739

DATA SOURCE: Handloader #107 Pg36

HISTORICAL DATA:

NOTES:

LOADING DATA:

BULLET WT./TYPE	POWDER WT./TYPE	VELOCITY ('/SEC)	SOURCE
170/FN	34.5/IMR3031	2240	Roberts

CASE PREPARATION: SHELLHOLDER (RCBS): 3

MAKE FROM: .30-30 Win. Fire factory ammo in "improved" chamber or fireform the case in the proper trim die.

PHYSICAL DATA (INCHES):

CASE TYPE: **Rimmed Bottleneck**

CASE LENGTH **A = 2.00**

HEAD DIAMETER **B = .422**

RIM DIAMETER **D = .506**

NECK DIAMETER **F = .330**

NECK LENGTH **H = .340**

SHOULDER LENGTH **K = .070**

BODY ANGLE (DEG'S/SIDE): **.370**

CASE CAPACITY **CC'S = 3.02**

LOADED LENGTH: **A/R**

BELT DIAMETER **C = N/A**

RIM THICKNESS **E = .063**

SHOULDER DIAMETER **G = .404**

LENGTH TO SHOULDER **J = 1.59**

SHOULDER ANGLE (DEG'S/SIDE): **27.85**

PRIMER: **L/R**

CASE CAPACITY (GR'S WATER): **46.63**

DIMENSIONAL DRAWING:

-NOT ACTUAL SIZE-
-DO NOT SCALE-

CARTRIDGE: .30-30 Wesson

OTHER NAMES:

DIA: .308

BALLISTEK NO: 308AA

NAI NO: RMS 21115/4.368

DATA SOURCE: COTW 4th Pg94

HISTORICAL DATA: By F. Wesson about 1880.

NOTES:

LOADING DATA:

BULLET WT./TYPE	POWDER WT./TYPE	VELOCITY ('/SEC)	SOURCE
165/RN	10.0/2400	1185	Barnes

CASE PREPARATION: SHELLHOLDER (RCBS): 6

MAKE FROM: .357 Max. F/L size .357 case in Wesson die. Trim & chamfer.

PHYSICAL DATA (INCHES):

CASE TYPE: **Rimmed Straight**

CASE LENGTH **A = 1.66**

HEAD DIAMETER **B = .380**

RIM DIAMETER **D = .440**

NECK DIAMETER **F = .329**

NECK LENGTH **H = N/A**

SHOULDER LENGTH **K = N/A**

BODY ANGLE (DEG'S/SIDE): **.907**

CASE CAPACITY **CC'S = 1.81**

LOADED LENGTH: **2.50**

BELT DIAMETER **C = N/A**

RIM THICKNESS **E = .05**

SHOULDER DIAMETER **G = N/A**

LENGTH TO SHOULDER **J = N/A**

SHOULDER ANGLE (DEG'S/SIDE): **N/A**

PRIMER: **S/P**

CASE CAPACITY (GR'S WATER): **28.06**

DIMENSIONAL DRAWING:

-NOT ACTUAL SIZE-
-DO NOT SCALE-

CARTRIDGE: .30-30 Winchester

OTHER NAMES:
7,62 x 51R

DIA: .308

BALLISTEK NO: 308E

NAI NO: RMB 22222/4.693

DATA SOURCE: Lyman Handbook 45th Pg91

HISTORICAL DATA: By Win. in 1895

NOTES:

LOADING DATA:

BULLET WT./TYPE	POWDER WT./TYPE	VELOCITY ('/SEC)	SOURCE
150/FN	29.4/IMR3031	2270	JJD

CASE PREPARATION: SHELLHOLDER (RCBS): 2

MAKE FROM: Factory or .375 Win. Anneal case and F/L size in .30-30 die. Trim to length and chamfer. 9,3 x 72R will also form in the same manner but require a bit more base swaging.

PHYSICAL DATA (INCHES):

CASE TYPE: **Rimmed Bottleneck**

CASE LENGTH **A = 1.976**

HEAD DIAMETER **B = .421**

RIM DIAMETER **D = .506**

NECK DIAMETER **F = .330**

NECK LENGTH **H = .426**

SHOULDER LENGTH **K = .132**

BODY ANGLE (DEG'S/SIDE): **.446**

CASE CAPACITY **CC'S = 2.88**

LOADED LENGTH: **2.55**

BELT DIAMETER **C = N/A**

RIM THICKNESS **E = .062**

SHOULDER DIAMETER **G = .402**

LENGTH TO SHOULDER **J = 1.418**

SHOULDER ANGLE (DEG'S/SIDE): **15.25**

PRIMER: **L/R**

CASE CAPACITY (GR'S WATER): **44.50**

DIMENSIONAL DRAWING:

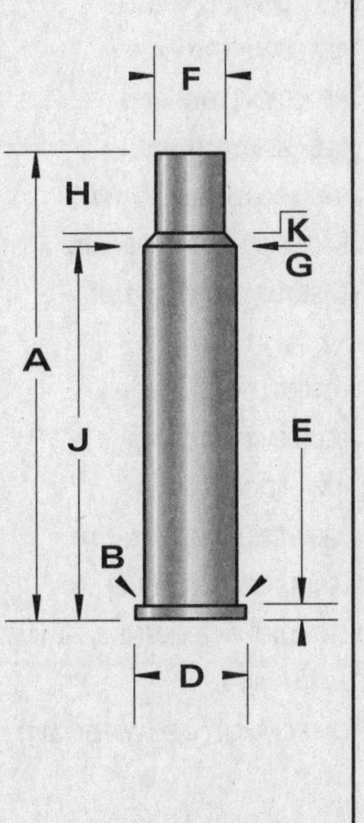

-NOT ACTUAL SIZE-
-DO NOT SCALE-

CARTRIDGE: .30-40 Krag

OTHER NAMES:
.30 Krag-Jorgensen
.30 U.S. Army

DIA: .308

BALLISTEK NO: 308F

NAI NO: RMB 22323/4.923

DATA SOURCE: Hornady 3rd Pg 225

HISTORICAL DATA: U.S. Mil. ctg. ca. 1892

NOTES:

LOADING DATA:

BULLET WT./TYPE	POWDER WT./TYPE	VELOCITY ('/SEC)	SOURCE
220/RN	43.0/IMR4350	2004	Lyman
190/HPBT	41.5/IMR4895	2500	JJD

CASE PREPARATION: SHELLHOLDER (RCBS): 7

MAKE FROM: Factory or .303 British. Size .303 case in .30-40 die. Square case mouth. F/L size. This case will be quite short, in the neck, but will work fine. BELL .405 win. basic will also work. Anneal and full-length size, trim and chamfer.

PHYSICAL DATA (INCHES):

CASE TYPE: **Rimmed Bottleneck**

CASE LENGTH **A = 2.25**

HEAD DIAMETER **B = .457**

RIM DIAMETER **D = .545**

NECK DIAMETER **F = .338**

NECK LENGTH **H = .42**

SHOULDER LENGTH **K = .100**

BODY ANGLE (DEG'S/SIDE): **.636**

CASE CAPACITY **CC'S = 3.653**

LOADED LENGTH: **3.09**

BELT DIAMETER **C = N/A**

RIM THICKNESS **E = .065**

SHOULDER DIAMETER **G = .423**

LENGTH TO SHOULDER **J = 1.73**

SHOULDER ANGLE (DEG'S/SIDE): **23.02**

PRIMER: **L/R**

CASE CAPACITY (GR'S WATER): **56.38**

DIMENSIONAL DRAWING:

-NOT ACTUAL SIZE-
-DO NOT SCALE-

CARTRIDGE: .30-40 Krag Improved (Ackley)

OTHER NAMES:

DIA: .308

BALLISTEK NO: 308AQ

NAI NO: RMB 22324/4.901

DATA SOURCE: Ackley Vol.1 Pg415

HISTORICAL DATA:

NOTES:

LOADING DATA:

BULLET WT./TYPE	POWDER WT./TYPE	VELOCITY ('/SEC)	SOURCE
180/Spire	54.0/4350	2740	Ackley

CASE PREPARATION: SHELLHOLDER (RCBS): 7

MAKE FROM: .30-40 Krag. Fire the factory ammo in the "improved" chamber.

PHYSICAL DATA (INCHES):

CASE TYPE: **Rimmed Bottleneck**

CASE LENGTH **A = 2.24**

HEAD DIAMETER **B = .457**

RIM DIAMETER **D = .545**

NECK DIAMETER **F = .338**

NECK LENGTH **H = .385**

SHOULDER LENGTH **K = .065**

BODY ANGLE (DEG'S/SIDE): **.522**

CASE CAPACITY **CC'S = 3.72**

LOADED LENGTH: **3.07**

BELT DIAMETER **C = N/A**

RIM THICKNESS **E = .063**

SHOULDER DIAMETER **G = .428**

LENGTH TO SHOULDER **J = 1.79**

SHOULDER ANGLE (DEG'S/SIDE): **34.69**

PRIMER: **L/R**

CASE CAPACITY (GR'S WATER): **57.35**

DIMENSIONAL DRAWING:

F H K G A J E B D

-NOT ACTUAL SIZE-
-DO NOT SCALE-

CARTRIDGE: .30-40 Wesson

OTHER NAMES:

DIA: .308

BALLISTEK NO: 308AB

NAI NO: RMB 13222/4.323

DATA SOURCE: COTW 4th Pg94

HISTORICAL DATA: From about 1880.

NOTES: As factory improvement over .30-30 Wesson (#308AA).

LOADING DATA:

BULLET WT./TYPE	POWDER WT./TYPE	VELOCITY ('/SEC)	SOURCE
165/FN	10.0/2400	1200	Barnes

CASE PREPARATION: SHELLHOLDER (RCBS): 6

MAKE FROM: .357 Max. F/L size case in .30-40 Wesson die. Square case mouth & chamfer. Case will be slightly short (.025") but, otherwise, okay.

PHYSICAL DATA (INCHES):

CASE TYPE: **Rimmed Bottleneck**

CASE LENGTH **A = 1.63**

HEAD DIAMETER **B = .381**

RIM DIAMETER **D = .436**

NECK DIAMETER **F = .329**

NECK LENGTH **H = .425**

SHOULDER LENGTH **K = .140**

BODY ANGLE (DEG'S/SIDE): **.132**

CASE CAPACITY **CC'S = 1.99**

LOADED LENGTH: **2.39**

BELT DIAMETER **C = N/A**

RIM THICKNESS **E = .048**

SHOULDER DIAMETER **G = .377**

LENGTH TO SHOULDER **J = 1.065**

SHOULDER ANGLE (DEG'S/SIDE): **10.52**

PRIMER: **S/P**

CASE CAPACITY (GR'S WATER): **30.77**

DIMENSIONAL DRAWING:

F H K G A J E B D

-NOT ACTUAL SIZE-
-DO NOT SCALE-

CARTRIDGE: .30-78 Single Shot

OTHER NAMES:

DIA: .310

BALLISTEK NO: 310H

NAI NO: RMS 31115/4.147

DATA SOURCE: Rifle #67 Pg6

HISTORICAL DATA: By G. Sengel.

NOTES:

LOADING DATA:

BULLET WT./TYPE	POWDER WT./TYPE	VELOCITY ('/SEC)	SOURCE
170/Lead	20.5/IMR4198	—	JJD

CASE PREPARATION: SHELLHOLDER (RCBS): 2

MAKE FROM: .30-30 Win. Trim case to length and F/L size. Use .310" dia. bullets or unsized .30 cal. bullets.

PHYSICAL DATA (INCHES):

CASE TYPE: **Rimmed Straight**
CASE LENGTH **A = 1.75**
HEAD DIAMETER **B = .422**
RIM DIAMETER **D = .506**
NECK DIAMETER **F = .331**
NECK LENGTH **H = .5**
SHOULDER LENGTH **K = N/A**
BODY ANGLE (DEG'S/SIDE): **1.54**
CASE CAPACITY **CC'S = .949**

LOADED LENGTH: **2.29**
BELT DIAMETER **C = N/A**
RIM THICKNESS **E = .062**
SHOULDER DIAMETER **G = N/A**
LENGTH TO SHOULDER **J = N/A**
SHOULDER ANGLE (DEG'S/SIDE): **N/A**
PRIMER: **L/R**
CASE CAPACITY (GR'S WATER): **14.65**

DIMENSIONAL DRAWING:

F
H
A
B
E
D

-NOT ACTUAL SIZE-
-DO NOT SCALE-

CARTRIDGE: .30-'06 Springfield

OTHER NAMES:
7,62 x 63mm
.30 M2 U.S.

DIA: .308

BALLISTEK NO: 308BJ

NAI NO: RXB 22332/5.306

DATA SOURCE: Hornady 3rd Pg228

HISTORICAL DATA: U.S. mil. ctg ca. 1906

NOTES: Followed the '03 version (same case).

LOADING DATA:

BULLET WT./TYPE	POWDER WT./TYPE	VELOCITY ('/SEC)	SOURCE
168/HPBT	45.0/IMR4064	2551	Lyman
190/HPBT	41.5/IMR4895	2500	JJD

CASE PREPARATION: SHELLHOLDER (RCBS): 3

MAKE FROM: Factory or .270 Win. Anneal case neck and taper expand to about .310-.315 I.D. F/L size in '06 die. Trim and chamfer.

PHYSICAL DATA (INCHES):

CASE TYPE: **Rimless Bottleneck**
CASE LENGTH **A = 2.494**
HEAD DIAMETER **B = .470**
RIM DIAMETER **D = .473**
NECK DIAMETER **F = .339**
NECK LENGTH **H = .383**
SHOULDER LENGTH **K = .162**
BODY ANGLE (DEG'S/SIDE): **.475**
CASE CAPACITY **CC'S = 4.40**

LOADED LENGTH: **3.34**
BELT DIAMETER **C = N/A**
RIM THICKNESS **E = .049**
SHOULDER DIAMETER **G = .441**
LENGTH TO SHOULDER **J = 1.949**
SHOULDER ANGLE (DEG'S/SIDE): **17.47**
PRIMER: **L/R**
CASE CAPACITY (GR'S WATER): **68.01**

DIMENSIONAL DRAWING:

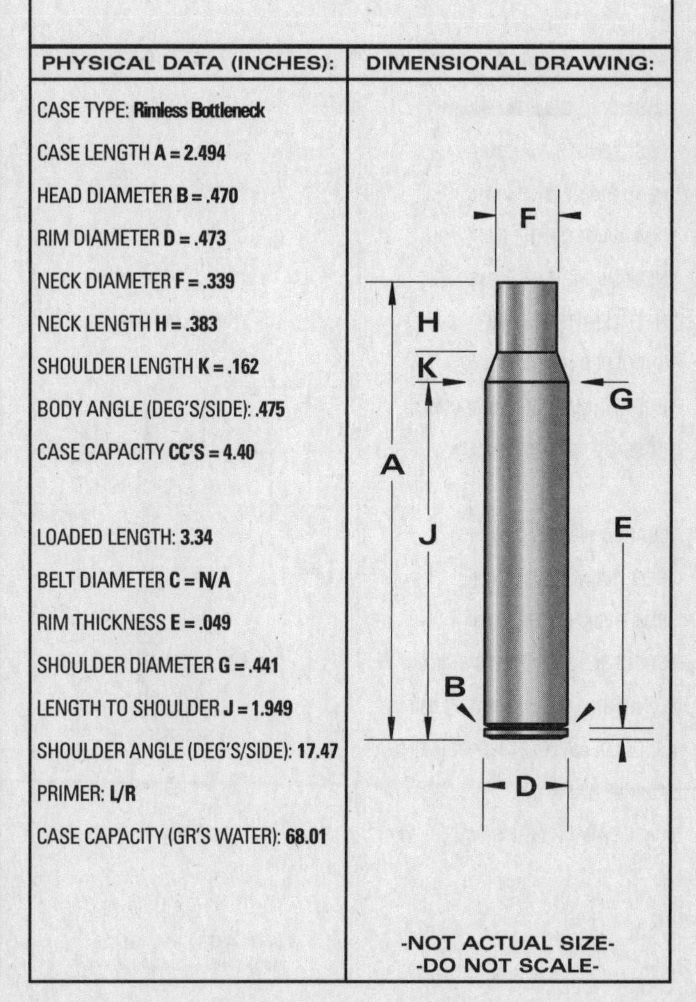

-NOT ACTUAL SIZE-
-DO NOT SCALE-

CARTRIDGE: .30/224 Weatherby

OTHER NAMES:

DIA: .308

BALLISTEK NO: 308BR

NAI NO: BEN 22243/4.626

DATA SOURCE: Ackley Vol.2 Pg186

HISTORICAL DATA:

NOTES:

LOADING DATA:

BULLET WT./TYPE	POWDER WT./TYPE	VELOCITY ('/SEC)	SOURCE
150/—	32.1/3031	2400	Ackley

CASE PREPARATION: SHELLHOLDER (RCBS): 27

MAKE FROM: .224 Weatherby. Anneal case neck and taper expand to .315" dia. F/L size, square case mouth and chamfer.

PHYSICAL DATA (INCHES):

CASE TYPE: **Belted Bottleneck**

CASE LENGTH **A = 1.920**

HEAD DIAMETER **B = .415**

RIM DIAMETER **D = .429**

NECK DIAMETER **F = .332**

NECK LENGTH **H = .300**

SHOULDER LENGTH **K = .086**

BODY ANGLE (DEG'S/SIDE): **.460**

CASE CAPACITY **CC'S = 2.74**

LOADED LENGTH: **2.55**

BELT DIAMETER **C = .429**

RIM THICKNESS **E = .049**

SHOULDER DIAMETER **G = .393**

LENGTH TO SHOULDER **J = 1.568**

SHOULDER ANGLE (DEG'S/SIDE): **N/A**

PRIMER: **L/R**

CASE CAPACITY (GR'S WATER): **42.41**

DIMENSIONAL DRAWING:

F H K G A J E B D C

-NOT ACTUAL SIZE-
-DO NOT SCALE-

CARTRIDGE: .30/284 Winchester

OTHER NAMES:

DIA: .308

BALLISTEK NO: 308N

NAI NO: RBB 22344/4.330

DATA SOURCE: Ackley Vol.2 Pg187

HISTORICAL DATA:

NOTES:

LOADING DATA:

BULLET WT./TYPE	POWDER WT./TYPE	VELOCITY ('/SEC)	SOURCE
180/Spire	51.0/4320	2750	Ackley

CASE PREPARATION: SHELLHOLDER (RCBS): 3

MAKE FROM: .284 Winchester. Taper expand the .284 case to .31" dia. Trim to length, chamfer and F/L size.

PHYSICAL DATA (INCHES):

CASE TYPE: **Rebated Bottleneck**

CASE LENGTH **A = 2.165**

HEAD DIAMETER **B = .500**

RIM DIAMETER **D = .473**

NECK DIAMETER **F = .342**

NECK LENGTH **H = .290**

SHOULDER LENGTH **K = .099**

BODY ANGLE (DEG'S/SIDE): **.454**

CASE CAPACITY **CC'S = 4.29**

LOADED LENGTH: **3.12**

BELT DIAMETER **C = N/A**

RIM THICKNESS **E = .05**

SHOULDER DIAMETER **G = .475**

LENGTH TO SHOULDER **J = 1.775**

SHOULDER ANGLE (DEG'S/SIDE): **33.62**

PRIMER: **L/R**

CASE CAPACITY (GR'S WATER): **66.20**

DIMENSIONAL DRAWING:

F H K G A J E B D

-NOT ACTUAL SIZE-
-DO NOT SCALE-

CARTRIDGE: .30/30 Carbine

OTHER NAMES:

DIA: .308
BALLISTEK NO: 308BQ
NAI NO: RXB 34344/2.851

DATA SOURCE: Ackley Vol.2 Pg185

HISTORICAL DATA:

NOTES: Similar to .30 Kurz (#308Y)

LOADING DATA:

BULLET WT./TYPE	POWDER WT./TYPE	VELOCITY ('/SEC)	SOURCE
110/RN	21.7/IMR4198	–	JJD

CASE PREPARATION: **SHELLHOLDER (RCBS):** 3

MAKE FROM: Any .30-06-type case head. You'll also need a form set. Anneal the case's shoulder area and run through form die set. Cut off to 1.40" and I.D. neck ream. Trim to length and F/L size. Not really worth the trouble unless you really can't find anything else to do!

PHYSICAL DATA (INCHES):

CASE TYPE: **Rimless Bottleneck**
CASE LENGTH **A = 1.34**
HEAD DIAMETER **B = .470**
RIM DIAMETER **D = .473**
NECK DIAMETER **F = .338**
NECK LENGTH **H = .30**
SHOULDER LENGTH **K = .06**
BODY ANGLE (DEG'S/SIDE): **1.46**
CASE CAPACITY **CC'S = 2.00**

LOADED LENGTH: **1.72**
BELT DIAMETER **C = N/A**
RIM THICKNESS **E = .050**
SHOULDER DIAMETER **G = .430**
LENGTH TO SHOULDER **J = .98**
SHOULDER ANGLE (DEG'S/SIDE): **37.17**
PRIMER: **L/R**
CASE CAPACITY (GR'S WATER): **30.91**

DIMENSIONAL DRAWING:

F
H
K
G
A
J
E
B
D

-NOT ACTUAL SIZE-
-DO NOT SCALE-

CARTRIDGE: .30/338 Winchester Magnum

OTHER NAMES:
.30 Belted Newton

DIA: .308
BALLISTEK NO: 308O
NAI NO: BEN 12343/4.873

DATA SOURCE: Sierra Manual 1985 Pg242

HISTORICAL DATA: By Fred Huntington (?) in 1958

NOTES: Excellent 1000 yard cartridge.

LOADING DATA:

BULLET WT./TYPE	POWDER WT./TYPE	VELOCITY ('/SEC)	SOURCE
190/HPBT	68.0/IMR4831	2950	JJD

CASE PREPARATION: **SHELLHOLDER (RCBS):** 4

MAKE FROM: .338 Winchester Mag. Anneal case neck. Size .338 case in the .30/338 F/L sizer. Trim and chamfer.

PHYSICAL DATA (INCHES):

CASE TYPE: **Belted Bottleneck**
CASE LENGTH **A = 2.500**
HEAD DIAMETER **B = .513**
RIM DIAMETER **D = .532**
NECK DIAMETER **F = .339**
NECK LENGTH **H = .299**
SHOULDER LENGTH **K = .161**
BODY ANGLE (DEG'S/SIDE): **.342**
CASE CAPACITY **CC'S = 5.47**

LOADED LENGTH: **3.34**
BELT DIAMETER **C = .532**
RIM THICKNESS **E = .049**
SHOULDER DIAMETER **G = .491**
LENGTH TO SHOULDER **J = 2.04**
SHOULDER ANGLE (DEG'S/SIDE): **25.26**
PRIMER: **L/R Mag**
CASE CAPACITY (GR'S WATER): **84.41**

DIMENSIONAL DRAWING:

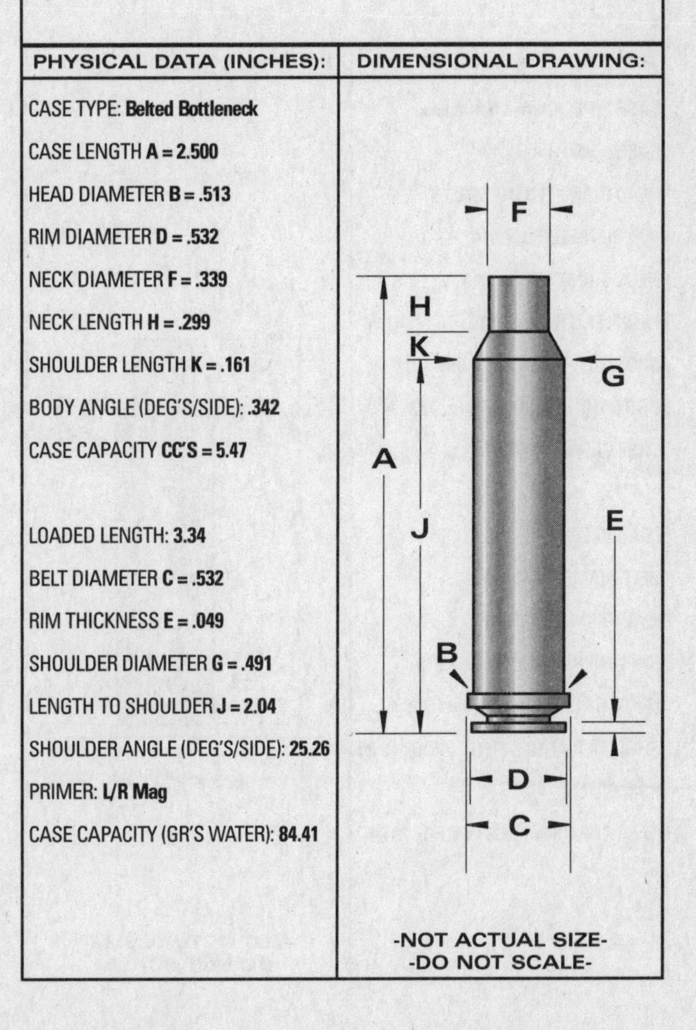

CARTRIDGE: .30/348 Winchester

OTHER NAMES:

DIA: .308

BALLISTEK NO: 308AR

NAI NO: RMB 23434/3.924

DATA SOURCE: Ackley Vol.1 Pg434

HISTORICAL DATA: One of B. Huton's.

NOTES:

LOADING DATA:

BULLET WT./TYPE	POWDER WT./TYPE	VELOCITY ('/SEC)	SOURCE
170/FN	60.0/4350	2710	Ackley

CASE PREPARATION: SHELLHOLDER (RCBS): 5

MAKE FROM: .348 Win. Form die set required. Anneal case, form and trim. Case will fireform (shoulder area) in the chamber.

PHYSICAL DATA (INCHES):

CASE TYPE: **Rimmed Bottleneck**

CASE LENGTH **A = 2.17**

HEAD DIAMETER **B = .553**

RIM DIAMETER **D = .610**

NECK DIAMETER **F = .340**

NECK LENGTH **H = .330**

SHOULDER LENGTH **K = .085**

BODY ANGLE (DEG'S/SIDE): **.792**

CASE CAPACITY **CC'S = 5.12**

LOADED LENGTH: **2.75**

BELT DIAMETER **C = N/A**

RIM THICKNESS **E = .07**

SHOULDER DIAMETER **G = .510**

LENGTH TO SHOULDER **J = 1.755**

SHOULDER ANGLE (DEG'S/SIDE): **45.00**

PRIMER: **L/R**

CASE CAPACITY (GR'S WATER): **79.02**

DIMENSIONAL DRAWING:

F H K G A J E B D

-NOT ACTUAL SIZE-
-DO NOT SCALE-

CARTRIDGE: .30/350 Remington Magnum

OTHER NAMES:

DIA: .308

BALLISTEK NO: 308AS

NAI NO: BEN 12343/4.220

DATA SOURCE: Ackley Vol.2 Pg189

HISTORICAL DATA:

NOTES: Very much overbore!

LOADING DATA:

BULLET WT./TYPE	POWDER WT./TYPE	VELOCITY ('/SEC)	SOURCE
180/—	52.0/4895	2780	Ackley

CASE PREPARATION: SHELLHOLDER (RCBS): 4

MAKE FROM: .300 H&H Mag. Anneal case and form in the F/L die with the expander removed. Trim to length. I.D. neck ream. F/L size. Fireform in the chamber.

PHYSICAL DATA (INCHES):

CASE TYPE: **Belted Bottleneck**

CASE LENGTH **A = 2.165**

HEAD DIAMETER **B = .513**

RIM DIAMETER **D = .532**

NECK DIAMETER **F = .340**

NECK LENGTH **H = .265**

SHOULDER LENGTH **K = .160**

BODY ANGLE (DEG'S/SIDE): **.335**

CASE CAPACITY **CC'S = 4.64**

LOADED LENGTH: **3.00**

BELT DIAMETER **C = .532**

RIM THICKNESS **E = .05**

SHOULDER DIAMETER **G = .495**

LENGTH TO SHOULDER **J = 1.74**

SHOULDER ANGLE (DEG'S/SIDE): **25.84**

PRIMER: **L/R Mag**

CASE CAPACITY (GR'S WATER): **71.55**

DIMENSIONAL DRAWING:

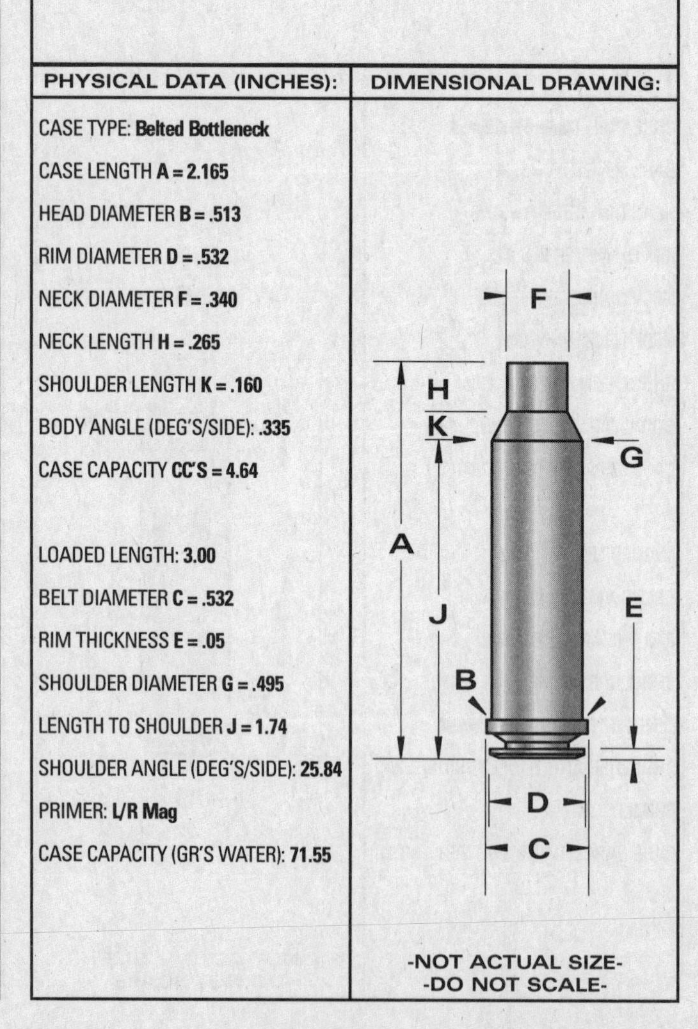

-NOT ACTUAL SIZE-
-DO NOT SCALE-

CARTRIDGE: .30/357 Paxton

OTHER NAMES:

DIA: .308

BALLISTEK NO: 308AC

NAI NO: RMB 14244/3.435

DATA SOURCE: Ackley Vol.2 Pg223

HISTORICAL DATA: By B. Paxton.

NOTES:

LOADING DATA:

BULLET WT./TYPE	POWDER WT./TYPE	VELOCITY ('/SEC)	SOURCE
110/RN	17.0/4227	1950	Paxton/ Ackley

CASE PREPARATION: SHELLHOLDER (RCBS): 6

MAKE FROM: .357 Mag. F/L size the .357 in the Paxton die. Square case mouth and chamfer.

PHYSICAL DATA (INCHES):

CASE TYPE: **Rimmed Bottleneck**

CASE LENGTH **A = 1.300**

HEAD DIAMETER **B = .379**

RIM DIAMETER **D = .440**

NECK DIAMETER **F = .330**

NECK LENGTH **H = .270**

SHOULDER LENGTH **K = .037**

BODY ANGLE (DEG'S/SIDE): **0**

CASE CAPACITY **CC'S = 1.54**

LOADED LENGTH: **1.725**

BELT DIAMETER **C = N/A**

RIM THICKNESS **E = .06**

SHOULDER DIAMETER **G = .379**

LENGTH TO SHOULDER **J = .995**

SHOULDER ANGLE (DEG'S/SIDE): **33.51**

PRIMER: **S/P**

CASE CAPACITY (GR'S WATER): **23.72**

DIMENSIONAL DRAWING:

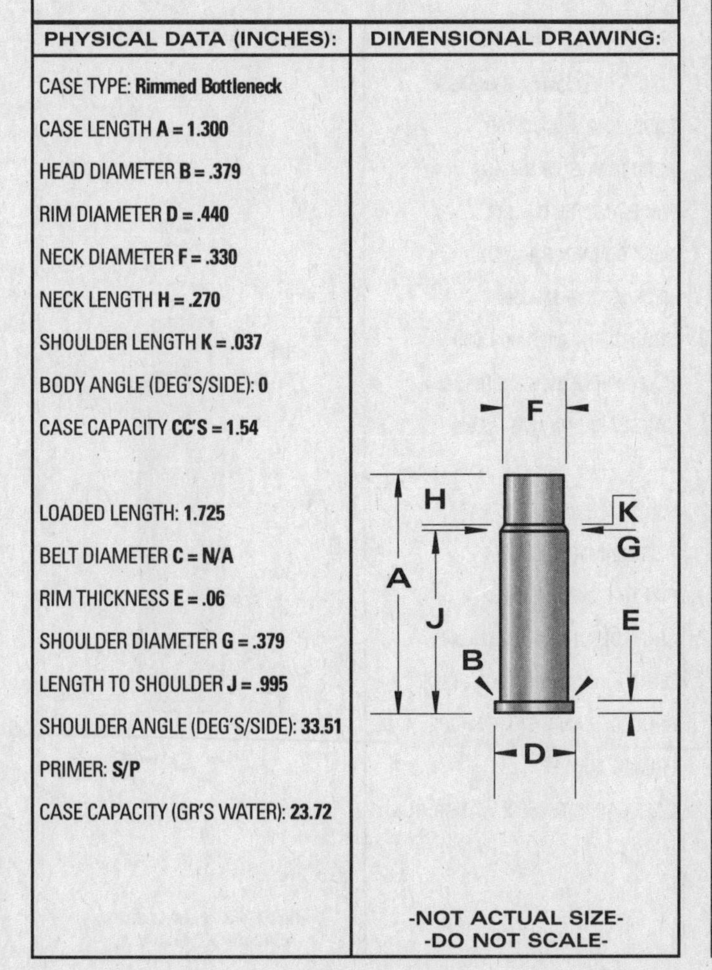

-NOT ACTUAL SIZE-
-DO NOT SCALE-

CARTRIDGE: .30/378 Arch

OTHER NAMES:

DIA: .308

BALLISTEK NO: 308AU

NAI NO: BEN 23434/3.866

DATA SOURCE: Ackley Vol.1 Pg442

HISTORICAL DATA: By Dr. E. Arch.

NOTES:

LOADING DATA:

BULLET WT./TYPE	POWDER WT./TYPE	VELOCITY ('/SEC)	SOURCE
250/RN	85.0/H570	2980	Ackley

CASE PREPARATION: SHELLHOLDER (RCBS): 14

MAKE FROM: .378 Weatherby. Form dies are required. Anneal the case neck <u>after</u> form die #1. Complete forming and trim to 2.3". F/L size, trim to length and chamfer. Shoulder area will fireform in chamber.

PHYSICAL DATA (INCHES):

CASE TYPE: **Belted Bottleneck**

CASE LENGTH **A = 2.25**

HEAD DIAMETER **B = .582**

RIM DIAMETER **D = .582**

NECK DIAMETER **F = .341**

NECK LENGTH **H = .380**

SHOULDER LENGTH **K = .120**

BODY ANGLE (DEG'S/SIDE): **.406**

CASE CAPACITY **CC'S = 6.01**

LOADED LENGTH: **3.05**

BELT DIAMETER **C = .603**

RIM THICKNESS **E = .062**

SHOULDER DIAMETER **G = .560**

LENGTH TO SHOULDER **J = 1.75**

SHOULDER ANGLE (DEG'S/SIDE): **42.38**

PRIMER: **L/R Mag**

CASE CAPACITY (GR'S WATER): **92.77**

DIMENSIONAL DRAWING:

F
H
K
G
A
J
E
B
D
C

-NOT ACTUAL SIZE-
-DO NOT SCALE-

CARTRIDGE: .30/378 Weatherby

OTHER NAMES:

DIA: .308
BALLISTEK NO: 308P
NAI NO: BEN 12434/4.991

\DATA SOURCE: Most current handloading manuals.
HISTORICAL DATA: Created by Roy Weatherby. In 1996 Weatherby introduced it as a factory cartridge.
NOTES: Cartridge was approved by SAAMI on 1/25/98. Older rifles and loading dies may vary slightly. Also, much of the old handloading data may be dangerously inaccurate. Use only current pressure tested data.

LOADING DATA:

BULLET WT./TYPE	POWDER WT./TYPE	VELOCITY ('/SEC)	SOURCE
180/Hornady	117.3/4VIHT 24N41	3300	Horn. #6
200/Nosler Partition	94/IMR-7828	3102	Nosler 5

CASE PREPARATION: SHELLHOLDER (RCBS): 14

Factory brass is readily available, or make from .378 Weatherby. Step down in Form die #1 (lube well). Anneal neck. Re-lube neck and step down in Form die #2. Expand neck with .30-caliber tapered expander. F/L Size, tumble clean, trim & chamfer. Check neck thickness, turn if necessary.

PHYSICAL DATA (INCHES):

CASE TYPE: **Belted Bottleneck**
CASE LENGTH **A = 2.913**
HEAD DIAMETER **B = .5817**
RIM DIAMETER **D = .579**
NECK DIAMETER **F = .3372**
NECK LENGTH **H = .344**
SHOULDER LENGTH **K = .224**
BODY ANGLE (DEG'S/SIDE): **.190**
CASE CAPACITY **CC'S = 9.05**

LOADED LENGTH: **3.648**
BELT DIAMETER **C = .6035**
RIM THICKNESS **E = .063**
SHOULDER DIAMETER **G = .561**
LENGTH TO SHOULDER **J = 2.345**
SHOULDER ANGLE (DEG'S/SIDE):
Front radius: **R .1512**
Rear radius: **R .130 - .005**
PRIMER: **L/R Mag**
CASE CAPACITY (GR'S WATER): **134.3**

DIMENSIONAL DRAWING:

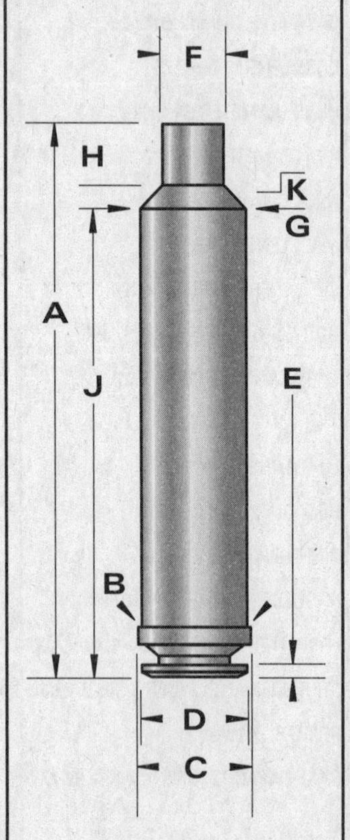

-NOT ACTUAL SIZE-
-DO NOT SCALE-

CARTRIDGE: .30/444 Marlin

OTHER NAMES:

DIA: .308
BALLISTEK NO: 308AV
NAI NO: RMB 22344/4.595

DATA SOURCE: Ackley Vol.2 Pg186
HISTORICAL DATA: Circa 1964.
NOTES:

LOADING DATA:

BULLET WT./TYPE	POWDER WT./TYPE	VELOCITY ('/SEC)	SOURCE
150/—	47.0/4320	2640	Ackley

CASE PREPARATION: SHELLHOLDER (RCBS): 28

MAKE FROM: .444 Marlin. Form die set required. Anneal case, form in 2-die set and trim to length. Chamfer and F/L size. This case can also be made by necking the annealed .444 case in a .357 AutoMag die (watch shoulder location!) and completing the neck reduction in a .308x1.5" F/L die. Best to get the proper form set.

PHYSICAL DATA (INCHES):

CASE TYPE: **Rimmed Bottleneck**
CASE LENGTH **A = 2.155**
HEAD DIAMETER **B = .469**
RIM DIAMETER **D = .514**
NECK DIAMETER **F = .332**
NECK LENGTH **H = .300**
SHOULDER LENGTH **K = .050**
BODY ANGLE (DEG'S/SIDE): **.464**
CASE CAPACITY **CC'S = 3.99**

LOADED LENGTH: **3.08**
BELT DIAMETER **C = N/A**
RIM THICKNESS **E = .062**
SHOULDER DIAMETER **G = .443**
LENGTH TO SHOULDER **J = 1.805**
SHOULDER ANGLE (DEG'S/SIDE): **47.98**
PRIMER: **L/R**
CASE CAPACITY (GR'S WATER): **61.63**

DIMENSIONAL DRAWING:

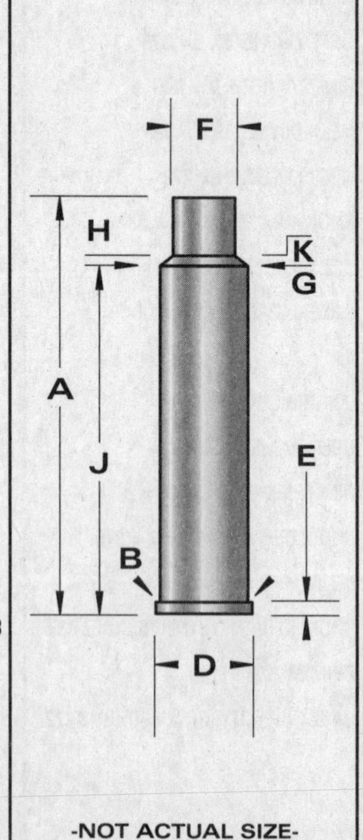

-NOT ACTUAL SIZE-
-DO NOT SCALE-

CARTRIDGE: .300 Ackley Magnum Improved

OTHER NAMES:

DIA: .308

BALLISTEK NO: 308AY

NAI NO: BEN 12334/5.068

DATA SOURCE: Ackley Vol.1 Pg440

HISTORICAL DATA:

NOTES:

LOADING DATA:

BULLET WT./TYPE	POWDER WT./TYPE	VELOCITY ('/SEC)	SOURCE
190/Spire	70.0/IMR4831	—	JJD

CASE PREPARATION: SHELLHOLDER (RCBS): 4

MAKE FROM: .300 H&H Mag. Size the H&H case in the "improved" die with the expander removed. Trim to 2.625". F/L size and final trim to length. Fireform in the Ackley chamber.

PHYSICAL DATA (INCHES):

CASE TYPE: **Belted Bottleneck**

CASE LENGTH **A = 2.600**

HEAD DIAMETER **B = .513**

RIM DIAMETER **D = .532**

NECK DIAMETER **F = .340**

NECK LENGTH **H = .355**

SHOULDER LENGTH **K = .069**

BODY ANGLE (DEG'S/SIDE): **.304**

CASE CAPACITY **CC'S = 5.70**

LOADED LENGTH: **A/R**

BELT DIAMETER **C = .532**

RIM THICKNESS **E = .05**

SHOULDER DIAMETER **G = .492**

LENGTH TO SHOULDER **J = 2.175**

SHOULDER ANGLE (DEG'S/SIDE): **47.35**

PRIMER: **L/R Mag**

CASE CAPACITY (GR'S WATER): **87.98**

DIMENSIONAL DRAWING:

F H K G A J E B D C

-NOT ACTUAL SIZE-
-DO NOT SCALE-

CARTRIDGE: .300 Apex Magnum

OTHER NAMES:

DIA: .308

BALLISTEK NO: 308AM

NAI NO: BEN 22333/5.097

DATA SOURCE: Ackley Vol.1 Pg 433

HISTORICAL DATA: By the Apex Rifle Co.

NOTES:

LOADING DATA:

BULLET WT./TYPE	POWDER WT./TYPE	VELOCITY ('/SEC)	SOURCE
180/Spire	69.0/4350	3213	Ackley

CASE PREPARATION: SHELLHOLDER (RCBS): 4

MAKE FROM: .300 H&H Mag. F/L size the H&H case in the Apex die with the expander removed. Trim to length & chamfer. Fireform in the Apex chamber.

PHYSICAL DATA (INCHES):

CASE TYPE: **Belted Bottleneck**

CASE LENGTH **A = 2.615**

HEAD DIAMETER **B = .513**

RIM DIAMETER **D = .532**

NECK DIAMETER **F = .340**

NECK LENGTH **H = .375**

SHOULDER LENGTH **K = .130**

BODY ANGLE (DEG'S/SIDE): **.359**

CASE CAPACITY **CC'S = 5.63**

LOADED LENGTH: **A/R**

BELT DIAMETER **C = .532**

RIM THICKNESS **E = .05**

SHOULDER DIAMETER **G = .489**

LENGTH TO SHOULDER **J = 2.11**

SHOULDER ANGLE (DEG'S/SIDE): **29.81**

PRIMER: **L/R Mag**

CASE CAPACITY (GR'S WATER): **86.93**

DIMENSIONAL DRAWING:

F H K G A J E B D C

-NOT ACTUAL SIZE-
-DO NOT SCALE-

CARTRIDGE: .300 Bennett Magnum

OTHER NAMES:

DIA: .308

BALLISTEK NO: 308BU

NAI NO: BEN 22324/4.893

DATA SOURCE: Ackley Vol.1 Pg430

HISTORICAL DATA: By B. Bennett

NOTES: Good case capacity for .308" dia. bore.

LOADING DATA:

BULLET WT./TYPE	POWDER WT./TYPE	VELOCITY ('/SEC)	SOURCE
180/Spire	61.6/IMR4350	—	JJD

CASE PREPARATION: SHELLHOLDER (RCBS): 4

MAKE FROM: .300 H&H Mag. F/L size the H&H case in the Bennett die with the expander removed. Trim to 2.60". F/L size and chamfer. Fireform in the Bennett chamber.

PHYSICAL DATA (INCHES):

CASE TYPE: **Belted Bottleneck**

CASE LENGTH **A = 2.51**

HEAD DIAMETER **B = .513**

RIM DIAMETER **D = .532**

NECK DIAMETER **F = .340**

NECK LENGTH **H = .385**

SHOULDER LENGTH **K = .085**

BODY ANGLE (DEG'S/SIDE): **.747**

CASE CAPACITY **CC'S = 5.11**

LOADED LENGTH: **3.50**

BELT DIAMETER **C = .532**

RIM THICKNESS **E = .05**

SHOULDER DIAMETER **G = .465**

LENGTH TO SHOULDER **J = 2.04**

SHOULDER ANGLE (DEG'S/SIDE): **36.32**

PRIMER: **L/R Mag**

CASE CAPACITY (GR'S WATER): **78.95**

DIMENSIONAL DRAWING:

-NOT ACTUAL SIZE-
-DO NOT SCALE-

CARTRIDGE: .300 Canadian Magnum

OTHER NAMES:

DIA: .308

BALLISTEK NO:

NAI NO:

DATA SOURCE:

HISTORICAL DATA: Developed about 1989 by North American Shooting Systems and is somewhat similar to the 300 Imperial Magnum.

NOTES:

LOADING DATA:

BULLET WT./TYPE	POWDER WT./TYPE	VELOCITY ('/SEC)	SOURCE
165	86.0/H-4831	3231	NASS
180	83.0/RL-22	3354	NASS
200	79.0/RL-22	4135	NASS

CASE PREPARATION: SHELL HOLDER (RCBS): 00

MAKE FROM: -

PHYSICAL DATA (INCHES):

CASE TYPE: **Rebated Bottleneck**

CASE LENGTH **A = 2.83**

HEAD DIAMETER **B = .544**

RIM DIAMETER **D = .532**

NECK DIAMETER **F = .340**

NECK LENGTH **H = .328**

SHOULDER LENGTH **K = .163**

BODY ANGLE (DEG'S/SIDE): **N/A**

CASE CAPACITY **CC'S = N/A**

LOADED LENGTH: **3.600**

BELT DIAMETER **C = N/A**

RIM THICKNESS **E = .05**

SHOULDER DIAMETER **G = .530**

LENGTH TO SHOULDER **J = 2.339**

SHOULDER ANGLE (DEG'S/SIDE): **N/A**

PRIMER: **L/R**

CASE CAPACITY (GR'S WATER):

DIMENSIONAL DRAWING:

-NOT ACTUAL SIZE-
-DO NOT SCALE-

CARTRIDGE: .300 Dakota

OTHER NAMES:	DIA: .308
	BALLISTEK NO:
	NAI NO:

DATA SOURCE:

HISTORICAL DATA: Based on the .404 Jeffrey case.

NOTES:

LOADING DATA:

BULLET WT./TYPE	POWDER WT./TYPE	VELOCITY ('/SEC)	SOURCE
165	77.0/IMR-4350	3247	Dakota
180	77.5/H-4831	3114	Dakota
200	77.5/H-4831	2965	Dakota

CASE PREPARATION: SHELL HOLDER (RCBS): SPL

MAKE FROM: 404 Jeffrey case shortened to create a 30-06 length cartridge (3.35").

PHYSICAL DATA (INCHES):

CASE TYPE: **Rimless Bottleneck**
CASE LENGTH **A = 2.550**
HEAD DIAMETER **B = .544**
RIM DIAMETER **D = .544**
NECK DIAMETER **F = .338**
NECK LENGTH **H = .300**
SHOULDER LENGTH **K = .150**
BODY ANGLE (DEG'S/SIDE): **N/A**
CASE CAPACITY **CC'S = N/A**

LOADED LENGTH: **3.33**
BELT DIAMETER **C = .N/A**
RIM THICKNESS **E = .050**
SHOULDER DIAMETER **G = .531**
LENGTH TO SHOULDER **J = 2.090**
SHOULDER ANGLE (DEG'S/SIDE): **32°**
PRIMER: **L/R**
CASE CAPACITY (GR'S WATER): **N/A**

DIMENSIONAL DRAWING:

-NOT ACTUAL SIZE-
-DO NOT SCALE-

CARTRIDGE: .300 Durham Magnum

OTHER NAMES:	DIA: .308
	BALLISTEK NO: 308AN
	NAI NO: BEN 22334/4.893

DATA SOURCE: Ackley Vol.1 Pg432

HISTORICAL DATA:

NOTES:

LOADING DATA:

BULLET WT./TYPE	POWDER WT./TYPE	VELOCITY ('/SEC)	SOURCE
180/Spire	71.0/4350	3090	Ackley

CASE PREPARATION: SHELLHOLDER (RCBS): 4

MAKE FROM: .300 H&H Mag. F/L size the H&H case, in the Durham die, with the expander removed. Trim to length and chamfer. Fireform in the Durham chamber.

PHYSICAL DATA (INCHES):

CASE TYPE: **Belted Bottleneck**
CASE LENGTH **A = 2.51**
HEAD DIAMETER **B = .513**
RIM DIAMETER **D = .532**
NECK DIAMETER **F = .340**
NECK LENGTH **H = .354**
SHOULDER LENGTH **K = .071**
BODY ANGLE (DEG'S/SIDE): **.471**
CASE CAPACITY **CC'S = 5.35**

LOADED LENGTH: **3.43**
BELT DIAMETER **C = .532**
RIM THICKNESS **E = .05**
SHOULDER DIAMETER **G = .482**
LENGTH TO SHOULDER **J = 2.085**
SHOULDER ANGLE (DEG'S/SIDE): **44.99**
PRIMER: **L/R Mag**
CASE CAPACITY (GR'S WATER): **82.58**

DIMENSIONAL DRAWING:

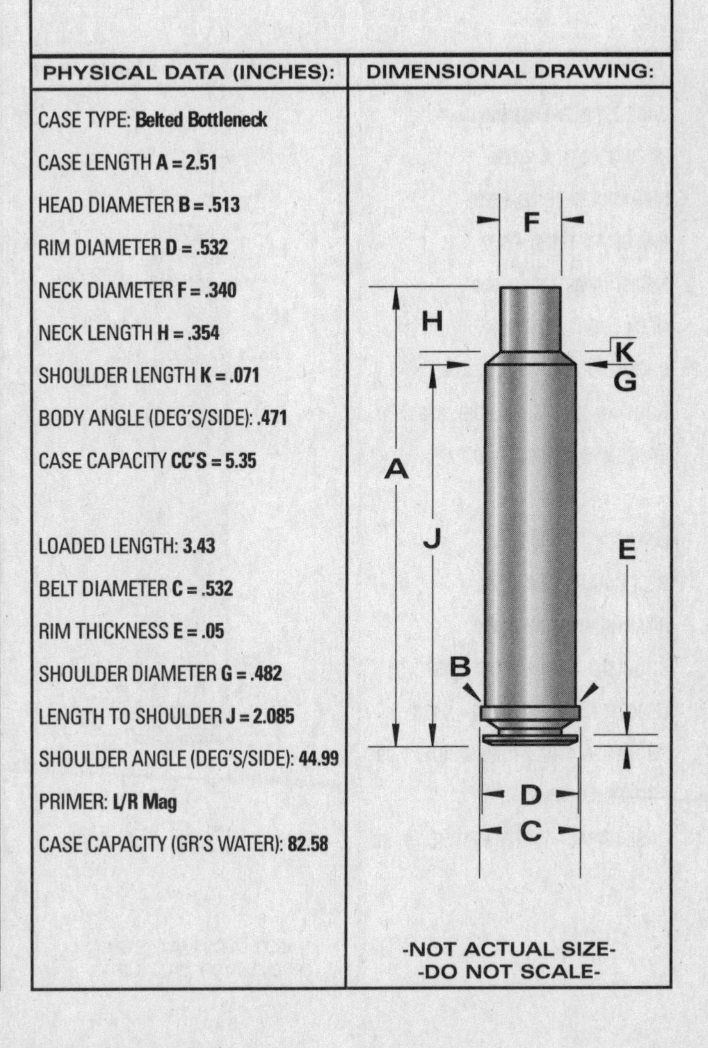

-NOT ACTUAL SIZE-
-DO NOT SCALE-

CARTRIDGE: .300 Holland & Holland Magnum

OTHER NAMES:
.300 H&H Magnum
7,63 x 72mm

DIA: .308

BALLISTEK NO: 308L

NAI NO: BEN 22341/5.555

DATA SOURCE: Hornady 3rd pg233

HISTORICAL DATA: By Holland & Holland about 1919.

NOTES:

LOADING DATA:

BULLET WT./TYPE	POWDER WT./TYPE	VELOCITY ('/SEC)	SOURCE
190/HPBT	68.0/IMR4350	3000	Horn.

CASE PREPARATION: **SHELLHOLDER (RCBS): 4**

MAKE FROM: Factory or .375 H&H. Anneal case neck. F/L size in .300 H&H die. Square case neck. Chamfer.

PHYSICAL DATA (INCHES):

CASE TYPE: **Belted Bottleneck**
CASE LENGTH **A = 2.85**
HEAD DIAMETER **B = .513**
RIM DIAMETER **D = .532**
NECK DIAMETER **F = .339**
NECK LENGTH **H = .30**
SHOULDER LENGTH **K = .437**
BODY ANGLE (DEG'S/SIDE): **.838**
CASE CAPACITY **CC'S = 5.755**

LOADED LENGTH: **3.65**
BELT DIAMETER **C = .532**
RIM THICKNESS **E = .05**
SHOULDER DIAMETER **G = .457**
LENGTH TO SHOULDER **J = 2.113**
SHOULDER ANGLE (DEG'S/SIDE): **7.69**
PRIMER: **L/R Mag**
CASE CAPACITY (GR'S WATER): **88.82**

DIMENSIONAL DRAWING:

-NOT ACTUAL SIZE-
-DO NOT SCALE-

CARTRIDGE: .300 ICL Grizzly Cub

OTHER NAMES:

DIA: .308

BALLISTEK NO: 308BD

NAI NO: BEN 12334/5.107

DATA SOURCE: Ackley Vol.1 Pg434

HISTORICAL DATA: Saturn Gun Shop design.

NOTES:

LOADING DATA:

BULLET WT./TYPE	POWDER WT./TYPE	VELOCITY ('/SEC)	SOURCE
180/Spire	69.0/4350	3105	Ackley

CASE PREPARATION: **SHELLHOLDER (RCBS): 4**

MAKE FROM: .300 H&H Mag. Size the H&H case in the ICL F/L die. Trim to length and chamfer. Fireform in the chamber.

PHYSICAL DATA (INCHES):

CASE TYPE: **Belted Bottleneck**
CASE LENGTH **A = 2.62**
HEAD DIAMETER **B = .513**
RIM DIAMETER **D = .532**
NECK DIAMETER **F = .339**
NECK LENGTH **H = .330**
SHOULDER LENGTH **K = .065**
BODY ANGLE (DEG'S/SIDE): **.311**
CASE CAPACITY **CC'S = 5.82**

LOADED LENGTH: **3.38**
BELT DIAMETER **C = .532**
RIM THICKNESS **E = .05**
SHOULDER DIAMETER **G = .491**
LENGTH TO SHOULDER **J = 2.225**
SHOULDER ANGLE (DEG'S/SIDE): **49.46**
PRIMER: **L/R Mag**
CASE CAPACITY (GR'S WATER): **89.77**

DIMENSIONAL DRAWING:

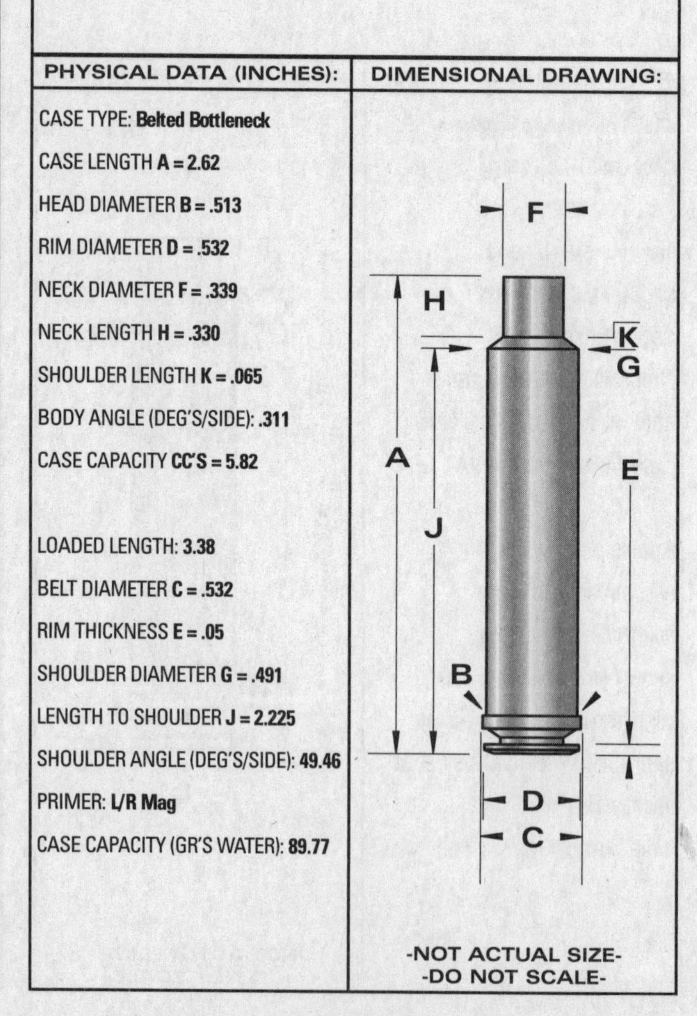

-NOT ACTUAL SIZE-
-DO NOT SCALE-

CARTRIDGE: .300 ICL Tornado

OTHER NAMES:

DIA: .308

BALLISTEK NO: 308BW

NAI NO: RXB 12334/4.734

DATA SOURCE: Ackley Vol.1 Pg424

HISTORICAL DATA:

NOTES:

LOADING DATA:

BULLET WT./TYPE	POWDER WT./TYPE	VELOCITY ('/SEC)	SOURCE
150/—	52.0/4350	2790	Ackley

CASE PREPARATION: SHELLHOLDER (RCBS): 11

MAKE FROM: .257 Roberts. Taper expand case to .315" dia. F/L size in the ICL chamber and fireform. Trim to proper length, chamfer and reload. These cases may also be fire-formed in the ICL trim die by firing with cornmeal.

PHYSICAL DATA (INCHES):

CASE TYPE: **Rimless Straight**

CASE LENGTH **A = 2.23**

HEAD DIAMETER **B = .471**

RIM DIAMETER **D = .473**

NECK DIAMETER **F = .339**

NECK LENGTH **H = .385**

SHOULDER LENGTH **K = .045**

BODY ANGLE (DEG'S/SIDE): **.196**

CASE CAPACITY **CC'S = 4.06**

LOADED LENGTH: **3.00**

BELT DIAMETER **C = N/A**

RIM THICKNESS **E = .049**

SHOULDER DIAMETER **G = .460**

LENGTH TO SHOULDER **J = 1.80**

SHOULDER ANGLE (DEG'S/SIDE): **53.35**

PRIMER: **L/R**

CASE CAPACITY (GR'S WATER): **62.72**

DIMENSIONAL DRAWING:

F H K G A E J B D

-NOT ACTUAL SIZE-
-DO NOT SCALE-

CARTRIDGE: .300 Mashburn Super

OTHER NAMES:

DIA: .308

BALLISTEK NO: 308AL

NAI NO: BEN 22333/5.458

DATA SOURCE: Ackley Vol.1 Pg439

HISTORICAL DATA:

NOTES:

LOADING DATA:

BULLET WT./TYPE	POWDER WT./TYPE	VELOCITY ('/SEC)	SOURCE
180/—	74.0/4350	3100	Ackley

CASE PREPARATION: SHELLHOLDER (RCBS): 4

MAKE FROM: .300 Weatherby. Fire the Weatherby ammo in the Mashburn chamber.

PHYSICAL DATA (INCHES):

CASE TYPE: **Belted Bottleneck**

CASE LENGTH **A = 2.800**

HEAD DIAMETER **B = .513**

RIM DIAMETER **D = .532**

NECK DIAMETER **F = .340**

NECK LENGTH **H = .333**

SHOULDER LENGTH **K = .137**

BODY ANGLE (DEG'S/SIDE): **.470**

CASE CAPACITY **CC'S = 6.03**

LOADED LENGTH: **3.52**

BELT DIAMETER **C = .532**

RIM THICKNESS **E = .05**

SHOULDER DIAMETER **G = .478**

LENGTH TO SHOULDER **J = 2.33**

SHOULDER ANGLE (DEG'S/SIDE): **26.73**

PRIMER: **L/R Mag**

CASE CAPACITY (GR'S WATER): **93.14**

DIMENSIONAL DRAWING:

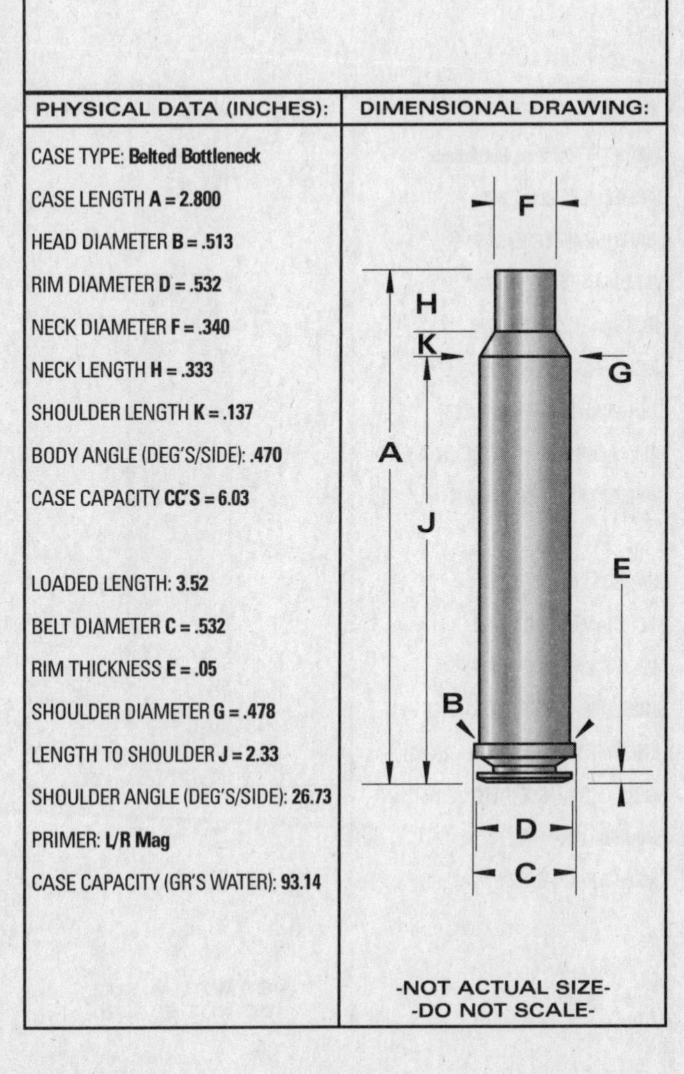

-NOT ACTUAL SIZE-
-DO NOT SCALE-

CARTRIDGE: 300 Pegasus

OTHER NAMES:

DIA: .308

BALLISTEK NO:

NAI NO:

DATA SOURCE:

HISTORICAL DATA: Created around 1994 and is based upon an entirely new case that features a 0.580" head size. The rim is essentially identical to the 378 Weatherby but there is no belt.

LOADING DATA:

BULLET WT./TYPE	POWDER WT./TYPE	VELOCITY ('/SEC)	SOURCE
150/Nosler BT	106/IMR-7828	3642	A-Square
180/Nosler BT	103.0/RL-22	3371	A-Square

CASE PREPARATION: SHELL HOLDER (RCBS): 00

MAKE FROM:

PHYSICAL DATA (INCHES):

CASE TYPE: **Rimless Bottleneck**

CASE LENGTH **A = 2.990**

HEAD DIAMETER **B = .580**

RIM DIAMETER **D = .580**

NECK DIAMETER **F = .339**

NECK LENGTH **H = .327**

SHOULDER LENGTH **K = .213**

BODY ANGLE (DEG'S/SIDE): **N/A**

CASE CAPACITY **CC'S = N/A**

LOADED LENGTH: **3.750**

BELT DIAMETER **C = N/A**

RIM THICKNESS **E = .065**

SHOULDER DIAMETER **G = .566**

LENGTH TO SHOULDER **J = 2.450**

SHOULDER ANGLE (DEG'S/SIDE): **N/A**

PRIMER: **L/R**

CASE CAPACITY (GR'S WATER):

DIMENSIONAL DRAWING:

F
H
K
G
A
J
E
B
D

-NOT ACTUAL SIZE-
-DO NOT SCALE-

CARTRIDGE: .300 PMVF

OTHER NAMES:
.300 CCC'

DIA: .308

BALLISTEK NO: 308BV

NAI NO: BEN 12334/5.493

DATA SOURCE: Ackley Vol.2 Pg197

HISTORICAL DATA: Designed by Hollywood Gun Shop.

NOTES: 'Controlled Combustion Chamberage!

LOADING DATA:

BULLET WT./TYPE	POWDER WT./TYPE	VELOCITY ('/SEC)	SOURCE
168/Spire	77.8/IMR4831	–	JJD

CASE PREPARATION: SHELLHOLDER (RCBS): 4

MAKE FROM: .300 H&H Mag. F/L size the Holland case in the PMVF die. Trim to length and chamfer. Fireform in the chamber. This case has a characteristic, inward, single radii as opposed to the Weatherby, double radius.

PHYSICAL DATA (INCHES):

CASE TYPE: **Belted Bottleneck**

CASE LENGTH **A = 2.818**

HEAD DIAMETER **B = .513**

RIM DIAMETER **D = .532**

NECK DIAMETER **F = .340**

NECK LENGTH **H = .332**

SHOULDER LENGTH **K = .061**

BODY ANGLE (DEG'S/SIDE): **.321**

CASE CAPACITY **CC'S = 6.27**

LOADED LENGTH: **3.675**

BELT DIAMETER **C = .532**

RIM THICKNESS **E = .05**

SHOULDER DIAMETER **G = .488**

LENGTH TO SHOULDER **J = 2.425**

SHOULDER ANGLE (DEG'S/SIDE): **N/A**

PRIMER: **L/R Mag**

CASE CAPACITY (GR'S WATER): **96.73**

DIMENSIONAL DRAWING:

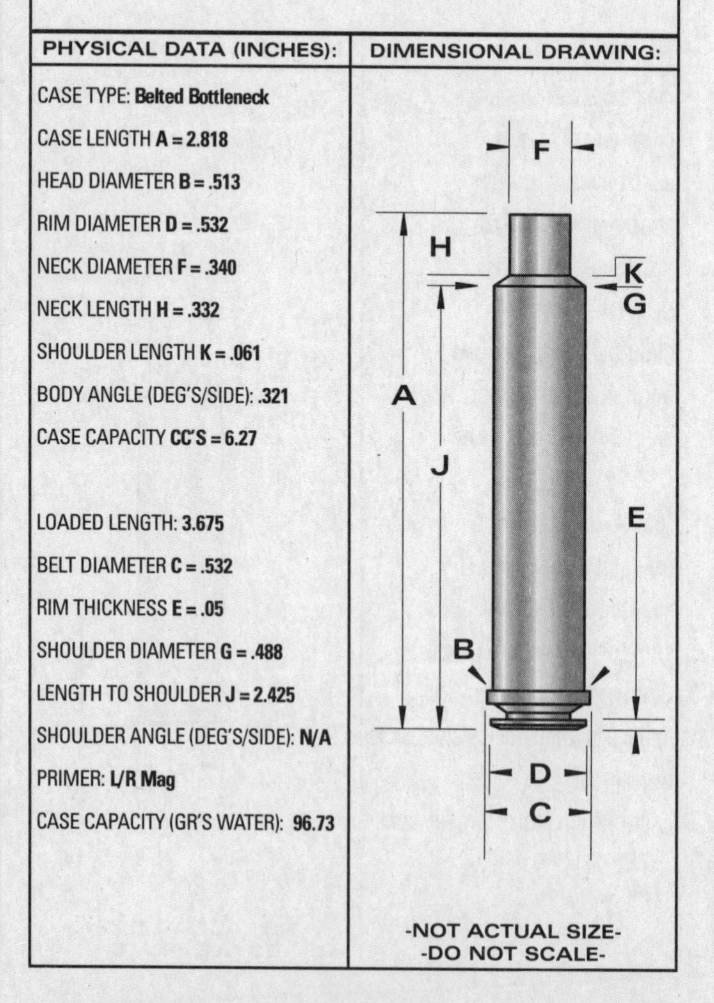

-NOT ACTUAL SIZE-
-DO NOT SCALE-

CARTRIDGE: .300 Remington Short Action Ultra Mag

OTHER NAMES:

DIA: .308

BALLISTEK NO:

NAI NO:

DATA SOURCE: Most current loading manuals.

HISTORICAL DATA: Introduced by Reminton in 2002 to increase power from a short action rifle. This case is a shorted version of the .300 Ultra Mag.

NOTES: While this case is similar to the .300 winches-ter Shrot Magnum the two are not inter-changeable.

LOADING DATA:

BULLET WT./TYPE	POWDER WT./TYPE	VELOCITY ('/SEC)	SOURCE
180/Hornady SP	63.1/Win	2,900	Horn. #6
180/Factory		2,960	Rem

CASE PREPARATION: **SHELL HOLDER (RCBS): 38**

MAKE FROM: Factory or neck up 7mm Reminton Short Action Ultra Mag.

PHYSICAL DATA (INCHES):

CASE TYPE: **Rebated Bottleneck**
CASE LENGTH **A = 2.015**
HEAD DIAMETER **B = .550**
RIM DIAMETER **D = .534**
NECK DIAMETER **F = .344**
NECK LENGTH **H = .311**
SHOULDER LENGTH **K = .166**
BODY ANGLE (DEG'S/SIDE):
CASE CAPACITY **CC'S =**
LOADED LENGTH: **2.825**
BELT DIAMETER **C =**
RIM THICKNESS **E = .050**
SHOULDER DIAMETER **G = .535**
LENGTH TO SHOULDER **J = 1.538**
SHOULDER ANGLE (DEG'S/SIDE): **30**
PRIMER: **L/R or L/R Mag**
CASE CAPACITY (GR'S WATER):

DIMENSIONAL DRAWING:

-NOT ACTUAL SIZE-
-DO NOT SCALE-

CARTRIDGE: .300 Remington Ultra Mag

OTHER NAMES:

DIA: .308

BALLISTEK NO:

NAI NO:

DATA SOURCE: Most current loading manuals.

HISTORICAL DATA: Introduced by Remington in 1999 to maxi-mize the performance from the 700 rifle. The case is made from the .404 Jeffery which was necked down and the body and shoulder blown out.

LOADING DATA:

BULLET WT./TYPE	POWDER WT./TYPE	VELOCITY ('/SEC)	SOURCE
200/Swift A-Frame	84.0/IMR-7828	2,989	Swift #1
180/Factory		3,250	Rem

CASE PREPARATION: **SHELL HOLDER (RCBS): 38**

MAKE FROM: Factory or neck up 7mm Remington Ultra Mag. Can be made from .404 Jeffery.

PHYSICAL DATA (INCHES):

CASE TYPE: **Rebated Bottleneck**
CASE LENGTH **A = 2.850**
HEAD DIAMETER **B = .550**
RIM DIAMETER **D = .534**
NECK DIAMETER **F = .344**
NECK LENGTH **H = .306**
SHOULDER LENGTH **K = .157**
BODY ANGLE (DEG'S/SIDE):
CASE CAPACITY **CC'S =**
LOADED LENGTH: **3.600**
BELT DIAMETER **C =**
RIM THICKNESS **E = .050**
SHOULDER DIAMETER **G = .525**
LENGTH TO SHOULDER **J = 2.387**
SHOULDER ANGLE (DEG'S/SIDE): **30**
PRIMER: **L/R Mag**
CASE CAPACITY (GR'S WATER): **113.8**

DIMENSIONAL DRAWING:

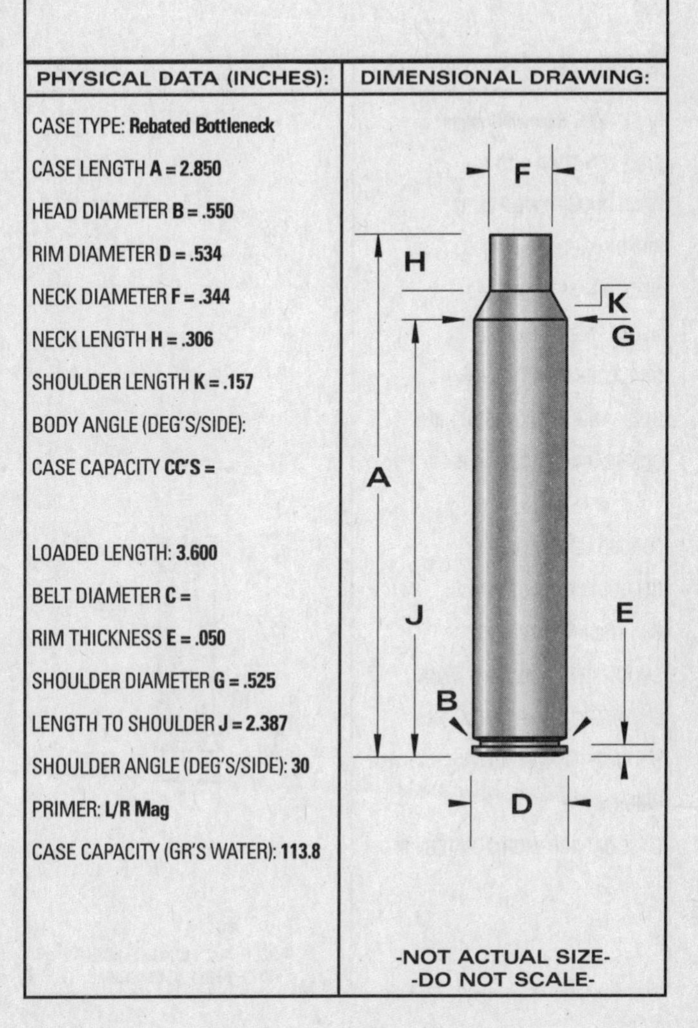

CARTRIDGE: .300 Rook Rifle

OTHER NAMES:
.295 Rook Rifle

DIA: .300

BALLISTEK NO: 300B

NAI NO: RMS 11115/3.667

DATA SOURCE: COTW 4th Pg229

HISTORICAL DATA: About 1880.

NOTES:

LOADING DATA:

BULLET WT./TYPE	POWDER WT./TYPE	VELOCITY ('/SEC)	SOURCE
90/RN	10.0/4198	1150	Barnes

CASE PREPARATION: **SHELLHOLDER (RCBS):** 16

MAKE FROM: .32 Colt Long. F/L size the Colt case to produce a "short" version of the Rook. Re-body a .30 Carbine case with 5/16" dia. tubing or lathe turn from solid brass to produce full length cases.

PHYSICAL DATA (INCHES):

CASE TYPE: **Rimmed Straight**
CASE LENGTH **A = 1.17**
HEAD DIAMETER **B = .319**
RIM DIAMETER **D = .369**
NECK DIAMETER **F = .317**
NECK LENGTH **H = N/A**
SHOULDER LENGTH **K = N/A**
BODY ANGLE (DEG'S/SIDE): **.051**
CASE CAPACITY **CC'S = .95**

LOADED LENGTH: **1.38**
BELT DIAMETER **C = N/A**
RIM THICKNESS **E = .052**
SHOULDER DIAMETER **G = N/A**
LENGTH TO SHOULDER **J = N/A**
SHOULDER ANGLE (DEG'S/SIDE): **N/A**
PRIMER: **S/P**
CASE CAPACITY (GR'S WATER): **14.66**

DIMENSIONAL DRAWING:

-NOT ACTUAL SIZE-
-DO NOT SCALE-

CARTRIDGE: .300 Savage

OTHER NAMES:

DIA: .308

BALLISTEK NO: 308C

NAI NO: RXB 22343/3.972

DATA SOURCE: Hornady Manual 3rd Pg212

HISTORICAL DATA: Designed by Savage Arms Co. about 1920.

NOTES:

LOADING DATA:

BULLET WT./TYPE	POWDER WT./TYPE	VELOCITY ('/SEC)	SOURCE
150/Spire	42.0/IMR4320	2590	Lyman

CASE PREPARATION: **SHELLHOLDER (RCBS):** 3

MAKE FROM: Factory or .30-06. Anneal case neck. Run into .300 Savage F/L die with expander removed. Trim to length. I.D. neck ream, if necessary. Chamfer. F/L size.

PHYSICAL DATA (INCHES):

CASE TYPE: **Rimless Bottleneck**
CASE LENGTH **A = 1.871**
HEAD DIAMETER **B = .471**
RIM DIAMETER **D = .473**
NECK DIAMETER **F = .339**
NECK LENGTH **H = .192**
SHOULDER LENGTH **K = .119**
BODY ANGLE (DEG'S/SIDE): **.463**
CASE CAPACITY **CC'S = 3.33**

LOADED LENGTH: **2.60**
BELT DIAMETER **C = N/A**
RIM THICKNESS **E = .049**
SHOULDER DIAMETER **G = .449**
LENGTH TO SHOULDER **J = 1.56**
SHOULDER ANGLE (DEG'S/SIDE): **24.80**
PRIMER: **L/R**
CASE CAPACITY (GR'S WATER): **51.42**

DIMENSIONAL DRAWING:

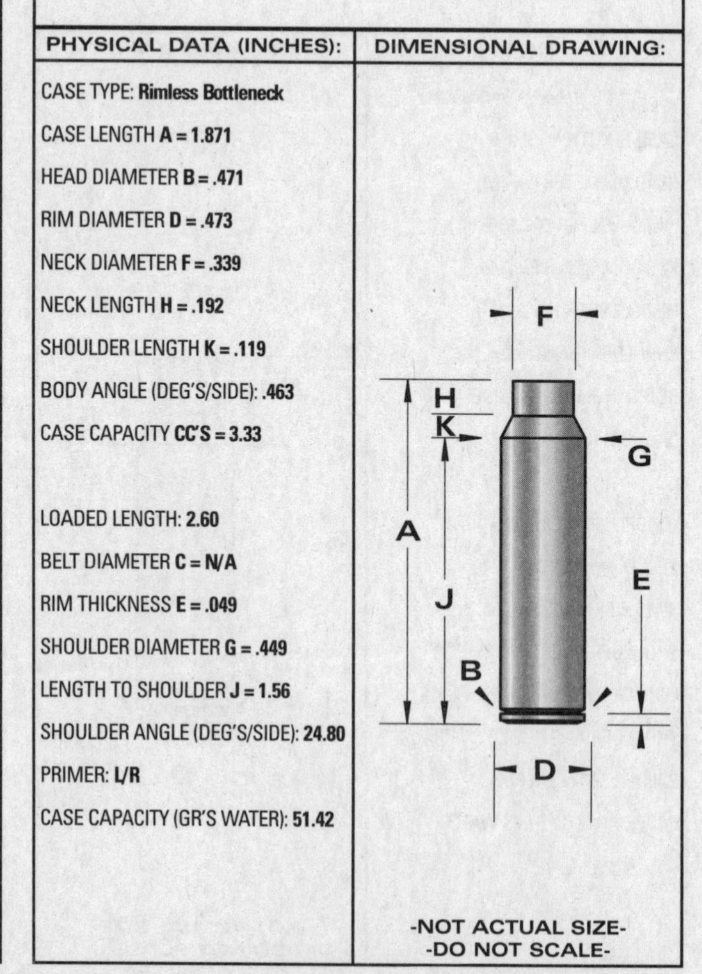

-NOT ACTUAL SIZE-
-DO NOT SCALE-

CARTRIDGE: .300 Sherwood

OTHER NAMES:

DIA: .300

BALLISTEK NO: 300C

NAI NO: RMS 11115/4.813

DATA SOURCE: COTW 4th Pg 230

HISTORICAL DATA: Introduced about 1900.

NOTES:

LOADING DATA:

BULLET WT./TYPE	POWDER WT./TYPE	VELOCITY ('/SEC)	SOURCE
150/Lead	14.0/4198	1400	Barnes

CASE PREPARATION: SHELLHOLDER (RCBS): 16

MAKE FROM: Turn from solid brass or re-body a .32 Colt Long case. Trim to length, chamfer and F/L size.

PHYSICAL DATA (INCHES):

CASE TYPE: **Rimmed Straight**

CASE LENGTH **A = 1.54**

HEAD DIAMETER **B = .320**

RIM DIAMETER **D = .370**

NECK DIAMETER **F = .318**

NECK LENGTH **H = N/A**

SHOULDER LENGTH **K = N/A**

BODY ANGLE (DEG'S/SIDE): **.038**

CASE CAPACITY **CC'S = 1.35**

LOADED LENGTH: **2.20**

BELT DIAMETER **C = N/A**

RIM THICKNESS **E = .055**

SHOULDER DIAMETER **G = N/A**

LENGTH TO SHOULDER **J = N/A**

SHOULDER ANGLE (DEG'S/SIDE): **N/A**

PRIMER: **S/P**

CASE CAPACITY (GR'S WATER): **20.89**

DIMENSIONAL DRAWING:

F A E B D

-NOT ACTUAL SIZE-
-DO NOT SCALE-

CARTRIDGE: .300 Super Flanged

OTHER NAMES:
.30 Super Flanged

DIA: .308

BALLISTEK NO: 308BG

NAI NO: RMB 22331/5.686

DATA SOURCE: COTW 4th Pg230

HISTORICAL DATA: Holland & Holland ctg.

NOTES:

LOADING DATA:

BULLET WT./TYPE	POWDER WT./TYPE	VELOCITY ('/SEC)	SOURCE
220/RN	72.6/MRP	2000	JJD

CASE PREPARATION: SHELLHOLDER (RCBS): 14

MAKE FROM: .375 Flanged (B.E.L.L.) Anneal case neck. Form case in form die. Trim to length, chamfer and F/L size.

PHYSICAL DATA (INCHES):

CASE TYPE: **Rimmed Bottleneck**

CASE LENGTH **A = 2.94**

HEAD DIAMETER **B = .517**

RIM DIAMETER **D = .572**

NECK DIAMETER **F = .338**

NECK LENGTH **H = .382**

SHOULDER LENGTH **K = .339**

BODY ANGLE (DEG'S/SIDE): **.950**

CASE CAPACITY **CC'S = 5.94**

LOADED LENGTH: **3.69**

BELT DIAMETER **C = N/A**

RIM THICKNESS **E = .062**

SHOULDER DIAMETER **G = .450**

LENGTH TO SHOULDER **J = 2.219**

SHOULDER ANGLE (DEG'S/SIDE): **9.38**

PRIMER: **L/R**

CASE CAPACITY (GR'S WATER): **91.63**

DIMENSIONAL DRAWING:

F H K G A J E B D

-NOT ACTUAL SIZE-
-DO NOT SCALE-

CARTRIDGE: .300 Wade Magnum

OTHER NAMES:

DIA: .308
BALLISTEK NO: 308BT
NAI NO: BEN 22323/5.029

DATA SOURCE: Ackley Vol.1 Pg433

HISTORICAL DATA: By F. Wade.

NOTES:

LOADING DATA:

BULLET WT./TYPE	POWDER WT./TYPE	VELOCITY ('/SEC)	SOURCE
180/Spire	57.0/4064	2875	Ackley

CASE PREPARATION: **SHELLHOLDER (RCBS): 4**

MAKE FROM: .300 Win. Mag. Use F/L die to set .300's shoulder back to proper location. Trim to length and fire-form in the Wade chamber to fill out shoulders.

PHYSICAL DATA (INCHES):

CASE TYPE: **Belted Bottleneck**
CASE LENGTH **A = 2.58**
HEAD DIAMETER **B = .513**
RIM DIAMETER **D = .532**
NECK DIAMETER **F = .340**
NECK LENGTH **H = .415**
SHOULDER LENGTH **K = .125**
BODY ANGLE (DEG'S/SIDE): **.358**
CASE CAPACITY **CC'S = 5.50**

LOADED LENGTH: **3.40**
BELT DIAMETER **C = .532**
RIM THICKNESS **E = .05**
SHOULDER DIAMETER **G = .490**
LENGTH TO SHOULDER **J = 2.04**
SHOULDER ANGLE (DEG'S/SIDE): **30.96**
PRIMER: **L/R Mag**
CASE CAPACITY (GR'S WATER): **84.86**

DIMENSIONAL DRAWING:

-NOT ACTUAL SIZE-
-DO NOT SCALE-

CARTRIDGE: .300 Weatherby Magnum

OTHER NAMES:

DIA: .308
BALLISTEK NO: 308K
NAI NO: BEN 12333/5.518

DATA SOURCE: Hornady 3rd Pg247

HISTORICAL DATA: By Roy Weatherby in 1944-45.

NOTES: Most popular Weatherby ctg.

LOADING DATA:

BULLET WT./TYPE	POWDER WT./TYPE	VELOCITY ('/SEC)	SOURCE
180/Spire	81.0/IMR4831	3100	JJD

CASE PREPARATION: **SHELLHOLDER (RCBS): 4**

MAKE FROM: Factory or .300 H&H. Fire the H&H ammo in the Weatherby chamber. Safer to fill the H&H cases with cornmeal and fireform in a .300 Weatherby trim die. Either way works well.

PHYSICAL DATA (INCHES):

CASE TYPE: **Belted Bottleneck**
CASE LENGTH **A = 2.82**
HEAD DIAMETER **B = .511**
RIM DIAMETER **D = .530**
NECK DIAMETER **F = .332**
NECK LENGTH **H = .325**
SHOULDER LENGTH **K = .175**
BODY ANGLE (DEG'S/SIDE): **.284**
CASE CAPACITY **CC'S = 6.50**

LOADED LENGTH: **3.56**
BELT DIAMETER **C = .530**
RIM THICKNESS **E = .05**
SHOULDER DIAMETER **G = .490**
LENGTH TO SHOULDER **J = 2.32**
SHOULDER ANGLE (DEG'S/SIDE): **N/A**
PRIMER: **L/R Mag**
CASE CAPACITY (GR'S WATER): **100.36**

DIMENSIONAL DRAWING:

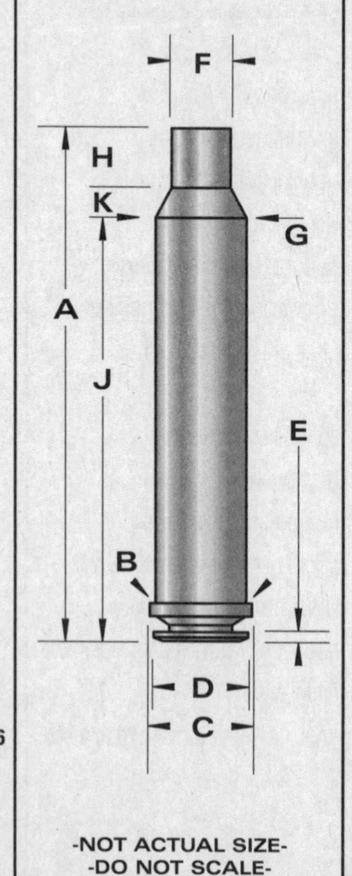

-NOT ACTUAL SIZE-
-DO NOT SCALE-

CARTRIDGE: 300 Whisper

OTHER NAMES:
List them here...

DIA: .308

BALLISTEK NO:

NAI NO:

DATA SOURCE:

HISTORICAL DATA: This cartridge is a new concept in the development of small case capacity, highly efficient cartridges combined with bullets of extreme efficiency.

NOTES:

LOADING DATA:

BULLET WT./TYPE	POWDER WT./TYPE	VELOCITY (*/SEC)	SOURCE
125/Nosler BT	20.6/H-110	2283	
150/Nosler BT	18.0/H-110	2073	
165	10.3/A-1680	1013	

CASE PREPARATION: SHELL HOLDER (RCBS): 10

MAKE FROM: 221 Remington case necked up to a 308 caliber.

PHYSICAL DATA (INCHES):

CASE TYPE: **Rimless Bottleneck**
CASE LENGTH **A = 1.400**
HEAD DIAMETER **B = .375**
RIM DIAMETER **D = .375**
NECK DIAMETER **F = .330**
NECK LENGTH **H = .298**
SHOULDER LENGTH **K = .031**
BODY ANGLE (DEG'S/SIDE): **.284**
CASE CAPACITY **CC'S = 6.50**

LOADED LENGTH: **2.575**
BELT DIAMETER **C = .N/A**
RIM THICKNESS **E = .045**
SHOULDER DIAMETER **G = .369**
LENGTH TO SHOULDER **J = 1.071**
SHOULDER ANGLE (DEG'S/SIDE): **23°**
PRIMER: **S/R**
CASE CAPACITY (GR'S WATER): **100.36**

DIMENSIONAL DRAWING:

-NOT ACTUAL SIZE-
-DO NOT SCALE-

CARTRIDGE: .300 Winchester Magnum

OTHER NAMES:

DIA: .308

BALLISTEK NO: 308J

NAI NO: BEN 12343/5.107

DATA SOURCE: Hornady 3rd Pg243

HISTORICAL DATA: By Win. in 1963.

NOTES:

LOADING DATA:

BULLET WT./TYPE	POWDER WT./TYPE	VELOCITY ('/SEC)	SOURCE
180/Spire	71.0/IMR4350	2950	JJD

CASE PREPARATION: SHELLHOLDER (RCBS): 4

MAKE FROM: .300 H&H for factory. Run .300 H&H case into .300 Win. die. Trim to length & chamfer. Fireform in chamber.

PHYSICAL DATA (INCHES):

CASE TYPE: **Belted Bottleneck**
CASE LENGTH **A = 2.620**
HEAD DIAMETER **B = .513**
RIM DIAMETER **D = .532**
NECK DIAMETER **F = .339**
NECK LENGTH **H = .264**
SHOULDER LENGTH **K = .142**
BODY ANGLE (DEG'S/SIDE): **.312**
CASE CAPACITY **CC'S = 5.85**

LOADED LENGTH: **3.34**
BELT DIAMETER **C = .532**
RIM THICKNESS **E = .05**
SHOULDER DIAMETER **G = .491**
LENGTH TO SHOULDER **J = 2.214**
SHOULDER ANGLE (DEG'S/SIDE): **28.15**
PRIMER: **L/R Mag**
CASE CAPACITY (GR'S WATER): **90.36**

DIMENSIONAL DRAWING:

-NOT ACTUAL SIZE-
-DO NOT SCALE-

CARTRIDGE: .300 Winchester Short Magnum

OTHER NAMES:

DIA: .308

BALLISTEK NO:

NAI NO:

DATA SOURCE: Most current loading manuals.

HISTORICAL DATA: Developed by Winchester in 2001 to extract the most performance possible from a .30-caliber cartridge chambered in current short action rifle designs.

NOTES: Performance is similar to the .300 Winchester Magnum.

LOADING DATA:

BULLET WT./TYPE	POWDER WT./TYPE	VELOCITY ('/SEC)	SOURCE
180/Swift Scirocco	63.2/WW 760	2,937	Lyman #48
150/Swift Scirroco	67.5/H414	3,223	Lyman #48

CASE PREPARATION: SHELL HOLDER (RCBS): 43

MAKE FROM: Factory. Can be made from .404 Jeffery. Shorten and die form to fit chamber. Fireform body to get the proper diameter, shoulder angle and head space. Then turn the rim to correct dimension. Ream neck if necessary. Square mouth and chamfer.

PHYSICAL DATA (INCHES):

CASE TYPE: **Rebated Bottleneck**

CASE LENGTH **A = 2.100**

HEAD DIAMETER **B = .5550**

RIM DIAMETER **D = .535**

NECK DIAMETER **F = .3440**

NECK LENGTH **H = .298**

SHOULDER LENGTH **K = .1381**

BODY ANGLE (DEG'S/SIDE):

CASE CAPACITY **CC'S =**

LOADED LENGTH: **2.860**

BELT DIAMETER **C =**

RIM THICKNESS **E = .54**

SHOULDER DIAMETER **G = .5381**

LENGTH TO SHOULDER **J = 1.664**

SHOULDER ANGLE (DEG'S/SIDE): **35**

PRIMER: **L/R or L/R Mag**

CASE CAPACITY (GR'S WATER): **79.2**

DIMENSIONAL DRAWING:

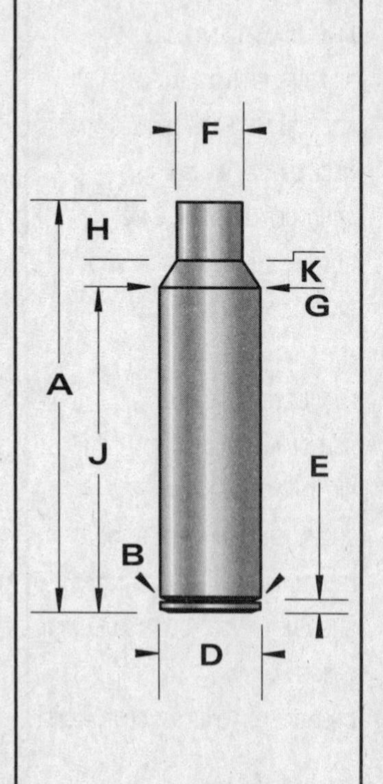

-NOT ACTUAL SIZE-
-DO NOT SCALE-

CARTRIDGE: .303 British

OTHER NAMES:
.303 MkII
7,7 x 56mm Enfield

DIA: .312

BALLISTEK NO: 311A

NAI NO: RMB 32332/4.895

DATA SOURCE: Hornady 3rd Pg253

HISTORICAL DATA: British mil. ctg. ca. 1888.

NOTES:

LOADING DATA:

BULLET WT./TYPE	POWDER WT./TYPE	VELOCITY ('/SEC)	SOURCE
174/RN	39.3/IMR4895	2300	Horn.

CASE PREPARATION: SHELLHOLDER (RCBS): 7

MAKE FROM: Factory or .30-40 Krag. Run Krag case into the .303 die with the expander removed. Trim to length. Chamfer. F/L size.

PHYSICAL DATA (INCHES):

CASE TYPE: **Rimmed Bottleneck**

CASE LENGTH **A = 2.207**

HEAD DIAMETER **B = .451**

RIM DIAMETER **D = .525**

NECK DIAMETER **F = .332**

NECK LENGTH **H = .304**

SHOULDER LENGTH **K = .117**

BODY ANGLE (DEG'S/SIDE): **1.048**

CASE CAPACITY **CC'S = 3.606**

LOADED LENGTH: **3.08**

BELT DIAMETER **C = N/A**

RIM THICKNESS **E = .06**

SHOULDER DIAMETER **G = .393**

LENGTH TO SHOULDER **J = 1.786**

SHOULDER ANGLE (DEG'S/SIDE): **14.61**

PRIMER: **L/R**

CASE CAPACITY (GR'S WATER): **55.65**

DIMENSIONAL DRAWING:

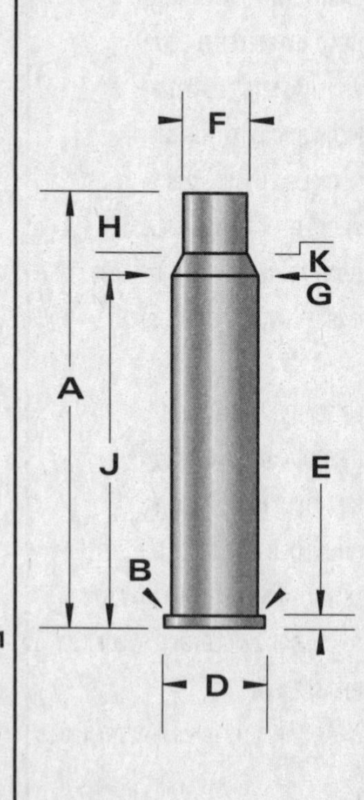

-NOT ACTUAL SIZE-
-DO NOT SCALE-

CARTRIDGE: .303 British Improved (Epps)

OTHER NAMES:

DIA: .311

BALLISTEK NO: 311D

NAI NO: RMB 12344/4.747

DATA SOURCE: Ackley Vol.1 Pg 447

HISTORICAL DATA: By E. Epps.

NOTES:

LOADING DATA:

BULLET WT./TYPE	POWDER WT./TYPE	VELOCITY ('/SEC)	SOURCE
175/Spire	46.7/IMR4320	—	JJD

CASE PREPARATION: SHELLHOLDER (RCBS): 7

MAKE FROM: .303 British. Taper expand the .303 case to about .325" dia. F/L size in the Epps die, square case mouth, chamfer and fireform. I believe it would be okay to fire factory ammo in the Epps chamber but, I have not done so.

PHYSICAL DATA (INCHES):

CASE TYPE: **Rimmed Bottleneck**

CASE LENGTH **A = 2.16**

HEAD DIAMETER **B = .455**

RIM DIAMETER **D = .540**

NECK DIAMETER **F = .339**

NECK LENGTH **H = .278**

SHOULDER LENGTH **K = .081**

BODY ANGLE (DEG'S/SIDE): **.136**

CASE CAPACITY **CC'S = 3.87**

LOADED LENGTH: **A/R**

BELT DIAMETER **C = N/A**

RIM THICKNESS **E = .064**

SHOULDER DIAMETER **G = .447**

LENGTH TO SHOULDER **J = 1.88**

SHOULDER ANGLE (DEG'S/SIDE): **42.87**

PRIMER: **L/R**

CASE CAPACITY (GR'S WATER): **59.73**

DIMENSIONAL DRAWING:

-NOT ACTUAL SIZE-
-DO NOT SCALE-

CARTRIDGE: .303 ICL Improved

OTHER NAMES:

DIA: .311

BALLISTEK NO: 311E

NAI NO: RMB 12335/4.729

DATA SOURCE: Ackley Vol.1 Pg446

HISTORICAL DATA:

NOTES:

LOADING DATA:

BULLET WT./TYPE	POWDER WT./TYPE	VELOCITY ('/SEC)	SOURCE
180/—	43.0/4320	2450	Ackley

CASE PREPARATION: SHELLHOLDER (RCBS): 7

MAKE FROM: .303 British. Fire factory ammo in ICL chamber. Trim to length, chamfer and F/L size. Reload.

PHYSICAL DATA (INCHES):

CASE TYPE: **Rimmed Bottleneck**

CASE LENGTH **A = 2.152**

HEAD DIAMETER **B = .455**

RIM DIAMETER **D = .540**

NECK DIAMETER **F = .339**

NECK LENGTH **H = .350**

SHOULDER LENGTH **K = .057**

BODY ANGLE (DEG'S/SIDE): **.105**

CASE CAPACITY **CC'S = 3.81**

LOADED LENGTH: **A/R**

BELT DIAMETER **C = N/A**

RIM THICKNESS **E = .064**

SHOULDER DIAMETER **G = .449**

LENGTH TO SHOULDER **J = 1.83**

SHOULDER ANGLE (DEG'S/SIDE): **45.01**

PRIMER: **L/R**

CASE CAPACITY (GR'S WATER): **58.91**

DIMENSIONAL DRAWING:

-NOT ACTUAL SIZE-
-DO NOT SCALE-

CARTRIDGE: .303 Magnum

OTHER NAMES:

DIA: .312

BALLISTEK NO: 312G

NAI NO: RMB 32332/4.415

DATA SOURCE: COTW 4th Pg231

HISTORICAL DATA: By Jeffery circa 1920.

NOTES:

LOADING DATA:

BULLET WT./TYPE	POWDER WT./TYPE	VELOCITY ('/SEC)	SOURCE
175/Spire	48.0/IMR4350	—	JJD

CASE PREPARATION: SHELLHOLDER (RCBS): 13

MAKE FROM: .280 Flanged (B.E.L.L.). Turn rim to .557" dia. and back chamfer. Anneal case and F/L size. Trim & chamfer. Fireform.

PHYSICAL DATA (INCHES):

CASE TYPE: **Rimmed Bottleneck**

CASE LENGTH **A = 2.34**

HEAD DIAMETER **B = .530**

RIM DIAMETER **D = .557**

NECK DIAMETER **F = .345**

NECK LENGTH **H = .350**

SHOULDER LENGTH **K = .200**

BODY ANGLE (DEG'S/SIDE): **1.26**

CASE CAPACITY **CC'S = 4.74**

LOADED LENGTH: **3.25**

BELT DIAMETER **C = N/A**

RIM THICKNESS **E = .052**

SHOULDER DIAMETER **G = .460**

LENGTH TO SHOULDER **J = 1.79**

SHOULDER ANGLE (DEG'S/SIDE): **16.04**

PRIMER: **L/R**

CASE CAPACITY (GR'S WATER): **73.18**

DIMENSIONAL DRAWING:

-NOT ACTUAL SIZE-
-DO NOT SCALE-

CARTRIDGE: .303 Savage

OTHER NAMES:

DIA: .308

BALLISTEK NO: 308D

NAI NO: RMB 23322/4.558

DATA SOURCE: Ackley Vol.1 Pg443

HISTORICAL DATA: By A. Savage about 1890.

NOTES:

LOADING DATA:

BULLET WT./TYPE	POWDER WT./TYPE	VELOCITY ('/SEC)	SOURCE
150/FN	35.0/4895	2275	Ackley

CASE PREPARATION: SHELLHOLDER (RCBS): 2

MAKE FROM: .220 Swift. Anneal case neck and taper expand to .310" dia. Re-anneal and F/L size in the .303 die with expanded removed. Trim to length, chamfer and F/L size again. Fireform in chamber. You may find that the .220's rim will need to be thickened to .060-.063".

PHYSICAL DATA (INCHES):

CASE TYPE: **Rimmed Bottleneck**

CASE LENGTH **A = 2.015**

HEAD DIAMETER **B = .442**

RIM DIAMETER **D = .505**

NECK DIAMETER **F = .333**

NECK LENGTH **H = .537**

SHOULDER LENGTH **K = .128**

BODY ANGLE (DEG'S/SIDE): **.722**

CASE CAPACITY **CC'S = 3.02**

LOADED LENGTH: **2.52**

BELT DIAMETER **C = N/A**

RIM THICKNESS **E = .063**

SHOULDER DIAMETER **G = .413**

LENGTH TO SHOULDER **J = 1.35**

SHOULDER ANGLE (DEG'S/SIDE): **17.35**

PRIMER: **L/R**

CASE CAPACITY (GR'S WATER): **46.58**

DIMENSIONAL DRAWING:

-NOT ACTUAL SIZE-
-DO NOT SCALE-

CARTRIDGE: .307 Winchester

OTHER NAMES:	DIA: .308 BALLISTEK NO: 308BZ NAI NO: RMB 22343/4.278

DATA SOURCE: Handloader #110 pg21

HISTORICAL DATA: By Win. in 1982.

NOTES:

LOADING DATA:

BULLET WT./TYPE	POWDER WT./TYPE	VELOCITY ('/SEC)	SOURCE
150/FNSP	39.0/IMR3031	2441	Waters

CASE PREPARATION: SHELLHOLDER (RCBS): 2

MAKE FROM: Factory or 7x57R. Turn rim to .506" dia. & back chamfer. Taper expand to about .315" dia. Trim to length, chamfer and F/L size. Fireform.

PHYSICAL DATA (INCHES):

CASE TYPE: **Rimmed Bottleneck**

CASE LENGTH **A = 2.015**

HEAD DIAMETER **B = .471**

RIM DIAMETER **D = .506**

NECK DIAMETER **F = .334**

NECK LENGTH **H = .303**

SHOULDER LENGTH **K = .152**

BODY ANGLE (DEG'S/SIDE): **.358**

CASE CAPACITY **CC'S = 3.67**

LOADED LENGTH: **2.56**

BELT DIAMETER **C = N/A**

RIM THICKNESS **E = .063**

SHOULDER DIAMETER **G = .454**

LENGTH TO SHOULDER **J = 1.56**

SHOULDER ANGLE (DEG'S/SIDE): **21.54**

PRIMER: **L/R**

CASE CAPACITY (GR'S WATER): **56.69**

DIMENSIONAL DRAWING:

-NOT ACTUAL SIZE-
-DO NOT SCALE-

CARTRIDGE: .308 B-J Express

OTHER NAMES:	DIA: .308 BALLISTEK NO: 308BX NAI NO: BEN 12334/5.477

DATA SOURCE: Ackley Vol.1 Pg449

HISTORICAL DATA: By Barnes & Johnson.

NOTES:

LOADING DATA:

BULLET WT./TYPE	POWDER WT./TYPE	VELOCITY ('/SEC)	SOURCE
190/Spire	77.4/IMR4831	—	JJD

CASE PREPARATION: SHELLHOLDER (RCBS): 4

MAKE FROM: .300 H&H Mag. F/L size the Holland case in the B-J die. Square case mouth and fireform. You may find it necessary, headspace-wise, to taper expand the Holland's neck to about .320" dia. to create a false shoulder (read: bulge) which will help center the case for fireforming.

PHYSICAL DATA (INCHES):

CASE TYPE: **Belted Bottleneck**

CASE LENGTH **A = 2.81**

HEAD DIAMETER **B = .513**

RIM DIAMETER **D = .532**

NECK DIAMETER **F = .339**

NECK LENGTH **H = .351**

SHOULDER LENGTH **K = .124**

BODY ANGLE (DEG'S/SIDE): **.282**

CASE CAPACITY **CC'S = 6.26**

LOADED LENGTH: **4.00**

BELT DIAMETER **C = .532**

RIM THICKNESS **E = .05**

SHOULDER DIAMETER **G = .492**

LENGTH TO SHOULDER **J = 2.335**

SHOULDER ANGLE (DEG'S/SIDE): **31.67**

PRIMER: **L/R Mag**

CASE CAPACITY (GR'S WATER): **96.65**

DIMENSIONAL DRAWING:

-NOT ACTUAL SIZE-
-DO NOT SCALE-

CARTRIDGE: .308 Barnes Supreme

OTHER NAMES:

DIA: .308

BALLISTEK NO: 308BY

NAI NO: BEN 12333/5.604

DATA SOURCE: Ackley Vol.1 Pg.449

HISTORICAL DATA: By F. Barnes.

NOTES:

LOADING DATA:

BULLET WT./TYPE	POWDER WT./TYPE	VELOCITY ('/SEC)	SOURCE
180/Spire	78.9/MRP	3075	JJD

CASE PREPARATION: SHELLHOLDER (RCBS): 4

MAKE FROM: .300 H&H Mag. Taper expand the Holland case to about .315" dia. F/L size in the Barnes die or fire-form in the chamber. This neck expansion is generally necessary to help center the case. Trim the cases and chamfer.

PHYSICAL DATA (INCHES):

CASE TYPE: **Belted Bottleneck**

CASE LENGTH **A = 2.875**

HEAD DIAMETER **B = .513**

RIM DIAMETER **D = .532**

NECK DIAMETER **F = .339**

NECK LENGTH **H = .345**

SHOULDER LENGTH **K = .185**

BODY ANGLE (DEG'S/SIDE): **.307**

CASE CAPACITY **CC'S = 6.37**

LOADED LENGTH: **4.05**

BELT DIAMETER **C = .532**

RIM THICKNESS **E = .05**

SHOULDER DIAMETER **G = .490**

LENGTH TO SHOULDER **J = 2.345**

SHOULDER ANGLE (DEG'S/SIDE): **22.20**

PRIMER: **L/R Mag**

CASE CAPACITY (GR'S WATER): **98.36**

DIMENSIONAL DRAWING:

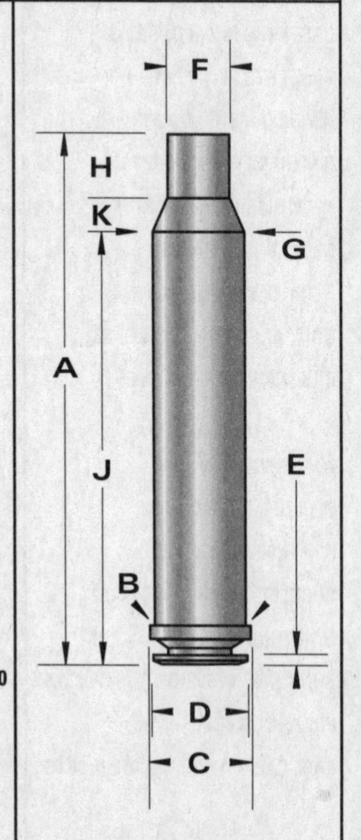

-NOT ACTUAL SIZE-
-DO NOT SCALE-

CARTRIDGE: .308 Norma Magnum

OTHER NAMES:

DIA: .308

BALLISTEK NO: 308M

NAI NO: BEN 12333/5.008

DATA SOURCE: Hornady Manual 3rd Pg 238

HISTORICAL DATA: By Norma in 1961.

NOTES:

LOADING DATA:

BULLET WT./TYPE	POWDER WT./TYPE	VELOCITY ('/SEC)	SOURCE
180/Spire	69.0/IMR4350	2987	JJD

CASE PREPARATION: SHELLHOLDER (RCBS): 4

MAKE FROM: Factory or .300 H&H. Run the H&H case into the .308 Norma die. This should just begin the shoulder. Trim to length and chamfer. Use a light load to fireform in the chamber.

PHYSICAL DATA (INCHES):

CASE TYPE: **Belted Bottleneck**

CASE LENGTH **A = 2.559**

HEAD DIAMETER **B = .511**

RIM DIAMETER **D = .530**

NECK DIAMETER **F = .339**

NECK LENGTH **H = .318**

SHOULDER LENGTH **K = .156**

BODY ANGLE (DEG'S/SIDE): **.334**

CASE CAPACITY **CC'S = 5.55**

LOADED LENGTH: **3.28**

BELT DIAMETER **C = .530**

RIM THICKNESS **E = .05**

SHOULDER DIAMETER **G = .489**

LENGTH TO SHOULDER **J = 2.085**

SHOULDER ANGLE (DEG'S/SIDE): **25.67**

PRIMER: **L/R Mag**

CASE CAPACITY (GR'S WATER): **85.78**

DIMENSIONAL DRAWING:

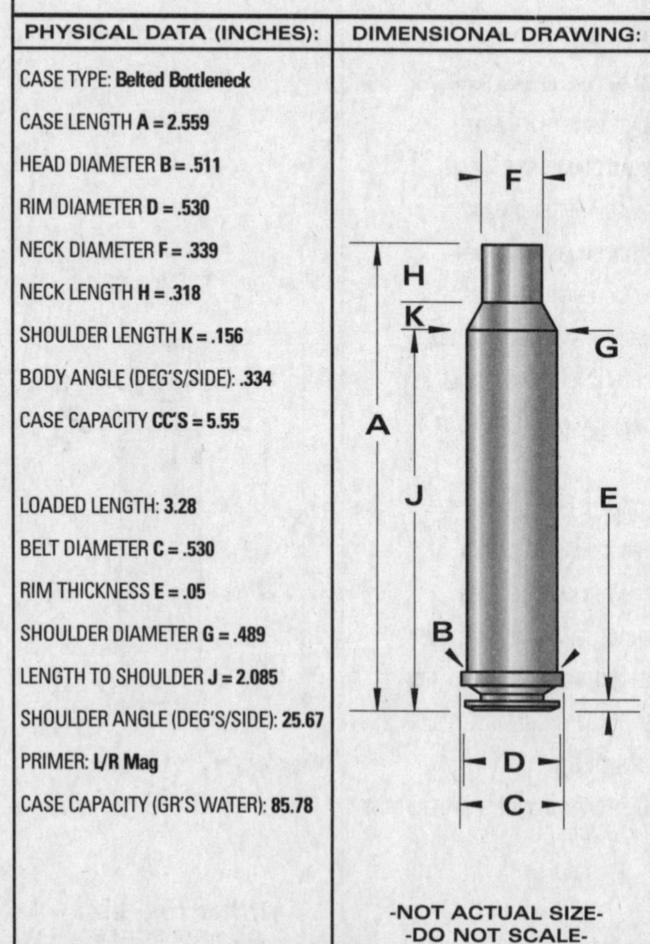

-NOT ACTUAL SIZE-
-DO NOT SCALE-

CARTRIDGE: .308 Winchester

OTHER NAMES:
7.62 x 51mm NATO
.30 Cal. T-65

DIA: .308

BALLISTEK NO: 308H

NAI NO: RXB
22342/4.322

DATA SOURCE: Hornady Manual 3rd Pg216

HISTORICAL DATA: Developed as mil. ctg. in early 1950s.

NOTES: Commercial by Winchester in 1952.

LOADING DATA:

BULLET WT./TYPE	POWDER WT./TYPE	VELOCITY ('/SEC)	SOURCE
168/HPBT	39.5/IMR4895	2500	Horn.
190/HPBT	41.5/IMR4895	2500	JJD

CASE PREPARATION: SHELLHOLDER (RCBS): 3

MAKE FROM: Factory or .30-06. Anneal case neck. Run into .308 F/L die with expander removed. Trim to length. I.D. (or O.D.) neck ream. Chamfer. F/L size.

PHYSICAL DATA (INCHES):

CASE TYPE: **Rimless Bottleneck**

CASE LENGTH **A = 2.010**

HEAD DIAMETER **B = .465**

RIM DIAMETER **D = .466**

NECK DIAMETER **F = .338**

NECK LENGTH **H = .266**

SHOULDER LENGTH **K = .186**

BODY ANGLE (DEG'S/SIDE): **.359**

CASE CAPACITY **CC'S = 3.50**

LOADED LENGTH: **2.80**

BELT DIAMETER **C = N/A**

RIM THICKNESS **E = .049**

SHOULDER DIAMETER **G = .488**

LENGTH TO SHOULDER **J = 1.558**

SHOULDER ANGLE (DEG'S/SIDE): **16.47**

PRIMER: **L/R**

CASE CAPACITY (GR'S WATER): **54.07**

DIMENSIONAL DRAWING:

-NOT ACTUAL SIZE-
-DO NOT SCALE-

CARTRIDGE: .308 x 1.5"

OTHER NAMES:

DIA: .308

BALLISTEK NO: 308G

NAI NO: RXB
24342/3.205

DATA SOURCE: NAI/Ballistek

HISTORICAL DATA: By F. Barnes in 1961.

NOTES:

LOADING DATA:

BULLET WT./TYPE	POWDER WT./TYPE	VELOCITY ('/SEC)	SOURCE
125/Spire	28.0/4198	2641	Barnes

CASE PREPARATION: SHELLHOLDER (RCBS): 3

MAKE FROM: .308 Win. (or any similar case). Form the case in a .308x1.5" trim die - cut off excess brass. Trim to length & chamfer. F/L size.

PHYSICAL DATA (INCHES):

CASE TYPE: **Rimless Bottleneck**

CASE LENGTH **A = 1.50**

HEAD DIAMETER **B = .468**

RIM DIAMETER **D = .470**

NECK DIAMETER **F = .338**

NECK LENGTH **H = .260**

SHOULDER LENGTH **K = .155**

BODY ANGLE (DEG'S/SIDE): **.582**

CASE CAPACITY **CC'S = 2.44**

LOADED LENGTH: **A/R**

BELT DIAMETER **C = N/A**

RIM THICKNESS **E = .05**

SHOULDER DIAMETER **G = .450**

LENGTH TO SHOULDER **J = 1.085**

SHOULDER ANGLE (DEG'S/SIDE): **19.86**

PRIMER: **L/R**

CASE CAPACITY (GR'S WATER): **37.61**

DIMENSIONAL DRAWING:

-NOT ACTUAL SIZE-
-DO NOT SCALE-

CARTRIDGE: .308 x 1.75"

OTHER NAMES:
.30 Doggie

DIA: .308

BALLISTEK NO: 308CC

NAI NO: RXB
23342/3.723

DATA SOURCE: NAI/Ballistek

HISTORICAL DATA:

NOTES:

LOADING DATA:

BULLET WT./TYPE	POWDER WT./TYPE	VELOCITY ('/SEC)	SOURCE
125/Spire	30.0/IMR3031	2450	JJD

CASE PREPARATION: SHELLHOLDER (RCBS): 3

MAKE FROM: .308 Winchester. Use .308 x 1.5" dies but adjust for the longer case length.

PHYSICAL DATA (INCHES):

CASE TYPE: **Rimless Bottleneck**

CASE LENGTH **A = 1.75**

HEAD DIAMETER **B = .47**

RIM DIAMETER **D = .473**

NECK DIAMETER **F = .338**

NECK LENGTH **H = .260**

SHOULDER LENGTH **K = .155**

BODY ANGLE (DEG'S/SIDE): **.504**

CASE CAPACITY **CC'S = 3.00**

LOADED LENGTH: **A/R**

BELT DIAMETER **C = N/A**

RIM THICKNESS **E = .049**

SHOULDER DIAMETER **G = .450**

LENGTH TO SHOULDER **J = 1.335**

SHOULDER ANGLE (DEG'S/SIDE): **19.86**

PRIMER: **L/R**

CASE CAPACITY (GR'S WATER): **46.30**

DIMENSIONAL DRAWING:

-NOT ACTUAL SIZE-
-DO NOT SCALE-

CARTRIDGE: .309 JDJ

OTHER NAMES:

DIA: .308

BALLISTEK NO: 308CD

NAI NO: RMB
12334/4.762

DATA SOURCE: SSK Industries

HISTORICAL DATA: By J.D. Jones.

NOTES:

LOADING DATA:

BULLET WT./TYPE	POWDER WT./TYPE	VELOCITY ('/SEC)	SOURCE
165/Nosler	—/—	2350	SSK

CASE PREPARATION: SHELLHOLDER (RCBS): 28

MAKE FROM: .444 Marlin. Available data says to F/L size the Marlin case and trim but, in my experience, going from .44 to .30 cal. will require a form die set.
I would obtain same, anneal the case, form as required, trim & chamfer and F/L size.

PHYSICAL DATA (INCHES):

CASE TYPE: **Rimmed Bottleneck**

CASE LENGTH **A = 2.205**

HEAD DIAMETER **B = .463**

RIM DIAMETER **D = .506**

NECK DIAMETER **F = .336**

NECK LENGTH **H = .320**

SHOULDER LENGTH **K = .067**

BODY ANGLE (DEG'S/SIDE): **.159**

CASE CAPACITY **CC'S = 4.02**

LOADED LENGTH: **3.045**

BELT DIAMETER **C = N/A**

RIM THICKNESS **E = .06**

SHOULDER DIAMETER **G = .454**

LENGTH TO SHOULDER **J = 1.818**

SHOULDER ANGLE (DEG'S/SIDE): **41.36**

PRIMER: **L/R**

CASE CAPACITY (GR'S WATER): **62.05**

DIMENSIONAL DRAWING:

-NOT ACTUAL SIZE-
-DO NOT SCALE-

CARTRIDGE: .310 Cadet Rifle

OTHER NAMES:
.310 Greener

DIA: .316

BALLISTEK NO: 316A

NAI NO: RMS 21115/2.889

DATA SOURCE: COTW 4th Pg232

HISTORICAL DATA: By Greener in 1900.

NOTES: Similar to .32-20 WCF.

LOADING DATA:

BULLET WT./TYPE	POWDER WT./TYPE	VELOCITY ('/SEC)	SOURCE
110/Lead	9.0/IMR4227	1400	JJD

CASE PREPARATION: SHELLHOLDER (RCBS): --

MAKE FROM: .32-20 WCF. Trim case to length & chamfer. F/L size in .310 die. Fireform.

PHYSICAL DATA (INCHES):

CASE TYPE: **Rimmed Straight**

CASE LENGTH **A = 1.02**

HEAD DIAMETER **B = .353**

RIM DIAMETER **D = .405**

NECK DIAMETER **F = .320**

NECK LENGTH **H = N/A**

SHOULDER LENGTH **K = N/A**

BODY ANGLE (DEG'S/SIDE): **.98**

CASE CAPACITY **CC'S = 1.00**

LOADED LENGTH: **1.59**

BELT DIAMETER **C = N/A**

RIM THICKNESS **E = .055**

SHOULDER DIAMETER **G = N/A**

LENGTH TO SHOULDER **J = N/A**

SHOULDER ANGLE (DEG'S/SIDE): **N/A**

PRIMER: **S/R**

CASE CAPACITY (GR'S WATER): **15.48**

DIMENSIONAL DRAWING:

F
A
E
B
D

-NOT ACTUAL SIZE-
-DO NOT SCALE-

CARTRIDGE: .318 Westley Richards Rimless

OTHER NAMES:
.318 Rimless Nitro Express

DIA: .330

BALLISTEK NO: 330B

NAI NO: RXB 12234/5.118

DATA SOURCE: COTW 4th Pg232

HISTORICAL DATA: By W. Richards in 1910.

NOTES:

LOADING DATA:

BULLET WT./TYPE	POWDER WT./TYPE	VELOCITY ('/SEC)	SOURCE
250/RN	58.0/4831	2400	Barnes

CASE PREPARATION: SHELLHOLDER (RCBS): 3

MAKE FROM: .30-06 Spgf. Taper expand neck to about .330" dia. F/L size in .318 die with expander removed. Trim to length. Neck I.D. may need reaming. F/L size and fire-form in chamber.

PHYSICAL DATA (INCHES):

CASE TYPE: **Rimless Bottleneck**

CASE LENGTH **A = 2.38**

HEAD DIAMETER **B = .465**

RIM DIAMETER **D = .465**

NECK DIAMETER **F = .358**

NECK LENGTH **H = .331**

SHOULDER LENGTH **K = .077**

BODY ANGLE (DEG'S/SIDE): **.226**

CASE CAPACITY **CC'S = 4.45**

LOADED LENGTH: **3.35**

BELT DIAMETER **C = N/A**

RIM THICKNESS **E = .05**

SHOULDER DIAMETER **G = .451**

LENGTH TO SHOULDER **J = 1.972**

SHOULDER ANGLE (DEG'S/SIDE): **31.12**

PRIMER: **L/R**

CASE CAPACITY (GR'S WATER): **68.72**

DIMENSIONAL DRAWING:

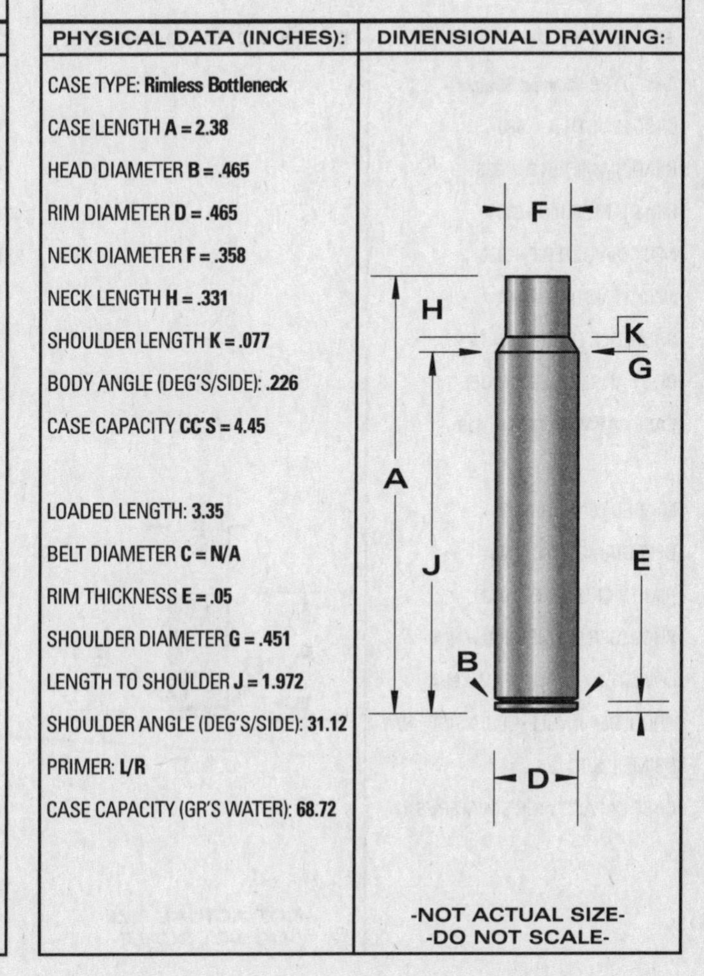

-NOT ACTUAL SIZE-
-DO NOT SCALE-

CARTRIDGE: .32 ACP

OTHER NAMES:
7,65 x 17mm
7,65 Browning SL

DIA: .309

BALLISTEK NO: 309B

NAI NO: RMS 11115/2.040

DATA SOURCE: COTW 4th Pg167

HISTORICAL DATA: By J. Browning about 1899.

NOTES: Projectile dia. is .309" (7,85mm).

LOADING DATA:

BULLET WT./TYPE	POWDER WT./TYPE	VELOCITY ('/SEC)	SOURCE
100/RN	2.8/Unique	875	Barnes

CASE PREPARATION: **SHELLHOLDER (RCBS): 17**

MAKE FROM: Factory or .32 S&W (short or long). Turn rim to .354" dia. and thin to .043" thick. Back chamfer. Cut new extractor groove. Trim to length and F/L size.

PHYSICAL DATA (INCHES):

CASE TYPE: **Rimmed Straight**
CASE LENGTH **A = .680**
HEAD DIAMETER **B = .336**
RIM DIAMETER **D = .354**
NECK DIAMETER **F = .336**
NECK LENGTH **H = N/A**
SHOULDER LENGTH **K = N/A**
BODY ANGLE (DEG'S/SIDE): **0**
CASE CAPACITY **CC'S = .419**

LOADED LENGTH: **1.03**
BELT DIAMETER **C = N/A**
RIM THICKNESS **E = .043**
SHOULDER DIAMETER **G = N/A**
LENGTH TO SHOULDER **J = N/A**
SHOULDER ANGLE (DEG'S/SIDE): **N/A**
PRIMER: **S/P**
CASE CAPACITY (GR'S WATER): **6.47**

DIMENSIONAL DRAWING:

F
A
B
E
D

-NOT ACTUAL SIZE-
-DO NOT SCALE-

CARTRIDGE: .32 Ballard Extra Long

OTHER NAMES:
.32 Extra Long

DIA: .317

BALLISTEK NO: 317C

NAI NO: RMS 11115/3.863

DATA SOURCE: COTW 4th Pg95

HISTORICAL DATA: About 1879.

NOTES:

LOADING DATA:

BULLET WT./TYPE	POWDER WT./TYPE	VELOCITY ('/SEC)	SOURCE
115/—	9.0/4198	1360	Barnes

CASE PREPARATION: **SHELLHOLDER (RCBS): 23**

MAKE FROM: Turn from solid brass. No known case will form into this one!

PHYSICAL DATA (INCHES):

CASE TYPE: **Rimmed Straight**
CASE LENGTH **A = 1.24**
HEAD DIAMETER **B = .321**
RIM DIAMETER **D = .369**
NECK DIAMETER **F = .319**
NECK LENGTH **H = N/A**
SHOULDER LENGTH **K = N/A**
BODY ANGLE (DEG'S/SIDE): **.048**
CASE CAPACITY **CC'S = 1.21**

LOADED LENGTH: **1.80**
BELT DIAMETER **C = N/A**
RIM THICKNESS **E = .068**
SHOULDER DIAMETER **G = N/A**
LENGTH TO SHOULDER **J = N/A**
SHOULDER ANGLE (DEG'S/SIDE): **N/A**
PRIMER: **S/R**
CASE CAPACITY (GR'S WATER): **18.63**

DIMENSIONAL DRAWING:

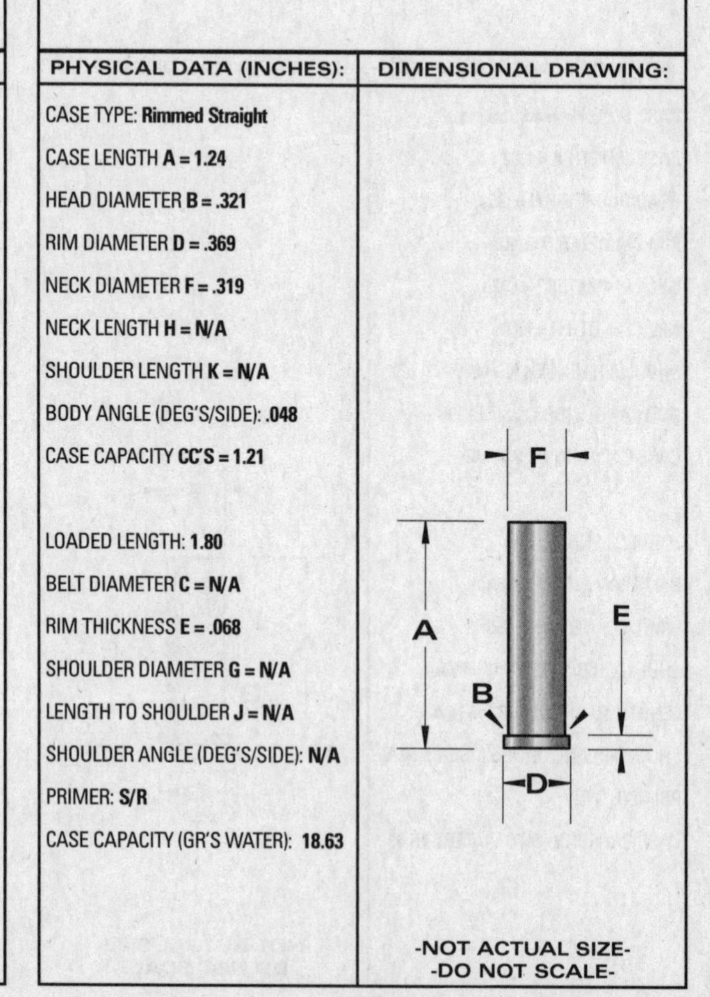

-NOT ACTUAL SIZE-
-DO NOT SCALE-

CARTRIDGE: .32 Colt Long

OTHER NAMES:

DIA: .313

BALLISTEK NO: 313D

NAI NO: RMS 11115/2.893

DATA SOURCE: COTW 4th Pg169

HISTORICAL DATA: By Colt in 1875.

NOTES:

LOADING DATA:

BULLET WT./TYPE	POWDER WT./TYPE	VELOCITY ('/SEC)	SOURCE
80/Lead	2.1/B'Eye	770	Barnes

CASE PREPARATION: SHELLHOLDER (RCBS): 10

MAKE FROM: Factory or re-body a .223 Rem. head with 5/16" tubing, F/L size and I.D. neck ream.

PHYSICAL DATA (INCHES):

CASE TYPE: **Rimmed Straight**

CASE LENGTH **A = .92**

HEAD DIAMETER **B = .318**

RIM DIAMETER **D = .374**

NECK DIAMETER **F = .313**

NECK LENGTH **H = N/A**

SHOULDER LENGTH **K = N/A**

BODY ANGLE (DEG'S/SIDE): **.164**

CASE CAPACITY **CC'S = .785**

LOADED LENGTH: **1.26**

BELT DIAMETER **C = N/A**

RIM THICKNESS **E = .05**

SHOULDER DIAMETER **G = N/A**

LENGTH TO SHOULDER **J = N/A**

SHOULDER ANGLE (DEG'S/SIDE): **N/A**

PRIMER: **S/P**

CASE CAPACITY (GR'S WATER): **12.12**

DIMENSIONAL DRAWING:

-NOT ACTUAL SIZE-
-DO NOT SCALE-

CARTRIDGE: .32 Colt Short

OTHER NAMES:

DIA: .313

BALLISTEK NO: 313C

NAI NO: RMS 11115/2.044

DATA SOURCE: COTW 4th Pg169

HISTORICAL DATA: By Colt in 1875.

NOTES:

LOADING DATA:

BULLET WT./TYPE	POWDER WT./TYPE	VELOCITY ('/SEC)	SOURCE
80/Lead	1.5/B'Eye	600	JJD

CASE PREPARATION: SHELLHOLDER (RCBS): 10

MAKE FROM: .32 Colt Long. Trim case to length and F/L size. .223 Rem. cases may be re-bodied, with 5/16" dia. tubing. F/L size and I.D. neck ream.

PHYSICAL DATA (INCHES):

CASE TYPE: **Rimmed Straight**

CASE LENGTH **A = .65**

HEAD DIAMETER **B = .318**

RIM DIAMETER **D = .374**

NECK DIAMETER **F = .313**

NECK LENGTH **H = N/A**

SHOULDER LENGTH **K = N/A**

BODY ANGLE (DEG'S/SIDE): **.239**

CASE CAPACITY **CC'S = .439**

LOADED LENGTH: **1.01**

BELT DIAMETER **C = N/A**

RIM THICKNESS **E = .052**

SHOULDER DIAMETER **G = N/A**

LENGTH TO SHOULDER **J = N/A**

SHOULDER ANGLE (DEG'S/SIDE): **N/A**

PRIMER: **S/P**

CASE CAPACITY (GR'S WATER): **6.78**

DIMENSIONAL DRAWING:

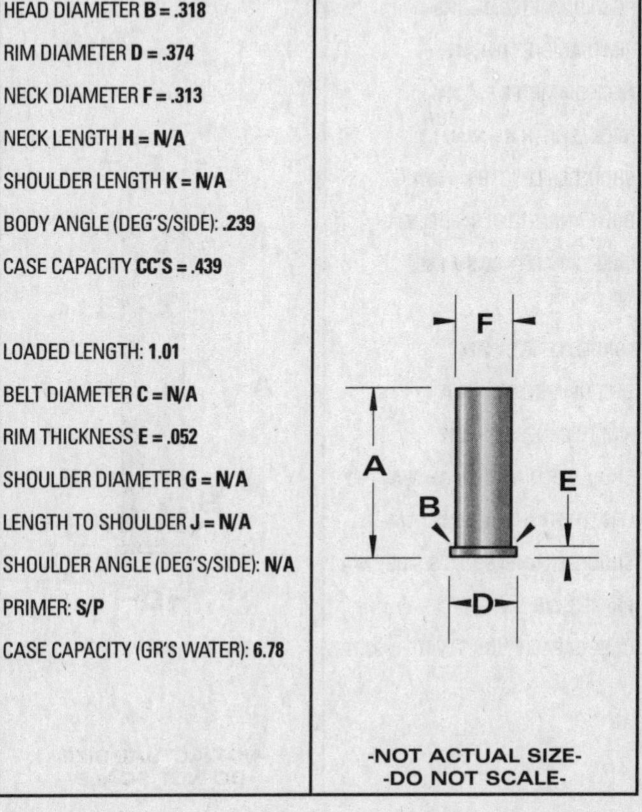

-NOT ACTUAL SIZE-
-DO NOT SCALE-

CARTRIDGE: .32 Ideal

OTHER NAMES:
.32-25-150 Ideal

DIA: .323

BALLISTEK NO: 323Z

NAI NO: RMS 11115/5.086

DATA SOURCE: COTW 4th Pg96

HISTORICAL DATA: Developed for Stevens M44 about 1903.

NOTES:

LOADING DATA:

BULLET WT./TYPE	POWDER WT./TYPE	VELOCITY ('/SEC)	SOURCE
150/Lead	12.0/4198	1330	Barnes

CASE PREPARATION: SHELLHOLDER (RCBS): 1

MAKE FROM: .32-20 WCF. Load the .32-20 case with 3.0 grs of Bullseye, corn meal and a wax plug. Fireform in the weapon or in a trim die to "blow-out" the case. F/L size and square case mouth. This case is quite short but, otherwise, fine.

PHYSICAL DATA (INCHES):

CASE TYPE: **Rimmed Straight**
CASE LENGTH **A = 1.77**
HEAD DIAMETER **B = .348**
RIM DIAMETER **D = .411**
NECK DIAMETER **F = .344**
NECK LENGTH **H = N/A**
SHOULDER LENGTH **K = N/A**
BODY ANGLE (DEG'S/SIDE): **.066**
CASE CAPACITY **CC'S = 1.86**
LOADED LENGTH: **2.25**
BELT DIAMETER **C = N/A**
RIM THICKNESS **E = .042**
SHOULDER DIAMETER **G = N/A**
LENGTH TO SHOULDER **J = N/A**
SHOULDER ANGLE (DEG'S/SIDE): **N/A**
PRIMER: **S/R**
CASE CAPACITY (GR'S WATER): **28.76**

DIMENSIONAL DRAWING:

F
A
B
E
D

-NOT ACTUAL SIZE-
-DO NOT SCALE-

CARTRIDGE: .32 Long Centerfire

OTHER NAMES:
.32 Long Rifle

DIA: .317

BALLISTEK NO: 317D

NAI NO: RMS 11115/2.554

DATA SOURCE: COTW 4th Pg95

HISTORICAL DATA: From about 1875.

NOTES: Similar to .32 S&W Long (#312C)

LOADING DATA:

BULLET WT./TYPE	POWDER WT./TYPE	VELOCITY ('/SEC)	SOURCE
85/Lead	3.4/2400	–	JJD

CASE PREPARATION: SHELLHOLDER (RCBS): 23

MAKE FROM: .32 Colt Long. Trim case to length and F/L size. Rim dia. may require reducing to .369" dia.

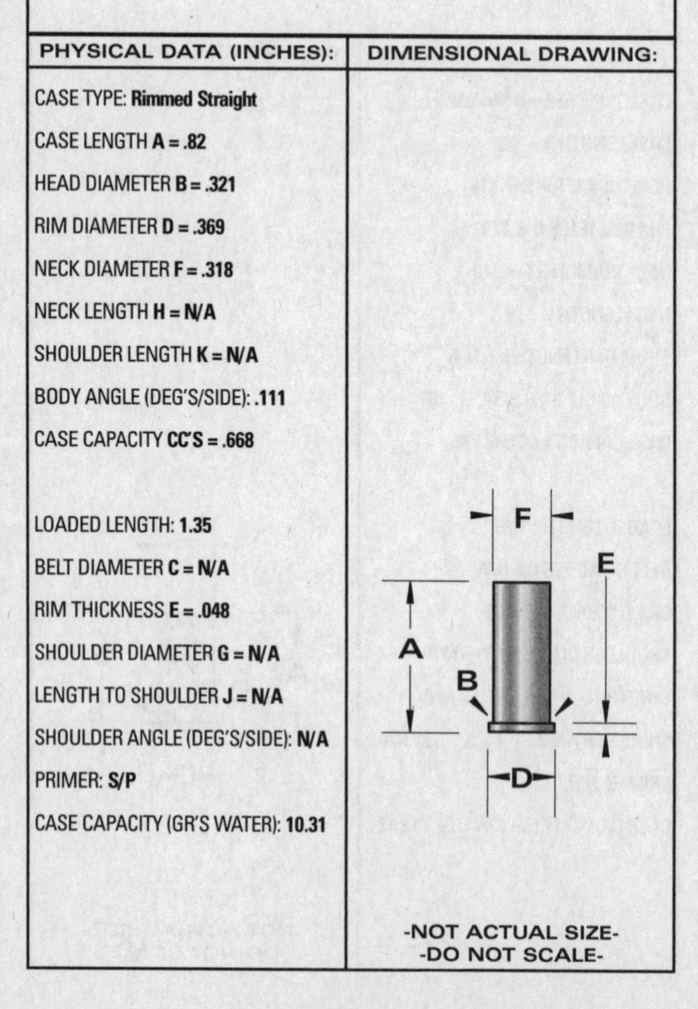

PHYSICAL DATA (INCHES):

CASE TYPE: **Rimmed Straight**
CASE LENGTH **A = .82**
HEAD DIAMETER **B = .321**
RIM DIAMETER **D = .369**
NECK DIAMETER **F = .318**
NECK LENGTH **H = N/A**
SHOULDER LENGTH **K = N/A**
BODY ANGLE (DEG'S/SIDE): **.111**
CASE CAPACITY **CC'S = .668**
LOADED LENGTH: **1.35**
BELT DIAMETER **C = N/A**
RIM THICKNESS **E = .048**
SHOULDER DIAMETER **G = N/A**
LENGTH TO SHOULDER **J = N/A**
SHOULDER ANGLE (DEG'S/SIDE): **N/A**
PRIMER: **S/P**
CASE CAPACITY (GR'S WATER): **10.31**

DIMENSIONAL DRAWING:

-NOT ACTUAL SIZE-
-DO NOT SCALE-

CARTRIDGE: .32 Remington

OTHER NAMES:

DIA: .320

BALLISTEK NO: 320B

NAI NO: RXB 22223/4.869

DATA SOURCE: COTW 4th Pg96

HISTORICAL DATA: By Rem. in 1906.

NOTES: Rimless version of .32 Win. Spl. (#321A)

LOADING DATA:

BULLET WT./TYPE	POWDER WT./TYPE	VELOCITY ('/SEC)	SOURCE
170/—	27.0/4198	2130	Barnes

CASE PREPARATION: SHELLHOLDER (RCBS): 25

MAKE FROM: .30-30 Win. Turn rim flush with body and cut extractor groove. Taper expand to .330" dia. and F/L size. Fireform in chamber. .30 Rem., if available, may be taper expanded and F/L sized.

PHYSICAL DATA (INCHES):

CASE TYPE: **Rimless Bottleneck**

CASE LENGTH **A = 2.05**

HEAD DIAMETER **B = .421**

RIM DIAMETER **D = .421**

NECK DIAMETER **F = .344**

NECK LENGTH **H = .489**

SHOULDER LENGTH **K = .063**

BODY ANGLE (DEG'S/SIDE): **.441**

CASE CAPACITY **CC'S = 3.01**

LOADED LENGTH: **2.53**

BELT DIAMETER **C = N/A**

RIM THICKNESS **E = .058**

SHOULDER DIAMETER **G = .401**

LENGTH TO SHOULDER **J = 1.498**

SHOULDER ANGLE (DEG'S/SIDE): **24.34**

PRIMER: **L/R**

CASE CAPACITY (GR'S WATER): **46.45**

DIMENSIONAL DRAWING:

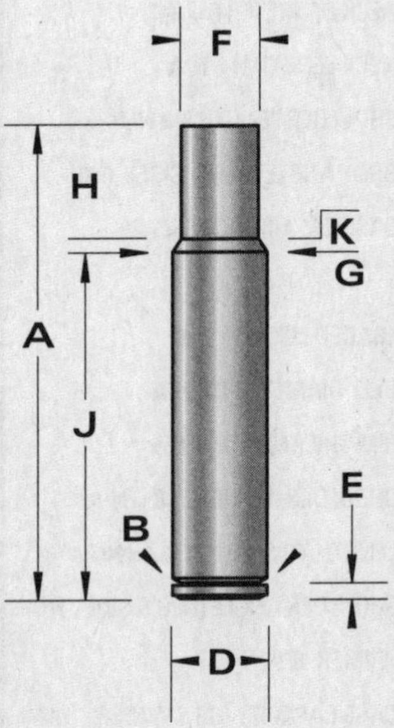

CARTRIDGE: .32 S&W Long

OTHER NAMES:
.32 Colt New Police
.32 Harrington & Richardson

DIA: .312

BALLISTEK NO: 312C

NAI NO: RMS 11115/2.746

DATA SOURCE: COTW 4th Pg168

HISTORICAL DATA: By S&W about 1896.

NOTES:

LOADING DATA:

BULLET WT./TYPE	POWDER WT./TYPE	VELOCITY ('/SEC)	SOURCE
100/Lead	2.0/B'Eye	750	JJD

CASE PREPARATION: SHELLHOLDER (RCBS): 10

MAKE FROM: Factory or use .32 S&W short, as is. It is possible to re-body a .223 Rem. case, with tubing but, this is, generally, not worth the effort as the factory brass is available.

PHYSICAL DATA (INCHES):

CASE TYPE: **Rimmed Straight**

CASE LENGTH **A = .92**

HEAD DIAMETER **B = .335**

RIM DIAMETER **D = .375**

NECK DIAMETER **F = .334**

NECK LENGTH **H = N/A**

SHOULDER LENGTH **K = N/A**

BODY ANGLE (DEG'S/SIDE): **.066**

CASE CAPACITY **CC'S = .704**

LOADED LENGTH: **1.28**

BELT DIAMETER **C = N/A**

RIM THICKNESS **E = .055**

SHOULDER DIAMETER **G = N/A**

LENGTH TO SHOULDER **J = N/A**

SHOULDER ANGLE (DEG'S/SIDE): **N/A**

PRIMER: **S/P**

CASE CAPACITY (GR'S WATER): **10.86**

DIMENSIONAL DRAWING:

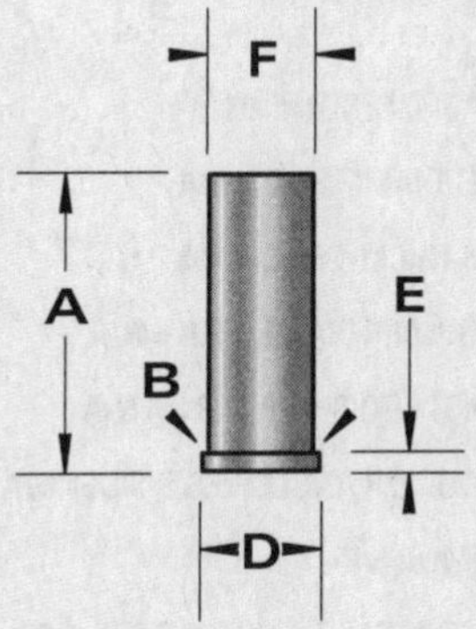

-NOT ACTUAL SIZE-
-DO NOT SCALE-

CARTRIDGE: .32 S&W Short

OTHER NAMES:
.32 Smith & Wesson

DIA: .312

BALLISTEK NO: 312B

NAI NO: RMS 11115/1.806

DATA SOURCE: COTW 4th Pg168

HISTORICAL DATA: In 1878 by S&W.

NOTES:

LOADING DATA:

BULLET WT./TYPE	POWDER WT./TYPE	VELOCITY ('/SEC)	SOURCE
85/Lead	1.2/B'Eye	725	JJD

CASE PREPARATION: SHELLHOLDER (RCBS): 10

MAKE FROM: .32 S&W Long. Cut case to .605" and chamfer. With moderate loads F/L dies are not always required but, they are a good idea!

PHYSICAL DATA (INCHES):

CASE TYPE: **Rimmed Straight**

CASE LENGTH **A = .605**

HEAD DIAMETER **B = .335**

RIM DIAMETER **D = .375**

NECK DIAMETER **F = .334**

NECK LENGTH **H = N/A**

SHOULDER LENGTH **K = N/A**

BODY ANGLE (DEG'S/SIDE): **.052**

CASE CAPACITY **CC'S = .348**

LOADED LENGTH: **.93**

BELT DIAMETER **C = N/A**

RIM THICKNESS **E = .055**

SHOULDER DIAMETER **G = N/A**

LENGTH TO SHOULDER **J = N/A**

SHOULDER ANGLE (DEG'S/SIDE): **N/A**

PRIMER: **S/P**

CASE CAPACITY (GR'S WATER): **5.38**

DIMENSIONAL DRAWING:

F E A B D

-NOT ACTUAL SIZE-
-DO NOT SCALE-

CARTRIDGE: .32 Winchester Self-Loading

OTHER NAMES:
.32 WCF SL

DIA: .320

BALLISTEK NO: 320D

NAI NO: RMS 11115/3.699

DATA SOURCE: COTW 4th pg96

HISTORICAL DATA: By Win. about 1905.

NOTES:

LOADING DATA:

BULLET WT./TYPE	POWDER WT./TYPE	VELOCITY ('/SEC)	SOURCE
165/Lead	12.5/4227	1440	Barnes

CASE PREPARATION: SHELLHOLDER (RCBS): 16

MAKE FROM: .32-20 WCF. Taper expand the Win. case and F/L size or fireform, with corn meal, in the SL trim die. Trim & chamfer. 11/32" dia. tubing can be used to re-body a .38 ACP case head.

PHYSICAL DATA (INCHES):

CASE TYPE: **Rimmed Straight**

CASE LENGTH **A = 1.28**

HEAD DIAMETER **B = .346**

RIM DIAMETER **D = .388**

NECK DIAMETER **F = .343**

NECK LENGTH **H = N/A**

SHOULDER LENGTH **K = N/A**

BODY ANGLE (DEG'S/SIDE): **.070**

CASE CAPACITY **CC'S = 1.20**

LOADED LENGTH: **1.88**

BELT DIAMETER **C = N/A**

RIM THICKNESS **E = .055**

SHOULDER DIAMETER **G = N/A**

LENGTH TO SHOULDER **J = N/A**

SHOULDER ANGLE (DEG'S/SIDE): **N/A**

PRIMER: **S/P**

CASE CAPACITY (GR'S WATER): **18.58**

DIMENSIONAL DRAWING:

F A E B D

-NOT ACTUAL SIZE-
-DO NOT SCALE-

CARTRIDGE: .32 Winchester Special

OTHER NAMES:	DIA: .321 BALLISTEK NO: 321A NAI NO: RMB 22222/4.684

DATA SOURCE: Hornady Manual 3rd Pg257

HISTORICAL DATA: By Win. in 1895.

NOTES:

LOADING DATA:

BULLET WT./TYPE	POWDER WT./TYPE	VELOCITY ('/SEC)	SOURCE
170/FN	31.5/IMR3031	2100	Horn.

CASE PREPARATION: SHELLHOLDER (RCBS): 2

MAKE FROM: Factory or .30-30 Win. Anneal neck and F/L size. Square case mouth. Chamfer.

PHYSICAL DATA (INCHES):

CASE TYPE: **Rimmed Bottleneck**

CASE LENGTH **A = 1.977**

HEAD DIAMETER **B = .422**

RIM DIAMETER **D = .506**

NECK DIAMETER **F = .343**

NECK LENGTH **H = .454**

SHOULDER LENGTH **K = .126**

BODY ANGLE (DEG'S/SIDE): **.478**

CASE CAPACITY **CC'S = 2.93**

LOADED LENGTH: **2.60**

BELT DIAMETER **C = N/A**

RIM THICKNESS **E = .063**

SHOULDER DIAMETER **G = .402**

LENGTH TO SHOULDER **J = 1.397**

SHOULDER ANGLE (DEG'S/SIDE): **13.17**

PRIMER: **L/R**

CASE CAPACITY (GR'S WATER): **45.23**

DIMENSIONAL DRAWING:

-NOT ACTUAL SIZE-
-DO NOT SCALE-

CARTRIDGE: .32-20 Winchester

OTHER NAMES: .32 Winchester .32 WCF	DIA: .312 BALLISTEK NO: 312H NAI NO: RMB 24231/3.725

DATA SOURCE: COTW 4th Pg60

HISTORICAL DATA: By Win. in 1882.

NOTES:

LOADING DATA:

BULLET WT./TYPE	POWDER WT./TYPE	VELOCITY ('/SEC)	SOURCE
100/Lead	12.0/2400	2220	Barnes

CASE PREPARATION: SHELLHOLDER (RCBS): 1

MAKE FROM: Factory or .25-20 WCF. Taper expand the .25-20 case to about .320" dia. and F/L size in the .32-20 die. Trim & chamfer.

PHYSICAL DATA (INCHES):

CASE TYPE: **Rimmed Bottleneck**

CASE LENGTH **A = 1.315**

HEAD DIAMETER **B = .353**

RIM DIAMETER **D = .408**

NECK DIAMETER **F = .326**

NECK LENGTH **H = .376**

SHOULDER LENGTH **K = .058**

BODY ANGLE (DEG'S/SIDE): **.467**

CASE CAPACITY **CC'S = 1.42**

LOADED LENGTH: **1.59**

BELT DIAMETER **C = N/A**

RIM THICKNESS **E = .064**

SHOULDER DIAMETER **G = .342**

LENGTH TO SHOULDER **J = .881**

SHOULDER ANGLE (DEG'S/SIDE): **7.85**

PRIMER: **S/R**

CASE CAPACITY (GR'S WATER): **21.89**

DIMENSIONAL DRAWING:

-NOT ACTUAL SIZE-
-DO NOT SCALE-

CARTRIDGE: .32-30 Remington

OTHER NAMES:

DIA: .312

BALLISTEK NO: 312D

NAI NO: RMB 22241/4.338

DATA SOURCE: COTW 4th Pg97

HISTORICAL DATA: About 1884.

NOTES: Similar to .32-20 WCF.

LOADING DATA:

BULLET WT./TYPE	POWDER WT./TYPE	VELOCITY ('/SEC)	SOURCE
115/Lead	13.5/IMR4198	1500	JJD

CASE PREPARATION: SHELLHOLDER (RCBS): 6

MAKE FROM: .357 Max. F/L size the .357 Max. case, square the case mouth and chamfer. Case is about .030" short but fine.

PHYSICAL DATA (INCHES):

CASE TYPE: **Rimmed Bottleneck**

CASE LENGTH **A = 1.64**

HEAD DIAMETER **B = .378**

RIM DIAMETER **D = .437**

NECK DIAMETER **F = .332**

NECK LENGTH **H = .250**

SHOULDER LENGTH **K = .115**

BODY ANGLE (DEG'S/SIDE): **.559**

CASE CAPACITY **CC'S = 1.98**

LOADED LENGTH: **2.01**

BELT DIAMETER **C = N/A**

RIM THICKNESS **E = .045**

SHOULDER DIAMETER **G = .357**

LENGTH TO SHOULDER **J = 1.275**

SHOULDER ANGLE (DEG'S/SIDE): **6.20**

PRIMER: **S/P**

CASE CAPACITY (GR'S WATER): **30.52**

DIMENSIONAL DRAWING:

-NOT ACTUAL SIZE-
-DO NOT SCALE-

CARTRIDGE: .32-35 Stevens & Maynard

OTHER NAMES:

DIA: .312

BALLISTEK NO: 312E

NAI NO: RMS 21115/4.676

DATA SOURCE: COTW 4th Pg97

HISTORICAL DATA: By J. Stevens Arms & Tool Co., about 1880.

NOTES:

LOADING DATA:

BULLET WT./TYPE	POWDER WT./TYPE	VELOCITY ('/SEC)	SOURCE
153/Lead	14.0/4198	1410	Barnes

CASE PREPARATION: SHELLHOLDER (RCBS): 21

MAKE FROM: 13/32" dia. tubing case, turn from solid brass or re-body a .30-30 case with 13/32 tubing. Anneal, trim to length and F/L size. Use light loads or black powder.

PHYSICAL DATA (INCHES):

CASE TYPE: **Rimmed Straight**

CASE LENGTH **A = 1.88**

HEAD DIAMETER **B = .402**

RIM DIAMETER **D = .503**

NECK DIAMETER **F = .339**

NECK LENGTH **H = N/A**

SHOULDER LENGTH **K = N/A**

BODY ANGLE (DEG'S/SIDE): **1.00**

CASE CAPACITY **CC'S = 2.21**

LOADED LENGTH: **2.29**

BELT DIAMETER **C = N/A**

RIM THICKNESS **E = .076**

SHOULDER DIAMETER **G = N/A**

LENGTH TO SHOULDER **J = N/A**

SHOULDER ANGLE (DEG'S/SIDE): **N/A**

PRIMER: **L/R**

CASE CAPACITY (GR'S WATER): **34.06**

DIMENSIONAL DRAWING:

-NOT ACTUAL SIZE-
-DO NOT SCALE-

CARTRIDGE: .32-40 Bullard

OTHER NAMES:

DIA: .315

BALLISTEK NO: 315A

NAI NO: RMB 34321/4.084

DATA SOURCE: COTW 4th Pg98

HISTORICAL DATA: Circa 1880.

NOTES: Smallest Bullard ctg.

LOADING DATA:

BULLET WT./TYPE	POWDER WT./TYPE	VELOCITY ('/SEC)	SOURCE
150/Lead	14.0/IMR4198	1325	JJD

CASE PREPARATION: SHELLHOLDER (RCBS): 2

MAKE FROM: .303 British. Taper expand case to .320" dia. Trim to 1.9". Anneal and chamfer. F/L size. Use unsized Lyman #311440 bullets.

(Note: In some rifles, you may find it necessary to reduce the .303's rim dia. and/or thicken the rim.)

PHYSICAL DATA (INCHES):

CASE TYPE: **Rimmed Bottleneck**

CASE LENGTH **A = 1.85**

HEAD DIAMETER **B = .453**

RIM DIAMETER **D = .510**

NECK DIAMETER **F = .332**

NECK LENGTH **H = .55**

SHOULDER LENGTH **K = .38**

BODY ANGLE (DEG'S/SIDE): **1.59**

CASE CAPACITY **CC'S = 2.76**

LOADED LENGTH: **2.26**

BELT DIAMETER **C = N/A**

RIM THICKNESS **E = .07**

SHOULDER DIAMETER **G = .413**

LENGTH TO SHOULDER **J = .92**

SHOULDER ANGLE (DEG'S/SIDE): **6.08**

PRIMER: **L/R**

CASE CAPACITY (GR'S WATER): **42.68**

DIMENSIONAL DRAWING:

-NOT ACTUAL SIZE-
-DO NOT SCALE-

CARTRIDGE: .32-40 Remington Hepburn

OTHER NAMES:
.32-40 Remington

DIA: .309

BALLISTEK NO: 309A

NAI NO: RMS 31115/4.702

DATA SOURCE: COTW 4th Pg97

HISTORICAL DATA: Circa 1871.

NOTES:

LOADING DATA:

BULLET WT./TYPE	POWDER WT./TYPE	VELOCITY ('/SEC)	SOURCE
150/Lead	14.0/IMR4198	1275	JJD

CASE PREPARATION: SHELLHOLDER (RCBS): 7

MAKE FROM: .303 British. Anneal case and F/L size in the .32-40 die. Trim to length, chamfer and fireform in the chamber.

PHYSICAL DATA (INCHES):

CASE TYPE: **Rimmed Straight**

CASE LENGTH **A = 2.13**

HEAD DIAMETER **B = .453**

RIM DIAMETER **D = .535**

NECK DIAMETER **F = .330**

NECK LENGTH **H = N/A**

SHOULDER LENGTH **K = N/A**

BODY ANGLE (DEG'S/SIDE): **1.70**

CASE CAPACITY **CC'S = 3.06**

LOADED LENGTH: **3.25**

BELT DIAMETER **C = N/A**

RIM THICKNESS **E = .065**

SHOULDER DIAMETER **G = N/A**

LENGTH TO SHOULDER **J = N/A**

SHOULDER ANGLE (DEG'S/SIDE): **N/A**

PRIMER: **L/R**

CASE CAPACITY (GR'S WATER): **47.25**

DIMENSIONAL DRAWING:

-NOT ACTUAL SIZE-
-DO NOT SCALE-

CARTRIDGE: .32-40 Winchester

OTHER NAMES:
.32-40 Ballard (or Marlin)
.32-40-165

DIA: .320

BALLISTEK NO: 320F

NAI NO: RMB
32225/5.023

DATA SOURCE: COTW 4th Pg98

HISTORICAL DATA: About 1884 for Ballard Union Hill Rifles.

NOTES:

LOADING DATA:

BULLET WT./TYPE	POWDER WT./TYPE	VELOCITY ('/SEC)	SOURCE
164/Lead	17.0/4198	1460	Barnes

CASE PREPARATION: SHELLHOLDER (RCBS): 2

MAKE FROM: Factory or .375 Win. Anneal case, cut to length & chamfer. F/L size. Size lead bullet. .320-.321" dia.

PHYSICAL DATA (INCHES):

CASE TYPE: **Rimmed Bottleneck**
CASE LENGTH **A = 2.13**
HEAD DIAMETER **B = .424**
RIM DIAMETER **D = .506**
NECK DIAMETER **F = .339**
NECK LENGTH **H = .485**
SHOULDER LENGTH **K = N/A**
BODY ANGLE (DEG'S/SIDE): **1.54**
CASE CAPACITY **CC'S = 2.93**

LOADED LENGTH: **2.50**
BELT DIAMETER **C = N/A**
RIM THICKNESS **E = .063**
SHOULDER DIAMETER **G = N/A**
LENGTH TO SHOULDER **J = N/A**
SHOULDER ANGLE (DEG'S/SIDE): **N/A**
PRIMER: **L/R**
CASE CAPACITY (GR'S WATER): **45.25**

DIMENSIONAL DRAWING:

-NOT ACTUAL SIZE-
-DO NOT SCALE-

CARTRIDGE: .32-44 Smith & Wesson

OTHER NAMES:

DIA: .321

BALLISTEK NO: 321B

NAI NO: RMS
11115/2.803

DATA SOURCE: Handloader #103 Pg14

HISTORICAL DATA: By Union Metallic Cgt. Co.

NOTES: Note: bullet is seated inside the case.

LOADING DATA:

BULLET WT./TYPE	POWDER WT./TYPE	VELOCITY ('/SEC)	SOURCE
85/Lead	1.4/B'Eye	—	JJD

CASE PREPARATION: SHELLHOLDER (RCBS): 1

MAKE FROM: .32-20 WCF. Taper expand case to straight case configuration (this is quite difficult) or cut Win. case to length and fireform (easier). I.D. neck ream so that case will accept .320-.321" dia. bullet or use undersize bullet.

PHYSICAL DATA (INCHES):

CASE TYPE: **Rimmed Straight**
CASE LENGTH **A = .970**
HEAD DIAMETER **B = .346**
RIM DIAMETER **D = .410**
NECK DIAMETER **F = .346**
NECK LENGTH **H = N/A**
SHOULDER LENGTH **K = N/A**
BODY ANGLE (DEG'S/SIDE): **0**
CASE CAPACITY **CC'S = .813**

LOADED LENGTH: **.985**
BELT DIAMETER **C = N/A**
RIM THICKNESS **E = .055**
SHOULDER DIAMETER **G = N/A**
LENGTH TO SHOULDER **J = N/A**
SHOULDER ANGLE (DEG'S/SIDE): **N/A**
PRIMER: **S/P**
CASE CAPACITY (GR'S WATER): **12.55**

DIMENSIONAL DRAWING:

-NOT ACTUAL SIZE-
-DO NOT SCALE-

CARTRIDGE: .320 Revolver

OTHER NAMES:

DIA: .317

BALLISTEK NO: 317B

NAI NO: RMS 11115/1.925

DATA SOURCE: COTW 4th Pg168

HISTORICAL DATA: British ctg. from about 1870

NOTES: Similar to .32 Colt Short (#313C)

LOADING DATA:

BULLET WT./TYPE	POWDER WT./TYPE	VELOCITY ('/SEC)	SOURCE
80/Lead	2.1/B'Eye	770	Barnes

CASE PREPARATION: SHELLHOLDER (RCBS): 12

MAKE FROM: .32 Colt Long. Turn rim to .350" dia. and back chamfer. Trim to .62" and chamfer. F/L size.

PHYSICAL DATA (INCHES):

CASE TYPE: **Rimmed Straight**

CASE LENGTH **A = .620**

HEAD DIAMETER **B = .322**

RIM DIAMETER **D = .350**

NECK DIAMETER **F = .320**

NECK LENGTH **H = N/A**

SHOULDER LENGTH **K = N/A**

BODY ANGLE (DEG'S/SIDE): **.089**

CASE CAPACITY **CC'S = .404**

LOADED LENGTH: **.90**

BELT DIAMETER **C = N/A**

RIM THICKNESS **E = .04**

SHOULDER DIAMETER **G = N/A**

LENGTH TO SHOULDER **J = N/A**

SHOULDER ANGLE (DEG'S/SIDE): **N/A**

PRIMER: **S/P**

CASE CAPACITY (GR'S WATER): **6.23**

DIMENSIONAL DRAWING:

F
A
B
E
D

-NOT ACTUAL SIZE-
-DO NOT SCALE-

CARTRIDGE: .323 Hollis

OTHER NAMES:

DIA: .323

BALLISTEK NO: 323W

NAI NO: BEN 12343/5.019

DATA SOURCE: Ackley Vol.2 Pg199

HISTORICAL DATA: By E. Hollis.

NOTES:

LOADING DATA:

BULLET WT./TYPE	POWDER WT./TYPE	VELOCITY ('/SEC)	SOURCE
175/—	71.0/4320	3474	Hollis/ Ackley

CASE PREPARATION: SHELLHOLDER (RCBS): 4

MAKE FROM: .308 Norma Mag. Taper expand neck to .330" dia. F/L size, square case mouth and chamfer.

PHYSICAL DATA (INCHES):

CASE TYPE: **Belted Bottleneck**

CASE LENGTH **A = 2.575**

HEAD DIAMETER **B = .513**

RIM DIAMETER **D = .532**

NECK DIAMETER **F = .354**

NECK LENGTH **H = .318**

SHOULDER LENGTH **K = .137**

BODY ANGLE (DEG'S/SIDE): **.313**

CASE CAPACITY **CC'S = 5.72**

LOADED LENGTH: **3.32**

BELT DIAMETER **C = .532**

RIM THICKNESS **E = .05**

SHOULDER DIAMETER **G = .492**

LENGTH TO SHOULDER **J = 2.12**

SHOULDER ANGLE (DEG'S/SIDE): **26.73**

PRIMER: **L/R Mag**

CASE CAPACITY (GR'S WATER): **88.28**

DIMENSIONAL DRAWING:

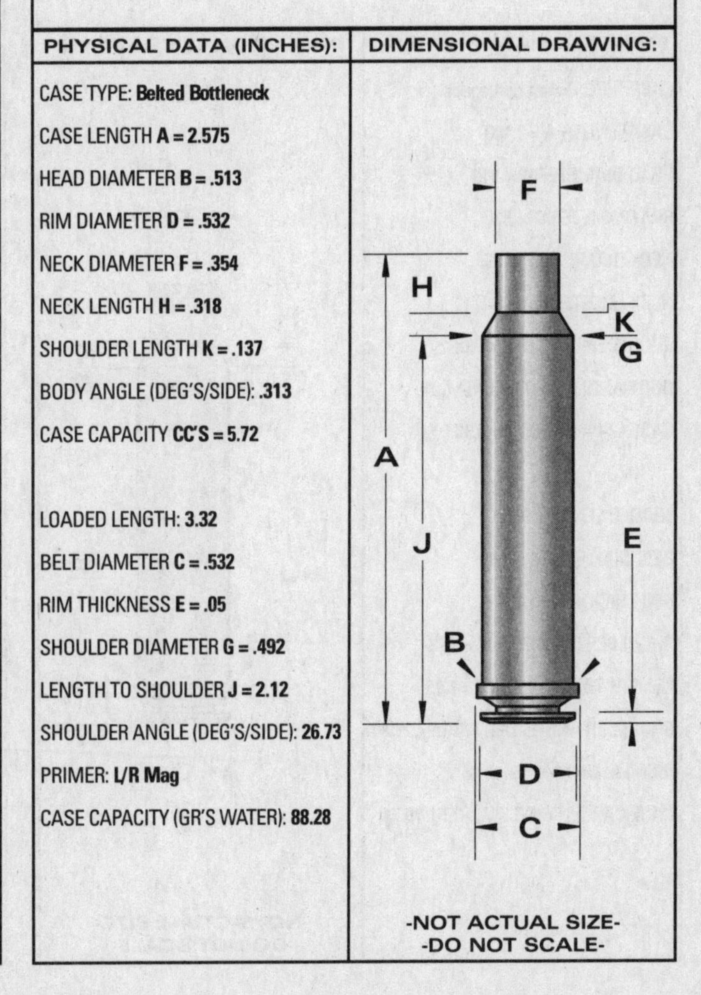

-NOT ACTUAL SIZE-
-DO NOT SCALE-

CARTRIDGE: .33 Belted Rimless (BSA)

OTHER NAMES:
.330 BSA

DIA: .338

BALLISTEK NO: 338L

NAI NO: BEN
32333/4.660

DATA SOURCE: COTW 4th Pg233

HISTORICAL DATA: By BSA about 1923.

NOTES:

LOADING DATA:

BULLET WT./TYPE	POWDER WT./TYPE	VELOCITY ('/SEC)	SOURCE
250/—	60.0/4350	2400	Barnes

CASE PREPARATION: SHELLHOLDER (RCBS): 4

MAKE FROM: .300 H&H Mag. Cut case to 2.5". Taper expand to .340" dia. F/L size, trim to length and chamfer. Fireform in chamber.

PHYSICAL DATA (INCHES):

CASE TYPE: **Belted Bottleneck**
CASE LENGTH A = **2.400**
HEAD DIAMETER B = **.515**
RIM DIAMETER D = **.530**
NECK DIAMETER F = **.369**
NECK LENGTH H = **.345**
SHOULDER LENGTH K = **.105**
BODY ANGLE (DEG'S/SIDE): **1.01**
CASE CAPACITY CC'S = **4.90**

LOADED LENGTH: **3.10**
BELT DIAMETER C = **.530**
RIM THICKNESS E = **.051**
SHOULDER DIAMETER G = **.453**
LENGTH TO SHOULDER J = **1.95**
SHOULDER ANGLE (DEG'S/SIDE): **21.80**
PRIMER: **L/R Mag**
CASE CAPACITY (GR'S WATER): **75.71**

DIMENSIONAL DRAWING:

-NOT ACTUAL SIZE-
-DO NOT SCALE-

CARTRIDGE: .33 Jeffery Rimmed

OTHER NAMES:
.333 Flanged Nitro Express

DIA: .333

BALLISTEK NO: 333C

NAI NO: RMB
22324/4.717

DATA SOURCE: COTW 4th Pg232

HISTORICAL DATA: By Jeffery in 1911.

NOTES:

LOADING DATA:

BULLET WT./TYPE	POWDER WT./TYPE	VELOCITY ('/SEC)	SOURCE
250/—	60.0/4350	2400	JJD

CASE PREPARATION: SHELLHOLDER (RCBS): 14

MAKE FROM: .280 Flanged (BELL). Form set required. Anneal case and form. Trim to length & chamfer. F/L size.

PHYSICAL DATA (INCHES):

CASE TYPE: **Rimmed Bottleneck**
CASE LENGTH A = **2.500**
HEAD DIAMETER B = **.530**
RIM DIAMETER D = **.625**
NECK DIAMETER F = **.356**
NECK LENGTH H = **.650**
SHOULDER LENGTH K = **.099**
BODY ANGLE (DEG'S/SIDE): **.85**
CASE CAPACITY CC'S = **5.46**

LOADED LENGTH: **3.43**
BELT DIAMETER C = **N/A**
RIM THICKNESS E = **.07**
SHOULDER DIAMETER G = **.484**
LENGTH TO SHOULDER J = **1.75**
SHOULDER ANGLE (DEG'S/SIDE): **32.62**
PRIMER: **L/R**
CASE CAPACITY (GR'S WATER): **84.25**

DIMENSIONAL DRAWING:

-NOT ACTUAL SIZE-
-DO NOT SCALE-

CARTRIDGE: .33 Newton

OTHER NAMES:	DIA: .333
	BALLISTEK NO: 333E
	NAI NO: RXB 12343/4.796

DATA SOURCE: NAI/Ballistek

HISTORICAL DATA: Designed by Charles Newton.

NOTES:

LOADING DATA:

BULLET WT./TYPE	POWDER WT./TYPE	VELOCITY ('/SEC)	SOURCE
275/Spire	61.0/IMR4350	2350	Nonte

CASE PREPARATION: SHELLHOLDER (RCBS): 4

MAKE FROM: .300 H&H. Turn belt to .520" dia. Turn rim to .520" dia. Cut new extractor groove. Anneal. Taper expand to .348" dia. F/L size with expander removed. Trim to length. Trim & chamfer. Fireform in chamber.

Note: case will fireform forward of belt to fill chamber.

PHYSICAL DATA (INCHES):

CASE TYPE: **Rimless Bottleneck**

CASE LENGTH **A = 2.494**

HEAD DIAMETER **B = .520**

RIM DIAMETER **D = .522**

NECK DIAMETER **F = .363**

NECK LENGTH **H = .325**

SHOULDER LENGTH **K = .169**

BODY ANGLE (DEG'S/SIDE): **.27**

CASE CAPACITY **CC'S = 5.74**

LOADED LENGTH: **3.35**

BELT DIAMETER **C = N/A**

RIM THICKNESS **E = .055**

SHOULDER DIAMETER **G = .503**

LENGTH TO SHOULDER **J = 2.00**

SHOULDER ANGLE (DEG'S/SIDE): **22.49**

PRIMER: **L/R**

CASE CAPACITY (GR'S WATER): **88.56**

DIMENSIONAL DRAWING:

-NOT ACTUAL SIZE-
-DO NOT SCALE-

CARTRIDGE: .33 Poacher's Pet

OTHER NAMES:	DIA: .338
	BALLISTEK NO: 338D
	NAI NO: RMS 21115/6.301

DATA SOURCE: Ackley Vol.2 Pg204

HISTORICAL DATA: By J. Wolford.

NOTES:

LOADING DATA:

BULLET WT./TYPE	POWDER WT./TYPE	VELOCITY ('/SEC)	SOURCE
250/—	66.0/4350	2600	Ackley

CASE PREPARATION: SHELLHOLDER (RCBS): 4

MAKE FROM: 9,3 x 74R. Anneal case and F/L size. Trim to length & chamfer. Fireform shoulder in chamber.

PHYSICAL DATA (INCHES):

CASE TYPE: **Rimmed Straight**

CASE LENGTH **A = 2.93**

HEAD DIAMETER **B = .465**

RIM DIAMETER **D = .524**

NECK DIAMETER **F = .369**

NECK LENGTH **H = N/A**

SHOULDER LENGTH **K = N/A**

BODY ANGLE (DEG'S/SIDE): **.958**

CASE CAPACITY **CC'S = 4.65**

LOADED LENGTH: **A/R**

BELT DIAMETER **C = N/A**

RIM THICKNESS **E = .06**

SHOULDER DIAMETER **G = N/A**

LENGTH TO SHOULDER **J = N/A**

SHOULDER ANGLE (DEG'S/SIDE): **N/A**

PRIMER: **L/R**

CASE CAPACITY (GR'S WATER): **71.73**

DIMENSIONAL DRAWING:

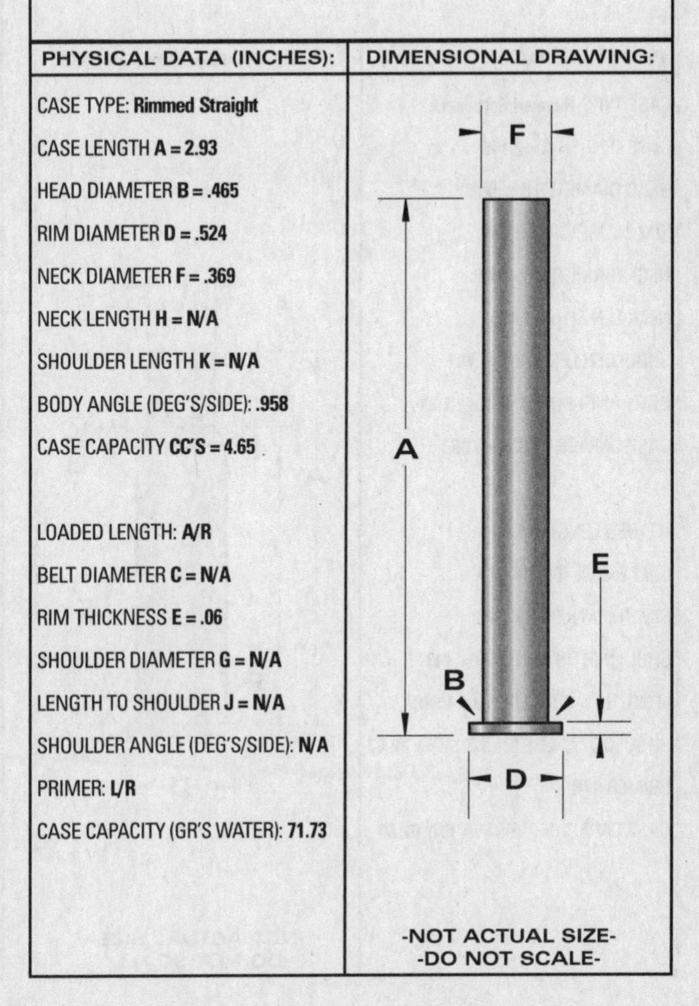

-NOT ACTUAL SIZE-
-DO NOT SCALE-

CARTRIDGE: .33 Winchester

OTHER NAMES:
.33 WCF

DIA: .338

BALLISTEK NO: 338A

NAI NO: RMB
33333/4.143

DATA SOURCE: Hornady Manual 3rd Pg270

HISTORICAL DATA: By Win. in 1902.

NOTES: Discontinued in 1940.

LOADING DATA:

BULLET WT./TYPE	POWDER WT./TYPE	VELOCITY ('/SEC)	SOURCE
200/FN	32.0/IMR4198	1925	JJD
200/FN	47.0/IMR4895	2520	Barnes

CASE PREPARATION: SHELLHOLDER (RCBS): 14

MAKE FROM: .45-70 Gov't. You will need a form die or an intermediate die such as the .40-65 Win. Anneal the case. Run into either the form die or the .40-65 F/L die. F/L size in .33 WCF die. Trim to length and chamfer.

PHYSICAL DATA (INCHES):

CASE TYPE: **Rimmed Bottleneck**
CASE LENGTH **A = 2.105**
HEAD DIAMETER **B = .508**
RIM DIAMETER **D = .610**
NECK DIAMETER **F = .366**
NECK LENGTH **H = .405**
SHOULDER LENGTH **K = .102**
BODY ANGLE (DEG'S/SIDE): **1.33**
CASE CAPACITY **CC'S = 4.05**
LOADED LENGTH: **2.70**
BELT DIAMETER **C = N/A**
RIM THICKNESS **E = .070**
SHOULDER DIAMETER **G = .443**
LENGTH TO SHOULDER **J = 1.598**
SHOULDER ANGLE (DEG'S/SIDE): **20.67**
PRIMER: **L/R**
CASE CAPACITY (GR'S WATER): **62.60**

DIMENSIONAL DRAWING:

-NOT ACTUAL SIZE-
-DO NOT SCALE-

CARTRIDGE: .33/308 Winchester

OTHER NAMES:

DIA: .338

BALLISTEK NO: 338E

NAI NO: RXB
12234/4.276

DATA SOURCE: NAI/Ballistek

HISTORICAL DATA:

NOTES:

LOADING DATA:

BULLET WT./TYPE	POWDER WT./TYPE	VELOCITY ('/SEC)	SOURCE
200/RN	41.5/IMR4895	—	JJD

CASE PREPARATION: SHELLHOLDER (RCBS): 3

MAKE FROM: .308 Win. Taper expand to .340" dia. F/L size, length trim & chamfer.

PHYSICAL DATA (INCHES):

CASE TYPE: **Rimless Bottleneck**
CASE LENGTH **A = 2.01**
HEAD DIAMETER **B = .470**
RIM DIAMETER **D = .473**
NECK DIAMETER **F = .369**
NECK LENGTH **H = .350**
SHOULDER LENGTH **K = .070**
BODY ANGLE (DEG'S/SIDE): **.309**
CASE CAPACITY **CC'S = 3.64**
LOADED LENGTH: **A/R**
BELT DIAMETER **C = N/A**
RIM THICKNESS **E = .05**
SHOULDER DIAMETER **G = .455**
LENGTH TO SHOULDER **J = 1.59**
SHOULDER ANGLE (DEG'S/SIDE): **31.56**
PRIMER: **L/R**
CASE CAPACITY (GR'S WATER): **56.21**

DIMENSIONAL DRAWING:

-NOT ACTUAL SIZE-
-DO NOT SCALE-

CARTRIDGE: 330 Dakota

OTHER NAMES:

DIA: .338

BALLISTEK NO:

NAI NO:

DATA SOURCE:

HISTORICAL DATA: This cartridge is based upon the 404 Jeffrey case. The 330 Dakota is dimensioned to function through a standard-length action (3.35").

LOADING DATA:

BULLET WT./TYPE	POWDER WT./TYPE	VELOCITY ('/SEC)	SOURCE
200	80.0 - IMR-4350	3082	Dakota
250	76.0 - IMR-4350	2853	Dakota

CASE PREPARATION: SHELL HOLDER (RCBS): SPL

MAKE FROM: 404 Jeffrey

PHYSICAL DATA (INCHES):

CASE TYPE: **Rimless Bottleneck**

CASE LENGTH **A = 2.540**

HEAD DIAMETER **B = .544**

RIM DIAMETER **D = .544**

NECK DIAMETER **F = .371**

NECK LENGTH **H = .340**

SHOULDER LENGTH **K = .130**

BODY ANGLE (DEG'S/SIDE): **N/A**

CASE CAPACITY **CC'S = N/A**

LOADED LENGTH: **3.32**

BELT DIAMETER **C = N/A**

RIM THICKNESS **E = .050**

SHOULDER DIAMETER **G = .530**

LENGTH TO SHOULDER **J = 2.070**

SHOULDER ANGLE (DEG'S/SIDE): **32°**

PRIMER: **L/R**

CASE CAPACITY (GR'S WATER): **N/A**

DIMENSIONAL DRAWING:

-NOT ACTUAL SIZE-
-DO NOT SCALE-

CARTRIDGE: .333 B-J Express

OTHER NAMES:

DIA: .333

BALLISTEK NO: 333G

NAI NO: BEN 22344/4.795

DATA SOURCE: Ackley Vol.1 Pg461

HISTORICAL DATA: By Barnes & Johnson.

NOTES:

LOADING DATA:

BULLET WT./TYPE	POWDER WT./TYPE	VELOCITY ('/SEC)	SOURCE
250/Spire	78.0/4831	2800	Ackley

CASE PREPARATION: SHELLHOLDER (RCBS): 4

MAKE FROM: .300 H&H Mag. Taper expand case neck to .340" dia. Cut-off at 2.5". F/L size in B-J die. Trim to length and chamfer. Case will fireform in chamber.

PHYSICAL DATA (INCHES):

CASE TYPE: **Belted Bottleneck**

CASE LENGTH **A = 2.46**

HEAD DIAMETER **B = .513**

RIM DIAMETER **D = .532**

NECK DIAMETER **F = .360**

NECK LENGTH **H = .270**

SHOULDER LENGTH **K = .110**

BODY ANGLE (DEG'S/SIDE): **.548**

CASE CAPACITY **CC'S = 5.23**

LOADED LENGTH: **A/R**

BELT DIAMETER **C = .532**

RIM THICKNESS **E = .05**

SHOULDER DIAMETER **G = .477**

LENGTH TO SHOULDER **J = 2.08**

SHOULDER ANGLE (DEG'S/SIDE): **44.61**

PRIMER: **L/R Mag**

CASE CAPACITY (GR'S WATER): **80.70**

DIMENSIONAL DRAWING:

-NOT ACTUAL SIZE-
-DO NOT SCALE-

CARTRIDGE: .333 Barnes Supreme

OTHER NAMES:

DIA: .333
BALLISTEK NO: 333H
NAI NO: BEN 22344/5.458

DATA SOURCE: Ackley Vol.1 Pg460

HISTORICAL DATA:

NOTES:

LOADING DATA:

BULLET WT./TYPE	POWDER WT./TYPE	VELOCITY ('/SEC)	SOURCE
300/Spire	79.0/IMR4831	—	JJD

CASE PREPARATION: **SHELLHOLDER (RCBS): 4**

MAKE FROM: .300 H&H Mag. Taper expand case neck to .340" dia. F/L size in Barnes die. Trim to length and fireform in chamber.

PHYSICAL DATA (INCHES):

CASE TYPE: **Belted Bottleneck**
CASE LENGTH **A = 2.80**
HEAD DIAMETER **B = .513**
RIM DIAMETER **D = .532**
NECK DIAMETER **F = .360**
NECK LENGTH **H = .330**
SHOULDER LENGTH **K = .095**
BODY ANGLE (DEG'S/SIDE): **.368**
CASE CAPACITY **CC'S = 6.44**

LOADED LENGTH: **A/R**
BELT DIAMETER **C = .532**
RIM THICKNESS **E = .05**
SHOULDER DIAMETER **G = .485**
LENGTH TO SHOULDER **J = 2.375**
SHOULDER ANGLE (DEG'S/SIDE): **33.34**
PRIMER: **L/R Mag**
CASE CAPACITY (GR'S WATER): **99.40**

DIMENSIONAL DRAWING:

F H K G A J E B D C

-NOT ACTUAL SIZE-
-DO NOT SCALE-

CARTRIDGE: .333 OKH

OTHER NAMES:

DIA: .333
BALLISTEK NO: 333A
NAI NO: RXB 22232/5.298

DATA SOURCE: Ackley Vol.1 Pg458

HISTORICAL DATA: By O'Neil.

NOTES:

LOADING DATA:

BULLET WT./TYPE	POWDER WT./TYPE	VELOCITY ('/SEC)	SOURCE
250/Spire	57.0/4064	2635	Ackley

CASE PREPARATION: **SHELLHOLDER (RCBS): 3**

MAKE FROM: .30-06. Taper expand '06 case to .340" dia. F/L size and square case mouth.

PHYSICAL DATA (INCHES):

CASE TYPE: **Rimless Bottleneck**
CASE LENGTH **A = 2.49**
HEAD DIAMETER **B = .470**
RIM DIAMETER **D = .473**
NECK DIAMETER **F = .360**
NECK LENGTH **H = .400**
SHOULDER LENGTH **K = .140**
BODY ANGLE (DEG'S/SIDE): **.491**
CASE CAPACITY **CC'S = 4.59**

LOADED LENGTH: **A/R**
BELT DIAMETER **C = N/A**
RIM THICKNESS **E = .05**
SHOULDER DIAMETER **G = .440**
LENGTH TO SHOULDER **J = 1.95**
SHOULDER ANGLE (DEG'S/SIDE): **15.94**
PRIMER: **L/R**
CASE CAPACITY (GR'S WATER): **70.88**

DIMENSIONAL DRAWING:

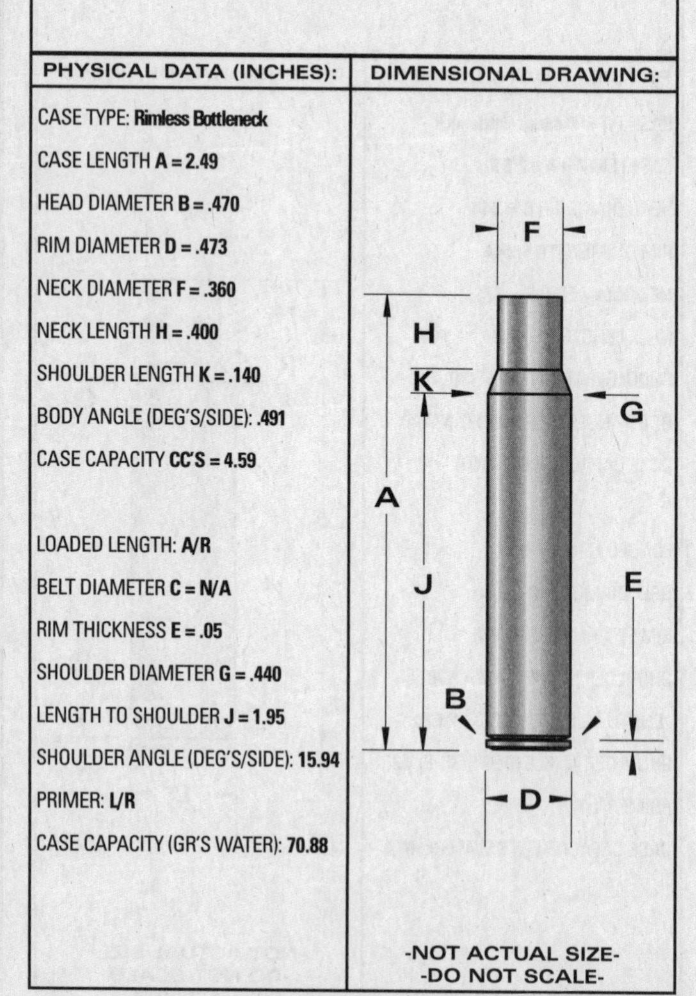

-NOT ACTUAL SIZE-
-DO NOT SCALE-

CARTRIDGE: .333 Rimless Nitro Express (Jeffrey)

OTHER NAMES:
.333 Jeffery

DIA: .333

BALLISTEK NO: 333D

NAI NO: RXB 22323/4.500

DATA SOURCE: COTW 4th Pg233

HISTORICAL DATA: By Jeffery in 1910.

NOTES:

LOADING DATA:

BULLET WT./TYPE	POWDER WT./TYPE	VELOCITY ('/SEC)	SOURCE
275/Spire	52.0/4350	2230	Barnes

CASE PREPARATION: **SHELLHOLDER (RCBS): 7**

MAKE FROM: .404 Jeffery (BELL). Anneal case and form in form die set. Trim to length & chamfer. F/L size.

PHYSICAL DATA (INCHES):

CASE TYPE: **Rimless Bottleneck**

CASE LENGTH **A = 2.43**

HEAD DIAMETER **B = .540**

RIM DIAMETER **D = .538**

NECK DIAMETER **F = .359**

NECK LENGTH **H = .50**

SHOULDER LENGTH **K = .13**

BODY ANGLE (DEG'S/SIDE): **.788**

CASE CAPACITY **CC'S = 5.58**

LOADED LENGTH: **3.50**

BELT DIAMETER **C = N/A**

RIM THICKNESS **E = .049**

SHOULDER DIAMETER **G = .496**

LENGTH TO SHOULDER **J = 1.80**

SHOULDER ANGLE (DEG'S/SIDE): **27.78**

PRIMER: **L/R**

CASE CAPACITY (GR'S WATER): **86.13**

DIMENSIONAL DRAWING:

-NOT ACTUAL SIZE-
-DO NOT SCALE-

CARTRIDGE: .333 Short Magnum (Ackley)

OTHER NAMES:

DIA: .333

BALLISTEK NO: 333I

NAI NO: BEN 22333/4.941

DATA SOURCE: Ackley Vol.1 Pg459

HISTORICAL DATA:

NOTES:

LOADING DATA:

BULLET WT./TYPE	POWDER WT./TYPE	VELOCITY ('/SEC)	SOURCE
275/—	51.0/4064	2307	Ackley

CASE PREPARATION: **SHELLHOLDER (RCBS): 4**

MAKE FROM: .338 Win. Mag. F/L size the .338 case in the .333 die. Trim.

PHYSICAL DATA (INCHES):

CASE TYPE: **Belted Bottleneck**

CASE LENGTH **A = 2.535**

HEAD DIAMETER **B = .513**

RIM DIAMETER **D = .532**

NECK DIAMETER **F = .360**

NECK LENGTH **H = .408**

SHOULDER LENGTH **K = .135**

BODY ANGLE (DEG'S/SIDE): **.67**

CASE CAPACITY **CC'S = 5.43**

LOADED LENGTH: **3.42**

BELT DIAMETER **C = .532**

RIM THICKNESS **E = .05**

SHOULDER DIAMETER **G = .471**

LENGTH TO SHOULDER **J = 1.992**

SHOULDER ANGLE (DEG'S/SIDE): **22.34**

PRIMER: **L/R Mag**

CASE CAPACITY (GR'S WATER): **83.82**

DIMENSIONAL DRAWING:

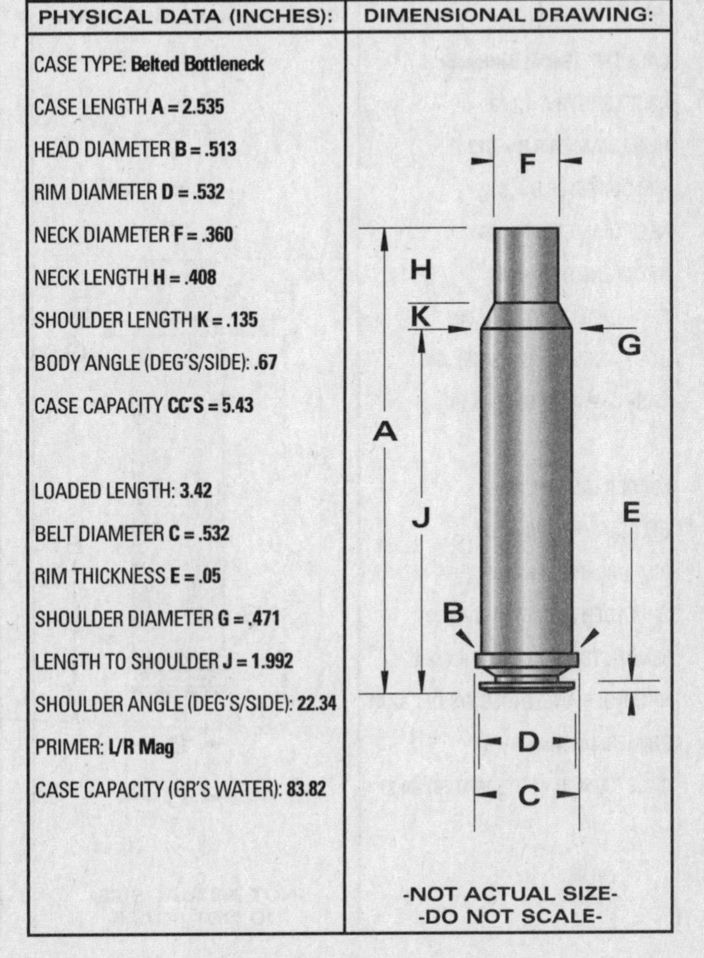

-NOT ACTUAL SIZE-
-DO NOT SCALE-

CARTRIDGE: .333 x 61mm Carlson Magnum

OTHER NAMES:

DIA: .333

BALLISTEK NO: 333B

NAI NO: BEN 22344/4.659

DATA SOURCE: Ackley Vol.1 Pg460

HISTORICAL DATA: By R. Carlson.

NOTES:

LOADING DATA:

BULLET WT./TYPE	POWDER WT./TYPE	VELOCITY ('/SEC)	SOURCE
275/--	64.0/4350	2610	Ackley

CASE PREPARATION: SHELLHOLDER (RCBS): 4

MAKE FROM: .300 H&H Mag. Taper expand case neck to about .340" dia. F/L size in the Carlson die, with the expander removed. Trim to length, chamfer and F/L size again. Fireform in the Carlson chamber.

PHYSICAL DATA (INCHES):

CASE TYPE: **Belted Bottleneck**

CASE LENGTH **A = 2.39**

HEAD DIAMETER **B = .513**

RIM DIAMETER **D = .532**

NECK DIAMETER **F = .360**

NECK LENGTH **H = .329**

SHOULDER LENGTH **K = .089**

BODY ANGLE (DEG'S/SIDE): **.646**

CASE CAPACITY **CC'S = 5.19**

LOADED LENGTH: **A/R**

BELT DIAMETER **C = .532**

RIM THICKNESS **E = .05**

SHOULDER DIAMETER **G = .473**

LENGTH TO SHOULDER **J = 1.972**

SHOULDER ANGLE (DEG'S/SIDE): **32.41**

PRIMER: **L/R Mag**

CASE CAPACITY (GR'S WATER): **80.21**

DIMENSIONAL DRAWING:

F H K G A J E B D C

-NOT ACTUAL SIZE-
-DO NOT SCALE-

CARTRIDGE: .334 OKH

OTHER NAMES:

DIA: .333

BALLISTEK NO: 333J

NAI NO: BEN 22334/5.468

DATA SOURCE: NAI/Ballistek

HISTORICAL DATA: By O'Neil, Keith & Hopkins.

NOTES:

LOADING DATA:

BULLET WT./TYPE	POWDER WT./TYPE	VELOCITY ('/SEC)	SOURCE
275/Spire	71.4/IMR4831	—	JJD

CASE PREPARATION: SHELLHOLDER (RCBS): 4

MAKE FROM: .300 H&H Mag. Taper expand case neck to .340" dia. F/L size in OKH die. Trim to length & chamfer. Fireform.

PHYSICAL DATA (INCHES):

CASE TYPE: **Belted Bottleneck**

CASE LENGTH **A = 2.80**

HEAD DIAMETER **B = .512**

RIM DIAMETER **D =.532**

NECK DIAMETER **F = .359**

NECK LENGTH **H = .340**

SHOULDER LENGTH **K = .048**

BODY ANGLE (DEG'S/SIDE): **.971**

CASE CAPACITY **CC'S = 5.86**

LOADED LENGTH: **3.84**

BELT DIAMETER **C = .532**

RIM THICKNESS **E = .05**

SHOULDER DIAMETER **G = .437**

LENGTH TO SHOULDER **J = 2.412**

SHOULDER ANGLE (DEG'S/SIDE): **39.09**

PRIMER: **L/R Mag**

CASE CAPACITY (GR'S WATER): **90.54**

DIMENSIONAL DRAWING:

F H K G A J E B D C

-NOT ACTUAL SIZE-
-DO NOT SCALE-

CARTRIDGE: .338 A&H Magnum

OTHER NAMES:

DIA: .338

BALLISTEK NO: 338G

NAI NO: RXB 12333/4.678

DATA SOURCE: Rifle #55 Pg36

HISTORICAL DATA: By W. Abe.

NOTES: Abe & Harris.

LOADING DATA:

BULLET WT./TYPE	POWDER WT./TYPE	VELOCITY ('/SEC)	SOURCE
275/Spire	78.0/H4831	2626	Hagel

CASE PREPARATION: SHELLHOLDER (RCBS): 7

MAKE FROM: .404 Jeffery (BELL). Form die set required. Anneal case. Form in form dies #1, #2 & #3. Trim to length. I.D. neck ream. F/L size and final fireform in the chamber.

PHYSICAL DATA (INCHES):

CASE TYPE: **Rimless Bottleneck**

CASE LENGTH **A = 2.55**

HEAD DIAMETER **B = .545**

RIM DIAMETER **D = .545**

NECK DIAMETER **F = .369**

NECK LENGTH **H = .375**

SHOULDER LENGTH **K = .175**

BODY ANGLE (DEG'S/SIDE): **.239**

CASE CAPACITY **CC'S = 6.49**

LOADED LENGTH: **3.37**

BELT DIAMETER **C = N/A**

RIM THICKNESS **E = .05**

SHOULDER DIAMETER **G = .530**

LENGTH TO SHOULDER **J = 2.00**

SHOULDER ANGLE (DEG'S/SIDE): **24.70**

PRIMER: **L/R Mag**

CASE CAPACITY (GR'S WATER): **98.76**

DIMENSIONAL DRAWING:

-NOT ACTUAL SIZE-
-DO NOT SCALE-

CARTRIDGE: .338 Excalibur

OTHER NAMES: .338 Excaliber

DIA: .338

BALLISTEK NO:

NAI NO:

DATA SOURCE:

HISTORICAL DATA: Approved by SAAMI 6/4/98. This cartridge is based on a new case. It is similar to the .378 Weatherby, but without the belt. It was designed to extract the highest velocity currently possible from .338-caliber hunting bullets.

LOADING DATA:

BULLET WT./TYPE	POWDER WT./TYPE	VELOCITY ('/SEC)	SOURCE
200/Nosler Bal. Tip	113.0/RL-22	3,202	Cart's of the World #9

CASE PREPARATION: SHELL HOLDER (RCBS): 14

MAKE FROM: Based on a new case. Use factory brass. May be possilbe to make from .338-378 Weatherby case. Turn belt off case in lathe, run through F/L sizer and fireform in chamber with cornmeal. The case will likely be short as parent case is .077-inch shorter. Square mouth, chamfer and deburr.

PHYSICAL DATA (INCHES):

CASE TYPE: **Rimless Bottleneck**

CASE LENGTH **A = 2.990**

HEAD DIAMETER **B = .5797**

RIM DIAMETER **D = .580**

NECK DIAMETER **F = .368**

NECK LENGTH **H = .357**

SHOULDER LENGTH **K = .183**

BODY ANGLE (DEG'S/SIDE):

CASE CAPACITY **CC'S =**

LOADED LENGTH: **3.750**

BELT DIAMETER **C =**

RIM THICKNESS **E = .065**

SHOULDER DIAMETER **G = .566**

LENGTH TO SHOULDER **J = 2.450**

SHOULDER ANGLE (DEG'S/SIDE): **28**

PRIMER: **L/R Mag**

CASE CAPACITY (GR'S WATER):

DIMENSIONAL DRAWING:

-NOT ACTUAL SIZE-
-DO NOT SCALE-

CARTRIDGE: .338 JDJ

OTHER NAMES:	DIA: .338 BALLISTEK NO: 338M NAI NO: RMB 22243/4.840

DATA SOURCE: SSK Industries

HISTORICAL DATA: By J.D. Jones.

NOTES:

LOADING DATA:

BULLET WT./TYPE	POWDER WT./TYPE	VELOCITY ('/SEC)	SOURCE
300/—	46.5/H4895	1931	SSK

CASE PREPARATION: SHELLHOLDER (RCBS): 7

MAKE FROM: .303 British. Taper expand case neck to hold .338" dia. bullet. Trim to length & chamfer. F/L size and fire-form, in the chamber, with moderate load.

PHYSICAL DATA (INCHES):

CASE TYPE: **Rimmed Bottleneck**
CASE LENGTH **A = 2.183**
HEAD DIAMETER **B = .451**
RIM DIAMETER **D = .524**
NECK DIAMETER **F = .358**
NECK LENGTH **H = .298**
SHOULDER LENGTH **K = .057**
BODY ANGLE (DEG'S/SIDE): **.475**
CASE CAPACITY **CC'S = 3.93**
LOADED LENGTH: **3.105**
BELT DIAMETER **C = N/A**
RIM THICKNESS **E = .06**
SHOULDER DIAMETER **G = .424**
LENGTH TO SHOULDER **J = 1.828**
SHOULDER ANGLE (DEG'S/SIDE): **30.06**
PRIMER: **L/R**
CASE CAPACITY (GR'S WATER): **60.54**

DIMENSIONAL DRAWING:

F H K G A J E B D

-NOT ACTUAL SIZE-
-DO NOT SCALE-

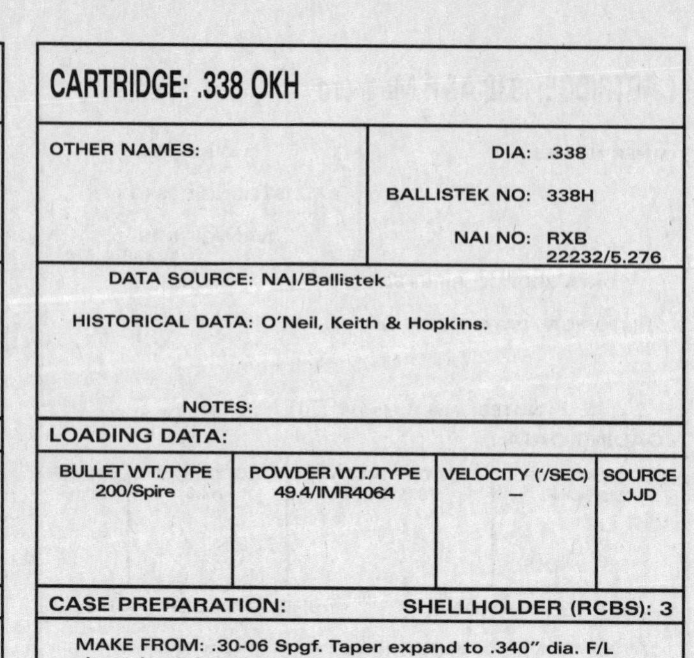

CARTRIDGE: .338 OKH

OTHER NAMES:	DIA: .338 BALLISTEK NO: 338H NAI NO: RXB 22232/5.276

DATA SOURCE: NAI/Ballistek

HISTORICAL DATA: O'Neil, Keith & Hopkins.

NOTES:

LOADING DATA:

BULLET WT./TYPE	POWDER WT./TYPE	VELOCITY ('/SEC)	SOURCE
200/Spire	49.4/IMR4064	—	JJD

CASE PREPARATION: SHELLHOLDER (RCBS): 3

MAKE FROM: .30-06 Spgf. Taper expand to .340" dia. F/L size, trim to length and chamfer.

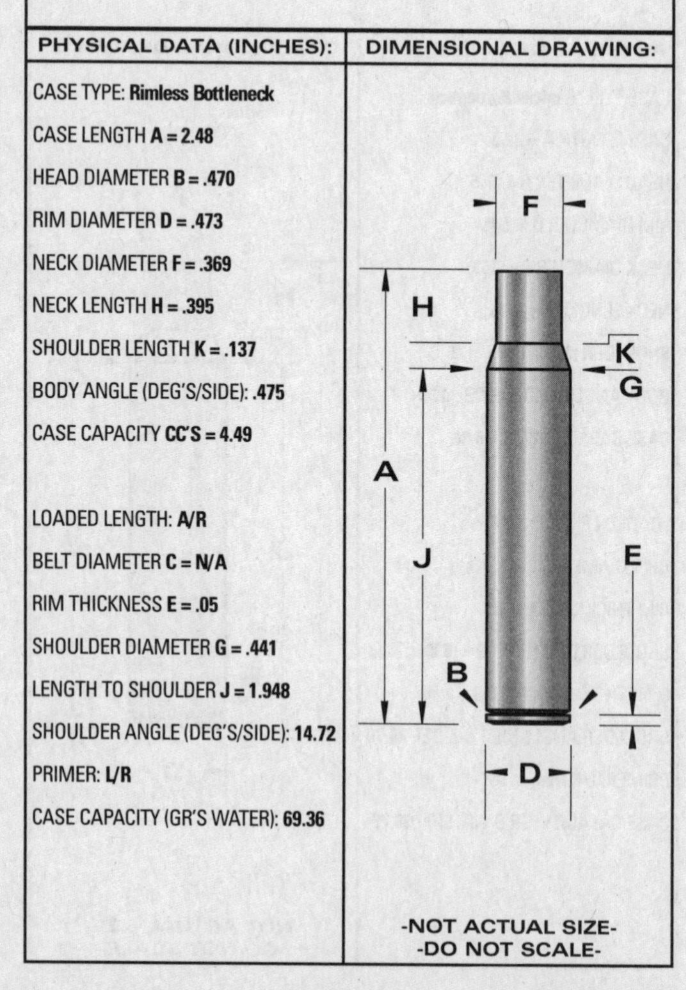

PHYSICAL DATA (INCHES):

CASE TYPE: **Rimless Bottleneck**
CASE LENGTH **A = 2.48**
HEAD DIAMETER **B = .470**
RIM DIAMETER **D = .473**
NECK DIAMETER **F = .369**
NECK LENGTH **H = .395**
SHOULDER LENGTH **K = .137**
BODY ANGLE (DEG'S/SIDE): **.475**
CASE CAPACITY **CC'S = 4.49**
LOADED LENGTH: **A/R**
BELT DIAMETER **C = N/A**
RIM THICKNESS **E = .05**
SHOULDER DIAMETER **G = .441**
LENGTH TO SHOULDER **J = 1.948**
SHOULDER ANGLE (DEG'S/SIDE): **14.72**
PRIMER: **L/R**
CASE CAPACITY (GR'S WATER): **69.36**

DIMENSIONAL DRAWING:

-NOT ACTUAL SIZE-
-DO NOT SCALE-

CARTRIDGE: .338 Remington Ultra Mag

OTHER NAMES:

DIA: .338

BALLISTEK NO:

NAI NO:

DATA SOURCE:

HISTORICAL DATA: Introduced by Remington in 2000. It followed the .300 RUM and was the second of the Ultra Mag family to be introduced.

NOTES: This is the only case of the Ultra Mag family that is a different length. The .338 RUM is 0.09-inch shorter than the other three Ultra Mag cartridges.

LOADING DATA:

BULLET WT./TYPE	POWDER WT./TYPE	VELOCITY ('/SEC)	SOURCE
225/Barnes XLC FB	95.0/RL-25	3,079	Barnes#3
250/Factory		2,860	Rem

CASE PREPARATION: **SHELL HOLDER (RCBS): 38**

MAKE FROM: Factory or neck up and shorten .300 Remington Ultra Mag. The shoulder must be pushed back. Can be made from .404 Jeffery.

PHYSICAL DATA (INCHES):

CASE TYPE: **Rebated Bottleneck**
CASE LENGTH **A = 2.760**
HEAD DIAMETER **B = .550**
RIM DIAMETER **D = .534**
NECK DIAMETER **F = .371**
NECK LENGTH **H = .330**
SHOULDER LENGTH **K = .134**
BODY ANGLE (DEG'S/SIDE):
CASE CAPACITY **CC'S =**
LOADED LENGTH: **3.600**
BELT DIAMETER **C =**
RIM THICKNESS **E = .050**
SHOULDER DIAMETER **G = .526**
LENGTH TO SHOULDER **J = 2.296**
SHOULDER ANGLE (DEG'S/SIDE): **30**
PRIMER: **L/R Mag**
CASE CAPACITY (GR'S WATER):

DIMENSIONAL DRAWING:

F H K G A J E B D

-NOT ACTUAL SIZE-
-DO NOT SCALE-

CARTRIDGE: 338 Whisper

OTHER NAMES:

DIA: .338

BALLISTEK NO:

NAI NO:

DATA SOURCE:

HISTORICAL DATA: Designed by J.D. Jones in the early 1990s, this chambering is the 7mm BR opened up to accept .338" bullets with no other changes.

NOTES:

LOADING DATA:

BULLET WT./TYPE	POWDER WT./TYPE	VELOCITY ('/SEC)	SOURCE
200 Nosler BT	11.5 - H-4227	1075	SSK
250 Nosler BT	9.6 - HP-38	585	SSK

CASE PREPARATION: **SHELL HOLDER (RCBS): 03**

MAKE FROM: 7mm BR opened up to accept .338" bullets with no other changes.

PHYSICAL DATA (INCHES):

CASE TYPE: **Rimless Bottleneck**
CASE LENGTH **A = 1.520**
HEAD DIAMETER **B = .471**
RIM DIAMETER **D = .473**
NECK DIAMETER **F = .361**
NECK LENGTH **H = .340**
SHOULDER LENGTH **K = .105**
BODY ANGLE (DEG'S/SIDE):
CASE CAPACITY **CC'S =**
LOADED LENGTH: **2.800**
BELT DIAMETER **C = N/A**
RIM THICKNESS **E = .054**
SHOULDER DIAMETER **G = N/A**
LENGTH TO SHOULDER **J = 1.075**
SHOULDER ANGLE (DEG'S/SIDE): **30°**
PRIMER: **SR**
CASE CAPACITY (GR'S WATER): **N/A**

DIMENSIONAL DRAWING:

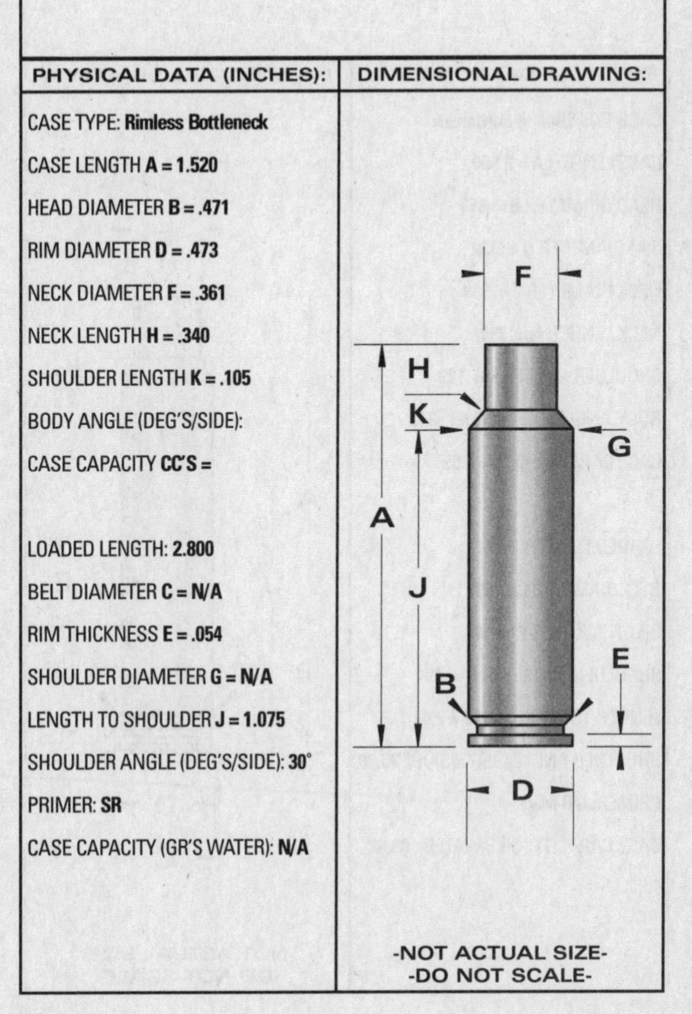

-NOT ACTUAL SIZE-
-DO NOT SCALE-

CARTRIDGE: .338 Winchester Magnum

OTHER NAMES:	DIA: .338 BALLISTEK NO: 338B NAI NO: BEN 12343/4.873

DATA SOURCE: Hornady Manual 3rd Pg271

HISTORICAL DATA: By Win. in 1958.

NOTES: .338" version of .30 Ackley #2 Magnum.

LOADING DATA:

BULLET WT./TYPE	POWDER WT./TYPE	VELOCITY ('/SEC)	SOURCE
200/–	78.0/IMR4831	3000	Barnes
250/RN	70.0/IMR4831	2600	Horn.

CASE PREPARATION: SHELLHOLDER (RCBS): 4

MAKE FROM: Factory or .375 H&H Mag. Run H&H case into .338 F/L die with expander removed. Trim to length & chamfer. Shoulder area will fireform upon ignition.

PHYSICAL DATA (INCHES):

CASE TYPE: **Belted Bottleneck**
CASE LENGTH **A = 2.500**
HEAD DIAMETER **B = .513**
RIM DIAMETER **D = .532**
NECK DIAMETER **F = .369**
NECK LENGTH **H = .331**
SHOULDER LENGTH **K = .129**
BODY ANGLE (DEG'S/SIDE): **.342**
CASE CAPACITY **CC'S = 5.55**

LOADED LENGTH: **3.34**
BELT DIAMETER **C = .532**
RIM THICKNESS **E = .049**
SHOULDER DIAMETER **G = .491**
LENGTH TO SHOULDER **J = 2.04**
SHOULDER ANGLE (DEG'S/SIDE): **25.30**
PRIMER: **L/R Mag**
CASE CAPACITY (GR'S WATER): **85.62**

DIMENSIONAL DRAWING:

-NOT ACTUAL SIZE-
-DO NOT SCALE-

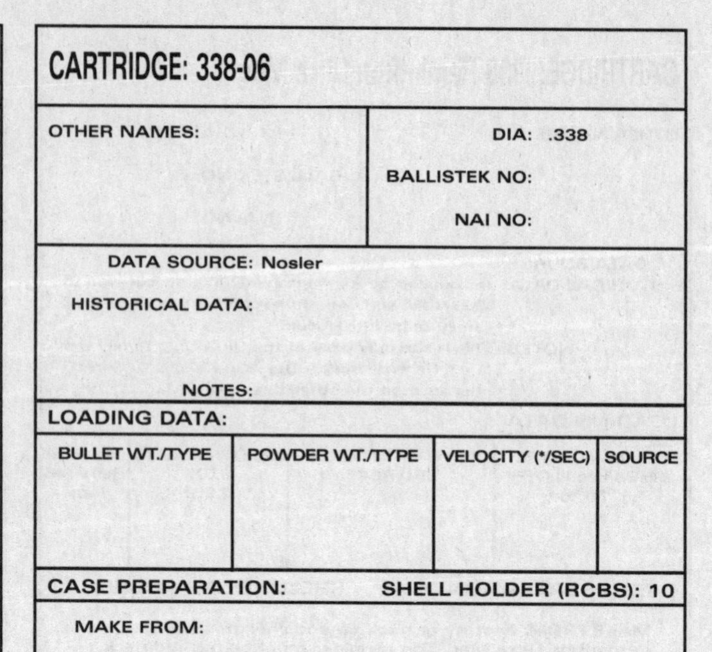

CARTRIDGE: 338-06

OTHER NAMES:	DIA: .338 BALLISTEK NO: NAI NO:

DATA SOURCE: Nosler

HISTORICAL DATA:

NOTES:

LOADING DATA:

BULLET WT./TYPE	POWDER WT./TYPE	VELOCITY ('/SEC)	SOURCE

CASE PREPARATION: SHELL HOLDER (RCBS): 10

MAKE FROM:

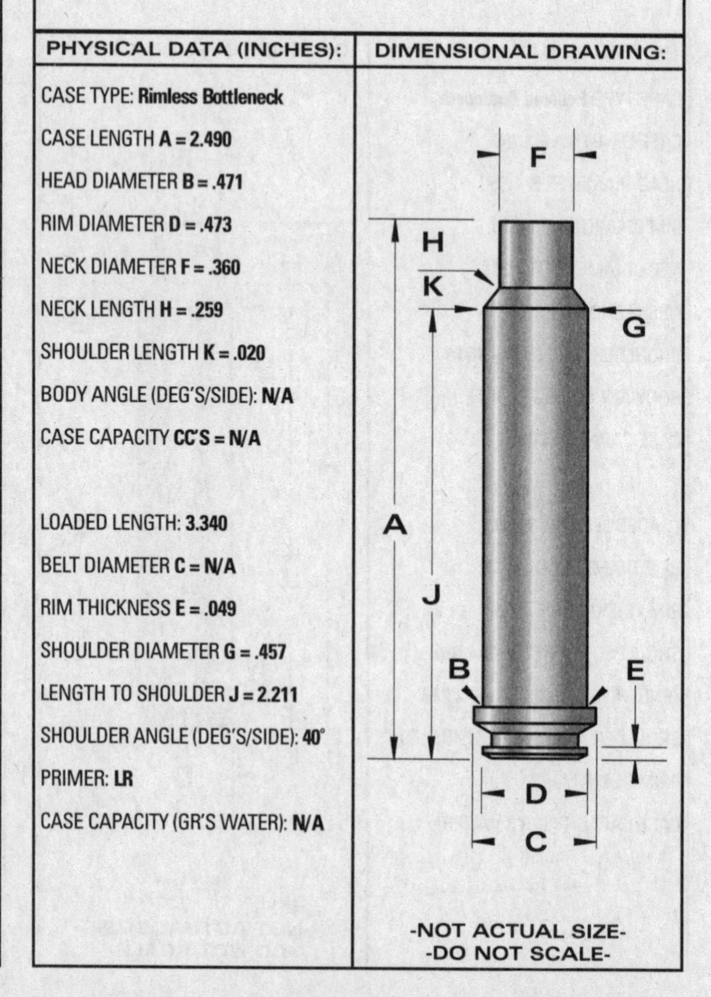

PHYSICAL DATA (INCHES):

CASE TYPE: **Rimless Bottleneck**
CASE LENGTH **A = 2.490**
HEAD DIAMETER **B = .471**
RIM DIAMETER **D = .473**
NECK DIAMETER **F = .360**
NECK LENGTH **H = .259**
SHOULDER LENGTH **K = .020**
BODY ANGLE (DEG'S/SIDE): **N/A**
CASE CAPACITY **CC'S = N/A**

LOADED LENGTH: **3.340**
BELT DIAMETER **C = N/A**
RIM THICKNESS **E = .049**
SHOULDER DIAMETER **G = .457**
LENGTH TO SHOULDER **J = 2.211**
SHOULDER ANGLE (DEG'S/SIDE): **40°**
PRIMER: **LR**
CASE CAPACITY (GR'S WATER): **N/A**

DIMENSIONAL DRAWING:

-NOT ACTUAL SIZE-
-DO NOT SCALE-

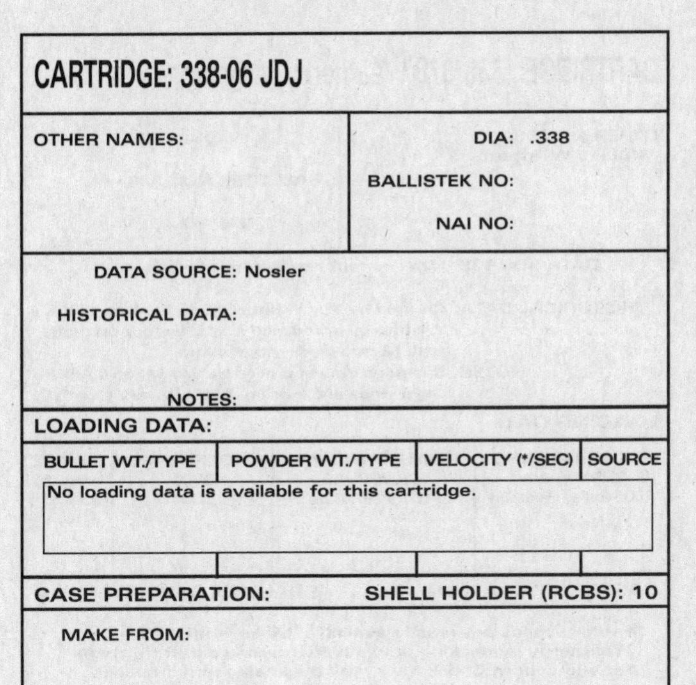

CARTRIDGE: 338-06 JDJ

OTHER NAMES:

DIA: .338

BALLISTEK NO:

NAI NO:

DATA SOURCE: Nosler

HISTORICAL DATA:

NOTES:

LOADING DATA:

BULLET WT./TYPE	POWDER WT./TYPE	VELOCITY ('/SEC)	SOURCE
No loading data is available for this cartridge.			

CASE PREPARATION: SHELL HOLDER (RCBS): 10

MAKE FROM:

PHYSICAL DATA (INCHES):

CASE TYPE: **Rimless Bottleneck**

CASE LENGTH **A = 2.490**

HEAD DIAMETER **B = .471**

RIM DIAMETER **D = .473**

NECK DIAMETER **F = .360**

NECK LENGTH **H = .259**

SHOULDER LENGTH **K = .020**

BODY ANGLE (DEG'S/SIDE): **N/A**

CASE CAPACITY **CC'S = N/A**

LOADED LENGTH: **3.340**

BELT DIAMETER **C = N/A**

RIM THICKNESS **E = .049**

SHOULDER DIAMETER **G = .457**

LENGTH TO SHOULDER **J = 2.211**

SHOULDER ANGLE (DEG'S/SIDE): **40°**

PRIMER: **LR**

CASE CAPACITY (GR'S WATER): **N/A**

DIMENSIONAL DRAWING:

F
H
K
G
A
J
E
B
D

-NOT ACTUAL SIZE-
-DO NOT SCALE-

CARTRIDGE: .338/06 A-Square

OTHER NAMES:

DIA: .338

BALLISTEK NO: 338I

NAI NO: RXB 22233/5.276

DATA SOURCE: Most modern handloading manuals.

HISTORICAL DATA: This long time wildcat was made "legitimate" when A-Square took it to SAAMI for approval in 1996. Dimensions are now standardized.

NOTES: Older rifles chambered as a wildcat may use different dimensions, always check with a chamber cast.

LOADING DATA:

BULLET WT./TYPE	POWDER WT./TYPE	VELOCITY ('/SEC)	SOURCE
225/Hornady	51.4/4Varget	2600	Horn. #6
250/Nosler Partition	55.5/H380	2418	Nosler #5

CASE PREPARATION: SHELLHOLDER (RCBS): 3

Run .30-06 into F/L sizing die with tapered expander. Trim, chamfer and deburr.

PHYSICAL DATA (INCHES):

CASE TYPE: **Rimless Bottleneck**

CASE LENGTH **A = 2.494**

HEAD DIAMETER **B = .4698**

RIM DIAMETER **D = .473**

NECK DIAMETER **F = .370**

NECK LENGTH **H = .438**

SHOULDER LENGTH **K = .111**

BODY ANGLE (DEG'S/SIDE): **.485**

CASE CAPACITY **CC'S = 4.74**

LOADED LENGTH: **3.440**

BELT DIAMETER **C = N/A**

RIM THICKNESS **E = .049**

SHOULDER DIAMETER **G = .4410**

LENGTH TO SHOULDER **J = 1.9480**

SHOULDER ANGLE (DEG'S/SIDE): **17 30'**

PRIMER: **L/R**

CASE CAPACITY (GR'S WATER): **71.4**

DIMENSIONAL DRAWING:

F
H
K
G
A
J
E
B
D

-NOT ACTUAL SIZE-
-DO NOT SCALE-

CARTRIDGE: .338/300 Holland & Holland Magnum

OTHER NAMES:

DIA: .338

BALLISTEK NO: 338F

NAI NO: BEN 22341/5.555

DATA SOURCE: NAI/Ballistek

HISTORICAL DATA:

NOTES:

LOADING DATA:

BULLET WT./TYPE	POWDER WT./TYPE	VELOCITY ('/SEC)	SOURCE
250/Spire	63.8/IMR4350	–	JJD

CASE PREPARATION: SHELLHOLDER (RCBS): 4

MAKE FROM: .300 H&H Mag. Anneal case neck and taper expand to .350" dia. F/L size, square case mouth & chamfer. Fireform in chamber.

PHYSICAL DATA (INCHES):

CASE TYPE: **Belted Bottleneck**

CASE LENGTH **A = 2.85**

HEAD DIAMETER **B = .513**

RIM DIAMETER **D = .532**

NECK DIAMETER **F = .368**

NECK LENGTH **H = .320**

SHOULDER LENGTH **K = .419**

BODY ANGLE (DEG'S/SIDE): **.839**

CASE CAPACITY **CC'S = 5.91**

LOADED LENGTH: **A/R**

BELT DIAMETER **C = .532**

RIM THICKNESS **E = .05**

SHOULDER DIAMETER **G = .457**

LENGTH TO SHOULDER **J = 2.11**

SHOULDER ANGLE (DEG'S/SIDE): **6.05**

PRIMER: **L/R Mag**

CASE CAPACITY (GR'S WATER): **91.27**

DIMENSIONAL DRAWING:

-NOT ACTUAL SIZE-
-DO NOT SCALE-

CARTRIDGE: .338/378 Weatherby Magnum

OTHER NAMES: .340/378 Weatherby

DIA: .338

BALLISTEK NO: 338J

NAI NO: BEN 12333/4.983

DATA SOURCE: Most current handloading manuals.

HISTORICAL DATA: Created by Roy Weatherby. In the late 1990's Weatherby introduced it as a factory cartridge with Norma loading the ammo.

NOTES: Cartridge was approved by SAAMI on 6/4/98. Older rifles and loading dies may vary slightly.

LOADING DATA:

BULLET WT./TYPE	POWDER WT./TYPE	VELOCITY ('/SEC)	SOURCE
225/Hornady	110.4/RL-25	3200	Horn. #6
250/Nosler Partition	101/IMR-7828	3034	Nosler #5

CASE PREPARATION: SHELLHOLDER (RCBS): 14

Factory cases are readily available. Make from .378 Weatherby by necking down in F/L resizing die. Trim, chamfer and deburr. Check neck wall thickness, turn if needed.

PHYSICAL DATA (INCHES):

CASE TYPE: **Belted Bottleneck**

CASE LENGTH **A = 2.913**

HEAD DIAMETER **B = .5817**

RIM DIAMETER **D = .579**

NECK DIAMETER **F = .369**

NECK LENGTH **H = .356**

SHOULDER LENGTH **K = .2117**

BODY ANGLE (DEG'S/SIDE): **.294**

CASE CAPACITY **CC'S = 8.73**

LOADED LENGTH: **3.763**

BELT DIAMETER **C = .6035**

RIM THICKNESS **E = .063**

SHOULDER DIAMETER **G = .561**

LENGTH TO SHOULDER **J = 2.345**

SHOULDER ANGLE (DEG'S/SIDE):
Front radius: **R .1531**
Rear radius: **R .130-0.13**

PRIMER: **L/R Mag**

CASE CAPACITY (GR'S WATER): **134.74**

DIMENSIONAL DRAWING:

-NOT ACTUAL SIZE-
-DO NOT SCALE-

CARTRIDGE: .340 Weatherby Magnum

OTHER NAMES:

DIA: .338

BALLISTEK NO: 338C

NAI NO: BEN 12332/5.518

DATA SOURCE: Hornady Manual 3rd Pg274

HISTORICAL DATA: Roy Weatherby development.

NOTES:

LOADING DATA:

BULLET WT./TYPE	POWDER WT./TYPE	VELOCITY ('/SEC)	SOURCE
200/–	80.0/4350	3075	Barnes
250/RN	81.0/IMR4831	2800	Horn.

CASE PREPARATION: SHELLHOLDER (RCBS): 4

MAKE FROM: Factory or .375 H&H. Run H&H case into .340 F/L die. Fireform in chamber with light load.

PHYSICAL DATA (INCHES):

CASE TYPE: **Belted Bottleneck**

CASE LENGTH A = **2.82**

HEAD DIAMETER B = **.511**

RIM DIAMETER D = **.530**

NECK DIAMETER F = **.361**

NECK LENGTH H = **.344**

SHOULDER LENGTH K = **.189**

BODY ANGLE (DEG'S/SIDE): **.288**

CASE CAPACITY CC'S = **6.61**

LOADED LENGTH: **3.68**

BELT DIAMETER C = **.530**

RIM THICKNESS E = **.05**

SHOULDER DIAMETER G = **.490**

LENGTH TO SHOULDER J = **2.287**

SHOULDER ANGLE (DEG'S/SIDE): **N/A**

PRIMER: **L/R Mag**

CASE CAPACITY (GR'S WATER): **102.05**

DIMENSIONAL DRAWING:

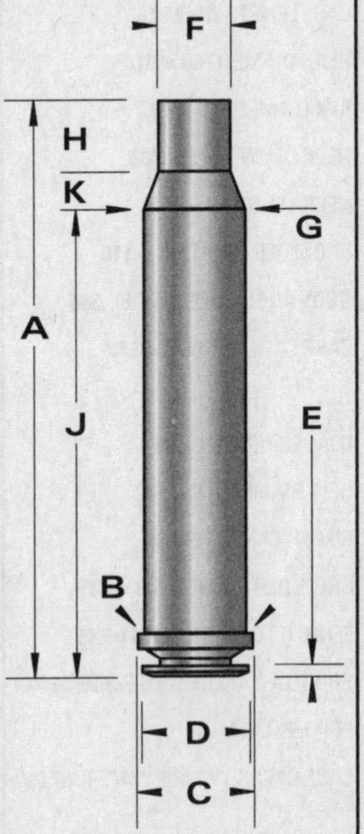

-NOT ACTUAL SIZE-
-DO NOT SCALE-

CARTRIDGE: .348 Winchester

OTHER NAMES:
.34 Winchester
.348 WCF

DIA: .348

BALLISTEK NO: 348A

NAI NO: RMB 33332/4.107

DATA SOURCE: Hornady Manual 3rd Pg276

HISTORICAL DATA: By Win. in 1933 for M71.

NOTES: Discontinued in 1958.

LOADING DATA:

BULLET WT./TYPE	POWDER WT./TYPE	VELOCITY ('/SEC)	SOURCE
200/FN	59.1/IMR4350	2400	Horn.
250/–	51.0/IMR4064	2300	Barnes

CASE PREPARATION: SHELLHOLDER (RCBS): 5

MAKE FROM: Factory or B.E.L.L. .450 NE. Cut case to 2.2". Anneal neck. Size in .348 F/L die. Turn rim to .610" dia. (if necessary). Rim thickness may also require building up to .070. Length trim and chamfer. F/L size.

PHYSICAL DATA (INCHES):

CASE TYPE: **Rimmed Bottleneck**

CASE LENGTH A = **2.246**

HEAD DIAMETER B = **.547**

RIM DIAMETER D = **.603**

NECK DIAMETER F = **.375**

NECK LENGTH H = **.420**

SHOULDER LENGTH K = **.085**

BODY ANGLE (DEG'S/SIDE): **1.572**

CASE CAPACITY CC'S = **4.95**

LOADED LENGTH: **2.83**

BELT DIAMETER C = **N/A**

RIM THICKNESS E = **.065**

SHOULDER DIAMETER G = **.467**

LENGTH TO SHOULDER J = **1.658**

SHOULDER ANGLE (DEG'S/SIDE): **15.31**

PRIMER: **L/R**

CASE CAPACITY (GR'S WATER): **76.45**

DIMENSIONAL DRAWING:

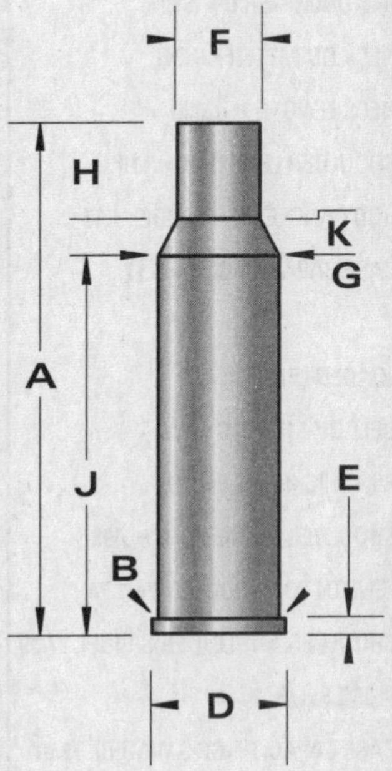

-NOT ACTUAL SIZE-
-DO NOT SCALE-

CARTRIDGE: .348 Winchester Improved

OTHER NAMES:

DIA: .348

BALLISTEK NO: 348B

NAI NO: RMB 33343/3.942

DATA SOURCE: Ackley Vol.1 Pg465

HISTORICAL DATA:

NOTES:

LOADING DATA:

BULLET WT./TYPE	POWDER WT./TYPE	VELOCITY ('/SEC)	SOURCE
250/—	58.0/4350	2245	Hutton/ Ackley

CASE PREPARATION: SHELLHOLDER (RCBS): 5

MAKE FROM: .348 Win. Fire factory ammo in "improved" chamber or fireform, with corn meal, in the proper trim die. This is a hard case to fireform in this manner and I generally fireform twice with 36 grs. of Unique, tissue paper or corn meal and a wax wad.

PHYSICAL DATA (INCHES):

CASE TYPE: **Rimmed Bottleneck**

CASE LENGTH **A = 2.18**

HEAD DIAMETER **B = .553**

RIM DIAMETER **D = .610**

NECK DIAMETER **F = .376**

NECK LENGTH **H = .330**

SHOULDER LENGTH **K = .110**

BODY ANGLE (DEG'S/SIDE): **1.17**

CASE CAPACITY **CC'S = 5.17**

LOADED LENGTH: **A/R**

BELT DIAMETER **C = N/A**

RIM THICKNESS **E = .07**

SHOULDER DIAMETER **G = .490**

LENGTH TO SHOULDER **J = 1.74**

SHOULDER ANGLE (DEG'S/SIDE): **27.39**

PRIMER: **L/R**

CASE CAPACITY (GR'S WATER): **79.87**

DIMENSIONAL DRAWING:

F H K G A J E B D

-NOT ACTUAL SIZE-
-DO NOT SCALE-

CARTRIDGE: .35 Ackley Magnum

OTHER NAMES:

DIA: .358

BALLISTEK NO: 358K

NAI NO: BEN 22233/4.980

DATA SOURCE: Ackley Vol.1 Pg475

HISTORICAL DATA: By P. Ackley in 1946.

NOTES: Similar to .358 Norma Mag. (#358H)

LOADING DATA:

BULLET WT./TYPE	POWDER WT./TYPE	VELOCITY ('/SEC)	SOURCE
250/RN	62.0/4350	2700	Ackley

CASE PREPARATION: SHELLHOLDER (RCBS): 4

MAKE FROM: .375 H&H Mag. Form in form die (one die) or reduce neck dia. in .358 Norma die. F/L size, with expander removed. Trim to length & chamfer. F/L size and fireform in the chamber.

PHYSICAL DATA (INCHES):

CASE TYPE: **Belted Bottleneck**

CASE LENGTH **A = 2.53**

HEAD DIAMETER **B = .513**

RIM DIAMETER **D = .532**

NECK DIAMETER **F = .385**

NECK LENGTH **H = .370**

SHOULDER LENGTH **K = .110**

BODY ANGLE (DEG'S/SIDE): **.356**

CASE CAPACITY **CC'S = 5.68**

LOADED LENGTH: **3.25**

BELT DIAMETER **C = .532**

RIM THICKNESS **E = .05**

SHOULDER DIAMETER **G = .485**

LENGTH TO SHOULDER **J = 2.05**

SHOULDER ANGLE (DEG'S/SIDE): **24.44**

PRIMER: **L/R Mag**

CASE CAPACITY (GR'S WATER): **87.65**

DIMENSIONAL DRAWING:

F H K G A J E B D C

-NOT ACTUAL SIZE-
-DO NOT SCALE-

CARTRIDGE: .35 Ackley Magnum Improved

OTHER NAMES:

DIA: .358

BALLISTEK NO: 358P

NAI NO: BEN 12332/5.477

DATA SOURCE: Ackley Vol.1 Pg479

HISTORICAL DATA:

NOTES:

LOADING DATA:

BULLET WT./TYPE	POWDER WT./TYPE	VELOCITY ('/SEC)	SOURCE
300/—	80.0/4350	2620	Ackley

CASE PREPARATION: SHELLHOLDER (RCBS): 4

MAKE FROM: .375 H&H Mag. F/L size the H&H case in the Ackley "improved" die. Square case mouth, chamfer and fireform in the chamber.

PHYSICAL DATA (INCHES):

CASE TYPE: **Belted Bottleneck**

CASE LENGTH **A = 2.81**

HEAD DIAMETER **B = .513**

RIM DIAMETER **D = .532**

NECK DIAMETER **F = .389**

NECK LENGTH **H = .400**

SHOULDER LENGTH **K = .080**

BODY ANGLE (DEG'S/SIDE): **.319**

CASE CAPACITY **CC'S = 6.35**

LOADED LENGTH: **3.70**

BELT DIAMETER **C = .532**

RIM THICKNESS **E = .05**

SHOULDER DIAMETER **G = .490**

LENGTH TO SHOULDER **J = 2.26**

SHOULDER ANGLE (DEG'S/SIDE): **36.20**

PRIMER: **L/R Mag**

CASE CAPACITY (GR'S WATER): **98.06**

DIMENSIONAL DRAWING:

F
H
K
G
A
J
E
B
D
C

-NOT ACTUAL SIZE-
-DO NOT SCALE-

CARTRIDGE: .35 Apex Magnum

OTHER NAMES:

DIA: .358

BALLISTEK NO: 358U

NAI NO: BEN 22334/4.932

DATA SOURCE: Ackley Vol.1 Pg478

HISTORICAL DATA: By Apex Rifle Co.

NOTES:

LOADING DATA:

BULLET WT./TYPE	POWDER WT./TYPE	VELOCITY ('/SEC)	SOURCE
275/Spire	59.7/IMR4320	—	JJD

CASE PREPARATION: SHELLHOLDER (RCBS): 4

MAKE FROM: .300 H&H Mag. Taper expand neck to .360" dia. Cut case off at 2.6" F/L size in Apex die, trim to length and chamfer. Fireform in chamber.

PHYSICAL DATA (INCHES):

CASE TYPE: **Belted Bottleneck**

CASE LENGTH **A = 2.525**

HEAD DIAMETER **B = .512**

RIM DIAMETER **D = .532**

NECK DIAMETER **F = .387**

NECK LENGTH **H = .395**

SHOULDER LENGTH **K = .070**

BODY ANGLE (DEG'S/SIDE): **.415**

CASE CAPACITY **CC'S = 5.64**

LOADED LENGTH: **3.21**

BELT DIAMETER **C = .532**

RIM THICKNESS **E = .05**

SHOULDER DIAMETER **G = .485**

LENGTH TO SHOULDER **J = 2.06**

SHOULDER ANGLE (DEG'S/SIDE): **34.99**

PRIMER: **L/R Mag**

CASE CAPACITY (GR'S WATER): **87.03**

DIMENSIONAL DRAWING:

F
H
K
G
A
J
E
B
D
C

-NOT ACTUAL SIZE-
-DO NOT SCALE-

CARTRIDGE: .35 Brown/Whelen

OTHER NAMES:

DIA: .358
BALLISTEK NO: 358R
NAI NO: RXB 12244/5.074

DATA SOURCE: Ackley Vol.1 Pg472

HISTORICAL DATA:

NOTES:

LOADING DATA:

BULLET WT./TYPE	POWDER WT./TYPE	VELOCITY ('/SEC)	SOURCE
250/—	71.0/4350	2700	Ackley

CASE PREPARATION: SHELLHOLDER (RCBS): 3

MAKE FROM: .30-06 Spgf. Taper expand case neck to .405" (or so) dia. F/L size in BW die. Square case mouth, chamfer and fireform.

PHYSICAL DATA (INCHES):

CASE TYPE: **Rimless Bottleneck**
CASE LENGTH **A = 2.385**
HEAD DIAMETER **B = .470**
RIM DIAMETER **D = .473**
NECK DIAMETER **F = .379**
NECK LENGTH **H = .235**
SHOULDER LENGTH **K = .060**
BODY ANGLE (DEG'S/SIDE): **.136**
CASE CAPACITY **CC'S = 4.98**
LOADED LENGTH: **A/R**
BELT DIAMETER **C = N/A**
RIM THICKNESS **E = .05**
SHOULDER DIAMETER **G = .461**
LENGTH TO SHOULDER **J = 2.09**
SHOULDER ANGLE (DEG'S/SIDE): **34.34**
PRIMER: **L/R**
CASE CAPACITY (GR'S WATER): **76.93**

DIMENSIONAL DRAWING:

F H K G A J E B D

-NOT ACTUAL SIZE-
-DO NOT SCALE-

CARTRIDGE: .35 G&H Magnum

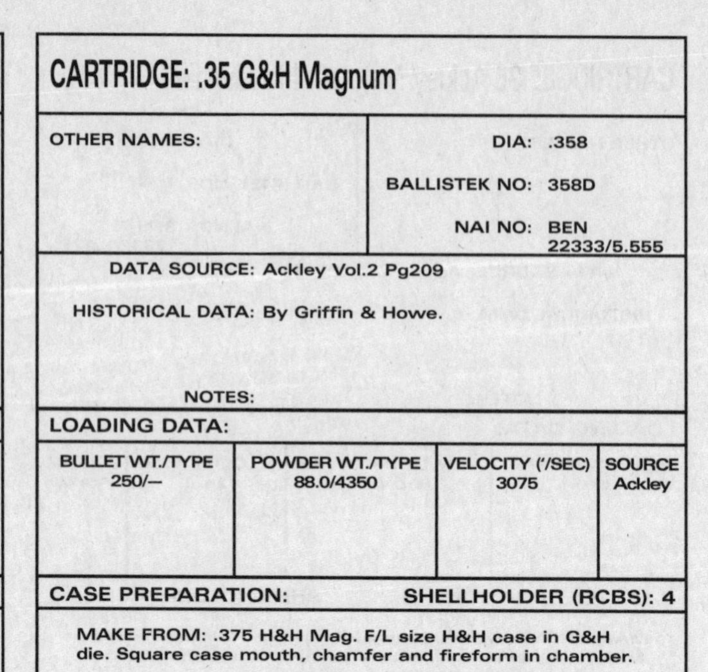

OTHER NAMES:

DIA: .358
BALLISTEK NO: 358D
NAI NO: BEN 22333/5.555

DATA SOURCE: Ackley Vol.2 Pg209

HISTORICAL DATA: By Griffin & Howe.

NOTES:

LOADING DATA:

BULLET WT./TYPE	POWDER WT./TYPE	VELOCITY ('/SEC)	SOURCE
250/—	88.0/4350	3075	Ackley

CASE PREPARATION: SHELLHOLDER (RCBS): 4

MAKE FROM: .375 H&H Mag. F/L size H&H case in G&H die. Square case mouth, chamfer and fireform in chamber.

PHYSICAL DATA (INCHES):

CASE TYPE: **Belted Bottleneck**
CASE LENGTH **A = 2.85**
HEAD DIAMETER **B = .513**
RIM DIAMETER **D = .532**
NECK DIAMETER **F = .390**
NECK LENGTH **H = .360**
SHOULDER LENGTH **K = .070**
BODY ANGLE (DEG'S/SIDE): **.683**
CASE CAPACITY **CC'S = 6.11**
LOADED LENGTH: **3.63**
BELT DIAMETER **C = .532**
RIM THICKNESS **E = .048**
SHOULDER DIAMETER **G = .460**
LENGTH TO SHOULDER **J = 2.42**
SHOULDER ANGLE (DEG'S/SIDE): **26.57**
PRIMER: **L/R Mag**
CASE CAPACITY (GR'S WATER): **94.41**

DIMENSIONAL DRAWING:

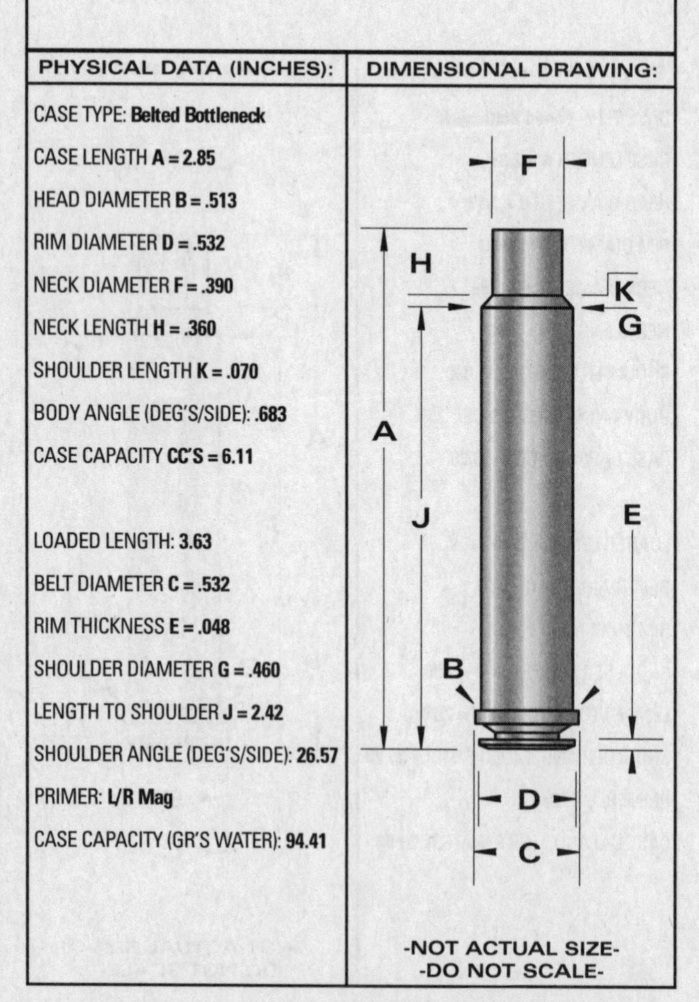

CARTRIDGE: .35 Lever Power

OTHER NAMES:

DIA: .358

BALLISTEK NO: 358V

NAI NO: RMB 12243/4.650

DATA SOURCE: Ackley Vol.1 Pg469

HISTORICAL DATA: By F. Wade.

NOTES:

LOADING DATA:

BULLET WT./TYPE	POWDER WT./TYPE	VELOCITY ('/SEC)	SOURCE
200/—	40.0/3031	2370	Ackley

CASE PREPARATION: **SHELLHOLDER (RCBS): 7**

MAKE FROM: .30-40 Krag. Turn rim to .500" dia. & back chamfer. Taper expand to .360" dia. (anneal as needed). Cut case off to 2.2" & chamfer. F/L size in LP die, trim to length and chamfer.

PHYSICAL DATA (INCHES):

CASE TYPE: **Rimmed Bottleneck**
CASE LENGTH **A = 2.125**
HEAD DIAMETER **B = .457**
RIM DIAMETER **D = .500**
NECK DIAMETER **F = .386**
NECK LENGTH **H = .346**
SHOULDER LENGTH **K = .059**
BODY ANGLE (DEG'S/SIDE): **.132**
CASE CAPACITY **CC'S = 3.90**

LOADED LENGTH: **2.74**
BELT DIAMETER **C = N/A**
RIM THICKNESS **E = .064**
SHOULDER DIAMETER **G = .450**
LENGTH TO SHOULDER **J = 1.72**
SHOULDER ANGLE (DEG'S/SIDE): **28.47**
PRIMER: **L/R**
CASE CAPACITY (GR'S WATER): **60.22**

DIMENSIONAL DRAWING:

-NOT ACTUAL SIZE-
-DO NOT SCALE-

CARTRIDGE: .35 Newton

OTHER NAMES:

DIA: .358

BALLISTEK NO: 358J

NAI NO: RXB 12344/4.796

DATA SOURCE: COTW 4th Pg100

HISTORICAL DATA: By Charles Newton in 1915.

NOTES:

LOADING DATA:

BULLET WT./TYPE	POWDER WT./TYPE	VELOCITY ('/SEC)	SOURCE
200/—	78.0/3031	3030	Barnes

CASE PREPARATION: **SHELLHOLDER (RCBS): 4**

MAKE FROM: .375 H&H. Turn belt to .520" dia. Turn rim to .525" dia. Cut new extractor groove. F/L size with expander removed. Trim to length. F/L size. Chamfer. Fireform in chamber.

PHYSICAL DATA (INCHES):

CASE TYPE: **Rimless Bottleneck**
CASE LENGTH **A = 2.494**
HEAD DIAMETER **B = .520**
RIM DIAMETER **D = .525**
NECK DIAMETER **F = .383**
NECK LENGTH **H = .329**
SHOULDER LENGTH **K = .065**
BODY ANGLE (DEG'S/SIDE): **.332**
CASE CAPACITY **CC'S = 5.99**

LOADED LENGTH: **3.35**
BELT DIAMETER **C = N/A**
RIM THICKNESS **E = .05**
SHOULDER DIAMETER **G = .498**
LENGTH TO SHOULDER **J = 2.10**
SHOULDER ANGLE (DEG'S/SIDE): **41.49**
PRIMER: **L/R**
CASE CAPACITY (GR'S WATER): **92.56**

DIMENSIONAL DRAWING:

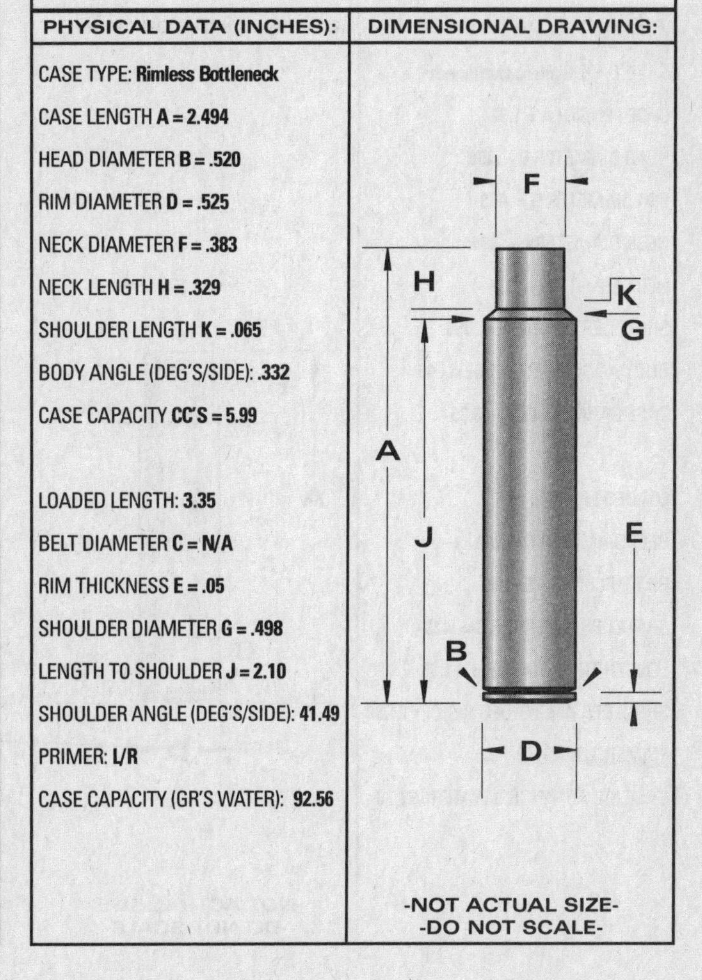

-NOT ACTUAL SIZE-
-DO NOT SCALE-

CARTRIDGE: .35 Remington

OTHER NAMES:
.35 Remington Auto

DIA: .358

BALLISTEK NO: 358B

NAI NO: RXB
22243/4.192

DATA SOURCE: Hornady Manual 3rd Pg280

HISTORICAL DATA: By Rem. in 1906 for M8.

NOTES:

LOADING DATA:

BULLET WT./TYPE	POWDER WT./TYPE	VELOCITY ('/SEC)	SOURCE
200/RN	37.0/IMR4895	1850	JJD
250/–	51.0/IMR4064	2300	Barnes

CASE PREPARATION: SHELLHOLDER (RCBS): 9

MAKE FROM: Factory or .303 British or .30-40 Krag. Trim case to 1.95". Turn rim flush with base and cut new extractor groove. F/L size case. Trim & chamfer.

PHYSICAL DATA (INCHES):

CASE TYPE: **Rimless Bottleneck**

CASE LENGTH **A = 1.92**

HEAD DIAMETER **B = .458**

RIM DIAMETER **D = .460**

NECK DIAMETER **F = .384**

NECK LENGTH **H = .335**

SHOULDER LENGTH **K = .055**

BODY ANGLE (DEG'S/SIDE): **.646**

CASE CAPACITY **CC'S = 3.35**

LOADED LENGTH: **2.51**

BELT DIAMETER **C = N/A**

RIM THICKNESS **E = .05**

SHOULDER DIAMETER **G = .428**

LENGTH TO SHOULDER **J = 1.53**

SHOULDER ANGLE (DEG'S/SIDE): **21.80**

PRIMER: **L/R**

CASE CAPACITY (GR'S WATER): **51.76**

DIMENSIONAL DRAWING:

F H K G A J E B D

-NOT ACTUAL SIZE-
-DO NOT SCALE-

CARTRIDGE: .35 Smith & Wesson Auto

OTHER NAMES:

DIA: .309

BALLISTEK NO: 309C

NAI NO: RXS
11115/1.936

DATA SOURCE: COTW 4th Pg169

HISTORICAL DATA: By S&W in 1913.

NOTES: .309" is correct dia.

LOADING DATA:

BULLET WT./TYPE	POWDER WT./TYPE	VELOCITY ('/SEC)	SOURCE
76/RN	1.6/B'Eye	806	Barnes

CASE PREPARATION: SHELLHOLDER (RCBS): 12

MAKE FROM: .25-20 Win. (or .32-20). Turn rim to .348" dia. and re-cut extractor groove. Cut case to length and I.D. neck ream. F/L size.

PHYSICAL DATA (INCHES):

CASE TYPE: **Rimless Straight**

CASE LENGTH **A = .670**

HEAD DIAMETER **B = .346**

RIM DIAMETER **D = .348**

NECK DIAMETER **F = .345**

NECK LENGTH **H = N/A**

SHOULDER LENGTH **K = N/A**

BODY ANGLE (DEG'S/SIDE): **.045**

CASE CAPACITY **CC'S = .397**

LOADED LENGTH: **.970**

BELT DIAMETER **C = N/A**

RIM THICKNESS **E = .04**

SHOULDER DIAMETER **G = N/A**

LENGTH TO SHOULDER **J = N/A**

SHOULDER ANGLE (DEG'S/SIDE): **N/A**

PRIMER: **S/P**

CASE CAPACITY (GR'S WATER): **6.13**

DIMENSIONAL DRAWING:

F A B E D

-NOT ACTUAL SIZE-
-DO NOT SCALE-

CARTRIDGE: .35 Whelen

OTHER NAMES:
.35/06 Springfield
.35/30-06 Spgf.

DIA: .358

BALLISTEK NO: 358C

NAI NO: RXB 22232/5.306

DATA SOURCE: Hornady Manual 3rd Pg286

HISTORICAL DATA: By Col. Whelen in 1922

NOTES:

LOADING DATA:

BULLET WT./TYPE	POWDER WT./TYPE	VELOCITY ('/SEC)	SOURCE
200/RN	54.0/IMR4895	2550	JJD

CASE PREPARATION: SHELLHOLDER (RCBS): 3

MAKE FROM: .30-06 Spfg. Anneal case neck. Taper expand to about .365" dia. F/L size in Whelen die. Square case mouth & chamfer.

PHYSICAL DATA (INCHES):

CASE TYPE: **Rimless Bottleneck**
CASE LENGTH **A = 2.494**
HEAD DIAMETER **B = .470**
RIM DIAMETER **D = .473**
NECK DIAMETER **F = .385**
NECK LENGTH **H = .434**
SHOULDER LENGTH **K = .112**
BODY ANGLE (DEG'S/SIDE): **.475**
CASE CAPACITY **CC'S = 4.70**
LOADED LENGTH: **3.26**
BELT DIAMETER **C = N/A**
RIM THICKNESS **E = .05**
SHOULDER DIAMETER **G = .441**
LENGTH TO SHOULDER **J = 1.948**
SHOULDER ANGLE (DEG'S/SIDE): **14.03**
PRIMER: **L/R**
CASE CAPACITY (GR'S WATER): **72.63**

DIMENSIONAL DRAWING:

-NOT ACTUAL SIZE-
-DO NOT SCALE-

CARTRIDGE: .35 Whelen Improved

OTHER NAMES:

DIA: .358

BALLISTEK NO: 358N

NAI NO: RXB 12224/5.298

DATA SOURCE: Ackley Vol.1 Pg472

HISTORICAL DATA:

NOTES:

LOADING DATA:

BULLET WT./TYPE	POWDER WT./TYPE	VELOCITY ('/SEC)	SOURCE
250/RN	53.0/4064	–	JJD

CASE PREPARATION: SHELLHOLDER (RCBS): 3

MAKE FROM: .30-06 Spgf. Taper expand '06 case to .330" dia. Anneal. Taper expand to .360" dia. F/L size in "improved" die. Square case mouth, chamfer and fireform in chamber.

PHYSICAL DATA (INCHES):

CASE TYPE: **Rimless Bottleneck**
CASE LENGTH **A = 2.49**
HEAD DIAMETER **B = .470**
RIM DIAMETER **D = .473**
NECK DIAMETER **F = .338**
NECK LENGTH **H = .465**
SHOULDER LENGTH **K = .049**
BODY ANGLE (DEG'S/SIDE): **.209**
CASE CAPACITY **CC'S = 6.31**
LOADED LENGTH: **3.20**
BELT DIAMETER **C = N/A**
RIM THICKNESS **E = .05**
SHOULDER DIAMETER **G = .457**
LENGTH TO SHOULDER **J = 1.976**
SHOULDER ANGLE (DEG'S/SIDE): **50.53**
PRIMER: **L/R**
CASE CAPACITY (GR'S WATER): **97.35**

DIMENSIONAL DRAWING:

-NOT ACTUAL SIZE-
-DO NOT SCALE-

CARTRIDGE: .35 Williams

OTHER NAMES:

DIA: .358

BALLISTEK NO: 358AA

NAI NO: BEN 12234/5.078

DATA SOURCE: Ackley Vol.1 Pg474

HISTORICAL DATA: By W. Williams in 1947.

NOTES: Similar to .35 Ackley Magnum (#358K).

LOADING DATA:

BULLET WT./TYPE	POWDER WT./TYPE	VELOCITY ('/SEC)	SOURCE
250/RN	63.0/4350	2550	Ackley

CASE PREPARATION: SHELLHOLDER (RCBS): 4

MAKE FROM: .300 H&H Mag. Taper expand case neck to .360" dia. Trim to 2.6". Chamfer and F/L size in Williams die. Fireform in chamber. .375 H&H brass may be necked down, trimmed, F/L sized and fireformed.

PHYSICAL DATA (INCHES):

CASE TYPE: **Belted Bottleneck**

CASE LENGTH **A = 2.600**

HEAD DIAMETER **B = .512**

RIM DIAMETER **D = .530**

NECK DIAMETER **F = .380**

NECK LENGTH **H = .402**

SHOULDER LENGTH **K = .063**

BODY ANGLE (DEG'S/SIDE): **.311**

CASE CAPACITY **CC'S = 6.14**

LOADED LENGTH: **3.40**

BELT DIAMETER **C = .530**

RIM THICKNESS **E = .05**

SHOULDER DIAMETER **G = .491**

LENGTH TO SHOULDER **J = 2.135**

SHOULDER ANGLE (DEG'S/SIDE): **41.37**

PRIMER: **L/R Mag**

CASE CAPACITY (GR'S WATER): **94.81**

DIMENSIONAL DRAWING:

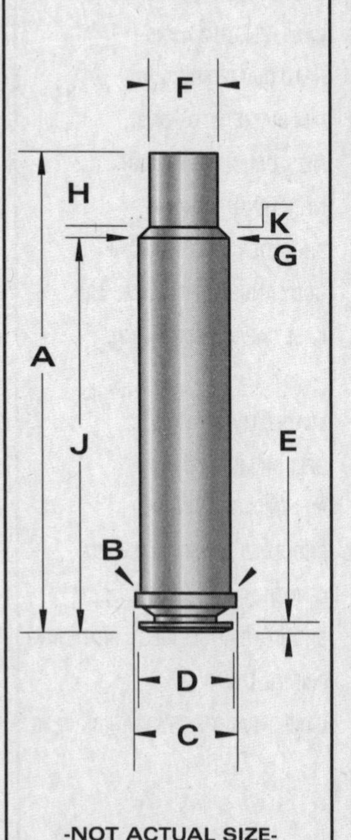

-NOT ACTUAL SIZE-
-DO NOT SCALE-

CARTRIDGE: .35 Winchester

OTHER NAMES:

DIA: .358

BALLISTEK NO: 358Q

NAI NO: RMB 22241/5.238

DATA SOURCE: Handloader #59 Pg66

HISTORICAL DATA: By Win. for M95 L/A rifle about 1903.

NOTES:

LOADING DATA:

BULLET WT./TYPE	POWDER WT./TYPE	VELOCITY ('/SEC)	SOURCE
250/RN	50.0/4895	2290	Barnes

CASE PREPARATION: SHELLHOLDER (RCBS): 7

MAKE FROM: .405 Win. Basic (BELL). Cut case to 2.5". Anneal. Reduce neck dia. in a .38-70 F/L die or use a proper form die. F/L size in .35 Win. die. Trim to final length and chamfer.

PHYSICAL DATA (INCHES):

CASE TYPE: **Rimmed Bottleneck**

CASE LENGTH **A = 2.415**

HEAD DIAMETER **B = .461**

RIM DIAMETER **D = .543**

NECK DIAMETER **F = .383**

NECK LENGTH **H = .284**

SHOULDER LENGTH **K = .130**

BODY ANGLE (DEG'S/SIDE): **.525**

CASE CAPACITY **CC'S = 4.45**

LOADED LENGTH: **3.16**

BELT DIAMETER **C = N/A**

RIM THICKNESS **E = .062**

SHOULDER DIAMETER **G = .428**

LENGTH TO SHOULDER **J = 2.00**

SHOULDER ANGLE (DEG'S/SIDE): **9.82**

PRIMER: **L/R**

CASE CAPACITY (GR'S WATER): **68.74**

DIMENSIONAL DRAWING:

F
H
K
G
A
J
E
B
D

-NOT ACTUAL SIZE-
-DO NOT SCALE-

CARTRIDGE: .35 Winchester Self-Loading

OTHER NAMES:
.35 Win. SL

DIA: .351

BALLISTEK NO: 351B

NAI NO: RMS 11115/3.000

DATA SOURCE: NAI/Ballistek

HISTORICAL DATA: By Win. in 1905 for M'05.

NOTES:

LOADING DATA:

BULLET WT./TYPE	POWDER WT./TYPE	VELOCITY ('/SEC)	SOURCE
180/—	12.9/2400	1420	Nonte

CASE PREPARATION: SHELLHOLDER (RCBS): 6

MAKE FROM: .357 Magnum. Turn rim to .405" dia. & back chamfer. Cut case to 1.15". F/L size in .35 SL die. Trim to length. Chamfer.

PHYSICAL DATA (INCHES):

CASE TYPE: **Rimmed Straight**

CASE LENGTH **A = 1.14**

HEAD DIAMETER **B = .380**

RIM DIAMETER **D = .405**

NECK DIAMETER **F = .374**

NECK LENGTH **H = N/A**

SHOULDER LENGTH **K = N/A**

BODY ANGLE (DEG'S/SIDE): **.157**

CASE CAPACITY **CC'S = 1.26**

LOADED LENGTH: **1.65**

BELT DIAMETER **C = N/A**

RIM THICKNESS **E = .05**

SHOULDER DIAMETER **G = N/A**

LENGTH TO SHOULDER **J = N/A**

SHOULDER ANGLE (DEG'S/SIDE): **N/A**

PRIMER: **S/R**

CASE CAPACITY (GR'S WATER): **19.43**

DIMENSIONAL DRAWING:

-NOT ACTUAL SIZE-
-DO NOT SCALE-

CARTRIDGE: .35-30 Maynard 1865

OTHER NAMES:

DIA: .370

BALLISTEK NO: 370A

NAI NO: RMS 11115/3.750

DATA SOURCE: NAI/Ballistek

HISTORICAL DATA: By Maynard in 1865.

NOTES:

LOADING DATA:

BULLET WT./TYPE	POWDER WT./TYPE	VELOCITY ('/SEC)	SOURCE
165/Lead	12.2/IMR4198	—	JJD

CASE PREPARATION: SHELLHOLDER (RCBS): Spl.

MAKE FROM: The only way to fabricate this case is to lathe turn the case heads and use 13/32" dia. tubing for the body. Anneal the case and F/L size.

PHYSICAL DATA (INCHES):

CASE TYPE: **Rimmed Straight**

CASE LENGTH **A = 1.53**

HEAD DIAMETER **B = .408**

RIM DIAMETER **D = .771**

NECK DIAMETER **F = .397**

NECK LENGTH **H = N/A**

SHOULDER LENGTH **K = N/A**

BODY ANGLE (DEG'S/SIDE): **.215**

CASE CAPACITY **CC'S = 2.07**

LOADED LENGTH: **1.98**

BELT DIAMETER **C = N/A**

RIM THICKNESS **E = .065**

SHOULDER DIAMETER **G = N/A**

LENGTH TO SHOULDER **J = N/A**

SHOULDER ANGLE (DEG'S/SIDE): **N/A**

PRIMER: **L/R**

CASE CAPACITY (GR'S WATER): **31.87**

DIMENSIONAL DRAWING:

-NOT ACTUAL SIZE-
-DO NOT SCALE-

CARTRIDGE: .35-30 Maynard 1873

OTHER NAMES:

DIA: .364

BALLISTEK NO: 363B

NAI NO: RMS 11115/4.044

DATA SOURCE: COTW 4th Pg 134

HISTORICAL DATA:

NOTES: Note the large rim diameter!

LOADING DATA:

BULLET WT./TYPE	POWDER WT./TYPE	VELOCITY ('/SEC)	SOURCE
125/Lead	13.2/IMR4198	—	JJD

CASE PREPARATION: SHELLHOLDER (RCBS): 577

MAKE FROM: Turn the large rim head and use 13/32" dia. tubing for the body. Anneal, trim to length and F/L size.

PHYSICAL DATA (INCHES):

CASE TYPE: **Rimmed Straight**

CASE LENGTH **A = 1.63**

HEAD DIAMETER **B = .403**

RIM DIAMETER **D = .765**

NECK DIAMETER **F = .397**

NECK LENGTH **H = N/A**

SHOULDER LENGTH **K = N/A**

BODY ANGLE (DEG'S/SIDE): **.109**

CASE CAPACITY **CC'S = 2.09**

LOADED LENGTH: **2.10**

BELT DIAMETER **C = N/A**

RIM THICKNESS **E = .063**

SHOULDER DIAMETER **G = N/A**

LENGTH TO SHOULDER **J = N/A**

SHOULDER ANGLE (DEG'S/SIDE): **N/A**

PRIMER: **L/R**

CASE CAPACITY (GR'S WATER): **32.33**

DIMENSIONAL DRAWING:

F

A

E

B

D

-NOT ACTUAL SIZE-
-DO NOT SCALE-

CARTRIDGE: .35-30 Maynard 1882

OTHER NAMES:

DIA: .359

BALLISTEK NO: 359B

NAI NO: RMS 11115/4.000

DATA SOURCE: COTW 4th Pg99

HISTORICAL DATA: From about 1882.

NOTES:

LOADING DATA:

BULLET WT./TYPE	POWDER WT./TYPE	VELOCITY ('/SEC)	SOURCE
165/Lead	16.0/4198	1320	Barnes

CASE PREPARATION: SHELLHOLDER (RCBS): 21

MAKE FROM: 13/32" dia. tubing case, turn from solid brass or re-body a .30-30 case head with 13/32 tubing. Anneal case, F/L size and trim to length.

PHYSICAL DATA (INCHES):

CASE TYPE: **Rimmed Straight**

CASE LENGTH **A = 1.600**

HEAD DIAMETER **B = .400**

RIM DIAMETER **D = .494**

NECK DIAMETER **F = .395**

NECK LENGTH **H = N/A**

SHOULDER LENGTH **K = N/A**

BODY ANGLE (DEG'S/SIDE): **.093**

CASE CAPACITY **CC'S = 1.96**

LOADED LENGTH: **2.03**

BELT DIAMETER **C = N/A**

RIM THICKNESS **E = .063**

SHOULDER DIAMETER **G = N/A**

LENGTH TO SHOULDER **J = N/A**

SHOULDER ANGLE (DEG'S/SIDE): **N/A**

PRIMER: **L/R**

CASE CAPACITY (GR'S WATER): **30.35**

DIMENSIONAL DRAWING:

F

A

E

B

D

-NOT ACTUAL SIZE-
-DO NOT SCALE-

CARTRIDGE: .35-40 Maynard 1882

OTHER NAMES:

DIA: .360

BALLISTEK NO: 360F

NAI NO: RMS 11115/5.150

DATA SOURCE: COTW 4th Pg99

HISTORICAL DATA: Longer version of .35-30 Maynard (#359B).

NOTES:

LOADING DATA:

BULLET WT./TYPE	POWDER WT./TYPE	VELOCITY ('/SEC)	SOURCE
165/Lead	18.0/4198	1400	Barnes

CASE PREPARATION: SHELLHOLDER (RCBS): 21

MAKE FROM: Turn head from brass and use 13/32" dia. tubing for body. Standard tubing case design. Anneal and F/L size. Trim & chamfer. Same tubing can be used with a 9,3 x 72R case head.

PHYSICAL DATA (INCHES):

CASE TYPE: **Rimmed Straight**

CASE LENGTH **A = 2.06**

HEAD DIAMETER **B = .400**

RIM DIAMETER **D = .492**

NECK DIAMETER **F = .390**

NECK LENGTH **H = N/A**

SHOULDER LENGTH **K = N/A**

BODY ANGLE (DEG'S/SIDE): **.143**

CASE CAPACITY **CC'S = 2.77**

LOADED LENGTH: **2.53**

BELT DIAMETER **C = N/A**

RIM THICKNESS **E = .065**

SHOULDER DIAMETER **G = N/A**

LENGTH TO SHOULDER **J = N/A**

SHOULDER ANGLE (DEG'S/SIDE): **N/A**

PRIMER: **L/R**

CASE CAPACITY (GR'S WATER): **42.70**

DIMENSIONAL DRAWING:

F

A

E

B

D

-NOT ACTUAL SIZE-
-DO NOT SCALE-

CARTRIDGE: .35/284 Winchester

OTHER NAMES:

DIA: .358

BALLISTEK NO: 358E

NAI NO: RBB 22244/4.320

DATA SOURCE: Ackley Vol.2 Pg205

HISTORICAL DATA:

NOTES:

LOADING DATA:

BULLET WT./TYPE	POWDER WT./TYPE	VELOCITY ('/SEC)	SOURCE
250/Spire	53.0/4064	2400	Ackley

CASE PREPARATION: SHELLHOLDER (RCBS): 3

MAKE FROM: .284 Win. Taper expand the .284 case to about .36" dia. F/L size, trim and chamfer.

PHYSICAL DATA (INCHES):

CASE TYPE: **Rebated Bottleneck**

CASE LENGTH **A = 2.16**

HEAD DIAMETER **B = .500**

RIM DIAMETER **D = .473**

NECK DIAMETER **F = .394**

NECK LENGTH **H = .300**

SHOULDER LENGTH **K = .06**

BODY ANGLE (DEG'S/SIDE): **.447**

CASE CAPACITY **CC'S = 4.38**

LOADED LENGTH: **2.94**

BELT DIAMETER **C = N/A**

RIM THICKNESS **E = .053**

SHOULDER DIAMETER **G = .475**

LENGTH TO SHOULDER **J = 1.80**

SHOULDER ANGLE (DEG'S/SIDE): **34.02**

PRIMER: **L/R**

CASE CAPACITY (GR'S WATER): **67.59**

DIMENSIONAL DRAWING:

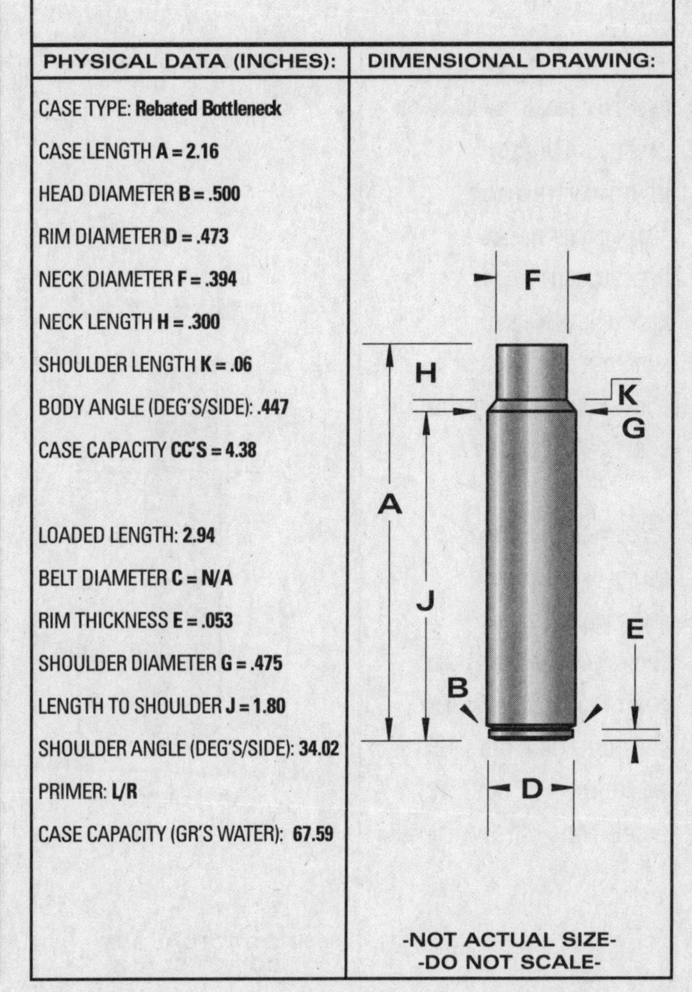

-NOT ACTUAL SIZE-
-DO NOT SCALE-

CARTRIDGE: .35/30-30 Winchester

OTHER NAMES:
.35-30 Winchester

DIA: .358

BALLISTEK NO: 358Z

NAI NO: RMB 22224/4.834

DATA SOURCE: COTW 4th Pg152

HISTORICAL DATA: Very old wildcat!

NOTES:

LOADING DATA:

BULLET WT./TYPE	POWDER WT./TYPE	VELOCITY ('/SEC)	SOURCE
200/Lead	25.0/4198	1893	Barnes

CASE PREPARATION: **SHELLHOLDER (RCBS): 2**

MAKE FROM: .30-30 Win. Anneal case neck. Taper expand to about .360" dia. F/L size, trim to length and chamfer.

PHYSICAL DATA (INCHES):

CASE TYPE: **Rimmed Bottleneck**
CASE LENGTH **A = 2.04**
HEAD DIAMETER **B = .422**
RIM DIAMETER **D = .506**
NECK DIAMETER **F = .378**
NECK LENGTH **H = .520**
SHOULDER LENGTH **K = .04**
BODY ANGLE (DEG'S/SIDE): **.59**
CASE CAPACITY **CC'S = 3.24**

LOADED LENGTH: **2.55**
BELT DIAMETER **C = N/A**
RIM THICKNESS **E = .062**
SHOULDER DIAMETER **G = .401**
LENGTH TO SHOULDER **J = 1.50**
SHOULDER ANGLE (DEG'S/SIDE): **28.67**
PRIMER: **L/R**
CASE CAPACITY (GR'S WATER): **50.06**

DIMENSIONAL DRAWING:

-NOT ACTUAL SIZE-
-DO NOT SCALE-

CARTRIDGE: .35/348 Winchester

OTHER NAMES:

DIA: .358

BALLISTEK NO: 358L

NAI NO: RMB 33333/3.942

DATA SOURCE: NAI/Ballistek

HISTORICAL DATA:

NOTES:

LOADING DATA:

BULLET WT./TYPE	POWDER WT./TYPE	VELOCITY ('/SEC)	SOURCE
250/RN	32.0/IMR4064	–	JJD

CASE PREPARATION: **SHELLHOLDER (RCBS): 5**

MAKE FROM: .348 Win. Taper expand .348 case to about .360" dia. Square case mouth and chamfer. F/L size.

PHYSICAL DATA (INCHES):

CASE TYPE: **Rimmed Bottleneck**
CASE LENGTH **A = 2.18**
HEAD DIAMETER **B = .553**
RIM DIAMETER **D = .610**
NECK DIAMETER **F = .386**
NECK LENGTH **H = .440**
SHOULDER LENGTH **K = .090**
BODY ANGLE (DEG'S/SIDE): **1.34**
CASE CAPACITY **CC'S = 5.03**

LOADED LENGTH: **A/R**
BELT DIAMETER **C = N/A**
RIM THICKNESS **E = .07**
SHOULDER DIAMETER **G = .485**
LENGTH TO SHOULDER **J = 1.65**
SHOULDER ANGLE (DEG'S/SIDE): **28.81**
PRIMER: **L/R**
CASE CAPACITY (GR'S WATER): **77.74**

DIMENSIONAL DRAWING:

-NOT ACTUAL SIZE-
-DO NOT SCALE-

CARTRIDGE: .350 Mashburn Short Magnum

OTHER NAMES:	
	DIA: .358
	BALLISTEK NO: 358T
	NAI NO: BEN 22334/4.824

DATA SOURCE: Ackley Vol.1 Pg478

HISTORICAL DATA:

NOTES: Similar to .35 Ackley Magnum (#358K)

LOADING DATA:

BULLET WT./TYPE	POWDER WT./TYPE	VELOCITY ('/SEC)	SOURCE
250/RN	58.0/IMR4320	–	JJD

CASE PREPARATION: SHELLHOLDER (RCBS): 4

MAKE FROM: .300 H&H Mag. Taper expand neck to .360" dia. Cut case to 2.5". Anneal neck, F/L size in Mashburn die, trim to length and chamfer.

PHYSICAL DATA (INCHES):

CASE TYPE: **Belted Bottleneck**

CASE LENGTH **A = 2.475**

HEAD DIAMETER **B = .513**

RIM DIAMETER **D = .532**

NECK DIAMETER **F = .389**

NECK LENGTH **H = .366**

SHOULDER LENGTH **K = .059**

BODY ANGLE (DEG'S/SIDE): **.511**

CASE CAPACITY **CC'S = 5.433**

LOADED LENGTH: **3.34**

BELT DIAMETER **C = .532**

RIM THICKNESS **E = .05**

SHOULDER DIAMETER **G = .480**

LENGTH TO SHOULDER **J = 2.05**

SHOULDER ANGLE (DEG'S/SIDE): **37.64**

PRIMER: **L/R**

CASE CAPACITY (GR'S WATER): **83.85**

DIMENSIONAL DRAWING:

-NOT ACTUAL SIZE-
-DO NOT SCALE-

CARTRIDGE: .350 Mashburn Super Magnum

OTHER NAMES:	
	DIA: .358
	BALLISTEK NO: 358Y
	NAI NO: BEN 22333/5.566

DATA SOURCE: Ackley Vol.1 Pg478

HISTORICAL DATA:

NOTES:

LOADING DATA:

BULLET WT./TYPE	POWDER WT./TYPE	VELOCITY ('/SEC)	SOURCE
250/Spire	67.0/H414	–	JJD

CASE PREPARATION: SHELLHOLDER (RCBS): 4

MAKE FROM: .300 H&H Mag. Taper expand case to .360" dia. F/L size in the Super Mashburn die, square case mouth, chamfer and fireform in the chamber.

PHYSICAL DATA (INCHES):

CASE TYPE: **Belted Bottleneck**

CASE LENGTH **A = 2.85**

HEAD DIAMETER **B = .512**

RIM DIAMETER **D = .532**

NECK DIAMETER **F = .389**

NECK LENGTH **H = .432**

SHOULDER LENGTH **K = .118**

BODY ANGLE (DEG'S/SIDE): **.504**

CASE CAPACITY **CC'S = 6.24**

LOADED LENGTH: **3.71**

BELT DIAMETER **C = .532**

RIM THICKNESS **E = .05**

SHOULDER DIAMETER **G = .475**

LENGTH TO SHOULDER **J = 2.30**

SHOULDER ANGLE (DEG'S/SIDE): **30.02**

PRIMER: **L/R Mag**

CASE CAPACITY (GR'S WATER): **96.36**

DIMENSIONAL DRAWING:

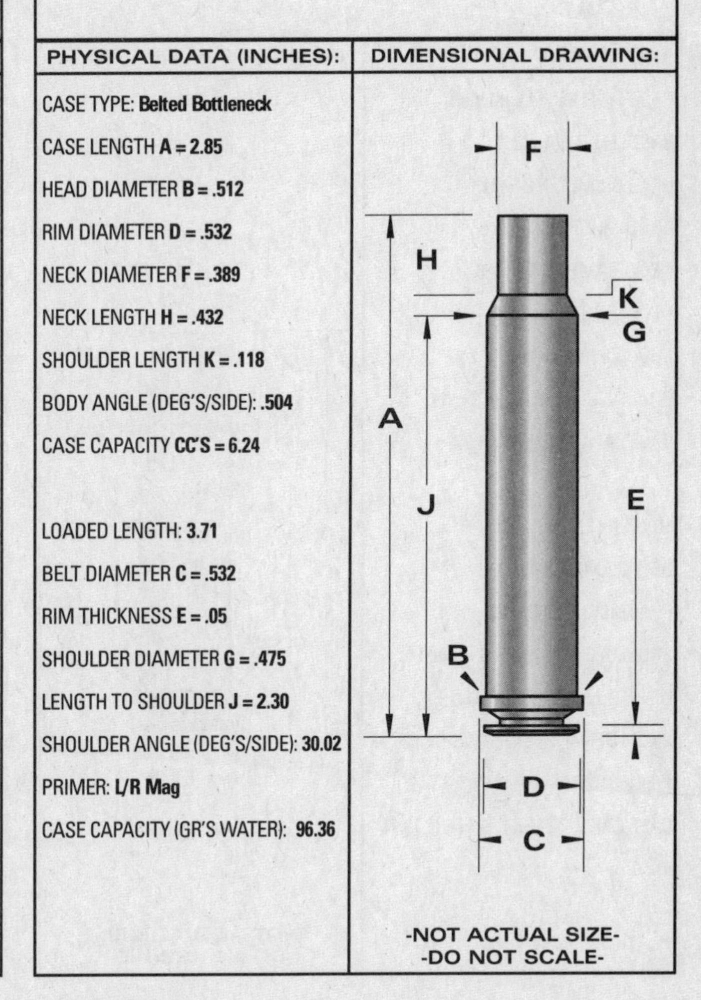

-NOT ACTUAL SIZE-
-DO NOT SCALE-

CARTRIDGE: .350 Remington Magnum

OTHER NAMES:	DIA: .358 BALLISTEK NO: 358F NAI NO: BEN 12343/4.230

DATA SOURCE: Hornady Manual 3rd Pg284

HISTORICAL DATA: By Rem. in 1955.

NOTES:

LOADING DATA:

BULLET WT./TYPE	POWDER WT./TYPE	VELOCITY ('/SEC)	SOURCE
250/RN	51.0/IMR4064	2270	JJD

CASE PREPARATION: SHELLHOLDER (RCBS): 4

MAKE FROM: Factory or .375 H&H. Anneal case neck. Cut-off at 2.25". Run into .350 F/L die. Trim & chamfer. Fireform in chamber.

PHYSICAL DATA (INCHES):

CASE TYPE: **Belted Bottleneck**
CASE LENGTH **A = 2.17**
HEAD DIAMETER **B = .513**
RIM DIAMETER **D = .532**
NECK DIAMETER **F = .388**
NECK LENGTH **H = .355**
SHOULDER LENGTH **K = .145**
BODY ANGLE (DEG'S/SIDE): **.331**
CASE CAPACITY **CC'S = 4.77**

LOADED LENGTH: **2.93**
BELT DIAMETER **C = .532**
RIM THICKNESS **E = .05**
SHOULDER DIAMETER **G = .496**
LENGTH TO SHOULDER **J = 1.67**
SHOULDER ANGLE (DEG'S/SIDE): **20.42**
PRIMER: **L/R Mag**
CASE CAPACITY (GR'S WATER): **73.74**

DIMENSIONAL DRAWING:

F H K G A J E B D C

-NOT ACTUAL SIZE-
-DO NOT SCALE-

CARTRIDGE: .350 Rigby Magnum

OTHER NAMES: .350 Rimless Magnum	DIA: .357 BALLISTEK NO: 357J NAI NO: RXB 32323/5.279

DATA SOURCE: COTW 4th Pg223

HISTORICAL DATA: By J. Rigby in 1908.

NOTES:

LOADING DATA:

BULLET WT./TYPE	POWDER WT./TYPE	VELOCITY ('/SEC)	SOURCE
225/RN	59.0/4320	2600	Barnes

CASE PREPARATION: SHELLHOLDER (RCBS): 26

MAKE FROM: .375 Flanged (BELL). Turn rim flush with case head and cut new extractor groove. Cut case to 2.8" and anneal. F/L size, trim to length & chamfer. Fireform in chamber.

PHYSICAL DATA (INCHES):

CASE TYPE: **Rimless Bottleneck**
CASE LENGTH **A = 2.74**
HEAD DIAMETER **B = .519**
RIM DIAMETER **D = .525**
NECK DIAMETER **F = .380**
NECK LENGTH **H = .485**
SHOULDER LENGTH **K = .075**
BODY ANGLE (DEG'S/SIDE): **1.09**
CASE CAPACITY **CC'S = 5.91**

LOADED LENGTH: **3.57**
BELT DIAMETER **C = N/A**
RIM THICKNESS **E = .057**
SHOULDER DIAMETER **G = .443**
LENGTH TO SHOULDER **J = 2.18**
SHOULDER ANGLE (DEG'S/SIDE): **22.78**
PRIMER: **L/R**
CASE CAPACITY (GR'S WATER): **91.25**

DIMENSIONAL DRAWING:

F H K G A J E B D

-NOT ACTUAL SIZE-
-DO NOT SCALE-

CARTRIDGE: .351 Winchester Self-Loading

OTHER NAMES:
.351SL

DIA: .351

BALLISTEK NO: 351A

NAI NO: RMS
11115/3.631

DATA SOURCE: COTW 4th Pg67

HISTORICAL DATA: By Win. in 1907 to replace the .35 SL.

NOTES:

LOADING DATA:

BULLET WT./TYPE	POWDER WT./TYPE	VELOCITY ('/SEC)	SOURCE
180/Spire	15.0/2400	1530	Nonte

CASE PREPARATION: SHELLHOLDER (RCBS): 6

MAKE FROM: .357 Max. Turn rim to .407" dia. Cut new extractor groove. Trim to 1.38" and F/L size. Chamfer.

PHYSICAL DATA (INCHES):

CASE TYPE: **Rimmed Straight**
CASE LENGTH **A = 1.38**
HEAD DIAMETER **B = .380**
RIM DIAMETER **D = .407**
NECK DIAMETER **F = .373**
NECK LENGTH **H = N/A**
SHOULDER LENGTH **K = N/A**
BODY ANGLE (DEG'S/SIDE): **.150**
CASE CAPACITY **CC'S = 1.63**

LOADED LENGTH: **1.91**
BELT DIAMETER **C = N/A**
RIM THICKNESS **E = .05**
SHOULDER DIAMETER **G = N/A**
LENGTH TO SHOULDER **J = N/A**
SHOULDER ANGLE (DEG'S/SIDE): **N/A**
PRIMER: **S/R**
CASE CAPACITY (GR'S WATER): **25.17**

DIMENSIONAL DRAWING:

F
A
E
B
D

-NOT ACTUAL SIZE-
-DO NOT SCALE-

CARTRIDGE: .357 Auto Mag

OTHER NAMES:
.357 AMP

DIA: .357

BALLISTEK NO: 357F

NAI NO: RXB
14243/2.762

DATA SOURCE: NAI/Ballistek

HISTORICAL DATA: Jurras development about 1970.

NOTES:

LOADING DATA:

BULLET WT./TYPE	POWDER WT./TYPE	VELOCITY ('/SEC)	SOURCE
158/JFP	16.0/B'Dot	1500	Barnes

CASE PREPARATION: SHELLHOLDER (RCBS): 3

MAKE FROM: .30-06 Spgf. (or any similar case head). Form dies are required. Form shoulder with form die. Trim to length. I.D. neck ream. Chamfer. F/L size.

PHYSICAL DATA (INCHES):

CASE TYPE: **Rimless Bottleneck**
CASE LENGTH **A = 1.298**
HEAD DIAMETER **B = .470**
RIM DIAMETER **D = .473**
NECK DIAMETER **F = .382**
NECK LENGTH **H = .260**
SHOULDER LENGTH **K = .099**
BODY ANGLE (DEG'S/SIDE): **.349**
CASE CAPACITY **CC'S = 2.26**

LOADED LENGTH: **1.60**
BELT DIAMETER **C = N/A**
RIM THICKNESS **E = .049**
SHOULDER DIAMETER **G = .461**
LENGTH TO SHOULDER **J = .938**
SHOULDER ANGLE (DEG'S/SIDE): **21.55**
PRIMER: **L/P**
CASE CAPACITY (GR'S WATER): **34.94**

DIMENSIONAL DRAWING:

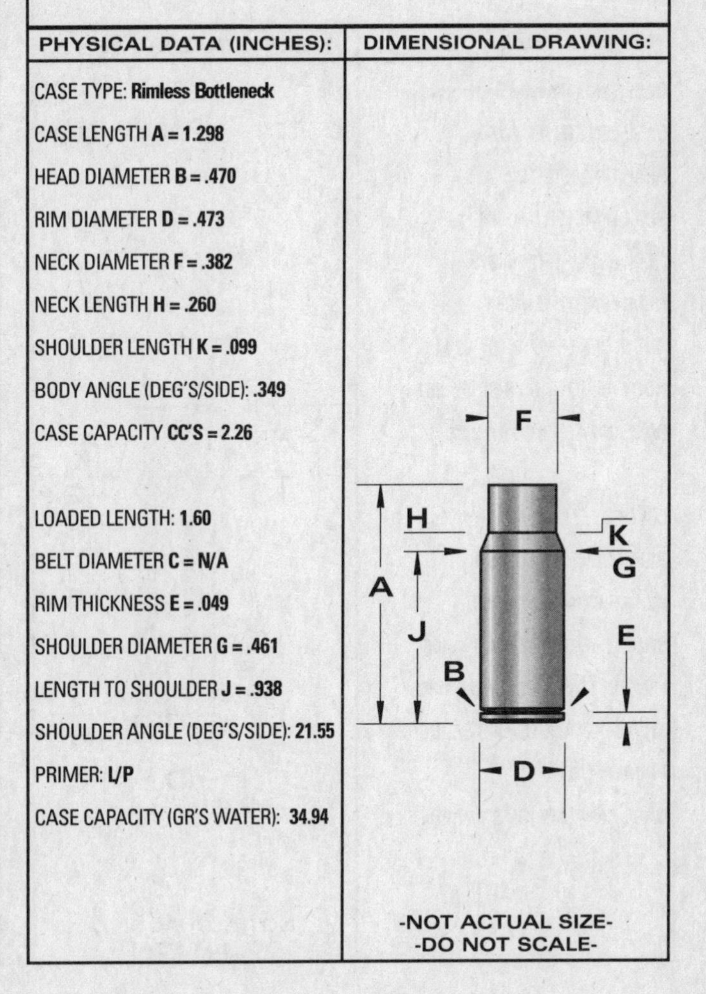

-NOT ACTUAL SIZE-
-DO NOT SCALE-

CARTRIDGE: .357 Herrett

OTHER NAMES:

DIA: .357

BALLISTEK NO: 357E

NAI NO: RMB 24234/4.156

DATA SOURCE: Hornady Manual 3rd Pg376

HISTORICAL DATA: By S. Herrett.

NOTES: Developed as a Contender ctg.

LOADING DATA:

BULLET WT./TYPE	POWDER WT./TYPE	VELOCITY ('/SEC)	SOURCE
158/JHP	24.3/IMR4227	1900	Horn.

CASE PREPARATION: SHELLHOLDER (RCBS): 2

MAKE FROM: .30-30 Win. Run case in F/L die with expander removed. Trim to length & chamfer. F/L size.

PHYSICAL DATA (INCHES):

CASE TYPE: **Rimmed Bottleneck**

CASE LENGTH **A = 1.750**

HEAD DIAMETER **B = .421**

RIM DIAMETER **D = .506**

NECK DIAMETER **F = .373**

NECK LENGTH **H = .358**

SHOULDER LENGTH **K = .024**

BODY ANGLE (DEG'S/SIDE): **.367**

CASE CAPACITY **CC'S = 2.82**

LOADED LENGTH: **2.06**

BELT DIAMETER **C = N/A**

RIM THICKNESS **E = .063**

SHOULDER DIAMETER **G = .406**

LENGTH TO SHOULDER **J = 1.368**

SHOULDER ANGLE (DEG'S/SIDE): **34.50**

PRIMER: **L/R**

CASE CAPACITY (GR'S WATER): **43.55**

DIMENSIONAL DRAWING:

-NOT ACTUAL SIZE-
-DO NOT SCALE-

CARTRIDGE: .357 Magnum

OTHER NAMES:
.357 Smith & Wesson Magnum

DIA: .357

BALLISTEK NO: 357B

NAI NO: RMS 11115/3.394

DATA SOURCE: Hornady Manual 3rd Pg344

HISTORICAL DATA: By Smith & Wesson in 1935.

NOTES: Developed by Winchester.

LOADING DATA:

BULLET WT./TYPE	POWDER WT./TYPE	VELOCITY ('/SEC)	SOURCE
158/JHP	14.0/2400	1200	Horn.
158/—	8.0/SR4756	1300	JJD

CASE PREPARATION: SHELLHOLDER (RCBS): 6

MAKE FROM: Factory or .357 Max. Trim the Max. case to 1.29" and chamfer. F/L size. Cases can also be made from .375 (3/8) tubing but, there should be no reason to expend this much effort. Tubing cases would require reduced loads.

PHYSICAL DATA (INCHES):

CASE TYPE: **Rimmed Straight**

CASE LENGTH **A = 1.29**

HEAD DIAMETER **B = .380**

RIM DIAMETER **D = .440**

NECK DIAMETER **F = .379**

NECK LENGTH **H = N/A**

SHOULDER LENGTH **K = N/A**

BODY ANGLE (DEG'S/SIDE): **1.34**

CASE CAPACITY **CC'S = 1.52**

LOADED LENGTH: **1.59**

BELT DIAMETER **C = N/A**

RIM THICKNESS **E = .06**

SHOULDER DIAMETER **G = N/A**

LENGTH TO SHOULDER **J = N/A**

SHOULDER ANGLE (DEG'S/SIDE): **N/A**

PRIMER: **S/P**

CASE CAPACITY (GR'S WATER): **23.47**

DIMENSIONAL DRAWING:

-NOT ACTUAL SIZE-
-DO NOT SCALE-

CARTRIDGE: .357 Maximum

OTHER NAMES:

DIA: .357

BALLISTEK NO: 357T

NAI NO: RMS 11115/4.234

DATA SOURCE: Sierra Manual 1985 Pg111

HISTORICAL DATA: By Rem. & Ruger in 1982.

NOTES:

LOADING DATA:

BULLET WT./TYPE	POWDER WT./TYPE	VELOCITY ('/SEC)	SOURCE
180/FMJ	22.9/W296	1800	Sierra

CASE PREPARATION: SHELLHOLDER (RCBS): 6

MAKE FROM: Factory. No other case will form to these dimensions. A low pressure case could be made from 3/8" tubing but, it would be worthless for the original loads. Rem. and Win. both offer brass.

PHYSICAL DATA (INCHES):

CASE TYPE: **Rimmed Straight**

CASE LENGTH **A = 1.605**

HEAD DIAMETER **B = .379**

RIM DIAMETER **D = .440**

NECK DIAMETER **F = .378**

NECK LENGTH **H = N/A**

SHOULDER LENGTH **K = N/A**

BODY ANGLE (DEG'S/SIDE): **.018**

CASE CAPACITY **CC'S = 2.014**

LOADED LENGTH: **1.99**

BELT DIAMETER **C = N/A**

RIM THICKNESS **E = .06**

SHOULDER DIAMETER **G = N/A**

LENGTH TO SHOULDER **J = N/A**

SHOULDER ANGLE (DEG'S/SIDE): **N/A**

PRIMER: **S/P**

CASE CAPACITY (GR'S WATER): **31.08**

DIMENSIONAL DRAWING:

F
A
E
B
D

-NOT ACTUAL SIZE-
-DO NOT SCALE-

CARTRIDGE: .357 SIG

OTHER NAMES:

DIA: .355

BALLISTEK NO:

NAI NO:

DATA SOURCE: Most reloading manuals.

HISTORICAL DATA: Developed jointly by Federal and SIG-Arms in 1994 to bring .357 Magnum revolver performance to semi-auto handguns.

NOTES: You must use 9mm .355 bullets and not the larger .357 bullets used in .357 Magnum. Cartridge headspaces on the case mouth.

LOADING DATA:

BULLET WT./TYPE	POWDER WT./TYPE	VELOCITY ('/SEC)	SOURCE
125/Speer HP	14.6/AA #9	1,437	Speer#13
147/Speer HP	8.8/Blue Dot	1,218	Speer#13

CASE PREPARATION: SHELL HOLDER (RCBS): 27

MAKE FROM: Cases can be formed from .40 S&W, but they will be as much as .020-inch short. Because the case headspaces off the case mouth this can result in problems with poor ignition and accuracy. If formed from 10mm Auto high pressure can result. This is because the 10mm case has a different internal configuration and uses a large primer. Also make by shortening .30 Remington case and fire forming, but loading caution is urged as capacity may be different and this case uses a large rifle primer.

PHYSICAL DATA (INCHES):

CASE TYPE: **Rimless Bottleneck**

CASE LENGTH **A = .865**

HEAD DIAMETER **B = .4240**

RIM DIAMETER **D = .424**

NECK DIAMETER **F = .381**

NECK LENGTH **H = .15**

SHOULDER LENGTH **K = .066**

BODY ANGLE (DEG'S/SIDE):

CASE CAPACITY **CC'S =**

LOADED LENGTH: **1.140**

BELT DIAMETER **C =**

RIM THICKNESS **E = .055**

SHOULDER DIAMETER **G = .424**

LENGTH TO SHOULDER **J = .649**

SHOULDER ANGLE (DEG'S/SIDE): **18**

PRIMER: **S/P**

CASE CAPACITY (GR'S WATER):

DIMENSIONAL DRAWING:

F
H
K
A
G
J
B
E
D

-NOT ACTUAL SIZE-
-DO NOT SCALE-

CARTRIDGE: .357/44 Bain & Davis

OTHER NAMES:	
.357/44 B&D	DIA: .357 BALLISTEK NO: 357G NAI NO: RMB 14241/2.800

DATA SOURCE: Hornady 3rd Pg374

HISTORICAL DATA: Developed about 1964.

NOTES:

LOADING DATA:

BULLET WT./TYPE	POWDER WT./TYPE	VELOCITY ('/SEC)	SOURCE
158/JHP	20.8/2400	1900	Horn.

CASE PREPARATION: SHELLHOLDER (RCBS): 18

MAKE FROM: .44 Magnum. Anneal case neck. Run into F/L die. Trim to length and chamfer.

PHYSICAL DATA (INCHES):

CASE TYPE: **Rimmed Bottleneck**
CASE LENGTH **A = 1.28**
HEAD DIAMETER **B = .457**
RIM DIAMETER **D = .514**
NECK DIAMETER **F = .384**
NECK LENGTH **H = .15**
SHOULDER LENGTH **K = .201**
BODY ANGLE (DEG'S/SIDE): **.078**
CASE CAPACITY **CC'S = 2.166**

LOADED LENGTH: **1.58**
BELT DIAMETER **C = N/A**
RIM THICKNESS **E = .06**
SHOULDER DIAMETER **G = .455**
LENGTH TO SHOULDER **J = .929**
SHOULDER ANGLE (DEG'S/SIDE): **10.01**
PRIMER: **L/P**
CASE CAPACITY (GR'S WATER): **33.43**

DIMENSIONAL DRAWING:

F
H
K
G
A
J
E
B
D

-NOT ACTUAL SIZE-
-DO NOT SCALE-

CARTRIDGE: .358 B-J Express

OTHER NAMES:	
	DIA: .358 BALLISTEK NO: 358AB NAI NO: BEN 22343/4.873

DATA SOURCE: Ackley Vol.1 Pg476

HISTORICAL DATA:

NOTES: Similar to .358 Norma Magnum (#358H).

LOADING DATA:

BULLET WT./TYPE	POWDER WT./TYPE	VELOCITY ('/SEC)	SOURCE
250/RN	78.0/4350	2900	Ackley

CASE PREPARATION: SHELLHOLDER (RCBS): 4

MAKE FROM: .375 H&H Mag. F/L size the H&H case in the B-J die with the expander removed. Trim the case to length and chamfer. Case will fireform in the chamber.

PHYSICAL DATA (INCHES):

CASE TYPE: **Belted Bottleneck**
CASE LENGTH **A = 2.51**
HEAD DIAMETER **B = .513**
RIM DIAMETER **D = .532**
NECK DIAMETER **F = .387**
NECK LENGTH **H = .292**
SHOULDER LENGTH **K = .078**
BODY ANGLE (DEG'S/SIDE): **.575**
CASE CAPACITY **CC'S = 5.93**

LOADED LENGTH: **3.40**
BELT DIAMETER **C = .532**
RIM THICKNESS **E = .05**
SHOULDER DIAMETER **G = .503**
LENGTH TO SHOULDER **J = 2.139**
SHOULDER ANGLE (DEG'S/SIDE): **36.5**
PRIMER: **L/R Mag**
CASE CAPACITY (GR'S WATER): **91.54**

DIMENSIONAL DRAWING:

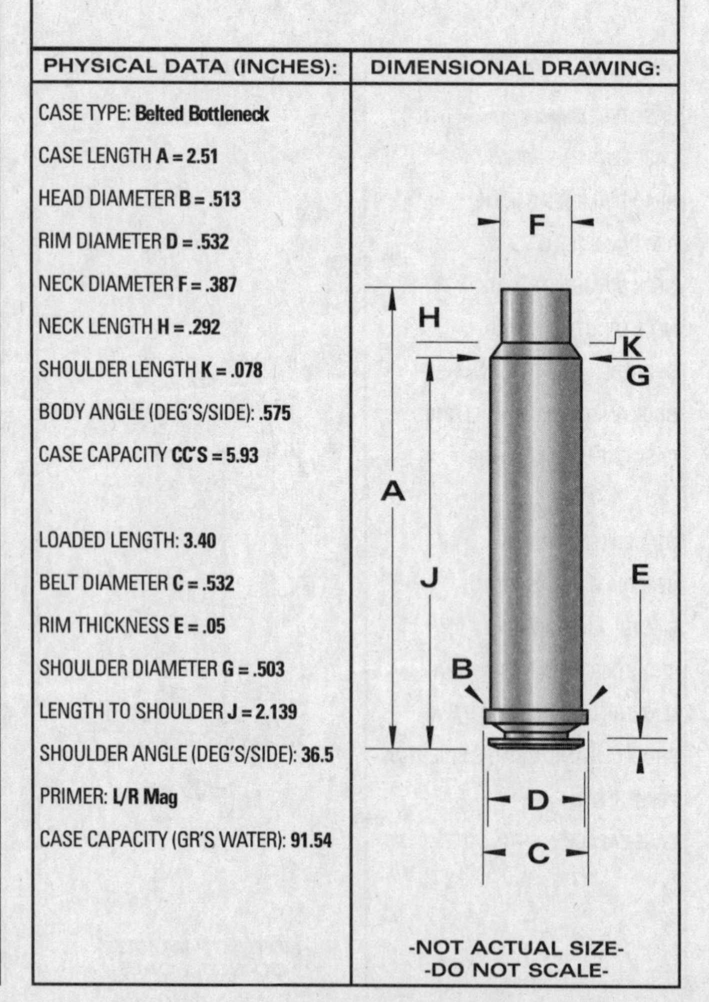

-NOT ACTUAL SIZE-
-DO NOT SCALE-

CARTRIDGE: .358 Barnes Supreme

OTHER NAMES:

DIA: .358

BALLISTEK NO: 358W

NAI NO: BEN 22333/5.567

DATA SOURCE: Ackley Vol.1 Pg480

HISTORICAL DATA:

NOTES:

LOADING DATA:

BULLET WT./TYPE	POWDER WT./TYPE	VELOCITY ('/SEC)	SOURCE
350/—	90.0/4350	3050	Ackley

CASE PREPARATION: SHELLHOLDER (RCBS): 4

MAKE FROM: .300 H&H Mag. Taper expand case neck to .360" dia. F/L size in the Barnes die. Trim to length & chamfer. Fireform in the chamber.

PHYSICAL DATA (INCHES):

CASE TYPE: **Belted Bottleneck**
CASE LENGTH **A = 2.856**
HEAD DIAMETER **B = .513**
RIM DIAMETER **D = .532**
NECK DIAMETER **F = .387**
NECK LENGTH **H = .380**
SHOULDER LENGTH **K = .091**
BODY ANGLE (DEG'S/SIDE): **.367**
CASE CAPACITY **CC'S = 6.54**
LOADED LENGTH: **4.00**
BELT DIAMETER **C = .532**
RIM THICKNESS **E = .049**
SHOULDER DIAMETER **G = .485**
LENGTH TO SHOULDER **J = 2.385**
SHOULDER ANGLE (DEG'S/SIDE): **28.30**
PRIMER: **L/R Mag**
CASE CAPACITY (GR'S WATER): **100.90**

DIMENSIONAL DRAWING:

-NOT ACTUAL SIZE-
-DO NOT SCALE-

CARTRIDGE: .358 JDJ

OTHER NAMES:

DIA: .358

BALLISTEK NO: 358AD

NAI NO: RMB 12242/4.798

DATA SOURCE: SSK Industries

HISTORICAL DATA: By J.D. Jones.

NOTES:

LOADING DATA:

BULLET WT./TYPE	POWDER WT./TYPE	VELOCITY ('/SEC)	SOURCE
250/—	44.0/H322	1920	SSK

CASE PREPARATION: SHELLHOLDER (RCBS): 28

MAKE FROM: .444 Marlin. F/L size the Marlin case, square the case mouth & chamfer.

PHYSICAL DATA (INCHES):

CASE TYPE: **Rimmed Bottleneck**
CASE LENGTH **A = 2.231**
HEAD DIAMETER **B = .465**
RIM DIAMETER **D = .511**
NECK DIAMETER **F = .381**
NECK LENGTH **H = .325**
SHOULDER LENGTH **K = .106**
BODY ANGLE (DEG'S/SIDE): **.250**
CASE CAPACITY **CC'S = 4.34**
LOADED LENGTH: **3.10**
BELT DIAMETER **C = N/A**
RIM THICKNESS **E = .061**
SHOULDER DIAMETER **G = .451**
LENGTH TO SHOULDER **J = 1.80**
SHOULDER ANGLE (DEG'S/SIDE): **18.27**
PRIMER: **L/R**
CASE CAPACITY (GR'S WATER): **66.94**

DIMENSIONAL DRAWING:

-NOT ACTUAL SIZE-
-DO NOT SCALE-

CARTRIDGE: .358 Lee Magnum

OTHER NAMES:

DIA: .358

BALLISTEK NO: 358X

NAI NO: RBB 22323/4.815

DATA SOURCE: Ackley Vol.1 Pg480

HISTORICAL DATA: By R. Lee.

NOTES:

LOADING DATA:

BULLET WT./TYPE	POWDER WT./TYPE	VELOCITY ('/SEC)	SOURCE
275/RN	80.0/4350	2630	Ackley

CASE PREPARATION: SHELLHOLDER (RCBS): 3

MAKE FROM: .425 Westley Richards. Form set required. Anneal case and form in form dies #1 & #2. Trim to length. F/L size. Fireform in chamber.

PHYSICAL DATA (INCHES):

CASE TYPE: **Rebated Bottleneck**

CASE LENGTH **A = 2.60**

HEAD DIAMETER **B = .540**

RIM DIAMETER **D = .467**

NECK DIAMETER **F = .385**

NECK LENGTH **H = .480**

SHOULDER LENGTH **K = .133**

BODY ANGLE (DEG'S/SIDE): **.48**

CASE CAPACITY **CC'S = 6.36**

LOADED LENGTH: **3.26**

BELT DIAMETER **C = N/A**

RIM THICKNESS **E = .045**

SHOULDER DIAMETER **G = .51**

LENGTH TO SHOULDER **J = 1.978**

SHOULDER ANGLE (DEG'S/SIDE): **25.17**

PRIMER: **L/R**

CASE CAPACITY (GR'S WATER): **98.19**

DIMENSIONAL DRAWING:

F
H
K
G
A
E
J
B
D

-NOT ACTUAL SIZE-
-DO NOT SCALE-

CARTRIDGE: .358 Norma Magnum

OTHER NAMES:

DIA: .358

BALLISTEK NO: 358H

NAI NO: BEN 12243/4.929

DATA SOURCE: Hornady Manual 3rd Pg288

HISTORICAL DATA: By Norma in 1959.

NOTES:

LOADING DATA:

BULLET WT./TYPE	POWDER WT./TYPE	VELOCITY ('/SEC)	SOURCE
250/RN	70.0/IMR4350	2650	JJD

CASE PREPARATION: SHELLHOLDER (RCBS): 4

MAKE FROM: Factory or .375 H&H. Anneal case neck. Run into .358 Norma F/L die with expander removed. Trim to 2.52". Chamfer. Fireform to establish Norma shoulder.

PHYSICAL DATA (INCHES):

CASE TYPE: **Belted Bottleneck**

CASE LENGTH **A = 2.519**

HEAD DIAMETER **B = .511**

RIM DIAMETER **D = .530**

NECK DIAMETER **F = .388**

NECK LENGTH **H = .328**

SHOULDER LENGTH **K = .106**

BODY ANGLE (DEG'S/SIDE): **.334**

CASE CAPACITY **CC'S = 5.67**

LOADED LENGTH: **3.28**

BELT DIAMETER **C = .530**

RIM THICKNESS **E = .049**

SHOULDER DIAMETER **G = .489**

LENGTH TO SHOULDER **J = 2.085**

SHOULDER ANGLE (DEG'S/SIDE): **25.47**

PRIMER: **L/R Mag**

CASE CAPACITY (GR'S WATER): **87.57**

DIMENSIONAL DRAWING:

F
H
K
G
A
J
E
B
D
C

-NOT ACTUAL SIZE-
-DO NOT SCALE-

CARTRIDGE: 358 STA

OTHER NAMES: 330 Shooting Times Alaskan

DIA: .358

BALLISTEK NO:

NAI NO:

DATA SOURCE:

HISTORICAL DATA: This wildcat chambering originated in 1990 by Layne Simpson. The original version was simply the 8mm Remington Magnum necked up with no other changes.

NOTES:

LOADING DATA:

BULLET WT./TYPE	POWDER WT./TYPE	VELOCITY ('/SEC)	SOURCE
125/Sierra JSP	91.0/H-4831	3046	A-Square
225/Sierra SBT	93.0/IMR-7828	3003	A-Square
275/A-Square Lion	90.0/RL-22	2835	A-Square

CASE PREPARATION: SHELL HOLDER (RCBS): 04

MAKE FROM: 8mm Remington Magnum

PHYSICAL DATA (INCHES):

CASE TYPE: **Belted Bottleneck**

CASE LENGTH **A = 2.845**

HEAD DIAMETER **B = .513**

RIM DIAMETER **D = .532**

NECK DIAMETER **F = .382**

NECK LENGTH **G = .495**

SHOULDER LENGTH **K = .081**

BODY ANGLE (DEG'S/SIDE): = **N/A**

CASE CAPACITY (CC'S):

LOADED LENGTH: **3.65**

BELT DIAMETER **C = .532**

RIM THICKNESS **E = .050**

SHOULDER DIAMETER **G = .495**

LENGTH TO SHOULDER **J = 2.405**

SHOULDER ANGLE (DEG'S/SIDE):

PRIMER: **L/R**

CASE CAPACITY (GR'S WATER):

DIMENSIONAL DRAWING:

-NOT ACTUAL SIZE-
-DO NOT SCALE-

CARTRIDGE: .358 Ultra Mag Towsley (UMT)

OTHER NAMES:

DIA: .358

BALLISTEK NO:

NAI NO:

DATA SOURCE: Barnes Bullets/Bryce M. Towsley

HISTORICAL DATA: Created in 1999 by Bryce Towsley. The 358 UMT shoots flatter than the 7mm Remington Magnum with 160-grain bullets and carries more energy at all ranges than the .416 Remington 400-grain.

NOTES: This cartridge requires the use of a high quality premium bullet. The Barnes X-bullets, notably the coated 225-grain, work very well.

LOADING DATA:

BULLET WT./TYPE	POWDER WT./TYPE	VELOCITY ('/SEC)	SOURCE
250/Barnes X-Bullet	90.0/RL-25	2,965	Barnes
225/Barnes XLC	98.0/IMR-7828	3,100	Barnes
250/Swift A-Frame	97.0/IMR-7828	3,114	Towsley

CASE PREPARATION: SHELL HOLDER (RCBS): 38

MAKE FROM: Factory .300 Ultra Mag cases. Expand neck with a tapered expander in full length resizing die. Fireform, trim, chamfer and deburr.

PHYSICAL DATA (INCHES):

CASE TYPE: **Rebated Bottleneck**

CASE LENGTH **A = 2.850**

HEAD DIAMETER **B = .551**

RIM DIAMETER **D = .534**

NECK DIAMETER **F = .386**

NECK LENGTH **H = .349**

SHOULDER LENGTH **K = .114**

BODY ANGLE (DEG'S/SIDE):

CASE CAPACITY **CC'S =**

LOADED LENGTH: **3.60**

BELT DIAMETER **C =**

RIM THICKNESS **E = .050**

SHOULDER DIAMETER **G = .525**

LENGTH TO SHOULDER **J = 2.387**

SHOULDER ANGLE (DEG'S/SIDE): **30**

PRIMER: **L/R Mag**

CASE CAPACITY (GR'S WATER): **116.5**

DIMENSIONAL DRAWING:

-NOT ACTUAL SIZE-
-DO NOT SCALE-

CARTRIDGE: .358 Winchester

OTHER NAMES: 8,8 x 51mm

DIA: .358

BALLISTEK NO: 358G

NAI NO: RXB 12233/4.287

DATA SOURCE: Hornady 3rd Pg282

HISTORICAL DATA: By Win. in 1955 for M70.

NOTES:

LOADING DATA:

BULLET WT./TYPE	POWDER WT./TYPE	VELOCITY ('/SEC)	SOURCE
250/RN	46.1/BL-C2	2200	Horn.

CASE PREPARATION: SHELLHOLDER (RCBS): 3

MAKE FROM: Factory .308 Win. Anneal case neck. Taper expand to .360" dia. F/L size in .358 die. Trim to length & chamfer.

PHYSICAL DATA (INCHES):

CASE TYPE: **Rimless Bottleneck**

CASE LENGTH **A = 2.015**

HEAD DIAMETER **B = .470**

RIM DIAMETER **D = .473**

NECK DIAMETER **F = .388**

NECK LENGTH **H = .365**

SHOULDER LENGTH **K = .092**

BODY ANGLE (DEG'S/SIDE): **.316**

CASE CAPACITY **CC'S = 3.73**

LOADED LENGTH: **2.78**

BELT DIAMETER **C = N/A**

RIM THICKNESS **E = .054**

SHOULDER DIAMETER **G = .455**

LENGTH TO SHOULDER **J = 1.558**

SHOULDER ANGLE (DEG'S/SIDE): **20.00**

PRIMER: **L/R**

CASE CAPACITY (GR'S WATER): **57.56**

DIMENSIONAL DRAWING:

-NOT ACTUAL SIZE-
-DO NOT SCALE-

CARTRIDGE: .360 #2 Nitro Express

OTHER NAMES:

DIA: .367

BALLISTEK NO: 367A

NAI NO: RMB 22322/5.547

DATA SOURCE: COTW 4th Pg235

HISTORICAL DATA: By Eley about 1905.

NOTES:

LOADING DATA:

BULLET WT./TYPE	POWDER WT./TYPE	VELOCITY ('/SEC)	SOURCE
320/RN	54.0/4320	2200	Barnes

CASE PREPARATION: SHELLHOLDER (RCBS): Spl.

MAKE FROM: .450 N.E. (BELL). Trim case to proper length, chamfer, anneal and F/L size in the .360 die.

PHYSICAL DATA (INCHES):

CASE TYPE: **Rimmed Bottleneck**

CASE LENGTH **A = 2.99**

HEAD DIAMETER **B = .539**

RIM DIAMETER **D = .631**

NECK DIAMETER **F = .393**

NECK LENGTH **H = .60**

SHOULDER LENGTH **K = .24**

BODY ANGLE (DEG'S/SIDE): **.455**

CASE CAPACITY **CC'S = 7.35**

LOADED LENGTH: **3.85**

BELT DIAMETER **C = N/A**

RIM THICKNESS **E = .063**

SHOULDER DIAMETER **G = .508**

LENGTH TO SHOULDER **J = 2.15**

SHOULDER ANGLE (DEG'S/SIDE): **13.47**

PRIMER: **L/R**

CASE CAPACITY (GR'S WATER): **113.56**

DIMENSIONAL DRAWING:

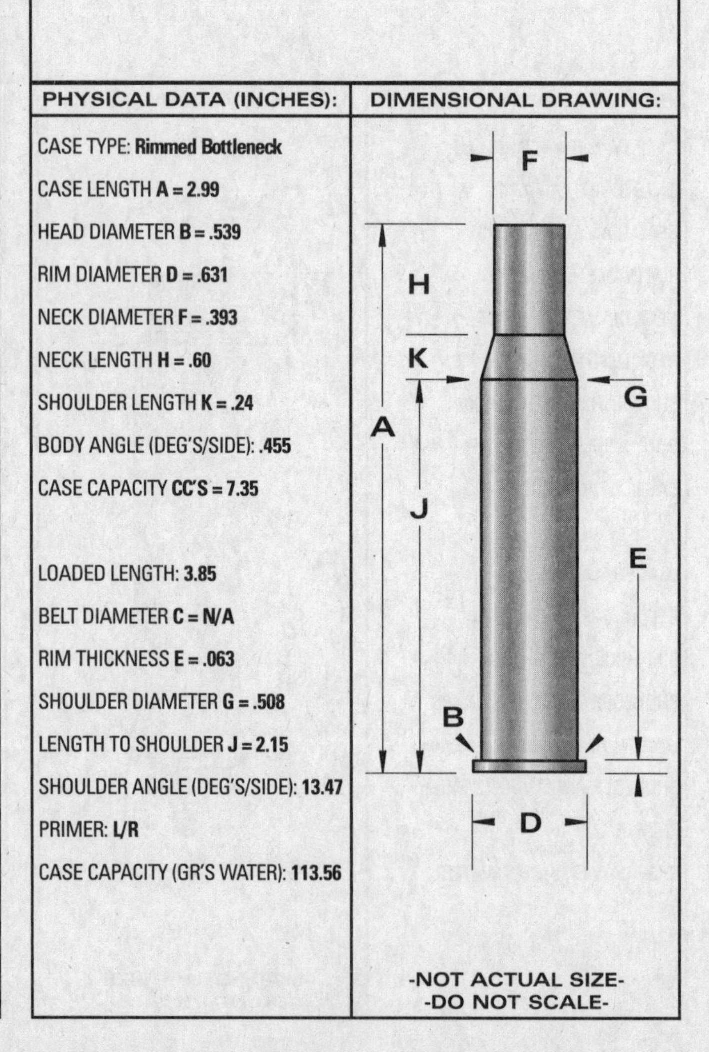

-NOT ACTUAL SIZE-
-DO NOT SCALE-

CARTRIDGE: .360 #5 Rook

OTHER NAMES:

DIA: .362

BALLISTEK NO: 362A

NAI NO: RMS 11115/2.763

DATA SOURCE: COTW 4th Pg234

HISTORICAL DATA: Black powder ctg. ca. 1875.

NOTES: Similar to .38 Colt Long (#357S).

LOADING DATA:

BULLET WT./TYPE	POWDER WT./TYPE	VELOCITY ('/SEC)	SOURCE
125/Lead	3.6/B'Eye	—	JJD

CASE PREPARATION: SHELLHOLDER (RCBS): 6

MAKE FROM: .357 Magnum. Cut the Mag. case to length, chamfer and F/L size. Good bullets are produced by drilling a "minie" in Lyman's #358430.

PHYSICAL DATA (INCHES):

CASE TYPE: **Rimmed Straight**

CASE LENGTH **A = 1.05**

HEAD DIAMETER **B = .380**

RIM DIAMETER **D = .432**

NECK DIAMETER **F = .375**

NECK LENGTH **H = N/A**

SHOULDER LENGTH **K = N/A**

BODY ANGLE (DEG'S/SIDE): **.143**

CASE CAPACITY **CC'S = 1.23**

LOADED LENGTH: **1.35**

BELT DIAMETER **C = N/A**

RIM THICKNESS **E = .05**

SHOULDER DIAMETER **G = N/A**

LENGTH TO SHOULDER **J = N/A**

SHOULDER ANGLE (DEG'S/SIDE): **N/A**

PRIMER: **S/R**

CASE CAPACITY (GR'S WATER): **18.94**

DIMENSIONAL DRAWING:

F
E
A
B
D

-NOT ACTUAL SIZE-
-DO NOT SCALE-

CARTRIDGE: .360 Nitro Express 2.25"

OTHER NAMES:

DIA: .357

BALLISTEK NO: 357N

NAI NO: RMS 21115/5.232

DATA SOURCE: COTW 4th Pg234

HISTORICAL DATA: Circa 1884 Kynoch ctg.

NOTES:

LOADING DATA:

BULLET WT./TYPE	POWDER WT./TYPE	VELOCITY ('/SEC)	SOURCE
250/Lead	26.0/4198	1700	Barnes

CASE PREPARATION: SHELLHOLDER (RCBS): 4

MAKE FROM: 9,3 x 72R. F/L size case with expander removed. Trim to length & chamfer. F/L size.

PHYSICAL DATA (INCHES):

CASE TYPE: **Rimmed Straight**

CASE LENGTH **A = 2.25**

HEAD DIAMETER **B = .430**

RIM DIAMETER **D = .480**

NECK DIAMETER **F = .384**

NECK LENGTH **H = N/A**

SHOULDER LENGTH **K = N/A**

BODY ANGLE (DEG'S/SIDE): **.598**

CASE CAPACITY **CC'S = 3.36**

LOADED LENGTH: **2.98**

BELT DIAMETER **C = N/A**

RIM THICKNESS **E = .048**

SHOULDER DIAMETER **G = N/A**

LENGTH TO SHOULDER **J = N/A**

SHOULDER ANGLE (DEG'S/SIDE): **N/A**

PRIMER: **L/R**

CASE CAPACITY (GR'S WATER): **51.88**

DIMENSIONAL DRAWING:

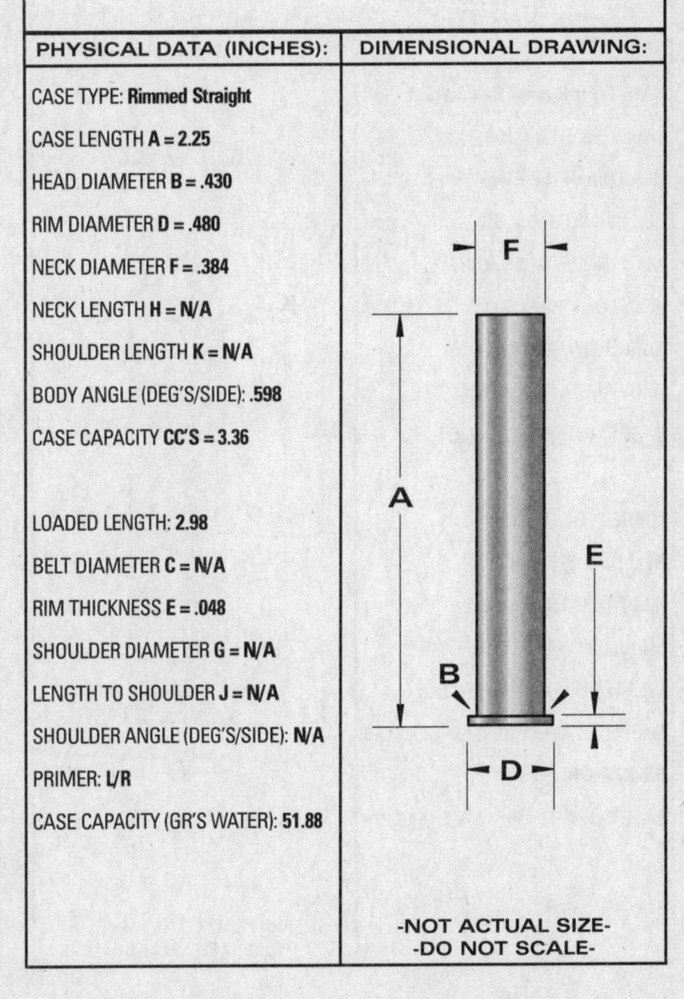

-NOT ACTUAL SIZE-
-DO NOT SCALE-

CARTRIDGE: .369 Purdey

OTHER NAMES:
.369 Nitro Express

DIA: .375

BALLISTEK NO: 375H

NAI NO: RMB 22342/4.954

DATA SOURCE: COTW 4th Pg236

HISTORICAL DATA: From about 1922.

NOTES: .375" is correct diameter.

LOADING DATA:

BULLET WT./TYPE	POWDER WT./TYPE	VELOCITY ('/SEC)	SOURCE
270/Spire	60.0/3031	2200	Nonte

CASE PREPARATION: SHELLHOLDER (RCBS): Spl.

MAKE FROM: .450 N.E. (BELL). Cut case to 2.7". Anneal and form in form die. F/L size, trim to length and chamfer. Fireform in chamber.

PHYSICAL DATA (INCHES):

CASE TYPE: **Rimmed Bottleneck**
CASE LENGTH **A = 2.69**
HEAD DIAMETER **B = .543**
RIM DIAMETER **D = .616**
NECK DIAMETER **F = .398**
NECK LENGTH **H = .300**
SHOULDER LENGTH **K = .190**
BODY ANGLE (DEG'S/SIDE): **.97**
CASE CAPACITY **CC'S = 6.63**

LOADED LENGTH: **3.59**
BELT DIAMETER **C = N/A**
RIM THICKNESS **E = .052**
SHOULDER DIAMETER **G = .475**
LENGTH TO SHOULDER **J = 2.20**
SHOULDER ANGLE (DEG'S/SIDE): **11.45**
PRIMER: **L/R**
CASE CAPACITY (GR'S WATER): **102.37**

DIMENSIONAL DRAWING:

F H K G A J E B D

-NOT ACTUAL SIZE-
-DO NOT SCALE-

CARTRIDGE: .375 Barnes Supreme

OTHER NAMES:

DIA: .375

BALLISTEK NO: 375AE

NAI NO: BEN 12243/5.546

DATA SOURCE: Ackley Vol.1 Pg482

HISTORICAL DATA:

NOTES:

LOADING DATA:

BULLET WT./TYPE	POWDER WT./TYPE	VELOCITY ('/SEC)	SOURCE
350/Spire	88.0/4831	2650	Ackley

CASE PREPARATION: SHELLHOLDER (RCBS): 4

MAKE FROM: .375 H&H Mag. Fireform factory ammo in the Barnes chamber.

PHYSICAL DATA (INCHES):

CASE TYPE: **Belted Bottleneck**
CASE LENGTH **A = 2.84**
HEAD DIAMETER **B = .512**
RIM DIAMETER **D = .532**
NECK DIAMETER **F = .404**
NECK LENGTH **H = .368**
SHOULDER LENGTH **K = .092**
BODY ANGLE (DEG'S/SIDE): **.079**
CASE CAPACITY **CC'S = 6.84**

LOADED LENGTH: **3.88**
BELT DIAMETER **C = .532**
RIM THICKNESS **E = .05**
SHOULDER DIAMETER **G = .506**
LENGTH TO SHOULDER **J = 2.38**
SHOULDER ANGLE (DEG'S/SIDE): **29.00**
PRIMER: **L/R Mag**
CASE CAPACITY (GR'S WATER): **105.56**

DIMENSIONAL DRAWING:

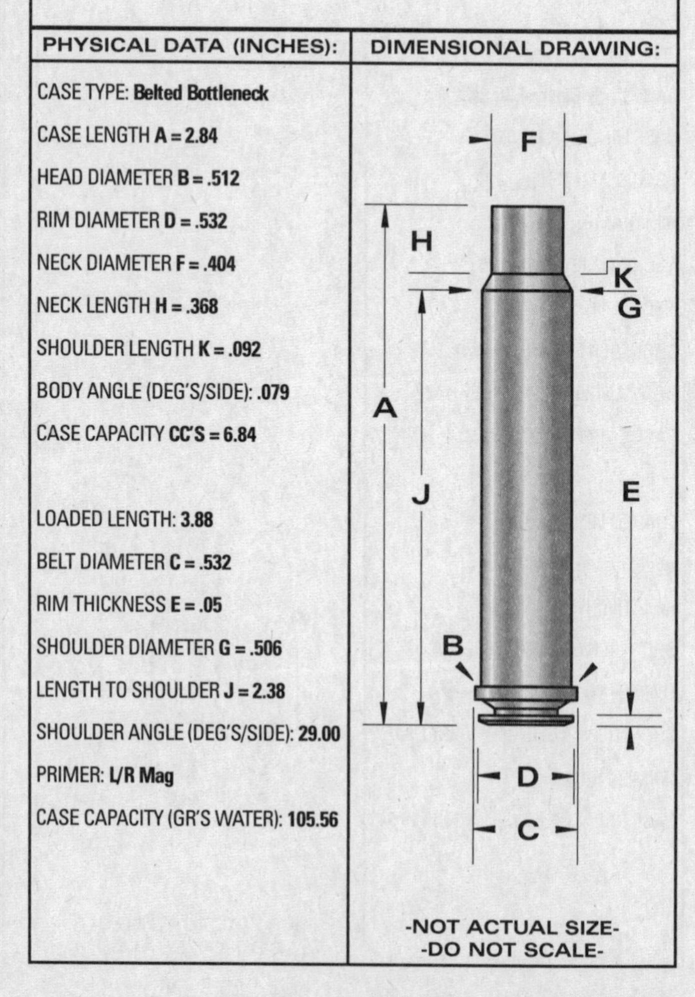

-NOT ACTUAL SIZE-
-DO NOT SCALE-

CARTRIDGE: .375 Dakota

OTHER NAMES:	DIA: .375
	BALLISTEK NO:
	NAI NO:

DATA SOURCE: Dakota Arms.
HISTORICAL DATA: Proprietary cartridge from Dakota Arms.
NOTES: Based on the .404 Jeffery case with the rim enlarged slightly to eliminate the rebated feature. The .375 Dakota is equal in all respects to the .375 H&H and is suitable for use in a standard length action.

LOADING DATA:

BULLET WT./TYPE	POWDER WT./TYPE	VELOCITY ('/SEC)	SOURCE
270	85.0/IMR-4350	2,895	Dakota Arms
300	78.0/H-4350	2,648	Dakota Arms

CASE PREPARATION: SHELL HOLDER (RCBS): 41

MAKE FROM: Factory or make by reforming and shortening .404 Jeffery.

PHYSICAL DATA (INCHES):

CASE TYPE: **Rimless Bottleneck**
CASE LENGTH **A = 2.570**
HEAD DIAMETER **B = .545**
RIM DIAMETER **D = .545**
NECK DIAMETER **F = .402**
NECK LENGTH **H = .400**
SHOULDER LENGTH **K = .109**
BODY ANGLE (DEG'S/SIDE):
CASE CAPACITY **CC'S =**
LOADED LENGTH: **3.320**
BELT DIAMETER **C =**
RIM THICKNESS **E = .050**
SHOULDER DIAMETER **G = .529**
LENGTH TO SHOULDER **J = 2.061**
SHOULDER ANGLE (DEG'S/SIDE): **30**
PRIMER: **L/R Mag**
CASE CAPACITY (GR'S WATER):

DIMENSIONAL DRAWING:

F
H
K
G
A
J
E
B
D

-NOT ACTUAL SIZE-
-DO NOT SCALE-

CARTRIDGE: .375 Durham Magnum

OTHER NAMES:	DIA: .375
	BALLISTEK NO: 375AH
	NAI NO: BEN 12244/5.039

DATA SOURCE: Ackley Vol.1 Pg483
HISTORICAL DATA:
NOTES:

LOADING DATA:

BULLET WT./TYPE	POWDER WT./TYPE	VELOCITY ('/SEC)	SOURCE
270/RN	80.0/H380	2725	Ackley

CASE PREPARATION: SHELLHOLDER (RCBS): 4

MAKE FROM: .338 Win. Mag. Taper expand the .338's neck to .380" dia. Cut to 2.6" and F/L size. Fireform in chamber. You can also taper expand the .375 H&H to .380" dia, trim to length and F/L size.

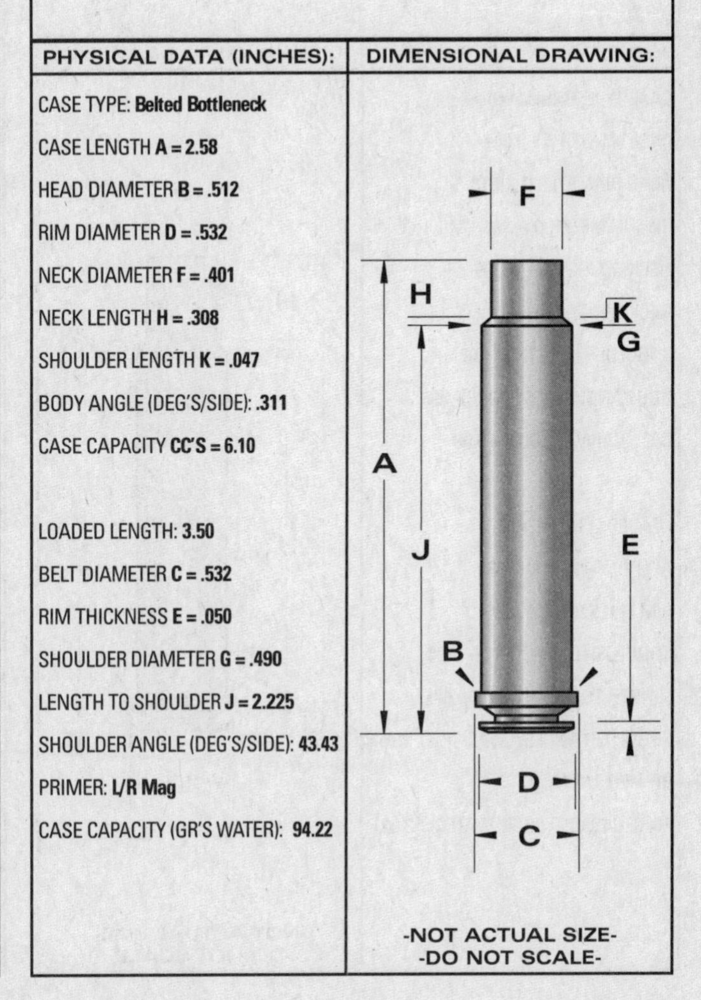

PHYSICAL DATA (INCHES):

CASE TYPE: **Belted Bottleneck**
CASE LENGTH **A = 2.58**
HEAD DIAMETER **B = .512**
RIM DIAMETER **D = .532**
NECK DIAMETER **F = .401**
NECK LENGTH **H = .308**
SHOULDER LENGTH **K = .047**
BODY ANGLE (DEG'S/SIDE): **.311**
CASE CAPACITY **CC'S = 6.10**
LOADED LENGTH: **3.50**
BELT DIAMETER **C = .532**
RIM THICKNESS **E = .050**
SHOULDER DIAMETER **G = .490**
LENGTH TO SHOULDER **J = 2.225**
SHOULDER ANGLE (DEG'S/SIDE): **43.43**
PRIMER: **L/R Mag**
CASE CAPACITY (GR'S WATER): **94.22**

DIMENSIONAL DRAWING:

-NOT ACTUAL SIZE-
-DO NOT SCALE-

CARTRIDGE: .375 Epstein Magnum

OTHER NAMES:	
	DIA: .375
	BALLISTEK NO: 375B
	NAI NO: BEN 22233/5.078

DATA SOURCE: Rifle #35 Pg16

HISTORICAL DATA: By M. Epstein about 1974.

NOTES:

LOADING DATA:

BULLET WT./TYPE	POWDER WT./TYPE	VELOCITY ('/SEC)	SOURCE
300/—	77.0/4350	2800	Epstein

CASE PREPARATION: SHELLHOLDER (RCBS): 4

MAKE FROM: .375 H&H Mag. Taper expand case neck to .390" dia. (to form a fireforming shoulder). F/L size in the Epstein die with the expander removed. Trim the case to 2.6" and chamfer. I.D. neck ream and F/L size again. Fireform in the chamber.

PHYSICAL DATA (INCHES):

CASE TYPE: **Belted Bottleneck**

CASE LENGTH **A = 2.605**

HEAD DIAMETER **B = .513**

RIM DIAMETER **D = .532**

NECK DIAMETER **F = .399**

NECK LENGTH **H = 460**

SHOULDER LENGTH **K = .094**

BODY ANGLE (DEG'S/SIDE): **.356**

CASE CAPACITY **CC'S = 6.08**

LOADED LENGTH: **A/R**

BELT DIAMETER **C = .532**

RIM THICKNESS **E = .044**

SHOULDER DIAMETER **G = .490**

LENGTH TO SHOULDER **J = 2.05**

SHOULDER ANGLE (DEG'S/SIDE): **25.59**

PRIMER: **L/R Mag**

CASE CAPACITY (GR'S WATER): **93.92**

DIMENSIONAL DRAWING:

F H K G A J E B D C

-NOT ACTUAL SIZE-
-DO NOT SCALE-

CARTRIDGE: .375 Express (Waters)

OTHER NAMES:	
	DIA: .375
	BALLISTEK NO: 375AL
	NAI NO: RMB 12243/4.775

DATA SOURCE: Rifle #66 Pg29

HISTORICAL DATA: By K. Waters about 1979.

NOTES:

LOADING DATA:

BULLET WT./TYPE	POWDER WT./TYPE	VELOCITY ('/SEC)	SOURCE
235/—	50.0/4895	2286	Waters

CASE PREPARATION: SHELLHOLDER (RCBS): 28

MAKE FROM: .444 Marlin. F/L size the Marlin case in the Waters die. Square case mouth, chamfer and fireform in the chamber.

PHYSICAL DATA (INCHES):

CASE TYPE: **Rimmed Bottleneck**

CASE LENGTH **A = 2.23**

HEAD DIAMETER **B = .467**

RIM DIAMETER **D = .511**

NECK DIAMETER **F = .399**

NECK LENGTH **H = .325**

SHOULDER LENGTH **K = .065**

BODY ANGLE (DEG'S/SIDE): **.192**

CASE CAPACITY **CC'S = 4.44**

LOADED LENGTH: **A/R**

BELT DIAMETER **C = N/A**

RIM THICKNESS **E = .06**

SHOULDER DIAMETER **G = .456**

LENGTH TO SHOULDER **J = 1.84**

SHOULDER ANGLE (DEG'S/SIDE): **23.67**

PRIMER: **L/R**

CASE CAPACITY (GR'S WATER): **68.55**

DIMENSIONAL DRAWING:

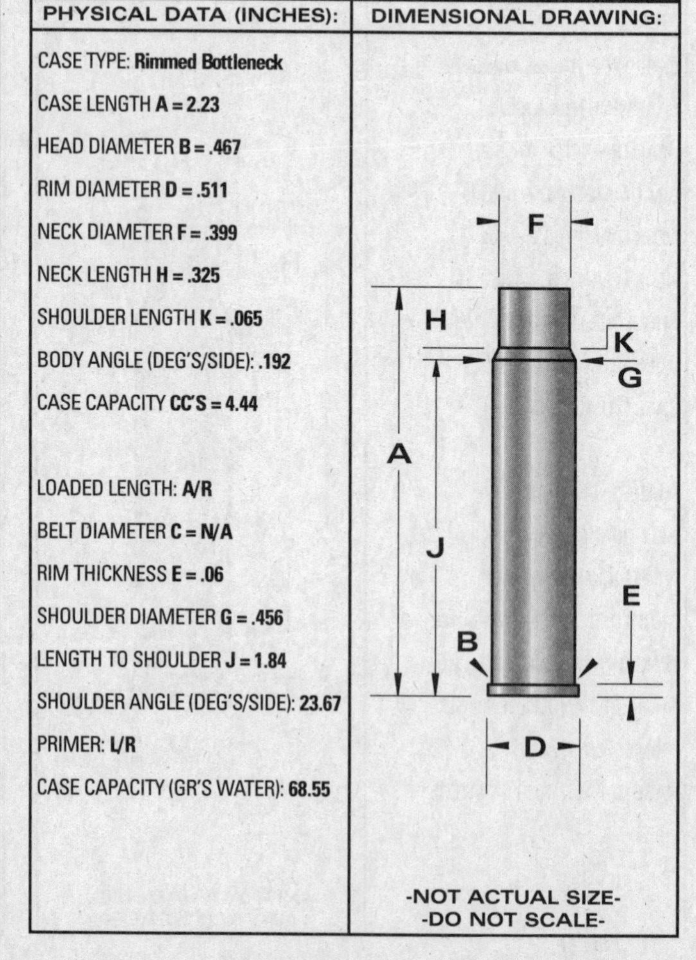

-NOT ACTUAL SIZE-
-DO NOT SCALE-

CARTRIDGE: .375 Express

OTHER NAMES:
.375 Payne Express

DIA: .375

BALLISTEK NO: 375AJ

NAI NO: RXB 22343/4.069

DATA SOURCE: Ackley Vol1 Pg482

HISTORICAL DATA: By R. Payne.

NOTES:

LOADING DATA:

BULLET WT./TYPE	POWDER WT./TYPE	VELOCITY ('/SEC)	SOURCE
300/RN	80.0/4350	2675	Ackley

CASE PREPARATION: SHELLHOLDER (RCBS): 13

MAKE FROM: .348 Winchester. Anneal case neck. Taper expand to .380" dia. F/L size in the Express die. Square case mouth. Turn rim flush with body and cut new extractor groove. Fireform in chamber.

PHYSICAL DATA (INCHES):

CASE TYPE: **Rimless Bottleneck**

CASE LENGTH **A = 2.25**

HEAD DIAMETER **B = .553**

RIM DIAMETER **D = .553**

NECK DIAMETER **F = .403**

NECK LENGTH **H = .335**

SHOULDER LENGTH **K = .115**

BODY ANGLE (DEG'S/SIDE): **.519**

CASE CAPACITY **CC'S = 5.82**

LOADED LENGTH: **3.05**

BELT DIAMETER **C = N/A**

RIM THICKNESS **E = .060**

SHOULDER DIAMETER **G = .524**

LENGTH TO SHOULDER **J = 1.80**

SHOULDER ANGLE (DEG'S/SIDE): **27.75**

PRIMER: **L/R**

CASE CAPACITY (GR'S WATER): **89.90**

DIMENSIONAL DRAWING:

F
H
K
G
A
J
E
B
D

-NOT ACTUAL SIZE-
-DO NOT SCALE-

CARTRIDGE: .375 Flanged Nitro Express

OTHER NAMES:

DIA: .375

BALLISTEK NO: 375V

NAI NO: RMS 21115/5.482

DATA SOURCE: COTW 4th Pg236

HISTORICAL DATA: Circa 1899.

NOTES:

LOADING DATA:

BULLET WT./TYPE	POWDER WT./TYPE	VELOCITY ('/SEC)	SOURCE
285/RN	46.0/3031	1890	Barnes

CASE PREPARATION: SHELLHOLDER (RCBS): 26

MAKE FROM: .405 Basic (BELL). Turn rim to .523" dia. and back chamfer. Thin to .061" thick. Cut case to 2.6". Anneal. F/L size, trim to length and chamfer. Fireform in chamber.

PHYSICAL DATA (INCHES):

CASE TYPE: **Rimmed Straight**

CASE LENGTH **A = 2.50**

HEAD DIAMETER **B = .456**

RIM DIAMETER **D = .523**

NECK DIAMETER **F = .397**

NECK LENGTH **H = N/A**

SHOULDER LENGTH **K = N/A**

BODY ANGLE (DEG'S/SIDE): **.692**

CASE CAPACITY **CC'S = 4.38**

LOADED LENGTH: **3.10**

BELT DIAMETER **C = N/A**

RIM THICKNESS **E = .061**

SHOULDER DIAMETER **G = N/A**

LENGTH TO SHOULDER **J = N/A**

SHOULDER ANGLE (DEG'S/SIDE): **N/A**

PRIMER: **L/R**

CASE CAPACITY (GR'S WATER): **67.55**

DIMENSIONAL DRAWING:

F
A
E
B
D

-NOT ACTUAL SIZE-
-DO NOT SCALE-

CARTRIDGE: .375 Flanged Magnum Nitro Express

OTHER NAMES:

DIA: .375

BALLISTEK NO: 375U

NAI NO: RMB 22232/5.753

DATA SOURCE: COTW 4th Pg236

HISTORICAL DATA: By H&H in 1912.

NOTES: Rimmed version of .375 H&H Mag. (#375A).

LOADING DATA:

BULLET WT./TYPE	POWDER WT./TYPE	VELOCITY ('/SEC)	SOURCE
300/RN	80.0/4350	2500	Barnes

CASE PREPARATION: SHELLHOLDER (RCBS): 14

MAKE FROM: .375 flanged (BELL). Trim case to length, chamfer, F/L size and square case mouth.

PHYSICAL DATA (INCHES):

CASE TYPE: **Rimmed Bottleneck**

CASE LENGTH **A = 2.94**

HEAD DIAMETER **B = .511**

RIM DIAMETER **D = .572**

NECK DIAMETER **F = .404**

NECK LENGTH **H = .444**

SHOULDER LENGTH **K = .096**

BODY ANGLE (DEG'S/SIDE): **.794**

CASE CAPACITY **CC'S = 6.30**

LOADED LENGTH: **3.80**

BELT DIAMETER **C = N/A**

RIM THICKNESS **E = .05**

SHOULDER DIAMETER **G = .450**

LENGTH TO SHOULDER **J = 2.40**

SHOULDER ANGLE (DEG'S/SIDE): **13.47**

PRIMER: **L/R**

CASE CAPACITY (GR'S WATER): **97.21**

DIMENSIONAL DRAWING:

F H K G A J E B D

-NOT ACTUAL SIZE-
-DO NOT SCALE-

CARTRIDGE: .375 H&H Magnum

OTHER NAMES:
.375 Holland & Holland
.375 Holland's Belted

DIA: .375

BALLISTEK NO: 375A

NAI NO: BEN 22241/5.555

DATA SOURCE: Hornady Manual 3rd Pg291

HISTORICAL DATA: By Holland & Holland in 1912.

NOTES:

LOADING DATA:

BULLET WT./TYPE	POWDER WT./TYPE	VELOCITY ('/SEC)	SOURCE
270/Spire	69.0/IMR4064	2550	JJD

CASE PREPARATION: SHELLHOLDER (RCBS): 4

MAKE FROM: Factory or .300 H&H. Anneal case neck. Taper expand neck (in several steps!) to about .380" I.D. F/L size and fireform for final shape.

PHYSICAL DATA (INCHES):

CASE TYPE: **Belted Bottleneck**

CASE LENGTH **A = 2.850**

HEAD DIAMETER **B = .513**

RIM DIAMETER **D = .532**

NECK DIAMETER **F = .405**

NECK LENGTH **H = .300**

SHOULDER LENGTH **K = .197**

BODY ANGLE (DEG'S/SIDE): **.705**

CASE CAPACITY **CC'S = 6.24**

LOADED LENGTH: **3.60**

BELT DIAMETER **C = .532**

RIM THICKNESS **E = .049**

SHOULDER DIAMETER **G = .460**

LENGTH TO SHOULDER **J = 2.353**

SHOULDER ANGLE (DEG'S/SIDE): **7.95**

PRIMER: **L/R Mag**

CASE CAPACITY (GR'S WATER): **96.37**

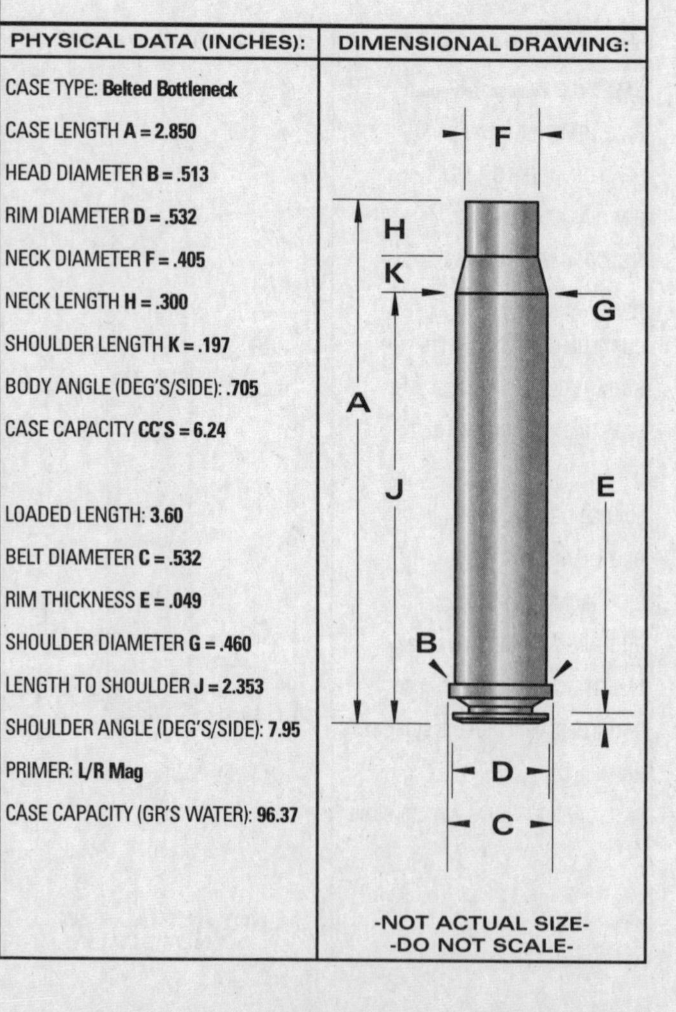

CARTRIDGE: .375 H&H Improved

OTHER NAMES:	DIA: .375 BALLISTEK NO: 375K NAI NO: BEN 12244/5.516

DATA SOURCE: Ackley Vol1 Pg487

HISTORICAL DATA:

NOTES:

LOADING DATA:

BULLET WT./TYPE	POWDER WT./TYPE	VELOCITY ('/SEC)	SOURCE
300/—	78.0/4064	2630	Ackley

CASE PREPARATION: SHELLHOLDER (RCBS): 4

MAKE FROM: .375 H&H Mag. Fire factory ammo in "improved" chamber. .375 H&H brass can be taper expanded to .385" dia. and F/L sized (fireforming is still necessary to fill-out the shoulders.)

PHYSICAL DATA (INCHES):

CASE TYPE: **Belted Bottleneck**
CASE LENGTH **A = 2.83**
HEAD DIAMETER **B = .513**
RIM DIAMETER **D = .532**
NECK DIAMETER **F = .404**
NECK LENGTH **H = .310**
SHOULDER LENGTH **K = .055**
BODY ANGLE (DEG'S/SIDE): **.189**
CASE CAPACITY **CC'S = 6.89**

LOADED LENGTH: **3.75**
BELT DIAMETER **C = .532**
RIM THICKNESS **E = .05**
SHOULDER DIAMETER **G = .498**
LENGTH TO SHOULDER **J = 2.465**
SHOULDER ANGLE (DEG'S/SIDE): **40.51**
PRIMER: **L/R Mag**
CASE CAPACITY (GR'S WATER): **104.77**

DIMENSIONAL DRAWING:

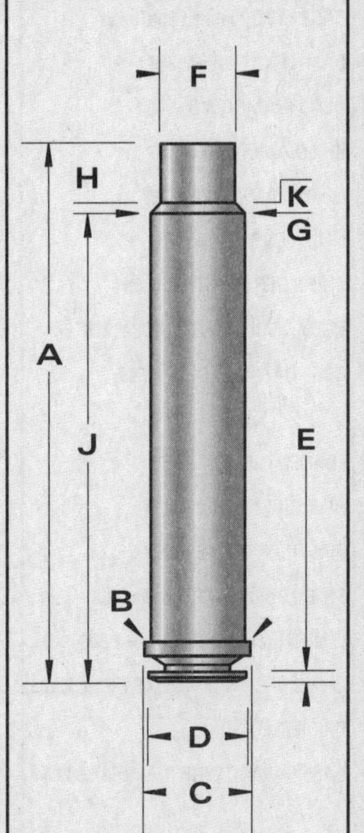

-NOT ACTUAL SIZE-
-DO NOT SCALE-

CARTRIDGE: 375 JRS

OTHER NAMES:	DIA: .375 BALLISTEK NO: NAI NO:

DATA SOURCE: Howell "Custom Cartridges"

HISTORICAL DATA: This cartridge was designed by Jon R. Sundra. Based upon the 8mm Remington Magnum case necked up to .375 with no other changes

NOTES:

LOADING DATA:

BULLET WT./TYPE	POWDER WT./TYPE	VELOCITY ('/SEC)	SOURCE
270	85.0/IMR-4350	2750	Sundra
300	83.0/IMR-4350	2700	Sundra

CASE PREPARATION: SHELL HOLDER (RCBS): 04

MAKE FROM: 8mm Remington Magnum case necked up to .375 with no other changes

PHYSICAL DATA (INCHES):

CASE TYPE: **Belted Bottleneck**
CASE LENGTH **A = 2.850**
HEAD DIAMETER **B = .513**
RIM DIAMETER **D = .532**
NECK DIAMETER **F = .402**
NECK LENGTH **H = .363**
SHOULDER LENGTH **K = .089**
BODY ANGLE (DEG'S/SIDE): **25°**
CASE CAPACITY **CC'S = N/A**

LOADED LENGTH: **3.60**
BELT DIAMETER **C = .532**
RIM THICKNESS **E = .050**
SHOULDER DIAMETER **G = .485**
LENGTH TO SHOULDER **J = 2.398**
SHOULDER ANGLE (DEG'S/SIDE):**N/A**
PRIMER: **L**
CASE CAPACITY (GR'S WATER): **N/A**

DIMENSIONAL DRAWING:

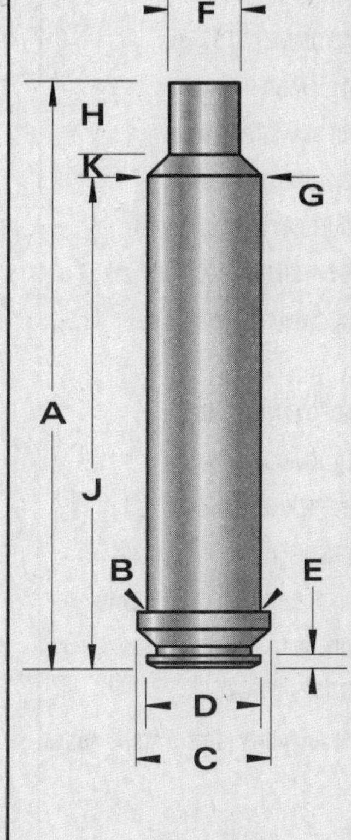

-NOT ACTUAL SIZE-
-DO NOT SCALE-

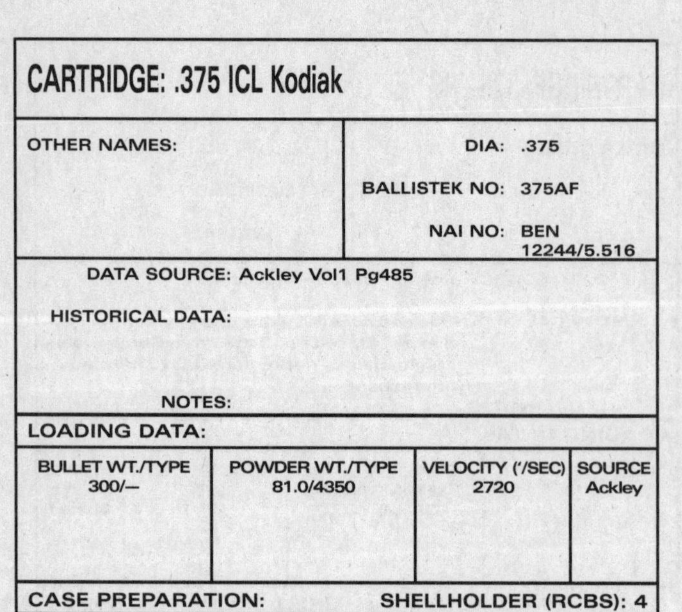

CARTRIDGE: .375 ICL Kodiak

OTHER NAMES:

DIA: .375

BALLISTEK NO: 375AF

NAI NO: BEN 12244/5.516

DATA SOURCE: Ackley Vol1 Pg485

HISTORICAL DATA:

NOTES:

LOADING DATA:

BULLET WT./TYPE	POWDER WT./TYPE	VELOCITY ('/SEC)	SOURCE
300/—	81.0/4350	2720	Ackley

CASE PREPARATION: SHELLHOLDER (RCBS): 4

MAKE FROM: .375 H&H Mag. Fire factory ammo in ICL chamber.

PHYSICAL DATA (INCHES):

CASE TYPE: **Belted Bottleneck**
CASE LENGTH **A = 2.83**
HEAD DIAMETER **B = .513**
RIM DIAMETER **D = .530**
NECK DIAMETER **F = .405**
NECK LENGTH **H = .350**
SHOULDER LENGTH **K = .065**
BODY ANGLE (DEG'S/SIDE): **.232**
CASE CAPACITY **CC'S = 6.66**

LOADED LENGTH: **A/R**
BELT DIAMETER **C = .530**
RIM THICKNESS **E = .05**
SHOULDER DIAMETER **G = .495**
LENGTH TO SHOULDER **J = 2.415**
SHOULDER ANGLE (DEG'S/SIDE): **34.69**
PRIMER: **L/R Mag**
CASE CAPACITY (GR'S WATER): **102.90**

DIMENSIONAL DRAWING:

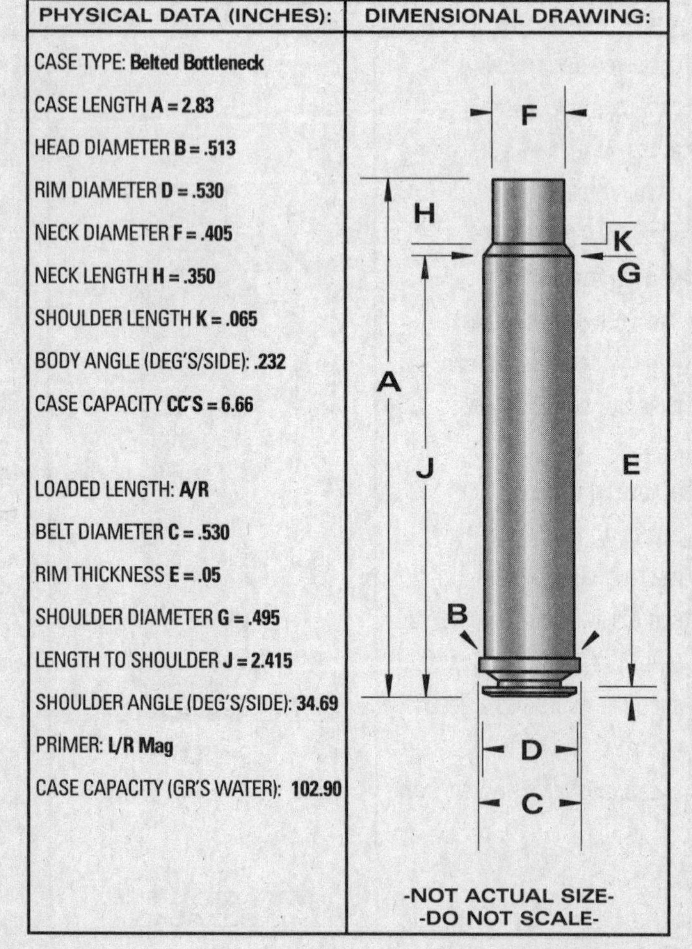

-NOT ACTUAL SIZE-
-DO NOT SCALE-

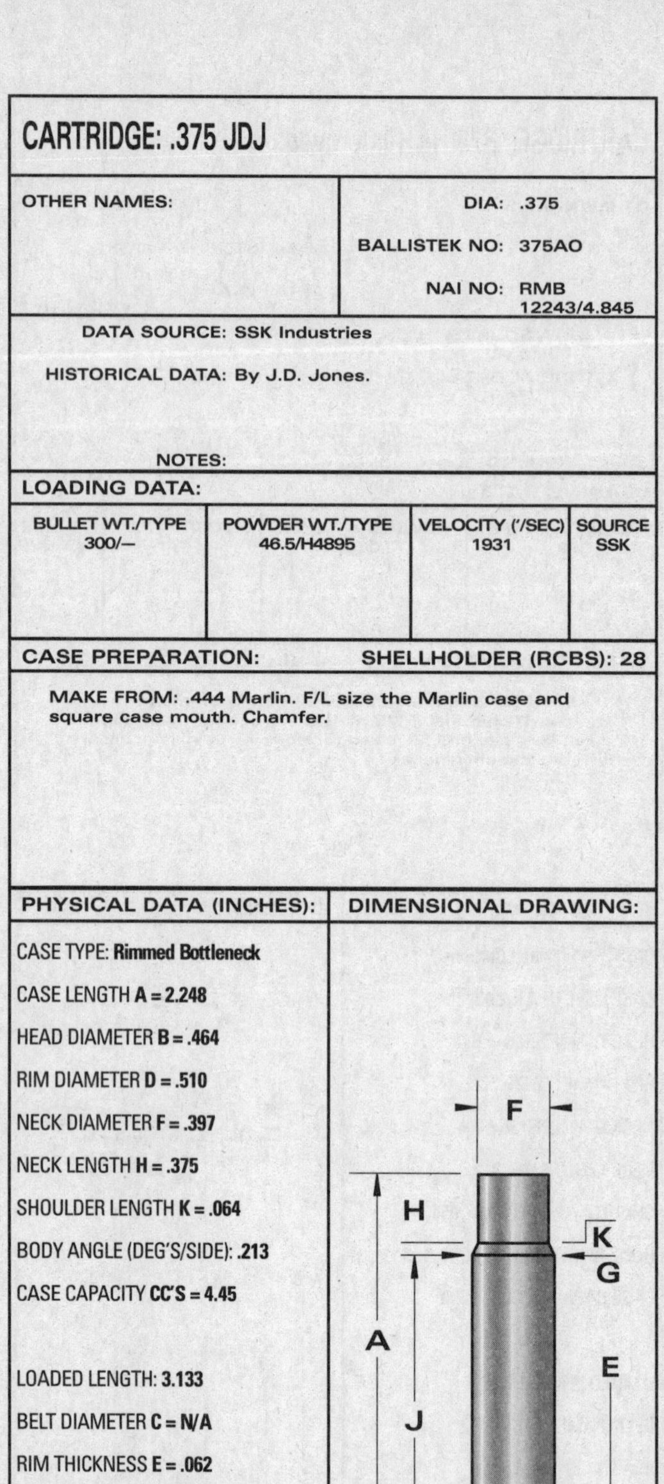

CARTRIDGE: .375 JDJ

OTHER NAMES:

DIA: .375

BALLISTEK NO: 375AO

NAI NO: RMB 12243/4.845

DATA SOURCE: SSK Industries

HISTORICAL DATA: By J.D. Jones.

NOTES:

LOADING DATA:

BULLET WT./TYPE	POWDER WT./TYPE	VELOCITY ('/SEC)	SOURCE
300/—	46.5/H4895	1931	SSK

CASE PREPARATION: SHELLHOLDER (RCBS): 28

MAKE FROM: .444 Marlin. F/L size the Marlin case and square case mouth. Chamfer.

PHYSICAL DATA (INCHES):

CASE TYPE: **Rimmed Bottleneck**
CASE LENGTH **A = 2.248**
HEAD DIAMETER **B = .464**
RIM DIAMETER **D = .510**
NECK DIAMETER **F = .397**
NECK LENGTH **H = .375**
SHOULDER LENGTH **K = .064**
BODY ANGLE (DEG'S/SIDE): **.213**
CASE CAPACITY **CC'S = 4.45**

LOADED LENGTH: **3.133**
BELT DIAMETER **C = N/A**
RIM THICKNESS **E = .062**
SHOULDER DIAMETER **G = .452**
LENGTH TO SHOULDER **J = 1.809**
SHOULDER ANGLE (DEG'S/SIDE): **23.25**
PRIMER: **L/R**
CASE CAPACITY (GR'S WATER): **68.66**

DIMENSIONAL DRAWING:

-NOT ACTUAL SIZE-
-DO NOT SCALE-

CARTRIDGE: .375 Jurras

OTHER NAMES:

DIA: .375

BALLISTEK NO: 375S

NAI NO: RMB 24323/3.605

DATA SOURCE: NAI/Ballistek

HISTORICAL DATA: Jurras/Contender ctg.

NOTES:

LOADING DATA:

BULLET WT./TYPE	POWDER WT./TYPE	VELOCITY ('/SEC)	SOURCE
200/RN	41.3/IMR4227	—	JJD

CASE PREPARATION: SHELLHOLDER (RCBS): 31

MAKE FROM: .500 N.E. (BELL). Form die set required (three dies). Form through dies #1, #2, & #3. Trim case to length and chamfer. F/L size.

PHYSICAL DATA (INCHES):

CASE TYPE: **Rimmed Bottleneck**

CASE LENGTH **A = 2.03**

HEAD DIAMETER **B = .563**

RIM DIAMETER **D = .648**

NECK DIAMETER **F = .407**

NECK LENGTH **H = .630**

SHOULDER LENGTH **K = .150**

BODY ANGLE (DEG'S/SIDE): **.764**

CASE CAPACITY **CC'S = 4.76**

LOADED LENGTH: **A/R**

BELT DIAMETER **C = N/A**

RIM THICKNESS **E = .054**

SHOULDER DIAMETER **G = .535**

LENGTH TO SHOULDER **J = 1.25**

SHOULDER ANGLE (DEG'S/SIDE): **23.10**

PRIMER: **L/R**

CASE CAPACITY (GR'S WATER): **73.53**

DIMENSIONAL DRAWING:

F
H
K
G
A
J
E
B
D

-NOT ACTUAL SIZE-
-DO NOT SCALE-

CARTRIDGE: .375 Mashburn Magnum Long

OTHER NAMES:

DIA: .375

BALLISTEK NO: 375AG

NAI NO: BEN 12243/5.507

DATA SOURCE: Ackley Vol1 Pg485

HISTORICAL DATA:

NOTES:

LOADING DATA:

BULLET WT./TYPE	POWDER WT./TYPE	VELOCITY ('/SEC)	SOURCE
270/—	81.0/4064	2825	Ackley

CASE PREPARATION: SHELLHOLDER (RCBS): 4

MAKE FROM: .375 H&H Mag. Fire factory ammo in Mashburn chamber.

PHYSICAL DATA (INCHES):

CASE TYPE: **Belted Bottleneck**

CASE LENGTH **A = 2.825**

HEAD DIAMETER **B = .513**

RIM DIAMETER **D = .532**

NECK DIAMETER **F = .404**

NECK LENGTH **H = .350**

SHOULDER LENGTH **K = .075**

BODY ANGLE (DEG'S/SIDE): **.325**

CASE CAPACITY **CC'S = 6.59**

LOADED LENGTH: **3.54**

BELT DIAMETER **C = .532**

RIM THICKNESS **E = .05**

SHOULDER DIAMETER **G = .488**

LENGTH TO SHOULDER **J = 2.40**

SHOULDER ANGLE (DEG'S/SIDE): **29.25**

PRIMER: **L/R Mag**

CASE CAPACITY (GR'S WATER): **101.80**

DIMENSIONAL DRAWING:

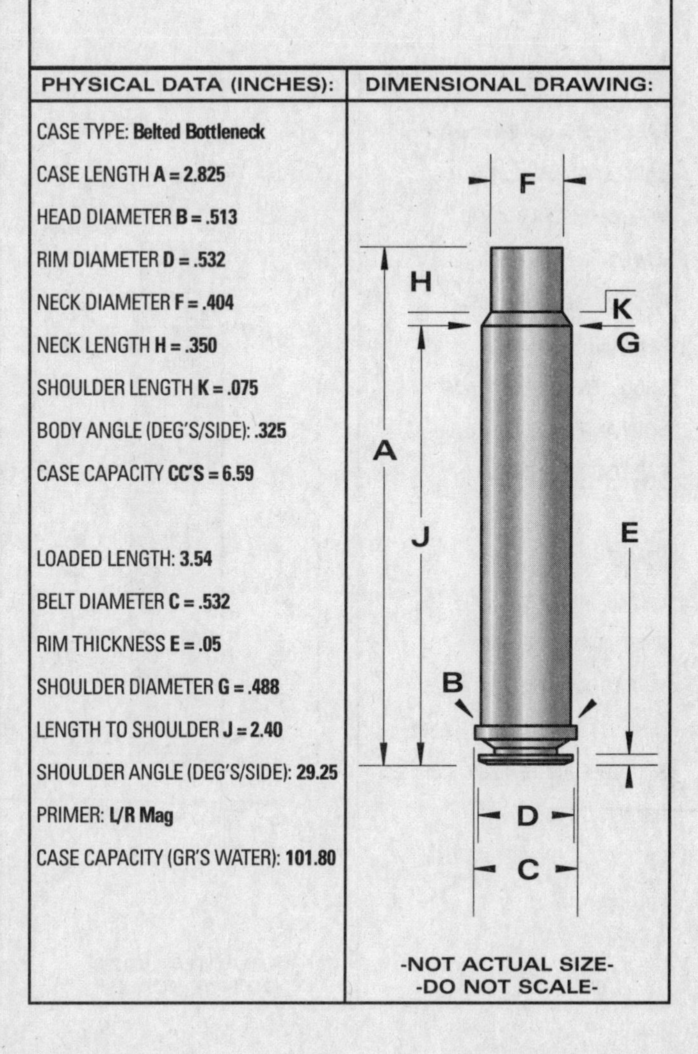

-NOT ACTUAL SIZE-
-DO NOT SCALE-

CARTRIDGE: .375 Remington Ultra Mag

OTHER NAMES:

DIA: .375

BALLISTEK NO:

NAI NO:

DATA SOURCE: Most current loading manuals.

HISTORICAL DATA: Introduced by Remington in 2001. This is the largest of the Ultra Mag family of cartridgees at the time of this writing. The case is made from the .404 Jeffery which was necked down and the body and shoulder blown out.

LOADING DATA:

BULLET WT./TYPE	POWDER WT./TYPE	VELOCITY ('/SEC)	SOURCE
300/Hornady BTSP	94.8/H-4831	2,700	Horn. #6
270/Factory		2,900	Rem

CASE PREPARATION: SHELL HOLDER (RCBS): 38

MAKE FROM: Factory or neck up .300 Remington Ultra Mag. Can be made from .404 Jeffery.

PHYSICAL DATA (INCHES):

CASE TYPE: **Rebated Bottleneck**

CASE LENGTH **A = 2.850**

HEAD DIAMETER **B = .550**

RIM DIAMETER **D = .534**

NECK DIAMETER **F = .405**

NECK LENGTH **H = .359**

SHOULDER LENGTH **K = .104**

BODY ANGLE (DEG'S/SIDE):

CASE CAPACITY **CC'S =**

LOADED LENGTH: **3.600**

BELT DIAMETER **C =**

RIM THICKNESS **E = .050**

SHOULDER DIAMETER **G = .525**

LENGTH TO SHOULDER **J = 2.387**

SHOULDER ANGLE (DEG'S/SIDE): **30**

PRIMER: **L/R Mag**

CASE CAPACITY (GR'S WATER):

DIMENSIONAL DRAWING:

F H K G A J E B D

-NOT ACTUAL SIZE-
-DO NOT SCALE-

CARTRIDGE: .375 Shannon

OTHER NAMES:

DIA: .375

BALLISTEK NO: 375M

NAI NO: RXB 24243/2.830

DATA SOURCE: Ackley Vol2 Pg211

HISTORICAL DATA: By J. Shannon.

NOTES:

LOADING DATA:

BULLET WT./TYPE	POWDER WT./TYPE	VELOCITY ('/SEC)	SOURCE
200/RN	22.0/2400	—	Shannon/ Ackley

CASE PREPARATION: SHELLHOLDER (RCBS): 3

MAKE FROM: .30-06 Spgf. Set back shoulder in Shannon form die. Cut to 1.35". I.D. neck ream. Trim to length and F/L size. Chamfer.

PHYSICAL DATA (INCHES):

CASE TYPE: **Rimless Bottleneck**

CASE LENGTH **A = 1.315**

HEAD DIAMETER **B = .470**

RIM DIAMETER **D = .473**

NECK DIAMETER **F = .419**

NECK LENGTH **H = .25**

SHOULDER LENGTH **K = .054**

BODY ANGLE (DEG'S/SIDE): **.420**

CASE CAPACITY **CC'S = 2.12**

LOADED LENGTH: **1.68**

BELT DIAMETER **C = N/A**

RIM THICKNESS **E = .05**

SHOULDER DIAMETER **G = .458**

LENGTH TO SHOULDER **J = 1.011**

SHOULDER ANGLE (DEG'S/SIDE): **20**

PRIMER: **L/R**

CASE CAPACITY (GR'S WATER): **32.76**

DIMENSIONAL DRAWING:

F H K G A J E B D

-NOT ACTUAL SIZE-
-DO NOT SCALE-

CARTRIDGE: .375 Van Horn

OTHER NAMES:

DIA: .375

BALLISTEK NO: 375N

NAI NO: BEN 12333/4.527

DATA SOURCE: Rifle #50 Pg40

HISTORICAL DATA: By G. Van Horn.

NOTES:

LOADING DATA:

BULLET WT./TYPE	POWDER WT./TYPE	VELOCITY ('/SEC)	SOURCE
300/PBT	98.0/H4831	2764	Hagel

CASE PREPARATION: **SHELLHOLDER (RCBS):** 14

MAKE FROM: .460 Weatherby. Run Weatherby case into form die (required). Trim to length. F/L size with expander removed to complete neck. I.D. neck ream. Chamfer I.D. and F/L size again. Fireform in chamber.

PHYSICAL DATA (INCHES):

CASE TYPE: **Belted Bottleneck**
CASE LENGTH **A = 2.635**
HEAD DIAMETER **B = .582**
RIM DIAMETER **D = .583**
NECK DIAMETER **F = .404**
NECK LENGTH **H = .424**
SHOULDER LENGTH **K = .161**
BODY ANGLE (DEG'S/SIDE): **.247**
CASE CAPACITY **CC'S = 7.74**

LOADED LENGTH: **3.33**
BELT DIAMETER **C = .603**
RIM THICKNESS **E = .062**
SHOULDER DIAMETER **G = .566**
LENGTH TO SHOULDER **J = 2.05**
SHOULDER ANGLE (DEG'S/SIDE): **26.71**
PRIMER: **L/R Mag**
CASE CAPACITY (GR'S WATER): **119.54**

DIMENSIONAL DRAWING:

F H K G A J E B D C

-NOT ACTUAL SIZE-
-DO NOT SCALE-

CARTRIDGE: .375 Weatherby

OTHER NAMES:

DIA: .375

BALLISTEK NO: 375AC

NAI NO: BEN 12233/5.575

DATA SOURCE: Ackley Vol1 Pg486

HISTORICAL DATA: By R. Weatherby about 1944.

NOTES:

LOADING DATA:

BULLET WT./TYPE	POWDER WT./TYPE	VELOCITY ('/SEC)	SOURCE
300/—	88.0/4350	2740	Barnes

CASE PREPARATION: **SHELLHOLDER (RCBS):** 4

MAKE FROM: .375 H&H Mag. Taper expand the H&H case to about .390" dia. F/L size in the Weatherby die. Square the case mouth and chamfer. Fireform in the proper chamber.

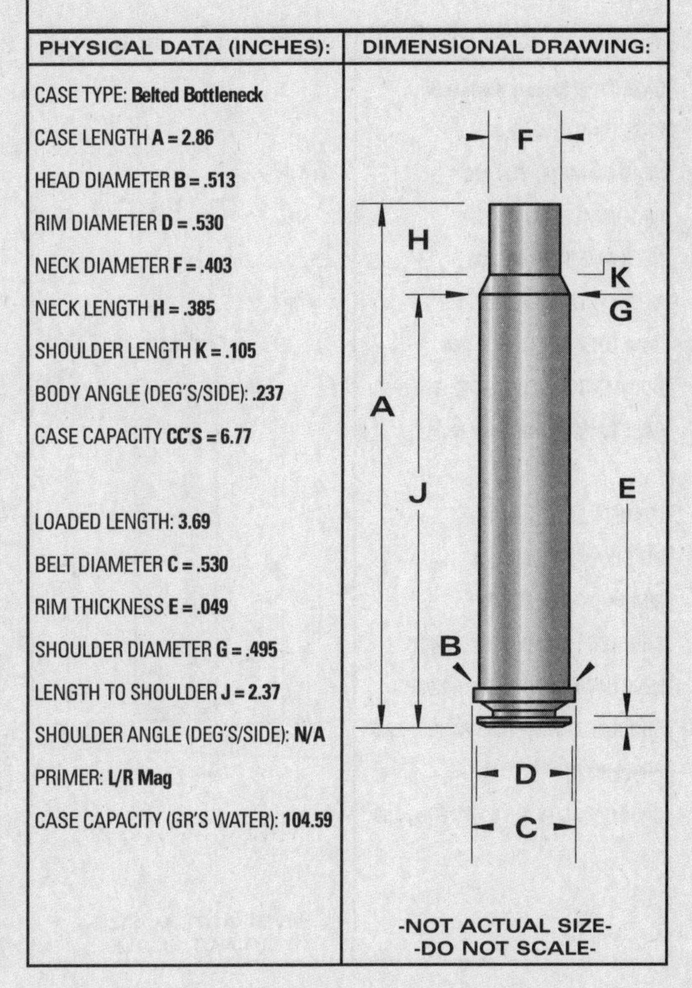

CASE TYPE: **Belted Bottleneck**
CASE LENGTH **A = 2.86**
HEAD DIAMETER **B = .513**
RIM DIAMETER **D = .530**
NECK DIAMETER **F = .403**
NECK LENGTH **H = .385**
SHOULDER LENGTH **K = .105**
BODY ANGLE (DEG'S/SIDE): **.237**
CASE CAPACITY **CC'S = 6.77**

LOADED LENGTH: **3.69**
BELT DIAMETER **C = .530**
RIM THICKNESS **E = .049**
SHOULDER DIAMETER **G = .495**
LENGTH TO SHOULDER **J = 2.37**
SHOULDER ANGLE (DEG'S/SIDE): **N/A**
PRIMER: **L/R Mag**
CASE CAPACITY (GR'S WATER): **104.59**

CARTRIDGE: .375 Whelen

OTHER NAMES:

DIA: .375

BALLISTEK NO: 375L

NAI NO: RXB 22235/5.298

DATA SOURCE: Ackley Vol1 Pg483

HISTORICAL DATA:

NOTES:

LOADING DATA:

BULLET WT./TYPE	POWDER WT./TYPE	VELOCITY ('/SEC)	SOURCE
350/RN	52.0/4895	2340	Ackley

CASE PREPARATION: SHELLHOLDER (RCBS): 3

MAKE FROM: .30-06 Spgf. Anneal case neck. Taper expand to about .385" dia. Square case mouth and F/L size. Trim to length and chamfer.

PHYSICAL DATA (INCHES):

CASE TYPE: **Rimless Bottleneck**

CASE LENGTH **A = 2.49**

HEAD DIAMETER **B = .470**

RIM DIAMETER **D = .473**

NECK DIAMETER **F = .402**

NECK LENGTH **H = .396**

SHOULDER LENGTH **K = .064**

BODY ANGLE (DEG'S/SIDE): **.391**

CASE CAPACITY **CC'S = 4.83**

LOADED LENGTH: **3.30**

BELT DIAMETER **C = N/A**

RIM THICKNESS **E = .05**

SHOULDER DIAMETER **G = .445**

LENGTH TO SHOULDER **J = 2.03**

SHOULDER ANGLE (DEG'S/SIDE): **32.05**

PRIMER: **L/R**

CASE CAPACITY (GR'S WATER): **74.50**

DIMENSIONAL DRAWING:

F H K G A J E B D

-NOT ACTUAL SIZE-
-DO NOT SCALE-

CARTRIDGE: .375 Whelen Improved

OTHER NAMES:

DIA: .375

BALLISTEK NO: 375R

NAI NO: RXB 12222/5.298

DATA SOURCE: Ackley Vol1 Pg484

HISTORICAL DATA:

NOTES:

LOADING DATA:

BULLET WT./TYPE	POWDER WT./TYPE	VELOCITY ('/SEC)	SOURCE
235/RN	53.5/IMR4895	2500	JJD

CASE PREPARATION: SHELLHOLDER (RCBS): 3

MAKE FROM: .30-06 Spgf. Taper expand (after annealing) to about .390" dia. F/L size in "improved" die. Square case mouth, chamfer and fireform. Standard .375 Whelen can, generally, be "improved" by firing in the chamber.

PHYSICAL DATA (INCHES):

CASE TYPE: **Rimless Bottleneck**

CASE LENGTH **A = 2.49**

HEAD DIAMETER **B = .470**

RIM DIAMETER **D = .473**

NECK DIAMETER **F = .402**

NECK LENGTH **H = .475**

SHOULDER LENGTH **K = .075**

BODY ANGLE (DEG'S/SIDE): **.230**

CASE CAPACITY **CC'S = 4.91**

LOADED LENGTH: **A/R**

BELT DIAMETER **C = N/A**

RIM THICKNESS **E = .05**

SHOULDER DIAMETER **G = .456**

LENGTH TO SHOULDER **J = 1.94**

SHOULDER ANGLE (DEG'S/SIDE): **19.80**

PRIMER: **L/R**

CASE CAPACITY (GR'S WATER): **75.79**

DIMENSIONAL DRAWING:

F H K G A J E B D

-NOT ACTUAL SIZE-
-DO NOT SCALE-

CARTRIDGE: .375/303 Westley Richards Express

OTHER NAMES:
.303 Axite

DIA: .330

BALLISTEK NO: 310F

NAI NO: RMB
22243/5.408

DATA SOURCE: COTW 4th Pg231

HISTORICAL DATA: By Westley Richards about 1906.

NOTES:

LOADING DATA:

BULLET WT./TYPE	POWDER WT./TYPE	VELOCITY ('/SEC)	SOURCE
200/RN	53.0/IMR4320	2700	JJD

CASE PREPARATION: SHELLHOLDER (RCBS): 21

MAKE FROM: 7 x 65R Brenneke. Turn rim to .505" dia. and back chamfer. Trim case to length and chamfer. F/L size and fireform. (Note: taper expand neck to .340" dia. before sizing.) .405 Win. basic (B.E.L.L.) brass will also work by turning the rim down. Anneal case, form in form die (two steps), cut to length and F/L size.

PHYSICAL DATA (INCHES):

CASE TYPE: **Rimmed Bottleneck**
CASE LENGTH **A = 2.488**
HEAD DIAMETER **B = .460**
RIM DIAMETER **D = .505**
NECK DIAMETER **F = .353**
NECK LENGTH **H = .280**
SHOULDER LENGTH **K = .065**
BODY ANGLE (DEG'S/SIDE): **.737**
CASE CAPACITY **CC'S = 4.42**

LOADED LENGTH: **3.48**
BELT DIAMETER **C = N/A**
RIM THICKNESS **E = .040**
SHOULDER DIAMETER **G = .410**
LENGTH TO SHOULDER **J = 2.143**
SHOULDER ANGLE (DEG'S/SIDE): **23.67**
PRIMER: **L/R**
CASE CAPACITY (GR'S WATER): **68.21**

DIMENSIONAL DRAWING:

F H K G A J E B D

-NOT ACTUAL SIZE-
-DO NOT SCALE-

CARTRIDGE: .375/338 Chatfield-Taylor

OTHER NAMES:
.375 Belted Newton
.375/338 Winchester Magnum

DIA: .375

BALLISTEK NO: 375O

NAI NO: BEN
22244/4.873

DATA SOURCE: Ackley Vol2 Pg212

HISTORICAL DATA: By R. Chatfield-Taylor.

NOTES:

LOADING DATA:

BULLET WT./TYPE	POWDER WT./TYPE	VELOCITY ('/SEC)	SOURCE
300/—	62.0/4064	2463	Ackley

CASE PREPARATION: SHELLHOLDER (RCBS): 4

MAKE FROM: .338 Win. Mag. Anneal case neck and taper expand to .385" dia. Trim to length and F/L size. .375 H&H brass can be cut off to length and fireformed (with corn meal) in the CT rim die.

PHYSICAL DATA (INCHES):

CASE TYPE: **Belted Bottleneck**
CASE LENGTH **A = 2.500**
HEAD DIAMETER **B = .513**
RIM DIAMETER **D = .532**
NECK DIAMETER **F = .405**
NECK LENGTH **H = .370**
SHOULDER LENGTH **K = .070**
BODY ANGLE (DEG'S/SIDE): **.354**
CASE CAPACITY **CC'S = 5.69**

LOADED LENGTH: **3.24**
BELT DIAMETER **C = .532**
RIM THICKNESS **E = .05**
SHOULDER DIAMETER **G = .490**
LENGTH TO SHOULDER **J = 2.06**
SHOULDER ANGLE (DEG'S/SIDE): **31.26**
PRIMER: **L/R Mag**
CASE CAPACITY (GR'S WATER): **87.92**

DIMENSIONAL DRAWING:

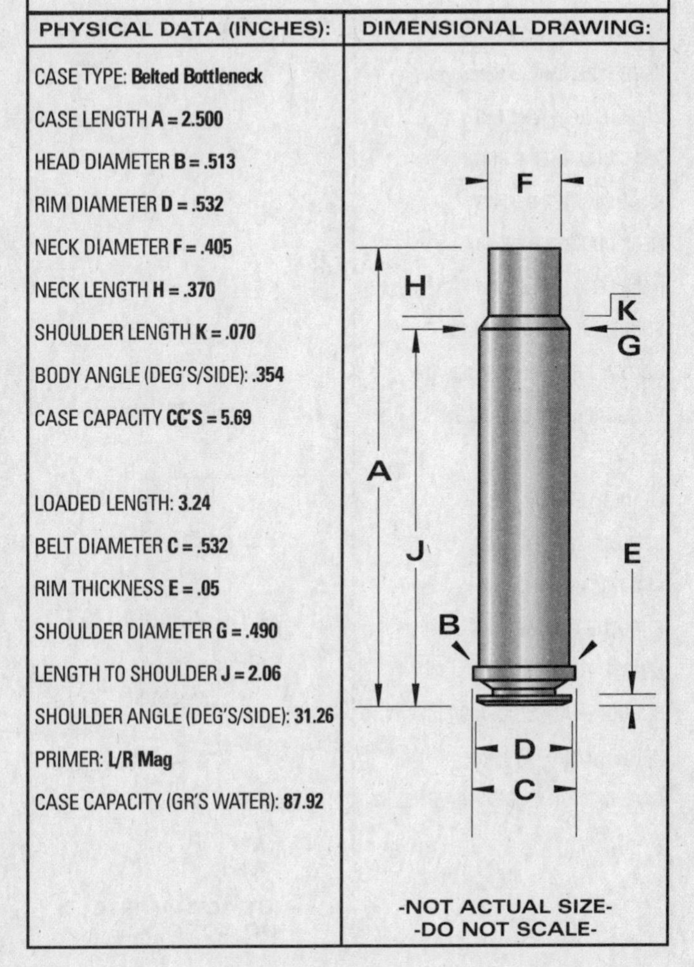

-NOT ACTUAL SIZE-
-DO NOT SCALE-

CARTRIDGE: .375/38-40 Rimless

OTHER NAMES:

DIA: .375

BALLISTEK NO: 375AK

NAI NO: RXB 24242/2.872

DATA SOURCE: Ackley Vol2 Pg211

HISTORICAL DATA: By G. Heubner.

NOTES: Similar to .38/40 (#400A)

LOADING DATA:

BULLET WT./TYPE	POWDER WT./TYPE	VELOCITY ('/SEC)	SOURCE
200/FP	20.0/2400	2400	Heubner/ Ackley

CASE PREPARATION: SHELLHOLDER (RCBS): 3

MAKE FROM: .30-06 Spgf. Cut '06 case off at 1.4" and F/L size with the expander removed. I.D. neck ream, trim to length, chamfer.

PHYSICAL DATA (INCHES):

CASE TYPE: **Rimless Bottleneck**

CASE LENGTH **A = 1.255**

HEAD DIAMETER **B = .470**

RIM DIAMETER **D = .473**

NECK DIAMETER **F = .407**

NECK LENGTH **H = .335**

SHOULDER LENGTH **K = .135**

BODY ANGLE (DEG'S/SIDE): **.421**

CASE CAPACITY **CC'S = 2.31**

LOADED LENGTH: **1.72**

BELT DIAMETER **C = N/A**

RIM THICKNESS **E = .05**

SHOULDER DIAMETER **G = .460**

LENGTH TO SHOULDER **J = .88**

SHOULDER ANGLE (DEG'S/SIDE): **11.10**

PRIMER: **L/P**

CASE CAPACITY (GR'S WATER): **35.63**

DIMENSIONAL DRAWING:

-NOT ACTUAL SIZE-
-DO NOT SCALE-

CARTRIDGE: .376 Steyr

OTHER NAMES:

DIA: .375

BALLISTEK NO:

NAI NO:

DATA SOURCE: Hornady #6

HISTORICAL DATA: Developed in a joint effort with Hornady and Steyr and introduced in 2000. It was designed for use in the short action Steyr Scout rifle while still giving performance that is adequate for large game.

NOTES: Accepted by SAAMI 6/13/01.

LOADING DATA:

BULLET WT./TYPE	POWDER WT./TYPE	VELOCITY ('/SEC)	SOURCE
270/Hornady	66.4/BL-C(2)	2,650	Horn. #6
300/Hornady	61.5/IMR-4895	2,500	Horn. #6

CASE PREPARATION: SHELL HOLDER (RCBS): 42

MAKE FROM: Based on a new case. Use factory cases from Hornady. May possibly be able to make cases from .284 Winchester or 9.3X64 Brenneke.

PHYSICAL DATA (INCHES):

CASE TYPE: **Rebated Bottleneck**

CASE LENGTH **A = 2.362**

HEAD DIAMETER **B = .506**

RIM DIAMETER **D = .495**

NECK DIAMETER **F = .403**

NECK LENGTH **H = .355**

SHOULDER LENGTH **K = .114**

BODY ANGLE (DEG'S/SIDE):

CASE CAPACITY **CC'S =**

LOADED LENGTH: **3.110**

BELT DIAMETER **C =**

RIM THICKNESS **E = .048**

SHOULDER DIAMETER **G = .473**

LENGTH TO SHOULDER **J = 1.893**

SHOULDER ANGLE (DEG'S/SIDE): **17.5**

PRIMER: **L/R**

CASE CAPACITY (GR'S WATER):

DIMENSIONAL DRAWING:

-NOT ACTUAL SIZE-
-DO NOT SCALE-

CARTRIDGE: .378 Weatherby

OTHER NAMES:

DIA: .375

BALLISTEK NO: 375D

NAI NO: BEN 12343/4.996

DATA SOURCE: Most current handloading manuals.

HISTORICAL DATA: Created by Roy Weatherby around 1953. Very powerful cartridge for use on the world's largest game.

NOTES: Cartridge was approved by SAAMI on 1/12/94. Older rifles and loading dies may vary slightly.

LOADING DATA:

BULLET WT./TYPE	POWDER WT./TYPE	VELOCITY ('/SEC)	SOURCE
270/Hornady	103.1/AA-4350	3100	Horn. #6
300/Nosler Partition	113/IMR-4831	2838	Nosler #5

CASE PREPARATION: SHELLHOLDER (RCBS): 14

Factory cases are readily available. Make from .338-378 Weatherby by necking up F/L resizing die with tapered expander. Trim, chamfer and deburr. Check neck wall thickness, turn if needed.

PHYSICAL DATA (INCHES):

CASE TYPE: **Belted Bottleneck**

CASE LENGTH **A = 2.913**

HEAD DIAMETER **B = .5817**

RIM DIAMETER **D = .579**

NECK DIAMETER **F = .403**

NECK LENGTH **H = .373**

SHOULDER LENGTH **K = .195**

BODY ANGLE (DEG'S/SIDE): **.294**

CASE CAPACITY **CC'S = 8.97**

LOADED LENGTH: **3.655**

BELT DIAMETER **C = .6035**

RIM THICKNESS **E = .063**

SHOULDER DIAMETER **G = .561**

LENGTH TO SHOULDER **J = 2.345**

SHOULDER ANGLE (DEG'S/SIDE):

Front radius: **R .1512**

Rear radius: **R .130 - .005**

PRIMER: **L/R Mag**

CASE CAPACITY (GR'S WATER): **138.45**

DIMENSIONAL DRAWING:

F H K G A J E B D C

-NOT ACTUAL SIZE-
-DO NOT SCALE-

CARTRIDGE: .38 ACP

OTHER NAMES: .38 Auto

DIA: .355

BALLISTEK NO: 355C

NAI NO: RMS 11115/2.337

DATA SOURCE: Hornady Manual 3rd Pg338

HISTORICAL DATA: Released in 1900.

NOTES: Do not confuse with .38 ACP Super (same case, different loads!)

LOADING DATA:

BULLET WT./TYPE	POWDER WT./TYPE	VELOCITY ('/SEC)	SOURCE
115/JHP	4.7/B'Eye	1100	Horn.

CASE PREPARATION: SHELLHOLDER (RCBS): 1

MAKE FROM: .38 Spl. Turn rim to .405" dia. & backchamfer. Cut new extractor groove. Cut off at .90" long & F/L size. Case is slightly undersize at base but will fireform to fill chamber.

PHYSICAL DATA (INCHES):

CASE TYPE: **Rimmed Straight**

CASE LENGTH **A = .900**

HEAD DIAMETER **B = .385**

RIM DIAMETER **D = .405**

NECK DIAMETER **F = .384**

NECK LENGTH **H = N/A**

SHOULDER LENGTH **K = N/A**

BODY ANGLE (DEG'S/SIDE): **.327**

CASE CAPACITY **CC'S = .889**

LOADED LENGTH: **1.11**

BELT DIAMETER **C = N/A**

RIM THICKNESS **E = .05**

SHOULDER DIAMETER **G = N/A**

LENGTH TO SHOULDER **J = N/A**

SHOULDER ANGLE (DEG'S/SIDE): **N/A**

PRIMER: **S/P**

CASE CAPACITY (GR'S WATER): **13.72**

DIMENSIONAL DRAWING:

F A E B D

-NOT ACTUAL SIZE-
-DO NOT SCALE-

CARTRIDGE: .38 ACP Super

OTHER NAMES:

DIA: .355

BALLISTEK NO: 355D

NAI NO: RMS 11115/2.337

DATA SOURCE: Hornady Manual 3rd Pg336

HISTORICAL DATA: By Colt in 1929.

NOTES: An improved .38 ACP.

LOADING DATA:

BULLET WT./TYPE	POWDER WT./TYPE	VELOCITY ('/SEC)	SOURCE
115/JHP	5.8/B'Eye	1300	Horn.

CASE PREPARATION: SHELLHOLDER (RCBS): 1

MAKE FROM: See instructions under .38 ACP (#355C).

PHYSICAL DATA (INCHES):

CASE TYPE: **Rimmed Straight**

CASE LENGTH **A = .900**

HEAD DIAMETER **B = .385**

RIM DIAMETER **D = .405**

NECK DIAMETER **F = .384**

NECK LENGTH **H = N/A**

SHOULDER LENGTH **K = N/A**

BODY ANGLE (DEG'S/SIDE): **.327**

CASE CAPACITY **CC'S = .889**

LOADED LENGTH: **1.11**

BELT DIAMETER **C = N/A**

RIM THICKNESS **E = .05**

SHOULDER DIAMETER **G = N/A**

LENGTH TO SHOULDER **J = N/A**

SHOULDER ANGLE (DEG'S/SIDE): **N/A**

PRIMER: **S/P**

CASE CAPACITY (GR'S WATER): **13.73**

DIMENSIONAL DRAWING:

F
A
B
E
D

-NOT ACTUAL SIZE-
-DO NOT SCALE-

CARTRIDGE: .38 Colt Long

OTHER NAMES:
.38 Colt Army
.38 Ball US Navy

DIA: .357

BALLISTEK NO: 357S

NAI NO: RMS 11115/2.727

DATA SOURCE: COTW 4th Pg176

HISTORICAL DATA: US mil ctg. 1892-1911.

NOTES:

LOADING DATA:

BULLET WT./TYPE	POWDER WT./TYPE	VELOCITY ('/SEC)	SOURCE
150/RN	3.0/B'Eye	810	Barnes

CASE PREPARATION: SHELLHOLDER (RCBS): 6

MAKE FROM: .38 Special. Trim to proper length. See remarks under .38 Colt Short (#357R).

PHYSICAL DATA (INCHES):

CASE TYPE: **Rimmed Straight**

CASE LENGTH **A = 1.031**

HEAD DIAMETER **B = .378**

RIM DIAMETER **D = .440**

NECK DIAMETER **F = .377**

NECK LENGTH **H = N/A**

SHOULDER LENGTH **K = N/A**

BODY ANGLE (DEG'S/SIDE): **.029**

CASE CAPACITY **CC'S = 1.13**

LOADED LENGTH: **1.36**

BELT DIAMETER **C = N/A**

RIM THICKNESS **E = .052**

SHOULDER DIAMETER **G = N/A**

LENGTH TO SHOULDER **J = N/A**

SHOULDER ANGLE (DEG'S/SIDE): **N/A**

PRIMER: **S/P**

CASE CAPACITY (GR'S WATER): **17.39**

DIMENSIONAL DRAWING:

F
A
B
E
D

-NOT ACTUAL SIZE-
-DO NOT SCALE-

CARTRIDGE: .38 Colt Short

OTHER NAMES:

DIA: .357

BALLISTEK NO: 357R

NAI NO: RMS 11115/2.016

DATA SOURCE: NAI/Ballistek

HISTORICAL DATA: By Colt about 1892.

NOTES:

LOADING DATA:

BULLET WT./TYPE	POWDER WT./TYPE	VELOCITY ('/SEC)	SOURCE
130/RN	2.1/B'Eye	700	JJD

CASE PREPARATION: SHELLHOLDER (RCBS): 6

MAKE FROM: .38 Special. Trim case to length, chamfer & F/L size. .38 Spl. or .357 Mag. dies may be used for sizing but, you'll need a seater die, for the colt, to crimp properly.

PHYSICAL DATA (INCHES):

CASE TYPE: **Rimmed Straight**

CASE LENGTH **A = .762**

HEAD DIAMETER **B = .378**

RIM DIAMETER **D = .433**

NECK DIAMETER **F = .377**

NECK LENGTH **H = N/A**

SHOULDER LENGTH **K = N/A**

BODY ANGLE (DEG'S/SIDE): **.04**

CASE CAPACITY **CC'S = .708**

LOADED LENGTH: **1.07**

BELT DIAMETER **C = N/A**

RIM THICKNESS **E = .052**

SHOULDER DIAMETER **G = N/A**

LENGTH TO SHOULDER **J = N/A**

SHOULDER ANGLE (DEG'S/SIDE): **N/A**

PRIMER: **S/P**

CASE CAPACITY (GR'S WATER): **10.94**

DIMENSIONAL DRAWING:

F
A
B
E
D

-NOT ACTUAL SIZE-
-DO NOT SCALE-

CARTRIDGE: .38 Ballard Extra Long

OTHER NAMES:
.38 Extra Long

DIA: .375

BALLISTEK NO: 375AB

NAI NO: RMS 11115/4.3001

DATA SOURCE: COTW 4th Pg101

HISTORICAL DATA: From about 1885.

NOTES:

LOADING DATA:

BULLET WT./TYPE	POWDER WT./TYPE	VELOCITY ('/SEC)	SOURCE
150/Lead	6.0/Unique	1160	Barnes

CASE PREPARATION: SHELLHOLDER (RCBS): 6

MAKE FROM: .357 Maximum. Use the Max. case as is.

PHYSICAL DATA (INCHES):

CASE TYPE: **Rimmed Straight**

CASE LENGTH **A = 1.63**

HEAD DIAMETER **B = .379**

RIM DIAMETER **D = .440**

NECK DIAMETER **F = .378**

NECK LENGTH **H = N/A**

SHOULDER LENGTH **K = N/A**

BODY ANGLE (DEG'S/SIDE): **N/A**

CASE CAPACITY **CC'S = 2.38**

LOADED LENGTH: **2.06**

BELT DIAMETER **C = N/A**

RIM THICKNESS **E = .052**

SHOULDER DIAMETER **G = N/A**

LENGTH TO SHOULDER **J = N/A**

SHOULDER ANGLE (DEG'S/SIDE): **N/A**

PRIMER: **S/P**

CASE CAPACITY (GR'S WATER): **36.80**

DIMENSIONAL DRAWING:

F
A
E
B
D

-NOT ACTUAL SIZE-
-DO NOT SCALE-

CARTRIDGE: .38 Long Centerfire

OTHER NAMES:

DIA: .376

BALLISTEK NO: 376B

NAI NO: RMS 11115/2.717

DATA SOURCE: COTW 4th Pg100

HISTORICAL DATA: From about 1875.

NOTES:

LOADING DATA:

BULLET WT./TYPE	POWDER WT./TYPE	VELOCITY ('/SEC)	SOURCE
150/Lead	3.2/B'Eye	—	JJD

CASE PREPARATION: SHELLHOLDER (RCBS): 6

MAKE FROM: .357 Magnum. Trim case to length and size in a .38 Spl. die.

PHYSICAL DATA (INCHES):

CASE TYPE: **Rimmed Straight**
CASE LENGTH **A = 1.03**
HEAD DIAMETER **B = .379**
RIM DIAMETER **D = .441**
NECK DIAMETER **F = .378**
NECK LENGTH **H = N/A**
SHOULDER LENGTH **K = N/A**
BODY ANGLE (DEG'S/SIDE): **.029**
CASE CAPACITY **CC'S = 1.31**
LOADED LENGTH: **1.45**
BELT DIAMETER **C = N/A**
RIM THICKNESS **E = .055**
SHOULDER DIAMETER **G = N/A**
LENGTH TO SHOULDER **J = N/A**
SHOULDER ANGLE (DEG'S/SIDE): **N/A**
PRIMER: **S/P**
CASE CAPACITY (GR'S WATER): **20.29**

DIMENSIONAL DRAWING:

F
A
E
B
D

-NOT ACTUAL SIZE-
-DO NOT SCALE-

CARTRIDGE: .38 Smith & Wesson

OTHER NAMES:
.380/200 British
.38 Colt New Police

DIA: .359

BALLISTEK NO: 359A

NAI NO: RMS 21115/2.021

DATA SOURCE: COTW 4th Pg177

HISTORICAL DATA: By S&W about 1877.

NOTES:

LOADING DATA:

BULLET WT./TYPE	POWDER WT./TYPE	VELOCITY ('/SEC)	SOURCE
150/Lead	4.7/Unique	890	Barnes

CASE PREPARATION: SHELLHOLDER (RCBS): 6

MAKE FROM: .38 Special. Trim case to length, chamfer and F/L size. Case body will fireform, slightly, in the chamber.

PHYSICAL DATA (INCHES):

CASE TYPE: **Rimmed Straight**
CASE LENGTH **A = .780**
HEAD DIAMETER **B = .386**
RIM DIAMETER **D = .433**
NECK DIAMETER **F = .386**
NECK LENGTH **H = N/A**
SHOULDER LENGTH **K = N/A**
BODY ANGLE (DEG'S/SIDE): **0**
CASE CAPACITY **CC'S = .728**
LOADED LENGTH: **1.20**
BELT DIAMETER **C = N/A**
RIM THICKNESS **E = .047**
SHOULDER DIAMETER **G = N/A**
LENGTH TO SHOULDER **J = N/A**
SHOULDER ANGLE (DEG'S/SIDE): **N/A**
PRIMER: **S/P**
CASE CAPACITY (GR'S WATER): **11.24**

DIMENSIONAL DRAWING:

F
A
E
B
D

-NOT ACTUAL SIZE-
-DO NOT SCALE-

CARTRIDGE: .38 Special

OTHER NAMES:
.38 S&W Special
.38 Colt Special

DIA: .357

BALLISTEK NO: 357A

NAI NO: RMS 11115/3.094

DATA SOURCE: NAI/Ballistek

HISTORICAL DATA: By Smith & Wesson in 1902.

NOTES:

LOADING DATA:

BULLET WT./TYPE	POWDER WT./TYPE	VELOCITY ('/SEC)	SOURCE
158/JHP	5.1/Unique	825	JJD
115/JHP	8.5/Bluedot	1200	Horn.

CASE PREPARATION: SHELLHOLDER (RCBS): 6

MAKE FROM: .357 Mag or .357 Max. Cut case to 1.555", chamfer & F/L size.

PHYSICAL DATA (INCHES):

CASE TYPE: **Rimmed Straight**
CASE LENGTH **A = 1.155**
HEAD DIAMETER **B = .380**
RIM DIAMETER **D = .440**
NECK DIAMETER **F = .379**
NECK LENGTH **H = N/A**
SHOULDER LENGTH **K = N/A**
BODY ANGLE (DEG'S/SIDE): **.327**
CASE CAPACITY **CC'S = 1.312**

LOADED LENGTH: **1.48**
BELT DIAMETER **C = N/A**
RIM THICKNESS **E = .058**
SHOULDER DIAMETER **G = N/A**
LENGTH TO SHOULDER **J = N/A**
SHOULDER ANGLE (DEG'S/SIDE): **N/A**
PRIMER: **L/P**
CASE CAPACITY (GR'S WATER): **20.25.**

DIMENSIONAL DRAWING:

F
A
E
B
D

-NOT ACTUAL SIZE-
-DO NOT SCALE-

CARTRIDGE: .38-35 Stevens

OTHER NAMES:
.38-33 Stevens

DIA: .375

BALLISTEK NO: 375AA

NAI NO: RMS 11115/4.020

DATA SOURCE: COTW 4th Pg101

HISTORICAL DATA: 1875 'Everlasting' ctg.

NOTES:

LOADING DATA:

BULLET WT./TYPE	POWDER WT./TYPE	VELOCITY ('/SEC)	SOURCE
200/Lead	13.7/IMR4198	—	JJD

CASE PREPARATION: SHELLHOLDER (RCBS): 21

MAKE FROM: Not much to work with here! 13/32" dia. tubing can be used to make tubing cases or the same tubing can be used to re-body a 41 S&W Mag. case head. Case has so little taper that F/L dies are not really needed to form.

PHYSICAL DATA (INCHES):

CASE TYPE: **Rimmed Straight**
CASE LENGTH **A = 1.62**
HEAD DIAMETER **B = .403**
RIM DIAMETER **D = .492**
NECK DIAMETER **F = .402**
NECK LENGTH **H = N/A**
SHOULDER LENGTH **K = N/A**
BODY ANGLE (DEG'S/SIDE): **.018**
CASE CAPACITY **CC'S = 2.21**

LOADED LENGTH: **2.43**
BELT DIAMETER **C = N/A**
RIM THICKNESS **E = .07**
SHOULDER DIAMETER **G = N/A**
LENGTH TO SHOULDER **J = N/A**
SHOULDER ANGLE (DEG'S/SIDE): **N/A**
PRIMER: **S/R**
CASE CAPACITY (GR'S WATER): **34.23**

DIMENSIONAL DRAWING:

F
A
E
B
D

-NOT ACTUAL SIZE-
-DO NOT SCALE-

CARTRIDGE: .38-40 Remington Hepburn

OTHER NAMES:
.38-40 Remington

DIA: .372

BALLISTEK NO: 372A

NAI NO: RMS 21115/3.899

DATA SOURCE: COTW 4th Pg101

HISTORICAL DATA: Circa 1875.

NOTES:

LOADING DATA:

BULLET WT./TYPE	POWDER WT./TYPE	VELOCITY ('/SEC)	SOURCE
250/Lead	15.0/2400	1300	Barnes

CASE PREPARATION: SHELLHOLDER (RCBS): 7

MAKE FROM: .303 British. Anneal case neck. Taper expand to .380" dia. Cut off at 1.8". F/L size, trim to length and chamfer. Fireform in R/H chamber.

PHYSICAL DATA (INCHES):

CASE TYPE: **Rimmed Straight**

CASE LENGTH **A = 1.77**

HEAD DIAMETER **B = .454**

RIM DIAMETER **D = .537**

NECK DIAMETER **F = .395**

NECK LENGTH **H = N/A**

SHOULDER LENGTH **K = N/A**

BODY ANGLE (DEG'S/SIDE): **.989**

CASE CAPACITY **CC'S = 2.87**

LOADED LENGTH: **2.32**

BELT DIAMETER **C = N/A**

RIM THICKNESS **E = .062**

SHOULDER DIAMETER **G = N/A**

LENGTH TO SHOULDER **J = N/A**

SHOULDER ANGLE (DEG'S/SIDE): **N/A**

PRIMER: **L/R**

CASE CAPACITY (GR'S WATER): **44.27**

DIMENSIONAL DRAWING:

F
A
E
B
D

-NOT ACTUAL SIZE-
-DO NOT SCALE-

CARTRIDGE: .38-40 Winchester

OTHER NAMES:

DIA: .401

BALLISTEK NO: 400A

NAI NO: RMS 32241/2.776

DATA SOURCE: COTW 4th Pg77

HISTORICAL DATA: By Win. in 1874.

NOTES:

LOADING DATA:

BULLET WT./TYPE	POWDER WT./TYPE	VELOCITY ('/SEC)	SOURCE
180/—	26.0/4227	1850	Barnes

CASE PREPARATION: SHELLHOLDER (RCBS): 21

MAKE FROM: 7 x 57R. Cut case off to 1.35". F/L size in .38-40 die with expander removed. I.D. neck ream and trim to length. Chamfer and F/L size. 9.3x74R brass may be used in much the same way.

PHYSICAL DATA (INCHES):

CASE TYPE: **Rimmed Bottleneck**

CASE LENGTH **A = 1.305**

HEAD DIAMETER **B = .470**

RIM DIAMETER **D = .525**

NECK DIAMETER **F = .416**

NECK LENGTH **H = .153**

SHOULDER LENGTH **K = .230**

BODY ANGLE (DEG'S/SIDE): **1.03**

CASE CAPACITY **CC'S = 2.47**

LOADED LENGTH: **1.59**

BELT DIAMETER **C = N/A**

RIM THICKNESS **E = .065**

SHOULDER DIAMETER **G = .444**

LENGTH TO SHOULDER **J = .922**

SHOULDER ANGLE (DEG'S/SIDE): **3.48**

PRIMER: **L/P**

CASE CAPACITY (GR'S WATER): **38.15**

DIMENSIONAL DRAWING:

F
H
K
G
A
J
E
B
D

-NOT ACTUAL SIZE-
-DO NOT SCALE-

CARTRIDGE: .38/45 ACP

OTHER NAMES:
.45-38

DIA: .357

BALLISTEK NO: 357H

NAI NO: RXB
14244/1.869

DATA SOURCE: Handloader #68 Pg28

HISTORICAL DATA: By B. Clerke about 1963.

NOTES:

LOADING DATA:

BULLET WT./TYPE	POWDER WT./TYPE	VELOCITY ('/SEC)	SOURCE
158/WC	3.5/B'Eye	834	Harvey

CASE PREPARATION: SHELLHOLDER (RCBS): 3

MAKE FROM: .45 ACP. Use only new brass and F/L size, in .45 ACP sizer, before you begin to form. This case is difficult to make and a form set is needed. Form in dies #1, #2, & #3. Trim to length. F/L size. Case will fireform, in chamber.

PHYSICAL DATA (INCHES):

CASE TYPE: **Rimless Bottleneck**

CASE LENGTH **A = .89**

HEAD DIAMETER **B = .476**

RIM DIAMETER **D = .480**

NECK DIAMETER **F = .381**

NECK LENGTH **H = .175**

SHOULDER LENGTH **K = .075**

BODY ANGLE (DEG'S/SIDE): **.130**

CASE CAPACITY **CC'S = 1.42**

LOADED LENGTH: **1.165**

BELT DIAMETER **C = N/A**

RIM THICKNESS **E = .05**

SHOULDER DIAMETER **G = .474**

LENGTH TO SHOULDER **J = .64**

SHOULDER ANGLE (DEG'S/SIDE): **31.79**

PRIMER: **L/P**

CASE CAPACITY (GR'S WATER): **22.00**

DIMENSIONAL DRAWING:

-NOT ACTUAL SIZE-
-DO NOT SCALE-

CARTRIDGE: .38-45 Bullard

OTHER NAMES:

DIA: .373

BALLISTEK NO: 373A

NAI NO: RMS
21115/4.053

DATA SOURCE: COTW 4th Pg102

HISTORICAL DATA: From about 1887 for Bullard LA rifles.

NOTES:

LOADING DATA:

BULLET WT./TYPE	POWDER WT./TYPE	VELOCITY ('/SEC)	SOURCE
250/Lead	10.0/Unique	1200	Barnes

CASE PREPARATION: SHELLHOLDER (RCBS): 7

MAKE FROM: .303 British. Turn rim to .526" dia. and back chamfer. Taper expand (after annealing) to .380" dia. cut off at 1.85". F/L size, trim to length and chamfer.

PHYSICAL DATA (INCHES):

CASE TYPE: **Rimmed Straight**

CASE LENGTH **A = 1.84**

HEAD DIAMETER **B = .454**

RIM DIAMETER **D = .526**

NECK DIAMETER **F = .397**

NECK LENGTH **H = N/A**

SHOULDER LENGTH **K = N/A**

BODY ANGLE (DEG'S/SIDE): **.914**

CASE CAPACITY **CC'S = 2.99**

LOADED LENGTH: **2.26**

BELT DIAMETER **C = N/A**

RIM THICKNESS **E = .055**

SHOULDER DIAMETER **G = N/A**

LENGTH TO SHOULDER **J = N/A**

SHOULDER ANGLE (DEG'S/SIDE): **N/A**

PRIMER: **L/R**

CASE CAPACITY (GR'S WATER): **46.27**

DIMENSIONAL DRAWING:

-NOT ACTUAL SIZE-
-DO NOT SCALE-

CARTRIDGE: .38-45 Stevens

OTHER NAMES:

DIA: .363

BALLISTEK NO: 363C

NAI NO: RMS 31115/3.868

DATA SOURCE: COTW 4th Pg102

HISTORICAL DATA: Stevens 'Everlasting' case, very rare.

NOTES:

LOADING DATA:

BULLET WT./TYPE	POWDER WT./TYPE	VELOCITY ('/SEC)	SOURCE
210/Lead	16.0/4198	1340	Barnes

CASE PREPARATION: SHELLHOLDER (RCBS): 7

MAKE FROM: .303 British. Turn rim to .522" dia. and back chamfer. Cut case to 1.8" taper expand to .370" dia. and F/L size. Trim to final length and fireform, in the chamber. Lyman bullet #358315 works well if paper patched.

PHYSICAL DATA (INCHES):

CASE TYPE: **Rimmed Straight**

CASE LENGTH **A = 1.76**

HEAD DIAMETER **B = .455**

RIM DIAMETER **D = .522**

NECK DIAMETER **F = .395**

NECK LENGTH **H = N/A**

SHOULDER LENGTH **K = N/A**

BODY ANGLE (DEG'S/SIDE): **1.01**

CASE CAPACITY **CC'S = 2.66**

LOADED LENGTH: **2.24**

BELT DIAMETER **C = N/A**

RIM THICKNESS **E = .06**

SHOULDER DIAMETER **G = N/A**

LENGTH TO SHOULDER **J = N/A**

SHOULDER ANGLE (DEG'S/SIDE): **N/A**

PRIMER: **L/R**

CASE CAPACITY (GR'S WATER): **41.07**

DIMENSIONAL DRAWING:

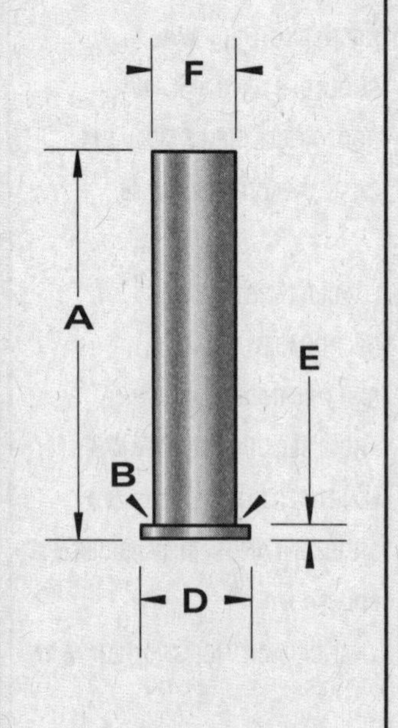

-NOT ACTUAL SIZE-
-DO NOT SCALE-

CARTRIDGE: .38-50 Ballard

OTHER NAMES:

DIA: .376

BALLISTEK NO: 376C

NAI NO: RMS 11115/4.706

DATA SOURCE: COTW 4th Pg103

HISTORICAL DATA: 1876 by Ballard as 'Everlasting' ctg.

NOTES: Dia's vary .376-.379. Slug the bore!

LOADING DATA:

BULLET WT./TYPE	POWDER WT./TYPE	VELOCITY ('/SEC)	SOURCE
250/Lead	17.0/4198	1350	Barnes

CASE PREPARATION: SHELLHOLDER (RCBS): 21

MAKE FROM: 9.3 x 72R. Cut case to length and F/L size in the Ballard die. Chamfer. I.D. neck if required (not usually).

PHYSICAL DATA (INCHES):

CASE TYPE: **Rimmed Straight**

CASE LENGTH **A = 2.00**

HEAD DIAMETER **B = .425**

RIM DIAMETER **D = .502**

NECK DIAMETER **F = .420**

NECK LENGTH **H = N/A**

SHOULDER LENGTH **K = N/A**

BODY ANGLE (DEG'S/SIDE): **.074**

CASE CAPACITY **CC'S = 2.77**

LOADED LENGTH: **2.72**

BELT DIAMETER **C = N/A**

RIM THICKNESS **E = .062**

SHOULDER DIAMETER **G = N/A**

LENGTH TO SHOULDER **J = N/A**

SHOULDER ANGLE (DEG'S/SIDE): **N/A**

PRIMER: **L/R**

CASE CAPACITY (GR'S WATER): **42.82**

DIMENSIONAL DRAWING:

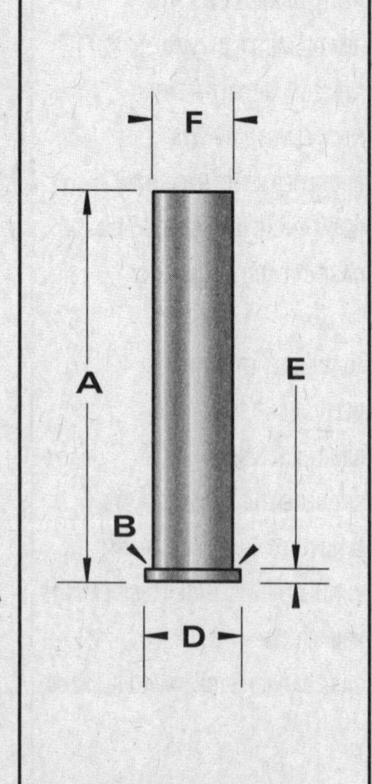

-NOT ACTUAL SIZE-
-DO NOT SCALE-

CARTRIDGE: .38-50 Maynard 1882

OTHER NAMES:

DIA: .375

BALLISTEK NO: 375X

NAI NO: RMS 11115/4.679

DATA SOURCE: COTW 4th Pg102

HISTORICAL DATA:

NOTES: Similar to .38-55 Win (#375P)

LOADING DATA:

BULLET WT./TYPE	POWDER WT./TYPE	VELOCITY ('/SEC)	SOURCE
250/LeadPP	16.0/4198	1320	Barnes

CASE PREPARATION: SHELLHOLDER (RCBS): 2

MAKE FROM: .375 Win. Build-up rim thickness to .075". Trim to length and F/L size. Chamfer. Rim may need turning to .500" dia. but, generally, not.

PHYSICAL DATA (INCHES):

CASE TYPE: **Rimmed Straight**

CASE LENGTH **A = 1.97**

HEAD DIAMETER **B = .421**

RIM DIAMETER **D = .500**

NECK DIAMETER **F = .415**

NECK LENGTH **H = N/A**

SHOULDER LENGTH **K = N/A**

BODY ANGLE (DEG'S/SIDE): **.091**

CASE CAPACITY **CC'S = 2.75**

LOADED LENGTH: **2.38**

BELT DIAMETER **C = N/A**

RIM THICKNESS **E = .075**

SHOULDER DIAMETER **G = N/A**

LENGTH TO SHOULDER **J = N/A**

SHOULDER ANGLE (DEG'S/SIDE): **N/A**

PRIMER: **L/R**

CASE CAPACITY (GR'S WATER): **42.41**

DIMENSIONAL DRAWING:

-NOT ACTUAL SIZE-
-DO NOT SCALE-

CARTRIDGE: .38-50 Remington

OTHER NAMES:

DIA: .376

BALLISTEK NO: 376E

NAI NO: RMS 21115/4.912

DATA SOURCE: COTW 4th Pg103

HISTORICAL DATA: By Rem. in 1883.

NOTES: Similar to .38-55 Win. (#375P).

LOADING DATA:

BULLET WT./TYPE	POWDER WT./TYPE	VELOCITY ('/SEC)	SOURCE
250/Lead	21.5/IMR4198	1400	JJD

CASE PREPARATION: SHELLHOLDER (RCBS): 7

MAKE FROM: .30-40 Krag. Turn rim to .535" dia. Anneal neck and taper expand to .380" dia. Trim to length and F/L size. Fireform in chamber.

PHYSICAL DATA (INCHES):

CASE TYPE: **Rimmed Straight**

CASE LENGTH **A = 2.23**

HEAD DIAMETER **B = .454**

RIM DIAMETER **D = .535**

NECK DIAMETER **F = .392**

NECK LENGTH **H = N/A**

SHOULDER LENGTH **K = N/A**

BODY ANGLE (DEG'S/SIDE): **.82**

CASE CAPACITY **CC'S = 3.95**

LOADED LENGTH: **3.07**

BELT DIAMETER **C = N/A**

RIM THICKNESS **E = .065**

SHOULDER DIAMETER **G = N/A**

LENGTH TO SHOULDER **J = N/A**

SHOULDER ANGLE (DEG'S/SIDE): **N/A**

PRIMER: **L/R**

CASE CAPACITY (GR'S WATER): **60.92**

DIMENSIONAL DRAWING:

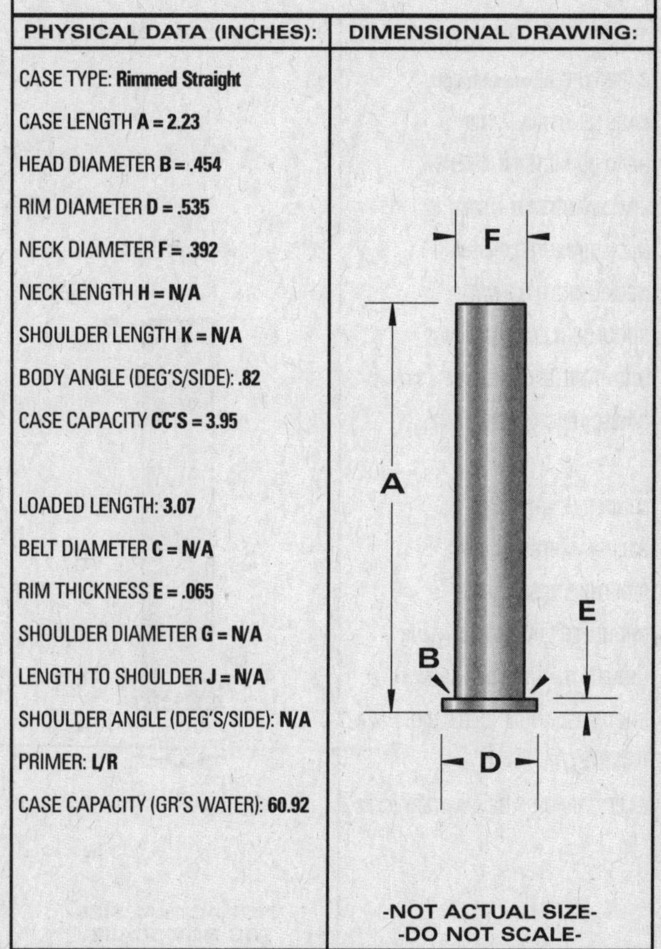

-NOT ACTUAL SIZE-
-DO NOT SCALE-

CARTRIDGE: .38-55 Winchester

OTHER NAMES:
.38-55 Ballard

DIA: .376

BALLISTEK NO: 376P

NAI NO: RMS 21115/5.057

DATA SOURCE: COTW 4th Pg75

HISTORICAL DATA: By Ballard as target ctg. about 1884.

NOTES:

LOADING DATA:

BULLET WT./TYPE	POWDER WT./TYPE	VELOCITY ('/SEC)	SOURCE
255/—	28.0/4895	1400	Barnes

CASE PREPARATION: SHELLHOLDER (RCBS): 2

MAKE FROM: .375 Win. Trim to correct length and F/L size. .30-30 cases may be fireformed, with corn meal, in the proper trim die for a short version of this case.

PHYSICAL DATA (INCHES):

CASE TYPE: **Rimmed Straight**

CASE LENGTH **A = 2.129**

HEAD DIAMETER **B = .421**

RIM DIAMETER **D = .506**

NECK DIAMETER **F = .393**

NECK LENGTH **H = N/A**

SHOULDER LENGTH **K = N/A**

BODY ANGLE (DEG'S/SIDE): **.387**

CASE CAPACITY **CC'S = 3.42**

LOADED LENGTH: **2.49**

BELT DIAMETER **C = N/A**

RIM THICKNESS **E = .061**

SHOULDER DIAMETER **G = N/A**

LENGTH TO SHOULDER **J = N/A**

SHOULDER ANGLE (DEG'S/SIDE): **N/A**

PRIMER: **L/R**

CASE CAPACITY (GR'S WATER): **52.72**

DIMENSIONAL DRAWING:

F
A
E
B
D

-NOT ACTUAL SIZE-
-DO NOT SCALE-

CARTRIDGE: .38-56 Winchester

OTHER NAMES:

DIA: .375

BALLISTEK NO: 375F

NAI NO: RMB 34221/4.168

DATA SOURCE: COTW 4th Pg103

HISTORICAL DATA: By Win. in (or about) 1885 for M86.

NOTES:

LOADING DATA:

BULLET WT./TYPE	POWDER WT./TYPE	VELOCITY ('/SEC)	SOURCE
250/Lead GC	25.0/IMR4198	—	JJD

CASE PREPARATION: SHELLHOLDER (RCBS): 14

MAKE FROM: .45-70 Gov't. Anneal case. Use form die set or first form in .40-65 WCF. F/L size in .38-56 die. Trim to length & chamfer.

PHYSICAL DATA (INCHES):

CASE TYPE: **Rimmed Bottleneck**

CASE LENGTH **A = 2.105**

HEAD DIAMETER **B = .505**

RIM DIAMETER **D = .606**

NECK DIAMETER **F = .402**

NECK LENGTH **H = .633**

SHOULDER LENGTH **K = .245**

BODY ANGLE (DEG'S/SIDE): **1.62**

CASE CAPACITY **CC'S = 4.04**

LOADED LENGTH: **2.60**

BELT DIAMETER **C = N/A**

RIM THICKNESS **E = .07**

SHOULDER DIAMETER **G = .447**

LENGTH TO SHOULDER **J = 1.227**

SHOULDER ANGLE (DEG'S/SIDE): **5.24**

PRIMER: **L/R**

CASE CAPACITY (GR'S WATER): **62.39**

DIMENSIONAL DRAWING:

F
H
K
G
A
J
E
B
D

-NOT ACTUAL SIZE-
-DO NOT SCALE-

CARTRIDGE: .38-70 Winchester

OTHER NAMES:

DIA: .376

BALLISTEK NO: 376A

NAI NO: RMB 32221/4.624

DATA SOURCE: COTW 4th Pg104

HISTORICAL DATA: By Win. in 1894.

NOTES:

LOADING DATA:

BULLET WT./TYPE	POWDER WT./TYPE	VELOCITY ('/SEC)	SOURCE
265/Lead	41.0/3031	1700	Nonte

CASE PREPARATION: SHELLHOLDER (RCBS): 14

MAKE FROM: .45 Basic. Form die required or use .40-65 form die. Anneal case and form. Trim to length & chamfer. F/L size & trim again.

PHYSICAL DATA (INCHES):

CASE TYPE: **Rimmed Bottleneck**

CASE LENGTH **A = 2.34**

HEAD DIAMETER **B = .506**

RIM DIAMETER **D = .600**

NECK DIAMETER **F = .403**

NECK LENGTH **H = .53**

SHOULDER LENGTH **K = .08**

BODY ANGLE (DEG'S/SIDE): **1.59**

CASE CAPACITY **CC'S = 4.54**

LOADED LENGTH: **2.73**

BELT DIAMETER **C = N/A**

RIM THICKNESS **E = .07**

SHOULDER DIAMETER **G = .421**

LENGTH TO SHOULDER **J = 1.73**

SHOULDER ANGLE (DEG'S/SIDE): **6.41**

PRIMER: **L/R**

CASE CAPACITY (GR'S WATER): **70.10**

DIMENSIONAL DRAWING:

-NOT ACTUAL SIZE-
-DO NOT SCALE-

CARTRIDGE: .38-72 Winchester

OTHER NAMES:

DIA: .378

BALLISTEK NO: 378A

NAI NO: RMB 22231/5.596

DATA SOURCE: COTW 4th Pg104

HISTORICAL DATA: By Win. in 1895.

NOTES:

LOADING DATA:

BULLET WT./TYPE	POWDER WT./TYPE	VELOCITY ('/SEC)	SOURCE
250/Lead	28.0/IMR4198	1400	JJD

CASE PREPARATION: SHELLHOLDER (RCBS): 18

MAKE FROM: .405 Basic (BELL). Turn rim to .519" dia. & back chamfer. Thin rim if necessary. Cut case to 2.6" & chamfer. F/L size in die. Trim to length.

PHYSICAL DATA (INCHES):

CASE TYPE: **Rimmed Bottleneck**

CASE LENGTH **A = 2.58**

HEAD DIAMETER **B = .461**

RIM DIAMETER **D = .519**

NECK DIAMETER **F = .397**

NECK LENGTH **H = .41**

SHOULDER LENGTH **K = .220**

BODY ANGLE (DEG'S/SIDE): **.556**

CASE CAPACITY **CC'S = 4.98**

LOADED LENGTH: **3.16**

BELT DIAMETER **C = N/A**

RIM THICKNESS **E = .061**

SHOULDER DIAMETER **G = .427**

LENGTH TO SHOULDER **J = 1.95**

SHOULDER ANGLE (DEG'S/SIDE): **3.90**

PRIMER: **L/R**

CASE CAPACITY (GR'S WATER): **76.86**

DIMENSIONAL DRAWING:

-NOT ACTUAL SIZE-
-DO NOT SCALE-

CARTRIDGE: .38-90 Winchester Express

OTHER NAMES:

DIA: .376

BALLISTEK NO: 376F

NAI NO: RMB 12221/6.813

DATA SOURCE: COTW 4th Pg105

HISTORICAL DATA: By Win. in 1886.

NOTES:

LOADING DATA:

BULLET WT./TYPE	POWDER WT./TYPE	VELOCITY ('/SEC)	SOURCE
225/Lead	23.0/IMR4198	—	JJD

CASE PREPARATION: SHELLHOLDER (RCBS): 13

MAKE FROM: .405 Win Basic (BELL). F/L size the .405 case in the .38-98 die. Load with about 10 grains of Bullseye and corn meal to fill the case. Fireform in a .38-90 trim die to 'blow-out' body. F/L size and trim case to length.

PHYSICAL DATA (INCHES):

CASE TYPE: **Rimmed Bottleneck**
CASE LENGTH **A = 3.25**
HEAD DIAMETER **B = .477**
RIM DIAMETER **D = .558**
NECK DIAMETER **F = .395**
NECK LENGTH **H = .650**
SHOULDER LENGTH **K = .240**
BODY ANGLE (DEG'S/SIDE): **.092**
CASE CAPACITY **CC'S = 7.10**

LOADED LENGTH: **3.70**
BELT DIAMETER **C = N/A**
RIM THICKNESS **E = .06**
SHOULDER DIAMETER **G = .470**
LENGTH TO SHOULDER **J = 2.36**
SHOULDER ANGLE (DEG'S/SIDE): **8.88**
PRIMER: **L/R**
CASE CAPACITY (GR'S WATER): **109.64**

DIMENSIONAL DRAWING:

-NOT ACTUAL SIZE-
-DO NOT SCALE-

CARTRIDGE: .380 Auto

OTHER NAMES:
.380 ACP
9mm Kurz (Corto) (Short)

DIA: .355

BALLISTEK NO: 355E

NAI NO: RXS 11115/1.823

DATA SOURCE: Hornady 3rd Pg331

HISTORICAL DATA: For Colt pistol about 1908.

NOTES: Designed by John Browning.

LOADING DATA:

BULLET WT./TYPE	POWDER WT./TYPE	VELOCITY ('/SEC)	SOURCE
90/JHP	3.8/B'Eye	1000	Horn.

CASE PREPARATION: SHELLHOLDER (RCBS): 10

MAKE FROM: .222 or .223 Rem. cut case off to .68" long. Inside neck ream for .355 projectile. F/L size. Not really worth the effort - commercial cases are easy to find.

PHYSICAL DATA (INCHES):

CASE TYPE: **Rimless Straight**
CASE LENGTH **A = .68**
HEAD DIAMETER **B = .373**
RIM DIAMETER **D = .374**
NECK DIAMETER **F = .372**
NECK LENGTH **H = N/A**
SHOULDER LENGTH **K = N/A**
BODY ANGLE (DEG'S/SIDE): **.327**
CASE CAPACITY **CC'S = .58**

LOADED LENGTH: **.98**
BELT DIAMETER **C = N/A**
RIM THICKNESS **E = .045**
SHOULDER DIAMETER **G = N/A**
LENGTH TO SHOULDER **J = N/A**
SHOULDER ANGLE (DEG'S/SIDE): **N/A**
PRIMER: **S/P**
CASE CAPACITY (GR'S WATER): **8.94**

DIMENSIONAL DRAWING:

-NOT ACTUAL SIZE-
-DO NOT SCALE-

CARTRIDGE: .380 Revolver

OTHER NAMES:

DIA: .375

BALLISTEK NO: 375AN

NAI NO: RMS 11115/1.842

DATA SOURCE: COTW 4th Pg176

HISTORICAL DATA: British ctg. from about 1868.

NOTES: Similar to .38 Colt Short.

LOADING DATA:

BULLET WT./TYPE	POWDER WT./TYPE	VELOCITY ('/SEC)	SOURCE
124/Lead	2.4/B'Eye	550	JJD

CASE PREPARATION: SHELLHOLDER (RCBS): 6

MAKE FROM: .38 Special. Trim case to length. This bullet was inside lubed which means that the bullet is about the same dia. as the case mouth. .375" dia. bullets can be turned to fit the case I.D. - leave the forward portion of the bullet full dia. - which leaves a step behind the crimping groove.

PHYSICAL DATA (INCHES):

CASE TYPE: **Rimmed Straight**

CASE LENGTH **A = .70**

HEAD DIAMETER **B = .380**

RIM DIAMETER **D = .426**

NECK DIAMETER **F = .377**

NECK LENGTH **H = N/A**

SHOULDER LENGTH **K = N/A**

BODY ANGLE (DEG'S/SIDE): **.134**

CASE CAPACITY **CC'S = .715**

LOADED LENGTH: **1.10**

BELT DIAMETER **C = N/A**

RIM THICKNESS **E = .062**

SHOULDER DIAMETER **G = N/A**

LENGTH TO SHOULDER **J = N/A**

SHOULDER ANGLE (DEG'S/SIDE): **N/A**

PRIMER: **S/P**

CASE CAPACITY (GR'S WATER): **11.04**

DIMENSIONAL DRAWING:

-NOT ACTUAL SIZE-
-DO NOT SCALE-

CARTRIDGE: .40-50 Sharps Necked

OTHER NAMES:
.40 - 1 11/16 Sharps

DIA: .403

BALLISTEK NO: 403P

NAI NO: RMB 24221/3.433

DATA SOURCE: COTW 4th Pg106

HISTORICAL DATA: Circa 1875.

NOTES:

LOADING DATA:

BULLET WT./TYPE	POWDER WT./TYPE	VELOCITY ('/SEC)	SOURCE
260/LeadPP	21.0/4198	1500	Barnes

CASE PREPARATION: SHELLHOLDER (RCBS): 14

MAKE FROM: .45-70 Gov't. Turn rim to .580" dia. Anneal case and form in form die. Trim to length and F/L size.

PHYSICAL DATA (INCHES):

CASE TYPE: **Rimmed Bottleneck**

CASE LENGTH **A = 1.72**

HEAD DIAMETER **B = .501**

RIM DIAMETER **D = .580**

NECK DIAMETER **F = .424**

NECK LENGTH **H = .560**

SHOULDER LENGTH **K = .193**

BODY ANGLE (DEG'S/SIDE): **.448**

CASE CAPACITY **CC'S = 3.64**

LOADED LENGTH: **2.37**

BELT DIAMETER **C = N/A**

RIM THICKNESS **E = .068**

SHOULDER DIAMETER **G = .489**

LENGTH TO SHOULDER **J = .967**

SHOULDER ANGLE (DEG'S/SIDE): **9.55**

PRIMER: **L/R**

CASE CAPACITY (GR'S WATER): **56.18**

DIMENSIONAL DRAWING:

-NOT ACTUAL SIZE-
-DO NOT SCALE-

CARTRIDGE: .40-50 Sharps Straight

OTHER NAMES:
.40 - 7/8 Sharps

DIA: .403

BALLISTEK NO: 403G

NAI NO: RMS
21115/4.141

DATA SOURCE: COTW 4th Pg105

HISTORICAL DATA: Circa 1879.

NOTES:

LOADING DATA:

BULLET WT./TYPE	POWDER WT./TYPE	VELOCITY ('/SEC)	SOURCE
260/LeadPP	21.0/4198	1450	Barnes

CASE PREPARATION: SHELLHOLDER (RCBS): 7

MAKE FROM: .303 British. Cut case to 1.9". Anneal case neck and taper expand to .40" dia. F/L size with expander removed. I.D. neck ream. Chamfer. Trim to length and F/L size again.

PHYSICAL DATA (INCHES):

CASE TYPE: **Rimmed Straight**

CASE LENGTH **A = 1.88**

HEAD DIAMETER **B = .454**

RIM DIAMETER **D = .554**

NECK DIAMETER **F = .421**

NECK LENGTH **H = N/A**

SHOULDER LENGTH **K = N/A**

BODY ANGLE (DEG'S/SIDE): **.519**

CASE CAPACITY **CC'S = 3.41**

LOADED LENGTH: **2.63**

BELT DIAMETER **C = N/A**

RIM THICKNESS **E = .06**

SHOULDER DIAMETER **G = N/A**

LENGTH TO SHOULDER **J = N/A**

SHOULDER ANGLE (DEG'S/SIDE): **N/A**

PRIMER: **L/R**

CASE CAPACITY (GR'S WATER): **52.72**

DIMENSIONAL DRAWING:

F
A
E
B
D

-NOT ACTUAL SIZE-
-DO NOT SCALE-

CARTRIDGE: .40-60 Marlin

OTHER NAMES:

DIA: .403

BALLISTEK NO: 403C

NAI NO: RMS
31115/4.186

DATA SOURCE: COTW 4th Pg107

HISTORICAL DATA: For M81 & M95 L/A rifles.

NOTES: Similar to .40-65 Win. (#406A).

LOADING DATA:

BULLET WT./TYPE	POWDER WT./TYPE	VELOCITY ('/SEC)	SOURCE
260/Lead	23.0/4198	1500	Barnes

CASE PREPARATION: SHELLHOLDER (RCBS): 14

MAKE FROM: .45-70 Gov't. Anneal case and F/L size. Square case mouth and chamfer. .40-65 WCF dies may be used; the cases are the same.

(Note: some rifles may require thinning the .45-70's rim to about .050".)

PHYSICAL DATA (INCHES):

CASE TYPE: **Rimmed Straight**

CASE LENGTH **A = 2.11**

HEAD DIAMETER **B = .504**

RIM DIAMETER **D = .604**

NECK DIAMETER **F = .425**

NECK LENGTH **H = N/A**

SHOULDER LENGTH **K = N/A**

BODY ANGLE (DEG'S/SIDE): **1.09**

CASE CAPACITY **CC'S = 4.33**

LOADED LENGTH: **2.55**

BELT DIAMETER **C = N/A**

RIM THICKNESS **E = .049**

SHOULDER DIAMETER **G = N/A**

LENGTH TO SHOULDER **J = N/A**

SHOULDER ANGLE (DEG'S/SIDE): **N/A**

PRIMER: **L/R**

CASE CAPACITY (GR'S WATER): **66.85**

DIMENSIONAL DRAWING:

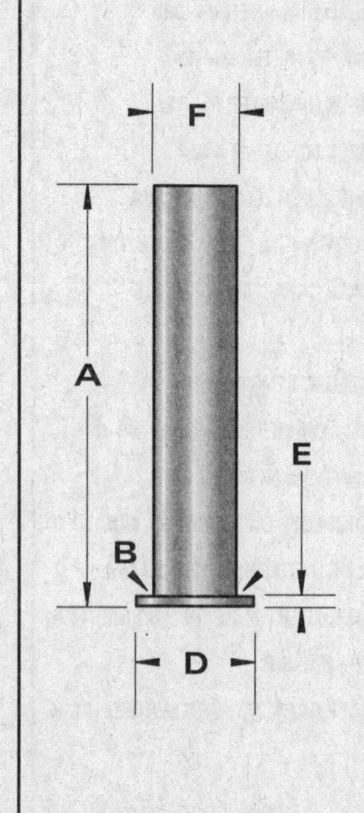

-NOT ACTUAL SIZE-
-DO NOT SCALE-

CARTRIDGE: .40-60 Winchester

OTHER NAMES:

DIA: .404

BALLISTEK NO: 404D

NAI NO: RMB 34245/3.695

DATA SOURCE: COTW 4th Pg106

HISTORICAL DATA: For M76 rifle about 1877.

NOTES:

LOADING DATA:

BULLET WT./TYPE	POWDER WT./TYPE	VELOCITY ('/SEC)	SOURCE
210/Lead	21.0/4198	1520	Barnes

CASE PREPARATION: SHELLHOLDER (RCBS): 14

MAKE FROM: .45-70 Gov't. Anneal case and form in form die. Trim to length, chamfer and F/L size. Use .403" dia. bullets.

PHYSICAL DATA (INCHES):

CASE TYPE: **Rimmed Bottleneck**

CASE LENGTH **A = 1.87**

HEAD DIAMETER **B = .506**

RIM DIAMETER **D = .63**

NECK DIAMETER **F = .425**

NECK LENGTH **H = .350**

SHOULDER LENGTH **K = .30**

BODY ANGLE (DEG'S/SIDE): **1.71**

CASE CAPACITY **CC'S = 3.87**

LOADED LENGTH: **2.10**

BELT DIAMETER **C = N/A**

RIM THICKNESS **E = .05**

SHOULDER DIAMETER **G = .445**

LENGTH TO SHOULDER **J = 1.22**

SHOULDER ANGLE (DEG'S/SIDE): **1.91**

PRIMER: **L/R**

CASE CAPACITY (GR'S WATER): **59.77**

DIMENSIONAL DRAWING:

-NOT ACTUAL SIZE-
-DO NOT SCALE-

CARTRIDGE: .40-65 Ballard Everlasting

OTHER NAMES:

DIA: .403

BALLISTEK NO: 403T

NAI NO: RMS 21115/4.685

DATA SOURCE: COTW 4th Pg107

HISTORICAL DATA: Circa 1876.

NOTES: Very rare!

LOADING DATA:

BULLET WT./TYPE	POWDER WT./TYPE	VELOCITY ('/SEC)	SOURCE
300/Lead	36.6/IMR3031	–	JJD

CASE PREPARATION: SHELLHOLDER (RCBS): 14

MAKE FROM: .45 Basic. Trim case to 2.4". Anneal and form in form die. F/L size, trim to length and chamfer.

PHYSICAL DATA (INCHES):

CASE TYPE: **Rimmed Straight**

CASE LENGTH **A = 2.38**

HEAD DIAMETER **B = .508**

RIM DIAMETER **D = .600**

NECK DIAMETER **F = .435**

NECK LENGTH **H = N/A**

SHOULDER LENGTH **K = N/A**

BODY ANGLE (DEG'S/SIDE): **.905**

CASE CAPACITY **CC'S = 4.79**

LOADED LENGTH: **2.95**

BELT DIAMETER **C = N/A**

RIM THICKNESS **E = .07**

SHOULDER DIAMETER **G = N/A**

LENGTH TO SHOULDER **J = N/A**

SHOULDER ANGLE (DEG'S/SIDE): **N/A**

PRIMER: **L/R**

CASE CAPACITY (GR'S WATER): **73.98**

DIMENSIONAL DRAWING:

-NOT ACTUAL SIZE-
-DO NOT SCALE-

CARTRIDGE: .40-65 Sharps Straight

OTHER NAMES:	
.40- 2 1/2 Sharps	DIA: .403
	BALLISTEK NO: 404C
	NAI NO: RMS 21115/5.519

DATA SOURCE: COTW 4th Pg109

HISTORICAL DATA: Circa 1879.

NOTES:

LOADING DATA:

BULLET WT./TYPE	POWDER WT./TYPE	VELOCITY ('/SEC)	SOURCE
330/LeadPP	36.9/IMR3031	–	JJD

CASE PREPARATION: SHELLHOLDER (RCBS): 7

MAKE FROM: .30-40 Krag. Turn rim to .533" dia. Anneal case and taper expand to .410" dia. Square case mouth. F/L size, trim to length, chamfer and fireform, in chamber.

PHYSICAL DATA (INCHES):

CASE TYPE: **Rimmed Straight**
CASE LENGTH **A = 2.50**
HEAD DIAMETER **B = .453**
RIM DIAMETER **D = .533**
NECK DIAMETER **F = .422**
NECK LENGTH **H = N/A**
SHOULDER LENGTH **K = N/A**
BODY ANGLE (DEG'S/SIDE): **.362**
CASE CAPACITY **CC'S = 4.72**

LOADED LENGTH: **3.18**
BELT DIAMETER **C = N/A**
RIM THICKNESS **E = .05**
SHOULDER DIAMETER **G = N/A**
LENGTH TO SHOULDER **J = N/A**
SHOULDER ANGLE (DEG'S/SIDE): **N/A**
PRIMER: **L/R**
CASE CAPACITY (GR'S WATER): **72.94**

DIMENSIONAL DRAWING:

F
A
E
B
D

-NOT ACTUAL SIZE-
-DO NOT SCALE-

CARTRIDGE: .40-65 Winchester

OTHER NAMES:	
.40-65 WCF	DIA: .406
	BALLISTEK NO: 406A
	NAI NO: RMS 31115/4.166

DATA SOURCE: COTW 4th Pg108

HISTORICAL DATA: By Win. in 1887.

NOTES:

LOADING DATA:

BULLET WT./TYPE	POWDER WT./TYPE	VELOCITY ('/SEC)	SOURCE
260/Lead	23.0/4198	1500	Barnes

CASE PREPARATION: SHELLHOLDER (RCBS): 14

MAKE FROM: .45-70 Gov't. Form case in form die. Square case mouth and F/L size.

PHYSICAL DATA (INCHES):

CASE TYPE: **Rimmed Straight**
CASE LENGTH **A = 2.10**
HEAD DIAMETER **B = .504**
RIM DIAMETER **D = .604**
NECK DIAMETER **F = .423**
NECK LENGTH **H = N/A**
SHOULDER LENGTH **K = N/A**
BODY ANGLE (DEG'S/SIDE): **1.14**
CASE CAPACITY **CC'S = 4.44**

LOADED LENGTH: **2.48**
BELT DIAMETER **C = N/A**
RIM THICKNESS **E = .06**
SHOULDER DIAMETER **G = N/A**
LENGTH TO SHOULDER **J = N/A**
SHOULDER ANGLE (DEG'S/SIDE): **N/A**
PRIMER: **L/R**
CASE CAPACITY (GR'S WATER): **68.52**

DIMENSIONAL DRAWING:

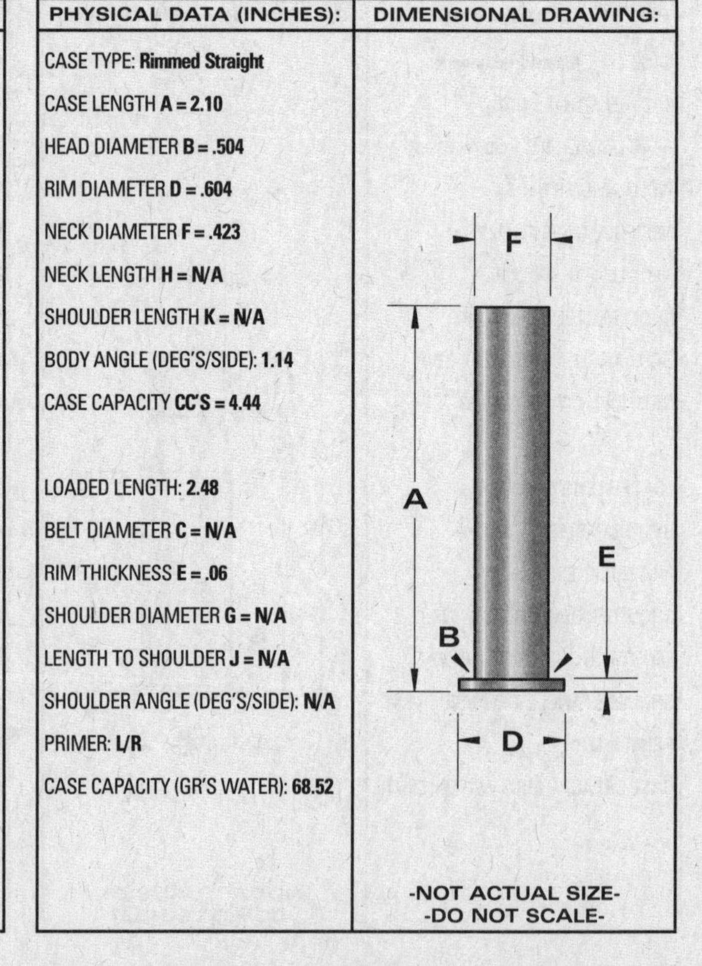

-NOT ACTUAL SIZE-
-DO NOT SCALE-

CARTRIDGE: .40-70 Ballard

OTHER NAMES:
.40-63 Ballard Everlasting

DIA: .403

BALLISTEK NO: 403L

NAI NO: RMS 21115/5.053

DATA SOURCE: COTW 4th Pg107

HISTORICAL DATA: Circa 1881.

NOTES:

LOADING DATA:

BULLET WT./TYPE	POWDER WT./TYPE	VELOCITY ('/SEC)	SOURCE
330/LeadPP	22.0/4198	1310	Barnes

CASE PREPARATION: SHELLHOLDER (RCBS): 21

MAKE FROM: 7 x 57R. Anneal case and taper expand to .410" dia. F/L size, square case mouth and chamfer. Rim is undersize but will headspace okay and case is slightly short.

PHYSICAL DATA (INCHES):

CASE TYPE: **Rimmed Straight**

CASE LENGTH **A = 2.38**

HEAD DIAMETER **B = .471**

RIM DIAMETER **D = .555**

NECK DIAMETER **F = .430**

NECK LENGTH **H = N/A**

SHOULDER LENGTH **K = N/A**

BODY ANGLE (DEG'S/SIDE): **.51**

CASE CAPACITY **CC'S = 4.49**

LOADED LENGTH: **3.07**

BELT DIAMETER **C = N/A**

RIM THICKNESS **E = .063**

SHOULDER DIAMETER **G = N/A**

LENGTH TO SHOULDER **J = N/A**

SHOULDER ANGLE (DEG'S/SIDE): **N/A**

PRIMER: **L/R**

CASE CAPACITY (GR'S WATER): **69.28**

DIMENSIONAL DRAWING:

-NOT ACTUAL SIZE-
-DO NOT SCALE-

CARTRIDGE: .40-70 Peabody What Cheer

OTHER NAMES:

DIA: .408

BALLISTEK NO: 408B

NAI NO: RMB 34241/3.029

DATA SOURCE: COTW 4th Pg108

HISTORICAL DATA: Circa 1877.

NOTES:

LOADING DATA:

BULLET WT./TYPE	POWDER WT./TYPE	VELOCITY ('/SEC)	SOURCE
330/LeadPP	22.0/4198	1350	Barnes

CASE PREPARATION: SHELLHOLDER (RCBS): Spl.

MAKE FROM: 11mm Beaumont (BELL). Cut case to 1.8". F/L size, trim to length and F/L size. Base may require lathe turning to .581" dia.

PHYSICAL DATA (INCHES):

CASE TYPE: **Rimmed Bottleneck**

CASE LENGTH **A = 1.76**

HEAD DIAMETER **B = .581**

RIM DIAMETER **D = .662**

NECK DIAMETER **F = .428**

NECK LENGTH **H = .380**

SHOULDER LENGTH **K = .392**

BODY ANGLE (DEG'S/SIDE): **1.09**

CASE CAPACITY **CC'S = 4.68**

LOADED LENGTH: **2.85**

BELT DIAMETER **C = N/A**

RIM THICKNESS **E = .071**

SHOULDER DIAMETER **G = .551**

LENGTH TO SHOULDER **J = .988**

SHOULDER ANGLE (DEG'S/SIDE): **8.92**

PRIMER: **L/R**

CASE CAPACITY (GR'S WATER): **72.29**

DIMENSIONAL DRAWING:

-NOT ACTUAL SIZE-
-DO NOT SCALE-

CARTRIDGE: .40-70 Remington

OTHER NAMES:

DIA: .405

BALLISTEK NO: 405B

NAI NO: RMB 13221/4.473

DATA SOURCE: COTW 4th Pg110

HISTORICAL DATA: Rem. version of .40-70 Sharps from about 1880.

NOTES: Some cases are 2-piece.

LOADING DATA:

BULLET WT./TYPE	POWDER WT./TYPE	VELOCITY ('/SEC)	SOURCE
330/LeadPP	27.0/4198	1450	Barnes

CASE PREPARATION: SHELLHOLDER (RCBS): 14

MAKE FROM: .45 Basic. Cut case to 2.3" and anneal. Form in form die and trim to length. Chamfer and F/L size.

PHYSICAL DATA (INCHES):

CASE TYPE: **Rimmed Bottleneck**

CASE LENGTH **A = 2.25**

HEAD DIAMETER **B = .503**

RIM DIAMETER **D = .595**

NECK DIAMETER **F = .434**

NECK LENGTH **H = .518**

SHOULDER LENGTH **K = .202**

BODY ANGLE (DEG'S/SIDE): **.064**

CASE CAPACITY **CC'S = 5.06**

LOADED LENGTH: **3.00**

BELT DIAMETER **C = N/A**

RIM THICKNESS **E = .065**

SHOULDER DIAMETER **G = .50**

LENGTH TO SHOULDER **J = 1.53**

SHOULDER ANGLE (DEG'S/SIDE): **9.27**

PRIMER: **L/R**

CASE CAPACITY (GR'S WATER): **78.17**

DIMENSIONAL DRAWING:

-NOT ACTUAL SIZE-
-DO NOT SCALE-

CARTRIDGE: .40-90 Sharps Necked

OTHER NAMES:
.40-100 Sharps Necked
.40 - 2 5/8 Sharps

DIA: .403

BALLISTEK NO: 403H

NAI NO: RMB 12231/5.197

DATA SOURCE: COTW 4th Pg112

HISTORICAL DATA: Circa 1876.

NOTES:

LOADING DATA:

BULLET WT./TYPE	POWDER WT./TYPE	VELOCITY ('/SEC)	SOURCE
370/LeadPP	28.0/4198	1450	Barnes

CASE PREPARATION: SHELLHOLDER (RCBS): 14

MAKE FROM: .45 Basic. Cut case to 2.7". Anneal neck and form in form die. Trim to length, chamfer and F/L size.

PHYSICAL DATA (INCHES):

CASE TYPE: **Rimmed Bottleneck**

CASE LENGTH **A = 2.63**

HEAD DIAMETER **B = .506**

RIM DIAMETER **D = .602**

NECK DIAMETER **F = .435**

NECK LENGTH **H = .50**

SHOULDER LENGTH **K = .33**

BODY ANGLE (DEG'S/SIDE): **.11**

CASE CAPACITY **CC'S = 5.99**

LOADED LENGTH: **3.44**

BELT DIAMETER **C = N/A**

RIM THICKNESS **E = .060**

SHOULDER DIAMETER **G = .50**

LENGTH TO SHOULDER **J = 1.80**

SHOULDER ANGLE (DEG'S/SIDE): **5.62**

PRIMER: **L/R**

CASE CAPACITY (GR'S WATER): **92.57**

DIMENSIONAL DRAWING:

-NOT ACTUAL SIZE-
-DO NOT SCALE-

CARTRIDGE: .40-70 Sharps Necked

OTHER NAMES:	DIA: .403 BALLISTEK NO: 403F NAI NO: RMB 13231/4.473

DATA SOURCE: COTW 4th Pg109

HISTORICAL DATA: From about 1876.

NOTES:

LOADING DATA:

BULLET WT./TYPE	POWDER WT./TYPE	VELOCITY ('/SEC)	SOURCE
300/LeadPP	27.0/4198	1450	Barnes

CASE PREPARATION: SHELLHOLDER (RCBS): 14

MAKE FROM: .45 Basic. Trim case to length and anneal neck. Form in form die. F/L size.

PHYSICAL DATA (INCHES):

CASE TYPE: **Rimmed Bottleneck**

CASE LENGTH **A = 2.25**

HEAD DIAMETER **B = .503**

RIM DIAMETER **D = .595**

NECK DIAMETER **F = .426**

NECK LENGTH **H = .445**

SHOULDER LENGTH **K = .215**

BODY ANGLE (DEG'S/SIDE): **.061**

CASE CAPACITY **CC'S = 5.24**

LOADED LENGTH: **3.02**

BELT DIAMETER **C = N/A**

RIM THICKNESS **E = .061**

SHOULDER DIAMETER **G = .500**

LENGTH TO SHOULDER **J = 1.59**

SHOULDER ANGLE (DEG'S/SIDE): **9.76**

PRIMER: **L/R**

CASE CAPACITY (GR'S WATER): **80.94**

DIMENSIONAL DRAWING:

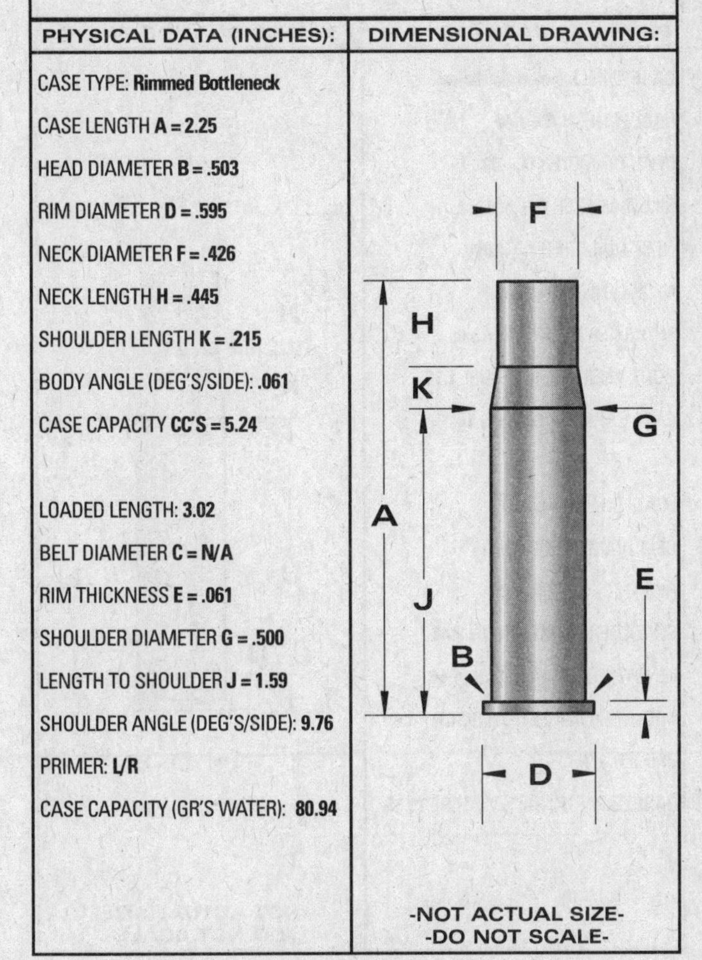

-NOT ACTUAL SIZE-
-DO NOT SCALE-

CARTRIDGE: .40-70 Winchester

OTHER NAMES:	DIA: .405 BALLISTEK NO: 405A NAI NO: RMB 12222/4.762

DATA SOURCE: COTW 4th Pg109

HISTORICAL DATA: Circa 1894.

NOTES:

LOADING DATA:

BULLET WT./TYPE	POWDER WT./TYPE	VELOCITY ('/SEC)	SOURCE
330/Lead	25.0/4198	1380	Barnes

CASE PREPARATION: SHELLHOLDER (RCBS): 14

MAKE FROM: .45 Basic. Cut case to 2.5", anneal and form in form die. Trim to length and F/L size.

PHYSICAL DATA (INCHES):

CASE TYPE: **Rimmed Bottleneck**

CASE LENGTH **A = 2.40**

HEAD DIAMETER **B = .504**

RIM DIAMETER **D = .604**

NECK DIAMETER **F = .430**

NECK LENGTH **H = .51**

SHOULDER LENGTH **K = .140**

BODY ANGLE (DEG'S/SIDE): **.147**

CASE CAPACITY **CC'S = 5.56**

LOADED LENGTH: **2.85**

BELT DIAMETER **C = N/A**

RIM THICKNESS **E = .059**

SHOULDER DIAMETER **G = .496**

LENGTH TO SHOULDER **J = 1.75**

SHOULDER ANGLE (DEG'S/SIDE): **13.26**

PRIMER: **L/R**

CASE CAPACITY (GR'S WATER): **85.94**

DIMENSIONAL DRAWING:

F
H
K
G
A
J
E
B
D

-NOT ACTUAL SIZE-
-DO NOT SCALE-

CARTRIDGE: .40-72 Winchester

OTHER NAMES:

DIA: .406

BALLISTEK NO: 406D

NAI NO: RMS 11115/5.652

DATA SOURCE: COTW 4th Pg110

HISTORICAL DATA: Early Winchester ctg.

NOTES:

LOADING DATA:

BULLET WT./TYPE	POWDER WT./TYPE	VELOCITY ('/SEC)	SOURCE
330/—	40.0/3031	1435	Barnes

CASE PREPARATION: SHELLHOLDER (RCBS): 24

MAKE FROM: .405 Win. Basic (BELL). Turn rim to .518" dia. and thin to .055". Cut case off to 2.65". F/L size with expander removed. I.D. neck ream if required. Chamfer, trim to length and F/L size.

PHYSICAL DATA (INCHES):

CASE TYPE: **Rimmed Straight**

CASE LENGTH **A = 2.60**

HEAD DIAMETER **B = .460**

RIM DIAMETER **D = .518**

NECK DIAMETER **F = .431**

NECK LENGTH **H = N/A**

SHOULDER LENGTH **K = N/A**

BODY ANGLE (DEG'S/SIDE): **.326**

CASE CAPACITY **CC'S = 4.92**

LOADED LENGTH: **3.15**

BELT DIAMETER **C = N/A**

RIM THICKNESS **E = .055**

SHOULDER DIAMETER **G = N/A**

LENGTH TO SHOULDER **J = N/A**

SHOULDER ANGLE (DEG'S/SIDE): **N/A**

PRIMER: **L/R**

CASE CAPACITY (GR'S WATER): **75.88**

DIMENSIONAL DRAWING:

-NOT ACTUAL SIZE-
-DO NOT SCALE-

CARTRIDGE: .40-82 Winchester

OTHER NAMES:
.40-82 WCF

DIA: .406

BALLISTEK NO: 406B

NAI NO: RMB 33235/4.780

DATA SOURCE: COTW 4th Pg111

HISTORICAL DATA: By Win. in 1885.

NOTES:

LOADING DATA:

BULLET WT./TYPE	POWDER WT./TYPE	VELOCITY ('/SEC)	SOURCE
260/Lead	28.0/4198	1425	Barnes

CASE PREPARATION: SHELLHOLDER (RCBS): 14

MAKE FROM: .45 Basic. Cut case to 2.45". Anneal and form in form die. Trim to length & chamfer. F/L size.

PHYSICAL DATA (INCHES):

CASE TYPE: **Rimmed Bottleneck**

CASE LENGTH **A = 2.40**

HEAD DIAMETER **B = .502**

RIM DIAMETER **D = .604**

NECK DIAMETER **F = .428**

NECK LENGTH **H = .45**

SHOULDER LENGTH **K = .350**

BODY ANGLE (DEG'S/SIDE): **1.10**

CASE CAPACITY **CC'S = 5.14**

LOADED LENGTH: **2.77**

BELT DIAMETER **C = N/A**

RIM THICKNESS **E = .06**

SHOULDER DIAMETER **G = .448**

LENGTH TO SHOULDER **J = 1.60**

SHOULDER ANGLE (DEG'S/SIDE): **1.63**

PRIMER: **L/R**

CASE CAPACITY (GR'S WATER): **79.38**

DIMENSIONAL DRAWING:

-NOT ACTUAL SIZE-
-DO NOT SCALE-

CARTRIDGE: .40-85 Ballard

OTHER NAMES:
.40-90 Ballard

DIA: .403

BALLISTEK NO: 403M

NAI NO: RMS 21115/6.164

DATA SOURCE: COTW 4th Pg111

HISTORICAL DATA: Circa 1878.

NOTES:

LOADING DATA:

BULLET WT./TYPE	POWDER WT./TYPE	VELOCITY ('/SEC)	SOURCE
370/LeadPP	28.0/4198	1400	Barnes

CASE PREPARATION: SHELLHOLDER (RCBS): 14

MAKE FROM: .45 Basic. Anneal case and form in form die. Trim to length and F/L size. Turn rim to .545" dia. and thin to .046".

PHYSICAL DATA (INCHES):

CASE TYPE: **Rimmed Straight**

CASE LENGTH **A = 2.94**

HEAD DIAMETER **B = .477**

RIM DIAMETER **D = .545**

NECK DIAMETER **F = .425**

NECK LENGTH **H = N/A**

SHOULDER LENGTH **K = N/A**

BODY ANGLE (DEG'S/SIDE): **.514**

CASE CAPACITY **CC'S = 5.93**

LOADED LENGTH: **3.81**

BELT DIAMETER **C = N/A**

RIM THICKNESS **E = .046**

SHOULDER DIAMETER **G = N/A**

LENGTH TO SHOULDER **J = N/A**

SHOULDER ANGLE (DEG'S/SIDE): **N/A**

PRIMER: **L/R**

CASE CAPACITY (GR'S WATER): **91.49**

DIMENSIONAL DRAWING:

-NOT ACTUAL SIZE-
-DO NOT SCALE-

CARTRIDGE: .40-90 Peabody "What Cheer"

OTHER NAMES:

DIA: .408

BALLISTEK NO: 408C

NAI NO: RMB 34331/3.413

DATA SOURCE: COTW 4th Pg113.

HISTORICAL DATA: Circa 1877.

NOTES:

LOADING DATA:

BULLET WT./TYPE	POWDER WT./TYPE	VELOCITY ('/SEC)	SOURCE
330/LeadPP	27.0/4198	1450	Barnes

CASE PREPARATION: SHELLHOLDER (RCBS): Spl.

MAKE FROM: 11mm Beaumont (BELL). Cut case to 2.1". size in F/L die with expander removed. Case may need annealing. Trim to length & chamfer. F/L size.

PHYSICAL DATA (INCHES):

CASE TYPE: **Rimmed Bottleneck**

CASE LENGTH **A = 2.00**

HEAD DIAMETER **B = .586**

RIM DIAMETER **D = .659**

NECK DIAMETER **F = .433**

NECK LENGTH **H = .430**

SHOULDER LENGTH **K = .325**

BODY ANGLE (DEG'S/SIDE): **1.09**

CASE CAPACITY **CC'S = 5.43**

LOADED LENGTH: **3.37**

BELT DIAMETER **C = N/A**

RIM THICKNESS **E = .07**

SHOULDER DIAMETER **G = .546**

LENGTH TO SHOULDER **J = 1.245**

SHOULDER ANGLE (DEG'S/SIDE): **9.86**

PRIMER: **L/R**

CASE CAPACITY (GR'S WATER): **83.75**

DIMENSIONAL DRAWING:

-NOT ACTUAL SIZE-
-DO NOT SCALE-

CARTRIDGE: .40-90 Sharps Straight

OTHER NAMES:

DIA: .403

BALLISTEK NO: 403K

NAI NO: RMS 21115/6.813

DATA SOURCE: COTW 4th Pg112

HISTORICAL DATA: Circa 1884.

NOTES: Similar to .40-90 Ballard (#403M).

LOADING DATA:

BULLET WT./TYPE	POWDER WT./TYPE	VELOCITY ('/SEC)	SOURCE
370/LeadPP	30.0/4198	1400	Barnes

CASE PREPARATION: SHELLHOLDER (RCBS): 14

MAKE FROM: .45 Basic. Anneal case and form in form die. Make certain that base is swaged (or turned) to .477" dia. Square case mouth, chamfer and F/L size. Turn rim to .546" dia. and back chamfer.

PHYSICAL DATA (INCHES):

CASE TYPE: **Rimmed Straight**

CASE LENGTH **A = 3.25**

HEAD DIAMETER **B = .477**

RIM DIAMETER **D = .546**

NECK DIAMETER **F = .425**

NECK LENGTH **H = N/A**

SHOULDER LENGTH **K = N/A**

BODY ANGLE (DEG'S/SIDE): **.467**

CASE CAPACITY **CC'S = 6.62**

LOADED LENGTH: **4.06**

BELT DIAMETER **C = N/A**

RIM THICKNESS **E = .065**

SHOULDER DIAMETER **G = N/A**

LENGTH TO SHOULDER **J = N/A**

SHOULDER ANGLE (DEG'S/SIDE): **N/A**

PRIMER: **L/R**

CASE CAPACITY (GR'S WATER): **102.25**

DIMENSIONAL DRAWING:

-NOT ACTUAL SIZE-
-DO NOT SCALE-

CARTRIDGE: .40-110 Winchester Express

OTHER NAMES:

DIA: .403

BALLISTEK NO: 403R

NAI NO: RMB 22221/5.985

DATA SOURCE: COTW 4th Pg113

HISTORICAL DATA: Circa 1886.

NOTES:

LOADING DATA:

BULLET WT./TYPE	POWDER WT./TYPE	VELOCITY ('/SEC)	SOURCE
260/Lead	32.0/4198	1617	Barnes

CASE PREPARATION: SHELLHOLDER (RCBS): 31

MAKE FROM: .450 N.E. (BELL). Anneal case and form in form die. Square case mouth. Build up rim thickness to .06". F/L size & chamfer.

PHYSICAL DATA (INCHES):

CASE TYPE: **Rimmed Bottleneck**

CASE LENGTH **A = 3.25**

HEAD DIAMETER **B = .543**

RIM DIAMETER **D = .651**

NECK DIAMETER **F = .428**

NECK LENGTH **H = .645**

SHOULDER LENGTH **K = .245**

BODY ANGLE (DEG'S/SIDE): **.769**

CASE CAPACITY **CC'S = 8.12**

LOADED LENGTH: **3.63**

BELT DIAMETER **C = N/A**

RIM THICKNESS **E = .06**

SHOULDER DIAMETER **G = .485**

LENGTH TO SHOULDER **J = 2.36**

SHOULDER ANGLE (DEG'S/SIDE): **6.63**

PRIMER: **L/R**

CASE CAPACITY (GR'S WATER): **125.33**

DIMENSIONAL DRAWING:

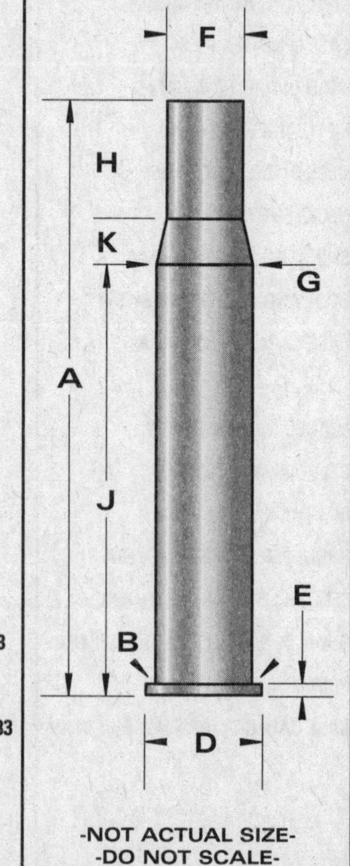

-NOT ACTUAL SIZE-
-DO NOT SCALE-

CARTRIDGE: .400 Nitro for Black Powder 3″

OTHER NAMES: .400 Purdey 3″

DIA: .395

BALLISTEK NO: 395A

NAI NO: RMS 21115/6.369

DATA SOURCE: COTW 4th Pg238

HISTORICAL DATA: Circa 1880.

NOTES:

LOADING DATA:

BULLET WT./TYPE	POWDER WT./TYPE	VELOCITY ('/SEC)	SOURCE
230/Lead	51.0/4895	2000	Barnes

CASE PREPARATION: SHELLHOLDER (RCBS): Spl.

MAKE FROM: 9.3 x 74R. Anneal case and taper expand to .400″ dia. Square case mouth and F/L size. Case will fire-form in the chamber.

PHYSICAL DATA (INCHES):

CASE TYPE: **Rimmed Straight**
CASE LENGTH **A = 3.00**
HEAD DIAMETER **B = .471**
RIM DIAMETER **D = .522**
NECK DIAMETER **F = .427**
NECK LENGTH **H = N/A**
SHOULDER LENGTH **K = N/A**
BODY ANGLE (DEG'S/SIDE): **.429**
CASE CAPACITY **CC'S = 5.58**

LOADED LENGTH: **3.55**
BELT DIAMETER **C = N/A**
RIM THICKNESS **E = .048**
SHOULDER DIAMETER **G = N/A**
LENGTH TO SHOULDER **J = N/A**
SHOULDER ANGLE (DEG'S/SIDE): **N/A**
PRIMER: **L/R**
CASE CAPACITY (GR'S WATER): **86.10**

DIMENSIONAL DRAWING:

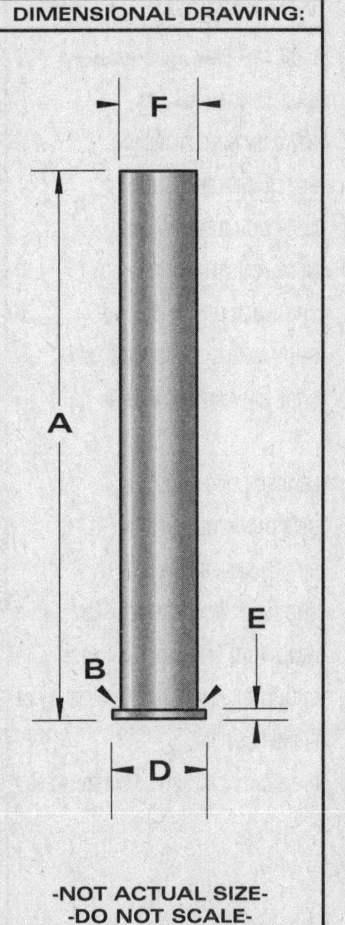

CARTRIDGE: .400 Williams

OTHER NAMES:

DIA: .411

BALLISTEK NO: 411B

NAI NO: BEN 22243/4.873

DATA SOURCE: Ackley Vol1 Pg490

HISTORICAL DATA: By W. Williams in 1944.

NOTES:

LOADING DATA:

BULLET WT./TYPE	POWDER WT./TYPE	VELOCITY ('/SEC)	SOURCE
250/RN	65.0/4198	2400	Ackley

CASE PREPARATION: SHELLHOLDER (RCBS): 4

MAKE FROM: .458 Win. Mag. F/L size the .458 case in the .400 die. Square case mouth and chamfer.

PHYSICAL DATA (INCHES):

CASE TYPE: **Belted Bottleneck**
CASE LENGTH **A = 2.500**
HEAD DIAMETER **B = .513**
RIM DIAMETER **D = .532**
NECK DIAMETER **F = .439**
NECK LENGTH **H = .360**
SHOULDER LENGTH **K = .05**
BODY ANGLE (DEG'S/SIDE): **.500**
CASE CAPACITY **CC'S = 5.80**

LOADED LENGTH: **3.31**
BELT DIAMETER **C = .532**
RIM THICKNESS **E = .05**
SHOULDER DIAMETER **G = .480**
LENGTH TO SHOULDER **J = 2.09**
SHOULDER ANGLE (DEG'S/SIDE): **22.29**
PRIMER: **L/R Mag**
CASE CAPACITY (GR'S WATER): **89.65**

DIMENSIONAL DRAWING:

-NOT ACTUAL SIZE-
-DO NOT SCALE-

CARTRIDGE: .400/350 Rigby Nitro Express

OTHER NAMES:	
.400/350 Nitro Express .350 Rigby	DIA: .357 BALLISTEK NO: 357P NAI NO: RMB 22222/5.851

DATA SOURCE: COTW 4th Pg234

HISTORICAL DATA: By J. Rigby in 1899.

NOTES:

LOADING DATA:

BULLET WT./TYPE	POWDER WT./TYPE	VELOCITY ('/SEC)	SOURCE
300/RN	58.0/4350	2180	Barnes

CASE PREPARATION: SHELLHOLDER (RCBS): 4

MAKE FROM: 9.3 x 74R. Trim case to length. F/L size. Case will fireform in the chamber.

PHYSICAL DATA (INCHES):

CASE TYPE: **Rimmed Bottleneck**
CASE LENGTH **A = 2.75**
HEAD DIAMETER **B = .470**
RIM DIAMETER **D = .520**
NECK DIAMETER **F = .380**
NECK LENGTH **H = .495**
SHOULDER LENGTH **K = .055**
BODY ANGLE (DEG'S/SIDE): **.787**
CASE CAPACITY **CC'S = 5.10**

LOADED LENGTH: **3.55**
BELT DIAMETER **C = N/A**
RIM THICKNESS **E = .047**
SHOULDER DIAMETER **G = .415**
LENGTH TO SHOULDER **J = 2.20**
SHOULDER ANGLE (DEG'S/SIDE): **17.65**
PRIMER: **L/R**
CASE CAPACITY (GR'S WATER): **78.73**

DIMENSIONAL DRAWING:

-NOT ACTUAL SIZE-
-DO NOT SCALE-

CARTRIDGE: .400/360 Nitro Express 2 3/4"

OTHER NAMES:	
.400/360 Purdey .400/360 Westley Richards	DIA: .367 BALLISTEK NO: 367B NAI NO: RMB 22222/5.652

DATA SOURCE: NAI/Ballistek

HISTORICAL DATA: Circa 1900.

NOTES:

LOADING DATA:

BULLET WT./TYPE	POWDER WT./TYPE	VELOCITY ('/SEC)	SOURCE
300/RN	40.0/IMR4320	–	JJD

CASE PREPARATION: SHELLHOLDER (RCBS): 13

MAKE FROM: .45 Colt + tubing extension or lathe turn from solid brass. Anneal case and form in form die. Trim to length, chamfer & F/L size.

PHYSICAL DATA (INCHES):

CASE TYPE: **Rimmed Bottleneck**
CASE LENGTH **A = 2.73**
HEAD DIAMETER **B = .467**
RIM DIAMETER **D = .522**
NECK DIAMETER **F = .387**
NECK LENGTH **H = .475**
SHOULDER LENGTH **K = .10**
BODY ANGLE (DEG'S/SIDE): **.674**
CASE CAPACITY **CC'S = 5.96**

LOADED LENGTH: **3.48**
BELT DIAMETER **C = N/A**
RIM THICKNESS **E = .047**
SHOULDER DIAMETER **G = .437**
LENGTH TO SHOULDER **J = 2.155**
SHOULDER ANGLE (DEG'S/SIDE): **17.22**
PRIMER: **L/R**
CASE CAPACITY (GR'S WATER): **92.06**

DIMENSIONAL DRAWING:

-NOT ACTUAL SIZE-
-DO NOT SCALE-

CARTRIDGE: .400/375 H&H Nitro Express

OTHER NAMES:
.400/375 Belted Nitro Express

DIA: .371

BALLISTEK NO: 371A

NAI NO: BEN 12221/5.513

DATA SOURCE: COTW 4th Pg235

HISTORICAL DATA: By Holland & Holland about 1905.

NOTES:

LOADING DATA:

BULLET WT./TYPE	POWDER WT./TYPE	VELOCITY ('/SEC)	SOURCE
270/Spire	42.0/3031	2000	Nonte

CASE PREPARATION: SHELLHOLDER (RCBS): 3

MAKE FROM: .240 Weatherby. Turn belt and rim to .466" dia. Back chamfer rim. Taper expand case neck (anneal first!) to .380" dia. F/L size, trim to length and chamfer.

PHYSICAL DATA (INCHES):

CASE TYPE: **Belted Bottleneck**
CASE LENGTH **A = 2.47**
HEAD DIAMETER **B = .448**
RIM DIAMETER **D = .466**
NECK DIAMETER **F = .397**
NECK LENGTH **H = .640**
SHOULDER LENGTH **K = .160**
BODY ANGLE (DEG'S/SIDE): **.058**
CASE CAPACITY **CC'S = 4.50**
LOADED LENGTH: **3.00**
BELT DIAMETER **C = .465**
RIM THICKNESS **E = .062**
SHOULDER DIAMETER **G = .445**
LENGTH TO SHOULDER **J = 1.67**
SHOULDER ANGLE (DEG'S/SIDE): **8.53**
PRIMER: **L/R**
CASE CAPACITY (GR'S WATER): **69.48**

DIMENSIONAL DRAWING:

-NOT ACTUAL SIZE-
-DO NOT SCALE-

CARTRIDGE: .401 Winchester Self Loading

OTHER NAMES:

DIA: .406

BALLISTEK NO: 406C

NAI NO: RMS 11115/3.496

DATA SOURCE: COTW 4th Pg113

HISTORICAL DATA: By Win. in 1910 for M10.

NOTES:

LOADING DATA:

BULLET WT./TYPE	POWDER WT./TYPE	VELOCITY ('/SEC)	SOURCE
200/—	31.0/4227	2100	Barnes

CASE PREPARATION: SHELLHOLDER (RCBS): 4

MAKE FROM: 9.3 x 72R. Turn rim to .457" dia. and back chamfer. Cut case to 1.55" and taper expand to .410" dia. Trim to length and size in .401 die with the expander removed. I.D. neck ream. Chamfer and F/L size. Fireform in chamber.

PHYSICAL DATA (INCHES):

CASE TYPE: **Rimmed Straight**
CASE LENGTH **A = 1.50**
HEAD DIAMETER **B = .429**
RIM DIAMETER **D = .457**
NECK DIAMETER **F = .428**
NECK LENGTH **H = N/A**
SHOULDER LENGTH **K = N/A**
BODY ANGLE (DEG'S/SIDE): **.019**
CASE CAPACITY **CC'S = 2.41**
LOADED LENGTH: **2.00**
BELT DIAMETER **C = N/A**
RIM THICKNESS **E = .042**
SHOULDER DIAMETER **G = N/A**
LENGTH TO SHOULDER **J = N/A**
SHOULDER ANGLE (DEG'S/SIDE): **N/A**
PRIMER: **L/R**
CASE CAPACITY (GR'S WATER): **37.13**

DIMENSIONAL DRAWING:

-NOT ACTUAL SIZE-
-DO NOT SCALE-

CARTRIDGE: .40 x 2″

OTHER NAMES:	
10.6 x 51mm .416 x 2″	DIA: .416 BALLISTEK NO: 416D NAI NO: BEN 13243/3.898

DATA SOURCE: Rifle #50 Pg36

HISTORICAL DATA: By O.A. Winters in 1976.

NOTES:

LOADING DATA:

BULLET WT./TYPE	POWDER WT./TYPE	VELOCITY ('/SEC)	SOURCE
300/RN	49.0/4198	2332	Winters

CASE PREPARATION: SHELLHOLDER (RCBS): 4

MAKE FROM: 7mm Rem. Mag. (or any similar magnum case). Form new case neck in form die. Trim to length and I.D. neck ream. Chamfer & F/L size.

PHYSICAL DATA (INCHES):

CASE TYPE: **Belted Bottleneck**
CASE LENGTH **A = 2.00**
HEAD DIAMETER **B = .513**
RIM DIAMETER **D = .532**
NECK DIAMETER **F = .446**
NECK LENGTH **H = .370**
SHOULDER LENGTH **K = .044**
BODY ANGLE (DEG'S/SIDE): **.31**
CASE CAPACITY **CC'S = 4.60**

LOADED LENGTH: **2.80**
BELT DIAMETER **C = .532**
RIM THICKNESS **E = .049**
SHOULDER DIAMETER **G = .498**
LENGTH TO SHOULDER **J = 1.586**
SHOULDER ANGLE (DEG'S/SIDE): **30.58**
PRIMER: **L/R Mag**
CASE CAPACITY (GR'S WATER): **70.94**

DIMENSIONAL DRAWING:

-NOT ACTUAL SIZE-
-DO NOT SCALE-

CARTRIDGE: .40-40 Maynard 1865

OTHER NAMES:	
	DIA: .423 BALLISTEK NO: 423C NAI NO: RMS 11115/3.821

DATA SOURCE: NAI/Ballistek

HISTORICAL DATA:

NOTES:

LOADING DATA:

BULLET WT./TYPE	POWDER WT./TYPE	VELOCITY ('/SEC)	SOURCE
230/Lead	19.3/IMR4198	–	JJD

CASE PREPARATION: SHELLHOLDER (RCBS): Spl.

MAKE FROM: Increase the rim dia. of a .30-40 Krag. Anneal case and taper expand to .430″ dia. Trim to length, I.D. neck ream and F/L size. This case can also be turned from solid brass.

PHYSICAL DATA (INCHES):

CASE TYPE: **Rimmed Straight**
CASE LENGTH **A = 1.75**
HEAD DIAMETER **B = .458**
RIM DIAMETER **D = .766**
NECK DIAMETER **F = .450**
NECK LENGTH **H = N/A**
SHOULDER LENGTH **K = N/A**
BODY ANGLE (DEG'S/SIDE): **.136**
CASE CAPACITY **CC'S = 3.18**

LOADED LENGTH: **2.24**
BELT DIAMETER **C = N/A**
RIM THICKNESS **E = .07**
SHOULDER DIAMETER **G = N/A**
LENGTH TO SHOULDER **J = N/A**
SHOULDER ANGLE (DEG'S/SIDE): **N/A**
PRIMER: **L/R**
CASE CAPACITY (GR'S WATER): **49.08**

DIMENSIONAL DRAWING:

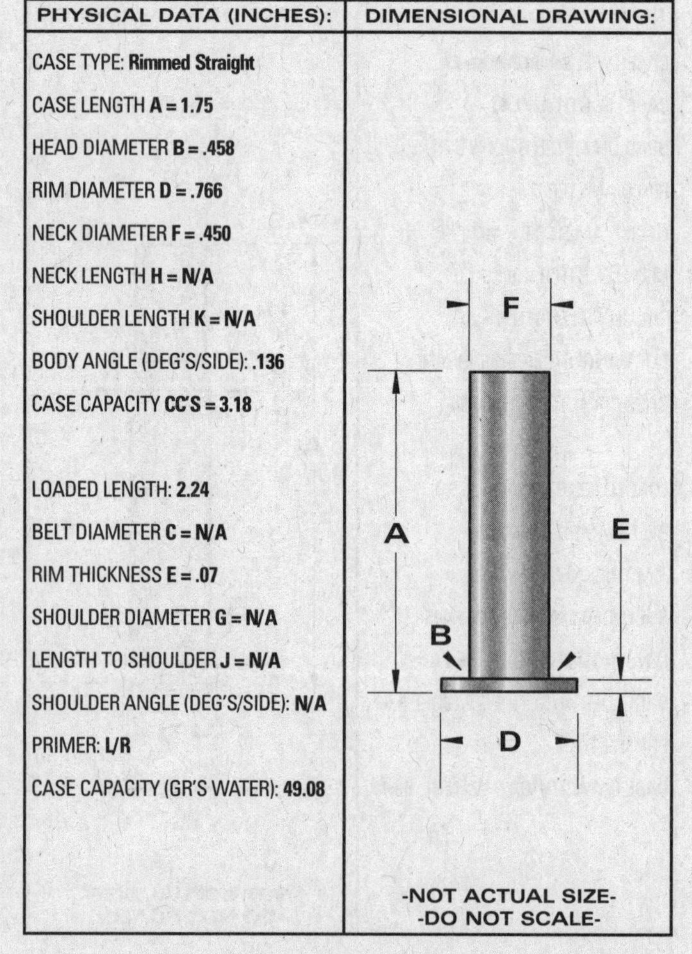

-NOT ACTUAL SIZE-
-DO NOT SCALE-

CARTRIDGE: .40-40 Maynard 1873

OTHER NAMES:

DIA: .422

BALLISTEK NO: 422A

NAI NO: RMS 11115/4.000

DATA SOURCE: NAI Ballistek

HISTORICAL DATA:

NOTES:

LOADING DATA:

BULLET WT./TYPE	POWDER WT./TYPE	VELOCITY ('/SEC)	SOURCE
200/Lead	21.4/IMR4198	—	JJD

CASE PREPARATION: SHELLHOLDER (RCBS): Spl.

MAKE FROM: Increase rim dia. of BELL's .405 Win. Basic or turn form solid brass. Trim case to length, chamfer and F/L size.

PHYSICAL DATA (INCHES):

CASE TYPE: **Rimmed Straight**

CASE LENGTH **A = 1.84**

HEAD DIAMETER **B = .460**

RIM DIAMETER **D = .743**

NECK DIAMETER **F = .450**

NECK LENGTH **H = N/A**

SHOULDER LENGTH **K = N/A**

BODY ANGLE (DEG'S/SIDE): **.161**

CASE CAPACITY **CC'S = 3.37**

LOADED LENGTH: **2.34**

BELT DIAMETER **C = N/A**

RIM THICKNESS **E = .07**

SHOULDER DIAMETER **G = N/A**

LENGTH TO SHOULDER **J = N/A**

SHOULDER ANGLE (DEG'S/SIDE): **N/A**

PRIMER: **L/R**

CASE CAPACITY (GR'S WATER): **52.02**

DIMENSIONAL DRAWING:

-NOT ACTUAL SIZE-
-DO NOT SCALE-

CARTRIDGE: .40-40 Maynard 1882

OTHER NAMES:

DIA: .415

BALLISTEK NO: 415C

NAI NO: RMS 11115/3.903

DATA SOURCE: COTW 4th Pg105

HISTORICAL DATA:

NOTES:

LOADING DATA:

BULLET WT./TYPE	POWDER WT./TYPE	VELOCITY ('/SEC)	SOURCE
260/Lead	24.0/4198	1400	Barnes

CASE PREPARATION: SHELLHOLDER (RCBS): 7

MAKE FROM: .303 British. Cut case to 1.8" and either taper expand to straight walled case or fireform, with corn meal, in the Maynard trim die. Trim to length, I.D. neck ream and F/L size.

PHYSICAL DATA (INCHES):

CASE TYPE: **Rimmed Straight**

CASE LENGTH **A = 1.78**

HEAD DIAMETER **B = .456**

RIM DIAMETER **D = .532**

NECK DIAMETER **F = .450**

NECK LENGTH **H = N/A**

SHOULDER LENGTH **K = N/A**

BODY ANGLE (DEG'S/SIDE): **.10**

CASE CAPACITY **CC'S = 3.04**

LOADED LENGTH: **2.32**

BELT DIAMETER **C = N/A**

RIM THICKNESS **E = .07**

SHOULDER DIAMETER **G = N/A**

LENGTH TO SHOULDER **J = N/A**

SHOULDER ANGLE (DEG'S/SIDE): **N/A**

PRIMER: **L/R**

CASE CAPACITY (GR'S WATER): **47.01**

DIMENSIONAL DRAWING:

-NOT ACTUAL SIZE-
-DO NOT SCALE-

CARTRIDGE: .40-60 Maynard 1882

OTHER NAMES:

DIA: .417

BALLISTEK NO: 415D

NAI NO: RMS 11115/4.845

DATA SOURCE: COTW 4th Pg106

HISTORICAL DATA:

NOTES:

LOADING DATA:

BULLET WT./TYPE	POWDER WT./TYPE	VELOCITY ('/SEC)	SOURCE
300/Lead	26.0/4198	1370	Barnes

CASE PREPARATION: SHELLHOLDER (RCBS): 7

MAKE FROM: .303 British. Either taper expand to straight walled case or fireform the .303 case, with corn meal, in the proper trim die. Square the case mouth, chamfer and F/L size.

PHYSICAL DATA (INCHES):

CASE TYPE: **Rimmed Straight**

CASE LENGTH **A = 2.20**

HEAD DIAMETER **B = .454**

RIM DIAMETER **D = .533**

NECK DIAMETER **F = .448**

NECK LENGTH **H = N/A**

SHOULDER LENGTH **K = N/A**

BODY ANGLE (DEG'S/SIDE): **.08**

CASE CAPACITY **CC'S = 3.99**

LOADED LENGTH: **2.75**

BELT DIAMETER **C = N/A**

RIM THICKNESS **E = .057**

SHOULDER DIAMETER **G = N/A**

LENGTH TO SHOULDER **J = N/A**

SHOULDER ANGLE (DEG'S/SIDE): **N/A**

PRIMER: **L/R**

CASE CAPACITY (GR'S WATER): **61.61**

DIMENSIONAL DRAWING:

-NOT ACTUAL SIZE-
-DO NOT SCALE-

CARTRIDGE: .40-70 Maynard 1882

OTHER NAMES:

DIA: .417

BALLISTEK NO: 417A

NAI NO: RMS 11115/5.366

DATA SOURCE: COTW 4th Pg108

HISTORICAL DATA:

NOTES:

LOADING DATA:

BULLET WT./TYPE	POWDER WT./TYPE	VELOCITY ('/SEC)	SOURCE
260/Lead	27.0/4198	1450	Barnes

CASE PREPARATION: SHELLHOLDER (RCBS): 7

MAKE FROM: .30-40 Krag. Turn rim to .535" dia. (not always necessary). Taper expand to .425" dia. and square case mouth. Chamfer and fireform, in chamber. This case is slightly short.

PHYSICAL DATA (INCHES):

CASE TYPE: **Rimmed Straight**

CASE LENGTH **A = 2.42**

HEAD DIAMETER **B =.451**

RIM DIAMETER **D = .535**

NECK DIAMETER **F = .450**

NECK LENGTH **H = N/A**

SHOULDER LENGTH **K = N/A**

BODY ANGLE (DEG'S/SIDE): **.012**

CASE CAPACITY **CC'S = 4.38**

LOADED LENGTH: **2.88**

BELT DIAMETER **C = N/A**

RIM THICKNESS **E = .065**

SHOULDER DIAMETER **G = N/A**

LENGTH TO SHOULDER **J = N/A**

SHOULDER ANGLE (DEG'S/SIDE): **N/A**

PRIMER: **L/R**

CASE CAPACITY (GR'S WATER): **67.59**

DIMENSIONAL DRAWING:

-NOT ACTUAL SIZE-
-DO NOT SCALE-

CARTRIDGE: .40-75 Bullard

OTHER NAMES:	
.40-60 Bullard	DIA: .413
	BALLISTEK NO: 413A
	NAI NO: RMS 31115/4.138

DATA SOURCE: COTW 4th Pg110

HISTORICAL DATA: Circa 1887.

NOTES:

LOADING DATA:

BULLET WT./TYPE	POWDER WT./TYPE	VELOCITY ('/SEC)	SOURCE
260/Lead	20.0/4198	1513	Barnes

CASE PREPARATION: SHELLHOLDER (RCBS): 14

MAKE FROM: .45-70 Gov't. Anneal case and form in form die. Trim to length and F/L size. Undersized bullets can be paper-patched to diameter.

PHYSICAL DATA (INCHES):

CASE TYPE: **Rimmed Straight**

CASE LENGTH **A = 2.09**

HEAD DIAMETER **B = .505**

RIM DIAMETER **D = .606**

NECK DIAMETER **F = .432**

NECK LENGTH **H = N/A**

SHOULDER LENGTH **K = N/A**

BODY ANGLE (DEG'S/SIDE): **1.04**

CASE CAPACITY **CC'S = 4.45**

LOADED LENGTH: **2.54**

BELT DIAMETER **C = N/A**

RIM THICKNESS **E = .085**

SHOULDER DIAMETER **G = N/A**

LENGTH TO SHOULDER **J = N/A**

SHOULDER ANGLE (DEG'S/SIDE): **N/A**

PRIMER: **L/R**

CASE CAPACITY (GR'S WATER): **68.73**

DIMENSIONAL DRAWING:

F

A

E

B

D

-NOT ACTUAL SIZE-
-DO NOT SCALE-

CARTRIDGE: .40-90 Bullard

OTHER NAMES:	
	DIA: .413
	BALLISTEK NO: 413B
	NAI NO: RMB 24241/3.585

DATA SOURCE: COTW 4th Pg111

HISTORICAL DATA: Circa 1886.

NOTES:

LOADING DATA:

BULLET WT./TYPE	POWDER WT./TYPE	VELOCITY ('/SEC)	SOURCE
300/Lead	29.0/4198	1450	Barnes

CASE PREPARATION: SHELLHOLDER (RCBS): Spl.

MAKE FROM: .50 Sharps (BELL). Turn rim to .622" dia. Cut case to 2.1" and anneal. Form in form die. Trim to length and chamfer. F/L size. Paper patch undersize lead bullets to proper diameter.

PHYSICAL DATA (INCHES):

CASE TYPE: **Rimmed Bottleneck**

CASE LENGTH **A = 2.04**

HEAD DIAMETER **B = .569**

RIM DIAMETER **D = .622**

NECK DIAMETER **F = .430**

NECK LENGTH **H = .380**

SHOULDER LENGTH **K = .330**

BODY ANGLE (DEG'S/SIDE): **.456**

CASE CAPACITY **CC'S = 5.78**

LOADED LENGTH: **2.55**

BELT DIAMETER **C = N/A**

RIM THICKNESS **E = .058**

SHOULDER DIAMETER **G = .551**

LENGTH TO SHOULDER **J = 1.33**

SHOULDER ANGLE (DEG'S/SIDE): **10.39**

PRIMER: **L/R**

CASE CAPACITY (GR'S WATER): **89.29.**

DIMENSIONAL DRAWING:

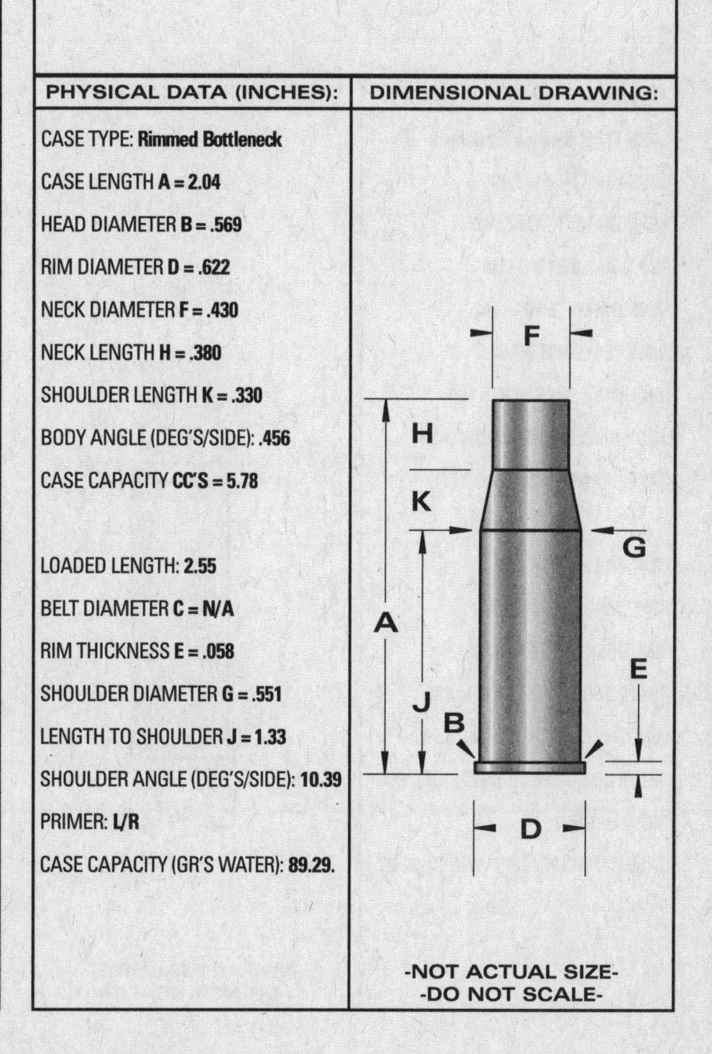

-NOT ACTUAL SIZE-
-DO NOT SCALE-

CARTRIDGE: .40/348 Winchester

OTHER NAMES:

DIA: .411

BALLISTEK NO: 411F

NAI NO: RMB 33222/4.068

DATA SOURCE: NAI/Ballistek

HISTORICAL DATA:

NOTES:

LOADING DATA:

BULLET WT./TYPE	POWDER WT./TYPE	VELOCITY ('/SEC)	SOURCE
300/RN	60.0/3031	2100	JJD

CASE PREPARATION: SHELLHOLDER (RCBS): 5

MAKE FROM: .348 Winchester. Anneal case neck and taper expand to .420" dia. Square case mouth, chamfer and F/L size.

PHYSICAL DATA (INCHES):

CASE TYPE: **Rimmed Bottleneck**

CASE LENGTH **A = 2.25**

HEAD DIAMETER **B = .553**

RIM DIAMETER **D = .610**

NECK DIAMETER **F = .442**

NECK LENGTH **H = .560**

SHOULDER LENGTH **K = .109**

BODY ANGLE (DEG'S/SIDE): **1.39**

CASE CAPACITY **CC'S = 5.33**

LOADED LENGTH: **A/R**

BELT DIAMETER **C = N/A**

RIM THICKNESS **E = .07**

SHOULDER DIAMETER **G = .486**

LENGTH TO SHOULDER **J = 1.58**

SHOULDER ANGLE (DEG'S/SIDE): **11.31**

PRIMER: **L/R**

CASE CAPACITY (GR'S WATER): **82.27**

DIMENSIONAL DRAWING:

F H K G A J E B D

-NOT ACTUAL SIZE-
-DO NOT SCALE-

CARTRIDGE: .40/348 Winchester Improved

OTHER NAMES:

DIA: .411

BALLISTEK NO: 411A

NAI NO: RMB 23244/4.069

DATA SOURCE: Ackley Vol1 Pg489

HISTORICAL DATA: Ackley's version; others exist.

NOTES:

LOADING DATA:

BULLET WT./TYPE	POWDER WT./TYPE	VELOCITY ('/SEC)	SOURCE
300/RN	59.0/4064	1960	Ackley

CASE PREPARATION: SHELLHOLDER (RCBS): 5

MAKE FROM: .348 Winchester. Anneal case neck and taper expand to about .420" dia. F/L size in "improved" die, square case mouth and fireform, in the chamber.

PHYSICAL DATA (INCHES):

CASE TYPE: **Rimmed Bottleneck**

CASE LENGTH **A = 2.25**

HEAD DIAMETER **B = .553**

RIM DIAMETER **D = .610**

NECK DIAMETER **F = .441**

NECK LENGTH **H = .389**

SHOULDER LENGTH **K = .068**

BODY ANGLE (DEG'S/SIDE): **.521**

CASE CAPACITY **CC'S = 5.88**

LOADED LENGTH: **3.05**

BELT DIAMETER **C = N/A**

RIM THICKNESS **E = .07**

SHOULDER DIAMETER **G = .524**

LENGTH TO SHOULDER **J = 1.793**

SHOULDER ANGLE (DEG'S/SIDE): **31.39**

PRIMER: **L/R**

CASE CAPACITY (GR'S WATER): **90.75**

DIMENSIONAL DRAWING:

F H K G A J E B D

-NOT ACTUAL SIZE-
-DO NOT SCALE-

CARTRIDGE: .400 Whelen

OTHER NAMES:	
	DIA: .412
	BALLISTEK NO: 410B
	NAI NO: RXS 21115/5.298

DATA SOURCE: Sharpe Pg398

HISTORICAL DATA: By Col. Whelen in early 1920's.

NOTES:

LOADING DATA:

BULLET WT./TYPE	POWDER WT./TYPE	VELOCITY ('/SEC)	SOURCE
350/Lead	53.0/3031	—	Hagel

CASE PREPARATION: SHELLHOLDER (RCBS): 3

MAKE FROM: .30-06 Spgf. Anneal case and taper expand to .420" dia. Square case mouth and F/L size. O'Connor Steelhead cases would work very well here as they are straight to begin with.

PHYSICAL DATA (INCHES):

CASE TYPE: **Rimless Straight**
CASE LENGTH **A = 2.49**
HEAD DIAMETER **B = .470**
RIM DIAMETER **D = .473**
NECK DIAMETER **F = .436**
NECK LENGTH **H = N/A**
SHOULDER LENGTH **K = N/A**
BODY ANGLE (DEG'S/SIDE): **.399**
CASE CAPACITY **CC'S = 4.89**

LOADED LENGTH: **A/R**
BELT DIAMETER **C = N/A**
RIM THICKNESS **E = .049**
SHOULDER DIAMETER **G = N/A**
LENGTH TO SHOULDER **J = N/A**
SHOULDER ANGLE (DEG'S/SIDE): **N/A**
PRIMER: **L/R**
CASE CAPACITY (GR'S WATER): **75.47**

DIMENSIONAL DRAWING:

F
A
E
B
D

-NOT ACTUAL SIZE-
-DO NOT SCALE-

CARTRIDGE: .404 B-J Express

OTHER NAMES:	
	DIA: .411
	BALLISTEK NO: 411G
	NAI NO: BEN 22242/5.243

DATA SOURCE: Ackley Vol1 Pg491

HISTORICAL DATA: By Barnes & Johnson.

NOTES:

LOADING DATA:

BULLET WT./TYPE	POWDER WT./TYPE	VELOCITY ('/SEC)	SOURCE
300/RN	85.0/4320	2810	Ackley

CASE PREPARATION: SHELLHOLDER (RCBS): 4

MAKE FROM: .375 H&H Mag. Anneal case neck and taper expand to .420" dia. F/L size with expander removed. Trim to length and chamfer. F/L size again.

PHYSICAL DATA (INCHES):

CASE TYPE: **Belted Bottleneck**
CASE LENGTH **A = 2.69**
HEAD DIAMETER **B = .513**
RIM DIAMETER **D = .532**
NECK DIAMETER **F = .442**
NECK LENGTH **H = .390**
SHOULDER LENGTH **K = .087**
BODY ANGLE (DEG'S/SIDE): **.398**
CASE CAPACITY **CC'S = 6.24**

LOADED LENGTH: **A/R**
BELT DIAMETER **C = .532**
RIM THICKNESS **E = .05**
SHOULDER DIAMETER **G = .485**
LENGTH TO SHOULDER **J = 2.213**
SHOULDER ANGLE (DEG'S/SIDE): **33.88**
PRIMER: **L/R Mag**
CASE CAPACITY (GR'S WATER): **96.38**

DIMENSIONAL DRAWING:

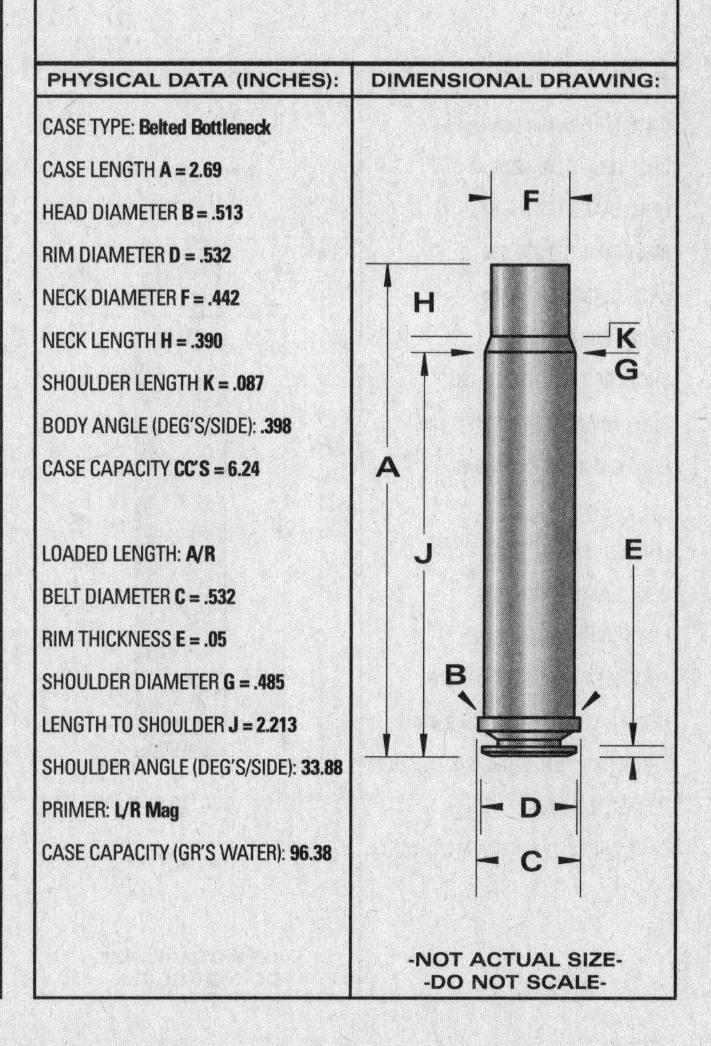

-NOT ACTUAL SIZE-
-DO NOT SCALE-

CARTRIDGE: .404 Barnes Supreme

OTHER NAMES:

DIA: .411
BALLISTEK NO: 411H
NAI NO: BEN 12243/5.555

DATA SOURCE: Ackley Vol1 Pg491

HISTORICAL DATA: By F. Barnes.

NOTES:

LOADING DATA:

BULLET WT./TYPE	POWDER WT./TYPE	VELOCITY ('/SEC)	SOURCE
400/RN	85.0/4064	2550	Ackley

CASE PREPARATION: SHELLHOLDER (RCBS): 4

MAKE FROM: .375 H&H Mag. Anneal case neck and taper expand to .420" dia. F/L size in the Barnes die, square the case mouth and chamfer. Fireform, in the chamber.

PHYSICAL DATA (INCHES):

CASE TYPE: **Belted Bottleneck**
CASE LENGTH **A = 2.85**
HEAD DIAMETER **B = .513**
RIM DIAMETER **D = .532**
NECK DIAMETER **F = .441**
NECK LENGTH **H = .365**
SHOULDER LENGTH **K = .061**
BODY ANGLE (DEG'S/SIDE): **.129**
CASE CAPACITY **CC'S = 6.96**

LOADED LENGTH: **3.96**
BELT DIAMETER **C = .532**
RIM THICKNESS **E = .049**
SHOULDER DIAMETER **G = .503**
LENGTH TO SHOULDER **J = 2.424**
SHOULDER ANGLE (DEG'S/SIDE): **26.94**
PRIMER: **L/R Mag**
CASE CAPACITY (GR'S WATER): **107.44**

DIMENSIONAL DRAWING:

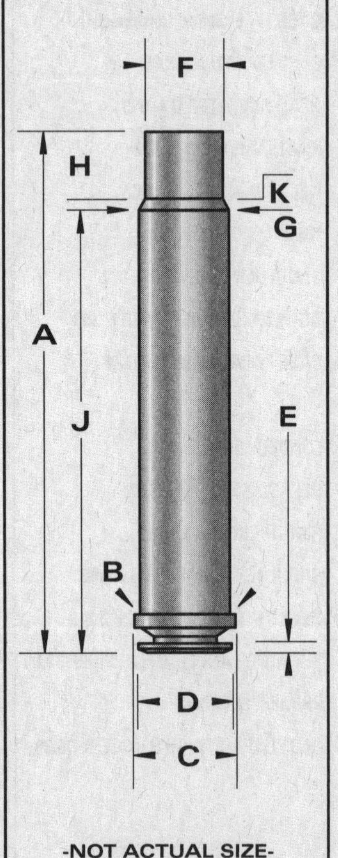

-NOT ACTUAL SIZE-
-DO NOT SCALE-

CARTRIDGE: .404 Jeffery Rimless

OTHER NAMES:
.404 Rimless Nitro Express
10.75 x 73mm

DIA: .423
BALLISTEK NO: 421A
NAI NO: RBB 22221/5.257

DATA SOURCE: COTW 4th Pg239

HISTORICAL DATA: Circa 1910.

NOTES:

LOADING DATA:

BULLET WT./TYPE	POWDER WT./TYPE	VELOCITY ('/SEC)	SOURCE
400/RN	65.0/3031	2150	Barnes

CASE PREPARATION: SHELLHOLDER (RCBS): Spl.

MAKE FROM: .404 Jeffery (BELL). The BELL brass is fully formed but I suggest that the cases be F/L sized, trimmed to square the mouth and chamfered. BELL's .450 N.E. brass would also work by turning off the rim, cutting a new extractor groove and forming.

PHYSICAL DATA (INCHES):

CASE TYPE: **Rebated Bottleneck**
CASE LENGTH **A = 2.86**
HEAD DIAMETER **B = .544**
RIM DIAMETER **D = .537**
NECK DIAMETER **F = .450**
NECK LENGTH **H = .543**
SHOULDER LENGTH **K = .247**
BODY ANGLE (DEG'S/SIDE): **.367**
CASE CAPACITY **CC'S = 7.51**

LOADED LENGTH: **3.53**
BELT DIAMETER **C = N/A**
RIM THICKNESS **E = .05**
SHOULDER DIAMETER **G = .520**
LENGTH TO SHOULDER **J = 2.07**
SHOULDER ANGLE (DEG'S/SIDE): **8.06**
PRIMER: **L/R Mag**
CASE CAPACITY (GR'S WATER): **115.88**

DIMENSIONAL DRAWING:

-NOT ACTUAL SIZE-
-DO NOT SCALE-

CARTRIDGE: .405 Winchester

OTHER NAMES:

DIA: .412
BALLISTEK NO: 412A
NAI NO: RMS 11115/5.596

DATA SOURCE: Hornady #6. Cartridges of the World #9.
HISTORICAL DATA: Introduced by Winchester in 1904 for use in the 1895 rifle. At that time the .405 Winchester was the most powerful cartridge ever developed for a lever action rifle. In 2002 Hornady reintroduced the cartridge and they are currently manufacturing ammo and brass.
NOTES: Accepted by SAAMI 1/24/98.

LOADING DATA:

BULLET WT./TYPE	POWDER WT./TYPE	VELOCITY ('/SEC)	SOURCE
300/Hornady FP	55.8/H-4895	2250	Horn. #6
300/SP	57.0/IMR-3031	2250	Cart.'s of the World #9

CASE PREPARATION: SHELLHOLDER (RCBS): 24

Factory brass is available from Hornady and Bertram. Can be made in a short version from .30-40 Krag brass. .405 Win. Basic (BELL) or 7 x 65R Brenneke. The Brenneke case can be annealed and taper expanded to .420" diameter, trimmed square, F/L sized and fireformed. Or the case can be fireformed with corn meal, in the .405 trim die. .30-40 Krag cases also work well in a short version of the .405 Win. Form as with the 7 x 65R case. However .405 Win. is unique and these cases are always a compromise and are not as satisfactory as factory brass.

PHYSICAL DATA (INCHES):

CASE TYPE: **Rimmed Straight**
CASE LENGTH **A = 2.583**
HEAD DIAMETER **B = .460**
RIM DIAMETER **D = .543**
NECK DIAMETER **F = .436**
NECK LENGTH **H = N/A**
SHOULDER LENGTH **K = N/A**
BODY ANGLE (DEG'S/SIDE): **.284**
CASE CAPACITY **CC'S = 4.98**

LOADED LENGTH: **3.175**
BELT DIAMETER **C = N/A**
RIM THICKNESS **E = .073**
SHOULDER DIAMETER **G = N/A**
LENGTH TO SHOULDER **J = N/A**
SHOULDER ANGLE (DEG'S/SIDE): **N/A**
PRIMER: **L/R**
CASE CAPACITY (GR'S WATER): **76.89**

DIMENSIONAL DRAWING:

-NOT ACTUAL SIZE-
-DO NOT SCALE-

CARTRIDGE: .408 Winchester

OTHER NAMES:

DIA: .410
BALLISTEK NO: 410F
NAI NO: BEN 22242/4.873

DATA SOURCE: Nonte

HISTORICAL DATA: Winchester (?) experimental ctg. No commercial production.

NOTES:

LOADING DATA:

BULLET WT./TYPE	POWDER WT./TYPE	VELOCITY ('/SEC)	SOURCE
275/RN	68.8/IMR4350	–	JJD

CASE PREPARATION: SHELLHOLDER (RCBS): 4

MAKE FROM: .338 Win. Mag. Anneal case neck and taper expand to .420" dia. Square case mouth & chamfer. F/L size.

PHYSICAL DATA (INCHES):

CASE TYPE: **Belted Bottleneck**
CASE LENGTH **A = 2.50**
HEAD DIAMETER **B = .513**
RIM DIAMETER **D = .532**
NECK DIAMETER **F = .435**
NECK LENGTH **H = .300**
SHOULDER LENGTH **K = .12**
BODY ANGLE (DEG'S/SIDE): **.358**
CASE CAPACITY **CC'S = 6.00**

LOADED LENGTH: **A/R**
BELT DIAMETER **C = .532**
RIM THICKNESS **E = .05**
SHOULDER DIAMETER **G = .490**
LENGTH TO SHOULDER **J = 2.04**
SHOULDER ANGLE (DEG'S/SIDE): **12.90**
PRIMER: **L/R Mag**
CASE CAPACITY (GR'S WATER): **92.71**

DIMENSIONAL DRAWING:

-NOT ACTUAL SIZE-
-DO NOT SCALE-

CARTRIDGE: .41 Avenger

OTHER NAMES:	DIA: .410 BALLISTEK NO: 410G NAI NO: RXB 34242/1.996

DATA SOURCE: Handloader #7 Pg28

HISTORICAL DATA: By J.D. Jones.

NOTES:

LOADING DATA:

BULLET WT./TYPE	POWDER WT./TYPE	VELOCITY ('/SEC)	SOURCE
170/JHC	10.0/B'Dot	980	Gaertner

CASE PREPARATION: SHELLHOLDER (RCBS): 3

MAKE FROM: .45 Win. Mag. Trim case to .97" and chamfer. F/L size. Trim to final length. Per Mr. Gaertner, a .41 Mag. carbide sizer was used to apply a light taper crimp.

PHYSICAL DATA (INCHES):

CASE TYPE: **Rimless Bottleneck**

CASE LENGTH **A = .950**

HEAD DIAMETER **B = .476**

RIM DIAMETER **D = .480**

NECK DIAMETER **F = .430**

NECK LENGTH **H = .275**

SHOULDER LENGTH **K = .050**

BODY ANGLE (DEG'S/SIDE): **1.07**

CASE CAPACITY **CC'S = 1.65**

LOADED LENGTH: **A/R**

BELT DIAMETER **C = N/A**

RIM THICKNESS **E = .049**

SHOULDER DIAMETER **G = .460**

LENGTH TO SHOULDER **J = .625**

SHOULDER ANGLE (DEG'S/SIDE): **16.69**

PRIMER: **L/P**

CASE CAPACITY (GR'S WATER): **25.46**

DIMENSIONAL DRAWING:

-NOT ACTUAL SIZE-
-DO NOT SCALE-

CARTRIDGE: .41 Colt Long

OTHER NAMES:	DIA: .386 BALLISTEK NO: 386A NAI NO: RMS 11115/2.769

DATA SOURCE: COTW 4th Pg179

HISTORICAL DATA: By Colt in 1877.

NOTES:

LOADING DATA:

BULLET WT./TYPE	POWDER WT./TYPE	VELOCITY ('/SEC)	SOURCE
200/Lead	4.5/Unique	820	Barnes

CASE PREPARATION: SHELLHOLDER (RCBS): 6

MAKE FROM: There is nothing that will form to this case. Our best results have been with 13/32" dia. tubing on a .357 Magnum (or .38 spl.) case head. With light loads; these work very well. Fabricate case, F/L size and trim to length. Use either undersize bullets that are 'minied' or I.D. neck ream as required.

PHYSICAL DATA (INCHES):

CASE TYPE: **Rimmed Straight**

CASE LENGTH **A = 1.13**

HEAD DIAMETER **B = .408**

RIM DIAMETER **D = .435**

NECK DIAMETER **F = .404**

NECK LENGTH **H = N/A**

SHOULDER LENGTH **K = N/A**

BODY ANGLE (DEG'S/SIDE): **.106**

CASE CAPACITY **CC'S = 1.52**

LOADED LENGTH: **1.39**

BELT DIAMETER **C = N/A**

RIM THICKNESS **E = .052**

SHOULDER DIAMETER **G = N/A**

LENGTH TO SHOULDER **J = N/A**

SHOULDER ANGLE (DEG'S/SIDE): **N/A**

PRIMER: **S/P**

CASE CAPACITY (GR'S WATER): **23.53**

DIMENSIONAL DRAWING:

-NOT ACTUAL SIZE-
-DO NOT SCALE-

CARTRIDGE: .41 Colt Short

OTHER NAMES:

DIA: .386

BALLISTEK NO: 386B

NAI NO: RMS 11115/1.593

DATA SOURCE: NAI/Ballistek

HISTORICAL DATA: By Colt about 1877.

NOTES:

LOADING DATA:

BULLET WT./TYPE	POWDER WT./TYPE	VELOCITY ('/SEC)	SOURCE
160/Lead	3.5/Unique	700	JJD

CASE PREPARATION: SHELLHOLDER (RCBS): 6

MAKE FROM: Use 13/32" dia. tubing to re-body a .38 Spl. or .357 Mag. case head. These cases work very well with all black powder loads. F/L size case and use 'minied' .375 dia. bullets (Lyman #375449 cut off to 150-160 grs.).

PHYSICAL DATA (INCHES):

CASE TYPE: **Rimmed Straight**

CASE LENGTH **A = .650**

HEAD DIAMETER **B = .408**

RIM DIAMETER **D = .435**

NECK DIAMETER **F = .404**

NECK LENGTH **H = N/A**

SHOULDER LENGTH **K = N/A**

BODY ANGLE (DEG'S/SIDE): **.191**

CASE CAPACITY **CC'S = .63**

LOADED LENGTH: **1.10**

BELT DIAMETER **C = N/A**

RIM THICKNESS **E = .052**

SHOULDER DIAMETER **G = N/A**

LENGTH TO SHOULDER **J = N/A**

SHOULDER ANGLE (DEG'S/SIDE): **N/A**

PRIMER: **S/P**

CASE CAPACITY (GR'S WATER): **9.83**

DIMENSIONAL DRAWING:

-NOT ACTUAL SIZE-
-DO NOT SCALE-

CARTRIDGE: .41 Jurras

OTHER NAMES:
.41 AMP

DIA: .410

BALLISTEK NO: 410D

NAI NO: RXB 24242/2.762

DATA SOURCE: NAI/Ballistek

HISTORICAL DATA: By L. Jurras.

NOTES: For Auto Mag. pistol about 1972-73.

LOADING DATA:

BULLET WT./TYPE	POWDER WT./TYPE	VELOCITY ('/SEC)	SOURCE
170/JHP	30.0/W296	2025	Jurras

CASE PREPARATION: SHELLHOLDER (RCBS): 3

MAKE FROM: .308 Win. (or any similar case head). Form case in form/rim die. Trim to length. I.D. neck ream & chamfer. F/L size.

PHYSICAL DATA (INCHES):

CASE TYPE: **Rimless Bottleneck**

CASE LENGTH **A = 1.298**

HEAD DIAMETER **B = .470**

RIM DIAMETER **D = .473**

NECK DIAMETER **F = .436**

NECK LENGTH **H = .284**

SHOULDER LENGTH **K = .064**

BODY ANGLE (DEG'S/SIDE): **.382**

CASE CAPACITY **CC'S = 2.40**

LOADED LENGTH: **A/R**

BELT DIAMETER **C = N/A**

RIM THICKNESS **E = .05**

SHOULDER DIAMETER **G = .460**

LENGTH TO SHOULDER **J = .95**

SHOULDER ANGLE (DEG'S/SIDE): **10.62**

PRIMER: **L/P**

CASE CAPACITY (GR'S WATER): **37.17**

DIMENSIONAL DRAWING:

-NOT ACTUAL SIZE-
-DO NOT SCALE-

CARTRIDGE: .41 S&W Magnum

OTHER NAMES:
.41 Smith & Wesson
.41 Magnum

DIA: .410

BALLISTEK NO: 410A

NAI NO: RMS 11115/2.965

DATA SOURCE: Hornady Manual 3rd Pg348

HISTORICAL DATA: By S&W in 1964 for M57.

NOTES:

LOADING DATA:

BULLET WT./TYPE	POWDER WT./TYPE	VELOCITY ('/SEC)	SOURCE
210/JHP	20.0/2400	1350	Horn.

CASE PREPARATION: SHELLHOLDER (RCBS): 30

MAKE FROM: Factory or .303 Savage. Turn rim to .492" dia. (may not be necessary). cut case off to 1.30". I.D. neck ream for .410" dia. bullet. F/L size case & chamfer. You can generally find the .41 Magnum cases.

PHYSICAL DATA (INCHES):

CASE TYPE: **Rimmed Straight**
CASE LENGTH **A = 1.29**
HEAD DIAMETER **B = .435**
RIM DIAMETER **D = .492**
NECK DIAMETER **F = .434**
NECK LENGTH **H = N/A**
SHOULDER LENGTH **K = N/A**
BODY ANGLE (DEG'S/SIDE): **.327**
CASE CAPACITY **CC'S = 2.013**

LOADED LENGTH: **1.60**
BELT DIAMETER **C = N/A**
RIM THICKNESS **E = .06**
SHOULDER DIAMETER **G = N/A**
LENGTH TO SHOULDER **J = N/A**
SHOULDER ANGLE (DEG'S/SIDE): **N/A**
PRIMER: **L/P**
CASE CAPACITY (GR'S WATER): **31.07**

DIMENSIONAL DRAWING:

-NOT ACTUAL SIZE-
-DO NOT SCALE-

CARTRIDGE: .41-44 CL

OTHER NAMES:

DIA: .410

BALLISTEK NO: 410H

NAI NO: RMS 21115/3.864

DATA SOURCE: Handloader #99 Pg36.

HISTORICAL DATA: By R. Gaertner in 1982.

NOTES:

LOADING DATA:

BULLET WT./TYPE	POWDER WT./TYPE	VELOCITY ('/SEC)	SOURCE
170/JHC	34.0/H110	1032	Gaertner

CASE PREPARATION: SHELLHOLDER (RCBS): 7

MAKE FROM: .30-40 Krag. Turn rim to .510" dia. Cut case off at 1.8". Taper expand to straight case configuration. F/L size in .44 Mag die. Neck size in .41 Mag die. Pass case mouth over a .407" dia expander. Turn necks to a thickness of .012" and trim to length. Bullets are seated backwards and totally inside case.

PHYSICAL DATA (INCHES):

CASE TYPE: **Rimmed Straight**
CASE LENGTH **A = 1.77**
HEAD DIAMETER **B = .458**
RIM DIAMETER **D = .510**
NECK DIAMETER **F = .435**
NECK LENGTH **H = approx. 1 cal. (.41)**
SHOULDER LENGTH **K = N/A**
BODY ANGLE (DEG'S/SIDE): **.383**
CASE CAPACITY **CC'S = 3.15**

LOADED LENGTH: **1.77**
BELT DIAMETER **C = N/A**
RIM THICKNESS **E = .054**
SHOULDER DIAMETER **G = N/A**
LENGTH TO SHOULDER **J = N/A**
SHOULDER ANGLE (DEG'S/SIDE): **N/A**
PRIMER: **L/P**
CASE CAPACITY (GR'S WATER): **48.67**

DIMENSIONAL DRAWING:

-NOT ACTUAL SIZE-
-DO NOT SCALE-

CARTRIDGE: .411 Magnum

OTHER NAMES:	DIA: .411 BALLISTEK NO: 411C NAI NO: BEN 22242/4.873

DATA SOURCE: Ackley Vol2 Pg213

HISTORICAL DATA: By L. Bowman.

NOTES:

LOADING DATA:

BULLET WT./TYPE	POWDER WT./TYPE	VELOCITY ('/SEC)	SOURCE
300/RN	69.0/4320	2500	Ackley

CASE PREPARATION: SHELLHOLDER (RCBS): 4

MAKE FROM: .458 Win. Mag. F/L size the .458 case in the .411 die, square the case mouth & chamfer. Original version used .338 Win. Mag. case.

PHYSICAL DATA (INCHES):

CASE TYPE: **Belted Bottleneck**
CASE LENGTH **A = 2.500**
HEAD DIAMETER **B = .513**
RIM DIAMETER **D = .532**
NECK DIAMETER **F = .431**
NECK LENGTH **H = .345**
SHOULDER LENGTH **K = .065**
BODY ANGLE (DEG'S/SIDE): **.545**
CASE CAPACITY **CC'S = 6.04**

LOADED LENGTH: **A/R**
BELT DIAMETER **C = .532**
RIM THICKNESS **E = .049**
SHOULDER DIAMETER **G = .477**
LENGTH TO SHOULDER **J = 2.09**
SHOULDER ANGLE (DEG'S/SIDE): **19.48**
PRIMER: **L/R Mag**
CASE CAPACITY (GR'S WATER): **93.28**

DIMENSIONAL DRAWING:

-NOT ACTUAL SIZE-
-DO NOT SCALE-

CARTRIDGE: .411 JDJ

OTHER NAMES:	DIA: .411 BALLISTEK NO: 411J NAI NO: RMB 12242/4.811

DATA SOURCE: SSK Industries

HISTORICAL DATA: By J.D. Jones

NOTES:

LOADING DATA:

BULLET WT./TYPE	POWDER WT./TYPE	VELOCITY ('/SEC)	SOURCE
210/—	40.0/4227	2150	SSK

CASE PREPARATION: SHELLHOLDER (RCBS): 28

MAKE FROM: .444 Marlin. F/L size case and trim to length.

PHYSICAL DATA (INCHES):

CASE TYPE: **Rimmed Bottleneck**
CASE LENGTH **A = 2.237**
HEAD DIAMETER **B = .465**
RIM DIAMETER **D = .511**
NECK DIAMETER **F = .430**
NECK LENGTH **H = .396**
SHOULDER LENGTH **K = .036**
BODY ANGLE (DEG'S/SIDE): **.303**
CASE CAPACITY **CC'S = 4.62**

LOADED LENGTH: **3.072**
BELT DIAMETER **C = N/A**
RIM THICKNESS **E = .06**
SHOULDER DIAMETER **G = .448**
LENGTH TO SHOULDER **J = 1.805**
SHOULDER ANGLE (DEG'S/SIDE): **14.04**
PRIMER: **L/R**
CASE CAPACITY (GR'S WATER): **71.32**

DIMENSIONAL DRAWING:

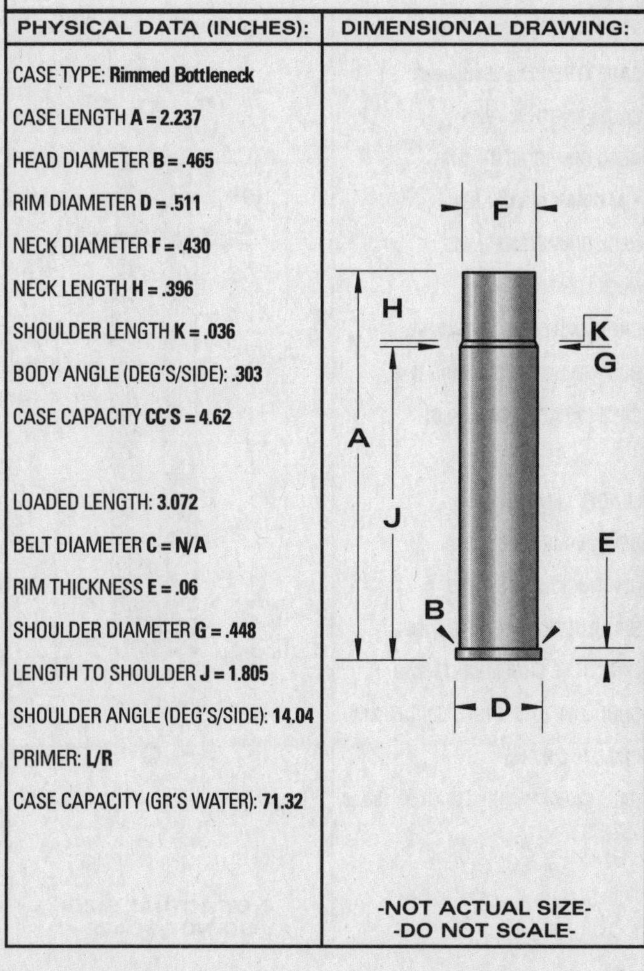

-NOT ACTUAL SIZE-
-DO NOT SCALE-

CARTRIDGE: .416 Barnes Supreme

OTHER NAMES:

DIA: .416

BALLISTEK NO: 416J

NAI NO: BEN 12243/5.555

DATA SOURCE: Ackley Vol.1 Pg. 492

HISTORICAL DATA: By F. Barnes.

NOTES:

LOADING DATA:

BULLET WT./TYPE	POWDER WT./TYPE	VELOCITY ('/SEC)	SOURCE
500/Spire	85.0/4831	2400	Ackley

CASE PREPARATION: SHELLHOLDER (RCBS): 4

MAKE FROM: .375 H&H Mag. Taper expand the H&H case to .420" dia. Square case mouth & chamfer. F/L size and fireform in the Barnes chamber.

PHYSICAL DATA (INCHES):

CASE TYPE: **Belted Bottleneck**

CASE LENGTH **A = 2.85**

HEAD DIAMETER **B = .513**

RIM DIAMETER **D = .532**

NECK DIAMETER **F = .445**

NECK LENGTH **H = .395**

SHOULDER LENGTH **K = .075**

BODY ANGLE (DEG'S/SIDE): **.118**

CASE CAPACITY **CC'S = 7.01**

LOADED LENGTH: **A/R**

BELT DIAMETER **C = .532**

RIM THICKNESS **E = .049**

SHOULDER DIAMETER **G = .504**

LENGTH TO SHOULDER **J = 2.38**

SHOULDER ANGLE (DEG'S/SIDE): **21.47**

PRIMER: **L/R Mag**

CASE CAPACITY (GR'S WATER): **108.28**

DIMENSIONAL DRAWING:

-NOT ACTUAL SIZE-
-DO NOT SCALE-

CARTRIDGE: .416 Dakota

OTHER NAMES:

DIA: .416

BALLISTEK NO:

NAI NO:

DATA SOURCE: Dakota Arms

HISTORICAL DATA: Proprietary cartridge from Dakota Arms.

NOTES: Based on the .404 Jeffery case with the rim enlarged slightly to eliminate the rebated feature. The .416 Dakota offers similar performance to the .416 Rigby and is suitable for use in a a standard length action.

LOADING DATA:

BULLET WT./TYPE	POWDER WT./TYPE	VELOCITY ('/SEC)	SOURCE
400	100.0/RL-19	2558	Dakota Arms

CASE PREPARATION: SHELL HOLDER (RCBS): 41

MAKE FROM: Factory or make by reforming .404 Jeffery.

PHYSICAL DATA (INCHES):

CASE TYPE: **Bottleneck**

CASE LENGTH **A = 2.850**

HEAD DIAMETER **B = .545**

RIM DIAMETER **D = .545**

NECK DIAMETER **F = .441**

NECK LENGTH **H = .50**

SHOULDER LENGTH **K = .074**

BODY ANGLE (DEG'S/SIDE):

CASE CAPACITY **CC'S =**

LOADED LENGTH: **3.645**

BELT DIAMETER **C = N/A**

RIM THICKNESS **E = .050**

SHOULDER DIAMETER **G = .527**

LENGTH TO SHOULDER **J = 2.276**

SHOULDER ANGLE (DEG'S/SIDE): **30**

PRIMER: **L/R Mag**

CASE CAPACITY (GR'S WATER): **116.0**

DIMENSIONAL DRAWING:

-NOT ACTUAL SIZE-
-DO NOT SCALE-

CARTRIDGE: .416 Howell

OTHER NAMES:

DIA: .416

BALLISTEK NO: 416L

NAI NO: RXB 22243/4.629

DATA SOURCE: Handloader #92 Pg55

HISTORICAL DATA: By K. Howell in 1980-81

NOTES:

LOADING DATA:

BULLET WT./TYPE	POWDER WT./TYPE	VELOCITY ('/SEC)	SOURCE
300/RN	85.0/W748	2750	Howell

CASE PREPARATION: SHELLHOLDER (RCBS): Spl.

MAKE FROM: .404 Jeffery (BELL). Cut case to 2.55", F/L size in Howell die, trim to length and chamfer.

PHYSICAL DATA (INCHES):

CASE TYPE: **Rimless Bottleneck**

CASE LENGTH **A = 2.500**

HEAD DIAMETER **B = .540**

RIM DIAMETER **D = .540**

NECK DIAMETER **F = .444**

NECK LENGTH **H = .345**

SHOULDER LENGTH **K = .086**

BODY ANGLE (DEG'S/SIDE): **.383**

CASE CAPACITY **CC'S = 6.53**

LOADED LENGTH: **A/R**

BELT DIAMETER **C = N/A**

RIM THICKNESS **E = .06**

SHOULDER DIAMETER **G = .515**

LENGTH TO SHOULDER **J = 2.069**

SHOULDER ANGLE (DEG'S/SIDE): **22.43**

PRIMER: **L/R**

CASE CAPACITY (GR'S WATER): **100.78**

DIMENSIONAL DRAWING:

F H K G A J E B D

-NOT ACTUAL SIZE-
-DO NOT SCALE-

CARTRIDGE: .416 Jurras

OTHER NAMES:

DIA: .416

BALLISTEK NO: 416H

NAI NO: RMB 34223/3.496

DATA SOURCE: NAI/Ballistek

HISTORICAL DATA: By L. Jurras

NOTES: Contender ctg.

LOADING DATA:

BULLET WT./TYPE	POWDER WT./TYPE	VELOCITY ('/SEC)	SOURCE
400/RN	43.2/IMR4227	–	JJD

CASE PREPARATION: SHELLHOLDER (RCBS): Spl.

MAKE FROM: .500 N.E. (BELL). Anneal case and form in form die set (required). Trim to length. I.D. neck ream, if necessary. F/L size.

PHYSICAL DATA (INCHES):

CASE TYPE: **Rimmed Bottleneck**

CASE LENGTH **A = 2.028**

HEAD DIAMETER **B = .580**

RIM DIAMETER **D = .660**

NECK DIAMETER **F = .440**

NECK LENGTH **H = .695**

SHOULDER LENGTH **K = .087**

BODY ANGLE (DEG'S/SIDE): **1.29**

CASE CAPACITY **CC'S = 5.28**

LOADED LENGTH: **A/R**

BELT DIAMETER **C = N/A**

RIM THICKNESS **E = .05**

SHOULDER DIAMETER **G = .533**

LENGTH TO SHOULDER **J = 1.246**

SHOULDER ANGLE (DEG'S/SIDE): **28.12**

PRIMER: **L/P**

CASE CAPACITY (GR'S WATER): **81.53**

DIMENSIONAL DRAWING:

F H K G A J E B D

-NOT ACTUAL SIZE-
-DO NOT SCALE-

CARTRIDGE: 416 Hoffman

OTHER NAMES:

DIA: .416

BALLISTEK NO:

NAI NO:

DATA SOURCE:

HISTORICAL DATA: Wildcat cartridge adopted by A-Square Co. In other words, brass cases, bullets, and loaded ammunition are available from A-Square. Originated in the 1970s with George L. Hoffman of Sonora, Texas.

LOADING DATA:

BULLET WT./TYPE	POWDER WT./TYPE	VELOCITY ('/SEC)	SOURCE
400/Sp	77/IMR-4064	2400	

CASE PREPARATION: SHELL HOLDER (RCBS): 04

MAKE FROM: 375 H&H Magnum case

PHYSICAL DATA (INCHES):

CASE TYPE: **Belted Bottleneck**

CASE LENGTH **A = 2.850**

HEAD DIAMETER **B = .513**

RIM DIAMETER **D = .532**

NECK DIAMETER **F = .447**

NECK LENGTH **H = .372**

SHOULDER LENGTH **K = .048**

BODY ANGLE (DEG'S/SIDE): **24°**

CASE CAPACITY **CC'S = N/A**

LOADED LENGTH: **3.60**

BELT DIAMETER **C = .532**

RIM THICKNESS **E = .050**

SHOULDER DIAMETER **G = .491**

LENGTH TO SHOULDER **J = 2.430**

SHOULDER ANGLE (DEG'S/SIDE): **N/A**

PRIMER: **L/R**

CASE CAPACITY (GR'S WATER): **N/A**

DIMENSIONAL DRAWING:

-NOT ACTUAL SIZE-
-DO NOT SCALE-

CARTRIDGE: .416 Rigby

OTHER NAMES:

DIA: .416

BALLISTEK NO: 416B

NAI NO: RXB 22232/4.923

DATA SOURCE: COTW 4th Pg240

HISTORICAL DATA: By Rigby in 1911.

NOTES:

LOADING DATA:

BULLET WT./TYPE	POWDER WT./TYPE	VELOCITY ('/SEC)	SOURCE
410/RN	95.0/4350	2300	Barnes

CASE PREPARATION: SHELLHOLDER (RCBS): Spl.

MAKE FROM: .416 Rigby (BELL) or 11mm Beaumont (BELL). Use the .416 brass as is or turn the rim off of the Beaumont case, cut a new extractor groove, anneal and form. Trim and F/L size.

PHYSICAL DATA (INCHES):

CASE TYPE: **Rimless Bottleneck**

CASE LENGTH **A = 2.90**

HEAD DIAMETER **B = .589**

RIM DIAMETER **D = .586**

NECK DIAMETER **F = .445**

NECK LENGTH **H = .460**

SHOULDER LENGTH **K = .130**

BODY ANGLE (DEG'S/SIDE): **.678**

CASE CAPACITY **CC'S = 8.59**

LOADED LENGTH: **3.72**

BELT DIAMETER **C = N/A**

RIM THICKNESS **E = .058**

SHOULDER DIAMETER **G = .539**

LENGTH TO SHOULDER **J = 2.31**

SHOULDER ANGLE (DEG'S/SIDE): **45**

PRIMER: **L/R**

CASE CAPACITY (GR'S WATER): **132.56**

DIMENSIONAL DRAWING:

-NOT ACTUAL SIZE-
-DO NOT SCALE-

CARTRIDGE: .416 Taylor

OTHER NAMES:

DIA: .416

BALLISTEK NO: 416C

NAI NO: BEN 22234/4.922

DATA SOURCE: Handloader #50 Pg32

HISTORICAL DATA: By R. Chatfield-Taylor.

NOTES:

LOADING DATA:

BULLET WT./TYPE	POWDER WT./TYPE	VELOCITY ('/SEC)	SOURCE
400/RN	70.0/IMR4320	2401	Waters

CASE PREPARATION: SHELLHOLDER (RCBS): 4

MAKE FROM: .458 Win. Mag. Form .458 case in form die. Square case mouth & chamfer. F/L size.

PHYSICAL DATA (INCHES):

CASE TYPE: **Belted Bottleneck**

CASE LENGTH A = **2.525**

HEAD DIAMETER B = **.513**

RIM DIAMETER D = **.532**

NECK DIAMETER F = **.440**

NECK LENGTH H = **.450**

SHOULDER LENGTH K = **.022**

BODY ANGLE (DEG'S/SIDE): **.402**

CASE CAPACITY CC'S = **6.06**

LOADED LENGTH: **3.375**

BELT DIAMETER C = **.532**

RIM THICKNESS E = **.049**

SHOULDER DIAMETER G = **.487**

LENGTH TO SHOULDER J = **2.053**

SHOULDER ANGLE (DEG'S/SIDE): **46.88**

PRIMER: **L/R Mag**

CASE CAPACITY (GR'S WATER): **93.65**

DIMENSIONAL DRAWING:

-NOT ACTUAL SIZE-
-DO NOT SCALE-

CARTRIDGE: .416 Weatherby

OTHER NAMES: .416/.378 Weatherby

DIA: .416

BALLISTEK NO:

NAI NO:

DATA SOURCE: Many current handloading manuals.

HISTORICAL DATA: Introduced by Weatherby in 1989. Very powerful cartridge for use on the world's largest game.

NOTES: Cartridge was approved by SAAMI on 1/12/94.

LOADING DATA:

BULLET WT./TYPE	POWDER WT./TYPE	VELOCITY ('/SEC)	SOURCE
400/Hornady	115.1/RL-19	2,700	Horn. #6
400/Nosler Partition	117/IMR-7828	2,704	Nosler #5

CASE PREPARATION: SHELL HOLDER (RCBS): 14

MAKE FROM: Factory is preferred. make from .378 Weatherby by necking up in F/L resizing die with tapered expander. Trim, chamfer and deburr. Check neck wall thick-ness, turn if needed. Fireform in chamber.

PHYSICAL DATA (INCHES):

CASE TYPE: **Belted Bottleneck**

CASE LENGTH A = **2.913**

HEAD DIAMETER B = **.5817**

RIM DIAMETER D = **.579**

NECK DIAMETER F = **.444**

NECK LENGTH H = **.392**

SHOULDER LENGTH K = **.176**

BODY ANGLE (DEG'S/SIDE):

CASE CAPACITY CC'S =

LOADED LENGTH: **3.750**

BELT DIAMETER C = **.6035**

RIM THICKNESS E = **.063**

SHOULDER DIAMETER G = **.561**

LENGTH TO SHOULDER J = **2.345**

SHOULDER ANGLE (DEG'S/SIDE):

PRIMER: **L/R Mag**

CASE CAPACITY (GR'S WATER): **140.8**

DIMENSIONAL DRAWING:

-NOT ACTUAL SIZE-
-DO NOT SCALE-

CARTRIDGE: .416/338 Winchester Magnum

OTHER NAMES:

DIA: .416

BALLISTEK NO: 416A

NAI NO: BEN 22242/4.873

DATA SOURCE: NAI/Ballistek

HISTORICAL DATA:

NOTES:

LOADING DATA:

BULLET WT./TYPE	POWDER WT./TYPE	VELOCITY ('/SEC)	SOURCE
500/RN	82.0/IMR4831	2300	JJD

CASE PREPARATION: SHELLHOLDER (RCBS): 4

MAKE FROM: .458 Win. Mag. F/L size the .458 case in the .416 die. Square case mouth, chamfer.

PHYSICAL DATA (INCHES):

CASE TYPE: **Belted Bottleneck**

CASE LENGTH **A = 2.500**

HEAD DIAMETER **B = .513**

RIM DIAMETER **D = .532**

NECK DIAMETER **F = .448**

NECK LENGTH **H = .38**

SHOULDER LENGTH **K = .080**

BODY ANGLE (DEG'S/SIDE): **.358**

CASE CAPACITY **CC'S = 5.79**

LOADED LENGTH: **A/R**

BELT DIAMETER **C = .532**

RIM THICKNESS **E = .049**

SHOULDER DIAMETER **G = .490**

LENGTH TO SHOULDER **J = 2.04**

SHOULDER ANGLE (DEG'S/SIDE): **14.70**

PRIMER: **L/R Mag**

CASE CAPACITY (GR'S WATER): **89.45**

DIMENSIONAL DRAWING:

-NOT ACTUAL SIZE-
-DO NOT SCALE-

CARTRIDGE: .416/348 Winchester

OTHER NAMES:

DIA: .416

BALLISTEK NO: 416K

NAI NO: RMB 33221/4.069

DATA SOURCE: NAI/Ballistek

HISTORICAL DATA:

NOTES:

LOADING DATA:

BULLET WT./TYPE	POWDER WT./TYPE	VELOCITY ('/SEC)	SOURCE
400/RN	55.0/IMR3031	–	JJD

CASE PREPARATION: SHELLHOLDER (RCBS): 5

MAKE FROM: .348 Winchester. Taper expand case neck to .420" dia. (anneal first). Square case mouth, chamfer and F/L size.

PHYSICAL DATA (INCHES):

CASE TYPE: **Rimmed Bottleneck**

CASE LENGTH **A = 2.25**

HEAD DIAMETER **B = .553**

RIM DIAMETER **D = .610**

NECK DIAMETER **F = .447**

NECK LENGTH **H = .565**

SHOULDER LENGTH **K = .105**

BODY ANGLE (DEG'S/SIDE): **1.41**

CASE CAPACITY **CC'S = 5.35**

LOADED LENGTH: **A/R**

BELT DIAMETER **C = N/A**

RIM THICKNESS **E = .07**

SHOULDER DIAMETER **G = .485**

LENGTH TO SHOULDER **J = 1.58**

SHOULDER ANGLE (DEG'S/SIDE): **10.25**

PRIMER: **L/R**

CASE CAPACITY (GR'S WATER): **82.62**

DIMENSIONAL DRAWING:

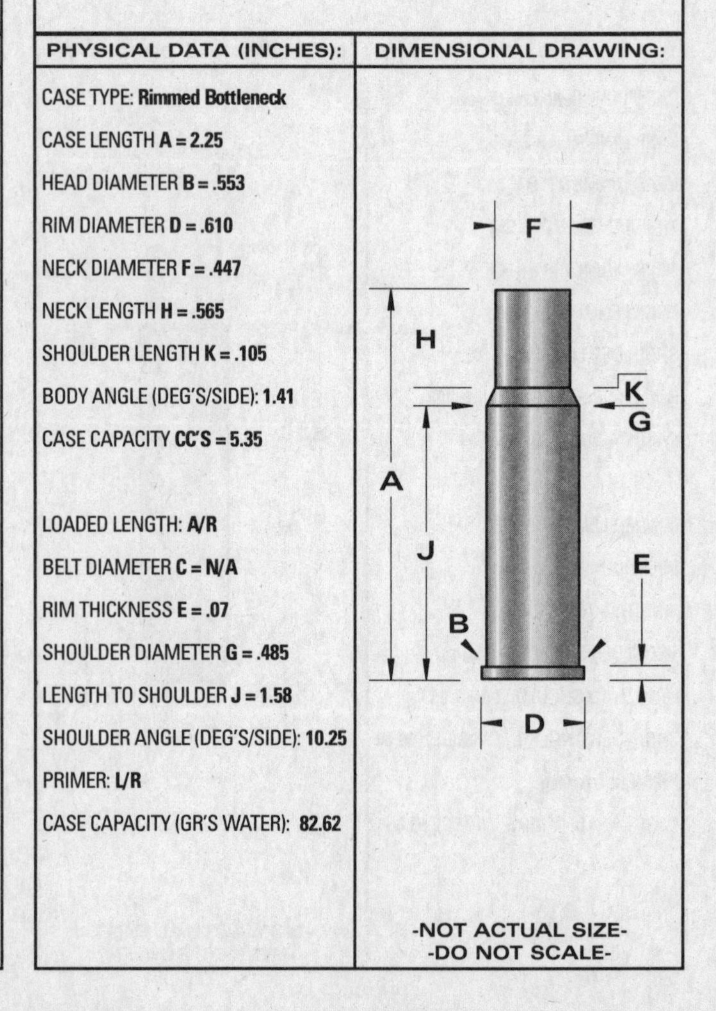

CARTRIDGE: .416/378 Weatherby Magnum

OTHER NAMES:

DIA: .416

BALLISTEK NO: 416F

NAI NO: BEN 12243/4.982

DATA SOURCE: NAI/Ballistek

HISTORICAL DATA:

NOTES:

LOADING DATA:

BULLET WT./TYPE	POWDER WT./TYPE	VELOCITY ('/SEC)	SOURCE
500/RN	86.0/H570	—	JJD

CASE PREPARATION: SHELLHOLDER (RCBS): 14

MAKE FROM: .378 Weatherby. Taper expand case neck to .420" dia. Square case mouth, chamfer and F/L size. .460 cases may be used but, you'll need a form die to reduce the neck diameter - I've had no luck reducing the neck with the F/L die.

PHYSICAL DATA (INCHES):

CASE TYPE: **Belted Bottleneck**
CASE LENGTH **A = 2.90**
HEAD DIAMETER **B = .582**
RIM DIAMETER **D = .582**
NECK DIAMETER **F = .439**
NECK LENGTH **H = .405**
SHOULDER LENGTH **K = .155**
BODY ANGLE (DEG'S/SIDE): **.294**
CASE CAPACITY **CC'S = 9.12**

LOADED LENGTH: **A/R**
BELT DIAMETER **C = .603**
RIM THICKNESS **E = .062**
SHOULDER DIAMETER **G = .560**
LENGTH TO SHOULDER **J = 2.34**
SHOULDER ANGLE (DEG'S/SIDE): **N/A**
PRIMER: **L/R Mag**
CASE CAPACITY (GR'S WATER): **140.80**

DIMENSIONAL DRAWING:

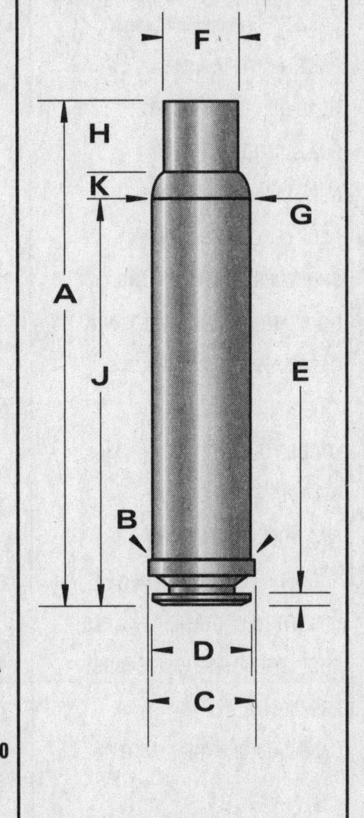

-NOT ACTUAL SIZE-
-DO NOT SCALE-

CARTRIDGE: .423 Van Horn

OTHER NAMES:

DIA: .423

BALLISTEK NO: 423D

NAI NO: RXB 12233/5.220

DATA SOURCE: Handloader #96 Pg28

HISTORICAL DATA: By G. Van Horn.

NOTES:

LOADING DATA:

BULLET WT./TYPE	POWDER WT./TYPE	VELOCITY ('/SEC)	SOURCE
400/RN	87.0/4064	2610	Hagel

CASE PREPARATION: SHELLHOLDER (RCBS): Spl.

MAKE FROM: .404 Jeffery (BELL). Taper expand the .404's case neck to about .450" dia. F/L size in the Van Horn die. Square case mouth and chamfer. Fireform in trim die with corn meal, or with moderate load, in chamber.

PHYSICAL DATA (INCHES):

CASE TYPE: **Rimless Bottleneck**
CASE LENGTH **A = 2.845**
HEAD DIAMETER **B = .545**
RIM DIAMETER **D = .543**
NECK DIAMETER **F = .447**
NECK LENGTH **H = .475**
SHOULDER LENGTH **K = .095**
BODY ANGLE (DEG'S/SIDE): **.345**
CASE CAPACITY **CC'S = 7.79**

LOADED LENGTH: **3.63**
BELT DIAMETER **C = N/A**
RIM THICKNESS **E = .06**
SHOULDER DIAMETER **G = .520**
LENGTH TO SHOULDER **J = 2.275**
SHOULDER ANGLE (DEG'S/SIDE): **21.02**
PRIMER: **L/R Mag**
CASE CAPACITY (GR'S WATER): **120.32**

DIMENSIONAL DRAWING:

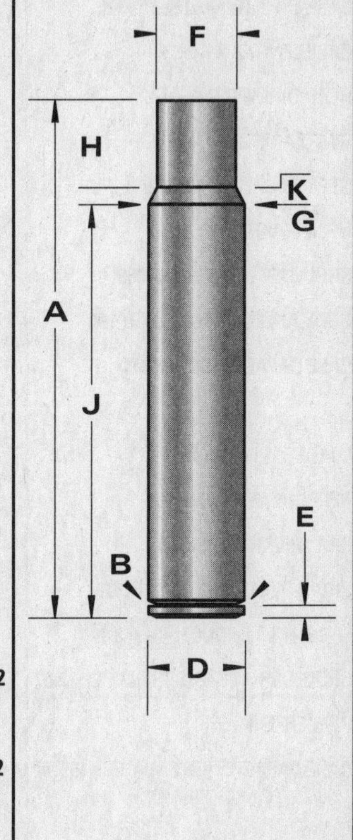

-NOT ACTUAL SIZE-
-DO NOT SCALE-

CARTRIDGE: .425 Lee Magnum

OTHER NAMES:

DIA: .435

BALLISTEK NO: 435D

NAI NO: RBB 12223/4.870

DATA SOURCE: Ackley Vol.1 Pg494

HISTORICAL DATA:

NOTES:

LOADING DATA:

BULLET WT./TYPE	POWDER WT./TYPE	VELOCITY ('/SEC)	SOURCE
410/RN	85.0/4895	2490	Ackley

CASE PREPARATION: SHELLHOLDER (RCBS): 3

MAKE FROM: .425 Westley Richards (BELL). Taper expand the BELL case to about .450" dia. F/L size in the Lee die, trim to length, chamfer and fireform in the chamber.

PHYSICAL DATA (INCHES):

CASE TYPE: **Rebated Bottleneck**

CASE LENGTH **A = 2.63**

HEAD DIAMETER **B = .54**

RIM DIAMETER **D = .467**

NECK DIAMETER **F = .449**

NECK LENGTH **H = .535**

SHOULDER LENGTH **K = .095**

BODY ANGLE (DEG'S/SIDE): **.190**

CASE CAPACITY **CC'S = 7.11**

LOADED LENGTH: **A/R**

BELT DIAMETER **C = N/A**

RIM THICKNESS **E = .045**

SHOULDER DIAMETER **G = .528**

LENGTH TO SHOULDER **J = 2.00**

SHOULDER ANGLE (DEG'S/SIDE): **22.57**

PRIMER: **L/R**

CASE CAPACITY (GR'S WATER): **109.78**

DIMENSIONAL DRAWING:

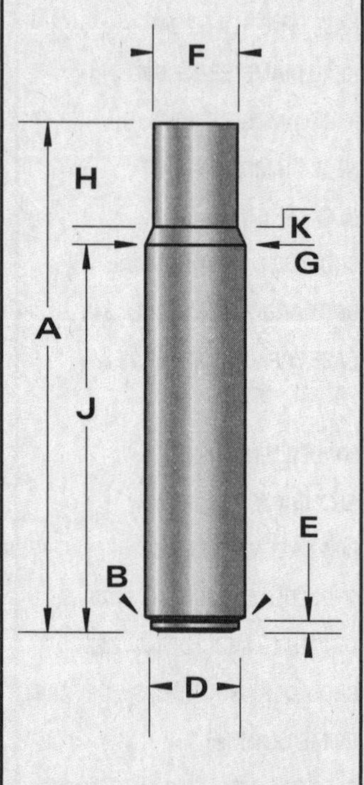

-NOT ACTUAL SIZE-
-DO NOT SCALE-

CARTRIDGE: 425 Express

OTHER NAMES:

DIA: .423

BALLISTEK NO:

NAI NO:

DATA SOURCE: A-Square Manual

HISTORICAL DATA: Based upon the 300 Winchester Magnum case shortened to allow for case stretching, then fireformed in the 425 chamber.

NOTES:

LOADING DATA:

BULLET WT./TYPE	POWDER WT./TYPE	VELOCITY ('/SEC)	SOURCE
350 400	79.0/IMR-4064 73.0/H-4895	2535 2420	Cameron Hopkins

CASE PREPARATION: SHELL HOLDER (RCBS): 10

MAKE FROM: 300 Winchester Magnum

PHYSICAL DATA (INCHES):

CASE TYPE: **Belted Bottleneck**

CASE LENGTH **A = 2.552**

HEAD DIAMETER **B = .513**

RIM DIAMETER **D = .532**

NECK DIAMETER **F = .448**

NECK LENGTH **H = .309**

SHOULDER LENGTH **K = .340**

BODY ANGLE (DEG'S/SIDE): **.N/A**

CASE CAPACITY **CC'S = N/A**

LOADED LENGTH: **3.34**

BELT DIAMETER **C = .532**

RIM THICKNESS **E = .050**

SHOULDER DIAMETER **G = .490**

LENGTH TO SHOULDER **J = 2.209**

SHOULDER ANGLE (DEG'S/SIDE):

PRIMER: **L/R**

CASE CAPACITY (GR'S WATER):

DIMENSIONAL DRAWING:

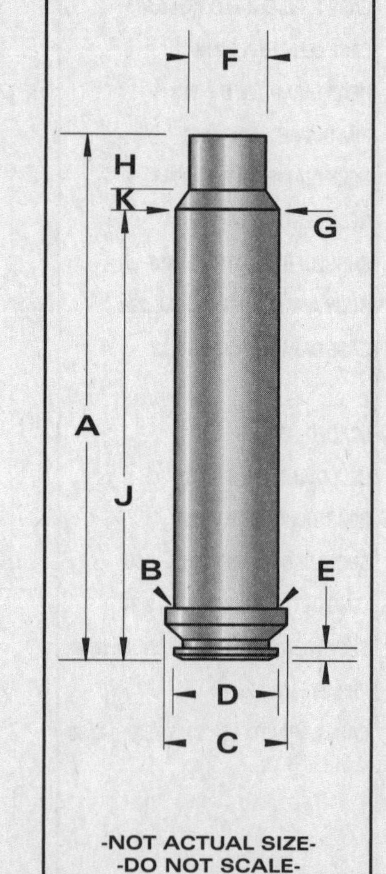

-NOT ACTUAL SIZE-
-DO NOT SCALE-

CARTRIDGE: .425 Westley Richards

OTHER NAMES:

DIA: .435

BALLISTEK NO: 435C

NAI NO: RBB 13222/4.862

DATA SOURCE: COTW 4th Pg240

HISTORICAL DATA: By W. Richards in 1909.

NOTES:

LOADING DATA:

BULLET WT./TYPE	POWDER WT./TYPE	VELOCITY ('/SEC)	SOURCE
410/—	80.0/4895	2410	Barnes

CASE PREPARATION: SHELLHOLDER (RCBS): 3

MAKE FROM: Factory (BELL) or .404 Jeffery (BELL). Turn rim to .467" dia. and cut new extractor groove. Anneal case and F/L size with the expander removed. Trim to length and chamfer. F/L size.

PHYSICAL DATA (INCHES):

CASE TYPE: **Rebated Bottleneck**

CASE LENGTH **A = 2.64**

HEAD DIAMETER **B = .543**

RIM DIAMETER **D = .467**

NECK DIAMETER **F = .456**

NECK LENGTH **H = .70**

SHOULDER LENGTH **K = .220**

BODY ANGLE (DEG'S/SIDE): **.056**

CASE CAPACITY **CC'S = 7.29**

LOADED LENGTH: **3.30**

BELT DIAMETER **C = N/A**

RIM THICKNESS **E = .057**

SHOULDER DIAMETER **G = .540**

LENGTH TO SHOULDER **J = 1.72**

SHOULDER ANGLE (DEG'S/SIDE): **10.80**

PRIMER: **L/R**

CASE CAPACITY (GR'S WATER): **112.53**

DIMENSIONAL DRAWING:

F
H
K
A
G
J
E
B
D

-NOT ACTUAL SIZE-
-DO NOT SCALE-

CARTRIDGE: .43 Dutch Beaumont M71/78

OTHER NAMES:
11 x 52R Netherlands Beaumont

DIA: .457

BALLISTEK NO: 457C

NAI NO: RMB 34231/3.541

DATA SOURCE: COTW 4th Pg216

HISTORICAL DATA:

NOTES:

LOADING DATA:

BULLET WT./TYPE	POWDER WT./TYPE	VELOCITY ('/SEC)	SOURCE
400/Lead	31.0/4198	1430	Barnes

CASE PREPARATION: SHELLHOLDER (RCBS): Spl.

MAKE FROM: 11mm Beaumont (BELL). Trim case to length, chamfer and F/L size. Fireform.

PHYSICAL DATA (INCHES):

CASE TYPE: **Rimmed Bottleneck**

CASE LENGTH **A = 2.04**

HEAD DIAMETER **B = .576**

RIM DIAMETER **D = .665**

NECK DIAMETER **F = .484**

NECK LENGTH **H = .500**

SHOULDER LENGTH **K = .18**

BODY ANGLE (DEG'S/SIDE): **1.18**

CASE CAPACITY **CC'S = 5.63**

LOADED LENGTH: **2.54**

BELT DIAMETER **C = N/A**

RIM THICKNESS **E = .062**

SHOULDER DIAMETER **G = .528**

LENGTH TO SHOULDER **J = 1.36**

SHOULDER ANGLE (DEG'S/SIDE): **6.97**

PRIMER: **L/R**

CASE CAPACITY (GR'S WATER): **86.95**

DIMENSIONAL DRAWING:

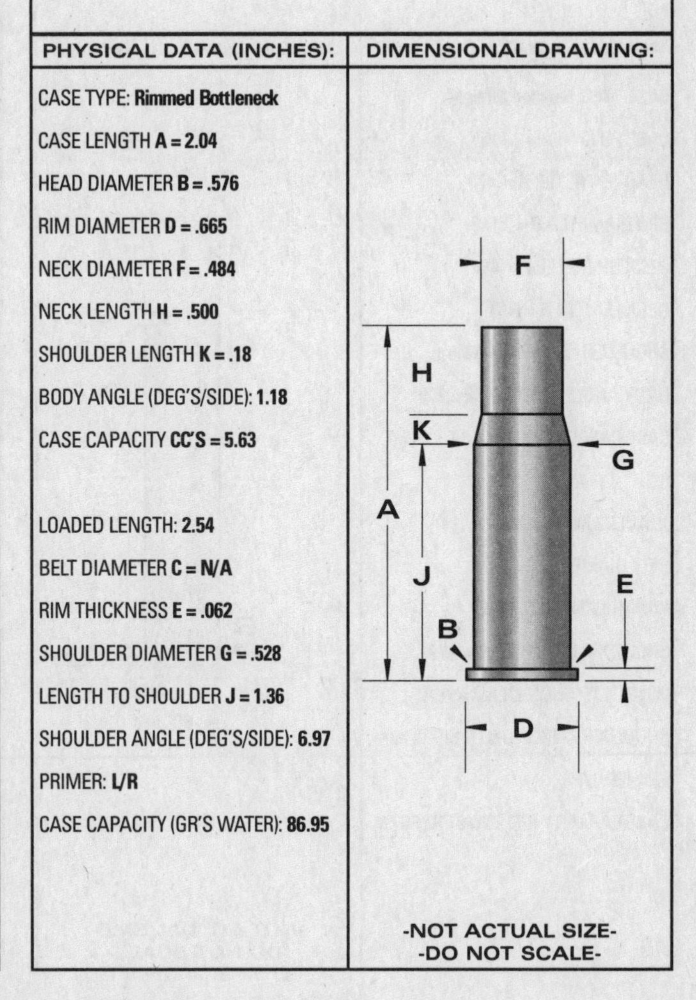

-NOT ACTUAL SIZE-
-DO NOT SCALE-

CARTRIDGE: .430 JDJ

OTHER NAMES:

DIA: .430

BALLISTEK NO: 430L

NAI NO: RMS 11115/4.500

DATA SOURCE: Handloader #94 Pg18

HISTORICAL DATA: By J.D. Jones about 1980.

NOTES:

LOADING DATA:

BULLET WT./TYPE	POWDER WT./TYPE	VELOCITY ('/SEC)	SOURCE
310/Lead	42.0/3031	1485	Jones/HL

CASE PREPARATION: SHELLHOLDER (RCBS): 28

MAKE FROM: .444 Marlin. Trim case to length, chamfer & F/L size.

PHYSICAL DATA (INCHES):

CASE TYPE: **Rimmed Straight**

CASE LENGTH A = **2.115**

HEAD DIAMETER B = **.470**

RIM DIAMETER D = **.514**

NECK DIAMETER F = **.450**

NECK LENGTH H = **N/A**

SHOULDER LENGTH K = **N/A**

BODY ANGLE (DEG'S/SIDE): **.279**

CASE CAPACITY CC'S = **4.30**

LOADED LENGTH: **2.56**

BELT DIAMETER C = **N/A**

RIM THICKNESS E = **.063**

SHOULDER DIAMETER G = **N/A**

LENGTH TO SHOULDER J = **N/A**

SHOULDER ANGLE (DEG'S/SIDE): **N/A**

PRIMER: **L/P**

CASE CAPACITY (GR'S WATER): **66.48**

DIMENSIONAL DRAWING:

F

A

E

B

D

-NOT ACTUAL SIZE-
-DO NOT SCALE-

CARTRIDGE: .44 Auto Mag

OTHER NAMES:
.44 AMP

DIA: .430

BALLISTEK NO: 430B

NAI NO: RXS 11115/2.761

DATA SOURCE: Hornady Manual 3rd Pg 355

HISTORICAL DATA: By H. Sanford in 1971.

NOTES: Commercial for L. Jurras.

LOADING DATA:

BULLET WT./TYPE	POWDER WT./TYPE	VELOCITY ('/SEC)	SOURCE
180/JHP	29.3/W296	1650	Sierra
240/JHP	20.8/2400	1350	Horn.

CASE PREPARATION: SHELLHOLDER (RCBS): 3

MAKE FROM: .308 Win. or .30-06. Cut case to 1.298". Inside neck ream to hold .429" dia. bullet. Chamfer & solvent clean. F/L size. Tumble. Chase chamfer & load.

PHYSICAL DATA (INCHES):

CASE TYPE: **Rimless Straight**

CASE LENGTH A = **1.298**

HEAD DIAMETER B = **.470**

RIM DIAMETER D = **.473**

NECK DIAMETER F = **.459**

NECK LENGTH H = **N/A**

SHOULDER LENGTH K = **N/A**

BODY ANGLE (DEG'S/SIDE): **.253**

CASE CAPACITY CC'S = **2.26**

LOADED LENGTH: **1.70**

BELT DIAMETER C = **N/A**

RIM THICKNESS E = **.054**

SHOULDER DIAMETER G = **N/A**

LENGTH TO SHOULDER J = **N/A**

SHOULDER ANGLE (DEG'S/SIDE): **N/A**

PRIMER: **L/R**

CASE CAPACITY (GR'S WATER): **34.96**

DIMENSIONAL DRAWING:

F

A

E

B

D

-NOT ACTUAL SIZE-
-DO NOT SCALE-

CARTRIDGE: .44 Ballard Extra Long

OTHER NAMES:
.44 Extra Long

DIA: .439

BALLISTEK NO: 439B

NAI NO: RMS 11115/3.696

DATA SOURCE: COTW 4th Pg115

HISTORICAL DATA: Circa 1876.

NOTES:

LOADING DATA:

BULLET WT./TYPE	POWDER WT./TYPE	VELOCITY ('/SEC)	SOURCE
250/LeadPP	17.7/IMR4198	–	JJD

CASE PREPARATION: SHELLHOLDER (RCBS): 2

MAKE FROM: .220 Swift. Cut case to 1.7". I.D. neck ream. Trim to length, chamfer and F/L size.

PHYSICAL DATA (INCHES):

CASE TYPE: **Rimmed Straight**
CASE LENGTH **A = 1.63**
HEAD DIAMETER **B = .441**
RIM DIAMETER **D = .506**
NECK DIAMETER **F = .441**
NECK LENGTH **H = N/A**
SHOULDER LENGTH **K = N/A**
BODY ANGLE (DEG'S/SIDE): **0**
CASE CAPACITY **CC'S = 3.27**

LOADED LENGTH: **2.10**
BELT DIAMETER **C = N/A**
RIM THICKNESS **E = .052**
SHOULDER DIAMETER **G = N/A**
LENGTH TO SHOULDER **J = N/A**
SHOULDER ANGLE (DEG'S/SIDE): **N/A**
PRIMER: **L/R**
CASE CAPACITY (GR'S WATER): **50.50**

DIMENSIONAL DRAWING:

F
A
B
E
D

-NOT ACTUAL SIZE-
-DO NOT SCALE-

CARTRIDGE: .44 Ballard Long

OTHER NAMES:
.44 Long Centerfire

DIA: .439

BALLISTEK NO: 439C

NAI NO: RMS 11115/2.472

DATA SOURCE: COTW 4th Pg115

HISTORICAL DATA: Circa 1875.

NOTES:

LOADING DATA:

BULLET WT./TYPE	POWDER WT./TYPE	VELOCITY ('/SEC)	SOURCE
227/Lead	35.0/FFg	1200	Barnes

CASE PREPARATION: SHELLHOLDER (RCBS): 2

MAKE FROM: .220 Swift. Cut case to 1.15". Trim to length. Run over .440" dia. expander. F/L size. I.D. neck ream & chamfer. F/L size.

PHYSICAL DATA (INCHES):

CASE TYPE: **Rimmed Straight**
CASE LENGTH **A = 1.09**
HEAD DIAMETER **B = .441**
RIM DIAMETER **D = .506**
NECK DIAMETER **F = .440**
NECK LENGTH **H = N/A**
SHOULDER LENGTH **K = N/A**
BODY ANGLE (DEG'S/SIDE): **.027**
CASE CAPACITY **CC'S = 1.94**

LOADED LENGTH: **1.65**
BELT DIAMETER **C = N/A**
RIM THICKNESS **E = .052**
SHOULDER DIAMETER **G = N/A**
LENGTH TO SHOULDER **J = N/A**
SHOULDER ANGLE (DEG'S/SIDE): **N/A**
PRIMER: **L/R**
CASE CAPACITY (GR'S WATER): **30.05**

DIMENSIONAL DRAWING:

F
A
B
E
D

-NOT ACTUAL SIZE-
-DO NOT SCALE-

CARTRIDGE: .44 Bull Dog

OTHER NAMES:

DIA: .440

BALLISTEK NO: 440C

NAI NO: RMS 11115/1.205

DATA SOURCE: COTW 4th Pg184

HISTORICAL DATA: From about 1880 for Webley Bull Dog revolver.

NOTES:

LOADING DATA:

BULLET WT./TYPE	POWDER WT./TYPE	VELOCITY ('/SEC)	SOURCE
170/Lead	3.5/B'Eye	—	JJD

CASE PREPARATION: SHELLHOLDER (RCBS): 8

MAKE FROM: .44 Magnum. Turn rim to .503" dia. and thin to .048". Trim to length. F/L size.

PHYSICAL DATA (INCHES):

CASE TYPE: **Rimmed Straight**

CASE LENGTH **A = .57**

HEAD DIAMETER **B = .455**

RIM DIAMETER **D = .503**

NECK DIAMETER **F = .470**

NECK LENGTH **H = N/A**

SHOULDER LENGTH **K = N/A**

BODY ANGLE (DEG'S/SIDE): **.164**

CASE CAPACITY **CC'S = .62**

LOADED LENGTH: **.95**

BELT DIAMETER **C = N/A**

RIM THICKNESS **E = .048**

SHOULDER DIAMETER **G = N/A**

LENGTH TO SHOULDER **J = N/A**

SHOULDER ANGLE (DEG'S/SIDE): **N/A**

PRIMER: **L/P**

CASE CAPACITY (GR'S WATER): **9.58**

DIMENSIONAL DRAWING:

-NOT ACTUAL SIZE-
-DO NOT SCALE-

CARTRIDGE: .44 Colt

OTHER NAMES:

DIA: .443

BALLISTEK NO: 443B

NAI NO: RMS 11115/2.412

DATA SOURCE: COTW 4th Pg183

HISTORICAL DATA: 1871 black powder ctg.

NOTES:

LOADING DATA:

BULLET WT./TYPE	POWDER WT./TYPE	VELOCITY ('/SEC)	SOURCE
210/Lead	7.5/Unique	—	JJD

CASE PREPARATION: SHELLHOLDER (RCBS): 30

MAKE FROM: .44 S&W Mag. Turn rim to .483" dia. and thin, if necessary, to .045-.047" thick. Trim to length. .44 Spl. dies can be used to size & load.

PHYSICAL DATA (INCHES):

CASE TYPE: **Rimmed Straight**

CASE LENGTH **A = 1.10**

HEAD DIAMETER **B = .456**

RIM DIAMETER **D = .483**

NECK DIAMETER **F = .450**

NECK LENGTH **H = N/A**

SHOULDER LENGTH **K = N/A**

BODY ANGLE (DEG'S/SIDE): **.163**

CASE CAPACITY **CC'S = 2.00**

LOADED LENGTH: **1.50**

BELT DIAMETER **C = N/A**

RIM THICKNESS **E = .047**

SHOULDER DIAMETER **G = N/A**

LENGTH TO SHOULDER **J = N/A**

SHOULDER ANGLE (DEG'S/SIDE): **N/A**

PRIMER: **L/P**

CASE CAPACITY (GR'S WATER): **30.93**

DIMENSIONAL DRAWING:

-NOT ACTUAL SIZE-
-DO NOT SCALE-

CARTRIDGE: .44 Evans Long

OTHER NAMES:
.44 Evans New Model

DIA: .419

BALLISTEK NO: 419A

NAI NO: RMS 11115/3.430

DATA SOURCE: COTW 4th Pg115

HISTORICAL DATA: Circa 1877.

NOTES:

LOADING DATA:

BULLET WT./TYPE	POWDER WT./TYPE	VELOCITY ('/SEC)	SOURCE
275/Lead	14.8/IMR4198	—	JJD

CASE PREPARATION: SHELLHOLDER (RCBS): 7

MAKE FROM: .303 British. Cut case to 1.6". Taper expand to .420" dia. Turn rim to .509" dia. & back chamfer. Trim case to length. I.D. neck ream. F/L size and fireform in the chamber.

PHYSICAL DATA (INCHES):

CASE TYPE: **Rimmed Straight**
CASE LENGTH A = **1.54**
HEAD DIAMETER B = **.449**
RIM DIAMETER D = **.509**
NECK DIAMETER F = **.434**
NECK LENGTH H = **N/A**
SHOULDER LENGTH K = **N/A**
BODY ANGLE (DEG'S/SIDE): **.289**
CASE CAPACITY CC'S = **2.79**
LOADED LENGTH: **2.00**
BELT DIAMETER C = **N/A**
RIM THICKNESS E = **.055**
SHOULDER DIAMETER G = **N/A**
LENGTH TO SHOULDER J = **N/A**
SHOULDER ANGLE (DEG'S/SIDE): **N/A**
PRIMER: **L/R**
CASE CAPACITY (GR'S WATER): **43.07**

DIMENSIONAL DRAWING:

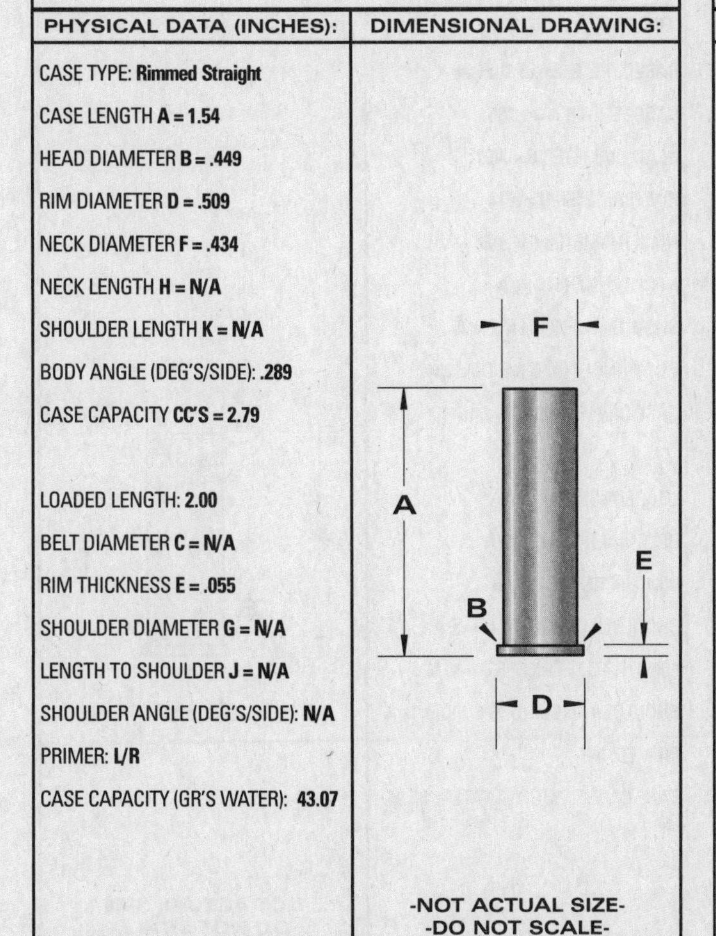

-NOT ACTUAL SIZE-
-DO NOT SCALE-

CARTRIDGE: .44 Evans Short

OTHER NAMES:
.44 Evans Old Model

DIA: .419

BALLISTEK NO: 419B

NAI NO: RMS 11115/2.250

DATA SOURCE: COTW 4th Pg114

HISTORICAL DATA: Circa 1875.

NOTES:

LOADING DATA:

BULLET WT./TYPE	POWDER WT./TYPE	VELOCITY ('/SEC)	SOURCE
215/Lead	10.0/IMR4198	—	JJD

CASE PREPARATION: SHELLHOLDER (RCBS): 21

MAKE FROM: .303 Savage. Cut case to 1.0". Taper expand to .425" dia. Trim to length. F/L size & I.D. neck ream.

PHYSICAL DATA (INCHES):

CASE TYPE: **Rimmed Straight**
CASE LENGTH A = **.99**
HEAD DIAMETER B = **.44**
RIM DIAMETER D = **.513**
NECK DIAMETER F = **.439**
NECK LENGTH H = **N/A**
SHOULDER LENGTH K = **N/A**
BODY ANGLE (DEG'S/SIDE): **.03**
CASE CAPACITY CC'S = **1.47**
LOADED LENGTH: **1.44**
BELT DIAMETER C = **N/A**
RIM THICKNESS E = **.055**
SHOULDER DIAMETER G = **N/A**
LENGTH TO SHOULDER J = **N/A**
SHOULDER ANGLE (DEG'S/SIDE): **N/A**
PRIMER: **L/R**
CASE CAPACITY (GR'S WATER): **22.82**

DIMENSIONAL DRAWING:

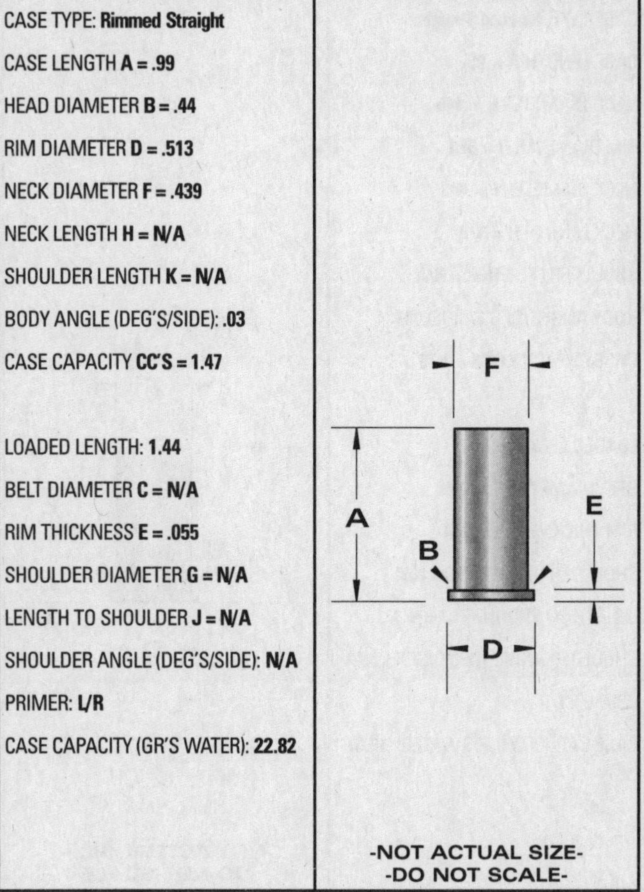

-NOT ACTUAL SIZE-
-DO NOT SCALE-

CARTRIDGE: .44 Henry

OTHER NAMES:	DIA: .423
	BALLISTEK NO: 423B
	NAI NO: RMS 11115/1.977

DATA SOURCE: COTW 4th Pg114

HISTORICAL DATA: Centerfire version of .44 Henry Flat.

NOTES:

LOADING DATA:

BULLET WT./TYPE	POWDER WT./TYPE	VELOCITY ('/SEC)	SOURCE
225/Lead	25.0/FFg	–	JJD

CASE PREPARATION: SHELLHOLDER (RCBS): 11

MAKE FROM: .220 Swift. Cut case to .90". I.D. neck ream & chamfer. Trim to length. F/L size. Rim is quite a bit under diameter but, enough to headspace case.

PHYSICAL DATA (INCHES):

CASE TYPE: **Rimmed Straight**

CASE LENGTH **A = .88**

HEAD DIAMETER **B = .445**

RIM DIAMETER **D = .523**

NECK DIAMETER **F = .443**

NECK LENGTH **H = N/A**

SHOULDER LENGTH **K = N/A**

BODY ANGLE (DEG'S/SIDE): **.07**

CASE CAPACITY **CC'S = 1.26**

LOADED LENGTH: **1.36**

BELT DIAMETER **C = N/A**

RIM THICKNESS **E = .064**

SHOULDER DIAMETER **G = N/A**

LENGTH TO SHOULDER **J = N/A**

SHOULDER ANGLE (DEG'S/SIDE): **N/A**

PRIMER: **L/R**

CASE CAPACITY (GR'S WATER): **19.57**

DIMENSIONAL DRAWING:

-NOT ACTUAL SIZE-
-DO NOT SCALE-

CARTRIDGE: .44 Remington Magnum

OTHER NAMES: .44 S&W Magnum	DIA: .430
	BALLISTEK NO: 430A
	NAI NO: RMS 11115/2.811

DATA SOURCE: Hornady 3rd Pg352

HISTORICAL DATA: By Rem. & S&W in 1955.

NOTES: Largely designed by Elmer Keith (per Keith).

LOADING DATA:

BULLET WT./TYPE	POWDER WT./TYPE	VELOCITY ('/SEC)	SOURCE
200/HP	27.8/IMR4227	1450	Horn.
250/HP	21.7/2400	1300	Horn.

CASE PREPARATION: SHELLHOLDER (RCBS): 18

MAKE FROM: Factory or .30-40 Krag or .405 Win. Basic (BELL). Case should be cut to length and the rim turned to the proper diameter. The .405 rim will require thinning to .06". I.D. neck ream to hold. .429 or .430" dia. bullets. F/L size. Fortunately, factory brass is widely available!

PHYSICAL DATA (INCHES):

CASE TYPE: **Rimmed Straight**

CASE LENGTH **A = 1.285**

HEAD DIAMETER **B = .457**

RIM DIAMETER **D = .514**

NECK DIAMETER **F = .456**

NECK LENGTH **H = N/A**

SHOULDER LENGTH **K = N/A**

BODY ANGLE (DEG'S/SIDE): **.294**

CASE CAPACITY **CC'S = 2.19**

LOADED LENGTH: **1.61**

BELT DIAMETER **C = N/A**

RIM THICKNESS **E = .06**

SHOULDER DIAMETER **G = N/A**

LENGTH TO SHOULDER **J = N/A**

SHOULDER ANGLE (DEG'S/SIDE): **N/A**

PRIMER: **L/P**

CASE CAPACITY (GR'S WATER): **33.94**

DIMENSIONAL DRAWING:

-NOT ACTUAL SIZE-
-DO NOT SCALE-

CARTRIDGE: .44 Russian (Smith & Wesson)

OTHER NAMES:

DIA: .429

BALLISTEK NO: 427C

NAI NO: RMS 11115/2.122

DATA SOURCE: COTW 4th Pg182

HISTORICAL DATA: By S&W in 1870 for Russian mil.

NOTES:

LOADING DATA:

BULLET WT./TYPE	POWDER WT./TYPE	VELOCITY ('/SEC)	SOURCE
246/—	3.6/B'Eye	700	Barnes

CASE PREPARATION: SHELLHOLDER (RCBS): 18

MAKE FROM: .44 Magnum (or Special). Trim case to length. Use .44 Spl. dies to size & seat.

PHYSICAL DATA (INCHES):

CASE TYPE: **Rimmed Straight**

CASE LENGTH **A = .97**

HEAD DIAMETER **B = .457**

RIM DIAMETER **D = .515**

NECK DIAMETER **F = .457**

NECK LENGTH **H = N/A**

SHOULDER LENGTH **K = N/A**

BODY ANGLE (DEG'S/SIDE): **0**

CASE CAPACITY **CC'S = 1.47**

LOADED LENGTH: **1.43**

BELT DIAMETER **C = N/A**

RIM THICKNESS **E = .059**

SHOULDER DIAMETER **G = N/A**

LENGTH TO SHOULDER **J = N/A**

SHOULDER ANGLE (DEG'S/SIDE): **N/A**

PRIMER: **L/P**

CASE CAPACITY (GR'S WATER): **22.76**

DIMENSIONAL DRAWING:

F
A
B
E
D

-NOT ACTUAL SIZE-
-DO NOT SCALE-

CARTRIDGE: .44 American (Smith & Wesson)

OTHER NAMES:

DIA: .434

BALLISTEK NO: 434A

NAI NO: RMS 11115/2.000

DATA SOURCE: COTW 4th Pg181

HISTORICAL DATA: By S&W about 1869.

NOTES:

LOADING DATA:

BULLET WT./TYPE	POWDER WT./TYPE	VELOCITY ('/SEC)	SOURCE
200/Lead	4.3/230P	740	Barnes

CASE PREPARATION: SHELLHOLDER (RCBS): 30

MAKE FROM: .41 S&W Magnum. Thin rim to .05" thick and trim case to length. F/L size.

PHYSICAL DATA (INCHES):

CASE TYPE: **Rimmed Straight**

CASE LENGTH **A = .88**

HEAD DIAMETER **B = .440**

RIM DIAMETER **D = .500**

NECK DIAMETER **F = .440**

NECK LENGTH **H = N/A**

SHOULDER LENGTH **K = N/A**

BODY ANGLE (DEG'S/SIDE): **0**

CASE CAPACITY **CC'S = 1.37**

LOADED LENGTH: **1.46**

BELT DIAMETER **C = N/A**

RIM THICKNESS **E = .05**

SHOULDER DIAMETER **G = N/A**

LENGTH TO SHOULDER **J = N/A**

SHOULDER ANGLE (DEG'S/SIDE): **N/A**

PRIMER: **L/P**

CASE CAPACITY (GR'S WATER): **21.22**

DIMENSIONAL DRAWING:

F
A
B
E
D

-NOT ACTUAL SIZE-
-DO NOT SCALE-

CARTRIDGE: .44 S&W Special

OTHER NAMES:
.44 Special

DIA: .430

BALLISTEK NO: 429C

NAI NO: RMS 11115/2.538

DATA SOURCE: Hornady Manual 3rd pg350

HISTORICAL DATA: By S&W in 1907.

NOTES:

LOADING DATA:

BULLET WT./TYPE	POWDER WT./TYPE	VELOCITY ('/SEC)	SOURCE
240/SWC	6.4/Unique	800	Horn.

CASE PREPARATION: SHELLHOLDER (RCBS): 18

MAKE FROM: Factory or .44 Mag. Cut case to 1.16" and F/L size. Chamfer case mouth.

PHYSICAL DATA (INCHES):

CASE TYPE: **Rimmed Straight**

CASE LENGTH **A = 1.16**

HEAD DIAMETER **B = .457**

RIM DIAMETER **D = .514**

NECK DIAMETER **F = .456**

NECK LENGTH **H = N/A**

SHOULDER LENGTH **K = N/A**

BODY ANGLE (DEG'S/SIDE): **.026**

CASE CAPACITY **CC'S = 1.92**

LOADED LENGTH: **1.50**

BELT DIAMETER **C = N/A**

RIM THICKNESS **E = .06**

SHOULDER DIAMETER **G = N/A**

LENGTH TO SHOULDER **J = N/A**

SHOULDER ANGLE (DEG'S/SIDE): **N/A**

PRIMER: **L/R**

CASE CAPACITY (GR'S WATER): **29.61**

DIMENSIONAL DRAWING:

F
A
E
B
D

-NOT ACTUAL SIZE-
-DO NOT SCALE-

CARTRIDGE: .44 Van Houten Super

OTHER NAMES:

DIA: .430

BALLISTEK NO: 430K

NAI NO: RMS 11115/4.376

DATA SOURCE: Ackley Vol.1 Pg494

HISTORICAL DATA: By Van Houten & Wade.

NOTES: For use in M94 Winchester.

LOADING DATA:

BULLET WT./TYPE	POWDER WT./TYPE	VELOCITY ('/SEC)	SOURCE
250/LeadGC	49.0/3031	1970	Ackley

CASE PREPARATION: SHELLHOLDER (RCBS): 2

MAKE FROM: .30-40 Krag. Anneal case neck and taper expand to .440" dia. Turn rim to .505" dia. Cut to 2.05". F/L size, trim to length and chamfer. Fireform in chamber.

PHYSICAL DATA (INCHES):

CASE TYPE: **Rimmed Straight**

CASE LENGTH **A = 2.00**

HEAD DIAMETER **B = .457**

RIM DIAMETER **D = .505**

NECK DIAMETER **F = .451**

NECK LENGTH **H = N/A**

SHOULDER LENGTH **K = N/A**

BODY ANGLE (DEG'S/SIDE): **.088**

CASE CAPACITY **CC'S = 3.89**

LOADED LENGTH: **2.54**

BELT DIAMETER **C = N/A**

RIM THICKNESS **E = .062**

SHOULDER DIAMETER **G = N/A**

LENGTH TO SHOULDER **J = N/A**

SHOULDER ANGLE (DEG'S/SIDE): **N/A**

PRIMER: **L/R**

CASE CAPACITY (GR'S WATER): **60.11**

DIMENSIONAL DRAWING:

F
A
E
B
D

-NOT ACTUAL SIZE-
-DO NOT SCALE-

CARTRIDGE: .44 Webley

OTHER NAMES:
.442 RIC
10,5 x 17R
.442 Revolver Centerfire

DIA: .436

BALLISTEK NO: 436B

NAI NO: RMS 11115/1.462

DATA SOURCE: COTW 4th Pg181

HISTORICAL DATA: British circa 1868.

NOTES:

LOADING DATA:

BULLET WT./TYPE	POWDER WT./TYPE	VELOCITY ('/SEC)	SOURCE
200/Lead	4.0/B'Eye	—	JJD

CASE PREPARATION: **SHELLHOLDER (RCBS): 21**

MAKE FROM: .44 Magnum. Turn rim to .503" dia. and thin to .042". Trim to length and F/L size.

PHYSICAL DATA (INCHES):

CASE TYPE: **Rimmed Straight**
CASE LENGTH **A = .69**
HEAD DIAMETER **B = .455**
RIM DIAMETER **D = .503**
NECK DIAMETER **F = .470**
NECK LENGTH **H = N/A**
SHOULDER LENGTH **K = N/A**
BODY ANGLE (DEG'S/SIDE): **.088**
CASE CAPACITY **CC'S = .874**

LOADED LENGTH: **1.10**
BELT DIAMETER **C = N/A**
RIM THICKNESS **E = .042**
SHOULDER DIAMETER **G = N/A**
LENGTH TO SHOULDER **J = N/A**
SHOULDER ANGLE (DEG'S/SIDE): **N/A**
PRIMER: **L/P**
CASE CAPACITY (GR'S WATER): **13.50**

DIMENSIONAL DRAWING:

-NOT ACTUAL SIZE-
-DO NOT SCALE-

CARTRIDGE: .44 Wesson Extra Long

OTHER NAMES:

DIA: .440

BALLISTEK NO: 440B

NAI NO: RMS 11115/3.696

DATA SOURCE: COTW 4th Pg116

HISTORICAL DATA: Similar to .44 Ballard Extra Long (#4398).

NOTES:

LOADING DATA:

BULLET WT./TYPE	POWDER WT./TYPE	VELOCITY ('/SEC)	SOURCE
250/Lead	17.8/IMR4198	—	JJD

CASE PREPARATION: **SHELLHOLDER (RCBS): 20**

MAKE FROM: .220 Swift. Cut case to 1.7". Pass over. 440" dia. expander. F/L size and I.D. neck ream. Trim to length, chamfer & F/L size again.

PHYSICAL DATA (INCHES):

CASE TYPE: **Rimmed Straight**
CASE LENGTH **A = 1.63**
HEAD DIAMETER **B = .441**
RIM DIAMETER **D = .510**
NECK DIAMETER **F = .441**
NECK LENGTH **H = N/A**
SHOULDER LENGTH **K = N/A**
BODY ANGLE (DEG'S/SIDE): **0**
CASE CAPACITY **CC'S = 3.29**

LOADED LENGTH: **2.19**
BELT DIAMETER **C = N/A**
RIM THICKNESS **E = .061**
SHOULDER DIAMETER **G = N/A**
LENGTH TO SHOULDER **J = N/A**
SHOULDER ANGLE (DEG'S/SIDE): **N/A**
PRIMER: **L/R**
CASE CAPACITY (GR'S WATER): **50.95**

DIMENSIONAL DRAWING:

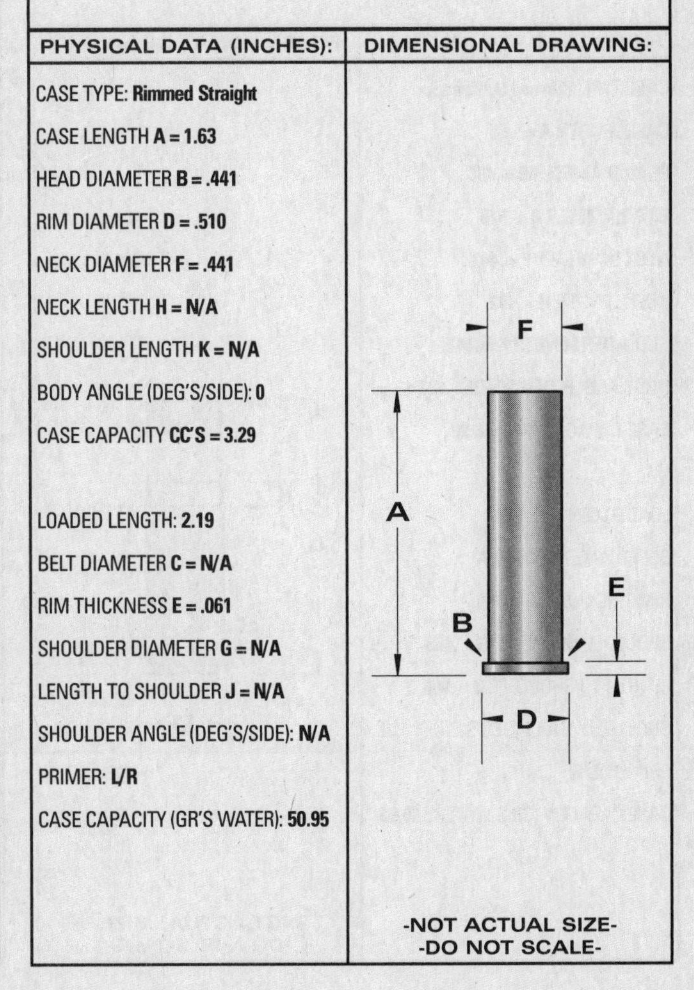

-NOT ACTUAL SIZE-
-DO NOT SCALE-

CARTRIDGE: .44-40 Extra Long

OTHER NAMES:

DIA: .428

BALLISTEK NO: 428A

NAI NO: RMB 14231/3.376

DATA SOURCE: COTW 4th Pg116

HISTORICAL DATA: .44-40 shot case with solid bullet(?).

NOTES:

LOADING DATA:

BULLET WT./TYPE	POWDER WT./TYPE	VELOCITY ('/SEC)	SOURCE
225/Lead	22.9/IMR4198	—	JJD

CASE PREPARATION: SHELLHOLDER (RCBS): 18

MAKE FROM: .444 Marlin. Thin rim to .050" thick (if necessary). Cut to 1.6". F/L size, trim to length and chamfer.

PHYSICAL DATA (INCHES):

CASE TYPE: **Rimmed Bottleneck**

CASE LENGTH A = **1.58**

HEAD DIAMETER B = **.468**

RIM DIAMETER D = **.515**

NECK DIAMETER F = **.442**

NECK LENGTH H = **.472**

SHOULDER LENGTH K = **.142**

BODY ANGLE (DEG'S/SIDE): **.187**

CASE CAPACITY CC'S = **3.30**

LOADED LENGTH: **1.96**

BELT DIAMETER C = **N/A**

RIM THICKNESS E = **.049**

SHOULDER DIAMETER G = **.463**

LENGTH TO SHOULDER J = **.966**

SHOULDER ANGLE (DEG'S/SIDE): **4.22**

PRIMER: **L/R**

CASE CAPACITY (GR'S WATER): **50.91**

DIMENSIONAL DRAWING:

F H K G A J E B D

-NOT ACTUAL SIZE-
-DO NOT SCALE-

CARTRIDGE: .44-40 Winchester

OTHER NAMES:

DIA: .427

BALLISTEK NO: 427A

NAI NO: RMB 14241/2.946

DATA SOURCE: COTW 4th Pg78

HISTORICAL DATA: By Win. about 1872.

NOTES:

LOADING DATA:

BULLET WT./TYPE	POWDER WT./TYPE	VELOCITY ('/SEC)	SOURCE
200/—	25.0/2400	1850	Barnes

CASE PREPARATION: SHELLHOLDER (RCBS): 6

MAKE FROM: Factory or .303 British. Turn rim to .525" dia., back chamfer and cut case to 1.35". Taper expand case mouth to .430" dia. F/L size. Trim to length. I.D. neck ream & chamfer. F/L size.

PHYSICAL DATA (INCHES):

CASE TYPE: **Rimmed Bottleneck**

CASE LENGTH A = **1.305**

HEAD DIAMETER B = **.471**

RIM DIAMETER D = **.525**

NECK DIAMETER F = **.443**

NECK LENGTH H = **.30**

SHOULDER LENGTH K = **.10**

BODY ANGLE (DEG'S/SIDE): **.568**

CASE CAPACITY CC'S = **2.48**

LOADED LENGTH: **1.55**

BELT DIAMETER C = **N/A**

RIM THICKNESS E = **.065**

SHOULDER DIAMETER G = **.457**

LENGTH TO SHOULDER J = **.905**

SHOULDER ANGLE (DEG'S/SIDE): **4.00**

PRIMER: **L/R**

CASE CAPACITY (GR'S WATER): **38.42**

DIMENSIONAL DRAWING:

F H K G A J E B D

-NOT ACTUAL SIZE-
-DO NOT SCALE-

CARTRIDGE: .44-60 Peabody "Creedmore"

OTHER NAMES:

DIA: .445

BALLISTEK NO: 445A

NAI NO: RMB 24231/3.648

DATA SOURCE: COTW 4th Pg117

HISTORICAL DATA: Circa 1877.

NOTES:

LOADING DATA:

BULLET WT./TYPE	POWDER WT./TYPE	VELOCITY ('/SEC)	SOURCE
350/Lead	23.5/IMR3031	—	JJD

CASE PREPARATION: SHELLHOLDER (RCBS): Spl.

MAKE FROM: .43 Span. Rem. Cut case to 1.9". Chamfer and F/L size. Trim to length.

PHYSICAL DATA (INCHES):

CASE TYPE: **Rimmed Bottleneck**

CASE LENGTH **A = 1.89**

HEAD DIAMETER **B = .518**

RIM DIAMETER **D = .628**

NECK DIAMETER **F = .464**

NECK LENGTH **H = .508**

SHOULDER LENGTH **K = .162**

BODY ANGLE (DEG'S/SIDE): **.449**

CASE CAPACITY **CC'S = 4.66**

LOADED LENGTH: **2.56**

BELT DIAMETER **C = N/A**

RIM THICKNESS **E = .073**

SHOULDER DIAMETER **G = .502**

LENGTH TO SHOULDER **J = 1.22**

SHOULDER ANGLE (DEG'S/SIDE): **6.67**

PRIMER: **L/R**

CASE CAPACITY (GR'S WATER): **71.84**

DIMENSIONAL DRAWING:

-NOT ACTUAL SIZE-
-DO NOT SCALE-

CARTRIDGE: .44-60 Sharps Necked

OTHER NAMES:
.44-60 Sharps Remington

DIA: .446

BALLISTEK NO: 446A

NAI NO: RMB 24222/3.641

DATA SOURCE: COTW 4th Pg117

HISTORICAL DATA: Circa 1875.

NOTES:

LOADING DATA:

BULLET WT./TYPE	POWDER WT./TYPE	VELOCITY ('/SEC)	SOURCE
315/Lead	24.0/4198	1300	Barnes

CASE PREPARATION: SHELLHOLDER (RCBS): Spl.

MAKE FROM: .43 Span. Rem. Thin rim to .068" thick (if necessary). Cut to 1.9". F/L size, trim to length and chamfer.

PHYSICAL DATA (INCHES):

CASE TYPE: **Rimmed Bottleneck**

CASE LENGTH **A = 1.875**

HEAD DIAMETER **B = .515**

RIM DIAMETER **D = .630**

NECK DIAMETER **F = .464**

NECK LENGTH **H = .61**

SHOULDER LENGTH **K = .095**

BODY ANGLE (DEG'S/SIDE): **.384**

CASE CAPACITY **CC'S = 4.59**

LOADED LENGTH: **2.55**

BELT DIAMETER **C = N/A**

RIM THICKNESS **E = .068**

SHOULDER DIAMETER **G = .502**

LENGTH TO SHOULDER **J = 1.17**

SHOULDER ANGLE (DEG'S/SIDE): **11.31**

PRIMER: **L/R**

CASE CAPACITY (GR'S WATER): **70.81**

DIMENSIONAL DRAWING:

-NOT ACTUAL SIZE-
-DO NOT SCALE-

CARTRIDGE: .44-77 Sharps & Remington

OTHER NAMES:

DIA: .446

BALLISTEK NO: 446C

NAI NO: RMB 13231/4.360

DATA SOURCE: COTW 4th Pg118

HISTORICAL DATA: Circa 1875.

NOTES: Two-piece, steel-head cases are reported.

LOADING DATA:

BULLET WT./TYPE	POWDER WT./TYPE	VELOCITY ('/SEC)	SOURCE
365/LeadPP	28.0/4198	1480	Barnes

CASE PREPARATION: SHELLHOLDER (RCBS): Spl.

MAKE FROM: .375 Flanged (BELL). Cut case to 2.3". F/L size, trim to length and chamfer.

PHYSICAL DATA (INCHES):

CASE TYPE: **Rimmed Bottleneck**

CASE LENGTH **A = 2.25**

HEAD DIAMETER **B = .516**

RIM DIAMETER **D = .625**

NECK DIAMETER **F = .467**

NECK LENGTH **H = .480**

SHOULDER LENGTH **K = .21**

BODY ANGLE (DEG'S/SIDE): **.294**

CASE CAPACITY **CC'S = 5.67**

LOADED LENGTH: **3.05**

BELT DIAMETER **C = N/A**

RIM THICKNESS **E = .065**

SHOULDER DIAMETER **G = .502**

LENGTH TO SHOULDER **J = 1.56**

SHOULDER ANGLE (DEG'S/SIDE): **4.76**

PRIMER: **L/R**

CASE CAPACITY (GR'S WATER): **87.52**

DIMENSIONAL DRAWING:

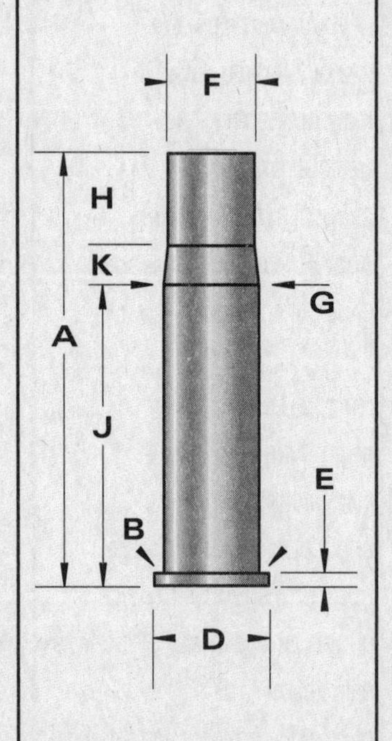

-NOT ACTUAL SIZE-
-DO NOT SCALE-

CARTRIDGE: .44-70 Maynard 1882

OTHER NAMES:

DIA: .445

BALLISTEK NO: 445B

NAI NO: RMS 21115/4.429

DATA SOURCE: COTW 4th Pg118

HISTORICAL DATA: For 1882 type Maynard rifles.

NOTES:

LOADING DATA:

BULLET WT./TYPE	POWDER WT./TYPE	VELOCITY ('/SEC)	SOURCE
470/Lead	26.0/4198	1300	Barnes

CASE PREPARATION: SHELLHOLDER (RCBS): 14

MAKE FROM: .45 Basic. Trim to length and F/L size.

PHYSICAL DATA (INCHES):

CASE TYPE: **Rimmed Straight**

CASE LENGTH **A = 2.21**

HEAD DIAMETER **B = .499**

RIM DIAMETER **D = .601**

NECK DIAMETER **F = .466**

NECK LENGTH **H = N/A**

SHOULDER LENGTH **K = N/A**

BODY ANGLE (DEG'S/SIDE): **.441**

CASE CAPACITY **CC'S = 4.98**

LOADED LENGTH: **2.87**

BELT DIAMETER **C = N/A**

RIM THICKNESS **E = .067**

SHOULDER DIAMETER **G = N/A**

LENGTH TO SHOULDER **J = N/A**

SHOULDER ANGLE (DEG'S/SIDE): **N/A**

PRIMER: **L/R**

CASE CAPACITY (GR'S WATER): **76.98**

DIMENSIONAL DRAWING:

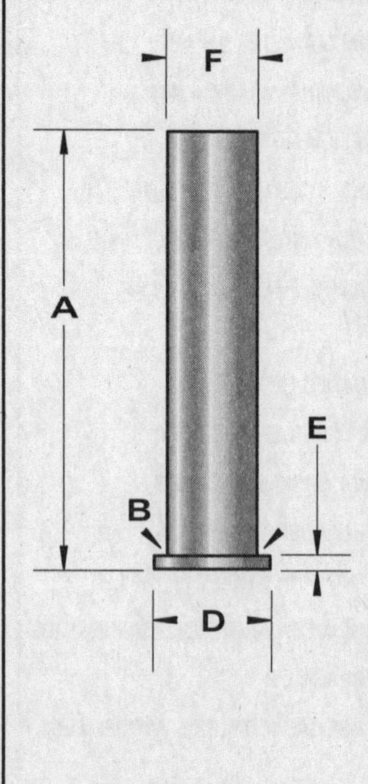

-NOT ACTUAL SIZE-
-DO NOT SCALE-

CARTRIDGE: .44-75 Ballard Everlasting

OTHER NAMES:

DIA: .445

BALLISTEK NO: 445F

NAI NO: RMS 11115/5.030

DATA SOURCE: COTW 4th Pg118

HISTORICAL DATA:

NOTES: Very rare.

LOADING DATA:

BULLET WT./TYPE	POWDER WT./TYPE	VELOCITY ('/SEC)	SOURCE
400/LeadPP	39.1/IMR3031	–	JJD

CASE PREPARATION: SHELLHOLDER (RCBS): 14

MAKE FROM: .45 Basic. Trim case to length and F/L size.

PHYSICAL DATA (INCHES):

CASE TYPE: **Rimmed Straight**

CASE LENGTH **A = 2.50**

HEAD DIAMETER **B = .497**

RIM DIAMETER **D = .603**

NECK DIAMETER **F = .487**

NECK LENGTH **H = N/A**

SHOULDER LENGTH **K = N/A**

BODY ANGLE (DEG'S/SIDE): **.118**

CASE CAPACITY **CC'S = 5.20**

LOADED LENGTH: **3.00**

BELT DIAMETER **C = N/A**

RIM THICKNESS **E = .072**

SHOULDER DIAMETER **G = N/A**

LENGTH TO SHOULDER **J = N/A**

SHOULDER ANGLE (DEG'S/SIDE): **N/A**

PRIMER: **L/R**

CASE CAPACITY (GR'S WATER): **80.29**

DIMENSIONAL DRAWING:

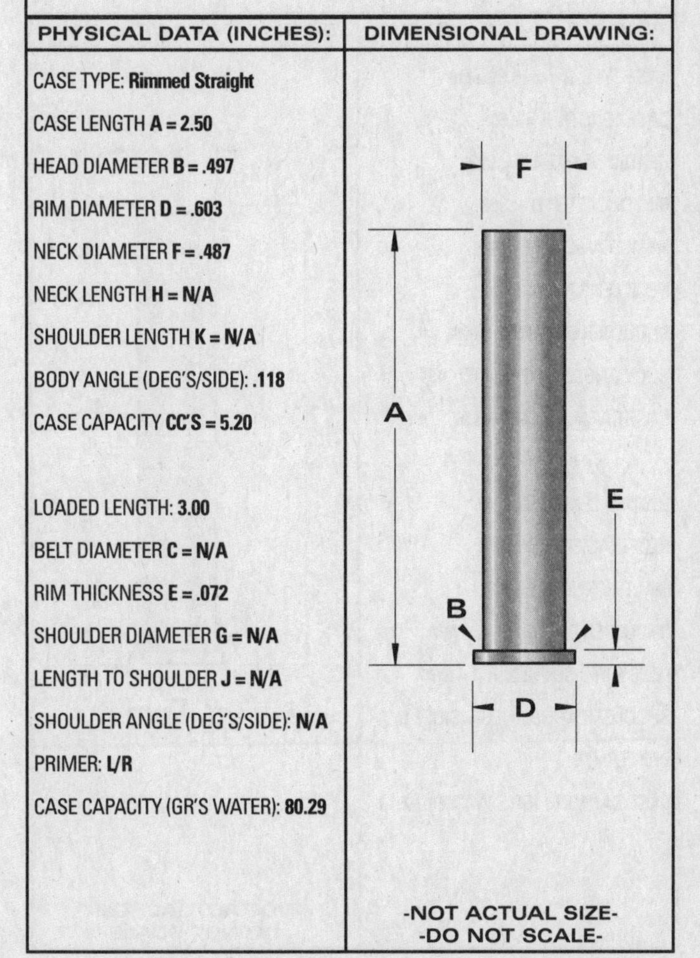

-NOT ACTUAL SIZE-
-DO NOT SCALE-

CARTRIDGE: .44-85 Wesson

OTHER NAMES:

DIA: .446

BALLISTEK NO: 446E

NAI NO: RMS 11115/6.101

DATA SOURCE: NAI/Ballistek

HISTORICAL DATA: Dim's by interpolation.

NOTES: Very rare.

LOADING DATA:

BULLET WT./TYPE	POWDER WT./TYPE	VELOCITY ('/SEC)	SOURCE
390/LeadPP	25.0/IMR4198	–	JJD

CASE PREPARATION: SHELLHOLDER (RCBS): 26

MAKE FROM: 9,3 x 74R. Anneal case neck and taper expand to hold .446" dia. bullets. Trim to length, chamfer and fireform, in the chamber. Use F/L die set to reload.

PHYSICAL DATA (INCHES):

CASE TYPE: **Rimmed Straight**

CASE LENGTH **A = 2.88**

HEAD DIAMETER **B = .472**

RIM DIAMETER **D = .530**

NECK DIAMETER **F = .460**

NECK LENGTH **H = N/A**

SHOULDER LENGTH **K = N/A**

BODY ANGLE (DEG'S/SIDE): **.121**

CASE CAPACITY **CC'S = 6.56**

LOADED LENGTH: **3.31**

BELT DIAMETER **C = N/A**

RIM THICKNESS **E = .05**

SHOULDER DIAMETER **G = N/A**

LENGTH TO SHOULDER **J = N/A**

SHOULDER ANGLE (DEG'S/SIDE): **N/A**

PRIMER: **L/R**

CASE CAPACITY (GR'S WATER): **101.30**

DIMENSIONAL DRAWING:

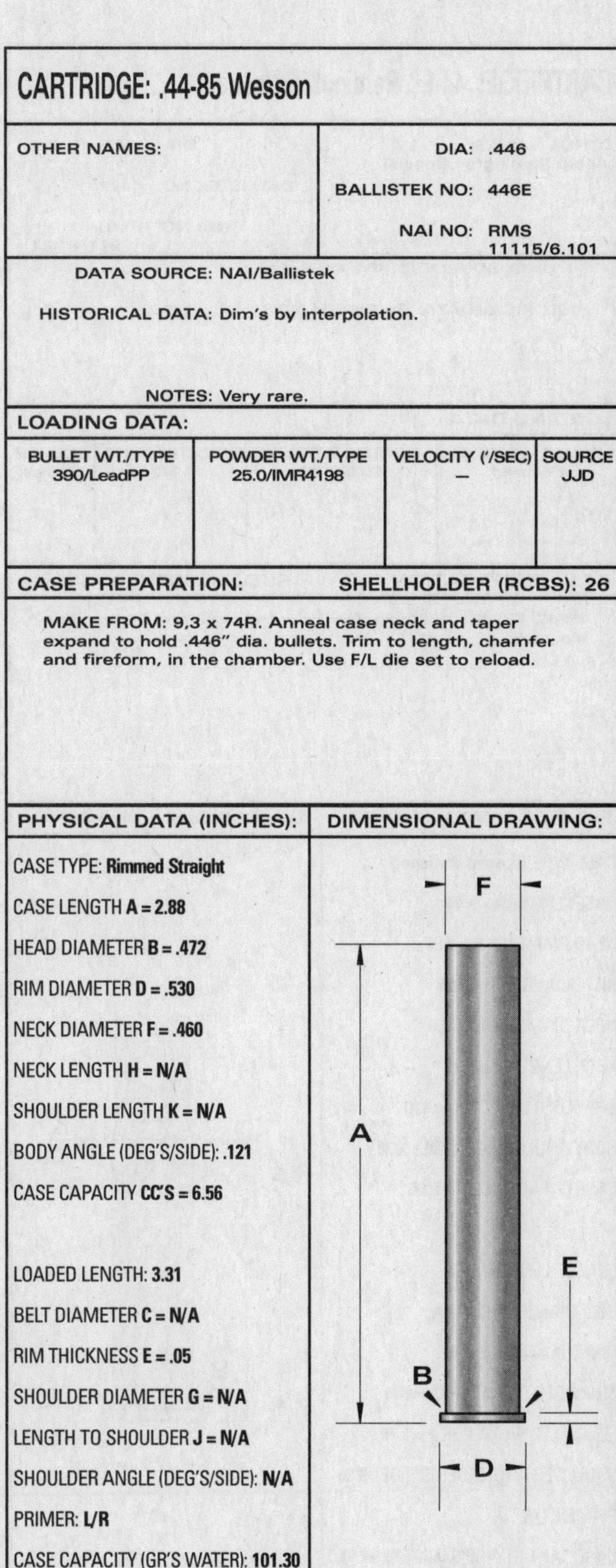

-NOT ACTUAL SIZE-
-DO NOT SCALE-

CARTRIDGE: .44-90 Remington Necked

OTHER NAMES:
.44-90 Remington Special

DIA: .442

BALLISTEK NO: 442A

NAI NO: RMB 12221/4.822

DATA SOURCE: COTW 4th Pg119

HISTORICAL DATA: By Rem. in 1873.

NOTES:

LOADING DATA:

BULLET WT./TYPE	POWDER WT./TYPE	VELOCITY ('/SEC)	SOURCE
470/Lead	30.0/4198	1270	Barnes

CASE PREPARATION: SHELLHOLDER (RCBS): 14

MAKE FROM: .45 Basic. Trim case to length and F/L size in die.

PHYSICAL DATA (INCHES):

CASE TYPE: **Rimmed Bottleneck**

CASE LENGTH **A = 2.44**

HEAD DIAMETER **B = .506**

RIM DIAMETER **D = .628**

NECK DIAMETER **F = .466**

NECK LENGTH **H = .630**

SHOULDER LENGTH **K = .110**

BODY ANGLE (DEG'S/SIDE): **.038**

CASE CAPACITY **CC'S = 6.00**

LOADED LENGTH: **3.08**

BELT DIAMETER **C = N/A**

RIM THICKNESS **E = .07**

SHOULDER DIAMETER **G = .504**

LENGTH TO SHOULDER **J = 1.70**

SHOULDER ANGLE (DEG'S/SIDE): **9.80**

PRIMER: **L/R**

CASE CAPACITY (GR'S WATER): **92.71**

DIMENSIONAL DRAWING:

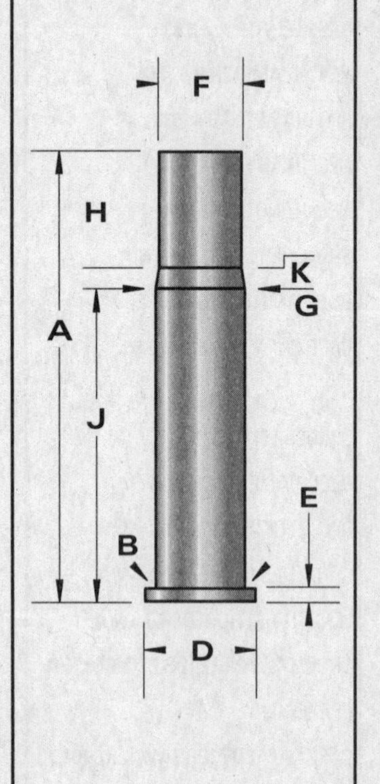

-NOT ACTUAL SIZE-
-DO NOT SCALE-

CARTRIDGE: .44-90 Remington Straight

OTHER NAMES:
.44-100 Remington Straight

DIA: .442

BALLISTEK NO: 442B

NAI NO: RMS 21115/5.169

DATA SOURCE: COTW 4th Pg120

HISTORICAL DATA:

NOTES:

LOADING DATA:

BULLET WT./TYPE	POWDER WT./TYPE	VELOCITY ('/SEC)	SOURCE
470/LeadPP	27.0/4198	1410	Barnes

CASE PREPARATION: SHELLHOLDER (RCBS): 14

MAKE FROM: .45 Basic. Trim case to length and F/L size. Turn rim to .568" dia. and back chamfer.

PHYSICAL DATA (INCHES):

CASE TYPE: **Rimmed Straight**

CASE LENGTH **A = 2.60**

HEAD DIAMETER **B = .503**

RIM DIAMETER **D = .568**

NECK DIAMETER **F = .465**

NECK LENGTH **H = N/A**

SHOULDER LENGTH **K = N/A**

BODY ANGLE (DEG'S/SIDE): **.43**

CASE CAPACITY **CC'S = 5.97**

LOADED LENGTH: **3.97**

BELT DIAMETER **C = N/A**

RIM THICKNESS **E = .07**

SHOULDER DIAMETER **G = N/A**

LENGTH TO SHOULDER **J = N/A**

SHOULDER ANGLE (DEG'S/SIDE): **N/A**

PRIMER: **L/R**

CASE CAPACITY (GR'S WATER): **92.13**

DIMENSIONAL DRAWING:

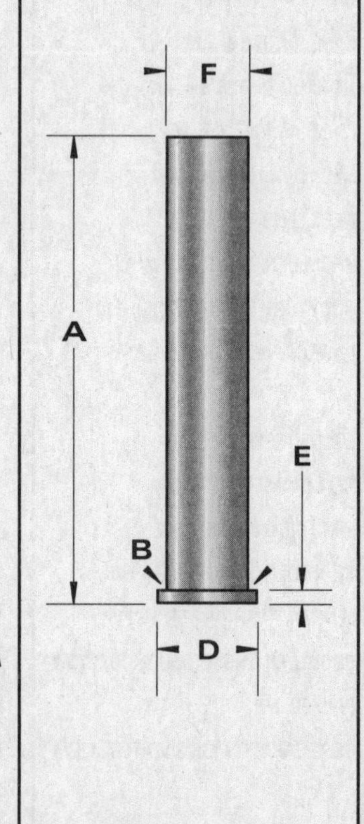

-NOT ACTUAL SIZE-
-DO NOT SCALE-

CARTRIDGE: .44-90 Sharps Necked

OTHER NAMES:
.44-100 Sharps
.44-105 Sharps

DIA: .446

BALLISTEK NO: 446D

NAI NO: RMB
12225/5.087

DATA SOURCE: COTW 4th Pg119

HISTORICAL DATA: Circa 1873.

NOTES:

LOADING DATA:

BULLET WT./TYPE	POWDER WT./TYPE	VELOCITY ('/SEC)	SOURCE
470/Lead	28.0/4198	1300	Barnes

CASE PREPARATION: SHELLHOLDER (RCBS): Spl.

MAKE FROM: .375 Flanged (BELL). Cut case to 2.7". F/L size, trim to length and chamfer.

PHYSICAL DATA (INCHES):

CASE TYPE: **Rimmed Bottleneck**

CASE LENGTH **A = 2.63**

HEAD DIAMETER **B = .517**

RIM DIAMETER **D = .625**

NECK DIAMETER **F = .468**

NECK LENGTH **H = .515**

SHOULDER LENGTH **K = .385**

BODY ANGLE (DEG'S/SIDE): **.219**

CASE CAPACITY **CC'S = 6.64**

LOADED LENGTH: **3.30**

BELT DIAMETER **C = N/A**

RIM THICKNESS **E = .06**

SHOULDER DIAMETER **G = .504**

LENGTH TO SHOULDER **J = 1.89**

SHOULDER ANGLE (DEG'S/SIDE): **2.67**

PRIMER: **L/R**

CASE CAPACITY (GR'S WATER): **102.57**

DIMENSIONAL DRAWING:

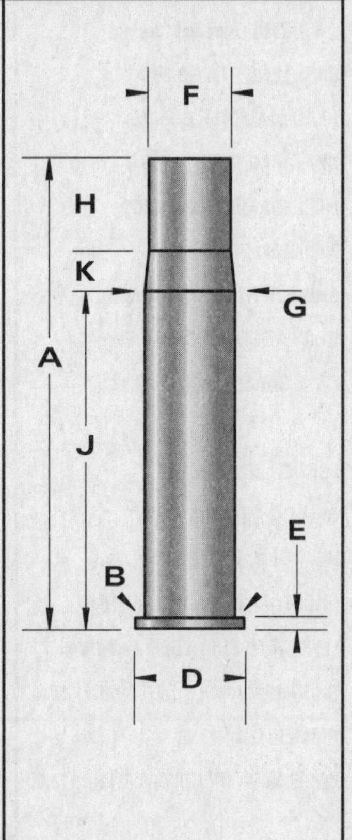

-NOT ACTUAL SIZE-
-DO NOT SCALE-

CARTRIDGE: .44-90 Peabody "What Cheer"

OTHER NAMES:
.44-100 Peabody

DIA: .443

BALLISTEK NO: 443A

NAI NO: RMB
24222/3.983

DATA SOURCE: COTW 4th Pg120

HISTORICAL DATA: Circa 1876.

NOTES:

LOADING DATA:

BULLET WT./TYPE	POWDER WT./TYPE	VELOCITY ('/SEC)	SOURCE
470/LeadPP	21.0/4759	1380	Barnes

CASE PREPARATION: SHELLHOLDER (RCBS): Spl.

MAKE FROM: 11mm Beaumont (BELL). F/L size the Beaumont case in the Peabody die with the expander removed. Trim to length, chamfer & F/L size.

PHYSICAL DATA (INCHES):

CASE TYPE: **Rimmed Bottleneck**

CASE LENGTH **A = 2.31**

HEAD DIAMETER **B = .580**

RIM DIAMETER **D = .670**

NECK DIAMETER **F = .465**

NECK LENGTH **H = .660**

SHOULDER LENGTH **K = .180**

BODY ANGLE (DEG'S/SIDE): **.677**

CASE CAPACITY **CC'S = 6.68**

LOADED LENGTH: **3.32**

BELT DIAMETER **C = N/A**

RIM THICKNESS **E = .062**

SHOULDER DIAMETER **G = .550**

LENGTH TO SHOULDER **J = 1.47**

SHOULDER ANGLE (DEG'S/SIDE): **13.28**

PRIMER: **L/R**

CASE CAPACITY (GR'S WATER): **103.11**

DIMENSIONAL DRAWING:

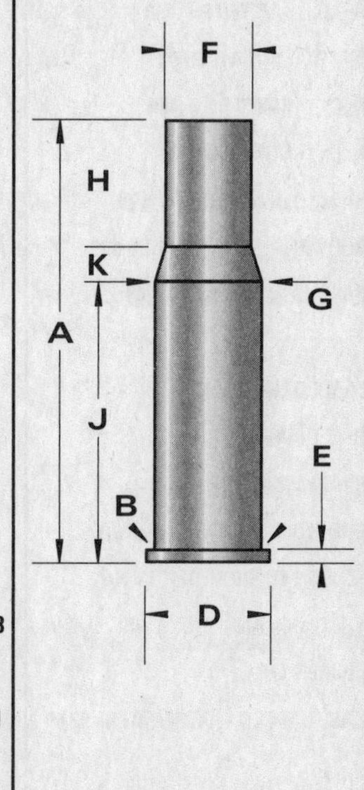

-NOT ACTUAL SIZE-
-DO NOT SCALE-

CARTRIDGE: .44-100 Ballard Everlasting

OTHER NAMES:

DIA: .445

BALLISTEK NO: 445H

NAI NO: RMS 11115/5.642

DATA SOURCE: COTW 4th Pg121

HISTORICAL DATA:

NOTES:

LOADING DATA:

BULLET WT./TYPE	POWDER WT./TYPE	VELOCITY ('/SEC)	SOURCE
365/LeadPP	26.0/4198	1500	Barnes

CASE PREPARATION: SHELLHOLDER (RCBS): 14

MAKE FROM: .45 Basic. Trim case to length and F/L size.

PHYSICAL DATA (INCHES):

CASE TYPE: **Rimmed Straight**

CASE LENGTH A = **2.81**

HEAD DIAMETER B = **.498**

RIM DIAMETER D = **.597**

NECK DIAMETER F = **.485**

NECK LENGTH H = **N/A**

SHOULDER LENGTH K = **N/A**

BODY ANGLE (DEG'S/SIDE): **.135**

CASE CAPACITY CC'S = **6.00**

LOADED LENGTH: **3.25**

BELT DIAMETER C = **N/A**

RIM THICKNESS E = **.065**

SHOULDER DIAMETER G = **N/A**

LENGTH TO SHOULDER J = **N/A**

SHOULDER ANGLE (DEG'S/SIDE): **N/A**

PRIMER: **L/R**

CASE CAPACITY (GR'S WATER): **92.69**

DIMENSIONAL DRAWING:

F A E B D

-NOT ACTUAL SIZE-
-DO NOT SCALE-

CARTRIDGE: .44-100 Wesson

OTHER NAMES:

DIA: .445

BALLISTEK NO: 445J

NAI NO: RMS 11115/6.969

DATA SOURCE: COTW 4th Pg120

HISTORICAL DATA: Circa 1881.

NOTES:

LOADING DATA:

BULLET WT./TYPE	POWDER WT./TYPE	VELOCITY ('/SEC)	SOURCE
550/Lead	31.9/IMR4198	—	JJD

CASE PREPARATION: SHELLHOLDER (RCBS): 14

MAKE FROM: .45 Basic. F/L size basic case and square case mouth. Fireform in chamber. Better cases can be turned from solid brass. The case volume is somewhat reduced but they work very well with full charges of Pyrodex.

PHYSICAL DATA (INCHES):

CASE TYPE: **Rimmed Straight**

CASE LENGTH A = **3.38**

HEAD DIAMETER B = **.485**

RIM DIAMETER D = **.615**

NECK DIAMETER F = **.480**

NECK LENGTH H = **N/A**

SHOULDER LENGTH K = **N/A**

BODY ANGLE (DEG'S/SIDE): **.043**

CASE CAPACITY CC'S = **7.32**

LOADED LENGTH: **3.85**

BELT DIAMETER C = **N/A**

RIM THICKNESS E = **.06**

SHOULDER DIAMETER G = **N/A**

LENGTH TO SHOULDER J = **N/A**

SHOULDER ANGLE (DEG'S/SIDE): **N/A**

PRIMER: **L/R**

CASE CAPACITY (GR'S WATER): **113.00**

DIMENSIONAL DRAWING:

F A E B D

-NOT ACTUAL SIZE-
-DO NOT SCALE-

CARTRIDGE: .44/06

OTHER NAMES:	
	DIA: .432
	BALLISTEK NO: 430D
	NAI NO: RMS 11115/2.812

DATA SOURCE: Handloader #67 Pg20

HISTORICAL DATA: By W. Blackwell in 1976.

NOTES:

LOADING DATA:

BULLET WT./TYPE	POWDER WT./TYPE	VELOCITY ('/SEC)	SOURCE
240/Lead	19.0/IMR4227	1168	Blackwell

CASE PREPARATION: SHELLHOLDER (RCBS): 3

MAKE FROM: .30-06 Spgf. (or any similar type case head). Trim case to length. Swage case base by F/L sizing in .44 Mag. die (case must be pushed all the way inside die to fully reduce base die.). I.D. neck ream. Anneal case. Peen rim to increase thickness to .060".

PHYSICAL DATA (INCHES):

CASE TYPE: **Rimmed Straight**

CASE LENGTH **A = 1.285**

HEAD DIAMETER **B = .457**

RIM DIAMETER **D = .473**

NECK DIAMETER **F = .456**

NECK LENGTH **H = N/A**

SHOULDER LENGTH **K = N/A**

BODY ANGLE (DEG'S/SIDE): **.023**

CASE CAPACITY **CC'S = 2.23**

LOADED LENGTH: **1.61**

BELT DIAMETER **C = N/A**

RIM THICKNESS **E = .06**

SHOULDER DIAMETER **G = N/A**

LENGTH TO SHOULDER **J = N/A**

SHOULDER ANGLE (DEG'S/SIDE): **N/A**

PRIMER: **L/P**

CASE CAPACITY (GR'S WATER): **34.43**

DIMENSIONAL DRAWING:

-NOT ACTUAL SIZE-
-DO NOT SCALE-

CARTRIDGE: .444 Marlin

OTHER NAMES:	
	DIA: .430
	BALLISTEK NO: 429B
	NAI NO: RMS 11115/4.734

DATA SOURCE: Hornady Manual 3rd pg297

HISTORICAL DATA: By Marlin in 1964.

NOTES:

LOADING DATA:

BULLET WT./TYPE	POWDER WT./TYPE	VELOCITY ('/SEC)	SOURCE
265/FP	42.0/IMR4198	2030	JJD

CASE PREPARATION: SHELLHOLDER (RCBS): 28

MAKE FROM: Factory or 9,3x74R. Turn rim to .514" dia. Cut-off at 2.220". Fireform in trim die or taper expand to hold .429" dia. lead bullets and fireform in chamber. 7x57R brass may also be used by fireforming in trim die. Finally, cases can be made from 15/32" dia. tubing.

PHYSICAL DATA (INCHES):

CASE TYPE: **Rimmed Straight**

CASE LENGTH **A = 2.225**

HEAD DIAMETER **B = .470**

RIM DIAMETER **D = .514**

NECK DIAMETER **F = .453**

NECK LENGTH **H = N/A**

SHOULDER LENGTH **K = N/A**

BODY ANGLE (DEG'S/SIDE): **.294**

CASE CAPACITY **CC'S = 4.50**

LOADED LENGTH: **2.60**

BELT DIAMETER **C = N/A**

RIM THICKNESS **E = .063**

SHOULDER DIAMETER **G = N/A**

LENGTH TO SHOULDER **J = N/A**

SHOULDER ANGLE (DEG'S/SIDE): **N/A**

PRIMER: **L/R**

CASE CAPACITY (GR'S WATER): **69.54**

DIMENSIONAL DRAWING:

-NOT ACTUAL SIZE-
-DO NOT SCALE-

CARTRIDGE: .45 ACP

OTHER NAMES:
.45 Colt Auto
11,25mm Auto

DIA: .451

BALLISTEK NO: 451C

NAI NO: RXS 11115/1.894

DATA SOURCE: Hornady 3rd Pg357

HISTORICAL DATA: By J. Browning in 1905.

NOTES:

LOADING DATA:

BULLET WT./TYPE	POWDER WT./TYPE	VELOCITY ('/SEC)	SOURCE
185/JHP	8.1/Unique	1000	Horn.

CASE PREPARATION: SHELLHOLDER (RCBS): 3

MAKE FROM: Factory or .45 Win. Mag. Cut case to length and F/L size. I.D. may require reaming, depending on bullet length. also, .308, .30-06, .270 or any similar case may be cut to length, I.D. reamed and sized. You should not have a problem finding commercial brass.

PHYSICAL DATA (INCHES):

CASE TYPE: **Rimless Straight**

CASE LENGTH **A = .898**

HEAD DIAMETER **B = .474**

RIM DIAMETER **D = .479**

NECK DIAMETER **F = .473**

NECK LENGTH **H = N/A**

SHOULDER LENGTH **K = N/A**

BODY ANGLE (DEG'S/SIDE): **.03**

CASE CAPACITY **CC'S = 1.48**

LOADED LENGTH: **1.24**

BELT DIAMETER **C = N/A**

RIM THICKNESS **E = .049**

SHOULDER DIAMETER **G = N/A**

LENGTH TO SHOULDER **J = N/A**

SHOULDER ANGLE (DEG'S/SIDE): **N/A**

PRIMER: **L/P**

CASE CAPACITY (GR'S WATER): **22.87**

DIMENSIONAL DRAWING:

-NOT ACTUAL SIZE-
-DO NOT SCALE-

CARTRIDGE: .45 Auto Rimmed

OTHER NAMES:
.45 AR

DIA: .451

BALLISTEK NO: 451D

NAI NO: RMS 11115/1.898

DATA SOURCE: Hornady 3rd Pg360

HISTORICAL DATA: By Peters in 1920.

NOTES: Rimmed version of .45 ACP (#451C).

LOADING DATA:

BULLET WT./TYPE	POWDER WT./TYPE	VELOCITY ('/SEC)	SOURCE
185/JHP	6.1/B'Eye	900	Horn.
250/—	6.5/Unique	830	Barnes

CASE PREPARATION: SHELLHOLDER (RCBS): 8

MAKE FROM: Factory or .45 Colt. Cut off case to .898". Build rim up to .09" thickness only if required. Turn I.D. of case to depth required for seating bullet. F/L size.

PHYSICAL DATA (INCHES):

CASE TYPE: **Rimmed Straight**

CASE LENGTH **A = .898**

HEAD DIAMETER **B = .473**

RIM DIAMETER **D = .516**

NECK DIAMETER **F = .472**

NECK LENGTH **H = N/A**

SHOULDER LENGTH **K = N/A**

BODY ANGLE (DEG'S/SIDE): **.035**

CASE CAPACITY **CC'S = 1.48**

LOADED LENGTH: **1.22**

BELT DIAMETER **C = N/A**

RIM THICKNESS **E = .09**

SHOULDER DIAMETER **G = N/A**

LENGTH TO SHOULDER **J = N/A**

SHOULDER ANGLE (DEG'S/SIDE): **N/A**

PRIMER: **L/R**

CASE CAPACITY (GR'S WATER): **22.92**

DIMENSIONAL DRAWING:

-NOT ACTUAL SIZE-
-DO NOT SCALE-

CARTRIDGE: .45 Boxer-Henry Long 1869

OTHER NAMES:

DIA: .450

BALLISTEK NO: 450A

NAI NO: RMS 21115/5.919

DATA SOURCE: Hoyem Vol.2 Pg97

HISTORICAL DATA: Ctg. for Martini-Henry rifle.

NOTES:

LOADING DATA:

BULLET WT./TYPE	POWDER WT./TYPE	VELOCITY ('/SEC)	SOURCE
480/LeadPP	32.0/IMR3031	—	JJD

CASE PREPARATION: SHELLHOLDER (RCBS): Spl.

MAKE FROM: .450 N.E. (BELL). F/L size the BELL case, square case mouth & chamfer.

PHYSICAL DATA (INCHES):

CASE TYPE: **Rimmed Straight**

CASE LENGTH **A = 3.22**

HEAD DIAMETER **B = .544**

RIM DIAMETER **D = .621**

NECK DIAMETER **F = .491**

NECK LENGTH **H = N/A**

SHOULDER LENGTH **K = N/A**

BODY ANGLE (DEG'S/SIDE): **.477**

CASE CAPACITY **CC'S = 7.80**

LOADED LENGTH: **3.75**

BELT DIAMETER **C = N/A**

RIM THICKNESS **E = .04**

SHOULDER DIAMETER **G = N/A**

LENGTH TO SHOULDER **J = N/A**

SHOULDER ANGLE (DEG'S/SIDE): **N/A**

PRIMER: **L/R**

CASE CAPACITY (GR'S WATER): **120.43**

DIMENSIONAL DRAWING:

F

A

E

B

D

-NOT ACTUAL SIZE-
-DO NOT SCALE-

CARTRIDGE: .45 Brown

OTHER NAMES:

DIA: .451

BALLISTEK NO: 451M

NAI NO: RMS 11115/5.148

DATA SOURCE: Nonte

HISTORICAL DATA:

NOTES: Similar to .45-100 Sharps.

LOADING DATA:

BULLET WT./TYPE	POWDER WT./TYPE	VELOCITY ('/SEC)	SOURCE
500/Lead	90/FFg	—	Nonte

CASE PREPARATION: SHELLHOLDER (RCBS): 14

MAKE FROM: .45 Basic. Trim the case to length and F/L size. .45-70 dies will work for sizing, if backed-off.

PHYSICAL DATA (INCHES):

CASE TYPE: **Rimmed Straight**

CASE LENGTH **A = 2.60**

HEAD DIAMETER **B = .505**

RIM DIAMETER **D = .601**

NECK DIAMETER **F = .475**

NECK LENGTH **H = N/A**

SHOULDER LENGTH **K = N/A**

BODY ANGLE (DEG'S/SIDE): **.403**

CASE CAPACITY **CC'S = 6.09**

LOADED LENGTH: **3.10**

BELT DIAMETER **C = N/A**

RIM THICKNESS **E = .071**

SHOULDER DIAMETER **G = N/A**

LENGTH TO SHOULDER **J = N/A**

SHOULDER ANGLE (DEG'S/SIDE): **N/A**

PRIMER: **L/R**

CASE CAPACITY (GR'S WATER): **93.89**

DIMENSIONAL DRAWING:

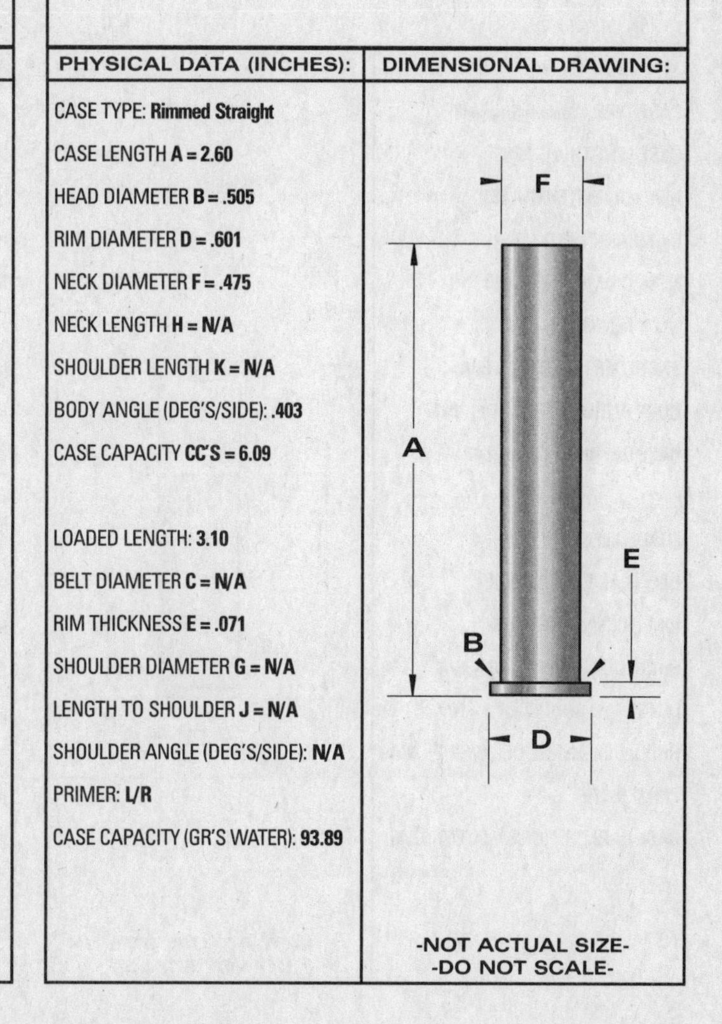

-NOT ACTUAL SIZE-
-DO NOT SCALE-

CARTRIDGE: .45 Colt Long

OTHER NAMES:
.45 Colt

DIA: .452

BALLISTEK NO: 452A

NAI NO: RMS 11115/2.671

DATA SOURCE: Hornady Manual 3rd Pg362

HISTORICAL DATA: Colt Development in 1873.

NOTES:

LOADING DATA:

BULLET WT./TYPE	POWDER WT./TYPE	VELOCITY ('/SEC)	SOURCE
250/JHP	16.4/2400	950	Horn.

CASE PREPARATION: SHELLHOLDER (RCBS): 20

MAKE FROM: Factory or 7x57R Mauser. Cut case to 1.3". Turn rim to .512" dia. I.D. neck ream to accept .451-.452" dia. bullet. F/L size. Trim to final length & chamfer.

PHYSICAL DATA (INCHES):

CASE TYPE: **Rimmed Straight**

CASE LENGTH **A = 1.285**

HEAD DIAMETER **B = .481**

RIM DIAMETER **D = .512**

NECK DIAMETER **F = .480**

NECK LENGTH **H = N/A**

SHOULDER LENGTH **K = N/A**

BODY ANGLE (DEG'S/SIDE): **.023**

CASE CAPACITY **CC'S = 2.43**

LOADED LENGTH: **1.60**

BELT DIAMETER **C = N/A**

RIM THICKNESS **E = .06**

SHOULDER DIAMETER **G = N/A**

LENGTH TO SHOULDER **J = N/A**

SHOULDER ANGLE (DEG'S/SIDE): **N/A**

PRIMER: **L/P**

CASE CAPACITY (GR'S WATER): **37.45**

DIMENSIONAL DRAWING:

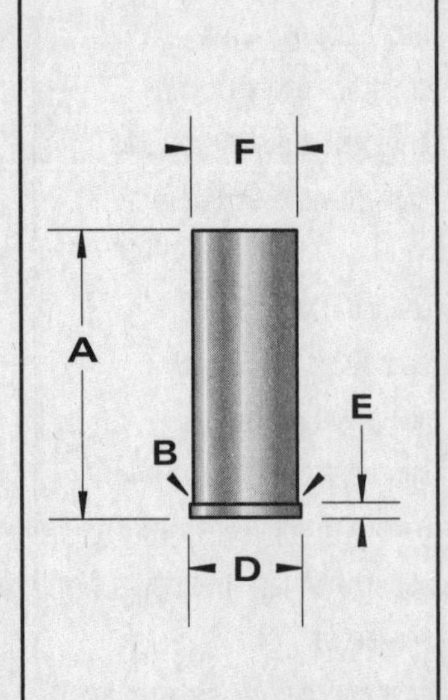

-NOT ACTUAL SIZE-
-DO NOT SCALE-

CARTRIDGE: 45 Glock Automatic Pistol

OTHER NAMES: 45 G.A.P.

DIA: .451

BALLISTEK NO:

NAI NO:

DATA SOURCE: CCI-Speer

HISTORICAL DATA: Designed by CCI-Speer and Glock to permit the use of compact frame/magazine (9mm Luger size) to accommodate a 45 caliber cartridge.

NOTES: Cases cannot be safely made by cutting down 45 Auto cases. Cut-down 45 Auto cases may cause feeding/ejection problems. 45 G.A.P. has a higher maximum pressure than 45 Auto; some 45 Auto cases may not handle the increased pressure.

LOADING DATA:

BULLET WT./TYPE	POWDER WT./TYPE	VELOCITY ('/SEC)	SOURCE

CASE PREPARATION: SHELL HOLDER (RCBS): 3

MAKE FROM: Use factory cases.

PHYSICAL DATA (INCHES):

CASE TYPE: **Rebated Rim Straight Taper**

CASE LENGTH **A = .760**

HEAD DIAMETER **B = .4760**

RIM DIAMETER **D = .470**

NECK DIAMETER **F = .4732**

NECK LENGTH **H =**

SHOULDER LENGTH **K =**

BODY ANGLE (DEG'S/SIDE):

CASE CAPACITY **CC'S = N/A**

LOADED LENGTH: **1.137 (w/ RN bullet)**

BELT DIAMETER **C = N/A**

RIM THICKNESS **E = .049**

SHOULDER DIAMETER **G = N/A**

LENGTH TO SHOULDER **J = N/A**

SHOULDER ANGLE (DEG'S/SIDE):

PRIMER: **Small Pistol**

CASE CAPACITY (GR'S WATER): **24.00**

DIMENSIONAL DRAWING:

F
A
B
E
D

-NOT ACTUAL SIZE-
-DO NOT SCALE-

CARTRIDGE: .45 New South Wales

OTHER NAMES:	DIA: .457
	BALLISTEK NO: 457K
	NAI NO: RMS 21115/4.503

DATA SOURCE: Hoyem Vol.2 pg97

HISTORICAL DATA: Ctg. for Martini-Henry rifles from about 1871.

NOTES:

LOADING DATA:

BULLET WT./TYPE	POWDER WT./TYPE	VELOCITY ('/SEC)	SOURCE
500/Lead	44.1/IMR3031	–	JJD

CASE PREPARATION: SHELLHOLDER (RCBS): Spl.

MAKE FROM: .450 N.E. (BELL). Cut case to 2.5". F/L size, trim to length and chamfer. Rather than I.D. neck reaming, use undersize bullets and 'minie', as required.

PHYSICAL DATA (INCHES):

CASE TYPE: **Rimmed Straight**

CASE LENGTH **A = 2.45**

HEAD DIAMETER **B = .544**

RIM DIAMETER **D = .621**

NECK DIAMETER **F = .491**

NECK LENGTH **H = N/A**

SHOULDER LENGTH **K = N/A**

BODY ANGLE (DEG'S/SIDE): **.630**

CASE CAPACITY **CC'S = 6.01**

LOADED LENGTH: **3.00**

BELT DIAMETER **C = N/A**

RIM THICKNESS **E = .042**

SHOULDER DIAMETER **G = N/A**

LENGTH TO SHOULDER **J = N/A**

SHOULDER ANGLE (DEG'S/SIDE): **N/A**

PRIMER: **L/R**

CASE CAPACITY (GR'S WATER): **92.75**

DIMENSIONAL DRAWING:

F
A
E
B
D

-NOT ACTUAL SIZE-
-DO NOT SCALE-

CARTRIDGE: .45 Schofield

OTHER NAMES: .45 Smith & Wesson	DIA: .454
	BALLISTEK NO: 454D
	NAI NO: RMS 11115/2.311

DATA SOURCE: COTW 4th Pg187

HISTORICAL DATA: Black powder ctg. ca. 1875.

NOTES:

LOADING DATA:

BULLET WT./TYPE	POWDER WT./TYPE	VELOCITY ('/SEC)	SOURCE
230/Lead	4.6/B'Eye	740	Barnes

CASE PREPARATION: SHELLHOLDER (RCBS): 20

MAKE FROM: .45 Colt. Trim case to length. Rim may need thinning to .05" (on some revolvers). F/L size.

PHYSICAL DATA (INCHES):

CASE TYPE: **Rimmed Straight**

CASE LENGTH **A = 1.10**

HEAD DIAMETER **B = .476**

RIM DIAMETER **D = .522**

NECK DIAMETER **F = .477**

NECK LENGTH **H = N/A**

SHOULDER LENGTH **K = N/A**

BODY ANGLE (DEG'S/SIDE): **.027**

CASE CAPACITY **CC'S = 2.00**

LOADED LENGTH: **1.43**

BELT DIAMETER **C = N/A**

RIM THICKNESS **E = .05**

SHOULDER DIAMETER **G = N/A**

LENGTH TO SHOULDER **J = N/A**

SHOULDER ANGLE (DEG'S/SIDE): **N/A**

PRIMER: **L/P**

CASE CAPACITY (GR'S WATER): **30.87**

DIMENSIONAL DRAWING:

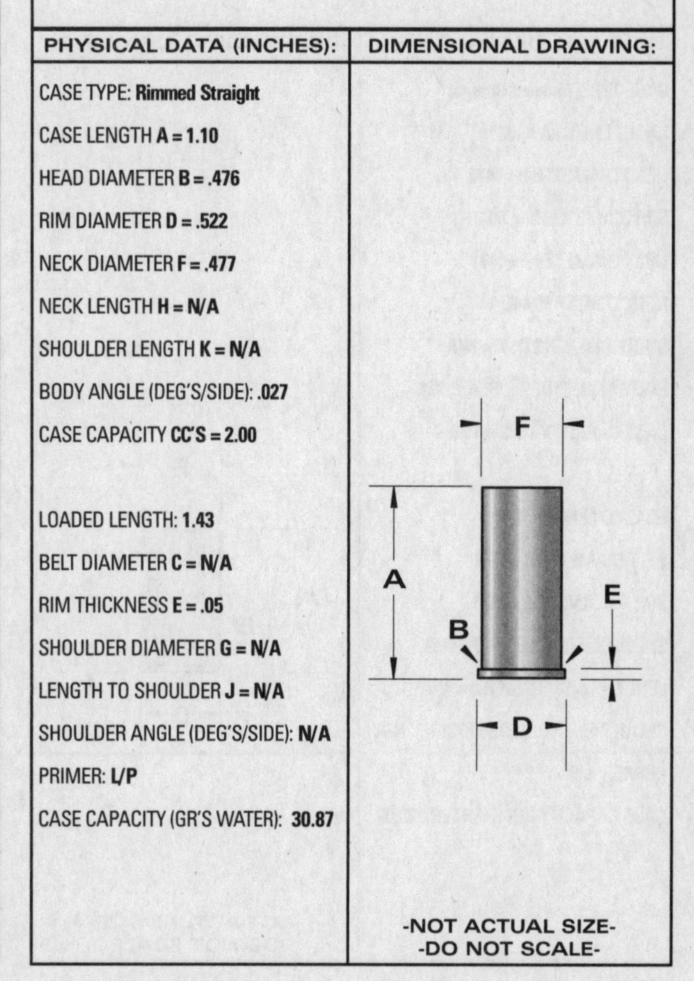

CARTRIDGE: .45 Webley

OTHER NAMES:

DIA: .452

BALLISTEK NO: 452B

NAI NO: RMS 11115/1.737

DATA SOURCE: COTW 4th Pg188

HISTORICAL DATA: Circa 1875 revolver ctg.

NOTES:

LOADING DATA:

BULLET WT./TYPE	POWDER WT./TYPE	VELOCITY ('/SEC)	SOURCE
230/Lead	4.4/B'Eye	—	JJD

CASE PREPARATION: **LOADING DATA:**

MAKE FROM: .45 Auto Rimmed. Turn rim to .504" dia. and thin to .036". Trim to length, chamfer and F/L size.

PHYSICAL DATA (INCHES):

CASE TYPE: **Rimmed Straight**

CASE LENGTH **A = .82**

HEAD DIAMETER **B = .472**

RIM DIAMETER **D = .504**

NECK DIAMETER **F = .471**

NECK LENGTH **H = N/A**

SHOULDER LENGTH **K = N/A**

BODY ANGLE (DEG'S/SIDE): **.036**

CASE CAPACITY **CC'S = 1.30**

LOADED LENGTH: **1.15**

BELT DIAMETER **C = N/A**

RIM THICKNESS **E = .036**

SHOULDER DIAMETER **G = N/A**

LENGTH TO SHOULDER **J = N/A**

SHOULDER ANGLE (DEG'S/SIDE): **N/A**

PRIMER: **L/P**

CASE CAPACITY (GR'S WATER): **20.09**

DIMENSIONAL DRAWING:

F
A
B
E
D

-NOT ACTUAL SIZE-
-DO NOT SCALE-

CARTRIDGE: .45 Winchester Magnum

OTHER NAMES:

DIA: .451

BALLISTEK NO: 451H

NAI NO: RXS 11115/2.516

DATA SOURCE: Hornady 3rd Pg382

HISTORICAL DATA: By Win. in 1979 for Wildey Pistol.

NOTES:

LOADING DATA:

BULLET WT./TYPE	POWDER WT./TYPE	VELOCITY ('/SEC)	SOURCE
230/FMJ	17.5/B'Dot	1600	Horn.

CASE PREPARATION: **SHELLHOLDER (RCBS): 3**

MAKE FROM: Factory or .308 Win, .30-06, .270 etc. Cut case to 1.2". I.D. neck ream to depth required. Trim to 1.198". F/L size & chamfer.

PHYSICAL DATA (INCHES):

CASE TYPE: **Rimless Straight**

CASE LENGTH **A = 1.198**

HEAD DIAMETER **B = .470**

RIM DIAMETER **D = .475**

NECK DIAMETER **F = .473**

NECK LENGTH **H = N/A**

SHOULDER LENGTH **K = N/A**

BODY ANGLE (DEG'S/SIDE): **.074**

CASE CAPACITY **CC'S = 2.24**

LOADED LENGTH: **1.51**

BELT DIAMETER **C = N/A**

RIM THICKNESS **E = .047**

SHOULDER DIAMETER **G = N/A**

LENGTH TO SHOULDER **J = N/A**

SHOULDER ANGLE (DEG'S/SIDE): **N/A**

PRIMER: **L/P**

CASE CAPACITY (GR'S WATER): **34.59**

DIMENSIONAL DRAWING:

F
A
B
E
D

-NOT ACTUAL SIZE-
-DO NOT SCALE-

CARTRIDGE: .45-50 Martini Sporting

OTHER NAMES:

DIA: .454

BALLISTEK NO: 454F

NAI NO: RMB 14241/2.984

DATA SOURCE: COTW 4th Pg122

HISTORICAL DATA: From about 1873. Win. was loading ammo in 1876.

NOTES:

LOADING DATA:

BULLET WT./TYPE	POWDER WT./TYPE	VELOCITY ('/SEC)	SOURCE
255/LeadPP	25.0/4198	1350	Barnes

CASE PREPARATION: SHELLHOLDER (RCBS): 14

MAKE FROM: .45-70 Gov't. Trim case to length, chamfer and F/L size. Fireform case in chamber.

PHYSICAL DATA (INCHES):

CASE TYPE: **Rimmed Bottleneck**
CASE LENGTH **A = 1.54**
HEAD DIAMETER **B = .516**
RIM DIAMETER **D = .634**
NECK DIAMETER **F = .478**
NECK LENGTH **H = .30**
SHOULDER LENGTH **K = .09**
BODY ANGLE (DEG'S/SIDE): **.241**
CASE CAPACITY **CC'S = 3.69**

LOADED LENGTH: **2.08**
BELT DIAMETER **C = N/A**
RIM THICKNESS **E = .065**
SHOULDER DIAMETER **G = .508**
LENGTH TO SHOULDER **J = 1.15**
SHOULDER ANGLE (DEG'S/SIDE): **9.46**
PRIMER: **L/R**
CASE CAPACITY (GR'S WATER): **56.98**

DIMENSIONAL DRAWING:

F
H
K
G
A
J
E
B
D

-NOT ACTUAL SIZE-
-DO NOT SCALE-

CARTRIDGE: .45-60 Winchester

OTHER NAMES:

DIA: .454

BALLISTEK NO: 454A

NAI NO: RMS 21115/3.720

DATA SOURCE: COTW 4th Pg122

HISTORICAL DATA: By Win. in 1879.

NOTES:

LOADING DATA:

BULLET WT./TYPE	POWDER WT./TYPE	VELOCITY ('/SEC)	SOURCE
300/Lead	25.0/4198	1450	Barnes

CASE PREPARATION: SHELLHOLDER (RCBS): 14

MAKE FROM: .45-70 Gov't. F/L size case, trim to length & chamfer.

PHYSICAL DATA (INCHES):

CASE TYPE: **Rimmed Straight**
CASE LENGTH **A = 1.89**
HEAD DIAMETER **B = .508**
RIM DIAMETER **D = .629**
NECK DIAMETER **F = .479**
NECK LENGTH **H = N/A**
SHOULDER LENGTH **K = N/A**
BODY ANGLE (DEG'S/SIDE): **.454**
CASE CAPACITY **CC'S = 4.24**

LOADED LENGTH: **2.15**
BELT DIAMETER **C = N/A**
RIM THICKNESS **E = .062**
SHOULDER DIAMETER **G = N/A**
LENGTH TO SHOULDER **J = N/A**
SHOULDER ANGLE (DEG'S/SIDE): **N/A**
PRIMER: **L/R**
CASE CAPACITY (GR'S WATER): **65.48**

DIMENSIONAL DRAWING:

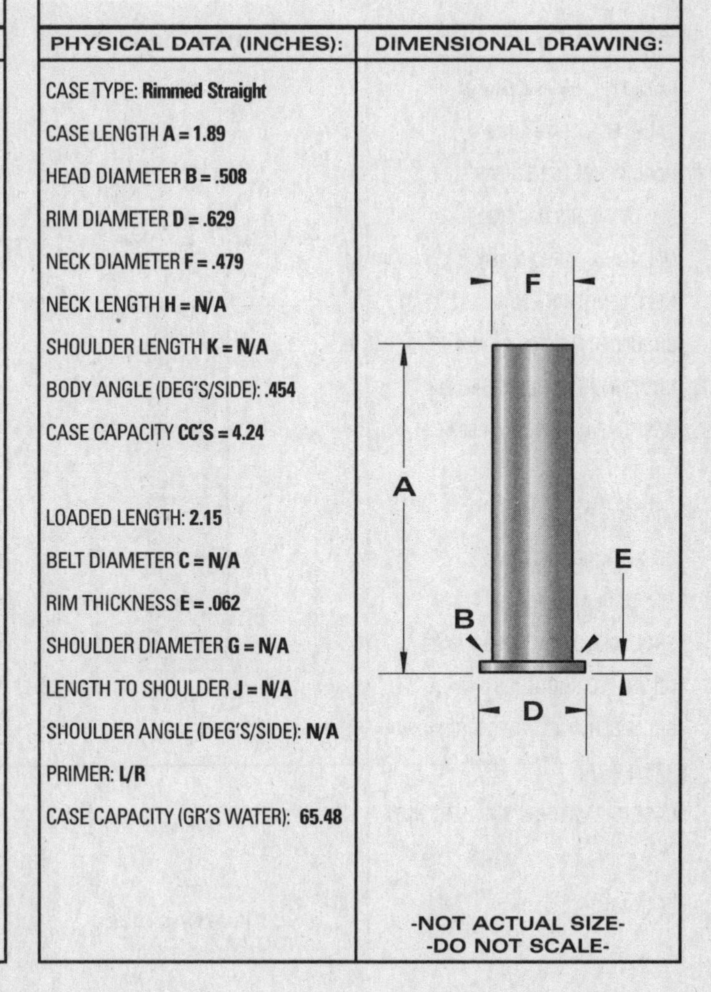

-NOT ACTUAL SIZE-
-DO NOT SCALE-

CARTRIDGE: .45-70 Government

OTHER NAMES:
.45-70-405
.45 U.S. Army M73

DIA: .458

BALLISTEK NO: 458A

NAI NO: RMS 21115/4.168

DATA SOURCE: Hornady Manual 3rd Pg300

HISTORICAL DATA: U.S. Mil. ctg. adopted 1973.

NOTES: Replaced by .30-40 Krag in 1892.

LOADING DATA:

BULLET WT./TYPE	POWDER WT./TYPE	VELOCITY ('/SEC)	SOURCE
300/JHP	53.0/Pyrodex	1425	JJD
350/RN	45.0/IMR4198	1800	Horn.

CASE PREPARATION: SHELLHOLDER (RCBS): 14

MAKE FROM: Factory or .45 Basic. Cut basic case to length, F/L size & chamfer. 1/2" O.D. tubing can be used for low pressure loads.

PHYSICAL DATA (INCHES):

CASE TYPE: **Rimmed Straight**
CASE LENGTH **A = 2.105**
HEAD DIAMETER **B = .505**
RIM DIAMETER **D = .608**
NECK DIAMETER **F = .480**
NECK LENGTH **H = N/A**
SHOULDER LENGTH **K = N/A**
BODY ANGLE (DEG'S/SIDE): **.294**
CASE CAPACITY **CC'S = 4.893**

LOADED LENGTH: **2.55**
BELT DIAMETER **C = N/A**
RIM THICKNESS **E = .07**
SHOULDER DIAMETER **G = N/A**
LENGTH TO SHOULDER **J = N/A**
SHOULDER ANGLE (DEG'S/SIDE): **N/A**
PRIMER: **L/R**
CASE CAPACITY (GR'S WATER): **75.51**

DIMENSIONAL DRAWING:

F A E B D

-NOT ACTUAL SIZE-
-DO NOT SCALE-

CARTRIDGE: .45-70 Van Choat

OTHER NAMES:

DIA: .457

BALLISTEK NO: 457H

NAI NO: RMS 11115/4.455

DATA SOURCE: NAI/Ballistek

HISTORICAL DATA: For experimental mil. rifle about 1872.

NOTES:

LOADING DATA:

BULLET WT./TYPE	POWDER WT./TYPE	VELOCITY ('/SEC)	SOURCE
420/LeadPP	36.4/4198	—	JJD

CASE PREPARATION: SHELLHOLDER (RCBS): 14

MAKE FROM: .45 Basic. Trim case to length and F/L size in .45-70 die backed off about .150". .45-70 can be used, as is, for a short version of this ctg.

PHYSICAL DATA (INCHES):

CASE TYPE: **Rimmed Straight**
CASE LENGTH **A = 2.25**
HEAD DIAMETER **B = .505**
RIM DIAMETER **D = .605**
NECK DIAMETER **F = .481**
NECK LENGTH **H = N/A**
SHOULDER LENGTH **K = N/A**
BODY ANGLE (DEG'S/SIDE): **.314**
CASE CAPACITY **CC'S = 5.22**

LOADED LENGTH: **2.91**
BELT DIAMETER **C = N/A**
RIM THICKNESS **E = .065**
SHOULDER DIAMETER **G = N/A**
LENGTH TO SHOULDER **J = N/A**
SHOULDER ANGLE (DEG'S/SIDE): **N/A**
PRIMER: **L/R**
CASE CAPACITY (GR'S WATER): **80.71**

DIMENSIONAL DRAWING:

F A E B D

-NOT ACTUAL SIZE-
-DO NOT SCALE-

CARTRIDGE: .45-75 Sharps Straight

OTHER NAMES:
.45-70 Sharps

DIA: .457

BALLISTEK NO: 457D

NAI NO: RMS 11115/4.158

DATA SOURCE: COTW 4th Pg123

HISTORICAL DATA: Same case as .45-70 Gov't.

NOTES:

LOADING DATA:

BULLET WT./TYPE	POWDER WT./TYPE	VELOCITY ('/SEC)	SOURCE
350/LeadPP	32.2/IMR4198	–	JJD

CASE PREPARATION: SHELLHOLDER (RCBS): 14

MAKE FROM: .45-70 Gov't. Use case as is. F/L size.

PHYSICAL DATA (INCHES):

CASE TYPE: **Rimmed Straight**
CASE LENGTH **A = 2.10**
HEAD DIAMETER **B = .505**
RIM DIAMETER **D = .600**
NECK DIAMETER **F = .481**
NECK LENGTH **H = N/A**
SHOULDER LENGTH **K = N/A**
BODY ANGLE (DEG'S/SIDE): **.338**
CASE CAPACITY **CC'S = 4.82**
LOADED LENGTH: **2.90**
BELT DIAMETER **C = N/A**
RIM THICKNESS **E = .066**
SHOULDER DIAMETER **G = N/A**
LENGTH TO SHOULDER **J = N/A**
SHOULDER ANGLE (DEG'S/SIDE): **N/A**
PRIMER: **L/R**
CASE CAPACITY (GR'S WATER): **74.48**

DIMENSIONAL DRAWING:

-NOT ACTUAL SIZE-
-DO NOT SCALE-

CARTRIDGE: .45/75 Winchester

OTHER NAMES:

DIA: .454

BALLISTEK NO: 454B

NAI NO: RMB 24231/3.384

DATA SOURCE: COTW 5th Pg126; Handloader #65

HISTORICAL DATA: For M1876 rifle. Loaded until 1937.

NOTES:

LOADING DATA:

BULLET WT./TYPE	POWDER WT./TYPE	VELOCITY ('/SEC)	SOURCE
350/Lead	24.0/4198	1380	Barnes

CASE PREPARATION: SHELLHOLDER (RCBS): Spl.

MAKE FROM: .50 Sharps (BELL). Cut case to 1.90". Anneal. Form in single form die. Trim to length. F/L size. Chamfer.

PHYSICAL DATA (INCHES):

CASE TYPE: **Rimmed Bottleneck**
CASE LENGTH **A = 1.895**
HEAD DIAMETER **B = .560**
RIM DIAMETER **D = .616**
NECK DIAMETER **F = .477**
NECK LENGTH **H = .55**
SHOULDER LENGTH **K = .205**
BODY ANGLE (DEG'S/SIDE): **.396**
CASE CAPACITY **CC'S = 5.18**
LOADED LENGTH: **2.25**
BELT DIAMETER **C = N/A**
RIM THICKNESS **E = .070**
SHOULDER DIAMETER **G = .547**
LENGTH TO SHOULDER **J = 1.14**
SHOULDER ANGLE (DEG'S/SIDE): **9.55**
PRIMER: **L/R**
CASE CAPACITY (GR'S WATER): **79.92**

DIMENSIONAL DRAWING:

-NOT ACTUAL SIZE-
-DO NOT SCALE-

CARTRIDGE: .45-78 Wolcott

OTHER NAMES: .45-78-475

DIA: .457
BALLISTEK NO: 457I
NAI NO: RMS 11115/4.583

DATA SOURCE: COTW 4th Pg123

HISTORICAL DATA: Variation of .45-70 Gov't from about 1873.

NOTES:

LOADING DATA:

BULLET WT./TYPE	POWDER WT./TYPE	VELOCITY ('/SEC)	SOURCE
475/LeadPP	38.4/IMR4198	–	JJD

CASE PREPARATION: **SHELLHOLDER (RCBS): 14**

MAKE FROM: .45 Basic. Trim to length and either F/L size or size in .45-70 dies backed-off .21".

PHYSICAL DATA (INCHES):

CASE TYPE: **Rimmed Straight**
CASE LENGTH **A = 2.31**
HEAD DIAMETER **B = .504**
RIM DIAMETER **D = .603**
NECK DIAMETER **F = .480**
NECK LENGTH **H = N/A**
SHOULDER LENGTH **K = N/A**
BODY ANGLE (DEG'S/SIDE): **.306**
CASE CAPACITY **CC'S = 5.40**

LOADED LENGTH: **3.19**
BELT DIAMETER **C = N/A**
RIM THICKNESS **E = .065**
SHOULDER DIAMETER **G = N/A**
LENGTH TO SHOULDER **J = N/A**
SHOULDER ANGLE (DEG'S/SIDE): **N/A**
PRIMER: **L/R**
CASE CAPACITY (GR'S WATER): **83.38**

DIMENSIONAL DRAWING:

-NOT ACTUAL SIZE-
-DO NOT SCALE-

CARTRIDGE: .45-80 Sharpshooter

OTHER NAMES:

DIA: .457
BALLISTEK NO: 457J
NAI NO: RMS 11115/4.752

DATA SOURCE: COTW 4th Pg123

HISTORICAL DATA: Another experimental US mil. ctg. from about 1874.

NOTES:

LOADING DATA:

BULLET WT./TYPE	POWDER WT./TYPE	VELOCITY ('/SEC)	SOURCE
500/LeadPP	41.1/IMR3031	–	JJD

CASE PREPARATION: **SHELLHOLDER (RCBS): 14**

MAKE FROM: .45 Basic. Cut case to 2.45". F/L size, trim to length & chamfer.

PHYSICAL DATA (INCHES):

CASE TYPE: **Rimmed Straight**
CASE LENGTH **A = 2.40**
HEAD DIAMETER **B = .505**
RIM DIAMETER **D = .605**
NECK DIAMETER **F = .480**
NECK LENGTH **H = N/A**
SHOULDER LENGTH **K = N/A**
BODY ANGLE (DEG'S/SIDE): **.306**
CASE CAPACITY **CC'S = 5.65**

LOADED LENGTH: **3.25**
BELT DIAMETER **C = N/A**
RIM THICKNESS **E = .065**
SHOULDER DIAMETER **G = N/A**
LENGTH TO SHOULDER **J = N/A**
SHOULDER ANGLE (DEG'S/SIDE): **N/A**
PRIMER: **L/R**
CASE CAPACITY (GR'S WATER): **87.31**

DIMENSIONAL DRAWING:

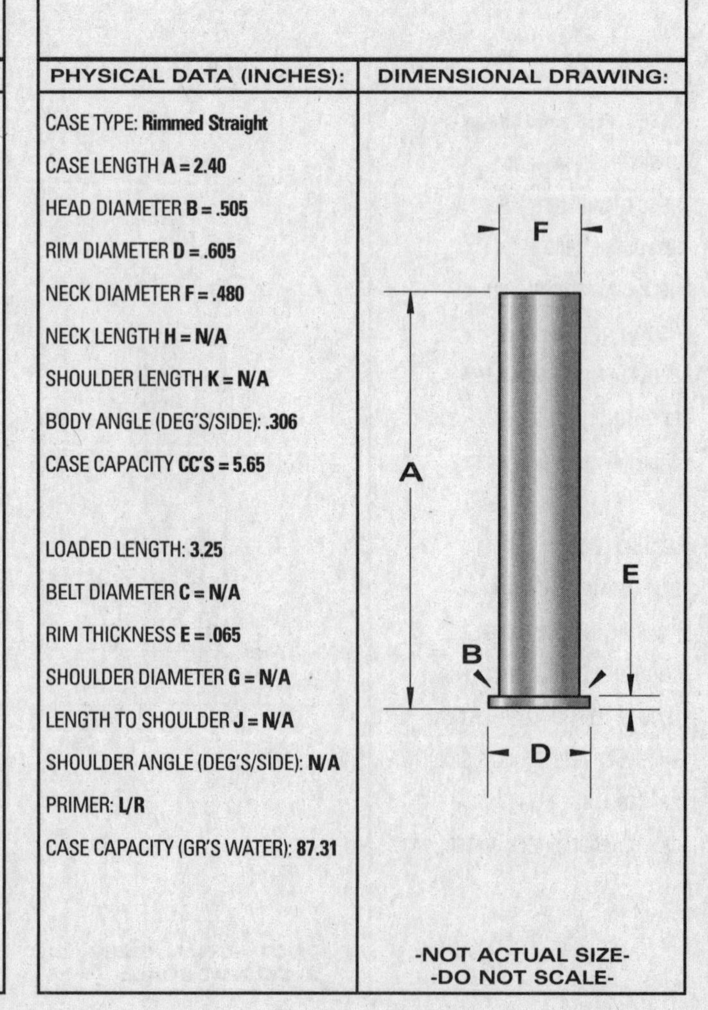

-NOT ACTUAL SIZE-
-DO NOT SCALE-

CARTRIDGE: .45-85 Ward Burton

OTHER NAMES:

DIA: .458

BALLISTEK NO: 458S

NAI NO: RMB 34332/2.759

DATA SOURCE: Hoyem Vol.2 Pg43

HISTORICAL DATA: Experimental ctg. for US mil. about 1875.

NOTES:

LOADING DATA:

BULLET WT./TYPE	POWDER WT./TYPE	VELOCITY ('/SEC)	SOURCE
400/LeadPP	21.7/IMR4198	—	JJD

CASE PREPARATION: SHELLHOLDER (RCBS): 577

MAKE FROM: .577 N.E. (BELL). Cut case to 1.9". Build rim to .092" thick. Anneal case neck and form in 2-die form set. Trim to length & chamfer. F/L size.

PHYSICAL DATA (INCHES):

CASE TYPE: **Rimmed Bottleneck**

CASE LENGTH **A = 1.81**

HEAD DIAMETER **B = .656**

RIM DIAMETER **D = .753**

NECK DIAMETER **F = .483**

NECK LENGTH **H = .480**

SHOULDER LENGTH **K = .215**

BODY ANGLE (DEG'S/SIDE): **1.34**

CASE CAPACITY **CC'S = 6.06**

LOADED LENGTH: **2.36**

BELT DIAMETER **C = N/A**

RIM THICKNESS **E = .092**

SHOULDER DIAMETER **G = .613**

LENGTH TO SHOULDER **J = 1.115**

SHOULDER ANGLE (DEG'S/SIDE): **16.82**

PRIMER: **L/R**

CASE CAPACITY (GR'S WATER): **93.44**

DIMENSIONAL DRAWING:

F H K G A J E B D

-NOT ACTUAL SIZE-
-DO NOT SCALE-

CARTRIDGE: .45-90 Sharps Straight

OTHER NAMES:

DIA: .451

BALLISTEK NO: 451L

NAI NO: RMS 11115/5.500

DATA SOURCE: COTW 4th Pg125

HISTORICAL DATA:

NOTES:

LOADING DATA:

BULLET WT./TYPE	POWDER WT./TYPE	VELOCITY ('/SEC)	SOURCE
485/LeadPP	36.0/IMR3031	—	JJD

CASE PREPARATION: SHELLHOLDER (RCBS): 14

MAKE FROM: .45 Basic. Cut case to 2.8" and F/L size.

PHYSICAL DATA (INCHES):

CASE TYPE: **Rimmed Straight**

CASE LENGTH **A = 2.75**

HEAD DIAMETER **B = .500**

RIM DIAMETER **D = .597**

NECK DIAMETER **F = .489**

NECK LENGTH **H = N/A**

SHOULDER LENGTH **K = N/A**

BODY ANGLE (DEG'S/SIDE): **.117**

CASE CAPACITY **CC'S = 6.02**

LOADED LENGTH: **3.30**

BELT DIAMETER **C = N/A**

RIM THICKNESS **E = .059**

SHOULDER DIAMETER **G = N/A**

LENGTH TO SHOULDER **J = N/A**

SHOULDER ANGLE (DEG'S/SIDE): **N/A**

PRIMER: **L/R**

CASE CAPACITY (GR'S WATER): **93.00**

DIMENSIONAL DRAWING:

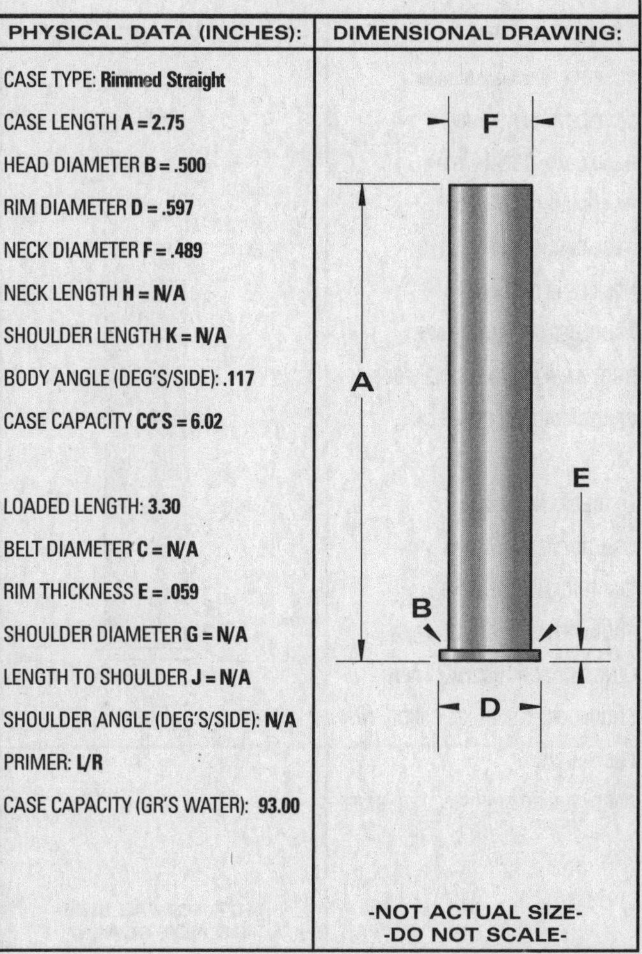

-NOT ACTUAL SIZE-
-DO NOT SCALE-

CARTRIDGE: .45-90 Winchester

OTHER NAMES:	
.45-82 Winchester .45-85 Winchester	DIA: .457 BALLISTEK NO: 457B NAI NO: RMS 11115/4.790

DATA SOURCE: COTW 4th Pg124

HISTORICAL DATA: By Win. ca. 1886.

NOTES:

LOADING DATA:

BULLET WT./TYPE	POWDER WT./TYPE	VELOCITY ('/SEC)	SOURCE
300/LeadPP	38.0/4198	1530	Barnes

CASE PREPARATION: SHELLHOLDER (RCBS): 14

MAKE FROM: .45 Basic. Cut case to 2.45". F/L size & trim to length. Chamfer.

PHYSICAL DATA (INCHES):

CASE TYPE: **Rimmed Straight**
CASE LENGTH **A = 2.40**
HEAD DIAMETER **B = .501**
RIM DIAMETER **D = .597**
NECK DIAMETER **F = .477**
NECK LENGTH **H = N/A**
SHOULDER LENGTH **K = N/A**
BODY ANGLE (DEG'S/SIDE): **.294**
CASE CAPACITY **CC'S = 5.68**
LOADED LENGTH: **2.88**
BELT DIAMETER **C = N/A**
RIM THICKNESS **E = .065**
SHOULDER DIAMETER **G = N/A**
LENGTH TO SHOULDER **J = N/A**
SHOULDER ANGLE (DEG'S/SIDE): **N/A**
PRIMER: **L/R**
CASE CAPACITY (GR'S WATER): **87.69**

DIMENSIONAL DRAWING:

-NOT ACTUAL SIZE-
-DO NOT SCALE-

CARTRIDGE: .45-100 Ballard

OTHER NAMES:	
	DIA: .454 BALLISTEK NO: 454C NAI NO: RMS 11115/5.642

DATA SOURCE: COTW 4th Pg124

HISTORICAL DATA: Circa 1878.

NOTES:

LOADING DATA:

BULLET WT./TYPE	POWDER WT./TYPE	VELOCITY ('/SEC)	SOURCE
500/LeadPP	22.0/4198	1400	Barnes

CASE PREPARATION: SHELLHOLDER (RCBS): 14

MAKE FROM: .45 Basic. Cut case to 2.9". F/L size & trim to length.

PHYSICAL DATA (INCHES):

CASE TYPE: **Rimmed Straight**
CASE LENGTH **A = 2.81**
HEAD DIAMETER **B = .498**
RIM DIAMETER **D = .597**
NECK DIAMETER **F = .487**
NECK LENGTH **H = N/A**
SHOULDER LENGTH **K = N/A**
BODY ANGLE (DEG'S/SIDE): **.114**
CASE CAPACITY **CC'S = 6.33**
LOADED LENGTH: **3.25**
BELT DIAMETER **C = N/A**
RIM THICKNESS **E = .065**
SHOULDER DIAMETER **G = N/A**
LENGTH TO SHOULDER **J = N/A**
SHOULDER ANGLE (DEG'S/SIDE): **N/A**
PRIMER: **L/R**
CASE CAPACITY (GR'S WATER): **97.69**

DIMENSIONAL DRAWING:

-NOT ACTUAL SIZE-
-DO NOT SCALE-

CARTRIDGE: .45-100 Remington

OTHER NAMES:

DIA: .452

BALLISTEK NO: 452C

NAI NO: RMB 12222/4.713

DATA SOURCE: COTW 4th Pg125

HISTORICAL DATA: By Rem. about 1880.

NOTES:

LOADING DATA:

BULLET WT./TYPE	POWDER WT./TYPE	VELOCITY ('/SEC)	SOURCE
475/LeadPP	25.9/IMR4198	—	JJD

CASE PREPARATION: SHELLHOLDER (RCBS): 31

MAKE FROM: .450 #2 N.E. (BELL). Cut case to 2.7". Thin rim to .062" thick. Anneal case, F/L size, trim to length & chamfer.

PHYSICAL DATA (INCHES):

CASE TYPE: **Rimmed Bottleneck**
CASE LENGTH **A = 2.63**
HEAD DIAMETER **B = .558**
RIM DIAMETER **D = .645**
NECK DIAMETER **F = .490**
NECK LENGTH **H = .640**
SHOULDER LENGTH **K = .14**
BODY ANGLE (DEG'S/SIDE): **.138**
CASE CAPACITY **CC'S = 7.21**

LOADED LENGTH: **3.26**
BELT DIAMETER **C = N/A**
RIM THICKNESS **E = .062**
SHOULDER DIAMETER **G = .550**
LENGTH TO SHOULDER **J = 1.85**
SHOULDER ANGLE (DEG'S/SIDE): **12.09**
PRIMER: **L/R**
CASE CAPACITY (GR'S WATER): **111.26**

DIMENSIONAL DRAWING:

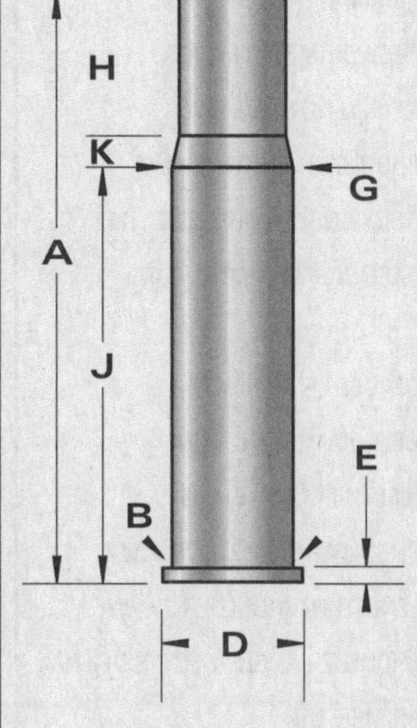

-NOT ACTUAL SIZE-
-DO NOT SCALE-

CARTRIDGE: .45-100 Sharps 2.4"

OTHER NAMES:

DIA: .451

BALLISTEK NO: 451A

NAI NO: RMS 11115/4.800

DATA SOURCE: COTW 4th Pg125

HISTORICAL DATA:

NOTES:

LOADING DATA:

BULLET WT./TYPE	POWDER WT./TYPE	VELOCITY ('/SEC)	SOURCE
485/LeadPP	37.7/IMR3031	—	JJD

CASE PREPARATION: SHELLHOLDER (RCBS): 14

MAKE FROM: .45 Basic. Trim case to 2.40" and F/L size.

PHYSICAL DATA (INCHES):

CASE TYPE: **Rimmed Straight**
CASE LENGTH **A = 2.40**
HEAD DIAMETER **B = .500**
RIM DIAMETER **D = .597**
NECK DIAMETER **F = .489**
NECK LENGTH **H = N/A**
SHOULDER LENGTH **K = N/A**
BODY ANGLE (DEG'S/SIDE): **.134**
CASE CAPACITY **CC'S = 5.16**

LOADED LENGTH: **2.85**
BELT DIAMETER **C = N/A**
RIM THICKNESS **E = .059**
SHOULDER DIAMETER **G = N/A**
LENGTH TO SHOULDER **J = N/A**
SHOULDER ANGLE (DEG'S/SIDE): **N/A**
PRIMER: **L/R**
CASE CAPACITY (GR'S WATER): **79.69**

DIMENSIONAL DRAWING:

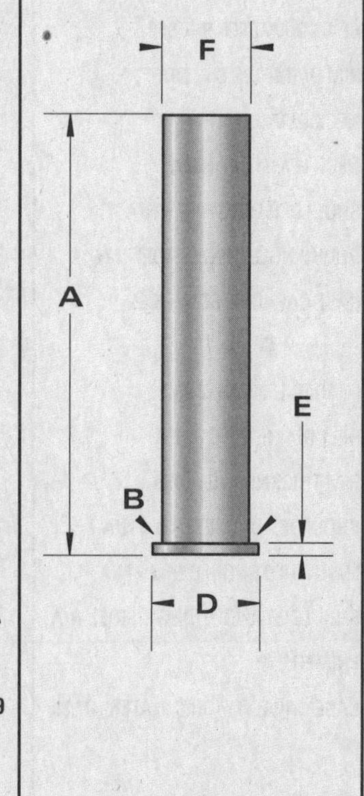

-NOT ACTUAL SIZE-
-DO NOT SCALE-

CARTRIDGE: .45-100 Sharps 2.6″

OTHER NAMES:

DIA: .451

BALLISTEK NO: 451E

NAI NO: RMS 11115/5.200

DATA SOURCE: COTW 4th Pg125

HISTORICAL DATA:

NOTES:

LOADING DATA:

BULLET WT./TYPE	POWDER WT./TYPE	VELOCITY ('/SEC)	SOURCE
485/LeadPP	37.7/IMR3031	—	JJD

CASE PREPARATION: SHELLHOLDER (RCBS): 14

MAKE FROM: .45 Basic. Trim case to 2.60″, chamfer & F/L size.

PHYSICAL DATA (INCHES):

CASE TYPE: **Rimmed Straight**

CASE LENGTH **A = 2.60**

HEAD DIAMETER **B = .500**

RIM DIAMETER **D = .597**

NECK DIAMETER **F = .489**

NECK LENGTH **H = N/A**

SHOULDER LENGTH **K = N/A**

BODY ANGLE (DEG'S/SIDE): **.124**

CASE CAPACITY **CC'S = 5.65**

LOADED LENGTH: **2.85**

BELT DIAMETER **C = N/A**

RIM THICKNESS **E = .059**

SHOULDER DIAMETER **G = N/A**

LENGTH TO SHOULDER **J = N/A**

SHOULDER ANGLE (DEG'S/SIDE): **N/A**

PRIMER: **L/R**

CASE CAPACITY (GR'S WATER): **87.30**

DIMENSIONAL DRAWING:

F

A

E

B

D

-NOT ACTUAL SIZE-
-DO NOT SCALE-

CARTRIDGE: .45-100 Sharps Straight

OTHER NAMES:
.45-100 Sharps 2 3/4″

DIA: .451

BALLISTEK NO: 451N

NAI NO: RMS 11115/5.500

DATA SOURCE: COTW 4th Pg125

HISTORICAL DATA:

NOTES:

LOADING DATA:

BULLET WT./TYPE	POWDER WT./TYPE	VELOCITY ('/SEC)	SOURCE
450/LeadPP	34.0/IMR3031	—	JJD

CASE PREPARATION: SHELLHOLDER (RCBS): 14

MAKE FROM: .45 Basic. Trim case to length and F/L size.

PHYSICAL DATA (INCHES):

CASE TYPE: **Rimmed Straight**

CASE LENGTH **A = 2.87**

HEAD DIAMETER **B = .500**

RIM DIAMETER **D = .597**

NECK DIAMETER **F = .489**

NECK LENGTH **H = N/A**

SHOULDER LENGTH **K = N/A**

BODY ANGLE (DEG'S/SIDE): **.117**

CASE CAPACITY **CC'S = 6.02**

LOADED LENGTH: **3.30**

BELT DIAMETER **C = N/A**

RIM THICKNESS **E = .06**

SHOULDER DIAMETER **G = N/A**

LENGTH TO SHOULDER **J = N/A**

SHOULDER ANGLE (DEG'S/SIDE): **N/A**

PRIMER: **L/R**

CASE CAPACITY (GR'S WATER): **92.99**

DIMENSIONAL DRAWING:

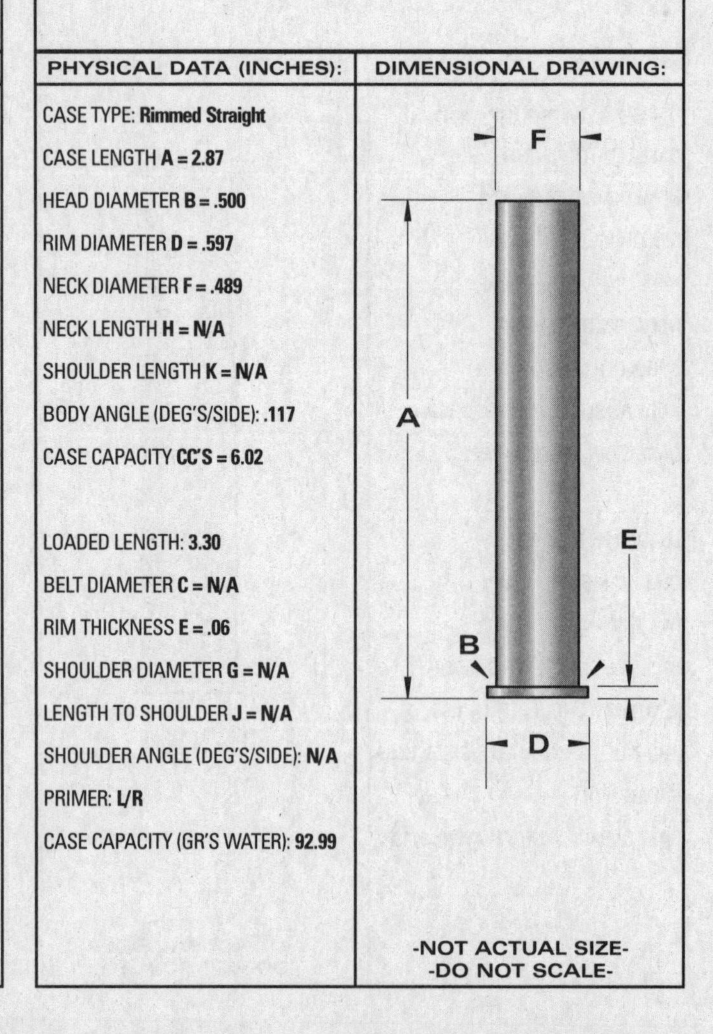

-NOT ACTUAL SIZE-
-DO NOT SCALE-

CARTRIDGE: .45-120 Sharps Straight

OTHER NAMES:

DIA: .451

BALLISTEK NO: 451B

NAI NO: RMS 11115/6.423

DATA SOURCE: COTW 4th Pg125

HISTORICAL DATA: Largest Sharps case.

NOTES:

LOADING DATA:

BULLET WT./TYPE	POWDER WT./TYPE	VELOCITY ('/SEC)	SOURCE
485/LeadPP	26.0/4198	1360	Barnes

CASE PREPARATION: SHELLHOLDER (RCBS): 14

MAKE FROM: .45 Basic. F/L size case & square case mouth.

PHYSICAL DATA (INCHES):

CASE TYPE: **Rimmed Straight**

CASE LENGTH **A = 3.25**

HEAD DIAMETER **B = .506**

RIM DIAMETER **D = .597**

NECK DIAMETER **F = .490**

NECK LENGTH **H = N/A**

SHOULDER LENGTH **K = N/A**

BODY ANGLE (DEG'S/SIDE): **.143**

CASE CAPACITY **CC'S = 7.32**

LOADED LENGTH: **4.16**

BELT DIAMETER **C = N/A**

RIM THICKNESS **E = .059**

SHOULDER DIAMETER **G = N/A**

LENGTH TO SHOULDER **J = N/A**

SHOULDER ANGLE (DEG'S/SIDE): **N/A**

PRIMER: **L/R**

CASE CAPACITY (GR'S WATER): **113.04**

DIMENSIONAL DRAWING:

F
A
E
B
D

-NOT ACTUAL SIZE-
-DO NOT SCALE-

CARTRIDGE: .45-125 Winchester Express

OTHER NAMES:

DIA: .456

BALLISTEK NO: 456A

NAI NO: RMB 12221/6.097

DATA SOURCE: COTW 4th Pg126

HISTORICAL DATA: By Win. in 1886.

NOTES:

LOADING DATA:

BULLET WT./TYPE	POWDER WT./TYPE	VELOCITY ('/SEC)	SOURCE
300/LeadPP	35.0/4198	1475	Barnes

CASE PREPARATION: SHELLHOLDER (RCBS): 14

MAKE FROM: .450 N.E. (BELL). Turn rim to .601" dia. and back chamfer. Build up rim to .065" thick. F/L size and lathe turn base to .533" dia. Square case mouth & chamfer.

PHYSICAL DATA (INCHES):

CASE TYPE: **Rimmed Bottleneck**

CASE LENGTH **A = 3.25**

HEAD DIAMETER **B = .533**

RIM DIAMETER **D = .601**

NECK DIAMETER **F = .470**

NECK LENGTH **H = .660**

SHOULDER LENGTH **K = .360**

BODY ANGLE (DEG'S/SIDE): **.169**

CASE CAPACITY **CC'S = 9.38**

LOADED LENGTH: **3.63**

BELT DIAMETER **C = N/A**

RIM THICKNESS **E = .065**

SHOULDER DIAMETER **G = .521**

LENGTH TO SHOULDER **J = 2.23**

SHOULDER ANGLE (DEG'S/SIDE): **4.05**

PRIMER: **L/R**

CASE CAPACITY (GR'S WATER): **144.86**

DIMENSIONAL DRAWING:

F
H
K
G
A
J
E
B
D

-NOT ACTUAL SIZE-
-DO NOT SCALE-

CARTRIDGE: .45-200-500

OTHER NAMES:

DIA: .457

BALLISTEK NO: 458Y

NAI NO: RMB 22343/4.594

DATA SOURCE: Hoyem Vol.2 Pg58

HISTORICAL DATA: Experimental Win. ctg. about 1884.

NOTES:

LOADING DATA:

BULLET WT./TYPE	POWDER WT./TYPE	VELOCITY ('/SEC)	SOURCE
500/Lead	200/FFg	—	Original

CASE PREPARATION: SHELLHOLDER (RCBS): Spl.

MAKE FROM: Turn case from solid brass. Anneal and form in 2-die form set. Trim to length and F/L size.

PHYSICAL DATA (INCHES):

CASE TYPE: **Rimmed Bottleneck**

CASE LENGTH **A = 3.11**

HEAD DIAMETER **B = .677**

RIM DIAMETER **D = .778**

NECK DIAMETER **F = .476**

NECK LENGTH **H = .17**

SHOULDER LENGTH **K = .17**

BODY ANGLE (DEG'S/SIDE): **.69**

CASE CAPACITY **CC'S = 13.91**

LOADED LENGTH: **3.80**

BELT DIAMETER **C = N/A**

RIM THICKNESS **E = .078**

SHOULDER DIAMETER **G = .615**

LENGTH TO SHOULDER **J = 2.77**

SHOULDER ANGLE (DEG'S/SIDE): **23.09**

PRIMER: **L/R**

CASE CAPACITY (GR'S WATER): **214.69**

DIMENSIONAL DRAWING:

-NOT ACTUAL SIZE-
-DO NOT SCALE-

CARTRIDGE: .450 Ackley Magnum

OTHER NAMES:

DIA: .458

BALLISTEK NO: 458K

NAI NO: BEN 12241/5.555

DATA SOURCE: Ackley Vol.1 Pg502

HISTORICAL DATA:

NOTES:

LOADING DATA:

BULLET WT./TYPE	POWDER WT./TYPE	VELOCITY ('/SEC)	SOURCE
500/RN	90.0/4320	2350	Ackley

CASE PREPARATION: SHELLHOLDER (RCBS): 4

MAKE FROM: .375 H&H Mag. Anneal case neck and taper expand to .460" dia. Square case mouth and F/L size. Fireform in chamber.

PHYSICAL DATA (INCHES):

CASE TYPE: **Belted Bottleneck**

CASE LENGTH **A = 2.85**

HEAD DIAMETER **B = .513**

RIM DIAMETER **D = .532**

NECK DIAMETER **F = .480**

NECK LENGTH **H = .345**

SHOULDER LENGTH **K = .145**

BODY ANGLE (DEG'S/SIDE): **.172**

CASE CAPACITY **CC'S = 7.44**

LOADED LENGTH: **3.75**

BELT DIAMETER **C = .532**

RIM THICKNESS **E = .05**

SHOULDER DIAMETER **G = .500**

LENGTH TO SHOULDER **J = 2.36**

SHOULDER ANGLE (DEG'S/SIDE): **3.94**

PRIMER: **L/R Mag**

CASE CAPACITY (GR'S WATER): **114.91**

DIMENSIONAL DRAWING:

-NOT ACTUAL SIZE-
-DO NOT SCALE-

CARTRIDGE: .450 Alaskan

OTHER NAMES:	
.450/348 Winchester	DIA: .458
	BALLISTEK NO: 458J
	NAI NO: RMB 23242/4.069

DATA SOURCE: Ackley Vol.1 Pg496

HISTORICAL DATA: By H. Johnson.

NOTES:

LOADING DATA:

BULLET WT./TYPE	POWDER WT./TYPE	VELOCITY ('/SEC)	SOURCE
350/RN	63.0/3031	2000	JJD

CASE PREPARATION: SHELLHOLDER (RCBS): 5

MAKE FROM: .348 Winchester. Anneal case neck and taper expand to .460" dia. Square case mouth, chamfer and F/L size. Fireform in the chamber.

PHYSICAL DATA (INCHES):

CASE TYPE: **Rimmed Bottleneck**
CASE LENGTH **A = 2.25**
HEAD DIAMETER **B = .553**
RIM DIAMETER **D = .610**
NECK DIAMETER **F = .480**
NECK LENGTH **H = .430**
SHOULDER LENGTH **K = .07**
BODY ANGLE (DEG'S/SIDE): **.59**
CASE CAPACITY **CC'S = 6.29**

LOADED LENGTH: **A/R**
BELT DIAMETER **C = N/A**
RIM THICKNESS **E = .07**
SHOULDER DIAMETER **G = .521**
LENGTH TO SHOULDER **J = 1.75**
SHOULDER ANGLE (DEG'S/SIDE): **16.32**
PRIMER: **L/R**
CASE CAPACITY (GR'S WATER): **97.09**

DIMENSIONAL DRAWING:

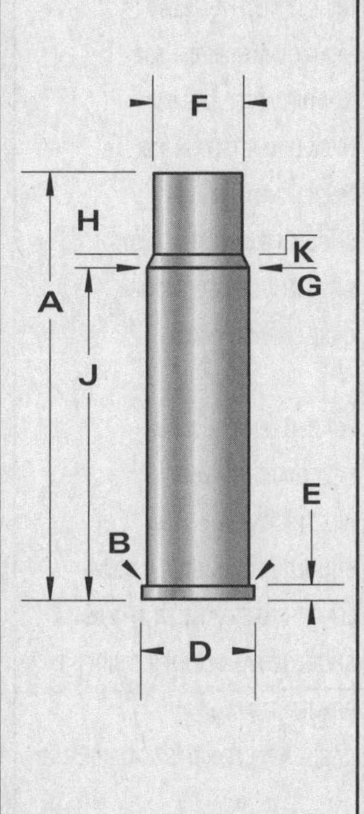

-NOT ACTUAL SIZE-
-DO NOT SCALE-

CARTRIDGE: .450 B-J Express

OTHER NAMES:	
	DIA: .458
	BALLISTEK NO: 458T
	NAI NO: BEN 12241/4.873

DATA SOURCE: Ackley Vol.1 Pg495

HISTORICAL DATA: By Barnes & Johnson

NOTES:

LOADING DATA:

BULLET WT./TYPE	POWDER WT./TYPE	VELOCITY ('/SEC)	SOURCE
500/RN	78.0/4320	2200	Ackley

CASE PREPARATION: SHELLHOLDER (RCBS): 4

MAKE FROM: .458 Win. Mag. Taper expand case neck to .475" dia. Square case mouth. F/L size in the B-J die, chamfer and fireform in the chamber.

PHYSICAL DATA (INCHES):

CASE TYPE: **Belted Bottleneck**
CASE LENGTH **A = 2.500**
HEAD DIAMETER **B = .513**
RIM DIAMETER **D = .532**
NECK DIAMETER **F = .481**
NECK LENGTH **H = .335**
SHOULDER LENGTH **K = .115**
BODY ANGLE (DEG'S/SIDE): **.185**
CASE CAPACITY **CC'S = 6.42**

LOADED LENGTH: **3.50**
BELT DIAMETER **C = .532**
RIM THICKNESS **E = .05**
SHOULDER DIAMETER **G = .501**
LENGTH TO SHOULDER **J = 2.05**
SHOULDER ANGLE (DEG'S/SIDE): **4.97**
PRIMER: **L/R Mag**
CASE CAPACITY (GR'S WATER): **99.04**

DIMENSIONAL DRAWING:

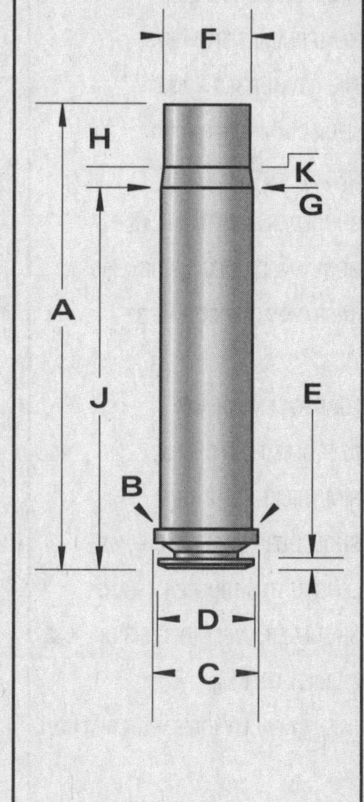

-NOT ACTUAL SIZE-
-DO NOT SCALE-

CARTRIDGE: .450 Barnes Supreme

OTHER NAMES:

DIA: .458

BALLISTEK NO: 458U

NAI NO: BEN 12241/5.516

DATA SOURCE: Ackley Vol.1 Pg497

HISTORICAL DATA: By F. Barnes.

NOTES:

LOADING DATA:

BULLET WT./TYPE	POWDER WT./TYPE	VELOCITY ('/SEC)	SOURCE
500/RN	95.0/4320	2440	Ackley

CASE PREPARATION: SHELLHOLDER (RCBS): 4

MAKE FROM: .375 H&H Mag. Anneal case neck and taper expand to about .465" dia. Square case mouth, F/L size and fireform, in the chamber.

PHYSICAL DATA (INCHES):

CASE TYPE: **Belted Bottleneck**

CASE LENGTH **A = 2.83**

HEAD DIAMETER **B = .513**

RIM DIAMETER **D = .532**

NECK DIAMETER **F = .481**

NECK LENGTH **H = .350**

SHOULDER LENGTH **K = .11**

BODY ANGLE (DEG'S/SIDE): **.198**

CASE CAPACITY **CC'S = 7.32**

LOADED LENGTH: **4.00**

BELT DIAMETER **C = .532**

RIM THICKNESS **E = .05**

SHOULDER DIAMETER **G = .498**

LENGTH TO SHOULDER **J = 2.37**

SHOULDER ANGLE (DEG'S/SIDE): **4.42**

PRIMER: **L/R Mag**

CASE CAPACITY (GR'S WATER): **113.01**

DIMENSIONAL DRAWING:

-NOT ACTUAL SIZE-
-DO NOT SCALE-

CARTRIDGE: .450 Dakota

OTHER NAMES:

DIA: .458

BALLISTEK NO:

NAI NO:

DATA SOURCE: Dakota Arms

HISTORICAL DATA: Proprietary cartridge from Dakota Arms. Based on the .416 Rigby case. The .450 Dakota offers performance suitable for use on the world's largest game.

NOTES:

LOADING DATA:

BULLET WT./TYPE	POWDER WT./TYPE	VELOCITY ('/SEC)	SOURCE
500	110.0/IMR-4350	2470	Dakota Arms

CASE PREPARATION: SHELL HOLDER (RCBS): 37

MAKE FROM: Factory or make fy reforming .416 Rigby case.

PHYSICAL DATA (INCHES):

CASE TYPE: **Rimless Bottleneck**

CASE LENGTH **A = 2.900**

HEAD DIAMETER **B = .582**

RIM DIAMETER **D = .582**

NECK DIAMETER **F = .485**

NECK LENGTH **H = .50**

SHOULDER LENGTH **K = .076**

BODY ANGLE (DEG'S/SIDE):

CASE CAPACITY **CC'S =**

LOADED LENGTH: **3.740**

BELT DIAMETER **C =**

RIM THICKNESS **E = .065**

SHOULDER DIAMETER **G = .560**

LENGTH TO SHOULDER **J = 2.324**

SHOULDER ANGLE (DEG'S/SIDE): **30**

PRIMER: **L/R Mag**

CASE CAPACITY (GR'S WATER): **145.0**

DIMENSIONAL DRAWING:

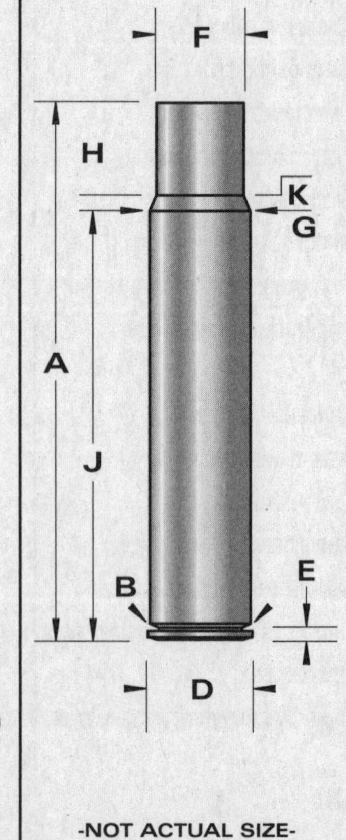

-NOT ACTUAL SIZE-
-DO NOT SCALE-

CARTRIDGE: .450 Fuller

OTHER NAMES:	
	DIA: .458
	BALLISTEK NO: 458V
	NAI NO: RMB 23232/4.086

DATA SOURCE: Ackley Vol.1 Pg497

HISTORICAL DATA: By H. Fuller.

NOTES:

LOADING DATA:

BULLET WT./TYPE	POWDER WT./TYPE	VELOCITY ('/SEC)	SOURCE
400/RN	66.0/3031	2045	Ackley

CASE PREPARATION: SHELLHOLDER (RCBS): 5

MAKE FROM: .348 Winchester. Anneal case neck and taper expand to .465" dia. Square case mouth, F/L size and fireform, in the chamber.

PHYSICAL DATA (INCHES):

CASE TYPE: **Rimmed Bottleneck**

CASE LENGTH **A = 2.26**

HEAD DIAMETER **B = .553**

RIM DIAMETER **D = .610**

NECK DIAMETER **F = .479**

NECK LENGTH **H = .460**

SHOULDER LENGTH **K = .06**

BODY ANGLE (DEG'S/SIDE): **.613**

CASE CAPACITY **CC'S = 6.33**

LOADED LENGTH: **2.90**

BELT DIAMETER **C = N/A**

RIM THICKNESS **E = .07**

SHOULDER DIAMETER **G = .520**

LENGTH TO SHOULDER **J = 1.74**

SHOULDER ANGLE (DEG'S/SIDE): **18.86**

PRIMER: **L/R**

CASE CAPACITY (GR'S WATER): **97.68**

DIMENSIONAL DRAWING:

-NOT ACTUAL SIZE-
-DO NOT SCALE-

CARTRIDGE: .450 Mashburn Magnum

OTHER NAMES:	
	DIA: .458
	BALLISTEK NO: 458W
	NAI NO: BES 11115/5.575

DATA SOURCE: Ackley Vol.1 Pg501

HISTORICAL DATA:

NOTES:

LOADING DATA:

BULLET WT./TYPE	POWDER WT./TYPE	VELOCITY ('/SEC)	SOURCE
500/RN	87.0/4320	2100	JJD

CASE PREPARATION: SHELLHOLDER (RCBS): 4

MAKE FROM: .375 H&H Mag. Anneal case neck. Taper expand to .465" dia. Square case mouth. F/L size, chamfer & fireform, in chamber.

PHYSICAL DATA (INCHES):

CASE TYPE: **Belted Straight**

CASE LENGTH **A = 2.86**

HEAD DIAMETER **B = .513**

RIM DIAMETER **D = .531**

NECK DIAMETER **F = .481**

NECK LENGTH **H = N/A**

SHOULDER LENGTH **K = N/A**

BODY ANGLE (DEG'S/SIDE): **.326**

CASE CAPACITY **CC'S = 7.04**

LOADED LENGTH: **3.65**

BELT DIAMETER **C = .530**

RIM THICKNESS **E = .049**

SHOULDER DIAMETER **G = N/A**

LENGTH TO SHOULDER **J = N/A**

SHOULDER ANGLE (DEG'S/SIDE): **N/A**

PRIMER: **L/R Mag**

CASE CAPACITY (GR'S WATER): **108.59**

DIMENSIONAL DRAWING:

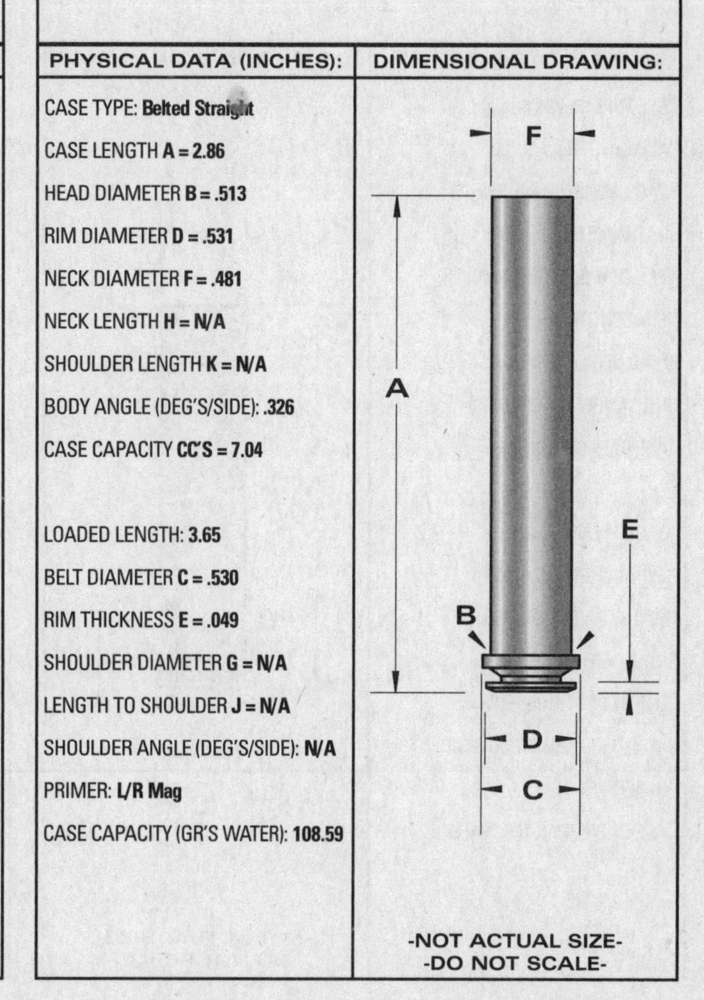

-NOT ACTUAL SIZE-
-DO NOT SCALE-

CARTRIDGE: .450 Marlin

OTHER NAMES:

DIA: .458

BALLISTEK NO:

NAI NO:

DATA SOURCE: Most current handloading manuals.

HISTORICAL DATA: Created Marlin and Hornady to create a cartridge for the 1895 Marlin rifle that would equal the performance of a handloaded .45-70. Similar in performance to .458X2-inch American wildcat. Accepted by SAAMI 1/10/01.

NOTES: The belt is thicker than on other magnum cases to prevent the catridge from chambering in other rifles.

LOADING DATA:

BULLET WT./TYPE	POWDER WT./TYPE	VELOCITY ('/SEC)	SOURCE
300/Hornady HP	57.1RL-7	2,200	Horn. #6
350/Hornady FP	55.3/AA-2015	2,000	Horn. #6

CASE PREPARATION: SHELL HOLDER (RCBS): 4

MAKE FROM: New case design. Do not make from other belted cases.

PHYSICAL DATA (INCHES):

CASE TYPE: **Belted Straight**

CASE LENGTH **A = 2.100**

HEAD DIAMETER **B = .512**

RIM DIAMETER **D = .532**

NECK DIAMETER **F = .4808**

NECK LENGTH **H =**

SHOULDER LENGTH **K =**

BODY ANGLE (DEG'S/SIDE):

CASE CAPACITY **CC'S =**

LOADED LENGTH: **2.550**

BELT DIAMETER **C = .532**

RIM THICKNESS **E = .050**

SHOULDER DIAMETER **G =**

LENGTH TO SHOULDER **J =**

SHOULDER ANGLE (DEG'S/SIDE):

PRIMER: **L/R**

CASE CAPACITY (GR'S WATER):

DIMENSIONAL DRAWING:

-NOT ACTUAL SIZE-
-DO NOT SCALE-

CARTRIDGE: .450 Nitro Express 3.25"

OTHER NAMES:

DIA: .458

BALLISTEK NO: 458R

NAI NO: RMS 21115/5.931

DATA SOURCE: COTW 4th Pg242

HISTORICAL DATA: By Rigby in 1898.

NOTES:

LOADING DATA:

BULLET WT./TYPE	POWDER WT./TYPE	VELOCITY ('/SEC)	SOURCE
485/–	35.0/4198	1450	Barnes

CASE PREPARATION: SHELLHOLDER (RCBS): Spl.

MAKE FROM: .450 N.E. (BELL). Square case mouth & F/L size.

PHYSICAL DATA (INCHES):

CASE TYPE: **Rimmed Straight**

CASE LENGTH **A = 3.25**

HEAD DIAMETER **B = .548**

RIM DIAMETER **D = .626**

NECK DIAMETER **F = .479**

NECK LENGTH **H = N/A**

SHOULDER LENGTH **K = N/A**

BODY ANGLE (DEG'S/SIDE): **.621**

CASE CAPACITY **CC'S = 8.81**

LOADED LENGTH: **3.85**

BELT DIAMETER **C = N/A**

RIM THICKNESS **E = .045**

SHOULDER DIAMETER **G = N/A**

LENGTH TO SHOULDER **J = N/A**

SHOULDER ANGLE (DEG'S/SIDE): **N/A**

PRIMER: **L/R**

CASE CAPACITY (GR'S WATER): **136.04**

DIMENSIONAL DRAWING:

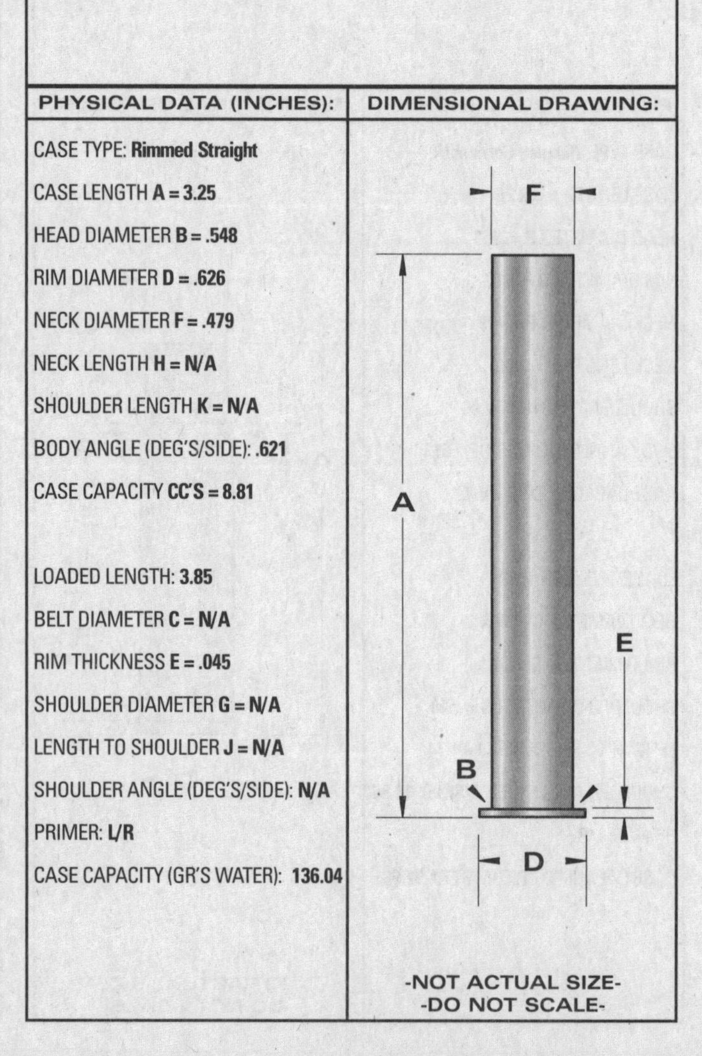

-NOT ACTUAL SIZE-
-DO NOT SCALE-

CARTRIDGE: .450 Revolver

OTHER NAMES:	
.450 Short .450 Adams	DIA: .455 BALLISTEK NO: 455E NAI NO: RMS 11115/1.446

DATA SOURCE: COTW 4th Pg188

HISTORICAL DATA: Early British revolver ctg. ca. 1868.

NOTES:

LOADING DATA:

BULLET WT./TYPE	POWDER WT./TYPE	VELOCITY ('/SEC)	SOURCE
255/Lead	2.7/B'Eye	–	JJD

CASE PREPARATION: SHELLHOLDER (RCBS): 20

MAKE FROM: .45 Auto Rimmed. Turn rim to .510" dia., back chamfer and thin to .040" thick. Trim to length and F/L size. .45 ACP dies may be used. See priming notes under .455 Webley MkI (#454J).

PHYSICAL DATA (INCHES):

CASE TYPE: **Rimmed Straight**
CASE LENGTH **A = .69**
HEAD DIAMETER **B = .477**
RIM DIAMETER **D = .510**
NECK DIAMETER **F = .475**
NECK LENGTH **H = N/A**
SHOULDER LENGTH **K = N/A**
BODY ANGLE (DEG'S/SIDE): **.088**
CASE CAPACITY **CC'S = .986**
LOADED LENGTH: **1.10**
BELT DIAMETER **C = N/A**
RIM THICKNESS **E = .04**
SHOULDER DIAMETER **G = N/A**
LENGTH TO SHOULDER **J = N/A**
SHOULDER ANGLE (DEG'S/SIDE): **N/A**
PRIMER: **L/P**
CASE CAPACITY (GR'S WATER): **15.22**

DIMENSIONAL DRAWING:

-NOT ACTUAL SIZE-
-DO NOT SCALE-

CARTRIDGE: .450 Rigby Match

OTHER NAMES:	
	DIA: .461 BALLISTEK NO: 461A NAI NO: RMS 21115/4.734

DATA SOURCE: COTW 4th Pg241

HISTORICAL DATA: Circa 1874 British ctg.

NOTES:

LOADING DATA:

BULLET WT./TYPE	POWDER WT./TYPE	VELOCITY ('/SEC)	SOURCE
480/LeadPP	85.0/FFg	1350	Original

CASE PREPARATION: SHELLHOLDER (RCBS): 14

MAKE FROM: .45 Basic. Trim case to length & F/L size.

PHYSICAL DATA (INCHES):

CASE TYPE: **Rimmed Straight**
CASE LENGTH **A = 2.40**
HEAD DIAMETER **B = .507**
RIM DIAMETER **D = .598**
NECK DIAMETER **F = .472**
NECK LENGTH **H = N/A**
SHOULDER LENGTH **K = N/A**
BODY ANGLE (DEG'S/SIDE): **.429**
CASE CAPACITY **CC'S = 6.06**
LOADED LENGTH: **3.70**
BELT DIAMETER **C = N/A**
RIM THICKNESS **E = .065**
SHOULDER DIAMETER **G = N/A**
LENGTH TO SHOULDER **J = N/A**
SHOULDER ANGLE (DEG'S/SIDE): **N/A**
PRIMER: **L/R**
CASE CAPACITY (GR'S WATER): **93.15**

DIMENSIONAL DRAWING:

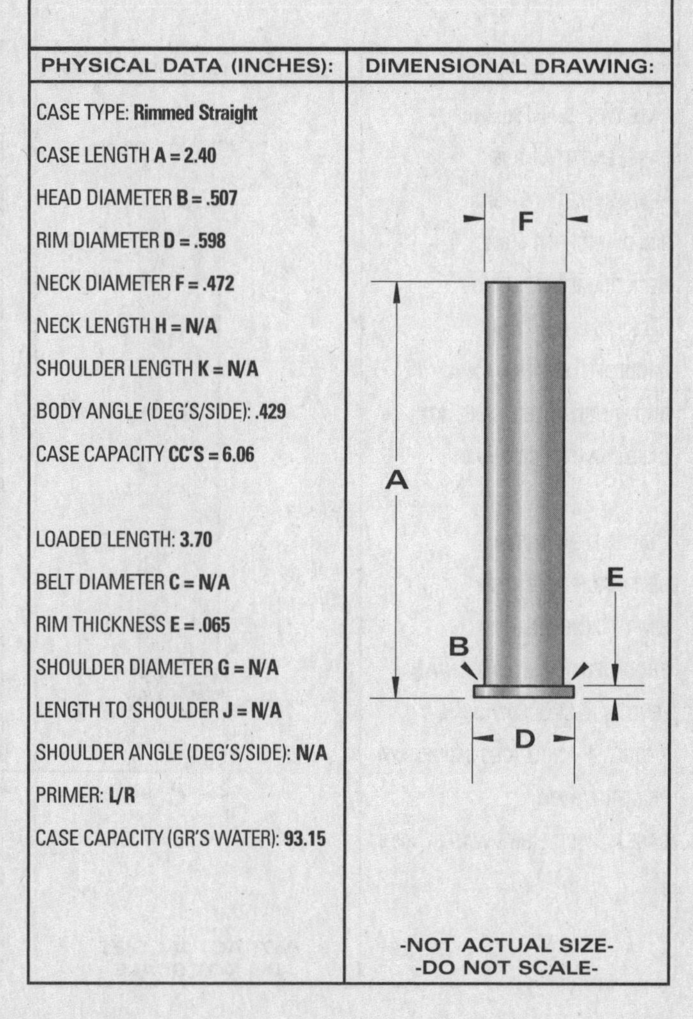

-NOT ACTUAL SIZE-
-DO NOT SCALE-

CARTRIDGE: .450 Watts Magnum

OTHER NAMES:

DIA: .458

BALLISTEK NO: 458I

NAI NO: BES 11115/5.555

DATA SOURCE: Ackley Vol.1 Pg501

HISTORICAL DATA: By Watts & Anderson.

NOTES:

LOADING DATA:

BULLET WT./TYPE	POWDER WT./TYPE	VELOCITY ('/SEC)	SOURCE
500/RN	90.0/3031	2470	Ackley

CASE PREPARATION: SHELLHOLDER (RCBS): 4

MAKE FROM: .375 H&H Mag. Anneal case neck and taper expand to .460" dia. Square case mouth and F/L size. Fireform in the chamber.

PHYSICAL DATA (INCHES):

CASE TYPE: **Belted Straight**
CASE LENGTH **A = 2.85**
HEAD DIAMETER **B = .513**
RIM DIAMETER **D = .532**
NECK DIAMETER **F = .480**
NECK LENGTH **H = N/A**
SHOULDER LENGTH **K = N/A**
BODY ANGLE (DEG'S/SIDE): **.337**
CASE CAPACITY **CC'S = 7.04**

LOADED LENGTH: **3.68**
BELT DIAMETER **C = .532**
RIM THICKNESS **E = .05**
SHOULDER DIAMETER **G = N/A**
LENGTH TO SHOULDER **J = N/A**
SHOULDER ANGLE (DEG'S/SIDE): **N/A**
PRIMER: **L/R Mag**
CASE CAPACITY (GR'S WATER): **108.63**

DIMENSIONAL DRAWING:

-NOT ACTUAL SIZE-
-DO NOT SCALE-

CARTRIDGE: .450/348 Winchester Improved (Ackley)

OTHER NAMES:

DIA: .458

BALLISTEK NO: 458C

NAI NO: RMB 13242/4.069

DATA SOURCE: Ackley Vol.1 Pg498

HISTORICAL DATA:

NOTES:

LOADING DATA:

BULLET WT./TYPE	POWDER WT./TYPE	VELOCITY ('/SEC)	SOURCE
350/RN	66.0/3031	2220	Ackley

CASE PREPARATION: SHELLHOLDER (RCBS): 5

MAKE FROM: .348 Winchester. Anneal case neck and taper expand to .460" dia. Square case mouth and F/L size in "improved" die. Fireform in chamber. This case fireforms quite well in the proper trim die.

PHYSICAL DATA (INCHES):

CASE TYPE: **Rimmed Bottleneck**
CASE LENGTH **A = 2.25**
HEAD DIAMETER **B = .553**
RIM DIAMETER **D = .610**
NECK DIAMETER **F = .488**
NECK LENGTH **H = .423**
SHOULDER LENGTH **K = .092**
BODY ANGLE (DEG'S/SIDE): **.223**
CASE CAPACITY **CC'S = 6.25**

LOADED LENGTH: **3.08**
BELT DIAMETER **C = N/A**
RIM THICKNESS **E = .07**
SHOULDER DIAMETER **G = .541**
LENGTH TO SHOULDER **J = 1.735**
SHOULDER ANGLE (DEG'S/SIDE): **16.06**
PRIMER: **L/R**
CASE CAPACITY (GR'S WATER): **96.56**

DIMENSIONAL DRAWING:

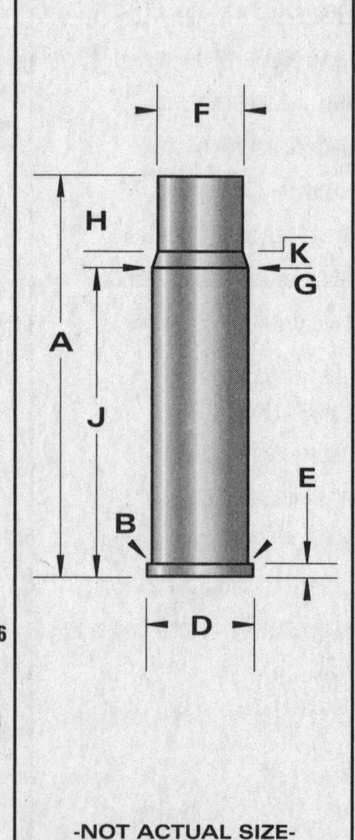

-NOT ACTUAL SIZE-
-DO NOT SCALE-

CARTRIDGE: .450/400 Nitro (2 3/8″ Black Powder)

OTHER NAMES:

DIA: .407

BALLISTEK NO: 407B

NAI NO: RMB 33221/4.338

DATA SOURCE: COTW 4th Pg238

HISTORICAL DATA: Circa 1880.

NOTES:

LOADING DATA:

BULLET WT./TYPE	POWDER WT./TYPE	VELOCITY ('/SEC)	SOURCE
270/Lead	35.2/IMR3031	–	JJD

CASE PREPARATION: SHELLHOLDER (RCBS): Spl.

MAKE FROM: .450 N.E. (BELL). Turn rim to .615″ dia. Cut case to 2.4″ & anneal. Form case in F/L die with expander removed. Trim to length, chamfer and fireform.

PHYSICAL DATA (INCHES):

CASE TYPE: **Rimmed Bottleneck**
CASE LENGTH **A = 2.36**
HEAD DIAMETER **B = .544**
RIM DIAMETER **D = .615**
NECK DIAMETER **F = .427**
NECK LENGTH **H = .545**
SHOULDER LENGTH **K = .165**
BODY ANGLE (DEG'S/SIDE): **1.14**
CASE CAPACITY **CC'S = 5.83**

LOADED LENGTH: **2.89**
BELT DIAMETER **C = N/A**
RIM THICKNESS **E = .052**
SHOULDER DIAMETER **G = .486**
LENGTH TO SHOULDER **J = 1.65**
SHOULDER ANGLE (DEG'S/SIDE): **10.13**
PRIMER: **L/R**
CASE CAPACITY (GR'S WATER): **89.90**

DIMENSIONAL DRAWING:

-NOT ACTUAL SIZE-
-DO NOT SCALE-

CARTRIDGE: .450/400 Magnum Nitro Express

OTHER NAMES:

DIA: .405

BALLISTEK NO: 405C

NAI NO: RMB 22222/5.974

DATA SOURCE: COTW 4th Pg238

HISTORICAL DATA: Early Kynoch sporting ctg.

NOTES:

LOADING DATA:

BULLET WT./TYPE	POWDER WT./TYPE	VELOCITY ('/SEC)	SOURCE
400/RN	52.7/H414	–	JJD

CASE PREPARATION: SHELLHOLDER (RCBS): Spl.

MAKE FROM: .450 N.E. (BELL). Turn rim to .615″ dia. Anneal case and form in form die. Square case mouth & chamfer. F/L size.

PHYSICAL DATA (INCHES):

CASE TYPE: **Rimmed Bottleneck**
CASE LENGTH **A = 3.25**
HEAD DIAMETER **B = .544**
RIM DIAMETER **D = .615**
NECK DIAMETER **F = .432**
NECK LENGTH **H = 1.00**
SHOULDER LENGTH **K = .170**
BODY ANGLE (DEG'S/SIDE): **.64**
CASE CAPACITY **CC'S = 8.04**

LOADED LENGTH: **3.90**
BELT DIAMETER **C = N/A**
RIM THICKNESS **E = .048**
SHOULDER DIAMETER **G = .502**
LENGTH TO SHOULDER **J = 2.08**
SHOULDER ANGLE (DEG'S/SIDE): **11.63**
PRIMER: **L/R**
CASE CAPACITY (GR'S WATER): **124.03**

DIMENSIONAL DRAWING:

-NOT ACTUAL SIZE-
-DO NOT SCALE-

CARTRIDGE: .450/400 3.25″ Magnum Nitro Express

OTHER NAMES:

DIA: .405
BALLISTEK NO: 405C
NAI NO: RMB 22222/5.974

DATA SOURCE: COTW 4th Pg238

HISTORICAL DATA: Early Kynoch sporting ctg.

NOTES:

LOADING DATA:

BULLET WT./TYPE	POWDER WT./TYPE	VELOCITY ('/SEC)	SOURCE
400/RN	52.7/H414	–	JJD

CASE PREPARATION: SHELLHOLDER (RCBS): Spl.

MAKE FROM: .450 N.E. (BELL). F/L size case and square case mouth. It may be necessary to reduce rim diameter.

PHYSICAL DATA (INCHES): DIMENSIONAL DRAWING:

CASE TYPE: **Rimmed Bottleneck**
CASE LENGTH **A = 3.25**
HEAD DIAMETER **B = .544**
RIM DIAMETER **D = .615**
NECK DIAMETER **F = .432**
NECK LENGTH **H = 1.00**
SHOULDER LENGTH **K = .17**
BODY ANGLE (DEG'S/SIDE): **.640**
CASE CAPACITY **CC'S = 8.04**

LOADED LENGTH: **3.90**
BELT DIAMETER **C = N/A**
RIM THICKNESS **E = .048**
SHOULDER DIAMETER **G = .502**
LENGTH TO SHOULDER **J = 2.08**
SHOULDER ANGLE (DEG'S/SIDE): **11.63**
PRIMER: **L/R**
CASE CAPACITY (GR'S WATER): **124.03**

-NOT ACTUAL SIZE-
-DO NOT SCALE-

CARTRIDGE: .454 Casull

OTHER NAMES:

DIA: .4525
BALLISTEK NO: 452F
NAI NO: RMS 11115/2.671

DATA SOURCE: Most reloading manuals

HISTORICAL DATA: Developed by Dick Casull and adapted to industry standards by SAAMI in 1997.

NOTES: Older editions of this book listed the .454 Casull with .45 Colt cases and triplex loads using three different powders. This is very dangerous! Use only new factory brass and single powder loads from current handloading manuals.

LOADING DATA:

BULLET WT./TYPE	POWDER WT./TYPE	VELOCITY ('/SEC)	SOURCE
300/Nosler Part. Gold	27.5/WW 296	1686	Nosler #5
260/Nosler Part. Gold	32.0/H-110	1942	Nosler #5

CASE PREPARATION: SHELLHOLDER (RCBS): 20

This is a very high-pressure revolver cartridge and only current factory brass should be used. Use only single powder loads. Bullets must be constructed for use in the .454 Casull. Due to the cartridge's high pressure (65,000 p.s.i. MAP) the use of .45-caliber bullets not designed for high-pressure use may result in damage to the firearm.

Warning: ***Information re .454 Casull in earlier editions of this book is now considered to be incorrect and dangerous.***

PHYSICAL DATA (INCHES): DIMENSIONAL DRAWING:

CASE TYPE: **Rimmed Straight**
CASE LENGTH **A = 1.383**
HEAD DIAMETER **B = .4775**
RIM DIAMETER **D = .512**
NECK DIAMETER **F = .4775**
NECK LENGTH **H = N/A**
SHOULDER LENGTH **K = N/A**
BODY ANGLE (DEG'S/SIDE): **.023**
CASE CAPACITY **CC'S = 2.42**

LOADED LENGTH: **1.765**
BELT DIAMETER **C = N/A**
RIM THICKNESS **E = .057**
SHOULDER DIAMETER **G = N/A**
LENGTH TO SHOULDER **J = N/A**
SHOULDER ANGLE (DEG'S/SIDE): **N/A**
PRIMER: **S/R**
CASE CAPACITY (GR'S WATER): **46.1**

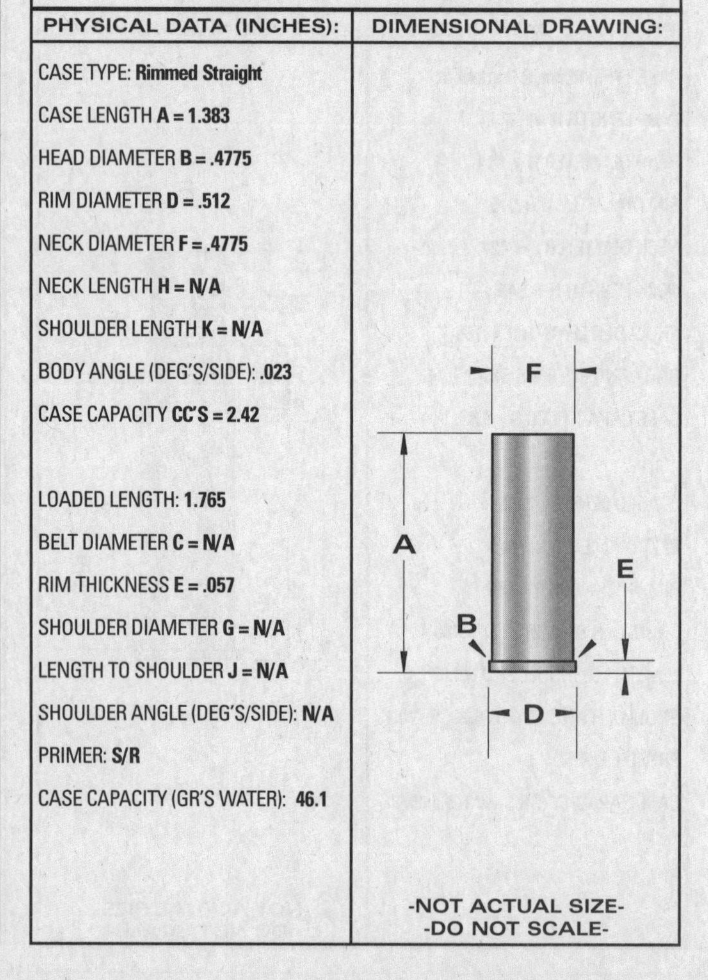

CARTRIDGE: .455 Webley Auto

OTHER NAMES:

DIA: .455

BALLISTEK NO: 455I

NAI NO: RMS 11115/1.962

DATA SOURCE: COTW 4th Pg188

HISTORICAL DATA: British naval ctg. from 1912.

NOTES:

LOADING DATA:

BULLET WT./TYPE	POWDER WT./TYPE	VELOCITY ('/SEC)	SOURCE
220/RN	5.5/Unique	700	JJD

CASE PREPARATION: SHELLHOLDER (RCBS): 21

MAKE FROM: .45 Colt. Turn rim to .500" dia. and cut new extractor groove. Cut to .95". F/L size in .45 ACP die adjusted for length. Trim to length & chamfer.

PHYSICAL DATA (INCHES):

CASE TYPE: **Rimmed Straight**
CASE LENGTH **A = .930**
HEAD DIAMETER **B = .474**
RIM DIAMETER **D = .500**
NECK DIAMETER **F = .473**
NECK LENGTH **H = N/A**
SHOULDER LENGTH **K = N/A**
BODY ANGLE (DEG'S/SIDE): **.032**
CASE CAPACITY **CC'S = 1.60**
LOADED LENGTH: **1.23**
BELT DIAMETER **C = N/A**
RIM THICKNESS **E = .048**
SHOULDER DIAMETER **G = N/A**
LENGTH TO SHOULDER **J = N/A**
SHOULDER ANGLE (DEG'S/SIDE): **N/A**
PRIMER: **L/P**
CASE CAPACITY (GR'S WATER): **24.77**

DIMENSIONAL DRAWING:

F
A
B
E
D

-NOT ACTUAL SIZE-
-DO NOT SCALE-

CARTRIDGE: .455 Webley MkI

OTHER NAMES:
.455 Colt-Eley
.455 Enfield

DIA: .455

BALLISTEK NO: 454J

NAI NO: RMS 11115/1.853

DATA SOURCE: COTW 4th Pg189

HISTORICAL DATA: British mil. ctg. ca. 1892.

NOTES:

LOADING DATA:

BULLET WT./TYPE	POWDER WT./TYPE	VELOCITY ('/SEC)	SOURCE
265/Lead	3.3/B'Eye	–	JJD

CASE PREPARATION: SHELLHOLDER (RCBS): 8

MAKE FROM: .45 Auto Rimmed. Thin rim to .039". Trim to length & chamfer. Use proper dies or .45 ACP to size & seat. You may also need to make up some .030" thick washers to slip over the case (to make up for the material removed in thinning) while priming. Some cases are strong enough to withstand priming, as is (i.e., at .039" thick), while others will shear off the rim.

PHYSICAL DATA (INCHES):

CASE TYPE: **Rimmed Straight**
CASE LENGTH **A = .886**
HEAD DIAMETER **B = .478**
RIM DIAMETER **D = .530**
NECK DIAMETER **F = .473**
NECK LENGTH **H = N/A**
SHOULDER LENGTH **K = N/A**
BODY ANGLE (DEG'S/SIDE): **.169**
CASE CAPACITY **CC'S = 1.50**
LOADED LENGTH: **1.46**
BELT DIAMETER **C = N/A**
RIM THICKNESS **E = .039**
SHOULDER DIAMETER **G = N/A**
LENGTH TO SHOULDER **J = N/A**
SHOULDER ANGLE (DEG'S/SIDE): **N/A**
PRIMER: **L/P**
CASE CAPACITY (GR'S WATER): **23.23**

DIMENSIONAL DRAWING:

F
A
B
E
D

-NOT ACTUAL SIZE-
-DO NOT SCALE-

CARTRIDGE: .455 Webley MkII

OTHER NAMES:
.455 Revolver Mk-2

DIA: .454

BALLISTEK NO: 454G

NAI NO: RMS 11115/1.604

DATA SOURCE: COTW 4th Pg189

HISTORICAL DATA: British revolver ctg. ca. 1879.

NOTES:

LOADING DATA:

BULLET WT./TYPE	POWDER WT./TYPE	VELOCITY ('/SEC)	SOURCE
250/Lead	3.5/B'Eye	710	Barnes

CASE PREPARATION: SHELLHOLDER (RCBS): 8

MAKE FROM: .45 Auto Rimmed. Thin rim to .045" thick. Trim case to length. Size & seat in proper die or use .45 ACP dies.

PHYSICAL DATA (INCHES):

CASE TYPE: **Rimmed Straight**

CASE LENGTH A = **.77**

HEAD DIAMETER B = **.480**

RIM DIAMETER D = **.535**

NECK DIAMETER F = **.476**

NECK LENGTH H = **N/A**

SHOULDER LENGTH K = **N/A**

BODY ANGLE (DEG'S/SIDE): **.158**

CASE CAPACITY CC'S = **1.18**

LOADED LENGTH: **1.23**

BELT DIAMETER C = **N/A**

RIM THICKNESS E = **.045**

SHOULDER DIAMETER G = **N/A**

LENGTH TO SHOULDER J = **N/A**

SHOULDER ANGLE (DEG'S/SIDE): **N/A**

PRIMER: **L/P**

CASE CAPACITY (GR'S WATER): **18.30**

DIMENSIONAL DRAWING:

-NOT ACTUAL SIZE-
-DO NOT SCALE-

CARTRIDGE: .457 Woods Magnum

OTHER NAMES:

DIA: .458

BALLISTEK NO: 458G

NAI NO: BEN 24245/3.479

DATA SOURCE: Guns Illustrated 1979

HISTORICAL DATA: By B. Woods in 1978.

NOTES:

LOADING DATA:

BULLET WT./TYPE	POWDER WT./TYPE	VELOCITY ('/SEC)	SOURCE
300/SP	42.0/4227	1763	Del Savio

CASE PREPARATION: SHELLHOLDER (RCBS): 4

MAKE FROM: .458 Win. Mag. Cut case to 1.8". Form in form die and I.D. neck ream. Trim to length, chamfer and F/L size.

PHYSICAL DATA (INCHES):

CASE TYPE: **Belted Bottleneck**

CASE LENGTH A = **1.785**

HEAD DIAMETER B = **.513**

RIM DIAMETER D = **.532**

NECK DIAMETER F = **.484**

NECK LENGTH H = **.32**

SHOULDER LENGTH K = **.305**

BODY ANGLE (DEG'S/SIDE): **.387**

CASE CAPACITY CC'S = **4.29**

LOADED LENGTH: **2.35**

BELT DIAMETER C = **.532**

RIM THICKNESS E = **.05**

SHOULDER DIAMETER G = **.50**

LENGTH TO SHOULDER J = **1.16**

SHOULDER ANGLE (DEG'S/SIDE): **1.50**

PRIMER: **L/P**

CASE CAPACITY (GR'S WATER): **66.24**

DIMENSIONAL DRAWING:

-NOT ACTUAL SIZE-
-DO NOT SCALE-

CARTRIDGE: .457 Woods Super Magnum

OTHER NAMES:

DIA: .458

BALLISTEK NO: 458H

NAI NO: BEN 24241/3.067

DATA SOURCE: Guns Illustrated 1979.

HISTORICAL DATA: By B. Woods in 1978.

NOTES:

LOADING DATA:

BULLET WT./TYPE	POWDER WT./TYPE	VELOCITY ('/SEC)	SOURCE
500/RN	38.0/4227	1340	Del Savio

CASE PREPARATION: SHELLHOLDER (RCBS): 14

MAKE FROM: .460 Weatherby (or .378 Weatherby). Cut case to 1.8". Form in form die and I.D. neck ream. F/L size. Trim to length & chamfer.

PHYSICAL DATA (INCHES):

CASE TYPE: **Belted Bottleneck**

CASE LENGTH **A = 1.785**

HEAD DIAMETER **B = .582**

RIM DIAMETER **D = .582**

NECK DIAMETER **F = .483**

NECK LENGTH **H = .320**

SHOULDER LENGTH **K = .305**

BODY ANGLE (DEG'S/SIDE): **.656**

CASE CAPACITY **CC'S = 5.19**

LOADED LENGTH: **2.35**

BELT DIAMETER **C = .603**

RIM THICKNESS **E = .062**

SHOULDER DIAMETER **G = .560**

LENGTH TO SHOULDER **J = 1.16**

SHOULDER ANGLE (DEG'S/SIDE): **7.19**

PRIMER: **L/P**

CASE CAPACITY (GR'S WATER): **80.09**

DIMENSIONAL DRAWING:

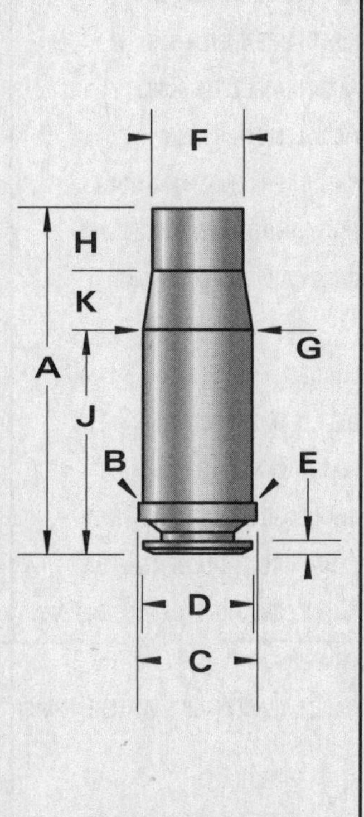

-NOT ACTUAL SIZE-
-DO NOT SCALE-

CARTRIDGE: .458 Canadian Magnum

OTHER NAMES:

DIA: .458

BALLISTEK NO:

NAI NO:

DATA SOURCE:

HISTORICAL DATA: North American Shooting Systems developed this cartridge around 1994.

NOTES:

LOADING DATA:

BULLET WT./TYPE	POWDER WT./TYPE	VELOCITY ('/SEC)	SOURCE
350	FL	2575	NASS
500	89.0/IMR-4064	2360	NASS

CASE PREPARATION: SHELL HOLDER (RCBS): 00

MAKE FROM:

PHYSICAL DATA (INCHES):

CASE TYPE: **Rebated Bottleneck**

CASE LENGTH **A = 2.830**

HEAD DIAMETER **B = .544**

RIM DIAMETER **D = 2.830**

NECK DIAMETER **F = .485**

NECK LENGTH **H = .460**

SHOULDER LENGTH **K = .031**

BODY ANGLE (DEG'S/SIDE): **.656**

CASE CAPACITY **CC'S = 5.19**

LOADED LENGTH: **3.600**

BELT DIAMETER **C = N/A**

RIM THICKNESS **E = .050**

SHOULDER DIAMETER **G = .530**

LENGTH TO SHOULDER **J = 2.339**

SHOULDER ANGLE (DEG'S/SIDE): **30°**

PRIMER: **L/R**

CASE CAPACITY (GR'S WATER): **80.09**

DIMENSIONAL DRAWING:

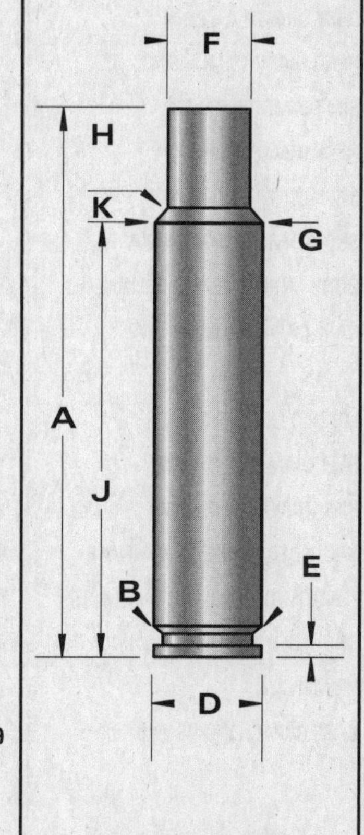

-NOT ACTUAL SIZE-
-DO NOT SCALE-

CARTRIDGE: .458 Lott

OTHER NAMES:

DIA: .458

BALLISTEK NO:

NAI NO:

DATA SOURCE:

HISTORICAL DATA: Originated by Jack Lott in 1971. Based on the blown out and shortened 375 H&H Magnum case.

NOTES:

LOADING DATA:

BULLET WT./TYPE	POWDER WT./TYPE	VELOCITY ('/SEC)	SOURCE
500 SP	85.0/IMR-4320	2330	

CASE PREPARATION: SHELL HOLDER (RCBS): 04

MAKE FROM: 375 H&H Magnum

PHYSICAL DATA (INCHES):

CASE TYPE: **Belted Straight**
CASE LENGTH **A = 2.800**
HEAD DIAMETER **B = .5013**
RIM DIAMETER **D = .532**
NECK DIAMETER **F = .481**
NECK LENGTH **H = N/A**
SHOULDER LENGTH **K = N/A**
BODY ANGLE (DEG'S/SIDE): **N/A**
CASE CAPACITY **CC'S = N/A**

LOADED LENGTH: **3.600**
BELT DIAMETER **C = .532**
RIM THICKNESS **E = .050**
SHOULDER DIAMETER **G = .N/A**
LENGTH TO SHOULDER **J = N/A**
SHOULDER ANGLE (DEG'S/SIDE): **N/A**
PRIMER: **L/R**
CASE CAPACITY (GR'S WATER):

DIMENSIONAL DRAWING:

-NOT ACTUAL SIZE-
-DO NOT SCALE-

CARTRIDGE: .458 RCBS

OTHER NAMES:

DIA: .458

BALLISTEK NO: 458Z

NAI NO: RMS 11115/5.500

DATA SOURCE: Handloader #56 Pg26

HISTORICAL DATA: By F. Huntington about 1974.

NOTES:

LOADING DATA:

BULLET WT./TYPE	POWDER WT./TYPE	VELOCITY ('/SEC)	SOURCE
500/RN	65.0/IMR4064	1875	Waters

CASE PREPARATION: SHELLHOLDER (RCBS): 14

MAKE FROM: .45 Basic. Trim case to length, chamfer & F/L size.

PHYSICAL DATA (INCHES):

CASE TYPE: **Rimmed Straight**
CASE LENGTH **A = 2.75**
HEAD DIAMETER **B = .500**
RIM DIAMETER **D = .595**
NECK DIAMETER **F = .483**
NECK LENGTH **H = N/A**
SHOULDER LENGTH **K = N/A**
BODY ANGLE (DEG'S/SIDE): **.181**
CASE CAPACITY **CC'S = 6.48**

LOADED LENGTH: **3.60**
BELT DIAMETER **C = N/A**
RIM THICKNESS **E = .065**
SHOULDER DIAMETER **G = N/A**
LENGTH TO SHOULDER **J = N/A**
SHOULDER ANGLE (DEG'S/SIDE): **N/A**
PRIMER: **L/R**
CASE CAPACITY (GR'S WATER): **100.13**

DIMENSIONAL DRAWING:

-NOT ACTUAL SIZE-
-DO NOT SCALE-

CARTRIDGE: 458 Whisper

OTHER NAMES:

DIA: .458

BALLISTEK NO:

NAI NO:

DATA SOURCE: J.D. Jones

HISTORICAL DATA: Developed by J.D. Jones in 1993 at SSK Industries, this chambering uses a shortened 458 Winchester Magnum case.

NOTES:

LOADING DATA:

BULLET WT./TYPE	POWDER WT./TYPE	VELOCITY ('/SEC)	SOURCE
500 H.T.	15.6 - W-231	1021	SSK
560 H.T.	18.0 - Blue Dot	1101	SSK

CASE PREPARATION: SHELL HOLDER (RCBS): 04

MAKE FROM: 458 Winchester Magnum

PHYSICAL DATA (INCHES):

CASE TYPE: **Belted Straight**

CASE LENGTH **A = 1.750**

HEAD DIAMETER **B = .506**

RIM DIAMETER **D = .525**

NECK DIAMETER **F = .485**

NECK LENGTH **H = N/A**

SHOULDER LENGTH **K = N/A**

BODY ANGLE (DEG'S/SIDE): **N/A**

CASE CAPACITY **CC'S =**

LOADED LENGTH: **2.800**

BELT DIAMETER **C = .532**

RIM THICKNESS **E = .050**

SHOULDER DIAMETER **G = N/A**

LENGTH TO SHOULDER **J = N/A**

SHOULDER ANGLE (DEG'S/SIDE): **N/A**

PRIMER: **LR**

CASE CAPACITY (GR'S WATER): **100.13**

DIMENSIONAL DRAWING:

-NOT ACTUAL SIZE-
-DO NOT SCALE-

CARTRIDGE: .458 Winchester Magnum

OTHER NAMES:

DIA: .458

BALLISTEK NO: 458E

NAI NO: BES 21115/4.873

DATA SOURCE: Hornady 3rd Pg304

HISTORICAL DATA: By Win. in 1956.

NOTES:

LOADING DATA:

BULLET WT./TYPE	POWDER WT./TYPE	VELOCITY ('/SEC)	SOURCE
300/FN	71.0/IMR4198	2730	Barnes
500/RN	70.0/IMR3031	2100	JJD

CASE PREPARATION: SHELLHOLDER (RCBS): 4

MAKE FROM: Factory or .375 H&H. Cut the H&H case to 2.50" and fireform, with cornmeal, in a .458 Win. trim die. Or, cases may be cut, taper expanded to hold .458" dia. bullets and fired in the chamber.

PHYSICAL DATA (INCHES):

CASE TYPE: **Belted Straight**

CASE LENGTH **A = 2.500**

HEAD DIAMETER **B = .513**

RIM DIAMETER **D = .532**

NECK DIAMETER **F = .481**

NECK LENGTH **H = N/A**

SHOULDER LENGTH **K = N/A**

BODY ANGLE (DEG'S/SIDE): **.294**

CASE CAPACITY **CC'S = 6.04**

LOADED LENGTH: **3.30**

BELT DIAMETER **C = .532**

RIM THICKNESS **E = .049**

SHOULDER DIAMETER **G = N/A**

LENGTH TO SHOULDER **J = N/A**

SHOULDER ANGLE (DEG'S/SIDE): **N/A**

PRIMER: **L/R Mag**

CASE CAPACITY (GR'S WATER): **93.29**

DIMENSIONAL DRAWING:

-NOT ACTUAL SIZE-
-DO NOT SCALE-

CARTRIDGE: .458 x 2" American

OTHER NAMES:	DIA: .458 BALLISTEK NO: 458D NAI NO: BES 11115/3.898

DATA SOURCE: COTW 4th Pg154

HISTORICAL DATA: By F. Barnes in 1962.

NOTES:

LOADING DATA:

BULLET WT./TYPE	POWDER WT./TYPE	VELOCITY ('/SEC)	SOURCE
350/JFN	40.0/4198	1825	Barnes

CASE PREPARATION: SHELLHOLDER (RCBS): 4

MAKE FROM: .458 Winchester Mag. Cut case to length & F/L size. .458 Mag. dies will work. I.D. cut to the proper length.

PHYSICAL DATA (INCHES):

CASE TYPE: **Belted Straight**
CASE LENGTH **A = 2.00**
HEAD DIAMETER **B = .513**
RIM DIAMETER **D = .532**
NECK DIAMETER **F = .478**
NECK LENGTH **H = N/A**
SHOULDER LENGTH **K = N/A**
BODY ANGLE (DEG'S/SIDE): **.514**
CASE CAPACITY **CC'S = 4.73**

LOADED LENGTH: **2.60**
BELT DIAMETER **C = .532**
RIM THICKNESS **E = .049**
SHOULDER DIAMETER **G = N/A**
LENGTH TO SHOULDER **J = N/A**
SHOULDER ANGLE (DEG'S/SIDE): **N/A**
PRIMER: **L/R**
CASE CAPACITY (GR'S WATER): **72.98**

DIMENSIONAL DRAWING:

-NOT ACTUAL SIZE-
-DO NOT SCALE-

CARTRIDGE: .460 G&A Magnum

OTHER NAMES:	DIA: .458 BALLISTEK NO: 458M NAI NO: RXB 12242/5.153

DATA SOURCE: NAI/Ballistek

HISTORICAL DATA: By Guns & Ammo magazine staff.

NOTES:

LOADING DATA:

BULLET WT./TYPE	POWDER WT./TYPE	VELOCITY ('/SEC)	SOURCE
500/FMJ	90.0/IMR4064	—	JJD

CASE PREPARATION: SHELLHOLDER (RCBS): 7

MAKE FROM: .404 Jeffery (BELL). Trim case to 2.78". F/L size. May be necessary to anneal case neck.

PHYSICAL DATA (INCHES):

CASE TYPE: **Rimless Bottleneck**
CASE LENGTH **A = 2.788**
HEAD DIAMETER **B = .544**
RIM DIAMETER **D = .541**
NECK DIAMETER **F = .485**
NECK LENGTH **H = .452**
SHOULDER LENGTH **K = .092**
BODY ANGLE (DEG'S/SIDE): **.182**
CASE CAPACITY **CC'S = 7.77**

LOADED LENGTH: **3.609**
BELT DIAMETER **C = N/A**
RIM THICKNESS **E = .049**
SHOULDER DIAMETER **G = .528**
LENGTH TO SHOULDER **J = 2.244**
SHOULDER ANGLE (DEG'S/SIDE): **13.15**
PRIMER: **L/R Mag**
CASE CAPACITY (GR'S WATER): **119.99**

DIMENSIONAL DRAWING:

-NOT ACTUAL SIZE-
-DO NOT SCALE-

CARTRIDGE: .460 Van Horn

OTHER NAMES:

DIA: .458

BALLISTEK NO: 458L

NAI NO: BEN 12243/4.295

DATA SOURCE: Rifle #43 Pg24

HISTORICAL DATA: By G. van Horn.

NOTES:

LOADING DATA:

BULLET WT./TYPE	POWDER WT./TYPE	VELOCITY ('/SEC)	SOURCE
500/FMJ	88.0/IMR4350	2280	Hagel

CASE PREPARATION: SHELLHOLDER (RCBS): 14

MAKE FROM: .460 Weatherby. Cut case to 2.55". F/L size with expander removed. Trim to length and chamfer.

PHYSICAL DATA (INCHES):

CASE TYPE: **Belted Bottleneck**

CASE LENGTH **A = 2.500**

HEAD DIAMETER **B = .582**

RIM DIAMETER **D = .582**

NECK DIAMETER **F = .481**

NECK LENGTH **H = .375**

SHOULDER LENGTH **K = .075**

BODY ANGLE (DEG'S/SIDE): **.185**

CASE CAPACITY **CC'S = 8.07**

LOADED LENGTH: **3.35**

BELT DIAMETER **C = .603**

RIM THICKNESS **E = .062**

SHOULDER DIAMETER **G = .570**

LENGTH TO SHOULDER **J = 2.05**

SHOULDER ANGLE (DEG'S/SIDE): **30.68**

PRIMER: **L/R Mag**

CASE CAPACITY (GR'S WATER): **124.60**

DIMENSIONAL DRAWING:

-NOT ACTUAL SIZE-
-DO NOT SCALE-

CARTRIDGE: .460 Weatherby Magnum

OTHER NAMES:

DIA: .458

BALLISTEK NO: 458F

NAI NO: BEN 12242/4.996

DATA SOURCE: Hornady Manual 3rd Pg306

HISTORICAL DATA: By Roy Weatherby in 1958.

NOTES:

LOADING DATA:

BULLET WT./TYPE	POWDER WT./TYPE	VELOCITY ('/SEC)	SOURCE
300/—	125.0/4350	2950	Barnes
500/RN	120.0/IMR4831	2530	JJD

CASE PREPARATION: SHELLHOLDER (RCBS): 14

MAKE FROM: Factory or .378 Weatherby. Anneal case neck. Taper expand to hold .458" dia. bullets. Fireform. See also .378 Weatherby Magnum.

PHYSICAL DATA (INCHES):

CASE TYPE: **Belted Bottleneck**

CASE LENGTH **A = 2.908**

HEAD DIAMETER **B = .582**

RIM DIAMETER **D = .583**

NECK DIAMETER **F = .481**

NECK LENGTH **H = .416**

SHOULDER LENGTH **K = .152**

BODY ANGLE (DEG'S/SIDE): **.294**

CASE CAPACITY **CC'S = 9.37**

LOADED LENGTH: **3.71**

BELT DIAMETER **C = .603**

RIM THICKNESS **E = .063**

SHOULDER DIAMETER **G = .560**

LENGTH TO SHOULDER **J = 2.34**

SHOULDER ANGLE (DEG'S/SIDE): **N/A**

PRIMER: **L/R Mag**

CASE CAPACITY (GR'S WATER): **144.67**

DIMENSIONAL DRAWING:

-NOT ACTUAL SIZE-
-DO NOT SCALE-

CARTRIDGE: .470 Capstick

OTHER NAMES:

DIA: .475

BALLISTEK NO:

NAI NO:

DATA SOURCE:

HISTORICAL DATA: Developed by Col. Arthur B. Alphin of A Square and named for the writer Peter Capstick. Very similar to the .475 Ackley Magnum.

NOTES: Approved by SAAMI 6/5/96.

LOADING DATA:

BULLET WT./TYPE	POWDER WT./TYPE	VELOCITY ('/SEC)	SOURCE
500/solid	85.0/H-4895	2,387	Cartrides of the World #9
500/solid	89.5/RL-12	2,410	

CASE PREPARATION: SHELL HOLDER (RCBS): 4

MAKE FROM: Factory or neck up .375 H&H brass with tapered expander and fireform. Trim, chamfer and deburr.

PHYSICAL DATA (INCHES):

CASE TYPE: **Belted Straight**

CASE LENGTH **A = 2.850**

HEAD DIAMETER **B = .5126**

RIM DIAMETER **D = .532**

NECK DIAMETER **F = .499**

NECK LENGTH **H =**

SHOULDER LENGTH **K =**

BODY ANGLE (DEG'S/SIDE):

CASE CAPACITY **CC'S =**

LOADED LENGTH: **3.600**

BELT DIAMETER **C = .532**

RIM THICKNESS **E = .050**

SHOULDER DIAMETER **G =**

LENGTH TO SHOULDER **J =**

SHOULDER ANGLE (DEG'S/SIDE):

PRIMER: **L/R Mag**

CASE CAPACITY (GR'S WATER):

DIMENSIONAL DRAWING:

F
A
E
B
D
C

-NOT ACTUAL SIZE-
-DO NOT SCALE-

CARTRIDGE: .470 Nitro Express

OTHER NAMES:

DIA: .483

BALLISTEK NO: 483A

NAI NO: RMB 22221/5.682

DATA SOURCE: COTW 4th Pg243

HISTORICAL DATA: Circa 1907.

NOTES:

LOADING DATA:

BULLET WT./TYPE	POWDER WT./TYPE	VELOCITY ('/SEC)	SOURCE
500/RN	75.1/IMR4320	2150	JJD

CASE PREPARATION: SHELLHOLDER (RCBS): Spl.

MAKE FROM: .470 N.E. (BELL). Use brass as is. F/L size & trim to length.

PHYSICAL DATA (INCHES):

CASE TYPE: **Rimmed Bottleneck**

CASE LENGTH **A = 3.25**

HEAD DIAMETER **B = .572**

RIM DIAMETER **D = .655**

NECK DIAMETER **F = .504**

NECK LENGTH **H = .75**

SHOULDER LENGTH **K = .1**

BODY ANGLE (DEG'S/SIDE): **.609**

CASE CAPACITY **CC'S = 10.15**

LOADED LENGTH: **3.86**

BELT DIAMETER **C = N/A**

RIM THICKNESS **E = .04**

SHOULDER DIAMETER **G = .531**

LENGTH TO SHOULDER **J = 2.40**

SHOULDER ANGLE (DEG'S/SIDE): **4.15**

PRIMER: **L/R**

CASE CAPACITY (GR'S WATER): **156.65**

DIMENSIONAL DRAWING:

F
H
K
G
A
J
E
B
D

-NOT ACTUAL SIZE-
-DO NOT SCALE-

CARTRIDGE: .475 #2 Jeffery Nitro Express

OTHER NAMES:

DIA: .483

BALLISTEK NO: 483B

NAI NO: RMB 12241/6.059

DATA SOURCE: COTW 4th Pg244

HISTORICAL DATA: British sporting ctg. ca. 1905.

NOTES:

LOADING DATA:

BULLET WT./TYPE	POWDER WT./TYPE	VELOCITY ('/SEC)	SOURCE
500/RN	97.5/IMR4350	—	JJD

CASE PREPARATION: SHELLHOLDER (RCBS): Spl.

MAKE FROM: .475 #2 N.E. (BELL). Use brass as is. Trim to length & F/L size.

PHYSICAL DATA (INCHES):

CASE TYPE: **Rimmed Bottleneck**

CASE LENGTH **A = 3.49**

HEAD DIAMETER **B = .576**

RIM DIAMETER **D = .665**

NECK DIAMETER **F = .510**

NECK LENGTH **H = .440**

SHOULDER LENGTH **K = .20**

BODY ANGLE (DEG'S/SIDE): **.313**

CASE CAPACITY **CC'S = 11.07**

LOADED LENGTH: **4.26**

BELT DIAMETER **C = N/A**

RIM THICKNESS **E = .09**

SHOULDER DIAMETER **G = .547**

LENGTH TO SHOULDER **J = 2.85**

SHOULDER ANGLE (DEG'S/SIDE): **5.28**

PRIMER: **L/R**

CASE CAPACITY (GR'S WATER): **170.90**

DIMENSIONAL DRAWING:

-NOT ACTUAL SIZE-
-DO NOT SCALE-

CARTRIDGE: .475 A&M Magnum

OTHER NAMES:

DIA: .475

BALLISTEK NO: 475A

NAI NO: BEN 12243/4.983

DATA SOURCE: Ackley Vol.1 Pg504

HISTORICAL DATA: By Atkinson & Marquart.

NOTES: About 1958

LOADING DATA:

BULLET WT./TYPE	POWDER WT./TYPE	VELOCITY ('/SEC)	SOURCE
500/RN	110.0/3031	2980	Barnes

CASE PREPARATION: SHELLHOLDER (RCBS): 14

MAKE FROM: .460 Weatherby. Taper expand case neck to about .480" dia. Square case mouth and F/L size.

PHYSICAL DATA (INCHES):

CASE TYPE: **Belted Bottleneck**

CASE LENGTH **A = 2.90**

HEAD DIAMETER **B = .582**

RIM DIAMETER **D = .583**

NECK DIAMETER **F = .502**

NECK LENGTH **H = .400**

SHOULDER LENGTH **K = .058**

BODY ANGLE (DEG'S/SIDE): **.281**

CASE CAPACITY **CC'S = 9.31**

LOADED LENGTH: **3.75**

BELT DIAMETER **C = .603**

RIM THICKNESS **E = .062**

SHOULDER DIAMETER **G = .560**

LENGTH TO SHOULDER **J = 2.442**

SHOULDER ANGLE (DEG'S/SIDE): **26.56**

PRIMER: **L/R Mag**

CASE CAPACITY (GR'S WATER): **143.74**

DIMENSIONAL DRAWING:

-NOT ACTUAL SIZE-
-DO NOT SCALE-

CARTRIDGE: .475 Ackley Magnum

OTHER NAMES:	
	DIA: .475
	BALLISTEK NO: 475B
	NAI NO: BES 11115/5.555

DATA SOURCE: Ackley Vol.1 Pg503

HISTORICAL DATA:

NOTES:

LOADING DATA:

BULLET WT./TYPE	POWDER WT./TYPE	VELOCITY ('/SEC)	SOURCE
600/RN	90.0/4320	2250	Ackley

CASE PREPARATION: SHELLHOLDER (RCBS): 4

MAKE FROM: .375 H&H Mag. Anneal case neck. Taper expand to .480" dia. (somewhat larger than the straight wall dia). Square case mouth and F/L size.

PHYSICAL DATA (INCHES):

CASE TYPE: **Belted Straight**
CASE LENGTH **A = 2.85**
HEAD DIAMETER **B = .513**
RIM DIAMETER **D = .532**
NECK DIAMETER **F = .497**
NECK LENGTH **H = N/A**
SHOULDER LENGTH **K = N/A**
BODY ANGLE (DEG'S/SIDE): **.163**
CASE CAPACITY **CC'S = 7.30**

LOADED LENGTH: **3.80**
BELT DIAMETER **C = .532**
RIM THICKNESS **E = .05**
SHOULDER DIAMETER **G = N/A**
LENGTH TO SHOULDER **J = N/A**
SHOULDER ANGLE (DEG'S/SIDE): **N/A**
PRIMER: **L/R Mag**
CASE CAPACITY (GR'S WATER): **112.65**

DIMENSIONAL DRAWING:

F
A
E
B
D
C

-NOT ACTUAL SIZE-
-DO NOT SCALE-

CARTRIDGE: .475 Barnes Supreme

OTHER NAMES:	
	DIA: .475
	BALLISTEK NO: 475D
	NAI NO: BES 11115/5.458

DATA SOURCE: Ackley Vol.1 Pg504

HISTORICAL DATA: By F. Barnes.

NOTES: Pretty close to Ackley's version (#475B).

LOADING DATA:

BULLET WT./TYPE	POWDER WT./TYPE	VELOCITY ('/SEC)	SOURCE
600/RN	90.0/4320	2250	Ackley

CASE PREPARATION: SHELLHOLDER (RCBS): 4

MAKE FROM: .375 H&H Mag. Anneal case neck. Taper expand to .480" dia. Square case mouth, F/L size & chamfer.

PHYSICAL DATA (INCHES):

CASE TYPE: **Belted Straight**
CASE LENGTH **A = 2.80**
HEAD DIAMETER **B = .513**
RIM DIAMETER **D = .532**
NECK DIAMETER **F = .496**
NECK LENGTH **H = N/A**
SHOULDER LENGTH **K = N/A**
BODY ANGLE (DEG'S/SIDE): **.177**
CASE CAPACITY **CC'S = 7.19**

LOADED LENGTH: **A/R**
BELT DIAMETER **C = .532**
RIM THICKNESS **E = .05**
SHOULDER DIAMETER **G = N/A**
LENGTH TO SHOULDER **J = N/A**
SHOULDER ANGLE (DEG'S/SIDE): **N/A**
PRIMER: **L/R Mag**
CASE CAPACITY (GR'S WATER): **110.91**

DIMENSIONAL DRAWING:

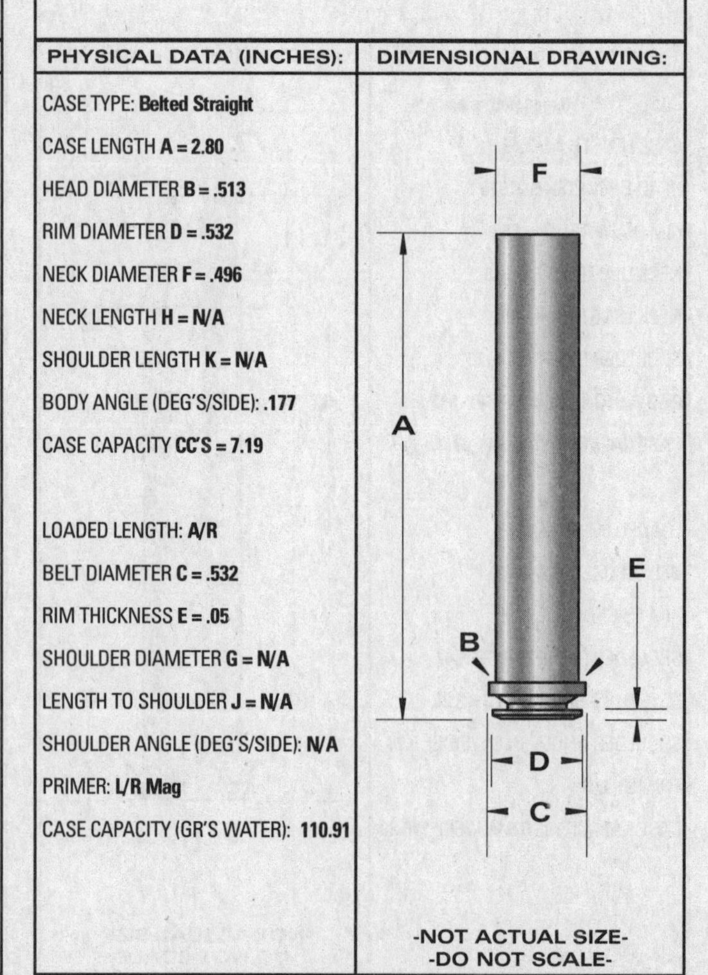

-NOT ACTUAL SIZE-
-DO NOT SCALE-

CARTRIDGE: .475 Jurras

OTHER NAMES:

DIA: .475

BALLISTEK NO: 475C

NAI NO: RMB 34221/3.535

DATA SOURCE: NAI/Ballistek

HISTORICAL DATA:

NOTES:

LOADING DATA:

BULLET WT./TYPE	POWDER WT./TYPE	VELOCITY ('/SEC)	SOURCE
350/RN	50.0/IMR4198	—	JJD

CASE PREPARATION: SHELLHOLDER (RCBS): Spl.

MAKE FROM: .500 N.E. (BELL). Form dies are required. Anneal case and form in form die. Trim to length & chamfer.

PHYSICAL DATA (INCHES):

CASE TYPE: **Rimmed Bottleneck**

CASE LENGTH **A = 2.029**

HEAD DIAMETER **B = .574**

RIM DIAMETER **D = .655**

NECK DIAMETER **F = .500**

NECK LENGTH **H = .695**

SHOULDER LENGTH **K = .104**

BODY ANGLE (DEG'S/SIDE): **1.22**

CASE CAPACITY **CC'S = 5.70**

LOADED LENGTH: **A/R**

BELT DIAMETER **C = N/A**

RIM THICKNESS **E = .04**

SHOULDER DIAMETER **G = .530**

LENGTH TO SHOULDER **J = 1.23**

SHOULDER ANGLE (DEG'S/SIDE): **8.21**

PRIMER: **L/P**

CASE CAPACITY (GR'S WATER): **87.95**

DIMENSIONAL DRAWING:

-NOT ACTUAL SIZE-
-DO NOT SCALE-

CARTRIDGE: .475 LTD

OTHER NAMES:

DIA: .475

BALLISTEK NO: 475E

NAI NO: RXB 22233/4.528

DATA SOURCE: Guns & Ammo Annual 1973 Pg14

HISTORICAL DATA: By J. Lott in 1972-73.

NOTES: LTD = Lott-Tanner Dinosaur.

LOADING DATA:

BULLET WT./TYPE	POWDER WT./TYPE	VELOCITY ('/SEC)	SOURCE
500/FMJ	125.0/3031	—	Lott

CASE PREPARATION: SHELLHOLDER (RCBS): Spl.

MAKE FROM: .577 N.E. (BELL). Turn rim flush with body and re-cut the extractor groove. Anneal case and form in form die. Trim to length & chamfer. F/L size.

PHYSICAL DATA (INCHES):

CASE TYPE: **Rimless Bottleneck**

CASE LENGTH **A = 2.98**

HEAD DIAMETER **B = .658**

RIM DIAMETER **D = .658**

NECK DIAMETER **F = .500**

NECK LENGTH **H = .500**

SHOULDER LENGTH **K = .112**

BODY ANGLE (DEG'S/SIDE): **.436**

CASE CAPACITY **CC'S = 11.85**

LOADED LENGTH: **3.95**

BELT DIAMETER **C = N/A**

RIM THICKNESS **E = .05**

SHOULDER DIAMETER **G = .625**

LENGTH TO SHOULDER **J = 2.368**

SHOULDER ANGLE (DEG'S/SIDE): **29.16**

PRIMER: **L/R Mag**

CASE CAPACITY (GR'S WATER): **182.83**

DIMENSIONAL DRAWING:

-NOT ACTUAL SIZE-
-DO NOT SCALE-

CARTRIDGE: .475 Linebaugh

OTHER NAMES:

DIA: .475

BALLISTEK NO:

NAI NO:

DATA SOURCE: Most current loading manuals.

HISTORICAL DATA: Created by John Linebaugh for his custom revolvers. Factory ammo first offered by Hornady in 2000.

NOTES: A very big and powerful handgun cartridge.

LOADING DATA:

BULLET WT./TYPE	POWDER WT./TYPE	VELOCITY ('/SEC)	SOURCE
325/Hornady XTP-MAG	31.6/Lil' Gun	1,550	Hornady #6
400/Hornady XTP-MAG	26.0/ WW 296	1,350	Hornady #6

CASE PREPARATION: SHELL HOLDER (RCBS): 40

MAKE FROM: Factory. Or make from .45-70 Government. The starting brass is the standard .45-70, preferably Winchester-Western brand, which is trimmed to 1.410", and it is ready to load. If any other brand of .45-70 brass is used, the process becomes more difficult as inside neck reaming is required.

PHYSICAL DATA (INCHES):

CASE TYPE: **Rimmed Straight**

CASE LENGTH **A = 1.40**

HEAD DIAMETER **B = .504**

RIM DIAMETER **D = .542**

NECK DIAMETER **F = .504**

NECK LENGTH **H = N/A**

SHOULDER LENGTH **K = N/A**

BODY ANGLE (DEG'S/SIDE):

CASE CAPACITY **CC'S =**

LOADED LENGTH: **1.765**

BELT DIAMETER **C = N/A**

RIM THICKNESS **E = .070**

SHOULDER DIAMETER **G = N/A**

LENGTH TO SHOULDER **J = N/A**

SHOULDER ANGLE (DEG'S/SIDE):

PRIMER: **L/R**

CASE CAPACITY (GR'S WATER):

DIMENSIONAL DRAWING:

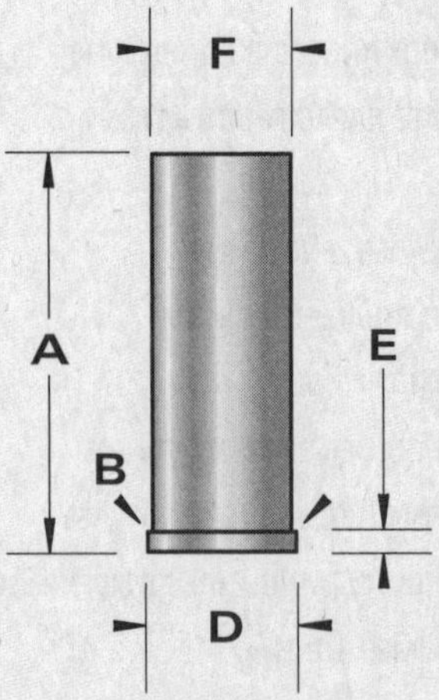

F
A
E
B
D

-NOT ACTUAL SIZE-
-DO NOT SCALE-

CARTRIDGE: .475 Nitro Express

OTHER NAMES:

DIA: .476

BALLISTEK NO: 476D

NAI NO: RMS 21115/6.055

DATA SOURCE: COTW 4th Pg244

HISTORICAL DATA: By Westley Richards about 1907.

NOTES:

LOADING DATA:

BULLET WT./TYPE	POWDER WT./TYPE	VELOCITY ('/SEC)	SOURCE
480/—	70.0/3031	2100	Barnes

CASE PREPARATION: SHELLHOLDER (RCBS): Spl.

MAKE FROM: .450 N.E. (BELL). Use brass as is. Square case mouth & F/L size.

PHYSICAL DATA (INCHES):

CASE TYPE: **Rimmed Straight**

CASE LENGTH **A = 3.30**

HEAD DIAMETER **B = .545**

RIM DIAMETER **D = .621**

NECK DIAMETER **F = .502**

NECK LENGTH **H = N/A**

SHOULDER LENGTH **K = N/A**

BODY ANGLE (DEG'S/SIDE): **.379**

CASE CAPACITY **CC'S = 9.05**

LOADED LENGTH: **3.82**

BELT DIAMETER **C = N/A**

RIM THICKNESS **E = .053**

SHOULDER DIAMETER **G = N/A**

LENGTH TO SHOULDER **J = N/A**

SHOULDER ANGLE (DEG'S/SIDE): **N/A**

PRIMER: **L/R**

CASE CAPACITY (GR'S WATER): **139.70**

DIMENSIONAL DRAWING:

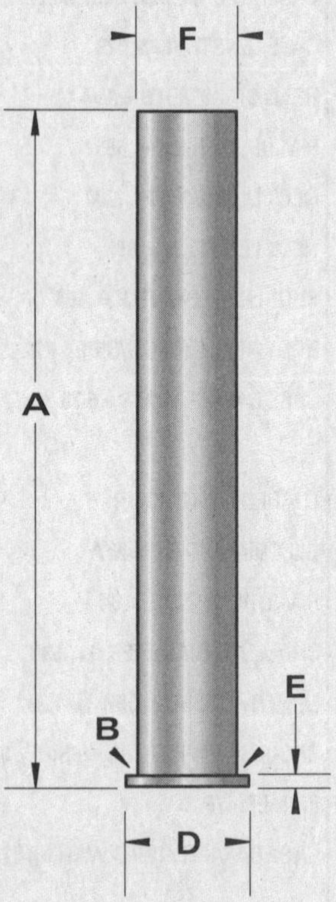

F
A
E
B
D

-NOT ACTUAL SIZE-
-DO NOT SCALE-

CARTRIDGE: .476 Enfield Revolver Mk-III

OTHER NAMES:
.476 Eley
.455/476 Revolver

DIA: .472
BALLISTEK NO: 472A
NAI NO: RMS 11115/1.820

DATA SOURCE: COTW 4th Pg190

HISTORICAL DATA: British mil. ctg. (1880's) before .455 Webley Mk-I (#454J).

NOTES:

LOADING DATA:

BULLET WT./TYPE	POWDER WT./TYPE	VELOCITY ('/SEC)	SOURCE
250/Lead	3.5/B'Eye	—	JJD

CASE PREPARATION: SHELLHOLDER (RCBS): 8

MAKE FROM: .45 Auto Rimmed. Trim case to proper length. Thin rim to .04" thick.

PHYSICAL DATA (INCHES):

CASE TYPE: **Rimmed Straight**
CASE LENGTH **A = .87**
HEAD DIAMETER **B = .478**
RIM DIAMETER **D = .530**
NECK DIAMETER **F = .474**
NECK LENGTH **H = N/A**
SHOULDER LENGTH **K = N/A**
BODY ANGLE (DEG'S/SIDE): **.138**
CASE CAPACITY **CC'S = 1.63**

LOADED LENGTH: **1.33**
BELT DIAMETER **C = N/A**
RIM THICKNESS **E = .04**
SHOULDER DIAMETER **G = N/A**
LENGTH TO SHOULDER **J = N/A**
SHOULDER ANGLE (DEG'S/SIDE): **N/A**
PRIMER: **L/P**
CASE CAPACITY (GR'S WATER): **25.13**

DIMENSIONAL DRAWING:

F
A
E
B
D

-NOT ACTUAL SIZE-
-DO NOT SCALE-

CARTRIDGE: .476 Westley Richards

OTHER NAMES:
.476 Nitro Express

DIA: .476
BALLISTEK NO: 476A
NAI NO: RMB 22221/5.263

DATA SOURCE: COTW 4th Pg244

HISTORICAL DATA: Circa 1907.

NOTES:

LOADING DATA:

BULLET WT./TYPE	POWDER WT./TYPE	VELOCITY ('/SEC)	SOURCE
500/LeadPP	68.6/4320	—	JJD

CASE PREPARATION: SHELLHOLDER (RCBS): Spl.

MAKE FROM: .470 N.E. (BELL). Trim case to length & F/L size.

PHYSICAL DATA (INCHES):

CASE TYPE: **Rimmed Bottleneck**
CASE LENGTH **A = 3.00**
HEAD DIAMETER **B = .570**
RIM DIAMETER **D = .643**
NECK DIAMETER **F = .508**
NECK LENGTH **H = .66**
SHOULDER LENGTH **K = .095**
BODY ANGLE (DEG'S/SIDE): **.56**
CASE CAPACITY **CC'S = 8.70**

LOADED LENGTH: **3.77**
BELT DIAMETER **C = N/A**
RIM THICKNESS **E = .055**
SHOULDER DIAMETER **G = .530**
LENGTH TO SHOULDER **J = 2.245**
SHOULDER ANGLE (DEG'S/SIDE): **6.60**
PRIMER: **L/R**
CASE CAPACITY (GR'S WATER): **134.34**

DIMENSIONAL DRAWING:

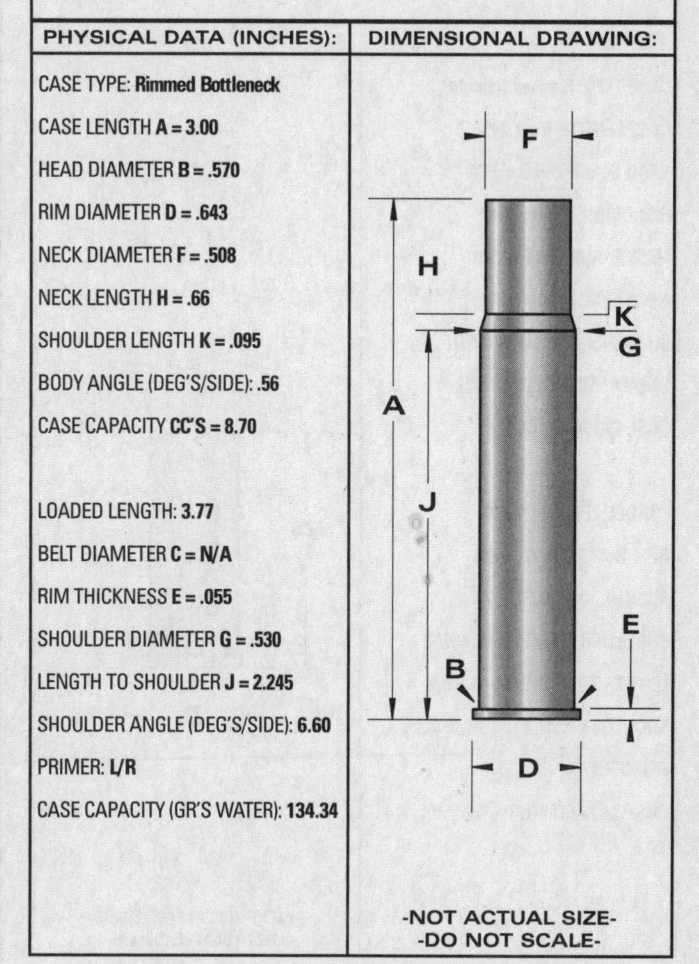

CARTRIDGE: .480 Ruger

OTHER NAMES:	DIA: .475 BALLISTEK NO: NAI NO:

DATA SOURCE: Most current handloading manuals.
HISTORICAL DATA: Developed in a joint effort between Sturm Ruger and Hornady in 2001. The goal was to produce a handgun cartridge that would exceed the performance level of the .44 Magnum, but without the stiff recoil of the .475 Linebaugh or .454 Casull.

LOADING DATA:

BULLET WT./TYPE	POWDER WT./TYPE	VELOCITY ('/SEC)	SOURCE
325/Hornady XTP Mag	H110 27.3-grains	1500	Hornady #6
400/Hornady XTP Mag.	Lil' Gun 19.6-grains	1250	Hornady #6

CASE PREPARATION: SHELL HOLDER (RCBS): 40

MAKE FROM: Factory or shorten .475 Linebaugh or .45-70 Winchester cases. Rim may require turning smaller.

PHYSICAL DATA (INCHES):

CASE TYPE: **Rimmed Straight**
CASE LENGTH **A = 1.285**
HEAD DIAMETER **B = .504**
RIM DIAMETER **D = .542**
NECK DIAMETER **F = .504**
NECK LENGTH **H = N/A**
SHOULDER LENGTH **K = N/A**
BODY ANGLE (DEG'S/SIDE):
CASE CAPACITY **CC'S =**
LOADED LENGTH: **1.650**
BELT DIAMETER **C = N/A**
RIM THICKNESS **E = .070**
SHOULDER DIAMETER **G = N/A**
LENGTH TO SHOULDER **J = N/A**
SHOULDER ANGLE (DEG'S/SIDE):
PRIMER: **L/P**
CASE CAPACITY (GR'S WATER):

DIMENSIONAL DRAWING:

F
A
E
B
D

-NOT ACTUAL SIZE-
-DO NOT SCALE-

CARTRIDGE: .50 Carbine

OTHER NAMES:	DIA: .515 BALLISTEK NO: 515A NAI NO: RMS 21115/2.371

DATA SOURCE: COTW 4th Pg126
HISTORICAL DATA: Circa 1870 for Springfield carbine.
NOTES: Short version of .50-70 Gov't (#513A).

LOADING DATA:

BULLET WT./TYPE	POWDER WT./TYPE	VELOCITY ('/SEC)	SOURCE
400/Lead	9.8/IMR4227	—	JJD

CASE PREPARATION: SHELLHOLDER (RCBS): Spl.

MAKE FROM: .50 Sharps (BELL). Cut case to 1.4". F/L size, I.D. neck ream, trim to length & chamfer.

PHYSICAL DATA (INCHES):

CASE TYPE: **Rimmed Straight**
CASE LENGTH **A = 1.34**
HEAD DIAMETER **B = .565**
RIM DIAMETER **D = .660**
NECK DIAMETER **F = .536**
NECK LENGTH **H = N/A**
SHOULDER LENGTH **K = N/A**
BODY ANGLE (DEG'S/SIDE): **.654**
CASE CAPACITY **CC'S = 3.59**
LOADED LENGTH: **1.80**
BELT DIAMETER **C = N/A**
RIM THICKNESS **E = .068**
SHOULDER DIAMETER **G = N/A**
LENGTH TO SHOULDER **J = N/A**
SHOULDER ANGLE (DEG'S/SIDE): **N/A**
PRIMER: **L/R**
CASE CAPACITY (GR'S WATER): **55.43**

DIMENSIONAL DRAWING:

F
A
E
B
D

-NOT ACTUAL SIZE-
-DO NOT SCALE-

CARTRIDGE: .50 Maynard 1865

OTHER NAMES:

DIA: .520

BALLISTEK NO: 520C

NAI NO: RMS 11115/2.275

DATA SOURCE: COTW 4th Pg135

HISTORICAL DATA:

NOTES:

LOADING DATA:

BULLET WT./TYPE	POWDER WT./TYPE	VELOCITY ('/SEC)	SOURCE
450/Lead	10.0/IMR4227	—	JJD

CASE PREPARATION: SHELLHOLDER (RCBS): Spl.

MAKE FROM: .450 N.E. (BELL). Cut case to 1.3". Increase rim dia. to .770". thin rim to .068". F/L size, trim to length & chamfer.

PHYSICAL DATA (INCHES):

CASE TYPE: **Rimmed Straight**

CASE LENGTH **A = 1.24**

HEAD DIAMETER **B = .545**

RIM DIAMETER **D = .770**

NECK DIAMETER **F = .543**

NECK LENGTH **H = N/A**

SHOULDER LENGTH **K = N/A**

BODY ANGLE (DEG'S/SIDE): **.049**

CASE CAPACITY **CC'S = 3.13**

LOADED LENGTH: **1.75**

BELT DIAMETER **C = N/A**

RIM THICKNESS **E = .068**

SHOULDER DIAMETER **G = N/A**

LENGTH TO SHOULDER **J = N/A**

SHOULDER ANGLE (DEG'S/SIDE): **N/A**

PRIMER: **L/R**

CASE CAPACITY (GR'S WATER): **48.27**

DIMENSIONAL DRAWING:

-NOT ACTUAL SIZE-
-DO NOT SCALE-

CARTRIDGE: .50 Peabody

OTHER NAMES:
.46 Peabody

DIA: .502

BALLISTEK NO: 502B

NAI NO: RMS 21115/2.870

DATA SOURCE: Hoyem Vol.2 Pg43

HISTORICAL DATA:

NOTES:

LOADING DATA:

BULLET WT./TYPE	POWDER WT./TYPE	VELOCITY ('/SEC)	SOURCE
450/Lead	19.3/IMR4198	—	JJD

CASE PREPARATION: SHELLHOLDER (RCBS): Spl.

MAKE FROM: .450 #2 N.E. (BELL). Cut case to 1.7" and thin rim to .052" thick. F/L size with expander removed. I.D. neck ream. Trim to length, F/L size & chamfer.

PHYSICAL DATA (INCHES):

CASE TYPE: **Rimmed Straight**

CASE LENGTH **A = 1.63**

HEAD DIAMETER **B = .568**

RIM DIAMETER **D = .650**

NECK DIAMETER **F = .544**

NECK LENGTH **H = N/A**

SHOULDER LENGTH **K = N/A**

BODY ANGLE (DEG'S/SIDE): **.435**

CASE CAPACITY **CC'S = 4.15**

LOADED LENGTH: **2.20**

BELT DIAMETER **C = N/A**

RIM THICKNESS **E = .052**

SHOULDER DIAMETER **G = N/A**

LENGTH TO SHOULDER **J = N/A**

SHOULDER ANGLE (DEG'S/SIDE): **N/A**

PRIMER: **L/R**

CASE CAPACITY (GR'S WATER): **64.02**

DIMENSIONAL DRAWING:

-NOT ACTUAL SIZE-
-DO NOT SCALE-

CARTRIDGE: .50 Remington Army M71

OTHER NAMES:

DIA: .508

BALLISTEK NO: 508A

NAI NO: RMS 31115/1.008

DATA SOURCE: COTW 4th Pg190

HISTORICAL DATA: Circa 1871 Army version of earlier .50 Rem. pistol.

NOTES:

LOADING DATA:

BULLET WT./TYPE	POWDER WT./TYPE	VELOCITY ('/SEC)	SOURCE
265/Lead	7.0/Unique	750	Barnes

CASE PREPARATION: SHELLHOLDER (RCBS): Spl.

MAKE FROM: .50 Sharps (BELL). Cut case to .60". I.D. neck ream. F/L size. Trim to length & chamfer.

PHYSICAL DATA (INCHES):

CASE TYPE: **Rimmed Straight**

CASE LENGTH A = **.570**

HEAD DIAMETER B = **.565**

RIM DIAMETER D = **.665**

NECK DIAMETER F = **.532**

NECK LENGTH H = **N/A**

SHOULDER LENGTH K = **N/A**

BODY ANGLE (DEG'S/SIDE): **1.89**

CASE CAPACITY CC'S = **.89**

LOADED LENGTH: **1.24**

BELT DIAMETER C = **N/A**

RIM THICKNESS E = **.072**

SHOULDER DIAMETER G = **N/A**

LENGTH TO SHOULDER J = **N/A**

SHOULDER ANGLE (DEG'S/SIDE): **N/A**

PRIMER: **L/P**

CASE CAPACITY (GR'S WATER): **13.87**

DIMENSIONAL DRAWING:

F
A
B
E
D

-NOT ACTUAL SIZE-
-DO NOT SCALE-

CARTRIDGE: .50 Springfield Cadet

OTHER NAMES:

DIA: .520

BALLISTEK NO: 520B

NAI NO: RMS 11115/2.459

DATA SOURCE: Hoyem Vol.2 Pg45

HISTORICAL DATA: Ctg. for Springfield carbine about 1868.

NOTES:

LOADING DATA:

BULLET WT./TYPE	POWDER WT./TYPE	VELOCITY ('/SEC)	SOURCE
400/Lead	10.0/IMR4227	–	JJD

CASE PREPARATION: SHELLHOLDER (RCBS): 5

MAKE FROM: .348 Winchester. Anneal case neck and taper expand to .530" dia. Trim to length & chamfer. F/L size.

PHYSICAL DATA (INCHES):

CASE TYPE: **Rimmed Straight**

CASE LENGTH A = **1.37**

HEAD DIAMETER B = **.557**

RIM DIAMETER D = **.664**

NECK DIAMETER F = **.556**

NECK LENGTH H = **N/A**

SHOULDER LENGTH K = **N/A**

BODY ANGLE (DEG'S/SIDE): **.022**

CASE CAPACITY CC'S = **3.46**

LOADED LENGTH: **1.78**

BELT DIAMETER C = **N/A**

RIM THICKNESS E = **.07**

SHOULDER DIAMETER G = **N/A**

LENGTH TO SHOULDER J = **N/A**

SHOULDER ANGLE (DEG'S/SIDE): **N/A**

PRIMER: **L/R**

CASE CAPACITY (GR'S WATER): **53.49**

DIMENSIONAL DRAWING:

F
A
E
B
D

-NOT ACTUAL SIZE-
-DO NOT SCALE-

CARTRIDGE: .50-50 Maynard 1882

OTHER NAMES:

DIA: .513

BALLISTEK NO: 513C

NAI NO: RMS 21115/2.433

DATA SOURCE: COTW 4th Pg126

HISTORICAL DATA:

NOTES: Similar to .50 Carbine (#515A).

LOADING DATA:

BULLET WT./TYPE	POWDER WT./TYPE	VELOCITY ('/SEC)	SOURCE
350/Lead	10.9/IMR4227	—	JJD

CASE PREPARATION: SHELLHOLDER (RCBS): Spl.

MAKE FROM: .450 #2 N.E. (BELL). Cut case to 1.4". Thin rim to .058" thick. F/L size, trim to length & chamfer.

PHYSICAL DATA (INCHES):

CASE TYPE: **Rimmed Straight**

CASE LENGTH **A = 1.37**

HEAD DIAMETER **B = .563**

RIM DIAMETER **D = .661**

NECK DIAMETER **F = .535**

NECK LENGTH **H = N/A**

SHOULDER LENGTH **K = N/A**

BODY ANGLE (DEG'S/SIDE): **.611**

CASE CAPACITY **CC'S = 3.65**

LOADED LENGTH: **1.91**

BELT DIAMETER **C = N/A**

RIM THICKNESS **E = .058**

SHOULDER DIAMETER **G = N/A**

LENGTH TO SHOULDER **J = N/A**

SHOULDER ANGLE (DEG'S/SIDE): **N/A**

PRIMER: **L/R**

CASE CAPACITY (GR'S WATER): **56.38**

DIMENSIONAL DRAWING:

-NOT ACTUAL SIZE-
-DO NOT SCALE-

CARTRIDGE: .50-70 Gatling

OTHER NAMES:

DIA: .510

BALLISTEK NO: 510H

NAI NO: RMS 21115/3.150

DATA SOURCE: Hoyem Vol.2 Pg73

HISTORICAL DATA: U.S. MG ctg. circa 1876.

NOTES:

LOADING DATA:

BULLET WT./TYPE	POWDER WT./TYPE	VELOCITY ('/SEC)	SOURCE
450/Lead	26.1/IMR3031	—	JJD

CASE PREPARATION: SHELLHOLDER (RCBS): Spl.

MAKE FROM: .450 #2 N.E. (BELL). Trim case to proper length and F/L size. Either I.D. neck ream for correct bullet diameter or use undersized, "minied" bullets.

PHYSICAL DATA (INCHES):

CASE TYPE: **Rimmed Straight**

CASE LENGTH **A = 1.78**

HEAD DIAMETER **B = .565**

RIM DIAMETER **D = .645**

NECK DIAMETER **F = .533**

NECK LENGTH **H = N/A**

SHOULDER LENGTH **K = N/A**

BODY ANGLE (DEG'S/SIDE): **.539**

CASE CAPACITY **CC'S = 5.03**

LOADED LENGTH: **2.19**

BELT DIAMETER **C = N/A**

RIM THICKNESS **E = .08**

SHOULDER DIAMETER **G = N/A**

LENGTH TO SHOULDER **J = N/A**

SHOULDER ANGLE (DEG'S/SIDE): **N/A**

PRIMER: **L/R**

CASE CAPACITY (GR'S WATER): **77.60**

DIMENSIONAL DRAWING:

-NOT ACTUAL SIZE-
-DO NOT SCALE-

CARTRIDGE: .50-70 Government

OTHER NAMES:	
.50-70 Musket .50 Government	DIA: .513 BALLISTEK NO: 513A NAI NO: RMS 21115/3.097

DATA SOURCE: COTW 4th Pg127

HISTORICAL DATA: U.S. mil. ctg. 1866-1873.

NOTES: Patterned after .50-60 Joslyn (rimfire).

LOADING DATA:

BULLET WT./TYPE	POWDER WT./TYPE	VELOCITY ('/SEC)	SOURCE
450/Lead	24.0/IMR4198	1300	JJD

CASE PREPARATION: SHELLHOLDER (RCBS): Spl.

MAKE FROM: .50 Sharps (BELL) or Dixie Gun Works. Cut BELL's brass to 1.8". F/L size, trim to length & chamfer. Use Dixie brass as is.

PHYSICAL DATA (INCHES):

CASE TYPE: **Rimmed Straight**

CASE LENGTH **A = 1.75**

HEAD DIAMETER **B = .565**

RIM DIAMETER **D = .660**

NECK DIAMETER **F = .535**

NECK LENGTH **H = N/A**

SHOULDER LENGTH **K = N/A**

BODY ANGLE (DEG'S/SIDE): .511

CASE CAPACITY **CC'S = 4.97**

LOADED LENGTH: **2.25**

BELT DIAMETER **C = N/A**

RIM THICKNESS **E = .07**

SHOULDER DIAMETER **G = N/A**

LENGTH TO SHOULDER **J = N/A**

SHOULDER ANGLE (DEG'S/SIDE): **N/A**

PRIMER: **L/R**

CASE CAPACITY (GR'S WATER): **76.77**

DIMENSIONAL DRAWING:

-NOT ACTUAL SIZE-
-DO NOT SCALE-

CARTRIDGE: .50-70 Maynard 1873

OTHER NAMES:	
	DIA: .514 BALLISTEK NO: 514A NAI NO: RMS 11115/3.406

DATA SOURCE: NAI/Ballistek

HISTORICAL DATA:

NOTES:

LOADING DATA:

BULLET WT./TYPE	POWDER WT./TYPE	VELOCITY ('/SEC)	SOURCE
450/Lead	27.9/IMR4198	—	JJD

CASE PREPARATION: SHELLHOLDER (RCBS): Spl.

MAKE FROM: .348 Winchester. Anneal case neck and taper expand to .520" dia. Cut to 1.9". Increase rim dia. to .760". F/L size case, trim to length & fireform. Also, 9/16" dia. tubing, on a .348 Win. case head, will work.

PHYSICAL DATA (INCHES):

CASE TYPE: **Rimmed Straight**

CASE LENGTH **A = 1.88**

HEAD DIAMETER **B = .552**

RIM DIAMETER **D = .760**

NECK DIAMETER **F = .547**

NECK LENGTH **H = N/A**

SHOULDER LENGTH **K = N/A**

BODY ANGLE (DEG'S/SIDE): **.079**

CASE CAPACITY **CC'S = 5.07**

LOADED LENGTH: **2.34**

BELT DIAMETER **C = N/A**

RIM THICKNESS **E = .07**

SHOULDER DIAMETER **G = N/A**

LENGTH TO SHOULDER **J = N/A**

SHOULDER ANGLE (DEG'S/SIDE): **N/A**

PRIMER: **L/R**

CASE CAPACITY (GR'S WATER): **78.28**

DIMENSIONAL DRAWING:

-NOT ACTUAL SIZE-
-DO NOT SCALE-

CARTRIDGE: .50-70 Rodman-Crispin

OTHER NAMES:

DIA: .510

BALLISTEK NO: 510G

NAI NO: RMS 21115/3.234

DATA SOURCE: Hoyem Vol.2 Pg47

HISTORICAL DATA: Ctg. for Springfield carbine about 1867.

NOTES: Original case was of brass (copper?) foil.

LOADING DATA:

BULLET WT./TYPE	POWDER WT./TYPE	VELOCITY ('/SEC)	SOURCE
450/Lead	20.7/IMR4198	—	JJD

CASE PREPARATION: SHELLHOLDER (RCBS): Spl.

MAKE FROM: .470 N.E. (BELL). Cut case to 1.9". F/L size, trim to length & chamfer.

PHYSICAL DATA (INCHES):

CASE TYPE: **Rimmed Straight**

CASE LENGTH **A = 1.85**

HEAD DIAMETER **B = .572**

RIM DIAMETER **D = .653**

NECK DIAMETER **F = .519**

NECK LENGTH **H = N/A**

SHOULDER LENGTH **K = N/A**

BODY ANGLE (DEG'S/SIDE): **.844**

CASE CAPACITY **CC'S = 5.63**

LOADED LENGTH: **2.18**

BELT DIAMETER **C = N/A**

RIM THICKNESS **E = .053**

SHOULDER DIAMETER **G = N/A**

LENGTH TO SHOULDER **J = N/A**

SHOULDER ANGLE (DEG'S/SIDE): **N/A**

PRIMER: **L/R**

CASE CAPACITY (GR'S WATER): **86.96**

DIMENSIONAL DRAWING:

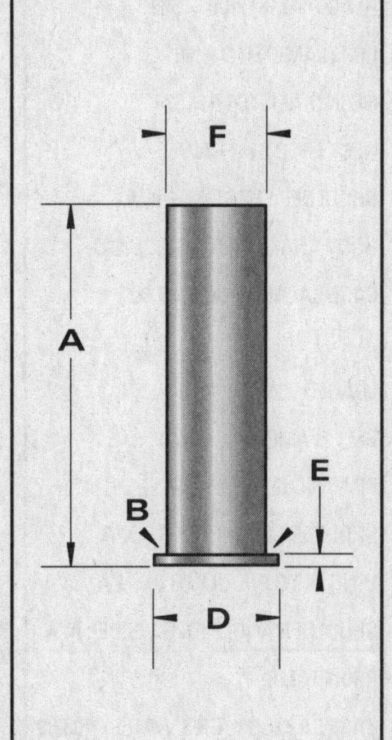

-NOT ACTUAL SIZE-
-DO NOT SCALE-

CARTRIDGE: .50-90 Sharps 2.5"

OTHER NAMES:
.50 - 2 1/2" Sharps
.50-100 (or -110) Sharps

DIA: .509

BALLISTEK NO: 509A

NAI NO: RMS 21115/4.425

DATA SOURCE: COTW 4th Pg127

HISTORICAL DATA: By Sharps in 1875.

NOTES: Winchester loaded ammo in 1878.

LOADING DATA:

BULLET WT./TYPE	POWDER WT./TYPE	VELOCITY ('/SEC)	SOURCE
465/LeadPP	30.0/4198	1320	Barnes

CASE PREPARATION: SHELLHOLDER (RCBS): 31

MAKE FROM: .50 Sharps (BELL). Trim case to proper length & F/L size.

PHYSICAL DATA (INCHES):

CASE TYPE: **Rimmed Straight**

CASE LENGTH **A = 2.50**

HEAD DIAMETER **B = .565**

RIM DIAMETER **D = .663**

NECK DIAMETER **F = .528**

NECK LENGTH **H = N/A**

SHOULDER LENGTH **K = N/A**

BODY ANGLE (DEG'S/SIDE): **.434**

CASE CAPACITY **CC'S = 7.58**

LOADED LENGTH: **3.20**

BELT DIAMETER **C = N/A**

RIM THICKNESS **E = .062**

SHOULDER DIAMETER **G = N/A**

LENGTH TO SHOULDER **J = N/A**

SHOULDER ANGLE (DEG'S/SIDE): **N/A**

PRIMER: **L/R**

CASE CAPACITY (GR'S WATER): **117.05**

DIMENSIONAL DRAWING:

F

A

B E

D

-NOT ACTUAL SIZE-
-DO NOT SCALE-

CARTRIDGE: .50-95 Winchester

OTHER NAMES:	DIA: .513
	BALLISTEK NO: 513B
	NAI NO: RMB 14242/3.416

DATA SOURCE: COTW 4th Pg128

HISTORICAL DATA: By Win. in 1879.

NOTES:

LOADING DATA:

BULLET WT./TYPE	POWDER WT./TYPE	VELOCITY ('/SEC)	SOURCE
350/Lead	23.0/4198	1350	Barnes

CASE PREPARATION: SHELLHOLDER (RCBS): Spl.

MAKE FROM: .450 #2 N.E. (BELL). Cut case to 2.0". Turn rim to .627" dia. and thin to .06" thick. F/L size. I.D. neck ream, if necessary. Trim to length & chamfer.

PHYSICAL DATA (INCHES):

CASE TYPE: **Rimmed Bottleneck**

CASE LENGTH **A = 1.92**

HEAD DIAMETER **B = .562**

RIM DIAMETER **D = .627**

NECK DIAMETER **F = .533**

NECK LENGTH **H = .430**

SHOULDER LENGTH **K = .050**

BODY ANGLE (DEG'S/SIDE): **.208**

CASE CAPACITY **CC'S = 5.92**

LOADED LENGTH: **2.26**

BELT DIAMETER **C = N/A**

RIM THICKNESS **E = .06**

SHOULDER DIAMETER **G = .553**

LENGTH TO SHOULDER **J = 1.440**

SHOULDER ANGLE (DEG'S/SIDE): **11.31**

PRIMER: **L/R**

CASE CAPACITY (GR'S WATER): **91.31**

DIMENSIONAL DRAWING:

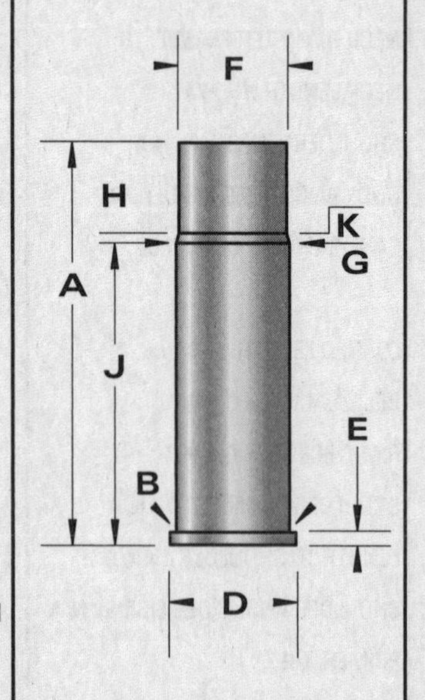

-NOT ACTUAL SIZE-
-DO NOT SCALE-

CARTRIDGE: .50-110 Winchester

OTHER NAMES:	DIA: .512
.50-100 Winchester	BALLISTEK NO: 512B
.50-105 Winchester	NAI NO: RMS 11115/4.355

DATA SOURCE: COTW 4th Pg128

HISTORICAL DATA: .50-110 by Win. in 1899.

NOTES: Cases are the same - loads vary.

LOADING DATA:

BULLET WT./TYPE	POWDER WT./TYPE	VELOCITY ('/SEC)	SOURCE
600/LeadPP	37.9/IMR4198	–	JJD

CASE PREPARATION: SHELLHOLDER (RCBS): 5

MAKE FROM: .450 N.E. (BELL). Cut case to 2.5". Turn rim to .607" dia. F/L size case. I.D. neck ream. F/L size again, trim to length & chamfer. Build up rim to .07" thick, if necessary.

PHYSICAL DATA (INCHES):

CASE TYPE: **Rimmed Straight**

CASE LENGTH **A = 2.40**

HEAD DIAMETER **B = .551**

RIM DIAMETER **D = .607**

NECK DIAMETER **F = .534**

NECK LENGTH **H = N/A**

SHOULDER LENGTH **K = N/A**

BODY ANGLE (DEG'S/SIDE): **.209**

CASE CAPACITY **CC'S = 7.00**

LOADED LENGTH: **2.75**

BELT DIAMETER **C = N/A**

RIM THICKNESS **E = .07**

SHOULDER DIAMETER **G = N/A**

LENGTH TO SHOULDER **J = N/A**

SHOULDER ANGLE (DEG'S/SIDE): **N/A**

PRIMER: **L/R**

CASE CAPACITY (GR'S WATER): **108.08**

DIMENSIONAL DRAWING:

F
A
E
B
D

-NOT ACTUAL SIZE-
-DO NOT SCALE-

CARTRIDGE: .50-115 Bullard

OTHER NAMES:	
	DIA: .512
	BALLISTEK NO: 512E
	NAI NO: RMB 13241/3.744

DATA SOURCE: COTW 4th Pg128

HISTORICAL DATA: Largest Bullard ctg. from about 1886.

NOTES:

LOADING DATA:

BULLET WT./TYPE	POWDER WT./TYPE	VELOCITY ('/SEC)	SOURCE
290/Lead	32.0/4198	1570	Barnes

CASE PREPARATION: SHELLHOLDER (RCBS): Spl.

MAKE FROM: 11mm Beaumont (BELL). Trim rim to .619" dia. and cut off to 2.2". F/L size, trim to length & chamfer.

PHYSICAL DATA (INCHES):

CASE TYPE: **Rimmed Bottleneck**
CASE LENGTH **A = 2.19**
HEAD DIAMETER **B = .585**
RIM DIAMETER **D = .619**
NECK DIAMETER **F = .547**
NECK LENGTH **H = .315**
SHOULDER LENGTH **K = .161**
BODY ANGLE (DEG'S/SIDE): **.151**
CASE CAPACITY **CC'S = 6.92**

LOADED LENGTH: **2.56**
BELT DIAMETER **C = N/A**
RIM THICKNESS **E = .068**
SHOULDER DIAMETER **G = .577**
LENGTH TO SHOULDER **J = 1.714**
SHOULDER ANGLE (DEG'S/SIDE): **5.32**
PRIMER: **L/R**
CASE CAPACITY (GR'S WATER): **106.87**

DIMENSIONAL DRAWING:

-NOT ACTUAL SIZE-
-DO NOT SCALE-

CARTRIDGE: .50-140 Sharps 3.25"

OTHER NAMES: .50 - 3 1/4" Sharps	
	DIA: .509
	BALLISTEK NO: 509B
	NAI NO: RMS 11115/5.752

DATA SOURCE: COTW 4th Pg129

HISTORICAL DATA: Circa 1882 by Sharps.

NOTES:

LOADING DATA:

BULLET WT./TYPE	POWDER WT./TYPE	VELOCITY ('/SEC)	SOURCE
465/LeadPP	33.0/4198	1450	Barnes

CASE PREPARATION: SHELLHOLDER (RCBS): Spl.

MAKE FROM: .50 Sharps (BELL). F/L size case & square case mouth.

PHYSICAL DATA (INCHES):

CASE TYPE: **Rimmed Straight**
CASE LENGTH **A = 3.25**
HEAD DIAMETER **B = .565**
RIM DIAMETER **D = .665**
NECK DIAMETER **F = .528**
NECK LENGTH **H = N/A**
SHOULDER LENGTH **K = N/A**
BODY ANGLE (DEG'S/SIDE): **.333**
CASE CAPACITY **CC'S = 10.17**

LOADED LENGTH: **3.94**
BELT DIAMETER **C = N/A**
RIM THICKNESS **E = .065**
SHOULDER DIAMETER **G = N/A**
LENGTH TO SHOULDER **J = N/A**
SHOULDER ANGLE (DEG'S/SIDE): **N/A**
PRIMER: **L/R**
CASE CAPACITY (GR'S WATER): **157.01**

DIMENSIONAL DRAWING:

-NOT ACTUAL SIZE-
-DO NOT SCALE-

CARTRIDGE: .50-140 Winchester Express

OTHER NAMES:

DIA: .512

BALLISTEK NO: 512D

NAI NO: RMS 11115/5.898

DATA SOURCE: COTW 4th Pg129

HISTORICAL DATA:

NOTES: Similar to .50-140 Sharps (#509B).

LOADING DATA:

BULLET WT./TYPE	POWDER WT./TYPE	VELOCITY ('/SEC)	SOURCE
465/LeadPP	33.0/4198	1450	Barnes

CASE PREPARATION: SHELLHOLDER (RCBS): Spl.

MAKE FROM: .450 N.E. (BELL). Build up rim to .07" thick. Square case mouth and F/L size. Fireform in the chamber.

PHYSICAL DATA (INCHES):

CASE TYPE: **Rimmed Straight**

CASE LENGTH **A = 3.25**

HEAD DIAMETER **B = .551**

RIM DIAMETER **D = .652**

NECK DIAMETER **F = .528**

NECK LENGTH **H = N/A**

SHOULDER LENGTH **K = N/A**

BODY ANGLE (DEG'S/SIDE): **.207**

CASE CAPACITY **CC'S = 10.08**

LOADED LENGTH: **3.95**

BELT DIAMETER **C = N/A**

RIM THICKNESS **E = .07**

SHOULDER DIAMETER **G = N/A**

LENGTH TO SHOULDER **J = N/A**

SHOULDER ANGLE (DEG'S/SIDE): **N/A**

PRIMER: **L/R**

CASE CAPACITY (GR'S WATER): **155.51**

DIMENSIONAL DRAWING:

-NOT ACTUAL SIZE-
-DO NOT SCALE-

CARTRIDGE: .500 A-Square

OTHER NAMES:

DIA: .510

BALLISTEK NO:

NAI NO:

DATA SOURCE:

HISTORICAL DATA: Developed by Col. Arthur B. Alphin of A-Square. The cartridge is based on a necked up .460 Weatherby.

NOTES: Approved by SAMMI 1/24/98.

LOADING DATA:

BULLET WT./TYPE	POWDER WT./TYPE	VELOCITY ('/SEC)	SOURCE
600/SP	116.5/IMR-4320	2,475	Cart.'s of the World #9

CASE PREPARATION: SHELL HOLDER (RCBS): 14

MAKE FROM: Factory or neck up .460 Weatherby brass with tapered expander and fireform. Trim, chamfer and deburr.

PHYSICAL DATA (INCHES):

CASE TYPE: **Belted Bottleneck**

CASE LENGTH **A = 2.900**

HEAD DIAMETER **B = .5818**

RIM DIAMETER **D = .579**

NECK DIAMETER **F = .536**

NECK LENGTH **H = .393**

SHOULDER LENGTH **K = .022**

BODY ANGLE (DEG'S/SIDE):

CASE CAPACITY **CC'S =**

LOADED LENGTH: **3.740**

BELT DIAMETER **C = .603**

RIM THICKNESS **E = .063**

SHOULDER DIAMETER **G = .5677**

LENGTH TO SHOULDER **J = 2.485**

SHOULDER ANGLE (DEG'S/SIDE): **35**

PRIMER: **L/R Mag**

CASE CAPACITY (GR'S WATER):

DIMENSIONAL DRAWING:

-NOT ACTUAL SIZE-
-DO NOT SCALE-

CARTRIDGE: .500 Jeffery Rimless

OTHER NAMES:
12,7 x 70 Schuler

DIA: .510

BALLISTEK NO: 510D

NAI NO: RBB
12242/4.455

DATA SOURCE: COTW 4th Pg278

HISTORICAL DATA: Exact origin seems cloudy.

NOTES:

LOADING DATA:

BULLET WT./TYPE	POWDER WT./TYPE	VELOCITY ('/SEC)	SOURCE
535/RN	100.0/3031	2400	Barnes

CASE PREPARATION: SHELLHOLDER (RCBS): Spl.

MAKE FROM: This case must be turned from solid brass rod. Turn the straight walled case, anneal, F/L size and trim to length.

PHYSICAL DATA (INCHES):

CASE TYPE: **Rebated Bottleneck**

CASE LENGTH **A = 2.74**

HEAD DIAMETER **B = .615**

RIM DIAMETER **D = .578**

NECK DIAMETER **F = .535**

NECK LENGTH **H = .30**

SHOULDER LENGTH **K = .140**

BODY ANGLE (DEG'S/SIDE): **0**

CASE CAPACITY **CC'S = 10.38**

LOADED LENGTH: **3.50**

BELT DIAMETER **C = N/A**

RIM THICKNESS **E = .06**

SHOULDER DIAMETER **G = .615**

LENGTH TO SHOULDER **J = 2.30**

SHOULDER ANGLE (DEG'S/SIDE): **15.94**

PRIMER: **L/R Mag**

CASE CAPACITY (GR'S WATER): **160.19**

DIMENSIONAL DRAWING:

-NOT ACTUAL SIZE-
-DO NOT SCALE-

CARTRIDGE: .500 Jurras

OTHER NAMES:

DIA: .505

BALLISTEK NO: 505C

NAI NO: RMB
34225/3.542

DATA SOURCE: NAI/Ballistek

HISTORICAL DATA: By L. Jurras.

NOTES:

LOADING DATA:

BULLET WT./TYPE	POWDER WT./TYPE	VELOCITY ('/SEC)	SOURCE
500/RN	54.5/IMR3031	–	JJD

CASE PREPARATION: SHELLHOLDER (RCBS): Spl.

MAKE FROM: .500 N.E. (BELL). Form dies are required. Cut case to 2.1" and form in form set. Trim to length and F/L size.

PHYSICAL DATA (INCHES):

CASE TYPE: **Rimmed Bottleneck**

CASE LENGTH **A = 2.033**

HEAD DIAMETER **B = .574**

RIM DIAMETER **D = .647**

NECK DIAMETER **F = .530**

NECK LENGTH **H = .708**

SHOULDER LENGTH **K = .071**

BODY ANGLE (DEG'S/SIDE): **1.00**

CASE CAPACITY **CC'S = 6.04**

LOADED LENGTH: **A/R**

BELT DIAMETER **C = N/A**

RIM THICKNESS **E = .054**

SHOULDER DIAMETER **G = .537**

LENGTH TO SHOULDER **J = 1.254**

SHOULDER ANGLE (DEG'S/SIDE): **2.82**

PRIMER: **L/P**

CASE CAPACITY (GR'S WATER): **93.34**

DIMENSIONAL DRAWING:

-NOT ACTUAL SIZE-
-DO NOT SCALE-

CARTRIDGE: .500 Linebaugh

OTHER NAMES:

DIA: .512

BALLISTEK NO:

NAI NO:

DATA SOURCE:

HISTORICAL DATA: Developed by Wyoming custom gunsmith John Linebaugh for use in his single action handguns.

NOTES:

LOADING DATA:

BULLET WT./TYPE	POWDER WT./TYPE	VELOCITY ('/SEC)	SOURCE
410/Cast	35.5/H110	1458	Linebaugh Custom Sixguns
468/Cast	29.5/Lil' Gun	1343	

CASE PREPARATION: SHELL HOLDER (RCBS): 5

MAKE FROM: Make from .348 Winchester brass. Cut on lathe to 1.435-inch. Trim to 1.400-inch Inside ream with RCBS Ream Die for .500 Linebaugh. Ream to slightly deeper than base of bullet. Case wall should be .014-inch. Chamfer and deburr. Use .510-inch expander plug. Cast bullets should be .512-inch, barrel is .510-inch.

PHYSICAL DATA (INCHES):

CASE TYPE: **Rimmed Straight**

CASE LENGTH **A = 1.40**

HEAD DIAMETER **B = .5474**

RIM DIAMETER **D = .610**

NECK DIAMETER **F = .543**

NECK LENGTH **H = N/A**

SHOULDER LENGTH **K = N/A**

BODY ANGLE (DEG'S/SIDE):

CASE CAPACITY **CC'S =**

LOADED LENGTH: **1.80**

BELT DIAMETER **C = N/A**

RIM THICKNESS **E = .070**

SHOULDER DIAMETER **G = N/A**

LENGTH TO SHOULDER **J = N/A**

SHOULDER ANGLE (DEG'S/SIDE):

PRIMER: **L/P**

CASE CAPACITY (GR'S WATER):

DIMENSIONAL DRAWING:

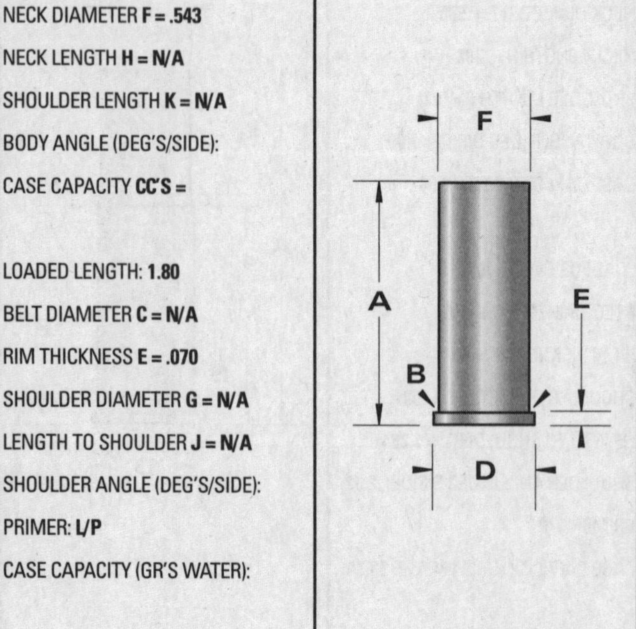

-NOT ACTUAL SIZE-
-DO NOT SCALE-

CARTRIDGE: .500 NAI Long Magnum

OTHER NAMES:

DIA: .510

BALLISTEK NO: 510J

NAI NO: BEN 12245/6.678

DATA SOURCE: NAI/Ballistek

HISTORICAL DATA: Designed by author in 1985.

NOTES:

LOADING DATA:

BULLET WT./TYPE	POWDER WT./TYPE	VELOCITY ('/SEC)	SOURCE
600/FMJ	150.0/H870	2925	JJD

CASE PREPARATION: SHELLHOLDER (RCBS): 14

MAKE FROM: Lathe-turned long version of the .460 Weatherby case head.

PHYSICAL DATA (INCHES):

CASE TYPE: **Belted Bottleneck**

CASE LENGTH **A = 3.76**

HEAD DIAMETER **B = .582**

RIM DIAMETER **D = .583**

NECK DIAMETER **F = .530**

NECK LENGTH **H = .465**

SHOULDER LENGTH **K = .099**

BODY ANGLE (DEG'S/SIDE): **.402**

CASE CAPACITY **CC'S = 12.24**

LOADED LENGTH: **4.755**

BELT DIAMETER **C = .603**

RIM THICKNESS **E = .06**

SHOULDER DIAMETER **G = .540**

LENGTH TO SHOULDER **J = 3.195**

SHOULDER ANGLE (DEG'S/SIDE): **2.86**

PRIMER: **Electronic Ignition**

CASE CAPACITY (GR'S WATER): **188.95**

DIMENSIONAL DRAWING:

-NOT ACTUAL SIZE-
-DO NOT SCALE-

CARTRIDGE: .500 NAI Short Magnum

OTHER NAMES:

DIA: .510

BALLISTEK NO: 510K

NAI NO: BEN 12245/5.338

DATA SOURCE: NAI/Ballistek

HISTORICAL DATA: Designed by author in 1985.

NOTES:

LOADING DATA:

BULLET WT./TYPE	POWDER WT./TYPE	VELOCITY ('/SEC)	SOURCE
600/FMJ	130.0/H870	2800	JJD

CASE PREPARATION: SHELLHOLDER (RCBS): 14

MAKE FROM: Lathe-turned long version of the .460 Weatherby case head.

PHYSICAL DATA (INCHES):

CASE TYPE: **Belted Bottleneck**

CASE LENGTH A = **3.000**

HEAD DIAMETER B = **.582**

RIM DIAMETER D = **.583**

NECK DIAMETER F = **.530**

NECK LENGTH H = **.465**

SHOULDER LENGTH K = **.099**

BODY ANGLE (DEG'S/SIDE): **.538**

CASE CAPACITY CC'S = **9.84**

LOADED LENGTH: **3.955**

BELT DIAMETER C = **.603**

RIM THICKNESS E = **.060**

SHOULDER DIAMETER G = **.540**

LENGTH TO SHOULDER J = **2.435**

SHOULDER ANGLE (DEG'S/SIDE): **2.86**

PRIMER: **Electronic Ignition**

CASE CAPACITY (GR'S WATER): **152.00**

DIMENSIONAL DRAWING:

-NOT ACTUAL SIZE-
-DO NOT SCALE-

CARTRIDGE: .500 Nitro Express 3"

OTHER NAMES:

DIA: .510

BALLISTEK NO: 510A

NAI NO: RMS 21115/5.189

DATA SOURCE: COTW 4th Pg246

HISTORICAL DATA: Circa 1896.

NOTES:

LOADING DATA:

BULLET WT./TYPE	POWDER WT./TYPE	VELOCITY ('/SEC)	SOURCE
440/LeadPP	41.0/4198	1560	Barnes

CASE PREPARATION: SHELLHOLDER (RCBS): Spl.

MAKE FROM: .500 N.E. (BELL). Trim case to proper length and F/L size.

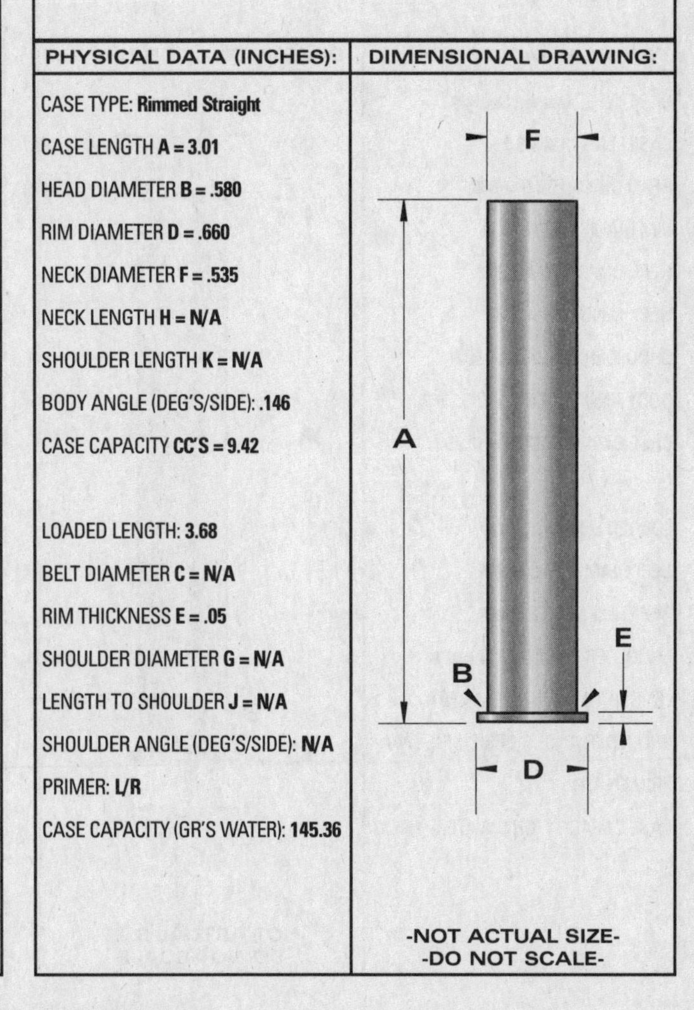

PHYSICAL DATA (INCHES):

CASE TYPE: **Rimmed Straight**

CASE LENGTH A = **3.01**

HEAD DIAMETER B = **.580**

RIM DIAMETER D = **.660**

NECK DIAMETER F = **.535**

NECK LENGTH H = **N/A**

SHOULDER LENGTH K = **N/A**

BODY ANGLE (DEG'S/SIDE): **.146**

CASE CAPACITY CC'S = **9.42**

LOADED LENGTH: **3.68**

BELT DIAMETER C = **N/A**

RIM THICKNESS E = **.05**

SHOULDER DIAMETER G = **N/A**

LENGTH TO SHOULDER J = **N/A**

SHOULDER ANGLE (DEG'S/SIDE): **N/A**

PRIMER: **L/R**

CASE CAPACITY (GR'S WATER): **145.36**

DIMENSIONAL DRAWING:

-NOT ACTUAL SIZE-
-DO NOT SCALE-

CARTRIDGE: .500 Nitro Express 3.25"

OTHER NAMES:

DIA: .510

BALLISTEK NO: 510F

NAI NO: RMS 21115/5.603

DATA SOURCE: COTW 4th Pg246

HISTORICAL DATA:

NOTES: Long version of the 3" ctg.

LOADING DATA:

BULLET WT./TYPE	POWDER WT./TYPE	VELOCITY ('/SEC)	SOURCE
440/LeadPP	43.0/IMR4198	1600	JJD

CASE PREPARATION: SHELLHOLDER (RCBS): Spl.

MAKE FROM: .500 N.E. (BELL). Use BELL's brass, as is. Square case mouth and F/L size.

PHYSICAL DATA (INCHES):

CASE TYPE: **Rimmed Straight**

CASE LENGTH **A = 3.25**

HEAD DIAMETER **B = .580**

RIM DIAMETER **D = .660**

NECK DIAMETER **F = .535**

NECK LENGTH **H = N/A**

SHOULDER LENGTH **K = N/A**

BODY ANGLE (DEG'S/SIDE): **.362**

CASE CAPACITY **CC'S = 10.25**

LOADED LENGTH: **3.62**

BELT DIAMETER **C = N/A**

RIM THICKNESS **E = .048**

SHOULDER DIAMETER **G = N/A**

LENGTH TO SHOULDER **J = N/A**

SHOULDER ANGLE (DEG'S/SIDE): **N/A**

PRIMER: **L/R**

CASE CAPACITY (GR'S WATER): **158.23**

DIMENSIONAL DRAWING:

-NOT ACTUAL SIZE-
-DO NOT SCALE-

CARTRIDGE: .500/450 #1 Express

OTHER NAMES:

DIA: .458

BALLISTEK NO: 458Q

NAI NO: RMB 23221/4.766

DATA SOURCE: COTW 4th Pg242

HISTORICAL DATA: Circa 1880.

NOTES:

LOADING DATA:

BULLET WT./TYPE	POWDER WT./TYPE	VELOCITY ('/SEC)	SOURCE
300/LeadPP	42.0/IMR3031	1500	JJD

CASE PREPARATION: SHELLHOLDER (RCBS): Spl.

MAKE FROM: .475 #2 N.E. (BELL). Turn rim to .660" dia. and thin to .062" thick. Cut case to 2.8". F/L size, trim to length & chamfer.

PHYSICAL DATA (INCHES):

CASE TYPE: **Rimmed Bottleneck**

CASE LENGTH **A = 2.75**

HEAD DIAMETER **B = .577**

RIM DIAMETER **D = .660**

NECK DIAMETER **F = .485**

NECK LENGTH **H = .760**

SHOULDER LENGTH **K = .160**

BODY ANGLE (DEG'S/SIDE): **.826**

CASE CAPACITY **CC'S = 7.90**

LOADED LENGTH: **3.25**

BELT DIAMETER **C = N/A**

RIM THICKNESS **E = .063**

SHOULDER DIAMETER **G = .530**

LENGTH TO SHOULDER **J = 1.83**

SHOULDER ANGLE (DEG'S/SIDE): **8.00**

PRIMER: **L/R**

CASE CAPACITY (GR'S WATER): **121.92**

DIMENSIONAL DRAWING:

-NOT ACTUAL SIZE-
-DO NOT SCALE-

CARTRIDGE: .500/450 #2 Musket

OTHER NAMES:	
	DIA: .451
	BALLISTEK NO: 451K
	NAI NO: RMB 24221/4.097

DATA SOURCE: COTW 4th Pg242

HISTORICAL DATA: Early Westley Richards black powder ctg.

NOTES:

LOADING DATA:

BULLET WT./TYPE	POWDER WT./TYPE	VELOCITY ('/SEC)	SOURCE
480/LeadPP	37.2/IMR3031	–	JJD

CASE PREPARATION: SHELLHOLDER (RCBS): Spl.

MAKE FROM: .475 #2 N.E. (BELL). Thin rim to .063". Cut case to 2.4". F/L size, trim to length and chamfer.

PHYSICAL DATA (INCHES):

CASE TYPE: **Rimmed Bottleneck**
CASE LENGTH **A = 2.36**
HEAD DIAMETER **B = .576**
RIM DIAMETER **D = .663**
NECK DIAMETER **F = .486**
NECK LENGTH **H = .585**
SHOULDER LENGTH **K = .255**
BODY ANGLE (DEG'S/SIDE): **.889**
CASE CAPACITY **CC'S = 6.43**

LOADED LENGTH: **2.99**
BELT DIAMETER **C = N/A**
RIM THICKNESS **E = .063**
SHOULDER DIAMETER **G = .535**
LENGTH TO SHOULDER **J = 1.52**
SHOULDER ANGLE (DEG'S/SIDE): **5.48**
PRIMER: **L/R**
CASE CAPACITY (GR'S WATER): **99.22**

DIMENSIONAL DRAWING:

F H K G A J E B D

-NOT ACTUAL SIZE-
-DO NOT SCALE-

CARTRIDGE: .500/450 Magnum Black Powder Express

OTHER NAMES:	
	DIA: .455
	BALLISTEK NO: 451P
	NAI NO: RMB 22225/5.702

DATA SOURCE: COTW 4th Pg241

HISTORICAL DATA: Circa 1880.

NOTES:

LOADING DATA:

BULLET WT./TYPE	POWDER WT./TYPE	VELOCITY ('/SEC)	SOURCE
325/Lead	51.0/IMR4064	–	JJD

CASE PREPARATION: SHELLHOLDER (RCBS): Spl.

MAKE FROM: .470 #2 N.E. (BELL). Thin rim to .042". Trim case to length. F/L size.

PHYSICAL DATA (INCHES):

CASE TYPE: **Rimmed Bottleneck**
CASE LENGTH **A = 3.25**
HEAD DIAMETER **B = .570**
RIM DIAMETER **D = .644**
NECK DIAMETER **F = .479**
NECK LENGTH **H = .595**
SHOULDER LENGTH **K = .305**
BODY ANGLE (DEG'S/SIDE): **.932**
CASE CAPACITY **CC'S = 9.19**

LOADED LENGTH: **3.91**
BELT DIAMETER **C = N/A**
RIM THICKNESS **E = .042**
SHOULDER DIAMETER **G = .50**
LENGTH TO SHOULDER **J = 2.35**
SHOULDER ANGLE (DEG'S/SIDE): **1.97**
PRIMER: **L/R**
CASE CAPACITY (GR'S WATER): **141.82**

DIMENSIONAL DRAWING:

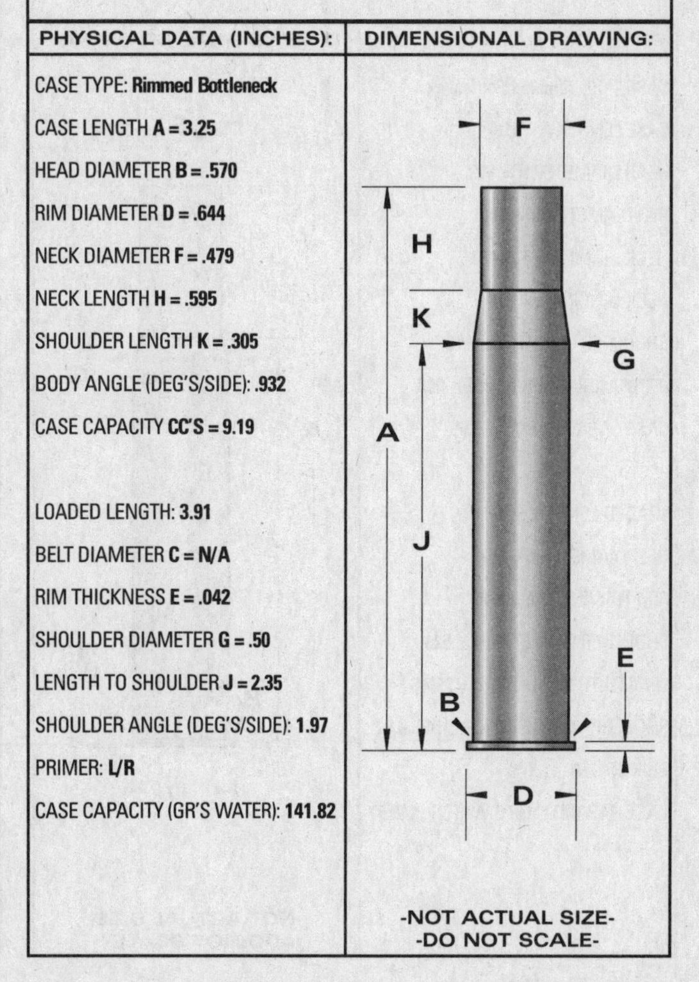

-NOT ACTUAL SIZE-
-DO NOT SCALE-

CARTRIDGE: .500/465 Nitro Express

OTHER NAMES:

DIA: .466

BALLISTEK NO: 466A

NAI NO: RMB 22221/5.654

DATA SOURCE: COTW 4th Pg243

HISTORICAL DATA: By Holland & Holland about 1907.

NOTES:

LOADING DATA:

BULLET WT./TYPE	POWDER WT./TYPE	VELOCITY ('/SEC)	SOURCE
480/—	75.0/3031	2100	Barnes

CASE PREPARATION: SHELLHOLDER (RCBS): Spl.

MAKE FROM: .470 N.E. (BELL). F/L size case and trim to length.

PHYSICAL DATA (INCHES):

CASE TYPE: **Rimmed Bottleneck**

CASE LENGTH A = **3.24**

HEAD DIAMETER B = **.573**

RIM DIAMETER D = **.650**

NECK DIAMETER F = **.488**

NECK LENGTH H = **.878**

SHOULDER LENGTH K = **.116**

BODY ANGLE (DEG'S/SIDE): **.686**

CASE CAPACITY CC'S = **9.65**

LOADED LENGTH: **3.89**

BELT DIAMETER C = **N/A**

RIM THICKNESS E = **.046**

SHOULDER DIAMETER G = **.524**

LENGTH TO SHOULDER J = **2.246**

SHOULDER ANGLE (DEG'S/SIDE): **8.82**

PRIMER: **L/R**

CASE CAPACITY (GR'S WATER): **148.97**

DIMENSIONAL DRAWING:

F H K G A J E B D

-NOT ACTUAL SIZE-
-DO NOT SCALE-

CARTRIDGE: .505 Barnes Supreme

OTHER NAMES:

DIA: .505

BALLISTEK NO: 505B

NAI NO: BES 21115/4.983

DATA SOURCE: Ackley Vol.2 Pg215

HISTORICAL DATA: By F. Barnes.

NOTES:

LOADING DATA:

BULLET WT./TYPE	POWDER WT./TYPE	VELOCITY ('/SEC)	SOURCE
600/FMJ	97.9/IMR4350	—	JJD

CASE PREPARATION: SHELLHOLDER (RCBS): 14

MAKE FROM: .460 Weatherby. Taper expand case neck to .510" dia. (you may need to anneal the case). F/L size in the Barnes die and trim to length.

PHYSICAL DATA (INCHES):

CASE TYPE: **Belted Straight**

CASE LENGTH A = **2.90**

HEAD DIAMETER B = **.583**

RIM DIAMETER D = **.582**

NECK DIAMETER F = **.528**

NECK LENGTH H = **N/A**

SHOULDER LENGTH K = **N/A**

BODY ANGLE (DEG'S/SIDE): **.545**

CASE CAPACITY CC'S = **9.05**

LOADED LENGTH: **4.00**

BELT DIAMETER C = **.603**

RIM THICKNESS E = **.062**

SHOULDER DIAMETER G = **N/A**

LENGTH TO SHOULDER J = **N/A**

SHOULDER ANGLE (DEG'S/SIDE): **N/A**

PRIMER: **L/R Mag**

CASE CAPACITY (GR'S WATER): **139.74**

DIMENSIONAL DRAWING:

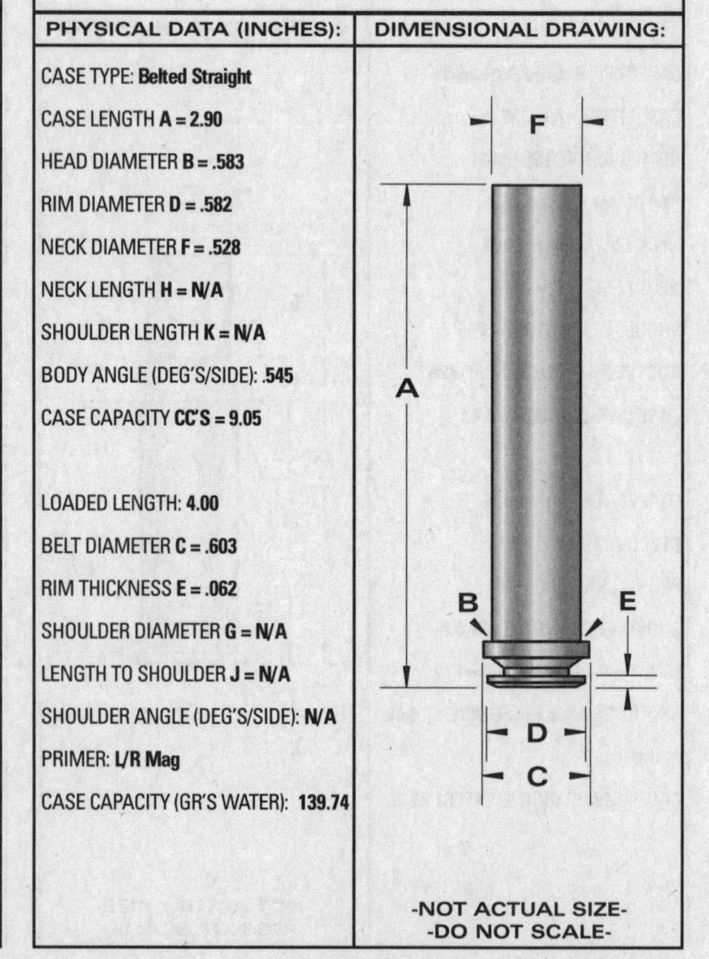

-NOT ACTUAL SIZE-
-DO NOT SCALE-

CARTRIDGE: .505 Gibbs Magnum

OTHER NAMES: .505 Rimless Magnum

DIA: .505

BALLISTEK NO: 505A

NAI NO: RXB 22232/4.961

DATA SOURCE: COTW 4th Pg247

HISTORICAL DATA: By Gibbs about 1910.

NOTES:

LOADING DATA:

BULLET WT./TYPE	POWDER WT./TYPE	VELOCITY ('/SEC)	SOURCE
525/RN	100.0/4064	2060	Barnes

CASE PREPARATION: SHELLHOLDER (RCBS): Spl.

MAKE FROM: There is no case that can be re-formed into the .505 Gibbs. It is possible to turn these cases from solid, 3/4" brass but, it's a lot of work. Anneal the formed case and F/L size. Trim to length.

PHYSICAL DATA (INCHES):

CASE TYPE: **Rimless Bottleneck**
CASE LENGTH **A = 3.15**
HEAD DIAMETER **B = .64**
RIM DIAMETER **D = .64**
NECK DIAMETER **F = .535**
NECK LENGTH **H = .66**
SHOULDER LENGTH **K = .03**
BODY ANGLE (DEG'S/SIDE): **.602**
CASE CAPACITY **CC'S = 11.67**

LOADED LENGTH: **3.75**
BELT DIAMETER **C = N/A**
RIM THICKNESS **E = .065**
SHOULDER DIAMETER **G = .600**
LENGTH TO SHOULDER **J = 2.46**
SHOULDER ANGLE (DEG'S/SIDE): **18.84**
PRIMER: **L/R Mag**
CASE CAPACITY (GR'S WATER): **180.17**

DIMENSIONAL DRAWING:

F H K G A J E B D

-NOT ACTUAL SIZE-
-DO NOT SCALE-

CARTRIDGE: .510 Wells

OTHER NAMES:

DIA: .510

BALLISTEK NO: 510B

NAI NO: BEN 12241/4.948

DATA SOURCE: NAI/Ballistek

HISTORICAL DATA:

NOTES:

LOADING DATA:

BULLET WT./TYPE	POWDER WT./TYPE	VELOCITY ('/SEC)	SOURCE
600/RN	118.0/IMR4320	–	JJD

CASE PREPARATION: SHELLHOLDER (RCBS): 14

MAKE FROM: .460 Weatherby. Taper expand case neck to .515" dia. Trim to length & chamfer. F/L size.

PHYSICAL DATA (INCHES):

CASE TYPE: **Belted Bottleneck**
CASE LENGTH **A = 2.88**
HEAD DIAMETER **B = .582**
RIM DIAMETER **D = .582**
NECK DIAMETER **F = .534**
NECK LENGTH **H = .430**
SHOULDER LENGTH **K = .11**
BODY ANGLE (DEG'S/SIDE): **.295**
CASE CAPACITY **CC'S = 9.55**

LOADED LENGTH: **3.644**
BELT DIAMETER **C = .603**
RIM THICKNESS **E = .062**
SHOULDER DIAMETER **G = .560**
LENGTH TO SHOULDER **J = 2.34**
SHOULDER ANGLE (DEG'S/SIDE): **6.74**
PRIMER: **L/R Mag**
CASE CAPACITY (GR'S WATER): **147.37**

DIMENSIONAL DRAWING:

F H K G A J E B D C

-NOT ACTUAL SIZE-
-DO NOT SCALE-

CARTRIDGE: .52-70 Sharps

OTHER NAMES:

DIA: .530

BALLISTEK NO: 530A

NAI NO: RMS 21115/2.788

DATA SOURCE: Hoyem Vol.2 Pg43

HISTORICAL DATA: Early Sharps ctg. for converted percussion rifles.

NOTES:

LOADING DATA:

BULLET WT./TYPE	POWDER WT./TYPE	VELOCITY ('/SEC)	SOURCE
325/Lead	19.2/IMR4198	—	JJD

CASE PREPARATION: SHELLHOLDER (RCBS): Spl.

MAKE FROM: .450 #2 N.E. (BELL). Cut case to 1.6". Thin rim to .063" thick. F/L size, trim to length & chamfer.

PHYSICAL DATA (INCHES):

CASE TYPE: **Rimmed Straight**

CASE LENGTH **A = 1.57**

HEAD DIAMETER **B = .563**

RIM DIAMETER **D = .657**

NECK DIAMETER **F = .531**

NECK LENGTH **H = N/A**

SHOULDER LENGTH **K = N/A**

BODY ANGLE (DEG'S/SIDE): **.608**

CASE CAPACITY **CC'S = 4.85**

LOADED LENGTH: **2.28**

BELT DIAMETER **C = N/A**

RIM THICKNESS **E = .063**

SHOULDER DIAMETER **G = N/A**

LENGTH TO SHOULDER **J = N/A**

SHOULDER ANGLE (DEG'S/SIDE): **N/A**

PRIMER: **L/R**

CASE CAPACITY (GR'S WATER): **74.86**

DIMENSIONAL DRAWING:

-NOT ACTUAL SIZE-
-DO NOT SCALE-

CARTRIDGE: .55-100 Maynard 1882

OTHER NAMES:

DIA: .551

BALLISTEK NO: 551A

NAI NO: RMS 11115/3.288

DATA SOURCE: COTW 4th Pg129

HISTORICAL DATA: By Maynard in 1882.

NOTES:

LOADING DATA:

BULLET WT./TYPE	POWDER WT./TYPE	VELOCITY ('/SEC)	SOURCE
530/Lead	22.0/IMR4198	—	JJD

CASE PREPARATION: SHELLHOLDER (RCBS): Spl.

MAKE FROM: 11mm Beaumont (BELL). Cut case to 2.0". Chamfer & F/L size. Trim to length. Rim will be quite a bit under diameter but enough to headspace.

PHYSICAL DATA (INCHES):

CASE TYPE: **Rimmed Straight**

CASE LENGTH **A = 1.94**

HEAD DIAMETER **B = .590**

RIM DIAMETER **D = .718**

NECK DIAMETER **F = .582**

NECK LENGTH **H = N/A**

SHOULDER LENGTH **K = N/A**

BODY ANGLE (DEG'S/SIDE): **.122**

CASE CAPACITY **CC'S = 6.13**

LOADED LENGTH: **2.56**

BELT DIAMETER **C = N/A**

RIM THICKNESS **E = .067**

SHOULDER DIAMETER **G = N/A**

LENGTH TO SHOULDER **J = N/A**

SHOULDER ANGLE (DEG'S/SIDE): **N/A**

PRIMER: **L/R**

CASE CAPACITY (GR'S WATER): **94.61**

DIMENSIONAL DRAWING:

-NOT ACTUAL SIZE-
-DO NOT SCALE-

CARTRIDGE: .56-56 Experimental

OTHER NAMES:

DIA: .520

BALLISTEK NO: 520D

NAI NO: RMS 21115/2.118

DATA SOURCE: Hoyem Vol.2 Pg44

HISTORICAL DATA: Experimental U.S. ctg from about 1865.

NOTES:

LOADING DATA:

BULLET WT./TYPE	POWDER WT./TYPE	VELOCITY ('/SEC)	SOURCE
300/Lead	11.0/IMR4227	—	JJD

CASE PREPARATION: SHELLHOLDER (RCBS): 5

MAKE FROM: .348 Winchester. Anneal case neck and taper expand to .530" dia. Cut case to 1.2". Thin rim to .047" thick. F/L size, trim to length & chamfer. Fireform in chamber.

PHYSICAL DATA (INCHES):

CASE TYPE: **Rimmed Straight**

CASE LENGTH **A = 1.18**

HEAD DIAMETER **B = .557**

RIM DIAMETER **D = .645**

NECK DIAMETER **F = .524**

NECK LENGTH **H = N/A**

SHOULDER LENGTH **K = N/A**

BODY ANGLE (DEG'S/SIDE): **.834**

CASE CAPACITY **CC'S = 3.22**

LOADED LENGTH: **1.59**

BELT DIAMETER **C = N/A**

RIM THICKNESS **E = .047**

SHOULDER DIAMETER **G = N/A**

LENGTH TO SHOULDER **J = N/A**

SHOULDER ANGLE (DEG'S/SIDE): **N/A**

PRIMER: **L/R**

CASE CAPACITY (GR'S WATER): **49.72**

DIMENSIONAL DRAWING:

F A B E D

-NOT ACTUAL SIZE-
-DO NOT SCALE-

CARTRIDGE: .577 Nitro Express 2.75"

OTHER NAMES:

DIA: .585

BALLISTEK NO: 585B

NAI NO: RMS 21115/4.166

DATA SOURCE: COTW 4th Pg248

HISTORICAL DATA: About 1880, first as black powder then smokeless.

NOTES:

LOADING DATA:

BULLET WT./TYPE	POWDER WT./TYPE	VELOCITY ('/SEC)	SOURCE
750/Lead	90.0/IMR3031	2000	JJD

CASE PREPARATION: SHELLHOLDER (RCBS): 577

MAKE FROM: .577 Basic (BELL). F/L size case & trim to length. Chamfer.

PHYSICAL DATA (INCHES):

CASE TYPE: **Rimmed Straight**

CASE LENGTH **A = 2.75**

HEAD DIAMETER **B = .660**

RIM DIAMETER **D = .748**

NECK DIAMETER **F = .608**

NECK LENGTH **H = N/A**

SHOULDER LENGTH **K = N/A**

BODY ANGLE (DEG'S/SIDE): **.556**

CASE CAPACITY **CC'S = 11.32**

LOADED LENGTH: **3.70**

BELT DIAMETER **C = N/A**

RIM THICKNESS **E = .072**

SHOULDER DIAMETER **G = N/A**

LENGTH TO SHOULDER **J = N/A**

SHOULDER ANGLE (DEG'S/SIDE): **N/A**

PRIMER: **L/R**

CASE CAPACITY (GR'S WATER): **174.69**

DIMENSIONAL DRAWING:

F A B E D

-NOT ACTUAL SIZE-
-DO NOT SCALE-

CARTRIDGE: .577 Nitro Express 3″

OTHER NAMES:

DIA: .585

BALLISTEK NO: 585A

NAI NO: RMS 21115/4.545

DATA SOURCE: COTW 4th Pg248

HISTORICAL DATA: Longer version of the 2.75″ case.

NOTES:

LOADING DATA:

BULLET WT./TYPE	POWDER WT./TYPE	VELOCITY ('/SEC)	SOURCE
750/Lead	100.0/3031	2000	Barnes

CASE PREPARATION: SHELLHOLDER (RCBS): 577

MAKE FROM: .577 N.E. (BELL). F/L size case & square case mouth.

PHYSICAL DATA (INCHES):

CASE TYPE: **Rimmed Straight**
CASE LENGTH **A = 3.00**
HEAD DIAMETER **B = .660**
RIM DIAMETER **D = .748**
NECK DIAMETER **F = .608**
NECK LENGTH **H = N/A**
SHOULDER LENGTH **K = N/A**
BODY ANGLE (DEG'S/SIDE): **.508**
CASE CAPACITY **CC'S = 12.48**
LOADED LENGTH: **4.00**
BELT DIAMETER **C = N/A**
RIM THICKNESS **E = .072**
SHOULDER DIAMETER **G = N/A**
LENGTH TO SHOULDER **J = N/A**
SHOULDER ANGLE (DEG'S/SIDE): **N/A**
PRIMER: **L/R**
CASE CAPACITY (GR'S WATER): **192.54**

DIMENSIONAL DRAWING:

-NOT ACTUAL SIZE-
-DO NOT SCALE-

CARTRIDGE: .577 REWA

OTHER NAMES:
.577/600 Nitro Express

DIA: .585

BALLISTEK NO: 585C

NAI NO: RMB 23221/4.304

DATA SOURCE: NAI/Ballistek

HISTORICAL DATA: Special ctg. for the Maharajah of Rewa.

NOTES: Ammo loaded by Kynoch.

LOADING DATA:

BULLET WT./TYPE	POWDER WT./TYPE	VELOCITY ('/SEC)	SOURCE
750/LeadPP	80.0/IMR4320	–	JJD

CASE PREPARATION: SHELLHOLDER (RCBS): Spl.

MAKE FROM: Turn .600 Nitro Express (#622A) case from solid brass. Anneal case neck and F/L size in REWA die. Trim to length & chamfer.

PHYSICAL DATA (INCHES):

CASE TYPE: **Rimmed Bottleneck**
CASE LENGTH **A = 3.000**
HEAD DIAMETER **B = .697**
RIM DIAMETER **D = .805**
NECK DIAMETER **F = .613**
NECK LENGTH **H = .75**
SHOULDER LENGTH **K = .290**
BODY ANGLE (DEG'S/SIDE): **.358**
CASE CAPACITY **CC'S = 13.88**
LOADED LENGTH: **3.70**
BELT DIAMETER **C = N/A**
RIM THICKNESS **E = .073**
SHOULDER DIAMETER **G = .675**
LENGTH TO SHOULDER **J = 1.96**
SHOULDER ANGLE (DEG'S/SIDE): **6.10**
PRIMER: **L/R Mag**
CASE CAPACITY (GR'S WATER): **214.30**

DIMENSIONAL DRAWING:

-NOT ACTUAL SIZE-
-DO NOT SCALE-

CARTRIDGE: .577 Snider

OTHER NAMES:

DIA: .570

BALLISTEK NO: 570A

NAI NO: RMS 21115/3.030

DATA SOURCE: COTW 4th Pg218

HISTORICAL DATA: British mil. ctg. ca. 1867.

NOTES:

LOADING DATA:

BULLET WT./TYPE	POWDER WT./TYPE	VELOCITY ('/SEC)	SOURCE
450/Lead	30.0/4198	1300	Barnes

CASE PREPARATION: SHELLHOLDER (RCBS): 577

MAKE FROM: .577 N.E. (BELL). Trim case to length & chamfer. Anneal. F/L size. Lyman bullet #575213 works very well.

PHYSICAL DATA (INCHES):

CASE TYPE: **Rimmed Straight**

CASE LENGTH A = **2.00**

HEAD DIAMETER B = **.660**

RIM DIAMETER D = **.747**

NECK DIAMETER F = **.602**

NECK LENGTH H = **N/A**

SHOULDER LENGTH K = **N/A**

BODY ANGLE (DEG'S/SIDE): **.858**

CASE CAPACITY CC'S = **7.42**

LOADED LENGTH: **2.45**

BELT DIAMETER C = **N/A**

RIM THICKNESS E = **.065**

SHOULDER DIAMETER G = **N/A**

LENGTH TO SHOULDER J = **N/A**

SHOULDER ANGLE (DEG'S/SIDE): **N/A**

PRIMER: **L/R**

CASE CAPACITY (GR'S WATER): **114.55**

DIMENSIONAL DRAWING:

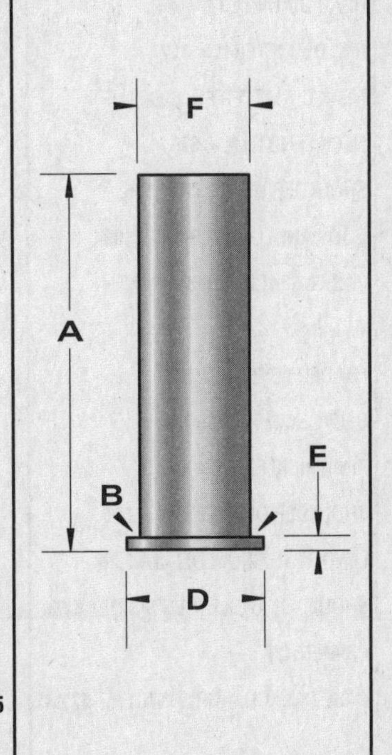

-NOT ACTUAL SIZE-
-DO NOT SCALE-

CARTRIDGE: .577/450 Martini Henry

OTHER NAMES: 11,43 x 60R (61R)

DIA: .455

BALLISTEK NO: 455A

NAI NO: RMB 24323/3.503

DATA SOURCE: COTW 4th Pg214

HISTORICAL DATA: British mil. ctg. ca. 1871.

NOTES: Early cases were of rolled brass foil with an iron base.

LOADING DATA:

BULLET WT./TYPE	POWDER WT./TYPE	VELOCITY ('/SEC)	SOURCE
400/LeadPP	Full case/Ctg. Pyrodex	1200	JJD
400/Lead	38.0/4198	1450	Barnes

CASE PREPARATION: SHELLHOLDER (RCBS): 577

MAKE FROM: .577 N.E. (BELL). Form die set is a must! Anneal case. Form in form dies #1 & #2 and in trim die. Trim to length & chamfer. F/L size. Slug the bore for bullet diameter; bores vary .452-.458" dia. 21/32" dia. tubing also works very well but, use only black powder or Pyrodex loads.

PHYSICAL DATA (INCHES):

CASE TYPE: **Rimmed Bottleneck**

CASE LENGTH A = **2.34**

HEAD DIAMETER B = **.668**

RIM DIAMETER D = **.746**

NECK DIAMETER F = **.487**

NECK LENGTH H = **.780**

SHOULDER LENGTH K = **.185**

BODY ANGLE (DEG'S/SIDE): **.975**

CASE CAPACITY CC'S = **7.98**

LOADED LENGTH: **3.12**

BELT DIAMETER C = **N/A**

RIM THICKNESS E = **.06**

SHOULDER DIAMETER G = **.628**

LENGTH TO SHOULDER J = **1.375**

SHOULDER ANGLE (DEG'S/SIDE): **20.86**

PRIMER: **L/R**

CASE CAPACITY (GR'S WATER): **123.20**

DIMENSIONAL DRAWING:

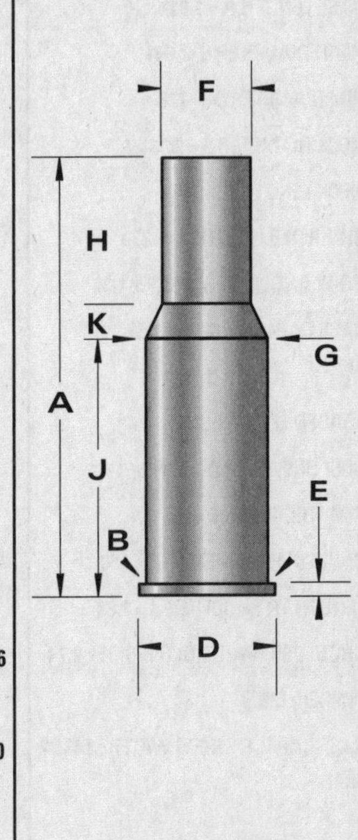

-NOT ACTUAL SIZE-
-DO NOT SCALE-

CARTRIDGE: .577/500 #2 Nitro Express

OTHER NAMES:
.500 #2 Express

DIA: .507

BALLISTEK NO: 507A

NAI NO: RMB 33225/4.384

DATA SOURCE: COTW 4th Pg245

HISTORICAL DATA: Black powder ctg. ca. 1883.

NOTES:

LOADING DATA:

BULLET WT./TYPE	POWDER WT./TYPE	VELOCITY ('/SEC)	SOURCE
360/LeadPP	37.0/4198	1700	Barnes

CASE PREPARATION: SHELLHOLDER (RCBS): 577

MAKE FROM: .577 N.E. (BELL). Trim case to length and anneal case neck. Build rim thickness up to .07". Turn rim to .726" dia. F/L size to form. Trim to length & chamfer.

PHYSICAL DATA (INCHES):

CASE TYPE: **Rimmed Bottleneck**

CASE LENGTH **A = 2.81**

HEAD DIAMETER **B = .641**

RIM DIAMETER **D = .726**

NECK DIAMETER **F = .538**

NECK LENGTH **H = .70**

SHOULDER LENGTH **K = .230**

BODY ANGLE (DEG'S/SIDE): **1.38**

CASE CAPACITY **CC'S = 9.58**

LOADED LENGTH: **3.40**

BELT DIAMETER **C = N/A**

RIM THICKNESS **E = .07**

SHOULDER DIAMETER **G = .560**

LENGTH TO SHOULDER **J = 1.88**

SHOULDER ANGLE (DEG'S/SIDE): **2.74**

PRIMER: **L/R**

CASE CAPACITY (GR'S WATER): **147.82**

DIMENSIONAL DRAWING:

-NOT ACTUAL SIZE-
-DO NOT SCALE-

CARTRIDGE: .577/500 Magnum Nitro Express

OTHER NAMES:

DIA: .500

BALLISTEK NO: 500A

NAI NO: RMB 22221/4.853

DATA SOURCE: COTW 4th Pg245

HISTORICAL DATA: Circa 1905.

NOTES:

LOADING DATA:

BULLET WT./TYPE	POWDER WT./TYPE	VELOCITY ('/SEC)	SOURCE
465/—	40.0/4198	1500	Barnes

CASE PREPARATION: SHELLHOLDER (RCBS): 577

MAKE FROM: .577 N.E. (BELL). Anneal case and form in form die. Square case mouth and F/L size. Turn rim to .717" dia. Lathe turn case base to .645" dia.

PHYSICAL DATA (INCHES):

CASE TYPE: **Rimmed Bottleneck**

CASE LENGTH **A = 3.13**

HEAD DIAMETER **B = .645**

RIM DIAMETER **D = .717**

NECK DIAMETER **F = .526**

NECK LENGTH **H = .675**

SHOULDER LENGTH **K = .195**

BODY ANGLE (DEG'S/SIDE): **.834**

CASE CAPACITY **CC'S = 11.49**

LOADED LENGTH: **3.74**

BELT DIAMETER **C = N/A**

RIM THICKNESS **E = .063**

SHOULDER DIAMETER **G = .585**

LENGTH TO SHOULDER **J = 2.26**

SHOULDER ANGLE (DEG'S/SIDE): **8.60**

PRIMER: **L/R**

CASE CAPACITY (GR'S WATER): **177.32**

DIMENSIONAL DRAWING:

-NOT ACTUAL SIZE-
-DO NOT SCALE-

CARTRIDGE: .58 Berdan

OTHER NAMES:

DIA: .579

BALLISTEK NO: 579A

NAI NO: RMS 21115/2.613

DATA SOURCE: Hoyem Vol.2 Pg41

HISTORICAL DATA: U.S. Berdan musket ctg. from about 1870.

NOTES: Several versions exist.

LOADING DATA:

BULLET WT./TYPE	POWDER WT./TYPE	VELOCITY ('/SEC)	SOURCE
530/Lead	24.5/IMR4198	–	JJD

CASE PREPARATION: SHELLHOLDER (RCBS): Spl.

MAKE FROM: .577 N.E. (BELL). Cut case to 1.7". Turn rim down to .71" dia. F/L size, trim to length. Turn base dia. to .643 after sizing.

PHYSICAL DATA (INCHES):

CASE TYPE: **Rimmed Straight**

CASE LENGTH **A = 1.68**

HEAD DIAMETER **B = .643**

RIM DIAMETER **D = .710**

NECK DIAMETER **F = .600**

NECK LENGTH **H = N/A**

SHOULDER LENGTH **K = N/A**

BODY ANGLE (DEG'S/SIDE): **.76**

CASE CAPACITY **CC'S = 6.17**

LOADED LENGTH: **2.15**

BELT DIAMETER **C = N/A**

RIM THICKNESS **E = .061**

SHOULDER DIAMETER **G = N/A**

LENGTH TO SHOULDER **J = N/A**

SHOULDER ANGLE (DEG'S/SIDE): **N/A**

PRIMER: **L/R**

CASE CAPACITY (GR'S WATER): **95.18**

DIMENSIONAL DRAWING:

F
A
B
E
D

-NOT ACTUAL SIZE-
-DO NOT SCALE-

CARTRIDGE: .58 Gatling 1875

OTHER NAMES:

DIA: .586

BALLISTEK NO: 586A

NAI NO: RMS 11115/2.206

DATA SOURCE: Hoyem Vol.2 Pg73

HISTORICAL DATA: Experimental U.S. MG ctg from about 1875.

NOTES:

LOADING DATA:

BULLET WT./TYPE	POWDER WT./TYPE	VELOCITY ('/SEC)	SOURCE
700/Lead	15.0/IMR4227	–	JJD

CASE PREPARATION: SHELLHOLDER (RCBS): Spl.

MAKE FROM: Solid brass. Turn from 3/4" dia. brass rod. Anneal case and F/L size. Trim to length & chamfer.

PHYSICAL DATA (INCHES):

CASE TYPE: **Rimmed Straight**

CASE LENGTH **A = 1.35**

HEAD DIAMETER **B = .612**

RIM DIAMETER **D = .709**

NECK DIAMETER **F = .598**

NECK LENGTH **H = N/A**

SHOULDER LENGTH **K = N/A**

BODY ANGLE (DEG'S/SIDE): **.315**

CASE CAPACITY **CC'S = 4.64**

LOADED LENGTH: **2.09**

BELT DIAMETER **C = N/A**

RIM THICKNESS **E = .08**

SHOULDER DIAMETER **G = N/A**

LENGTH TO SHOULDER **J = N/A**

SHOULDER ANGLE (DEG'S/SIDE): **N/A**

PRIMER: **L/R**

CASE CAPACITY (GR'S WATER): **71.62**

DIMENSIONAL DRAWING:

F
A
B
E
D

-NOT ACTUAL SIZE-
-DO NOT SCALE-

CARTRIDGE: .58 Remington Carbine

OTHER NAMES:	DIA: .610 BALLISTEK NO: 610A NAI NO: RMS 11115/1.905

DATA SOURCE: Hoyem Vol.2 Pg40

HISTORICAL DATA: Circa 1870.

NOTES: Several variations exist.

LOADING DATA:

BULLET WT./TYPE	POWDER WT./TYPE	VELOCITY ('/SEC)	SOURCE
900/Lead	20.0/IMR4198	–	JJD

CASE PREPARATION: SHELLHOLDER (RCBS): Spl.

MAKE FROM: Best method is to use 5/8" dia. tubing on a .577 N.E. case head. Fabricate case, anneal, F/L size, trim to length & chamfer.

PHYSICAL DATA (INCHES):

CASE TYPE: **Rimmed Straight**

CASE LENGTH **A = 1.20**

HEAD DIAMETER **B = .630**

RIM DIAMETER **D = .745**

NECK DIAMETER **F = .622**

NECK LENGTH **H = N/A**

SHOULDER LENGTH **K = N/A**

BODY ANGLE (DEG'S/SIDE): **.199**

CASE CAPACITY **CC'S = 4.26**

LOADED LENGTH: **1.73**

BELT DIAMETER **C = N/A**

RIM THICKNESS **E = .052**

SHOULDER DIAMETER **G = N/A**

LENGTH TO SHOULDER **J = N/A**

SHOULDER ANGLE (DEG'S/SIDE): **N/A**

PRIMER: **L/R**

CASE CAPACITY (GR'S WATER): **65.83**

DIMENSIONAL DRAWING:

F
A
B
E
D

-NOT ACTUAL SIZE-
-DO NOT SCALE-

CARTRIDGE: .58 Roberts

OTHER NAMES:	DIA: .612 BALLISTEK NO: 612A NAI NO: RMS 21115/2.088

DATA SOURCE: Hoyem Vol.2 Pg42

HISTORICAL DATA: Circa 1870 for Roberts conversion of percussion rifle.

NOTES:

LOADING DATA:

BULLET WT./TYPE	POWDER WT./TYPE	VELOCITY ('/SEC)	SOURCE
900/Lead	10.0/IMR4227	–	JJD

CASE PREPARATION: SHELLHOLDER (RCBS): 577

MAKE FROM: .577 N.E. (BELL). Cut case to length. F/L size & chamfer.

PHYSICAL DATA (INCHES):

CASE TYPE: **Rimmed Straight**

CASE LENGTH **A = 1.37**

HEAD DIAMETER **B = .656**

RIM DIAMETER **D = .741**

NECK DIAMETER **F = .613**

NECK LENGTH **H = N/A**

SHOULDER LENGTH **K = N/A**

BODY ANGLE (DEG'S/SIDE): **.943**

CASE CAPACITY **CC'S = 5.50**

LOADED LENGTH: **1.90**

BELT DIAMETER **C = N/A**

RIM THICKNESS **E = .064**

SHOULDER DIAMETER **G = N/A**

LENGTH TO SHOULDER **J = N/A**

SHOULDER ANGLE (DEG'S/SIDE): **N/A**

PRIMER: **L/R**

CASE CAPACITY (GR'S WATER): **84.92**

DIMENSIONAL DRAWING:

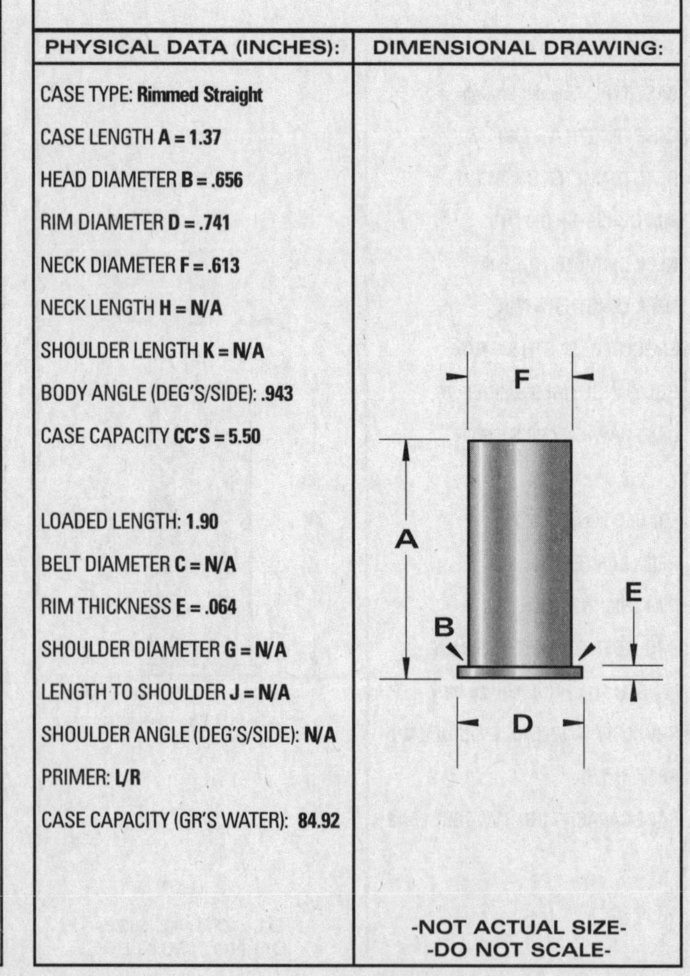

-NOT ACTUAL SIZE-
-DO NOT SCALE-

CARTRIDGE: .58 U.S. Musket

OTHER NAMES:

DIA: .605

BALLISTEK NO: 605A

NAI NO: RMS 21115/1.858

DATA SOURCE: Hoyem Vol.2 Pg40

HISTORICAL DATA: Ctg. for U.S. Musket converted to C/F.

NOTES:

LOADING DATA:

BULLET WT./TYPE	POWDER WT./TYPE	VELOCITY ('/SEC)	SOURCE
700/Lead	10.0/IMR4227	—	JJD

CASE PREPARATION: SHELLHOLDER (RCBS): 577

MAKE FROM: .577 N.E. (BELL). Cut case to 1.3". F/L size. Turn base to .651" dia. Trim to length & chamfer.

PHYSICAL DATA (INCHES):

CASE TYPE: **Rimmed Straight**

CASE LENGTH **A = 1.21**

HEAD DIAMETER **B = .651**

RIM DIAMETER **D = .743**

NECK DIAMETER **F = .613**

NECK LENGTH **H = N/A**

SHOULDER LENGTH **K = N/A**

BODY ANGLE (DEG'S/SIDE): **.952**

CASE CAPACITY **CC'S = 4.48**

LOADED LENGTH: **1.74**

BELT DIAMETER **C = N/A**

RIM THICKNESS **E = .067**

SHOULDER DIAMETER **G = N/A**

LENGTH TO SHOULDER **J = N/A**

SHOULDER ANGLE (DEG'S/SIDE): **N/A**

PRIMER: **L/R**

CASE CAPACITY (GR'S WATER): **69.23**

DIMENSIONAL DRAWING:

F
A
E
B
D

-NOT ACTUAL SIZE-
-DO NOT SCALE-

CARTRIDGE: .600 Nitro Express

OTHER NAMES:

DIA: .621

BALLISTEK NO: 622A

NAI NO: RMS 21115/4.276

DATA SOURCE: NAI/Ballistek

HISTORICAL DATA: Largest British sporting ctg. by Jeffery.

NOTES:

LOADING DATA:

BULLET WT./TYPE	POWDER WT./TYPE	VELOCITY ('/SEC)	SOURCE
900/RN	98.0/IMR3031	—	JJD

CASE PREPARATION: SHELLHOLDER (RCBS): Spl.

MAKE FROM: Turn case from solid 7/8" dia. brass rod. Trim to proper length, chamfer and anneal case. F/L size.

PHYSICAL DATA (INCHES):

CASE TYPE: **Rimmed Straight**

CASE LENGTH **A = 2.985**

HEAD DIAMETER **B = .698**

RIM DIAMETER **D = .802**

NECK DIAMETER **F = .649**

NECK LENGTH **H = N/A**

SHOULDER LENGTH **K = N/A**

BODY ANGLE (DEG'S/SIDE): **.479**

CASE CAPACITY **CC'S = 13.77**

LOADED LENGTH: **3.596**

BELT DIAMETER **C = N/A**

RIM THICKNESS **E = .06**

SHOULDER DIAMETER **G = N/A**

LENGTH TO SHOULDER **J = N/A**

SHOULDER ANGLE (DEG'S/SIDE): **N/A**

PRIMER: **L/R Mag**

CASE CAPACITY (GR'S WATER): **212.45**

DIMENSIONAL DRAWING:

F
A
E
B
D

-NOT ACTUAL SIZE-
-DO NOT SCALE-

CARTRIDGE: .70-150 Winchester

OTHER NAMES:

DIA: .705

BALLISTEK NO: 705A

NAI NO: RMB 24241/2.945

DATA SOURCE: NAI/Ballistek

HISTORICAL DATA: Likely a gimmick by Winchester. No commercial production.

NOTES:

LOADING DATA:

BULLET WT./TYPE	POWDER WT./TYPE	VELOCITY ('/SEC)	SOURCE
900/Lead	95.0/IMR3031	—	JJD

CASE PREPARATION: **SHELLHOLDER (RCBS):** Spl.

MAKE FROM: Cases can be made from brass, 12 ga. shotgun shells or turned from solid 1" brass rod. Either way, a form die set is required. Anneal the case and form in form set (three dies). Trim to length and F/L size.

PHYSICAL DATA (INCHES):

CASE TYPE: **Rimmed Bottleneck**

CASE LENGTH **A = 2.38**

HEAD DIAMETER **B = .808**

RIM DIAMETER **D = .872**

NECK DIAMETER **F = .735**

NECK LENGTH **H = .425**

SHOULDER LENGTH **K = .255**

BODY ANGLE (DEG'S/SIDE): **.439**

CASE CAPACITY **CC'S = 15.06**

LOADED LENGTH: **2.78**

BELT DIAMETER **C = N/A**

RIM THICKNESS **E = .058**

SHOULDER DIAMETER **G = .785**

LENGTH TO SHOULDER **J = 1.70**

SHOULDER ANGLE (DEG'S/SIDE): **5.59**

PRIMER: **L/R**

CASE CAPACITY (GR'S WATER): **232.45**

DIMENSIONAL DRAWING:

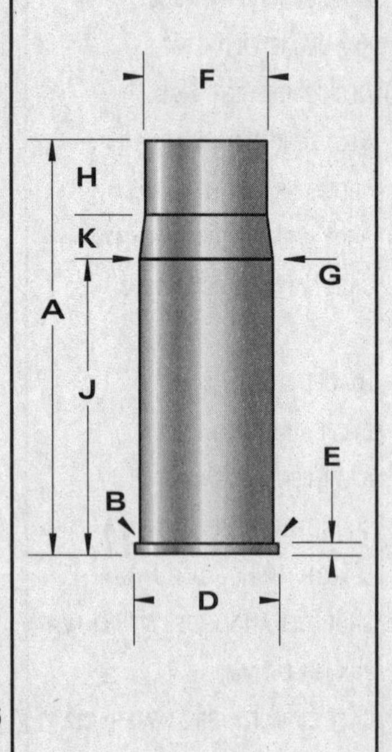

-NOT ACTUAL SIZE-
-DO NOT SCALE-

Metric Cartridges

CARTRIDGE: 4,85 U.K.

OTHER NAMES:
4,85 British

DIA: .191

BALLISTEK NO: 191A

NAI NO: RXB
12422/5.329

DATA SOURCE: MSAA Pg21

HISTORICAL DATA: Experimental British ctg.

NOTES:

LOADING DATA:

BULLET WT./TYPE	POWDER WT./TYPE	VELOCITY ('/SEC)	SOURCE
55/Spire	Original Load	2740	MSAA

CASE PREPARATION: SHELLHOLDER (RCBS): 23

MAKE FROM: .222 Rem. Mag. Anneal case neck. F/L size in 4,85 UK sizer. Square case mouth. Case will be about .08" short but OK. Base will fireform in chamber.

PHYSICAL DATA (INCHES):

CASE TYPE: **Rimless Bottleneck**
CASE LENGTH **A = 1.94**
HEAD DIAMETER **B = .364**
RIM DIAMETER **D = .368**
NECK DIAMETER **F = .225**
NECK LENGTH **H = .340**
SHOULDER LENGTH **K = .18**
BODY ANGLE (DEG'S/SIDE): **.211**
CASE CAPACITY **CC'S = 1.75**

LOADED LENGTH: **2.46**
BELT DIAMETER **C = N/A**
RIM THICKNESS **E = .045**
SHOULDER DIAMETER **G = .355**
LENGTH TO SHOULDER **J = 1.42**
SHOULDER ANGLE (DEG'S/SIDE): **19.85**
PRIMER: **S/R**
CASE CAPACITY (GR'S WATER): **27.05**

DIMENSIONAL DRAWING:

F
H
K
G
A
J
E
B
D

-NOT ACTUAL SIZE-
-DO NOT SCALE-

CARTRIDGE: 5mm Bergmann

OTHER NAMES:

DIA: .203

BALLISTEK NO: 203A

NAI NO: RXS
31115/2.161

DATA SOURCE: COTW 4th Pg160

HISTORICAL DATA: Circa 1894.

NOTES:

LOADING DATA:

BULLET WT./TYPE	POWDER WT./TYPE	VELOCITY ('/SEC)	SOURCE
35/FMJ	Factory Load	600	COTW

CASE PREPARATION: SHELLHOLDER (RCBS): 29

MAKE FROM: .25 ACP. Anneal case. F/L size in Bergmann die. Trim to length. Chamfer. Turn rim to .274" dia & back chamfer.

PHYSICAL DATA (INCHES):

CASE TYPE: **Rimless Straight**
CASE LENGTH **A = .590**
HEAD DIAMETER **B = .273**
RIM DIAMETER **D = .274**
NECK DIAMETER **F = .230**
NECK LENGTH **H = N/A**
SHOULDER LENGTH **K = N/A**
BODY ANGLE (DEG'S/SIDE):
CASE CAPACITY **CC'S = .16**

LOADED LENGTH: **.96**
BELT DIAMETER **C = N/A**
RIM THICKNESS **E = .04**
SHOULDER DIAMETER **G = N/A**
LENGTH TO SHOULDER **J = N/A**
SHOULDER ANGLE (DEG'S/SIDE): **N/A**
PRIMER: **S/P**
CASE CAPACITY (GR'S WATER): **2.47**

DIMENSIONAL DRAWING:

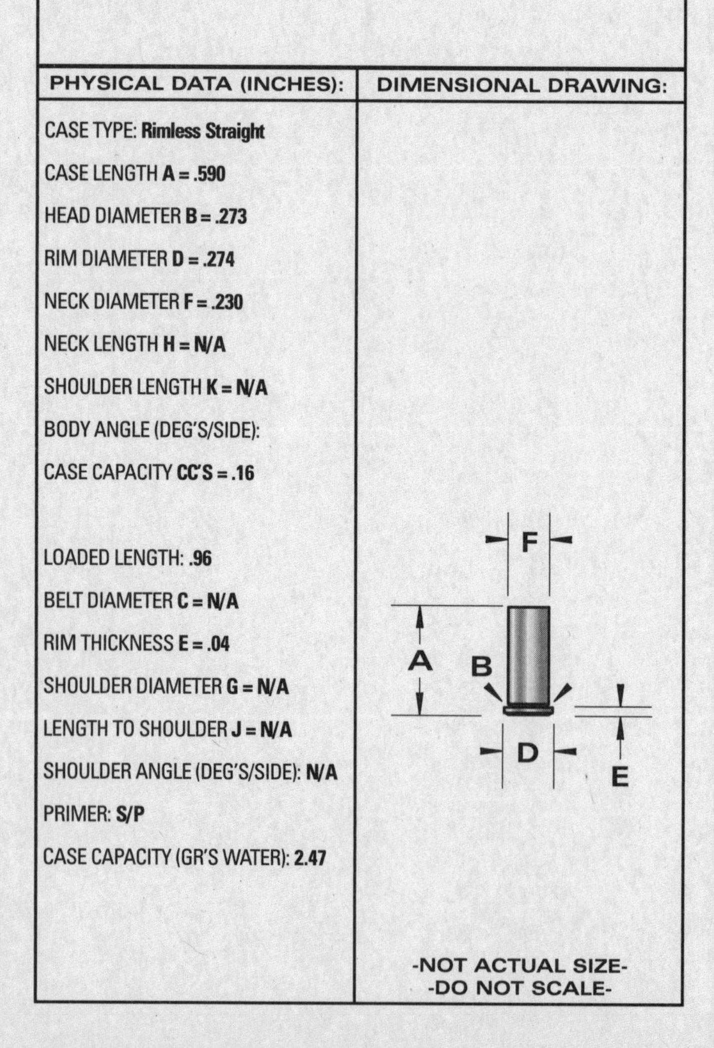

-NOT ACTUAL SIZE-
-DO NOT SCALE-

CARTRIDGE: 5mm Clement Auto

OTHER NAMES:

DIA: .202

BALLISTEK NO: 202A

NAI NO: RXB 14234/2.535

DATA SOURCE: COTW 4th Pg160

HISTORICAL DATA:

NOTES:

LOADING DATA:

BULLET WT./TYPE	POWDER WT./TYPE	VELOCITY ('/SEC)	SOURCE
36/FMJ		1030	Barnes

CASE PREPARATION: SHELLHOLDER (RCBS): 29

MAKE FROM: .25 ACP. Turn rim to .281" dia. Size .25 ACP case in 5mm F/L die. Square case mouth. Neck will be about .1" short but will headspace Okay. Case will fireform in chamber.

PHYSICAL DATA (INCHES):

CASE TYPE: **Rimless Bottleneck**

CASE LENGTH **A = .710**

HEAD DIAMETER **B = .280**

RIM DIAMETER **D = .281**

NECK DIAMETER **F = .223**

NECK LENGTH **H = .208**

SHOULDER LENGTH **K = .032**

BODY ANGLE (DEG'S/SIDE): **.318**

CASE CAPACITY **CC'S = .305**

LOADED LENGTH: **1.01**

BELT DIAMETER **C = N/A**

RIM THICKNESS **E = .04**

SHOULDER DIAMETER **G = .277**

LENGTH TO SHOULDER **J = .47**

SHOULDER ANGLE (DEG'S/SIDE): **40.14**

PRIMER: **S/P**

CASE CAPACITY (GR'S WATER): **4.71**

DIMENSIONAL DRAWING:

F H K A J B G D E

-NOT ACTUAL SIZE-
-DO NOT SCALE-

CARTRIDGE: 5mm/223 Remington

OTHER NAMES:

DIA: .204

BALLISTEK NO: 204B

NAI NO: RXB 22444/4.667

DATA SOURCE: Ackley Vol.2 Pg115

HISTORICAL DATA:

NOTES:

LOADING DATA:

BULLET WT./TYPE	POWDER WT./TYPE	VELOCITY ('/SEC)	SOURCE
30/—	20.0/4198	3410	Ackley

CASE PREPARATION: SHELLHOLDER (RCBS): 10

MAKE FROM: .223 Rem. Neck case down in 5mm/223 trim die. chamfer (to excess on I.D. to aid bullet seating). Use 5mm bullets pulled from 5mm Rimfire ammo. (Drill small [.050-.060" dia.] hole in rimfire case. Cut in half with jewelers saw. Very carefully, knock out bullet [from inside out]. Seat bullet with .222 Rem. seater die.)

PHYSICAL DATA (INCHES):

CASE TYPE: **Rimless Bottleneck**

CASE LENGTH **A = 1.755**

HEAD DIAMETER **B = .376**

RIM DIAMETER **D = .378**

NECK DIAMETER **F = .221**

NECK LENGTH **H = .17**

SHOULDER LENGTH **K = .085**

BODY ANGLE (DEG'S/SIDE): **.793**

CASE CAPACITY **CC'S = 1.95**

LOADED LENGTH: **2.29**

BELT DIAMETER **C = N/A**

RIM THICKNESS **E = .045**

SHOULDER DIAMETER **G = .340**

LENGTH TO SHOULDER **J = 1.50**

SHOULDER ANGLE (DEG'S/SIDE): **34.99**

PRIMER: **S/R**

CASE CAPACITY (GR'S WATER): **30.13**

DIMENSIONAL DRAWING:

F H K G A J E B D

-NOT ACTUAL SIZE-
-DO NOT SCALE-

CARTRIDGE: 5,5mm Velo Dog Revolver

OTHER NAMES:

DIA: .225

BALLISTEK NO: 225C

NAI NO: RMS 11115/4.427

DATA SOURCE: COTW 4th Pg160

HISTORICAL DATA: Circa 1894.

NOTES: Similar to .22 Maynard XL.

LOADING DATA:

BULLET WT./TYPE	POWDER WT./TYPE	VELOCITY ('/SEC)	SOURCE
45/—	4.7/Unique	650	Barnes

CASE PREPARATION: SHELLHOLDER (RCBS): Spl.

MAKE FROM: There is nothing that will form to this case. Cases can be turned from solid brass, annealed, trimmed to length and F/L sized. Also, we have made cases by using 1/4" tubing to re-body a .25 ACP case head.

PHYSICAL DATA (INCHES):

CASE TYPE: **Rimmed Straight**
CASE LENGTH **A = 1.12**
HEAD DIAMETER **B = .253**
RIM DIAMETER **D = .308**
NECK DIAMETER **F = .248**
NECK LENGTH **H = N/A**
SHOULDER LENGTH **K = N/A**
BODY ANGLE (DEG'S/SIDE): **.14**
CASE CAPACITY **CC'S = .487**
LOADED LENGTH: **1.35**
BELT DIAMETER **C = N/A**
RIM THICKNESS **E = .07**
SHOULDER DIAMETER **G = N/A**
LENGTH TO SHOULDER **J = N/A**
SHOULDER ANGLE (DEG'S/SIDE): **N/A**
PRIMER: **S/P**
CASE CAPACITY (GR'S WATER): **7.52**

DIMENSIONAL DRAWING:

-NOT ACTUAL SIZE-
-DO NOT SCALE-

CARTRIDGE: 5,6 x 33 Rook

OTHER NAMES:

DIA: .222

BALLISTEK NO: 222C

NAI NO: RXB 13342/4.030

DATA SOURCE: COTW 4th Pg254

HISTORICAL DATA: Ca. 1900.

NOTES: Similar to 5,7mm Johnson.

LOADING DATA:

BULLET WT./TYPE	POWDER WT./TYPE	VELOCITY ('/SEC)	SOURCE
60/RN	5.0/2400	1600	Barnes

CASE PREPARATION: SHELLHOLDER (RCBS): 17

MAKE FROM: .25-20 Win. Turn rim to .327" dia. and cut new extractor groove. F/L size which will swage head. Square case mouth & chamfer. Case is slightly short but, otherwise, fine. (Note: if swaging throws up a large "lump," just forward of the extractor groove, turn off with lathe.)

PHYSICAL DATA (INCHES):

CASE TYPE: **Rimless Bottleneck**
CASE LENGTH **A = 1.31**
HEAD DIAMETER **B = .325**
RIM DIAMETER **D = .327**
NECK DIAMETER **F = .247**
NECK LENGTH **H = .20**
SHOULDER LENGTH **K = .11**
BODY ANGLE (DEG'S/SIDE): **.250**
CASE CAPACITY **CC'S = 1.00**
LOADED LENGTH: **1.62**
BELT DIAMETER **C = N/A**
RIM THICKNESS **E = .062**
SHOULDER DIAMETER **G = .318**
LENGTH TO SHOULDER **J = 1.00**
SHOULDER ANGLE (DEG'S/SIDE): **17.88**
PRIMER: **S/R**
CASE CAPACITY (GR'S WATER): **15.48**

DIMENSIONAL DRAWING:

-NOT ACTUAL SIZE-
-DO NOT SCALE-

CARTRIDGE: 5,6 x 33R Rook

OTHER NAMES:

DIA: .222

BALLISTEK NO: 222B

NAI NO: RMB 13342/4.033

DATA SOURCE: COTW 4th Pg254

HISTORICAL DATA: Ca. 1900. Obsolete since 1936.

NOTES: Similar to .22 Hornet.

LOADING DATA:

BULLET WT./TYPE	POWDER WT./TYPE	VELOCITY ('/SEC)	SOURCE
60/RN	5.0/2400	1600	Barnes

CASE PREPARATION: SHELLHOLDER (RCBS): 1

MAKE FROM: .32-20 Win. Turn rim to .366" dia. F/L size in 5,6x33R die which will swage base. Turn off excess material with lathe. Case is slightly short but, otherwise, fine. Fireform.

PHYSICAL DATA (INCHES):

CASE TYPE: **Rimmed Bottleneck**

CASE LENGTH **A = 1.311**

HEAD DIAMETER **B = .325**

RIM DIAMETER **D = .366**

NECK DIAMETER **F = .247**

NECK LENGTH **H = .200**

SHOULDER LENGTH **K = .111**

BODY ANGLE (DEG'S/SIDE): **.25**

CASE CAPACITY **CC'S = 1.00**

LOADED LENGTH: **1.64**

BELT DIAMETER **C = N/A**

RIM THICKNESS **E = .045**

SHOULDER DIAMETER **G = .318**

LENGTH TO SHOULDER **J = 1.00**

SHOULDER ANGLE (DEG'S/SIDE): **17.73**

PRIMER: **S/R**

CASE CAPACITY (GR'S WATER): **15.49**

DIMENSIONAL DRAWING:

-NOT ACTUAL SIZE-
-DO NOT SCALE-

CARTRIDGE: 5,6 x 50mm Magnum

OTHER NAMES:

DIA: .224

BALLISTEK NO: 224BF

NAI NO: RXB 22333/5.234

DATA SOURCE: Hornady Manual 3rd Pg80

HISTORICAL DATA: By DWM in 1966.

NOTES:

LOADING DATA:

BULLET WT./TYPE	POWDER WT./TYPE	VELOCITY ('/SEC)	SOURCE
50/Spire	26.3/IMR3031	3300	Horn.

CASE PREPARATION: SHELLHOLDER (RCBS): 10

MAKE FROM: RWS factory. Cases could be made by blowing out (fireforming) .222 Rem. Mag. brass but, neck would be very short (approx. .16") long. May work, however.

PHYSICAL DATA (INCHES):

CASE TYPE: **Rimless Bottleneck**

CASE LENGTH **A = 1.968**

HEAD DIAMETER **B = .376**

RIM DIAMETER **D = .378**

NECK DIAMETER **F = .255**

NECK LENGTH **H = .266**

SHOULDER LENGTH **K = .116**

BODY ANGLE (DEG'S/SIDE): **.454**

CASE CAPACITY **CC'S = 2.03**

LOADED LENGTH: **2.45**

BELT DIAMETER **C = N/A**

RIM THICKNESS **E = .045**

SHOULDER DIAMETER **G = .354**

LENGTH TO SHOULDER **J = 1.586**

SHOULDER ANGLE (DEG'S/SIDE): **23.11**

PRIMER: **L/R**

CASE CAPACITY (GR'S WATER): **31.38**

DIMENSIONAL DRAWING:

-NOT ACTUAL SIZE-
-DO NOT SCALE-

CARTRIDGE: 5,6 x 57mm RWS

OTHER NAMES:

DIA: .224

BALLISTEK NO: 224BG

NAI NO: RXB 22432/4.759

DATA SOURCE: Hornady 3rd Pg102

HISTORICAL DATA: By RWS in 1963.

NOTES:

LOADING DATA:

BULLET WT./TYPE	POWDER WT./TYPE	VELOCITY ('/SEC)	SOURCE
55/Spire	41.3/IMR4350	3700	Horn.

CASE PREPARATION: SHELLHOLDER (RCBS): 3

MAKE FROM: .270 Win. Anneal. F/L size case. Trim to length. Chamfer.

PHYSICAL DATA (INCHES):

CASE TYPE: **Rimless Bottleneck**
CASE LENGTH **A = 2.232**
HEAD DIAMETER **B = .469**
RIM DIAMETER **D = .470**
NECK DIAMETER **F = .280**
NECK LENGTH **H = .247**
SHOULDER LENGTH **K = .233**
BODY ANGLE (DEG'S/SIDE): **.701**
CASE CAPACITY **CC'S = 3.03**

LOADED LENGTH: **2.70**
BELT DIAMETER **C = N/A**
RIM THICKNESS **E = .05**
SHOULDER DIAMETER **G = .431**
LENGTH TO SHOULDER **J = 1.752**
SHOULDER ANGLE (DEG'S/SIDE): **17.95**
PRIMER: **L/R**
CASE CAPACITY (GR'S WATER): **46.79**

DIMENSIONAL DRAWING:

F H K G A J E B D

-NOT ACTUAL SIZE-
-DO NOT SCALE-

CARTRIDGE: 5,6 x 61mm Vom Hofe Super Express

OTHER NAMES:

DIA: .228

BALLISTEK NO: 228P

NAI NO: RXB 12422/5.042

DATA SOURCE: COTW 4th Pg255

HISTORICAL DATA: By E. Vom Hofe in 1937.

NOTES: Bores vary .227-.228".

LOADING DATA:

BULLET WT./TYPE	POWDER WT./TYPE	VELOCITY ('/SEC)	SOURCE
70/Spire	35.0/IMR4895	—	JJD

CASE PREPARATION: SHELLHOLDER (RCBS): 3

MAKE FROM: .25-06 Rem. Anneal case neck and size F/L in Vom Hofe die. Trim to length & chamfer.

PHYSICAL DATA (INCHES):

CASE TYPE: **Rimless Bottleneck**
CASE LENGTH **A = 2.400**
HEAD DIAMETER **B = .476**
RIM DIAMETER **D = .478**
NECK DIAMETER **F = .258**
NECK LENGTH **H = .315**
SHOULDER LENGTH **K = .355**
BODY ANGLE (DEG'S/SIDE): **.318**
CASE CAPACITY **CC'S = 4.13**

LOADED LENGTH: **3.14**
BELT DIAMETER **C = N/A**
RIM THICKNESS **E = .049**
SHOULDER DIAMETER **G = .459**
LENGTH TO SHOULDER **J = 1.73**
SHOULDER ANGLE (DEG'S/SIDE): **15.80**
PRIMER: **L/R**
CASE CAPACITY (GR'S WATER): **63.76**

DIMENSIONAL DRAWING:

F H K G A J E B D

-NOT ACTUAL SIZE-
-DO NOT SCALE-

CARTRIDGE: 5,6 x 61R Vom Hofe Super Express

OTHER NAMES:

DIA: .228

BALLISTEK NO: 228Q

NAI NO: RMB 12422/5.000

DATA SOURCE: COTW 4th Pg255

HISTORICAL DATA: By E. Vom Hofe in 1937.

NOTES: Similar to .22 Savage High Power.

LOADING DATA:

BULLET WT./TYPE	POWDER WT./TYPE	VELOCITY ('/SEC)	SOURCE
70/Spire	44.0/4350	3000	Nonte

CASE PREPARATION: SHELLHOLDER (RCBS): 4

MAKE FROM: 9,3 x 74R. Cut case to 2.4". Anneal. Size in .30-06 die to establish length-to-shoulder dim. (1.73"). Neck size in .270 and .243 die to reduce neck dia. Trim to length. Chamfer. F/L size and I.D. neck ream, if necessary.

PHYSICAL DATA (INCHES):

CASE TYPE: **Rimmed Bottleneck**

CASE LENGTH **A = 2.39**

HEAD DIAMETER **B = .478**

RIM DIAMETER **D = .533**

NECK DIAMETER **F = .260**

NECK LENGTH **H = .315**

SHOULDER LENGTH **K = .345**

BODY ANGLE (DEG'S/SIDE): **.337**

CASE CAPACITY **CC'S = 4.09**

LOADED LENGTH: **3.18**

BELT DIAMETER **C = N/A**

RIM THICKNESS **E = .05**

SHOULDER DIAMETER **G = .460**

LENGTH TO SHOULDER **J = 1.73**

SHOULDER ANGLE (DEG'S/SIDE): **16.16**

PRIMER: **L/R**

CASE CAPACITY (GR'S WATER): **63.14**

DIMENSIONAL DRAWING:

F H K G A J E B D

-NOT ACTUAL SIZE-
-DO NOT SCALE-

CARTRIDGE: 5.7mm Johnson

OTHER NAMES:
MMJ 5,7 Johnson
.22/30 Carbine

DIA: .224

BALLISTEK NO: 224M

NAI NO: RXB 24343/3.654

DATA SOURCE: COTW 4th Pg138

HISTORICAL DATA: By M. Johnson in 1963 for Johnson carbine.

NOTES:

LOADING DATA:

BULLET WT./TYPE	POWDER WT./TYPE	VELOCITY ('/SEC)	SOURCE
40/Spire	14.0/2400	2850	Nonte
50/Spire	14.0/IMR4198	2700	Barnes

CASE PREPARATION: SHELLHOLDER (RCBS): 17

MAKE FROM: .30 Carbine. Anneal case. Form in form dies #1 and #20. Trim to length. F/L size & chamfer. Clean.

PHYSICAL DATA (INCHES):

CASE TYPE: **Rimless Bottleneck**

CASE LENGTH **A = 1.29**

HEAD DIAMETER **B = .353**

RIM DIAMETER **D = .356**

NECK DIAMETER **F = .252**

NECK LENGTH **H = .22**

SHOULDER LENGTH **K = .096**

BODY ANGLE (DEG'S/SIDE): **.85**

CASE CAPACITY **CC'S = 1.06**

LOADED LENGTH: **1.65**

BELT DIAMETER **C = N/A**

RIM THICKNESS **E = .05**

SHOULDER DIAMETER **G = .33**

LENGTH TO SHOULDER **J = .974**

SHOULDER ANGLE (DEG'S/SIDE): **22.10**

PRIMER: **S/R**

CASE CAPACITY (GR'S WATER): **16.47**

DIMENSIONAL DRAWING:

F H K G A J E B D

-NOT ACTUAL SIZE-
-DO NOT SCALE-

CARTRIDGE: 6mm Arch

OTHER NAMES:

DIA: .243

BALLISTEK NO: 243AB

NAI NO: RXB 22434/4.548

DATA SOURCE: Ackley Vol.1 Pg311

HISTORICAL DATA: By Dr. E. Arch.

NOTES:

LOADING DATA:

BULLET WT./TYPE	POWDER WT./TYPE	VELOCITY ('/SEC)	SOURCE
60/Spire	50.0/4320	4055(?)	Ackley

CASE PREPARATION: SHELLHOLDER (RCBS): 2

MAKE FROM: 6,5 x 55 Swedish. Run case into Arch die to F/L size. Trim & chamfer. Use light loads to finish forming in chamber.

PHYSICAL DATA (INCHES):

CASE TYPE: **Rimless Bottleneck**

CASE LENGTH **A = 2.165**

HEAD DIAMETER **B = .476**

RIM DIAMETER **D = .476**

NECK DIAMETER **F = .265**

NECK LENGTH **H = .300**

SHOULDER LENGTH **K = .100**

BODY ANGLE (DEG'S/SIDE): **.512**

CASE CAPACITY **CC'S = 3.97**

LOADED LENGTH: **3.05**

BELT DIAMETER **C = N/A**

RIM THICKNESS **E = .06**

SHOULDER DIAMETER **G = .448**

LENGTH TO SHOULDER **J = 1.765**

SHOULDER ANGLE (DEG'S/SIDE): **42.45**

PRIMER: **L/R**

CASE CAPACITY (GR'S WATER): **61.37**

DIMENSIONAL DRAWING:

F
H
K
G
A
J
E
B
D

-NOT ACTUAL SIZE-
-DO NOT SCALE-

CARTRIDGE: 6mm Atlas

OTHER NAMES:

DIA: .243

BALLISTEK NO: 243X

NAI NO: BEN 22433/4.824

DATA SOURCE: Ackley Vol.1 Pg325

HISTORICAL DATA: By H. Baker.

NOTES: Badly overbore!

LOADING DATA:

BULLET WT./TYPE	POWDER WT./TYPE	VELOCITY ('/SEC)	SOURCE
100/Spire	78.0/H570	3760	Baker

CASE PREPARATION: SHELLHOLDER (RCBS): 4

MAKE FROM: .264 Win. Mag. F/L size the magnum case in the Atlas die. Trim to length and chamfer. Fireform in the chamber.

PHYSICAL DATA (INCHES):

CASE TYPE: **Belted Bottleneck**

CASE LENGTH **A = 2.475**

HEAD DIAMETER **B = .513**

RIM DIAMETER **D = .532**

NECK DIAMETER **F = .269**

NECK LENGTH **H = .255**

SHOULDER LENGTH **K = .195**

BODY ANGLE (DEG'S/SIDE): **.486**

CASE CAPACITY **CC'S = 5.33**

LOADED LENGTH: **3.02**

BELT DIAMETER **C = .532**

RIM THICKNESS **E = .05**

SHOULDER DIAMETER **G = .482**

LENGTH TO SHOULDER **J = 2.025**

SHOULDER ANGLE (DEG'S/SIDE): **28.64**

PRIMER: **L/R Mag**

CASE CAPACITY (GR'S WATER): **82.22**

DIMENSIONAL DRAWING:

F
H
K
G
A
J
E
B
D
C

-NOT ACTUAL SIZE-
-DO NOT SCALE-

CARTRIDGE: 6mm Belted Express Ackley

OTHER NAMES:	DIA: .243 BALLISTEK NO: 243BG NAI NO: BEN 12424/5.177

DATA SOURCE: Ackley Vol.1 Pg308

HISTORICAL DATA: Ackley ctg. with formed belt.

NOTES: G. Nonte sold cases back in the 1960's.

LOADING DATA:

BULLET WT./TYPE	POWDER WT./TYPE	VELOCITY ('/SEC)	SOURCE
100/Spire	40.0/4320	3030	Ackley

CASE PREPARATION: SHELLHOLDER (RCBS): 3

MAKE FROM: .25-06 Rem. Swage on or solder on a magnum-type belt. You must first reduce the base dia. to .450". Anneal the case and size in the F/L die. Trim and chamfer. F/L size again. Seldom worth the trouble.

PHYSICAL DATA (INCHES):

CASE TYPE: **Belted Bottleneck**
CASE LENGTH **A = 2.33**
HEAD DIAMETER **B = .450**
RIM DIAMETER **D = .473**
NECK DIAMETER **F = .265**
NECK LENGTH **H = .315**
SHOULDER LENGTH **K = .145**
BODY ANGLE (DEG'S/SIDE): **.171**
CASE CAPACITY **CC'S = 3.99**
LOADED LENGTH: **3.05**
BELT DIAMETER **C = .473**
RIM THICKNESS **E = .05**
SHOULDER DIAMETER **G = .440**
LENGTH TO SHOULDER **J = 1.87**
SHOULDER ANGLE (DEG'S/SIDE): **31.11**
PRIMER: **L/R**
CASE CAPACITY (GR'S WATER): **61.71**

DIMENSIONAL DRAWING:

-NOT ACTUAL SIZE-
-DO NOT SCALE-

CARTRIDGE: 6mm Benchrest

OTHER NAMES:	DIA: .243 BALLISTEK NO: 243AS NAI NO: RXB 24433/3.241

DATA SOURCE: Rifle #63 Pg34.

HISTORICAL DATA: Remington development in 1978.

NOTES:

LOADING DATA:

BULLET WT./TYPE	POWDER WT./TYPE	VELOCITY ('/SEC)	SOURCE
120/Spire	29.8/IMR4320	–	JJD

CASE PREPARATION: SHELLHOLDER (RCBS): 3

MAKE FROM: Rem. BR. Form dies required. Form in order and trim to length. Chamfer and F/L size. Generally, annealing is not necessary.

PHYSICAL DATA (INCHES):

CASE TYPE: **Rimless Bottleneck**
CASE LENGTH **A = 1.52**
HEAD DIAMETER **B = .469**
RIM DIAMETER **D = .473**
NECK DIAMETER **F = .265**
NECK LENGTH **H = .270**
SHOULDER LENGTH **K = .170**
BODY ANGLE (DEG'S/SIDE): **.358**
CASE CAPACITY **CC'S = 2.45**
LOADED LENGTH: **A/R**
BELT DIAMETER **C = N/A**
RIM THICKNESS **E = .05**
SHOULDER DIAMETER **G = .458**
LENGTH TO SHOULDER **J = 1.08**
SHOULDER ANGLE (DEG'S/SIDE): **29.58**
PRIMER: **S/R**
CASE CAPACITY (GR'S WATER): **37.91**

DIMENSIONAL DRAWING:

-NOT ACTUAL SIZE-
-DO NOT SCALE-

CARTRIDGE: 6mm Donaldson International

OTHER NAMES:

DIA: .243

BALLISTEK NO: 243Y

NAI NO: RXB 23424/3.808

DATA SOURCE: Ackley Vol.1 Pg298

HISTORICAL DATA: By H. Donaldson.

NOTES:

LOADING DATA:

BULLET WT./TYPE	POWDER WT./TYPE	VELOCITY ('/SEC)	SOURCE
60/SHP	32.0/3031	3650	Ackley

CASE PREPARATION: SHELLHOLDER (RCBS): 3

MAKE FROM: .25-06 Rem. Anneal shoulder area. Size in F/L die with expander removed. Trim to length & chamfer. Ream neck I.D. F/L size. Case will final form on ignition.

PHYSICAL DATA (INCHES):

CASE TYPE: **Rimless Bottleneck**

CASE LENGTH **A = 1.79**

HEAD DIAMETER **B = .470**

RIM DIAMETER **D = .473**

NECK DIAMETER **F = .269**

NECK LENGTH **H = .318**

SHOULDER LENGTH **K = .122**

BODY ANGLE (DEG'S/SIDE): **.971**

CASE CAPACITY **CC'S = 2.81**

LOADED LENGTH: **2.26**

BELT DIAMETER **C = N/A**

RIM THICKNESS **E = .05**

SHOULDER DIAMETER **G = .431**

LENGTH TO SHOULDER **J = 1.35**

SHOULDER ANGLE (DEG'S/SIDE): **33.58**

PRIMER: **L/R**

CASE CAPACITY (GR'S WATER): **43.49**

DIMENSIONAL DRAWING:

F H K G A J E B D

-NOT ACTUAL SIZE-
-DO NOT SCALE-

CARTRIDGE: 6mm Durham International

OTHER NAMES:

DIA: .243

BALLISTEK NO: 243Z

NAI NO: RXB 23424/3.989

DATA SOURCE: Ackley Vol.1 Pg300

HISTORICAL DATA:

NOTES: Similar to 6mm Donaldson Int'l (#243Y).

LOADING DATA:

BULLET WT./TYPE	POWDER WT./TYPE	VELOCITY ('/SEC)	SOURCE
100/Spire	41.0/4350	3031	Ackley

CASE PREPARATION: SHELLHOLDER (RCBS): 3

MAKE FROM: .243 Win. Size the Winchester case in the Durham die. Some cases have required I.D. neck reaming. Trim to length & chamfer.

PHYSICAL DATA (INCHES):

CASE TYPE: **Rimless Bottleneck**

CASE LENGTH **A = 1.875**

HEAD DIAMETER **B = .470**

RIM DIAMETER **D = .473**

NECK DIAMETER **F = .268**

NECK LENGTH **H = .350**

SHOULDER LENGTH **K = .115**

BODY ANGLE (DEG'S/SIDE): **.568**

CASE CAPACITY **CC'S = 3.08**

LOADED LENGTH: **2.53**

BELT DIAMETER **C = N/A**

RIM THICKNESS **E = .05**

SHOULDER DIAMETER **G = .446**

LENGTH TO SHOULDER **J = 1.41**

SHOULDER ANGLE (DEG'S/SIDE): **37.73**

PRIMER: **L/R**

CASE CAPACITY (GR'S WATER): **74.60**

DIMENSIONAL DRAWING:

F H K G A J E B D

-NOT ACTUAL SIZE-
-DO NOT SCALE-

CARTRIDGE: 6mm Express Junior

OTHER NAMES:	DIA: .243
	BALLISTEK NO: 243BH
	NAI NO: RBB 22433/4.340

DATA SOURCE: Ackley Vol.2 Pg147

HISTORICAL DATA: By R. Payne.

NOTES: Shoulders are of the familiar Weatherby type.

LOADING DATA:

BULLET WT./TYPE	POWDER WT./TYPE	VELOCITY ('/SEC)	SOURCE
100/Spire	50.0/4350	3353	JJD

CASE PREPARATION: SHELLHOLDER (RCBS): 3

MAKE FROM: .284 Winchester. Anneal case and F/L size in the Express die. Trim & chamfer. Shoulders will fireform upon ignition.

PHYSICAL DATA (INCHES):

CASE TYPE: **Rebated Bottleneck**

CASE LENGTH **A = 2.17**

HEAD DIAMETER **B = .500**

RIM DIAMETER **D = .473**

NECK DIAMETER **F = .270**

NECK LENGTH **H = .265**

SHOULDER LENGTH **K = .185**

BODY ANGLE (DEG'S/SIDE): **.716**

CASE CAPACITY **CC'S = 4.16**

LOADED LENGTH: **2.78**

BELT DIAMETER **C = N/A**

RIM THICKNESS **E = .05**

SHOULDER DIAMETER **G = .462**

LENGTH TO SHOULDER **J = 1.72**

SHOULDER ANGLE (DEG'S/SIDE): **N/A**

PRIMER: **L/R**

CASE CAPACITY (GR'S WATER): **64.31**

DIMENSIONAL DRAWING:

-NOT ACTUAL SIZE-
-DO NOT SCALE-

CARTRIDGE: 6mm Halger Magnum

OTHER NAMES:	DIA: .243
.244 Halger Magnum	BALLISTEK NO: 243BI
	NAI NO: RMB 22432/4.818

DATA SOURCE: COTW 4th Pg256

HISTORICAL DATA: By Halger Arms Co. (Gr.) in 1920's.

NOTES: A rimmed version also exists.

LOADING DATA:

BULLET WT./TYPE	POWDER WT./TYPE	VELOCITY ('/SEC)	SOURCE
105/RN	41.0/4350	2900	JJD

CASE PREPARATION: SHELLHOLDER (RCBS):

MAKE FROM: .444 Marlin. Turn rim to .519" dia. and re-cut extractor groove. Anneal case. Use form set or .33 WCF die to reduce neck die. Size again in 7mm/08 die being careful to locate shoulder properly. Anneal again. F/L size with expander removed. Trim & chamfer. I.D. neck ream if required. F/L size.

PHYSICAL DATA (INCHES):

CASE TYPE: **Rimmed Bottleneck**

CASE LENGTH **A = 2.25**

HEAD DIAMETER **B = .467**

RIM DIAMETER **D = .519**

NECK DIAMETER **F = .287**

NECK LENGTH **H = .275**

SHOULDER LENGTH **K = .208**

BODY ANGLE (DEG'S/SIDE): **.585**

CASE CAPACITY **CC'S = 3.38**

LOADED LENGTH: **3.04**

BELT DIAMETER **C = N/A**

RIM THICKNESS **E = .062**

SHOULDER DIAMETER **G = .435**

LENGTH TO SHOULDER **J = 1.767**

SHOULDER ANGLE (DEG'S/SIDE): **19.58**

PRIMER: **L/R**

CASE CAPACITY (GR'S WATER): **52.26**

DIMENSIONAL DRAWING:

-NOT ACTUAL SIZE-
-DO NOT SCALE-

CARTRIDGE: 6mm HLS

OTHER NAMES:

DIA: .243
BALLISTEK NO: 243AC
NAI NO: RXB 23423/4.246

DATA SOURCE: Ackley Vol.1 Pg297

HISTORICAL DATA: By Consolidated Armslube Co.

NOTES:

LOADING DATA:

BULLET WT./TYPE	POWDER WT./TYPE	VELOCITY ('/SEC)	SOURCE
75/—	39.7/4320	3360	Hutton

CASE PREPARATION: SHELLHOLDER (RCBS): 3

MAKE FROM: 6mm Rem. Size Rem. case in HLS die or a 6mm Rem. die can be ground off to produce the 2.00" ctg. length.

PHYSICAL DATA (INCHES):

CASE TYPE: **Rimless Bottleneck**
CASE LENGTH **A = 2.000**
HEAD DIAMETER **B = .471**
RIM DIAMETER **D = .472**
NECK DIAMETER **F = .269**
NECK LENGTH **H = .365**
SHOULDER LENGTH **K = .135**
BODY ANGLE (DEG'S/SIDE): **.903**
CASE CAPACITY **CC'S = 3.20**
LOADED LENGTH: **2.57**
BELT DIAMETER **C = N/A**
RIM THICKNESS **E = .048**
SHOULDER DIAMETER **G = .430**
LENGTH TO SHOULDER **J = 1.50**
SHOULDER ANGLE (DEG'S/SIDE): **30.80**
PRIMER: **L/R**
CASE CAPACITY (GR'S WATER): **49.35**

DIMENSIONAL DRAWING:

-NOT ACTUAL SIZE-
-DO NOT SCALE-

CARTRIDGE: 6mm International

OTHER NAMES:
6mm/250 Savage
6mm Remington International

DIA: .243
BALLISTEK NO: 243AI
NAI NO: RXB 33423/4.059

DATA SOURCE: Sierra Manual 1985 Pg133

HISTORICAL DATA: Developed in 1955.

NOTES:

LOADING DATA:

BULLET WT./TYPE	POWDER WT./TYPE	VELOCITY ('/SEC)	SOURCE
75/HP	34.9/IMR4895	3200	Sierra
100/—	34.5/IMR4064	3065	Ackley

CASE PREPARATION: SHELLHOLDER (RCBS): 3

MAKE FROM: .250 Savage or 6mm Rem. Size case in 6mm Int'l. die to set the shoulder back. Annealing not always necessary but will increase case life. Case will fireform in chamber.

PHYSICAL DATA (INCHES):

CASE TYPE: **Rimless Bottleneck**
CASE LENGTH **A = 1.912**
HEAD DIAMETER **B = .471**
RIM DIAMETER **D = .472**
NECK DIAMETER **F = .272**
NECK LENGTH **H = .372**
SHOULDER LENGTH **K = .150**
BODY ANGLE (DEG'S/SIDE): **1.15**
CASE CAPACITY **CC'S = 2.85**
LOADED LENGTH: **2.45**
BELT DIAMETER **C = N/A**
RIM THICKNESS **E = .048**
SHOULDER DIAMETER **G = .419**
LENGTH TO SHOULDER **J = 1.39**
SHOULDER ANGLE (DEG'S/SIDE): **26.10**
PRIMER: **L/R**
CASE CAPACITY (GR'S WATER): **44.06**

DIMENSIONAL DRAWING:

-NOT ACTUAL SIZE-
-DO NOT SCALE-

CARTRIDGE: 6mm JDJ #2

OTHER NAMES:

DIA: .243

BALLISTEK NO: 243BT

NAI NO: RMB 12344/4.561

DATA SOURCE: SSK Industries

HISTORICAL DATA: By J.D. Jones.

NOTES:

LOADING DATA:

BULLET WT./TYPE	POWDER WT./TYPE	VELOCITY ('/SEC)	SOURCE
50/Hornaday	32.0/IMR 3031	2864	SSK
60/Hornaday SX	35.0/H414	2732	SSK
63/Sierra	38.5/H4831	2831	SSK

CASE PREPARATION: SHELLHOLDER (RCBS): 11

MAKE FROM: .225 Winchester. Taper expand neck to hold .243" dia. bullets and fireform with a moderate load.

PHYSICAL DATA (INCHES):

CASE TYPE: **Rimmed Bottleneck**

CASE LENGTH **A = 1.902**

HEAD DIAMETER **B = .417**

RIM DIAMETER **D = .467**

NECK DIAMETER **F = .270**

NECK LENGTH **H = .192**

SHOULDER LENGTH **K = .090**

BODY ANGLE (DEG'S/SIDE): **.161**

CASE CAPACITY **CC'S = 2.72**

LOADED LENGTH: **2.562**

BELT DIAMETER **C = N/A**

RIM THICKNESS **E = .046**

SHOULDER DIAMETER **G = .409**

LENGTH TO SHOULDER **J = 1.62**

SHOULDER ANGLE (DEG'S/SIDE): **37.67**

PRIMER: **L/R**

CASE CAPACITY (GR'S WATER): **41.92**

DIMENSIONAL DRAWING:

F H K G A J E B D

-NOT ACTUAL SIZE-
-DO NOT SCALE-

CARTRIDGE: 6mm Krag Improved (Ackley)

OTHER NAMES:

DIA: .243

BALLISTEK NO: 243BJ

NAI NO: RMB 12424/4.858

DATA SOURCE: Ackley Vol.1 Pg310

HISTORICAL DATA:

NOTES: About max. case capacity for .243" dia. bore.

LOADING DATA:

BULLET WT./TYPE	POWDER WT./TYPE	VELOCITY ('/SEC)	SOURCE
75/Spire	51.0/4350	3809	Ackley

CASE PREPARATION: SHELLHOLDER (RCBS): 7

MAKE FROM: .30-40 Krag. Anneal case and form in form die. Trim to length & chamfer. F/L size. Fireform case in chamber.

PHYSICAL DATA (INCHES):

CASE TYPE: **Rimmed Bottleneck**

CASE LENGTH **A = 2.22**

HEAD DIAMETER **B = .457**

RIM DIAMETER **D = .545**

NECK DIAMETER **F = .272**

NECK LENGTH **H = .330**

SHOULDER LENGTH **K = .08**

BODY ANGLE (DEG'S/SIDE): **.338**

CASE CAPACITY **CC'S = 3.65**

LOADED LENGTH: **3.09**

BELT DIAMETER **C = N/A**

RIM THICKNESS **E = .064**

SHOULDER DIAMETER **G = .438**

LENGTH TO SHOULDER **J = 1.81**

SHOULDER ANGLE (DEG'S/SIDE): **46.05**

PRIMER: **L/R**

CASE CAPACITY (GR'S WATER): **56.33**

DIMENSIONAL DRAWING:

F H K G A J E B D

-NOT ACTUAL SIZE-
-DO NOT SCALE-

CARTRIDGE: 6mm Krag Short (Ackley)

OTHER NAMES:

DIA: .243

BALLISTEK NO: 243AA

NAI NO: RMB 12423/4.781

DATA SOURCE: Ackley Vol.1 Pg. 309

HISTORICAL DATA:

NOTES:

LOADING DATA:

BULLET WT./TYPE	POWDER WT./TYPE	VELOCITY ('/SEC)	SOURCE
90/Spire	40.0/4350	–	Ackley

CASE PREPARATION: SHELLHOLDER (RCBS): 7

MAKE FROM: .30-40 Krag. Three-die forming set required. Form in order, anneal between 1st and 2nd reduction. Trim to length. F/L size & chamfer.

PHYSICAL DATA (INCHES):

CASE TYPE: **Rimmed Bottleneck**

CASE LENGTH **A = 2.185**

HEAD DIAMETER **B = .457**

RIM DIAMETER **D = .545**

NECK DIAMETER **F = .274**

NECK LENGTH **H = .308**

SHOULDER LENGTH **K = .147**

BODY ANGLE (DEG'S/SIDE): **.187**

CASE CAPACITY **CC'S = 3.58**

LOADED LENGTH: **2.96**

BELT DIAMETER **C = N/A**

RIM THICKNESS **E = .063**

SHOULDER DIAMETER **G = .447**

LENGTH TO SHOULDER **J = 1.73**

SHOULDER ANGLE (DEG'S/SIDE): **30.47**

PRIMER: **L/R**

CASE CAPACITY (GR'S WATER): **55.37**

DIMENSIONAL DRAWING:

F H K G A J E B D

-NOT ACTUAL SIZE-
-DO NOT SCALE-

CARTRIDGE: 6mm Lee Navy

OTHER NAMES:
.236 Navy
.236 Lee Remington

DIA: .243

BALLISTEK NO: 243B

NAI NO: RXB 22422/5.327

DATA SOURCE: NAI/Ballistek

HISTORICAL DATA: U.S. Navy ctg. ca. 1895.

NOTES:

LOADING DATA:

BULLET WT./TYPE	POWDER WT./TYPE	VELOCITY ('/SEC)	SOURCE
75/Spire	35.0/3031	3150	Nonte
100/–	34.0/4895	2680	Barnes

CASE PREPARATION: SHELLHOLDER (RCBS): 11

MAKE FROM: .220 Swift. Run case neck over .22-.243" tapered expander. Run neck over a .257" dia. expander. F/L size. Turn rim to .448" dia and cut new extractor groove. Case will be .12" short but will function.

PHYSICAL DATA (INCHES):

CASE TYPE: **Rimless Bottleneck**

CASE LENGTH **A = 2.36**

HEAD DIAMETER **B = .443**

RIM DIAMETER **D = .448**

NECK DIAMETER **F = .276**

NECK LENGTH **H = .426**

SHOULDER LENGTH **K = .174**

BODY ANGLE (DEG'S/SIDE): **.753**

CASE CAPACITY **CC'S = 3.25**

LOADED LENGTH: **3.11**

BELT DIAMETER **C = N/A**

RIM THICKNESS **E = .05**

SHOULDER DIAMETER **G = .402**

LENGTH TO SHOULDER **J = 1.76**

SHOULDER ANGLE (DEG'S/SIDE): **19.90**

PRIMER: **L/R**

CASE CAPACITY (GR'S WATER): **50.13**

DIMENSIONAL DRAWING:

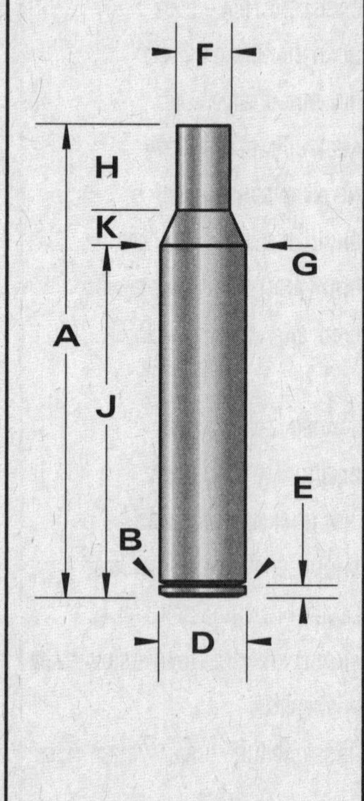

-NOT ACTUAL SIZE-
-DO NOT SCALE-

CARTRIDGE: 6mm Magnum (Ackley)

OTHER NAMES:

DIA: .243

BALLISTEK NO: 243BM

NAI NO: BEN 22423/4.824

DATA SOURCE: Ackley Vol.1 Pg323

HISTORICAL DATA: This cartridge is not recommended; even by Ackley!

NOTES: Hard to produce and overbore!

LOADING DATA:

BULLET WT./TYPE	POWDER WT./TYPE	VELOCITY ('/SEC)	SOURCE
100/Spire	50.0/4350	—	JJD

CASE PREPARATION: SHELLHOLDER (RCBS): 4

MAKE FROM: .300 H&H. Form set is required. Anneal case and run through the form die series. Trim to length. Chamfer and F/L size.

PHYSICAL DATA (INCHES):

CASE TYPE: **Belted Bottleneck**

CASE LENGTH **A = 2.475**

HEAD DIAMETER **B = .513**

RIM DIAMETER **D = .532**

NECK DIAMETER **F = .276**

NECK LENGTH **H = .333**

SHOULDER LENGTH **K = .242**

BODY ANGLE (DEG'S/SIDE): **.387**

CASE CAPACITY **CC'S = 5.00**

LOADED LENGTH: **3.30**

BELT DIAMETER **C = .532**

RIM THICKNESS **E = .049**

SHOULDER DIAMETER **G = .490**

LENGTH TO SHOULDER **J = 1.90**

SHOULDER ANGLE (DEG'S/SIDE): **23.85**

PRIMER: **L/R Mag**

CASE CAPACITY (GR'S WATER): **77.16**

DIMENSIONAL DRAWING:

-NOT ACTUAL SIZE-
-DO NOT SCALE-

CARTRIDGE: 6mm Nieomiller

OTHER NAMES:

DIA: .243

BALLISTEK NO: 243BL

NAI NO: RXB 22423/4.404

DATA SOURCE: Ackley Vol2 Pg143

HISTORICAL DATA: By H. Nieomiller.

NOTES: Long neck design for increased grip on bullet.

LOADING DATA:

BULLET WT./TYPE	POWDER WT./TYPE	VELOCITY ('/SEC)	SOURCE
85/—	44.0/4350	3249	Ackley

CASE PREPARATION: SHELLHOLDER (RCBS): 3

MAKE FROM: .243 Win. A standard .243 F/L die can be modified, by grinding away the base, to produce this cartridge. Remove 0.100" of stock from the die. Anneal the .243 case and use the modified die to push the shoulder back.

PHYSICAL DATA (INCHES):

CASE TYPE: **Rimless Bottleneck**

CASE LENGTH **A = 2.07**

HEAD DIAMETER **B = .47**

RIM DIAMETER **D = .473**

NECK DIAMETER **F = .261**

NECK LENGTH **H = .370**

SHOULDER LENGTH **K = .16**

BODY ANGLE (DEG'S/SIDE): **.790**

CASE CAPACITY **CC'S = 3.51**

LOADED LENGTH: **2.64**

BELT DIAMETER **C = N/A**

RIM THICKNESS **E = .05**

SHOULDER DIAMETER **G = .433**

LENGTH TO SHOULDER **J = 1.54**

SHOULDER ANGLE (DEG'S/SIDE): **28.25**

PRIMER: **L/R**

CASE CAPACITY (GR'S WATER): **54.21**

DIMENSIONAL DRAWING:

-NOT ACTUAL SIZE-
-DO NOT SCALE-

CARTRIDGE: 6mm PPC

OTHER NAMES:

DIA: .243

BALLISTEK NO: 243T

NAI NO: RXB 14433/3.458

DATA SOURCE: NAI/Ballistek

HISTORICAL DATA: By Palmisano & Pindell in 1975.

NOTES:

LOADING DATA:

BULLET WT./TYPE	POWDER WT./TYPE	VELOCITY ('/SEC)	SOURCE
70/Spire	25.0/IMR4198	3320	JJD

CASE PREPARATION: SHELLHOLDER (RCBS): 6

MAKE FROM: .220 Russian (Sako). Taper expand the .220 case to .243" + dia. F/L size in PPC die. Trim & chamfer.

PHYSICAL DATA (INCHES):

CASE TYPE: **Rimless Bottleneck**

CASE LENGTH **A = 1.525**

HEAD DIAMETER **B = .441**

RIM DIAMETER **D = .445**

NECK DIAMETER **F = .265**

NECK LENGTH **H = .272**

SHOULDER LENGTH **K = .178**

BODY ANGLE (DEG'S/SIDE): **.327**

CASE CAPACITY **CC'S = 2.19**

LOADED LENGTH: **2.10**

BELT DIAMETER **C = N/A**

RIM THICKNESS **E = .05**

SHOULDER DIAMETER **G = .431**

LENGTH TO SHOULDER **J = 1.075**

SHOULDER ANGLE (DEG'S/SIDE): **24.99**

PRIMER: **S/R**

CASE CAPACITY (GR'S WATER): **33.80**

DIMENSIONAL DRAWING:

F H K G A J E B D

-NOT ACTUAL SIZE-
-DO NOT SCALE-

CARTRIDGE: 6mm Remington

OTHER NAMES:
.244 Remington

DIA: .243

BALLISTEK NO: 243A

NAI NO: RXB 22423/4.740

DATA SOURCE: Hornady Manual 3rd Pg112

HISTORICAL DATA: By Rem. in 1963.

NOTES: .244 version originated in 1955.

LOADING DATA:

BULLET WT./TYPE	POWDER WT./TYPE	VELOCITY ('/SEC)	SOURCE
100/Spire	38.0/IMR4064	2975	JJD

CASE PREPARATION: SHELLHOLDER (RCBS): 3

MAKE FROM: Factory or .270 Win. Anneal neck. F/L size with expander removed. Trim to length. Check neck I.D. & O.D. Ream if necessary. Clean. F/L size. Chamfer.

PHYSICAL DATA (INCHES):

CASE TYPE: **Rimless Bottleneck**

CASE LENGTH **A = 2.233**

HEAD DIAMETER **B = .471**

RIM DIAMETER **D = .472**

NECK DIAMETER **F = .276**

NECK LENGTH **H = .351**

SHOULDER LENGTH **K = .174**

BODY ANGLE (DEG'S/SIDE): **.760**

CASE CAPACITY **CC'S = 3.53**

LOADED LENGTH: **2.83**

BELT DIAMETER **C = N/A**

RIM THICKNESS **E = .05**

SHOULDER DIAMETER **G = .431**

LENGTH TO SHOULDER **J = 1.708**

SHOULDER ANGLE (DEG'S/SIDE): **24.00**

PRIMER: **L/R**

CASE CAPACITY (GR'S WATER): **54.57**

DIMENSIONAL DRAWING:

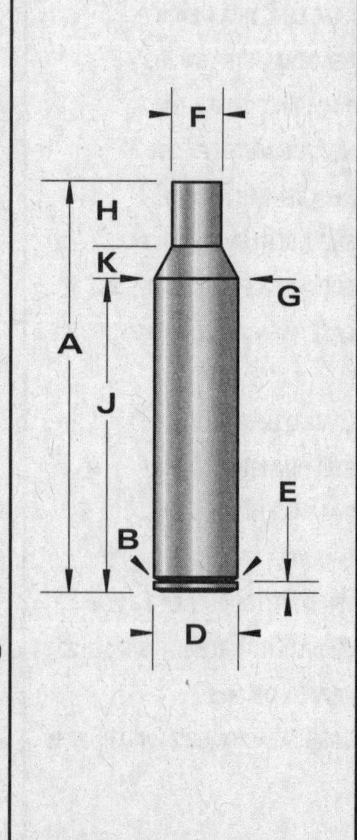

-NOT ACTUAL SIZE-
-DO NOT SCALE-

CARTRIDGE: 6mm Shipley Pipsqueak

OTHER NAMES:
6mm/222 Remington Magnum

DIA: .243

BALLISTEK NO: 243BK

NAI NO: RXB 12343/4.866

DATA SOURCE: Ackley Vol.2 Pg144

HISTORICAL DATA: By H. Shipley.

NOTES: Similar to 6 x 47mm.

LOADING DATA:

BULLET WT./TYPE	POWDER WT./TYPE	VELOCITY ('/SEC)	SOURCE
60/Spire	30.0/BL-C2	3160	Ackley/ Saylor

CASE PREPARATION: SHELLHOLDER (RCBS): 10

MAKE FROM: .222 Rem. Mag. Taper expand neck to .250" dia. and F/L size. Fireform in chamber. Ackley says that factory ammo may be fired in the chamber by seating the bullet out and lubing the case.

PHYSICAL DATA (INCHES):

CASE TYPE: **Rimless Bottleneck**

CASE LENGTH **A = 1.825**

HEAD DIAMETER **B = .375**

RIM DIAMETER **D = .378**

NECK DIAMETER **F = .264**

NECK LENGTH **H = .215**

SHOULDER LENGTH **K = .095**

BODY ANGLE (DEG'S/SIDE): **.217**

CASE CAPACITY **CC'S = 2.14**

LOADED LENGTH: **2.45**

BELT DIAMETER **C = N/A**

RIM THICKNESS **E = .045**

SHOULDER DIAMETER **G = .365**

LENGTH TO SHOULDER **J = 1.515**

SHOULDER ANGLE (DEG'S/SIDE): **27.99**

PRIMER: **L/R**

CASE CAPACITY (GR'S WATER): **33.04**

DIMENSIONAL DRAWING:

-NOT ACTUAL SIZE-
-DO NOT SCALE-

CARTRIDGE: 6mm Swift

OTHER NAMES:

DIA: .243

BALLISTEK NO: 243W

NAI NO: RMB 22423/4.900

DATA SOURCE: Rifle #47 Pg35

HISTORICAL DATA: Early wildcat with many variations.

NOTES:

LOADING DATA:

BULLET WT./TYPE	POWDER WT./TYPE	VELOCITY ('/SEC)	SOURCE
75/SHP	40.0/4064	3527	Hanson

CASE PREPARATION: SHELLHOLDER (RCBS): 11

MAKE FROM: .220 Swift. Taper expand Swift case to .250" dia. F/L size in 6mm Swift die. Trim to length and chamfer.

PHYSICAL DATA (INCHES):

CASE TYPE: **Rimmed Bottleneck**

CASE LENGTH **A = 2.205**

HEAD DIAMETER **B = .450**

RIM DIAMETER **D = .473**

NECK DIAMETER **F = .276**

NECK LENGTH **H = .310**

SHOULDER LENGTH **K = .170**

BODY ANGLE (DEG'S/SIDE): **.90**

CASE CAPACITY **CC'S = 3.14**

LOADED LENGTH: **A/R**

BELT DIAMETER **C = N/A**

RIM THICKNESS **E = .05**

SHOULDER DIAMETER **G = .402**

LENGTH TO SHOULDER **J = 1.725**

SHOULDER ANGLE (DEG'S/SIDE): **20.33**

PRIMER: **L/R**

CASE CAPACITY (GR'S WATER): **48.50**

DIMENSIONAL DRAWING:

-NOT ACTUAL SIZE-
-DO NOT SCALE-

CARTRIDGE: 6mm/06 Springfield

OTHER NAMES:

DIA: .243

BALLISTEK NO: 243AP

NAI NO: RXB 22442/5.298

DATA SOURCE: NAI/Ballistek

HISTORICAL DATA:

NOTES:

LOADING DATA:

BULLET WT./TYPE	POWDER WT./TYPE	VELOCITY ('/SEC)	SOURCE
75/Spire	39.7/IMR4320	—	JJD

CASE PREPARATION: SHELLHOLDER (RCBS): 3

MAKE FROM: .25-06 Rem. Form the .25 cal. case in the 6mm die. Trim to length and chamfer.

PHYSICAL DATA (INCHES):

CASE TYPE: **Rimless Bottleneck**
CASE LENGTH **A = 2.49**
HEAD DIAMETER **B = .470**
RIM DIAMETER **D = .473**
NECK DIAMETER **F = .276**
NECK LENGTH **H = .240**
SHOULDER LENGTH **K = .31**
BODY ANGLE (DEG'S/SIDE): **.493**
CASE CAPACITY **CC'S = 4.20**

LOADED LENGTH: **A/R**
BELT DIAMETER **C = N/A**
RIM THICKNESS **E = .05**
SHOULDER DIAMETER **G = .440**
LENGTH TO SHOULDER **J = 1.94**
SHOULDER ANGLE (DEG'S/SIDE): **14.81**
PRIMER: **L/R**
CASE CAPACITY (GR'S WATER): **64.82**

DIMENSIONAL DRAWING:

-NOT ACTUAL SIZE-
-DO NOT SCALE-

CARTRIDGE: 6mm/222 Remington

OTHER NAMES:

DIA: .243

BALLISTEK NO: 243AF

NAI NO: RXB 22323/4.521

DATA SOURCE: Rifle #38 Pg32

HISTORICAL DATA:

NOTES:

LOADING DATA:

BULLET WT./TYPE	POWDER WT./TYPE	VELOCITY ('/SEC)	SOURCE
80/—	25.0/BL-C2	2700	JJD

CASE PREPARATION: SHELLHOLDER (RCBS): 10

MAKE FROM: .222 Rem. Anneal case neck and taper expand to about .250" dia. F/L size and chamfer.

PHYSICAL DATA (INCHES):

CASE TYPE: **Rimless Bottleneck**
CASE LENGTH **A = 1.70**
HEAD DIAMETER **B = .376**
RIM DIAMETER **D = .376**
NECK DIAMETER **F = .276**
NECK LENGTH **H = .330**
SHOULDER LENGTH **K = .106**
BODY ANGLE (DEG'S/SIDE): **.511**
CASE CAPACITY **CC'S = 1.677**

LOADED LENGTH: **A/R**
BELT DIAMETER **C = N/A**
RIM THICKNESS **E = .045**
SHOULDER DIAMETER **G = .357**
LENGTH TO SHOULDER **J = 1.264**
SHOULDER ANGLE (DEG'S/SIDE): **20.90**
PRIMER: **S/R**
CASE CAPACITY (GR'S WATER): **24.88**

DIMENSIONAL DRAWING:

-NOT ACTUAL SIZE-
-DO NOT SCALE-

CARTRIDGE: 6mm/224 Weatherby

OTHER NAMES:
6mm Tom Cat
.243/224 Weatherby

DIA: .243

BALLISTEK NO: 243AU

NAI NO: BEN 22333/4.703

DATA SOURCE: Ackley Vol.2 Pg145

HISTORICAL DATA:

NOTES:

LOADING DATA:

BULLET WT./TYPE	POWDER WT./TYPE	VELOCITY ('/SEC)	SOURCE
100/—	31.0/4064	2940	Ackley

CASE PREPARATION: **SHELLHOLDER (RCBS): 27**

MAKE FROM: .224 Weatherby. Taper expand the Weatherby brass to .250" dia. Chamfer & F/L size. Fireform in chamber.

PHYSICAL DATA (INCHES):

CASE TYPE: **Belted Bottleneck**

CASE LENGTH **A = 1.952**

HEAD DIAMETER **B = .415**

RIM DIAMETER **D = .429**

NECK DIAMETER **F = .275**

NECK LENGTH **H = .285**

SHOULDER LENGTH **K = .109**

BODY ANGLE (DEG'S/SIDE): **/422**

CASE CAPACITY **CC'S = 2.50**

LOADED LENGTH: **2.51**

BELT DIAMETER **C = .429**

RIM THICKNESS **E = .049**

SHOULDER DIAMETER **G = .395**

LENGTH TO SHOULDER **J = 1.558**

SHOULDER ANGLE (DEG'S/SIDE): **N/A**

PRIMER: **L/R**

CASE CAPACITY (GR'S WATER): **38.58**

DIMENSIONAL DRAWING:

-NOT ACTUAL SIZE-
-DO NOT SCALE-

CARTRIDGE: 6mm/225 Winchester

OTHER NAMES:
.243/225 Winchester

DIA: .243

BALLISTEK NO: 243AH

NAI NO: RMB 22343/4.695

DATA SOURCE: Ackley Vol.2 Pg146

HISTORICAL DATA:

NOTES:

LOADING DATA:

BULLET WT./TYPE	POWDER WT./TYPE	VELOCITY ('/SEC)	SOURCE
90/—	30.0/3031	3020	Ackley

CASE PREPARATION: **SHELLHOLDER (RCBS): 11**

MAKE FROM: .225 Winchester. Anneal case neck. Taper expand to .250" dia. Trim & chamfer. F/L size.

PHYSICAL DATA (INCHES):

CASE TYPE: **Rimmed Bottleneck**

CASE LENGTH **A = 1.986**

HEAD DIAMETER **B = .423**

RIM DIAMETER **D = .473**

NECK DIAMETER **F = .266**

NECK LENGTH **H = .235**

SHOULDER LENGTH **K = .141**

BODY ANGLE (DEG'S/SIDE): **.629**

CASE CAPACITY **CC'S = 2.79**

LOADED LENGTH: **2.54**

BELT DIAMETER **C = N/A**

RIM THICKNESS **E = .05**

SHOULDER DIAMETER **G = .392**

LENGTH TO SHOULDER **J = 1.61**

SHOULDER ANGLE (DEG'S/SIDE): **24.07**

PRIMER: **L/R**

CASE CAPACITY (GR'S WATER): **43.12**

DIMENSIONAL DRAWING:

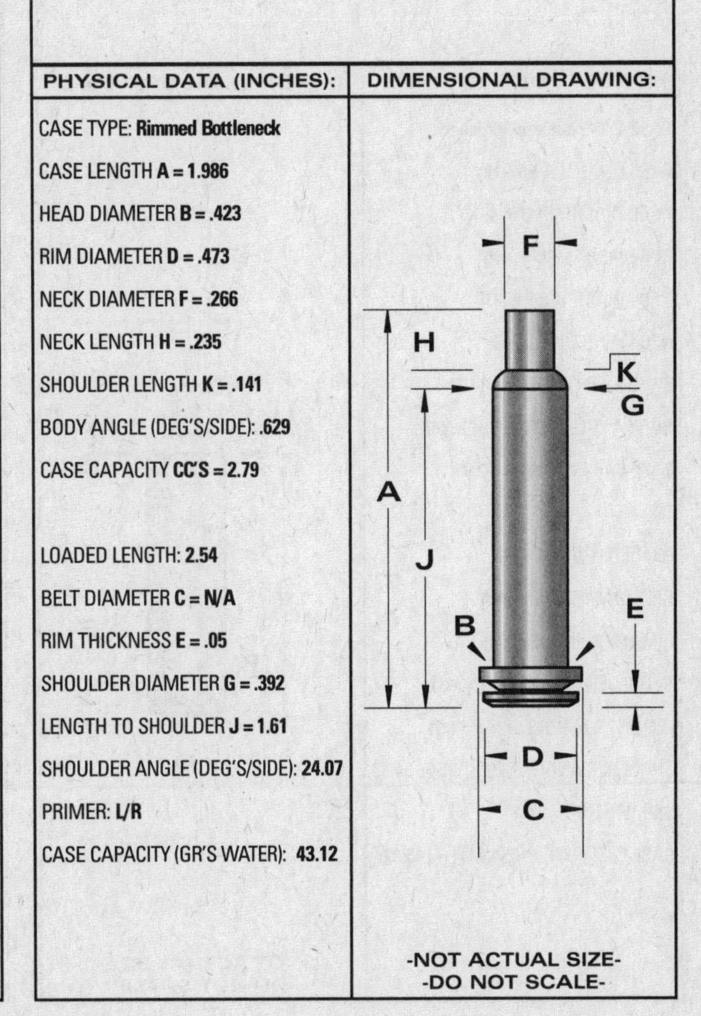

-NOT ACTUAL SIZE-
-DO NOT SCALE-

CARTRIDGE: 6mm/250 Walker

OTHER NAMES:
6mm-250 Savage

DIA: .243

BALLISTEK NO: 243AV

NAI NO: RXB
23434/4.094

DATA SOURCE: Ackley Vol.1 Pg297

HISTORICAL DATA:

NOTES:

LOADING DATA:

BULLET WT./TYPE	POWDER WT./TYPE	VELOCITY ('/SEC)	SOURCE
75/Spire	32.0/IMR3031	–	JJD

CASE PREPARATION: SHELLHOLDER (RCBS): 3

MAKE FROM: .22-250 Rem. Anneal case neck and taper expand to .250" dia. F/L size. .250 Savage brass may also be annealed and necked down in the F/L die.

PHYSICAL DATA (INCHES):

CASE TYPE: **Rimless Bottleneck**

CASE LENGTH **A = 1.912**

HEAD DIAMETER **B = .467**

RIM DIAMETER **D = .470**

NECK DIAMETER **F = .276**

NECK LENGTH **H = .290**

SHOULDER LENGTH **K = .112**

BODY ANGLE (DEG'S/SIDE): **.787**

CASE CAPACITY **CC'S = 2.96**

LOADED LENGTH: **A/R**

BELT DIAMETER **C = N/A**

RIM THICKNESS **E = .055**

SHOULDER DIAMETER **G = .431**

LENGTH TO SHOULDER **J = 1.51**

SHOULDER ANGLE (DEG'S/SIDE): **34.68**

PRIMER: **L/R**

CASE CAPACITY (GR'S WATER): **45.82**

DIMENSIONAL DRAWING:

F H K G A J E B D

-NOT ACTUAL SIZE-
-DO NOT SCALE-

CARTRIDGE: 6mm/284 Winchester

OTHER NAMES:

DIA: .243

BALLISTEK NO: 243AJ

NAI NO: RBB
22434/4.340

DATA SOURCE: Hornady Manual 3rd Pg116

HISTORICAL DATA:

NOTES:

LOADING DATA:

BULLET WT./TYPE	POWDER WT./TYPE	VELOCITY ('/SEC)	SOURCE
87/Spire	50.0/IMR4831	3300	JJD

CASE PREPARATION: SHELLHOLDER (RCBS): 3

MAKE FROM: .284 Winchester. Anneal case neck. Size in F/L die. Square case mouth. Chamfer.

PHYSICAL DATA (INCHES):

CASE TYPE: **Rebated Bottleneck**

CASE LENGTH **A = 2.17**

HEAD DIAMETER **B = .500**

RIM DIAMETER **D = .473**

NECK DIAMETER **F = .320**

NECK LENGTH **H = .285**

SHOULDER LENGTH **K = .110**

BODY ANGLE (DEG'S/SIDE): **.455**

CASE CAPACITY **CC'S = 3.09**

LOADED LENGTH: **2.83**

BELT DIAMETER **C = N/A**

RIM THICKNESS **E = .055**

SHOULDER DIAMETER **G = .475**

LENGTH TO SHOULDER **J = 1.775**

SHOULDER ANGLE (DEG'S/SIDE): **35.16**

PRIMER: **L/R**

CASE CAPACITY (GR'S WATER): **47.75**

DIMENSIONAL DRAWING:

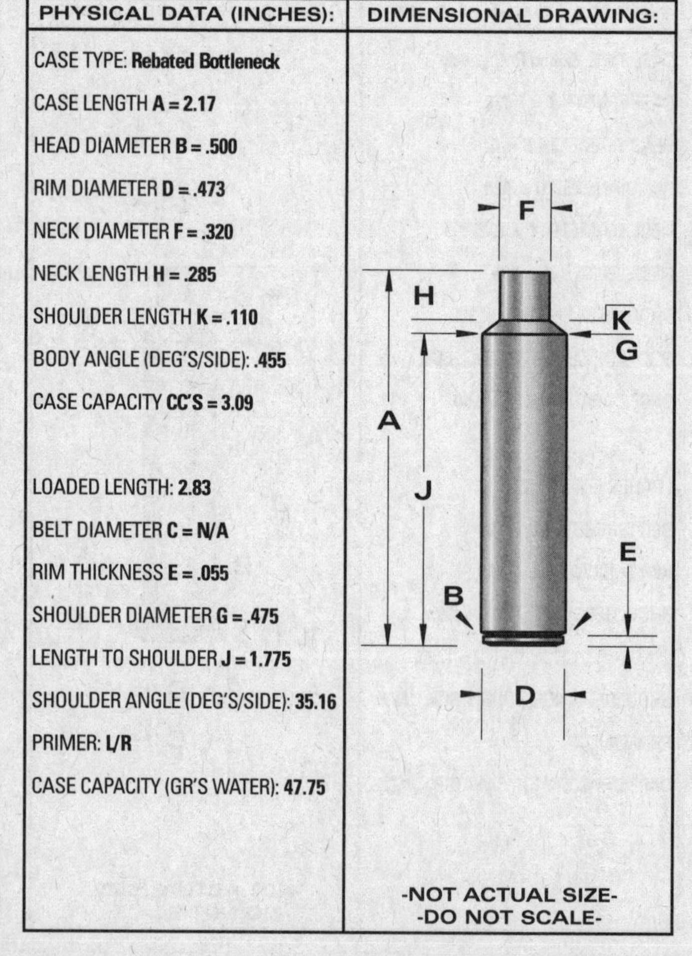

-NOT ACTUAL SIZE-
-DO NOT SCALE-

CARTRIDGE: 6mm/30-30 Winchester

OTHER NAMES:
.243/30-30 Winchester

DIA: .243

BALLISTEK NO: 243AM

NAI NO: RMB
12324/5.142

DATA SOURCE: Ackley Vol.1 Pg296

HISTORICAL DATA:

NOTES: Ideal case capacity for .243" dia. bore.

LOADING DATA:

BULLET WT./TYPE	POWDER WT./TYPE	VELOCITY ('/SEC)	SOURCE
75/Spire	36.0/3031	3430	Ackley

CASE PREPARATION: SHELLHOLDER (RCBS): 2

MAKE FROM: .30-30 Win. Form case in form die (1-die). Trim to length, chamfer and F/L size.

PHYSICAL DATA (INCHES):

CASE TYPE: **Rimmed Bottleneck**

CASE LENGTH A = **2.165**

HEAD DIAMETER B = **.421**

RIM DIAMETER D = **.506**

NECK DIAMETER F = **.272**

NECK LENGTH H = **.328**

SHOULDER LENGTH K = **.102**

BODY ANGLE (DEG'S/SIDE): **.093**

CASE CAPACITY CC'S = **3.16**

LOADED LENGTH: **2.885**

BELT DIAMETER C = **N/A**

RIM THICKNESS E = **.063**

SHOULDER DIAMETER G = **.416**

LENGTH TO SHOULDER J = **1.735**

SHOULDER ANGLE (DEG'S/SIDE): **37.05**

PRIMER: **L/R**

CASE CAPACITY (GR'S WATER): **48.72**

DIMENSIONAL DRAWING:

F H K G A J E B D

-NOT ACTUAL SIZE-
-DO NOT SCALE-

CARTRIDGE: 6mm/30-30 Winchester Improved

OTHER NAMES:
.243/30-30 Improved

DIA: .243

BALLISTEK NO: 243K

NAI NO: RMB
23322/4.703

DATA SOURCE: NAI/Ballistek

HISTORICAL DATA:

NOTES:

LOADING DATA:

BULLET WT./TYPE	POWDER WT./TYPE	VELOCITY ('/SEC)	SOURCE
70/Spire	25.0/IMR4895	–	JJD

CASE PREPARATION: SHELLHOLDER (RCBS): 2

MAKE FROM: .25-35 Win. Anneal case neck and F/L size. Trim to length & chamfer. Fireform (with light load) in chamber.

PHYSICAL DATA (INCHES):

CASE TYPE: **Rimmed Bottleneck**

CASE LENGTH A = **1.98**

HEAD DIAMETER B = **.421**

RIM DIAMETER D = **.506**

NECK DIAMETER F = **.265**

NECK LENGTH H = **.415**

SHOULDER LENGTH K = **.215**

BODY ANGLE (DEG'S/SIDE): **.523**

CASE CAPACITY CC'S = **2.61**

LOADED LENGTH: **A/R**

BELT DIAMETER C = **N/A**

RIM THICKNESS E = **.062**

SHOULDER DIAMETER G = **.400**

LENGTH TO SHOULDER J = **1.35**

SHOULDER ANGLE (DEG'S/SIDE): **17.43**

PRIMER: **L/R**

CASE CAPACITY (GR'S WATER): **40.31**

DIMENSIONAL DRAWING:

F H K G A J E B D

-NOT ACTUAL SIZE-
-DO NOT SCALE-

CARTRIDGE: 6mm/303 British

OTHER NAMES:
.243/303 British

DIA: .243

BALLISTEK NO: 243AL

NAI NO: RMB 32422/5.055

DATA SOURCE: Ackley Vol.1 Pg309

HISTORICAL DATA: Australian design.

NOTES:

LOADING DATA:

BULLET WT./TYPE	POWDER WT./TYPE	VELOCITY ('/SEC)	SOURCE
75/Spire	30.0/3031	3000	JJD

CASE PREPARATION: SHELLHOLDER (RCBS): 7

MAKE FROM: .303 British. Reduce neck in 7mm/08 sizer (or, use 6mm/303 form die). Anneal. F/L size, trim & chamfer.

PHYSICAL DATA (INCHES):

CASE TYPE: **Rimmed Bottleneck**

CASE LENGTH **A = 2.300**

HEAD DIAMETER **B = .455**

RIM DIAMETER **D = .540**

NECK DIAMETER **F = .263**

NECK LENGTH **H = .320**

SHOULDER LENGTH **K = .183**

BODY ANGLE (DEG'S/SIDE): **1.11**

CASE CAPACITY **CC'S = 3.57**

LOADED LENGTH: **2.87**

BELT DIAMETER **C = N/A**

RIM THICKNESS **E = .063**

SHOULDER DIAMETER **G = .393**

LENGTH TO SHOULDER **J = 1.797**

SHOULDER ANGLE (DEG'S/SIDE): **19.55**

PRIMER: **L/R**

CASE CAPACITY (GR'S WATER): **55.14**

DIMENSIONAL DRAWING:

-NOT ACTUAL SIZE-
-DO NOT SCALE-

CARTRIDGE: 6mm/303 Epps

OTHER NAMES:

DIA: .243

BALLISTEK NO: 243AW

NAI NO: RMB 22424/5.011

DATA SOURCE: Ackley Vol.2 Pg148

HISTORICAL DATA: By E. Epps.

NOTES:

LOADING DATA:

BULLET WT./TYPE	POWDER WT./TYPE	VELOCITY ('/SEC)	SOURCE
75/Spire	50.0/4831	3300	JJD

CASE PREPARATION: SHELLHOLDER (RCBS): 7

MAKE FROM: .303 British. Run case into 7mm/08 F/L die to reduce neck dia. Anneal case. F/L size in Epps die with expander removed. Trim to length & chamfer. F/L size.

PHYSICAL DATA (INCHES):

CASE TYPE: **Rimmed Bottleneck**

CASE LENGTH **A = 2.28**

HEAD DIAMETER **B = .455**

RIM DIAMETER **D = .540**

NECK DIAMETER **F = .263**

NECK LENGTH **H = .323**

SHOULDER LENGTH **K = .097**

BODY ANGLE (DEG'S/SIDE): **.552**

CASE CAPACITY **CC'S = 3.85**

LOADED LENGTH: **2.82**

BELT DIAMETER **C = N/A**

RIM THICKNESS **E = .064**

SHOULDER DIAMETER **G = .423**

LENGTH TO SHOULDER **J = 1.86**

SHOULDER ANGLE (DEG'S/SIDE): **39.51**

PRIMER: **L/R**

CASE CAPACITY (GR'S WATER): **59.50**

DIMENSIONAL DRAWING:

-NOT ACTUAL SIZE-
-DO NOT SCALE-

CARTRIDGE: 6mm/350 Remington Magnum

OTHER NAMES:

DIA: .243

BALLISTEK NO: 243AQ

NAI NO: BEN 22433/4.327

DATA SOURCE: Ackley Vol.2 Pg148

HISTORICAL DATA:

NOTES: Overbore for .243" dia.

LOADING DATA:

BULLET WT./TYPE	POWDER WT./TYPE	VELOCITY ('/SEC)	SOURCE
100/Spire	60.0/4831	3444	Ackley

CASE PREPARATION: SHELLHOLDER (RCBS): 4

MAKE FROM: 6,5 Rem. Mag. F/L size case in 6,5/350 die to reduce neck dia. Trim to length & chamfer.

PHYSICAL DATA (INCHES):

CASE TYPE: **Belted Bottleneck**

CASE LENGTH **A = 2.22**

HEAD DIAMETER **B = .513**

RIM DIAMETER **D = .532**

NECK DIAMETER **F = .268**

NECK LENGTH **H = .246**

SHOULDER LENGTH **K = .214**

BODY ANGLE (DEG'S/SIDE): **.587**

CASE CAPACITY **CC'S = 4.65**

LOADED LENGTH: **2.75**

BELT DIAMETER **C = .532**

RIM THICKNESS **E = .05**

SHOULDER DIAMETER **G = .481**

LENGTH TO SHOULDER **J = 1.76**

SHOULDER ANGLE (DEG'S/SIDE): **26.45**

PRIMER: **L/R**

CASE CAPACITY (GR'S WATER): **71.80**

DIMENSIONAL DRAWING:

-NOT ACTUAL SIZE-
-DO NOT SCALE-

CARTRIDGE: 6 x 29.5mm Stahl

OTHER NAMES:

DIA: .243

BALLISTEK NO: 243BD

NAI NO: RMB 24242/3.625

DATA SOURCE: COTW 4th Pg255

HISTORICAL DATA: Circa 1880.

NOTES:

LOADING DATA:

BULLET WT./TYPE	POWDER WT./TYPE	VELOCITY ('/SEC)	SOURCE
85/Lead	7.0/2400	1460	Barnes

CASE PREPARATION: SHELLHOLDER (RCBS): 16

MAKE FROM: .22 Hornet. Turn a brass ring, .300" I.D. x .320" O.D. x .12" long. Slip over the Hornet's body and hold in place, against the rim, with super glue. Use 2.0 grs. of Bullseye to fireform the case in either the Stahl trim die or the actual chamber. The brass, forward of the ring, will expand and hold the ring in place. In place of turning rings, short sections of 5/16 O.D. thin wall tubing may be used but, they are harder to push on the case. Trim the formed case.

PHYSICAL DATA (INCHES):

CASE TYPE: **Rimmed Bottleneck**

CASE LENGTH **A = 1.16**

HEAD DIAMETER **B = .320**

RIM DIAMETER **D = .370**

NECK DIAMETER **F = .262**

NECK LENGTH **H = .240**

SHOULDER LENGTH **K = .070**

BODY ANGLE (DEG'S/SIDE): **.837**

CASE CAPACITY **CC'S = .861**

LOADED LENGTH: **1.44**

BELT DIAMETER **C = N/A**

RIM THICKNESS **E = .05**

SHOULDER DIAMETER **G = .301**

LENGTH TO SHOULDER **J = .85**

SHOULDER ANGLE (DEG'S/SIDE): **15.5**

PRIMER: **S/R**

CASE CAPACITY (GR'S WATER): **13.29**

DIMENSIONAL DRAWING:

-NOT ACTUAL SIZE-
-DO NOT SCALE-

CARTRIDGE: 6 x 39mm

OTHER NAMES:

DIA: .243

BALLISTEK NO: 243U

NAI NO: RXB 24434/3.248

DATA SOURCE: Rifle #53 Pg38

HISTORICAL DATA: By King & Ferguson about 1977.

NOTES:

LOADING DATA:

BULLET WT./TYPE	POWDER WT./TYPE	VELOCITY ('/SEC)	SOURCE
85/HPBT	36.8/W760	3000	JJD

CASE PREPARATION: SHELLHOLDER (RCBS): 3

MAKE FROM: .308 Win. Form in form dies #1 and #2. Run into trim die and trim to length. Chamfer. F/L size in 6x39 die. Case will fireform, slightly, in the chamber. This case can be made without the actual forming dies but, the intermediate dies are almost as exotic as the 6x39 forming dies!

PHYSICAL DATA (INCHES):

CASE TYPE: **Rimless Bottleneck**

CASE LENGTH **A = 1.53**

HEAD DIAMETER **B = .471**

RIM DIAMETER **D = .473**

NECK DIAMETER **F = .276**

NECK LENGTH **H = .303**

SHOULDER LENGTH **K = .087**

BODY ANGLE (DEG'S/SIDE): **.640**

CASE CAPACITY **CC'S = 2.30**

LOADED LENGTH: **A/R**

BELT DIAMETER **C = N/A**

RIM THICKNESS **E = .05**

SHOULDER DIAMETER **G = .450**

LENGTH TO SHOULDER **J = 1.14**

SHOULDER ANGLE (DEG'S/SIDE): **45.00**

PRIMER: **L/R**

CASE CAPACITY (GR'S WATER): **35.51**

DIMENSIONAL DRAWING:

-NOT ACTUAL SIZE-
-DO NOT SCALE-

CARTRIDGE: 6 x 47mm Remington

OTHER NAMES:
6mm/222 Remington Magnum

DIA: .243

BALLISTEK NO: 243V

NAI NO: RXB 22333/4.933

DATA SOURCE: Hornady 3rd Pg106

HISTORICAL DATA: Accredited to M. Walker.

NOTES:

LOADING DATA:

BULLET WT./TYPE	POWDER WT./TYPE	VELOCITY ('/SEC)	SOURCE
75/Spire	23.0/IMR4198	2990	JJD

CASE PREPARATION: SHELLHOLDER (RCBS): 10

MAKE FROM: .222 Rem. Mag. Taper neck expand to .243" dia. F/L size. Square case mouth and chamfer.

PHYSICAL DATA (INCHES):

CASE TYPE: **Rimless Bottleneck**

CASE LENGTH **A = 1.850**

HEAD DIAMETER **B = .375**

RIM DIAMETER **D = .378**

NECK DIAMETER **F = .266**

NECK LENGTH **H = .280**

SHOULDER LENGTH **K = .106**

BODY ANGLE (DEG'S/SIDE): **.407**

CASE CAPACITY **CC'S = 2.05**

LOADED LENGTH: **2.48**

BELT DIAMETER **C = N/A**

RIM THICKNESS **E = .045**

SHOULDER DIAMETER **G = .357**

LENGTH TO SHOULDER **J = 1.464**

SHOULDER ANGLE (DEG'S/SIDE): **23.23**

PRIMER: **S/R**

CASE CAPACITY (GR'S WATER): **31.63**

DIMENSIONAL DRAWING:

-NOT ACTUAL SIZE-
-DO NOT SCALE-

CARTRIDGE: 6 x 57 Mauser

OTHER NAMES:

DIA: .243

BALLISTEK NO: 243BO

NAI NO: RXB 32432/4.694

DATA SOURCE: COTW 4th Pg255

HISTORICAL DATA: German ctg. ca. 1895.

NOTES:

LOADING DATA:

BULLET WT./TYPE	POWDER WT./TYPE	VELOCITY ('/SEC)	SOURCE
140/—	40.8/4831	—	JJD

CASE PREPARATION: SHELLHOLDER (RCBS): 3

MAKE FROM: 7 x 57 Mauser by necking down, in 6x57 die. Or, from .25-06 Rem. by annealing and forming in 6x57 die with expander removed. Trim the case and I.D. neck ream, if needed. F/L size.

PHYSICAL DATA (INCHES):

CASE TYPE: **Rimless Bottleneck**

CASE LENGTH **A = 2.23**

HEAD DIAMETER **B = .475**

RIM DIAMETER **D = .476**

NECK DIAMETER **F = .284**

NECK LENGTH **H = .300**

SHOULDER LENGTH **K = .230**

BODY ANGLE (DEG'S/SIDE): **1.05**

CASE CAPACITY **CC'S = 3.30**

LOADED LENGTH: **2.95**

BELT DIAMETER **C = N/A**

RIM THICKNESS **E = .05**

SHOULDER DIAMETER **G = .420**

LENGTH TO SHOULDER **J = 1.70**

SHOULDER ANGLE (DEG'S/SIDE): **16.47**

PRIMER: **L/R**

CASE CAPACITY (GR'S WATER): **51.06**

DIMENSIONAL DRAWING:

F
H
K
A
G
J
E
B
D

-NOT ACTUAL SIZE-
-DO NOT SCALE-

CARTRIDGE: 6 x 58mm Forster

OTHER NAMES:

DIA: .243

BALLISTEK NO: 243BE

NAI NO: RXB 22423/4.655

DATA SOURCE: COTW 4th Pg256

HISTORICAL DATA: German ctg from about 1905.

NOTES:

LOADING DATA:

BULLET WT./TYPE	POWDER WT./TYPE	VELOCITY ('/SEC)	SOURCE
100/Spire	22.0/IMR3031	2600	JJD

CASE PREPARATION: SHELLHOLDER (RCBS): 3

MAKE FROM: .25-06 Rem. Anneal case neck and F/L size in the 6x58 die. Trim to length and chamfer. F/L size.

PHYSICAL DATA (INCHES):

CASE TYPE: **Rimless Bottleneck**

CASE LENGTH **A = 2.188**

HEAD DIAMETER **B = .470**

RIM DIAMETER **D = .468**

NECK DIAMETER **F = .285**

NECK LENGTH **H = .340**

SHOULDER LENGTH **K = .168**

BODY ANGLE (DEG'S/SIDE): **.639**

CASE CAPACITY **CC'S = 3.30**

LOADED LENGTH: **3.06**

BELT DIAMETER **C = N/A**

RIM THICKNESS **E = .062**

SHOULDER DIAMETER **G = .437**

LENGTH TO SHOULDER **J = 1.68**

SHOULDER ANGLE (DEG'S/SIDE): **24.34**

PRIMER: **L/R**

CASE CAPACITY (GR'S WATER): **50.91**

DIMENSIONAL DRAWING:

F
H
K
A
G
J
E
B
D

-NOT ACTUAL SIZE-
-DO NOT SCALE-

CARTRIDGE: 6 x 58R Forster

OTHER NAMES:

DIA: .243

BALLISTEK NO: 243BF

NAI NO: RMB 22423/4.645

DATA SOURCE: COTW 4th Pg256

HISTORICAL DATA: German ctg. from about 1904.

NOTES:

LOADING DATA:

BULLET WT./TYPE	POWDER WT./TYPE	VELOCITY ('/SEC)	SOURCE
100/Spire	22.0/IMR3031	2800	JJD

CASE PREPARATION: SHELLHOLDER (RCBS): 4

MAKE FROM: 7x57R or 9,3x74R. Trim case to length & chamfer mouth. Anneal. Form in 6x58R die. Check length and trim, if necessary. Neck ream if required. F/L size.

PHYSICAL DATA (INCHES):

CASE TYPE: **Rimmed Bottleneck**

CASE LENGTH **A = 2.188**

HEAD DIAMETER **B = .471**

RIM DIAMETER **D = .532**

NECK DIAMETER **F = .284**

NECK LENGTH **H = .340**

SHOULDER LENGTH **K = .168**

BODY ANGLE (DEG'S/SIDE): **.658**

CASE CAPACITY **CC'S = 3.33**

LOADED LENGTH: **3.06**

BELT DIAMETER **C = N/A**

RIM THICKNESS **E = .062**

SHOULDER DIAMETER **G = .437**

LENGTH TO SHOULDER **J = 1.68**

SHOULDER ANGLE (DEG'S/SIDE): **24.48**

PRIMER: **L/R**

CASE CAPACITY (GR'S WATER): **51.36**

DIMENSIONAL DRAWING:

-NOT ACTUAL SIZE-
-DO NOT SCALE-

CARTRIDGE: 6 x 61 Sharpe & Hart

OTHER NAMES:

DIA: .243

BALLISTEK NO: 243BS

NAI NO: BEN 22434/4.658

DATA SOURCE: NAI/Ballistek

HISTORICAL DATA:

NOTES:

LOADING DATA:

BULLET WT./TYPE	POWDER WT./TYPE	VELOCITY ('/SEC)	SOURCE
87/Spire	44.5/IMR4350	3000	JJD

CASE PREPARATION: SHELLHOLDER (RCBS): 4

MAKE FROM: 7 x 61 S&H. F/L size the 7mm case or form the 7x61 case from any suitable magnum and re-form, again, to the 6 x 61 case.

PHYSICAL DATA (INCHES):

CASE TYPE: **Belted Bottleneck**

CASE LENGTH **A = 2.425**

HEAD DIAMETER **B = .513**

RIM DIAMETER **D = .532**

NECK DIAMETER **F = .281**

NECK LENGTH **H = .323**

SHOULDER LENGTH **K = .102**

BODY ANGLE (DEG'S/SIDE): **.538**

CASE CAPACITY **CC'S = 4.73**

LOADED LENGTH: **A/R**

BELT DIAMETER **C = .532**

RIM THICKNESS **E = .05**

SHOULDER DIAMETER **G = .477**

LENGTH TO SHOULDER **J = 2.00**

SHOULDER ANGLE (DEG'S/SIDE): **44**

PRIMER: **L/R Mag**

CASE CAPACITY (GR'S WATER): **73.10**

DIMENSIONAL DRAWING:

-NOT ACTUAL SIZE-
-DO NOT SCALE-

CARTRIDGE: 6,3 x 53R Finnish

OTHER NAMES:

DIA: .257

BALLISTEK NO: 257BF

NAI NO: RMB 23423/4.294

DATA SOURCE: COTW 4th Pg256

HISTORICAL DATA: Finnish sporting ctg.

NOTES:

LOADING DATA:

BULLET WT./TYPE	POWDER WT./TYPE	VELOCITY ('/SEC)	SOURCE
140/Spire	38.0/IMR4831	—	JJD

CASE PREPARATION: SHELLHOLDER (RCBS): 13

MAKE FROM: 7,62 x 54R Russian. Anneal case. Size in 6,3x53R die with expander removed. Trim and chamfer. F/L size. Fireform in chamber.

PHYSICAL DATA (INCHES):

CASE TYPE: **Rimmed Bottleneck**

CASE LENGTH **A = 2.087**

HEAD DIAMETER **B = .486**

RIM DIAMETER **D = .565**

NECK DIAMETER **F = .285**

NECK LENGTH **H = .330**

SHOULDER LENGTH **K = .211**

BODY ANGLE (DEG'S/SIDE): **.489**

CASE CAPACITY **CC'S = 3.74**

LOADED LENGTH: **2.51**

BELT DIAMETER **C = N/A**

RIM THICKNESS **E = .064**

SHOULDER DIAMETER **G = .463**

LENGTH TO SHOULDER **J = 1.546**

SHOULDER ANGLE (DEG'S/SIDE): **22.87**

PRIMER: **L/R**

CASE CAPACITY (GR'S WATER): **57.80**

DIMENSIONAL DRAWING:

-NOT ACTUAL SIZE-
-DO NOT SCALE-

CARTRIDGE: 6,35 x 47mm

OTHER NAMES:
.25/222 Remington Magnum

DIA: .257

BALLISTEK NO: 257AP

NAI NO: RXB 22333/4.920

DATA SOURCE: Handloader #85 Pg31

HISTORICAL DATA: By W. Blackwell.

NOTES: Designed as a Contender ctg.

LOADING DATA:

BULLET WT./TYPE	POWDER WT./TYPE	VELOCITY ('/SEC)	SOURCE
75/SHP	25.3/IMR3031	2272	Blackwell

CASE PREPARATION: SHELLHOLDER (RCBS): 10

MAKE FROM: .222 Rem. Mag. Fire factory .222 Rem. Mag. ammo in 6,35x47 chamber. The Mag. case can also be taper expanded to about .260" dia. and F/L sized down.

PHYSICAL DATA (INCHES):

CASE TYPE: **Rimless Bottleneck**

CASE LENGTH **A = 1.85**

HEAD DIAMETER **B = .376**

RIM DIAMETER **D = .378**

NECK DIAMETER **F = .280**

NECK LENGTH **H = .319**

SHOULDER LENGTH **K = .068**

BODY ANGLE (DEG'S/SIDE): **.430**

CASE CAPACITY **CC'S = 2.07**

LOADED LENGTH: **A/R**

BELT DIAMETER **C = N/A**

RIM THICKNESS **E = .045**

SHOULDER DIAMETER **G = .357**

LENGTH TO SHOULDER **J = 1.463**

SHOULDER ANGLE (DEG'S/SIDE): **29.51**

PRIMER: **S/R**

CASE CAPACITY (GR'S WATER): **32.01**

DIMENSIONAL DRAWING:

-NOT ACTUAL SIZE-
-DO NOT SCALE-

CARTRIDGE: 6,35 x 59mm

OTHER NAMES:

DIA: .257

BALLISTEK NO: 257AQ

NAI NO: RXB 22423/4.942

DATA SOURCE: Rifle #53 Pg40

HISTORICAL DATA: By O. Winters.

NOTES: Version of .25-06 Improved.

LOADING DATA:

BULLET WT./TYPE	POWDER WT./TYPE	VELOCITY ('/SEC)	SOURCE
100/—	52.0/4350	3356	Unknown

CASE PREPARATION: SHELLHOLDER (RCBS): 3

MAKE FROM: .270 Win. Form dies are required. Form in form die, trim in trim die and ream in ream die. Chamfer and F/L size. Fireform in chamber. You may also just F/L size .25-06 (with fireforming) and I do not know why Mr. Winters did not go this route. If you do use the .25-06 case, watch headspacing!

PHYSICAL DATA (INCHES):

CASE TYPE: **Rimless Bottleneck**

CASE LENGTH **A = 2.323**

HEAD DIAMETER **B = .470**

RIM DIAMETER **D = .473**

NECK DIAMETER **F = .290**

NECK LENGTH **H = .340**

SHOULDER LENGTH **K = .133**

BODY ANGLE (DEG'S/SIDE): **.486**

CASE CAPACITY **CC'S = 3.89**

LOADED LENGTH: **3.08**

BELT DIAMETER **C = N/A**

RIM THICKNESS **E = .05**

SHOULDER DIAMETER **G = .442**

LENGTH TO SHOULDER **J = 1.85**

SHOULDER ANGLE (DEG'S/SIDE): **29.74**

PRIMER: **L/R**

CASE CAPACITY (GR'S WATER): **60.12**

DIMENSIONAL DRAWING:

-NOT ACTUAL SIZE-
-DO NOT SCALE-

CARTRIDGE: 6,5 Apex Magnum

OTHER NAMES:

DIA: .264

BALLISTEK NO: 264V

NAI NO: BEN 22424/4.922

DATA SOURCE: Ackley Vol.1 Pg373

HISTORICAL DATA: By Apex Rifle Co.

NOTES:

LOADING DATA:

BULLET WT./TYPE	POWDER WT./TYPE	VELOCITY ('/SEC)	SOURCE
140/—	56.0/4350	2940	Ackley

CASE PREPARATION: SHELLHOLDER (RCBS): 4

MAKE FROM: 7mm Rem. Mag. F/L size the Rem. case in the Apex die. Fireform in the Apex trim die and size, again, in the F/L sizer. Form dies are available.

PHYSICAL DATA (INCHES):

CASE TYPE: **Belted Bottleneck**

CASE LENGTH **A = 2.52**

HEAD DIAMETER **B = .512**

RIM DIAMETER **D = .532**

NECK DIAMETER **F = .290**

NECK LENGTH **H = .335**

SHOULDER LENGTH **K = .125**

BODY ANGLE (DEG'S/SIDE): **.877**

CASE CAPACITY **CC'S = 5.11**

LOADED LENGTH: **A/R**

BELT DIAMETER **C = .532**

RIM THICKNESS **E = .049**

SHOULDER DIAMETER **G = .455**

LENGTH TO SHOULDER **J = 2.06**

SHOULDER ANGLE (DEG'S/SIDE): **33.42**

PRIMER: **L/R Mag**

CASE CAPACITY (GR'S WATER): **78.94**

DIMENSIONAL DRAWING:

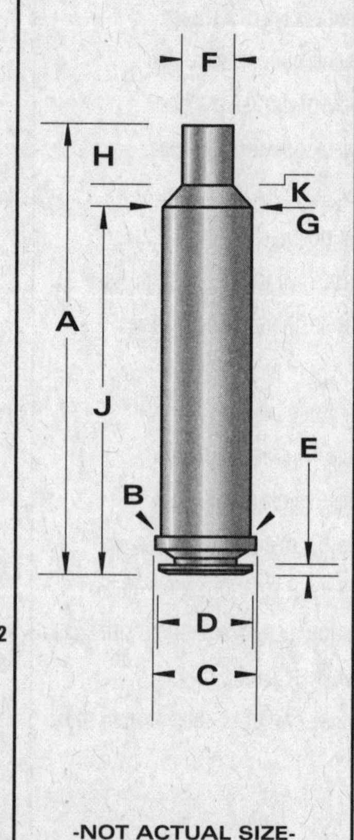

-NOT ACTUAL SIZE-
-DO NOT SCALE-

CARTRIDGE: 6,5mm Bergmann

OTHER NAMES:

DIA: .264

BALLISTEK NO: 264N

NAI NO: RXB 34243/2.370

DATA SOURCE: COTW 4th Pg163

HISTORICAL DATA: By T. Bergmann about 1894.

NOTES: Early versions have no extractor groove.

LOADING DATA:

BULLET WT./TYPE	POWDER WT./TYPE	VELOCITY ('/SEC)	SOURCE
85/Lead	7.0/2400	—	JJD

CASE PREPARATION: SHELLHOLDER (RCBS): 10

MAKE FROM: .223 Rem. Cut case off to .89". Turn base and rim to .368" dia. (some pistols will not need this; .223 cases can be used as is, after sizing). Deepen extractor groove. F/L size and I.D. neck ream. Chamfer.

PHYSICAL DATA (INCHES):

CASE TYPE: **Rimless Bottleneck**

CASE LENGTH **A = .870**

HEAD DIAMETER **B = .367**

RIM DIAMETER **D = .370**

NECK DIAMETER **F = .289**

NECK LENGTH **H = .175**

SHOULDER LENGTH **K = .035**

BODY ANGLE (DEG'S/SIDE): **2.61**

CASE CAPACITY **CC'S = .698**

LOADED LENGTH: **1.23**

BELT DIAMETER **C = N/A**

RIM THICKNESS **E = .040**

SHOULDER DIAMETER **G = .325**

LENGTH TO SHOULDER **J = .66**

SHOULDER ANGLE (DEG'S/SIDE): **27.21**

PRIMER: **S/R**

CASE CAPACITY (GR'S WATER): **10.78**

DIMENSIONAL DRAWING:

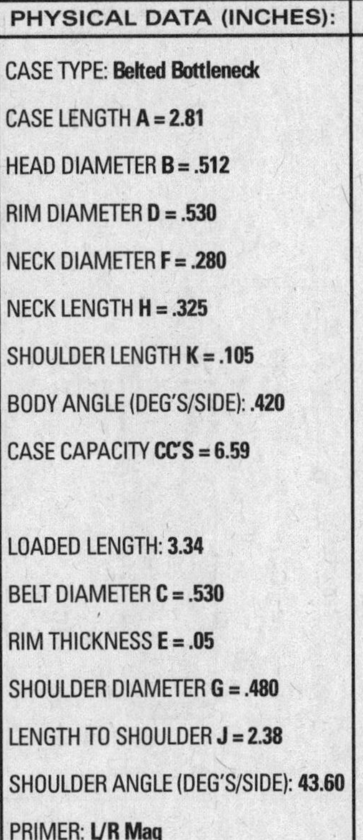

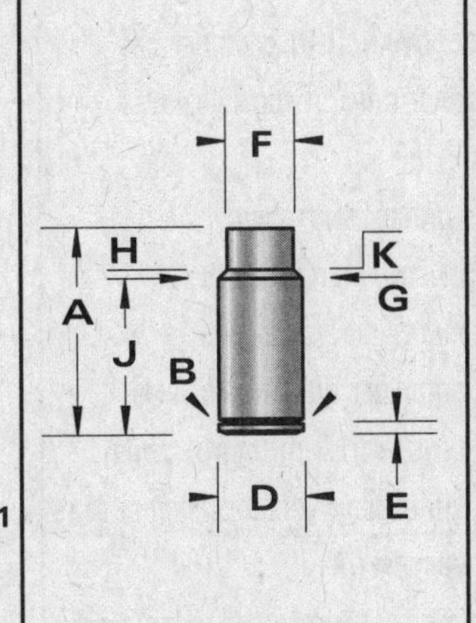

-NOT ACTUAL SIZE-
-DO NOT SCALE-

CARTRIDGE: 6,5 Critser Express

OTHER NAMES:

DIA: .264

BALLISTEK NO: 264S

NAI NO: BEN 22434/5.488

DATA SOURCE: Ackley Vol.1 Pg374

HISTORICAL DATA:

NOTES: About max volume for .300 H&H case. (per Ackley)

LOADING DATA:

BULLET WT./TYPE	POWDER WT./TYPE	VELOCITY ('/SEC)	SOURCE
140/Spire	65.0/H570	2910	Ackley

CASE PREPARATION: SHELLHOLDER (RCBS): 4

MAKE FROM: .300 H&H Mag. Reduce case neck in 7mm Rem. Mag. F/L size in Critser die. Trim, chamfer and fire-form in chamber of trim die.

PHYSICAL DATA (INCHES):

CASE TYPE: **Belted Bottleneck**

CASE LENGTH **A = 2.81**

HEAD DIAMETER **B = .512**

RIM DIAMETER **D = .530**

NECK DIAMETER **F = .280**

NECK LENGTH **H = .325**

SHOULDER LENGTH **K = .105**

BODY ANGLE (DEG'S/SIDE): **.420**

CASE CAPACITY **CC'S = 6.59**

LOADED LENGTH: **3.34**

BELT DIAMETER **C = .530**

RIM THICKNESS **E = .05**

SHOULDER DIAMETER **G = .480**

LENGTH TO SHOULDER **J = 2.38**

SHOULDER ANGLE (DEG'S/SIDE): **43.60**

PRIMER: **L/R Mag**

CASE CAPACITY (GR'S WATER):

101.84

DIMENSIONAL DRAWING:

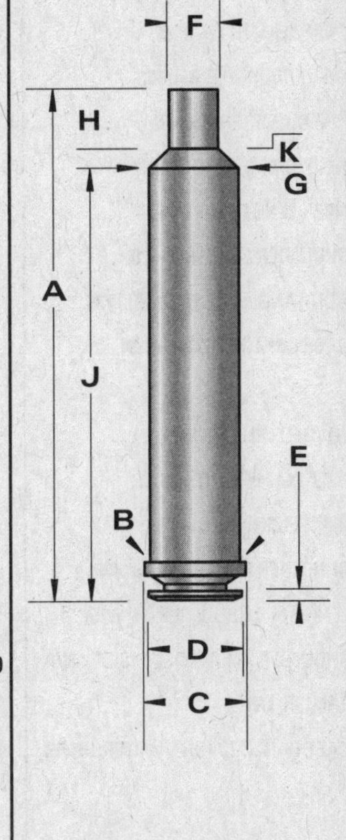

-NOT ACTUAL SIZE-
-DO NOT SCALE-

CARTRIDGE: 6,5mm Express Jr.

OTHER NAMES:

DIA: .264

BALLISTEK NO: 264BD

NAI NO: RBB 22433/4.460

DATA SOURCE: Ackley Vol.2 Pg158

HISTORICAL DATA: By R. Payne.

NOTES:

LOADING DATA:

BULLET WT./TYPE	POWDER WT./TYPE	VELOCITY ('/SEC)	SOURCE
140/—	52.0/4350	3080	Payne/ Ackley

CASE PREPARATION: SHELLHOLDER (RCBS): 3

MAKE FROM: .284 Winchester. F/L size Win. case in 6,5mm die and fireform in chamber to produce typical Payne (read: Weatherby) shoulder. Square case mouth, chamfer and reload.

PHYSICAL DATA (INCHES):

CASE TYPE: **Rebated Bottleneck**

CASE LENGTH **A = 2.23**

HEAD DIAMETER **B = .500**

RIM DIAMETER **D = .473**

NECK DIAMETER **F = .289**

NECK LENGTH **H = .316**

SHOULDER LENGTH **K = .164**

BODY ANGLE (DEG'S/SIDE): **.739**

CASE CAPACITY **CC'S = 4.34**

LOADED LENGTH: **3.06**

BELT DIAMETER **C = N/A**

RIM THICKNESS **E = .054**

SHOULDER DIAMETER **G = .460**

LENGTH TO SHOULDER **J = 1.75**

SHOULDER ANGLE (DEG'S/SIDE): **N/A**

PRIMER: **L/R**

CASE CAPACITY (GR'S WATER): **66.98**

DIMENSIONAL DRAWING:

F H K G A J E B D

-NOT ACTUAL SIZE-
-DO NOT SCALE-

CARTRIDGE: 6,5 Gibbs

OTHER NAMES:
6,5 Gibbs Magnum

DIA: .264

BALLISTEK NO: 264BI

NAI NO: RXB 12434/5.351

DATA SOURCE: Ackley Vol.2 Pg164

HISTORICAL DATA: By G. Gibbs.

NOTES:

LOADING DATA:

BULLET WT./TYPE	POWDER WT./TYPE	VELOCITY ('/SEC)	SOURCE
140/RN	61.0/4831	3010	Gibbs/ Ackley

CASE PREPARATION: SHELLHOLDER (RCBS): 3

MAKE FROM: .270 Win. F/L size Win. case in Gibbs die. Square case mouth, chamfer and fireformm in chamber.

PHYSICAL DATA (INCHES):

CASE TYPE: **Rimless Bottleneck**

CASE LENGTH **A = 2.515**

HEAD DIAMETER **B = .470**

RIM DIAMETER **D = .473**

NECK DIAMETER **F = .288**

NECK LENGTH **H = .270**

SHOULDER LENGTH **K = .049**

BODY ANGLE (DEG'S/SIDE): **.301**

CASE CAPACITY **CC'S = 4.82**

LOADED LENGTH: **A/R**

BELT DIAMETER **C = N/A**

RIM THICKNESS **E = .049**

SHOULDER DIAMETER **G = .449**

LENGTH TO SHOULDER **J = 2.195**

SHOULDER ANGLE (DEG'S/SIDE): **58.15**

PRIMER: **L/R**

CASE CAPACITY (GR'S WATER): **74.41**

DIMENSIONAL DRAWING:

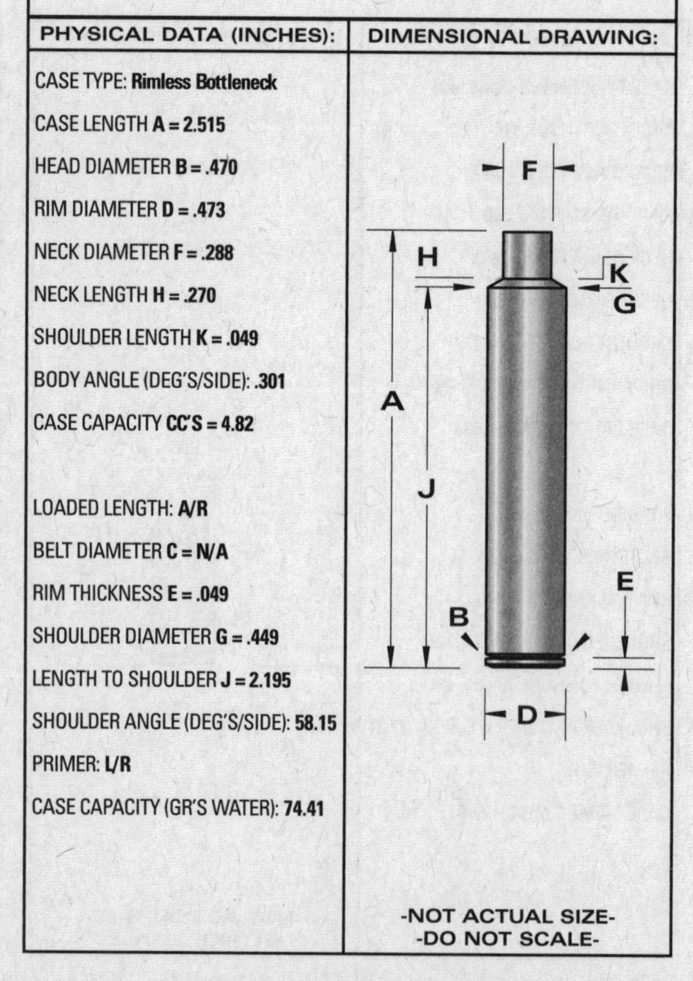

-NOT ACTUAL SIZE-
-DO NOT SCALE-

CARTRIDGE: 6,5 ICL Boar

OTHER NAMES:

DIA: .264

BALLISTEK NO: 264AY

NAI NO: RXB 12424/5.298

DATA SOURCE: Ackley Vol.1 Pg365

HISTORICAL DATA:

NOTES:

LOADING DATA:

BULLET WT./TYPE	POWDER WT./TYPE	VELOCITY ('/SEC)	SOURCE
150/Spire	54.0/4350	2885	Ackley

CASE PREPARATION: SHELLHOLDER (RCBS): 3

MAKE FROM: .270 Winchester. F/L size the .270 case in the ICL chamber. Fireform in chamber.

PHYSICAL DATA (INCHES):

CASE TYPE: **Rimless Bottleneck**

CASE LENGTH **A = 2.49**

HEAD DIAMETER **B = .470**

RIM DIAMETER **D = .473**

NECK DIAMETER **F = .295**

NECK LENGTH **H = .390**

SHOULDER LENGTH **K = .10**

BODY ANGLE (DEG'S/SIDE): **.239**

CASE CAPACITY **CC'S = 4.41**

LOADED LENGTH: **3.23**

BELT DIAMETER **C = N/A**

RIM THICKNESS **E = .05**

SHOULDER DIAMETER **G = .455**

LENGTH TO SHOULDER **J = 2.00**

SHOULDER ANGLE (DEG'S/SIDE): **38.66**

PRIMER: **L/R**

CASE CAPACITY (GR'S WATER): **68.13**

DIMENSIONAL DRAWING:

F H K G A J E B D

-NOT ACTUAL SIZE-
-DO NOT SCALE-

CARTRIDGE: 6,5 ICL Magnum

OTHER NAMES:

DIA: .264

BALLISTEK NO: 264BE

NAI NO: BEN 12434/4.834

DATA SOURCE: Ackley Vol.2 Pg163

HISTORICAL DATA:

NOTES: Reported as minimum body taper and 45° shoulder.

LOADING DATA:

BULLET WT./TYPE	POWDER WT./TYPE	VELOCITY ('/SEC)	SOURCE
140/Spire	65.5/4831	3145	Ackley

CASE PREPARATION: SHELLHOLDER (RCBS): 4

MAKE FROM: 7mm Rem. Mag. Size Rem. case in F/L die. Trim & chamfer. Fireform in ICL chamber.

PHYSICAL DATA (INCHES):

CASE TYPE: **Belted Bottleneck**

CASE LENGTH **A = 2.48**

HEAD DIAMETER **B = .513**

RIM DIAMETER **D = .532**

NECK DIAMETER **F = .291**

NECK LENGTH **H = .320**

SHOULDER LENGTH **K = .096**

BODY ANGLE (DEG'S/SIDE): **.184**

CASE CAPACITY **CC'S = 5.55**

LOADED LENGTH: **3.25**

BELT DIAMETER **C = .532**

RIM THICKNESS **E = .049**

SHOULDER DIAMETER **G = .501**

LENGTH TO SHOULDER **J = 2.063**

SHOULDER ANGLE (DEG'S/SIDE): **47.26**

PRIMER: **L/R Mag**

CASE CAPACITY (GR'S WATER): **85.74**

DIMENSIONAL DRAWING:

F H K G A J B E D C

-NOT ACTUAL SIZE-
-DO NOT SCALE-

CARTRIDGE: 6,5 JDJ

OTHER NAMES:

DIA: .264

BALLISTEK NO: 264BN

NAI NO: RMB 12344/4.540

DATA SOURCE: SSK Industries

HISTORICAL DATA: By J.D. Jones.

NOTES:

LOADING DATA:

BULLET WT./TYPE	POWDER WT./TYPE	VELOCITY ('/SEC)	SOURCE
140/—	35.0/4350	2141	SSK

CASE PREPARATION: SHELLHOLDER (RCBS): 11

MAKE FROM: .225 Winchester. Taper expand neck to hold .264" dia. bullets and fireform with a 10% reduced load.

PHYSICAL DATA (INCHES):

CASE TYPE: **Rimmed Bottleneck**

CASE LENGTH **A = 1.907**

HEAD DIAMETER **B = .420**

RIM DIAMETER **D = .469**

NECK DIAMETER **F = .294**

NECK LENGTH **H = .196**

SHOULDER LENGTH **K = .071**

BODY ANGLE (DEG'S/SIDE): **.219**

CASE CAPACITY **CC'S = 2.72**

LOADED LENGTH: **2.880**

BELT DIAMETER **C = N/A**

RIM THICKNESS **E = .048**

SHOULDER DIAMETER **G = .409**

LENGTH TO SHOULDER **J = 1.640**

SHOULDER ANGLE (DEG'S/SIDE): **39.00**

PRIMER: **L/R**

CASE CAPACITY (GR'S WATER): **42.01**

DIMENSIONAL DRAWING:

F H K G A J E B D

-NOT ACTUAL SIZE-
-DO NOT SCALE-

CARTRIDGE: 6,5 Remington Magnum

OTHER NAMES:

DIA: .264

BALLISTEK NO: 264C

NAI NO: BEN 12443/4.23

DATA SOURCE: Hornady Manual 3rd Pg162

HISTORICAL DATA: By Rem. in 1966.

NOTES:

LOADING DATA:

BULLET WT./TYPE	POWDER WT./TYPE	VELOCITY ('/SEC)	SOURCE
129/Spire	47.7/IMR4350	2900	Horn.

CASE PREPARATION: SHELLHOLDER (RCBS): 4

MAKE FROM: Factory or .300 H&H Mag. Run Holland case into 6,5 Rem. die. You may need to anneal. Trim to length. Chamfer. F/L size.

PHYSICAL DATA (INCHES):

CASE TYPE: **Belted Bottleneck**

CASE LENGTH **A = 2.17**

HEAD DIAMETER **B = .513**

RIM DIAMETER **D = .532**

NECK DIAMETER **F = .298**

NECK LENGTH **H = .261**

SHOULDER LENGTH **K = .212**

BODY ANGLE (DEG'S/SIDE): **.325**

CASE CAPACITY **CC'S = 4.44**

LOADED LENGTH: **3.05**

BELT DIAMETER **C = .532**

RIM THICKNESS **E = .05**

SHOULDER DIAMETER **G = .496**

LENGTH TO SHOULDER **J = 1.697**

SHOULDER ANGLE (DEG'S/SIDE): **25.03**

PRIMER: **L/R Mag**

CASE CAPACITY (GR'S WATER): **68.64**

DIMENSIONAL DRAWING:

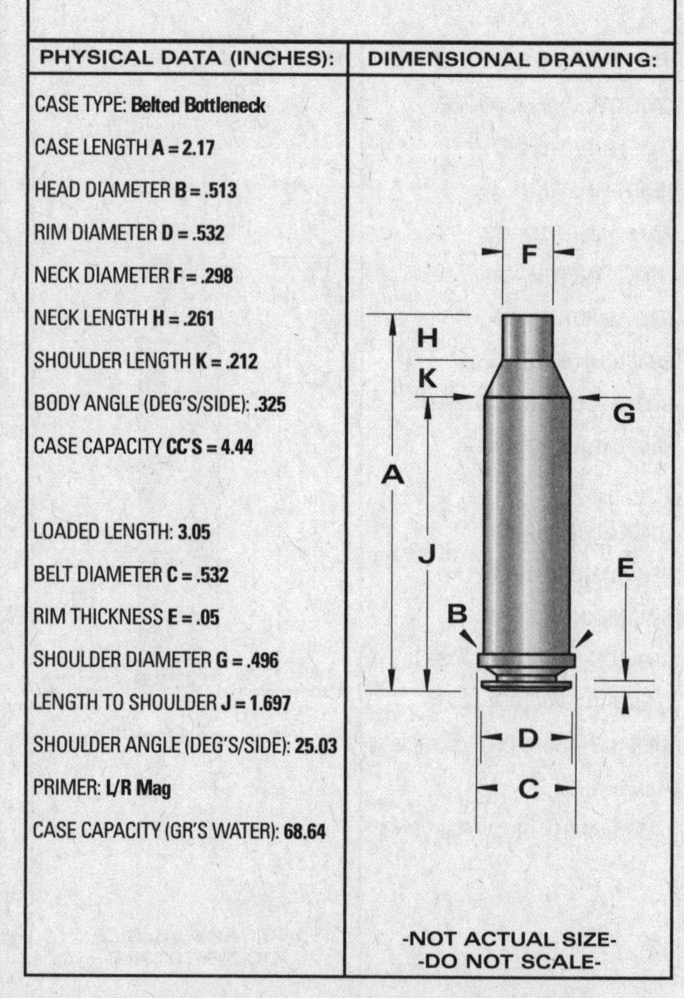

-NOT ACTUAL SIZE-
-DO NOT SCALE-

CARTRIDGE: 6,5 Spence Special

OTHER NAMES:

DIA: .264

BALLISTEK NO: 264AZ

NAI NO: RXB 22433/4.734

DATA SOURCE: Ackley Vol.1 Pg355

HISTORICAL DATA: By G. Spence.

NOTES: Similar to 6,5x57 Mauser.

LOADING DATA:

BULLET WT./TYPE	POWDER WT./TYPE	VELOCITY ('/SEC)	SOURCE
120/Spire	38.0/H380	2480	Ackley

CASE PREPARATION: SHELLHOLDER (RCBS): 11

MAKE FROM: 7 x 57 Mauser. F/L size Mauser case in F/L die. .257 Roberts brass bay also be taper expanded to 6,5 cal. (.264) and fireformed in the chamber.

PHYSICAL DATA (INCHES):

CASE TYPE: **Rimless Bottleneck**

CASE LENGTH **A = 2.23**

HEAD DIAMETER **B = .471**

RIM DIAMETER **D = .473**

NECK DIAMETER **F = .295**

NECK LENGTH **H = .320**

SHOULDER LENGTH **K = .180**

BODY ANGLE (DEG'S/SIDE): **.767**

CASE CAPACITY **CC'S = 3.66**

LOADED LENGTH: **2.85**

BELT DIAMETER **C = N/A**

RIM THICKNESS **E = .05**

SHOULDER DIAMETER **G = .430**

LENGTH TO SHOULDER **J = 1.73**

SHOULDER ANGLE (DEG'S/SIDE): **20.55**

PRIMER: **L/R**

CASE CAPACITY (GR'S WATER): **56.55**

DIMENSIONAL DRAWING:

-NOT ACTUAL SIZE-
-DO NOT SCALE-

CARTRIDGE: 6,5 Super Mashburn

OTHER NAMES:

DIA: .264

BALLISTEK NO: 264U

NAI NO: BEN 22424/5.126

DATA SOURCE: Ackley Vol.1 Pg374

HISTORICAL DATA:

NOTES: Similar to 6,5 Apex (#264V).

LOADING DATA:

BULLET WT./TYPE	POWDER WT./TYPE	VELOCITY ('/SEC)	SOURCE
140/RN	64.0/4831	3345	Ackley

CASE PREPARATION: SHELLHOLDER (RCBS): 4

MAKE FROM: .300 H&H Mag. Reduce case neck in 7mm Weatherby Mag. F/L die and F/L size in Mashburn die, with expander removed. Trim, chamfer and F/L size again. Fireform in chamber. You can also obtain a form set for the ctg.

PHYSICAL DATA (INCHES):

CASE TYPE: **Belted Bottleneck**

CASE LENGTH **A = 2.63**

HEAD DIAMETER **B = .513**

RIM DIAMETER **D = .532**

NECK DIAMETER **F = .294**

NECK LENGTH **H = .405**

SHOULDER LENGTH **K = .155**

BODY ANGLE (DEG'S/SIDE): **.382**

CASE CAPACITY **CC'S = 5.50**

LOADED LENGTH: **3.52**

BELT DIAMETER **C = .532**

RIM THICKNESS **E = .05**

SHOULDER DIAMETER **G = .488**

LENGTH TO SHOULDER **J = 2.07**

SHOULDER ANGLE (DEG'S/SIDE): **32.03**

PRIMER: **L/R Mag**

CASE CAPACITY (GR'S WATER): **84.95**

DIMENSIONAL DRAWING:

-NOT ACTUAL SIZE-
-DO NOT SCALE-

CARTRIDGE: 6.5mm Whisper

OTHER NAMES:

DIA: .264

BALLISTEK NO:

NAI NO:

DATA SOURCE: J.D. Jones

HISTORICAL DATA: The design intent was application in sound-suppressed M-15s, bolt-action rifles and T/C Contenders.

NOTES:

LOADING DATA:

BULLET WT./TYPE	POWDER WT./TYPE	VELOCITY (°/SEC)	SOURCE
100 Hornady	19.0 - H-110	2300	SSK
120 Nosler BT	19.0 - A-1680	2150	SSK
155	8.3 - H-110	970	SSK

CASE PREPARATION: SHELL HOLDER (RCBS): 10

MAKE FROM: 221 Remington. Neck up with expander ball.

PHYSICAL DATA (INCHES):

CASE TYPE: **Rimless Bottleneck**

CASE LENGTH **A = 1.36**

HEAD DIAMETER **B = .376**

RIM DIAMETER **D = .378**

NECK DIAMETER **F = .286**

NECK LENGTH **H = .278**

SHOULDER LENGTH **K = .051**

BODY ANGLE (DEG'S/SIDE): **N/A**

CASE CAPACITY **CC'S**

LOADED LENGTH: **2.26**

BELT DIAMETER **C = N/A**

RIM THICKNESS **E = .045**

SHOULDER DIAMETER **G = .361**

LENGTH TO SHOULDER **J = 1.071**

SHOULDER ANGLE (DEG'S/SIDE):

PRIMER: **S/R**

CASE CAPACITY (GR'S WATER):

DIMENSIONAL DRAWING:

F H K G A J B E D

-NOT ACTUAL SIZE-
-DO NOT SCALE-

CARTRIDGE: 6,5 x 27R

OTHER NAMES:
6,5mm Einzelladerbuchse

DIA: .257

BALLISTEK NO: 257L

NAI NO: RMB 34342/2.796

DATA SOURCE: COTW 4th Pg257

HISTORICAL DATA: Early German ctg. for single-shot rifles.

NOTES:

LOADING DATA:

BULLET WT./TYPE	POWDER WT./TYPE	VELOCITY (°/SEC)	SOURCE
85/Lead	7.0/2400	1460	Barnes

CASE PREPARATION: SHELLHOLDER (RCBS): 6

MAKE FROM: .357 Magnum. Anneal case and trim to length. Chamfer and form in F/L die.

PHYSICAL DATA (INCHES):

CASE TYPE: **Rimmed Bottleneck**

CASE LENGTH **A = 1.06**

HEAD DIAMETER **B = .379**

RIM DIAMETER **D = .428**

NECK DIAMETER **F = .284**

NECK LENGTH **H = .220**

SHOULDER LENGTH **K = .090**

BODY ANGLE (DEG'S/SIDE): **1.61**

CASE CAPACITY **CC'S = .96**

LOADED LENGTH: **1.54**

BELT DIAMETER **C = N/A**

RIM THICKNESS **E = .051**

SHOULDER DIAMETER **G = .348**

LENGTH TO SHOULDER **J = .750**

SHOULDER ANGLE (DEG'S/SIDE): **19.57**

PRIMER: **S/P**

CASE CAPACITY (GR'S WATER): **14.82**

DIMENSIONAL DRAWING:

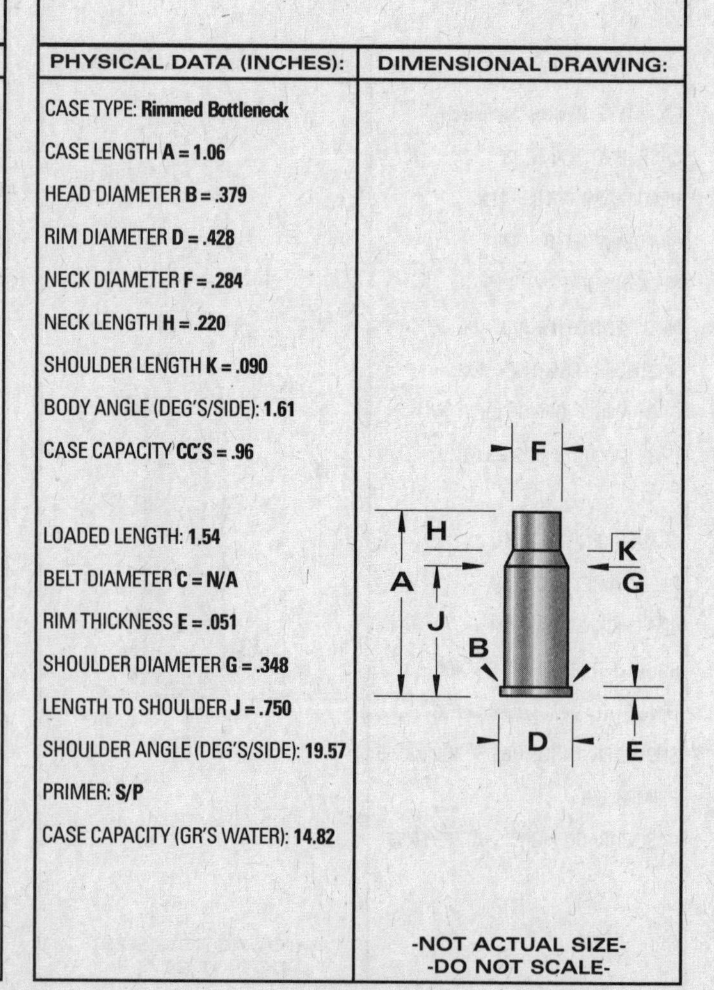

-NOT ACTUAL SIZE-
-DO NOT SCALE-

CARTRIDGE: 6,5 x 38 Ballistek

OTHER NAMES:	
6,5 Ballistek	DIA: .264 BALLISTEK NO: 264AM NAI NO: RXB 24344/3.232

DATA SOURCE: NAI/Ballistek

HISTORICAL DATA: By author in 1980.

NOTES: Experimental assault rifle cartridge.

LOADING DATA:

BULLET WT./TYPE	POWDER WT./TYPE	VELOCITY ('/SEC)	SOURCE
140/Spire	30.0/BL-C2	2250	JJD

CASE PREPARATION: SHELLHOLDER (RCBS): 3

MAKE FROM: Rem. B-R. Anneal case neck. Run into 6,5x38 form die. Trim to length & chamfer. Clean. F/L size. Final shoulder will fireform on ignition.

PHYSICAL DATA (INCHES):

CASE TYPE: **Rimless Bottleneck**

CASE LENGTH **A = 1.500**

HEAD DIAMETER **B = .464**

RIM DIAMETER **D = .468**

NECK DIAMETER **F = .290**

NECK LENGTH **H = .263**

SHOULDER LENGTH **K = .108**

BODY ANGLE (DEG'S/SIDE): **.709**

CASE CAPACITY **CC'S = 2.34**

LOADED LENGTH: **2.285**

BELT DIAMETER **C = N/A**

RIM THICKNESS **E = .049**

SHOULDER DIAMETER **G = .441**

LENGTH TO SHOULDER **J = 1.129**

SHOULDER ANGLE (DEG'S/SIDE): **34.95**

PRIMER: **S/R**

CASE CAPACITY (GR'S WATER): **36.17**

DIMENSIONAL DRAWING:

-NOT ACTUAL SIZE-
-DO NOT SCALE-

CARTRIDGE: 6,5 x 40R Sauer

OTHER NAMES:	
	DIA: .250 BALLISTEK NO: 250C NAI NO: RMS 31115/3.989

DATA SOURCE: COTW 4th Pg257

HISTORICAL DATA: Early German ctg. for single-shot rifles.

NOTES:

LOADING DATA:

BULLET WT./TYPE	POWDER WT./TYPE	VELOCITY ('/SEC)	SOURCE
100/Lead	5.0/2400	1200	Barnes

CASE PREPARATION: SHELLHOLDER (RCBS): Spl.

MAKE FROM: Fabricate this case from brass solid stock. Anneal, trim to length and F/L size.

PHYSICAL DATA (INCHES):

CASE TYPE: **Rimmed Straight**

CASE LENGTH **A = 1.58**

HEAD DIAMETER **B = .396**

RIM DIAMETER **D = .451**

NECK DIAMETER **F = .290**

NECK LENGTH **H = N/A**

SHOULDER LENGTH **K = N/A**

BODY ANGLE (DEG'S/SIDE): **2.01**

CASE CAPACITY **CC'S = 1.31**

LOADED LENGTH: **2.07**

BELT DIAMETER **C = N/A**

RIM THICKNESS **E = .07**

SHOULDER DIAMETER **G = N/A**

LENGTH TO SHOULDER **J = N/A**

SHOULDER ANGLE (DEG'S/SIDE): **N/A**

PRIMER: **S/R**

CASE CAPACITY (GR'S WATER): **20.32**

DIMENSIONAL DRAWING:

-NOT ACTUAL SIZE-
-DO NOT SCALE-

CARTRIDGE: 6,5 x 48R Sauer

OTHER NAMES:

DIA: .260

BALLISTEK NO: 260B

NAI NO: RMS 31115/4.342

DATA SOURCE: COTW 4th Pg257

HISTORICAL DATA:

NOTES:

LOADING DATA:

BULLET WT./TYPE	POWDER WT./TYPE	VELOCITY ('/SEC)	SOURCE
120/Lead	12.0/4198	1260	Barnes

CASE PREPARATION: SHELLHOLDER (RCBS): 21

MAKE FROM: .303 Savage. Anneal case neck and F/L size. Trim & chamfer.

PHYSICAL DATA (INCHES):

CASE TYPE: **Rimmed Straight**
CASE LENGTH **A = 1.88**
HEAD DIAMETER **B = .433**
RIM DIAMETER **D = .495**
NECK DIAMETER **F = .284**
NECK LENGTH **H = N/A**
SHOULDER LENGTH **K = N/A**
BODY ANGLE (DEG'S/SIDE): **2.33**
CASE CAPACITY **CC'S = 2.13**
LOADED LENGTH: **2.43**
BELT DIAMETER **C = N/A**
RIM THICKNESS **E = .051**
SHOULDER DIAMETER **G = N/A**
LENGTH TO SHOULDER **J = N/A**
SHOULDER ANGLE (DEG'S/SIDE): **N/A**
PRIMER: **L/R**
CASE CAPACITY (GR'S WATER): **32.88**

DIMENSIONAL DRAWING:

F A B E D

-NOT ACTUAL SIZE-
-DO NOT SCALE-

CARTRIDGE: 6,5 x 50mm Arisaka

OTHER NAMES:
6,5mm Japanese
6,5 x 51 Arisaka

DIA: .264

BALLISTEK NO: 264D

NAI NO: RMB 22333/4.438

DATA SOURCE: Hornady Manual 3rd Pg147

HISTORICAL DATA: Japanese mil. ctg. ca. 1905.

NOTES:

LOADING DATA:

BULLET WT./TYPE	POWDER WT./TYPE	VELOCITY ('/SEC)	SOURCE
140/Spire	33.0/IMR4895	2450	JJD

CASE PREPARATION: SHELLHOLDER (RCBS): 15

MAKE FROM: .270 Win. Swage base to .447" dia. Size in 6,5 Japenese die with expander removed. Trim to length. I.D. neck ream. F/L size. Chamfer.

PHYSICAL DATA (INCHES):

CASE TYPE: **Rimmed Bottleneck**
CASE LENGTH **A = 1.984**
HEAD DIAMETER **B = .447**
RIM DIAMETER **D = .466**
NECK DIAMETER **F = .288**
NECK LENGTH **H = .281**
SHOULDER LENGTH **K = .172**
BODY ANGLE (DEG'S/SIDE): **.645**
CASE CAPACITY **CC'S = 3.09**
LOADED LENGTH: **2.85**
BELT DIAMETER **C = N/A**
RIM THICKNESS **E = .05**
SHOULDER DIAMETER **G = .417**
LENGTH TO SHOULDER **J = 1.531**
SHOULDER ANGLE (DEG'S/SIDE): **20.55**
PRIMER: **L/R**
CASE CAPACITY (GR'S WATER): **47.67**

DIMENSIONAL DRAWING:

F H K G A J E B D

-NOT ACTUAL SIZE-
-DO NOT SCALE-

CARTRIDGE: 6,5 x 52mm Mannlicher-Carcano

OTHER NAMES:
6,5 Carcano
6,5 Italian

DIA: .264

BALLISTEK NO: 264E

NAI NO: RXB
22333/4.63

DATA SOURCE: Hornady Manual 3rd Pg150

HISTORICAL DATA: Italian mil. ctg. adopted in 1891.

NOTES:

LOADING DATA:

BULLET WT./TYPE	POWDER WT./TYPE	VELOCITY ('/SEC)	SOURCE
140/Spire	31.0/IMR4895	2125	JJD

CASE PREPARATION: SHELLHOLDER (RCBS): 9

MAKE FROM: .220 Swift. Cut off rim and deepen extractor groove. Size in 6,5 M/C sizer. Trim to length. Chamfer. F/L size. .270 Win. may also be used by swaging base to .446" dia., annealing, sizing and I.D. neck reaming.

PHYSICAL DATA (INCHES):

CASE TYPE: **Rimless Bottleneck**

CASE LENGTH **A = 2.065**

HEAD DIAMETER **B = .446**

RIM DIAMETER **D = .449**

NECK DIAMETER **F = .295**

NECK LENGTH **H = .290**

SHOULDER LENGTH **K = .157**

BODY ANGLE (DEG'S/SIDE): **.424**

CASE CAPACITY **CC'S = 3.155**

LOADED LENGTH: **2.94**

BELT DIAMETER **C = N/A**

RIM THICKNESS **E = .04**

SHOULDER DIAMETER **G = .425**

LENGTH TO SHOULDER **J = 1.618**

SHOULDER ANGLE (DEG'S/SIDE): **22.49**

PRIMER: **L/R**

CASE CAPACITY (GR'S WATER): **48.70**

DIMENSIONAL DRAWING:

F H K G A J E B D

-NOT ACTUAL SIZE-
-DO NOT SCALE-

CARTRIDGE: 6,5 x 53.5R Daudeteau

OTHER NAMES:
6,5 x 54R Daudeteau

DIA: .263

BALLISTEK NO: 264P

NAI NO: RMB
23434/4.265

DATA SOURCE: COTW 4th Pg257

HISTORICAL DATA: French naval ctg. ca. 1895.

NOTES:

LOADING DATA:

BULLET WT./TYPE	POWDER WT./TYPE	VELOCITY ('/SEC)	SOURCE
140/RN	44.0/IMR4350	—	JJD

CASE PREPARATION: SHELLHOLDER (RCBS): 13

MAKE FROM: 7,62 x 54R Russian. Turn rim to .524" dia. & back chamfer. Anneal case (may not always be necessary) and F/L size. Trim & chamfer.

PHYSICAL DATA (INCHES):

CASE TYPE: **Rimmed Bottleneck**

CASE LENGTH **A = 2.09**

HEAD DIAMETER **B = .490**

RIM DIAMETER **D = .524**

NECK DIAMETER **F = .298**

NECK LENGTH **H = .360**

SHOULDER LENGTH **K = .17**

BODY ANGLE (DEG'S/SIDE): **.505**

CASE CAPACITY **CC'S = 3.65**

LOADED LENGTH: **3.02**

BELT DIAMETER **C = N/A**

RIM THICKNESS **E = .062**

SHOULDER DIAMETER **G = .466**

LENGTH TO SHOULDER **J = 1.56**

SHOULDER ANGLE (DEG'S/SIDE): **26.29**

PRIMER: **L/R**

CASE CAPACITY (GR'S WATER): **56.39**

DIMENSIONAL DRAWING:

F H K G A J E B D

-NOT ACTUAL SIZE-
-DO NOT SCALE-

CARTRIDGE: 6,5 x 53R Dutch Mannlicher

OTHER NAMES:	
6,5mm Mannlicher .256 Mannlicher	DIA: .263 BALLISTEK NO: 263A NAI NO: RMB 22332/4.666

DATA SOURCE: COTW 4th Pg197

HISTORICAL DATA: By Mannlicher about 1890.

NOTES:

LOADING DATA:

BULLET WT./TYPE	POWDER WT./TYPE	VELOCITY ('/SEC)	SOURCE
160/RN	32.0/IMR3031	2100	JJD

CASE PREPARATION: SHELLHOLDER (RCBS): 7

MAKE FROM: .303 British. Turn rim to .526" dia. & back chamfer. Thin rim if necessary for chambering. Trim to length, anneal and F/L size.

PHYSICAL DATA (INCHES):

CASE TYPE: **Rimmed Bottleneck**

CASE LENGTH **A = 2.10**

HEAD DIAMETER **B = .450**

RIM DIAMETER **D = .526**

NECK DIAMETER **F = .297**

NECK LENGTH **H = .320**

SHOULDER LENGTH **K = .180**

BODY ANGLE (DEG'S/SIDE): **.552**

CASE CAPACITY **CC'S = 3.12**

LOADED LENGTH: **3.03**

BELT DIAMETER **C = N/A**

RIM THICKNESS **E = .049**

SHOULDER DIAMETER **G = .423**

LENGTH TO SHOULDER **J = 1.60**

SHOULDER ANGLE (DEG'S/SIDE): **19.29**

PRIMER: **L/R**

CASE CAPACITY (GR'S WATER): **48.23**

DIMENSIONAL DRAWING:

-NOT ACTUAL SIZE-
-DO NOT SCALE-

CARTRIDGE: 6,5 x 54mm Mannlicher-Schoeneauer

OTHER NAMES:	
	DIA: .264 BALLISTEK NO: 264F NAI NO: RXB 22333/4.720

DATA SOURCE: Hornady Manual 3rd Pg153

HISTORICAL DATA: Greek mil. ctg. ca. 1903.

NOTES:

LOADING DATA:

BULLET WT./TYPE	POWDER WT./TYPE	VELOCITY ('/SEC)	SOURCE
160/RN	39.0/IMR4831	2100	JJD

CASE PREPARATION: SHELLHOLDER (RCBS): 9

MAKE FROM: .220 Swift. Cut off rim and deepen extractor groove. F/L size in 6,5 M/S die. Trim to length. Chamfer.

PHYSICAL DATA (INCHES):

CASE TYPE: **Rimless Bottleneck**

CASE LENGTH **A = 2.11**

HEAD DIAMETER **B = .447**

RIM DIAMETER **D = .453**

NECK DIAMETER **F = .288**

NECK LENGTH **H = .320**

SHOULDER LENGTH **K = .172**

BODY ANGLE (DEG'S/SIDE): **.383**

CASE CAPACITY **CC'S = 3.39**

LOADED LENGTH: **2.99**

BELT DIAMETER **C = N/A**

RIM THICKNESS **E = .055**

SHOULDER DIAMETER **G = .428**

LENGTH TO SHOULDER **J = 1.618**

SHOULDER ANGLE (DEG'S/SIDE): **22.14**

PRIMER: **L/R**

CASE CAPACITY (GR'S WATER): **52.26**

DIMENSIONAL DRAWING:

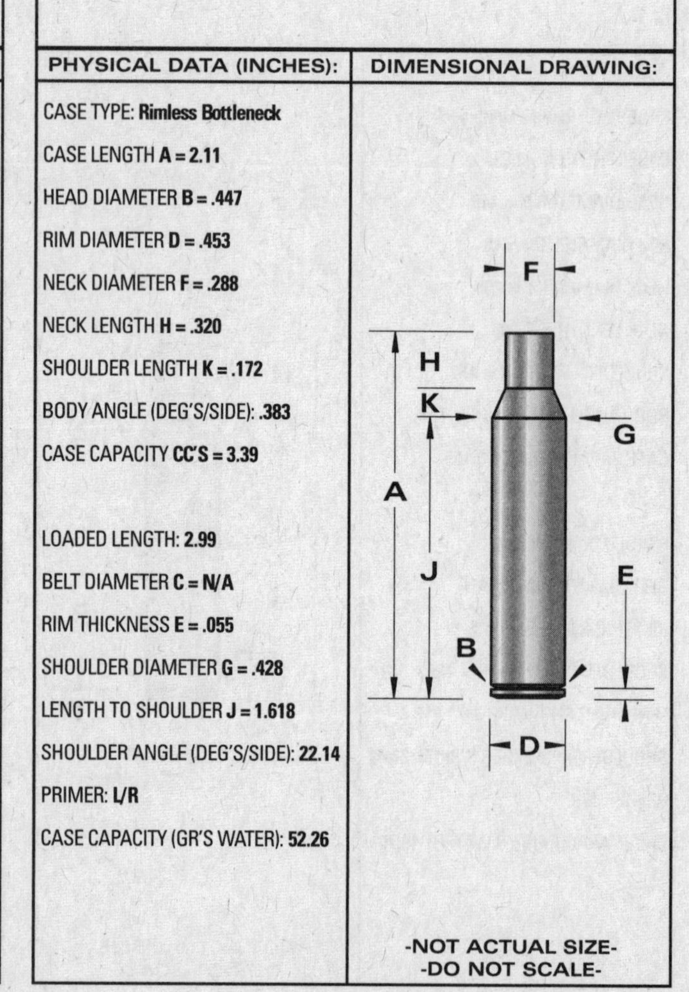

-NOT ACTUAL SIZE-
-DO NOT SCALE-

CARTRIDGE: 6,5 x 54mm Mauser

OTHER NAMES:
6.5 x 54K
6.5mm Kurz

DIA: .262

BALLISTEK NO: 262A

NAI NO: RXB
22422/4.560

DATA SOURCE: NAI/Ballistek

HISTORICAL DATA: By Mauser about 1900.

NOTES:

LOADING DATA:

BULLET WT./TYPE	POWDER WT./TYPE	VELOCITY ('/SEC)	SOURCE
120/RN	36.0/4895	2500	Barnes

CASE PREPARATION: SHELLHOLDER (RCBS): 3

MAKE FROM: .270 Win. Size the .270 case in the 6.5x54 F/L die with the expander removed. Trim to length and chamfer. F/L size. Ream neck I.D. in neck diameter (with seated bullet) exceeds .296" dia. In some weapons, standard .264 bullets will work fine and in others, the bullet should be swaged to .262. Slug your bore.

PHYSICAL DATA (INCHES):

CASE TYPE: **Rimless Bottleneck**
CASE LENGTH **A = 2.125**
HEAD DIAMETER **B = .466**
RIM DIAMETER **D = .466**
NECK DIAMETER **F = .296**
NECK LENGTH **H = .385**
SHOULDER LENGTH **K = .21**
BODY ANGLE (DEG'S/SIDE): **.624**
CASE CAPACITY **CC'S = 3.29**

LOADED LENGTH: **2.67**
BELT DIAMETER **C = N/A**
RIM THICKNESS **E = .05**
SHOULDER DIAMETER **G = .437**
LENGTH TO SHOULDER **J = 1.53**
SHOULDER ANGLE (DEG'S/SIDE): **18.55**
PRIMER: **L/R**
CASE CAPACITY (GR'S WATER): **50.86**

DIMENSIONAL DRAWING:

F
H
K
G
A
J
E
B
D

-NOT ACTUAL SIZE-
-DO NOT SCALE-

CARTRIDGE: 6,5 x 55mm Swedish Mauser

OTHER NAMES:
6,5 Swedish
6,5 Norwegian

DIA: .264

BALLISTEK NO: 264G

NAI NO: RXB
22423/4.548

DATA SOURCE: Hornady Manual 3rd Pg156

HISTORICAL DATA: Swedish mil. ctg. ca. 1894.

NOTES:

LOADING DATA:

BULLET WT./TYPE	POWDER WT./TYPE	VELOCITY ('/SEC)	SOURCE
140/Spire	37.7/IMR4895	2400	Horn.

CASE PREPARATION: SHELLHOLDER (RCBS): 2

MAKE FROM: .270 Win. Anneal case. Run into 6,5x55 F/L die with expander removed. Trim to length. Chamfer. F/L size. Base will fireform on ignition.

PHYSICAL DATA (INCHES):

CASE TYPE: **Rimless Bottleneck**
CASE LENGTH **A = 2.165**
HEAD DIAMETER **B = .476**
RIM DIAMETER **D = .476**
NECK DIAMETER **F = .297**
NECK LENGTH **H = .332**
SHOULDER LENGTH **K = .149**
BODY ANGLE (DEG'S/SIDE): **.791**
CASE CAPACITY **CC'S = 3.566**

LOADED LENGTH: **3.03**
BELT DIAMETER **C = N/A**
RIM THICKNESS **E = .06**
SHOULDER DIAMETER **G = .435**
LENGTH TO SHOULDER **J = 1.684**
SHOULDER ANGLE (DEG'S/SIDE): **24.84**
PRIMER: **L/R**
CASE CAPACITY (GR'S WATER): **55.04**

DIMENSIONAL DRAWING:

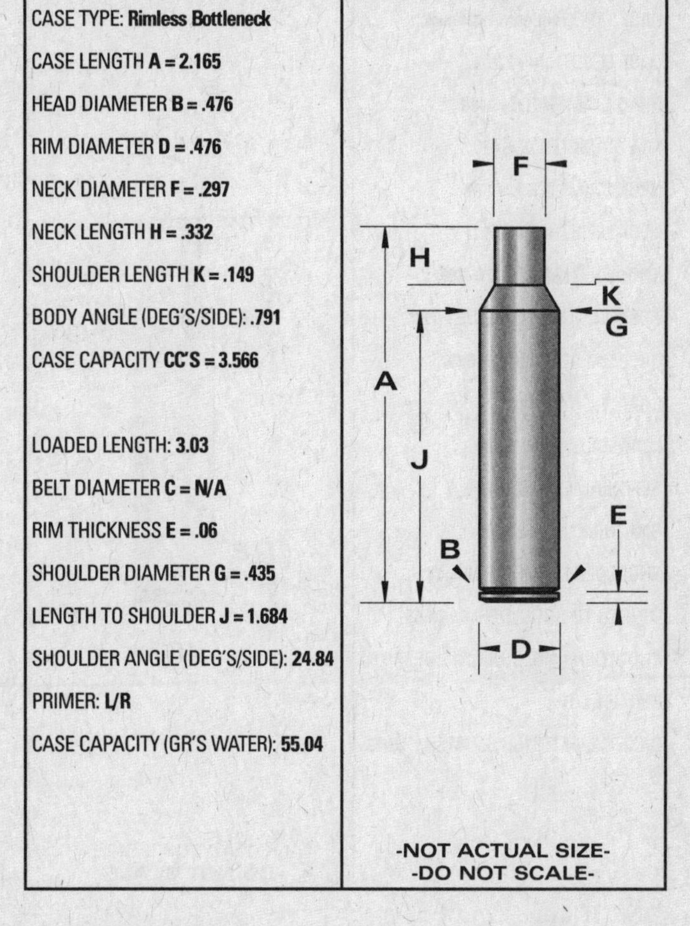

-NOT ACTUAL SIZE-
-DO NOT SCALE-

CARTRIDGE: 6,5 x 57mm Mauser

OTHER NAMES:
6,5/7 x 57 Mauser
6,5/257 Roberts

DIA: .264

BALLISTEK NO: 264R

NAI NO: RXB
22432/4.769

DATA SOURCE: Hornady 3rd Pg159

HISTORICAL DATA: circa. 1893.

NOTES:

LOADING DATA:

BULLET WT./TYPE	POWDER WT./TYPE	VELOCITY ('/SEC)	SOURCE
140/RN	42.4/IMR4350	2600	Horn.

CASE PREPARATION: SHELLHOLDER (RCBS): 3

MAKE FROM: 7 x 57 Mauser. Anneal neck and size in 6,5 x 57 sizer. Trim and chamfer.

PHYSICAL DATA (INCHES):

CASE TYPE: **Rimless Bottleneck**
CASE LENGTH **A = 2.232**
HEAD DIAMETER **B = .468**
RIM DIAMETER **D = .470**
NECK DIAMETER **F = .301**
NECK LENGTH **H = .291**
SHOULDER LENGTH **K = .286**
BODY ANGLE (DEG'S/SIDE): **.728**
CASE CAPACITY **CC'S = 3.48**

LOADED LENGTH: **3.02**
BELT DIAMETER **C = N/A**
RIM THICKNESS **E = .05**
SHOULDER DIAMETER **G = .431**
LENGTH TO SHOULDER **J = 1.655**
SHOULDER ANGLE (DEG'S/SIDE): **18.80**
PRIMER: **L/R**
CASE CAPACITY (GR'S WATER): **53.73**

DIMENSIONAL DRAWING:

F H K G A J E B D

-NOT ACTUAL SIZE-
-DO NOT SCALE-

CARTRIDGE: 6,5 x 57R Mauser

OTHER NAMES:

DIA: .264

BALLISTEK NO: 264AL

NAI NO: RMB
22432/4.766

DATA SOURCE: COTW 4th Pg259

HISTORICAL DATA: Circa 1893.

NOTES:

LOADING DATA:

BULLET WT./TYPE	POWDER WT./TYPE	VELOCITY ('/SEC)	SOURCE
120/Spire	39.0/IMR4320	2500	JJD

CASE PREPARATION: SHELLHOLDER (RCBS): 21

MAKE FROM: 7 x 57R Mauser. F/L size case, trim to length & chamfer. It is also possible to use .444 Marlin brass but, a form die set is required. Anneal case, form, F/L size, trim & chamfer.

PHYSICAL DATA (INCHES):

CASE TYPE: **Rimmed Bottleneck**
CASE LENGTH **A = 2.24**
HEAD DIAMETER **B = .470**
RIM DIAMETER **D = .521**
NECK DIAMETER **F = .292**
NECK LENGTH **H = .275**
SHOULDER LENGTH **K = .215**
BODY ANGLE (DEG'S/SIDE): **.739**
CASE CAPACITY **CC'S = 3.78**

LOADED LENGTH: **3.18**
BELT DIAMETER **C = N/A**
RIM THICKNESS **E = .05**
SHOULDER DIAMETER **G = .430**
LENGTH TO SHOULDER **J = 1.75**
SHOULDER ANGLE (DEG'S/SIDE): **17.79**
PRIMER: **L/R**
CASE CAPACITY (GR'S WATER): **58.37**

DIMENSIONAL DRAWING:

F H K G A J E B D

-NOT ACTUAL SIZE-
-DO NOT SCALE-

CARTRIDGE: 6,5 x 58 Mauser

OTHER NAMES:
6,5 x 58 Portuguese Vergueiro
6,5 x 58 Mauser (P)

DIA: .264

BALLISTEK NO: 264BH

NAI NO: RXB
22432/4.855

DATA SOURCE: COTW 4th Pg199

HISTORICAL DATA: Portuguese mil. ctg. ca. 1937.

NOTES:

LOADING DATA:

BULLET WT./TYPE	POWDER WT./TYPE	VELOCITY ('/SEC)	SOURCE
160/RN	43.0/4350	2510	Barnes

CASE PREPARATION: SHELLHOLDER (RCBS): 3

MAKE FROM: .270 Win. Anneal case neck. Run into F/L sizer with expander removed. Trim to length and chamfer. F/L size. I.D. neck ream if seated bullet expands neck beyond .293" dia.

PHYSICAL DATA (INCHES):

CASE TYPE: **Rimless Bottleneck**

CASE LENGTH **A = 2.272**

HEAD DIAMETER **B = .468**

RIM DIAMETER **D = .465**

NECK DIAMETER **F = .293**

NECK LENGTH **H = .300**

SHOULDER LENGTH **K = .272**

BODY ANGLE (DEG'S/SIDE): **.802**

CASE CAPACITY **CC'S = 3.70**

LOADED LENGTH: **3.22**

BELT DIAMETER **C = N/A**

RIM THICKNESS **E = .052**

SHOULDER DIAMETER **G = .426**

LENGTH TO SHOULDER **J = 1.70**

SHOULDER ANGLE (DEG'S/SIDE): **13.73**

PRIMER: **L/R**

CASE CAPACITY (GR'S WATER): **57.18**

DIMENSIONAL DRAWING:

-NOT ACTUAL SIZE-
-DO NOT SCALE-

CARTRIDGE: 6,5 x 58R Krag-Jorgensen

OTHER NAMES:
6,5 Krag

DIA: .264

BALLISTEK NO: 264BG

NAI NO: RMB
22432/4.570

DATA SOURCE: NAI/Ballistek

HISTORICAL DATA: Danish target ctg. ca. 1933.

NOTES:

LOADING DATA:

BULLET WT./TYPE	POWDER WT./TYPE	VELOCITY ('/SEC)	SOURCE
140/Spire	46.0/4350	2500	Barnes

CASE PREPARATION: SHELLHOLDER (RCBS): 14

MAKE FROM: .45 Basic. A form set, for this ctg., is a must! Anneal case and run through form dies #1 & #2. Trim in trim die (rough trim). F/L size. Neck I.D. will need reaming. F/L size, trim & chamfer.

PHYSICAL DATA (INCHES):

CASE TYPE: **Rimmed Bottleneck**

CASE LENGTH **A = 2.285**

HEAD DIAMETER **B = .500**

RIM DIAMETER **D = .575**

NECK DIAMETER **F = .300**

NECK LENGTH **H = .30**

SHOULDER LENGTH **K = .30**

BODY ANGLE (DEG'S/SIDE): **.771**

CASE CAPACITY **CC'S = 4.09**

LOADED LENGTH: **3.25**

BELT DIAMETER **C = N/A**

RIM THICKNESS **E = .062**

SHOULDER DIAMETER **G = .460**

LENGTH TO SHOULDER **J = 1.685**

SHOULDER ANGLE (DEG'S/SIDE): **14.90**

PRIMER: **L/R**

CASE CAPACITY (GR'S WATER): **63.24**

DIMENSIONAL DRAWING:

-NOT ACTUAL SIZE-
-DO NOT SCALE-

CARTRIDGE: 6,5 x 58R Sauer

OTHER NAMES:	
6.6 x 58.5R Sauer	DIA: .260 BALLISTEK NO: 260A NAI NO: RMS 31115/5.311

DATA SOURCE: COTW 4th Pg259

HISTORICAL DATA: Ctg. for Sauer single-shot rifle.

NOTES:

LOADING DATA:

BULLET WT./TYPE	POWDER WT./TYPE	VELOCITY ('/SEC)	SOURCE
120/—	15.0/4198	1480	Barnes

CASE PREPARATION: SHELLHOLDER (RCBS): 30

MAKE FROM: 9.3 x 72R. Size 9.3 case in 6,5x58R F/L die with expander removed. Trim and chamfer. F/L size.

PHYSICAL DATA (INCHES):

CASE TYPE: **Rimmed Straight**
CASE LENGTH **A = 2.300**
HEAD DIAMETER **B = .433**
RIM DIAMETER **D = .500**
NECK DIAMETER **F = .290**
NECK LENGTH **H = N/A**
SHOULDER LENGTH **K = N/A**
BODY ANGLE (DEG'S/SIDE): **1.82**
CASE CAPACITY **CC'S = 2.59**

LOADED LENGTH: **3.08**
BELT DIAMETER **C = N/A**
RIM THICKNESS **E = .05**
SHOULDER DIAMETER **G = N/A**
LENGTH TO SHOULDER **J = N/A**
SHOULDER ANGLE (DEG'S/SIDE): **N/A**
PRIMER: **L/R**
CASE CAPACITY (GR'S WATER): **40.06**

DIMENSIONAL DRAWING:

-NOT ACTUAL SIZE-
-DO NOT SCALE-

CARTRIDGE: 6,5 x 61mm Mauser

OTHER NAMES:	
	DIA: .264 BALLISTEK NO: 264BJ NAI NO: RXB 22423/5.031

DATA SOURCE: COTW 4th Pg260

HISTORICAL DATA: DWM version of .256 Newton (per P. Sharpe).

NOTES:

LOADING DATA:

BULLET WT./TYPE	POWDER WT./TYPE	VELOCITY ('/SEC)	SOURCE
140/Spire	50.0/4350	2640	Barnes

CASE PREPARATION: SHELLHOLDER (RCBS): 3

MAKE FROM: .270 Winchester. Anneal case and F/L size. Trim to length and chamfer. Fireform in chamber.

PHYSICAL DATA (INCHES):

CASE TYPE: **Rimless Bottleneck**
CASE LENGTH **A = 2.400**
HEAD DIAMETER **B = .477**
RIM DIAMETER **D = .479**
NECK DIAMETER **F = .297**
NECK LENGTH **H = .340**
SHOULDER LENGTH **K = .140**
BODY ANGLE (DEG'S/SIDE): **.416**
CASE CAPACITY **CC'S = 4.23**

LOADED LENGTH: **3.55**
BELT DIAMETER **C = N/A**
RIM THICKNESS **E = .052**
SHOULDER DIAMETER **G = .452**
LENGTH TO SHOULDER **J = 1.92**
SHOULDER ANGLE (DEG'S/SIDE): **28.97**
PRIMER: **L/R**
CASE CAPACITY (GR'S WATER): **65.31**

DIMENSIONAL DRAWING:

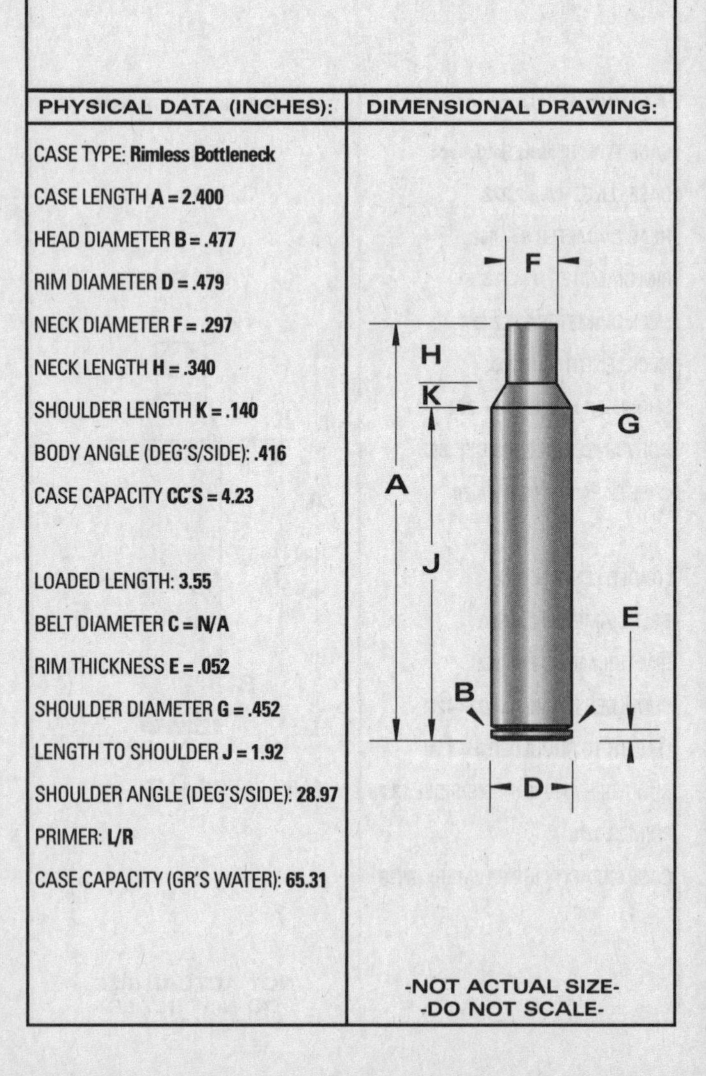

-NOT ACTUAL SIZE-
-DO NOT SCALE-

CARTRIDGE: 6,5 x 61R Mauser

OTHER NAMES:	DIA: .264 BALLISTEK NO: 264BK NAI NO: RMB 22423/5.031

DATA SOURCE: COTW 4th Pg260

HISTORICAL DATA: DWM version of a rimmed .256 Newton.

NOTES:

LOADING DATA:

BULLET WT./TYPE	POWDER WT./TYPE	VELOCITY ('/SEC)	SOURCE
140/–	50.0/4350	2640	Barnes

CASE PREPARATION: SHELLHOLDER (RCBS): 4

MAKE FROM: 9,3 x 74R. Anneal case neck. Run into .308 F/L die to reduce neck dia. Size again in 7mm/08 die. Watch location of shoulder! Trim to length. F/L size and chamfer. Fireform.

PHYSICAL DATA (INCHES):

CASE TYPE: **Rimmed Bottleneck**
CASE LENGTH **A = 2.400**
HEAD DIAMETER **B = .477**
RIM DIAMETER **D = .532**
NECK DIAMETER **F = .296**
NECK LENGTH **H = .340**
SHOULDER LENGTH **K = .140**
BODY ANGLE (DEG'S/SIDE): **.416**
CASE CAPACITY **CC'S = 4.25**
LOADED LENGTH: **3.55**
BELT DIAMETER **C = N/A**
RIM THICKNESS **E = .055**
SHOULDER DIAMETER **G = .452**
LENGTH TO SHOULDER **J = 1.92**
SHOULDER ANGLE (DEG'S/SIDE): **29.12**
PRIMER: **L/R**
CASE CAPACITY (GR'S WATER): **65.74**

DIMENSIONAL DRAWING:

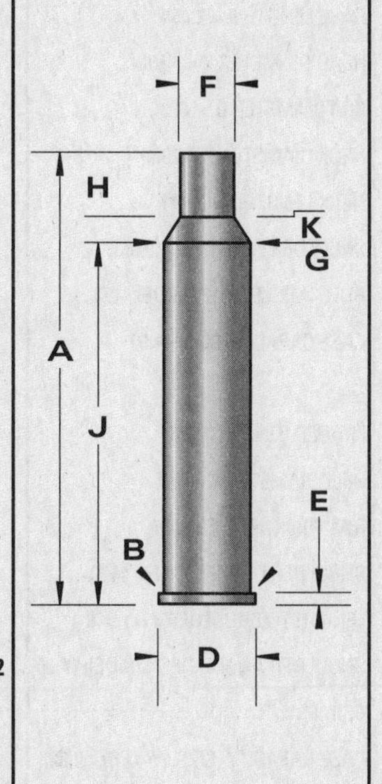

-NOT ACTUAL SIZE-
-DO NOT SCALE-

CARTRIDGE: 6,5 x 68 Schuler

OTHER NAMES:	DIA: .264 BALLISTEK NO: 264Q NAI NO: RBB 22432/5.115

DATA SOURCE: COTW 4th Pg260

HISTORICAL DATA: By RWS about 1938.

NOTES: Similar to .264 Win. Mag.

LOADING DATA:

BULLET WT./TYPE	POWDER WT./TYPE	VELOCITY ('/SEC)	SOURCE
87/Spire	65.0/4350	3750	Barnes

CASE PREPARATION: SHELLHOLDER (RCBS): 4

MAKE FROM: .300 H&H Mag. Trun belt to .520" dia. Reduce neck dia. in 7mm Rem. Mag. die (neck only!). Anneal. Trim & chamfer. F/L size in Schuler die. Fireform in chamber.

PHYSICAL DATA (INCHES):

CASE TYPE: **Rebated Bottleneck**
CASE LENGTH **A = 2.660**
HEAD DIAMETER **B = .520**
RIM DIAMETER **D = .510**
NECK DIAMETER **F = .297**
NECK LENGTH **H = .280**
SHOULDER LENGTH **K = .340**
BODY ANGLE (DEG'S/SIDE): **.607**
CASE CAPACITY **CC'S = 5.52**
LOADED LENGTH: **3.40**
BELT DIAMETER **C = N/A**
RIM THICKNESS **E = .05**
SHOULDER DIAMETER **G = .481**
LENGTH TO SHOULDER **J = 2.04**
SHOULDER ANGLE (DEG'S/SIDE): **15.14**
PRIMER: **L/R**
CASE CAPACITY (GR'S WATER): **85.16**

DIMENSIONAL DRAWING:

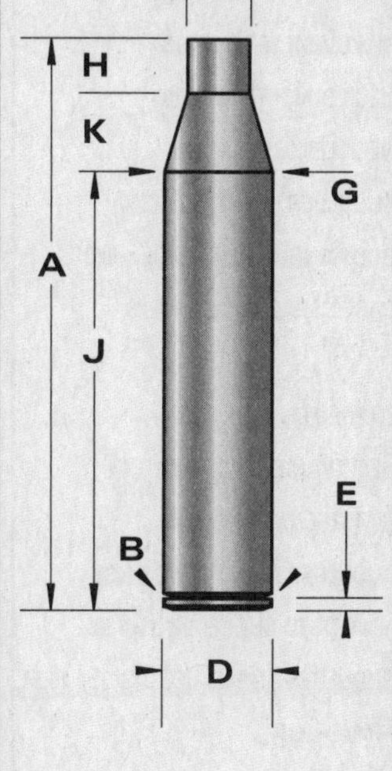

-NOT ACTUAL SIZE-
-DO NOT SCALE-

CARTRIDGE: 6,5/06 Improved (Ackley)

OTHER NAMES:	
.256/06 Improved	DIA: .264
	BALLISTEK NO: 264X
	NAI NO: RXB 22432/5.298

DATA SOURCE: Ackley Vol.1 Pg363

HISTORICAL DATA: Early wildcat.

NOTES: Very underrated cartridge!

LOADING DATA:

BULLET WT./TYPE	POWDER WT./TYPE	VELOCITY ('/SEC)	SOURCE
140/Spire	51.0/4350	2920	Ackley

CASE PREPARATION: SHELLHOLDER (RCBS): 3

MAKE FROM: .270 Win. Neck size the .270 case in a 6,5/06 Imp. die. Trim & chamfer. Fireform in the "improved" chamber or in the proper trim die.

PHYSICAL DATA (INCHES):

CASE TYPE: **Rimless Bottleneck**

CASE LENGTH **A = 2.49**

HEAD DIAMETER **B = .470**

RIM DIAMETER **D = .473**

NECK DIAMETER **F = .293**

NECK LENGTH **H = .286**

SHOULDER LENGTH **K = .244**

BODY ANGLE (DEG'S/SIDE): **.520**

CASE CAPACITY **CC'S = 4.33**

LOADED LENGTH: **3.25**

BELT DIAMETER **C = N/A**

RIM THICKNESS **E = .05**

SHOULDER DIAMETER **G = .438**

LENGTH TO SHOULDER **J = 1.96**

SHOULDER ANGLE (DEG'S/SIDE): **16.54**

PRIMER: **L/R**

CASE CAPACITY (GR'S WATER): **66.93**

DIMENSIONAL DRAWING:

F H K G A J E B D

-NOT ACTUAL SIZE-
-DO NOT SCALE-

CARTRIDGE: 6,5-06 A-Square

OTHER NAMES:	
6,5-06 Springfield	DIA: .264
	BALLISTEK NO: 264W
	NAI NO: RXB 22423/5.297

DATA SOURCE: Many current reloading manuals.

HISTORICAL DATA: This long time wildcat was made "legitimate when A-Square took it to SAAMI for approval in 1997. Dimensions are now standardized.

NOTES: Older rifles chambered as a wildcat may use different dimensions, always check with a chamber cast.

LOADING DATA:

BULLET WT./TYPE	POWDER WT./TYPE	VELOCITY ('/SEC)	SOURCE
140/Hornady	51.6/RL-19	2900	Horn. #6
160/Hornady	52.9/AA3100	2800	Horn. #6

CASE PREPARATION: SHELLHOLDER (RCBS): 3

Run .270 Winchester run .30-06 case through 6,5-06 A-Square F/L sizing die. Fireform, trim, chamfer and deburr. Check neck wall thickness and turn if needed.

PHYSICAL DATA (INCHES):

CASE TYPE: **Rimless Bottleneck**

CASE LENGTH **A = 2.494**

HEAD DIAMETER **B = .4698**

RIM DIAMETER **D = .473**

NECK DIAMETER **F = .296**

NECK LENGTH **H = .3177**

SHOULDER LENGTH **K = .2283**

BODY ANGLE (DEG'S/SIDE): **.474**

CASE CAPACITY **CC'S = 4.07**

LOADED LENGTH: **3.240**

BELT DIAMETER **C = N/A**

RIM THICKNESS **E = .049**

SHOULDER DIAMETER **G = .4410**

LENGTH TO SHOULDER **J = 1.948**

SHOULDER ANGLE (DEG'S/SIDE): **17° 30'**

PRIMER: **L/R**

CASE CAPACITY (GR'S WATER): **62.82**

DIMENSIONAL DRAWING:

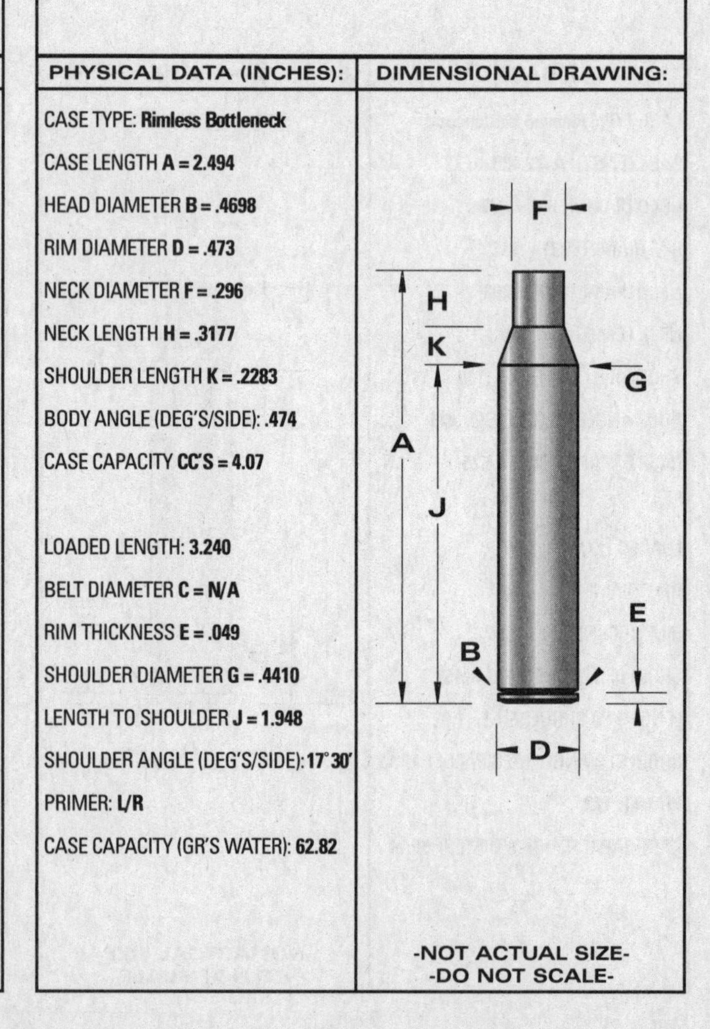

-NOT ACTUAL SIZE-
-DO NOT SCALE-

CARTRIDGE: 6,5/257 Roberts

OTHER NAMES:

DIA: .264

BALLISTEK NO: 264Y

NAI NO: RXB 22433/4.713

DATA SOURCE: NAI/Ballistek

HISTORICAL DATA:

NOTES:

LOADING DATA:

BULLET WT./TYPE	POWDER WT./TYPE	VELOCITY ('/SEC)	SOURCE
140/Spire	32.0/IMR4320	—	JJD

CASE PREPARATION: SHELLHOLDER (RCBS): 11

MAKE FROM: .257 Roberts. Taper expand case to .270" dia. F/L size, trim to length and chamfer.

PHYSICAL DATA (INCHES):

CASE TYPE: **Rimless Bottleneck**

CASE LENGTH **A = 2.22**

HEAD DIAMETER **B = .471**

RIM DIAMETER **D = .473**

NECK DIAMETER **F = .298**

NECK LENGTH **H = .320**

SHOULDER LENGTH **K = .170**

BODY ANGLE (DEG'S/SIDE): **.767**

CASE CAPACITY **CC'S = 3.57**

LOADED LENGTH: **A/R**

BELT DIAMETER **C = N/A**

RIM THICKNESS **E = .049**

SHOULDER DIAMETER **G = .430**

LENGTH TO SHOULDER **J = 1.73**

SHOULDER ANGLE (DEG'S/SIDE): **21.22**

PRIMER: **L/R**

CASE CAPACITY (GR'S WATER): **55.23**

DIMENSIONAL DRAWING:

-NOT ACTUAL SIZE-
-DO NOT SCALE-

CARTRIDGE: 6,5/280 RCBS Improved

OTHER NAMES:

DIA: .264

BALLISTEK NO: 264Z

NAI NO: RXB 22424/5.361

DATA SOURCE: Ackley Vol.2 Pg164

HISTORICAL DATA: Huntington design.

NOTES:

LOADING DATA:

BULLET WT./TYPE	POWDER WT./TYPE	VELOCITY ('/SEC)	SOURCE
150/Spire	50.0/4350	2838	Ackley

CASE PREPARATION: SHELLHOLDER (RCBS): 3

MAKE FROM: .280 Remington. F/L size Rem. case in "improved" die. Trim, chamfer and fireform. .270 Win. cases will work in the same manner as will .25-06 cases, if they are taper expanded.

PHYSICAL DATA (INCHES):

CASE TYPE: **Rimless Bottleneck**

CASE LENGTH **A = 2.52**

HEAD DIAMETER **B = .470**

RIM DIAMETER **D = .473**

NECK DIAMETER **F = .297**

NECK LENGTH **H = .360**

SHOULDER LENGTH **K = .136**

BODY ANGLE (DEG'S/SIDE): **.402**

CASE CAPACITY **CC'S = 4.93**

LOADED LENGTH: **3.42**

BELT DIAMETER **C = N/A**

RIM THICKNESS **E = .048**

SHOULDER DIAMETER **G = .443**

LENGTH TO SHOULDER **J = 2.02**

SHOULDER ANGLE (DEG'S/SIDE): **33.99**

PRIMER: **L/R**

CASE CAPACITY (GR'S WATER): **76.11**

DIMENSIONAL DRAWING:

-NOT ACTUAL SIZE-
-DO NOT SCALE-

CARTRIDGE: 6,5/284 Winchester

OTHER NAMES:

DIA: .264

BALLISTEK NO: 264J

NAI NO: RBB 22434/4.210

DATA SOURCE: Ackley Vol.2 Pg160

HISTORICAL DATA:

NOTES:

LOADING DATA:

BULLET WT./TYPE	POWDER WT./TYPE	VELOCITY ('/SEC)	SOURCE
140/Spire	52.5/4350	2929	Ackley

CASE PREPARATION:

SHELLHOLDER (RCBS): 3

MAKE FROM: .284 Winchester. F/L size case in proper die. Trim & chamfer.

PHYSICAL DATA (INCHES):

CASE TYPE: **Rebated Bottleneck**

CASE LENGTH **A = 2.16**

HEAD DIAMETER **B = .500**

RIM DIAMETER **D = .473**

NECK DIAMETER **F = .293**

NECK LENGTH **H = .280**

SHOULDER LENGTH **K = .080**

BODY ANGLE (DEG'S/SIDE): **.483**

CASE CAPACITY **CC'S = 4.31**

LOADED LENGTH: **2.95**

BELT DIAMETER **C = N/A**

RIM THICKNESS **E = .05**

SHOULDER DIAMETER **G = .473**

LENGTH TO SHOULDER **J = 1.80**

SHOULDER ANGLE (DEG'S/SIDE): **48.36**

PRIMER: **L/R**

CASE CAPACITY (GR'S WATER): **66.54**

DIMENSIONAL DRAWING:

F H K G A J E B D

-NOT ACTUAL SIZE-
-DO NOT SCALE-

CARTRIDGE: 6,5/300 Weatherby

OTHER NAMES:
6,5 x 300 W.W.H.

DIA: .264

BALLISTEK NO: 264K

NAI NO: BEN 12423/5.506

DATA SOURCE: Sierra Manual 1985 Pg171

HISTORICAL DATA: By P.J. Wright.

NOTES: Overbore but very accurate.

LOADING DATA:

BULLET WT./TYPE	POWDER WT./TYPE	VELOCITY ('/SEC)	SOURCE
139/Spire	84.0/H870	3400	Ackley
140/HPBT	83.0/H870	–	Sierra

CASE PREPARATION:

SHELLHOLDER (RCBS): 4

MAKE FROM: .300 Weatherby. DO NOT ANNEAL CASE! Run .300 Weatherby case into a 6,5/300 Weatherby trim die & trim to length (Note: trim length will be about 2.815"). F/L size. Tumble clean and case chamfer.

PHYSICAL DATA (INCHES):

CASE TYPE: **Belted Bottleneck**

CASE LENGTH **A = 2.825**

HEAD DIAMETER **B = .513**

RIM DIAMETER **D = .532**

NECK DIAMETER **F = .292**

NECK LENGTH **H = .343**

SHOULDER LENGTH **K = .184**

BODY ANGLE (DEG'S/SIDE): **.286**

CASE CAPACITY **CC'S = 6.23**

LOADED LENGTH: **3.67**

BELT DIAMETER **C = .532**

RIM THICKNESS **E = .05**

SHOULDER DIAMETER **G = .492**

LENGTH TO SHOULDER **J = 2.298**

SHOULDER ANGLE (DEG'S/SIDE): **N/A**

PRIMER: **L/R Mag**

CASE CAPACITY (GR'S WATER): **96.21**

DIMENSIONAL DRAWING:

F H K G A J E B D C

-NOT ACTUAL SIZE-
-DO NOT SCALE-

CARTRIDGE: 6,5/300 Winchester Magnum

OTHER NAMES:

DIA: .264

BALLISTEK NO: 264AQ

NAI NO: BEN 12444/5.078

DATA SOURCE: NAI/Ballistek

HISTORICAL DATA: Several versions exist.

NOTES:

LOADING DATA:

BULLET WT./TYPE	POWDER WT./TYPE	VELOCITY ('/SEC)	SOURCE
129/Spire	59.0/IMR4831	3100	JJD

CASE PREPARATION: SHELLHOLDER (RCBS): 4

MAKE FROM: .300 Win. Mag. Anneal case neck and F/L size. Trim & chamfer. Fireform in chamber.

PHYSICAL DATA (INCHES):

CASE TYPE: **Belted Bottleneck**

CASE LENGTH **A = 2.605**

HEAD DIAMETER **B = .513**

RIM DIAMETER **D = .532**

NECK DIAMETER **F = .289**

NECK LENGTH **H = .255**

SHOULDER LENGTH **K = .145**

BODY ANGLE (DEG'S/SIDE): **.185**

CASE CAPACITY **CC'S = 6.04**

LOADED LENGTH: **A/R**

BELT DIAMETER **C = .532**

RIM THICKNESS **E = .049**

SHOULDER DIAMETER **G = .500**

LENGTH TO SHOULDER **J = 2.205**

SHOULDER ANGLE (DEG'S/SIDE): **36.04**

PRIMER: **L/R Mag**

CASE CAPACITY (GR'S WATER): **93.24**

DIMENSIONAL DRAWING:

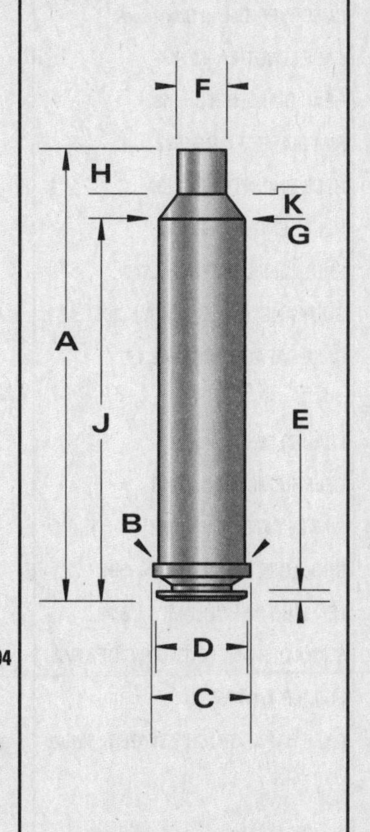

-NOT ACTUAL SIZE-
-DO NOT SCALE-

CARTRIDGE: 6,5/303 Epps

OTHER NAMES:

DIA: .264

BALLISTEK NO: 264AR

NAI NO: RMB 22334/4.879

DATA SOURCE: Ackley Vol.2 Pg157

HISTORICAL DATA: By E. Epps.

NOTES:

LOADING DATA:

BULLET WT./TYPE	POWDER WT./TYPE	VELOCITY ('/SEC)	SOURCE
120/Spire	43.7/IMR4350	—	JJD

CASE PREPARATION: SHELLHOLDER (RCBS): 7

MAKE FROM: .303 British. Run case into any short, 7mm die to reduce neck dia. F/L size in Epps die, trim, chamfer and fireform.

PHYSICAL DATA (INCHES):

CASE TYPE: **Rimmed Bottleneck**

CASE LENGTH **A = 2.22**

HEAD DIAMETER **B = .455**

RIM DIAMETER **D = .540**

NECK DIAMETER **F = .290**

NECK LENGTH **H = .318**

SHOULDER LENGTH **K = .092**

BODY ANGLE (DEG'S/SIDE): **.694**

CASE CAPACITY **CC'S = 3.57**

LOADED LENGTH: **3.10**

BELT DIAMETER **C = N/A**

RIM THICKNESS **E = .064**

SHOULDER DIAMETER **G = .416**

LENGTH TO SHOULDER **J = 1.81**

SHOULDER ANGLE (DEG'S/SIDE): **34.40**

PRIMER: **L/R**

CASE CAPACITY (GR'S WATER): **55.24**

DIMENSIONAL DRAWING:

F
H
K
G
A
J
E
B
D

-NOT ACTUAL SIZE-
-DO NOT SCALE-

CARTRIDGE: 6,5/350 Remington Magnum

OTHER NAMES:	
	DIA: .264
	BALLISTEK NO: 264AB
	NAI NO: BEN 12443/4.269

DATA SOURCE: Ackley Vol.2 Pg162.

HISTORICAL DATA: Designed for short M600 actions.

NOTES:

LOADING DATA:

BULLET WT./TYPE	POWDER WT./TYPE	VELOCITY ('/SEC)	SOURCE
120/—	56.2/4350	2940	Ackley

CASE PREPARATION: SHELLHOLDER (RCBS): 4

MAKE FROM: .350 Rem. Mag. You really need a form set for this cartridge! Form, trim, chamfer and fireform in chamber. This is nothing more or less than the 6,5mm Rem. Mag. Any difference is not worth the effort!

PHYSICAL DATA (INCHES):

CASE TYPE: **Belted Bottleneck**

CASE LENGTH **A = 2.19**

HEAD DIAMETER **B = .513**

RIM DIAMETER **D = .532**

NECK DIAMETER **F = .291**

NECK LENGTH **H = .242**

SHOULDER LENGTH **K = .198**

BODY ANGLE (DEG'S/SIDE): **.295**

CASE CAPACITY **CC'S = 4.75**

LOADED LENGTH: **2.80**

BELT DIAMETER **C = .532**

RIM THICKNESS **E = .05**

SHOULDER DIAMETER **G = .497**

LENGTH TO SHOULDER **J = 1.75**

SHOULDER ANGLE (DEG'S/SIDE): **27.48**

PRIMER: **L/R Mag**

CASE CAPACITY (GR'S WATER): **73.31**

DIMENSIONAL DRAWING:

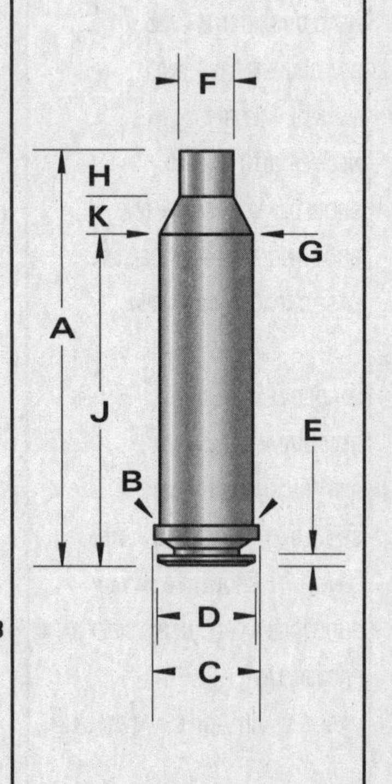

-NOT ACTUAL SIZE-
-DO NOT SCALE-

CARTRIDGE: 6,5/378 Weatherby

OTHER NAMES:	
	DIA: .264
	BALLISTEK NO: 264L
	NAI NO: BEN 22424/4.983

DATA SOURCE: NAI/Ballistek

HISTORICAL DATA:

NOTES: Overbore!

LOADING DATA:

BULLET WT./TYPE	POWDER WT./TYPE	VELOCITY ('/SEC)	SOURCE
140/Spire	114.0/MRP	3300	JJD

CASE PREPARATION: SHELLHOLDER (RCBS): 14

MAKE FROM: .378 Weatherby. Form die set required! Do not anneal. Form in form dies #1 and #2. Anneal. Form in form die #3. Trim to length. F/L size & chamfer.

PHYSICAL DATA (INCHES):

CASE TYPE: **Belted Bottleneck**

CASE LENGTH **A = 2.900**

HEAD DIAMETER **B = .582**

RIM DIAMETER **D = .582**

NECK DIAMETER **F = .294**

NECK LENGTH **H = .370**

SHOULDER LENGTH **K = .140**

BODY ANGLE (DEG'S/SIDE): **.366**

CASE CAPACITY **CC'S = 8.177**

LOADED LENGTH: **3.82**

BELT DIAMETER **C = .603**

RIM THICKNESS **E = .062**

SHOULDER DIAMETER **G = .554**

LENGTH TO SHOULDER **J = 2.39**

SHOULDER ANGLE (DEG'S/SIDE): **N/A**

PRIMER: **L/R Mag**

CASE CAPACITY (GR'S WATER): **126.20**

DIMENSIONAL DRAWING:

F
H
K
G
A
J
E
B
D
C

-NOT ACTUAL SIZE-
-DO NOT SCALE-

CARTRIDGE: 6.53 Lazzaroni Scramjet

OTHER NAMES:

DIA: .257

BALLISTEK NO:

NAI NO:

DATA SOURCE:

HISTORICAL DATA: Created by John Lazzaroni.

NOTES: This is the .25 caliber in John Lazzaroni's line of long action beltless magnums. The case however is slightly different than the other long action cases.

LOADING DATA:

BULLET WT./TYPE	POWDER WT./TYPE	VELOCITY ('/SEC)	SOURCE
100/Nosler Partition	92.0/AA8700	3,754	Lazzaroni
115/Nosler Partititon	74.0/IMR7828	3,501	Lazzaroni

CASE PREPARATION: SHELL HOLDER (RCBS): 4

MAKE FROM:

PHYSICAL DATA (INCHES):

CASE TYPE: **Rimless Bottleneck**

CASE LENGTH **A = 2.810**

HEAD DIAMETER **B = .5314**

RIM DIAMETER **D = .532**

NECK DIAMETER **F = .294**

NECK LENGTH **H = .300**

SHOULDER LENGTH **K = .188**

BODY ANGLE (DEG'S/SIDE):

CASE CAPACITY **CC'S =**

LOADED LENGTH: **3.460**

BELT DIAMETER **C = N/A**

RIM THICKNESS **E = .060**

SHOULDER DIAMETER **G = .512**

LENGTH TO SHOULDER **J = 2.322**

SHOULDER ANGLE (DEG'S/SIDE): **30**

PRIMER: **L/R Mag**

CASE CAPACITY (GR'S WATER):

DIMENSIONAL DRAWING:

-NOT ACTUAL SIZE-
-DO NOT SCALE-

CARTRIDGE: 6,7 x 60mm

OTHER NAMES: 6,5mm E.G.M.

DIA: .264

BALLISTEK NO: 264BC

NAI NO: BEN 12334/5.248

DATA SOURCE: Ackley Vol.2 Pg158

HISTORICAL DATA: By Dr. Eichhorn.

NOTES: EGM = Eichhorn Gamlakarleby Magnum.

LOADING DATA:

BULLET WT./TYPE	POWDER WT./TYPE	VELOCITY ('/SEC)	SOURCE
139/—	49.5/4831	2650	Ackley

CASE PREPARATION: SHELLHOLDER (RCBS): 3

MAKE FROM: .270 Win. Swage the case base to .450" dia. and silver solder a rough magnum-type belt in place. Turn the belt to .473" dia. A set of belt forming dies may also be used for this operation. F/L size and fireform shoulder in chamber or trim die. Trim case, chamfer and reload.

PHYSICAL DATA (INCHES):

CASE TYPE: **Belted Bottleneck**

CASE LENGTH **A = 2.362**

HEAD DIAMETER **B = .450**

RIM DIAMETER **D = .473**

NECK DIAMETER **F = .290**

NECK LENGTH **H = .327**

SHOULDER LENGTH **K = .065**

BODY ANGLE (DEG'S/SIDE): **.235**

CASE CAPACITY **CC'S = 4.04**

LOADED LENGTH: **3.25**

BELT DIAMETER **C = .473**

RIM THICKNESS **E = .049**

SHOULDER DIAMETER **G = .435**

LENGTH TO SHOULDER **J = 1.97**

SHOULDER ANGLE (DEG'S/SIDE): **45.14**

PRIMER: **L/R**

CASE CAPACITY (GR'S WATER): **62.32**

DIMENSIONAL DRAWING:

-NOT ACTUAL SIZE-
-DO NOT SCALE-

CARTRIDGE: 6,8 x 57mm Chinese Mauser

OTHER NAMES:

DIA: .277

BALLISTEK NO: 277E

NAI NO: RXB 22332/4.734

DATA SOURCE: NAI/Ballistek

HISTORICAL DATA:

NOTES:

LOADING DATA:

BULLET WT./TYPE	POWDER WT./TYPE	VELOCITY ('/SEC)	SOURCE
150/RN	42.6/4350	—	JJD

CASE PREPARATION: **SHELLHOLDER (RCBS): 3**

MAKE FROM: .270 Win. Size .270 case in 6,8x57 F/L die with expander removed. Trim to length. Chamfer. I.D. neck ream. F/L size. Fireform in chamber.

PHYSICAL DATA (INCHES):

CASE TYPE: **Rimless Bottleneck**

CASE LENGTH **A = 2.192**

HEAD DIAMETER **B = .463**

RIM DIAMETER **D = .468**

NECK DIAMETER **F = .311**

NECK LENGTH **H = .298**

SHOULDER LENGTH **K = .185**

BODY ANGLE (DEG'S/SIDE): **.645**

CASE CAPACITY **CC'S = 3.50**

LOADED LENGTH: **2.95**

BELT DIAMETER **C = N/A**

RIM THICKNESS **E = .047**

SHOULDER DIAMETER **G = .429**

LENGTH TO SHOULDER **J = 1.709**

SHOULDER ANGLE (DEG'S/SIDE): **17.68**

PRIMER: **L/R**

CASE CAPACITY (GR'S WATER): **54.09**

DIMENSIONAL DRAWING:

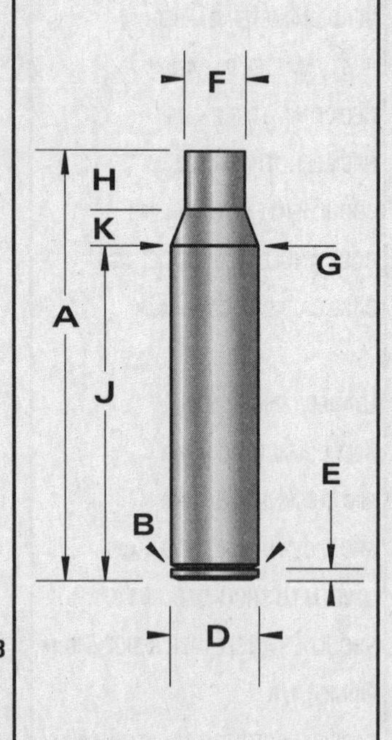

-NOT ACTUAL SIZE-
-DO NOT SCALE-

CARTRIDGE: 7mm Ackley Magnum

OTHER NAMES:

DIA: .284

BALLISTEK NO: 284AE

NAI NO: BEN 22424/4.814

DATA SOURCE: Ackley Vol.1 Pg403

HISTORICAL DATA:

NOTES:

LOADING DATA:

BULLET WT./TYPE	POWDER WT./TYPE	VELOCITY ('/SEC)	SOURCE
174/—	60.0/4350	2950	Ackley

CASE PREPARATION: **SHELLHOLDER (RCBS): 4**

MAKE FROM: 7mm Rem. Mag. Anneal neck and shoulder area. F/L size in Ackley die with expander removed. Trim and chamfer. I.D. may need neck reaming. F/L size and fireform in chamber.

PHYSICAL DATA (INCHES):

CASE TYPE: **Belted Bottleneck**

CASE LENGTH **A = 2.47**

HEAD DIAMETER **B = .513**

RIM DIAMETER **D = .532**

NECK DIAMETER **F = .315**

NECK LENGTH **H = .375**

SHOULDER LENGTH **K = .135**

BODY ANGLE (DEG'S/SIDE): **.374**

CASE CAPACITY **CC'S = 5.20**

LOADED LENGTH: **3.22**

BELT DIAMETER **C = .532**

RIM THICKNESS **E = .05**

SHOULDER DIAMETER **G = .490**

LENGTH TO SHOULDER **J = 1.96**

SHOULDER ANGLE (DEG'S/SIDE): **32.95**

PRIMER: **L/R Mag**

CASE CAPACITY (GR'S WATER): **80.32**

DIMENSIONAL DRAWING:

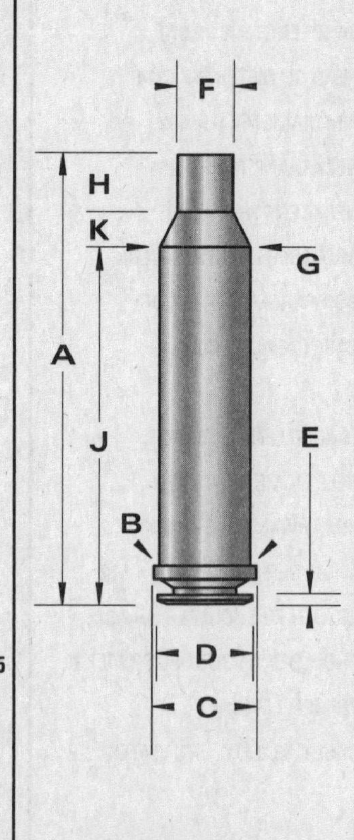

-NOT ACTUAL SIZE-
-DO NOT SCALE-

CARTRIDGE: 7mm Dakota

OTHER NAMES:

DIA: .284

BALLISTEK NO:

NAI NO:

DATA SOURCE: Dakota

HISTORICAL DATA: Based on the 404 Jeffrey case. This case is long enough to create standard- or magnum-length cases.

NOTES:

LOADING DATA:

BULLET WT./TYPE	POWDER WT./TYPE	VELOCITY ('/SEC)	SOURCE
140	73.0/IMR-4831	3495	Dakota
160	68.0/IMR-4831	3645	Dakota

CASE PREPARATION: SHELL HOLDER (RCBS): SPL

MAKE FROM: 404 Jeffery case.

PHYSICAL DATA (INCHES):

CASE TYPE: **Rimless Bottleneck**

CASE LENGTH **A = 2.50**

HEAD DIAMETER **B = .544**

RIM DIAMETER **D = .544**

NECK DIAMETER **F = .314**

NECK LENGTH **H = .333**

SHOULDER LENGTH **K = .187**

BODY ANGLE (DEG'S/SIDE): **N/A**

CASE CAPACITY **CC'S**

LOADED LENGTH: **3.33**

BELT DIAMETER **C = N/A**

RIM THICKNESS **E = .050**

SHOULDER DIAMETER **G = .531**

LENGTH TO SHOULDER **J = 1.980**

SHOULDER ANGLE (DEG'S/SIDE):**30°**

PRIMER: **L/R**

CASE CAPACITY (GR'S WATER):

DIMENSIONAL DRAWING:

F

H

K

G

A

J

E

B

D

-NOT ACTUAL SIZE-
-DO NOT SCALE-

CARTRIDGE: 7mm Canadian Magnum

OTHER NAMES:

DIA: .284

BALLISTEK NO:

NAI NO:

DATA SOURCE: Aubrey White

HISTORICAL DATA: This cartridge was developed around 1989 by North American Shooting Systems (NASS) and is similar to the 7mm Imperial Magnum.

NOTES:

LOADING DATA:

BULLET WT./TYPE	POWDER WT./TYPE	VELOCITY ('/SEC)	SOURCE
140	82.0 - H-4831	3426	NASS
160	82.0 - RL22	3264	NASS
175	79.0 - IMR-7828	3018	NASS

CASE PREPARATION: SHELL HOLDER (RCBS): 10

MAKE FROM: .404 Jeffery

PHYSICAL DATA (INCHES):

CASE TYPE: **Rebated Bottleneck**

CASE LENGTH **A = 2.830**

HEAD DIAMETER **B = .544**

RIM DIAMETER **D = .532**

NECK DIAMETER **F = .322**

NECK LENGTH **H = .311**

SHOULDER LENGTH **K = .180**

BODY ANGLE (DEG'S/SIDE): **N/A**

CASE CAPACITY **CC'S**

LOADED LENGTH: **3.600**

BELT DIAMETER **C = N/A**

RIM THICKNESS **E = .050**

SHOULDER DIAMETER **G = .530**

LENGTH TO SHOULDER **J = 2.339**

SHOULDER ANGLE (DEG'S/SIDE):**30°**

PRIMER: **L/R**

CASE CAPACITY (GR'S WATER):

DIMENSIONAL DRAWING:

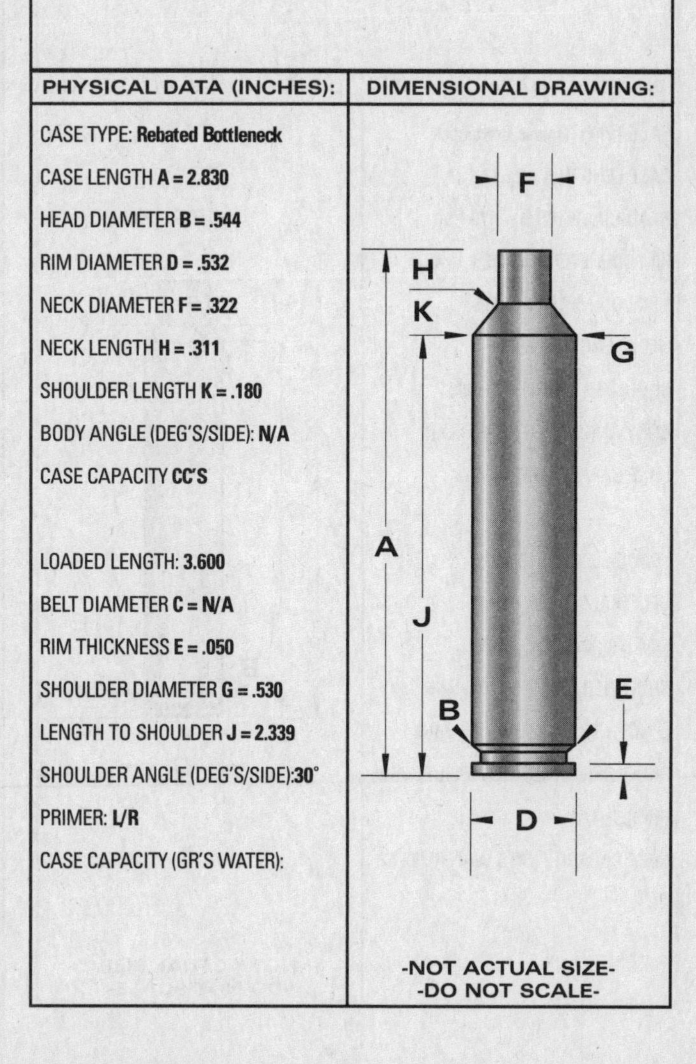

-NOT ACTUAL SIZE-
-DO NOT SCALE-

CARTRIDGE: 7mm Gibbs

OTHER NAMES:

DIA: .284

BALLISTEK NO: 284T

NAI NO: RXB 12344/5.355

DATA SOURCE: Ackley Vol.2 Pg178

HISTORICAL DATA: By R. Gibbs.

NOTES:

LOADING DATA:

BULLET WT./TYPE	POWDER WT./TYPE	VELOCITY ('/SEC)	SOURCE
160/—	64.0/4831	3106	Ackley

CASE PREPARATION: SHELLHOLDER (RCBS): 3

MAKE FROM: .280 Remington. Fireform the Rem. case (with cornmeal) in the Gibbs trim die. Trim & size. Also, .30-06 cases may be F/L sized and fireformed in the weapon chamber. Do not fireform the .280 case, in the chamber, due to headspace problems.

PHYSICAL DATA (INCHES):

CASE TYPE: **Rimless Bottleneck**

CASE LENGTH **A = 2.517**

HEAD DIAMETER **B = .470**

RIM DIAMETER **D = .473**

NECK DIAMETER **F = .315**

NECK LENGTH **H = .255**

SHOULDER LENGTH **K = .099**

BODY ANGLE (DEG'S/SIDE): **.219**

CASE CAPACITY **CC'S = 4.70**

LOADED LENGTH: **3.34**

BELT DIAMETER **C = N/A**

RIM THICKNESS **E = .049**

SHOULDER DIAMETER **G = .455**

LENGTH TO SHOULDER **J = 2.163**

SHOULDER ANGLE (DEG'S/SIDE): **35.26**

PRIMER: **L/R**

CASE CAPACITY (GR'S WATER): **72.52**

DIMENSIONAL DRAWING:

-NOT ACTUAL SIZE-
-DO NOT SCALE-

CARTRIDGE: 7mm ICL Tortilla

OTHER NAMES:

DIA: .284

BALLISTEK NO: 284BB

NAI NO: RXB 22323/4.713

DATA SOURCE: Ackley Vol.1 Pg389

HISTORICAL DATA:

NOTES: Another "improved" 7 x 57mm Mauser.

LOADING DATA:

BULLET WT./TYPE	POWDER WT./TYPE	VELOCITY ('/SEC)	SOURCE
140/Spire	55.0/4831	3150	Ackley

CASE PREPARATION: SHELLHOLDER (RCBS):

MAKE FROM: .30-06 Spgf. F/L size the '06 case in the ICL die with the expander removed. Trim & chamfer. F/L size. Fireform in chamber.

PHYSICAL DATA (INCHES):

CASE TYPE: **Rimless Bottleneck**

CASE LENGTH **A = 2.220**

HEAD DIAMETER **B = .471**

RIM DIAMETER **D = .473**

NECK DIAMETER **F = .310**

NECK LENGTH **H = .360**

SHOULDER LENGTH **K = .120**

BODY ANGLE (DEG'S/SIDE): **.613**

CASE CAPACITY **CC'S = 3.88**

LOADED LENGTH: **2.94**

BELT DIAMETER **C = N/A**

RIM THICKNESS **E = .049**

SHOULDER DIAMETER **G = .438**

LENGTH TO SHOULDER **J = 1.74**

SHOULDER ANGLE (DEG'S/SIDE): **28.07**

PRIMER: **L/R**

CASE CAPACITY (GR'S WATER): **59.95**

DIMENSIONAL DRAWING:

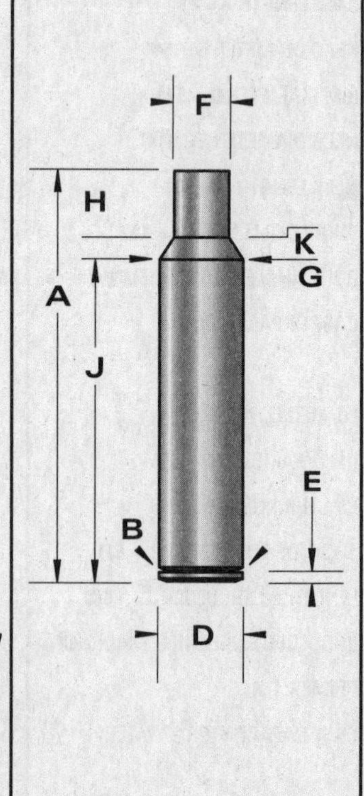

-NOT ACTUAL SIZE-
-DO NOT SCALE-

CARTRIDGE: 7mm ICL Wapiti

OTHER NAMES:

DIA: .284

BALLISTEK NO: 284BC

NAI NO: BEN 12424/4.971

DATA SOURCE: Ackley Vol.1 Pg405

HISTORICAL DATA:

NOTES:

LOADING DATA:

BULLET WT./TYPE	POWDER WT./TYPE	VELOCITY ('/SEC)	SOURCE
160/Spire	70.0/4831	3125	Ackley

CASE PREPARATION: SHELLHOLDER (RCBS): 4

MAKE FROM: .300 H&H Mag. Trim case to 2.60" and chamfer. F/L size in the ICL die. Fireform in the ICL chamber.

PHYSICAL DATA (INCHES):

CASE TYPE: **Belted Bottleneck**

CASE LENGTH **A = 2.55**

HEAD DIAMETER **B = .513**

RIM DIAMETER **D = .532**

NECK DIAMETER **F = .317**

NECK LENGTH **H = .375**

SHOULDER LENGTH **K = .085**

BODY ANGLE (DEG'S/SIDE): **.348**

CASE CAPACITY **CC'S = 5.39**

LOADED LENGTH: **A/R**

BELT DIAMETER **C = .532**

RIM THICKNESS **E = .05**

SHOULDER DIAMETER **G = .490**

LENGTH TO SHOULDER **J = 2.09**

SHOULDER ANGLE (DEG'S/SIDE): **45.50**

PRIMER: **L/R Mag**

CASE CAPACITY (GR'S WATER): **83.26**

DIMENSIONAL DRAWING:

F H K G A J E B D C

-NOT ACTUAL SIZE-
-DO NOT SCALE-

CARTRIDGE: 7mm JDJ

OTHER NAMES:

DIA: .284

BALLISTEK NO: 284BR

NAI NO: RMB 12344/4.497

DATA SOURCE: SSK Industries

HISTORICAL DATA: By J.D. Jones.

NOTES:

LOADING DATA:

BULLET WT./TYPE	POWDER WT./TYPE	VELOCITY ('/SEC)	SOURCE
154/—	34.0/4320	2107	SSK

CASE PREPARATION: SHELLHOLDER (RCBS): 11

MAKE FROM: .225 Winchester. Taper expand case to hold .284" dia. bullets and fireform with a 10% reduced load.

PHYSICAL DATA (INCHES):

CASE TYPE: **Rimmed Bottleneck**

CASE LENGTH **A = 1.889**

HEAD DIAMETER **B = .420**

RIM DIAMETER **D = .468**

NECK DIAMETER **F = .311**

NECK LENGTH **H = .187**

SHOULDER LENGTH **K = .077**

BODY ANGLE (DEG'S/SIDE): **.201**

CASE CAPACITY **CC'S = 2.79**

LOADED LENGTH: **2.886**

BELT DIAMETER **C = N/A**

RIM THICKNESS **E = .047**

SHOULDER DIAMETER **G = .410**

LENGTH TO SHOULDER **J = 1.625**

SHOULDER ANGLE (DEG'S/SIDE): **32.73**

PRIMER: **L/R**

CASE CAPACITY (GR'S WATER): **43.14**

DIMENSIONAL DRAWING:

F H K G A J E B D

-NOT ACTUAL SIZE-
-DO NOT SCALE-

CARTRIDGE: 7mm JDJ #2

OTHER NAMES:

DIA: .284

BALLISTEK NO: 284BS

NAI NO: RMB 12344/4.281

DATA SOURCE: SSK Industries

HISTORICAL DATA: By J.D. Jones.

NOTES:

LOADING DATA:

BULLET WT./TYPE	POWDER WT./TYPE	VELOCITY ('/SEC)	SOURCE
140/Nosler	–/–	2370	SSK

CASE PREPARATION: SHELLHOLDER (RCBS): 2

MAKE FROM: .307 Winchester. F/L size the factory case, square case mouth, chamfer and fireform with moderate load.

PHYSICAL DATA (INCHES):

CASE TYPE: **Rimmed Bottleneck**

CASE LENGTH **A = 1.995**

HEAD DIAMETER **B = .466**

RIM DIAMETER **D = .501**

NECK DIAMETER **F = .318**

NECK LENGTH **H = .250**

SHOULDER LENGTH **K = .095**

BODY ANGLE (DEG'S/SIDE): **.316**

CASE CAPACITY **CC'S = 3.41**

LOADED LENGTH: **2.927**

BELT DIAMETER **C = N/A**

RIM THICKNESS **E = .058**

SHOULDER DIAMETER **G = .450**

LENGTH TO SHOULDER **J = 1.650**

SHOULDER ANGLE (DEG'S/SIDE): **34.79**

PRIMER: **L/R**

CASE CAPACITY (GR'S WATER): **52.58**

DIMENSIONAL DRAWING:

F H K G A J E B D

-NOT ACTUAL SIZE-
-DO NOT SCALE-

CARTRIDGE: 7mm JRS

OTHER NAMES:

DIA: .284

BALLISTEK NO:

NAI NO:

DATA SOURCE: Jon Sundra

HISTORICAL DATA: Based on the 280/7mm Express Remington case, but is more than an Improved 280 in that it cannot be made by fireforming 280 Remington ammo in a 7mm JRS chamber.

NOTES:

LOADING DATA:

BULLET WT./TYPE	POWDER WT./TYPE	VELOCITY ('/SEC)	SOURCE
145	63.0/RL22	3130	Sundra
154	61.5/RL22	3020	Sundra

CASE PREPARATION: SHELL HOLDER (RCBS): 03

MAKE FROM: .280 Remington
Seat bullet into lands from .280 Rem. case with reduced load and fireform.

PHYSICAL DATA (INCHES):

CASE TYPE: **Rimless Bottleneck**

CASE LENGTH **A = 2.540**

HEAD DIAMETER **B = .470**

RIM DIAMETER **D = .473**

NECK DIAMETER **F = .315**

NECK LENGTH **H = .300**

SHOULDER LENGTH **K = .110**

BODY ANGLE (DEG'S/SIDE): **N/A**

CASE CAPACITY **CC'S = 3.41**

LOADED LENGTH: **3.330**

BELT DIAMETER **C = N/A**

RIM THICKNESS **E = .049**

SHOULDER DIAMETER **G = .455**

LENGTH TO SHOULDER **J = 2.130**

SHOULDER ANGLE (DEG'S/SIDE): **35°**

PRIMER: **L/R**

CASE CAPACITY (GR'S WATER): **52.58**

DIMENSIONAL DRAWING:

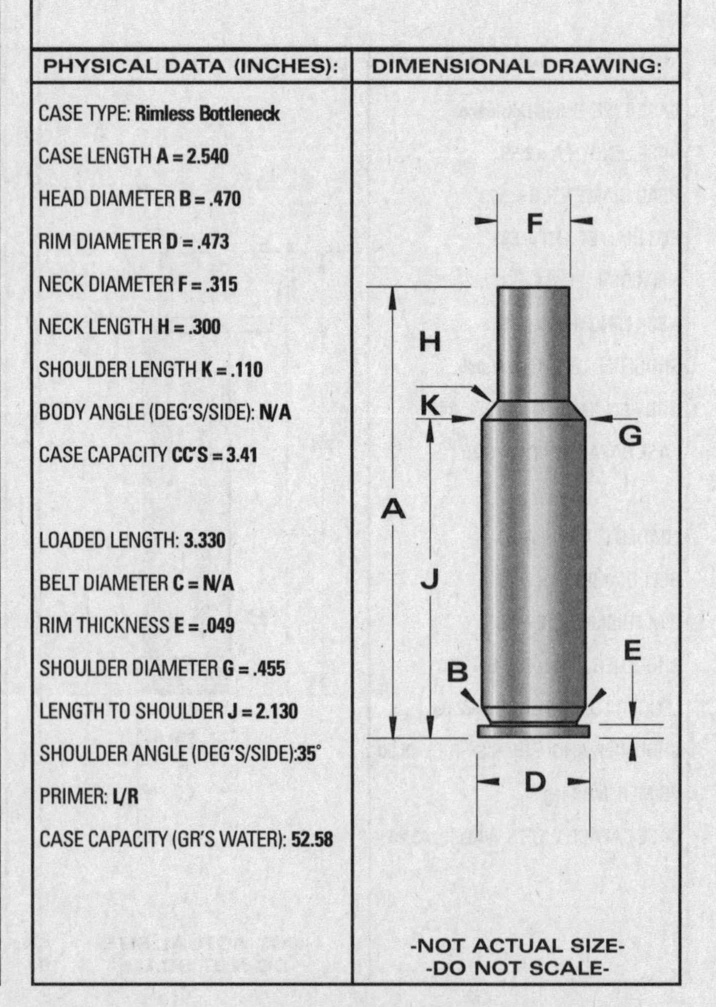

-NOT ACTUAL SIZE-
-DO NOT SCALE-

CARTRIDGE: 7mm Krag Improved

OTHER NAMES:

DIA: .284

BALLISTEK NO: 284V

NAI NO: RMB 22323/5.054

DATA SOURCE: NAI/Ballistek

HISTORICAL DATA: Ballistek's version. Others exist.

NOTES:

LOADING DATA:

BULLET WT./TYPE	POWDER WT./TYPE	VELOCITY ('/SEC)	SOURCE
140/Spire	44.8/H380	—	JJD

CASE PREPARATION: SHELLHOLDER (RCBS): 7

MAKE FROM: .30-40 Krag. Size the Krag case in the "improved" die and fireform. Trim and chamfer. Resize and reload.

PHYSICAL DATA (INCHES):

CASE TYPE: **Rimmed Bottleneck**

CASE LENGTH **A = 2.31**

HEAD DIAMETER **B = .457**

RIM DIAMETER **D = .545**

NECK DIAMETER **F = .316**

NECK LENGTH **H = .370**

SHOULDER LENGTH **K = .160**

BODY ANGLE (DEG'S/SIDE): **.398**

CASE CAPACITY **CC'S = 3.75**

LOADED LENGTH: **2.245**

BELT DIAMETER **C = N/A**

RIM THICKNESS **E = .064**

SHOULDER DIAMETER **G = .435**

LENGTH TO SHOULDER **J = 1.78**

SHOULDER ANGLE (DEG'S/SIDE): **20.39**

PRIMER: **L/R**

CASE CAPACITY (GR'S WATER): **58.01**

DIMENSIONAL DRAWING:

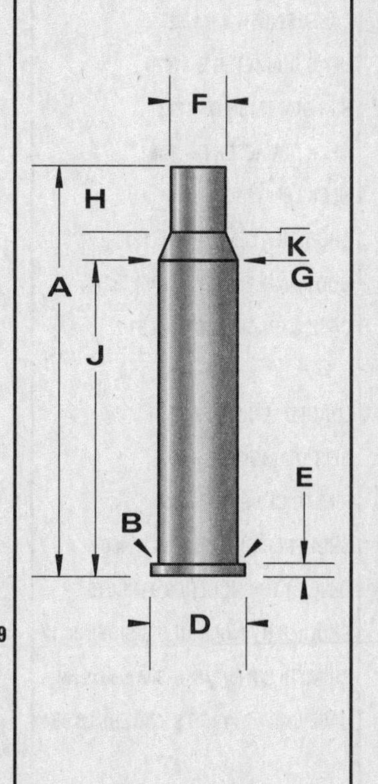

-NOT ACTUAL SIZE-
-DO NOT SCALE-

CARTRIDGE: 7mm Nambu

OTHER NAMES:
7mm Baby Nambu

DIA: .280

BALLISTEK NO: 280A

NAI NO: RXB 34242/2.222

DATA SOURCE: COTW 4th Pg163

HISTORICAL DATA: Japanese mil. ctg. ca. 1920.

NOTES:

LOADING DATA:

BULLET WT./TYPE	POWDER WT./TYPE	VELOCITY ('/SEC)	SOURCE
60/Lead	1.7/B'Eye	1000	JJD

CASE PREPARATION: SHELLHOLDER (RCBS): 17 (ext)

MAKE FROM: .30 Carbine. Swage base to .351" dia. or turn in lathe. Use extended shell holder #17 to run case into form die (required!). Cut case off to .79". I.D. ream (also required). F/L size. Old Lyman mold #280473 (unsized) makes fine bullets if cut off to weigh 60 grains.

PHYSICAL DATA (INCHES):

CASE TYPE: **Rimless Bottleneck**

CASE LENGTH **A = .78**

HEAD DIAMETER **B = .351**

RIM DIAMETER **D = .360**

NECK DIAMETER **F = .296**

NECK LENGTH **H = .120**

SHOULDER LENGTH **K = .077**

BODY ANGLE (DEG'S/SIDE): **1.34**

CASE CAPACITY **CC'S = .647**

LOADED LENGTH: **1.06**

BELT DIAMETER **C = N/A**

RIM THICKNESS **E = .045**

SHOULDER DIAMETER **G = .333**

LENGTH TO SHOULDER **J = .583**

SHOULDER ANGLE (DEG'S/SIDE): **13.50**

PRIMER: **L/R**

CASE CAPACITY (GR'S WATER): **9.98**

DIMENSIONAL DRAWING:

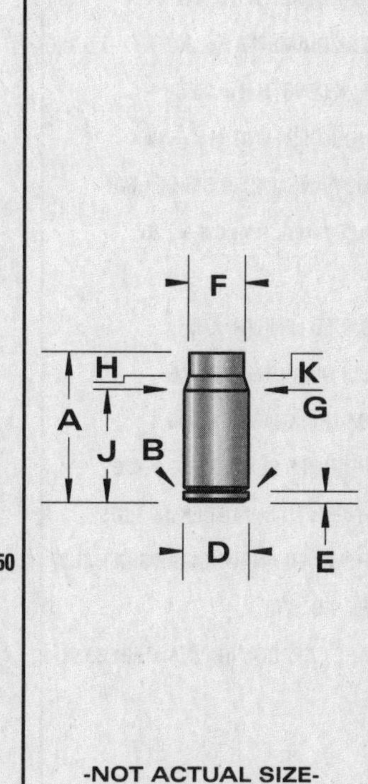

-NOT ACTUAL SIZE-
-DO NOT SCALE-

CARTRIDGE: 7mm PPC

OTHER NAMES:

DIA: .284

BALLISTEK NO: 284AD

NAI NO: RXB 24333/3.447

DATA SOURCE: NAI/Ballistek

HISTORICAL DATA: 7mm version of original 6mm PPC ctg.

NOTES:

LOADING DATA:

BULLET WT./TYPE	POWDER WT./TYPE	VELOCITY ('/SEC)	SOURCE
120/Spire	25.0/IMR4198	—	JJD

CASE PREPARATION: SHELLHOLDER (RCBS): 6

MAKE FROM: .220 Russian (Sako). Anneal case and taper expand to about .290" dia. F/L size, trim and chamfer.

PHYSICAL DATA (INCHES):

CASE TYPE: **Rimless Bottleneck**

CASE LENGTH **A = 1.520**

HEAD DIAMETER **B = .441**

RIM DIAMETER **D = .442**

NECK DIAMETER **F = .307**

NECK LENGTH **H = .305**

SHOULDER LENGTH **K = .140**

BODY ANGLE (DEG'S/SIDE): **.687**

CASE CAPACITY **CC'S = 2.20**

LOADED LENGTH: **A/R**

BELT DIAMETER **C = N/A**

RIM THICKNESS **E = .048**

SHOULDER DIAMETER **G = .420**

LENGTH TO SHOULDER **J = 1.075**

SHOULDER ANGLE (DEG'S/SIDE): **21.97**

PRIMER: **S/R**

CASE CAPACITY (GR'S WATER): **33.98**

DIMENSIONAL DRAWING:

F H K G A J E B D

-NOT ACTUAL SIZE-
-DO NOT SCALE-

CARTRIDGE: 7mm Remington BR

OTHER NAMES:
7mm Benchrest (Rem.)

DIA: .284

BALLISTEK NO: 284AP

NAI NO: RXB 14333/3.234

DATA SOURCE: Hornady 3rd Pg326

HISTORICAL DATA: By Rem. in 1980.

NOTES: BR cases use small rifle primers!

LOADING DATA:

BULLET WT./TYPE	POWDER WT./TYPE	VELOCITY ('/SEC)	SOURCE
139/Spire	30.0/BL-C2	2000	JJD

CASE PREPARATION: SHELLHOLDER (RCBS): 3

MAKE FROM: Rem. BR or .308 or .270 Win. Anneal case neck. Run into form die #1 to set shoulder back. Form die #2 will reduce neck dia. Trim case to length. F/L size, chamfer and fireform in chamber.

PHYSICAL DATA (INCHES):

CASE TYPE: **Rimless Bottleneck**

CASE LENGTH **A = 1.52**

HEAD DIAMETER **B = .470**

RIM DIAMETER **D = .473**

NECK DIAMETER **F = .308**

NECK LENGTH **H = .314**

SHOULDER LENGTH **K = .131**

BODY ANGLE (DEG'S/SIDE): **.327**

CASE CAPACITY **CC'S = 2.51**

LOADED LENGTH: **2.33**

BELT DIAMETER **C = N/A**

RIM THICKNESS **E = .054**

SHOULDER DIAMETER **G = .460**

LENGTH TO SHOULDER **J = 1.075**

SHOULDER ANGLE (DEG'S/SIDE): **30.12**

PRIMER: **S/R (or L/R in .308 type case)**

CASE CAPACITY (GR'S WATER): **38.80**

DIMENSIONAL DRAWING:

F H K G A J E B D

-NOT ACTUAL SIZE-
-DO NOT SCALE-

CARTRIDGE: 7mm Remington Magnum

OTHER NAMES:

DIA: .284

BALLISTEK NO: 284E

NAI NO: BEN 12443/4.873

DATA SOURCE: Hornady Manual 3rd Pg196

HISTORICAL DATA: By Rem. in 1962.

NOTES:

LOADING DATA:

BULLET WT./TYPE	POWDER WT./TYPE	VELOCITY ('/SEC)	SOURCE
168/Spire	65.0/IMR4831	2479	JJD

CASE PREPARATION: **SHELLHOLDER (RCBS): 4**

MAKE FROM: Factory or .300 H&H Mag. Anneal case neck and size in 7mm Mag. sizer. Trim to length. Check neck wall thickness and ream if necessary. F/L size, chamfer and fireform in chamber.

PHYSICAL DATA (INCHES):

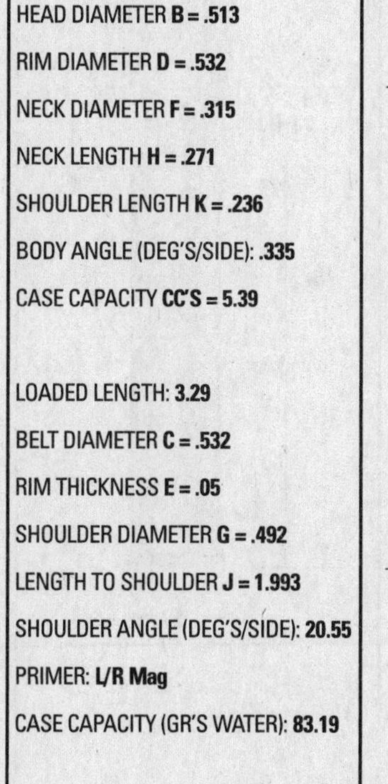

CASE TYPE: **Belted Bottleneck**

CASE LENGTH **A = 2.500**

HEAD DIAMETER **B = .513**

RIM DIAMETER **D = .532**

NECK DIAMETER **F = .315**

NECK LENGTH **H = .271**

SHOULDER LENGTH **K = .236**

BODY ANGLE (DEG'S/SIDE): **.335**

CASE CAPACITY **CC'S = 5.39**

LOADED LENGTH: **3.29**

BELT DIAMETER **C = .532**

RIM THICKNESS **E = .05**

SHOULDER DIAMETER **G = .492**

LENGTH TO SHOULDER **J = 1.993**

SHOULDER ANGLE (DEG'S/SIDE): **20.55**

PRIMER: **L/R Mag**

CASE CAPACITY (GR'S WATER): **83.19**

DIMENSIONAL DRAWING:

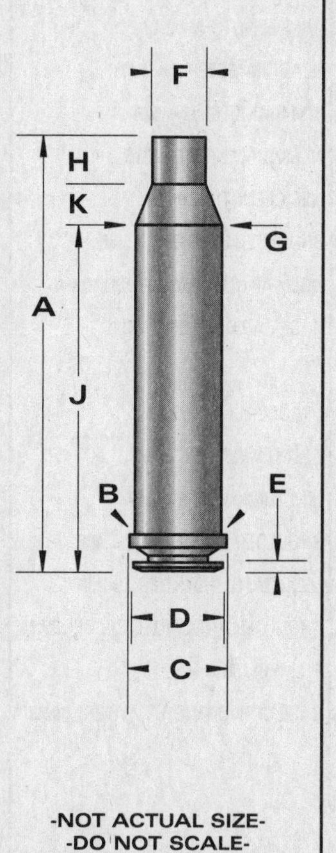

-NOT ACTUAL SIZE-
-DO NOT SCALE-

CARTRIDGE: 7mm Remington Short Action Ultra Mag

OTHER NAMES:

DIA: .284

BALLISTEK NO:

NAI NO:

DATA SOURCE: Most current loading manuals.

HISTORICAL DATA: Introduced by Remington in 2002 to increase power from a short action rifle. This is a shortened version of the 7mm Ultra Mag cartridge.

NOTES: While similar to the 7mm Winchester Short Magnum the two are not interchangeable.

LOADING DATA:

BULLET WT./TYPE	POWDER WT./TYPE	VELOCITY ('/SEC)	SOURCE
139/Hornady SP	64.9/RL-19	3,200	Horn. #6
150/Factory		3,110	Rem

CASE PREPARATION: **SHELL HOLDER (RCBS): 38**

MAKE FROM: Factory or make from .300 Remington Short Action Ultra Mag by necking down to .284. Trim, chamfer and deburr. Check neck wall thickness and turn if necessary.

PHYSICAL DATA (INCHES):

CASE TYPE: **Rebated B/N**

CASE LENGTH **A = 2.035**

HEAD DIAMETER **B = .550**

RIM DIAMETER **D = .534**

NECK DIAMETER **F = .320**

NECK LENGTH **H = .3108**

SHOULDER LENGTH **K = .186**

BODY ANGLE (DEG'S/SIDE):

CASE CAPACITY **CC'S =**

LOADED LENGTH: **2.825**

BELT DIAMETER **C =**

RIM THICKNESS **E = .050**

SHOULDER DIAMETER **G = .535**

LENGTH TO SHOULDER **J = 1.538**

SHOULDER ANGLE (DEG'S/SIDE): **30**

PRIMER: **L/R**

CASE CAPACITY (GR'S WATER):

DIMENSIONAL DRAWING:

-NOT ACTUAL SIZE-
-DO NOT SCALE-

CARTRIDGE: 7mm Remington Ultra Mag

OTHER NAMES:

DIA: .284

BALLISTEK NO:

NAI NO:

DATA SOURCE: Most current loading manuals.

HISTORICAL DATA: Introduced by Remington in 2001. The case is made from the .404 Jeffery which was necked down and the body and shoulder blown out.

NOTES:

LOADING DATA:

BULLET WT./TYPE	POWDER WT./TYPE	VELOCITY ('/SEC)	SOURCE
175/Swift A-Frame	98.0/H-50BMG	3,014	Swift #1
160/Factory		3,200	Rem

CASE PREPARATION: SHELL HOLDER (RCBS): 38

MAKE FROM: Factory or neck down .300 Remington Ultra Mag. Can be made from .404 Jeffery.

PHYSICAL DATA (INCHES):

CASE TYPE: **Rebated Bottleneck**

CASE LENGTH **A = 2.850**

HEAD DIAMETER **B = .550**

RIM DIAMETER **D = .534**

NECK DIAMETER **F = .322**

NECK LENGTH **H = .287**

SHOULDER LENGTH **K = .176**

BODY ANGLE (DEG'S/SIDE):

CASE CAPACITY **CC'S =**

LOADED LENGTH: **3.600**

BELT DIAMETER **C =**

RIM THICKNESS **E = .050**

SHOULDER DIAMETER **G = .525**

LENGTH TO SHOULDER **J = 2.387**

SHOULDER ANGLE (DEG'S/SIDE): **30**

PRIMER: **L/R Mag**

CASE CAPACITY (GR'S WATER):

DIMENSIONAL DRAWING:

-NOT ACTUAL SIZE-
-DO NOT SCALE-

CARTRIDGE: 7mm Rigby Magnum

OTHER NAMES:
.275 #2 Magnum

DIA: .287

BALLISTEK NO: 287C

NAI NO: RMB 22334/5.298

DATA SOURCE: COTW 4th Pg228

HISTORICAL DATA: By Rigby about 1927.

NOTES:

LOADING DATA:

BULLET WT./TYPE	POWDER WT./TYPE	VELOCITY ('/SEC)	SOURCE
140/LeadPP	27.7/IMR4064	—	JJD

CASE PREPARATION: SHELLHOLDER (RCBS): 26

MAKE FROM: .405 Win. Basic (BELL). Turn rim to .528" dia. and cut off to 2.5". Anneal case neck and form in form die set (2-dies). Trim to length, chamfer and F/L size. Fireform in chamber.

PHYSICAL DATA (INCHES):

CASE TYPE: **Rimmed Bottleneck**

CASE LENGTH **A = 2.49**

HEAD DIAMETER **B = .470**

RIM DIAMETER **D = .528**

NECK DIAMETER **F = .315**

NECK LENGTH **H = .35**

SHOULDER LENGTH **K = .060**

BODY ANGLE (DEG'S/SIDE): **.975**

CASE CAPACITY **CC'S = 4.16**

LOADED LENGTH: **3.25**

BELT DIAMETER **C = N/A**

RIM THICKNESS **E = .047**

SHOULDER DIAMETER **G = .406**

LENGTH TO SHOULDER **J = 2.08**

SHOULDER ANGLE (DEG'S/SIDE): **37.17**

PRIMER: **L/R**

CASE CAPACITY (GR'S WATER): **64.23**

DIMENSIONAL DRAWING:

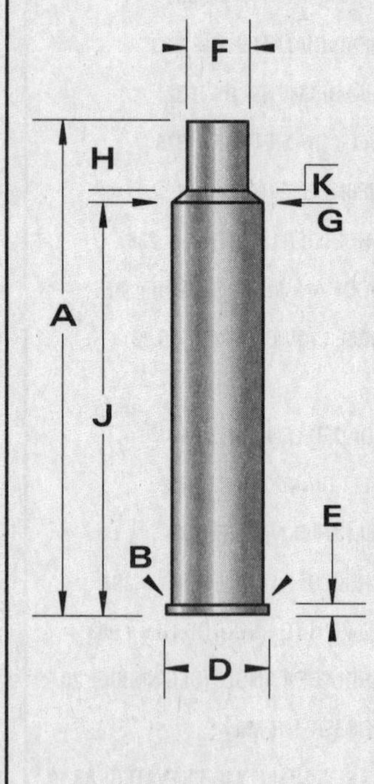

-NOT ACTUAL SIZE-
-DO NOT SCALE-

CARTRIDGE: 7mm Rimmed Magnum (Holland & Holland)

OTHER NAMES:

DIA: .287

BALLISTEK NO: 287D

NAI NO: RMB 22323/5.333

DATA SOURCE: COTW 4th Pg227

HISTORICAL DATA: Old H&H ctg.

NOTES: Very rare.

LOADING DATA:

BULLET WT./TYPE	POWDER WT./TYPE	VELOCITY ('/SEC)	SOURCE
140/Spire	40.0/4064	—	Barnes

CASE PREPARATION: SHELLHOLDER (RCBS): 7

MAKE FROM: 9,3 x 75R. Trim case to length and anneal neck. F/L size with expander removed. Case may need I.D. neck reaming. F/L size again.

PHYSICAL DATA (INCHES):

CASE TYPE: **Rimmed Bottleneck**

CASE LENGTH **A = 2.48**

HEAD DIAMETER **B = .465**

RIM DIAMETER **D = .532**

NECK DIAMETER **F = .319**

NECK LENGTH **H = .410**

SHOULDER LENGTH **K = .100**

BODY ANGLE (DEG'S/SIDE): **.97**

CASE CAPACITY **CC'S = 3.90**

LOADED LENGTH: **3.22**

BELT DIAMETER **C = N/A**

RIM THICKNESS **E = .055**

SHOULDER DIAMETER **G = .405**

LENGTH TO SHOULDER **J = 1.97**

SHOULDER ANGLE (DEG'S/SIDE): **23.26**

PRIMER: **L/R**

CASE CAPACITY (GR'S WATER): **60.21**

DIMENSIONAL DRAWING:

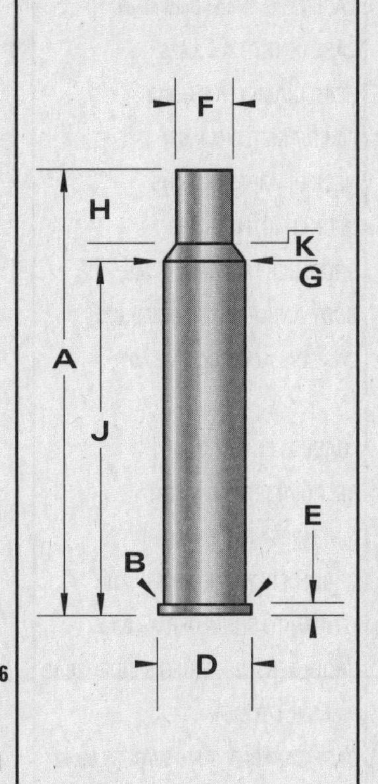

-NOT ACTUAL SIZE-
-DO NOT SCALE-

CARTRIDGE: 7mm Smith Magnum

OTHER NAMES:

DIA: .284

BALLISTEK NO: 284BD

NAI NO: BEN 22423/4.629

DATA SOURCE: Ackley Vol.2 Pg179

HISTORICAL DATA: By B. Smith.

NOTES:

LOADING DATA:

BULLET WT./TYPE	POWDER WT./TYPE	VELOCITY ('/SEC)	SOURCE
160/Spire	65.0/4350	3174	Ackley

CASE PREPARATION: SHELLHOLDER (RCBS): 4

MAKE FROM: 7mm Rem. Mag. F/L size the Rem. case in the Smith die. Trim to length and chamfer. Fireform in Smith chamber.

PHYSICAL DATA (INCHES):

CASE TYPE: **Belted Bottleneck**

CASE LENGTH **A = 2.375**

HEAD DIAMETER **B = .513**

RIM DIAMETER **D = .532**

NECK DIAMETER **F = .317**

NECK LENGTH **H = .400**

SHOULDER LENGTH **K = .153**

BODY ANGLE (DEG'S/SIDE): **.494**

CASE CAPACITY **CC'S = 4.79**

LOADED LENGTH: **3.32**

BELT DIAMETER **C = .532**

RIM THICKNESS **E = .05**

SHOULDER DIAMETER **G = .485**

LENGTH TO SHOULDER **J = 1.822**

SHOULDER ANGLE (DEG'S/SIDE): **28.77**

PRIMER: **L/R Mag**

CASE CAPACITY (GR'S WATER): **73.90**

DIMENSIONAL DRAWING:

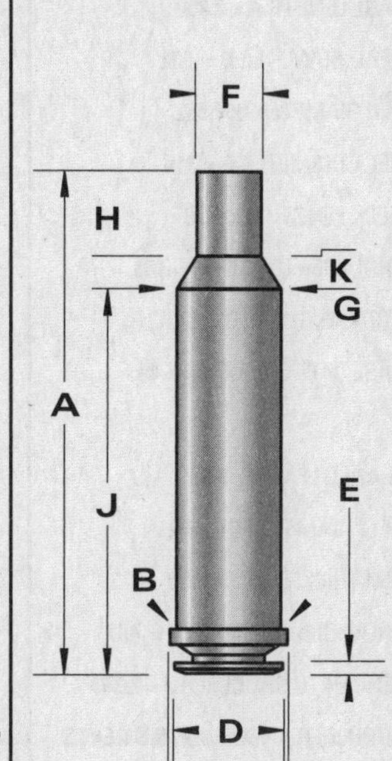

-NOT ACTUAL SIZE-
-DO NOT SCALE-

CARTRIDGE: 7mm STW

OTHER NAMES:
7mm Shooting Times Westerner

DIA: .284

BALLISTEK NO:

NAI NO:

DATA SOURCE: Nosler

HISTORICAL DATA: This Wildcat chambering was designed in 1989 by Layne Simpson. This design includes a slight body taper compared to the original 8mm Remington Magnum case.

NOTES:

LOADING DATA:

BULLET WT./TYPE	POWDER WT./TYPE	VELOCITY (*/SEC)	SOURCE
140 Nosler BT	75.0/H-4831	3234	A-Square
160 Nosler Part.	75.0/RL-22	3084	A-Square

CASE PREPARATION: **SHELL HOLDER (RCBS): 4**

MAKE FROM: 8mm Remington Magnum case.

PHYSICAL DATA (INCHES):

CASE TYPE: **Belted Bottleneck**
CASE LENGTH A = **2.850**
HEAD DIAMETER B = **.532**
RIM DIAMETER D = **.532**
NECK DIAMETER F = **.316**
NECK LENGTH H = **.278**
SHOULDER LENGTH K = **.183**
BODY ANGLE (DEG'S/SIDE):
CASE CAPACITY CC'S = **5.40**

LOADED LENGTH: **3.65**
BELT DIAMETER C = **.532**
RIM THICKNESS E = **.050**
SHOULDER DIAMETER G = **.487**
LENGTH TO SHOULDER J = **2.389**
SHOULDER ANGLE (DEG'S/SIDE): **25°**
PRIMER: **L/R**
CASE CAPACITY (GR'S WATER):

DIMENSIONAL DRAWING:

-NOT ACTUAL SIZE-
-DO NOT SCALE-

CARTRIDGE: 7mm Super Mashburn Magnum

OTHER NAMES:

DIA: .284

BALLISTEK NO: 284AF

NAI NO: BEN 22433/5.127

DATA SOURCE: Ackley Vol.1 Pg406

HISTORICAL DATA:

NOTES:

LOADING DATA:

BULLET WT./TYPE	POWDER WT./TYPE	VELOCITY ('/SEC)	SOURCE
160/—	68.0/4350	3165	Ackley

CASE PREPARATION: **SHELLHOLDER (RCBS): 4**

MAKE FROM: .300 H&H Mag. Begin reducing the .300's neck dia. in a 7mm Rem. Mag. sizer (watch shoulder location!). Size F/L in the Mashburn die. Trim and chamfer.

PHYSICAL DATA (INCHES):

CASE TYPE: **Belted Bottleneck**
CASE LENGTH A = **2.625**
HEAD DIAMETER B = **.512**
RIM DIAMETER D = **.531**
NECK DIAMETER F = **.315**
NECK LENGTH H = **.350**
SHOULDER LENGTH K = **.145**
BODY ANGLE (DEG'S/SIDE): **.623**
CASE CAPACITY CC'S = **5.40**

LOADED LENGTH: **3.36**
BELT DIAMETER C = **.531**
RIM THICKNESS E = **.05**
SHOULDER DIAMETER G = **.470**
LENGTH TO SHOULDER J = **2.13**
SHOULDER ANGLE (DEG'S/SIDE): **28.12**
PRIMER: **L/R Mag**
CASE CAPACITY (GR'S WATER): **83.42**

DIMENSIONAL DRAWING:

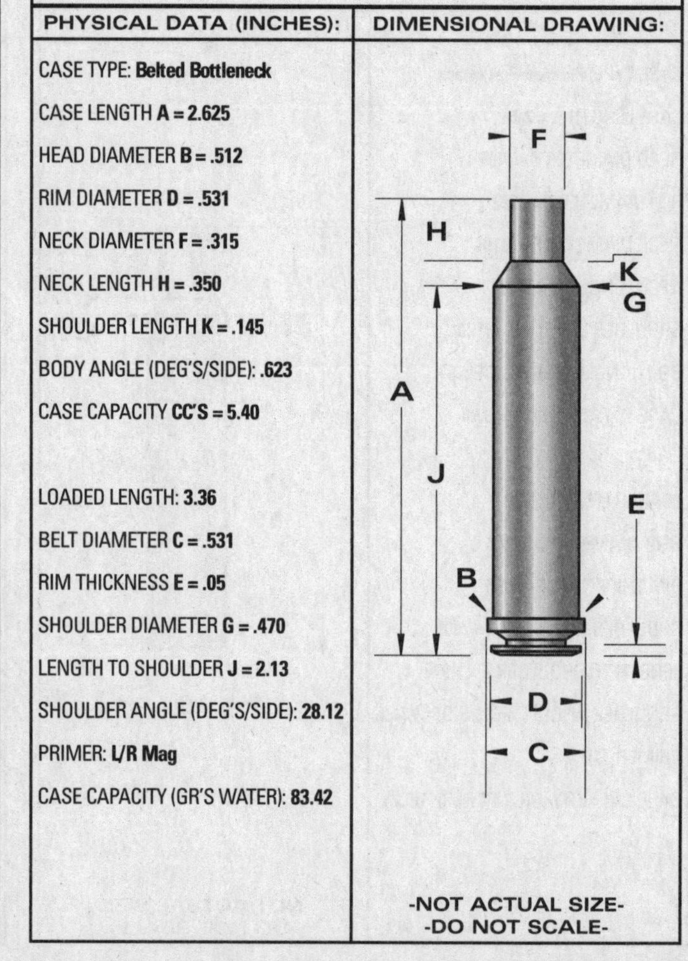

-NOT ACTUAL SIZE-
-DO NOT SCALE-

CARTRIDGE: 7mm TC/U

OTHER NAMES:
7mm/223 Remington

DIA: .284

BALLISTEK NO: 284AR

NAI NO: RXB
12243/4.680

DATA SOURCE: Handloader #96

HISTORICAL DATA: Designed by W. Ugalde.

NOTES: Commercial by Federal.

LOADING DATA:

BULLET WT./TYPE	POWDER WT./TYPE	VELOCITY ('/SEC)	SOURCE
130/Spire	26.0/IMR3031	1885	Blackwell

CASE PREPARATION: SHELLHOLDER (RCBS): 10

MAKE FROM: .223 Rem. Taper expand .223 case to .285-.290" dia. F/L size in TC/U die. Trim & chamfer. Fireform in chamber.

PHYSICAL DATA (INCHES):

CASE TYPE: **Rimless Bottleneck**

CASE LENGTH **A = 1.76**

HEAD DIAMETER **B = .376**

RIM DIAMETER **D = .378**

NECK DIAMETER **F = .310**

NECK LENGTH **H = .253**

SHOULDER LENGTH **K = .057**

BODY ANGLE (DEG'S/SIDE): **.16**

CASE CAPACITY **CC'S = 2.05**

LOADED LENGTH: **2.70**

BELT DIAMETER **C = N/A**

RIM THICKNESS **E = .045**

SHOULDER DIAMETER **G = .369**

LENGTH TO SHOULDER **J = 1.45**

SHOULDER ANGLE (DEG'S/SIDE): **40**

PRIMER: **S/R**

CASE CAPACITY (GR'S WATER): **31.77**

DIMENSIONAL DRAWING:

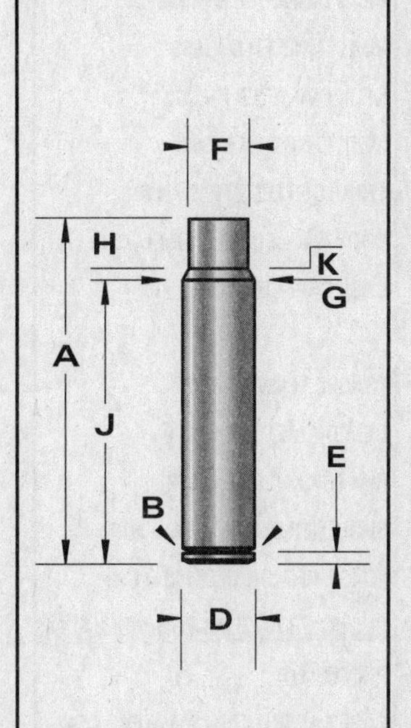

-NOT ACTUAL SIZE-
-DO NOT SCALE-

CARTRIDGE: 7mm Venturan

OTHER NAMES:

DIA: .284

BALLISTEK NO: 284U

NAI NO: BEN
22433/4.356

DATA SOURCE: Ackley Vol.1 Pg398

HISTORICAL DATA: By R. Payne.

NOTES: Weatherby-type shoulders.

LOADING DATA:

BULLET WT./TYPE	POWDER WT./TYPE	VELOCITY ('/SEC)	SOURCE
160/RN	64.0/4350	3235	Ackley

CASE PREPARATION: SHELLHOLDER (RCBS): 4

MAKE FROM: 7mm Rem. Mag. F/L size the Rem. case in the Venturan die with the expander removed. Some cases may need annealing. Trim to length and chamfer. F/L size and I.D. neck ream, if necessary. Fireform in the chamber.

PHYSICAL DATA (INCHES):

CASE TYPE: **Belted Bottleneck**

CASE LENGTH **A = 2.235**

HEAD DIAMETER **B = .513**

RIM DIAMETER **D = .532**

NECK DIAMETER **F = .314**

NECK LENGTH **H = .305**

SHOULDER LENGTH **K = .178**

BODY ANGLE (DEG'S/SIDE): **.424**

CASE CAPACITY **CC'S = 4.68**

LOADED LENGTH: **3.09**

BELT DIAMETER **C = .532**

RIM THICKNESS **E = .05**

SHOULDER DIAMETER **G = .490**

LENGTH TO SHOULDER **J = 1.752**

SHOULDER ANGLE (DEG'S/SIDE): **N/A**

PRIMER: **L/R Mag**

CASE CAPACITY (GR'S WATER): **72.27**

DIMENSIONAL DRAWING:

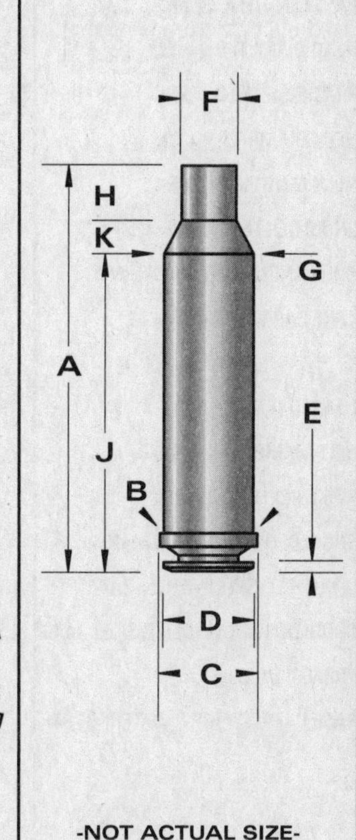

-NOT ACTUAL SIZE-
-DO NOT SCALE-

CARTRIDGE: 7mm Vom Hofe Belted

OTHER NAMES:	
7 x 73mm Vom Hofe	DIA: .284
	BALLISTEK NO: 284AH
	NAI NO: BEN 22422/5.446

DATA SOURCE: COTW 4th Pg203

HISTORICAL DATA: By Vom Hofe in 1931.

NOTES:

LOADING DATA:

BULLET WT./TYPE	POWDER WT./TYPE	VELOCITY ('/SEC)	SOURCE
175/Spire	75.0/H870	–	JJD

CASE PREPARATION: SHELLHOLDER (RCBS): 4

MAKE FROM: .300 H&H Mag. Turn belt to .527" dia. F/L size in Vom Hofe die. Square case mouth. Fireform in chamber with moderate load. Case will expand, forward of the belt, to fill the chamber. You'll need to be quite careful with cases made in this manner as the normal headspacing, on the belt, will not exist. Instead, the case will headspace on the neck. We have good luck fireforming these cases in a trim die with cornmeal.

PHYSICAL DATA (INCHES):

CASE TYPE: **Belted Bottleneck**

CASE LENGTH **A = 2.87**

HEAD DIAMETER **B = .527**

RIM DIAMETER **D = .533**

NECK DIAMETER **F = .315**

NECK LENGTH **H = .405**

SHOULDER LENGTH **K = .325**

BODY ANGLE (DEG'S/SIDE): **.650**

CASE CAPACITY **CC'S = 6.12**

LOADED LENGTH: **3.68**

BELT DIAMETER **C = .545**

RIM THICKNESS **E = .058**

SHOULDER DIAMETER **G = .483**

LENGTH TO SHOULDER **J = 2.14**

SHOULDER ANGLE (DEG'S/SIDE): **14.49**

PRIMER: **L/R Mag**

CASE CAPACITY (GR'S WATER): **94.54**

DIMENSIONAL DRAWING:

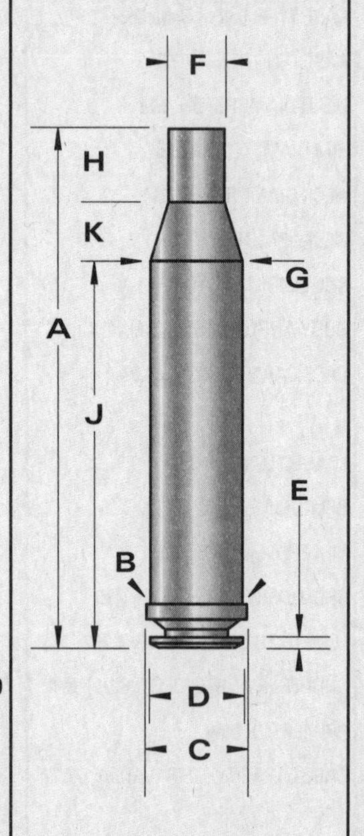

-NOT ACTUAL SIZE-
-DO NOT SCALE-

CARTRIDGE: 7mm Wade Super

OTHER NAMES:	
	DIA: .284
	BALLISTEK NO: 284BK
	NAI NO: RXB 23424/3.945

DATA SOURCE: Ackley Vol.1 Pg396

HISTORICAL DATA: By F. Wade.

NOTES:

LOADING DATA:

BULLET WT./TYPE	POWDER WT./TYPE	VELOCITY ('/SEC)	SOURCE
160/Spire	74.0/H570	3000	Ackley

CASE PREPARATION: SHELLHOLDER (RCBS): 5

MAKE FROM: .348 Winchester. Turn the .348's rim flush with the base and cut a new extractor groove. A form set is required to reduce the neck dia. Anneal and form. Trim to length and chamfer. F/L size and fireform.

PHYSICAL DATA (INCHES):

CASE TYPE: **Rimless Bottleneck**

CASE LENGTH **A = 2.17**

HEAD DIAMETER **B = .550**

RIM DIAMETER **D = .551**

NECK DIAMETER **F = .313**

NECK LENGTH **H = .360**

SHOULDER LENGTH **K = .110**

BODY ANGLE (DEG'S/SIDE): **.763**

CASE CAPACITY **CC'S = 5.02**

LOADED LENGTH: **2.95**

BELT DIAMETER **C = N/A**

RIM THICKNESS **E = .052**

SHOULDER DIAMETER **G = .510**

LENGTH TO SHOULDER **J = 1.70**

SHOULDER ANGLE (DEG'S/SIDE): **41.84**

PRIMER: **L/R**

CASE CAPACITY (GR'S WATER): **77.47**

DIMENSIONAL DRAWING:

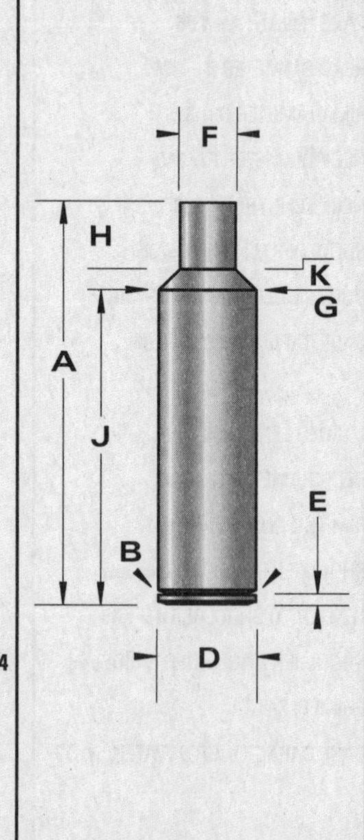

-NOT ACTUAL SIZE-
-DO NOT SCALE-

CARTRIDGE: 7mm Weatherby Magnum

OTHER NAMES:

DIA: .284

BALLISTEK NO: 284AQ

NAI NO: BEN 12433/4.980

DATA SOURCE: Hornady Manual 3rd Pg200

HISTORICAL DATA: By Roy Weatherby.

NOTES:

LOADING DATA:

BULLET WT./TYPE	POWDER WT./TYPE	VELOCITY ('/SEC)	SOURCE
120/Spire	73.0/IMR4831	3400	JJD
162/HPBT	69.0/IMR4831	2910	JJD

CASE PREPARATION: SHELLHOLDER (RCBS): 4

MAKE FROM: Factory or .300 H&H Mag. Size the H&H case in the Weatherby die. Trim to length & chamfer. Case will fireform upon ignition.

PHYSICAL DATA (INCHES):

CASE TYPE: **Belted Bottleneck**

CASE LENGTH **A = 2.545**

HEAD DIAMETER **B = .511**

RIM DIAMETER **D = .530**

NECK DIAMETER **F = .307**

NECK LENGTH **H = .350**

SHOULDER LENGTH **K = .157**

BODY ANGLE (DEG'S/SIDE): **.327**

CASE CAPACITY **CC'S = 5.66**

LOADED LENGTH: **3.26**

BELT DIAMETER **C = .530**

RIM THICKNESS **E = .05**

SHOULDER DIAMETER **G = .490**

LENGTH TO SHOULDER **J = 2.038**

SHOULDER ANGLE (DEG'S/SIDE): **N/A**

PRIMER: **L/R Mag**

CASE CAPACITY (GR'S WATER): **87.46**

DIMENSIONAL DRAWING:

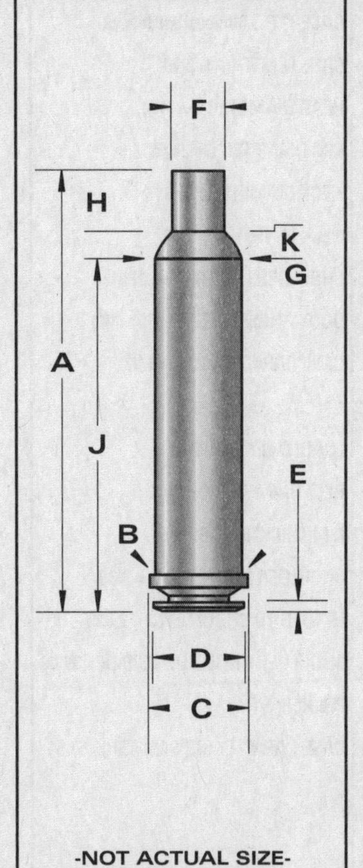

-NOT ACTUAL SIZE-
-DO NOT SCALE-

CARTRIDGE: 7mm Whisper

OTHER NAMES:

DIA: .284

BALLISTEK NO:

NAI NO:

DATA SOURCE: J.D. Jones

HISTORICAL DATA: The design intent was application in sound-suppressed M-15s, bolt-action rifles and T/C Contenders.

NOTES:

LOADING DATA:

BULLET WT./TYPE	POWDER WT./TYPE	VELOCITY ('/SEC)	SOURCE
120	20 - A-1680	2250	SSK
140/Nosler BT	18.5 - A-1680	2060	SSK
168	9.5 - A-1680	1056	SSK

CASE PREPARATION: SHELL HOLDER (RCBS): 10

MAKE FROM: 221 Remington
Neck up with expander ball.

PHYSICAL DATA (INCHES):

CASE TYPE: **Rimless Bottleneck**

CASE LENGTH **A = 1.400**

HEAD DIAMETER **B = .376**

RIM DIAMETER **D = .378**

NECK DIAMETER **F = .306**

NECK LENGTH **H = .288**

SHOULDER LENGTH **K = .041**

BODY ANGLE (DEG'S/SIDE):

CASE CAPACITY **CC'S =**

LOADED LENGTH: **2.26**

BELT DIAMETER **C = N/A**

RIM THICKNESS **E = .045**

SHOULDER DIAMETER **G = .361**

LENGTH TO SHOULDER **J = 1.071**

SHOULDER ANGLE (DEG'S/SIDE): **23°**

PRIMER: **S/R**

CASE CAPACITY (GR'S WATER):

DIMENSIONAL DRAWING:

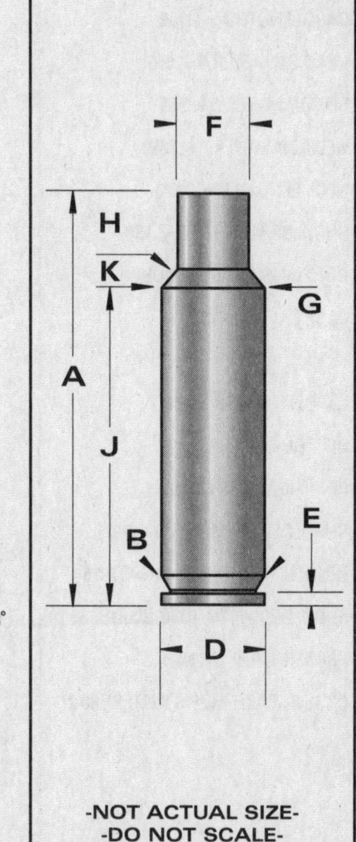

-NOT ACTUAL SIZE-
-DO NOT SCALE-

CARTRIDGE: 7mm Winchester Short Magnum

OTHER NAMES:	DIA: .284
	BALLISTEK NO:
	NAI NO:

DATA SOURCE: Most current loading manuals.

HISTORICAL DATA: Developed by Winchester in 2002. This is the only case in the current WSM family that uses a different datum line.

LOADING DATA:

BULLET WT./TYPE	POWDER WT./TYPE	VELOCITY ('/SEC)	SOURCE
140/Nosler Ball. Tip	66.7/Ramshot Hunter	3221	Ramshot
162/Hornady	75.0/Ramshot Mag.	3085	#3

CASE PREPARATION: SHELL HOLDER (RCBS): 43

MAKE FROM: Factory or make from .270 WSM by fireforming. The body on the 7mm WSM is .0378-inch longer than the .270 or .300 WSM. This moves the shoulder forward to prevent the 7mm WSM from chambering in a .270 WSM chambered rifle.
Can be made from .404 Jeffery. Shorten and die form to fit chamber. Fireform body to get the proper diameter, shoulder angle and head space. Then turn the rim to correct dimension. Ream neck if necessary. Square mouth and chamfer.

PHYSICAL DATA (INCHES):

CASE TYPE: **Rebated Bottleneck**
CASE LENGTH **A = 2.100**
HEAD DIAMETER **B = .5550**
RIM DIAMETER **D = .535**
NECK DIAMETER **F = .3210**
NECK LENGTH **H = .2435**
SHOULDER LENGTH **K = .1547**
BODY ANGLE (DEG'S/SIDE):
CASE CAPACITY **CC'S =**
LOADED LENGTH: **2.860**
BELT DIAMETER **C =**
RIM THICKNESS **E = .54**
SHOULDER DIAMETER **G = .5377**
LENGTH TO SHOULDER **J = 1.7018**
SHOULDER ANGLE (DEG'S/SIDE): **35**
PRIMER: **L/R or L/R Mag**
CASE CAPACITY (GR'S WATER): **80.2*-**

DIMENSIONAL DRAWING:

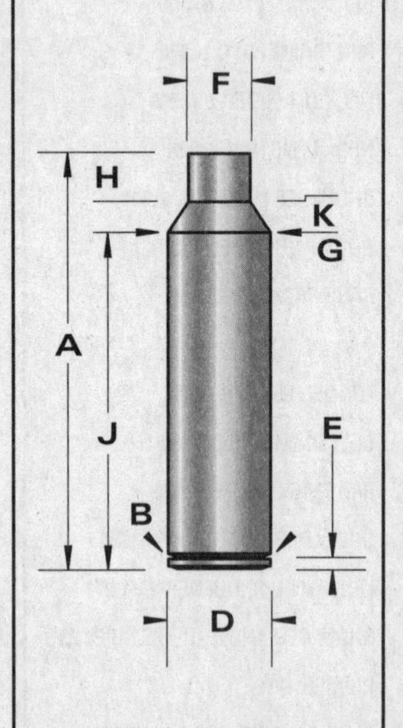

-NOT ACTUAL SIZE-
-DO NOT SCALE-

CARTRIDGE: 7mm/06 Improved (Ackley)

OTHER NAMES:	DIA: .284
	BALLISTEK NO: 284H
	NAI NO: RXB 22324/5.298

DATA SOURCE: Ackley Vol.1 Pg393

HISTORICAL DATA:

NOTES:

LOADING DATA:

BULLET WT./TYPE	POWDER WT./TYPE	VELOCITY ('/SEC)	SOURCE
160/Spire	57.0/4350	3010	Ackley

CASE PREPARATION: SHELLHOLDER (RCBS): 3

MAKE FROM: .30-06 Spgf. F/L size in either 7mm/06 or 7mm/06 "improved" die. Square case mouth, chamfer and fireform.

PHYSICAL DATA (INCHES):

CASE TYPE: **Rimless Bottleneck**
CASE LENGTH **A = 2.49**
HEAD DIAMETER **B = .470**
RIM DIAMETER **D = .473**
NECK DIAMETER **F = .314**
NECK LENGTH **H = .380**
SHOULDER LENGTH **K = .075**
BODY ANGLE (DEG'S/SIDE): **.483**
CASE CAPACITY **CC'S = 4.37**
LOADED LENGTH: **3.41**
BELT DIAMETER **C = N/A**
RIM THICKNESS **E = .05**
SHOULDER DIAMETER **G = .439**
LENGTH TO SHOULDER **J = 2.035**
SHOULDER ANGLE (DEG'S/SIDE): **39.80**
PRIMER: **L/R**
CASE CAPACITY (GR'S WATER): **67.51**

DIMENSIONAL DRAWING:

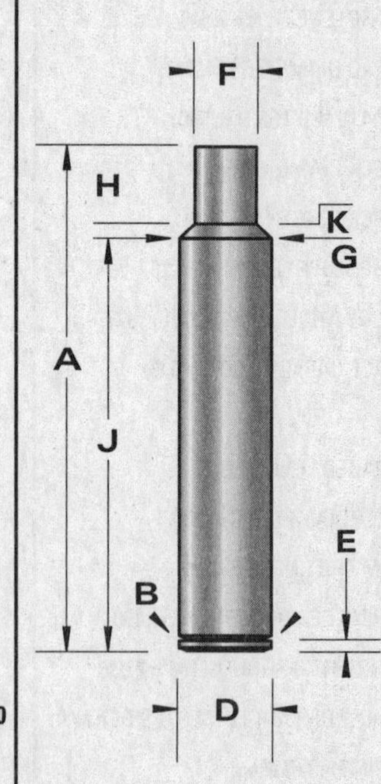

-NOT ACTUAL SIZE-
-DO NOT SCALE-

CARTRIDGE: 7mm-30 JDJ

OTHER NAMES:	DIA: .284
	BALLISTEK NO:
	NAI NO:

DATA SOURCE:

HISTORICAL DATA: This JDJ cartridge is the 7-30 Waters improved. The purpose is to meet demand for improved performance with readily obtained cases.

NOTES:

LOADING DATA:

BULLET WT./TYPE	POWDER WT./TYPE	VELOCITY ('/SEC)	SOURCE
–	–	–	–

CASE PREPARATION: SHELL HOLDER (RCBS): 02

MAKE FROM: .30-30 Winchester

PHYSICAL DATA (INCHES):

CASE TYPE: **Rimmed Bottleneck**
CASE LENGTH A = **2.03**
HEAD DIAMETER B = **.419**
RIM DIAMETER D = **.497**
NECK DIAMETER F = **.306**
NECK LENGTH H =
SHOULDER LENGTH K =
BODY ANGLE (DEG'S/SIDE):
CASE CAPACITY CC'S
LOADED LENGTH:
BELT DIAMETER C =
RIM THICKNESS E =
SHOULDER DIAMETER G = **.409**
LENGTH TO SHOULDER J =
SHOULDER ANGLE (DEG'S/SIDE):
PRIMER: **L/R**
CASE CAPACITY (GR'S WATER):

DIMENSIONAL DRAWING:

F
H
K
G
A
J
E
B
D

-NOT ACTUAL SIZE-
-DO NOT SCALE-

CARTRIDGE: 7mm-30 Waters

OTHER NAMES:	DIA: .284
	BALLISTEK NO: 284BM
	NAI NO: RMB 22335/4.106

DATA SOURCE: Handloader #113 Pg23

HISTORICAL DATA: By K. Waters in 1983-84.

NOTES:

LOADING DATA:

BULLET WT./TYPE	POWDER WT./TYPE	VELOCITY ('/SEC)	SOURCE
120/Spire	35.5/IMR4064	2698	Waters

CASE PREPARATION: SHELLHOLDER (RCBS): 2

MAKE FROM: Factory (Federal) or .30-30. F/L size the .30-30 case in the Waters die. Trim, chamfer and fireform.

PHYSICAL DATA (INCHES):

CASE TYPE: **Rimmed Bottleneck**
CASE LENGTH A = **2.04 (max)**
HEAD DIAMETER B = **.422**
RIM DIAMETER D = **.506**
NECK DIAMETER F = **.307**
NECK LENGTH H = **.307**
SHOULDER LENGTH K = **.153**
BODY ANGLE (DEG'S/SIDE): **.475**
CASE CAPACITY CC'S = **2.96**
LOADED LENGTH: **2.55**
BELT DIAMETER C = **N/A**
RIM THICKNESS E = **.062**
SHOULDER DIAMETER G = **.399**
LENGTH TO SHOULDER J = **1.58**
SHOULDER ANGLE (DEG'S/SIDE): **16.73**
PRIMER: **L/R**
CASE CAPACITY (GR'S WATER): **45.72**

DIMENSIONAL DRAWING:

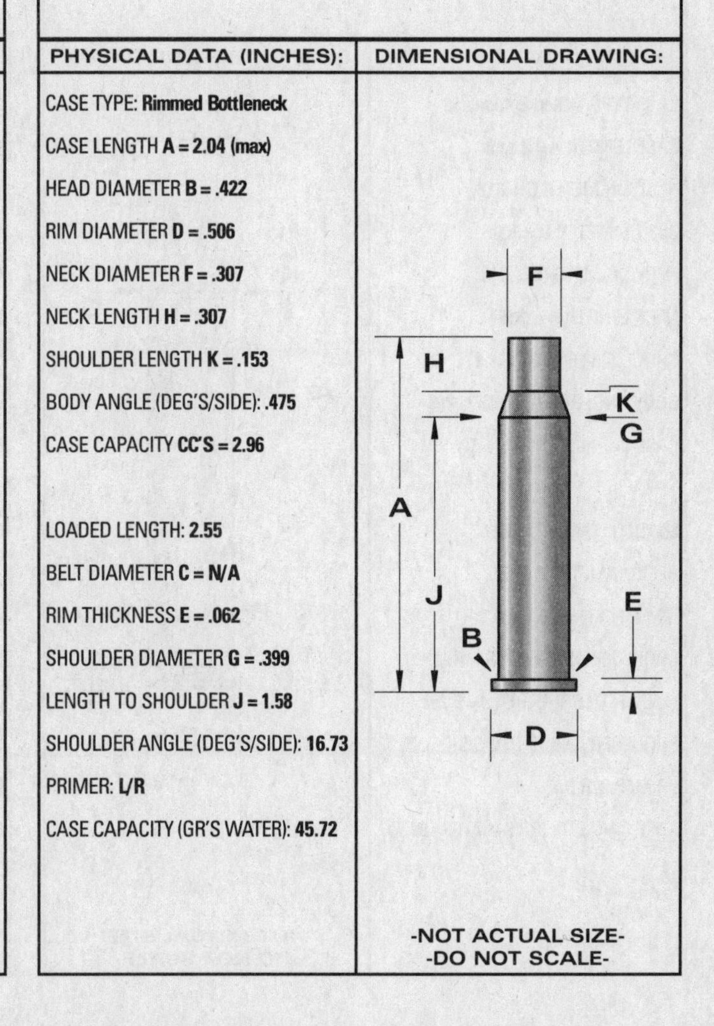

-NOT ACTUAL SIZE-
-DO NOT SCALE-

CARTRIDGE: 7mm/300 Weatherby Magnum

OTHER NAMES:

DIA: .284

BALLISTEK NO: 284I

NAI NO: BEN 12433/5.506

DATA SOURCE: Sierra Manual 1985 Pg199

HISTORICAL DATA: By H. Wolfe in 1960.

NOTES: Author's choice for 1,000 yard shooting.

LOADING DATA:

BULLET WT./TYPE	POWDER WT./TYPE	VELOCITY ('/SEC)	SOURCE
162/HPBT	85.0/H870	3210	JJD
140/HPBT	83.0/H870	–	Sierra

CASE PREPARATION: SHELLHOLDER (RCBS): 4

MAKE FROM: .300 Weatherby. For best results, you'll need a ream die & reamer for this ctg. Run .300 Weatherby case into ream die & ream. F/L size. Chamfer case mouth I.D. & O.D. Solvent clean.

PHYSICAL DATA (INCHES):

CASE TYPE: **Belted Bottleneck**
CASE LENGTH **A = 2.825**
HEAD DIAMETER **B = .513**
RIM DIAMETER **D = .532**
NECK DIAMETER **F = .310**
NECK LENGTH **H = .316**
SHOULDER LENGTH **K = .211**
BODY ANGLE (DEG'S/SIDE): **.286**
CASE CAPACITY **CC'S = 6.39**
LOADED LENGTH: **3.722**
BELT DIAMETER **C = .532**
RIM THICKNESS **E = .05**
SHOULDER DIAMETER **G = .492**
LENGTH TO SHOULDER **J = 2.298**
SHOULDER ANGLE (DEG'S/SIDE): **23.33**
PRIMER: **L/R Mag**
CASE CAPACITY (GR'S WATER): **98.70**

DIMENSIONAL DRAWING:

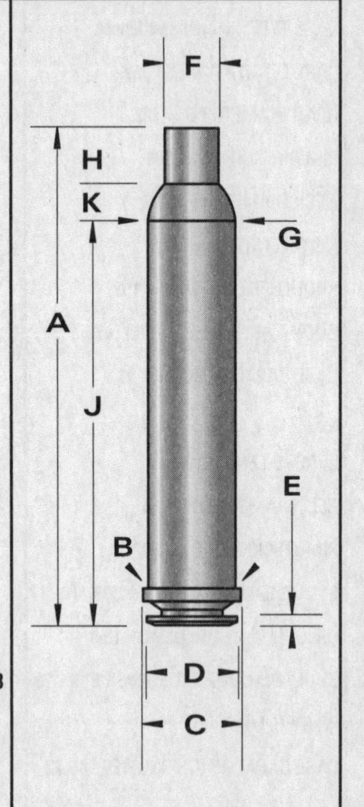

-NOT ACTUAL SIZE-
-DO NOT SCALE-

CARTRIDGE: 7mm/303 British

OTHER NAMES:

DIA: .284

BALLISTEK NO: 284X

NAI NO: RMB 22334/4.747

DATA SOURCE: NAI/Ballistek

HISTORICAL DATA:

NOTES:

LOADING DATA:

BULLET WT./TYPE	POWDER WT./TYPE	VELOCITY ('/SEC)	SOURCE
150/RN	41.8/IMR4350	–	JJD

CASE PREPARATION: SHELLHOLDER (RCBS): 7

MAKE FROM: .303 British. F/L size the .303 case in the 7mm/303 die. Trim and chamfer.

PHYSICAL DATA (INCHES):

CASE TYPE: **Rimmed Bottleneck**
CASE LENGTH **A = 2.16**
HEAD DIAMETER **B = .455**
RIM DIAMETER **D = .540**
NECK DIAMETER **F = .310**
NECK LENGTH **H = .325**
SHOULDER LENGTH **K = .065**
BODY ANGLE (DEG'S/SIDE): **.784**
CASE CAPACITY **CC'S = 3.48**
LOADED LENGTH: **A/R**
BELT DIAMETER **C = N/A**
RIM THICKNESS **E = .064**
SHOULDER DIAMETER **G = .412**
LENGTH TO SHOULDER **J = 1.77**
SHOULDER ANGLE (DEG'S/SIDE): **38.11**
PRIMER: **L/R**
CASE CAPACITY (GR'S WATER): **53.80**

DIMENSIONAL DRAWING:

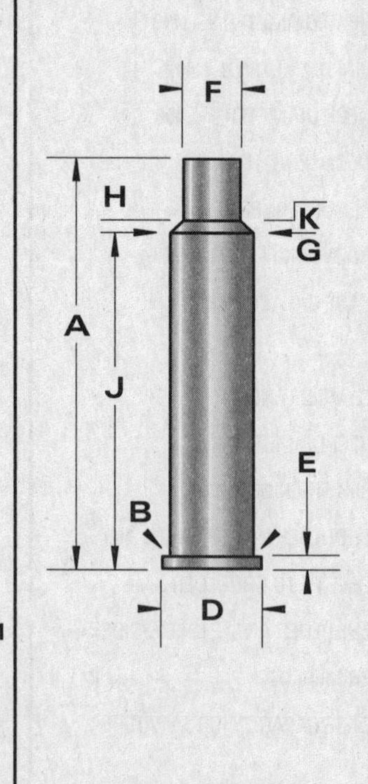

-NOT ACTUAL SIZE-
-DO NOT SCALE-

CARTRIDGE: 7mm/303 British Improved

OTHER NAMES:
7mm/303 Epps

DIA: .284

BALLISTEK NO: 284AB

NAI NO: RMB
22333/4.747

DATA SOURCE: NAI/Ballistek

HISTORICAL DATA:

NOTES:

LOADING DATA:

BULLET WT./TYPE	POWDER WT./TYPE	VELOCITY ('/SEC)	SOURCE
150/Spire	43.6/W760	–	JJD

CASE PREPARATION: SHELLHOLDER (RCBS): 7

MAKE FROM: .303 British. Size the .303 case in the 7mm/303 F/L die. Trim and chamfer. Fireform in chamber.

PHYSICAL DATA (INCHES):

CASE TYPE: **Rimmed Bottleneck**
CASE LENGTH **A = 2.16**
HEAD DIAMETER **B = .455**
RIM DIAMETER **D = .540**
NECK DIAMETER **F = .311**
NECK LENGTH **H = .325**
SHOULDER LENGTH **K = .115**
BODY ANGLE (DEG'S/SIDE): **.377**
CASE CAPACITY **CC'S = 3.62**

LOADED LENGTH: **A/R**
BELT DIAMETER **C = N/A**
RIM THICKNESS **E = .063**
SHOULDER DIAMETER **G = .435**
LENGTH TO SHOULDER **J = 1.72**
SHOULDER ANGLE (DEG'S/SIDE): **28.33**
PRIMER: **L/R**
CASE CAPACITY (GR'S WATER): **55.91**

DIMENSIONAL DRAWING:

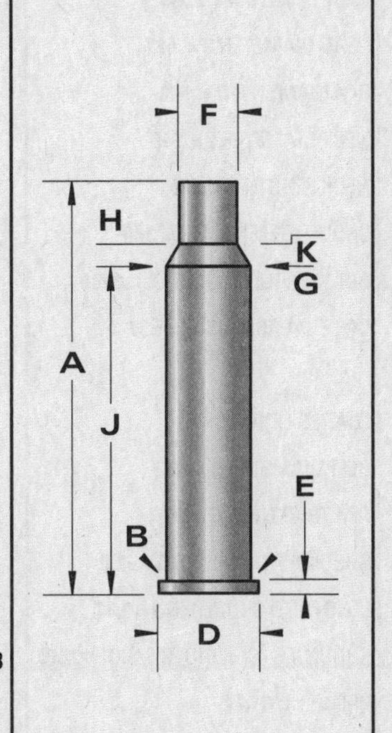

-NOT ACTUAL SIZE-
-DO NOT SCALE-

CARTRIDGE: 7mm/308 Durham

OTHER NAMES:

DIA: .284

BALLISTEK NO: 285Y

NAI NO: RXB
12333/4.276

DATA SOURCE: Ackley Vol.2 Pg171

HISTORICAL DATA:

NOTES: Several versions exist.

LOADING DATA:

BULLET WT./TYPE	POWDER WT./TYPE	VELOCITY ('/SEC)	SOURCE
140/Spire	45.0/4064	3032	Ackley

CASE PREPARATION: SHELLHOLDER (RCBS): 3

MAKE FROM: .308 Winchester. F/L size the .308 case in Durham die to set the shoulder back. Trim and chamfer. Fireform.

PHYSICAL DATA (INCHES):

CASE TYPE: **Rimless Bottleneck**
CASE LENGTH **A = 2.01**
HEAD DIAMETER **B = .470**
RIM DIAMETER **D = .473**
NECK DIAMETER **F = .313**
NECK LENGTH **H = .320**
SHOULDER LENGTH **K = .130**
BODY ANGLE (DEG'S/SIDE): **.316**
CASE CAPACITY **CC'S = 4.14**

LOADED LENGTH: **A/R**
BELT DIAMETER **C = N/A**
RIM THICKNESS **E = .05**
SHOULDER DIAMETER **G = .455**
LENGTH TO SHOULDER **J = 1.56**
SHOULDER ANGLE (DEG'S/SIDE): **28.64**
PRIMER: **L/R**
CASE CAPACITY (GR'S WATER): **63.87**

DIMENSIONAL DRAWING:

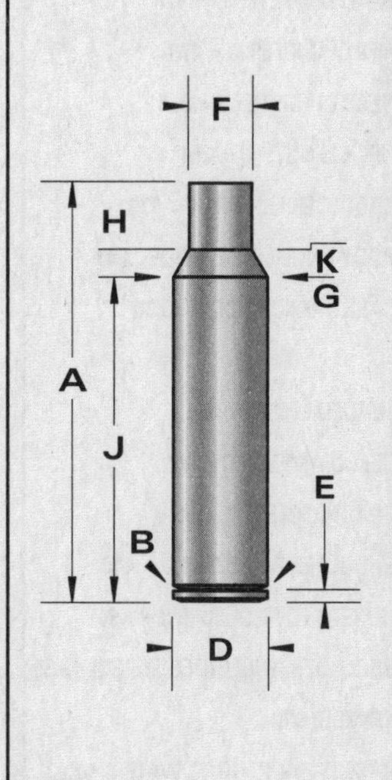

-NOT ACTUAL SIZE-
-DO NOT SCALE-

CARTRIDGE: 7mm/308 Winchester

OTHER NAMES:

DIA: .284

BALLISTEK NO: 284B

NAI NO: RXB 22332/4.276

DATA SOURCE: Ackley Vol.1 Pg390

HISTORICAL DATA: Early wildcat (1950's), now commercial, by Remington.

NOTES: Several other versions exist.

LOADING DATA:

BULLET WT./TYPE	POWDER WT./TYPE	VELOCITY ('/SEC)	SOURCE
160/Spire	39.0/3031	2595	Ackley

CASE PREPARATION: SHELLHOLDER (RCBS): 3

MAKE FROM: Factory or .308 Win. F/L size the .308 case in the 7mm/08 die. Trim and chamfer.

PHYSICAL DATA (INCHES):

CASE TYPE: **Rimless Bottleneck**

CASE LENGTH **A = 2.01**

HEAD DIAMETER **B = .470**

RIM DIAMETER **D = .473**

NECK DIAMETER **F = .314**

NECK LENGTH **H = .291**

SHOULDER LENGTH **K = .179**

BODY ANGLE (DEG'S/SIDE): **.641**

CASE CAPACITY **CC'S = 3.38**

LOADED LENGTH: **A/R**

BELT DIAMETER **C = N/A**

RIM THICKNESS **E = .05**

SHOULDER DIAMETER **G = .440**

LENGTH TO SHOULDER **J = 1.54**

SHOULDER ANGLE (DEG'S/SIDE): **19.39**

PRIMER: **L/R**

CASE CAPACITY (GR'S WATER): **52.23**

DIMENSIONAL DRAWING:

F H K G A J E B D

-NOT ACTUAL SIZE-
-DO NOT SCALE-

CARTRIDGE: 7mm/338 Winchester Magnum

OTHER NAMES:

DIA: .284

BALLISTEK NO: 284J

NAI NO: BEN 22434/4.873

DATA SOURCE: NAI/Ballistek

HISTORICAL DATA:

NOTES: This is pretty much a 7mm Rem. Mag.

LOADING DATA:

BULLET WT./TYPE	POWDER WT./TYPE	VELOCITY ('/SEC)	SOURCE
139/Spire	62.0/IMR4831	3000	JJD

CASE PREPARATION: SHELLHOLDER (RCBS): 4

MAKE FROM: 7mm Rem. Mag. F/L size the Rem. case in the 7mm/338 die and fireform in the chamber. This case has a slightly greater shoulder angle than does the Rem. case.

PHYSICAL DATA (INCHES):

CASE TYPE: **Belted Bottleneck**

CASE LENGTH **A = 2.500**

HEAD DIAMETER **B = .513**

RIM DIAMETER **D = .532**

NECK DIAMETER **F = .315**

NECK LENGTH **H = .320**

SHOULDER LENGTH **K = .140**

BODY ANGLE (DEG'S/SIDE): **.358**

CASE CAPACITY **CC'S = 5.37**

LOADED LENGTH: **A/R**

BELT DIAMETER **C = .532**

RIM THICKNESS **E = .05**

SHOULDER DIAMETER **G = .490**

LENGTH TO SHOULDER **J = 2.04**

SHOULDER ANGLE (DEG'S/SIDE): **32.00**

PRIMER: **L/R Mag**

CASE CAPACITY (GR'S WATER): **82.87**

DIMENSIONAL DRAWING:

F H K G A J E B D C

-NOT ACTUAL SIZE-
-DO NOT SCALE-

CARTRIDGE: 7mm/350 Remington Magnum

OTHER NAMES:

DIA: .284

BALLISTEK NO: 284Z

NAI NO: BEN 12443/4.220

DATA SOURCE: Ackley Vol.2 Pg178

HISTORICAL DATA:

NOTES: Badly overbore for .284" dia.!

LOADING DATA:

BULLET WT./TYPE	POWDER WT./TYPE	VELOCITY ('/SEC)	SOURCE
175/—	55.5/4350	2695	Ackley

CASE PREPARATION: SHELLHOLDER (RCBS): 4

MAKE FROM: 7mm Rem. Mag. F/L size the Rem. case. Shoulder area may need to be annealed. Trim & chamfer. F/L size, again, with expander in place. Fireform.

PHYSICAL DATA (INCHES):

CASE TYPE: **Belted Bottleneck**

CASE LENGTH **A = 2.165**

HEAD DIAMETER **B = .513**

RIM DIAMETER **D = .532**

NECK DIAMETER **F = .316**

NECK LENGTH **H = .275**

SHOULDER LENGTH **K = .180**

BODY ANGLE (DEG'S/SIDE): **.341**

CASE CAPACITY **CC'S = 4.54**

LOADED LENGTH: **2.98**

BELT DIAMETER **C = .532**

RIM THICKNESS **E = .049**

SHOULDER DIAMETER **G = .495**

LENGTH TO SHOULDER **J = 1.71**

SHOULDER ANGLE (DEG'S/SIDE): **26.44**

PRIMER: **L/R Mag**

CASE CAPACITY (GR'S WATER): **70.02**

DIMENSIONAL DRAWING:

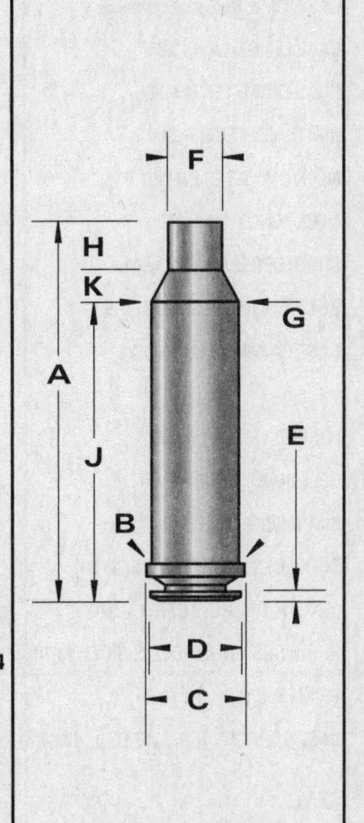

-NOT ACTUAL SIZE-
-DO NOT SCALE-

CARTRIDGE: 7mm/350 Rigby Rimmed

OTHER NAMES:
.350/7mm Rigby
7mm/400 Rigby

DIA: .280

BALLISTEK NO: 280B

NAI NO: RMB 22322/5.872

DATA SOURCE: NAI/Ballistek

HISTORICAL DATA:

NOTES:

LOADING DATA:

BULLET WT./TYPE	POWDER WT./TYPE	VELOCITY ('/SEC)	SOURCE
160/RN	51.5/4350	—	JJD

CASE PREPARATION: SHELLHOLDER (RCBS): 4

MAKE FROM: 9,3 x 74R. Thin rim to .050" (may not be necessary). Cut case off to 2.80". Anneal. F/L size and trim to length. Fireform.

PHYSICAL DATA (INCHES):

CASE TYPE: **Rimmed Bottleneck**

CASE LENGTH **A = 2.76**

HEAD DIAMETER **B = .470**

RIM DIAMETER **D = .528**

NECK DIAMETER **F = .315**

NECK LENGTH **H = .430**

SHOULDER LENGTH **K = .163**

BODY ANGLE (DEG'S/SIDE): **.946**

CASE CAPACITY **CC'S = 4.32**

LOADED LENGTH: **3.25**

BELT DIAMETER **C = N/A**

RIM THICKNESS **E = .05**

SHOULDER DIAMETER **G = .405**

LENGTH TO SHOULDER **J = 2.167**

SHOULDER ANGLE (DEG'S/SIDE): **15.43**

PRIMER: **L/R**

CASE CAPACITY (GR'S WATER): **66.74**

DIMENSIONAL DRAWING:

F
H
K
G
A
J
E
B
D

-NOT ACTUAL SIZE-
-DO NOT SCALE-

CARTRIDGE: 7 x 30 JDJ

OTHER NAMES:

DIA: .284

BALLISTEK NO:

NAI NO:

DATA SOURCE: J.D. Jones

HISTORICAL DATA: An improved 7-30 Waters

NOTES:

LOADING DATA:

BULLET WT./TYPE	POWDER WT./TYPE	VELOCITY ('/SEC)	SOURCE

CASE PREPARATION: **SHELLHOLDER (RCBS):** 02

MAKE FROM: 7-30 Waters
Fireform 7-30 Waters with reduced load in JDJ Chamber

PHYSICAL DATA (INCHES):

CASE TYPE: **Rimmed Bottleneck**

CASE LENGTH **A = 2.040**

HEAD DIAMETER **B = .421**

RIM DIAMETER **D = .506**

NECK DIAMETER **F = .306**

NECK LENGTH **H = .280**

SHOULDER LENGTH **K = .070**

BODY ANGLE (DEG'S/SIDE):

CASE CAPACITY **CC'S**

LOADED LENGTH: **2.800**

BELT DIAMETER **C = N/A**

RIM THICKNESS **E = .063**

SHOULDER DIAMETER **G = .415**

LENGTH TO SHOULDER **J = .1.690**

SHOULDER ANGLE (DEG'S/SIDE):

PRIMER: **L/R**

CASE CAPACITY (GR'S WATER): **18.66**

DIMENSIONAL DRAWING:

F H K G A J E B D

-NOT ACTUAL SIZE-
-DO NOT SCALE-

CARTRIDGE: 7 x 33mm Ballistek

OTHER NAMES:
7mm Ballistek
7mm/30 Carbine

DIA: .284

BALLISTEK NO: 284AV

NAI NO: RXB 24232/3.606

DATA SOURCE: NAI/Ballistek

HISTORICAL DATA: Designed by author in 1981.

NOTES: For use in a 7mm Rem. Mag. chamber insert.

LOADING DATA:

BULLET WT./TYPE	POWDER WT./TYPE	VELOCITY ('/SEC)	SOURCE
120/Spire	13.0/IMR4227	1650	JJD

CASE PREPARATION: **SHELLHOLDER (RCBS):** 17

MAKE FROM: .30 Carbine. Run .30 Carbine case into 7x33 F/L sizer with expander removed. Trim to length. Chamfer I.D. & O.D. F/L size.

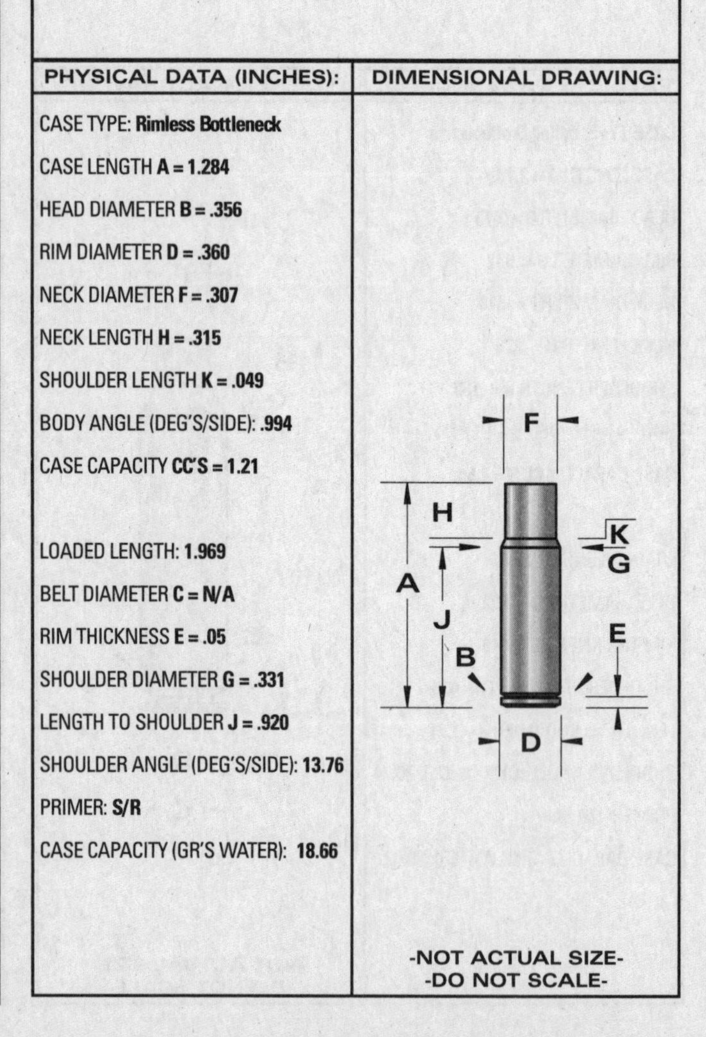

PHYSICAL DATA (INCHES):

CASE TYPE: **Rimless Bottleneck**

CASE LENGTH **A = 1.284**

HEAD DIAMETER **B = .356**

RIM DIAMETER **D = .360**

NECK DIAMETER **F = .307**

NECK LENGTH **H = .315**

SHOULDER LENGTH **K = .049**

BODY ANGLE (DEG'S/SIDE): **.994**

CASE CAPACITY **CC'S = 1.21**

LOADED LENGTH: **1.969**

BELT DIAMETER **C = N/A**

RIM THICKNESS **E = .05**

SHOULDER DIAMETER **G = .331**

LENGTH TO SHOULDER **J = .920**

SHOULDER ANGLE (DEG'S/SIDE): **13.76**

PRIMER: **S/R**

CASE CAPACITY (GR'S WATER): **18.66**

DIMENSIONAL DRAWING:

-NOT ACTUAL SIZE-
-DO NOT SCALE-

CARTRIDGE: 7 x 33 Sako

OTHER NAMES:
7 x 33 Finnish Rimless

DIA: .284

BALLISTEK NO: 284BJ

NAI NO: RXB 24242/3.355

DATA SOURCE: Rifle #82 Pg18

HISTORICAL DATA: Finnish development about 1945.

NOTES: Correct dia. is .286".

LOADING DATA:

BULLET WT./TYPE	POWDER WT./TYPE	VELOCITY ('/SEC)	SOURCE
120/Spire	17.3/IMR4227	—	JJD

CASE PREPARATION: SHELLHOLDER (RCBS): 16

MAKE FROM: Factory (Sako) or 9mm Win. Mag. Anneal the case and F/L size in the 7x33 die. Square case mouth. Case is .150" short which does not leave much of a neck but, it will work. It is also possible to re-body a 9mm Luger case with 13/32" dia. tubing, trimming and sizing. This case is okay for light loads.

PHYSICAL DATA (INCHES):

CASE TYPE: **Rimless Bottleneck**
CASE LENGTH **A = 1.312**
HEAD DIAMETER **B = .391**
RIM DIAMETER **D = .394**
NECK DIAMETER **F = .313**
NECK LENGTH **H = .173**
SHOULDER LENGTH **K = .096**
BODY ANGLE (DEG'S/SIDE): **.543**
CASE CAPACITY **CC'S = 1.48**

LOADED LENGTH: **A/R**
BELT DIAMETER **C = N/A**
RIM THICKNESS **E = .05**
SHOULDER DIAMETER **G = .375**
LENGTH TO SHOULDER **J = 1.043**
SHOULDER ANGLE (DEG'S/SIDE): **17.89**
PRIMER: **S/R**
CASE CAPACITY (GR'S WATER): **22.91**

DIMENSIONAL DRAWING:

F
H
K
G
A
J
E
B
D

-NOT ACTUAL SIZE-
-DO NOT SCALE-

CARTRIDGE: 7 x 45 Ingram

OTHER NAMES:
7mm/223 Rem. Improved

DIA: .284

BALLISTEK NO: 284AZ

NAI NO: RXB 12243/4.666

DATA SOURCE: Hornady Manual 3rd pg324

HISTORICAL DATA: By D. Ingram.

NOTES:

LOADING DATA:

BULLET WT./TYPE	POWDER WT./TYPE	VELOCITY ('/SEC)	SOURCE
139/Spire	31.0/BL-C2	2125	JJD

CASE PREPARATION: SHELLHOLDER (RCBS): 10

MAKE FROM: .223 Rem. Anneal case neck. Taper expand to .290" dia. F/L size in 7x45 die. Shoulder will fireform upon ignition in chamber.

PHYSICAL DATA (INCHES):

CASE TYPE: **Rimless Bottleneck**
CASE LENGTH **A = 1.75**
HEAD DIAMETER **B = .375**
RIM DIAMETER **D = .378**
NECK DIAMETER **F = .304**
NECK LENGTH **H = .219**
SHOULDER LENGTH **K = .061**
BODY ANGLE (DEG'S/SIDE): **.225**
CASE CAPACITY **CC'S = 2.126**

LOADED LENGTH: **2.69**
BELT DIAMETER **C = N/A**
RIM THICKNESS **E = .045**
SHOULDER DIAMETER **G = .365**
LENGTH TO SHOULDER **J = 1.47**
SHOULDER ANGLE (DEG'S/SIDE): **45.56**
PRIMER: **S/R**
CASE CAPACITY (GR'S WATER): **32.82**

DIMENSIONAL DRAWING:

F
H
K
G
A
J
E
B
D

-NOT ACTUAL SIZE-
-DO NOT SCALE-

CARTRIDGE: 7 x 49mm Medium

OTHER NAMES:
7mm Medium
7mm Second Optimum

DIA: .284
BALLISTEK NO: 284BN
NAI NO: RXB 23332/4.163

DATA SOURCE: MSAA Pg29

HISTORICAL DATA: British mil. Experimental ctg. ca. 1952.

NOTES:

LOADING DATA:

BULLET WT./TYPE	POWDER WT./TYPE	VELOCITY ('/SEC)	SOURCE
140/Spire	38.1/IMR4320	2750	JJD

CASE PREPARATION: **SHELLHOLDER (RCBS): 3**

MAKE FROM: 7 x 64 Brenneke. F/L size in the 7x49 die with the expander removed. Trim to 1.95" & chamfer. I.D. neck ream. F/L size and trim to length.

PHYSICAL DATA (INCHES):

CASE TYPE: **Rimless Bottleneck**
CASE LENGTH **A = 1.936**
HEAD DIAMETER **B = .465**
RIM DIAMETER **D = .466**
NECK DIAMETER **F = .315**
NECK LENGTH **H = .315**
SHOULDER LENGTH **K = .197**
BODY ANGLE (DEG'S/SIDE): **.374**
CASE CAPACITY **CC'S = 3.20**

LOADED LENGTH: **2.88**
BELT DIAMETER **C = N/A**
RIM THICKNESS **E = .05**
SHOULDER DIAMETER **G = .449**
LENGTH TO SHOULDER **J = 1.424**
SHOULDER ANGLE (DEG'S/SIDE): **18.78**
PRIMER: **L/R**
CASE CAPACITY (GR'S WATER): **49.39**

DIMENSIONAL DRAWING:

F
H
K
G
A
J
E
B
D

-NOT ACTUAL SIZE-
-DO NOT SCALE-

CARTRIDGE: 7 x 57mm Improved (Ackley)

OTHER NAMES:

DIA: .284
BALLISTEK NO: 284AA
NAI NO: RXB 12324/4.744

DATA SOURCE: Ackley Vol.1 Pg389

HISTORICAL DATA:

NOTES:

LOADING DATA:

BULLET WT./TYPE	POWDER WT./TYPE	VELOCITY ('/SEC)	SOURCE
160/Spire	47.5/IMR4350	–	JJD

CASE PREPARATION: **SHELLHOLDER (RCBS): 3**

MAKE FROM: 7 x 57mm Mauser or .30-06 Spgf. Fire factory ammo in the "improved" chamber or F/L size the '06 case, trim and chamfer.

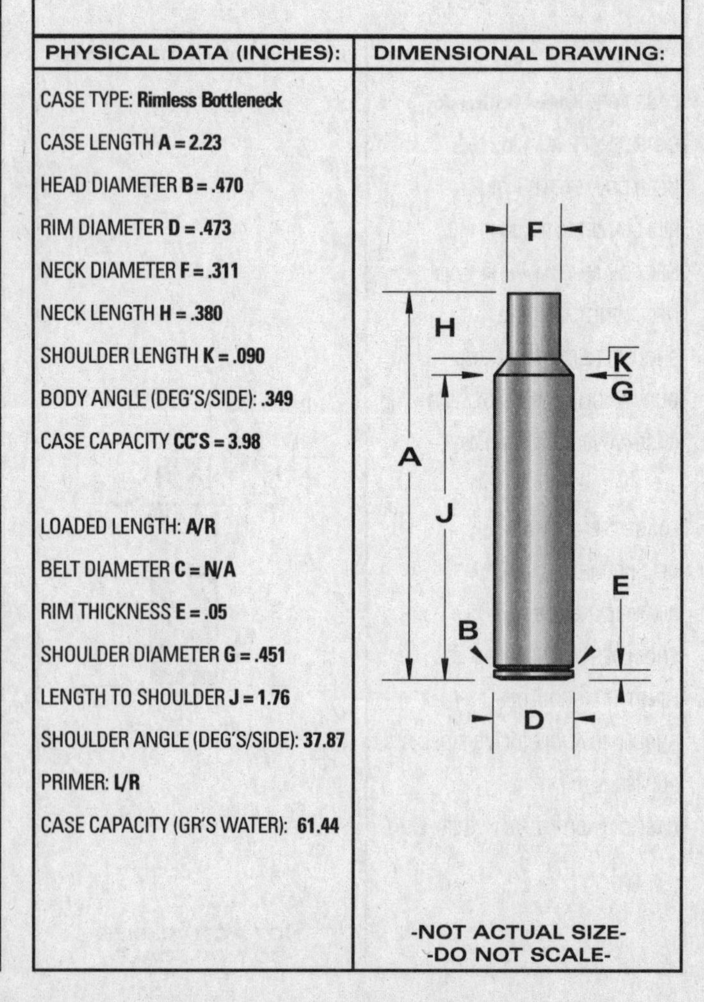

CARTRIDGE: 7 x 57 Mauser

OTHER NAMES: 7mm Mauser

DIA: .284

BALLISTEK NO: 284D

NAI NO: RXB 22333/4.745

DATA SOURCE: Hornady Manual 3rd Pg178

HISTORICAL DATA: By Mauser about 1890.

NOTES: Near perfect case capacity for .284" dia. bore.

LOADING DATA:

BULLET WT./TYPE	POWDER WT./TYPE	VELOCITY ('/SEC)	SOURCE
154/Spire	40.0/IMR4895	2500	JJD

CASE PREPARATION: SHELLHOLDER (RCBS): 3

MAKE FROM: Factory or .270 Win. Anneal case neck. Run into 7x57 sizer with expander removed. Trim to length. Check neck wall thickness and I.D. ream, if required. F/L size, chamfer and final fireform in chamber.

PHYSICAL DATA (INCHES):

CASE TYPE: **Rimless Bottleneck**
CASE LENGTH **A = 2.235**
HEAD DIAMETER **B = .471**
RIM DIAMETER **D = .473**
NECK DIAMETER **F = .321**
NECK LENGTH **H = .340**
SHOULDER LENGTH **K = .143**
BODY ANGLE (DEG'S/SIDE): **.738**
CASE CAPACITY **CC'S = 3.599**

LOADED LENGTH: **3.02**
BELT DIAMETER **C = N/A**
RIM THICKNESS **E = .049**
SHOULDER DIAMETER **G = .431**
LENGTH TO SHOULDER **J = 1.752**
SHOULDER ANGLE (DEG'S/SIDE): **21.04**
PRIMER: **L/R**
CASE CAPACITY (GR'S WATER): **55.55**

DIMENSIONAL DRAWING:

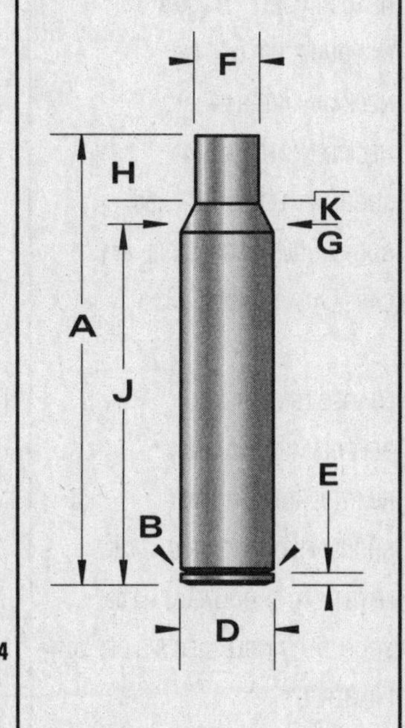

-NOT ACTUAL SIZE-
-DO NOT SCALE-

CARTRIDGE: 7 x 57R Mauser

OTHER NAMES:

DIA: .284

BALLISTEK NO: 284AU

NAI NO: RMB 22322/4.744

DATA SOURCE: NAI/Ballistek

HISTORICAL DATA:

NOTES:

LOADING DATA:

BULLET WT./TYPE	POWDER WT./TYPE	VELOCITY ('/SEC)	SOURCE
139/Spire	40.5/IMR4895	2500	JJD

CASE PREPARATION: SHELLHOLDER (RCBS): 26

MAKE FROM: Factory (Norma) or 7 x 65R Brenneke (RWS). F/L size the Brenneke case with expander removed. Trim and chamfer. F/L size. .444 Marlin in brass will also work but, a form set is required.

PHYSICAL DATA (INCHES):

CASE TYPE: **Rimmed Bottleneck**
CASE LENGTH **A = 2.23**
HEAD DIAMETER **B = .470**
RIM DIAMETER **D = .521**
NECK DIAMETER **F = .320**
NECK LENGTH **H = .365**
SHOULDER LENGTH **K = .140**
BODY ANGLE (DEG'S/SIDE): **.939**
CASE CAPACITY **CC'S = 3.49**

LOADED LENGTH: **3.07**
BELT DIAMETER **C = N/A**
RIM THICKNESS **E = .055**
SHOULDER DIAMETER **G = .420**
LENGTH TO SHOULDER **J = 1.725**
SHOULDER ANGLE (DEG'S/SIDE): **19.65**
PRIMER: **L/R**
CASE CAPACITY (GR'S WATER): **53.91**

DIMENSIONAL DRAWING:

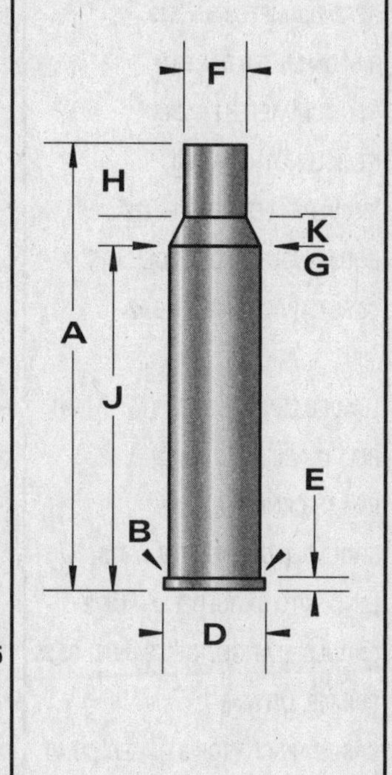

-NOT ACTUAL SIZE-
-DO NOT SCALE-

CARTRIDGE: 7 x 61 Sharpe & Hart

OTHER NAMES:

DIA: .284
BALLISTEK NO: 284F
NAI NO: BEN 22434/4.666

DATA SOURCE: Hornady Manual 3rd Pg192

HISTORICAL DATA: By Phil Sharpe about 1946.

NOTES:

LOADING DATA:

BULLET WT./TYPE	POWDER WT./TYPE	VELOCITY ('/SEC)	SOURCE
150/Spire	57.0/IMR4350	2865	Lyman

CASE PREPARATION: **SHELLHOLDER (RCBS):** 4

MAKE FROM: Factory or .300 H&H Mag. Do Not Anneal! Run into 7x61 form die. Trim to 2.4". Chamfer. F/L size. Fireform in chamber. I have also had very good luck making this case from .300 Win. Mag. Run into a 7mm/338 form die, cut to 2.4" and F/L size with expander removed. Final trim, chamfer and resize, with expander in place. Fireform.

PHYSICAL DATA (INCHES):

CASE TYPE: **Belted Bottleneck**
CASE LENGTH A = **2.394**
HEAD DIAMETER B = **.513**
RIM DIAMETER D = **.532**
NECK DIAMETER F = **.313**
NECK LENGTH H = **.300**
SHOULDER LENGTH K = **.122**
BODY ANGLE (DEG'S/SIDE): **.646**
CASE CAPACITY CC'S = **5.02**
LOADED LENGTH: **3.28**
BELT DIAMETER C = **.532**
RIM THICKNESS E = **.05**
SHOULDER DIAMETER G = **.473**
LENGTH TO SHOULDER J = **1.972**
SHOULDER ANGLE (DEG'S/SIDE): **33.25**
PRIMER: **L/R Mag**
CASE CAPACITY (GR'S WATER): **77.40**

DIMENSIONAL DRAWING:

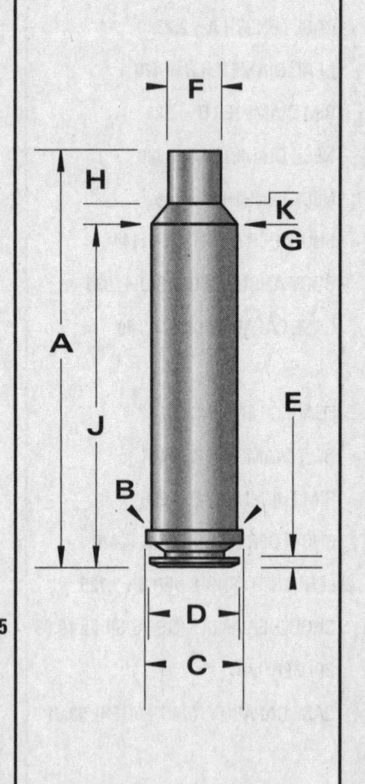

-NOT ACTUAL SIZE-
-DO NOT SCALE-

CARTRIDGE: 7 x 64 Brenneke

OTHER NAMES:

DIA: .284
BALLISTEK NO: 284N
NAI NO: RXB 22333/5.408

DATA SOURCE: Handloader #68 Pg53

HISTORICAL DATA: By Brenneke about 1916.

NOTES: Old DWM #557.

LOADING DATA:

BULLET WT./TYPE	POWDER WT./TYPE	VELOCITY ('/SEC)	SOURCE
140/Spire	48.8/IMR4350	2725	JJD

CASE PREPARATION: **SHELLHOLDER (RCBS):** 3

MAKE FROM: .30-06 Spgf. F/L size in Brenneke die. Watch headspace! Square case mouth. This case will be slightly short but, otherwise, fine.

PHYSICAL DATA (INCHES):

CASE TYPE: **Rimless Bottleneck**
CASE LENGTH A = **2.52**
HEAD DIAMETER B = **.466**
RIM DIAMETER D = **.470**
NECK DIAMETER F = **.313**
NECK LENGTH H = **.340**
SHOULDER LENGTH K = **.150**
BODY ANGLE (DEG'S/SIDE): **.641**
CASE CAPACITY CC'S = **4.29**
LOADED LENGTH: **3.30**
BELT DIAMETER C = **N/A**
RIM THICKNESS E = **.051**
SHOULDER DIAMETER G = **.425**
LENGTH TO SHOULDER J = **2.03**
SHOULDER ANGLE (DEG'S/SIDE): **20.47**
PRIMER: **L/R**
CASE CAPACITY (GR'S WATER): **66.22**

DIMENSIONAL DRAWING:

F H K G A J E B D

-NOT ACTUAL SIZE-
-DO NOT SCALE-

CARTRIDGE: 7 x 65R Brenneke

OTHER NAMES:

DIA: .284

BALLISTEK NO: 284AY

NAI NO: RMB 22323/5.467

DATA SOURCE: Hornady Manual 3rd Pg189

HISTORICAL DATA: By Wm. Brenneke in 1917.

NOTES: Similar to .280 Rem.

LOADING DATA:

BULLET WT./TYPE	POWDER WT./TYPE	VELOCITY ('/SEC)	SOURCE
154/Spire	46.3/IMR4064	2700	Horn.

CASE PREPARATION: SHELLHOLDER (RCBS): 26

MAKE FROM: Factory or 9,3 x 74R. Run case into form die (or, a .33 Win. sizer just far enough to set shoulder). F/L size. Trim and chamfer. Case will fireform slightly.

PHYSICAL DATA (INCHES):

CASE TYPE: **Rimmed Bottleneck**
CASE LENGTH **A = 2.559**
HEAD DIAMETER **B = .468**
RIM DIAMETER **D = .525**
NECK DIAMETER **F = .313**
NECK LENGTH **H = .380**
SHOULDER LENGTH **K = .151**
BODY ANGLE (DEG'S/SIDE): **.674**
CASE CAPACITY **CC'S = 4.35**

LOADED LENGTH: **3.35**
BELT DIAMETER **C = N/A**
RIM THICKNESS **E = .055**
SHOULDER DIAMETER **G = .425**
LENGTH TO SHOULDER **J = 2.028**
SHOULDER ANGLE (DEG'S/SIDE): **20.34**
PRIMER: **L/R**
CASE CAPACITY (GR'S WATER): **67.10**

DIMENSIONAL DRAWING:

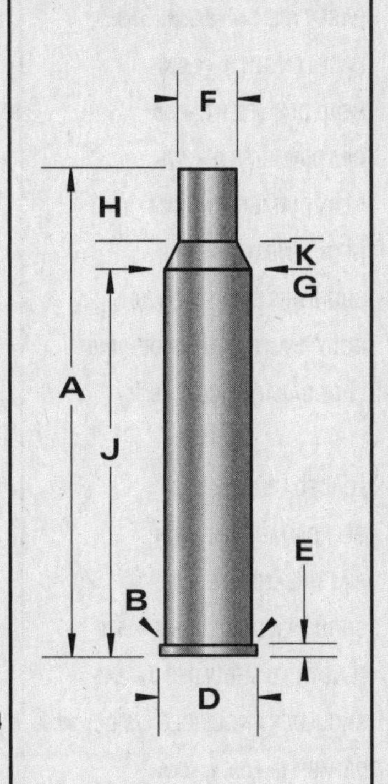

-NOT ACTUAL SIZE-
-DO NOT SCALE-

CARTRIDGE: 7 x 66 Vom Hofe SE

OTHER NAMES:
7mm Vom Hofe Express
7,6 x 66mm Vom Hofe Super

DIA: .284

BALLISTEK NO: 284AT

NAI NO: RBB 22442/4.751

DATA SOURCE: COTW 4th Pg262

HISTORICAL DATA: German ctg. ca. 1956.

NOTES:

LOADING DATA:

BULLET WT./TYPE	POWDER WT./TYPE	VELOCITY ('/SEC)	SOURCE
175/–	60.0/4350	2900	Barnes

CASE PREPARATION: SHELLHOLDER (RCBS): 18

MAKE FROM: .404 Jeffery (B.E.L.L.) Turn rim to .510" dia. (rebated) and cut new extractor groove. You'll need a form set for this case. Anneal and run case through the form set (three dies). Trim and F/L size.

PHYSICAL DATA (INCHES):

CASE TYPE: **Rebated Bottleneck**
CASE LENGTH **A = 2.58**
HEAD DIAMETER **B = .543**
RIM DIAMETER **D = .510**
NECK DIAMETER **F = .316**
NECK LENGTH **H = .215**
SHOULDER LENGTH **K = .280**
BODY ANGLE (DEG'S/SIDE): **.881**
CASE CAPACITY **CC'S = 5.86**

LOADED LENGTH: **3.25**
BELT DIAMETER **C = N/A**
RIM THICKNESS **E = .05**
SHOULDER DIAMETER **G = .485**
LENGTH TO SHOULDER **J = 2.085**
SHOULDER ANGLE (DEG'S/SIDE): **16.79**
PRIMER: **L/R**
CASE CAPACITY (GR'S WATER): **90.53**

DIMENSIONAL DRAWING:

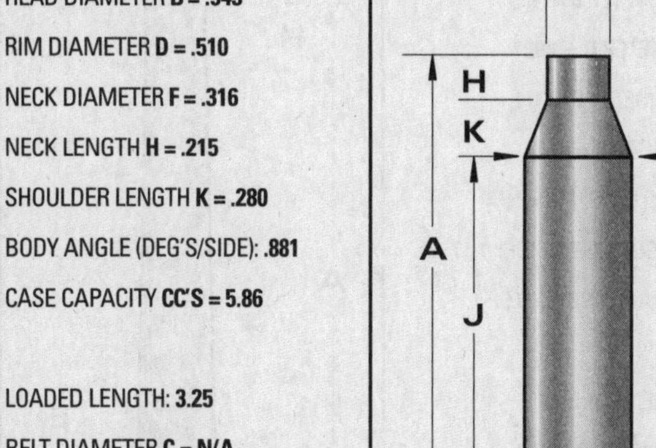

-NOT ACTUAL SIZE-
-DO NOT SCALE-

CARTRIDGE: 7 x 72R

OTHER NAMES:

DIA: .284

BALLISTEK NO: 284BH

NAI NO: RMS 31115/6.682

DATA SOURCE: COTW 4th Pg263

HISTORICAL DATA: By RWS ca. 1932.

NOTES: Reported as both .284" and .285" dia.

LOADING DATA:

BULLET WT./TYPE	POWDER WT./TYPE	VELOCITY ('/SEC)	SOURCE
160/FN	27.0/4895	1810	Barnes

CASE PREPARATION: SHELLHOLDER (RCBS): 30

MAKE FROM: 9,3 x 72R. Anneal case and F/L size in the 7x72R die. Square case mouth and chamfer.

PHYSICAL DATA (INCHES):

CASE TYPE: **Rimmed Straight**
CASE LENGTH **A = 2.84**
HEAD DIAMETER **B = .425**
RIM DIAMETER **D = .482**
NECK DIAMETER **F = .311**
NECK LENGTH **H = N/A**
SHOULDER LENGTH **K = N/A**
BODY ANGLE (DEG'S/SIDE): **1.17**
CASE CAPACITY **CC'S = 3.52**

LOADED LENGTH: **3.48**
BELT DIAMETER **C = N/A**
RIM THICKNESS **E = .06**
SHOULDER DIAMETER **G = N/A**
LENGTH TO SHOULDER **J = N/A**
SHOULDER ANGLE (DEG'S/SIDE): **N/A**
PRIMER: **L/R**
CASE CAPACITY (GR'S WATER): **54.28**

DIMENSIONAL DRAWING:

F
H
A
E
B
D

-NOT ACTUAL SIZE-
-DO NOT SCALE-

CARTRIDGE: 7 x 74 NAI/USC

OTHER NAMES:
7mm Ultra Sniper

DIA: .284

BALLISTEK NO: 284AX

NAI NO: BEN 12434/4.991

DATA SOURCE: NAI/Ballistek

HISTORICAL DATA: By author, in 1984, for mil. evaluation.

NOTES: Uses flash-tubes, electronic ignition & 40" barrel. Special weapon.

LOADING DATA:

BULLET WT./TYPE	POWDER WT./TYPE	VELOCITY ('/SEC)	SOURCE
175/Rebated Boattail	100/H870	3715	JJD

CASE PREPARATION: SHELLHOLDER (RCBS): 14

MAKE FROM: Case is made from .378 Weatherby in 7x74 form die set. 1.2" long flash-tubes are installed through primer pocket. Primer uses electronic spark for ignition. Details available from NAI.

PHYSICAL DATA (INCHES):

CASE TYPE: **Belted Bottleneck**
CASE LENGTH **A = 2.900**
HEAD DIAMETER **B = .581**
RIM DIAMETER **D = .576**
NECK DIAMETER **F = .314**
NECK LENGTH **H = .29**
SHOULDER LENGTH **K = .16**
BODY ANGLE (DEG'S/SIDE): **.140**
CASE CAPACITY **CC'S = 8.65**

LOADED LENGTH: **3.82**
BELT DIAMETER **C = .601**
RIM THICKNESS **E = .06**
SHOULDER DIAMETER **G = .570**
LENGTH TO SHOULDER **J = 2.45**
SHOULDER ANGLE (DEG'S/SIDE): **38.66**
PRIMER: **Electric Ignition**
CASE CAPACITY (GR'S WATER): **133.56**

DIMENSIONAL DRAWING:

F
H
K
G
A
E
J
B
D
C

-NOT ACTUAL SIZE-
-DO NOT SCALE-

CARTRIDGE: 7 x 75R Vom Hofe Rimmed

OTHER NAMES:
7 x 75R Vom Hofe SE

DIA: .284

BALLISTEK NO: 284AS

NAI NO: RMB 22322/6.303

DATA SOURCE: COTW 4th Pg262

HISTORICAL DATA: By Vom Hofe in 1939.

NOTES:

LOADING DATA:

BULLET WT./TYPE	POWDER WT./TYPE	VELOCITY ('/SEC)	SOURCE
175/RN	59.0/MRP	–	JJD

CASE PREPARATION: SHELLHOLDER (RCBS): 4

MAKE FROM: 9,3 x 74R. Turn rim dia. to .519", if necessary. Anneal case and size F/L in the 7 x 75R die. If case will not form, size in .33 WCF and .30-30 dies to reduce neck dia. (or, obtain form set).

PHYSICAL DATA (INCHES):

CASE TYPE: **Rimmed Bottleneck**

CASE LENGTH **A = 2.95**

HEAD DIAMETER **B = .468**

RIM DIAMETER **D = .519**

NECK DIAMETER **F = .318**

NECK LENGTH **H = .365**

SHOULDER LENGTH **K = .245**

BODY ANGLE (DEG'S/SIDE): **.696**

CASE CAPACITY **CC'S = 4.85**

LOADED LENGTH: **3.68**

BELT DIAMETER **C = N/A**

RIM THICKNESS **E = .052**

SHOULDER DIAMETER **G = .416**

LENGTH TO SHOULDER **J = 2.34**

SHOULDER ANGLE (DEG'S/SIDE): **11.31**

PRIMER: **L/R**

CASE CAPACITY (GR'S WATER): **74.90**

DIMENSIONAL DRAWING:

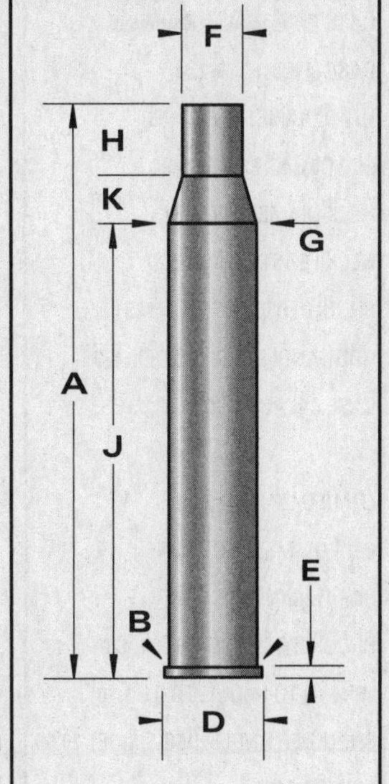

-NOT ACTUAL SIZE-
-DO NOT SCALE-

CARTRIDGE: 7.21 Lazzaroni Firebird

OTHER NAMES:

DIA: .284

BALLISTEK NO:

NAI NO:

DATA SOURCE:

HISTORICAL DATA: Created by John Lazzaroni, cartridge drawing dated 8/27/01.

NOTES: This is a 7mm cartridge in Lazzaroni's line of long action beltless magnums.

LOADING DATA:

BULLET WT./TYPE	POWDER WT./TYPE	VELOCITY ('/SEC)	SOURCE
140/Nosler Partition	113.0/AA 8700	3,729	Lazzaroni
160/Nosler Partition	97/RL-25	3,507	Lazzaroni

CASE PREPARATION: SHELL HOLDER (RCBS): 14

MAKE FROM:

PHYSICAL DATA (INCHES):

CASE TYPE: **Rimless Bottleneck**

CASE LENGTH **A = 2.81**

HEAD DIAMETER **B = .579**

RIM DIAMETER **D = .580**

NECK DIAMETER **F = .321**

NECK LENGTH **H = .292**

SHOULDER LENGTH **K = .205**

BODY ANGLE (DEG'S/SIDE):

CASE CAPACITY **CC'S =**

LOADED LENGTH: **3.630**

BELT DIAMETER **C =**

RIM THICKNESS **E = .065**

SHOULDER DIAMETER **G = .560**

LENGTH TO SHOULDER **J = 2.313**

SHOULDER ANGLE (DEG'S/SIDE): **30**

PRIMER: **L/R Mag**

CASE CAPACITY (GR'S WATER):

DIMENSIONAL DRAWING:

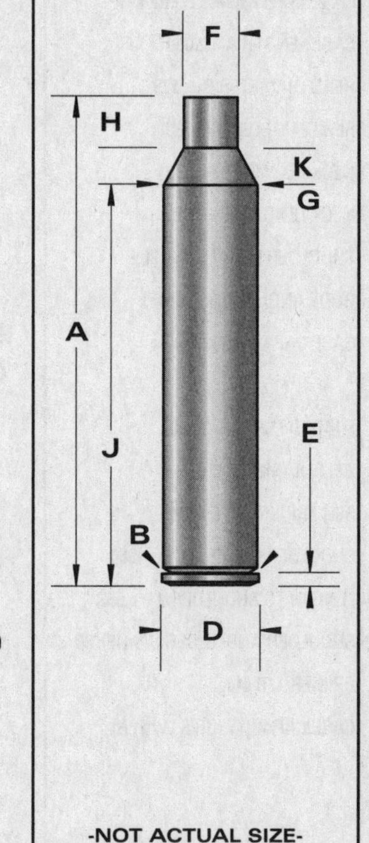

-NOT ACTUAL SIZE-
-DO NOT SCALE-

CARTRIDGE: 7.21 Lazzaroni Tomahawk

OTHER NAMES:

DIA: .284

BALLISTEK NO:

NAI NO:

DATA SOURCE:

HISTORICAL DATA: Created by John Lazzaroni, cartridge drawing dated 8/23/01.

NOTES: This is the 7mm cartridge in the Lazzaroni short action magnum lineup.

LOADING DATA:

BULLET WT./TYPE	POWDER WT./TYPE	VELOCITY ('/SEC)	SOURCE
140/Nosler Partition	67/H4350	3,373	Lazzaroni
160/Nosler Partition	67/RL-19	3,215	Lazzaroni

CASE PREPARATION: SHELL HOLDER (RCBS): 14

MAKE FROM:

PHYSICAL DATA (INCHES):

CASE TYPE: **Rimless Bottleneck**

CASE LENGTH **A = 2.050**

HEAD DIAMETER **B = .579**

RIM DIAMETER **D = .580**

NECK DIAMETER **F = .321**

NECK LENGTH **H = .292**

SHOULDER LENGTH **K = .205**

BODY ANGLE (DEG'S/SIDE):

CASE CAPACITY **CC'S =**

LOADED LENGTH: **2.800**

BELT DIAMETER **C =**

RIM THICKNESS **E = .065**

SHOULDER DIAMETER **G = .560**

LENGTH TO SHOULDER **J = 1.553**

SHOULDER ANGLE (DEG'S/SIDE): **30**

PRIMER: **L/R Mag**

CASE CAPACITY (GR'S WATER):

DIMENSIONAL DRAWING:

F H K G A J E B D

-NOT ACTUAL SIZE-
-DO NOT SCALE-

CARTRIDGE: 7.35mm Carcano

OTHER NAMES:
7,35 Italian
7,35 Mannlicher Carcano

DIA: .300

BALLISTEK NO: 300A

NAI NO: RXB 22344/4.517

DATA SOURCE: COTW 4th Pg201

HISTORICAL DATA: Italian mil. ctg. ca. 1938.

NOTES:

LOADING DATA:

BULLET WT./TYPE	POWDER WT./TYPE	VELOCITY ('/SEC)	SOURCE
145/Spire	40.0/4320	2550	Barnes

CASE PREPARATION: SHELLHOLDER (RCBS): 9

MAKE FROM: 6,5 x 54 M/S. Taper expand case to .30 cal. Trim & chamfer. F/L size and fireform. I have had some luck obtaining old military cases from The Old Western Scrounger. Bullets are produced by lathe turning .30 cal., 220 gr. RN bullets to .300-.301" dia. and cutting to equal 150 gr. (approx.).

PHYSICAL DATA (INCHES):

CASE TYPE: **Rimless Bottleneck**

CASE LENGTH **A = 2.01**

HEAD DIAMETER **B = .445**

RIM DIAMETER **D = .449**

NECK DIAMETER **F = .323**

NECK LENGTH **H = .252**

SHOULDER LENGTH **K = .143**

BODY ANGLE (DEG'S/SIDE): **.477**

CASE CAPACITY **CC'S = 3.35**

LOADED LENGTH: **2.90**

BELT DIAMETER **C = N/A**

RIM THICKNESS **E = .05**

SHOULDER DIAMETER **G = .420**

LENGTH TO SHOULDER **J = 1.70**

SHOULDER ANGLE (DEG'S/SIDE): **29.90**

PRIMER: **L/R**

CASE CAPACITY (GR'S WATER): **51.68**

DIMENSIONAL DRAWING:

F H K G A J E B D

-NOT ACTUAL SIZE-
-DO NOT SCALE-

CARTRIDGE: 7,5mm Schmidt Rubin

OTHER NAMES:
7,5mm Swiss

7,5 x 54.5 Schmidt Rubin
M90/03

DIA: .308

BALLISTEK NO: 308R

NAI NO: RXB 22343/4.340

DATA SOURCE: Hornady 3rd Pg204

HISTORICAL DATA: Swiss mil. Ctg. ca. 1889.

NOTES:

LOADING DATA:

BULLET WT./TYPE	POWDER WT./TYPE	VELOCITY ('/SEC)	SOURCE
150/Spire	46.0/IMR4895	2775	JJD
168/HPBT	65.0/IMR4831	2810	JJD

CASE PREPARATION: SHELLHOLDER (RCBS): 2

MAKE FROM: Factory or 7,62x54R Russian. Turn rim to .495" dia. Cut new extractor groove. F/L size in Schmidt Rubin die. Case will be about .090" short but, otherwise, fine. Final forming in chamber.

PHYSICAL DATA (INCHES):

CASE TYPE: **Rimless Bottleneck**

CASE LENGTH **A = 2.14**

HEAD DIAMETER **B = .493**

RIM DIAMETER **D = .495**

NECK DIAMETER **F = .337**

NECK LENGTH **H = .290**

SHOULDER LENGTH **K = .099**

BODY ANGLE (DEG'S/SIDE): **.794**

CASE CAPACITY **CC'S = 4.07**

LOADED LENGTH: **3.03**

BELT DIAMETER **C = N/A**

RIM THICKNESS **E = .06**

SHOULDER DIAMETER **G = .45**

LENGTH TO SHOULDER **J = 1.75**

SHOULDER ANGLE (DEG'S/SIDE): **29.46**

PRIMER: **L/R**

CASE CAPACITY (GR'S WATER): **62.89**

DIMENSIONAL DRAWING:

-NOT ACTUAL SIZE-
-DO NOT SCALE-

CARTRIDGE: 7,5mm Swedish Nagant Revolver

OTHER NAMES:
7,5mm Swedish Revolver

DIA: .325

BALLISTEK NO: 325A

NAI NO: RMS 21115/2.543

DATA SOURCE: COTW 4th Pg170

HISTORICAL DATA: By Swiss mil. about 1882.

NOTES:

LOADING DATA:

BULLET WT./TYPE	POWDER WT./TYPE	VELOCITY ('/SEC)	SOURCE
100/Lead	2.0/B'Eye	720	Barnes

CASE PREPARATION: SHELLHOLDER (RCBS): 1

MAKE FROM: .25-20 or .32-20 Win. Cut case to length & chamfer. Use Lyman bullet #323470 cut to .52" long. Minie bullet with .250" dia. drill x 3/16 deep. Case will fireform upon ignition and undersize bullet will expand to "take" rifling.

PHYSICAL DATA (INCHES):

CASE TYPE: **Rimmed Straight**

CASE LENGTH **A = .890**

HEAD DIAMETER **B = .350**

RIM DIAMETER **D = .406**

NECK DIAMETER **F = .328**

NECK LENGTH **H = N/A**

SHOULDER LENGTH **K = N/A**

BODY ANGLE (DEG'S/SIDE): **.757**

CASE CAPACITY **CC'S = .84**

LOADED LENGTH: **1.35**

BELT DIAMETER **C = N/A**

RIM THICKNESS **E = .058**

SHOULDER DIAMETER **G = N/A**

LENGTH TO SHOULDER **J = N/A**

SHOULDER ANGLE (DEG'S/SIDE): **N/A**

PRIMER: **S/P**

CASE CAPACITY (GR'S WATER): **12.97**

DIMENSIONAL DRAWING:

-NOT ACTUAL SIZE-
-DO NOT SCALE-

CARTRIDGE: 7,5mm Swiss Army

OTHER NAMES:

DIA: .317

BALLISTEK NO: 317A

NAI NO: RMS 11115/2.579

DATA SOURCE: COTW 4th Pg170

HISTORICAL DATA: Adopted by Swiss mil. in 1882.

NOTES: Similar to 7,5mm Swedish Nagant (#325A).

LOADING DATA:

BULLET WT./TYPE	POWDER WT./TYPE	VELOCITY ('/SEC)	SOURCE
110/Lead	1.7/B'Eye	650	JJD

CASE PREPARATION: SHELLHOLDER (RCBS): 1

MAKE FROM: .32-20 WCF. Cut case to length, chamfer and F/L size. Neck may require I.D. reaming. Clean & resize. Fireform.

PHYSICAL DATA (INCHES):

CASE TYPE: **Rimmed Straight**

CASE LENGTH **A = .89**

HEAD DIAMETER **B = .345**

RIM DIAMETER **D = .407**

NECK DIAMETER **F = .335**

NECK LENGTH **H = N/A**

SHOULDER LENGTH **K = N/A**

BODY ANGLE (DEG'S/SIDE): **.347**

CASE CAPACITY **CC'S = .735**

LOADED LENGTH: **1.29**

BELT DIAMETER **C = N/A**

RIM THICKNESS **E = .065**

SHOULDER DIAMETER **G = N/A**

LENGTH TO SHOULDER **J = N/A**

SHOULDER ANGLE (DEG'S/SIDE): **N/A**

PRIMER: **S/P**

CASE CAPACITY (GR'S WATER): **11.35**

DIMENSIONAL DRAWING:

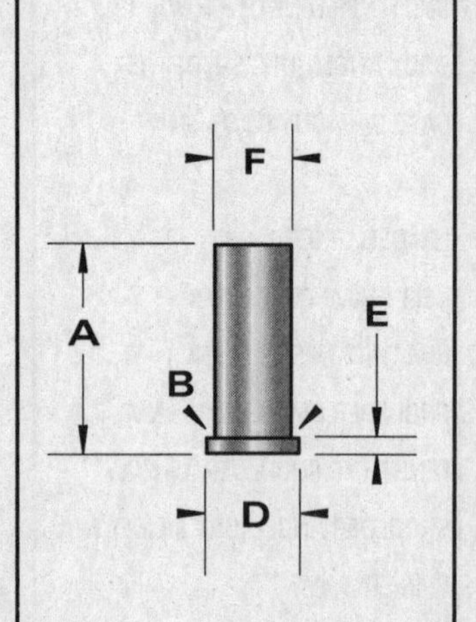

-NOT ACTUAL SIZE-
-DO NOT SCALE-

CARTRIDGE: 7,5 x 53.5 French MAS

OTHER NAMES:
7,5 x 54 M29 French

DIA: .308

BALLISTEK NO: 308AZ

NAI NO: RXB 22333/4.395

DATA SOURCE: COTW 4th Pg202

HISTORICAL DATA: French mil. ctg. ca. 1925.

NOTES: MAS = Manufacture d' Armes de Saint Etiene.

LOADING DATA:

BULLET WT./TYPE	POWDER WT./TYPE	VELOCITY ('/SEC)	SOURCE
150/Spire	54.0/4831	2680	Barnes

CASE PREPARATION: SHELLHOLDER (RCBS): 30

MAKE FROM: 7,62 x 54R Russian. Lathe turn base to .480" dia. and rim to .482" dia. Cut new extractor groove. F/L size case in 7,5 x 53.5mm die. Trim and chamfer. Fireform in chamber.

PHYSICAL DATA (INCHES):

CASE TYPE: **Rimless Bottleneck**

CASE LENGTH **A = 2.11**

HEAD DIAMETER **B = .480**

RIM DIAMETER **D = .482**

NECK DIAMETER **F = .340**

NECK LENGTH **H = .320**

SHOULDER LENGTH **K = .100**

BODY ANGLE (DEG'S/SIDE): **.750**

CASE CAPACITY **CC'S = 3.72**

LOADED LENGTH: **2.99**

BELT DIAMETER **C = N/A**

RIM THICKNESS **E = .05**

SHOULDER DIAMETER **G = .441**

LENGTH TO SHOULDER **J = 1.69**

SHOULDER ANGLE (DEG'S/SIDE): **26.79**

PRIMER: **L/R**

CASE CAPACITY (GR'S WATER): **57.44**

DIMENSIONAL DRAWING:

-NOT ACTUAL SIZE-
-DO NOT SCALE-

CARTRIDGE: 7,5 x 53.5R Rubin

OTHER NAMES:

DIA: .312

BALLISTEK NO: 312M

NAI NO: RMS 21115/4.617

DATA SOURCE: Hoyem Vol.2 Pg185

HISTORICAL DATA: Circa 1885.

NOTES: A rimless version also exists.

LOADING DATA:

BULLET WT./TYPE	POWDER WT./TYPE	VELOCITY ('/SEC)	SOURCE
160/RN	17.1/IMR4198	—	JJD

CASE PREPARATION: **SHELLHOLDER (RCBS): 7**

MAKE FROM: .303 British. turn rim to .528" dia. Taper expand case neck to .320" dia. F/L size with expander removed. Trim to length & chamfer. F/L size. Fireform in chamber.

PHYSICAL DATA (INCHES):

CASE TYPE: **Rimmed Straight**

CASE LENGTH **A = 2.11**

HEAD DIAMETER **B = .457**

RIM DIAMETER **D = .528**

NECK DIAMETER **F = .393**

NECK LENGTH **H = N/A**

SHOULDER LENGTH **K = N/A**

BODY ANGLE (DEG'S/SIDE): **.895**

CASE CAPACITY **CC'S = 2.14**

LOADED LENGTH: **3.08**

BELT DIAMETER **C = N/A**

RIM THICKNESS **E = .062**

SHOULDER DIAMETER **G = N/A**

LENGTH TO SHOULDER **J = N/A**

SHOULDER ANGLE (DEG'S/SIDE): **N/A**

PRIMER: **L/R**

CASE CAPACITY (GR'S WATER): **33.00**

DIMENSIONAL DRAWING:

F
A
E
B
D

-NOT ACTUAL SIZE-
-DO NOT SCALE-

CARTRIDGE: 7,62mm Russian Nagant

OTHER NAMES:
7,62mm Nagant Revolver
7,62 x 38R
7,5mm Russian Nagant

DIA: .306

BALLISTEK NO: 306A

NAI NO: RMS 31115/4.248

DATA SOURCE: COTW 4th Pg165

HISTORICAL DATA: Gas-seal revolver - bullet is seated inside case.

NOTES: Russian ctg. ca. 1890.

LOADING DATA:

BULLET WT./TYPE	POWDER WT./TYPE	VELOCITY ('/SEC)	SOURCE
100/Lead	3.8/B'Eye	900	JJD

CASE PREPARATION: **SHELLHOLDER (RCBS): 16**

MAKE FROM: .32-20 Winchester. Taper expand to .300" dia. F/L size. Square case mouth. Chamfer and fireform. Case is short but okay. You may also re-body a .38 ACP with 11/32" dia. tubing and F/L size. Use any .30 Cal. mold to produce 150 grain projectiles and cut to equal 100-105 grains. Seat bullet completely inside the case and apply roll crimp to case mouth. It may be necessary to turn the .32-20's rim to the .388" dia.

PHYSICAL DATA (INCHES):

CASE TYPE: **Rimmed Straight**

CASE LENGTH **A = 1.52**

HEAD DIAMETER **B = .358**

RIM DIAMETER **D = .388**

NECK DIAMETER **F = .296**

NECK LENGTH **H = N/A**

SHOULDER LENGTH **K = N/A**

BODY ANGLE (DEG'S/SIDE): **1.2**

CASE CAPACITY **CC'S = 1.83**

LOADED LENGTH: **1.52**

BELT DIAMETER **C = N/A**

RIM THICKNESS **E = .042**

SHOULDER DIAMETER **G = N/A**

LENGTH TO SHOULDER **J = N/A**

SHOULDER ANGLE (DEG'S/SIDE): **N/A**

PRIMER: **S/P**

CASE CAPACITY (GR'S WATER): **28.3**

DIMENSIONAL DRAWING:

F
A
E
B
D

-NOT ACTUAL SIZE-
-DO NOT SCALE-

CARTRIDGE: 7,62mm Tokarev

OTHER NAMES:
7,62mm Russian

DIA: .307

BALLISTEK NO: 307A

NAI NO: RMB 24242/2.552

DATA SOURCE: COTW 4th Pg164

HISTORICAL DATA: Russian pistol ctg. about 1930.

NOTES:

LOADING DATA:

BULLET WT./TYPE	POWDER WT./TYPE	VELOCITY ('/SEC)	SOURCE
86/RN	5.0/B'Eye	1390	Barnes

CASE PREPARATION: SHELLHOLDER (RCBS): 10

MAKE FROM: .38 Special. Turn rim to .390" dia. Trim to .98". F/L size. Final trim to length. .223 Rem. brass may also be used, by cutting to length and F/L sizing but, you'll need to I.D. neck ream.

PHYSICAL DATA (INCHES):

CASE TYPE: **Rimmed Straight**

CASE LENGTH **A = .970**

HEAD DIAMETER **B = .380**

RIM DIAMETER **D = .390**

NECK DIAMETER **F = .33**

NECK LENGTH **H = .125**

SHOULDER LENGTH **K = .085**

BODY ANGLE (DEG'S/SIDE): **.511**

CASE CAPACITY **CC'S = 1.04**

LOADED LENGTH: **1.35**

BELT DIAMETER **C = N/A**

RIM THICKNESS **E = .052**

SHOULDER DIAMETER **G = .370**

LENGTH TO SHOULDER **J = .76**

SHOULDER ANGLE (DEG'S/SIDE): **13.24**

PRIMER: **S/P**

CASE CAPACITY (GR'S WATER): **15.98**

DIMENSIONAL DRAWING:

F H K A J G B E D

-NOT ACTUAL SIZE-
-DO NOT SCALE-

CARTRIDGE: 7,62 Micro-Whisper

OTHER NAMES:

DIA: .308

BALLISTEK NO:

NAI NO:

DATA SOURCE: J.D. Jones, Frank Barnes

HISTORICAL DATA: Designed by J.D. Jones in the early 1990s, this is simply the .30 Luger case adapted to use a 30-caliber rifle bullet.

NOTES:

LOADING DATA:

BULLET WT./TYPE	POWDER WT./TYPE	VELOCITY ('/SEC)	SOURCE
93 Norma SP	11.5/A#9	1762	SSK
150 Hornady FMJ	7.0/H-110	1018	SSK
168 Hornady Match	7.1/A#9	1096	SSK

CASE PREPARATION: SHELL HOLDER (RCBS): 01

MAKE FROM: .30 Luger
Fireform .30 Luger with bullet-engaging lands.

PHYSICAL DATA (INCHES):

CASE TYPE: **Rimless Bottleneck**

CASE LENGTH **A = .810**

HEAD DIAMETER **B = .384**

RIM DIAMETER **D = .390**

NECK DIAMETER **F = .314**

NECK LENGTH **H = 1.66**

SHOULDER LENGTH **K = .035**

BODY ANGLE (DEG'S/SIDE):

CASE CAPACITY **CC'S = 1.04**

LOADED LENGTH: **1.700**

BELT DIAMETER **C = N/A**

RIM THICKNESS **E = .050**

SHOULDER DIAMETER **G = .378**

LENGTH TO SHOULDER **J = .609**

SHOULDER ANGLE (DEG'S/SIDE): **~20°**

PRIMER: **S/P**

CASE CAPACITY (GR'S WATER):

DIMENSIONAL DRAWING:

F H K G A B J E D

-NOT ACTUAL SIZE-
-DO NOT SCALE-

CARTRIDGE: 7,62 Mini-Whisper

OTHER NAMES:

DIA: .308

BALLISTEK NO:

NAI NO:

DATA SOURCE: J.D. Jones

HISTORICAL DATA: Designed by J.D. Jones in the early 1990s, this is simply the .30 Mauser case adapted to use a 30-caliber rifle bullet.

NOTES:

LOADING DATA:

BULLET WT./TYPE	POWDER WT./TYPE	VELOCITY ('/SEC)	SOURCE
93/Norma	7.5/Clays	1727	SSK
110/Speer Carb	7.5/Clays	1588	SSK
150/Hornady FMJ	4.5/A#2	350	SSK

CASE PREPARATION: SHELL HOLDER (RCBS): 16

MAKE FROM: 7.63 Mauser
Fireform 7.63 Mauser with bullet-engaging lands.

PHYSICAL DATA (INCHES):

CASE TYPE: **Rimless Bottleneck**

CASE LENGTH **A = .990**

HEAD DIAMETER **B = .386**

RIM DIAMETER **D = .393**

NECK DIAMETER **F = .329**

NECK LENGTH **H = .175**

SHOULDER LENGTH **K = .035**

BODY ANGLE (DEG'S/SIDE):

CASE CAPACITY **CC'S**

LOADED LENGTH: **1.85**

BELT DIAMETER **C = N/A**

RIM THICKNESS **E = .050**

SHOULDER DIAMETER **G = .376**

LENGTH TO SHOULDER **J = .078**

SHOULDER ANGLE (DEG'S/SIDE): **~20°**

PRIMER: **S/P**

CASE CAPACITY (GR'S WATER):

DIMENSIONAL DRAWING:

F H K G A J B E D

-NOT ACTUAL SIZE-
-DO NOT SCALE-

CARTRIDGE: 7,62 x 39 (M43)

OTHER NAMES:
7,62mm Russian Short
7,62mm Kalashnikov

DIA: .310

BALLISTEK NO: 310A

NAI NO: RXB 34242/3.444

DATA SOURCE: Sierra Manual 1985 Pg202

HISTORICAL DATA: Soviet mil. ctg. adopted 1943.

NOTES: .308" projectiles work OK but, .310 is correct dia.

LOADING DATA:

BULLET WT./TYPE	POWDER WT./TYPE	VELOCITY ('/SEC)	SOURCE
150/Spire	21.5/IMR4227	2000	JJD

CASE PREPARATION: SHELLHOLDER (RCBS): 6

MAKE FROM: Factory (Sako) or .220 Russian (Sako). Taper neck expand to .284" dia. Anneal neck. Further taper expand to .315. F/L size and fireform in chamber. .220 Swift cases may also be used by turning the rim flush with the body and cutting a new extractor groove. Anneal neck and set back shoulder in 7,62x39 die. Trim to length, chamfer and fireform in the chamber.

PHYSICAL DATA (INCHES):

CASE TYPE: **Rimless Bottleneck**

CASE LENGTH **A = 1.519**

HEAD DIAMETER **B = .441**

RIM DIAMETER **D = .443**

NECK DIAMETER **F = .334**

NECK LENGTH **H = .217**

SHOULDER LENGTH **K = .096**

BODY ANGLE (DEG'S/SIDE): **1.48**

CASE CAPACITY **CC'S = 2.20**

LOADED LENGTH: **2.19**

BELT DIAMETER **C = N/A**

RIM THICKNESS **E = .058**

SHOULDER DIAMETER **G = .389**

LENGTH TO SHOULDER **J = 1.205**

SHOULDER ANGLE (DEG'S/SIDE): **15.82**

PRIMER: **S/R**

CASE CAPACITY (GR'S WATER): **33.98**

DIMENSIONAL DRAWING:

F H K G A J E B D

-NOT ACTUAL SIZE-
-DO NOT SCALE-

CARTRIDGE: 7,62 x 45 Czech Short M52

OTHER NAMES:
7,62mm M52 Czech

DIA: .309

BALLISTEK NO: 309E

NAI NO: RXB
23243/4.000

DATA SOURCE: MSAA Pg32

HISTORICAL DATA: The earlier version (1945) was 7,5mm.

NOTES:

LOADING DATA:

BULLET WT./TYPE	POWDER WT./TYPE	VELOCITY ('/SEC)	SOURCE
130/Spire	32.4/IMR3031	2450	JJD

CASE PREPARATION: SHELLHOLDER (RCBS): 6

MAKE FROM: .220 Swift. Turn rim to .441" dia. and cut new extractor groove. Anneal case neck and taper expand to about .315" dia. F/L size to set shoulder back, with expander removed. Trim to length. I.D. neck ream, if necessary. F/L size and fireform.

PHYSICAL DATA (INCHES):

CASE TYPE: **Rimless Bottleneck**
CASE LENGTH **A = 1.764**
HEAD DIAMETER **B = .441**
RIM DIAMETER **D = .441**
NECK DIAMETER **F = .337**
NECK LENGTH **H = .263**
SHOULDER LENGTH **K = .101**
BODY ANGLE (DEG'S/SIDE): **.358**
CASE CAPACITY **CC'S = 2.76**
LOADED LENGTH: **2.45**
BELT DIAMETER **C = N/A**
RIM THICKNESS **E = .06**
SHOULDER DIAMETER **G = .426**
LENGTH TO SHOULDER **J = 1.40**
SHOULDER ANGLE (DEG'S/SIDE): **23.77**
PRIMER: **L/R**
CASE CAPACITY (GR'S WATER): **42.67**

DIMENSIONAL DRAWING:

F
H
K
G
A
J
E
B
D

-NOT ACTUAL SIZE-
-DO NOT SCALE-

CARTRIDGE: 7.62 x 54R Russian

OTHER NAMES:
7,62mm Russian
7,62 Mosin-Nagant M91

DIA: .308

BALLISTEK NO: 310B

NAI NO: RMB
23333/4.192

DATA SOURCE: Hornady Manual 3rd Pg221

HISTORICAL DATA: Russian mil. ctg. ca. 1891.

NOTES: Loaded in U.S. until about 1950.

LOADING DATA:

BULLET WT./TYPE	POWDER WT./TYPE	VELOCITY ('/SEC)	SOURCE
165/Spire	46.2/IMR4895	2600	Horn.

CASE PREPARATION: SHELLHOLDER (RCBS): 13

MAKE FROM: Factory or .45-70 Govt. Swage base to .490" dia. Turn rim to .570 & back chamfer. Anneal case. Use 7.62 form die set or form .40-65 then .33 Win. necks. Trim to length. F/L size.

PHYSICAL DATA (INCHES):

CASE TYPE: **Rimmed Bottleneck**
CASE LENGTH **A = 2.05**
HEAD DIAMETER **B = .489**
RIM DIAMETER **D = .570**
NECK DIAMETER **F = .336**
NECK LENGTH **H = .350**
SHOULDER LENGTH **K = .155**
BODY ANGLE (DEG'S/SIDE): **.660**
CASE CAPACITY **CC'S = 3.817**
LOADED LENGTH: **3.01**
BELT DIAMETER **C = N/A**
RIM THICKNESS **E = .065**
SHOULDER DIAMETER **G = .458**
LENGTH TO SHOULDER **J = 1.545**
SHOULDER ANGLE (DEG'S/SIDE): **21.48**
PRIMER: **L/R**
CASE CAPACITY (GR'S WATER): **58.91**

DIMENSIONAL DRAWING:

F
H
K
G
A
J
E
B
D

-NOT ACTUAL SIZE-
-DO NOT SCALE-

CARTRIDGE: 7,63mm Mannlicher

OTHER NAMES:
7,65mm Mannlicher

DIA: .308

BALLISTEK NO: 308BK

NAI NO: RXS
11115/2.530

DATA SOURCE: COTW 4th Pg166

HISTORICAL DATA: Mil. auto ctg. ca. 1900.

NOTES:

LOADING DATA:

BULLET WT./TYPE	POWDER WT./TYPE	VELOCITY ('/SEC)	SOURCE
85/RN	3.2/Unique	1000	Barnes

CASE PREPARATION: SHELLHOLDER (RCBS): 12

MAKE FROM: .32 S&W Long. Turn rim flush with body dia. Trim case to .84" and F/L size. Chamfer. (It is also necessary to cut a new extractor groove!)

PHYSICAL DATA (INCHES):

CASE TYPE: **Rimless Straight**

CASE LENGTH **A = .84**

HEAD DIAMETER **B = .332**

RIM DIAMETER **D = .334**

NECK DIAMETER **F = .331**

NECK LENGTH **H = N/A**

SHOULDER LENGTH **K = N/A**

BODY ANGLE (DEG'S/SIDE): **.036**

CASE CAPACITY **CC'S = .605**

LOADED LENGTH: **1.12**

BELT DIAMETER **C = N/A**

RIM THICKNESS **E = .045**

SHOULDER DIAMETER **G = N/A**

LENGTH TO SHOULDER **J = N/A**

SHOULDER ANGLE (DEG'S/SIDE): **N/A**

PRIMER: **S/P**

CASE CAPACITY (GR'S WATER): **9.34**

DIMENSIONAL DRAWING:

F
A
E
B
D

-NOT ACTUAL SIZE-
-DO NOT SCALE-

CARTRIDGE: 7,63mm Mauser

OTHER NAMES:
.30 Mauser

DIA: .308

BALLISTEK NO: 308U

NAI NO: RMB
24241/2.598

DATA SOURCE: COTW 4th Pg165

HISTORICAL DATA: By H. Borchardt in 1893.

NOTES: Same case as .30 Borchardt (#307B).

LOADING DATA:

BULLET WT./TYPE	POWDER WT./TYPE	VELOCITY ('/SEC)	SOURCE
86/RN	5.0/B'Eye	1300	JJD

CASE PREPARATION: SHELLHOLDER (RCBS): 6

MAKE FROM: .38 Special. Turn rim to .390" dia. Thin, if necessary, to .045". Cut case to 1.00" and F/L size. .223 Rem. brass may also be used by cutting to 1.00", F/L sizing, I.D. neck reaming and resizing.

PHYSICAL DATA (INCHES):

CASE TYPE: **Rimmed Bottleneck**

CASE LENGTH **A = .99**

HEAD DIAMETER **B = .381**

RIM DIAMETER **D = .390**

NECK DIAMETER **F = .332**

NECK LENGTH **H = .130**

SHOULDER LENGTH **K = .125**

BODY ANGLE (DEG'S/SIDE): **.589**

CASE CAPACITY **CC'S = 1.05**

LOADED LENGTH: **1.36**

BELT DIAMETER **C = N/A**

RIM THICKNESS **E = .045**

SHOULDER DIAMETER **G = .370**

LENGTH TO SHOULDER **J = .735**

SHOULDER ANGLE (DEG'S/SIDE): **8.64**

PRIMER: **S/P**

CASE CAPACITY (GR'S WATER): **16.30**

DIMENSIONAL DRAWING:

F
H
K
G
A
J
E
B
D

-NOT ACTUAL SIZE-
-DO NOT SCALE-

CARTRIDGE: 7,65mm French Long MAS

OTHER NAMES:	
7,65mm Longue 7,65mm MAS Auto	DIA: .309 BALLISTEK NO: 309D NAI NO: RXS 11115/2.309

DATA SOURCE: COTW 4th Pg166

HISTORICAL DATA: French mil. ctg.

NOTES: Similar to .30 Pedersen.

LOADING DATA:

BULLET WT./TYPE	POWDER WT./TYPE	VELOCITY ('/SEC)	SOURCE
85/Lead	2.3/B'Eye	1000	Barnes

CASE PREPARATION: SHELLHOLDER (RCBS): 12

MAKE FROM: .32 S&W Long. Turn rim flush with body dia. Cut new extractor groove. Trim case to length and F/L size. Use any .30 cal. cast lead bullet, unsized and cut off to equal 85-90 grains.

PHYSICAL DATA (INCHES):

CASE TYPE: **Rimless Straight**
CASE LENGTH **A = .776**
HEAD DIAMETER **B = .336**
RIM DIAMETER **D = .335**
NECK DIAMETER **F = .335**
NECK LENGTH **H = N/A**
SHOULDER LENGTH **K = N/A**
BODY ANGLE (DEG'S/SIDE): **.039**
CASE CAPACITY **CC'S = .53**

LOADED LENGTH: **1.198**
BELT DIAMETER **C = N/A**
RIM THICKNESS **E = .046**
SHOULDER DIAMETER **G = N/A**
LENGTH TO SHOULDER **J = N/A**
SHOULDER ANGLE (DEG'S/SIDE): **N/A**
PRIMER: **S/P**
CASE CAPACITY (GR'S WATER): **8.19**

DIMENSIONAL DRAWING:

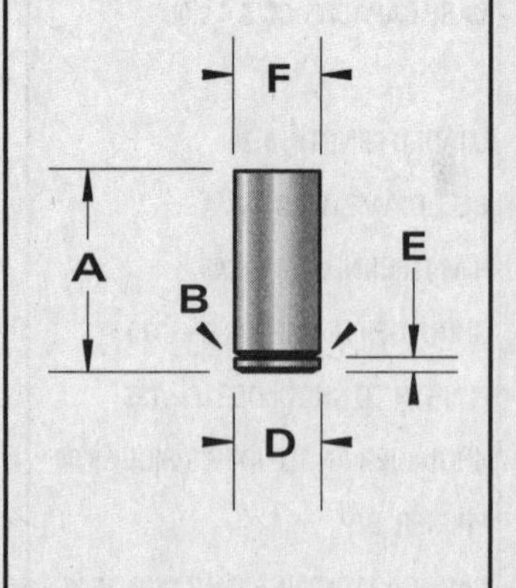

-NOT ACTUAL SIZE-
-DO NOT SCALE-

CARTRIDGE: 7,65 x 54 Mauser

OTHER NAMES:	
7,65mm Argentine Mauser 7,65 x 53 Belgian Mauser	DIA: .313 BALLISTEK NO: 312F NAI NO: RXB 22343/4.466

DATA SOURCE: COTW 4th Pg206

HISTORICAL DATA: Widely manufactured mil. ctg. ca. 1889.

NOTES:

LOADING DATA:

BULLET WT./TYPE	POWDER WT./TYPE	VELOCITY ('/SEC)	SOURCE
175/RN	49.0/4350	2560	Barnes

CASE PREPARATION: SHELLHOLDER (RCBS): 3

MAKE FROM: Factory or .30-06 Spgf. Trim the case to 2.1", anneal and F/L size in the Mauser die. Trim to length and chamfer.

PHYSICAL DATA (INCHES):

CASE TYPE: **Rimless Bottleneck**
CASE LENGTH **A = 2.09**
HEAD DIAMETER **B = .468**
RIM DIAMETER **D = .470**
NECK DIAMETER **F = .338**
NECK LENGTH **H = .270**
SHOULDER LENGTH **K = .125**
BODY ANGLE (DEG'S/SIDE): **.747**
CASE CAPACITY **CC'S = 3.70**

LOADED LENGTH: **2.95**
BELT DIAMETER **C = N/A**
RIM THICKNESS **E = .06**
SHOULDER DIAMETER **G = .429**
LENGTH TO SHOULDER **J = 1.695**
SHOULDER ANGLE (DEG'S/SIDE): **20.00**
PRIMER: **L/R**
CASE CAPACITY (GR'S WATER): **57.11**

DIMENSIONAL DRAWING:

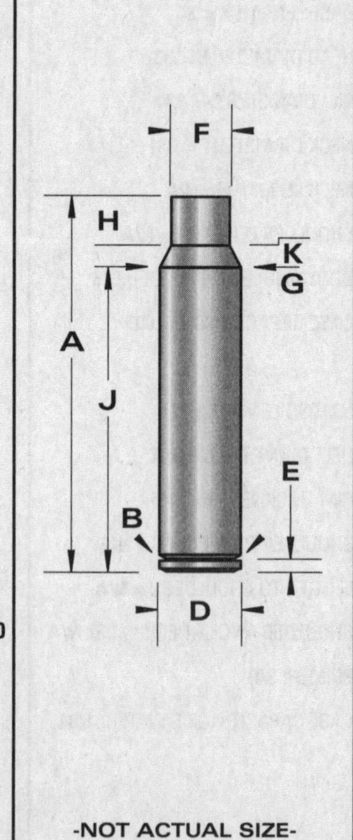

-NOT ACTUAL SIZE-
-DO NOT SCALE-

CARTRIDGE: 7,65mm Roth Sauer

OTHER NAMES: 7,65mm Frommer

DIA: .301
BALLISTEK NO: 301A
NAI NO: RXS 11115/1.522

DATA SOURCE: COTW 4th Pg167

HISTORICAL DATA: About 1900.

NOTES:

LOADING DATA:

BULLET WT./TYPE	POWDER WT./TYPE	VELOCITY ('/SEC)	SOURCE
75/Lead	2.0/B'Eye	800	JJD

CASE PREPARATION: **SHELLHOLDER (RCBS):** 17

MAKE FROM: .32 ACP. Trim to .52" and chamfer. Turn rim flush with body dia. F/L size and trim. Use Lyman mold #311440 to cast bullets and cut to equal 75-80 grs. Swage bullets to .301-302" dia.

PHYSICAL DATA (INCHES):

CASE TYPE: **Rimless Straight**
CASE LENGTH **A = .510**
HEAD DIAMETER **B = .335**
RIM DIAMETER **D = .335**
NECK DIAMETER **F = .332**
NECK LENGTH **H = N/A**
SHOULDER LENGTH **K = N/A**
BODY ANGLE (DEG'S/SIDE): **.18**
CASE CAPACITY **CC'S = .215**
LOADED LENGTH: **.840**
BELT DIAMETER **C = N/A**
RIM THICKNESS **E = .044**
SHOULDER DIAMETER **G = N/A**
LENGTH TO SHOULDER **J = N/A**
SHOULDER ANGLE (DEG'S/SIDE): **N/A**
PRIMER: **S/P**
CASE CAPACITY (GR'S WATER): **3.32**

DIMENSIONAL DRAWING:

F
A
B
E
D

-NOT ACTUAL SIZE-
-DO NOT SCALE-

CARTRIDGE: 7,7 x 60R

OTHER NAMES:

DIA: .304
BALLISTEK NO: 304A
NAI NO: RMB 13322/4.816

DATA SOURCE: COTW 4th Pg263

HISTORICAL DATA:

NOTES:

LOADING DATA:

BULLET WT./TYPE	POWDER WT./TYPE	VELOCITY ('/SEC)	SOURCE
180/RN	57.8/IMR4350	—	JJD

CASE PREPARATION: **SHELLHOLDER (RCBS):** 14

MAKE FROM: Mauser "A" base (BELL). Thin rim to .07" thick. Anneal case and form through 2-die form set. Trim to length, F/L size & chamfer.

PHYSICAL DATA (INCHES):

CASE TYPE: **Rimmed Bottleneck**
CASE LENGTH **A = 2.49**
HEAD DIAMETER **B = .517**
RIM DIAMETER **D = .607**
NECK DIAMETER **F = .336**
NECK LENGTH **H = .61**
SHOULDER LENGTH **K = .255**
BODY ANGLE (DEG'S/SIDE): **.201**
CASE CAPACITY **CC'S = 5.05**
LOADED LENGTH: **3.20**
BELT DIAMETER **C = N/A**
RIM THICKNESS **E = .07**
SHOULDER DIAMETER **G = .507**
LENGTH TO SHOULDER **J = 1.625**
SHOULDER ANGLE (DEG'S/SIDE): **18.54**
PRIMER: **L/R**
CASE CAPACITY (GR'S WATER): **78.06**

DIMENSIONAL DRAWING:

F
H
K
G
A
J
E
B
D

-NOT ACTUAL SIZE-
-DO NOT SCALE-

CARTRIDGE: 7,7 x 58mm Arisaka

OTHER NAMES:
7,7mm Japanese
7,7mm Jap T-99
.31 Jap

DIA: .312

BALLISTEK NO: 311C

NAI NO: RXB 22343/4.819

DATA SOURCE: Hornady 3rd Pg255

HISTORICAL DATA: Japanese mil. ctg. ca. 1939.

NOTES: Replaced 6,5 Arisaka as service ctg.

LOADING DATA:

BULLET WT./TYPE	POWDER WT./TYPE	VELOCITY ('/SEC)	SOURCE
150/Spire	43.0/IMR4064	2267	Lyman

CASE PREPARATION: SHELLHOLDER (RCBS): 3

MAKE FROM: Factory or .30-06. Run '06 case into 7,7 die with expander removed. Trim to length & chamfer. F/L size.

PHYSICAL DATA (INCHES):

CASE TYPE: **Rimless Bottleneck**
CASE LENGTH **A = 2.27**
HEAD DIAMETER **B = .471**
RIM DIAMETER **D = .471**
NECK DIAMETER **F = .337**
NECK LENGTH **H = .302**
SHOULDER LENGTH **K = .096**
BODY ANGLE (DEG'S/SIDE): **.719**
CASE CAPACITY **CC'S = 4.10**
LOADED LENGTH: **3.15**
BELT DIAMETER **C = N/A**
RIM THICKNESS **E = .049**
SHOULDER DIAMETER **G = .429**
LENGTH TO SHOULDER **J = 1.872**
SHOULDER ANGLE (DEG'S/SIDE): **25.60**
PRIMER: **L/R**
CASE CAPACITY (GR'S WATER): **63.29**

DIMENSIONAL DRAWING:

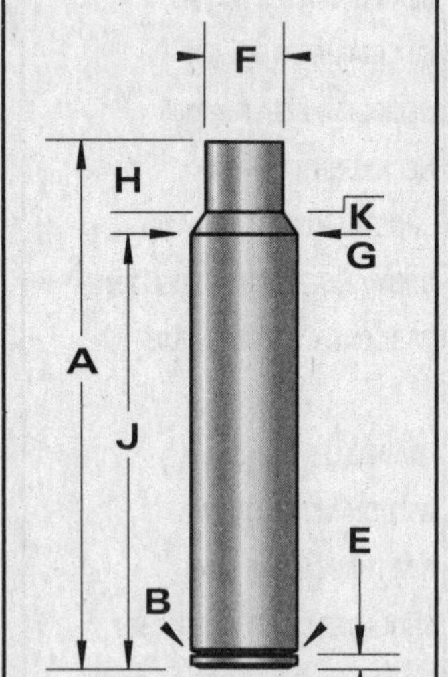

-NOT ACTUAL SIZE-
-DO NOT SCALE-

CARTRIDGE: 7.82 Lazzaroni Patriot

OTHER NAMES:

DIA: .308

BALLISTEK NO:

NAI NO:

DATA SOURCE:

HISTORICAL DATA: Created by John Lazzaroni, cartridge drawing dated 9/11/01.

NOTES: This is the .308 caliber cartridge in the Lazzaroni short action beltless magnum line.

LOADING DATA:

BULLET WT./TYPE	POWDER WT./TYPE	VELOCITY ('/SEC)	SOURCE
150/Nosler Ball. Tip	74/RL-19	3382	Lazzaroni
180/Nosler Ball. Tip	68/IMR 4831	3120	Lazzaroni

CASE PREPARATION: SHELL HOLDER (RCBS): 14

MAKE FROM:

PHYSICAL DATA (INCHES):

CASE TYPE: **Rimless Bottleneck**
CASE LENGTH **A = 2.050**
HEAD DIAMETER **B = .579**
RIM DIAMETER **D = .580**
NECK DIAMETER **F = .344**
NECK LENGTH **H = .312**
SHOULDER LENGTH **K = .185**
BODY ANGLE (DEG'S/SIDE):
CASE CAPACITY **CC'S =**
LOADED LENGTH: **2.800**
BELT DIAMETER **C =**
RIM THICKNESS **E = .065**
SHOULDER DIAMETER **G = .560**
LENGTH TO SHOULDER **J = 1.553**
SHOULDER ANGLE (DEG'S/SIDE): **30**
PRIMER: **L/R or L/R Mag**
CASE CAPACITY (GR'S WATER): **82.7**

DIMENSIONAL DRAWING:

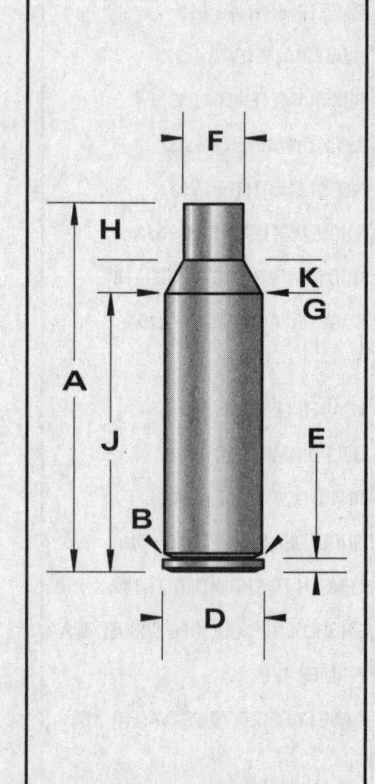

-NOT ACTUAL SIZE-
-DO NOT SCALE-

CARTRIDGE: 7.82 Lazzaroni Warbird

OTHER NAMES:

DIA: .308

BALLISTEK NO:

NAI NO:

DATA SOURCE:

HISTORICAL DATA: Created by John Lazzaroni, cartridge drawing dated 4/5/99.

NOTES: This is the 30 caliber in John Lazzaroni's line of long action beltless magnums.

LOADING DATA:

BULLET WT./TYPE	POWDER WT./TYPE	VELOCITY ('/SEC)	SOURCE
150/Nosler Ball. Tip	103.5/IMR 7828	3745	Lazzaroni
200/Nosler Partition	100.0/RL-25	3358	Lazzaroni

CASE PREPARATION: SHELL HOLDER (RCBS): 14

MAKE FROM:

PHYSICAL DATA (INCHES):

CASE TYPE: **Rimless Bottleneck**

CASE LENGTH A = **2.81**

HEAD DIAMETER B = **.579**

RIM DIAMETER D = **.580**

NECK DIAMETER F = **.345**

NECK LENGTH H = **.312**

SHOULDER LENGTH K = **.185**

BODY ANGLE (DEG'S/SIDE):

CASE CAPACITY CC'S =

LOADED LENGTH: **3.630**

BELT DIAMETER C =

RIM THICKNESS E = **.065**

SHOULDER DIAMETER G = **.560**

LENGTH TO SHOULDER J = **2.313**

SHOULDER ANGLE (DEG'S/SIDE): **30**

PRIMER: **L/R Mag**

CASE CAPACITY (GR'S WATER):

DIMENSIONAL DRAWING:

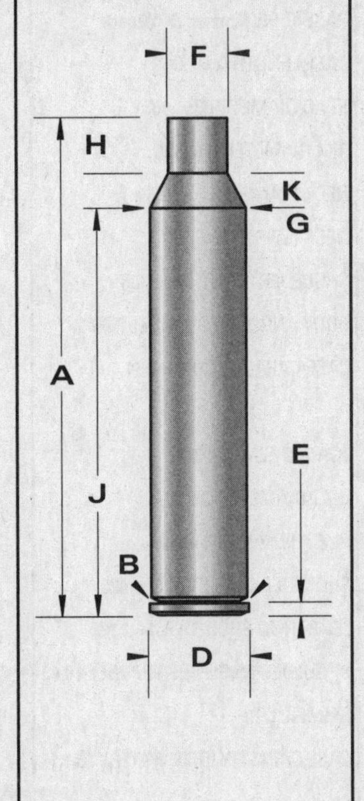

-NOT ACTUAL SIZE-
-DO NOT SCALE-

CARTRIDGE: 7,92mm CETME

OTHER NAMES: 7,92mm Spanish

DIA: .312

BALLISTEK NO: 312J

NAI NO: RXB 23342/3.557

DATA SOURCE: Nonte Pg197

HISTORICAL DATA: By Vorgrimmler for C.E.T.M.E. arsenal.

NOTES:

LOADING DATA:

BULLET WT./TYPE	POWDER WT./TYPE	VELOCITY ('/SEC)	SOURCE
170/Spire	30.0/IMR3031	—	JJD

CASE PREPARATION: SHELLHOLDER (RCBS): 3

MAKE FROM: .308 Win. F/L size Win. case, in 7,92 die, with expander removed. Trim to length & chamfer. I.D. neck ream and clean. F/L size.

PHYSICAL DATA (INCHES):

CASE TYPE: **Rimless Bottleneck**

CASE LENGTH A = **1.665**

HEAD DIAMETER B = **.468**

RIM DIAMETER D = **.466**

NECK DIAMETER F = **.359**

NECK LENGTH H = **.186**

SHOULDER LENGTH K = **.139**

BODY ANGLE (DEG'S/SIDE): **.829**

CASE CAPACITY CC'S = **2.50**

LOADED LENGTH: **3.01**

BELT DIAMETER C = **N/A**

RIM THICKNESS E = **.044**

SHOULDER DIAMETER G = **.435**

LENGTH TO SHOULDER J = **1.34**

SHOULDER ANGLE (DEG'S/SIDE): **15.29**

PRIMER: **L/R**

CASE CAPACITY (GR'S WATER): **38.56**

DIMENSIONAL DRAWING:

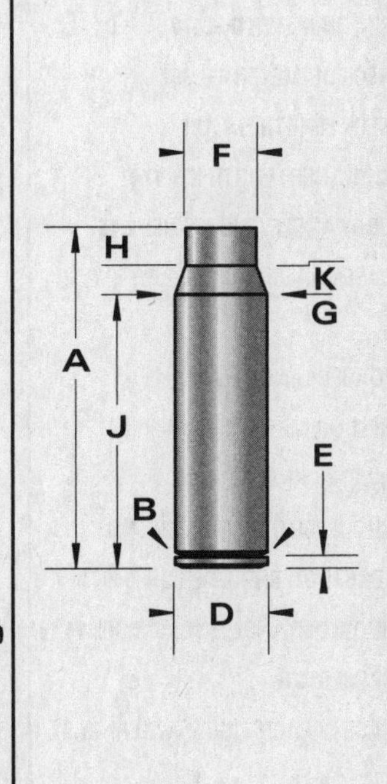

-NOT ACTUAL SIZE-
-DO NOT SCALE-

CARTRIDGE: 7,92 x 33 Kurz

OTHER NAMES:
7,9 x 33mm
7,92mm Pistolen Patrone

DIA: .323

BALLISTEK NO: 323K

NAI NO: RXB
34342/2.766

DATA SOURCE: COTW 4th Pg205

HISTORICAL DATA: 1940 German assault weapon ctg.

NOTES:

LOADING DATA:

BULLET WT./TYPE	POWDER WT./TYPE	VELOCITY ('/SEC)	SOURCE
125/Spire	21.5/IMR4198	2100	JJD

CASE PREPARATION: SHELLHOLDER (RCBS): 3

MAKE FROM: .308 Win. Taper expand neck to .330" dia. Anneal. Form in form die (one die) and cut to length. I.D. neck ream. Chamfer and F/L size.

PHYSICAL DATA (INCHES):

CASE TYPE: **Rimless Bottleneck**
CASE LENGTH **A = 1.300**
HEAD DIAMETER **B = .470**
RIM DIAMETER **D = .470**
NECK DIAMETER **F = .352**
NECK LENGTH **H = .181**
SHOULDER LENGTH **K = .174**
BODY ANGLE (DEG'S/SIDE): **1.15**
CASE CAPACITY **CC'S = 2.07**
LOADED LENGTH: **1.88**
BELT DIAMETER **C = N/A**
RIM THICKNESS **E = .05**
SHOULDER DIAMETER **G = .440**
LENGTH TO SHOULDER **J = .945**
SHOULDER ANGLE (DEG'S/SIDE): **14.19**
PRIMER: **L/R**
CASE CAPACITY (GR'S WATER): **31.97**

DIMENSIONAL DRAWING:

-NOT ACTUAL SIZE-
-DO NOT SCALE-

CARTRIDGE: 7,92 x 57R Schwarzlose

OTHER NAMES:
7,92mm Scherpe Patrone

DIA: .312

BALLISTEK NO: 312K

NAI NO: RMB
22332/5.054

DATA SOURCE: MSAA Pg45

HISTORICAL DATA: Adopted by Dutch in 1908 for Schwarzlose MG.

NOTES:

LOADING DATA:

BULLET WT./TYPE	POWDER WT./TYPE	VELOCITY ('/SEC)	SOURCE
195/Spire	42.5/IMR4320	—	JJD

CASE PREPARATION: SHELLHOLDER (RCBS): 7

MAKE FROM: 7 x 65R Brenneke. Trim case to length & chamfer. Taper expand to about .320" dia. F/L size and fire-form.

PHYSICAL DATA (INCHES):

CASE TYPE: **Rimmed Bottleneck**
CASE LENGTH **A = 2.34**
HEAD DIAMETER **B = .463**
RIM DIAMETER **D = .535**
NECK DIAMETER **F = .355**
NECK LENGTH **H = .350**
SHOULDER LENGTH **K = .105**
BODY ANGLE (DEG'S/SIDE): **.646**
CASE CAPACITY **CC'S = 3.64**
LOADED LENGTH: **3.15**
BELT DIAMETER **C = N/A**
RIM THICKNESS **E = .055**
SHOULDER DIAMETER **G = .425**
LENGTH TO SHOULDER **J = 1.885**
SHOULDER ANGLE (DEG'S/SIDE): **18.43**
PRIMER: **L/R**
CASE CAPACITY (GR'S WATER): **56.19**

DIMENSIONAL DRAWING:

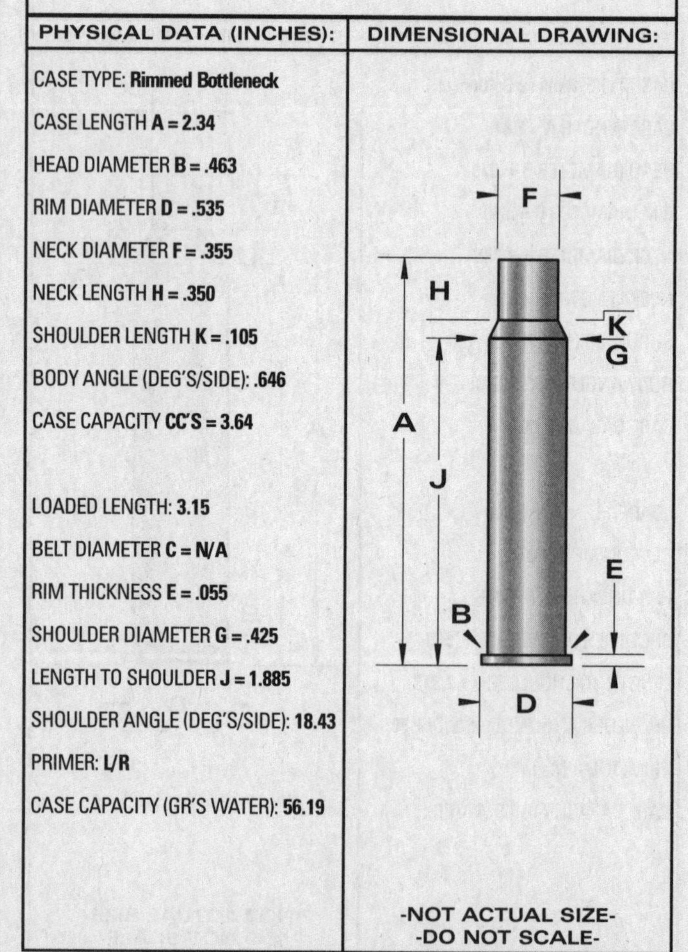

-NOT ACTUAL SIZE-
-DO NOT SCALE-

CARTRIDGE: 7,92 x 61 Norwegian Long

OTHER NAMES: 7,92 x 61 Colt

DIA: .312

BALLISTEK NO: 312L

NAI NO: RBB 22343/5.154

DATA SOURCE: MSAA Pg45

HISTORICAL DATA: For Norwegian M17 Browning MG.

NOTES: About 1938.

LOADING DATA:

BULLET WT./TYPE	POWDER WT./TYPE	VELOCITY ('/SEC)	SOURCE
220/Spire	50.4/H414	—	JJD

CASE PREPARATION: **SHELLHOLDER (RCBS):** 3

MAKE FROM: .30-06 Spgf. F/L size case and trim to length. This could well be nothing more than a European version of the .30-06.

PHYSICAL DATA (INCHES):

CASE TYPE: **Rebated Bottleneck**

CASE LENGTH A = **2.438**

HEAD DIAMETER B = **.473**

RIM DIAMETER D = **.467**

NECK DIAMETER F = **.354**

NECK LENGTH H = **.273**

SHOULDER LENGTH K = **.102**

BODY ANGLE (DEG'S/SIDE): **.492**

CASE CAPACITY CC'S = **4.14**

LOADED LENGTH: **3.35**

BELT DIAMETER C = **N/A**

RIM THICKNESS E = **.052**

SHOULDER DIAMETER G = **.441**

LENGTH TO SHOULDER J = **2.063**

SHOULDER ANGLE (DEG'S/SIDE): **23.09**

PRIMER: **L/R**

CASE CAPACITY (GR'S WATER): **63.91**

DIMENSIONAL DRAWING:

-NOT ACTUAL SIZE-
-DO NOT SCALE-

CARTRIDGE: 8mm Gibbs

OTHER NAMES:

DIA: .323

BALLISTEK NO: 323Q

NAI NO: RXB 12344/5.255

DATA SOURCE: Ackley Vol.2 Pg199

HISTORICAL DATA:

NOTES:

LOADING DATA:

BULLET WT./TYPE	POWDER WT./TYPE	VELOCITY ('/SEC)	SOURCE
170/Spire	67.0/4350	3120	Gibbs/ Ackley

CASE PREPARATION: **SHELLHOLDER (RCBS):** 3

MAKE FROM: .30-06 Spgf. Taper expand neck to .330" dia. F/L size in Gibbs die. Square case mouth, chamfer and fire-form in Gibbs chamber.

PHYSICAL DATA (INCHES):

CASE TYPE: **Rimless Bottleneck**

CASE LENGTH A = **2.47**

HEAD DIAMETER B = **.470**

RIM DIAMETER D = **.473**

NECK DIAMETER F = **.351**

NECK LENGTH H = **.270**

SHOULDER LENGTH K = **.070**

BODY ANGLE (DEG'S/SIDE): **.252**

CASE CAPACITY CC'S = **4.76**

LOADED LENGTH: **3.16**

BELT DIAMETER C = **N/A**

RIM THICKNESS E = **.049**

SHOULDER DIAMETER G = **.453**

LENGTH TO SHOULDER J = **2.13**

SHOULDER ANGLE (DEG'S/SIDE): **36.07**

PRIMER: **L/R**

CASE CAPACITY (GR'S WATER): **73.45**

DIMENSIONAL DRAWING:

-NOT ACTUAL SIZE-
-DO NOT SCALE-

CARTRIDGE: 8mm JDJ

OTHER NAMES:

DIA: .323

BALLISTEK NO:

NAI NO:

DATA SOURCE: J.D. Jones, Frank Barnes

HISTORICAL DATA: Designed by J.D. Jones around 1980, this chambering is the 444 Marlin case necked down to 8mm with no other changes.

NOTES:

LOADING DATA:

BULLET WT./TYPE	POWDER WT./TYPE	VELOCITY ('/SEC)	SOURCE
150 Hornady	47.5 - IMR-4320	2286	SSK
170 Hornady	47.5 - IMR-4320	2254	SSK

CASE PREPARATION: SHELL HOLDER (RCBS): 28

MAKE FROM: 444 Marlin
Neck down 444 Marlin cases.

PHYSICAL DATA (INCHES):

CASE TYPE: **Semi-rimmed Bottleneck**
CASE LENGTH **A = 2.225**
HEAD DIAMETER **B = .470**
RIM DIAMETER **D = .514**
NECK DIAMETER **F = .347**
NECK LENGTH **H = .343**
SHOULDER LENGTH **K = .081**
BODY ANGLE (DEG'S/SIDE):
CASE CAPACITY **CC'S = .946**
LOADED LENGTH: **3.340**
BELT DIAMETER **C = N/A**
RIM THICKNESS **E = .063**
SHOULDER DIAMETER **G = .456**
LENGTH TO SHOULDER **J = 1.801**
SHOULDER ANGLE (DEG'S/SIDE): **40°**
PRIMER: **L/R**
CASE CAPACITY (GR'S WATER):

DIMENSIONAL DRAWING:

F H K G A J E B D

-NOT ACTUAL SIZE-
-DO NOT SCALE-

CARTRIDGE: 8mm Nambu

OTHER NAMES:
8mm Type 14

DIA: .320

BALLISTEK NO: 320A

NAI NO: RMB 24242/2.103

DATA SOURCE: NAI/Ballistek

HISTORICAL DATA: Japanese mil. auto pistol ctg. from about 1904.

NOTES:

LOADING DATA:

BULLET WT./TYPE	POWDER WT./TYPE	VELOCITY ('/SEC)	SOURCE
85/Lead	2.7/Unique	900	JJD

CASE PREPARATION: SHELLHOLDER (RCBS): 25

MAKE FROM: .30 Rem. Cut off case to .840". Use #25 shell holder extension and form in form die. Clean. F/L size & chamfer. I.D. neck ream. F/L size again. Use Lyman bullet #32362 sized .321". (Note: Midway Arms has offered formed cases and jacketed bullets. You may find some around.)

PHYSICAL DATA (INCHES):

CASE TYPE: **Rimmed Bottleneck**
CASE LENGTH **A = .856**
HEAD DIAMETER **B = .407**
RIM DIAMETER **D = .417**
NECK DIAMETER **F = .347**
NECK LENGTH **H = .161**
SHOULDER LENGTH **K = .090**
BODY ANGLE (DEG'S/SIDE): **.990**
CASE CAPACITY **CC'S = .946**
LOADED LENGTH: **1.244**
BELT DIAMETER **C = N/A**
RIM THICKNESS **E = .042**
SHOULDER DIAMETER **G = .393**
LENGTH TO SHOULDER **J = .605**
SHOULDER ANGLE (DEG'S/SIDE): **14.33**
PRIMER: **L/P**
CASE CAPACITY (GR'S WATER): **14.60**

DIMENSIONAL DRAWING:

F H K A G J B E D

-NOT ACTUAL SIZE-
-DO NOT SCALE-

CARTRIDGE: 8mm PMM

OTHER NAMES:

DIA: .323

BALLISTEK NO: 323R

NAI NO: BEN 12342/4.873

DATA SOURCE: Ackley Vol.2 Pg200

HISTORICAL DATA: By J. Ochocki.

NOTES: PMM = Poor Man's Magnum

LOADING DATA:

BULLET WT./TYPE	POWDER WT./TYPE	VELOCITY ('/SEC)	SOURCE
170/Spire	75.0/4350	3000	Ackley

CASE PREPARATION: SHELLHOLDER (RCBS): 4

MAKE FROM: 7mm Rem. Mag. Taper expand to .330" dia. F/L size in PMM die. Square case mouth, chamfer and fire-form.

PHYSICAL DATA (INCHES):

CASE TYPE: **Belted Bottleneck**
CASE LENGTH **A = 2.500**
HEAD DIAMETER **B = .513**
RIM DIAMETER **D = .532**
NECK DIAMETER **F = .354**
NECK LENGTH **H = .290**
SHOULDER LENGTH **K = .217**
BODY ANGLE (DEG'S/SIDE): **.335**
CASE CAPACITY **CC'S = 5.50**

LOADED LENGTH: **A/R**
BELT DIAMETER **C = .532**
RIM THICKNESS **E = .05**
SHOULDER DIAMETER **G = .492**
LENGTH TO SHOULDER **J = 1.993**
SHOULDER ANGLE (DEG'S/SIDE): **17.64**
PRIMER: **L/R Mag**
CASE CAPACITY (GR'S WATER): **84.92**

DIMENSIONAL DRAWING:

-NOT ACTUAL SIZE-
-DO NOT SCALE-

CARTRIDGE: 8mm Rast-Gasser

OTHER NAMES:
8mm Gasser

DIA: .320

BALLISTEK NO: 320C

NAI NO: RMS 11115/3.095

DATA SOURCE: COTW 4th Pg172

HISTORICAL DATA: Austrian mil. revolver ctg. ca. 1873.

NOTES:

LOADING DATA:

BULLET WT./TYPE	POWDER WT./TYPE	VELOCITY ('/SEC)	SOURCE
95/Lead	2.0/B'Eye	800	Barnes

CASE PREPARATION: SHELLHOLDER (RCBS): 10

MAKE FROM: .32 S&W Long. Taper expand case to hold .319-.320" dia bullet. Cases, as such, are short but okay.

PHYSICAL DATA (INCHES):

CASE TYPE: **Rimmed Straight**
CASE LENGTH **A = 1.037**
HEAD DIAMETER **B = .335**
RIM DIAMETER **D = .376**
NECK DIAMETER **F = .332**
NECK LENGTH **H = N/A**
SHOULDER LENGTH **K = N/A**
BODY ANGLE (DEG'S/SIDE): **.086**
CASE CAPACITY **CC'S = .935**

LOADED LENGTH: **1.39**
BELT DIAMETER **C = N/A**
RIM THICKNESS **E = .048**
SHOULDER DIAMETER **G = N/A**
LENGTH TO SHOULDER **J = N/A**
SHOULDER ANGLE (DEG'S/SIDE): **N/A**
PRIMER: **S/P**
CASE CAPACITY (GR'S WATER): **14.44**

DIMENSIONAL DRAWING:

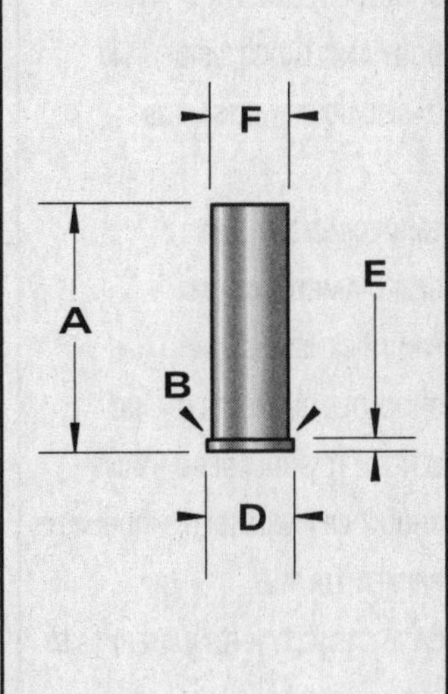

-NOT ACTUAL SIZE-
-DO NOT SCALE-

CARTRIDGE: 8mm Remington Magnum

OTHER NAMES:

DIA: .323

BALLISTEK NO: 323J

NAI NO: BEN 12343/5.555

DATA SOURCE: Hornady Manual 3rd Pg207

HISTORICAL DATA: By Rem. in 1978.

NOTES:

LOADING DATA:

BULLET WT./TYPE	POWDER WT./TYPE	VELOCITY ('/SEC)	SOURCE
170/RN	71.0/IMR4320	3050	JJD

CASE PREPARATION: SHELLHOLDER (RCBS): 4

MAKE FROM: Factory or .375 H&H. Anneal case neck and size in 8mm Mag. sizer with expander removed. Trim to length & chamfer. Shoulder area will fireform upon ignition.

PHYSICAL DATA (INCHES):

CASE TYPE: **Belted Bottleneck**
CASE LENGTH **A = 2.85**
HEAD DIAMETER **B = .513**
RIM DIAMETER **D = .532**
NECK DIAMETER **F = .354**
NECK LENGTH **H = .320**
SHOULDER LENGTH **K = .141**
BODY ANGLE (DEG'S/SIDE): **.340**
CASE CAPACITY **CC'S = 6.38**

LOADED LENGTH: **3.60**
BELT DIAMETER **C = .532**
RIM THICKNESS **E = .05**
SHOULDER DIAMETER **G = .487**
LENGTH TO SHOULDER **J = 2.389**
SHOULDER ANGLE (DEG'S/SIDE): **25.25**
PRIMER: **L/R Mag**
CASE CAPACITY (GR'S WATER): **98.53**

DIMENSIONAL DRAWING:

-NOT ACTUAL SIZE-
-DO NOT SCALE-

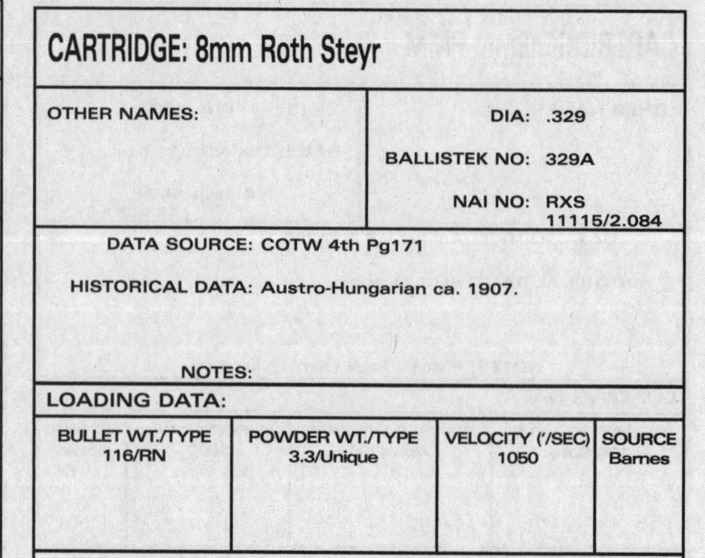

CARTRIDGE: 8mm Roth Steyr

OTHER NAMES:

DIA: .329

BALLISTEK NO: 329A

NAI NO: RXS 11115/2.084

DATA SOURCE: COTW 4th Pg171

HISTORICAL DATA: Austro-Hungarian ca. 1907.

NOTES:

LOADING DATA:

BULLET WT./TYPE	POWDER WT./TYPE	VELOCITY ('/SEC)	SOURCE
116/RN	3.3/Unique	1050	Barnes

CASE PREPARATION: SHELLHOLDER (RCBS): 17

MAKE FROM: .30 Carbine. Cut case to .75". I.D. neck ream. F/L size, length trim & chamfer. Use .323 unsized bullets or minie with .250" dia. drill.

PHYSICAL DATA (INCHES):

CASE TYPE: **Rimless Straight**
CASE LENGTH **A = .740**
HEAD DIAMETER **B = .355**
RIM DIAMETER **D = .356**
NECK DIAMETER **F = .353**
NECK LENGTH **H = N/A**
SHOULDER LENGTH **K = N/A**
BODY ANGLE (DEG'S/SIDE): **.083**
CASE CAPACITY **CC'S = .564**

LOADED LENGTH: **1.14**
BELT DIAMETER **C = N/A**
RIM THICKNESS **E = .046**
SHOULDER DIAMETER **G = N/A**
LENGTH TO SHOULDER **J = N/A**
SHOULDER ANGLE (DEG'S/SIDE): **N/A**
PRIMER: **S/P**
CASE CAPACITY (GR'S WATER): **8.71**

DIMENSIONAL DRAWING:

-NOT ACTUAL SIZE-
-DO NOT SCALE-

CARTRIDGE: 8mm/06 Springfield

OTHER NAMES:

DIA: .323

BALLISTEK NO: 323E

NAI NO: RXB 22323/5.306

DATA SOURCE: Hornady 3rd Pg261

HISTORICAL DATA: Early wildcat.

NOTES: Similar to 8 x 64(S).

LOADING DATA:

BULLET WT./TYPE	POWDER WT./TYPE	VELOCITY ('/SEC)	SOURCE
150/Spire	53.0/IMR4064	2850	JJD

CASE PREPARATION: SHELLHOLDER (RCBS): 3

MAKE FROM: .30-06 Spgf. Taper expand to .330-.335" dia. F/L size. Trim to length & chamfer.

PHYSICAL DATA (INCHES):

CASE TYPE: **Rimless Bottleneck**

CASE LENGTH **A = 2.494**

HEAD DIAMETER **B = .470**

RIM DIAMETER **D = .473**

NECK DIAMETER **F = .351**

NECK LENGTH **H = .427**

SHOULDER LENGTH **K = .119**

BODY ANGLE (DEG'S/SIDE): **.475**

CASE CAPACITY **CC'S = 4.53**

LOADED LENGTH: **3.18**

BELT DIAMETER **C = N/A**

RIM THICKNESS **E = .049**

SHOULDER DIAMETER **G = .441**

LENGTH TO SHOULDER **J = 1.948**

SHOULDER ANGLE (DEG'S/SIDE): **20.71**

PRIMER: **L/R**

CASE CAPACITY (GR'S WATER): **69.90**

DIMENSIONAL DRAWING:

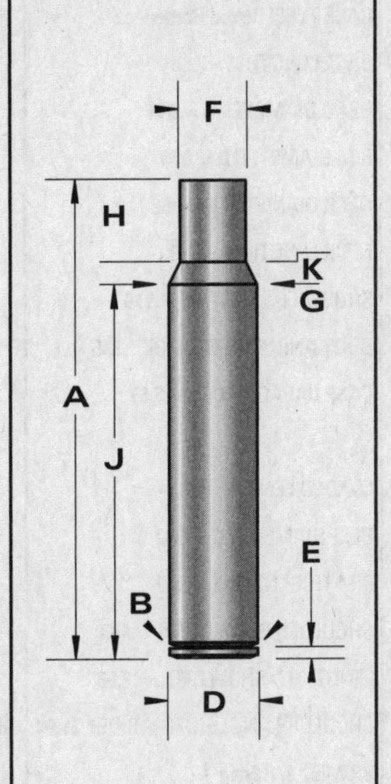

-NOT ACTUAL SIZE-
-DO NOT SCALE-

CARTRIDGE: 8mm/06 Springfield Improved

OTHER NAMES:

DIA: .323

BALLISTEK NO: 323S

NAI NO: RXB 12334/5.255

DATA SOURCE: NAI/Ballistek

HISTORICAL DATA: This is Ackley's version.

NOTES:

LOADING DATA:

BULLET WT./TYPE	POWDER WT./TYPE	VELOCITY ('/SEC)	SOURCE
150/Spire	55.0/IMR4064	–	JJD

CASE PREPARATION: SHELLHOLDER (RCBS): 3

MAKE FROM: .30-06 Spgf. Taper expand case neck to .330" dia. F/L size in "improved" die, square case mouth and fire-form.

PHYSICAL DATA (INCHES):

CASE TYPE: **Rimless Bottleneck**

CASE LENGTH **A = 2.47**

HEAD DIAMETER **B = .470**

RIM DIAMETER **D = .473**

NECK DIAMETER **F = .350**

NECK LENGTH **H = .380**

SHOULDER LENGTH **K = .07**

BODY ANGLE (DEG'S/SIDE): **.314**

CASE CAPACITY **CC'S = 4.66**

LOADED LENGTH: **3.25**

BELT DIAMETER **C = N/A**

RIM THICKNESS **E = .05**

SHOULDER DIAMETER **G = .450**

LENGTH TO SHOULDER **J = 2.02**

SHOULDER ANGLE (DEG'S/SIDE): **35.54**

PRIMER: **L/R**

CASE CAPACITY (GR'S WATER): **71.87**

DIMENSIONAL DRAWING:

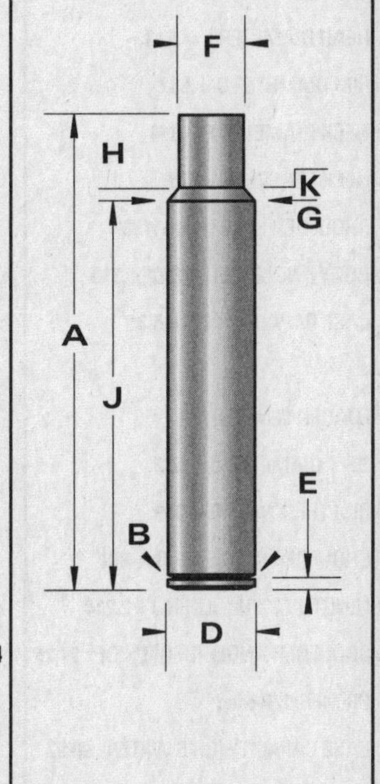

-NOT ACTUAL SIZE-
-DO NOT SCALE-

CARTRIDGE: 8mm/300 Winchester Magnum

OTHER NAMES:

DIA: .323

BALLISTEK NO: 323T

NAI NO: BEN 12343/5.107

DATA SOURCE: NAI/Ballistek

HISTORICAL DATA:

NOTES:

LOADING DATA:

BULLET WT./TYPE	POWDER WT./TYPE	VELOCITY ('/SEC)	SOURCE
220/RN	73.3/IMR4831	—	JJD

CASE PREPARATION: SHELLHOLDER (RCBS): 4

MAKE FROM: 8mm Rem. Mag. F/L size in 8mm/300 die to set shoulder back. Trim to length, chamfer.

PHYSICAL DATA (INCHES):

CASE TYPE: **Belted Bottleneck**
CASE LENGTH **A = 2.62**
HEAD DIAMETER **B = .513**
RIM DIAMETER **D = .532**
NECK DIAMETER **F = .354**
NECK LENGTH **H = .273**
SHOULDER LENGTH **K = .133**
BODY ANGLE (DEG'S/SIDE): **.313**
CASE CAPACITY **CC'S = 5.89**

LOADED LENGTH: **A/R**
BELT DIAMETER **C = .532**
RIM THICKNESS **E = .049**
SHOULDER DIAMETER **G = .491**
LENGTH TO SHOULDER **J = 2.214**
SHOULDER ANGLE (DEG'S/SIDE): **27.25**
PRIMER: **L/R Mag**
CASE CAPACITY (GR'S WATER): **90.92**

DIMENSIONAL DRAWING:

F H K G A J E B D C

-NOT ACTUAL SIZE-
-DO NOT SCALE-

CARTRIDGE: 8mm/308 Norma Magnum

OTHER NAMES:

DIA: .323

BALLISTEK NO: 323U

NAI NO: BEN 12343/5.009

DATA SOURCE: NAI/Ballistek

HISTORICAL DATA:

NOTES: Pretty much an 8mm Rem. Mag.

LOADING DATA:

BULLET WT./TYPE	POWDER WT./TYPE	VELOCITY ('/SEC)	SOURCE
220/Spire	69.8/IMR4831	—	JJD

CASE PREPARATION: SHELLHOLDER (RCBS): 4

MAKE FROM: 8mm Rem. Mag. Size the Rem. case in the 8mm/308 F/L die, with the expander removed. Trim to length, chamfer and F/L size again.

PHYSICAL DATA (INCHES):

CASE TYPE: **Belted Bottleneck**
CASE LENGTH **A = 2.56**
HEAD DIAMETER **B = .511**
RIM DIAMETER **D = .530**
NECK DIAMETER **F = .352**
NECK LENGTH **H = .310**
SHOULDER LENGTH **K = .170**
BODY ANGLE (DEG'S/SIDE): **.335**
CASE CAPACITY **CC'S = 5.67**

LOADED LENGTH: **A/R**
BELT DIAMETER **C = .530**
RIM THICKNESS **E = .049**
SHOULDER DIAMETER **G = .489**
LENGTH TO SHOULDER **J = 2.08**
SHOULDER ANGLE (DEG'S/SIDE): **21.94**
PRIMER: **L/R Mag**
CASE CAPACITY (GR'S WATER): **87.63**

DIMENSIONAL DRAWING:

F H K G A J E B D C

-NOT ACTUAL SIZE-
-DO NOT SCALE-

CARTRIDGE: 8mm/338 Winchester Magnum

OTHER NAMES:

DIA: .323

BALLISTEK NO: 323G

NAI NO: BEN 22333/4.873

DATA SOURCE: NAI/Ballistek

HISTORICAL DATA:

NOTES:

LOADING DATA:

BULLET WT./TYPE	POWDER WT./TYPE	VELOCITY ('/SEC)	SOURCE
220/RN	67.0/IMR4350	—	JJD

CASE PREPARATION: SHELLHOLDER (RCBS): 4

MAKE FROM: .338 Win. Mag. F/L size Win. case and trim to length. You may also set the shoulder back on the 8mm Rem. Mag. case.

PHYSICAL DATA (INCHES):

CASE TYPE: **Belted Bottleneck**

CASE LENGTH **A = 2.500**

HEAD DIAMETER **B = .513**

RIM DIAMETER **D = .532**

NECK DIAMETER **F = .354**

NECK LENGTH **H = .325**

SHOULDER LENGTH **K = .135**

BODY ANGLE (DEG'S/SIDE): **.358**

CASE CAPACITY **CC'S = 5.489**

LOADED LENGTH: **A/R**

BELT DIAMETER **C = .532**

RIM THICKNESS **E = .049**

SHOULDER DIAMETER **G = .490**

LENGTH TO SHOULDER **J = 2.04**

SHOULDER ANGLE (DEG'S/SIDE): **26.73**

PRIMER: **L/R Mag**

CASE CAPACITY (GR'S WATER): **84.71**

DIMENSIONAL DRAWING:

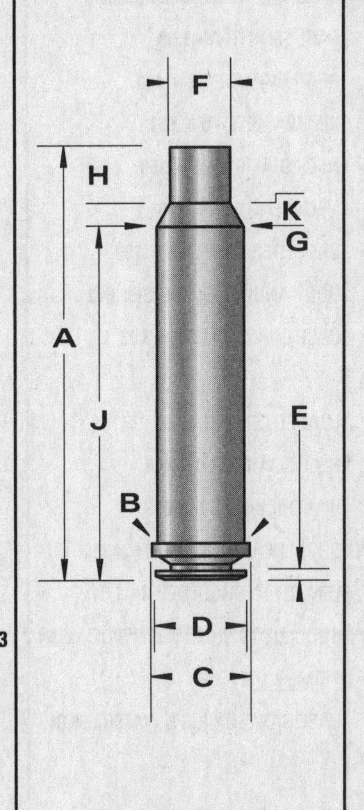

-NOT ACTUAL SIZE-
-DO NOT SCALE-

CARTRIDGE: 8 x 42R

OTHER NAMES:

DIA: .318

BALLISTEK NO: 318G

NAI NO: RMB 34333/3.547

DATA SOURCE: COTW 4th Pg264

HISTORICAL DATA: Circa 1888.

NOTES:

LOADING DATA:

BULLET WT./TYPE	POWDER WT./TYPE	VELOCITY ('/SEC)	SOURCE
175/—	22.0/4227	1580	Barnes

CASE PREPARATION: SHELLHOLDER (RCBS): 26

MAKE FROM: 9,3 x 74R. Cut case to 1.7". Anneal and F/L size with the expander removed. I.D. neck ream. Trim to length, chamfer and F/L size again.

PHYSICAL DATA (INCHES):

CASE TYPE: **Rimmed Bottleneck**

CASE LENGTH **A = 1.66**

HEAD DIAMETER **B = .468**

RIM DIAMETER **D = .525**

NECK DIAMETER **F = .347**

NECK LENGTH **H = .350**

SHOULDER LENGTH **K = .09**

BODY ANGLE (DEG'S/SIDE): **1.26**

CASE CAPACITY **CC'S = 2.63**

LOADED LENGTH: **2.28**

BELT DIAMETER **C = N/A**

RIM THICKNESS **E = .054**

SHOULDER DIAMETER **G = .423**

LENGTH TO SHOULDER **J = 1.22**

SHOULDER ANGLE (DEG'S/SIDE): **22.89**

PRIMER: **L/R**

CASE CAPACITY (GR'S WATER): **40.59**

DIMENSIONAL DRAWING:

F
H
K
G
A
J
E
B
D

-NOT ACTUAL SIZE-
-DO NOT SCALE-

CARTRIDGE: 8 x 48R Sauer

OTHER NAMES:

DIA: .316

BALLISTEK NO: 316B

NAI NO: RMS 31115/4.351

DATA SOURCE: COTW 4th Pg264

HISTORICAL DATA: Early black powder ctg.

NOTES:

LOADING DATA:

BULLET WT./TYPE	POWDER WT./TYPE	VELOCITY ('/SEC)	SOURCE
150/Lead	16.5/IMR4198	1330	JJD

CASE PREPARATION: SHELLHOLDER (RCBS): 21

MAKE FROM: .303 Savage. Taper expand the Savage case to about .325" dia. Trim to 1.9", chamfer and F/L size. Finish trim. Use paper patched .30 cal. lead bullets.

PHYSICAL DATA (INCHES):

CASE TYPE: **Rimmed Straight**

CASE LENGTH **A = 1.88**

HEAD DIAMETER **B = .432**

RIM DIAMETER **D = .500**

NECK DIAMETER **F = .344**

NECK LENGTH **H = N/A**

SHOULDER LENGTH **K = N/A**

BODY ANGLE (DEG'S/SIDE): **1.38**

CASE CAPACITY **CC'S = 2.43**

LOADED LENGTH: **2.58**

BELT DIAMETER **C = N/A**

RIM THICKNESS **E = .06**

SHOULDER DIAMETER **G = N/A**

LENGTH TO SHOULDER **J = N/A**

SHOULDER ANGLE (DEG'S/SIDE): **N/A**

PRIMER: **L/R**

CASE CAPACITY (GR'S WATER): **37.51**

DIMENSIONAL DRAWING:

-NOT ACTUAL SIZE-
-DO NOT SCALE-

CARTRIDGE: 8 x 50R Austrian Mannlicher M93

OTHER NAMES:

DIA: .323

BALLISTEK NO: 323BL

NAI NO: RMB 23342/4.041

DATA SOURCE: Smith Pg102

HISTORICAL DATA: Austrian mil. ctg. circa 1893.

NOTES:

LOADING DATA:

BULLET WT./TYPE	POWDER WT./TYPE	VELOCITY ('/SEC)	SOURCE
224/Lead	27.9/IMR4320	–	JJD

CASE PREPARATION: SHELLHOLDER (RCBS): 13

MAKE FROM: 7,62 x 54R Russian. Turn rim to .551" dia. and back chamfer. Thin rim to .047" thick. Taper expand to .330" dia. Trim to length & chamfer. F/L size. Fireform in chamber.

PHYSICAL DATA (INCHES):

CASE TYPE: **Rimmed Bottleneck**

CASE LENGTH **A = 1.98**

HEAD DIAMETER **B = .490**

RIM DIAMETER **D = .551**

NECK DIAMETER **F = .350**

NECK LENGTH **H = .270**

SHOULDER LENGTH **K = .199**

BODY ANGLE (DEG'S/SIDE): **.612**

CASE CAPACITY **CC'S = 3.82**

LOADED LENGTH: **2.99**

BELT DIAMETER **C = N/A**

RIM THICKNESS **E = .047**

SHOULDER DIAMETER **G = .462**

LENGTH TO SHOULDER **J = 1.51**

SHOULDER ANGLE (DEG'S/SIDE): **15.64**

PRIMER: **L/R**

CASE CAPACITY (GR'S WATER): **59.01**

DIMENSIONAL DRAWING:

-NOT ACTUAL SIZE-
-DO NOT SCALE-

CARTRIDGE: 8 x 50R Lebel

OTHER NAMES:
8mm M86 French

DIA: .323

BALLISTEK NO: 323A

NAI NO: RMB 34333/4.694

DATA SOURCE: COTW 4th Pg206

HISTORICAL DATA: First smallbore, smokeless, mil. cartridge.

NOTES: French development.

LOADING DATA:

BULLET WT./TYPE	POWDER WT./TYPE	VELOCITY ('/SEC)	SOURCE
170/Spire	47.5/IMR4895	2410	JJD

CASE PREPARATION: **SHELLHOLDER (RCBS): 5**

MAKE FROM: .348 Win. Do not anneal brass! Form in 8x50R tirm die. Cut to length. Lathe turn base to uniform .536" dia. F/L resize & chamfer.

PHYSICAL DATA (INCHES):

CASE TYPE: **Rimmed Bottleneck**

CASE LENGTH **A = 1.98**

HEAD DIAMETER **B = .536**

RIM DIAMETER **D = .621**

NECK DIAMETER **F = .347**

NECK LENGTH **H = .345**

SHOULDER LENGTH **K = .156**

BODY ANGLE (DEG'S/SIDE): **1.18**

CASE CAPACITY **CC'S = 4.33**

LOADED LENGTH: **2.75**

BELT DIAMETER **C = N/A**

RIM THICKNESS **E = .065**

SHOULDER DIAMETER **G = .483**

LENGTH TO SHOULDER **J = 1.479**

SHOULDER ANGLE (DEG'S/SIDE): **23.55**

PRIMER: **L/R**

CASE CAPACITY (GR'S WATER): **66.93**

DIMENSIONAL DRAWING:

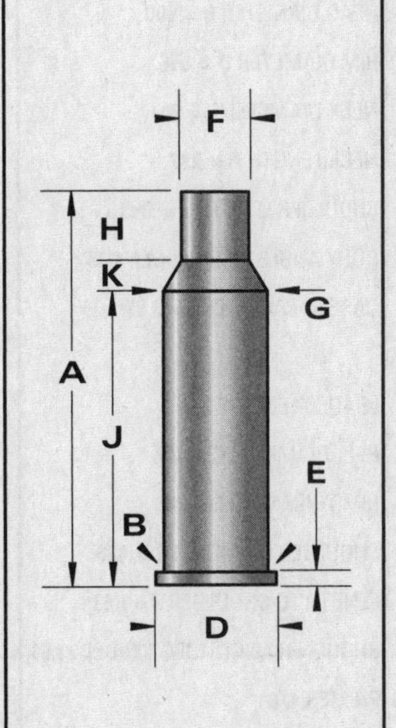

-NOT ACTUAL SIZE-
-DO NOT SCALE-

CARTRIDGE: 8 x 51mm Mauser

OTHER NAMES:

DIA: .316

BALLISTEK NO: 316C

NAI NO: RXB 23342/4.239

DATA SOURCE: COTW 4th Pg264

HISTORICAL DATA: For short-action Mausers about 1890.

NOTES:

LOADING DATA:

BULLET WT./TYPE	POWDER WT./TYPE	VELOCITY ('/SEC)	SOURCE
150/—	41.0/4064	2350	Barnes

CASE PREPARATION: **SHELLHOLDER (RCBS): 3**

MAKE FROM: .308 Win. Taper expand to .325" dia. Trim to length, chamfer and F/L size.

PHYSICAL DATA (INCHES):

CASE TYPE: **Rimless Bottleneck**

CASE LENGTH **A = 1.98**

HEAD DIAMETER **B = .467**

RIM DIAMETER **D = .467**

NECK DIAMETER **F = .344**

NECK LENGTH **H = .305**

SHOULDER LENGTH **K = .165**

BODY ANGLE (DEG'S/SIDE): **.678**

CASE CAPACITY **CC'S = 3.40**

LOADED LENGTH: **2.67**

BELT DIAMETER **C = N/A**

RIM THICKNESS **E = .052**

SHOULDER DIAMETER **G = .436**

LENGTH TO SHOULDER **J = 1.51**

SHOULDER ANGLE (DEG'S/SIDE): **15.58**

PRIMER: **L/R**

CASE CAPACITY (GR'S WATER): **52.50**

DIMENSIONAL DRAWING:

-NOT ACTUAL SIZE-
-DO NOT SCALE-

CARTRIDGE: 8 x 51R Mauser

OTHER NAMES:

DIA: .316

BALLISTEK NO: 316D

NAI NO: RMB 23342/4.239

DATA SOURCE: COTW 4th Pg264

HISTORICAL DATA: Rimmed version of 8x51 Mauser (#316C) from about 1890.

NOTES:

LOADING DATA:

BULLET WT./TYPE	POWDER WT./TYPE	VELOCITY ('/SEC)	SOURCE
150/—	41.0/4064	2350	Barnes

CASE PREPARATION: SHELLHOLDER (RCBS): 18

MAKE FROM: 9,3 x 74R. Turn rim to .515" dia. and back chamfer. Anneal case and trim to 2.0". F/L size, length trim and chamfer. Fireform.

PHYSICAL DATA (INCHES):

CASE TYPE: **Rimmed Bottleneck**

CASE LENGTH **A = 1.98**

HEAD DIAMETER **B = .467**

RIM DIAMETER **D = .515**

NECK DIAMETER **F = .344**

NECK LENGTH **H = .305**

SHOULDER LENGTH **K = .165**

BODY ANGLE (DEG'S/SIDE): **.678**

CASE CAPACITY **CC'S = 3.40**

LOADED LENGTH: **2.68**

BELT DIAMETER **C = N/A**

RIM THICKNESS **E = .055**

SHOULDER DIAMETER **G = .436**

LENGTH TO SHOULDER **J = 1.51**

SHOULDER ANGLE (DEG'S/SIDE): **15.57**

PRIMER: **L/R**

CASE CAPACITY (GR'S WATER): **52.50**

DIMENSIONAL DRAWING:

F H K G A J E B D

-NOT ACTUAL SIZE-
-DO NOT SCALE-

CARTRIDGE: 8 x 52R Siamese T66

OTHER NAMES:

DIA: .323

BALLISTEK NO: 323F

NAI NO: RMB 32343/4.040

DATA SOURCE: MSAA Pg36

HISTORICAL DATA: Designed for Siamese mil. T66 bolt rifle.

NOTES: Manufactured in Thailand and England.

LOADING DATA:

BULLET WT./TYPE	POWDER WT./TYPE	VELOCITY ('/SEC)	SOURCE
180/Spire	40.0/IMR4320	2000	JJD

CASE PREPARATION: SHELLHOLDER (RCBS): 14

MAKE FROM: .45-70 Gov't. Anneal case and run into F/L .40-65 WCF and .33 WCF dies (a form set will do the same job). Turn rim to .556" dia. and back chamfer. F/L size in 8x52R die. Case may (usually) need I.D. neck reaming. F/L size again.

PHYSICAL DATA (INCHES):

CASE TYPE: **Rimmed Bottleneck**

CASE LENGTH **A = 2.02**

HEAD DIAMETER **B = .500**

RIM DIAMETER **D = .556**

NECK DIAMETER **F = .347**

NECK LENGTH **H = .297**

SHOULDER LENGTH **K = .093**

BODY ANGLE (DEG'S/SIDE): **1.48**

CASE CAPACITY **CC'S = 3.81**

LOADED LENGTH: **2.95**

BELT DIAMETER **C = N/A**

RIM THICKNESS **E = .065**

SHOULDER DIAMETER **G = .426**

LENGTH TO SHOULDER **J = 1.63**

SHOULDER ANGLE (DEG'S/SIDE): **23.01**

PRIMER: **L/R**

CASE CAPACITY (GR'S WATER): **58.80**

DIMENSIONAL DRAWING:

F H K G A J E B D

-NOT ACTUAL SIZE-
-DO NOT SCALE-

CARTRIDGE: 8 x 53R Murata

OTHER NAMES:
8mm-20 Meiji

DIA: .329

BALLISTEK NO: 329C

NAI NO: RMB
13333/4.187

DATA SOURCE: COTW 4th Pg209

HISTORICAL DATA: Japanese mil. ctg. from 1889 to 1897.

NOTES:

LOADING DATA:

BULLET WT./TYPE	POWDER WT./TYPE	VELOCITY ('/SEC)	SOURCE
250/RN	47.4/IMR4320	–	JJD

CASE PREPARATION: SHELLHOLDER (RCBS): 13

MAKE FROM: 7,62 x 54R Russian. Turn rim to .558" dia. & back chamfer. Taper expand to .335" dia. and cut case to 2.1". F/L size in 8x53R die, trim to length and fireform.

PHYSICAL DATA (INCHES):

CASE TYPE: **Rimmed Bottleneck**

CASE LENGTH **A = 2.06**

HEAD DIAMETER **B = .492**

RIM DIAMETER **D = .558**

NECK DIAMETER **F = .361**

NECK LENGTH **H = .360**

SHOULDER LENGTH **K = .128**

BODY ANGLE (DEG'S/SIDE): **.146**

CASE CAPACITY **CC'S = 4.07**

LOADED LENGTH: **2.90**

BELT DIAMETER **C = N/A**

RIM THICKNESS **E = .07**

SHOULDER DIAMETER **G = .485**

LENGTH TO SHOULDER **J = 1.572**

SHOULDER ANGLE (DEG'S/SIDE): **25.84**

PRIMER: **L/R**

CASE CAPACITY (GR'S WATER): **62.84**

DIMENSIONAL DRAWING:

-NOT ACTUAL SIZE-
-DO NOT SCALE-

CARTRIDGE: 8 x 54 Krag-Jorgensen

OTHER NAMES:

DIA: .323

BALLISTEK NO: 323BB

NAI NO: RXB
22333/4.435

DATA SOURCE: COTW 4th Pg266

HISTORICAL DATA:

NOTES:

LOADING DATA:

BULLET WT./TYPE	POWDER WT./TYPE	VELOCITY ('/SEC)	SOURCE
150/–	52.0/3031	2850	Barnes

CASE PREPARATION: SHELLHOLDER (RCBS): 3

MAKE FROM: 6,5 x 55mm. Taper expand case to .330" dia. (anneal at .290-.300" dia.). Trim to length, chamfer and F/L size. Fireform.

PHYSICAL DATA (INCHES):

CASE TYPE: **Rimless Bottleneck**

CASE LENGTH **A = 2.12**

HEAD DIAMETER **B = .478**

RIM DIAMETER **D = .478**

NECK DIAMETER **F = .351**

NECK LENGTH **H = .343**

SHOULDER LENGTH **K = .077**

BODY ANGLE (DEG'S/SIDE): **.82**

CASE CAPACITY **CC'S = 3.81**

LOADED LENGTH: **2.85**

BELT DIAMETER **C = N/A**

RIM THICKNESS **E = .051**

SHOULDER DIAMETER **G = .435**

LENGTH TO SHOULDER **J = 1.70**

SHOULDER ANGLE (DEG'S/SIDE): **28.61**

PRIMER: **L/R**

CASE CAPACITY (GR'S WATER): **58.81**

DIMENSIONAL DRAWING:

-NOT ACTUAL SIZE-
-DO NOT SCALE-

CARTRIDGE: 8 x 56 Mannlicher-Schoenauer

OTHER NAMES:

DIA: .323

BALLISTEK NO: 323BC

NAI NO: RXB 22334/4.752

DATA SOURCE: COTW 4th Pg267

HISTORICAL DATA: From about 1908.

NOTES:

LOADING DATA:

BULLET WT./TYPE	POWDER WT./TYPE	VELOCITY ('/SEC)	SOURCE
200/RN	40.0/3031	2050	Barnes

CASE PREPARATION: SHELLHOLDER (RCBS): 3

MAKE FROM: 9,3 x 57mm. F/L size the 9,3mm case in the 8x56 die. Trim to length and chamfer.

PHYSICAL DATA (INCHES):

CASE TYPE: **Rimless Bottleneck**

CASE LENGTH **A = 2.21**

HEAD DIAMETER **B = .465**

RIM DIAMETER **D = .470**

NECK DIAMETER **F = .347**

NECK LENGTH **H = .375**

SHOULDER LENGTH **K = .035**

BODY ANGLE (DEG'S/SIDE): **.734**

CASE CAPACITY **CC'S = 3.90**

LOADED LENGTH: **2.97**

BELT DIAMETER **C = N/A**

RIM THICKNESS **E = .05**

SHOULDER DIAMETER **G = .424**

LENGTH TO SHOULDER **J = 1.80**

SHOULDER ANGLE (DEG'S/SIDE): **47.72**

PRIMER: **L/R**

CASE CAPACITY (GR'S WATER): **60.23**

DIMENSIONAL DRAWING:

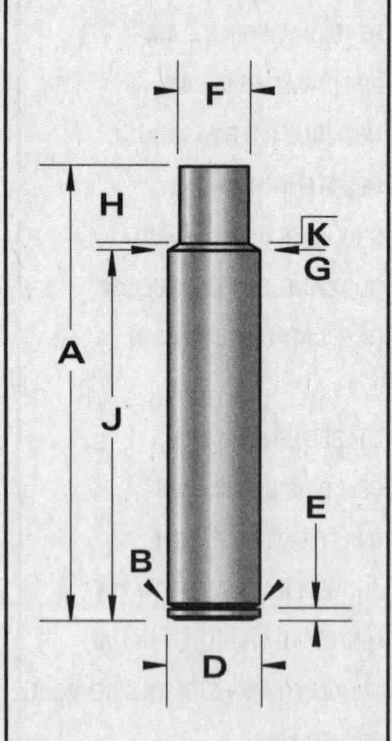

-NOT ACTUAL SIZE-
-DO NOT SCALE-

CARTRIDGE: 8 x 56R Hungarian Mannlicher

OTHER NAMES:
8mm M31 Hungarian
8mm M30 Austrian

DIA: .328

BALLISTEK NO: 329B

NAI NO: RMB 23341/4.501

DATA SOURCE: MSAA Pg37

HISTORICAL DATA: Hungarian mil. ctg. about 1930.

NOTES:

LOADING DATA:

BULLET WT./TYPE	POWDER WT./TYPE	VELOCITY ('/SEC)	SOURCE
205/—	45.0/3031	2300	Barnes

CASE PREPARATION: SHELLHOLDER (RCBS): 13

MAKE FROM: 7,62 x 54R Russian. Turn rim to .554" dia. & back chamfer. Taper expand to about .335" dia. and cut to 2.25". Chamfer. F/L szie in 8x56R die. Trim to length & chamfer. Fireform in chamber.

PHYSICAL DATA (INCHES):

CASE TYPE: **Rimmed Bottleneck**

CASE LENGTH **A = 2.21**

HEAD DIAMETER **B = .491**

RIM DIAMETER **D = .554**

NECK DIAMETER **F = .365**

NECK LENGTH **H = .300**

SHOULDER LENGTH **K = .370**

BODY ANGLE (DEG'S/SIDE): **.384**

CASE CAPACITY **CC'S = 4.15**

LOADED LENGTH: **3.02**

BELT DIAMETER **C = N/A**

RIM THICKNESS **E = .056**

SHOULDER DIAMETER **G = .473**

LENGTH TO SHOULDER **J = 1.54**

SHOULDER ANGLE (DEG'S/SIDE): **8.30**

PRIMER: **L/R**

CASE CAPACITY (GR'S WATER): **64.01**

DIMENSIONAL DRAWING:

-NOT ACTUAL SIZE-
-DO NOT SCALE-

CARTRIDGE: 8 x 56R Kropatchek

OTHER NAMES:

DIA: .323

BALLISTEK NO: 323M

NAI NO: RMB 33322/4.122

DATA SOURCE: Hoyem Vol.2

HISTORICAL DATA: For Kropatchek tubular mag. rifle about 1886.

NOTES:

LOADING DATA:

BULLET WT./TYPE	POWDER WT./TYPE	VELOCITY ('/SEC)	SOURCE
247/RN	56.0/IMR4350	–	JJD

CASE PREPARATION: SHELLHOLDER (RCBS): 5

MAKE FROM: .450 N.E. (BELL). Trim case to 2.25" and anneal. Form in form die set or 8x50R Lebel form die set (watch shoulder location). F/L size and I.D. neck ream. Chamfer and F/L size again. Increase rim thickness if necessary.

PHYSICAL DATA (INCHES):

CASE TYPE: **Rimmed Bottleneck**

CASE LENGTH **A = 2.23**

HEAD DIAMETER **B = .541**

RIM DIAMETER **D = .618**

NECK DIAMETER **F = .349**

NECK LENGTH **H = .445**

SHOULDER LENGTH **K = .235**

BODY ANGLE (DEG'S/SIDE): **1.18**

CASE CAPACITY **CC'S = 4.84**

LOADED LENGTH: **3.19**

BELT DIAMETER **C = N/A**

RIM THICKNESS **E = .075**

SHOULDER DIAMETER **G = .485**

LENGTH TO SHOULDER **J = 1.55**

SHOULDER ANGLE (DEG'S/SIDE): **16.13**

PRIMER: **L/R**

CASE CAPACITY (GR'S WATER): **74.71**

DIMENSIONAL DRAWING:

F H K G A J E B D

-NOT ACTUAL SIZE-
-DO NOT SCALE-

CARTRIDGE: 8 x 57 Mauser

OTHER NAMES:
8 x 57(s) Mauser
8mm (7.92mm) Mauser

DIA: .323

BALLISTEK NO: 323B

NAI NO: RXB 22343/4.776

DATA SOURCE: Hornady Manual 3rd Pg258

HISTORICAL DATA: German mil. ctg. ca. 1888.

NOTES: Watch bore dia.! S&JS=.323; J=.318. When in doubt, slug the bore!

LOADING DATA:

BULLET WT./TYPE	POWDER WT./TYPE	VELOCITY ('/SEC)	SOURCE
125/Spire	52.0/IMR4895	2890	Barnes

CASE PREPARATION: SHELLHOLDER (RCBS): 3

MAKE FROM: .30-06 or factory. Taper expand the '06 case to .325-.330" dia. Run into 8.57 die with expander removed (this will set shoulder back). Trim to 2.24" and F/L size, with expander in place. .270 Win. cases may be used in the same manner. Further, you can cut the rim off from .444 Marlin, cut new extractor groove, anneal and F/L size (form die may be required).

PHYSICAL DATA (INCHES):

CASE TYPE: **Rimless Bottleneck**

CASE LENGTH **A = 2.240**

HEAD DIAMETER **B = .469**

RIM DIAMETER **D = .473**

NECK DIAMETER **F = .349**

NECK LENGTH **H = .307**

SHOULDER LENGTH **K = .102**

BODY ANGLE (DEG'S/SIDE): **.614**

CASE CAPACITY **CC'S = 4.06**

LOADED LENGTH: **2.99**

BELT DIAMETER **C = N/A**

RIM THICKNESS **E = .049**

SHOULDER DIAMETER **G = .434**

LENGTH TO SHOULDER **J = 1.831**

SHOULDER ANGLE (DEG'S/SIDE): **22.62**

PRIMER: **L/R**

CASE CAPACITY (GR'S WATER): **62.68**

DIMENSIONAL DRAWING:

F H K G A J E B D

-NOT ACTUAL SIZE-
-DO NOT SCALE-

CARTRIDGE: 8 x 57.5R Rubin

OTHER NAMES:	DIA: .323 BALLISTEK NO: 323BR NAI NO: RMB 34231/3.969

DATA SOURCE: Hoyem Vol.2 Pg185

HISTORICAL DATA: Swiss-Rubin ctg. from about 1883.

NOTES: Similar to 8x60R Guedes (#326A).

LOADING DATA:

BULLET WT./TYPE	POWDER WT./TYPE	VELOCITY ('/SEC)	SOURCE
250/RN	21.0/IMR4198	—	JJD

CASE PREPARATION: SHELLHOLDER (RCBS): Spl.

MAKE FROM: .500 N.E. (BELL). Turn rim to .639" dia. and build up to .062" thick. Cut case to 2.3" and anneal. Form in form die set (2-dies) and trim to length. F/L size.

PHYSICAL DATA (INCHES):

CASE TYPE: **Rimmed Bottleneck**
CASE LENGTH **A = 2.27**
HEAD DIAMETER **B = .572**
RIM DIAMETER **D = .639**
NECK DIAMETER **F = .358**
NECK LENGTH **H = .350**
SHOULDER LENGTH **K = .590**
BODY ANGLE (DEG'S/SIDE): **1.14**
CASE CAPACITY **CC'S = 5.20**

LOADED LENGTH: **3.16**
BELT DIAMETER **C = N/A**
RIM THICKNESS **E = .062**
SHOULDER DIAMETER **G = .527**
LENGTH TO SHOULDER **J = 1.33**
SHOULDER ANGLE (DEG'S/SIDE): **8.15**
PRIMER: **L/R**
CASE CAPACITY (GR'S WATER): **80.27**

DIMENSIONAL DRAWING:

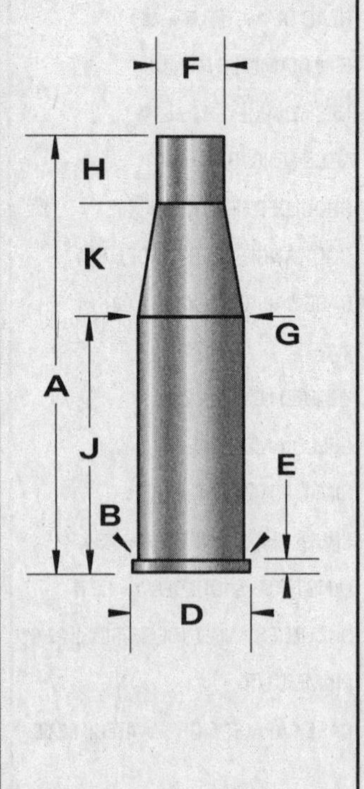

-NOT ACTUAL SIZE-
-DO NOT SCALE-

CARTRIDGE: 8 x 59mm Breda

OTHER NAMES: 8mm Italian M35	DIA: .323 BALLISTEK NO: 323BJ NAI NO: RBB 22344/4.734

DATA SOURCE: MSAA Pg37

HISTORICAL DATA:

NOTES:

LOADING DATA:

BULLET WT./TYPE	POWDER WT./TYPE	VELOCITY ('/SEC)	SOURCE
208/Spire	49.6/IMR3031	2600	JJD

CASE PREPARATION: SHELLHOLDER (RCBS): 5

MAKE FROM: 10,75 x 68. Increase rim's rebate by turning to .469-.470" dia. and deepening the extractor groove. F/L size case with expander removed. Trim to length. Chamfer and fireform.

PHYSICAL DATA (INCHES):

CASE TYPE: **Rebated Bottleneck**
CASE LENGTH **A = 2.32**
HEAD DIAMETER **B = .490**
RIM DIAMETER **D = .469**
NECK DIAMETER **F = .360**
NECK LENGTH **H = .308**
SHOULDER LENGTH **K = .062**
BODY ANGLE (DEG'S/SIDE): **.90**
CASE CAPACITY **CC'S = 4.15**

LOADED LENGTH: **3.29**
BELT DIAMETER **C = N/A**
RIM THICKNESS **E = .058**
SHOULDER DIAMETER **G = .435**
LENGTH TO SHOULDER **J = 1.95**
SHOULDER ANGLE (DEG'S/SIDE): **31.17**
PRIMER: **L/R**
CASE CAPACITY (GR'S WATER): **64.09**

DIMENSIONAL DRAWING:

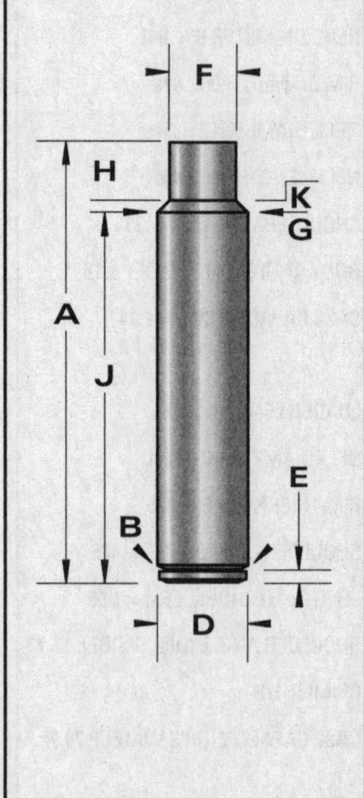

-NOT ACTUAL SIZE-
-DO NOT SCALE-

CARTRIDGE: 8 x 60 Magnum

OTHER NAMES:
8 x 60 (S) or (J)

DIA: .323'

BALLISTEK NO: 323BD

NAI NO: RXB 22333/4.974

DATA SOURCE: COTW 4th Pg266

HISTORICAL DATA: The Magnum Bombe is an 8x60(S) with a rather hot load.

NOTES: ' (S)=.323" dia. (J)=.318" dia.

LOADING DATA:

BULLET WT./TYPE	POWDER WT./TYPE	VELOCITY ('/SEC)	SOURCE
170/—	50.0/4320	—	Nonte

CASE PREPARATION: SHELLHOLDER (RCBS): 3

MAKE FROM: .30-06 Spgf. Anneal case neck and taper expand to .330" dia. Size in F/L die with expander removed. Trim to length & chamfer. F/L size.

PHYSICAL DATA (INCHES):

CASE TYPE: **Rimless Bottleneck**

CASE LENGTH **A = 2.338**

HEAD DIAMETER **B = .470**

RIM DIAMETER **D = .468**

NECK DIAMETER **F = .348**

NECK LENGTH **H = .330**

SHOULDER LENGTH **K = .073**

BODY ANGLE (DEG'S/SIDE): **.627**

CASE CAPACITY **CC'S = 4.28**

LOADED LENGTH: **3.11**

BELT DIAMETER **C = N/A**

RIM THICKNESS **E = .049**

SHOULDER DIAMETER **G = .432**

LENGTH TO SHOULDER **J = 1.935**

SHOULDER ANGLE (DEG'S/SIDE): **29.91**

PRIMER: **L/R**

CASE CAPACITY (GR'S WATER): **66.18**

DIMENSIONAL DRAWING:

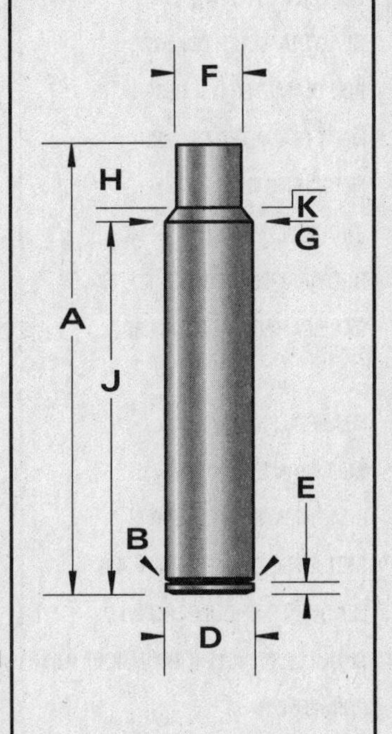

-NOT ACTUAL SIZE-
-DO NOT SCALE-

CARTRIDGE: 8 x 60 Mauser

OTHER NAMES:

DIA: .323

BALLISTEK NO: 323C

NAI NO: RXB 22333/4.974

DATA SOURCE: COTW 4th Pg266

HISTORICAL DATA:

NOTES:

LOADING DATA:

BULLET WT./TYPE	POWDER WT./TYPE	VELOCITY ('/SEC)	SOURCE
160/Spire	47.0/IMR3031	2700	JJD

CASE PREPARATION: SHELLHOLDER (RCBS): 3

MAKE FROM: .30-06 Spgf. Taper expand neck to about .330" dia. Trim case to 2.35". F/L size in 8x60 die. Trim & chamfer.

PHYSICAL DATA (INCHES):

CASE TYPE: **Rimless Bottleneck**

CASE LENGTH **A = 2.338**

HEAD DIAMETER **B = .470**

RIM DIAMETER **D = .468**

NECK DIAMETER **F = .348**

NECK LENGTH **H = .330**

SHOULDER LENGTH **K = .073**

BODY ANGLE (DEG'S/SIDE): **.627**

CASE CAPACITY **CC'S = 4.29**

LOADED LENGTH: **3.11**

BELT DIAMETER **C = N/A**

RIM THICKNESS **E = .049**

SHOULDER DIAMETER **G = .432**

LENGTH TO SHOULDER **J = 1.935**

SHOULDER ANGLE (DEG'S/SIDE): **29.91**

PRIMER: **L/R**

CASE CAPACITY (GR'S WATER): **66.18**

DIMENSIONAL DRAWING:

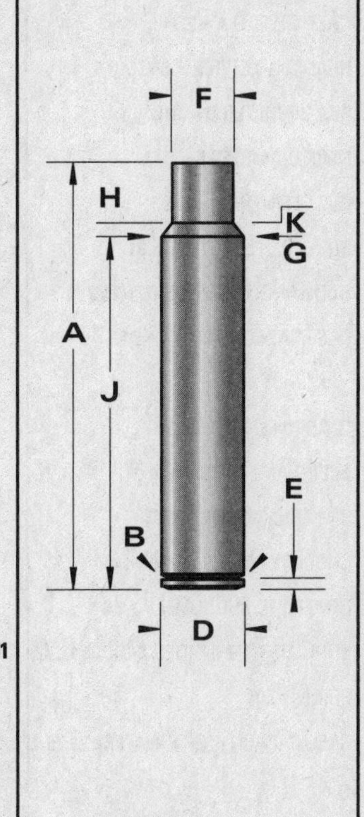

-NOT ACTUAL SIZE-
-DO NOT SCALE-

CARTRIDGE: 8 x 60R Kropatchek

OTHER NAMES:
8 x 60R Guedes

DIA: .318

BALLISTEK NO: 318D

NAI NO: RMB 34322/4.309

DATA SOURCE: Hoyem Vol.2

HISTORICAL DATA: Portugal circa 1885.

NOTES: The Guedes rounds is of .323" dia. and has a .621" dia. rim.

LOADING DATA:

BULLET WT./TYPE	POWDER WT./TYPE	VELOCITY ('/SEC)	SOURCE
150/RN	30/IMR4198	1700	JJD

CASE PREPARATION: SHELLHOLDER (RCBS): Spl.

MAKE FROM: .450 N.E. (BELL). Turn rim to .618" dia., if necessary, and build up rim to .065" thick. Anneal case and form in 2-die form set. Trim to length, F/L size and chamfer.

PHYSICAL DATA (INCHES):

CASE TYPE: **Rimmed Bottleneck**
CASE LENGTH **A = 2.34**
HEAD DIAMETER **B = .543**
RIM DIAMETER **D = .618**
NECK DIAMETER **F = .347**
NECK LENGTH **H = .560**
SHOULDER LENGTH **K = .31**
BODY ANGLE (DEG'S/SIDE): **1.26**
CASE CAPACITY **CC'S = 4.84**
LOADED LENGTH: **3.23**
BELT DIAMETER **C = N/A**
RIM THICKNESS **E = .065**
SHOULDER DIAMETER **G = .487**
LENGTH TO SHOULDER **J = 1.47**
SHOULDER ANGLE (DEG'S/SIDE): **12.72**
PRIMER: **L/R**
CASE CAPACITY (GR'S WATER): **74.75**

DIMENSIONAL DRAWING:

-NOT ACTUAL SIZE-
-DO NOT SCALE-

CARTRIDGE: 8 x 61R Rubin

OTHER NAMES:

DIA: .323

BALLISTEK NO: 323BQ

NAI NO: RMB 22331/4.668

DATA SOURCE: Hoyem Vol.2 Pg184

HISTORICAL DATA: By Rubin about 1880.

NOTES:

LOADING DATA:

BULLET WT./TYPE	POWDER WT./TYPE	VELOCITY ('/SEC)	SOURCE
175/Lead	35.1/IMR4320	–	JJD

CASE PREPARATION: SHELLHOLDER (RCBS): Spl.

MAKE FROM: .43 Span. Rem. turn rim to .586" dia. Form through two-die form set. Trim to length, chamfer and F/L size.

PHYSICAL DATA (INCHES):

CASE TYPE: **Rimmed Bottleneck**
CASE LENGTH **A = 2.39**
HEAD DIAMETER **B = .512**
RIM DIAMETER **D = .586**
NECK DIAMETER **F = .353**
NECK LENGTH **H = .34**
SHOULDER LENGTH **K = .340**
BODY ANGLE (DEG'S/SIDE): **.77**
CASE CAPACITY **CC'S = 4.86**
LOADED LENGTH: **3.34**
BELT DIAMETER **C = N/A**
RIM THICKNESS **E = .090**
SHOULDER DIAMETER **G = .471**
LENGTH TO SHOULDER **J = 1.71**
SHOULDER ANGLE (DEG'S/SIDE): **9.84**
PRIMER: **L/R**
CASE CAPACITY (GR'S WATER): **75.01**

DIMENSIONAL DRAWING:

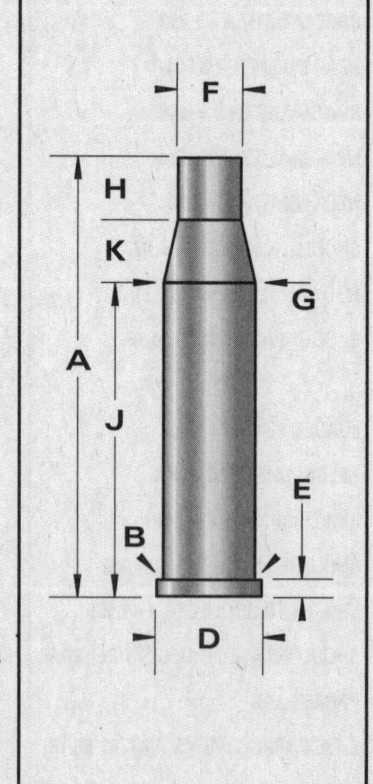

-NOT ACTUAL SIZE-
-DO NOT SCALE-

CARTRIDGE: 8 x 62 Durham Magnum

OTHER NAMES:

DIA: .323

BALLISTEK NO: 323BE

NAI NO: BEN 22344/4.990

DATA SOURCE: Ackley Vol.1 Pg456

HISTORICAL DATA:

NOTES:

LOADING DATA:

BULLET WT./TYPE	POWDER WT./TYPE	VELOCITY ('/SEC)	SOURCE
175/RN	71.0/4320	3467	Ackley

CASE PREPARATION: SHELLHOLDER (RCBS): 4

MAKE FROM: 8mm Rem. Mag. Set back shoulder in Durham F/L die with expander removed. Trim to length and chamfer. Neck I.D.'s generally need a little reaming. F/L size and fireform in the chamber.

PHYSICAL DATA (INCHES):

CASE TYPE: **Belted Bottleneck**

CASE LENGTH **A = 2.56**

HEAD DIAMETER **B = .513**

RIM DIAMETER **D = .532**

NECK DIAMETER **F = .352**

NECK LENGTH **H = .280**

SHOULDER LENGTH **K = .090**

BODY ANGLE (DEG'S/SIDE): **.375**

CASE CAPACITY **CC'S = 5.77**

LOADED LENGTH: **3.50**

BELT DIAMETER **C = .532**

RIM THICKNESS **E = .05**

SHOULDER DIAMETER **G = .487**

LENGTH TO SHOULDER **J = 2.19**

SHOULDER ANGLE (DEG'S/SIDE): **36.87**

PRIMER: **L/R Mag**

CASE CAPACITY (GR'S WATER): **89.09**

DIMENSIONAL DRAWING:

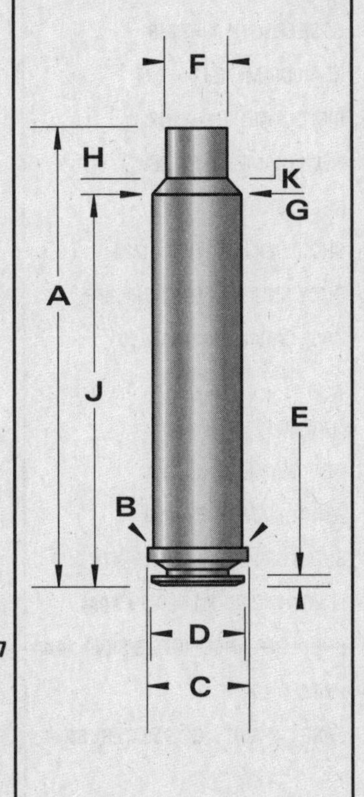

-NOT ACTUAL SIZE-
-DO NOT SCALE-

CARTRIDGE: 8 x 63mm Swedish

OTHER NAMES:

DIA: .323

BALLISTEK NO: 323BF

NAI NO: RBB 22343/5.082

DATA SOURCE: COTW 4th Pg267

HISTORICAL DATA: Swedish mil. ctg. ca. 1932.

NOTES:

LOADING DATA:

BULLET WT./TYPE	POWDER WT./TYPE	VELOCITY ('/SEC)	SOURCE
170/–	57.0/4320	2820	Barnes

CASE PREPARATION: SHELLHOLDER (RCBS): 5

MAKE FROM: 10,75 x 68mm. F/L size the 10,75mm case in the 8x63 die, with the expander removed. Trim to length and chamfer. F/L size.

PHYSICAL DATA (INCHES):

CASE TYPE: **Rebated Bottleneck**

CASE LENGTH **A = 2.48**

HEAD DIAMETER **B = .488**

RIM DIAMETER **D = .479**

NECK DIAMETER **F = .356**

NECK LENGTH **H = .318**

SHOULDER LENGTH **K = .132**

BODY ANGLE (DEG'S/SIDE): **.500**

CASE CAPACITY **CC'S = 4.76**

LOADED LENGTH: **3.35**

BELT DIAMETER **C = N/A**

RIM THICKNESS **E = .052**

SHOULDER DIAMETER **G = .456**

LENGTH TO SHOULDER **J = 2.03**

SHOULDER ANGLE (DEG'S/SIDE): **20.74**

PRIMER: **L/R**

CASE CAPACITY (GR'S WATER): **73.55**

DIMENSIONAL DRAWING:

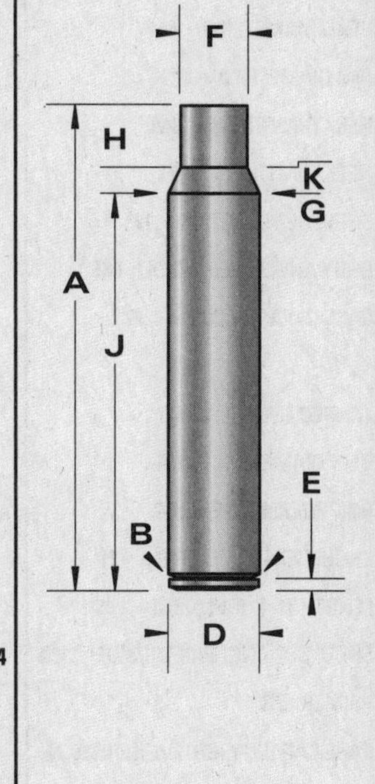

-NOT ACTUAL SIZE-
-DO NOT SCALE-

CARTRIDGE: 8 x 65R(S) Brenneke

OTHER NAMES:

DIA: .323

BALLISTEK NO: 323BG

NAI NO: RMB 22332/5.517

DATA SOURCE: COTW 4th Pg268

HISTORICAL DATA:

NOTES:

LOADING DATA:

BULLET WT./TYPE	POWDER WT./TYPE	VELOCITY ('/SEC)	SOURCE
170/—	52.0/4064	2710	Barnes

CASE PREPARATION: SHELLHOLDER (RCBS): 26

MAKE FROM: 7 x 65R Brenneke. Taper expand to .330" dia., F/L size, trim and chamfer. Cases can also be made from 9,3x74R by turning off the rim, cutting a new extractor groove, cutting to 2.6", annealing, F/L sizing and trimming to length.

PHYSICAL DATA (INCHES):

CASE TYPE: **Rimmed Bottleneck**

CASE LENGTH **A = 2.56**

HEAD DIAMETER **B = .464**

RIM DIAMETER **D = .520**

NECK DIAMETER **F = .348**

NECK LENGTH **H = .335**

SHOULDER LENGTH **K = .175**

BODY ANGLE (DEG'S/SIDE): **.665**

CASE CAPACITY **CC'S = 4.55**

LOADED LENGTH: **3.65**

BELT DIAMETER **C = N/A**

RIM THICKNESS **E = .054**

SHOULDER DIAMETER **G = .421**

LENGTH TO SHOULDER **J = 2.05**

SHOULDER ANGLE (DEG'S/SIDE): **11.78**

PRIMER: **L/R**

CASE CAPACITY (GR'S WATER): **70.26**

DIMENSIONAL DRAWING:

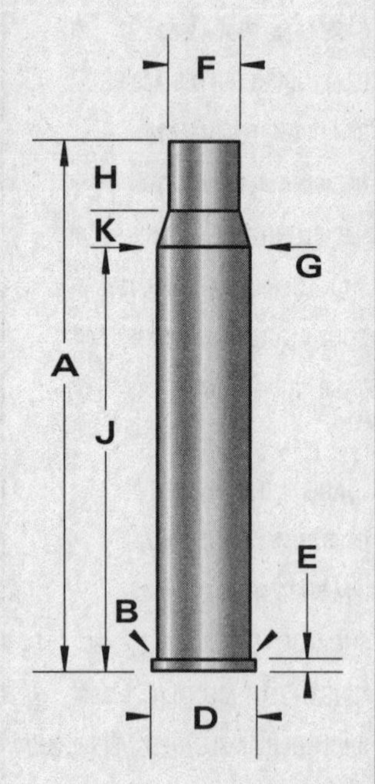

-NOT ACTUAL SIZE-
-DO NOT SCALE-

CARTRIDGE: 8 x 68(S) Magnum

OTHER NAMES:

DIA: .323

BALLISTEK NO: 323P

NAI NO: RBB 22332/5.072

DATA SOURCE: Hornady 3rd Pg264

HISTORICAL DATA: By RWS in 1940.

NOTES:

LOADING DATA:

BULLET WT./TYPE	POWDER WT./TYPE	VELOCITY ('/SEC)	SOURCE
170/RN	70.0/IMR4350	3000	JJD

CASE PREPARATION: SHELLHOLDER (RCBS): 4

MAKE FROM: Factory or .280 Ross (BELL). Swage base to .524" dia. Turn off rim and cut new extractor groove. Taper expand Ross case to .330. F/L size in 8 x 68 die. Chamfer. Fireform in chamber.

PHYSICAL DATA (INCHES):

CASE TYPE: **Rebated Bottleneck**

CASE LENGTH **A = 2.658**

HEAD DIAMETER **B = .524**

RIM DIAMETER **D = .512**

NECK DIAMETER **F = .360**

NECK LENGTH **H = .335**

SHOULDER LENGTH **K = .229**

BODY ANGLE (DEG'S/SIDE): **.695**

CASE CAPACITY **CC'S = 5.60**

LOADED LENGTH: **3.40**

BELT DIAMETER **C = N/A**

RIM THICKNESS **E = .055**

SHOULDER DIAMETER **G = .478**

LENGTH TO SHOULDER **J = 2.094**

SHOULDER ANGLE (DEG'S/SIDE): **14.44**

PRIMER: **L/R**

CASE CAPACITY (GR'S WATER): **86.54**

DIMENSIONAL DRAWING:

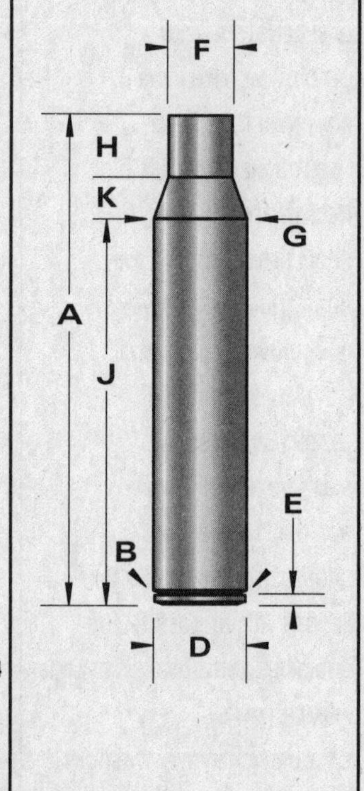

-NOT ACTUAL SIZE-
-DO NOT SCALE-

CARTRIDGE: 8 x 71mm Peterlongo

OTHER NAMES:

DIA: .318

BALLISTEK NO: 318A

NAI NO: RXB 22333/6.060

DATA SOURCE: COTW 4th Pg265

HISTORICAL DATA: By J. Peterlongo.

NOTES:

LOADING DATA:

BULLET WT./TYPE	POWDER WT./TYPE	VELOCITY ('/SEC)	SOURCE
200/RN	54.0/IMR4350	–	JJD

CASE PREPARATION: SHELLHOLDER (RCBS): 24

MAKE FROM: .405 Win. basic (BELL). Turn rim flush with head and cut new extractor groove. Cut case to 2.8" and anneal. F/L size with expander removed. I.D. neck ream. Re-chamfer, F/L size and fireform.

PHYSICAL DATA (INCHES):

CASE TYPE: **Rimless Bottleneck**

CASE LENGTH **A = 2.80**

HEAD DIAMETER **B = .462**

RIM DIAMETER **D = .468**

NECK DIAMETER **F = .349**

NECK LENGTH **H = .355**

SHOULDER LENGTH **K = .095**

BODY ANGLE (DEG'S/SIDE): **.532**

CASE CAPACITY **CC'S = 4.85**

LOADED LENGTH: **3.28**

BELT DIAMETER **C = N/A**

RIM THICKNESS **E = .05**

SHOULDER DIAMETER **G = .422**

LENGTH TO SHOULDER **J = 2.35**

SHOULDER ANGLE (DEG'S/SIDE): **21.02**

PRIMER: **L/R**

CASE CAPACITY (GR'S WATER): **74.91**

DIMENSIONAL DRAWING:

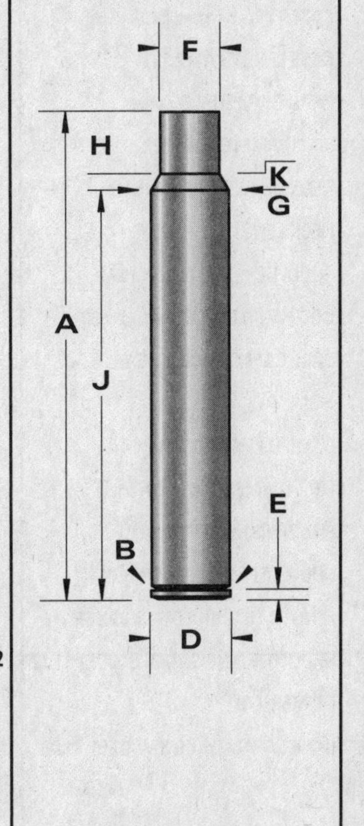

-NOT ACTUAL SIZE-
-DO NOT SCALE-

CARTRIDGE: 8 x 72R Sauer

OTHER NAMES:
8 x 72R S&S
8 x 72R/360

DIA: .324

BALLISTEK NO: 324A

NAI NO: RMS 21115/6.620

DATA SOURCE: COTW 4th Pg269

HISTORICAL DATA: Sauer & Son about 1910.

NOTES:

LOADING DATA:

BULLET WT./TYPE	POWDER WT./TYPE	VELOCITY ('/SEC)	SOURCE
225/FN	38.0/3031	1910	Barnes

CASE PREPARATION: SHELLHOLDER (RCBS): 4

MAKE FROM: 9,3 x 72R. Anneal case and F/L size in the 8x72R die. Square case mouth.

PHYSICAL DATA (INCHES):

CASE TYPE: **Rimmed Straight**

CASE LENGTH **A = 2.84**

HEAD DIAMETER **B = .429**

RIM DIAMETER **D = .483**

NECK DIAMETER **F = .344**

NECK LENGTH **H = N/A**

SHOULDER LENGTH **K = N/A**

BODY ANGLE (DEG'S/SIDE): **.873**

CASE CAPACITY **CC'S = 4.15**

LOADED LENGTH: **3.40**

BELT DIAMETER **C = N/A**

RIM THICKNESS **E = .053**

SHOULDER DIAMETER **G = N/A**

LENGTH TO SHOULDER **J = N/A**

SHOULDER ANGLE (DEG'S/SIDE): **N/A**

PRIMER: **L/R**

CASE CAPACITY (GR'S WATER): **64.14**

DIMENSIONAL DRAWING:

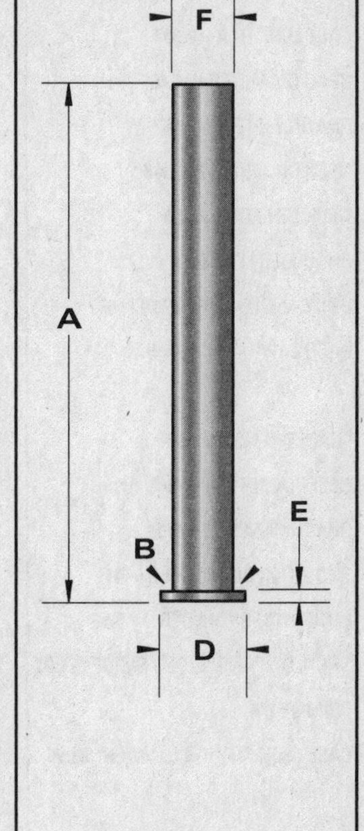

-NOT ACTUAL SIZE-
-DO NOT SCALE-

CARTRIDGE: 8 x 75mm

OTHER NAMES:	DIA: .318 BALLISTEK NO: 318B NAI NO: RXB 22342/6.303

DATA SOURCE: COTW 4th Pg265

HISTORICAL DATA: Circa 1910.

NOTES: Both .318 and .323 diameter bores exist!

LOADING DATA:

BULLET WT./TYPE	POWDER WT./TYPE	VELOCITY ('/SEC)	SOURCE
220/Spire	65.7/IMR4831	—	JJD

CASE PREPARATION: SHELLHOLDER (RCBS): 4

MAKE FROM: 9,3 x 74R. Turn rim flush with body diameter and re-cut the extractor groove. F/L size in the 8x75 die (annealing may be required), with the expander removed. Trim to length. F/L size and chamfer.

PHYSICAL DATA (INCHES):

CASE TYPE: **Rimless Bottleneck**
CASE LENGTH **A = 2.937**
HEAD DIAMETER **B = .466**
RIM DIAMETER **D = .467**
NECK DIAMETER **F = .344**
NECK LENGTH **H = .205**
SHOULDER LENGTH **K = .172**
BODY ANGLE (DEG'S/SIDE): **.667**
CASE CAPACITY **CC'S = 5.30**

LOADED LENGTH: **3.50**
BELT DIAMETER **C = N/A**
RIM THICKNESS **E = .05**
SHOULDER DIAMETER **G = .411**
LENGTH TO SHOULDER **J = 2.56**
SHOULDER ANGLE (DEG'S/SIDE): **11.02**
PRIMER: **L/R**
CASE CAPACITY (GR'S WATER): **81.76**

DIMENSIONAL DRAWING:

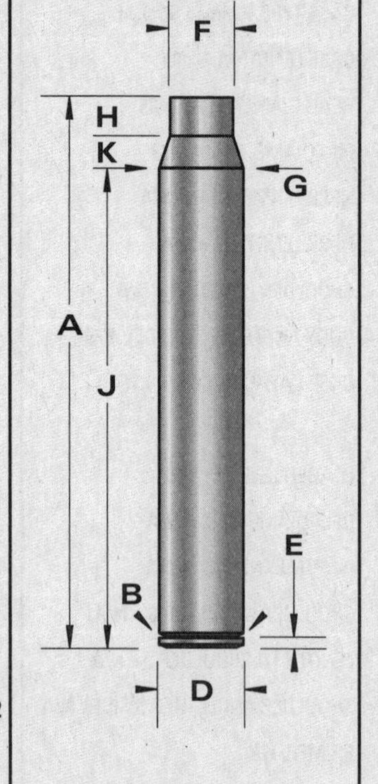

-NOT ACTUAL SIZE-
-DO NOT SCALE-

CARTRIDGE: 8 x 75R

OTHER NAMES:	DIA: .318 BALLISTEK NO: 318C NAI NO: RMB 22342/6.302

DATA SOURCE: COTW 4th Pg265

HISTORICAL DATA: Circa 1910.

NOTES: .318" and .323" dia. bores exist. Check diameter!

LOADING DATA:

BULLET WT./TYPE	POWDER WT./TYPE	VELOCITY ('/SEC)	SOURCE
220/Spire	61.0/IMR4350	—	JJD

CASE PREPARATION: SHELLHOLDER (RCBS): 26

MAKE FROM: 9,3 x 74R. Anneal case and F/L size. Trim to length.

PHYSICAL DATA (INCHES):

CASE TYPE: **Rimmed Bottleneck**
CASE LENGTH **A = 2.937**
HEAD DIAMETER **B = .466**
RIM DIAMETER **D = .522**
NECK DIAMETER **F = .345**
NECK LENGTH **H = .205**
SHOULDER LENGTH **K = .167**
BODY ANGLE (DEG'S/SIDE): **.666**
CASE CAPACITY **CC'S = 5.26**

LOADED LENGTH: **3.51**
BELT DIAMETER **C = N/A**
RIM THICKNESS **E = .052**
SHOULDER DIAMETER **G = .411**
LENGTH TO SHOULDER **J = 2.565**
SHOULDER ANGLE (DEG'S/SIDE): **11.17**
PRIMER: **L/R**
CASE CAPACITY (GR'S WATER): **81.23**

DIMENSIONAL DRAWING:

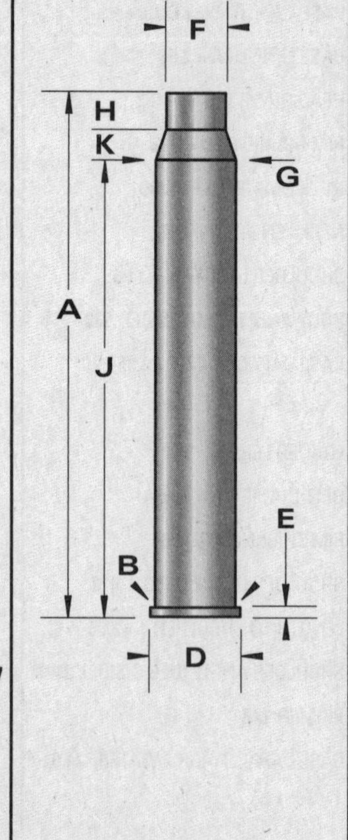

-NOT ACTUAL SIZE-
-DO NOT SCALE-

CARTRIDGE: 8,15 x 46R

OTHER NAMES:
8,2 x 46.5R
8,2 Patrone

DIA: .316

BALLISTEK NO: 302E

NAI NO: RMB 34221/4.323

DATA SOURCE: COTW 4th Pg269

HISTORICAL DATA:

NOTES: Bores vary - slug for actual dia.

LOADING DATA:

BULLET WT./TYPE	POWDER WT./TYPE	VELOCITY ('/SEC)	SOURCE
150/Lead	27.0/IMR3031	1675	JJD

CASE PREPARATION: SHELLHOLDER (RCBS): 30

MAKE FROM: .30-30 Win. Anneal case and F/L size with expander removed. Trim to length, chamfer and F/L size, again.

PHYSICAL DATA (INCHES):

CASE TYPE: **Rimmed Bottleneck**
CASE LENGTH **A = 1.82**
HEAD DIAMETER **B = .421**
RIM DIAMETER **D = .484**
NECK DIAMETER **F = .346**
NECK LENGTH **H = .485**
SHOULDER LENGTH **K = .220**
BODY ANGLE (DEG'S/SIDE): **1.34**
CASE CAPACITY **CC'S = 2.34**
LOADED LENGTH: **2.28**
BELT DIAMETER **C = N/A**
RIM THICKNESS **E = .063**
SHOULDER DIAMETER **G = .378**
LENGTH TO SHOULDER **J = 1.115**
SHOULDER ANGLE (DEG'S/SIDE): **4.16**
PRIMER: **L/R**
CASE CAPACITY (GR'S WATER): **36.13**

DIMENSIONAL DRAWING:

-NOT ACTUAL SIZE-
-DO NOT SCALE-

CARTRIDGE: 8,3 x 46.5R Rubin

OTHER NAMES:

DIA: .327

BALLISTEK NO: 327A

NAI NO: RMB 34343/3.233

DATA SOURCE: Hoyem Vol.2 Pg185

HISTORICAL DATA: Experimental Rubin (Swiss) ctg. circa 1880.

NOTES:

LOADING DATA:

BULLET WT./TYPE	POWDER WT./TYPE	VELOCITY ('/SEC)	SOURCE
180/RN	26.5/IMR4198	–	JJD

CASE PREPARATION: SHELLHOLDER (RCBS): Spl.

MAKE FROM: .450 #2 N.E. (BELL). Turn rim down to .624" dia. and thin to .057" thick. Cut case to 1.9" and anneal. Form in form die set (3-dies), trim to length and I.D. neck ream. F/L size.

PHYSICAL DATA (INCHES):

CASE TYPE: **Rimmed Bottleneck**
CASE LENGTH **A = 1.82**
HEAD DIAMETER **B = .563**
RIM DIAMETER **D = .624**
NECK DIAMETER **F = .355**
NECK LENGTH **H = .290**
SHOULDER LENGTH **K = .160**
BODY ANGLE (DEG'S/SIDE): **1.42**
CASE CAPACITY **CC'S = 4.24**
LOADED LENGTH: **2.75**
BELT DIAMETER **C = N/A**
RIM THICKNESS **E = .057**
SHOULDER DIAMETER **G = .505**
LENGTH TO SHOULDER **J = 1.37**
SHOULDER ANGLE (DEG'S/SIDE): **25.11**
PRIMER: **L/R**
CASE CAPACITY (GR'S WATER): **65.49**

DIMENSIONAL DRAWING:

-NOT ACTUAL SIZE-
-DO NOT SCALE-

CARTRIDGE: 8.59 Lazzaroni Galaxy

OTHER NAMES:

DIA: .338

BALLISTEK NO:

NAI NO:

DATA SOURCE:

HISTORICAL DATA: Created by John Lazzroni, cartridge drawing dated 10/02/01.

NOTES: This is the .338 cartridge in the Lazzaroni short action magnum lineup

LOADING DATA:

BULLET WT./TYPE	POWDER WT./TYPE	VELOCITY ('/SEC)	SOURCE
200/Nosler Ball. Tip	66/RL-15	3214	Lazzaroni
250/Nosler Partition	66/H4350	2743	Lazzaroni

CASE PREPARATION:

SHELL HOLDER (RCBS): 14

MAKE FROM:

PHYSICAL DATA (INCHES):

CASE TYPE: **Rimless Bottleneck**

CASE LENGTH **A = 2.050**

HEAD DIAMETER **B = .579**

RIM DIAMETER **D = .580**

NECK DIAMETER **F = .373**

NECK LENGTH **H = .336**

SHOULDER LENGTH **K = .161**

BODY ANGLE (DEG'S/SIDE):

CASE CAPACITY **CC'S =**

LOADED LENGTH: **2.800**

BELT DIAMETER **C =**

RIM THICKNESS **E = .065**

SHOULDER DIAMETER **G = .560**

LENGTH TO SHOULDER **J = 1.553**

SHOULDER ANGLE (DEG'S/SIDE): **30**

PRIMER: **L/R or L/R Mag**

CASE CAPACITY (GR'S WATER):

DIMENSIONAL DRAWING:

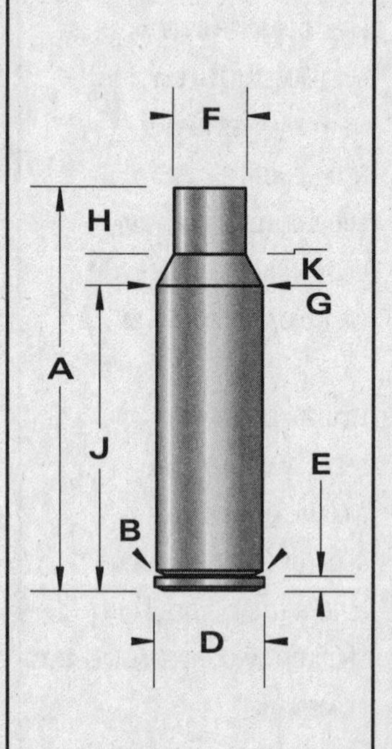

-NOT ACTUAL SIZE-
-DO NOT SCALE-

CARTRIDGE: 8.59 Lazzaroni Titan

OTHER NAMES:

DIA: .338

BALLISTEK NO:

NAI NO:

DATA SOURCE:

HISTORICAL DATA: Created by John Lazzaroni, cartridge drawing dated 4/11/01.

NOTES: This is the .338 caliber in John Lazzaroni's line of long action beltless magnums.

LOADING DATA:

BULLET WT./TYPE	POWDER WT./TYPE	VELOCITY ('/SEC)	SOURCE
225/Nosler Partition	100.0/IMR-7828	3225	Lazzaroni
250/Swift A-Frame	103.0/RL-25	3154	Lazzaroni

CASE PREPARATION:

SHELL HOLDER (RCBS): 14

MAKE FROM:

PHYSICAL DATA (INCHES):

CASE TYPE: **Rimless Bottleneck**

CASE LENGTH **A = 2.81**

HEAD DIAMETER **B = .579**

RIM DIAMETER **D = .580**

NECK DIAMETER **F = .374**

NECK LENGTH **H = .337**

SHOULDER LENGTH **K = .160**

BODY ANGLE (DEG'S/SIDE):

CASE CAPACITY **CC'S =**

LOADED LENGTH: **3.630**

BELT DIAMETER **C =**

RIM THICKNESS **E = .065**

SHOULDER DIAMETER **G = .560**

LENGTH TO SHOULDER **J = 2.313**

SHOULDER ANGLE (DEG'S/SIDE): **30**

PRIMER: **L/R Mag**

CASE CAPACITY (GR'S WATER):

DIMENSIONAL DRAWING:

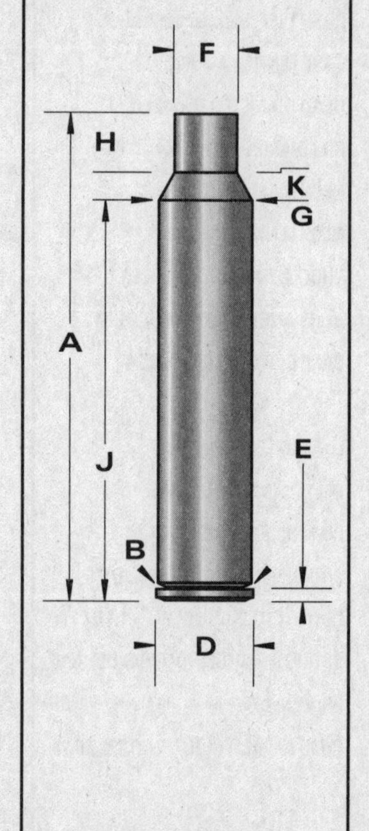

-NOT ACTUAL SIZE-
-DO NOT SCALE-

CARTRIDGE: 9mm Bayard Long

OTHER NAMES:
9mm Bergmann-Bayard #6
9mm Danish M10

DIA: .355

BALLISTEK NO: 355K

NAI NO: RXS
21115/2.333

DATA SOURCE: COTW 4th Pg174

HISTORICAL DATA: For Bergmann-Bayard auto.

NOTES:

LOADING DATA:

BULLET WT./TYPE	POWDER WT./TYPE	VELOCITY ('/SEC)	SOURCE
116/RN	7.0/Unique	1280	Barnes

CASE PREPARATION: **SHELLHOLDER (RCBS): 1**

MAKE FROM: 9mm Win. Mag. Trim case to .91" & chamfer. Size forward 1/4" of case mouth in .39 Spl. sizer, to reduce diameter. For quality ammo, purchase a F/L sizer because of the .375" dia. neck dim.

PHYSICAL DATA (INCHES):

CASE TYPE: **Rimless Straight**
CASE LENGTH **A = .91**
HEAD DIAMETER **B = .390**
RIM DIAMETER **D = .392**
NECK DIAMETER **F = .375**
NECK LENGTH **H = N/A**
SHOULDER LENGTH **K = N/A**
BODY ANGLE (DEG'S/SIDE): **.497**
CASE CAPACITY **CC'S = .966**
LOADED LENGTH: **1.32**
BELT DIAMETER **C = N/A**
RIM THICKNESS **E = .046**
SHOULDER DIAMETER **G = N/A**
LENGTH TO SHOULDER **J = N/A**
SHOULDER ANGLE (DEG'S/SIDE): **N/A**
PRIMER: **S/P**
CASE CAPACITY (GR'S WATER): **14.91**

DIMENSIONAL DRAWING:

F
A
E
B
D

-NOT ACTUAL SIZE-
-DO NOT SCALE-

CARTRIDGE: 9mm Browning Long

OTHER NAMES:

DIA: .355

BALLISTEK NO: 355J

NAI NO: RMS
21115/2.083

DATA SOURCE: COTW 4th Pg174

HISTORICAL DATA: Introduced in 1903.

NOTES:

LOADING DATA:

BULLET WT./TYPE	POWDER WT./TYPE	VELOCITY ('/SEC)	SOURCE
115/JHP	4.5/Unique	900	JJD

CASE PREPARATION: **SHELLHOLDER (RCBS): 6**

MAKE FROM: .38 ACP (or ACP Super). Trim case to .80" & chamfer. Use the .38 ACP dies for sizing & loading.

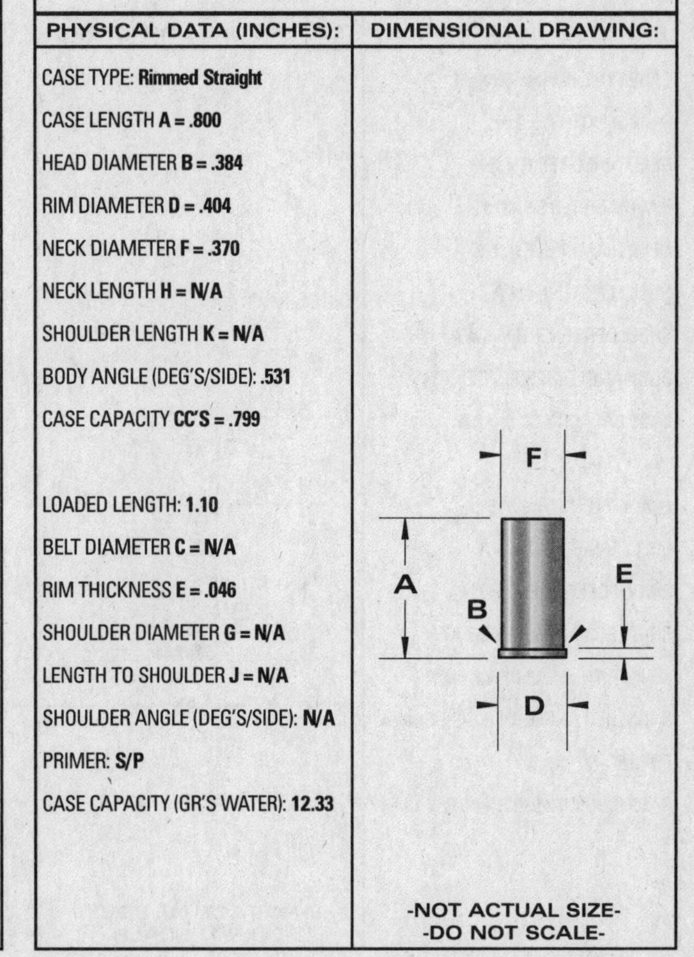

PHYSICAL DATA (INCHES):

CASE TYPE: **Rimmed Straight**
CASE LENGTH **A = .800**
HEAD DIAMETER **B = .384**
RIM DIAMETER **D = .404**
NECK DIAMETER **F = .370**
NECK LENGTH **H = N/A**
SHOULDER LENGTH **K = N/A**
BODY ANGLE (DEG'S/SIDE): **.531**
CASE CAPACITY **CC'S = .799**
LOADED LENGTH: **1.10**
BELT DIAMETER **C = N/A**
RIM THICKNESS **E = .046**
SHOULDER DIAMETER **G = N/A**
LENGTH TO SHOULDER **J = N/A**
SHOULDER ANGLE (DEG'S/SIDE): **N/A**
PRIMER: **S/P**
CASE CAPACITY (GR'S WATER): **12.33**

DIMENSIONAL DRAWING:

-NOT ACTUAL SIZE-
-DO NOT SCALE-

CARTRIDGE: 9mm Devel

OTHER NAMES:

DIA: .355

BALLISTEK NO: 355M

NAI NO: RXS 11115/2.342

DATA SOURCE: NAI/Ballistek

HISTORICAL DATA: Developed for early Devel prototype.

NOTES: By author in 1975.

LOADING DATA:

BULLET WT./TYPE	POWDER WT./TYPE	VELOCITY ('/SEC)	SOURCE
115/JHP	5.7/B'Eye	1300	JJD

CASE PREPARATION: SHELLHOLDER (RCBS): 1

MAKE FROM: .38 Super. Turn rim flush (or slightly below) body dieameter. Back chamfer and size in normal .38 ACP dies. This rim reduction was to help feeding in auto pistols.

PHYSICAL DATA (INCHES):

CASE TYPE: **Rimless Straight**

CASE LENGTH **A = .89**

HEAD DIAMETER **B = .380**

RIM DIAMETER **D = .379**

NECK DIAMETER **F = .379**

NECK LENGTH **H = N/A**

SHOULDER LENGTH **K = N/A**

BODY ANGLE (DEG'S/SIDE): **.034**

CASE CAPACITY **CC'S = .877**

LOADED LENGTH: **1.116**

BELT DIAMETER **C = N/A**

RIM THICKNESS **E = .047**

SHOULDER DIAMETER **G = N/A**

LENGTH TO SHOULDER **J = N/A**

SHOULDER ANGLE (DEG'S/SIDE): **N/A**

PRIMER: **S/P**

CASE CAPACITY (GR'S WATER): **13.53**

DIMENSIONAL DRAWING:

F
A
E
B
D

-NOT ACTUAL SIZE-
-DO NOT SCALE-

CARTRIDGE: 9mm Japanese Revolver T26

OTHER NAMES:

DIA: .360

BALLISTEK NO: 360A

NAI NO: RMS 21115/2.227

DATA SOURCE: Handloader #68 Pg36

HISTORICAL DATA: Japanese mil. ctg.

NOTES: Note the very thin rim.

LOADING DATA:

BULLET WT./TYPE	POWDER WT./TYPE	VELOCITY ('/SEC)	SOURCE
150/Lead	4.7/Unique	880	Nonte

CASE PREPARATION: SHELLHOLDER (RCBS): 6

MAKE FROM: .38 Special. Use lathe to thin rim to .028-.030". Deburr edges with fine file. Trim case to length. Fabricate thin brass or alum. washers (.440" O.D. x .390" I.D. x .030" thick) to slip over the case while priming. If these rings are not used, the rim may tear off, while seating the primer. Our best bullet has been a 158 gr. lead SWC, unsized.

PHYSICAL DATA (INCHES):

CASE TYPE: **Rimmed Straight**

CASE LENGTH **A = .862**

HEAD DIAMETER **B = .387**

RIM DIAMETER **D = .435**

NECK DIAMETER **F = .373**

NECK LENGTH **H = N/A**

SHOULDER LENGTH **K = N/A**

BODY ANGLE (DEG'S/SIDE): **.482**

CASE CAPACITY **CC'S = .93**

LOADED LENGTH: **1.181**

BELT DIAMETER **C = N/A**

RIM THICKNESS **E = .03**

SHOULDER DIAMETER **G = N/A**

LENGTH TO SHOULDER **J = N/A**

SHOULDER ANGLE (DEG'S/SIDE): **N/A**

PRIMER: **S/P**

CASE CAPACITY (GR'S WATER): **14.35**

DIMENSIONAL DRAWING:

F
A
E
B
D

-NOT ACTUAL SIZE-
-DO NOT SCALE-

CARTRIDGE: 9mm Luger (Parabellum)

OTHER NAMES:
9 x 19mm
9mm Glisenti

DIA: .355

BALLISTEK NO: 355A

NAI NO: RXS 21115/1.923

DATA SOURCE: Hornady Manual 3rd Pg333

HISTORICAL DATA: By G. Luger in 1902.

NOTES:

LOADING DATA:

BULLET WT./TYPE	POWDER WT./TYPE	VELOCITY ('/SEC)	SOURCE
90/FMJ	6.7/Unique	1350	Sierra
115/JHP	8.5/Bluedot	1200	Horn.

CASE PREPARATION: **SHELLHOLDER (RCBS): 16**

MAKE FROM: .38 Super. Cut case to length & chamfer. F/L size in Luger die. Carbide dies work best as some amount of base swaging will take place when sizing.

PHYSICAL DATA (INCHES):

CASE TYPE: **Rimless Straight**

CASE LENGTH **A = .754**

HEAD DIAMETER **B = .392**

RIM DIAMETER **D = .393**

NECK DIAMETER **F = .38**

NECK LENGTH **H = N/A**

SHOULDER LENGTH **K = N/A**

BODY ANGLE (DEG'S/SIDE): **.327**

CASE CAPACITY **CC'S = .70**

LOADED LENGTH: **1.10**

BELT DIAMETER **C = N/A**

RIM THICKNESS **E = .05**

SHOULDER DIAMETER **G = N/A**

LENGTH TO SHOULDER **J = N/A**

SHOULDER ANGLE (DEG'S/SIDE): **N/A**

PRIMER: **S/P**

CASE CAPACITY (GR'S WATER): **10.80**

DIMENSIONAL DRAWING:

F
A
B
E
D

-NOT ACTUAL SIZE-
-DO NOT SCALE-

CARTRIDGE: 9mm Makarov

OTHER NAMES:

DIA: .363

BALLISTEK NO: 363A

NAI NO: RMS 11115/1.839

DATA SOURCE: COTW 4th Pg174

HISTORICAL DATA: Current Soviet mil. pistol ctg.

NOTES:

LOADING DATA:

BULLET WT./TYPE	POWDER WT./TYPE	VELOCITY ('/SEC)	SOURCE
100/JFN	5.0/Unique	1000	JJD

CASE PREPARATION: **SHELLHOLDER (RCBS): 16**

MAKE FROM: .38 ACP (or ACP Super). Turn rim to .396" dia. and back chamfer. Cut case to length, F/L size in 9mm Luger die. Standard .355" dia. (9mm) bullets may be used or lead bullets can be "minied," for better accuracy.

Note: Nonte said that 9mm Luger brass could be used, as is, by trimming to length but, you'll find the Luger's bases need to be swaged to .386 dia. or they will not always chamber properly.

PHYSICAL DATA (INCHES):

CASE TYPE: **Rimmed Straight**

CASE LENGTH **A = .70**

HEAD DIAMETER **B = .39**

RIM DIAMETER **D = .396**

NECK DIAMETER **F = .388**

NECK LENGTH **H = N/A**

SHOULDER LENGTH **K = N/A**

BODY ANGLE (DEG'S/SIDE): **.086**

CASE CAPACITY **CC'S = .63**

LOADED LENGTH: **.97**

BELT DIAMETER **C = N/A**

RIM THICKNESS **E = .048**

SHOULDER DIAMETER **G = N/A**

LENGTH TO SHOULDER **J = N/A**

SHOULDER ANGLE (DEG'S/SIDE): **N/A**

PRIMER: **S/P**

CASE CAPACITY (GR'S WATER): **9.86**

DIMENSIONAL DRAWING:

F
A
B
E
D

-NOT ACTUAL SIZE-
-DO NOT SCALE-

CARTRIDGE: 9mm Mauser Pistol

OTHER NAMES:

DIA: .355

BALLISTEK NO: 355N

NAI NO: RXS 21115/2.519

DATA SOURCE: COTW 4th Pg173

HISTORICAL DATA: About 1908.

NOTES: DWM #487.

LOADING DATA:

BULLET WT./TYPE	POWDER WT./TYPE	VELOCITY ('/SEC)	SOURCE
115/JHP	9.0/B'Dot	1000	JJD

CASE PREPARATION: SHELLHOLDER (RCBS): 16

MAKE FROM: 9mm Win. Mag. Trim the Win. case to length and chamfer. F/L size in proper die or reduce case mouth dia. with .38 Spl. sizer.

PHYSICAL DATA (INCHES):

CASE TYPE: **Rimless Straight**

CASE LENGTH **A = .98**

HEAD DIAMETER **B = .389**

RIM DIAMETER **D = .39**

NECK DIAMETER **F = .376**

NECK LENGTH **H = N/A**

SHOULDER LENGTH **K = N/A**

BODY ANGLE (DEG'S/SIDE): **.400**

CASE CAPACITY **CC'S = 1.07**

LOADED LENGTH: **1.38**

BELT DIAMETER **C = N/A**

RIM THICKNESS **E = .049**

SHOULDER DIAMETER **G = N/A**

LENGTH TO SHOULDER **J = N/A**

SHOULDER ANGLE (DEG'S/SIDE): **N/A**

PRIMER: **S/P**

CASE CAPACITY (GR'S WATER): **16.49**

DIMENSIONAL DRAWING:

F
A
E
B
D

-NOT ACTUAL SIZE-
-DO NOT SCALE-

CARTRIDGE: 9mm Steyr

OTHER NAMES:
9mm Austrian M12

DIA: .355

BALLISTEK NO: 355H

NAI NO: RXS 11115/2.368

DATA SOURCE: NAI/Ballistek

HISTORICAL DATA: Austrian mil. ctg.

NOTES: Similar to 9mm Bayard.

LOADING DATA:

BULLET WT./TYPE	POWDER WT./TYPE	VELOCITY ('/SEC)	SOURCE
115/JHP	6.3/Unique	1000	JJD

CASE PREPARATION: SHELLHOLDER (RCBS): 16

MAKE FROM: .38 ACP (or Super ACP). Turn rim to .381" dia. & back chamfer. Use the .38 ACP dies to F/L size and load.

PHYSICAL DATA (INCHES):

CASE TYPE: **Rimless Straight**

CASE LENGTH **A = .900**

HEAD DIAMETER **B = .380**

RIM DIAMETER **D = .381**

NECK DIAMETER **F = .380**

NECK LENGTH **H = N/A**

SHOULDER LENGTH **K = N/A**

BODY ANGLE (DEG'S/SIDE): **0**

CASE CAPACITY **CC'S = .897**

LOADED LENGTH: **1.30**

BELT DIAMETER **C = N/A**

RIM THICKNESS **E = .049**

SHOULDER DIAMETER **G = N/A**

LENGTH TO SHOULDER **J = N/A**

SHOULDER ANGLE (DEG'S/SIDE): **N/A**

PRIMER: **S/P**

CASE CAPACITY (GR'S WATER): **13.84**

DIMENSIONAL DRAWING:

F
A
E
B
D

-NOT ACTUAL SIZE-
-DO NOT SCALE-

CARTRIDGE: 9mm Super Cooper

OTHER NAMES:
9mm Cooper

DIA: .355

BALLISTEK NO: 355B

NAI NO: RXS 11115/2.367

DATA SOURCE: NAI/Ballistek

HISTORICAL DATA: Concept by J. Cooper & I. Stone in 1974.

NOTES: Original ammo by author.

LOADING DATA:

BULLET WT./TYPE	POWDER WT./TYPE	VELOCITY ('/SEC)	SOURCE
90/JHP	9.6/SR4756	2000	JJD

CASE PREPARATION: SHELLHOLDER (RCBS): 10

MAKE FROM: .223 Rem. Cut case off to .90". Taper expand, slightly, to remove .223's normal taper. F/L size in 9mm Cooper die. I.D. ream deep enough for bullet.

Note: Special barrels are required for this cartridge.

PHYSICAL DATA (INCHES):

CASE TYPE: **Rimless Straight**

CASE LENGTH **A = .89**

HEAD DIAMETER **B = .376**

RIM DIAMETER **D = .375**

NECK DIAMETER **F = .374**

NECK LENGTH **H = N/A**

SHOULDER LENGTH **K = N/A**

BODY ANGLE (DEG'S/SIDE): **.067**

CASE CAPACITY **CC'S = .90**

LOADED LENGTH: **1.136**

BELT DIAMETER **C = N/A**

RIM THICKNESS **E = .042**

SHOULDER DIAMETER **G = N/A**

LENGTH TO SHOULDER **J = N/A**

SHOULDER ANGLE (DEG'S/SIDE): **N/A**

PRIMER: **S/P**

CASE CAPACITY (GR'S WATER): **13.93**

DIMENSIONAL DRAWING:

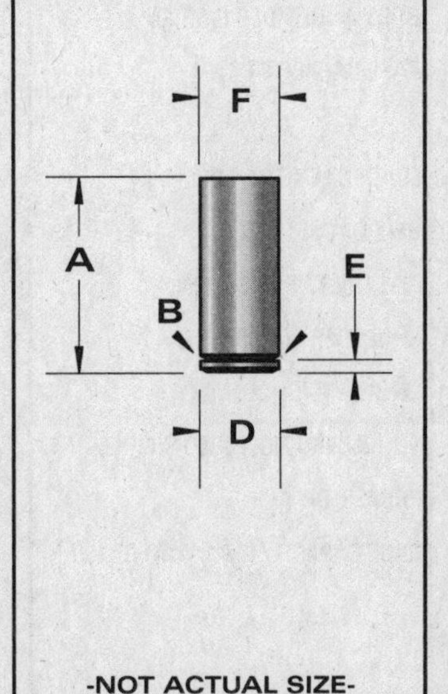

-NOT ACTUAL SIZE-
-DO NOT SCALE-

CARTRIDGE: 9mm Winchester Magnum

OTHER NAMES:

DIA: .355

BALLISTEK NO: 355L

NAI NO: RXS 11115/2.959

DATA SOURCE: COTW 4th Pg175

HISTORICAL DATA: By Win. about 1976.

NOTES: May still be in limbo!

LOADING DATA:

BULLET WT./TYPE	POWDER WT./TYPE	VELOCITY ('/SEC)	SOURCE
115/JHP	7.0/Herco	1200	JJD

CASE PREPARATION: SHELLHOLDER (RCBS): 16

MAKE FROM: Only way to come up with this case is to use 13/32" dia. tubing to re-body a standard 9mm Luger case head. Naturally, this defeats the purpose of the "magnum" so, let's hope the factory cases are available soon.

PHYSICAL DATA (INCHES):

CASE TYPE: **Rimless Straight**

CASE LENGTH **A = 1.160**

HEAD DIAMETER **B = .392**

RIM DIAMETER **D = .394**

NECK DIAMETER **F = .379**

NECK LENGTH **H = N/A**

SHOULDER LENGTH **K = N/A**

BODY ANGLE (DEG'S/SIDE): **.335**

CASE CAPACITY **CC'S = 1.34**

LOADED LENGTH: **1.550**

BELT DIAMETER **C = N/A**

RIM THICKNESS **E = .05**

SHOULDER DIAMETER **G = N/A**

LENGTH TO SHOULDER **J = N/A**

SHOULDER ANGLE (DEG'S/SIDE): **N/A**

PRIMER: **S/P**

CASE CAPACITY (GR'S WATER): **20.72**

DIMENSIONAL DRAWING:

F
A
E
B
D

-NOT ACTUAL SIZE-
-DO NOT SCALE-

CARTRIDGE: 9 x 18 Ultra

OTHER NAMES:
9mm Ultra
9mm Police

DIA: .355

BALLISTEK NO: 355G

NAI NO: RBS 11115/1.818

DATA SOURCE: Handloader #68 Pg18

HISTORICAL DATA: For Luftwaffe during WWII.

NOTES: Case has rebated rim.

LOADING DATA:

BULLET WT./TYPE	POWDER WT./TYPE	VELOCITY ('/SEC)	SOURCE
100/FMJ	6.0/Unique	—	JJD

CASE PREPARATION: SHELLHOLDER (RCBS): 1

MAKE FROM: .38 ACP (or ACP Super). Rebate rim by turning to .367" dia. and deepen extractor groove. Trim case to length. F/L size. I.D. neck ream as is necessary for the bullet being used.

PHYSICAL DATA (INCHES):

CASE TYPE: **Rebated Straight**

CASE LENGTH **A = .708**

HEAD DIAMETER **B = .385**

RIM DIAMETER **D = .370**

NECK DIAMETER **F = .378**

NECK LENGTH **H = N/A**

SHOULDER LENGTH **K = N/A**

BODY ANGLE (DEG'S/SIDE): **.306**

CASE CAPACITY **CC'S = .610**

LOADED LENGTH: **.99**

BELT DIAMETER **C = N/A**

RIM THICKNESS **E = .046**

SHOULDER DIAMETER **G = N/A**

LENGTH TO SHOULDER **J = N/A**

SHOULDER ANGLE (DEG'S/SIDE): **N/A**

PRIMER: **S/P**

CASE CAPACITY (GR'S WATER): **9.42**

DIMENSIONAL DRAWING:

F
A
B
E
D

-NOT ACTUAL SIZE-
-DO NOT SCALE-

CARTRIDGE: 9 X 23 Winchester

OTHER NAMES:

DIA: .355

BALLISTEK NO:

NAI NO:

DATA SOURCE:

HISTORICAL DATA: Introduced by Winchester in 1997 to provide a safe alternative to 38 auto handloads for competition.

NOTES: Do not confuse with the 9x23 Largo. The 9x23 Winchester operates at much higher pressure.

LOADING DATA:

BULLET WT./TYPE	POWDER WT./TYPE	VELOCITY ('/SEC)	SOURCE
125/Silvertip hollow point	Factory Load	1,450	Win

CASE PREPARATION: SHELL HOLDER (RCBS): 16

MAKE FROM: The 9x23 was designed for pressure of 55,000 PSI and it is not recommended that any other case be used to create 9x23 cases for this reason. It is not recommended that 9mm Win Mag brass be used to make this case.

PHYSICAL DATA (INCHES):

CASE TYPE: Rimless Straight

CASE LENGTH **A = .900**

HEAD DIAMETER **B = .3911**

RIM DIAMETER **D = .394**

NECK DIAMETER **F = .381**

NECK LENGTH **H =**

SHOULDER LENGTH **K =**

BODY ANGLE (DEG'S/SIDE):

CASE CAPACITY **CC'S =**

LOADED LENGTH: **1.30**

BELT DIAMETER **C =**

RIM THICKNESS **E = .050**

SHOULDER DIAMETER **G =**

LENGTH TO SHOULDER **J =**

SHOULDER ANGLE (DEG'S/SIDE):

PRIMER: SP

CASE CAPACITY (GR'S WATER): 15.7

DIMENSIONAL DRAWING:

F
A
B
E
D

-NOT ACTUAL SIZE-
-DO NOT SCALE-

CARTRIDGE: 9 x 51.5R Rubin

OTHER NAMES:

DIA: .364

BALLISTEK NO: 364A

NAI NO: RMB 24331/3.745

DATA SOURCE: Hoyem Vol.2 Pg184

HISTORICAL DATA: By Rubin in 1880.

NOTES: Generally found as rimfire version.

LOADING DATA:

BULLET WT./TYPE	POWDER WT./TYPE	VELOCITY ('/SEC)	SOURCE
200/LeadPP	33.4/IMR3031	—	JJD

CASE PREPARATION: SHELLHOLDER (RCBS): Spl.

MAKE FROM: .450 N.E. (BELL). Turn rim to .635" dia. Cut to 2.05". Anneal case and form in two-die form set. Trim to length, I.D. neck ream and F/L size.

PHYSICAL DATA (INCHES):

CASE TYPE: **Rimmed Bottleneck**

CASE LENGTH **A = 2.03**

HEAD DIAMETER **B = .542**

RIM DIAMETER **D = .635**

NECK DIAMETER **F = .394**

NECK LENGTH **H = .380**

SHOULDER LENGTH **K = .350**

BODY ANGLE (DEG'S/SIDE): **.781**

CASE CAPACITY **CC'S = 4.61**

LOADED LENGTH: **3.03**

BELT DIAMETER **C = N/A**

RIM THICKNESS **E = .077**

SHOULDER DIAMETER **G = .512**

LENGTH TO SHOULDER **J = 1.30**

SHOULDER ANGLE (DEG'S/SIDE): **9.57**

PRIMER: **L/R**

CASE CAPACITY (GR'S WATER): **71.24**

DIMENSIONAL DRAWING:

-NOT ACTUAL SIZE-
-DO NOT SCALE-

CARTRIDGE: 9 x 56 Mannlicher Schoenauer

OTHER NAMES:

DIA: .356

BALLISTEK NO: 356B

NAI NO: RXB 22244/4.784

DATA SOURCE: COTW 4th Pg270

HISTORICAL DATA: Circa 1905.

NOTES:

LOADING DATA:

BULLET WT./TYPE	POWDER WT./TYPE	VELOCITY ('/SEC)	SOURCE
200/Spire	40.0/3031	2110	Barnes

CASE PREPARATION: SHELLHOLDER (RCBS): 3

MAKE FROM: .30-06 Spgf. Anneal case neck and taper expand to .360" dia. F/L size in 9x56 die with the expander removed. Finish turn base to .464" dia. Trim to length, chamfer and F/L size again. Fireform in chamber.

PHYSICAL DATA (INCHES):

CASE TYPE: **Rimless Bottleneck**

CASE LENGTH **A = 2.22**

HEAD DIAMETER **B = .464**

RIM DIAMETER **D = .464**

NECK DIAMETER **F = .378**

NECK LENGTH **H = .350**

SHOULDER LENGTH **K = .02**

BODY ANGLE (DEG'S/SIDE): **.972**

CASE CAPACITY **CC'S = 3.96**

LOADED LENGTH: **3.56**

BELT DIAMETER **C = N/A**

RIM THICKNESS **E = .055**

SHOULDER DIAMETER **G = .408**

LENGTH TO SHOULDER **J = 1.85**

SHOULDER ANGLE (DEG'S/SIDE): **36.86**

PRIMER: **L/R**

CASE CAPACITY (GR'S WATER): **61.15**

DIMENSIONAL DRAWING:

-NOT ACTUAL SIZE-
-DO NOT SCALE-

CARTRIDGE: 9 x 57 Mauser

OTHER NAMES:	DIA: .356 BALLISTEK NO: 356A NAI NO: RXB 22242/4.732

DATA SOURCE: COTW 4th Pg270

HISTORICAL DATA:

NOTES:

LOADING DATA:

BULLET WT./TYPE	POWDER WT./TYPE	VELOCITY ('/SEC)	SOURCE
250/—	44.0/3031	2260	Barnes

CASE PREPARATION: SHELLHOLDER (RCBS): 3

MAKE FROM: .30-06 Spgf. Taper expand, trim to length and F/L size. 8x57 Mauser brass will form in the same way.

PHYSICAL DATA (INCHES):

CASE TYPE: **Rimless Bottleneck**
CASE LENGTH **A = 2.21**
HEAD DIAMETER **B = .467**
RIM DIAMETER **D = .468**
NECK DIAMETER **F = .380**
NECK LENGTH **H = .350**
SHOULDER LENGTH **K = .080**
BODY ANGLE (DEG'S/SIDE): **.707**
CASE CAPACITY **CC'S = 4.06**

LOADED LENGTH: **3.10**
BELT DIAMETER **C = N/A**
RIM THICKNESS **E = .05**
SHOULDER DIAMETER **G = .428**
LENGTH TO SHOULDER **J = 1.78**
SHOULDER ANGLE (DEG'S/SIDE): **16.69**
PRIMER: **L/R**
CASE CAPACITY (GR'S WATER): **62.74**

DIMENSIONAL DRAWING:

F
H
K
G
A
J
E
B
D

-NOT ACTUAL SIZE-
-DO NOT SCALE-

CARTRIDGE: 9 x 57R Mauser

OTHER NAMES:	DIA: .356 BALLISTEK NO: 356C NAI NO: RMB 22242 /4.732

DATA SOURCE: COTW 4th Pg270

HISTORICAL DATA: About 1905.

NOTES:

LOADING DATA:

BULLET WT./TYPE	POWDER WT./TYPE	VELOCITY ('/SEC)	SOURCE
250/—	44.0/3031	2260	Barnes

CASE PREPARATION: SHELLHOLDER (RCBS): 28

MAKE FROM: .444 Marlin. Do not anneal. Trim case to length, chamfer and F/L size in 9x57R die.

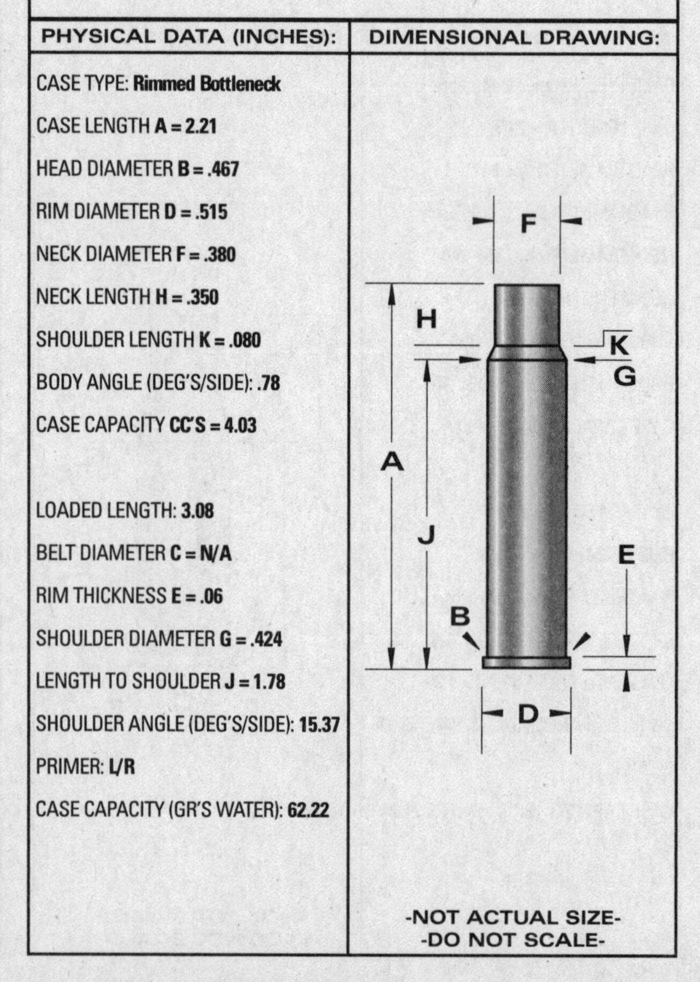

PHYSICAL DATA (INCHES):

CASE TYPE: **Rimmed Bottleneck**
CASE LENGTH **A = 2.21**
HEAD DIAMETER **B = .467**
RIM DIAMETER **D = .515**
NECK DIAMETER **F = .380**
NECK LENGTH **H = .350**
SHOULDER LENGTH **K = .080**
BODY ANGLE (DEG'S/SIDE): **.78**
CASE CAPACITY **CC'S = 4.03**

LOADED LENGTH: **3.08**
BELT DIAMETER **C = N/A**
RIM THICKNESS **E = .06**
SHOULDER DIAMETER **G = .424**
LENGTH TO SHOULDER **J = 1.78**
SHOULDER ANGLE (DEG'S/SIDE): **15.37**
PRIMER: **L/R**
CASE CAPACITY (GR'S WATER): **62.22**

DIMENSIONAL DRAWING:

-NOT ACTUAL SIZE-
-DO NOT SCALE-

CARTRIDGE: 9 x 57R Rubin

OTHER NAMES:

DIA: .362
BALLISTEK NO: 362B
NAI NO: RMB 13231/4.375

DATA SOURCE: Hoyem Vol.2 Pg184

HISTORICAL DATA: By Rubin in 1880.

NOTES:

LOADING DATA:

BULLET WT./TYPE	POWDER WT./TYPE	VELOCITY ('/SEC)	SOURCE
225/LeadPP	38.3/IMR4064	—	JJD

CASE PREPARATION: **SHELLHOLDER (RCBS):** Spl.

MAKE FROM: .43 Span. Rem. (BELL). Turn rim to .586" dia. Anneal case and form in two-die form set. Trim to length & chamfer. F/L size.

PHYSICAL DATA (INCHES):

CASE TYPE: **Rimmed Bottleneck**
CASE LENGTH **A = 2.24**
HEAD DIAMETER **B = .512**
RIM DIAMETER **D = .586**
NECK DIAMETER **F = .394**
NECK LENGTH **H = .39**
SHOULDER LENGTH **K = .39**
BODY ANGLE (DEG'S/SIDE): **.068**
CASE CAPACITY **CC'S = 4.89**

LOADED LENGTH: **3.21**
BELT DIAMETER **C = N/A**
RIM THICKNESS **E = .086**
SHOULDER DIAMETER **G = .509**
LENGTH TO SHOULDER **J = 1.46**
SHOULDER ANGLE (DEG'S/SIDE): **8.39**
PRIMER: **L/R**
CASE CAPACITY (GR'S WATER): **75.54**

DIMENSIONAL DRAWING:

F H K G A J E B D

-NOT ACTUAL SIZE-
-DO NOT SCALE-

CARTRIDGE: 9 x 63mm

OTHER NAMES:

DIA: .358
BALLISTEK NO: 357K
NAI NO: RXB 22232/5.310

DATA SOURCE: COTW 4th Pg271

HISTORICAL DATA: About 1905.

NOTES:

LOADING DATA:

BULLET WT./TYPE	POWDER WT./TYPE	VELOCITY ('/SEC)	SOURCE
231/—	57.0/4350	2510	Barnes

CASE PREPARATION: **SHELLHOLDER (RCBS):** 3

MAKE FROM: .30-06 Spgf. Taper expand to .360" dia. F/L size and trim to length. Fireform in chamber.

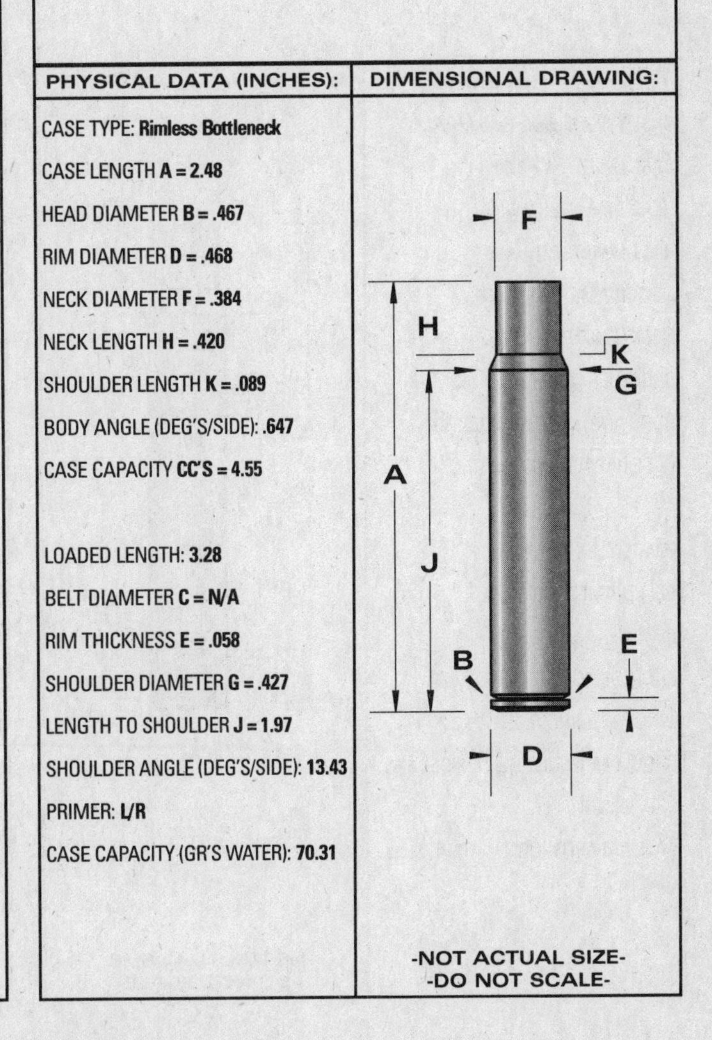

PHYSICAL DATA (INCHES):

CASE TYPE: **Rimless Bottleneck**
CASE LENGTH **A = 2.48**
HEAD DIAMETER **B = .467**
RIM DIAMETER **D = .468**
NECK DIAMETER **F = .384**
NECK LENGTH **H = .420**
SHOULDER LENGTH **K = .089**
BODY ANGLE (DEG'S/SIDE): **.647**
CASE CAPACITY **CC'S = 4.55**

LOADED LENGTH: **3.28**
BELT DIAMETER **C = N/A**
RIM THICKNESS **E = .058**
SHOULDER DIAMETER **G = .427**
LENGTH TO SHOULDER **J = 1.97**
SHOULDER ANGLE (DEG'S/SIDE): **13.43**
PRIMER: **L/R**
CASE CAPACITY (GR'S WATER): **70.31**

DIMENSIONAL DRAWING:

-NOT ACTUAL SIZE-
-DO NOT SCALE-

CARTRIDGE: 9 x 71mm Peterlongo

OTHER NAMES:	
	DIA: .350
	BALLISTEK NO: 350A
	NAI NO: RXB 22231/6.034

DATA SOURCE: COTW 4th Pg270

HISTORICAL DATA: By J. Peterlongo.

NOTES:

LOADING DATA:

BULLET WT./TYPE	POWDER WT./TYPE	VELOCITY ('/SEC)	SOURCE
250/RN	40.4/IMR4350	—	JJD

CASE PREPARATION: SHELLHOLDER (RCBS): 3

MAKE FROM: 9,3 x 74R. Turn rim flush with body dia. Cut new extracor groove. F/L size case in die with expander removed. Trim to length and chamfer.

PHYSICAL DATA (INCHES):

CASE TYPE: **Rimless Bottleneck**
CASE LENGTH **A = 2.800**
HEAD DIAMETER **B = .464**
RIM DIAMETER **D = .466**
NECK DIAMETER **F = .386**
NECK LENGTH **H = .365**
SHOULDER LENGTH **K = .145**
BODY ANGLE (DEG'S/SIDE): **.603**
CASE CAPACITY **CC'S = 4.78**

LOADED LENGTH: **3.28**
BELT DIAMETER **C = N/A**
RIM THICKNESS **E = .06**
SHOULDER DIAMETER **G = .420**
LENGTH TO SHOULDER **J = 2.29**
SHOULDER ANGLE (DEG'S/SIDE): **6.69**
PRIMER: **L/R**
CASE CAPACITY (GR'S WATER): **73.91**

DIMENSIONAL DRAWING:

-NOT ACTUAL SIZE-
-DO NOT SCALE-

CARTRIDGE: 9,1 x 40R

OTHER NAMES:	
	DIA: .374
	BALLISTEK NO: 374A
	NAI NO: RMS 21115/3.960

DATA SOURCE: COTW 4th Pg271

HISTORICAL DATA: Rare European ctg. ca. 1900.

NOTES:

LOADING DATA:

BULLET WT./TYPE	POWDER WT./TYPE	VELOCITY ('/SEC)	SOURCE
140/Lead	14.5/2400	—	JJD

CASE PREPARATION: SHELLHOLDER (RCBS): 6

MAKE FROM: 13/32" dia. tubing case or use the same tubing to re-body a .30-30 case head. Trim to length, chamfer and F/L size.

PHYSICAL DATA (INCHES):

CASE TYPE: **Rimmed Straight**
CASE LENGTH **A = 1.60**
HEAD DIAMETER **B = .404**
RIM DIAMETER **D = .446**
NECK DIAMETER **F = .385**
NECK LENGTH **H = N/A**
SHOULDER LENGTH **K = N/A**
BODY ANGLE (DEG'S/SIDE): **.353**
CASE CAPACITY **CC'S = 2.38**

LOADED LENGTH: **2.00**
BELT DIAMETER **C = N/A**
RIM THICKNESS **E = .062**
SHOULDER DIAMETER **G = N/A**
LENGTH TO SHOULDER **J = N/A**
SHOULDER ANGLE (DEG'S/SIDE): **N/A**
PRIMER: **L/R**
CASE CAPACITY (GR'S WATER): **36.74**

DIMENSIONAL DRAWING:

-NOT ACTUAL SIZE-
-DO NOT SCALE-

CARTRIDGE: 9,3 x 48R

OTHER NAMES:	DIA: .365 BALLISTEK NO: 365G NAI NO: RMS 21115/4.364

DATA SOURCE: COTW 4th Pg272

HISTORICAL DATA:

NOTES:

LOADING DATA:

BULLET WT./TYPE	POWDER WT./TYPE	VELOCITY ('/SEC)	SOURCE
160/Lead	21.0/IMR4198	—	JJD

CASE PREPARATION: SHELLHOLDER (RCBS): 21

MAKE FROM: .303 Savage. Taper expand (after annealing) to a straight case configuration. Trim to length and F/L size.

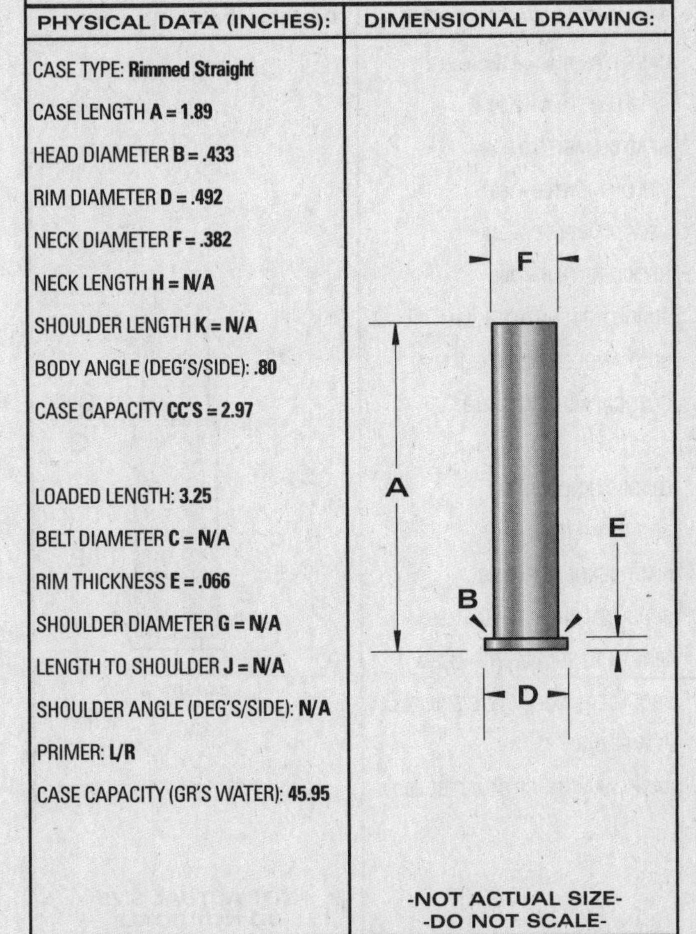

PHYSICAL DATA (INCHES):

CASE TYPE: **Rimmed Straight**

CASE LENGTH **A = 1.89**

HEAD DIAMETER **B = .433**

RIM DIAMETER **D = .492**

NECK DIAMETER **F = .382**

NECK LENGTH **H = N/A**

SHOULDER LENGTH **K = N/A**

BODY ANGLE (DEG'S/SIDE): **.80**

CASE CAPACITY **CC'S = 2.97**

LOADED LENGTH: **3.25**

BELT DIAMETER **C = N/A**

RIM THICKNESS **E = .066**

SHOULDER DIAMETER **G = N/A**

LENGTH TO SHOULDER **J = N/A**

SHOULDER ANGLE (DEG'S/SIDE): **N/A**

PRIMER: **L/R**

CASE CAPACITY (GR'S WATER): **45.95**

DIMENSIONAL DRAWING:

-NOT ACTUAL SIZE-
-DO NOT SCALE-

CARTRIDGE: 9,3 JDJ

OTHER NAMES:	DIA: .366 BALLISTEK NO: NAI NO:

DATA SOURCE: J.D. Jones

HISTORICAL DATA: Designed by J.D. Jones at SSk Industries, this chambering is the 444 Marlin case necked down to a 9.3mm with no other changes.

NOTES:

LOADING DATA:

BULLET WT./TYPE	POWDER WT./TYPE	VELOCITY ('/SEC)	SOURCE
270 Speer	52.0 - IMR-4064	1974	SSK

CASE PREPARATION: SHELL HOLDER (RCBS): 28

MAKE FROM: 444 Marlin

PHYSICAL DATA (INCHES):

CASE TYPE: **Rimless Bottleneck**

CASE LENGTH **A = 2.22**

HEAD DIAMETER **B = .492**

RIM DIAMETER **D = .506**

NECK DIAMETER **F = .389**

NECK LENGTH **H = .765**

SHOULDER LENGTH **K = .048**

BODY ANGLE (DEG'S/SIDE): **1.02**

CASE CAPACITY **CC'S = 3.99**

LOADED LENGTH: **2.80**

BELT DIAMETER **C = N/A**

RIM THICKNESS **E = .058**

SHOULDER DIAMETER **G = .455**

LENGTH TO SHOULDER **J = 1.297**

SHOULDER ANGLE (DEG'S/SIDE): **33.69**

PRIMER: **L/R**

CASE CAPACITY (GR'S WATER): **61.57**

DIMENSIONAL DRAWING:

F
H
K
G
A
J
E
B
D

-NOT ACTUAL SIZE-
-DO NOT SCALE-

CARTRIDGE: 9,3 x 53 Swiss M25

OTHER NAMES:

DIA: .365

BALLISTEK NO: 365E

NAI NO: RXB 34224/4.288

DATA SOURCE: COTW 4th Pg272

HISTORICAL DATA: Popular Swiss ctg. ca. 1925.

NOTES:

LOADING DATA:

BULLET WT./TYPE	POWDER WT./TYPE	VELOCITY ('/SEC)	SOURCE
200/Spire	23.0/IMR4227	—	JJD

CASE PREPARATION: SHELLHOLDER (RCBS): 2

MAKE FROM: 7,5 x 55mm Swiss. Taper expand (after annealing) to .370" dia. Trim to length, chamfer and F/L size. Fireform in chamber.

PHYSICAL DATA (INCHES):

CASE TYPE: **Rimless Bottleneck**

CASE LENGTH A = **2.11**

HEAD DIAMETER B = **.492**

RIM DIAMETER D = **.491**

NECK DIAMETER F = **.389**

NECK LENGTH H = **.765**

SHOULDER LENGTH K = **.048**

BODY ANGLE (DEG'S/SIDE): **1.02**

CASE CAPACITY CC'S = **3.99**

LOADED LENGTH: **2.80**

BELT DIAMETER C = **N/A**

RIM THICKNESS E = **.052**

SHOULDER DIAMETER G = **.453**

LENGTH TO SHOULDER J = **1.297**

SHOULDER ANGLE (DEG'S/SIDE): **33.69**

PRIMER: **L/R**

CASE CAPACITY (GR'S WATER): **61.57**

DIMENSIONAL DRAWING:

F H K G A J B E D

-NOT ACTUAL SIZE-
-DO NOT SCALE-

CARTRIDGE: 9,3 x 53R Swiss M26

OTHER NAMES:

DIA: .365

BALLISTEK NO: 365F

NAI NO: RMB 34224/4.231

DATA SOURCE: COTW 4th Pg272

HISTORICAL DATA:

NOTES:

LOADING DATA:

BULLET WT./TYPE	POWDER WT./TYPE	VELOCITY ('/SEC)	SOURCE
200/Spire	23.0/IMR4227	—	JJD

CASE PREPARATION: SHELLHOLDER (RCBS): 14

MAKE FROM: .45-70 Gov't. Reduce case neck in .40-65 WCF sizer or in proper form die. Anneal nech area. F/L size, trim to length and chamfer. Fireform in chamber. Turn rim to .563" dia., if required.

PHYSICAL DATA (INCHES):

CASE TYPE: **Rimmed Bottleneck**

CASE LENGTH A = **2.09**

HEAD DIAMETER B = **.494**

RIM DIAMETER D = **.563**

NECK DIAMETER F = **.391**

NECK LENGTH H = **.765**

SHOULDER LENGTH K = **.054**

BODY ANGLE (DEG'S/SIDE): **1.02**

CASE CAPACITY CC'S = **3.94**

LOADED LENGTH: **2.83**

BELT DIAMETER C = **N/A**

RIM THICKNESS E = **.068**

SHOULDER DIAMETER G = **.455**

LENGTH TO SHOULDER J = **1.298**

SHOULDER ANGLE (DEG'S/SIDE): **49.84**

PRIMER: **L/R**

CASE CAPACITY (GR'S WATER): **60.77**

DIMENSIONAL DRAWING:

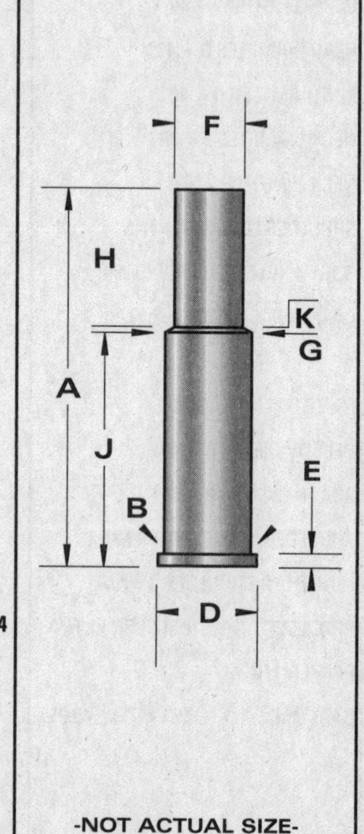

-NOT ACTUAL SIZE-
-DO NOT SCALE-

CARTRIDGE: 9,3 x 54R Hebler

OTHER NAMES:

DIA: .369

BALLISTEK NO: 369A

NAI NO: RMB 23241/4.380

DATA SOURCE: COTW 4th Pg274

HISTORICAL DATA: Rare, experimental ctg.

NOTES: Original had screw-on case head.

LOADING DATA:

BULLET WT./TYPE	POWDER WT./TYPE	VELOCITY ('/SEC)	SOURCE
240/RN	33.8/IMR3031	—	JJD

CASE PREPARATION: SHELLHOLDER (RCBS): 13

MAKE FROM: 7,62 x 54R Russian. Turn rim to .550" dia. and back chamfer. Taper expand to .370" dia. F/L size in Hebler die, square case mouth, chamfer and fireform in the chamber.

PHYSICAL DATA (INCHES):

CASE TYPE: **Rimmed Bottleneck**

CASE LENGTH **A = 2.12**

HEAD DIAMETER **B = .484**

RIM DIAMETER **D = .550**

NECK DIAMETER **F = .398**

NECK LENGTH **H = .330**

SHOULDER LENGTH **K = .340**

BODY ANGLE (DEG'S/SIDE): **.504**

CASE CAPACITY **CC'S = 4.14**

LOADED LENGTH: **2.92**

BELT DIAMETER **C = N/A**

RIM THICKNESS **E = .064**

SHOULDER DIAMETER **G = .462**

LENGTH TO SHOULDER **J = 1.45**

SHOULDER ANGLE (DEG'S/SIDE): **1.45**

PRIMER: **L/R**

CASE CAPACITY (GR'S WATER): **63.97**

DIMENSIONAL DRAWING:

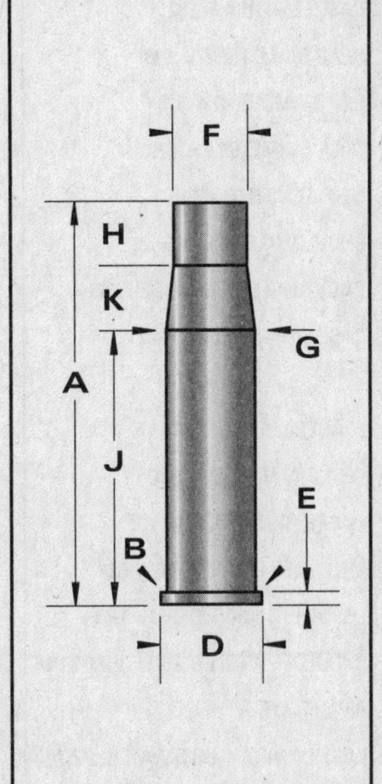

-NOT ACTUAL SIZE-
-DO NOT SCALE-

CARTRIDGE: 9,3 x 57mm Mauser

OTHER NAMES:
9,3/8x57 Mauser
9,2mm Mauser

DIA: .365

BALLISTEK NO: 365H

NAI NO: RXB 22222/4.786

DATA SOURCE: COTW 4th Pg272

HISTORICAL DATA:

NOTES:

LOADING DATA:

BULLET WT./TYPE	POWDER WT./TYPE	VELOCITY ('/SEC)	SOURCE
286/RN	43.0/3031	2070	Barnes

CASE PREPARATION: SHELLHOLDER (RCBS): 3

MAKE FROM: Factory or 8x57 Mauser. Taper expand the 8mm case to .370" dia. (anneal first!). Square case mouth, chamfer and F/L size. .30-06 brass may be used in much the same way but, trim to 2.25" after expanding.

PHYSICAL DATA (INCHES):

CASE TYPE: **Rimless Bottleneck**

CASE LENGTH **A = 2.24**

HEAD DIAMETER **B = .468**

RIM DIAMETER **D = .469**

NECK DIAMETER **F = .389**

NECK LENGTH **H = .365**

SHOULDER LENGTH **K = .075**

BODY ANGLE (DEG'S/SIDE): **.716**

CASE CAPACITY **CC'S = 4.17**

LOADED LENGTH: **3.23**

BELT DIAMETER **C = N/A**

RIM THICKNESS **E = .048**

SHOULDER DIAMETER **G = .428**

LENGTH TO SHOULDER **J = 1.80**

SHOULDER ANGLE (DEG'S/SIDE): **14.57**

PRIMER: **L/R**

CASE CAPACITY (GR'S WATER): **64.29**

DIMENSIONAL DRAWING:

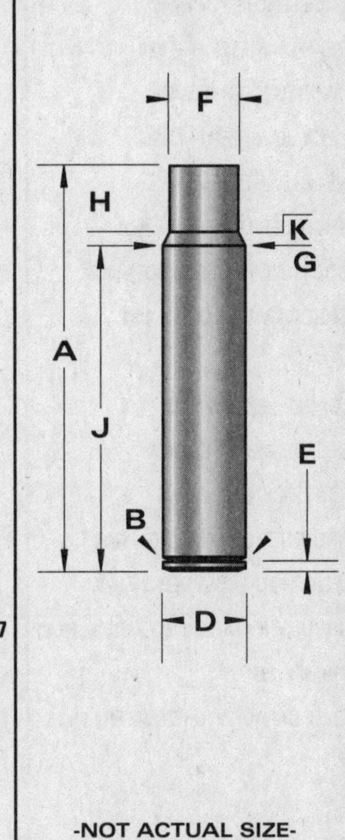

-NOT ACTUAL SIZE-
-DO NOT SCALE-

CARTRIDGE: 9,3 x 57R

OTHER NAMES:
9,3 x 57R/360
.360 Nitro Express

DIA: .365

BALLISTEK NO: 365D

NAI NO: RMS 21115/5.245

DATA SOURCE: COTW 4th Pg272

HISTORICAL DATA: European ctg. circa 1890.

NOTES: This is not a rimmed version of the 9.3 x 57 (#365H).

LOADING DATA:

BULLET WT./TYPE	POWDER WT./TYPE	VELOCITY ('/SEC)	SOURCE
190/Lead	22.0/3031	2000	JJD

CASE PREPARATION: SHELLHOLDER (RCBS): 13

MAKE FROM: .45-70 Gov't. Form die set required. Anneal case and form in form set (2-dies). Trim to length. F/L size.

PHYSICAL DATA (INCHES):

CASE TYPE: **Rimmed Bottleneck**
CASE LENGTH **A = 2.245**
HEAD DIAMETER **B = .428**
RIM DIAMETER **D = .486**
NECK DIAMETER **F = .390**
NECK LENGTH **H = N/A**
SHOULDER LENGTH **K = N/A**
BODY ANGLE (DEG'S/SIDE): **.498**
CASE CAPACITY **CC'S = 4.31**

LOADED LENGTH: **2.83**
BELT DIAMETER **C = N/A**
RIM THICKNESS **E = .06**
SHOULDER DIAMETER **G = N/A**
LENGTH TO SHOULDER **J = N/A**
SHOULDER ANGLE (DEG'S/SIDE): **N/A**
PRIMER: **L/R**
CASE CAPACITY (GR'S WATER): **53.16**

DIMENSIONAL DRAWING:

-NOT ACTUAL SIZE-
-DO NOT SCALE-

CARTRIDGE: 9,3 x 58R Koeffler

OTHER NAMES:

DIA: .367

BALLISTEK NO: 367E

NAI NO: RMB 14322/4.272

DATA SOURCE: Hoyem Vol.2 Pg177

HISTORICAL DATA: Werndl-Mannlicher ctg. about 1878.

NOTES:

LOADING DATA:

BULLET WT./TYPE	POWDER WT./TYPE	VELOCITY ('/SEC)	SOURCE
200/LeadPP	33.6/IMR3031	–	JJD

CASE PREPARATION: SHELLHOLDER (RCBS): Spl.

MAKE FROM: .450 N.E. (BELL). Cut case to 2.4". Build up rim to .08" thick. Form in form die set (two-dies). Trim to length, chamfer and F/L size.

PHYSICAL DATA (INCHES):

CASE TYPE: **Rimmed Bottleneck**
CASE LENGTH **A = 2.32**
HEAD DIAMETER **B = .543**
RIM DIAMETER **D = .618**
NECK DIAMETER **F = .428**
NECK LENGTH **H = .68**
SHOULDER LENGTH **K = .21**
BODY ANGLE (DEG'S/SIDE): **.209**
CASE CAPACITY **CC'S = 4.71**

LOADED LENGTH: **3.15**
BELT DIAMETER **C = N/A**
RIM THICKNESS **E = .08**
SHOULDER DIAMETER **G = .534**
LENGTH TO SHOULDER **J = 1.43**
SHOULDER ANGLE (DEG'S/SIDE): **14.16**
PRIMER: **L/R**
CASE CAPACITY (GR'S WATER): **72.70**

DIMENSIONAL DRAWING:

-NOT ACTUAL SIZE-
-DO NOT SCALE-

CARTRIDGE: 9,3 x 62 Mauser

OTHER NAMES:

DIA: .365

BALLISTEK NO: 365A

NAI NO: RXB 22242/5.116

DATA SOURCE: COTW 4th Pg273

HISTORICAL DATA: About 1905 by O. Bock.

NOTES:

LOADING DATA:

BULLET WT./TYPE	POWDER WT./TYPE	VELOCITY ('/SEC)	SOURCE
286/—	53.0/3031	2360	Barnes

CASE PREPARATION: SHELLHOLDER (RCBS): 3

MAKE FROM: Factory or .30-06 Spgf. Taper expand case neck (after annealing) to .370" dia. F/L size in 9.3x62 die. Trim to length and chamfer.

PHYSICAL DATA (INCHES):

CASE TYPE: **Rimless Bottleneck**

CASE LENGTH **A = 2.42**

HEAD DIAMETER **B = .473**

RIM DIAMETER **D = .470**

NECK DIAMETER **F = .390**

NECK LENGTH **H = .290**

SHOULDER LENGTH **K = .12**

BODY ANGLE (DEG'S/SIDE): **.411**

CASE CAPACITY **CC'S = 4.85**

LOADED LENGTH: **3.29**

BELT DIAMETER **C = N/A**

RIM THICKNESS **E = .049**

SHOULDER DIAMETER **G = .447**

LENGTH TO SHOULDER **J = 2.01**

SHOULDER ANGLE (DEG'S/SIDE): **13.81**

PRIMER: **L/R**

CASE CAPACITY (GR'S WATER): **74.87**

DIMENSIONAL DRAWING:

-NOT ACTUAL SIZE-
-DO NOT SCALE-

CARTRIDGE: 9,3 x 62R Mauser

OTHER NAMES:

DIA: .365

BALLISTEK NO: 365K

NAI NO: RMB 22242/5.116

DATA SOURCE: COTW 4th Pg273

HISTORICAL DATA: Rimmed version of 9,3 x 62 Mauser (#365A).

NOTES:

LOADING DATA:

BULLET WT./TYPE	POWDER WT./TYPE	VELOCITY ('/SEC)	SOURCE
232/RN	62.0/4320	2640	JJD

CASE PREPARATION: SHELLHOLDER (RCBS): 4

MAKE FROM: 9,3 x 74R. F/L size case in the 62R die with the expander removed. Trim to length & chamfer. F/L size again. Fireform in chamber.

PHYSICAL DATA (INCHES):

CASE TYPE: **Rimmed Bottleneck**

CASE LENGTH **A = 2.42**

HEAD DIAMETER **B = .473**

RIM DIAMETER **D = .520**

NECK DIAMETER **F = .388**

NECK LENGTH **H = .295**

SHOULDER LENGTH **K = .095**

BODY ANGLE (DEG'S/SIDE): **.407**

CASE CAPACITY **CC'S = 4.85**

LOADED LENGTH: **3.30**

BELT DIAMETER **C = N/A**

RIM THICKNESS **E = .061**

SHOULDER DIAMETER **G = .447**

LENGTH TO SHOULDER **J = 2.03**

SHOULDER ANGLE (DEG'S/SIDE): **17.25**

PRIMER: **L/R**

CASE CAPACITY (GR'S WATER): **74.95**

DIMENSIONAL DRAWING:

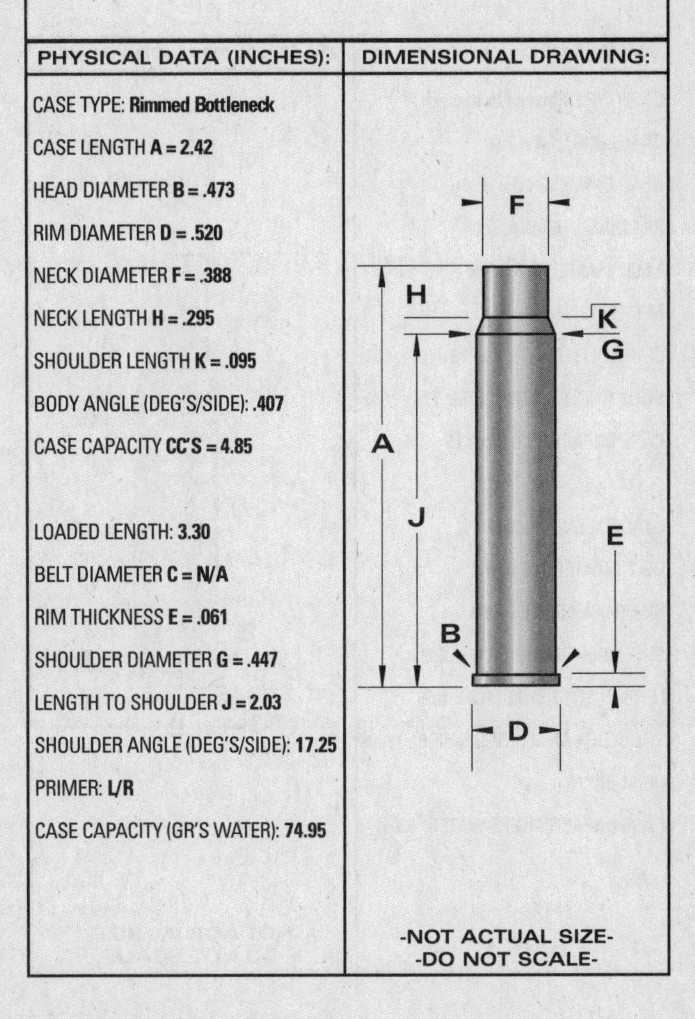

-NOT ACTUAL SIZE-
-DO NOT SCALE-

CARTRIDGE: 9,3 x 63.5R Koeffler

OTHER NAMES:

DIA: .367

BALLISTEK NO: 367F

NAI NO: RMB 22222/5.020

DATA SOURCE: Hoyem Vol.2 Pg179

HISTORICAL DATA: Ctg. for M86 Mannlicher from about 1878.

NOTES:

LOADING DATA:

BULLET WT./TYPE	POWDER WT./TYPE	VELOCITY ('/SEC)	SOURCE
200/LeadPP	34.5/IMR3031	—	JJD

CASE PREPARATION: SHELLHOLDER (RCBS): 14

MAKE FROM: .45 Basic. Cut case to 2.55" and anneal. Form in two-die set. Trim to length and F/L size. Build rim up to .085" thick.

PHYSICAL DATA (INCHES):

CASE TYPE: **Rimmed Bottleneck**

CASE LENGTH **A = 2.50**

HEAD DIAMETER **B = .498**

RIM DIAMETER **D = .572**

NECK DIAMETER **F = .388**

NECK LENGTH **H = .67**

SHOULDER LENGTH **K = .18**

BODY ANGLE (DEG'S/SIDE): **.77**

CASE CAPACITY **CC'S = 5.15**

LOADED LENGTH: **3.40**

BELT DIAMETER **C = N/A**

RIM THICKNESS **E = .085**

SHOULDER DIAMETER **G = .459**

LENGTH TO SHOULDER **J = 1.65**

SHOULDER ANGLE (DEG'S/SIDE): **11.15**

PRIMER: **L/R**

CASE CAPACITY (GR'S WATER): **79.51**

DIMENSIONAL DRAWING:

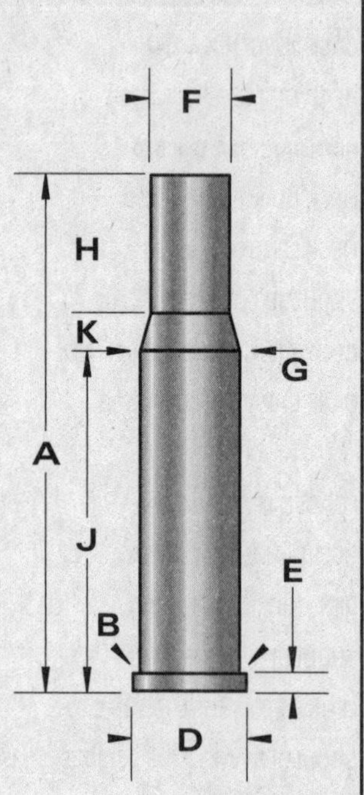

-NOT ACTUAL SIZE-
-DO NOT SCALE-

CARTRIDGE: 9,3 x 64 Brenneke

OTHER NAMES:

DIA: .365

BALLISTEK NO: 366A

NAI NO: RBB 22242/5.000

DATA SOURCE: COTW 4th Pg273

HISTORICAL DATA: By W. Brenneke.

NOTES:

LOADING DATA:

BULLET WT./TYPE	POWDER WT./TYPE	VELOCITY ('/SEC)	SOURCE
286/—	58.0/4064	2400	Barnes

CASE PREPARATION: SHELLHOLDER (RCBS): 21

MAKE FROM: Factory or .45 Basic. Rebate rim by turning to .492" dia. and cut extractor groove. Cut case to 2.6", anneal and F/L size (some cases need neck reducing in our .40-65 WCF die before F/L sizing). Trim and chamfer.

PHYSICAL DATA (INCHES):

CASE TYPE: **Rebated Bottleneck**

CASE LENGTH **A = 2.52**

HEAD DIAMETER **B = .504**

RIM DIAMETER **D = .492**

NECK DIAMETER **F = .391**

NECK LENGTH **H = .340**

SHOULDER LENGTH **K = .140**

BODY ANGLE (DEG'S/SIDE): **.358**

CASE CAPACITY **CC'S = 5.64**

LOADED LENGTH: **3.43**

BELT DIAMETER **C = N/A**

RIM THICKNESS **E = .061**

SHOULDER DIAMETER **G = .481**

LENGTH TO SHOULDER **J = 2.04**

SHOULDER ANGLE (DEG'S/SIDE): **17.82**

PRIMER: **L/R**

CASE CAPACITY (GR'S WATER): **87.04**

DIMENSIONAL DRAWING:

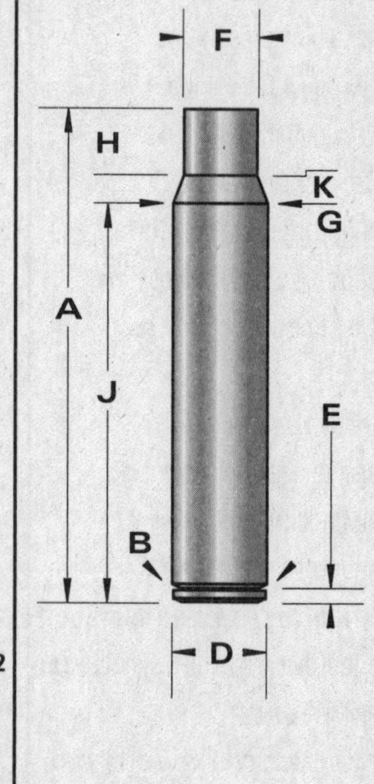

-NOT ACTUAL SIZE-
-DO NOT SCALE-

CARTRIDGE: 9,3 x 65R Collath

OTHER NAMES:

DIA: .367

BALLISTEK NO: 367C

NAI NO: RMB 12241/5.779

DATA SOURCE: COTW 4th Pg273

HISTORICAL DATA: By Collath in early 1900's.

NOTES:

LOADING DATA:

BULLET WT./TYPE	POWDER WT./TYPE	VELOCITY ('/SEC)	SOURCE
193/FN	41.5/BL-C2	—	JJD

CASE PREPARATION: SHELLHOLDER (RCBS): 4

MAKE FROM: 9.3 x 72R. Trim the 9,3mm case to 2.6". F/L size in the Collath die. Fireform in the Collath trim die, with cornmeal, to expand the body. F/L size again and trim to length.

PHYSICAL DATA (INCHES):

CASE TYPE: **Rimmed Bottleneck**

CASE LENGTH **A = 2.56**

HEAD DIAMETER **B = .443**

RIM DIAMETER **D = .508**

NECK DIAMETER **F = .384**

NECK LENGTH **H = .273**

SHOULDER LENGTH **K = .197**

BODY ANGLE (DEG'S/SIDE): **.348**

CASE CAPACITY **CC'S = 4.76**

LOADED LENGTH: **3.01**

BELT DIAMETER **C = N/A**

RIM THICKNESS **E = .053**

SHOULDER DIAMETER **G = .420**

LENGTH TO SHOULDER **J = 2.09**

SHOULDER ANGLE (DEG'S/SIDE): **5.22**

PRIMER: **L/R**

CASE CAPACITY (GR'S WATER): **73.51**

DIMENSIONAL DRAWING:

F
H
K
G
A
J
E
B
D

-NOT ACTUAL SIZE-
-DO NOT SCALE-

CARTRIDGE: 9,3 x 70R

OTHER NAMES:

DIA: .365

BALLISTEK NO: 365J

NAI NO: RMS 21115/6.440

DATA SOURCE: COTW 4th Pg272

HISTORICAL DATA: Circa 1890 German black powder ctg.

NOTES:

LOADING DATA:

BULLET WT./TYPE	POWDER WT./TYPE	VELOCITY ('/SEC)	SOURCE
300/Lead	36.8/IMR4320	—	JJD

CASE PREPARATION: SHELLHOLDER (RCBS): 30

MAKE FROM: 9,3 x 73R. Trim case to length and F/L size.

PHYSICAL DATA (INCHES):

CASE TYPE: **Rimmed Straight**

CASE LENGTH **A = 2.75**

HEAD DIAMETER **B = .427**

RIM DIAMETER **D = .482**

NECK DIAMETER **F = .387**

NECK LENGTH **H = N/A**

SHOULDER LENGTH **K = N/A**

BODY ANGLE (DEG'S/SIDE): **.425**

CASE CAPACITY **CC'S = 4.40**

LOADED LENGTH: **3.45**

BELT DIAMETER **C = N/A**

RIM THICKNESS **E = .06**

SHOULDER DIAMETER **G = N/A**

LENGTH TO SHOULDER **J = N/A**

SHOULDER ANGLE (DEG'S/SIDE): **N/A**

PRIMER: **L/R**

CASE CAPACITY (GR'S WATER): **67.93**

DIMENSIONAL DRAWING:

F
A
E
B
D

-NOT ACTUAL SIZE-
-DO NOT SCALE-

CARTRIDGE: 9,3 x 72R

OTHER NAMES:

DIA: .365

BALLISTEK NO: 365B

NAI NO: RMS 21115/6.651

DATA SOURCE: COTW 4th Pg272

HISTORICAL DATA:

NOTES: Do not confuse with 9,3 x 72R Sauer (#365M)

LOADING DATA:

BULLET WT./TYPE	POWDER WT./TYPE	VELOCITY ('/SEC)	SOURCE
193/RN	39.0/BL-C2	—	JJD

CASE PREPARATION: SHELLHOLDER (RCBS): 30

MAKE FROM: Factory or re-body a .30-30 case head with 7/16" dia. tubing. It is also possible to solder a tubing extension, on the .30-30 case but, if you are going to solder, the re-bodied case is much better.

Fabricate the case, turn the rim to diameter & back chamfer, anneal, F/L size, trim to length and chamfer.

PHYSICAL DATA (INCHES):

CASE TYPE: **Rimmed Straight**

CASE LENGTH **A = 2.84**

HEAD DIAMETER **B = .427**

RIM DIAMETER **D = .482**

NECK DIAMETER **F = .385**

NECK LENGTH **H = N/A**

SHOULDER LENGTH **K = N/A**

BODY ANGLE (DEG'S/SIDE): **.432**

CASE CAPACITY **CC'S = 4.61**

LOADED LENGTH: **3.27**

BELT DIAMETER **C = N/A**

RIM THICKNESS **E = .056**

SHOULDER DIAMETER **G = N/A**

LENGTH TO SHOULDER **J = N/A**

SHOULDER ANGLE (DEG'S/SIDE): **N/A**

PRIMER: **L/R**

CASE CAPACITY (GR'S WATER): **71.19**

DIMENSIONAL DRAWING:

F
A
E
B
D

-NOT ACTUAL SIZE-
-DO NOT SCALE-

CARTRIDGE: 9,3 x 72R Sauer

OTHER NAMES:

DIA: .365

BALLISTEK NO: 365M

NAI NO: RMB 22232/5.983

DATA SOURCE: COTW 4th Pg274

HISTORICAL DATA: By Sauer & Son

NOTES: Do not confuse with 9,3 x 72R (#365B).

LOADING DATA:

BULLET WT./TYPE	POWDER WT./TYPE	VELOCITY ('/SEC)	SOURCE
186/FN	42.2/BL-C2	—	JJD

CASE PREPARATION: SHELLHOLDER (RCBS): 18

MAKE FROM: A short version of this case can be made from 7x57R brass by taper expanding and F/L sizing. Also, you may re-body a .44 Mag. case head with 15/32" dia. tubing. Fabricate case, trim to length, anneal, F/L size and trim.

PHYSICAL DATA (INCHES):

CASE TYPE: **Rimmed Bottleneck**

CASE LENGTH **A = 2.83**

HEAD DIAMETER **B = .473**

RIM DIAMETER **D = .510**

NECK DIAMETER **F = .391**

NECK LENGTH **H = .450**

SHOULDER LENGTH **K = .060**

BODY ANGLE (DEG'S/SIDE): **.69**

CASE CAPACITY **CC'S = 5.34**

LOADED LENGTH: **3.34**

BELT DIAMETER **C = N/A**

RIM THICKNESS **E = .047**

SHOULDER DIAMETER **G = .422**

LENGTH TO SHOULDER **J = 2.32**

SHOULDER ANGLE (DEG'S/SIDE): **14.48**

PRIMER: **L/R**

CASE CAPACITY (GR'S WATER): **82.43**

DIMENSIONAL DRAWING:

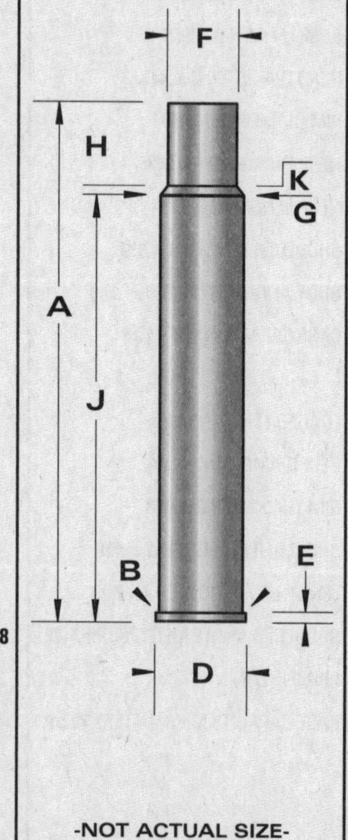

-NOT ACTUAL SIZE-
-DO NOT SCALE-

CARTRIDGE: 9,3 x 74R

OTHER NAMES:

DIA: .365

BALLISTEK NO: 365C

NAI NO: RMB 22221/6.290

DATA SOURCE: COTW 4th Pg274

HISTORICAL DATA: Popular German ctg. since 1900.

NOTES:

LOADING DATA:

BULLET WT./TYPE	POWDER WT./TYPE	VELOCITY ('/SEC)	SOURCE
286/RN	57.0/4064	2360	Barnes

CASE PREPARATION: SHELLHOLDER (RCBS): 26

MAKE FROM: Factory or .444 Marlin. Anneal the Marlin case and F/L size for a short version of the actual case. The Marlin case head can be re-bodied with 15/32" dia. tubing for a full-length (but, weaker) case. Fabricate case, trim to length, anneal, F/L size, trim to final length.

PHYSICAL DATA (INCHES):

CASE TYPE: **Rimmed Bottleneck**

CASE LENGTH **A = 2.925**

HEAD DIAMETER **B = .465**

RIM DIAMETER **D = .524**

NECK DIAMETER **F = .387**

NECK LENGTH **H = .496**

SHOULDER LENGTH **K = .129**

BODY ANGLE (DEG'S/SIDE): **.695**

CASE CAPACITY **CC'S = 5.46**

LOADED LENGTH: **3.47**

BELT DIAMETER **C = N/A**

RIM THICKNESS **E = .065**

SHOULDER DIAMETER **G = .414**

LENGTH TO SHOULDER **J = 2.30**

SHOULDER ANGLE (DEG'S/SIDE): **5.97**

PRIMER: **L/R**

CASE CAPACITY (GR'S WATER): **84.27**

DIMENSIONAL DRAWING:

F
H
K
G
A
J
B
E
D

-NOT ACTUAL SIZE-
-DO NOT SCALE-

CARTRIDGE: 9,3 x 80R

OTHER NAMES:

DIA: .365

BALLISTEK NO: 365L

NAI NO: RMS 21115/7.302

DATA SOURCE: COTW 4th Pg272

HISTORICAL DATA:

NOTES: Similar to 9,3x72R (#365B).

LOADING DATA:

BULLET WT./TYPE	POWDER WT./TYPE	VELOCITY ('/SEC)	SOURCE
225/FN	45.0/IMR4064	—	JJD

CASE PREPARATION: SHELLHOLDER (RCBS): 30

MAKE FROM: 9,3 x 72R. Use the 72mm case, as is, for a "short" version or re-body a .30-30 case head with 7/16" dia. tubing and turn the rim to .485" dia. Anneal, F/L size and trim to length.

PHYSICAL DATA (INCHES):

CASE TYPE: **Rimmed Straight**

CASE LENGTH **A = 3.14**

HEAD DIAMETER **B = .430**

RIM DIAMETER **D = .485**

NECK DIAMETER **F = .386**

NECK LENGTH **H = N/A**

SHOULDER LENGTH **K = N/A**

BODY ANGLE (DEG'S/SIDE): **.409**

CASE CAPACITY **CC'S = 5.17**

LOADED LENGTH: **3.50**

BELT DIAMETER **C = N/A**

RIM THICKNESS **E = .06**

SHOULDER DIAMETER **G = N/A**

LENGTH TO SHOULDER **J = N/A**

SHOULDER ANGLE (DEG'S/SIDE): **N/A**

PRIMER: **L/R**

CASE CAPACITY (GR'S WATER): **79.83**

DIMENSIONAL DRAWING:

F
A
E
B
D

-NOT ACTUAL SIZE-
-DO NOT SCALE-

CARTRIDGE: 9,3 x 82R

OTHER NAMES:	DIA: .365 BALLISTEK NO: 365I NAI NO: RMS 21115/7.465

DATA SOURCE: COTW 4th Pg272

HISTORICAL DATA:

NOTES:

LOADING DATA:

BULLET WT./TYPE	POWDER WT./TYPE	VELOCITY ('/SEC)	SOURCE
225/Lead	40.0/BL-C2	—	JJD

CASE PREPARATION: SHELLHOLDER (RCBS): 21

MAKE FROM: Use 9,3x72R brass, as is, or fabricate a 7/16" dia. tubing case. It is also possible to re-body the 9,3x72R case head or a '06-type case head if full-length cases are needed.

PHYSICAL DATA (INCHES):

CASE TYPE: **Rimmed Straight**
CASE LENGTH **A = 3.21**
HEAD DIAMETER **B = .430**
RIM DIAMETER **D = .485**
NECK DIAMETER **F = .386**
NECK LENGTH **H = N/A**
SHOULDER LENGTH **K = N/A**
BODY ANGLE (DEG'S/SIDE): **.40**
CASE CAPACITY **CC'S = 5.30**

LOADED LENGTH: **3.72**
BELT DIAMETER **C = N/A**
RIM THICKNESS **E = .062**
SHOULDER DIAMETER **G = N/A**
LENGTH TO SHOULDER **J = N/A**
SHOULDER ANGLE (DEG'S/SIDE): **N/A**
PRIMER: **L/R**
CASE CAPACITY (GR'S WATER): **81.80**

DIMENSIONAL DRAWING:

F
A
E
B
D

-NOT ACTUAL SIZE-
-DO NOT SCALE-

CARTRIDGE: 9,4mm Dutch East Indies

OTHER NAMES:	DIA: .380 BALLISTEK NO: 380C NAI NO: RMS 21115/2.488

DATA SOURCE: MSAA Pg42

HISTORICAL DATA: Long version of 9,4mm Dutch Revolver (#380B).

NOTES:

LOADING DATA:

BULLET WT./TYPE	POWDER WT./TYPE	VELOCITY ('/SEC)	SOURCE
150/Lead	1.3/B'Eye	—	JJD

CASE PREPARATION: SHELLHOLDER (RCBS): 21

MAKE FROM: .303 Savage. Cut case to 1.07". Turn rim to .480" dia. & back chamfer. I.D. neck ream, chamfer and F/L size.

PHYSICAL DATA (INCHES):

CASE TYPE: **Rimmed Straight**
CASE LENGTH **A = 1.07**
HEAD DIAMETER **B = .430**
RIM DIAMETER **D = .480**
NECK DIAMETER **F = .410**
NECK LENGTH **H = N/A**
SHOULDER LENGTH **K = N/A**
BODY ANGLE (DEG'S/SIDE): **.559**
CASE CAPACITY **CC'S = 1.38**

LOADED LENGTH: **1.29**
BELT DIAMETER **C = N/A**
RIM THICKNESS **E = .045**
SHOULDER DIAMETER **G = N/A**
LENGTH TO SHOULDER **J = N/A**
SHOULDER ANGLE (DEG'S/SIDE): **N/A**
PRIMER: **L/P**
CASE CAPACITY (GR'S WATER): **21.35**

DIMENSIONAL DRAWING:

F
A
E
B
D

-NOT ACTUAL SIZE-
-DO NOT SCALE-

CARTRIDGE: 9,4mm Dutch Revolver

OTHER NAMES:
9,4mm Netherlands M73
9,4mm Scherpe Patrone

DIA: .380

BALLISTEK NO: 380B

NAI NO: RMS 21115/1.884

DATA SOURCE: MSAA Pg42

HISTORICAL DATA: 1873 for Chamelot-Delvigne revolver.

NOTES: This is the "Home Service" version.

LOADING DATA:

BULLET WT./TYPE	POWDER WT./TYPE	VELOCITY ('/SEC)	SOURCE
150/Lead	1.2/B'Eye	600	JJD

CASE PREPARATION: SHELLHOLDER (RCBS): 21

MAKE FROM: .303 Savage. Cut case to .81". Turn rim to .490" dia. & back chamfer. I.D. neck ream, chamfer, F/L size.

PHYSICAL DATA (INCHES):

CASE TYPE: **Rimmed Straight**

CASE LENGTH **A = .81**

HEAD DIAMETER **B = .430**

RIM DIAMETER **D = .490**

NECK DIAMETER **F = .410**

NECK LENGTH **H = N/A**

SHOULDER LENGTH **K = N/A**

BODY ANGLE (DEG'S/SIDE): **.745**

CASE CAPACITY **CC'S = .91**

LOADED LENGTH: **1.14**

BELT DIAMETER **C = N/A**

RIM THICKNESS **E = .042**

SHOULDER DIAMETER **G = N/A**

LENGTH TO SHOULDER **J = N/A**

SHOULDER ANGLE (DEG'S/SIDE): **N/A**

PRIMER: **L/P**

CASE CAPACITY (GR'S WATER): **14.09**

DIMENSIONAL DRAWING:

F
E
A
B
D

-NOT ACTUAL SIZE-
-DO NOT SCALE-

CARTRIDGE: 9,5 x 47R

OTHER NAMES:
9,5 x 47R Martini
9,5 x 47R Deutsche Schuetzen

DIA: .375

BALLISTEK NO: 375Y

NAI NO: RMB 24221/3.606

DATA SOURCE: Handloader #113 Pg12

HISTORICAL DATA: Black powder target ctg. ca. 1880.

NOTES: Nonte reported dia. as .348-.350". I believe that .370-.375" is more likely.

LOADING DATA:

BULLET WT./TYPE	POWDER WT./TYPE	VELOCITY ('/SEC)	SOURCE
175/Lead	25.5/IMR3031	–	JJD

CASE PREPARATION: SHELLHOLDER (RCBS): Spl.

MAKE FROM: .43 Span. Rem. (BELL). Turn rim to .583" dia. & back chamfer. Cut case to 1.9" and anneal. Form in form die set (2 dies). Trim to length. May require I.D. neck reaming for proper bullet. F/L size.

PHYSICAL DATA (INCHES):

CASE TYPE: **Rimmed Bottleneck**

CASE LENGTH **A = 1.85**

HEAD DIAMETER **B = .513**

RIM DIAMETER **D = .583**

NECK DIAMETER **F = .409**

NECK LENGTH **H = .52**

SHOULDER LENGTH **K = .335**

BODY ANGLE (DEG'S/SIDE): **.576**

CASE CAPACITY **CC'S = 3.69**

LOADED LENGTH: **2.37**

BELT DIAMETER **C = N/A**

RIM THICKNESS **E = .09**

SHOULDER DIAMETER **G = .497**

LENGTH TO SHOULDER **J = .995**

SHOULDER ANGLE (DEG'S/SIDE): **7.48**

PRIMER: **L/R**

CASE CAPACITY (GR'S WATER): **56.98**

DIMENSIONAL DRAWING:

F
H
K
G
A
J
E
B
D

-NOT ACTUAL SIZE-
-DO NOT SCALE-

CARTRIDGE: 9,5 x 60R Mauser

OTHER NAMES:

DIA: .382
BALLISTEK NO: 382A
NAI NO: RMB 23221/4.581

DATA SOURCE: Hoyem Vol.2 Pg220

HISTORICAL DATA: Ctg. for Turkish-Peabody-Martini circa 1887.

NOTES:

LOADING DATA:

BULLET WT./TYPE	POWDER WT./TYPE	VELOCITY ('/SEC)	SOURCE
284/Lead	35.8/IMR3031	–	JJD

CASE PREPARATION: SHELLHOLDER (RCBS): Spl.

MAKE FROM: .43 Span. Rem. (BELL). Turn rim dia. down to .586" dia. and thin, if necessary, to .070" thick. Anneal case and form in form die set (2-dies). Trim to length, chamfer and F/L size.

PHYSICAL DATA (INCHES):

CASE TYPE: **Rimmed Bottleneck**
CASE LENGTH **A = 2.35**
HEAD DIAMETER **B = .513**
RIM DIAMETER **D = .586**
NECK DIAMETER **F = .410**
NECK LENGTH **H = .600**
SHOULDER LENGTH **K = .250**
BODY ANGLE (DEG'S/SIDE): **.617**
CASE CAPACITY **CC'S = 5.07**

LOADED LENGTH: **2.99**
BELT DIAMETER **C = N/A**
RIM THICKNESS **E = .07**
SHOULDER DIAMETER **G = .485**
LENGTH TO SHOULDER **J = 1.50**
SHOULDER ANGLE (DEG'S/SIDE): **8.53**
PRIMER: **L/R**
CASE CAPACITY (GR'S WATER): **78.23**

DIMENSIONAL DRAWING:

-NOT ACTUAL SIZE-
-DO NOT SCALE-

CARTRIDGE: 9,5 x 73 Miller Greiss

OTHER NAMES:

DIA: .375
BALLISTEK NO: 375T
NAI NO: RXB 12321/5.257

DATA SOURCE: COTW 4th Pg275

HISTORICAL DATA:

NOTES: Reported diameters vary .372-.375".

LOADING DATA:

BULLET WT./TYPE	POWDER WT./TYPE	VELOCITY ('/SEC)	SOURCE
270/RN	79.0/H414	–	JJD

CASE PREPARATION: SHELLHOLDER (RCBS): 7

MAKE FROM: .404 Jeffery (BELL). Form die is required. Anneal case and form. Square case mouth & chamfer. F/L size.

PHYSICAL DATA (INCHES):

CASE TYPE: **Rimless Bottleneck**
CASE LENGTH **A = 2.855**
HEAD DIAMETER **B = .543**
RIM DIAMETER **D = .541**
NECK DIAMETER **F = .402**
NECK LENGTH **H = .475**
SHOULDER LENGTH **K = .365**
BODY ANGLE (DEG'S/SIDE): **.189**
CASE CAPACITY **CC'S = 7.39**

LOADED LENGTH: **3.50**
BELT DIAMETER **C = N/A**
RIM THICKNESS **E = .05**
SHOULDER DIAMETER **G = .531**
LENGTH TO SHOULDER **J = 2.015**
SHOULDER ANGLE (DEG'S/SIDE): **10.02**
PRIMER: **L/R**
CASE CAPACITY (GR'S WATER): **114.10**

DIMENSIONAL DRAWING:

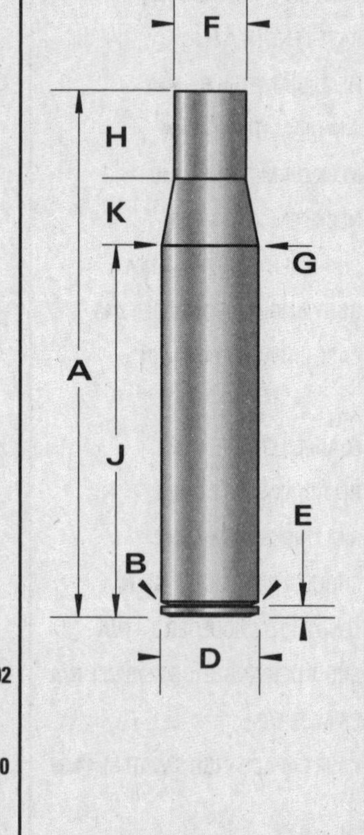

-NOT ACTUAL SIZE-
-DO NOT SCALE-

CARTRIDGE: 10mm Bren

OTHER NAMES:

DIA: .400

BALLISTEK NO: 400C

NAI NO: RXS 11115/2.369

DATA SOURCE: NAI/Ballistek

HISTORICAL DATA: By Dornaus & Dixon for Bren 10 auto pistol.

NOTES: Some of the original ammo was manufactured by the author.

LOADING DATA:

BULLET WT./TYPE	POWDER WT./TYPE	VELOCITY ('/SEC)	SOURCE
200/JTC	10.0/B'Dot	1000	D&D

CASE PREPARATION: SHELLHOLDER (RCBS): 19

MAKE FROM: Factory brass may be available by the time this is published. Otherwise, use .30 Rem. brass and cut to 1.01". Taper expand to .41" dia. F/L size. I.D. Neck ream. F/L size again & chamfer.

PHYSICAL DATA (INCHES):

CASE TYPE: **Rimless Straight**

CASE LENGTH **A = 1.00**

HEAD DIAMETER **B = .422**

RIM DIAMETER **D = .422**

NECK DIAMETER **F = .421**

NECK LENGTH **H = N/A**

SHOULDER LENGTH **K = N/A**

BODY ANGLE (DEG'S/SIDE): **.030**

CASE CAPACITY **CC'S = .50**

LOADED LENGTH: **1.275**

BELT DIAMETER **C = N/A**

RIM THICKNESS **E = .04**

SHOULDER DIAMETER **G = N/A**

LENGTH TO SHOULDER **J = N/A**

SHOULDER ANGLE (DEG'S/SIDE): **N/A**

PRIMER: **L/P**

CASE CAPACITY (GR'S WATER): **7.73**

DIMENSIONAL DRAWING:

F
A
E
B
D

-NOT ACTUAL SIZE-
-DO NOT SCALE-

CARTRIDGE: 10 x 49R Hotchkiss Auxiliary

OTHER NAMES:

DIA: .408

BALLISTEK NO: 408D

NAI NO: RMB 14221/3.587

DATA SOURCE: Hoyem Vol.2 Pg234

HISTORICAL DATA: Used with insert in 37mm Hotchkiss revolving cannon.

NOTES: Quite rare.

LOADING DATA:

BULLET WT./TYPE	POWDER WT./TYPE	VELOCITY ('/SEC)	SOURCE
340/Lead	30.9/RE#7	–	JJD

CASE PREPARATION: SHELLHOLDER (RCBS): Spl.

MAKE FROM: .280 Flanged (BELL). Cut case to 2.0". build up rim to .081" thickness. Taper expand case to .410" dia. and trim to length. F/L size with the expander removed and I.D. neck ream. F/L size and fireform in the chamber. .406" dia. bullets work fine.

PHYSICAL DATA (INCHES):

CASE TYPE: **Rimmed Bottleneck**

CASE LENGTH **A = 1.93**

HEAD DIAMETER **B = .538**

RIM DIAMETER **D = .660**

NECK DIAMETER **F = .431**

NECK LENGTH **H = .582**

SHOULDER LENGTH **K = .324**

BODY ANGLE (DEG'S/SIDE): **.243**

CASE CAPACITY **CC'S = 4.64**

LOADED LENGTH: **2.85**

BELT DIAMETER **C = N/A**

RIM THICKNESS **E = .081**

SHOULDER DIAMETER **G = .531**

LENGTH TO SHOULDER **J = 1.024**

SHOULDER ANGLE (DEG'S/SIDE): **8.77**

PRIMER: **L/R**

CASE CAPACITY (GR'S WATER): **71.53**

DIMENSIONAL DRAWING:

F
H
K
G
A
J
B
E
D

-NOT ACTUAL SIZE-
-DO NOT SCALE-

CARTRIDGE: 10,15 x 54R Jarman

OTHER NAMES:

DIA: .408

BALLISTEK NO: 408E

NAI NO: RMB 14233/3.888

DATA SOURCE: Hoyem Vol.2 Pg230

HISTORICAL DATA: Norwegian ctg. for Jarman rifle about 1881.

NOTES:

LOADING DATA:

BULLET WT./TYPE	POWDER WT./TYPE	VELOCITY ('/SEC)	SOURCE
250/Spire	28.2/IMR4198	—	JJD

CASE PREPARATION: SHELLHOLDER (RCBS): Spl.

MAKE FROM: .280 Flanged (BELL). Cut case to 2.15". Build rim up to .083" thickness. F/L size & chamfer.

PHYSICAL DATA (INCHES):

CASE TYPE: **Rimmed Bottleneck**

CASE LENGTH **A = 2.10**

HEAD DIAMETER **B = .540**

RIM DIAMETER **D = .613**

NECK DIAMETER **F = .434**

NECK LENGTH **H = .490**

SHOULDER LENGTH **K = .180**

BODY ANGLE (DEG'S/SIDE): **.256**

CASE CAPACITY **CC'S = 5.28**

LOADED LENGTH: **3.14**

BELT DIAMETER **C = N/A**

RIM THICKNESS **E = .083**

SHOULDER DIAMETER **G = .529**

LENGTH TO SHOULDER **J = 1.43**

SHOULDER ANGLE (DEG'S/SIDE): **14.78**

PRIMER: **L/R**

CASE CAPACITY (GR'S WATER): **81.49**

DIMENSIONAL DRAWING:

F H K G A J E B D

-NOT ACTUAL SIZE-
-DO NOT SCALE-

CARTRIDGE: 10,15 x 61R Jarman

OTHER NAMES:

DIA: .403

BALLISTEK NO: 403S

NAI NO: RMB 14222/4.379

DATA SOURCE: COTW 4th Pg209

HISTORICAL DATA: Norwegian mil. ctg. about 1887.

NOTES:

LOADING DATA:

BULLET WT./TYPE	POWDER WT./TYPE	VELOCITY ('/SEC)	SOURCE
290/LeadPP	32.0/4198	1430	Barnes

CASE PREPARATION: SHELLHOLDER (RCBS): 5

MAKE FROM: .348 Win. Taper expand (after annealing) to .410" dia. Square case mouth & chamfer. F/L size.

PHYSICAL DATA (INCHES):

CASE TYPE: **Rimmed Bottleneck**

CASE LENGTH **A = 2.40**

HEAD DIAMETER **B = .548**

RIM DIAMETER **D = .615**

NECK DIAMETER **F = .430**

NECK LENGTH **H = .78**

SHOULDER LENGTH **K = .170**

BODY ANGLE (DEG'S/SIDE): **.183**

CASE CAPACITY **CC'S = 6.03**

LOADED LENGTH: **3.06**

BELT DIAMETER **C = N/A**

RIM THICKNESS **E = .075**

SHOULDER DIAMETER **G = .54**

LENGTH TO SHOULDER **J = 1.45**

SHOULDER ANGLE (DEG'S/SIDE): **17.93**

PRIMER: **L/R**

CASE CAPACITY (GR'S WATER): **93.02**

DIMENSIONAL DRAWING:

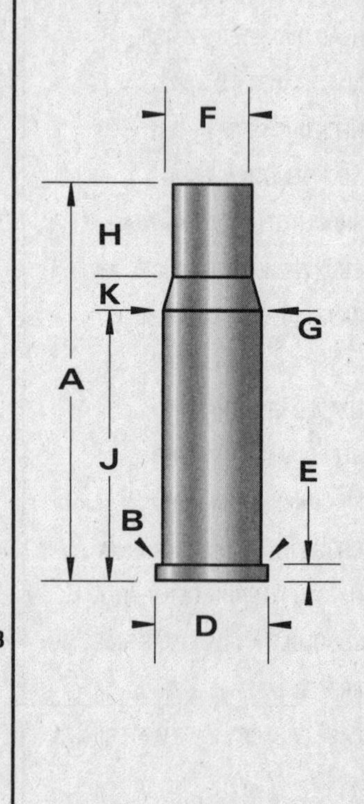

-NOT ACTUAL SIZE-
-DO NOT SCALE-

CARTRIDGE: 10,25 x 69R Hunting Express

OTHER NAMES:

DIA: .404

BALLISTEK NO: 404A

NAI NO: RMB 33221/4.954

DATA SOURCE: COTW 4th Pg275

HISTORICAL DATA: Circa 1885.

NOTES:

LOADING DATA:

BULLET WT./TYPE	POWDER WT./TYPE	VELOCITY ('/SEC)	SOURCE
235/Lead	43.0/IMR3031	—	JJD

CASE PREPARATION: SHELLHOLDER (RCBS): Spl.

MAKE FROM: .450 N.E. (BELL). Cut case to 2.8" and anneal. Form in form die. Trim to length, F/L size and chamfer. Use .403" dia. bullets.

PHYSICAL DATA (INCHES):

CASE TYPE: **Rimmed Bottleneck**

CASE LENGTH **A = 2.72**

HEAD DIAMETER **B = .549**

RIM DIAMETER **D = .630**

NECK DIAMETER **F = .415**

NECK LENGTH **H = .78**

SHOULDER LENGTH **K = .270**

BODY ANGLE (DEG'S/SIDE): **1.34**

CASE CAPACITY **CC'S = 6.86**

LOADED LENGTH: **3.17**

BELT DIAMETER **C = N/A**

RIM THICKNESS **E = .059**

SHOULDER DIAMETER **G = .480**

LENGTH TO SHOULDER **J = 1.67**

SHOULDER ANGLE (DEG'S/SIDE): **6.86**

PRIMER: **L/R**

CASE CAPACITY (GR'S WATER): **105.84**

DIMENSIONAL DRAWING:

F H K G A J E B D

-NOT ACTUAL SIZE-
-DO NOT SCALE-

CARTRIDGE: 10,3 x 60R Swiss

OTHER NAMES:

DIA: .415

BALLISTEK NO: 415A

NAI NO: RMB 33221/4.314

DATA SOURCE: COTW 4th Pg276

HISTORICAL DATA: Swiss target ctg.

NOTES:

LOADING DATA:

BULLET WT./TYPE	POWDER WT./TYPE	VELOCITY ('/SEC)	SOURCE
330/Lead	41.4/IMR3031	—	JJD

CASE PREPARATION: SHELLHOLDER (RCBS): Spl.

MAKE FROM: .450 N.E. (BELL). Cut case to 2.4". Anneal and form in form die. Trim to length and chamfer. F/L size.

PHYSICAL DATA (INCHES):

CASE TYPE: **Rimmed Bottleneck**

CASE LENGTH **A = 2.36**

HEAD DIAMETER **B = .547**

RIM DIAMETER **D = .619**

NECK DIAMETER **F = .440**

NECK LENGTH **H = .56**

SHOULDER LENGTH **K = .20**

BODY ANGLE (DEG'S/SIDE): **1.00**

CASE CAPACITY **CC'S = 5.85**

LOADED LENGTH: **3.08**

BELT DIAMETER **C = N/A**

RIM THICKNESS **E = .041**

SHOULDER DIAMETER **G = .498**

LENGTH TO SHOULDER **J = 1.60**

SHOULDER ANGLE (DEG'S/SIDE): **8.25**

PRIMER: **L/R**

CASE CAPACITY (GR'S WATER): **90.35**

DIMENSIONAL DRAWING:

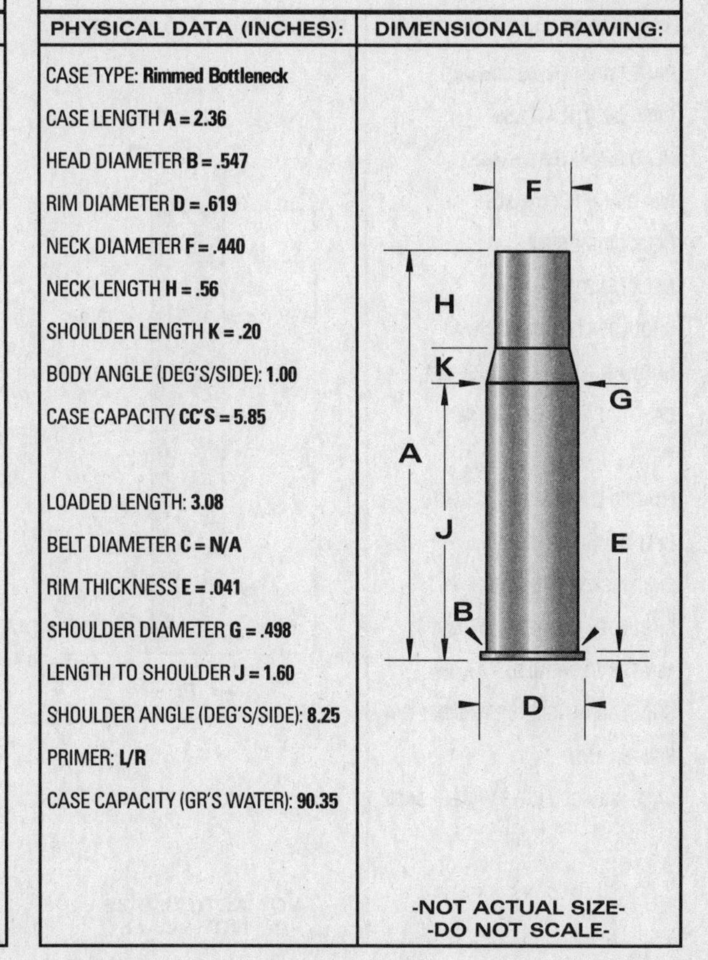

-NOT ACTUAL SIZE-
-DO NOT SCALE-

CARTRIDGE: 10,3 x 65R Baenziger

OTHER NAMES:

DIA: .423

BALLISTEK NO: 423A

NAI NO: RMS 21115/5.541

DATA SOURCE: COTW 4th Pg276

HISTORICAL DATA:

NOTES: DWM #164. Same as brass .410 shotshell.

LOADING DATA:

BULLET WT./TYPE	POWDER WT./TYPE	VELOCITY ('/SEC)	SOURCE
290/Lead	43.0/3031	1625	Barnes

CASE PREPARATION: SHELLHOLDER (RCBS): Spl.

MAKE FROM: 7 x 65R Brenneke. Turn rim to .505" dia. Taper expand case neck to .430" dia. Square case mouth, chamfer and F/L size. Fireform in chamber. Build rim up to .1" thick by super gluing .040" thick shim/washer in place.

PHYSICAL DATA (INCHES):

CASE TYPE: **Rimmed Straight**

CASE LENGTH A = **2.56**

HEAD DIAMETER B = **.462**

RIM DIAMETER D = **.505**

NECK DIAMETER F = **.431**

NECK LENGTH H = **N/A**

SHOULDER LENGTH K = **N/A**

BODY ANGLE (DEG'S/SIDE): **.361**

CASE CAPACITY CC'S = **5.48**

LOADED LENGTH: **3.15**

BELT DIAMETER C = **N/A**

RIM THICKNESS E = **.10**

SHOULDER DIAMETER G = **N/A**

LENGTH TO SHOULDER J = **N/A**

SHOULDER ANGLE (DEG'S/SIDE): **N/A**

PRIMER: **L/R**

CASE CAPACITY (GR'S WATER): **84.60**

DIMENSIONAL DRAWING:

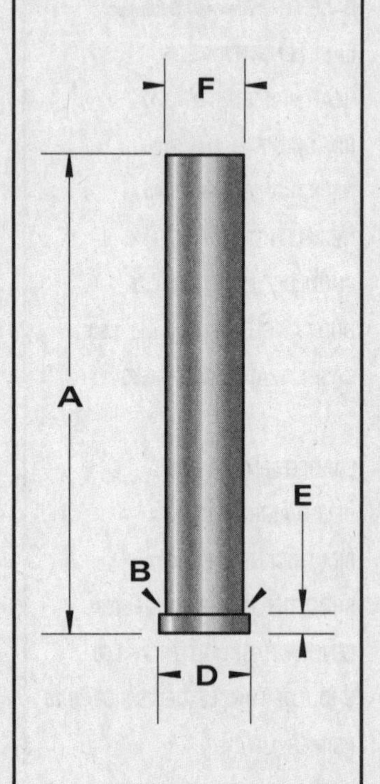

-NOT ACTUAL SIZE-
-DO NOT SCALE-

CARTRIDGE: 10,4mm Italian Revolver

OTHER NAMES:
10,35mm Italian Revolver
10,35mm Glisenti

DIA: .422

BALLISTEK NO: 422C

NAI NO: RMS 11115/1.973

DATA SOURCE: COTW 4th Pg181

HISTORICAL DATA: For M74 Italian service revolver.

NOTES:

LOADING DATA:

BULLET WT./TYPE	POWDER WT./TYPE	VELOCITY ('/SEC)	SOURCE
177/Lead	4.0/B'Eye	–	JJD

CASE PREPARATION: SHELLHOLDER (RCBS): 18

MAKE FROM: .44 Mag. (or Special). Turn rim to .505" dia. Cut case off to .90". Trim to length. Size in F/L die or .240 Weatherby sizer.

PHYSICAL DATA (INCHES):

CASE TYPE: **Rimmed Straight**

CASE LENGTH A = **.89**

HEAD DIAMETER B = **.451**

RIM DIAMETER D = **.505**

NECK DIAMETER F = **.444**

NECK LENGTH H = **N/A**

SHOULDER LENGTH K = **N/A**

BODY ANGLE (DEG'S/SIDE): **.243**

CASE CAPACITY CC'S = **1.29**

LOADED LENGTH: **1.25**

BELT DIAMETER C = **N/A**

RIM THICKNESS E = **.066**

SHOULDER DIAMETER G = **N/A**

LENGTH TO SHOULDER J = **N/A**

SHOULDER ANGLE (DEG'S/SIDE): **N/A**

PRIMER: **L/P**

CASE CAPACITY (GR'S WATER): **19.96**

DIMENSIONAL DRAWING:

F
A
E
B
D

-NOT ACTUAL SIZE-
-DO NOT SCALE-

CARTRIDGE: 10,4 x 38R Swiss Vetterli

OTHER NAMES:
.41 Swiss M69/81

DIA: .415

BALLISTEK NO: 415B

NAI NO: RMB 24244/2.685

DATA SOURCE: COTW 4th Pg210

HISTORICAL DATA: Swiss mil. ctg. ca. 1896.

NOTES:

LOADING DATA:

BULLET WT./TYPE	POWDER WT./TYPE	VELOCITY ('/SEC)	SOURCE
300/LeadPP	22.9/IMR4198	—	JJD

CASE PREPARATION: SHELLHOLDER (RCBS): Spl.

MAKE FROM: .280 Flanged (BELL). Cut case to 1.5". Build rim up to .08" thick. Anneal case and form in form die. Trim to length and F/L size. Fireform.

PHYSICAL DATA (INCHES):

CASE TYPE: **Rimmed Bottleneck**

CASE LENGTH **A = 1.45**

HEAD DIAMETER **B = .540**

RIM DIAMETER **D = .630**

NECK DIAMETER **F = .437**

NECK LENGTH **H = .366**

SHOULDER LENGTH **K = .014**

BODY ANGLE (DEG'S/SIDE): **.724**

CASE CAPACITY **CC'S = 3.46**

LOADED LENGTH: **2.20**

BELT DIAMETER **C = N/A**

RIM THICKNESS **E = .08**

SHOULDER DIAMETER **G = .518**

LENGTH TO SHOULDER **J = 1.07**

SHOULDER ANGLE (DEG'S/SIDE): **70.93**

PRIMER: **L/R**

CASE CAPACITY (GR'S WATER): **53.45**

DIMENSIONAL DRAWING:

-NOT ACTUAL SIZE-
-DO NOT SCALE-

CARTRIDGE: 10,4 x 42R Vetterli

OTHER NAMES:

DIA: .422

BALLISTEK NO: 422D

NAI NO: RMB 24241/2.783

DATA SOURCE: Hoyem Vol.2 Pg181

HISTORICAL DATA: Swiss ctg. ca. 1871.

NOTES:

LOADING DATA:

BULLET WT./TYPE	POWDER WT./TYPE	VELOCITY ('/SEC)	SOURCE
250/LeadPP	28.9/IMR4198	—	JJD

CASE PREPARATION: SHELLHOLDER (RCBS): 14

MAKE FROM: .280 Flanged (BELL). Cut case to 1.55". Anneal and taper expand to .430" dia. Trim to length, chamfer and F/L size. Fireform in chamber.

PHYSICAL DATA (INCHES):

CASE TYPE: **Rimmed Bottleneck**

CASE LENGTH **A = 1.50**

HEAD DIAMETER **B = .539**

RIM DIAMETER **D = .615**

NECK DIAMETER **F = .441**

NECK LENGTH **H = .25**

SHOULDER LENGTH **K = .27**

BODY ANGLE (DEG'S/SIDE): **.514**

CASE CAPACITY **CC'S = 3.72**

LOADED LENGTH: **2.21**

BELT DIAMETER **C = N/A**

RIM THICKNESS **E = .06**

SHOULDER DIAMETER **G = .525**

LENGTH TO SHOULDER **J = .98**

SHOULDER ANGLE (DEG'S/SIDE): **8.84**

PRIMER: **L/R**

CASE CAPACITY (GR'S WATER): **57.48**

DIMENSIONAL DRAWING:

-NOT ACTUAL SIZE-
-DO NOT SCALE-

CARTRIDGE: 10,4 x 47R Vetterli-Vitali

OTHER NAMES:
10.4 x 47R Italian M70

DIA: .430

BALLISTEK NO: 430H

NAI NO: RMB 24241/3.463

DATA SOURCE: COTW 4th Pg210

HISTORICAL DATA: Italian ctg. from about 1870.

NOTES:

LOADING DATA:

BULLET WT./TYPE	POWDER WT./TYPE	VELOCITY ('/SEC)	SOURCE
250/Lead	27.0/4198	1300	Barnes

CASE PREPARATION: **SHELLHOLDER (RCBS):** Spl.

MAKE FROM: Cut case to 1.9" and build rim up to .084" thick. Chamfer. F/L size, trim to length and chamfer.

PHYSICAL DATA (INCHES):

CASE TYPE: **Rimmed Bottleneck**

CASE LENGTH **A = 1.87**

HEAD DIAMETER **B = .540**

RIM DIAMETER **D = .634**

NECK DIAMETER **F = .437**

NECK LENGTH **H = .175**

SHOULDER LENGTH **K = .335**

BODY ANGLE (DEG'S/SIDE): **.567**

CASE CAPACITY **CC'S = 5.22**

LOADED LENGTH: **2.46**

BELT DIAMETER **C = N/A**

RIM THICKNESS **E = .084**

SHOULDER DIAMETER **G = .517**

LENGTH TO SHOULDER **J = 1.36**

SHOULDER ANGLE (DEG'S/SIDE): **6.81**

PRIMER: **L/R**

CASE CAPACITY (GR'S WATER): **80.64**

DIMENSIONAL DRAWING:

F H K G A J E B D

-NOT ACTUAL SIZE-
-DO NOT SCALE-

CARTRIDGE: 10,5 x 47R

OTHER NAMES:

DIA: .419

BALLISTEK NO: 419C

NAI NO: RMB 24221/3.606

DATA SOURCE: COTW 4th Pg276

HISTORICAL DATA:

NOTES:

LOADING DATA:

BULLET WT./TYPE	POWDER WT./TYPE	VELOCITY ('/SEC)	SOURCE
275/Lead	28.9/IMR4198	–	JJD

CASE PREPARATION: **SHELLHOLDER (RCBS):** Spl.

MAKE FROM: .43 Span. Rem. (BELL). Cut case to 1.9". Turn rim to .591" dia & back chamfer. I.D. neck ream. F/L size & trim to length.

PHYSICAL DATA (INCHES):

CASE TYPE: **Rimmed Bottleneck**

CASE LENGTH **A = 1.85**

HEAD DIAMETER **B = .513**

RIM DIAMETER **D = .591**

NECK DIAMETER **F = .445**

NECK LENGTH **H = .53**

SHOULDER LENGTH **K = .385**

BODY ANGLE (DEG'S/SIDE): **.66**

CASE CAPACITY **CC'S = 4.10**

LOADED LENGTH: **2.40**

BELT DIAMETER **C = N/A**

RIM THICKNESS **E = .09**

SHOULDER DIAMETER **G = .496**

LENGTH TO SHOULDER **J = .935**

SHOULDER ANGLE (DEG'S/SIDE): **3.79**

PRIMER: **L/R**

CASE CAPACITY (GR'S WATER): **63.33**

DIMENSIONAL DRAWING:

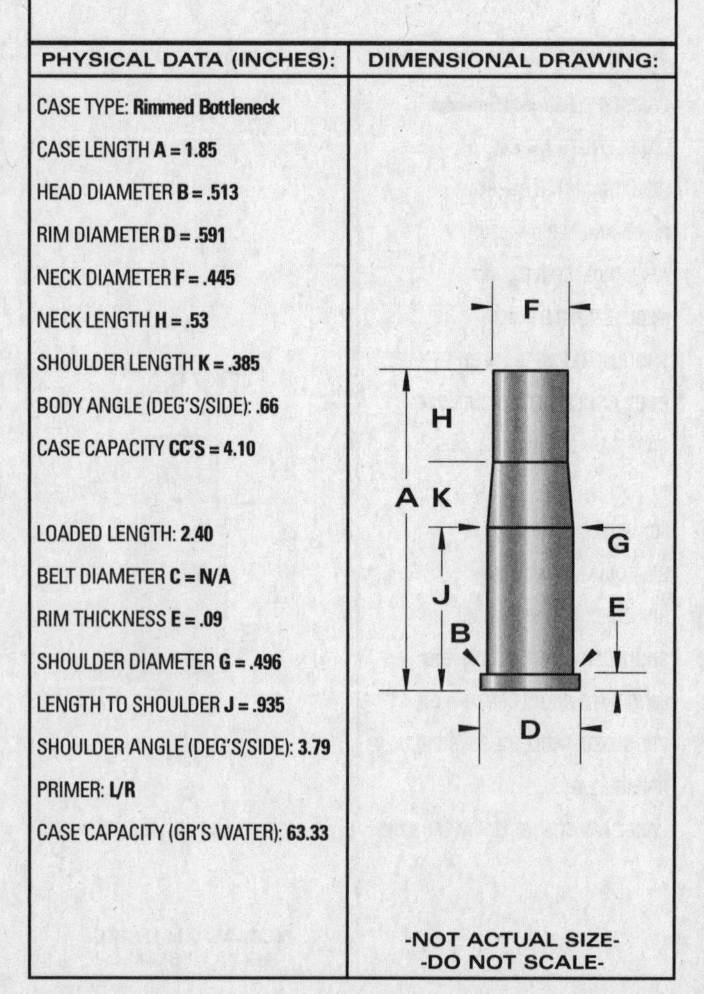

-NOT ACTUAL SIZE-
-DO NOT SCALE-

CARTRIDGE: 10,66 x 57.5R Gatling Russian

OTHER NAMES:

DIA: .434

BALLISTEK NO: 434B

NAI NO: RMB 13232/4.369

DATA SOURCE: Hoyem Vol.2 Pg193

HISTORICAL DATA: Russian Gatling gun ctg. circa 1869.

NOTES:

LOADING DATA:

BULLET WT./TYPE	POWDER WT./TYPE	VELOCITY ('/SEC)	SOURCE
380/RN	44.4/BL-C2	–	JJD

CASE PREPARATION: SHELLHOLDER (RCBS): Spl.

MAKE FROM: .43 Span. Rem. (BELL). Thin rim to .067" thick. F/L size with expander removed. Trim to length & chamfer.

PHYSICAL DATA (INCHES):

CASE TYPE: **Rimmed Bottleneck**

CASE LENGTH **A = 2.25**

HEAD DIAMETER **B = .515**

RIM DIAMETER **D = .634**

NECK DIAMETER **F = .448**

NECK LENGTH **H = .505**

SHOULDER LENGTH **K = .155**

BODY ANGLE (DEG'S/SIDE): **.082**

CASE CAPACITY **CC'S = 5.85**

LOADED LENGTH: **2.97**

BELT DIAMETER **C = N/A**

RIM THICKNESS **E = .067**

SHOULDER DIAMETER **G = .511**

LENGTH TO SHOULDER **J = 1.59**

SHOULDER ANGLE (DEG'S/SIDE): **11.49**

PRIMER: **L/R**

CASE CAPACITY (GR'S WATER): **90.29**

DIMENSIONAL DRAWING:

F H K G A J E B D

-NOT ACTUAL SIZE-
-DO NOT SCALE-

CARTRIDGE: 10,66 x 57R Berdan #2

OTHER NAMES:

DIA: .431

BALLISTEK NO: 431A

NAI NO: RMB 13232/4.369

DATA SOURCE: Hoyem Vol.2 Pg191

HISTORICAL DATA: Another 1066mm Russian ctg. for Berdan rifles.

NOTES: About 1870.

LOADING DATA:

BULLET WT./TYPE	POWDER WT./TYPE	VELOCITY ('/SEC)	SOURCE
370/Lead	44.0/IMR3031	–	JJD

CASE PREPARATION: SHELLHOLDER (RCBS): Spl.

MAKE FROM: .375 Flanged (BELL). Cut case to 2.3", anneal and F/L size. Trim to length & chamfer.

PHYSICAL DATA (INCHES):

CASE TYPE: **Rimmed Bottleneck**

CASE LENGTH **A = 2.25**

HEAD DIAMETER **B = .515**

RIM DIAMETER **D = .635**

NECK DIAMETER **F = .447**

NECK LENGTH **H = .507**

SHOULDER LENGTH **K = .163**

BODY ANGLE (DEG'S/SIDE): **.06**

CASE CAPACITY **CC'S = 5.78**

LOADED LENGTH: **2.96**

BELT DIAMETER **C = N/A**

RIM THICKNESS **E = .061**

SHOULDER DIAMETER **G = .512**

LENGTH TO SHOULDER **J = 1.58**

SHOULDER ANGLE (DEG'S/SIDE): **11.27**

PRIMER: **L/R**

CASE CAPACITY (GR'S WATER): **89.26**

DIMENSIONAL DRAWING:

F H K G A J E B D

-NOT ACTUAL SIZE-
-DO NOT SCALE-

CARTRIDGE: 10.57 Lazzaroni Maverick

OTHER NAMES:

DIA: .416

BALLISTEK NO:

NAI NO:

DATA SOURCE:

HISTORICAL DATA: Created by John Lazzaroni, cartridge drawing dated 4/3/02.

NOTES: This is the .416 caliber cartridge in the Lazzaroni short action magnum lineup.

LOADING DATA:

BULLET WT./TYPE	POWDER WT./TYPE	VELOCITY ('/SEC)	SOURCE
350/Swift A-Frame	72.0/RL-15	2572	Lazzaroni
400/Swift A-Frame	69.0/RL-15	2434	Lazzaroni

CASE PREPARATION: SHELL HOLDER (RCBS): 14

MAKE FROM:

PHYSICAL DATA (INCHES):

CASE TYPE: **Rimless Bottleneck**
CASE LENGTH **A = 2.050**
HEAD DIAMETER **B = .579**
RIM DIAMETER **D = .580**
NECK DIAMETER **F = .449**
NECK LENGTH **H = .403**
SHOULDER LENGTH **K = .094**
BODY ANGLE (DEG'S/SIDE):
CASE CAPACITY **CC'S =**
LOADED LENGTH: **2.800**
BELT DIAMETER **C =**
RIM THICKNESS **E = .065**
SHOULDER DIAMETER **G = .560**
LENGTH TO SHOULDER **J = 1.553**
SHOULDER ANGLE (DEG'S/SIDE): **30**
PRIMER: **L/R Mag**
CASE CAPACITY (GR'S WATER):

DIMENSIONAL DRAWING:

F H K G A J E B D

-NOT ACTUAL SIZE-
-DO NOT SCALE-

CARTRIDGE: 10.57 Lazzaroni Meteor

OTHER NAMES:

DIA: .416

BALLISTEK NO:

NAI NO:

DATA SOURCE:

HISTORICAL DATA: Created by John Lazzaroni, cartrige drawing dated 4/11/01.

NOTES: This is the Lazzaroni long action beltless magnum cartridge in .416.

LOADING DATA:

BULLET WT./TYPE	POWDER WT./TYPE	VELOCITY ('/SEC)	SOURCE
350/Swift A-Frame	113.0/RL-19	2965	
400/Swift A-Frame or Speer Monolithic Solid	108.0/RL-19	2770 A-Frame 2752 Solid	

CASE PREPARATION: SHELL HOLDER (RCBS): 14

MAKE FROM:

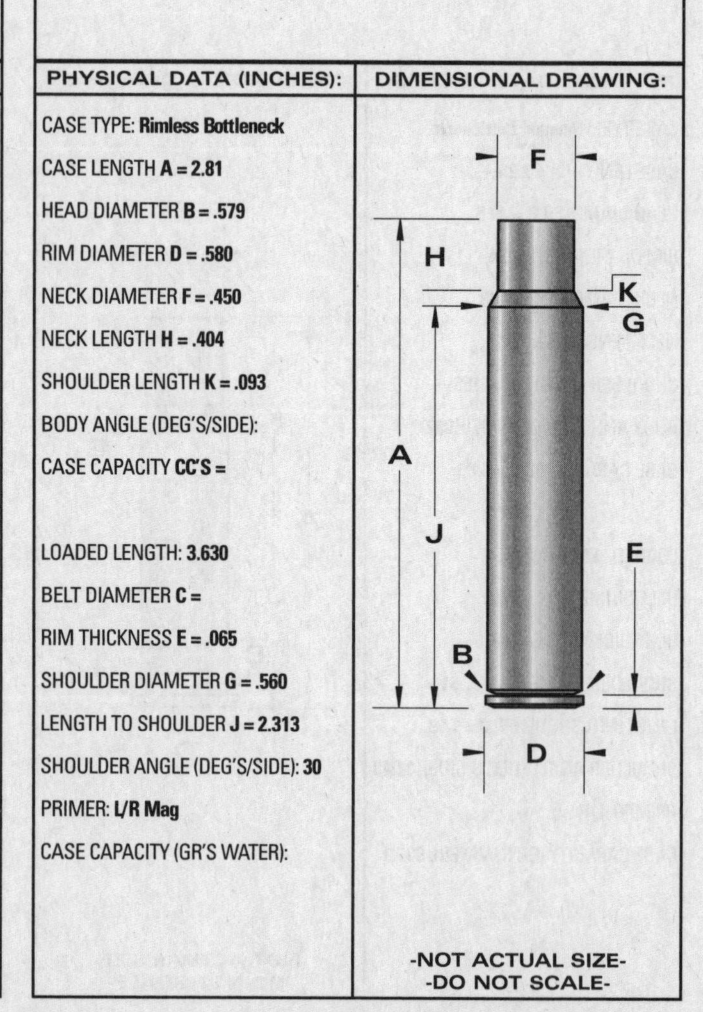

PHYSICAL DATA (INCHES):

CASE TYPE: **Rimless Bottleneck**
CASE LENGTH **A = 2.81**
HEAD DIAMETER **B = .579**
RIM DIAMETER **D = .580**
NECK DIAMETER **F = .450**
NECK LENGTH **H = .404**
SHOULDER LENGTH **K = .093**
BODY ANGLE (DEG'S/SIDE):
CASE CAPACITY **CC'S =**
LOADED LENGTH: **3.630**
BELT DIAMETER **C =**
RIM THICKNESS **E = .065**
SHOULDER DIAMETER **G = .560**
LENGTH TO SHOULDER **J = 2.313**
SHOULDER ANGLE (DEG'S/SIDE): **30**
PRIMER: **L/R Mag**
CASE CAPACITY (GR'S WATER):

DIMENSIONAL DRAWING:

-NOT ACTUAL SIZE-
-DO NOT SCALE-

CARTRIDGE: 10,7 x 57R Krag Petersen

OTHER NAMES:

DIA: .425

BALLISTEK NO: 425B

NAI NO: RMB 23231/4.395

DATA SOURCE: Hoyem Vol.2 Pg230

HISTORICAL DATA: Swedish mil. ctg. for Remington rolling block rifle.

NOTES:

LOADING DATA:

BULLET WT./TYPE	POWDER WT./TYPE	VELOCITY ('/SEC)	SOURCE
350/Lead	44.2/BL-C2	–	JJD

CASE PREPARATION: SHELLHOLDER (RCBS): 4

MAKE FROM: .375 H&H Mag. Turn belt to .520-.521" dia. - leave rim full diameter. Cut case to 2.3" & anneal. Taper expand to .430" dia. F/L size and trim to length. Chamfer and fireform in the chamber.

PHYSICAL DATA (INCHES):

CASE TYPE: **Rimmed Bottleneck**

CASE LENGTH **A = 2.29**

HEAD DIAMETER **B = .521**

RIM DIAMETER **D = .632**

NECK DIAMETER **F = .447**

NECK LENGTH **H = .460**

SHOULDER LENGTH **K = .190**

BODY ANGLE (DEG'S/SIDE): **.417**

CASE CAPACITY **CC'S = 5.66**

LOADED LENGTH: **2.93**

BELT DIAMETER **C = N/A**

RIM THICKNESS **E = .052**

SHOULDER DIAMETER **G = .50**

LENGTH TO SHOULDER **J = 1.64**

SHOULDER ANGLE (DEG'S/SIDE): **7.94**

PRIMER: **L/R**

CASE CAPACITY (GR'S WATER): **87.42**

DIMENSIONAL DRAWING:

F H K G A J E B D

-NOT ACTUAL SIZE-
-DO NOT SCALE-

CARTRIDGE: 10,75 x 57mm Mannlicher

OTHER NAMES:

DIA: .424

BALLISTEK NO: 424A

NAI NO: RXB 12241/4.786

DATA SOURCE: COTW 4th Pg277

HISTORICAL DATA: Circa 1900.

NOTES:

LOADING DATA:

BULLET WT./TYPE	POWDER WT./TYPE	VELOCITY ('/SEC)	SOURCE
350/RN	44.9/BL-C2	–	JJD

CASE PREPARATION: SHELLHOLDER (RCBS): 3

MAKE FROM: 9,3 x 57mm. Taper expand the case to .430" dia. Square case mouth and F/L size. Cases can be made from .308 Win. in mcuh the same manner.

PHYSICAL DATA (INCHES):

CASE TYPE: **Rimless Bottleneck**

CASE LENGTH **A = 2.24**

HEAD DIAMETER **B = .468**

RIM DIAMETER **D = .468**

NECK DIAMETER **F = .449**

NECK LENGTH **H = .265**

SHOULDER LENGTH **K = .055**

BODY ANGLE (DEG'S/SIDE): **.05**

CASE CAPACITY **CC'S = 4.71**

LOADED LENGTH: **3.05**

BELT DIAMETER **C = N/A**

RIM THICKNESS **E = .05**

SHOULDER DIAMETER **G = .465**

LENGTH TO SHOULDER **J = 1.92**

SHOULDER ANGLE (DEG'S/SIDE): **8.27**

PRIMER: **L/R**

CASE CAPACITY (GR'S WATER): **72.76**

DIMENSIONAL DRAWING:

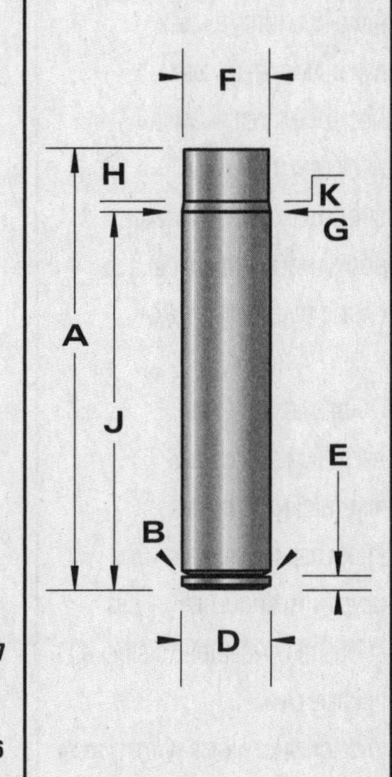

-NOT ACTUAL SIZE-
-DO NOT SCALE-

CARTRIDGE: 10,75 x 58R Russian Berdan

OTHER NAMES:

DIA: .430

BALLISTEK NO: 430J

NAI NO: RMB 34231/3.950

DATA SOURCE: COTW 4th Pg211

HISTORICAL DATA: Russian mil. ctg. ca. 1871.

NOTES:

LOADING DATA:

BULLET WT./TYPE	POWDER WT./TYPE	VELOCITY ('/SEC)	SOURCE
370/Lead	31.0/4198	1410	Barnes

CASE PREPARATION: SHELLHOLDER (RCBS): Spl.

MAKE FROM: .50 Sharps (BELL). Turn rim to .637" dia. and cut case off to 2.3". Anneal and F/L size. Trim to length and chamfer.

PHYSICAL DATA (INCHES):

CASE TYPE: **Rimmed Bottleneck**

CASE LENGTH **A = 2.24**

HEAD DIAMETER **B = .567**

RIM DIAMETER **D = .637**

NECK DIAMETER **F = .449**

NECK LENGTH **H = .470**

SHOULDER LENGTH **K = .240**

BODY ANGLE (DEG'S/SIDE): **1.31**

CASE CAPACITY **CC'S = 6.04**

LOADED LENGTH: **2.95**

BELT DIAMETER **C = N/A**

RIM THICKNESS **E = .07**

SHOULDER DIAMETER **G = .506**

LENGTH TO SHOULDER **J = 1.53**

SHOULDER ANGLE (DEG'S/SIDE): **6.77**

PRIMER: **L/R**

CASE CAPACITY (GR'S WATER): **93.24**

DIMENSIONAL DRAWING:

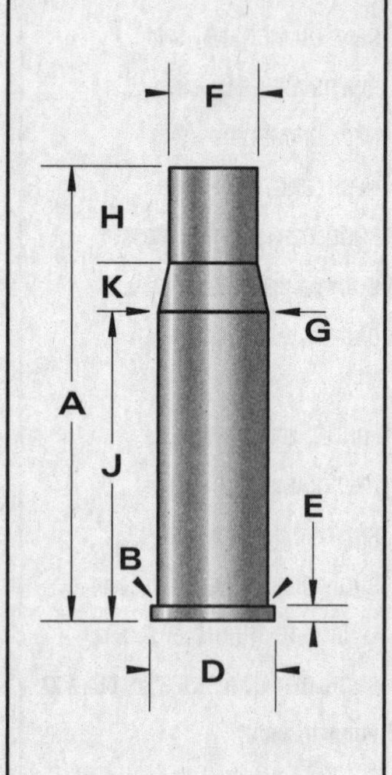

-NOT ACTUAL SIZE-
-DO NOT SCALE-

CARTRIDGE: 10,75 x 63 Mauser

OTHER NAMES:
10,75 x 62 Mauser

DIA: .424

BALLISTEK NO: 424C

NAI NO: RBB 12242/5.010

DATA SOURCE: COTW 4th Pg276

HISTORICAL DATA: Circa 1910.

NOTES: DWM #515.

LOADING DATA:

BULLET WT./TYPE	POWDER WT./TYPE	VELOCITY ('/SEC)	SOURCE
350/RN	65.0/IMR3031	–	JJD

CASE PREPARATION: SHELLHOLDER (RCBS): 5

MAKE FROM: 10.75 x 68 RWS. Rebate rim by turning to .467" dia. Re-cut extractor groove. Expand neck up to .430" dia. and size back down, in 10,75x63mm sizer. Trim to length, chamfer and fireform in chamber.

PHYSICAL DATA (INCHES):

CASE TYPE: **Rebated Bottleneck**

CASE LENGTH **A = 2.47**

HEAD DIAMETER **B = .493**

RIM DIAMETER **D = .467**

NECK DIAMETER **F = .447**

NECK LENGTH **H = .361**

SHOULDER LENGTH **K = .049**

BODY ANGLE (DEG'S/SIDE): **.215**

CASE CAPACITY **CC'S = 5.72**

LOADED LENGTH: **3.22**

BELT DIAMETER **C = N/A**

RIM THICKNESS **E = .06**

SHOULDER DIAMETER **G = .479**

LENGTH TO SHOULDER **J = 2.06**

SHOULDER ANGLE (DEG'S/SIDE): **18.08**

PRIMER: **L/R**

CASE CAPACITY (GR'S WATER): **88.29**

DIMENSIONAL DRAWING:

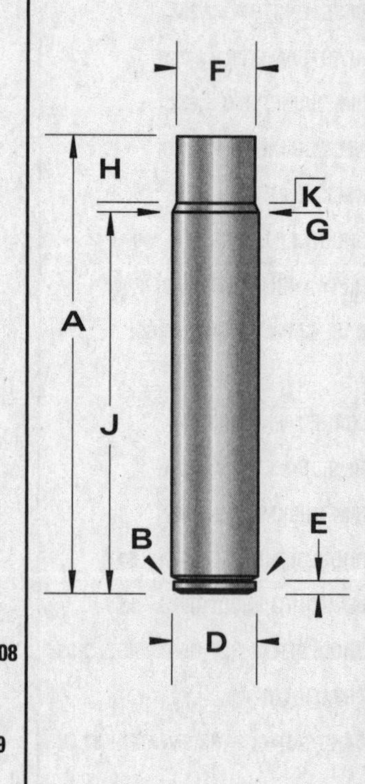

-NOT ACTUAL SIZE-
-DO NOT SCALE-

CARTRIDGE: 10,75 x 65R Collath

OTHER NAMES:

DIA: .424

BALLISTEK NO: 424D

NAI NO: RMS 21115/5.256

DATA SOURCE: COTW 4th Pg277

HISTORICAL DATA:

NOTES: Very rare cartridge.

LOADING DATA:

BULLET WT./TYPE	POWDER WT./TYPE	VELOCITY ('/SEC)	SOURCE
305/Lead	41.3/IMR3031	—	JJD

CASE PREPARATION: SHELLHOLDER (RCBS): 7

MAKE FROM: .45 Basic. Turn rim to .542" dia. and back chamfer. Cut case to 2.6" and anneal. Form in F/L die. Trim to length and chamfer. Fireform case in chamber.

PHYSICAL DATA (INCHES):

CASE TYPE: **Rimmed Straight**

CASE LENGTH **A = 2.56**

HEAD DIAMETER **B = .487**

RIM DIAMETER **D = .542**

NECK DIAMETER **F = .451**

NECK LENGTH **H = N/A**

SHOULDER LENGTH **K = N/A**

BODY ANGLE (DEG'S/SIDE): **.410**

CASE CAPACITY **CC'S = 5.33**

LOADED LENGTH: **3.02**

BELT DIAMETER **C = N/A**

RIM THICKNESS **E = .047**

SHOULDER DIAMETER **G = N/A**

LENGTH TO SHOULDER **J = N/A**

SHOULDER ANGLE (DEG'S/SIDE): **N/A**

PRIMER: **L/R**

CASE CAPACITY (GR'S WATER): **82.25**

DIMENSIONAL DRAWING:

-NOT ACTUAL SIZE-
-DO NOT SCALE-

CARTRIDGE: 10,75 x 68 Mauser

OTHER NAMES:

DIA: .424

BALLISTEK NO: 424E

NAI NO: RXB 12231/5.427

DATA SOURCE: COTW 4th Pg277

HISTORICAL DATA: From the early 1920's.

NOTES:

LOADING DATA:

BULLET WT./TYPE	POWDER WT./TYPE	VELOCITY ('/SEC)	SOURCE
255/—	36.0/3031	1750	Barnes

CASE PREPARATION: SHELLHOLDER (RCBS): 5

MAKE FROM: Factory (RWS). I have never done this but, it looks like 1/2" tubing could be used to re-body an 8x68S case head.

PHYSICAL DATA (INCHES):

CASE TYPE: **Rimless Bottleneck**

CASE LENGTH **A = 2.67**

HEAD DIAMETER **B = .492**

RIM DIAMETER **D = .488**

NECK DIAMETER **F = .445**

NECK LENGTH **H = .490**

SHOULDER LENGTH **K = .155**

BODY ANGLE (DEG'S/SIDE): **.345**

CASE CAPACITY **CC'S = 6.14**

LOADED LENGTH: **3.16**

BELT DIAMETER **C = N/A**

RIM THICKNESS **E = .05**

SHOULDER DIAMETER **G = .470**

LENGTH TO SHOULDER **J = 2.025**

SHOULDER ANGLE (DEG'S/SIDE): **4.61**

PRIMER: **L/R**

CASE CAPACITY (GR'S WATER): **94.82**

DIMENSIONAL DRAWING:

-NOT ACTUAL SIZE-
-DO NOT SCALE-

CARTRIDGE: 11mm French Revolver

OTHER NAMES:

DIA: .425

BALLISTEK NO: 425A

NAI NO: RMS 21115/1.543

DATA SOURCE: COTW 4th Pg182

HISTORICAL DATA: Early French mil. ctg.

NOTES:

LOADING DATA:

BULLET WT./TYPE	POWDER WT./TYPE	VELOCITY ('/SEC)	SOURCE
180/Lead	3.0/B'Eye	675	JJD

CASE PREPARATION: SHELLHOLDER (RCBS): 24

MAKE FROM: .44 Magnum (or Special). Cut case to length. Chamfer. Turn rim to .491" dia. Use .429" dia. lead bullets & seat with .44 Spl. die.

PHYSICAL DATA (INCHES):

CASE TYPE: **Rimmed Straight**
CASE LENGTH **A = .71**
HEAD DIAMETER **B = .460**
RIM DIAMETER **D = .491**
NECK DIAMETER **F = .449**
NECK LENGTH **H = N/A**
SHOULDER LENGTH **K = N/A**
BODY ANGLE (DEG'S/SIDE): **.471**
CASE CAPACITY **CC'S = .913**

LOADED LENGTH: **1.18**
BELT DIAMETER **C = N/A**
RIM THICKNESS **E = .041**
SHOULDER DIAMETER **G = N/A**
LENGTH TO SHOULDER **J = N/A**
SHOULDER ANGLE (DEG'S/SIDE): **N/A**
PRIMER: **L/P**
CASE CAPACITY (GR'S WATER): **14.09**

DIMENSIONAL DRAWING:

F
A
B
E
D

-NOT ACTUAL SIZE-
-DO NOT SCALE-

CARTRIDGE: 11mm German Service

OTHER NAMES:
10,6mm German Revolver
10,8mm German Revolver

DIA: .426

BALLISTEK NO: 426A

NAI NO: RMS 11115/2.119

DATA SOURCE: COTW 4th Pg183

HISTORICAL DATA: German mil. ctg. ca. 1879.

NOTES:

LOADING DATA:

BULLET WT./TYPE	POWDER WT./TYPE	VELOCITY ('/SEC)	SOURCE
262/Lead	5.0/B'Eye	—	JJD

CASE PREPARATION: SHELLHOLDER (RCBS): 18

MAKE FROM: .44 Magnum (or Special). Trim case to length. Use .429" dia. lead bullets.

PHYSICAL DATA (INCHES):

CASE TYPE: **Rimmed Straight**
CASE LENGTH **A = .96**
HEAD DIAMETER **B = .453**
RIM DIAMETER **D = .509**
NECK DIAMETER **F = .449**
NECK LENGTH **H = N/A**
SHOULDER LENGTH **K = N/A**
BODY ANGLE (DEG'S/SIDE): **.127**
CASE CAPACITY **CC'S = 1.46**

LOADED LENGTH: **1.21**
BELT DIAMETER **C = N/A**
RIM THICKNESS **E = .06**
SHOULDER DIAMETER **G = N/A**
LENGTH TO SHOULDER **J = N/A**
SHOULDER ANGLE (DEG'S/SIDE): **N/A**
PRIMER: **L/P**
CASE CAPACITY (GR'S WATER): **22.57**

DIMENSIONAL DRAWING:

F
A
B
E
D

-NOT ACTUAL SIZE-
-DO NOT SCALE-

CARTRIDGE: 11mm Murata

OTHER NAMES: 11mm Meiji-13

DIA: .432

BALLISTEK NO: 432A

NAI NO: RMB 24221/4.354

DATA SOURCE: COTW 4th Pg212

HISTORICAL DATA: Japanese mil. ctg. ca. 1881.

NOTES:

LOADING DATA:

BULLET WT./TYPE	POWDER WT./TYPE	VELOCITY ('/SEC)	SOURCE
420/LeadPP	41.6/BL-C2	–	JJD

CASE PREPARATION: SHELLHOLDER (RCBS): Spl.

MAKE FROM: .450 N.E. (BELL). Cut case to 2.4" and anneal. Build rim up to .09" thick. F/L size in the 11mm die. Trim to length.

PHYSICAL DATA (INCHES):

CASE TYPE: **Rimmed Bottleneck**

CASE LENGTH **A = 2.36**

HEAD DIAMETER **B = .542**

RIM DIAMETER **D = .632**

NECK DIAMETER **F = .465**

NECK LENGTH **H = .69**

SHOULDER LENGTH **K = .17**

BODY ANGLE (DEG'S/SIDE): **.374**

CASE CAPACITY **CC'S = 5.91**

LOADED LENGTH: **3.13**

BELT DIAMETER **C = N/A**

RIM THICKNESS **E = .093**

SHOULDER DIAMETER **G = .525**

LENGTH TO SHOULDER **J = 1.50**

SHOULDER ANGLE (DEG'S/SIDE): **10.00**

PRIMER: **L/R**

CASE CAPACITY (GR'S WATER): **91.24**

DIMENSIONAL DRAWING:

-NOT ACTUAL SIZE-
-DO NOT SCALE-

CARTRIDGE: 11 x 48R Remington Carbine

OTHER NAMES:

DIA: .433

BALLISTEK NO: 433A

NAI NO: RMB 24245/3.733

DATA SOURCE:

HISTORICAL DATA:

NOTES:

LOADING DATA:

BULLET WT./TYPE	POWDER WT./TYPE	VELOCITY ('/SEC)	SOURCE
375/LeadPP	34.1/IMR3031	–	JJD

CASE PREPARATION: SHELLHOLDER (RCBS): Spl.

MAKE FROM: .43 Span. Rem. (BELL). Cut case to 2.0". F/L size, trim to length and chamfer.

PHYSICAL DATA (INCHES):

CASE TYPE: **Rimmed Bottleneck**

CASE LENGTH **A = 1.90**

HEAD DIAMETER **B = .509**

RIM DIAMETER **D = .632**

NECK DIAMETER **F = .483**

NECK LENGTH **H = .381**

SHOULDER LENGTH **K = .219**

BODY ANGLE (DEG'S/SIDE): **.494**

CASE CAPACITY **CC'S = 3.90**

LOADED LENGTH: **2.43**

BELT DIAMETER **C = N/A**

RIM THICKNESS **E = .072**

SHOULDER DIAMETER **G = .49**

LENGTH TO SHOULDER **J = 1.30**

SHOULDER ANGLE (DEG'S/SIDE): **.915**

PRIMER: **L/R**

CASE CAPACITY (GR'S WATER): **60.26**

DIMENSIONAL DRAWING:

-NOT ACTUAL SIZE-
-DO NOT SCALE-

CARTRIDGE: 11 x 50.5R Comblain M71

OTHER NAMES:

DIA: .449

BALLISTEK NO: 449B

NAI NO: RMB 24241/3.473

DATA SOURCE: Hoyem Vol.2 Pg155

HISTORICAL DATA: Ctg. for Terssen rifle, ca. 1868.

NOTES:

LOADING DATA:

BULLET WT./TYPE	POWDER WT./TYPE	VELOCITY ('/SEC)	SOURCE
350/LeadPP	45.7/IMR3031	—	JJD

CASE PREPARATION: **SHELLHOLDER (RCBS):** Spl.

MAKE FROM: .470 #2 N.E. (BELL). Cut case to 2.0" and F/L size. Trim to length.

PHYSICAL DATA (INCHES):

CASE TYPE: **Rimmed Bottleneck**
CASE LENGTH **A = 1.99**
HEAD DIAMETER **B = .573**
RIM DIAMETER **D = .677**
NECK DIAMETER **F = .476**
NECK LENGTH **H = .245**
SHOULDER LENGTH **K = .145**
BODY ANGLE (DEG'S/SIDE): **.90**
CASE CAPACITY **CC'S = 5.60**

LOADED LENGTH: **2.64**
BELT DIAMETER **C = N/A**
RIM THICKNESS **E = .075**
SHOULDER DIAMETER **G = .529**
LENGTH TO SHOULDER **J = 1.60**
SHOULDER ANGLE (DEG'S/SIDE): **10.35**
PRIMER: **L/R**
CASE CAPACITY (GR'S WATER): **86.52**

DIMENSIONAL DRAWING:

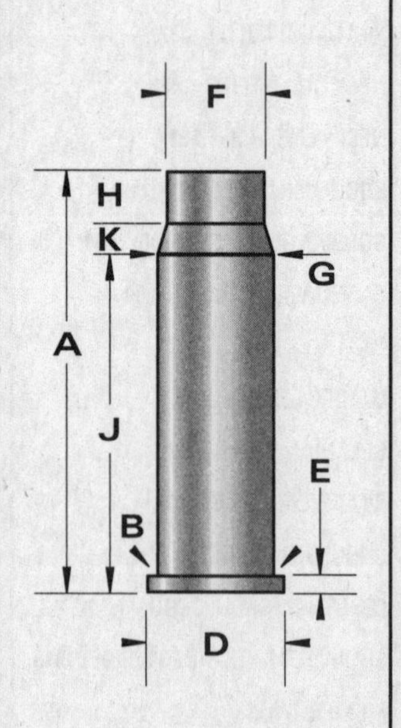

-NOT ACTUAL SIZE-
-DO NOT SCALE-

CARTRIDGE: 11 x 50R Belgian Albini M67/72

OTHER NAMES:

DIA: .435

BALLISTEK NO: 435B

NAI NO: RMB 23242/3.448

DATA SOURCE: COTW 4th Pg211

HISTORICAL DATA: Belgian ctg. ca. 1867.

NOTES: Quite rare.

LOADING DATA:

BULLET WT./TYPE	POWDER WT./TYPE	VELOCITY ('/SEC)	SOURCE
370/LeadPP	29.0/4198	1350	Barnes

CASE PREPARATION: **SHELLHOLDER (RCBS):** Spl.

MAKE FROM: 11mm Beaumont (BELL). Trim case to length & F/L size.

PHYSICAL DATA (INCHES):

CASE TYPE: **Rimmed Bottleneck**
CASE LENGTH **A = 2.00**
HEAD DIAMETER **B = .580**
RIM DIAMETER **D = .678**
NECK DIAMETER **F = .472**
NECK LENGTH **H = .165**
SHOULDER LENGTH **K = .155**
BODY ANGLE (DEG'S/SIDE): **.87**
CASE CAPACITY **CC'S = 5.49**

LOADED LENGTH: **2.60**
BELT DIAMETER **C = N/A**
RIM THICKNESS **E = .075**
SHOULDER DIAMETER **G = .535**
LENGTH TO SHOULDER **J = 1.68**
SHOULDER ANGLE (DEG'S/SIDE): **11.48**
PRIMER: **L/R**
CASE CAPACITY (GR'S WATER): **84.79**

DIMENSIONAL DRAWING:

-NOT ACTUAL SIZE-
-DO NOT SCALE-

CARTRIDGE: 11 x 53R Belgian Comblain

OTHER NAMES:

DIA: .436

BALLISTEK NO: 436A

NAI NO: RMB 24242/3.652

DATA SOURCE: COTW 4th Pg211

HISTORICAL DATA: Circa 1871.

NOTES: Do not confuse with Brazilian Comblain (#452E).

LOADING DATA:

BULLET WT./TYPE	POWDER WT./TYPE	VELOCITY ('/SEC)	SOURCE
370/LeadPP	32.0/4198	1460	Barnes

CASE PREPARATION: SHELLHOLDER (RCBS): Spl.

MAKE FROM: .475 #2 N.E. (BELL). Cut case to 2.15" & chamfer. Anneal and F/L size. Trim to length.

PHYSICAL DATA (INCHES):

CASE TYPE: **Rimmed Bottleneck**

CASE LENGTH **A = 2.10**

HEAD DIAMETER **B = .575**

RIM DIAMETER **D = .673**

NECK DIAMETER **F = .460**

NECK LENGTH **H = .380**

SHOULDER LENGTH **K = .190**

BODY ANGLE (DEG'S/SIDE): **.926**

CASE CAPACITY **CC'S = 5.90**

LOADED LENGTH: **2.76**

BELT DIAMETER **C = N/A**

RIM THICKNESS **E = .082**

SHOULDER DIAMETER **G = .532**

LENGTH TO SHOULDER **J = 1.53**

SHOULDER ANGLE (DEG'S/SIDE): **10.73**

PRIMER: **L/R**

CASE CAPACITY (GR'S WATER): **91.18**

DIMENSIONAL DRAWING:

F H K G A J E B D

-NOT ACTUAL SIZE-
-DO NOT SCALE-

CARTRIDGE: 11 x 57R Remington

OTHER NAMES:

DIA: .433

BALLISTEK NO: 433B

NAI NO: RMB 14231/4.316

DATA SOURCE: Hoyem Vol.2 Pg236

HISTORICAL DATA: First ctg. for Spanish Rem. with bottle necked case.

NOTES:

LOADING DATA:

BULLET WT./TYPE	POWDER WT./TYPE	VELOCITY ('/SEC)	SOURCE
375/LeadPP	41.8/IMR3031	–	JJD

CASE PREPARATION: SHELLHOLDER (RCBS): Spl.

MAKE FROM: .43 Span. Rem. (BELL). Trim case to length and F/L size.

PHYSICAL DATA (INCHES):

CASE TYPE: **Rimmed Bottleneck**

CASE LENGTH **A = 2.24**

HEAD DIAMETER **B = .519**

RIM DIAMETER **D = .629**

NECK DIAMETER **F = .470**

NECK LENGTH **H = .515**

SHOULDER LENGTH **K = .275**

BODY ANGLE (DEG'S/SIDE): **.275**

CASE CAPACITY **CC'S = 5.21**

LOADED LENGTH: **2.87**

BELT DIAMETER **C = N/A**

RIM THICKNESS **E = .084**

SHOULDER DIAMETER **G = .507**

LENGTH TO SHOULDER **J = 1.45**

SHOULDER ANGLE (DEG'S/SIDE): **3.84**

PRIMER: **L/R**

CASE CAPACITY (GR'S WATER): **80.46**

DIMENSIONAL DRAWING:

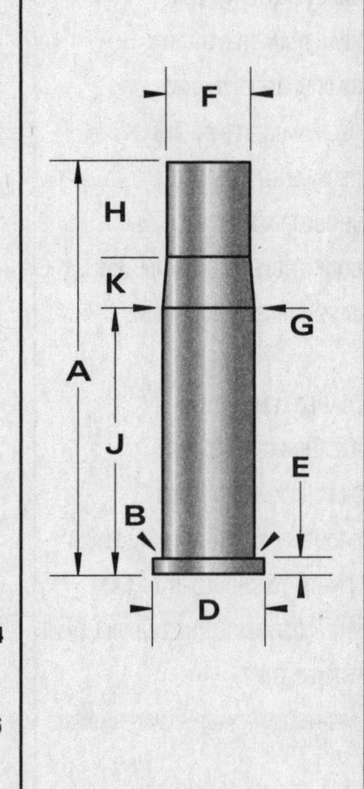

-NOT ACTUAL SIZE-
-DO NOT SCALE-

CARTRIDGE: 11 x 59mm Gras

OTHER NAMES:

DIA: .445

BALLISTEK NO: 445C

NAI NO: RMB 14222/4.301

DATA SOURCE: COTW 4th Pg213

HISTORICAL DATA: French mil. ctg. ca. 1874.

NOTES:

LOADING DATA:

BULLET WT./TYPE	POWDER WT./TYPE	VELOCITY ('/SEC)	SOURCE
385/LeadPP	33.0/4198	1400	Barnes

CASE PREPARATION: SHELLHOLDER (RCBS): 5

MAKE FROM: .348 Winchester. Anneal case neck and taper expand to .450" dia. Lathe turn base to .544" dia. Build rim thickness to .08". F/L size. Square case mouth & chamfer. Fireform.

PHYSICAL DATA (INCHES):

CASE TYPE: **Rimmed Bottleneck**

CASE LENGTH **A = 2.34**

HEAD DIAMETER **B = .544**

RIM DIAMETER **D = .667**

NECK DIAMETER **F = .468**

NECK LENGTH **H = .78**

SHOULDER LENGTH **K = .09**

BODY ANGLE (DEG'S/SIDE): **.293**

CASE CAPACITY **CC'S = 6.24**

LOADED LENGTH: **3.00**

BELT DIAMETER **C = N/A**

RIM THICKNESS **E = .083**

SHOULDER DIAMETER **G = .531**

LENGTH TO SHOULDER **J = 1.47**

SHOULDER ANGLE (DEG'S/SIDE): **19.29**

PRIMER: **L/R**

CASE CAPACITY (GR'S WATER): **96.36**

DIMENSIONAL DRAWING:

-NOT ACTUAL SIZE-
-DO NOT SCALE-

CARTRIDGE: 11 x 59R Vickers

OTHER NAMES:

DIA: .442

BALLISTEK NO: 442C

NAI NO: RMB 13221/4.360

DATA SOURCE: Hoyem Vol.2 Pg147

HISTORICAL DATA: Vickers M/G cartridge.

NOTES:

LOADING DATA:

BULLET WT./TYPE	POWDER WT./TYPE	VELOCITY ('/SEC)	SOURCE
300/RN	45.3/IMR3031	—	JJD

CASE PREPARATION: SHELLHOLDER (RCBS): 31

MAKE FROM: .280 Flanged (BELL). Trim case to length, chamfer and F/L size.

PHYSICAL DATA (INCHES):

CASE TYPE: **Rimmed Bottleneck**

CASE LENGTH **A = 2.35**

HEAD DIAMETER **B = .539**

RIM DIAMETER **D = .659**

NECK DIAMETER **F = .473**

NECK LENGTH **H = .60**

SHOULDER LENGTH **K = .19**

BODY ANGLE (DEG'S/SIDE): **.294**

CASE CAPACITY **CC'S = 6.04**

LOADED LENGTH: **2.99**

BELT DIAMETER **C = N/A**

RIM THICKNESS **E = .072**

SHOULDER DIAMETER **G = .525**

LENGTH TO SHOULDER **J = 1.56**

SHOULDER ANGLE (DEG'S/SIDE): **7.79**

PRIMER: **L/R**

CASE CAPACITY (GR'S WATER): **93.17**

DIMENSIONAL DRAWING:

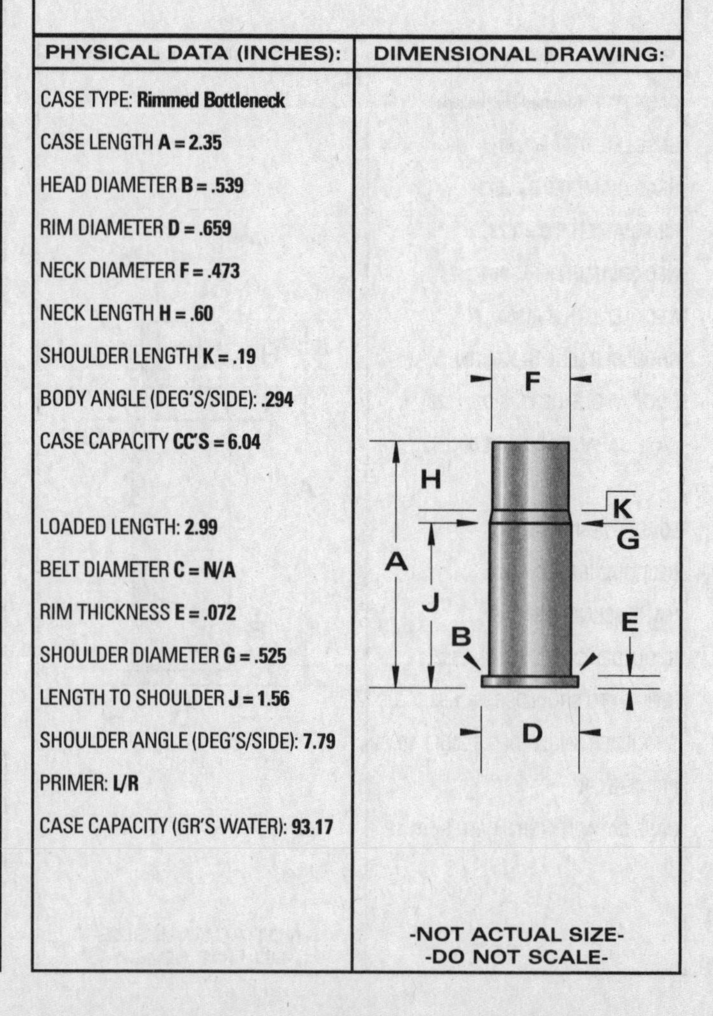

-NOT ACTUAL SIZE-
-DO NOT SCALE-

CARTRIDGE: 11,15 x 36R Fruhwirth Carbine

OTHER NAMES:

DIA: .439

BALLISTEK NO: 439F

NAI NO: RMB 24241/2.883

DATA SOURCE: Hoyem Vol.2 Pg177

HISTORICAL DATA:

NOTES:

LOADING DATA:

BULLET WT./TYPE	POWDER WT./TYPE	VELOCITY ('/SEC)	SOURCE
225/Lead	21.6/IMR4198	–	JJD

CASE PREPARATION: SHELLHOLDER (RCBS): Spl.

MAKE FROM: 7,62 x 54R Russian. Cut case to 1.5". Anneal and taper expand to .440" dia. Trim to length and F/L size. Fireform in chamber.

PHYSICAL DATA (INCHES):

CASE TYPE: **Rimmed Bottleneck**

CASE LENGTH **A = 1.41**

HEAD DIAMETER **B = .489**

RIM DIAMETER **D = .577**

NECK DIAMETER **F = .456**

NECK LENGTH **H = .39**

SHOULDER LENGTH **K = .075**

BODY ANGLE (DEG'S/SIDE): **.422**

CASE CAPACITY **CC'S = 3.05**

LOADED LENGTH: **2.10**

BELT DIAMETER **C = N/A**

RIM THICKNESS **E = .06**

SHOULDER DIAMETER **G = .478**

LENGTH TO SHOULDER **J = .945**

SHOULDER ANGLE (DEG'S/SIDE): **8.34**

PRIMER: **L/R**

CASE CAPACITY (GR'S WATER): **47.13**

DIMENSIONAL DRAWING:

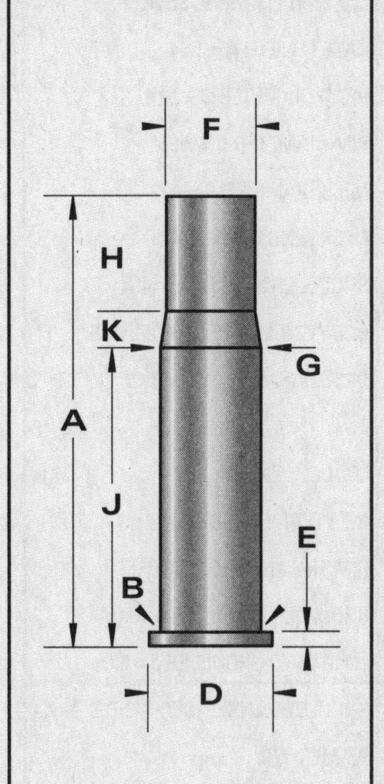

-NOT ACTUAL SIZE-
-DO NOT SCALE-

CARTRIDGE: 11,15 x 36R Werndl Carbine

OTHER NAMES:

DIA: .442

BALLISTEK NO: 442E

NAI NO: RMS 21115/2.872

DATA SOURCE: Hoyem Vol.2 Pg174

HISTORICAL DATA: Austrian ctg. from about 1867.

NOTES:

LOADING DATA:

BULLET WT./TYPE	POWDER WT./TYPE	VELOCITY ('/SEC)	SOURCE
275/Lead	12.0/IMR4227	–	JJD

CASE PREPARATION: SHELLHOLDER (RCBS): 13

MAKE FROM: 7,62 x 54R Russian. Cut case to 1.5", anneal and taper expand to .450" dia. Trim to length and F/L size.

PHYSICAL DATA (INCHES):

CASE TYPE: **Rimmed Straight**

CASE LENGTH **A = 1.41**

HEAD DIAMETER **B = .491**

RIM DIAMETER **D = .574**

NECK DIAMETER **F = .473**

NECK LENGTH **H = N/A**

SHOULDER LENGTH **K = N/A**

BODY ANGLE (DEG'S/SIDE): **.386**

CASE CAPACITY **CC'S = 2.69**

LOADED LENGTH: **1.84**

BELT DIAMETER **C = N/A**

RIM THICKNESS **E = .076**

SHOULDER DIAMETER **G = N/A**

LENGTH TO SHOULDER **J = N/A**

SHOULDER ANGLE (DEG'S/SIDE): **N/A**

PRIMER: **L/R**

CASE CAPACITY (GR'S WATER): **41.64**

DIMENSIONAL DRAWING:

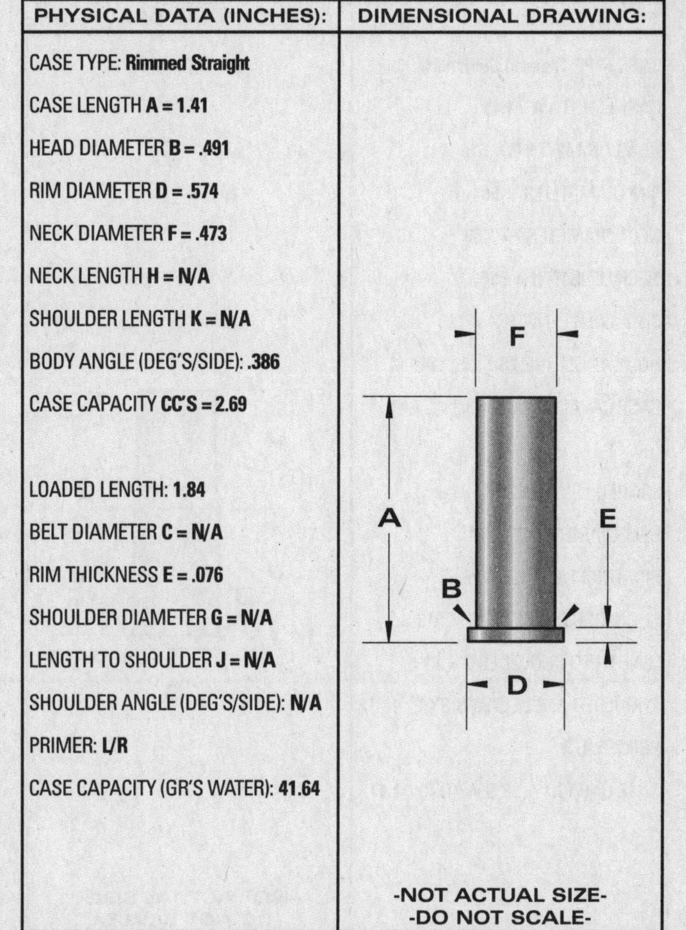

-NOT ACTUAL SIZE-
-DO NOT SCALE-

CARTRIDGE: 11,15 x 37R Grensaufsher

OTHER NAMES:	DIA: .439 BALLISTEK NO: 439E NAI NO: RMB 14221/2.887

DATA SOURCE: Hoyem Vol.2 Pg209

HISTORICAL DATA: For M71/84 Mauser.

NOTES:

LOADING DATA:

BULLET WT./TYPE	POWDER WT./TYPE	VELOCITY ('/SEC)	SOURCE
250/Lead	20.2/IMR4198	–	JJD

CASE PREPARATION: SHELLHOLDER (RCBS): Spl.

MAKE FROM: .43 Span. Rem. (BELL). Cut case to 1.5". Turn rim to .584" dia. Trim to length, chamfer and F/L size.

PHYSICAL DATA (INCHES):

CASE TYPE: **Rimmed Bottleneck**

CASE LENGTH **A = 1.49**

HEAD DIAMETER **B = .516**

RIM DIAMETER **D = .584**

NECK DIAMETER **F = .468**

NECK LENGTH **H = .554**

SHOULDER LENGTH **K = .126**

BODY ANGLE (DEG'S/SIDE): **.140**

CASE CAPACITY **CC'S = 3.33**

LOADED LENGTH: **2.10**

BELT DIAMETER **C = N/A**

RIM THICKNESS **E = .075**

SHOULDER DIAMETER **G = .513**

LENGTH TO SHOULDER **J = .81**

SHOULDER ANGLE (DEG'S/SIDE): **10.12**

PRIMER: **L/R**

CASE CAPACITY (GR'S WATER): **51.47**

DIMENSIONAL DRAWING:

-NOT ACTUAL SIZE-
-DO NOT SCALE-

CARTRIDGE: 11,15 x 42R Werndl M67

OTHER NAMES:	DIA: .442 BALLISTEK NO: 442D NAI NO: RMS 21115/3.190

DATA SOURCE: Hoyem Vol.2 Pg174

HISTORICAL DATA: Austrian, circa1867.

NOTES:

LOADING DATA:

BULLET WT./TYPE	POWDER WT./TYPE	VELOCITY ('/SEC)	SOURCE
313/Lead	18.6/IMR4227	–	JJD

CASE PREPARATION: SHELLHOLDER (RCBS): Spl.

MAKE FROM: .43 Span. Rem. (BELL). Turn rim to .590" dia. Cut case to 1.7". Trim to length, chamfer and F/L size.

PHYSICAL DATA (INCHES):

CASE TYPE: **Rimmed Straight**

CASE LENGTH **A = 1.64**

HEAD DIAMETER **B = .514**

RIM DIAMETER **D = .590**

NECK DIAMETER **F = .473**

NECK LENGTH **H = N/A**

SHOULDER LENGTH **K = N/A**

BODY ANGLE (DEG'S/SIDE): **.746**

CASE CAPACITY **CC'S = 3.43**

LOADED LENGTH: **2.37**

BELT DIAMETER **C = N/A**

RIM THICKNESS **E = .067**

SHOULDER DIAMETER **G = N/A**

LENGTH TO SHOULDER **J = N/A**

SHOULDER ANGLE (DEG'S/SIDE): **N/A**

PRIMER: **L/R**

CASE CAPACITY (GR'S WATER): **52.99**

DIMENSIONAL DRAWING:

-NOT ACTUAL SIZE-
-DO NOT SCALE-

CARTRIDGE: 11,15 x 58 Werndl M77

OTHER NAMES:

DIA: .441

BALLISTEK NO: 441A

NAI NO: RMB 14222/4.165

DATA SOURCE: COTW 4th Pg213

HISTORICAL DATA: Austrian ctg. ca. 1886.

NOTES:

LOADING DATA:

BULLET WT./TYPE	POWDER WT./TYPE	VELOCITY ('/SEC)	SOURCE
370/Lead	32.0/4198	1360	Barnes

CASE PREPARATION: SHELLHOLDER (RCBS): Spl.

MAKE FROM: .450 N.E. (BELL). Turn rim to .617" dia. & back chamfer. Cut to 2.3" Anneal case, F/L size, trim to length & chamfer. Build rim up to .09" thick.

PHYSICAL DATA (INCHES):

CASE TYPE: **Rimmed Bottleneck**

CASE LENGTH **A = 2.27**

HEAD DIAMETER **B = .545**

RIM DIAMETER **D = .617**

NECK DIAMETER **F = .466**

NECK LENGTH **H = .715**

SHOULDER LENGTH **K = .168**

BODY ANGLE (DEG'S/SIDE): **.217**

CASE CAPACITY **CC'S = 6.00**

LOADED LENGTH: **3.02**

BELT DIAMETER **C = N/A**

RIM THICKNESS **E = .09**

SHOULDER DIAMETER **G = .536**

LENGTH TO SHOULDER **J = 1.387**

SHOULDER ANGLE (DEG'S/SIDE): **11.76**

PRIMER: **L/R**

CASE CAPACITY (GR'S WATER): **92.64**

DIMENSIONAL DRAWING:

F

H

K

G

A

J

E

B

D

-NOT ACTUAL SIZE-
-DO NOT SCALE-

CARTRIDGE: 11,15 x 58R Spanish Remington

OTHER NAMES:

DIA: .439

BALLISTEK NO: 439A

NAI NO: RMB 13222/4.360

DATA SOURCE: COTW 4th Pg212

HISTORICAL DATA: Spanish mil. ctg. ca. 1871.

NOTES:

LOADING DATA:

BULLET WT./TYPE	POWDER WT./TYPE	VELOCITY ('/SEC)	SOURCE
387/Lead	32.0/4198	1360	Barnes

CASE PREPARATION: SHELLHOLDER (RCBS): Spl.

MAKE FROM: .43 Span. Rem. (BELL). Thin rim to .065" thick. Trim to length, chamfer and F/L size.

PHYSICAL DATA (INCHES):

CASE TYPE: **Rimmed Bottleneck**

CASE LENGTH **A = 2.25**

HEAD DIAMETER **B = .516**

RIM DIAMETER **D = .635**

NECK DIAMETER **F = .455**

NECK LENGTH **H = .567**

SHOULDER LENGTH **K = .123**

BODY ANGLE (DEG'S/SIDE): **.084**

CASE CAPACITY **CC'S = 5.81**

LOADED LENGTH: **2.82**

BELT DIAMETER **C = N/A**

RIM THICKNESS **E = .065**

SHOULDER DIAMETER **G = .512**

LENGTH TO SHOULDER **J = 1.56**

SHOULDER ANGLE (DEG'S/SIDE): **13.05**

PRIMER: **L/R**

CASE CAPACITY (GR'S WATER): **89.72**

DIMENSIONAL DRAWING:

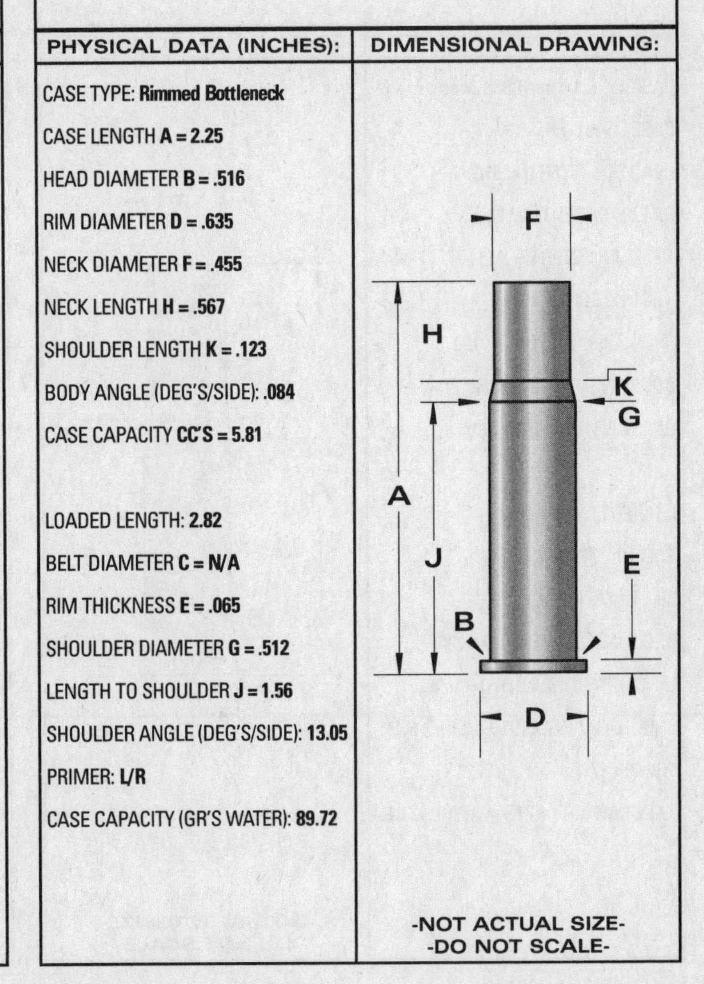

-NOT ACTUAL SIZE-
-DO NOT SCALE-

CARTRIDGE: 11,15 x 60R Mauser

OTHER NAMES:
.43 Mauser

DIA: .446

BALLISTEK NO: 446B

NAI NO: RMB 13221/4.593

DATA SOURCE: COTW 4th Pg213

HISTORICAL DATA: By P. Mauser about 1871.

NOTES:

LOADING DATA:

BULLET WT./TYPE	POWDER WT./TYPE	VELOCITY ('/SEC)	SOURCE
387/LeadPP	32.0/4198	1335	Barnes

CASE PREPARATION: SHELLHOLDER (RCBS): Spl.

MAKE FROM: 11mm "A" base (BELL). Anneal case and form in the F/L die with the expander removed. Form die may be required but, generally, cases will form okay in F/L die. Trim to length & chamfer. F/L size again.

PHYSICAL DATA (INCHES):

CASE TYPE: **Rimmed Bottleneck**

CASE LENGTH **A = 2.37**

HEAD DIAMETER **B = .516**

RIM DIAMETER **D = .586**

NECK DIAMETER **F = .465**

NECK LENGTH **H = .690**

SHOULDER LENGTH **K = .199**

BODY ANGLE (DEG'S/SIDE): **.134**

CASE CAPACITY **CC'S = 6.06**

LOADED LENGTH: **3.00**

BELT DIAMETER **C = N/A**

RIM THICKNESS **E = .091**

SHOULDER DIAMETER **G = .510**

LENGTH TO SHOULDER **J = 1.48**

SHOULDER ANGLE (DEG'S/SIDE): **6.45**

PRIMER: **L/R**

CASE CAPACITY (GR'S WATER): **93.61**

DIMENSIONAL DRAWING:

-NOT ACTUAL SIZE-
-DO NOT SCALE-

CARTRIDGE: 11,2 x 60 Mauser (Schuler)

OTHER NAMES:
11.15 x 59.8 Schuler

DIA: .440

BALLISTEK NO: 440A

NAI NO: RXB 13222/4.581

DATA SOURCE: COTW 4th Pg278

HISTORICAL DATA: German sporting ctg.

NOTES:

LOADING DATA:

BULLET WT./TYPE	POWDER WT./TYPE	VELOCITY ('/SEC)	SOURCE
370/—	45.0/3031	1500	Barnes

CASE PREPARATION: SHELLHOLDER (RCBS): 4

MAKE FROM: .375 H&H Mag. Turn belt flush with body. Turn rim to .465" dia. and re-cut extractor groove. Anneal neck and taper expand to .450" dia. Cut to 2.4". F/L size, trim to length and chamfer.

PHYSICAL DATA (INCHES):

CASE TYPE: **Rebated Bottleneck**

CASE LENGTH **A = 2.35**

HEAD DIAMETER **B = .513**

RIM DIAMETER **D = .465**

NECK DIAMETER **F = .465**

NECK LENGTH **H = .80**

SHOULDER LENGTH **K = .09**

BODY ANGLE (DEG'S/SIDE): **.023**

CASE CAPACITY **CC'S = 5.77**

LOADED LENGTH: **2.86**

BELT DIAMETER **C = N/A**

RIM THICKNESS **E = .056**

SHOULDER DIAMETER **G = .512**

LENGTH TO SHOULDER **J = 1.46**

SHOULDER ANGLE (DEG'S/SIDE): **14.63**

PRIMER: **L/R**

CASE CAPACITY (GR'S WATER): **89.11**

DIMENSIONAL DRAWING:

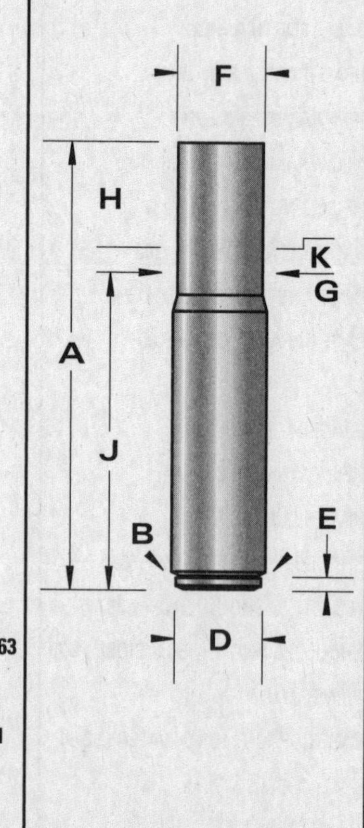

-NOT ACTUAL SIZE-
-DO NOT SCALE-

CARTRIDGE: 11,2 x 72 Schuler

OTHER NAMES:

DIA: .439

BALLISTEK NO: 439D

NAI NO: RBB 22241/5.224

DATA SOURCE: COTW 4th Pg278

HISTORICAL DATA: German ctg. ca. 1920.

NOTES:

LOADING DATA:

BULLET WT./TYPE	POWDER WT./TYPE	VELOCITY ('/SEC)	SOURCE
400/—	10.3/4350	2360	Barnes

CASE PREPARATION: SHELLHOLDER (RCBS): Spl.

MAKE FROM: .404 Jeffery (BELL). Rebate rim by turning to .469" dia. Re-cut extractor groove. F/L size in Schuler die. Trim to length & chamfer.

PHYSICAL DATA (INCHES):

CASE TYPE: **Rebated Bottleneck**

CASE LENGTH **A = 2.80**

HEAD DIAMETER **B = .536**

RIM DIAMETER **D = .469**

NECK DIAMETER **F = .465**

NECK LENGTH **H = .35**

SHOULDER LENGTH **K = .21**

BODY ANGLE (DEG'S/SIDE): **.365**

CASE CAPACITY **CC'S = 7.46**

LOADED LENGTH: **3.85**

BELT DIAMETER **C = N/A**

RIM THICKNESS **E = .06**

SHOULDER DIAMETER **G = .51**

LENGTH TO SHOULDER **J = 2.24**

SHOULDER ANGLE (DEG'S/SIDE): **6.11**

PRIMER: **L/R**

CASE CAPACITY (GR'S WATER): **115.16**

DIMENSIONAL DRAWING:

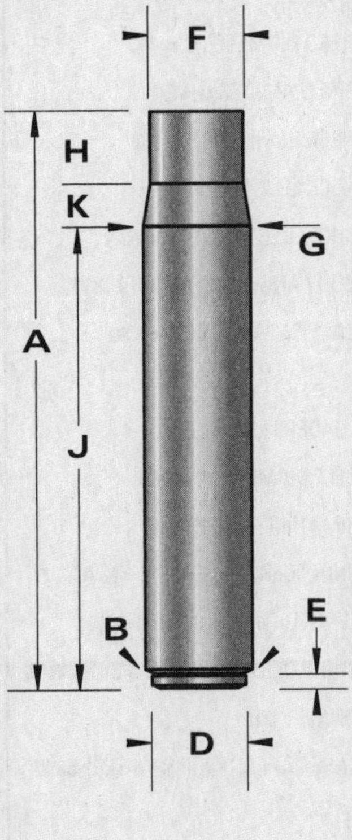

-NOT ACTUAL SIZE-
-DO NOT SCALE-

CARTRIDGE: 11,35mm Madsen

OTHER NAMES:

DIA: .460

BALLISTEK NO: 460C

NAI NO: RXB 23232/3.847

DATA SOURCE: MSAA Pg46

HISTORICAL DATA: Ctg. for Madsen MG circa 1932.

NOTES: Made in England and Argentina.

LOADING DATA:

BULLET WT./TYPE	POWDER WT./TYPE	VELOCITY ('/SEC)	SOURCE
300/Spire	80.0/IMR3031	—	JJD

CASE PREPARATION: SHELLHOLDER (RCBS): Spl.

MAKE FROM: Turn case from solid 5/8" dia. brass. Trim to 2.5". Anneal. F/L size, trim to length and chamfer.

PHYSICAL DATA (INCHES):

CASE TYPE: **Rimless Bottleneck**

CASE LENGTH **A = 2.42**

HEAD DIAMETER **B = .629**

RIM DIAMETER **D = .625**

NECK DIAMETER **F = .504**

NECK LENGTH **H = .466**

SHOULDER LENGTH **K = .134**

BODY ANGLE (DEG'S/SIDE): **.707**

CASE CAPACITY **CC'S = 7.68**

LOADED LENGTH: **3.28**

BELT DIAMETER **C = N/A**

RIM THICKNESS **E = .067**

SHOULDER DIAMETER **G = .589**

LENGTH TO SHOULDER **J = 1.82**

SHOULDER ANGLE (DEG'S/SIDE): **17.59**

PRIMER: **L/R**

CASE CAPACITY (GR'S WATER): **118.64**

DIMENSIONAL DRAWING:

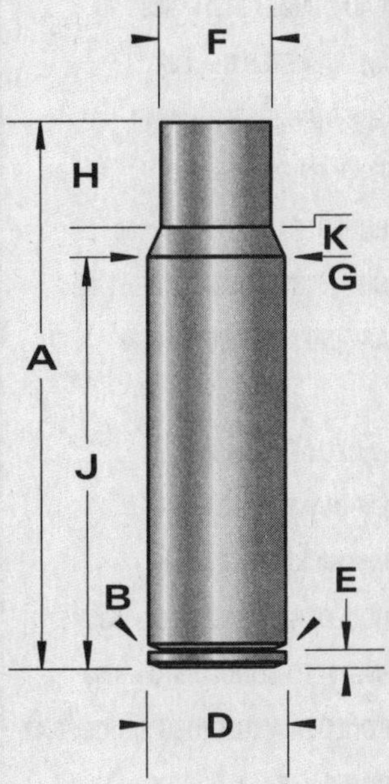

-NOT ACTUAL SIZE-
-DO NOT SCALE-

CARTRIDGE: 11,4 x 50R Brazilian Comblain M74

OTHER NAMES:
11.4 x 53R Brazilian M74

DIA: .452

BALLISTEK NO: 452E

NAI NO: RMB 33241/3.435

DATA SOURCE: COTW 4th Pg215

HISTORICAL DATA:

NOTES:

LOADING DATA:

BULLET WT./TYPE	POWDER WT./TYPE	VELOCITY ('/SEC)	SOURCE
485/LeadPP	27.0/4198	1280	Barnes

CASE PREPARATION: SHELLHOLDER (RCBS): Spl.

MAKE FROM: 11mm Beaumont (BELL). Cut case to 2.1". Build rim thickness to .085". F/L size case, trim to length.

PHYSICAL DATA (INCHES):

CASE TYPE: **Rimmed Bottleneck**
CASE LENGTH **A = 2.02**
HEAD DIAMETER **B = .588**
RIM DIAMETER **D = .682**
NECK DIAMETER **F = .494**
NECK LENGTH **H = .170**
SHOULDER LENGTH **K = .160**
BODY ANGLE (DEG'S/SIDE): **1.11**
CASE CAPACITY **CC'S = 5.48**

LOADED LENGTH: **2.62**
BELT DIAMETER **C = N/A**
RIM THICKNESS **E = .085**
SHOULDER DIAMETER **G = .530**
LENGTH TO SHOULDER **J = 1.69**
SHOULDER ANGLE (DEG'S/SIDE): **6.42**
PRIMER: **L/R**
CASE CAPACITY (GR'S WATER): **84.58**

DIMENSIONAL DRAWING:

-NOT ACTUAL SIZE-
-DO NOT SCALE-

CARTRIDGE: 11.4 x 50R Werndl M73

OTHER NAMES:

DIA: .449

BALLISTEK NO: 449A

NAI NO: RMS 11115/3.996

DATA SOURCE: COTW 4th Pg214

HISTORICAL DATA: Austrian mil. ctg. ca. 1873.

NOTES:

LOADING DATA:

BULLET WT./TYPE	POWDER WT./TYPE	VELOCITY ('/SEC)	SOURCE
465/Lead	29.0/4198	1300	Barnes

CASE PREPARATION: SHELLHOLDER (RCBS): 14

MAKE FROM: .45-70 Gov't. Turn rim to .570" dia. and back chamfer. Trim to length and F/L size.

PHYSICAL DATA (INCHES):

CASE TYPE: **Rimmed Straight**
CASE LENGTH **A = 1.97**
HEAD DIAMETER **B = .493**
RIM DIAMETER **D = .571**
NECK DIAMETER **F = .472**
NECK LENGTH **H = N/A**
SHOULDER LENGTH **K = N/A**
BODY ANGLE (DEG'S/SIDE): **.314**
CASE CAPACITY **CC'S = 4.30**

LOADED LENGTH: **2.55**
BELT DIAMETER **C = N/A**
RIM THICKNESS **E = .06**
SHOULDER DIAMETER **G = N/A**
LENGTH TO SHOULDER **J = N/A**
SHOULDER ANGLE (DEG'S/SIDE): **N/A**
PRIMER: **L/R**
CASE CAPACITY (GR'S WATER): **66.39**

DIMENSIONAL DRAWING:

-NOT ACTUAL SIZE-
-DO NOT SCALE-

CARTRIDGE: 11.43 x 41 Peabody-Martini

OTHER NAMES:

DIA: .451

BALLISTEK NO: 451R

NAI NO: RMB 24241/2.788

DATA SOURCE: Hoyem Vol.2 Pg220

HISTORICAL DATA: For Turkish Martini rifles.

NOTES:

LOADING DATA:

BULLET WT./TYPE	POWDER WT./TYPE	VELOCITY ('/SEC)	SOURCE
405/LeadPP	21.5/IMR4198	—	JJD

CASE PREPARATION: SHELLHOLDER (RCBS): Spl.

MAKE FROM: .475 #2 N.E. (BELL). Cut case to 1.7" and thin rim to .062" thick. Chamfer. F/L size. I.D. neck reaming will be necessary to seat correct dia. bullet, however, slightly undersized bullets may be "minied," for general use.

PHYSICAL DATA (INCHES):

CASE TYPE: **Rimmed Bottleneck**

CASE LENGTH **A = 1.62**

HEAD DIAMETER **B = .581**

RIM DIAMETER **D = .665**

NECK DIAMETER **F = .474**

NECK LENGTH **H = .412**

SHOULDER LENGTH **K = .268**

BODY ANGLE (DEG'S/SIDE): **.735**

CASE CAPACITY **CC'S = 4.50**

LOADED LENGTH: **2.32**

BELT DIAMETER **C = N/A**

RIM THICKNESS **E = .062**

SHOULDER DIAMETER **G = .562**

LENGTH TO SHOULDER **J = .94**

SHOULDER ANGLE (DEG'S/SIDE): **9.32**

PRIMER: **L/R**

CASE CAPACITY (GR'S WATER): **69.60**

DIMENSIONAL DRAWING:

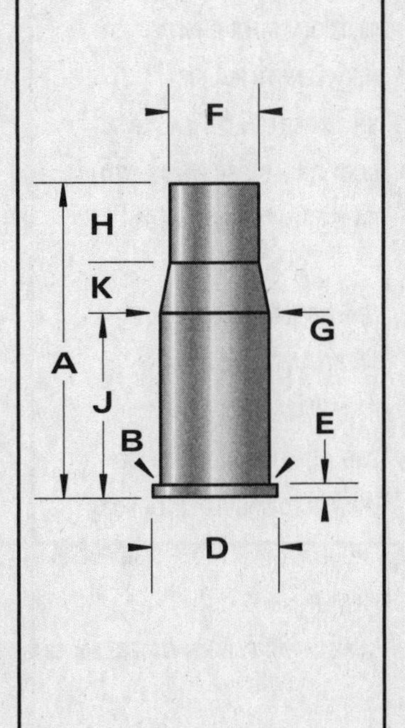

-NOT ACTUAL SIZE-
-DO NOT SCALE-

CARTRIDGE: 11,43 x 49R Romanian Peabody

OTHER NAMES:

DIA: .459

BALLISTEK NO: 459A

NAI NO: RMB 24243/3.416

DATA SOURCE: Hoyem Vol.2 Pg197

HISTORICAL DATA: Romanian ctg. for Peabody M67 circa 1868.

NOTES:

LOADING DATA:

BULLET WT./TYPE	POWDER WT./TYPE	VELOCITY ('/SEC)	SOURCE
380/LeadPP	39.8/IMR3031	—	JJD

CASE PREPARATION: SHELLHOLDER (RCBS): Spl.

MAKE FROM: .450 #2 N.E. (BELL). Thin rim to .06". Cut case to 2.0". F/L size, trim to length & chamfer.

PHYSICAL DATA (INCHES):

CASE TYPE: **Rimmed Bottleneck**

CASE LENGTH **A = 1.92**

HEAD DIAMETER **B = .562**

RIM DIAMETER **D = .657**

NECK DIAMETER **F = .484**

NECK LENGTH **H = .450**

SHOULDER LENGTH **K = .06**

BODY ANGLE (DEG'S/SIDE): **.449**

CASE CAPACITY **CC'S = 5.38**

LOADED LENGTH: **2.46**

BELT DIAMETER **C = N/A**

RIM THICKNESS **E = .06**

SHOULDER DIAMETER **G = .543**

LENGTH TO SHOULDER **J = 1.41**

SHOULDER ANGLE (DEG'S/SIDE): **26.18**

PRIMER: **L/R**

CASE CAPACITY (GR'S WATER): **83.11**

DIMENSIONAL DRAWING:

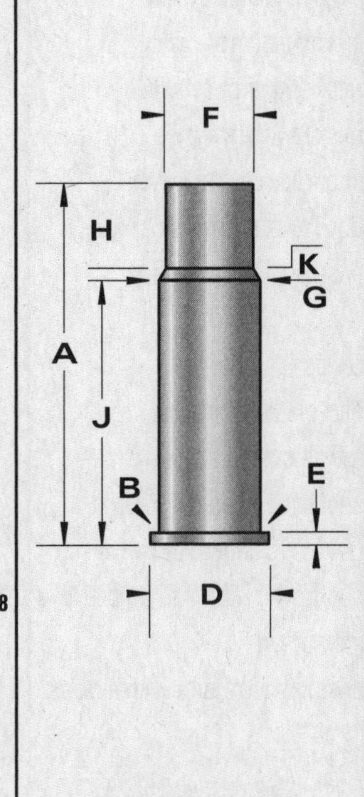

-NOT ACTUAL SIZE-
-DO NOT SCALE-

CARTRIDGE: 11,43 x 50R Egyptian Remington

OTHER NAMES:
11mm Egyptian

DIA: .448

BALLISTEK NO: 448A

NAI NO: RMB 34222/3.339

DATA SOURCE: COTW 4th Pg215

HISTORICAL DATA: Egyptian mil. ctg. about 1870.

NOTES:

LOADING DATA:

BULLET WT./TYPE	POWDER WT./TYPE	VELOCITY ('/SEC)	SOURCE
465/LeadPP	34.0/4198	1440	Barnes

CASE PREPARATION: SHELLHOLDER (RCBS): Spl.

MAKE FROM: .475 #2 N.E. (BELL). Cut case to 2.0" and thin rim to .065". F/L size, trim to length and chamfer.

PHYSICAL DATA (INCHES):

CASE TYPE: **Rimmed Bottleneck**
CASE LENGTH **A = 1.94**
HEAD DIAMETER **B = .581**
RIM DIAMETER **D = .670**
NECK DIAMETER **F = .480**
NECK LENGTH **H = .570**
SHOULDER LENGTH **K = .14**
BODY ANGLE (DEG'S/SIDE): **1.08**
CASE CAPACITY **CC'S = 5.23**
LOADED LENGTH: **2.73**
BELT DIAMETER **C = N/A**
RIM THICKNESS **E = .065**
SHOULDER DIAMETER **G = .542**
LENGTH TO SHOULDER **J = 1.23**
SHOULDER ANGLE (DEG'S/SIDE): **12.48**
PRIMER: **L/R**
CASE CAPACITY (GR'S WATER): **80.75**

DIMENSIONAL DRAWING:

F H K G A J E B D

-NOT ACTUAL SIZE-
-DO NOT SCALE-

CARTRIDGE: 11,43 x 55R Turkish

OTHER NAMES:
.450 Turkish
.45 Peabody Martini

DIA: .447

BALLISTEK NO: 447D

NAI NO: RMB 24221/3.952

DATA SOURCE: COTW 4th Pg219

HISTORICAL DATA: Turkish mil. ctg. ca. 1874.

NOTES:

LOADING DATA:

BULLET WT./TYPE	POWDER WT./TYPE	VELOCITY ('/SEC)	SOURCE
465/LeadPP	36.0/4198	1410	Barnes

CASE PREPARATION: SHELLHOLDER (RCBS): Spl.

MAKE FROM: 11mm Beaumont (BELL). Trim case to length and F/L size.

PHYSICAL DATA (INCHES):

CASE TYPE: **Rimmed Bottleneck**
CASE LENGTH **A = 2.30**
HEAD DIAMETER **B = .582**
RIM DIAMETER **D = .668**
NECK DIAMETER **F = .474**
NECK LENGTH **H = .678**
SHOULDER LENGTH **K = .299**
BODY ANGLE (DEG'S/SIDE): **.561**
CASE CAPACITY **CC'S = 6.58**
LOADED LENGTH: **3.12**
BELT DIAMETER **C = N/A**
RIM THICKNESS **E = .064**
SHOULDER DIAMETER **G = .560**
LENGTH TO SHOULDER **J = 1.323**
SHOULDER ANGLE (DEG'S/SIDE): **8.18**
PRIMER: **L/R**
CASE CAPACITY (GR'S WATER): **101.53**

DIMENSIONAL DRAWING:

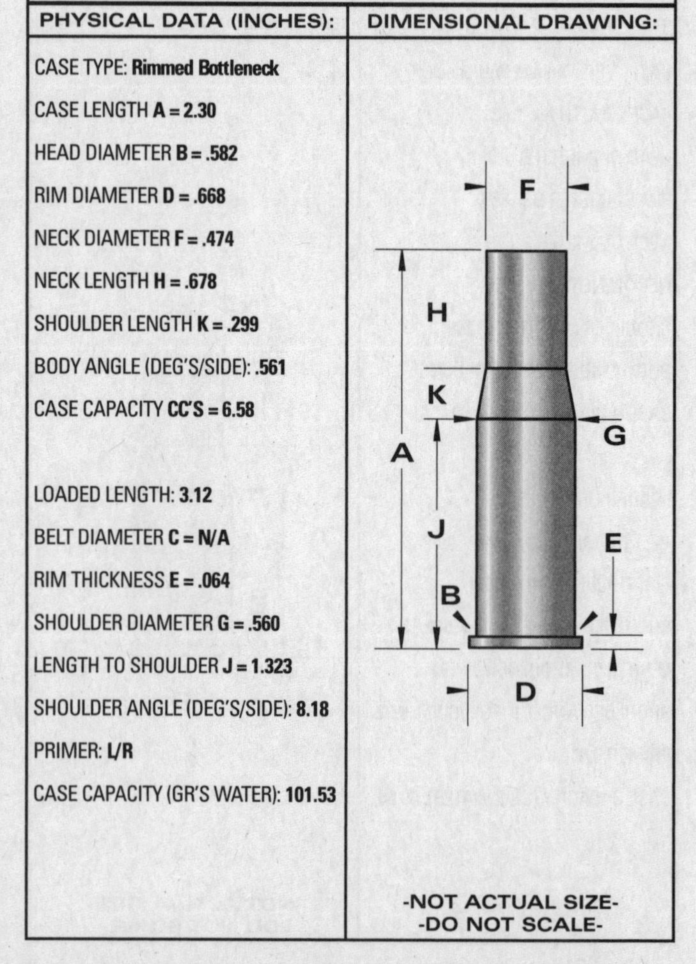

-NOT ACTUAL SIZE-
-DO NOT SCALE-

CARTRIDGE: 11,43 x 59R Turkish Mauser

OTHER NAMES:

DIA: .456

BALLISTEK NO: 456B

NAI NO: RMB 24221/3.993

DATA SOURCE: Hoyem Vol.2 Pg220

HISTORICAL DATA: Turkish mil. ctg. ca. 1874.

NOTES:

LOADING DATA:

BULLET WT./TYPE	POWDER WT./TYPE	VELOCITY ('/SEC)	SOURCE
480/LeadPP	27.2/IMR4198	—	JJD

CASE PREPARATION: **SHELLHOLDER (RCBS):** Spl.

MAKE FROM: 11mm Beaumont (BELL). Trim case to length and F/L size.

PHYSICAL DATA (INCHES):

CASE TYPE: **Rimmed Bottleneck**
CASE LENGTH **A = 2.32**
HEAD DIAMETER **B = .581**
RIM DIAMETER **D = .668**
NECK DIAMETER **F = .474**
NECK LENGTH **H = .675**
SHOULDER LENGTH **K = .285**
BODY ANGLE (DEG'S/SIDE): **.494**
CASE CAPACITY **CC'S = 6.97**

LOADED LENGTH: **3.14**
BELT DIAMETER **C = N/A**
RIM THICKNESS **E = .063**
SHOULDER DIAMETER **G = .561**
LENGTH TO SHOULDER **J = 1.36**
SHOULDER ANGLE (DEG'S/SIDE): **8.68**
PRIMER: **L/R**
CASE CAPACITY (GR'S WATER): **107.65**

DIMENSIONAL DRAWING:

F H K G A J E B D

-NOT ACTUAL SIZE-
-DO NOT SCALE-

CARTRIDGE: 11,5 x 50R Werder M69

OTHER NAMES:

DIA: .460

BALLISTEK NO: 460B

NAI NO: RMB 14241/3.794

DATA SOURCE: Hoyem Vol.2 Pg202

HISTORICAL DATA:

NOTES:

LOADING DATA:

BULLET WT./TYPE	POWDER WT./TYPE	VELOCITY ('/SEC)	SOURCE
329/Lead	31.8/IMR3031	—	JJD

CASE PREPARATION: **SHELLHOLDER (RCBS):** Spl.

MAKE FROM: .43 Span. Rem. (BELL). Trim case to length and F/L size. May be necessary to reduce rim dia. to .622".

PHYSICAL DATA (INCHES):

CASE TYPE: **Rimmed Bottleneck**
CASE LENGTH **A = 1.95**
HEAD DIAMETER **B = .514**
RIM DIAMETER **D = .622**
NECK DIAMETER **F = .477**
NECK LENGTH **H = .380**
SHOULDER LENGTH **K = .30**
BODY ANGLE (DEG'S/SIDE): **.026**
CASE CAPACITY **CC'S = 5.06**

LOADED LENGTH: **2.55**
BELT DIAMETER **C = N/A**
RIM THICKNESS **E = .074**
SHOULDER DIAMETER **G = .513**
LENGTH TO SHOULDER **J = 1.27**
SHOULDER ANGLE (DEG'S/SIDE): **3.43**
PRIMER: **L/R**
CASE CAPACITY (GR'S WATER): **78.12**

DIMENSIONAL DRAWING:

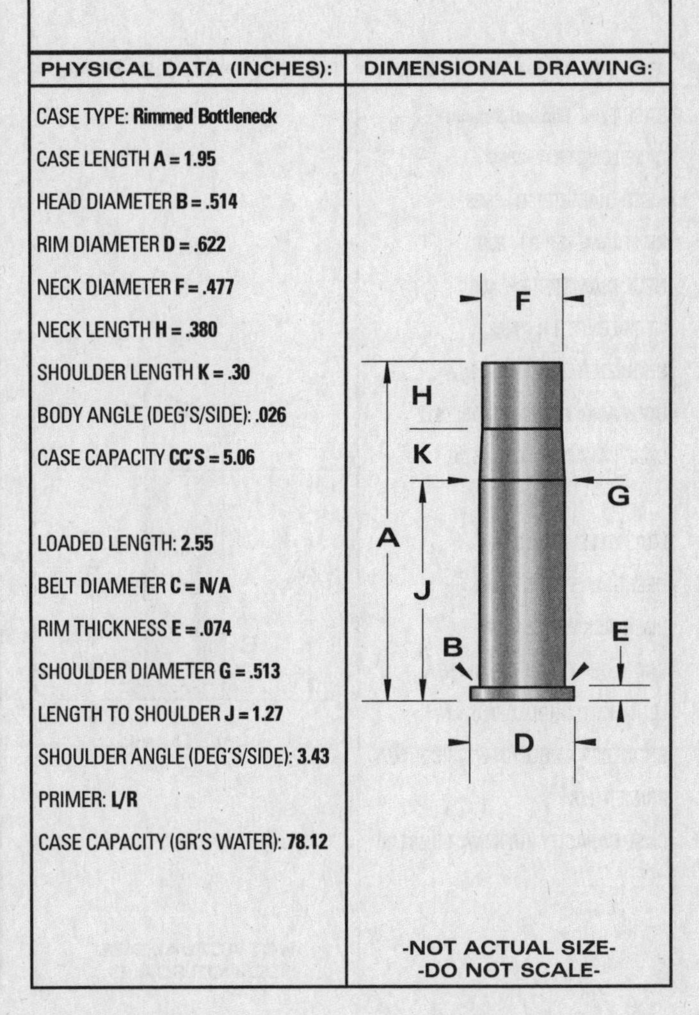

-NOT ACTUAL SIZE-
-DO NOT SCALE-

CARTRIDGE: 11,5 x 57R Spanish Reformado

OTHER NAMES:

DIA: .454

BALLISTEK NO: 454H

NAI NO: RMS 21115/4.304

DATA SOURCE: COTW 4th Pg216

HISTORICAL DATA: Early Spanish mil. ctg. ca. 1867.

NOTES:

LOADING DATA:

BULLET WT./TYPE	POWDER WT./TYPE	VELOCITY ('/SEC)	SOURCE
250/Lead	32.0/4198	1220	Barnes

CASE PREPARATION: SHELLHOLDER (RCBS): Spl.

MAKE FROM: .43 Span. Rem. (BELL). Trim case to length. F/L size. Case will fireform in the chamber.

PHYSICAL DATA (INCHES):

CASE TYPE: **Rimmed Straight**

CASE LENGTH **A = 2.26**

HEAD DIAMETER **B = .525**

RIM DIAMETER **D = .630**

NECK DIAMETER **F = .486**

NECK LENGTH **H = N/A**

SHOULDER LENGTH **K = N/A**

BODY ANGLE (DEG'S/SIDE): **.512**

CASE CAPACITY **CC'S = 5.26**

LOADED LENGTH: **3.06**

BELT DIAMETER **C = N/A**

RIM THICKNESS **E = .08**

SHOULDER DIAMETER **G = N/A**

LENGTH TO SHOULDER **J = N/A**

SHOULDER ANGLE (DEG'S/SIDE): **N/A**

PRIMER: **L/R**

CASE CAPACITY (GR'S WATER): **81.30**

DIMENSIONAL DRAWING:

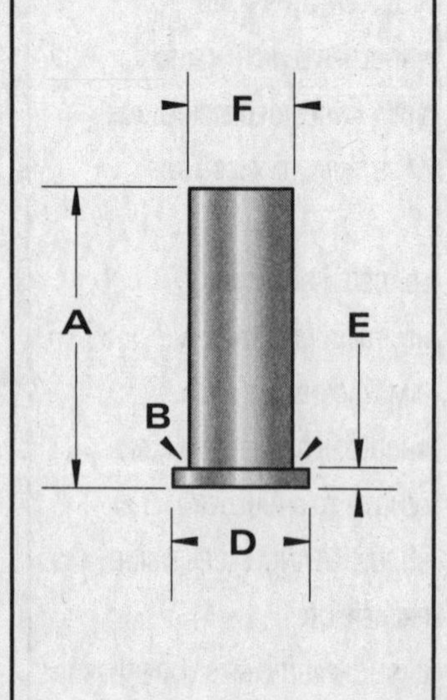

-NOT ACTUAL SIZE-
-DO NOT SCALE-

CARTRIDGE: 11,7 x 41.5R Danish Remington

OTHER NAMES:

DIA: .465

BALLISTEK NO: 465A

NAI NO: RMS 21115/3.183

DATA SOURCE: Hoyem Vol.2 Pg233

HISTORICAL DATA: Danish mil. ctg. about 1875.

NOTES:

LOADING DATA:

BULLET WT./TYPE	POWDER WT./TYPE	VELOCITY ('/SEC)	SOURCE
250/Lead	19.3/IMR4227	—	JJD

CASE PREPARATION: SHELLHOLDER (RCBS): Spl.

MAKE FROM: .375 Flanged (BELL). Cut case to 1.7". F/L size, trim to length & chamfer.

PHYSICAL DATA (INCHES):

CASE TYPE: **Rimmed Straight**

CASE LENGTH **A = 1.63**

HEAD DIAMETER **B = .512**

RIM DIAMETER **D = .574**

NECK DIAMETER **F = .483**

NECK LENGTH **H = N/A**

SHOULDER LENGTH **K = N/A**

BODY ANGLE (DEG'S/SIDE): **.531**

CASE CAPACITY **CC'S = 3.77**

LOADED LENGTH: **2.32**

BELT DIAMETER **C = N/A**

RIM THICKNESS **E = .066**

SHOULDER DIAMETER **G = N/A**

LENGTH TO SHOULDER **J = N/A**

SHOULDER ANGLE (DEG'S/SIDE): **N/A**

PRIMER: **L/R**

CASE CAPACITY (GR'S WATER): **58.29**

DIMENSIONAL DRAWING:

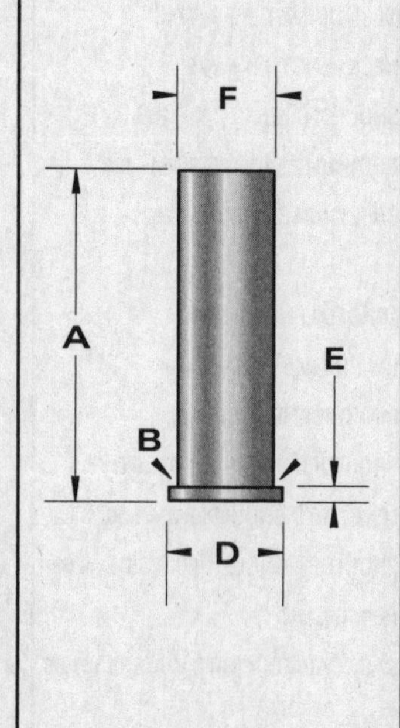

-NOT ACTUAL SIZE-
-DO NOT SCALE-

CARTRIDGE: 11,7 x 45.5R Danish Remington

OTHER NAMES:

DIA: .469

BALLISTEK NO: 469A

NAI NO: RMS 21115/3.552

DATA SOURCE: Hoyem Vol.2 Pg233

HISTORICAL DATA: Civilian version of 1880's Danish mil. ctg.

NOTES:

LOADING DATA:

BULLET WT./TYPE	POWDER WT./TYPE	VELOCITY ('/SEC)	SOURCE
378/LeadPP	24.4/IMR4198	–	JJD

CASE PREPARATION: SHELLHOLDER (RCBS): 14

MAKE FROM: .45-70 Gov't. Trim case to length & chamfer. Turn rim to .576" dia. and back chamfer. F/L size.

PHYSICAL DATA (INCHES):

CASE TYPE: **Rimmed Straight**

CASE LENGTH **A = 1.79**

HEAD DIAMETER **B = .504**

RIM DIAMETER **D = .576**

NECK DIAMETER **F = .481**

NECK LENGTH **H = N/A**

SHOULDER LENGTH **K = N/A**

BODY ANGLE (DEG'S/SIDE): **.379**

CASE CAPACITY **CC'S = 4.30**

LOADED LENGTH: **2.51**

BELT DIAMETER **C = N/A**

RIM THICKNESS **E = .052**

SHOULDER DIAMETER **G = N/A**

LENGTH TO SHOULDER **J = N/A**

SHOULDER ANGLE (DEG'S/SIDE): **N/A**

PRIMER: **L/R**

CASE CAPACITY (GR'S WATER): **66.44**

DIMENSIONAL DRAWING:

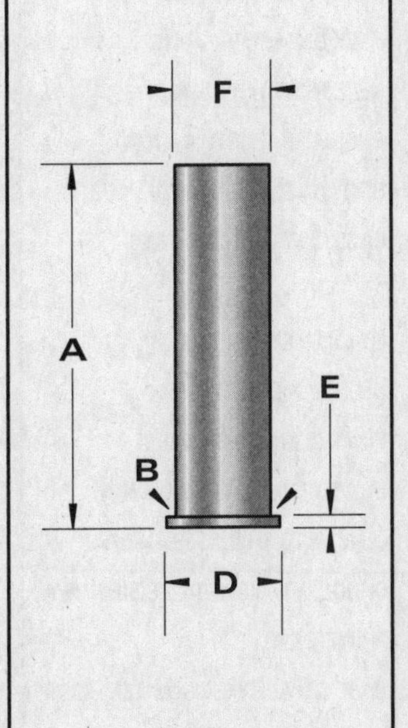

-NOT ACTUAL SIZE-
-DO NOT SCALE-

CARTRIDGE: 11,7 x 51.6R Remington

OTHER NAMES:

DIA: .463

BALLISTEK NO: 463A

NAI NO: RMS 21115/3.957

DATA SOURCE: Hoyem Vol.2 Pg233

HISTORICAL DATA: Danish mil. ctg. about 1875.

NOTES:

LOADING DATA:

BULLET WT./TYPE	POWDER WT./TYPE	VELOCITY ('/SEC)	SOURCE
385/LeadPP	20.9/IMR4198	–	JJD

CASE PREPARATION: SHELLHOLDER (RCBS): 14

MAKE FROM: .43 Span. Rem. (BELL). Cut case to 2.1". Turn rim to .581" dia. and thin to .06". F/L size, trim to length and chamfer.

PHYSICAL DATA (INCHES):

CASE TYPE: **Rimmed Straight**

CASE LENGTH **A = 2.03**

HEAD DIAMETER **B = .513**

RIM DIAMETER **D = .581**

NECK DIAMETER **F = .486**

NECK LENGTH **H = N/A**

SHOULDER LENGTH **K = N/A**

BODY ANGLE (DEG'S/SIDE): **.392**

CASE CAPACITY **CC'S = 4.80**

LOADED LENGTH: **2.75**

BELT DIAMETER **C = N/A**

RIM THICKNESS **E = .06**

SHOULDER DIAMETER **G = N/A**

LENGTH TO SHOULDER **J = N/A**

SHOULDER ANGLE (DEG'S/SIDE): **N/A**

PRIMER: **L/R**

CASE CAPACITY (GR'S WATER): **74.11**

DIMENSIONAL DRAWING:

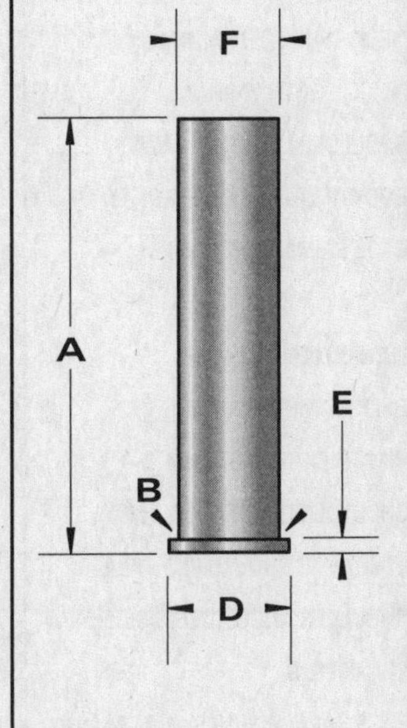

-NOT ACTUAL SIZE-
-DO NOT SCALE-

CARTRIDGE: 11,7 x 51R Danish Remington

OTHER NAMES: .45 Danish Remington

DIA: .455

BALLISTEK NO: 455C

NAI NO: RMS 21115/3.910

DATA SOURCE: COTW 4th Pg215

HISTORICAL DATA: Danish mil. ctg. about 1867.

NOTES:

LOADING DATA:

BULLET WT./TYPE	POWDER WT./TYPE	VELOCITY ('/SEC)	SOURCE
300/Lead	34.0/4198	1480	Barnes

CASE PREPARATION: SHELLHOLDER (RCBS): Spl.

MAKE FROM: .375 Flanged (BELL). Cut case to 2.1". F/L size & chamfer.

PHYSICAL DATA (INCHES):

CASE TYPE: **Rimmed Straight**
CASE LENGTH **A = 2.01**
HEAD DIAMETER **B = .514**
RIM DIAMETER **D = .579**
NECK DIAMETER **F = .486**
NECK LENGTH **H = N/A**
SHOULDER LENGTH **K = N/A**
BODY ANGLE (DEG'S/SIDE): **.41**
CASE CAPACITY **CC'S = 4.51**

LOADED LENGTH: **2.45**
BELT DIAMETER **C = N/A**
RIM THICKNESS **E = .057**
SHOULDER DIAMETER **G = N/A**
LENGTH TO SHOULDER **J = N/A**
SHOULDER ANGLE (DEG'S/SIDE): **N/A**
PRIMER: **L/R**
CASE CAPACITY (GR'S WATER): **69.67**

DIMENSIONAL DRAWING:

-NOT ACTUAL SIZE-
-DO NOT SCALE-

CARTRIDGE: 11,7 x 56R Danish Remington

OTHER NAMES:

DIA: .485

BALLISTEK NO: 485A

NAI NO: RMS 21115/4.288

DATA SOURCE: Hoyem Vol.2 Pg234

HISTORICAL DATA: Danish mil. ctg. Largest of the Danish-Remington line.

NOTES: Bullet featured 7mm hole axially through bullet.

LOADING DATA:

BULLET WT./TYPE	POWDER WT./TYPE	VELOCITY ('/SEC)	SOURCE
325/LeadPP	39.6/IMR3031	–	JJD

CASE PREPARATION: SHELLHOLDER (RCBS): Spl.

MAKE FROM: .43 Span. Rem. (BELL). Turn rim to .580" dia. and thin to .048". Cut case to 2.25". F/L size, trim to length & chamfer.

PHYSICAL DATA (INCHES):

CASE TYPE: **Rimmed Straight**
CASE LENGTH **A = 2.20**
HEAD DIAMETER **B = .513**
RIM DIAMETER **D = .580**
NECK DIAMETER **F = .485**
NECK LENGTH **H = N/A**
SHOULDER LENGTH **K = N/A**
BODY ANGLE (DEG'S/SIDE): **.373**
CASE CAPACITY **CC'S = 6.07**

LOADED LENGTH: **2.98**
BELT DIAMETER **C = N/A**
RIM THICKNESS **E = .048**
SHOULDER DIAMETER **G = N/A**
LENGTH TO SHOULDER **J = N/A**
SHOULDER ANGLE (DEG'S/SIDE): **N/A**
PRIMER: **L/R**
CASE CAPACITY (GR'S WATER): **93.79**

DIMENSIONAL DRAWING:

-NOT ACTUAL SIZE-
-DO NOT SCALE-

CARTRIDGE: 11,7 x 57R Berdan

OTHER NAMES:

DIA: .461

BALLISTEK NO: 461B

NAI NO: RMS 21115/4.409

DATA SOURCE: Hoyem Vol.2 Pg191

HISTORICAL DATA: Russian mil. ctg.

NOTES:

LOADING DATA:

BULLET WT./TYPE	POWDER WT./TYPE	VELOCITY ('/SEC)	SOURCE
350/LeadPP	38.0/IMR4198	—	JJD

CASE PREPARATION: SHELLHOLDER (RCBS): 14

MAKE FROM: .45 Basic. Cut case to 2.25". F/L size, trim to length & chamfer.

PHYSICAL DATA (INCHES):

CASE TYPE: **Rimmed Straight**

CASE LENGTH **A = 2.20**

HEAD DIAMETER **B = .499**

RIM DIAMETER **D = .600**

NECK DIAMETER **F = .464**

NECK LENGTH **H = N/A**

SHOULDER LENGTH **K = N/A**

BODY ANGLE (DEG'S/SIDE): **.471**

CASE CAPACITY **CC'S = 5.55**

LOADED LENGTH: **3.05**

BELT DIAMETER **C = N/A**

RIM THICKNESS **E = .074**

SHOULDER DIAMETER **G = N/A**

LENGTH TO SHOULDER **J = N/A**

SHOULDER ANGLE (DEG'S/SIDE): **N/A**

PRIMER: **L/R**

CASE CAPACITY (GR'S WATER): **85.70**

DIMENSIONAL DRAWING:

F

A

E

B

D

-NOT ACTUAL SIZE-
-DO NOT SCALE-

CARTRIDGE: 11,75 x 36 Montenegrin (Short)

OTHER NAMES:
11mm Austrian Gasser (Short)
11,25 x 36 Montenegrin

DIA: .445

BALLISTEK NO: 445L

NAI NO: RMS 21115/2.898

DATA SOURCE: COTW 4th Pg185

HISTORICAL DATA: Montenegrin revolver ctg. ca. 1870.

NOTES:

LOADING DATA:

BULLET WT./TYPE	POWDER WT./TYPE	VELOCITY ('/SEC)	SOURCE
280/Lead	4.0/B'Eye	—	JJD

CASE PREPARATION: SHELLHOLDER (RCBS): 13

MAKE FROM: 7,62 x 54R Russian. Turn rim to .555" dia. Cut case to 1.5". F/L size with the expander removed. I.D. neck ream. Trim to length, chamfer and F/L size.

PHYSICAL DATA (INCHES):

CASE TYPE: **Rimmed Straight**

CASE LENGTH **A = 1.42**

HEAD DIAMETER **B = .490**

RIM DIAMETER **D = .555**

NECK DIAMETER **F = .472**

NECK LENGTH **H = N/A**

SHOULDER LENGTH **K = N/A**

BODY ANGLE (DEG'S/SIDE): **.698**

CASE CAPACITY **CC'S = 2.78**

LOADED LENGTH: **1.73**

BELT DIAMETER **C = N/A**

RIM THICKNESS **E = .06**

SHOULDER DIAMETER **G = N/A**

LENGTH TO SHOULDER **J = N/A**

SHOULDER ANGLE (DEG'S/SIDE): **N/A**

PRIMER: **L/P**

CASE CAPACITY (GR'S WATER): **42.98**

DIMENSIONAL DRAWING:

F

A

E

B

D

-NOT ACTUAL SIZE-
-DO NOT SCALE-

CARTRIDGE: 11,75 x 51 Montenegrin (Long)

OTHER NAMES:
11mm Austrian Gasser (Long)
11,25 x 51 Montenegrin

DIA: .445

BALLISTEK NO: 445D

NAI NO: RMS 11115/4 .082

DATA SOURCE: Smith Pg533

HISTORICAL DATA: Circa 1870 for Montenegrin revolvers.

NOTES:

LOADING DATA:

BULLET WT./TYPE	POWDER WT./TYPE	VELOCITY ('/SEC)	SOURCE
300/Lead	3.5/B'Eye	—	JJD

CASE PREPARATION: SHELLHOLDER (RCBS): 13

MAKE FROM: 7,62 x 54R Russian. Turn rim to .555" dia. Cut case to 2.1". F/L size with expander removed. I.D. neck ream. Trim to length, chamfer & F/L size.

PHYSICAL DATA (INCHES):

CASE TYPE: **Rimmed Straight**
CASE LENGTH **A = 2.00**
HEAD DIAMETER **B = .490**
RIM DIAMETER **D = .555**
NECK DIAMETER **F = .472**
NECK LENGTH **H = N/A**
SHOULDER LENGTH **K = N/A**
BODY ANGLE (DEG'S/SIDE): **.698**
CASE CAPACITY **CC'S = 1.40**

LOADED LENGTH: **2.37**
BELT DIAMETER **C = N/A**
RIM THICKNESS **E = .06**
SHOULDER DIAMETER **G = N/A**
LENGTH TO SHOULDER **J = N/A**
SHOULDER ANGLE (DEG'S/SIDE): **N/A**
PRIMER: **L/P**
CASE CAPACITY (GR'S WATER): **21.69**

DIMENSIONAL DRAWING:

F
A
E
B
D

-NOT ACTUAL SIZE-
-DO NOT SCALE-

CARTRIDGE: 12,7 x 70R Mitrailleuse

OTHER NAMES:

DIA: .515

BALLISTEK NO: 515C

NAI NO: RMS 21115/5.676

DATA SOURCE: Hoyem Vol.2 Pg144

HISTORICAL DATA: Ctg. for French volley gun, ca. 1869.

NOTES:

LOADING DATA:

BULLET WT./TYPE	POWDER WT./TYPE	VELOCITY ('/SEC)	SOURCE
400/LeadPP	43.0/IMR4198	—	JJD

CASE PREPARATION: SHELLHOLDER (RCBS): Spl.

MAKE FROM: A section of 19/32" dia. tubing on a .475 #2 N.E. (BELL). Case head is about the only way to make this case. Anneal, F/L size and trim to length. Build up rim to .092" thick.

PHYSICAL DATA (INCHES):

CASE TYPE: **Rimmed Straight**
CASE LENGTH **A = 3.40**
HEAD DIAMETER **B = .599**
RIM DIAMETER **D = .673**
NECK DIAMETER **F = .556**
NECK LENGTH **H = N/A**
SHOULDER LENGTH **K = N/A**
BODY ANGLE (DEG'S/SIDE): **.372**
CASE CAPACITY **CC'S = 10.61**

LOADED LENGTH: **4.00**
BELT DIAMETER **C = N/A**
RIM THICKNESS **E = .092**
SHOULDER DIAMETER **G = N/A**
LENGTH TO SHOULDER **J = N/A**
SHOULDER ANGLE (DEG'S/SIDE): **N/A**
PRIMER: **L/R**
CASE CAPACITY (GR'S WATER): **163.75**

DIMENSIONAL DRAWING:

F
A
E
B
D

-NOT ACTUAL SIZE-
-DO NOT SCALE-

CARTRIDGE: 12,8 x 45R Papal Remington

OTHER NAMES:

DIA: .503

BALLISTEK NO: 503A

NAI NO: RMS 21115/3.238

DATA SOURCE: Hoyem Vol.2 Pg171

HISTORICAL DATA: Ctg. for Vetterli rifle about 1870.

NOTES:

LOADING DATA:

BULLET WT./TYPE	POWDER WT./TYPE	VELOCITY ('/SEC)	SOURCE
430/Lead	26.5/IMR4198	–	JJD

CASE PREPARATION: SHELLHOLDER (RCBS): Spl.

MAKE FROM: .450 #2 N.E. (BELL). Cut case to 1.9". Trim to final length and F/L size.

PHYSICAL DATA (INCHES):

CASE TYPE: **Rimmed Straight**

CASE LENGTH **A = 1.82**

HEAD DIAMETER **B = .562**

RIM DIAMETER **D = .667**

NECK DIAMETER **F = .535**

NECK LENGTH **H = N/A**

SHOULDER LENGTH **K = N/A**

BODY ANGLE (DEG'S/SIDE): **.445**

CASE CAPACITY **CC'S = 4.88**

LOADED LENGTH: **2.40**

BELT DIAMETER **C = N/A**

RIM THICKNESS **E = .082**

SHOULDER DIAMETER **G = N/A**

LENGTH TO SHOULDER **J = N/A**

SHOULDER ANGLE (DEG'S/SIDE): **N/A**

PRIMER: **L/R**

CASE CAPACITY (GR'S WATER): **75.43**

DIMENSIONAL DRAWING:

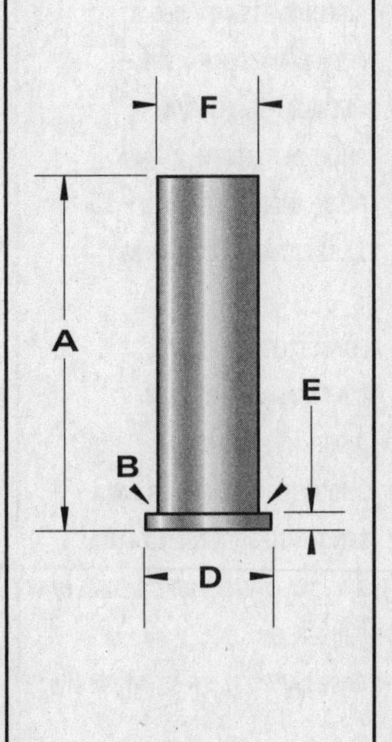

-NOT ACTUAL SIZE-
-DO NOT SCALE-

CARTRIDGE: 13 x 87R Mitrailleuse

OTHER NAMES:

DIA: .523

BALLISTEK NO: 523A

NAI NO: RMS 11115/5.826

DATA SOURCE: Hoyem Vol.2 Pg144

HISTORICAL DATA: French MG ctg.

NOTES: Case has tapered rim - tapers from .669" dia. in, towards case head.

LOADING DATA:

BULLET WT./TYPE	POWDER WT./TYPE	VELOCITY ('/SEC)	SOURCE
770/LeadPP	40.0/IMR4198	–	JJD

CASE PREPARATION: SHELLHOLDER (RCBS): Spl.

MAKE FROM: You'll have to make this case from scratch. Lathe turn tapered rim from brass. Use 19/32" dia. tubing for body. Standard tubing case design. Anneal and F/L size.

PHYSICAL DATA (INCHES):

CASE TYPE: **Rimmed Straight**

CASE LENGTH **A = 3.49**

HEAD DIAMETER **B = .599**

RIM DIAMETER **D = .669**

NECK DIAMETER **F = .589**

NECK LENGTH **H = N/A**

SHOULDER LENGTH **K = N/A**

BODY ANGLE (DEG'S/SIDE): **.085**

CASE CAPACITY **CC'S = 10.05**

LOADED LENGTH: **4.63**

BELT DIAMETER **C = N/A**

RIM THICKNESS **E = .125**

SHOULDER DIAMETER **G = N/A**

LENGTH TO SHOULDER **J = N/A**

SHOULDER ANGLE (DEG'S/SIDE): **N/A**

PRIMER: **L/R**

CASE CAPACITY (GR'S WATER): **155.25**

DIMENSIONAL DRAWING:

F

A

E

B

D

-NOT ACTUAL SIZE-
-DO NOT SCALE-

CARTRIDGE: 14,4 x 41R Madrid Berdan

OTHER NAMES:	
14,5 Berdan Spain	DIA: .587
	BALLISTEK NO: 587A
	NAI NO: RMS 21115/2.432

DATA SOURCE: Hoyem Vol.2 Pg236

HISTORICAL DATA: Spanish mil. ctg. about 1867.

NOTES:

LOADING DATA:

BULLET WT./TYPE	POWDER WT./TYPE	VELOCITY ('/SEC)	SOURCE
500/Lead	19.3/IMR4198	—	JJD

CASE PREPARATION: SHELLHOLDER (RCBS): 577

MAKE FROM: .577 N.E. (BELL). Cut case to 1.7" and build rim up to .072" thick. F/L size, trim to length & chamfer.

PHYSICAL DATA (INCHES):

CASE TYPE: **Rimmed Straight**
CASE LENGTH **A = 1.60**
HEAD DIAMETER **B = .658**
RIM DIAMETER **D = .759**
NECK DIAMETER **F = .631**
NECK LENGTH **H = N/A**
SHOULDER LENGTH **K = N/A**
BODY ANGLE (DEG'S/SIDE): **.506**
CASE CAPACITY **CC'S = 5.59**

LOADED LENGTH: **2.04**
BELT DIAMETER **C = N/A**
RIM THICKNESS **E = .072**
SHOULDER DIAMETER **G = N/A**
LENGTH TO SHOULDER **J = N/A**
SHOULDER ANGLE (DEG'S/SIDE): **N/A**
PRIMER: **L/R**
CASE CAPACITY (GR'S WATER): **86.19**

DIMENSIONAL DRAWING:

F
A
E
B
D

-NOT ACTUAL SIZE-
-DO NOT SCALE-

CARTRIDGE: 14,5 x 33R Wanzl

OTHER NAMES:	
	DIA: .561
	BALLISTEK NO: 561A
	NAI NO: RMS 11115/2.166

DATA SOURCE: Hoyem Vol.2 Pg174

HISTORICAL DATA: Austro-Hungarian ctg. from about 1870.

NOTES: Both c/f and r/f versions exist.

LOADING DATA:

BULLET WT./TYPE	POWDER WT./TYPE	VELOCITY ('/SEC)	SOURCE
458/Lead	10.0/IMR4198	—	JJD

CASE PREPARATION: SHELLHOLDER (RCBS): Spl.

MAKE FROM: This one must be a 19/32" dia. tubing case with the head turned from solid, 3/4" dia. stock. Fabricate case, anneal and F/L size. Trim to length and chamfer. Use "minied" .555-556" bullets.

PHYSICAL DATA (INCHES):

CASE TYPE: **Rimmed Straight**
CASE LENGTH **A = 1.30**
HEAD DIAMETER **B = .600**
RIM DIAMETER **D = .686**
NECK DIAMETER **F = .586**
NECK LENGTH **H = N/A**
SHOULDER LENGTH **K = N/A**
BODY ANGLE (DEG'S/SIDE): **.324**
CASE CAPACITY **CC'S = 3.95**

LOADED LENGTH: **1.97**
BELT DIAMETER **C = N/A**
RIM THICKNESS **E = .065**
SHOULDER DIAMETER **G = N/A**
LENGTH TO SHOULDER **J = N/A**
SHOULDER ANGLE (DEG'S/SIDE): **N/A**
PRIMER: **L/R**
CASE CAPACITY (GR'S WATER): **61.07**

DIMENSIONAL DRAWING:

F
A
E
B
D

-NOT ACTUAL SIZE-
-DO NOT SCALE-

CARTRIDGE: 14,66 x 35R Peabody 1870

OTHER NAMES:
14.66 Serbian Peabody

DIA: .592

BALLISTEK NO: 592A

NAI NO: RMS 21115/2.123

DATA SOURCE: Hoyem Vol.2 Pg225

HISTORICAL DATA: Serbian mil. ctg. about 1870.

NOTES:

LOADING DATA:

BULLET WT./TYPE	POWDER WT./TYPE	VELOCITY ('/SEC)	SOURCE
500/Lead	10.0/IMR4227	—	JJD

CASE PREPARATION: SHELLHOLDER (RCBS): 577

MAKE FROM: .577 N.E. (BELL). Cut case to 1.4" and build rim up to .08" thick. F/L size, trim to length & chamfer.

PHYSICAL DATA (INCHES):

CASE TYPE: **Rimmed Straight**

CASE LENGTH **A = 1.38**

HEAD DIAMETER **B = .650**

RIM DIAMETER **D = .757**

NECK DIAMETER **F = .618**

NECK LENGTH **H = N/A**

SHOULDER LENGTH **K = N/A**

BODY ANGLE (DEG'S/SIDE): **.705**

CASE CAPACITY **CC'S = 4.90**

LOADED LENGTH: **1.97**

BELT DIAMETER **C = N/A**

RIM THICKNESS **E = .08**

SHOULDER DIAMETER **G = N/A**

LENGTH TO SHOULDER **J = N/A**

SHOULDER ANGLE (DEG'S/SIDE): **N/A**

PRIMER: **L/R**

CASE CAPACITY (GR'S WATER): **75.70**

DIMENSIONAL DRAWING:

F
A
E
B
D

-NOT ACTUAL SIZE-
-DO NOT SCALE-

CARTRIDGE: 14,7 x 40R Snider

OTHER NAMES:

DIA: .570

BALLISTEK NO: 570C

NAI NO: RMS 31115/2.404

DATA SOURCE: Hoyem Vol.2 Pg218

HISTORICAL DATA: Circa 1875. Used in Turkish mil. rifle.

NOTES: A short version of the .577 Snider.

LOADING DATA:

BULLET WT./TYPE	POWDER WT./TYPE	VELOCITY ('/SEC)	SOURCE
350/Lead	20.2/IMR4198	—	JJD

CASE PREPARATION: SHELLHOLDER (RCBS): 577

MAKE FROM: .577 N.E. (BELL). Cut case to 1.6". F/L size, trim to length & chamfer.

PHYSICAL DATA (INCHES):

CASE TYPE: **Rimmed Straight**

CASE LENGTH **A = 1.57**

HEAD DIAMETER **B = .653**

RIM DIAMETER **D = .736**

NECK DIAMETER **F = .586**

NECK LENGTH **H = N/A**

SHOULDER LENGTH **K = N/A**

BODY ANGLE (DEG'S/SIDE): **1.26**

CASE CAPACITY **CC'S = 5.78**

LOADED LENGTH: **2.00**

BELT DIAMETER **C = N/A**

RIM THICKNESS **E = .05**

SHOULDER DIAMETER **G = N/A**

LENGTH TO SHOULDER **J = N/A**

SHOULDER ANGLE (DEG'S/SIDE): **N/A**

PRIMER: **L/R**

CASE CAPACITY (GR'S WATER): **89.26**

DIMENSIONAL DRAWING:

F
A
E
B
D

-NOT ACTUAL SIZE-
-DO NOT SCALE-

CARTRIDGE: 15,2 x 28R Gevelot

OTHER NAMES:

DIA: .598

BALLISTEK NO: 598A

NAI NO: RMS 21115/1.833

DATA SOURCE: Hoyem Vol.2 Pg197

HISTORICAL DATA: Ctg. for Peabody rifle about 1867.

NOTES:

LOADING DATA:

BULLET WT./TYPE	POWDER WT./TYPE	VELOCITY ('/SEC)	SOURCE
600/RN	10.0/IMR4227	–	JJD

CASE PREPARATION: SHELLHOLDER (RCBS): Spl.

MAKE FROM: Lathe turn solid. Trim to length & anneal. F/L size.

PHYSICAL DATA (INCHES):

CASE TYPE: **Rimmed Straight**

CASE LENGTH **A = 1.12**

HEAD DIAMETER **B = .611**

RIM DIAMETER **D = .697**

NECK DIAMETER **F = .598**

NECK LENGTH **H = N/A**

SHOULDER LENGTH **K = N/A**

BODY ANGLE (DEG'S/SIDE): **.140**

CASE CAPACITY **CC'S = 3.84**

LOADED LENGTH: **1.65**

BELT DIAMETER **C = N/A**

RIM THICKNESS **E = .092**

SHOULDER DIAMETER **G = N/A**

LENGTH TO SHOULDER **J = N/A**

SHOULDER ANGLE (DEG'S/SIDE): **N/A**

PRIMER: **L/R**

CASE CAPACITY (GR'S WATER): **59.27**

DIMENSIONAL DRAWING:

F

A

E

B

D

-NOT ACTUAL SIZE-
-DO NOT SCALE-

CARTRIDGE:

OTHER NAMES:

DIA:

BALLISTEK NO:

NAI NO:

DATA SOURCE:

HISTORICAL DATA:

NOTES:

LOADING DATA:

BULLET WT./TYPE	POWDER WT./TYPE	VELOCITY ('/SEC)	SOURCE

CASE PREPARATION: SHELL HOLDER (RCBS): 10

MAKE FROM:

PHYSICAL DATA (INCHES):

CASE TYPE:

CASE LENGTH **A =**

HEAD DIAMETER **B =**

RIM DIAMETER **D =**

NECK DIAMETER **F =**

NECK LENGTH **H =**

SHOULDER LENGTH **K =**

BODY ANGLE (DEG'S/SIDE):

CASE CAPACITY **CC'S =**

LOADED LENGTH:

BELT DIAMETER **C =**

RIM THICKNESS **E =**

SHOULDER DIAMETER **G =**

LENGTH TO SHOULDER **J =**

SHOULDER ANGLE (DEG'S/SIDE):

PRIMER:

CASE CAPACITY (GR'S WATER):

DIMENSIONAL DRAWING:

-NOT ACTUAL SIZE-
-DO NOT SCALE-

Appendix A

LONGEST MEMBER OF THAT FAMILY IS LISTED.

CASE	TYPE	BASE (inches)	RIM (inches)	TH'K (inches)	LENGTH (inches)	MFR	SH
.25 ACP	RX/S	.278	.302	.043	.615	C	29
.22 HORNET	RM/B	.299	.350	.065	1.403	C	12
5/16 TUBING	T/C	.313	A/R	A/R	A/R	B	A/R
.32 LONG COLT	RM/S	.318	.374	.05	.92	D	23
.32 S&W LONG	RM/S	.335	.375	.052	.93	D	23
.32 ACP	RX/S	.336	.354	.043	.68	O	17
11/32 TUBING	T/C	.344	A/R	A/R	A/R	B	A/R
.25-20 WCF	RM/B	.349	.408	.065	1.330	O	1
.30 CARBINE	RX/S	.356	.360	.050	1.29	C	17
.380 ACP	RX/S	.373	.374	.045	.68	C	10
3/8 TUBING	T/C	.375	A/R	A/R	A/R	B	A/R
.222 REM MAG	RX/B	.376	.378	.045	1.850	C	10
.357 MAXIMUM	RM/S	.376	.433	.056	1.605	D	6
.38 ACP	RX/S	.384	.406	.049	.90	D	1
9MM WIN MAG	RX/S	.392	.394	.05	1.160	C	16
13/32 TUBING	T/C	.406	A/R	A/R	A/R	B	A/R
.224 WEB'Y	BE/B	.415	.429	.05	1.92	W	27
5,6 × 52R	RM/B	.418	.492	.063	2.047	N	2
.375 WIN.	RM/S	.420	.506	.063	2.02	O	2
.30 REM.	RX/B	.421	.422	.05	2.03	D	19
.225 WIN.	RM/B	.422	.473	.05	1.93	O	11
9,3 × 72R	RM/S	.427	.482	.06	2.84	R	4
.303 SAVAGE	RM/B	.433	.500	.063	2.00	O	21
.41 S&W MAG	RM/S	.435	.492	.06	1.29	D	30
7/16 TUBING	T/C	.438	A/R	A/R	A/R	B	A/R
.220 RUSSIAN	RX/B	.441	.444	.058	1.518	S	6
.220 SWIFT	RM/B	.445	.473	.05	2.205	D	11
6,5 × 54 M/S	RX/B	.447	.453	.055	2.100	R	9
.240 WEB'Y	BE/B	.453	.473	.05	2.500	W	3
.303 BRITISH	RM/B	.455	.540	.065	2.160	C	7
.30-40 KRAG	RM/B	.457	.545	.065	2.314	C	7
.35 REM.	RX/B	.457	.460	.05	1.920	D	9
.44 S&WMAG	RM/S	.457	.514	.06	1.285	D	18
.405 WIN.	RM/S	.460	.539	.073	3.22	A	24
7 × 64 BREN.	RX/B	.463	.468	.055	2.51	R	3
7 × 65R BREN.	RM/B	.463	.521	.06	2.53	R	26
9,3 × 74R	RM/S	.465	.524	.06	2.93	R	4
9,3 × 57	RX/B	.468	.469	.055	2.24	N	3
.444 MARLIN	RM/S	.469	.514	.063	2.225	D	28

CASE	TYPE	BASE (inches)	RIM (inches)	TH'K (inches)	LENGTH (inches)	MFR	SH
15/32 TUBING	T/C	.469	A/R	A/R	A/R	B	A/R
.270 WIN.	RX/B	.470	.473	.05	2.540	C	3
.45 MAGNUM	RX/S	.470	.475	.047	1.194	O	3
7 × 57R	RM/B	.470	.521	.063	2.23	N	26
7.7 JAP	RX/B	.472	.470	.049	2.27	N	3
9,3 × 62	RX/B	.473	.470	.065	2.42	N	3
.45 A/R	RM/S	.476	.516	.089	.898	D	8
6,5 × 55	RX/B	.476	.476	.059	2.160	N	2
.45 COLT	RM/S	.480	.512	.058	1.285	O	20
7,62 × 54 RUS	RM/B	.489	.570	.064	2.114	N	13
10,75 × 68	RB/B	.492	.488	.053	2.67	R	5
7,5 × 55 SWIS	RX/B	.494	.496	.058	2.14	N	2
1/2 TUBING	T/C	.500	A/R	A/R	A/R	B	A/R
.45 BASIC	RM/S	.500	.590	.07	3.25	A	14
9,3 × 64 BREN.	RX/B	.504	.492	.062	2.52	R	21
.375 H&H	BE/B	.513	.532	.05	2.85	C	4
.43 SPAN REM	RM/S	.513	.627	.08	2.622	A	SPL
.375 FLANGED	RM/S	.516	.571	.06	3.25	A	SPL
11MM A BASE	RM/S	.516	.576	.09	2.662	A	SPL
8 × 685	RX/B	.524	.512	.055	2.66	R	28
17/32 TUBING	T/C	.531	A/R	A/R	A/R	B	A/R
.280 ROSS	RM/S	.535	.556	.06	2.6	A	SPL
.280 FLANGED	RM/B	.535	.607	.06	2.61	A	14
.425 W/R	RB/S	.540	.467	.045	2.64	A	3
.404 JEFFREY	RX/S	.544	.543	.05	2.86	A	7
.450 N/E	RM/S	.545	.624	.042	3.25	A	SPL
.348 WIN.	RM/B	.553	.610	.07	2.255	O	5
9/16 TUBING	T/C	.563	A/R	A/R	A/R	B	A/R
.450 #2	RM/S	.564	.643	.08	3.50	A	SPL
.50 SHARPS	RM/S	.565	.665	.062	3.25	A	31
.470 N/E 3.25	RM/B	.572	.644	.04	3.25	A	31
.500 N/E	RM/S	.574	.655	.04	3.00	A	SPL
.475 #2 3.5	RM/S	.576	.670	.08	3.50	A	SPL
.460 WEB'Y	BE/B	.582	.603	.062	2.908	W	14
11MM BEAU.	RM/S	.588	.660	.064	2.90	A	SPL
.416 RIGBY	RX/B	.589	.586	.065	2.875	A	SPL
19/32 TUBING	T/C	.594	A/R	A/R	A/R	B	A/R
5/8 TUBING	T/C	.625	A/R	A/R	A/R	B	A/R
.577 N/E	RM/S	.658	.745	.05	3.00	A	577

SUPPLIER CODE

S = SAKO
A = B.E.L.L.
C = GENERAL
B = NAI/BALLISTEK
R = RWS
N = NORMA
O = OLIN/WINCHESTER
D = DUPONT/REMINGTON

TYPE CODE

A/R = Auto Rimmed
T/C = Tubing Case
RX/s = Rimless Straight
RM/B = Rimmed Bottleneck
RM/S = Rimmed Straight
BE/B = Belted Straight
RX/B = Rimless Bottleneck

Appendix B

THE FOLLOWING LIST INCLUDES ALL OMARK/RCBS SHELLHOLDERS IN ORDER OF THEIR RIM DIAMETER AND THICKNESS:

DIAMETER (inches)	THICKNESS (inches)	SHELLHOLDER NUMBER
.298	.040	29
.350	.063	12
.360	.050	17
.370	.050	23
.378	.045	10
.385	.050	16
.408	.065	1
.422	.049	25
.440	.060	6
.460	.050	9
.473	.054	3
.475	.049	11
.488	.058	30
.505	.063	21
.506	.063	2
.510	.088	8
.510	.055	20
.514	.063	18
.525	.065	26
.545	.064	7
.532	.050	4
.564	.060	13
.608	.070	31
.610	.070	5
.660	.063	31
.750	.055	577

Appendix C

ACCURATE ARMS CO., INC
5891 HWY. 230 W.
MC EWEN, TN 37101
http://www.accuratepowder.com/
SMOKELESS POWDER

ALLIANT POWDER
ROUTE 114 P.O.BOX 6 BLDG. 229
RADFORD, VA 24143-0096
http://www.alliant_powder.com
SMOKELESS POWDER

ALPHA BULLET CO.
104 DUNLAP ST.
HARVARD, IL 60033
1-815-943-1004
http://www.alphabullet.com
COMPONETS, CAST LEAD BULLETS

AMERICAN CUSTOM AMMUNITION
5750 PLUNKETT ST. #5
HOLLYWOOD, FL 33023
http://www.americancustom.tripod.com
AMMUNITION & COMPONENTS

AMERICAN GAS & CHEMICAL
220 PEGASUS AVE NORTHVALE, NJ 07647
1-800-526-1008
http://www.amgas.com
LUBRICANTS

ARMFIELD CUSTOM BULLETS
4775 CAROLINE DR.
SAN DIEGO, CA 92115
1-619-582-7188
BULLETS

BALD EAGLE PRECISION MACHINE CO
101-D ALLISON ST.
LOCK HAVEN, PA 17745
1-717-748-6772
http://www.baldeaglemachine.com
PRECISION DIE BLANKS, RELOADING PRESSES,

BALLISTI-CAST INC
BOX 1057
MINOT, ND 58702
1-701-497-3333
http://www.ballisti-cast.com/
BULLET CASTING, PROPELANTS
RELOADING PRESSES

BARNES BULLETS
P.O. BOX 215
AMERICAN FORK, UT 84003
http://www.barnesbullets.com/
PREMIUM JACKETED BULLETS

BATTENFELD TECHNOLOGIES INC
P.O.BOX 1035 COLUMBIA, MO 65205
1-573-446-3857
http://www.battenfeldtechnologies.com
RELOADING SUPPLIES

BELL BRASS
MAST TECHNOLOGY
14555 US HWY 95 SOUTH
BOULDER CITY, NV 89005
http://www.bellammo.com/
BRASS CASES

BERGER BULLETS
5342 W CAMELBACK RD
GLENDALE, AZ 85301

1-714-447-5456 FAX: 1-714-447-5407
http://www.bergerbullets.com
AMMUNITION

BERRY'S MFG.
401 N. 3050 E. ST
ST.GEORGE, UT 84790
1-800-269-7373
http://www.berrysmfg.com
TUMBLERS

BLACKPOWDER PRODUCTS INC.
5988 PEACHTREE CORNERS EAST
NORCROSS, GA 30071
http://www.bpiguns.com/
CVA & WINCHESTER BLACKPOWDER

BLUE STAR CARTRIDGE & BRASS
915 EAST LINCOLN ST.
SEARCY, AZ 72143-7417
501-268-6443
http://www.blue-star-inc.com/
BRASS CASES & AMMUNITION

BROWNELL'S INC.
ROUTE 2, BOX 1
MONTEZUMA, IA 50171
1-800-741-0015
http://www.brownells.com
TOOLS AND SUPPLIES

B-SQUARE ENGINEERING CO.
BOX 11281
FT. WORTH, TX 76110
1-800-433-2909
www.b-square.com
RELOADING EQUIPMENT

BUFFALO ARMS CO
660 VERMEER COURT
PONDERAY, IDAHO 83852
208-263-6953
http://www.buffaloarms.com
RELOADING SUPPLIES, BRASS CASES

CAMDEX, INC.
2330 ALGER
TROY, MI 48083-2001
1-810-528-2300
http://www.camdexloader.com
RELOADING PRESSES,

CCI/SPEER INC.
2299 SNAKE RIVER AVE.
BOX 956
LEWISTON, ID 83501
http://www.cci-ammunition.com
AMMO & BULLETS

C-H TOOL & DIE COMPANY
4-D CUSTOM DIE CO.
711 N. SANDUSKY STREET BOX 889
MT. VERNON, OHIO 43050
740-397-7214
http://www.ch4d.com
RELOADING PRESSES, DIES, ACCESSORIES

CLYMER MFG. CO.
1645 WEST HAMLIN ROAD
ROCHESTER HILLS, MI 48309
1-248-853-5555
http://www.clymertool.com
REAMERS & OTHER TOOLS

CORBIN MANUFACTURING AND SUPPLY, INC.
600 INDUSTRIAL CIRCLE
WHITE CITY, OR 97503
1-541-826-5211
http://www.swage.com
SWAGE DIES, RELOADING PRESSES
JACKET MAKING DIES, LEAD EXTRUDERS

DENVER BULLET CO.
1600 W. 13TH AVE.
DENVER, CO 80204
1-303-893-3146
http::/ /www.denbullets@aol.com
BULLETS

DILLON PRECISION INC.
7442 E BUTHERUS DR.
SCOTTSDALE, AZ 85260
1-800-421-7632
http://www.reloader.com
RELOADING EQUIPMENT, PRESSES, DIES,

DIXIE GUN WORKS
1412 P.O. BOX 130
REELFOOT AVE.
HIGHWAY 51 SOUTH

UNION CITY, TN 38261
1-731-885-0700
http://www.dixiegun.com/
COMPONENTS

DYNAMIT NOBEL RWS
81 RUCKMAN ROAD
CLOSTER, NJ 07624
1-201-767-7971
www.dnrws.com
SMOKELESS POWDER, AMMUNITION

EFFICIENT MACHINERY CO.
12878 N.E. 15TH PL
BELLEVUE, WA 98005
1-800-375-8554
http://www.sturdybench.com
RELOADING BENCHES

BLUE RIDGE MACHINERY & TOOLS
BOX 536
HURRICANE, WV 25526
1- 800-872-6500
http://www.blueridgemachinery.com/
EMCO – COMPAC 5 LATHE

E-Z WAY SYSTEMS
BOX 4310
NEWARK, OH 43058
1-740-345-6645
http://www.infinet.com/~ezway
SIZING DIE WAX, RELOADING ACCESSORIES

FEDERAL CARTRIDGE CO.
900 EHLEN DRIVE
ANOKA, MN 55303-7503
1-800-322-2342
http://www.federalcartridge.com/
AMMUNITION

FLITZ INTERNATIONAL
821 MOHR AVE
WATERFORD, WI 53185
1-800-558-8611
http://www.flitz.com
METAL CLEANERS, RELOADING GEAR

FORSTER PRODUCTS INC.
310 EAST LANARK AVE.
LANARK, IL 61046
(815) 493-6360

http://www.forsterproducts.com/
RELOADING EQUIPMENT

HARRELL BROTHERS
5756 HICKORY DRIVE
SALEM, VA 24153
1-540-380-2683
http://www.harrellsprec.com/
COMPACT RELOADING PRESSES
POWDER MEASURES

HAWK BULLETS, INC.
849 HAWKS BRIDGE RD.
SALEM, NJ 08079
1-856-299-2800
http//:www.hawkbullets.com
CUSTOM BULLETS

HERCULES INC.
910 MARKET STREET
WILMINGTON, DE 19899
SMOKELESS POWDER

HODGDON POWDER CO., INC.
BOX 2932
SHAWNEE MISSION, KS 66201
913-362-9455
www. hodgdon.com
PROPELLANTS

HORNADY MFG. CO.
BOX 1848
GRAND ISLAND, NE 68802-1848
1-800-338-3220
http://www.hornady.com
AMMUNITION, RELOADING SUPPLIES

HUNTINGTON DIE SPECIALTIES
601 ORO DAM BLVD., BOX 991
OROVILLE, CA 95965
1-530-534-1210
http://www.huntingtons.com
RELOADING EQUIPMENT,

IMR, INC.
P.O. BOX 2932
SHAWNEE MISSION, KS 66201
http://www.imrpowder.com

SMOKELESS POWDERS

JGS PRECISION TOOL MFG., L.L.C.
60819 SELANDER RD.
COOS BAY, OR 97420
1-541-267-4331
http://www.jgstools.com/
REAMERS & TOOLS

J.T. HUNTER & CO.
BOX 288
OSTEGO, MI 49078
1-800-376-3166
http://www.jthunter.com/
RELOADING MACHINE BASE

K&M SERVICES
5430 SALMON RUN RD
DOVER, PA 17315
1-717-292-3175
RELOADING TOOLS

LEE PRECISION, INC.
4275 HWY U
HARTFORD, WI 53027
1-262-673-3075
http://www.leeprecision.com
RELOADING EQUIPMENT

LYMAN PRODUCTS CORP.
475 SMITH STREET
MIDDLETOWN, CT 06457
1-800-225-9626
http://www.lymanproducts.com
RELOADING EQUIPMENT, TECH. MANUALS
BLACK POWDER SUPPLIES

MAGMA ENGINEERING CO.
BOX 61
QUEEN CREEK, AZ 85242
1-480-987-9008
http://www.magmaengr.com/
AUTOMATIC BULLET CASTING MACHINES

MAYVILLE ENGINEERING CO, INC.Ï
MEC RELOADERS
715 SOUTH STREET
MAYVILLE, WI 53050
1-800-797-4632
http://www.mayvl.com
SHOTGUN RELOADING ACCESSORIES

MIDWAY USA
5875 WEST VAN HORN TAVERN ROAD
COLUMBIA, MO 65203
1-800-243-3220
http://www.midwayusa.com/
RELOADING SUPPLIES

NATIONAL RELOADING MANUFACTURERS
ONE CENTERPOINTE DR SUITE 300
LAKE OSWEGO, OR 97035
http://www.reload-nrma.com
RELOADERS ASSN.

NECO
BOX 427
LAFAYETTE, CA 94549
1-800-451-3550
http://www.neconos.com
PRESSURE LAPPING KIT
RUNOUT GUAGE, RELOADING TOOLS,

NEIL JONES CUSTOM PRODUCTS
R.D.1 BOX 483 A 17217 BROOKHOUSE
SAEGERTOWN, PA 16433
1-814-763-2769
http://www.neiljones.com
RELOADING PRESSSES, DIES,

NOSLER INC
BOX 671 BEND, OR 97709
1-503-382-3921
http://www.nosler.com
BULLETS

OHAUS SCALE CORP.
19 A CHAPIN ROAD, Box 900
PINE BROOK, NJ 07058
1-973-377-9000
http://www.ohaus.com/
SCALES & WEIGHTS

OLD WESTERN SCROUNGER, INC.
12924 HIGHWAY A-12
MONTAGUE, CA 96064
OBSOLITE AMMO & BRASS, BERDAN PRIMERS

PROMETHEUS TOOL CORP
WASHINGTON
1-425-239-9100

http://www.askfirst.com/prometheus/
SCALES, DISPENSERS, POWDER MEASURES

QUINETICS CORP.
235 WEST TURBO
SAN ANTONIO, TX 78216
1-210-308-0886
http://www.quinetics.com
BULLET PULLER, POWDER MEASURES,
AUTOMATIC SHELL HOLDERS

RAMSHOT POWDERS INC
P.O.BOX 158
MILES CITY, MT 59301
1-800-497-1007
http://www.ramshot.com
PROPELLANTS, RELOADING

RCBS
605 ORO DAM BLVD
OROVILLE, CA 95965
1-800-533-5000
http://www.rcbs.com
RELOADING TOOLS & ACCESSORIES,

RCE CO.
4090 COLVER RD.
PHOENIX, OR 97535
1-541-512-0440
http://www.rceco.com
BULLET SWAGING EQUIPMENT & SUPPLIES

REDDING RELOADING EQUIPMENT
1089 STARR RD
CORTLAND, NY 13045
1-607-753-3331
http://www.redding-reloading.com
RELOADING ACCESSORIES, MOLDS, SHELL HOLDERS,

RELOADING MFG ASSOCIATION
ONE CENTERPOINTE DR
LAKE OSWEGO, OR 97035
RELOADING ASSOCIATION

REMINGTON ARMS CO., INC.
870 REMINGTON DRIVE
MADISON, NC 27025-0700
1-800-243-9700;
www.remington.com
COMPONENTS

ST. MARKS POWDER
GENERAL DYNAMICS
7121 COASTAL HIGHWAY
CRAWFORDVILLE, FL 32327
1-850-925-6111
http://www.gd.com /st_marks_powder.htm
BALL POWDER PROPELLANT

SIERRA BULLET CO.
BOX 818
SEDALIA, MO 65302-0818
1-800-223-8799
http://www.sierrabullets.com/
BULLETS & RELOADING MANUALS

SINCLAIR INTERNATIONAL
2330 WAYNE HAVEN ST.
FT.WAYNE, IN 46803
1-219-493-1858
http://www.sinclairintl.com
ACCESSORIES, PRESSES,
CASE & PRIMER POCKET TOOLS,

SPEER/CCI
BLOUNT INTERNATIONAL, INC.
BOX 856
LEWISTON, ID 83501;
1-800-627-3640
http://www.speer-bullets.com

SPOLAR POWER LOAD INC.
2273 S VISTA B-2
BLOOMINGTON, CA 92335
1-800-227-9667
http://www.spolargold.com
RELOADING PRESS

SINCLAIR INTERNATIONAL
2330 WAYNE-HAVEN ST.
FORT WAYNE, IN 46803
1-219-493-1858
http: //www.sinclairintl.com

RELOADING TOOLS

SSK INDUSTRIES
590 WOODVUE LANE
WINTERSVILLE, OH 43953
1-740-264-0176
http://www.sskindustries.com/
COMPONENTS

STAR MACHINE WORKS
BOX 1872
PIONEER, CA 95666
1-209-295-5000
http://www.starmachineworks.com
RELOADING TOOLS

STONEY POINT PRODUCTS
BOX 234, 1815 N. SPRING ST.
NEW ULM, MN 56073-0234
1-507-354-3360
http://www.stoneypoint.com
PRECISION GUAGES
SWIFT BULLET CO.
201 MAIN STREET, BOX 27
QUINTER, KS 67752
1-785-754-3959
http://www.swiftbullet.com/
BULLETS

THOMPSON BULLET LUBE
BOX 472343
GARLAND, TX 75047
1-972-271-8063
http://www.thompsonbulletlube.com
BULLET LUBE

TRU-SQUARE METAL PROD.INC.
BOX 585
AUBURN, WA 98071
1-253-833-2310
RELOADING ACCESSORIES
TUMBLERS

WEATHERBY, INC.
3100 EL CAMINO REAL
ATASCADERO, CA 93422
1-805-466-1767
http://www.weatherby.com
AMMUNITION

WESTERN POWDERS, INC.
BOX 158 / TOP OF YELLOWSTONE HILL
MILES CITY, MT, 59301
1-800-497-1007
http://www.westernpowders.com
PROPELLANTS

WINCHESTER AMMUNITION
ATT: PRODUCT SERVICES
427 NORTH SHAMROCK ST.
EAST ALTON, IL 62024-1174
1-618-258-2000
http://www.winchester.com/
AMMO & COMPONENTS

ZERO BULLET CO
BOX 1188
CULLMAN, AL 35056
1-800-545-9376
http://www.zerobullets.com/
AMMUNITION & COMPONENTS

Z-HAT CUSTOM DIES
4010A S.POPLAR ST
CASPER, WY 82601
http://www.z-hat.com
CUSTOM RELOADING DIES,

Index

Part One

Part Two

ENGLISH AND AMERICAN CARTRIDGES

Part Two

METRIC CARTIDGES

CARTRIDGE:

OTHER NAMES:
9mm Ultra
9mm Police

DATA SOURCE:

HISTORICAL DATA:

NOTES:

LOADING DATA:

BULLET WT./TYPE	POWDER WT./TYPE	VELOCITY ('/SEC)	SOURCE

CASE PREPARATION: **SHELLHOLDER (RCBS): 10**

MAKE FROM:

PHYSICAL DATA (INCHES): **DIMENSIONAL DRAWING:**

CASE TYPE:
CASE LENGTH:
HEAD DIAMETER:
RIM DIAMETER:
NECK DIAMETER:
NECK LENGTH:
SHOULDER LENGTH:
BODY ANGLE (DEG'S/SIDE):
CASE CAPACITY:

LOADED LENGTH:
BELT DIAMETER:
RIM THICKNESS:
SHOULDER DIAMETER:
LENGTH TO SHOULDER:
SHOULDER ANGLE (DEG'S/SIDE):
PRIMER:
CASE CAPACITY (GR'S WATER):

CARTRIDGE:

OTHER NAMES:
9mm Ultra
9mm Police

DATA SOURCE:

HISTORICAL DATA:

NOTES:

LOADING DATA:

BULLET WT./TYPE	POWDER WT./TYPE	VELOCITY ('/SEC)	SOURCE

CASE PREPARATION: **SHELLHOLDER (RCBS): 10**

MAKE FROM:

PHYSICAL DATA (INCHES): **DIMENSIONAL DRAWING:**

CASE TYPE:
CASE LENGTH:
HEAD DIAMETER:
RIM DIAMETER:
NECK DIAMETER:
NECK LENGTH:
SHOULDER LENGTH:
BODY ANGLE (DEG'S/SIDE):
CASE CAPACITY:

LOADED LENGTH:
BELT DIAMETER:
RIM THICKNESS:
SHOULDER DIAMETER:
LENGTH TO SHOULDER:
SHOULDER ANGLE (DEG'S/SIDE):
PRIMER:
CASE CAPACITY (GR'S WATER):

CARTRIDGE:

OTHER NAMES:

DATA SOURCE:

HISTORICAL DATA:

NOTES:

LOADING DATA:

BULLET WT./TYPE	POWDER WT./TYPE	VELOCITY ('/SEC)	SOURCE

CASE PREPARATION: SHELLHOLDER (RCBS): 10

MAKE FROM:

PHYSICAL DATA (INCHES): DIMENSIONAL DRAWING:

CASE TYPE:
CASE LENGTH:
HEAD DIAMETER:
RIM DIAMETER:
NECK DIAMETER:
NECK LENGTH:
SHOULDER LENGTH:
BODY ANGLE (DEG'S/SIDE):
CASE CAPACITY:
LOADED LENGTH:
BELT DIAMETER:
RIM THICKNESS:
SHOULDER DIAMETER:
LENGTH TO SHOULDER:
SHOULDER ANGLE (DEG'S/SIDE):
PRIMER:
CASE CAPACITY (GR'S WATER):

CARTRIDGE:

OTHER NAMES:

DATA SOURCE:

HISTORICAL DATA:

NOTES:

LOADING DATA:

BULLET WT./TYPE	POWDER WT./TYPE	VELOCITY ('/SEC)	SOURCE

CASE PREPARATION: SHELLHOLDER (RCBS): 10

MAKE FROM:

PHYSICAL DATA (INCHES): DIMENSIONAL DRAWING:

CASE TYPE:
CASE LENGTH:
HEAD DIAMETER:
RIM DIAMETER:
NECK DIAMETER:
NECK LENGTH:
SHOULDER LENGTH:
BODY ANGLE (DEG'S/SIDE):
CASE CAPACITY:
LOADED LENGTH:
BELT DIAMETER:
RIM THICKNESS:
SHOULDER DIAMETER:
LENGTH TO SHOULDER:
SHOULDER ANGLE (DEG'S/SIDE):
PRIMER:
CASE CAPACITY (GR'S WATER):

CARTRIDGE:

OTHER NAMES:
9mm Ultra
9mm Police

DATA SOURCE:

HISTORICAL DATA:

NOTES:

LOADING DATA:

BULLET WT./TYPE	POWDER WT./TYPE	VELOCITY ('/SEC)	SOURCE

CASE PREPARATION: SHELLHOLDER (RCBS): 10

MAKE FROM:

PHYSICAL DATA (INCHES): DIMENSIONAL DRAWING:

CASE TYPE:
CASE LENGTH:
HEAD DIAMETER:
RIM DIAMETER:
NECK DIAMETER:
NECK LENGTH:
SHOULDER LENGTH:
BODY ANGLE (DEG'S/SIDE):
CASE CAPACITY:

LOADED LENGTH:
BELT DIAMETER:
RIM THICKNESS:
SHOULDER DIAMETER:
LENGTH TO SHOULDER:
SHOULDER ANGLE (DEG'S/SIDE):
PRIMER:
CASE CAPACITY (GR'S WATER):

CARTRIDGE:

OTHER NAMES:
9mm Ultra
9mm Police

DATA SOURCE:

HISTORICAL DATA:

NOTES:

LOADING DATA:

BULLET WT./TYPE	POWDER WT./TYPE	VELOCITY ('/SEC)	SOURCE

CASE PREPARATION: SHELLHOLDER (RCBS): 10

MAKE FROM:

PHYSICAL DATA (INCHES): DIMENSIONAL DRAWING:

CASE TYPE:
CASE LENGTH:
HEAD DIAMETER:
RIM DIAMETER:
NECK DIAMETER:
NECK LENGTH:
SHOULDER LENGTH:
BODY ANGLE (DEG'S/SIDE):
CASE CAPACITY:

LOADED LENGTH:
BELT DIAMETER:
RIM THICKNESS:
SHOULDER DIAMETER:
LENGTH TO SHOULDER:
SHOULDER ANGLE (DEG'S/SIDE):
PRIMER:
CASE CAPACITY (GR'S WATER):

CARTRIDGE:

OTHER NAMES:

DATA SOURCE:

HISTORICAL DATA:

NOTES:

LOADING DATA:

BULLET WT./TYPE	POWDER WT./TYPE	VELOCITY ('/SEC)	SOURCE

CASE PREPARATION: SHELLHOLDER (RCBS): 10

MAKE FROM:

PHYSICAL DATA (INCHES):

CASE TYPE:
CASE LENGTH:
HEAD DIAMETER:
RIM DIAMETER:
NECK DIAMETER:
NECK LENGTH:
SHOULDER LENGTH:
BODY ANGLE (DEG'S/SIDE):
CASE CAPACITY:
LOADED LENGTH:
BELT DIAMETER:
RIM THICKNESS:
SHOULDER DIAMETER:
LENGTH TO SHOULDER:
SHOULDER ANGLE (DEG'S/SIDE):
PRIMER:
CASE CAPACITY (GR'S WATER):

DIMENSIONAL DRAWING:

CARTRIDGE:

OTHER NAMES:

DATA SOURCE:

HISTORICAL DATA:

NOTES:

LOADING DATA:

BULLET WT./TYPE	POWDER WT./TYPE	VELOCITY ('/SEC)	SOURCE

CASE PREPARATION: SHELLHOLDER (RCBS): 10

MAKE FROM:

PHYSICAL DATA (INCHES):

CASE TYPE:
CASE LENGTH:
HEAD DIAMETER:
RIM DIAMETER:
NECK DIAMETER:
NECK LENGTH:
SHOULDER LENGTH:
BODY ANGLE (DEG'S/SIDE):
CASE CAPACITY:
LOADED LENGTH:
BELT DIAMETER:
RIM THICKNESS:
SHOULDER DIAMETER:
LENGTH TO SHOULDER:
SHOULDER ANGLE (DEG'S/SIDE):
PRIMER:
CASE CAPACITY (GR'S WATER):

DIMENSIONAL DRAWING:

CARTRIDGE:

OTHER NAMES:
9mm Ultra
9mm Police

DATA SOURCE:

HISTORICAL DATA:

NOTES:

LOADING DATA:

BULLET WT./TYPE	POWDER WT./TYPE	VELOCITY ('/SEC)	SOURCE

CASE PREPARATION: SHELLHOLDER (RCBS): 10

MAKE FROM:

PHYSICAL DATA (INCHES):

CASE TYPE:
CASE LENGTH:
HEAD DIAMETER:
RIM DIAMETER:
NECK DIAMETER:
NECK LENGTH:
SHOULDER LENGTH:
BODY ANGLE (DEG'S/SIDE):
CASE CAPACITY:

LOADED LENGTH:
BELT DIAMETER:
RIM THICKNESS:
SHOULDER DIAMETER:
LENGTH TO SHOULDER:
SHOULDER ANGLE (DEG'S/SIDE):
PRIMER:
CASE CAPACITY (GR'S WATER):

DIMENSIONAL DRAWING:

CARTRIDGE:

OTHER NAMES:
9mm Ultra
9mm Police

DATA SOURCE:

HISTORICAL DATA:

NOTES:

LOADING DATA:

BULLET WT./TYPE	POWDER WT./TYPE	VELOCITY ('/SEC)	SOURCE

CASE PREPARATION: SHELLHOLDER (RCBS): 10

MAKE FROM:

PHYSICAL DATA (INCHES):

CASE TYPE:
CASE LENGTH:
HEAD DIAMETER:
RIM DIAMETER:
NECK DIAMETER:
NECK LENGTH:
SHOULDER LENGTH:
BODY ANGLE (DEG'S/SIDE):
CASE CAPACITY:

LOADED LENGTH:
BELT DIAMETER:
RIM THICKNESS:
SHOULDER DIAMETER:
LENGTH TO SHOULDER:
SHOULDER ANGLE (DEG'S/SIDE):
PRIMER:
CASE CAPACITY (GR'S WATER):

DIMENSIONAL DRAWING:

CARTRIDGE:

OTHER NAMES:

DATA SOURCE:

HISTORICAL DATA:

NOTES:

LOADING DATA:

BULLET WT./TYPE	POWDER WT./TYPE	VELOCITY ('/SEC)	SOURCE

CASE PREPARATION: SHELLHOLDER (RCBS): 10

MAKE FROM:

PHYSICAL DATA (INCHES):	DIMENSIONAL DRAWING:

CARTRIDGE:

OTHER NAMES:

DATA SOURCE:

HISTORICAL DATA:

NOTES:

LOADING DATA:

BULLET WT./TYPE	POWDER WT./TYPE	VELOCITY ('/SEC)	SOURCE

CASE PREPARATION: SHELLHOLDER (RCBS): 10

MAKE FROM:

PHYSICAL DATA (INCHES):	DIMENSIONAL DRAWING:

CARTRIDGE:

OTHER NAMES:
9mm Ultra
9mm Police

DATA SOURCE:

HISTORICAL DATA:

NOTES:

LOADING DATA:

BULLET WT./TYPE	POWDER WT./TYPE	VELOCITY ('/SEC)	SOURCE

CASE PREPARATION: **SHELLHOLDER (RCBS): 10**

MAKE FROM:

PHYSICAL DATA (INCHES): **DIMENSIONAL DRAWING:**

CARTRIDGE:

OTHER NAMES:
9mm Ultra
9mm Police

DATA SOURCE:

HISTORICAL DATA:

NOTES:

LOADING DATA:

BULLET WT./TYPE	POWDER WT./TYPE	VELOCITY ('/SEC)	SOURCE

CASE PREPARATION: **SHELLHOLDER (RCBS): 10**

MAKE FROM:

PHYSICAL DATA (INCHES): **DIMENSIONAL DRAWING:**

CARTRIDGE:

OTHER NAMES:

DIA:

BALLISTEK NO:

NAI NO:

DATA SOURCE:

HISTORICAL DATA:

NOTES:

LOADING DATA:

BULLET WT./TYPE	POWDER WT./TYPE	VELOCITY ('/SEC)	SOURCE

CASE PREPARATION: SHELL HOLDER (RCBS): 10

MAKE FROM:

PHYSICAL DATA (INCHES):	DIMENSIONAL DRAWING:

CARTRIDGE:

OTHER NAMES:

DIA:

BALLISTEK NO:

NAI NO:

DATA SOURCE:

HISTORICAL DATA:

NOTES:

LOADING DATA:

BULLET WT./TYPE	POWDER WT./TYPE	VELOCITY ('/SEC)	SOURCE

CASE PREPARATION: SHELL HOLDER (RCBS): 10

MAKE FROM:

PHYSICAL DATA (INCHES):	DIMENSIONAL DRAWING: